AN INTRODUCTION TO MATHEMATICAL ANALYSIS

FOR

ECONOMIC THEORY AND ECONOMETRICS

AN INTRODUCTION TO MATHEMATICAL ANALYSIS

FOR

ECONOMIC THEORY AND ECONOMETRICS

Dean Corbae

Maxwell B. Stinchcombe

Juraj Zeman

PRINCETON UNIVERSITY PRESS

Princeton and Oxford

Published by Princeton University Press, 41 William Street, Princeton,
New Jersey 08540

In the United Kingdom: Princeton University Press, 6 Oxford Street, Woodstock,
Oxfordshire OX20 1TW

Library of Congress Cataloging-in-Publication Data

Corbae, Dean.
An introduction to mathematical analysis for economic theory and econometrics /
Dean Corbae, Maxwell B. Stinchcombe, Juraj Zeman.
p.   cm.
Includes bibliographical references and index.
ISBN 978-0-691-11867-3 (hardcover : alk. paper)
1. Economics, Mathematical.   2. Mathematical analysis.   3. Econometrics.
I. Stinchcombe, Maxwell.   II. Zeman, Juraj, 1957–   III. Title.
HB135.C657   2009
330.01′5195—dc22   2008047711

British Library Cataloging-in-Publication Data is available

This book was composed in Times Roman and Bembo using ZzTEX
by Princeton Editorial Associates, Inc., Scottsdale, Arizona.

Printed on acid-free paper.   ∞

press.princeton.edu

Printed in the United States of America

1 3 5 7 9 10 8 6 4 2

To my family: those who put up with me in the past—Josephine and Phil—and especially those who put up with me in the present—Margaret, Bethany, Paul, and Elena.

    D.C.

To Rigo and Beth, who sometimes tolerated, sometimes encouraged, and sometimes rather disliked my absorption with this project.

    M.B.S.

To my family.

    J.Z.

Contents

Chapter 7 ◆ Measure Spaces and Probability

Chapter 8 ◆ The $L^p(\Omega, \mathcal{F}, P)$ and ℓ^p Spaces, $p \in [1, \infty]$

Chapter 9 ◆ Probabilities on Metric Spaces

Chapter 10 ◆ Infinite-Dimensional Convex Analysis   595

Chapter 11 ◆ Expanded Spaces   627

Preface

The objective of this volume is to provide a simple introduction to mathematical analysis with applications in economic theory and econometrics. There is increasing use of real and functional analysis in economics, but few books cover that material at an introductory level. Our rationale for writing this one is to bridge the gap between basic mathematical economics books (which deal with calculus, linear algebra, and constrained optimization) and advanced economics texts, such as Stokey and Lucas's *Recursive Methods in Economic Dynamics,* that presume a working knowledge of functional analysis and measure theory.

We take a unified approach to understanding basic and advanced spaces through the application of the metric completion theorem. This is the concept by which, for example, the real numbers complete the rational numbers and measure spaces complete fields of measurable sets. There are three other major innovations in this work relative to mathematics textbooks: (i) we have gathered material from very different areas in mathematics, from lattices and convex analysis to measure theory and functional analysis, because they are useful for economists doing regression analysis, working on both static and dynamic choice problems, and analyzing both strategic and competitive equilibria; (ii) we try to use concepts familiar to economists—approximation and existence of a solution—to understand analysis and measure theory; and (iii) pedagogically, we provide extensive simple examples drawn from economic theory and econometrics to provide the intuition necessary for grasping difficult ideas. It is important to emphasize that while we aim to make this material as accessible as possible, we have not excluded demanding mathematical concepts used by economists and that, aside from examples assuming an undergraduate background in economics, the book is self-contained (i.e., almost any theorem used in proving a given result is itself proved in these pages as well).

We taught the first half of this manuscript in the first-semester Ph.D. core sequence at the University of Pittsburgh and the University of Texas. Those programs begin with an intensive, one-month remedial summer math class that focuses on calculus and linear algebra. Much of Chapters 1 through 6 was taught in a basic class. We also believe that parts of those chapters are appropriate for an upper-level undergraduate mathematical economics class. The remainder of the book was taught as a graduate mathematical economics class. While we used the manuscript in a classroom, we hope it will be useful to researchers; for instance, anyone who reads a book like Stokey and Lucas's *Recursive Methods* must understand the background concepts in our text. In fact, it was because one of the authors found that his students were ill prepared to understand Stokey and Lucas in his upper-level macroeconomics class that this project began.

Throughout, we have sought developments of the mathematics that are directly tied to economics. In some cases, this means that there are more direct ways to reach the mathematical results. We hope that the added motivation of the subjects is worth the costs that the reader bears.

User's Guide

We intend this text to cover the advanced undergraduate and beginning graduate mathematics that we think most economics Ph.D. students ought to know. As our intended audience is economists, we have kept the economic applications as much in the foreground as possible and have emphasized approximation.

In this roadmap, we lay out how we think this whole enterprise is supposed to work. We first give a rough overview of the contents of the chapters and the main points. We believe that Chapters 1 through 5 can form the basis for an advanced undergraduate course in mathematical economics. Our experience is that adding Sections 6.1 through 6.5 to the undergraduate course provides the basis for a good, though very full, semester-long *graduate* Math for Economists class. The rest of the book is suitable for a second, advanced course designed for students who plan to be econometricians or theoreticians of any stripe.

Outline of Chapters

Chapter 1. Logic
This chapter goes about systematically reducing logical statements to statements about relations between sets. For instance, we show how "if A, then B" statements are the same as "A is a subset of B" statements and "there exists an x in A such that B is true about it" statements are the same as "A and B have a nonempty intersection." If the book succeeds as we hope, students will always be able to parse the statement of a result in the economics literature so as to see the subset relation that underlies it.

Chapter 2. Set Theory
Here we explain how to manipulate and compare sets. We begin with relations, which are simply subsets of a given set. Functions, correspondences, and equivalence relations are simply special cases of relations. We also show how rational choice theory can be formulated as conditions (such as completeness and transitivity) on preference relations. Relations can be used to order sets in different ways, such as lattices. Using lattices we introduce monotone comparative statics and a "monotone" fixed-point theorem due to Tarski. We then turn to the study of the sizes of sets and the distinction between countable and uncountable infinities. Having introduced the notion of infinity, we move onto ways to handle it.

The key assumption here is known as the axiom of choice, which is equivalent to the existence of certain kinds of maximal elements, a statement known as Zorn's lemma.

Chapter 3. The Space of Real Numbers

The one-dimensional real line (denoted $\mathbb{R}$) is the most basic space in mathematics (even elementary school students are exposed to it). It is what we use to measure quantities. Rather than follow the standard axiomatic approach to the real numbers, we emphasize how rational numbers provide an arbitrarily good system for measurements of quantities, a property called "denseness," yet the space of rational numbers has "holes" in it. In the space of rational numbers, we measure distance in the usual way, known to Euclid, by the absolute value of the difference, $|q - q'|$. We then show how to add new points to the space of rational numbers to represent the results of arbitrarily fine approximations, a property called "completeness." Completeness is absolutely critical for representing the solutions to problems. In particular, one can construct a sequence of approximate solutions that get closer and closer together (i.e., a Cauchy sequence), and, provided the space is complete, the limit of this sequence *exists* and is the solution of the original problem. Another particularly important property of completeness of the real numbers for economics is the existence of maximal elements.

Chapter 4. The Finite-Dimensional Metric Space of Real Vectors

This is the first of two workhorse chapters for young economists. The underlying set is the set of length-ℓ vectors of the real numbers constructed in Chapter 3 (denoted $\mathbb{R}^\ell$). Here there are many "usual ways" to measure the distance between points, and their underlying commonality, a "metric" or a distance function, introduces a new issue that we explore throughout the rest of the book—how to measure distances. Apart from the ℓ-dimensional analogue of Euclidean distance, the square root of the sum of squares of component differences, we cover a number of other distance functions or metrics; "the sum of absolute values of component differences," known as the L^1 metric; and the "maximum of absolute values of component differences," known as the sup norm metric. In addition to completeness, which was emphasized in Chapter 3, this chapter focuses on compactness and continuity, which are essential for guaranteeing the existence of a solution in both static and dynamic optimization models used by economists. Many of the existence results can be understood in terms of a simple application of the intermediate value theorem.

Chapter 5. Finite-Dimensional Convex Analysis

This is the second of the two workhorse chapters for young economists. Much of economics is about finding the solution to a well-posed optimum problem characterized by a system of equations. The optimum problem may be written as maximizing a continuous function (e.g., a utility function) on a compact convex set (e.g., a budget set). We begin with basic separation theorems (sufficient to prove the second welfare theorem) that talk about ways of separating disjoint convex sets from each other using linear functions that lie in a "dual space." We also apply these separation results to a proof of the Kuhn-Tucker theorem in optimization and cover Kakutani's fixed-point theorem, which is the major mathematical tool for proving the existence of equilibria in economics.

Chapter 6. Metric Spaces

Chapters 6 and 7 form the core of the book for Ph.D. students in economics. Chapter 6 introduces the general notion of a metric space—simply an underlying set M and a distance function, also known as a metric. In this chapter we introduce and study the distance between sets, a crucial ingredient in the study of the behavior of sets of maxima in comparative statics problems; the distance between cumulative distribution functions, which forms the basis for asymptotic analysis in econometrics and for the theory of choice under uncertainty; and the distance between continuous functions, which we use in the study of value functions for dynamic optimization problems. After we present the fundamental approximation results in the study of function spaces (the algebraic and the lattice theoretic versions of the Stone-Weierstrass theorem), we give an extended presentation of regression analysis as approximation; regression analysis; the distance between infinite sequences, which economists use to model both sequences of plays in repeated games and strategy sets for repeated games; sequences of outcomes in stochastic process theory and other dynamical systems; and rewards in dynamic optimization problems. After this, we turn to two classes of extension theorems for continuous functions. These give us conditions under which continuous functions with useful properties exist, and this in turn tells us that continuous functions have very powerful approximation capabilities. Finally, given the importance of completeness for the existence of solutions, we provide the generalization of the process by which we completed the rationals, the metric completion theorem, which tells us how to add a minimal set of elements to any metric space so as to make it complete.

Chapter 7. Measure Spaces and Probability

A fruitful generalization of continuous functions on metric spaces is measurable functions on measure spaces. Metric spaces are defined as a set and a special class of sets, the open sets, closed under arbitrary union and finite intersection. Measure spaces are an underlying set and a class of sets, the measurable sets, that are closed under countable union, countable intersection, and complementation. Probabilities are functions from the class of measurable sets to the interval [0, 1]. Measurable functions, also known as random variables, are those with the inverse image of every open set being measurable. These are the tools used to model random phenomena.

The first results that we prove using these tools involve the behavior of sequences of random variables, the weak law of large numbers (WLLN), the Borel-Cantelli lemma, the strong law of large numbers (SLLN), and Kolmogorov's 0-1 law. All of these are results about "large numbers" of random variables, and the mathematics of talking about large numbers uses sequences and convergence.

A probability is nonatomic if one can partition the measure space arbitrarily finely. Without this most versatile and interesting class of spaces, we do not have models of sequences of independent random variables. After developing the limit theorems, we return to basics and give a careful proof of the existence of these nonatomic spaces.

Our approach to the existence of nonatomic probabilities is to begin with a probability on a class of sets closed under finite unions, finite intersections, and complementation. We define the distance between simple sets as the probability of their symmetric difference. In keeping with the approximation theme that runs through this book, we define the measurable sets as the completion of this small class of sets. Probabilities can be extended from the small class of sets to the larger one using denseness and continuity. This is not the approach taken by most texts, which sandwich every set between an outer approximation and an inner approximation and define the measurable ones as those whose inner and outer approximations differ by an arbitrarily small amount; an appendix to the chapter gives a brief sketch of this approach.

Chapter 8. The $L^p(\Omega, \mathcal{F}, P)$ and ℓ^p Spaces, $p \in [1, \infty]$

Beyond their uses in statistics and econometrics, spaces of random variables with pth moments, the L^p spaces, are used to model commodities such as securities or stocks or bonds. In order to understand the basic properties of these spaces, we use the results on measurability and integrability from Chapter 7, as well as geometric intuitions from Chapter 5 on convex analysis.

Prices are continuous linear functions on the space of commodities. When the space of commodities is an L^p space, this leads us to the so-called "dual space" of L^p. L^2 is the space of random variables with finite variance, and in this setting the dual space is the original space, something that makes proofs and intuitions much easier. This is also the setting in which one does regression analysis, here viewed as the process of finding the conditional expectation of a random variable, Y, given the value of a random vector, X. Conditional expectation is widely useful outside econometrics as well. It turns out to be a projection in L^2 spaces, and we carefully study the geometry of these projections.

Models of perfect competition and of anonymous interactions of large groups of people typically assume that each agent is a negligibly small part of the economy. One way to do this is to assume that the space of agents is a nonatomic probability space. This is one of the most important uses of nonatomic probabilities, and the existence of equilibria relies on results about the structures of L^p spaces.

Chapter 9. Probabilities on Metric Spaces

Probabilities on metric spaces of outcomes are the foundation of models of choice under uncertainty, and probabilities on metric spaces of time paths are the foundation of models of stochastic processes. This chapter develops the basic results on the structure of probabilities on metric spaces. We begin with the Prokhorov metric, which generalizes the Levy metric on probabilities on $\mathbb{R}$ given in Chapter 6, and give five equivalent formulations of convergence in this metric. These lead to the continuous mapping theorem, a crucial tool in econometrics, and the denseness of finitely supported probabilities, a crucial interpretational tool for probabilities. We then turn to continuous expected utility preferences, showing how they can be represented as Prokhorov continuous, linear functions on the set of probabilities. The Riesz representation theorem delivers the essential duality result: probabilities on complete separable metric spaces are equivalent to continuous $\mathbb{R}$-valued linear functions on the set of continuous bounded functions. The requisite continuity is closely related to the countable additivity of probabilities. Sets of probabilities are tight if they are "almost" compactly supported, and this property characterizes compact sets of probabilities. This characterization is a crucial part of the analysis of the behavior of sums of very many independent random variables. Probabilities can also be identified with continuous linear operators on spaces of continuous functions, and we present a fairly simple, operator theoretic proof of the central

limit theorem, which concerns the distribution of the sums of very many independent random variables, all of which are small.

Chapter 10. Infinite-Dimensional Convex Analysis
The results in convex analysis in vector space $\mathbb{R}^{\ell}$ are the basic tools of economics.
Over time, we have been driven to the use of larger vector spaces in our models.
This chapter begins with a brief introduction to general topological spaces before
focusing on Hausdorff locally convex topological vector spaces, which generalize
the Banach spaces of Chapter 8 and form the broadest class of vector spaces
regularly used by economists. The Hahn-Banach theorem reappears here to show
that the duals of these vector spaces are sufficiently rich to perform the separation of
convex sets from each other that is at the core of a great deal of economic analysis.
We then turn to convergence and compactness in topological spaces and give
Tychonoff's theorem, which yields the crucial criterion for weak* compactness
in topological vector spaces. We end with Schauder's fixed-point theorem for
compact convex subsets of locally convex topological vector spaces.

Chapter 11. Expanded Spaces
Despite the fact that no conceivable system of physical measurements could distinguish between rational and irrational quantities, we added a great many new
numbers to the rationals in order to exactly, rather than approximately, represent
solutions to equations. This gave us $\mathbb{R}$, which is the usual continuous model of
quantities and probabilities. There are modeling situations with large numbers of
individuals or actions in which one more level of expansion has proven itself very
useful, which we cover in this chapter. We show that this new level of expansion en-
compasses all of the compactifications that appeared in Chapter 6. We also show
how to represent the effects of independent shocks on every member of a large
population. This model of uncountably many uncorrelated random variables has
an associated law of large numbers and is most closely approximated by a large
finite number of random variables. The expansion turns the approximation into
exact mathematics.

Week	Advanced undergraduate	First-year graduate
1	Ch. 1, 2.1–4	Ch. 1, 2.1–4
2	Ch. 2.5–7	Ch. 2.5–7
3	Ch. 2.8	Ch. 2.8–9
4	Ch. 2.10	Ch. 2.10, 3.1–3
5	Ch. 3.1–3	Ch. 3.4–8
6	Ch. 3.4–7	Ch. 4.1–2
7	Ch. 4.1–2	Ch. 4.3, 5.1–4
8	Ch. 4.3, 5.1–2	Ch. 4.4–6
9	Ch. 5.3–4	Ch. 4.7
10	Ch. 4.4	Ch. 4.8–9
11	Ch. 4.5–7	Ch. 5.5–9
12	Ch. 4.8	Ch. 4.10–12
13	Ch. 5.5–6	Ch. 5.11–12
14	Ch. 5.7–8	Ch. 6.1–3
15	Ch. 5.8–10	Ch. 6.4–5

Week	Graduate theory	Graduate econometrics
1	Ch. 2.10–13	Same
2	Ch. 6.1–5 (review)	Same
3	Ch. 6.6	Same
4	Ch. 6.7–8	Same
5	Ch. 6.9, 7.1	Same
6	Ch. 7.1–2	Same
7	Ch. 7.3–4	Same
8	Ch. 7.5–6	Same
9	Ch. 7.7–9	Same
10	Ch. 7.10	Same
11	Ch. 8.1–3	Same
12	Ch. 8.4	Same
13	Ch. 8.5–6	9.1–3
14	Ch. 8.7–8	9.6–8
15	Ch. 8.9–10	9.8–10

Notation

We assume that the reader/student has had a calculus-based course in micro-economics as well as a course in linear algebra. We do not use linear algebra until Chapter 4, and then only the basics such as matrix multiplication and the concept of dimension. Chapter 5 uses it much more seriously in the study of the derivative characterizations of concavity and convexity. However, well before we introduce them formally, we use the real numbers in examples. We have tried to be scrupulously careful that no important result is given before all of its pieces are formally developed. Scrupulous care does not guarantee complete success. Both for reference purposes and because we may mistakenly use something before we define it, we include here the main notations we use, their meaning, and the pages on which they appear.

Spaces, Sets, and Classes of Sets

- $\mathbb{N} = \{1, 2, 3, \ldots\}$, the natural or "counting" numbers (p. 21).
- $\mathbb{Z} = \{\ldots, -2, -1, 0, 1, 2, \ldots\}$, the integers and $\mathbb{Z}_+ = \{0, 1, 2, \ldots\}$, the nonnegative integers (p. 21).
- $\mathbb{Q} = \{\frac{m}{n} : m, n \in \mathbb{Z}, n \neq 0\}$, the quotients, or rational numbers (p. 21).
- $\mathfrak{C}(\mathbb{Q})$, the set of Cauchy sequences of rational numbers (p. 76).
- $\mathbb{R}$, the set of "real numbers" that we construct in Chapter 3 as a set of equivalence classes of Cauchy sequences of rationals, that is, by adding the irrational numbers (p. 79).
- $\mathbb{R}^\ell$, the set of length-ℓ vectors of real numbers (Chapter 4).
- (M, d), a metric space, M a nonempty set, d a distance function on M (Chapter 6).
- $C(M)$ and $C_b(M)$, the set of continuous and the set of continuous bounded $\mathbb{R}$-valued functions on a metric space (M, d) (p. 272).
- $\mathcal{P}(A)$, the power set of A, that is, the set of all subsets of A (p. 20).
- $\mathcal{D}(\mathbb{R})$, the set of cumulative distribution functions on $\mathbb{R}$ (p. 275).
- $\mathfrak{X}$, a vector space over $\mathbb{R}$ (p. 115).
- $\mathbb{H}$, a Hilbert space (p. 475).
- $\mathcal{F}^\circ \subset \mathcal{P}(\Omega)$, a field of subsets (p. 20 and p. 335).

- $\mathcal{F} \subset \mathcal{P}(\Omega)$, a σ-field of subsets (p. 335).
- $(\Omega, \mathcal{F})$, a measure space (p. 335 and p. 356).
- $(\Omega, \mathcal{F}, P)$, a probability space (p. 335 and p. 356).
- $L^p(\Omega, \mathcal{F}, P)$ and ℓ^p, $p \in [1, \infty]$, the space of random variables with pth moments and the space of sequences with absolutely summable pth powers (Chapter 8).
- $\Delta(S)$, the set of probabilities on a finite set S (p. 155).
- $\Delta(M)$, the set of countably additive Borel probabilities on a metric space (M, d) (Chapter 9).
- $(\varphi, (M', d'))$, a compactification of a metric space (p. 153).
- $(f, (X', d'))$, the compact embedding of a metric space (p. 328).
- $(f, (M', d'))$ or $(\widehat{M}, \widehat{d})$, the completion of a metric space (p. 334).

Functions and Relations

- $1_A(x)$, the indicator function of a set A (p. 4).
- $\{x \in X : \mathbb{A}(x)\}$, the set of all x in X such that $\mathbb{A}(x)$ is true (Chapter 1).
- $A \times B$, the Cartesian product of the sets A and B (p. 22).
- $\mathbf{x} \le \mathbf{y}$, "less than or equal to" for vectors, $x_i \le y_i$ for $i = 1, \ldots, \ell$ for $\mathbf{x}, \mathbf{y} \in \mathbb{R}^\ell$.
- $\mathbf{x} \ll \mathbf{y}$, strictly less than for vectors, $x_i < y_i$ for $i = 1, \ldots, \ell$ for $\mathbf{x}, \mathbf{y} \in \mathbb{R}^\ell$.
- $\succsim$, a binary relation (p. 23), common properties of binary relations given in Table 2.7 (p. 39).
- $f : A \to B$ or $a \mapsto f(a)$, a function from the set A to the set B (p. 24).
- $\Gamma : A \twoheadrightarrow B$, a correspondence from the set A to the set B (p. 25).
- $f^{-1} : B \twoheadrightarrow A$, the inverse of a function $f : A \to B$ (p. 35).
- $\Gamma^+ : B \twoheadrightarrow A$ and $\Gamma^- : B \twoheadrightarrow A$, the upper and lower inverses of a correspondence $\Gamma : A \twoheadrightarrow B$ (p. 266).
- $g \circ f$, the composition of a function $f : A \to B$ and a function $g : B \to C$ (p. 38).
- $n \mapsto x_n$, $(x_n)_{n \in \mathbb{N}}$, $(x_n)_{n=1}^\infty$, or (x_n), a sequence (p. 75).
- $k \mapsto x_{n_k}$, a subsequence of the sequence $n \mapsto x_n$ (p. 76).
- $x_n \to x$, the convergence of a sequence x_n to a point x in $\mathbb{R}$ (p. 84), in (M, d) (p. 108).
- $X_n \to_w X$, the weak convergence of a sequence in L^p (p. 518).
- $\lim_n x_n$, the limit of a sequence x_n in $\mathbb{R}$ (p. 84), and in (M, d) (p. 108).
- $\sum_{n=0}^N r_n$, the sum of the numbers or vectors $r_1, r_2, \ldots, r_N$ (p. 94).
- $\pm\infty$, infinity
- $\sum_{n=0}^\infty r_n$, the limit, if it exists, of the sequence $s_N = \sum_{n=0}^N r_n$ (p. 95).
- $\mathbb{A}_n$ i.o. and $\mathbb{A}_n$ a.a., infinitely often and almost always (p. 263).

- $\arg\max_{x \in X} f(x)$, the set of solutions to the problem "max $f(x)$ subject to x being an element of X" (p. 31).
- $\mathbf{x} \cdot \mathbf{y} = \sum_{i=1}^{\ell} x_i y_i$, the inner product of vectors $\mathbf{x}$, $\mathbf{y}$ in $\mathbb{R}^{\ell}$.
- $\mathrm{cl}(E)$, ∂E, the closure and the boundary of a subset E of a metric space (p. 124).
- $\mathrm{co}(E)$, $\overline{\mathrm{co}}(E)$, the convex hull and the closed convex hull of a set E (p. 188).
- $\mathcal{A}(S)$, the algebra generated by a set of functions S (p. 299).
- $x \vee y$ and $x \wedge y$, the least upper bound and the greatest lower bound of the points x, y in a lattice, also denoted $\sup(\{x, y\})$ and $\inf(\{x, y\})$ (p. 39).
- $\sup(E)$ and $\inf(E)$, the supremum and infimum of a set of real numbers E (p. 90).
- $\limsup_n x_n$ and $\liminf_n x_n$ (p. 101).
- $\|\mathbf{x}\|$, the norm of a vector in $\mathbb{R}^{\ell}$ (p. 116) and in more general vector spaces (p. 117).
- $X = Y$ a.e., almost everywhere equality of random variables (p. 361).
- $\alpha(X, Y)$, the Ky Fan pseudometric on random variables (p. 361).
- $L^0 = L^0(\Omega, \mathcal{F}, P)$, the vector space of $\mathbb{R}$-valued random variables (p. 338 and p. 360).
- $M_s = M_s(\Omega, \mathcal{F}, P) \subset L^0$, the vector space of simple $\mathbb{R}$-valued random variables (p. 362).

AN INTRODUCTION TO
MATHEMATICAL ANALYSIS
FOR
ECONOMIC THEORY
AND ECONOMETRICS

CHAPTER 1

Logic

The building blocks of modern economics are based on logical reasoning to prove the validity of a conclusion, $\mathbb{B}$, from well-defined premises, $\mathbb{A}$. In general, statements such as $\mathbb{A}$ and/or $\mathbb{B}$ can be represented using sets, and a "proof" is constructed by applying, sometimes ingeniously, a fixed set of rules to establish that the statement $\mathbb{B}$ is true whenever $\mathbb{A}$ is true. We begin with examples of how we represent statements as sets, then turn to the rules that allow us to form more and more complex statements, and then give a taxonomy of the major types of proofs that we use in this book.

1.1 ◆ Statements, Sets, Subsets, and Implication

The idea of a set (of things), or group, or collection is a "primitive," one that we use without being able to clearly define it. The idea of belonging to a set (group, collection) is primitive in exactly the same sense. Our first step is to give the allowable rules by which we evaluate whether statements about sets are true.

We begin by fixing a set X of things that we might have an interest in. When talking about demand behavior, the set X has to include, at the very least, prices, incomes, affordable consumption sets, preference relations, and preference-optimal sets. The set X varies with the context, and is often not mentioned at all.

We express the primitive notion of membership by "$\in$," so that "$x \in A$" means that "x is an element of the set A" and "$y \notin A$" means that "y is not an element of A."

Notation Alert 1.1.A *Capitalized letters are usually reserved for sets and smaller letters for points/things in the set that we are studying. Sometimes, several levels of analysis are present simultaneously, and we cannot do this. Consider the study of utility functions, u, on a set of options, X. A function u is a set of pairs of the form $(x, u(x))$, with x an option and $u(x)$ the number representing its utility. However, in our study of demand behavior, we want to see what happens as u varies. From this perspective, u is a point in the set of possible utility functions.*

Membership allows us to define subsets. We say "A is a subset of B," written "$A \subset B$," if every $x \in A$ satisfies $x \in B$. Thus, subsets are defined in terms of the primitive relation "$\in$." We write $A = B$ if $A \subset B$ and $B \subset A$, and $A \neq B$ otherwise.

We usually specify the elements of a set explicitly by saying "The set A is the set of all elements x in X such that each x has the property $\mathbb{A}$, that is, that $\mathbb{A}(x)$ is true," and write "$A = \{x \in X : \mathbb{A}(x)\}$" as a shorter version of this. For example, with $X = \mathbb{R}^\ell$, the statement "$\mathbf{x} \geq 0$" is identified with the set $\mathbb{R}^\ell_+ = \{\mathbf{x} \in X : \mathbf{x} \geq 0\}$. In this way, we identify a statement with the set of elements of X for which the statement is true. There are deep issues in logic and the foundations of mathematics relating to the question of whether or not all sets can be identified by "properties." Fortunately, these issues rarely impinge on the mathematics that economists need. Chapter 2 is more explicit about these issues.

We are very often interested in establishing the truth of statements of the form "If $\mathbb{A}$, then $\mathbb{B}$." There are many equivalent ways of writing such a statement: "$\mathbb{A} \Rightarrow \mathbb{B}$," "$\mathbb{A}$ implies $\mathbb{B}$," "$\mathbb{A}$ only if $\mathbb{B}$," "$\mathbb{A}$ is sufficient for $\mathbb{B}$," or "$\mathbb{B}$ is necessary for $\mathbb{A}$." To remember the sufficiency and necessity, it may help to subvocalize them as "$\mathbb{A}$ is sufficiently strong to guarantee $\mathbb{B}$" and "$\mathbb{B}$ is necessarily true if $\mathbb{A}$ is true."

The logical relation of implication is a subset relation. If $A = \{x \in X : \mathbb{A}(x)\}$ and $B = \{x \in X : \mathbb{B}(x)\}$, then "$\mathbb{A} \Rightarrow \mathbb{B}$" is the same as "$A \subset B$."

Example 1.1.1 *Let X be the set of numbers, $\mathbb{A}(x)$ the statement "$x^2 < 1$," and $\mathbb{B}(x)$ the statement "$|x| \leq 1$." Now, $\mathbb{A} \Rightarrow \mathbb{B}$. In terms of sets, $A = \{x \in X : \mathbb{A}(x)\}$ is the set of numbers strictly between -1 and $+1$, $B = \{x \in X : \mathbb{B}(x)\}$ is the set of numbers greater than or equal to -1 and less than or equal to $+1$, and $A \subset B$.*

The statements of interest can be quite complex to write out in their entirety. If X is the set of allocations in a model $\mathcal{E}$ of an economy and $\mathbb{A}(x)$ is the statement "x is a Walrasian equilibrium allocated for the economy $\mathcal{E}$," then a complete specification of the statement takes a great deal of work. Presuming some familiarity with general equilibrium models, we offer the following.

Example 1.1.2 *Let X be the set of allocations in a model $\mathcal{E}$ of an economy; let $\mathbb{A}(x)$ be the statement "x is a Walrasian equilibrium allocation"; and $\mathbb{B}(x)$ be the statement "x is Pareto efficient for $\mathcal{E}$." The first fundamental theorem of welfare economics is $\mathbb{A} \Rightarrow \mathbb{B}$. In terms of the definition of subsets, this is expressed as, "Every Walrasian equilibrium allocation is Pareto efficient."*

In other cases, we are interested in the truth of statements of the form "$\mathbb{A}$ if and only if $\mathbb{B}$," often written "$\mathbb{A}$ iff $\mathbb{B}$." Equivalently, such a statement can be written: "$\mathbb{A} \Rightarrow \mathbb{B}$ and $\mathbb{B} \Rightarrow \mathbb{A}$," which is often shortened to "$\mathbb{A} \Leftrightarrow \mathbb{B}$." Other frequently used formulations are: "$\mathbb{A}$ implies $\mathbb{B}$ and $\mathbb{B}$ implies $\mathbb{A}$," "$\mathbb{A}$ is necessary and sufficient for $\mathbb{B}$," or "$\mathbb{A}$ is equivalent to $\mathbb{B}$." In terms of the corresponding sets A and B, these are all different ways of writing "$A = B$."

Example 1.1.3 *Let X be the set of numbers, $\mathbb{A}(x)$ the statement "$0 \leq x \leq 1$," and $\mathbb{B}(x)$ the statement "$x^2 \leq x$." From high school algebra, $\mathbb{A} \Leftrightarrow \mathbb{B}$. In terms of sets, $A = \{x \in X : \mathbb{A}(x)\}$ and $B = \{x \in X : \mathbb{B}(x)\}$ are both the sets of numbers greater than or equal to 0 and less than or equal to 1.*

1.2 ◆ Statements and Their Truth Values

Note that a statement of the form "$\mathbb{A} \Rightarrow \mathbb{B}$" is simply a construct of two simple statements connected by "$\Rightarrow$." This is one of seven ways of constructing new statements that we use. In this section, we cover the first five of them: ands, ors, nots, implies, and equivalence. Repeated applications of these seven ways of constructing statements yield more and more elaboration and complication.

We begin with the simplest three methods, which construct new sets directly from a set or pair of sets that we start with. We then turn to the statements that are about relations between sets and introduce another formulation in terms of indicator functions. Later we give the other two methods, which involve the logical quantifiers "for all" and "there exists." Throughout, interest focuses on methods of establishing the truth or falsity of statements, that is, on methods of proof.

1.2.a Ands/Ors/Nots as Intersections/Unions/Complements

The simplest three ways of constructing new statements from other ones are using the connectives "and" or "or," or by "not," which is negation. Notationally: "$\mathbb{A} \wedge \mathbb{B}$" means "$\mathbb{A}$ and $\mathbb{B}$," "$\mathbb{A} \vee \mathbb{B}$" means "$\mathbb{A}$ or $\mathbb{B}$," and "$\neg \mathbb{A}$" means "not $\mathbb{A}$."

In terms of the corresponding sets: "$\mathbb{A} \wedge \mathbb{B}$" is $A \cap B$, the intersection of A and B, that is, the set of all points that belong to both A *and* B; "$\mathbb{A} \vee \mathbb{B}$" is $A \cup B$, the union of A and B, that is, the set of all points that belong to A *or* belong to B; and "$\neg \mathbb{A}$" is $A^c = \{x \in X : x \notin A\}$, the complement of A, is the set of all elements of X that do *not* belong to A.

The meanings of these new statements, $\neg \mathbb{A}$, $\mathbb{A} \wedge \mathbb{B}$, and $\mathbb{A} \vee \mathbb{B}$, are given by a *truth table*, Table 1.a. The corresponding Table 1.b gives the corresponding set versions of the new statements.

	Table 1.a			
$\mathbb{A}$	$\mathbb{B}$	$\neg \mathbb{A}$	$\mathbb{A} \wedge \mathbb{B}$	$\mathbb{A} \vee \mathbb{B}$
T	*T*	*F*	*T*	*T*
T	*F*	*F*	*F*	*T*
F	*T*	*T*	*F*	*T*
F	*F*	*T*	*F*	*F*

	Table 1.b			
A	B	A^c	$A \cap B$	$A \cup B$
$x \in A$	$x \in B$	$x \notin A^c$	$x \in A \cap B$	$x \in A \cup B$
$x \in A$	$x \notin B$	$x \notin A^c$	$x \notin A \cap B$	$x \in A \cup B$
$x \notin A$	$x \in B$	$x \in A^c$	$x \notin A \cap B$	$x \in A \cup B$
$x \notin A$	$x \notin B$	$x \in A^c$	$x \notin A \cap B$	$x \notin A \cup B$

The first two columns of Table 1.a give possible truth values for the statements $\mathbb{A}$ and $\mathbb{B}$. The last three columns give the truth values for $\neg \mathbb{A}$, $\mathbb{A} \wedge \mathbb{B}$, and $\mathbb{A} \vee \mathbb{B}$ as a function of the truth values of $\mathbb{A}$ and $\mathbb{B}$. The first two columns of Table 1.b give the corresponding membership properties of an element x, and the last three columns give the corresponding membership properties of x in the sets A^c, $A \cap B$, and $A \cup B$.

Consider the second rows of both tables, the row where $\mathbb{A}$ is true and $\mathbb{B}$ is false. This corresponds to discussing an x with the properties that it belongs to A and does not belong to B. The statement "not $\mathbb{A}$," that is, $\neg \mathbb{A}$, is false, which corresponds to x not belonging to A^c, $x \notin A^c$. The statement "$\mathbb{A}$ and $\mathbb{B}$," that is, "$\mathbb{A} \wedge \mathbb{B}$," is also false. This is sensible: since $\mathbb{B}$ is false, it is not the case that both $\mathbb{A}$ and $\mathbb{B}$ are true. This corresponds to x not being in the intersection of A and B, that is, $x \notin A \cap B$.

The statement "$\mathbb{A}$ or $\mathbb{B}$," that is, "$\mathbb{A} \vee \mathbb{B}$," is true. This is sensible: since $\mathbb{A}$ is true, it is the case that at least one of $\mathbb{A}$ and $\mathbb{B}$ is true, corresponding to x being in the union of A and B.

It is important to note that we use the word "or" in its nonexclusive sense. When we describe someone as "tall or red-headed," we mean to allow tall red-headed people. We do not mean "or" in the exclusive sense that the person is either tall or red-headed but not both. One sees this by considering the last columns in the two tables, the ones with the patterns $TTTF$ and $\in\in\in\notin$. "$\mathbb{A}$ or $\mathbb{B}$" is true as long as at least one of $\mathbb{A}$ and $\mathbb{B}$ is true, and we do not exclude the possibility that both are true. The exclusive "or" is defined by $(\mathbb{A} \vee \mathbb{B}) \wedge (\neg(\mathbb{A} \wedge \mathbb{B}))$, which has the truth pattern $FTTF$. In terms of sets, the exclusive "or" is $(A \cup B) \cap (A \cap B)^c$, which has the corresponding membership pattern $\notin\in\in\notin$.

1.2.b Implies/Equivalence as Subset/Equality

Two of the remaining four ways of constructing new statements are: "$\mathbb{A} \Rightarrow \mathbb{B}$," which means "$\mathbb{A}$ implies $\mathbb{B}$" and "$\mathbb{A} \Leftrightarrow \mathbb{B}$," which means "$\mathbb{A}$ is equivalent to $\mathbb{B}$." In terms of sets, these are "$A \subset B$" and "$A = B$." These are statements about relations between subsets of X.

Indicator functions are a very useful way to talk about the relations between subsets. For each $x \in X$ and $A \subset X$, define the **indicator of the set** A by

$$1_A(x) := \begin{cases} 1 & \text{if } x \in A, \\ 0 & \text{if } x \notin A. \end{cases} \tag{1.1}$$

Remember, a proposition, $\mathbb{A}$, is a statement about elements $x \in X$ that can be either true or false. When it is true, we write $\mathbb{A}(x)$. The corresponding set A is $\{x \in X : \mathbb{A}(x)\}$. The indicator of A takes on the value 1 for exactly those x for which $\mathbb{A}$ is true and takes on the value 0 for those x for which $\mathbb{A}$ is false.

Indicator functions are ordered pointwise; that is, $1_A \leq 1_B$ when $1_A(x) \leq 1_B(x)$ for every point x in the set X. Saying "$1_A \leq 1_B$" is the same as saying that "$A \subset B$." It is easy to give sets A and B that satisfy neither $A \subset B$ nor $B \subset A$. Therefore, unlike pairs of numbers r and s, for which it is always true that either $r \leq s$ or $s \leq r$, pairs of indicator functions may not be ranked by "$\leq$."

Example 1.2.1 *If X is the three-point set $\{a, b, c\}$, $A = \{a, b\}$, $B = \{b, c\}$, and $C = \{c\}$, then $1_A \leq 1_X$, $1_B \leq 1_X$, $1_C \leq 1_B$, $\neg(1_A \leq 1_B)$, and $\neg(1_B \leq 1_A)$.*

Proving statements of the form $\mathbb{A} \Rightarrow \mathbb{B}$ and $\mathbb{A} \Leftrightarrow \mathbb{B}$ is the essential part of mathematical reasoning. For the first, we take the truth of $\mathbb{A}$ as given and then establish logically that the truth of $\mathbb{B}$ follows. For the second, we take the additional step of taking the truth of $\mathbb{B}$ as given and then establish logically that the truth of $\mathbb{A}$ follows. In terms of sets, for proving the first, we take a point, x, assume only that $x \in A$, and establish that this implies that $x \in B$, thus proving that $A \subset B$. For proving the second, we take the additional step of taking a point, x, assume only that $x \in B$, and establish that this implies that $x \in A$. Here is the truth table for $\Rightarrow$ and $\Leftrightarrow$, both for statements and for indicator functions.

<table>
<tr><td colspan="3">Table 1.c</td></tr>
</table>

$\mathbb{A}$ $\mathbb{B}$	$\mathbb{A} \Rightarrow \mathbb{B}$	$\mathbb{A} \Leftrightarrow \mathbb{B}$
T T	T	T
T F	F	F
F T	T	F
F F	T	T

Table 1.d

$x \in A$ $x \in B$	$1_A(x) \le 1_B(x)$	$1_A(x) = 1_B(x)$
$x \in A$ $x \in B$	T	T
$x \in A$ $x \notin B$	F	F
$x \notin A$ $x \in B$	T	F
$x \notin A$ $x \notin B$	T	T

1.2.c The Empty Set and Vacuously True Statements

We now come to the idea of something that is vacuously true, and a substantial proportion of people find this idea tricky or annoying, or both. The idea that we are after is that starting from false premises, one can establish anything. In Table 1.c, if $\mathbb{A}$ is false, then the statement $\mathbb{A} \Rightarrow \mathbb{B}$ is true, whether $\mathbb{B}$ is true or false.

A statement that is false for all $x \in X$ corresponds to having an indicator function with the property that for all $x \in X$, $1_A(x) = 0$. In terms of sets, the notation for this is $A = \emptyset$, where we read "$\emptyset$" as the **empty set**, that is, the vacuous set, the one that contains no elements. No matter what the set B is, if $A = \emptyset$, then $1_A(x) \le 1_B(x)$ for all $x \in X$.

Definition 1.2.2   *The statement* $\mathbb{A} \Rightarrow \mathbb{B}$ *is **vacuously true** if* $A = \emptyset$.

This definition follows the convention that we use throughout: we show the term or terms being defined in boldface type.

In terms of sets, this is the observation that for all B, $\emptyset \subset B$, that is, that every element of $\emptyset$ belongs to B. What many people find distasteful is that "every element of $\emptyset$ belongs to B" suggests that there is an element of $\emptyset$, and since there is no such element, the statement feels wrong to them. There is nothing to be done except to get over the feeling.

1.2.d Indicators and Ands/Ors/Nots

Indicator functions can also be used to capture ands, ors, and nots. Often this makes proofs simpler.

The pointwise minimum of a pair of indicator functions, 1_A and 1_B, is written as "$1_A \wedge 1_B$," and is defined by $(1_A \wedge 1_B)(x) = \min\{1_A(x), 1_B(x)\}$. Now, $1_A(x)$ and $1_B(x)$ are equal either to 0 or to 1. Since the minimum of 1 and 1 is 1, the minimum of 0 and 1 is 0, and the minimum of 0 and 0 is 0, $1_{A \cap B} = 1_A \wedge 1_B$. This means that the indicator associated with the statement "$\mathbb{A} \wedge \mathbb{B}$" is $1_A \wedge 1_B$. By checking cases, we note that for all $x \in X$, $(1_A \wedge 1_B)(x) = 1_A(x) \cdot 1_B(x)$. As a result, $1_A \wedge 1_B$ is often written as $1_A \cdot 1_B$.

In a similar way, the pointwise maximum of a pair of indicator functions, 1_A and 1_B, is written as "$1_A \vee 1_B$" and defined by $(1_A \vee 1_B)(x) = \max\{1_A(x), 1_B(x)\}$. Here, $1_{A \cup B} = 1_A \vee 1_B$, and the indicator associated with the statement "$\mathbb{A} \vee \mathbb{B}$" is $1_A \vee 1_B$. Basic properties of numbers say that for all x, $(1_A \vee 1_B)(x) = 1_A(x) + 1_B(x) - 1_A(x) \cdot 1_B(x)$, so $1_A \vee 1_B$ could be defined as $1_A + 1_B - 1_A \cdot 1_B$.

For complements, we define "1" to be the indicator of X, that is, the function that is equal to 1 everywhere on X, and we note that $1_{A^c} = 1 - 1_A$.

1.3 ◆ Proofs, a First Look

Let us define a few terms. A *theorem* or *proposition* is a statement that we prove to be true. A *lemma* is a theorem that we use to prove another theorem. This seems to indicate that lemmas are less important than theorems. However, some lemmas are used to prove many theorems, and by this measure are more important than any of the theorems we use them for.

A *corollary* is a theorem whose proof is (supposed to) follow directly from the previous theorem. A *definition* is a statement that is true by interpreting one of its terms in such a way as to make the statement true. An *axiom* or *assumption* is a statement that is taken to be true without proof. A *tautology* is a statement that is true without assumptions (e.g., $x = x$). A *contradiction* is a statement that cannot be true (e.g., $\mathbb{A}$ is true and $\mathbb{A}$ is false).

We now have the tools to prove the validity of the basic forms of arguments that are used throughout economic and statistical theory. For example, we show how to use a truth table to prove what is called a distributive law. It is analogous to the distributive law you learned in elementary school for any numbers a, b, c, $a * (b + c) = (a * b) + (a * c)$, except that we replace $*$'s by $\vee$'s and $+$'s by $\wedge$'s.

Theorem 1.3.1 $(\mathbb{A} \vee (\mathbb{B} \wedge \mathbb{C})) \Leftrightarrow ((\mathbb{A} \vee \mathbb{B}) \wedge (\mathbb{A} \vee \mathbb{C}))$.

We give two different proofs of this result, both in the format we use throughout: the beginning of the arguments are marked by "***Proof***," the end of the arguments by "∎."

Proof. From Table 1.c, for any statements $\mathbb{D}$ and $\mathbb{E}$, establishing $\mathbb{D} \Leftrightarrow \mathbb{E}$ involves showing that the statements are either true together or false together. In this case, $\mathbb{D}$ is the constructed statement $(\mathbb{A} \vee (\mathbb{B} \wedge \mathbb{C}))$ and $\mathbb{E}$ is the constructed statement $((\mathbb{A} \vee \mathbb{B}) \wedge (\mathbb{A} \vee \mathbb{C}))$. The truth or falsity of these statements depends on the truth or falsity of the statements $\mathbb{A}$, $\mathbb{B}$, and $\mathbb{C}$. The left three columns of the following exhaustively list all of the possibilities.

Table 1.e

$\mathbb{A}$	$\mathbb{B}$	$\mathbb{C}$	$\mathbb{B} \wedge \mathbb{C}$	$\mathbb{A} \vee (\mathbb{B} \wedge \mathbb{C})$	$\mathbb{A} \vee \mathbb{B}$	$\mathbb{A} \vee \mathbb{C}$	$(\mathbb{A} \vee \mathbb{B}) \wedge (\mathbb{A} \vee \mathbb{C})$
T	*T*	*T*	*T*	*T*	*T*	*T*	*T*
T	*T*	*F*	*F*	*T*	*T*	*T*	*T*
T	*F*	*T*	*F*	*T*	*T*	*T*	*T*
T	*F*	*F*	*F*	*T*	*T*	*T*	*T*
F	*T*	*T*	*T*	*T*	*T*	*T*	*T*
F	*T*	*F*	*F*	*F*	*T*	*F*	*F*
F	*F*	*T*	*F*	*F*	*F*	*T*	*F*
F	*F*	*F*	*F*	*F*	*F*	*F*	*F*

The truth values in the fourth column, $\mathbb{B} \wedge \mathbb{C}$, are formed using the rules for $\wedge$ and the truth values of $\mathbb{B}$ and $\mathbb{C}$. The truth values in the next column, $\mathbb{A} \vee (\mathbb{B} \wedge \mathbb{C})$, are formed using the rules for $\vee$ and the truth values of the $\mathbb{A}$ column and the just-derived truth values of the $\mathbb{B} \wedge \mathbb{C}$ column. The truth values in the next three columns are derived analogously. Since the truth values in the column $\mathbb{A} \vee (\mathbb{B} \wedge \mathbb{C})$ match those in the column $(\mathbb{A} \vee \mathbb{B}) \wedge (\mathbb{A} \vee \mathbb{C})$, the two statements are equivalent.

■

Another Proof of Theorem 1.3.1.   In terms of indicator functions,

$$1_{A \cup (B \cap C)} = 1_A \cdot (1_B + 1_C - 1_B \cdot 1_C)$$
$$= 1_A \cdot 1_B + 1_A \cdot 1_C - 1_A \cdot 1_B \cdot 1_C, \tag{1.2}$$

while

$$1_{(A \cap B) \cup (A \cap C)} = 1_A \cdot 1_B + 1_A \cdot 1_C + (1_A \cdot 1_B) \cdot (1_B \cdot 1_C). \tag{1.3}$$

Since indicators take only the values 0 and 1, for any set, for example, B, $1_B \cdot 1_B = 1_B$. Therefore $(1_A \cdot 1_B) \cdot (1_B \cdot 1_C) = 1_A \cdot 1_{B} \cdot 1_C$.   ■

The following contains the commutative, associative, and distributive laws. To prove them, one can simply generate the appropriate truth table.

Theorem 1.3.2   *Let $\mathbb{A}$, $\mathbb{B}$, and $\mathbb{C}$ be any statements. Then*

1. *commutativity holds, $(\mathbb{A} \vee \mathbb{B}) \Leftrightarrow (\mathbb{B} \vee \mathbb{A})$ and $(\mathbb{A} \wedge \mathbb{B}) \Leftrightarrow (\mathbb{B} \wedge \mathbb{A})$,*
2. *associativity holds, $((\mathbb{A} \wedge \mathbb{B}) \wedge \mathbb{C}) \Leftrightarrow (\mathbb{A} \wedge (\mathbb{B} \wedge \mathbb{C}))$, $((\mathbb{A} \vee \mathbb{B}) \vee \mathbb{C}) \Leftrightarrow (\mathbb{A} \vee (\mathbb{B} \vee \mathbb{C}))$, and*
3. *the distributive laws hold, $(\mathbb{A} \wedge (\mathbb{B} \vee \mathbb{C})) \Leftrightarrow ((\mathbb{A} \wedge \mathbb{B}) \vee (\mathbb{A} \wedge \mathbb{C}))$, $(\mathbb{A} \vee (\mathbb{B} \wedge \mathbb{C})) \Leftrightarrow ((\mathbb{A} \vee \mathbb{B}) \wedge (\mathbb{A} \vee \mathbb{C}))$.*

Exercise 1.3.3   Restate Theorem 1.3.2 in terms of sets and in terms of indicator functions. Then complete the proof of Theorem 1.3.2 both by generating the appropriate truth tables and by using indicator functions.

We now prove two results, Lemma 1.3.4 and Theorem 1.3.6, that form the basis for the methods of logical reasoning we pursue in this book. The following is used so many times that it is at least as important as a theorem.

Lemma 1.3.4   *$\mathbb{A}$ implies $\mathbb{B}$ iff $\mathbb{A}$ is false or $\mathbb{B}$ is true,*

$$(\mathbb{A} \Rightarrow \mathbb{B}) \Leftrightarrow ((\neg \mathbb{A}) \vee \mathbb{B}), \tag{1.4}$$

and a double negative makes a positive,

$$\neg(\neg \mathbb{A}) \Leftrightarrow \mathbb{A}. \tag{1.5}$$

Proof.   In terms of indicator functions, (1.4) is $1_A(x) \le 1_B(x)$ iff $1_A(x) = 0$ or $1_B(x) = 1$, which is true because $1_A(x)$ and $1_B(x)$ can only take on the values 0 and 1. (1.5) is simpler; it says that $1 - (1 - 1_A) = 1_A$.   ■

An alternative method of proving (1.4) in Lemma 1.3.4 is to construct the truth table as follows.

Table 1.f

A	B	A $\Rightarrow$ B	¬A	(¬A $\vee$ B)
T	T	T	F	T
T	F	F	F	F
F	T	T	T	T
F	T	T	T	T

Since the third and the fifth columns are identical, we have $(A \Rightarrow B) \Leftrightarrow ((\neg A) \vee B)$.

Exercise 1.3.5 Complete the proof of Lemma 1.3.4 by generating the appropriate truth tables and restate (1.5) in terms of sets.

The next result, Theorem 1.3.6, forms the basis for most of the logical reasoning in this book. The first (direct) approach (1.6) is the *syllogism*, which says that "if A is true and 'A implies B' is true, then B is true." The second (indirect) approach (1.7) is the *contradiction*, which says in words that "if not A leads to a false statement of the form B and not B, then A is true. That is, one way to prove A is to hypothesize ¬A and show that this leads to a contradiction. Another (indirect) approach (1.8) is the *contrapositive*, which says that "A implies B is the same as "whenever B is false, A is false." In terms of sets, this last is "$[A \subset B] \Leftrightarrow [B^c \subset A^c]$."

Theorem 1.3.6 *If A is true and A implies B, then B is true,*

$$(A \wedge (A \Rightarrow B)) \Rightarrow B. \tag{1.6}$$

If A being false implies a contradiction, then A is true,

$$((\neg A) \Rightarrow (B \wedge (\neg B))) \Rightarrow A. \tag{1.7}$$

A implies B iff whenever B is false, A is false,

$$(A \Rightarrow B) \Leftrightarrow ((\neg B) \Rightarrow (\neg A)). \tag{1.8}$$

Proof. In the case of (1.6), $1_A(x) = 1$ and $1_A(x) \leq 1_B(x)$ imply that $1_B(x) = 1$.
In the case of (1.7), note that $(1_B(x) \wedge (1 - 1_B(x))) = 0$ for all x. Hence $(1 - 1_A(x)) \leq (1_B(x) \vee (1 - 1_B(x)))$ implies that $1 - 1_A(x) = 0$, that is, $1_A(x) = 1$.
In the case of (1.8), $1_A(x) \leq 1_B(x)$ iff $-1_B(x) \leq -1_A(x)$ iff $1 - 1_B(x) \leq 1 - 1_A(x)$. ∎

Exercise 1.3.7 Give an alternative proof of Theorem 1.3.6 using truth tables and another using the distributive laws and Lemma 1.3.4.

As we are mostly interested in proving statements of the form "A implies B," it is worth belaboring the notion of the contrapositive in (1.8). "A implies B" is the same as "whenever A is true, we know that B is true." There is one and only one way that this last statement could be false—if there is an $x \in X$ such that $\neg\mathbb{B}(x)$ while $\mathbb{A}(x)$. Therefore, "$\mathbb{A}$ implies $\mathbb{B}$" is equivalent to $(\neg\mathbb{B}) \Rightarrow (\neg\mathbb{A})$.

In terms of sets, we are saying that $A \subset B$ is equivalent to $B^c \subset A^c$. Often it is easier to pick a point y, assume only that y does *not* belong to B, and establish that this implies that y does *not* belong to A.

Example 1.3.8   *Let X be the set of humans, let $\mathbb{A}(x)$ be the statement "x has a Ph.D. in economics," and let $\mathbb{B}(x)$ be the statement "x is literate in at least one language." Showing that $\mathbb{A} \Rightarrow \mathbb{B}$ is the same as showing that there are no completely illiterate economics Ph.D.s. Which method of proving the statement one would want to use depends on whether or not it is easier to check all the economics Ph.D.s in X for literacy or to check all the illiterates in X for Ph.D.s in economics.*

A final note: the contrapositive of "$\mathbb{A} \Rightarrow \mathbb{B}$" is "$(\neg\mathbb{B}) \Rightarrow (\neg\mathbb{A})$." This is not the same as the *converse* of "$\mathbb{A} \Rightarrow \mathbb{B}$," which is "$\mathbb{B} \Rightarrow \mathbb{A}$."

1.4 ◆ Logical Quantifiers

The last two of our seven ways to construct statements use the two *quantifiers*, "$\exists$," read as "there exists," and "$\forall$," read as "for all." More specifically, "$(\exists x \in A)[\mathbb{B}(x)]$" means "there exists an x in the set A such that $\mathbb{B}(x)$" and "$(\forall x \in A)[\mathbb{B}(x)]$" means "for all x in the set A, $\mathbb{B}(x)$." Our discussion of indicator functions has already used these quantifiers; for example, $1_A \leq 1_B$ was defined as $(\forall x \in X)[1_A(x) \leq 1_B(x)]$. We now formalize the ways in which we use the quantifiers.

Quantifiers should be understood as statements about the relations between sets, and here the empty set, $\emptyset$, is again useful. In terms of sets, "$(\exists x \in A)[\mathbb{B}(x)]$" is the statement $(A \cap B) \neq \emptyset$, while "$(\forall x \in A)[\mathbb{B}(x)]$" is the statement $A \subset B$.

Notation Alert 1.4.A   *Following common usage, when the set A is supposed to be clear from context, we often write $(\exists x)[\mathbb{B}(x)]$ for $(\exists x \in A)[\mathbb{B}(x)]$. If A is not in fact clear from context, we run the risk of leaving the intended set A undefined.*

The two crucial properties of quantifiers are contained in the following, which gives the relationship among quantifiers, negations, and complements.

Theorem 1.4.1   *There is no x in A such that $\mathbb{B}(x)$ iff for all x in A, it is not the case that $\mathbb{B}(x)$,*

$$\neg(\exists x \in A)[\mathbb{B}(x)] \Leftrightarrow (\forall x \in A)[\neg\mathbb{B}(x)], \tag{1.9}$$

and it is not the case that for all x in A we have $\mathbb{B}(x)$ iff there is some x in A for which $\mathbb{B}(x)$ fails,

$$\neg(\forall x \in A)[\mathbb{B}(x)] \Leftrightarrow (\exists x \in A)[\neg\mathbb{B}(x)]. \tag{1.10}$$

Proof.   In terms of sets, (1.9) is $[A \cap B = \emptyset] \Leftrightarrow [A \subset B^c]$. In terms of indicators, letting 0 be the function identically equal to 0, it is $1_A \cdot 1_B = 0$ iff $1_A \leq (1 - 1_B)$.

In terms of sets, (1.10) is $\neg[A \subset B] \Leftrightarrow [A \cap B^c \neq \emptyset]$. In terms of indicators, the left-hand side of (1.10) is $\neg[1_A \leq 1_B]$, which is true iff for some x in X, $1_A(x) > 1_B(x)$. This happens iff for some x, $1_A(x) = 1$ and $1_B(x) = 0$, that is, iff for some x, $1_A(x) \cdot (1 - 1_B(x)) = 1$, which is the right-hand side of (1.10). ■

The second tautology in Theorem 1.4.1 is important since it illustrates the concept of a *counterexample*. In particular, (1.10) states: "If it is not true that $\mathbb{B}(x)$ for all x in A, then there must exist a counterexample (i.e., an x satisfying $\neg\mathbb{B}(x)$), and vice versa." Counterexamples are important tools, since knowing that $x \in A$ and $x \in B$ for hundreds and hundreds of x's does not prove that $A \subset B$, but a single counterexample shows that $\neg[A \subset B]$.

Often, one can profitably apply the rules in (1.9) and (1.10) time after time. The following anticipates material from the topics of convergence and continuity that we cover extensively later.

Example 1.4.2 *A sequence of numbers is a list $(x_1, x_2, x_3, \ldots)$, one x_n for each counting number $n \in \mathbb{N} = \{1, 2, 3, \ldots\}$, where the "$\ldots$" indicates "keep going in this fashion." Let $\mathbb{R}_{++}$ denote the set of strictly positive numbers; for any $N \in \mathbb{N}$, let $N_\geq$ be the set of integers greater than or equal to N; and let $\mathbb{A}(\epsilon, x)$ be the statement that $|x| < \epsilon$. We say that a sequence converges to 0 if*

$$(\forall \epsilon \in \mathbb{R}_{++})(\exists N \in \mathbb{N})(\forall n \in N_\geq)[\mathbb{A}(\epsilon, x_n)],$$

which is more much conveniently, and just as precisely, written as

$$(\forall \epsilon > 0)(\exists N \in \mathbb{N})(\forall n \geq N)[|x_n| < \epsilon].$$

This captures the idea that the numbers in the sequence become and stay arbitrarily small as we move further and further out in the sequence. A verbal shorthand for this is that "for all positive ϵ (no matter how small), $|x_n|$ is smaller than ϵ for large n."

*Applying (1.9) and (1.10) repeatedly shows that the following are all equivalent to the sequence **not** converging to 0:*

$$\neg(\forall \epsilon > 0)(\exists N \in \mathbb{N})(\forall n \geq N)[|x_n| < \epsilon],$$

$$(\exists \epsilon > 0)\neg(\exists N \in \mathbb{N})(\forall n \geq N)[|x_n| < \epsilon],$$

$$(\exists \epsilon > 0)(\forall N \in \mathbb{N})\neg(\forall n \geq N)[|x_n| < \epsilon],$$

$$(\exists \epsilon > 0)(\forall N \in \mathbb{N})(\exists n \geq N)\neg[|x_n| < \epsilon], \text{ and}$$

$$(\exists \epsilon > 0)(\forall N \in \mathbb{N})(\exists n \geq N)[|x_n| \geq \epsilon].$$

Thus, the statement "a sequence fails to converge to 0" is equivalent to "for some strictly positive ϵ, for all N (no matter how large), there is an even larger n such that $|x_n| \geq \epsilon$."

One should also note that the commutative and distributive laws we found with "$\vee$" and "$\wedge$" in them may break down with quantifiers. While

$$(\exists x)[\mathbb{A}(x) \vee \mathbb{B}(x)] \Leftrightarrow (\exists x)[\mathbb{A}(x)] \vee (\exists x)[\mathbb{B}(x)], \tag{1.11}$$

$$(\exists x)[\mathbb{A}(x) \wedge \mathbb{B}(x)] \Rightarrow (\exists x)[\mathbb{A}(x)] \wedge (\exists x)[\mathbb{B}(x)]. \tag{1.12}$$

Example 1.4.3   *To see why* (1.12) *cannot hold as an "if and only if" statement, suppose x is the set of countries in the world,* $\mathbb{A}(x)$ *is the property that x has a gross domestic product strictly above average, and* $\mathbb{B}(x)$ *is the property that x has a gross domestic product strictly below average. There will be at least one country above the mean and at least one country below the mean. That is,* $(\exists x)[\mathbb{A}(x)] \wedge (\exists x)[\mathbb{B}(x)]$ *is true, but clearly there cannot be a country that is both above and below the mean,* $\neg(\exists x)[\mathbb{A}(x) \wedge \mathbb{B}(x)]$.

In terms of sets, (1.11) can be rewritten as $[(A \cup B) \neq \emptyset] \Leftrightarrow [(A \neq \emptyset) \vee (B \neq \emptyset)]$. The set form of (1.12) is $[A \cap B \neq \emptyset] \Rightarrow [(A \neq \emptyset) \wedge (B \neq \emptyset)]$. Hopefully this formulation makes the reason we do not have an "if and only if" relation in (1.12) even clearer.

We can also make increasingly complex statements by adding more variables. For example, statements of the form $\mathbb{A}(x, y)$ as x and y both vary across X. One can always view this as a statement about a pair (x, y) and change X to contain pairs, but this may not mitigate the additional complexity.

Example 1.4.4   *When X is the set of numbers and* $\mathbb{A}(x, y)$ *states that "y that is larger than x," where x and y are numbers, the statement* $(\forall x)(\exists y)(x < y)$ *says "for every x there is a y that is larger than x." The statement* $(\exists y)(\forall x)(x < y)$ *says "there is a y that is larger than every x." The former statement is true, but the latter is false.*

1.5 ◆ Taxonomy of Proofs

We now discuss broadly the methodology of proofs you will frequently encounter in economics. The most intuitive is the *direct proof* in the form of "$\mathbb{A} \Rightarrow \mathbb{B}$," discussed in (1.6). The work is to fill in the intermediate steps so that $\mathbb{A} \Rightarrow \mathbb{A}_1$, $\mathbb{A}_1 \Rightarrow \mathbb{A}_2$, and $\ldots \mathbb{A}_n \Rightarrow \mathbb{B}$ are all tautologies. In terms of sets, this involves constructing n sets $A_1, \ldots, A_n$ such that $A \subset A_1 \subset \cdots \subset A_n \subset B$.

Notation Alert 1.5.A   *The "$\ldots$" indicates* A_2 *through* A_{n-1} *in the first list. The "$\cdots$" indicates the same sets in the second list, but we also mean to indicate that the subset relation holds for all the intermediate pairs.*

In some cases, the sets $A_1, \ldots, A_n$ arise from splitting $\mathbb{B}$ into cases. If we find $\mathbb{B}_1, \mathbb{B}_2$ such that $[\mathbb{B}_1 \vee \mathbb{B}_2] \Rightarrow \mathbb{B}$ and can show that $\mathbb{A} \Rightarrow [\mathbb{B}_1 \vee \mathbb{B}_2]$, then we are done.

In other cases it may be simpler to split $\mathbb{A}$ into cases. That is, sometimes it is easier to find $\mathbb{A}_1$ and $\mathbb{A}_2$ for which $\mathbb{A} \Rightarrow [\mathbb{A}^1$ and $\mathbb{A}^2]$ and then to show that $[\mathbb{A}_1 \Rightarrow \mathbb{B}] \vee [\mathbb{A}_2 \Rightarrow \mathbb{B}]$.

Another direct method of proof, called *induction*, works only for the natural numbers $\mathbb{N} = \{1, 2, 3, \ldots\}$. Suppose we wish to show that $(\forall n \in \mathbb{N}) \, \mathbb{A}(n)$ is true. This is equivalent to proving $\mathbb{A}(1) \wedge (\forall n \in \mathbb{N}) (\mathbb{A}(n) \Rightarrow \mathbb{A}(n + 1))$. This works since $\mathbb{A}(1)$ is true and $\mathbb{A}(1) \Rightarrow \mathbb{A}(2)$ and $\mathbb{A}(2) \Rightarrow \mathbb{A}(3)$ and so on. In Chapter 2 we show why induction works.

Proofs by contradiction are also known as indirect proofs. They may, initially, seem less natural than direct proofs. To help you on your way to becoming fluent in indirect proofs, we now give the exceedingly simple indirect proof of the first fundamental theorem of welfare economics. This is perhaps one of the most important things you will learn in all of economics, and finding a direct proof seems rather difficult. Again, we presume some familiarity with general equilibrium models.

Definition 1.5.1 *An **exchange economy model** is a triple, $\mathcal{E} = (I, \mathbf{y}_i, \succ_i)$, where I is a finite set (meant to represent the people in the economy), $\mathbf{y}_i \in \mathbb{R}_+^\ell$ is i **endowment** of the ℓ goods that are available in the model of the economy, and $\succ_i$ is i's **preference relation** over his or her own consumption.*

There are two things to note here: first, we did not say what a preference relation is and we do it, in detail, in Chapter 2; and second, we assumed that preferences are defined only over own consumption, which is a very strong assumption, and we discuss it further.

Definition 1.5.2 *An **allocation** is a list of vectors, written $(\mathbf{x}_i)_{i \in I}$, where $\mathbf{x}_i \in \mathbb{R}_+^\ell$ for each $i \in I$. An allocation $(\mathbf{x}_i)_{i \in I}$ is **feasible for** $\mathcal{E}$ if for each good k, $k = 1, \ldots, \ell$,*

$$\sum_i x_{i,k} \leq \sum_i y_{i,k}, \tag{1.13}$$

where the summation is over all of the individuals in the economy.

Definition 1.5.3 *A feasible allocation $(\mathbf{x}_i)_{i \in I}$ is **Pareto efficient** if there is no feasible allocation $(\mathbf{x}_i')_{i \in I}$ such that all agents prefer $\mathbf{x}_i'$ to $\mathbf{x}_i$.*

Definition 1.5.4 *A **price** $\mathbf{p}$ is a nonzero vector in $\mathbb{R}_+^\ell$. An allocation-price pair $((\mathbf{x}_i)_{i \in I}, \mathbf{p})$ is a **Walrasian equilibrium for** $\mathcal{E}$ if it is feasible, and if $\mathbf{x}_i'$ is preferred by i to $\mathbf{x}_i$, then i cannot afford $\mathbf{x}_i'$, that is,*

$$\sum_k p_k x_{i,k}' > \sum_k p_k y_{i,k}. \tag{1.14}$$

Theorem 1.5.5 (First Fundamental Theorem of Welfare Economics)
If $((\mathbf{x}_i)_{i \in I}, \mathbf{p})$ is a Walrasian equilibrium, then $(\mathbf{x}_i)_{i \in I}$ is Pareto efficient.

Let $A = \{(\mathbf{x}_i)_{i \in I} : (\exists \mathbf{p})[((\mathbf{x}_i)_{i \in I}, \mathbf{p}) \text{ is a Walrasian equilibrium}]\}$, and $B = \{(\mathbf{x}_i)_{i \in I} : (\mathbf{x}_i)_{i \in I} \text{ is Pareto efficient}\}$. In terms of sets, Theorem 1.5.5 states that $A \subset B$.

Proof. $\mathbb{A}$ is the statement "$((\mathbf{x}_i)_{i \in I}, \mathbf{p})$ is a Walrasian equilibrium" and $\mathbb{B}$ is the statement "$(\mathbf{x}_i)_{i \in I}$ is Pareto efficient." A proof by contradiction assumes $\mathbb{A} \wedge \neg \mathbb{B}$, and shows that this leads to a contradiction, $\mathbb{C} \wedge \neg \mathbb{C}$. In words, suppose that $((\mathbf{x}_i)_{i \in I}, \mathbf{p})$ is a Walrasian equilibrium but that $(\mathbf{x}_i)_{i \in I}$ is not Pareto efficient. We have to show that this leads to a contradiction.

By the definition of Pareto efficiency, failing to be Pareto efficient means that there exists a feasible allocation, $(\mathbf{x}_i')_{i \in I}$, that has the property that all agents prefer

$\mathbf{x}'_i$ to $\mathbf{x}_i$. By the definition of Walrasian equlibrium, we can sum (1.14) across all individuals to obtain

$$\sum_i \left(\sum_k p_k x'_{i,k} \right) > \sum_i \left(\sum_k p_k y_{i,k} \right). \tag{1.15}$$

Rearranging the summations in (1.15) gives

$$\sum_k \sum_i p_k x'_{i,k} > \sum_k \sum_i p_k y_{i,k}, \quad \text{equivalently}$$

$$\sum_k p_k \left(\sum_i x'_{i,k} \right) > \sum_k p_k \left(\sum_i y_{i,k} \right). \tag{1.16}$$

Since $(\mathbf{x}'_i)_{i \in I}$ is a feasible allocation, multiplying each term in (1.13) by the nonnegative number p_k and then summing yields

$$\sum_k p_k \left(\sum_i x'_{i,k} \right) \le \sum_k p_k \left(\sum_i p_k y_{i,k} \right). \tag{1.17}$$

Let r be the number $\sum_k p_k \left(\sum_i y_{i,k} \right)$ and let s be the number $\sum_k p_k \left(\sum_i x'_{i,k} \right)$. Equation (1.16) is the statement, $\mathbb{C}$, that $s > r$, whereas (1.17) is the statement $\neg\mathbb{C}$ that $s \le r$. We have derived the contradiction $[\mathbb{C} \wedge \neg\mathbb{C}]$, which we know to be false, from the supposition $[\mathbb{A} \wedge \neg\mathbb{B}]$. From this, we conclude that $[\mathbb{A} \Rightarrow \mathbb{B}]$. ∎

As one becomes more accustomed to the patterns of logical arguments, details of the arguments are suppressed. Here is a shorthand, three-sentence version of the foregoing proof.

Proof. If $(\mathbf{x}_i)_{i \in I}$ is not Pareto efficient, $\exists (\mathbf{x}'_i)_{i \in I}$ feasible and unanimously preferred to $(\mathbf{x}_i)_{i \in I}$. Summing (1.14) across individuals yields $\sum_k \sum_i p_k x'_{i,k} > \sum_k \sum_i p_k y_{i,k}$. Since $(\mathbf{x}'_i)_{i \in I}$ is feasible, summing (1.13) over goods, we have $\sum_k \sum_i p_k x'_{i,k} \le \sum_k \sum_i p_k y_{i,k}$. ∎

Just as "$7x^2 + 9x < 3$" is a shorter and clearer version of "seven times the square of a number plus nine times that number adds to a number less than three," the shortening of proofs is mostly meant to help. It can, however, feel like a diabolically designed code, one meant to obfuscate rather than elucidate.

Some decoding hints:

1. Looking at the statement of Theorem 1.5.5, we see that it ends in "then $(\mathbf{x}_i)_{i \in I}$ is Pareto efficient." Since the shortened proof starts with the sentence "If $(\mathbf{x}_i)_{i \in I}$ is not Pareto efficient," you should conclude that we are offering a proof by contradiction. This means that you should be looking for a conclusion that is always false. Reaching such a falsity completes the proof.

2. Despite what it says, the second sentence in the shortened proof does more than sum (1.14); it rearranges the summation. Your job as a reader is to

look at (1.14) and see that it leads to what is claimed. If the requisite rearrangement is tricky, then it should be given explicitly. Like beauty, trickiness is in the eye of the beholder.

3. The third sentence probably compresses too many steps. Sometimes, this will happen.

Throughout most of Chapter 2, we try to be explicit about the strategy of proof being used. As we get further and further into the book, we shorten proofs more and more. Hopefully, the early practice with proofs will help render our shortenings transparent.

CHAPTER 2

Set Theory

In the foundations of economic theory, one worries about the existence of optima for single-person decision problems and about the existence of simultaneous optima for linked, multiple-person-decision problems. The simultaneous optima are called equilibria. Often more interesting than the study of existence questions is the study of the changes in these optima and equilibria as aspects of the economic environment change, which is called "comparative statics." Since a change in one person's behavior can result in a change in another's optimal choices when the problems are linked, the comparative statics of equilibria will typically be a more complicated undertaking.

The early sections of this chapter cover notation, product spaces, relations, and functions. This is sufficient background for the foundational results in rational choice theory: conditions on preferences that guarantee the existence of optimal choices in finite contexts; representations of the optimal choices as solutions to utility maximization problems; and some elementary comparative statics results.

An introduction to weak orders, partial orders, and lattices provides sufficient background for the basics of monotone comparative statics based on supermodularity. It is also sufficient background for Tarski's fixed-point theorem, the first of the fixed point theorems we cover. Fixed-point theorems are often the tool used to show the existence of equilibria. Tarski's theorem also gives information useful for comparative statics, and we apply it to study the existence and properties of the set of stable matchings.

Whether or not the universe is infinite or finite but very large seems to be unanswerable. However, the mathematics of infinite sets often turns out to be much, much easier than finite mathematics. Imagine trying to study planar geometry under the simplifying assumption that the plane contains 293 million (or so) points. At the end of this chapter we deal with the basic results concerning infinite sets, results that we use extensively in our study of models of prices, quantities, and time, all of which begin in Chapter 3.

2.1 ◆ Some Simple Questions

Simple questions often have very complicated answers. In economics, a simple question with this property is, "What is money?" In mathematics, one can ask, "What is a set?" Intuitively, it seems that such simple questions ought to have answers of comparable simplicity. The existence of book-length treatments of both questions is an indication that these intuitions are wrong.

It would be wonderful if we could *always* associate with a property $\mathbb{A}$ a set $A = \{x \in X : \mathbb{A}(x)\}$ consisting of all objects having property $\mathbb{A}$. Bertrand Russell taught us that we cannot.

Example 2.1.1 (Russell's Paradox)   *Let $\mathbb{A}$ be the property "is a set and does not belong to itself." Suppose there is a set A of all sets with property $\mathbb{A}$. If A belongs to itself, then it does not belong to itself—it is a set and it belongs to the set of sets that do not belong to themselves. But, if A does not belong to itself, then it does.*

Our way around this difficulty is to limit ourselves, ahead of time, to a smaller group of sets and objects, X, that we talk about. This gives a correspondingly smaller notion of membership in that group. To that end, in what follows, we *fix* a given universe (or space) X and consider only sets (or groups) whose elements (or members) are elements of X. This limits the properties that we can talk about. The limits are deep, complicated, and fascinating. They are also irrelevant to essentially everything we do with mathematics in the study of economics because the limits are loose enough to allow everything we use.

In the study of consumer demand behavior, X would have to contain, at a minimum, the positive orthant ($\mathbb{R}^\ell_+$, as a consumption set), the set of preference relations on the positive orthant, and the set of functions from price-income pairs to the positive orthant. Suppose we wish to discuss a result of the form, "The demand functions of all smooth preference relations with indifference curves lying inside the strictly positive orthant are themselves smooth." This means that X has to include subsets of the positive orthant (e.g., the strictly positive orthant), subsets of the preference relations, and subsets of the possible demand functions.

The smaller group of objects that we talk about is called a **superstructure**. Superstructures are formally defined in §2.13 at the end of this chapter. The essential idea is that one starts with a set S. We have to start with some kind of primitive, we agree that S is a set, and we agree that none of the elements of S contains any elements. We then begin an inductive process, adding to S the class of all subsets of S, then the class of all subsets of everything we have so far, and so on and so on. As we will see, this allows us to construct and work with all of the spaces of functions, probabilities, preferences, stochastic process models, dynamic programming problems, equilibrium models, and so on, that we need to study economics. It also keeps us safely out of trouble by avoiding situations like Russell's example and allows us to identify our restricted class of sets with the properties that they have.

2.2 ◆ Notation and Other Basics

As in Chapter 1, we express the notion of membership by "$\in$" so that "$x \in A$" means "x is an element of the set A" and "$x \notin A$" means "x is not an element of A." We usually specify the elements of a set explicitly by saying "A is the set of all x in X having the property $\mathbb{A}$," and write $A = \{x \in X : \mathbb{A}(x)\}$. When the space X is understood, we may abbreviate this as $A = \{x : \mathbb{A}(x)\}$.

Example 2.2.1 *If $\mathbb{A}(\mathbf{x})$ is the property "is affordable at prices $\mathbf{p}$ and income w" and $X = \mathbb{R}_+^\ell$, then the Walrasian budget set, denoted $B(\mathbf{p}, w)$, is defined by $B(\mathbf{p}, w) = \{\mathbf{x} \in X : \mathbb{A}(\mathbf{x})\}$. With more detail about the statement $\mathbb{A}$, this is $B(\mathbf{p}, w) = \{\mathbf{x} \in \mathbb{R}_+^\ell : \mathbf{p} \cdot \mathbf{x} \le w\}$.*

Definition 2.2.2 *For A and B subsets of X, we define:*

1. *$A \cap B$, the **intersection of A and B**, by $A \cap B = \{x \in X : [x \in A] \wedge [x \in B]\}$,*

2. *$A \cup B$, the **union of A and B**, by $A \cup B = \{x \in X : [x \in A] \vee [x \in B]\}$,*

3. *$A \subset B$, A **is a subset of** B, or B **contains** A, if $[x \in A] \Rightarrow [x \in B]$,*

4. *$A = B$, A **is equal to** B, if $[A \subset B] \wedge [B \subset A]$,*

5. *$A \ne B$, A **is not equal to** B, if $\neg[A = B]$,*

6. *$A \subsetneq B$, A **is a proper subset** of B, if $[A \subset B] \wedge [A \ne B]$,*

7. *$A \setminus B$, the **difference between A and B**, by $A \setminus B = \{x \in A : x \notin B\}$,*

8. *$A \triangle B$, the **symmetric difference between A and B**, by $A \triangle B = (A \setminus B) \cup (B \setminus A)$,*

9. *A^c, the **complement of** A, by $A^c = \{x \in X : x \notin A\}$,*

10. *$\emptyset$, the **empty set**, by $\emptyset = X^c$, and*

11. *A and B to be **disjoint** if $A \cap B = \emptyset$.*

These definitions can be visualized using Venn diagrams as in Figure 2.2.2.

Example 2.2.3 *If $X = \{1, 2, \ldots, 10\}$, the counting numbers between 1 and 10, $A = \{$even numbers in $X\}$, $B = \{$odd numbers in $X\}$, $C = \{$powers of 2 in $X\}$, and $D = \{$primes in $X\}$, then $A \cap B = \emptyset$, $A \cap D = \{2\}$, $A \setminus C = \{6, 10\}$, $C \subsetneq A$, $B \ne C$, $C \cup D = \{2, 3, 4, 5, 7, 8\}$, and $B \triangle D = \{2, 9\}$.*

There is a purpose to the notational choices made in defining "$\cap$" using "$\wedge$" and defining "$\cup$" using "$\vee$." Being in $A \cap B$ requires being in $A \wedge$ being in B, being in $A \cup B$ requires being in $A \vee$ being in B. The definitions of unions and intersections can easily be extended to arbitrary collections of sets. Let I be an index set, for example, $I = \mathbb{N} = \{1, 2, 3, \ldots\}$ as in Example 1.4.2 (p. 10), and let A_i, $i \in I$ be subsets of X. Then $\cup_{i \in I} A_i = \{x \in X : (\exists i \in I)[x \in A_i]\}$ and $\cap_{i \in I} A_i = \{x \in X : (\forall i \in I)[x \in A_i]\}$.

We have seen the following commutative, associative, and distributive properties before in Theorem 1.3.2 (p. 7), and they are easily checked using Venn diagrams.

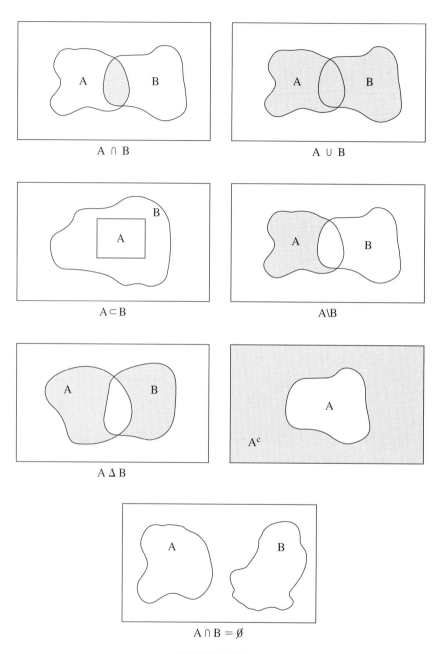

FIGURE 2.2.2

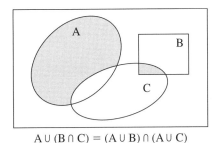
$A \cup (B \cap C) = (A \cup B) \cap (A \cup C)$

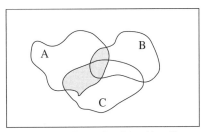
$A \cap (B \cup C) = (A \cap B) \cup (A \cap C)$

FIGURE 2.2.4

Theorem 2.2.4   *For sets A, B, and C,*

1. $A \cap B = B \cap A$, $A \cup B = B \cup A$;
2. $(A \cap B) \cap C = A \cap (B \cap C)$, $(A \cup B) \cup C = A \cup (B \cup C)$; *and*
3. $A \cap (B \cup C) = (A \cap B) \cup (A \cap C)$, $A \cup (B \cap C) = (A \cup B) \cap (A \cup C)$.

Exercise 2.2.5   Prove Theorem 2.2.4 from Theorem 1.3.2. [See Figure 2.2.4. The proof amounts to applying the logical connectives and above definitions: to show $A \cap B = B \cap A$, it is sufficient to note that $x \in A \cap B \Leftrightarrow (x \in A) \wedge (x \in B) \Leftrightarrow (x \in B) \wedge (x \in A) \Leftrightarrow x \in B \cap A$.]

The following properties are used extensively in probability theory and are easily checked in a Venn diagram. [See Figure 2.2.6.]

Theorem 2.2.6 (DeMorgan's Laws)   *If A, B, and C are any sets, then*

1. $A \backslash (B \cup C) = (A \backslash B) \cap (A \backslash C)$, *and*
2. $A \backslash (B \cap C) = (A \backslash B) \cup (A \backslash C)$.

In particular, taking $A = X$, $(B \cup C)^c = B^c \cap C^c$ and $(B \cap C)^c = B^c \cup C^c$.

The last two equalities are "the complement of a union is the intersection of the complements" and "the complement of an intersection is the union of the complements." When we think of B and C as statements, $(B \cup C)^c$ is "not B or C," which is equivalent to, "neither B nor C," which is equivalent to, "not B and not C," and this is $B^c \cap C^c$. In the same way, $(B \cap C)^c$ is "not both B and C," which is equivalent to "either not B or not C," and this is $B^c \cup C^c$.

Proof.   For (1) we show that $A \backslash (B \cup C) \subset (A \backslash B) \cap (A \backslash C)$, and $A \backslash (B \cup C) \supset (A \backslash B) \cap (A \backslash C)$.

($\subset$) Suppose $x \in A \backslash (B \cup C)$. Then $x \in A$ and $x \notin (B \cup C)$. Thus $x \in A$ and $(x \notin B$ and $x \notin C)$. This implies $x \in A \backslash B$ and $x \in A \backslash C$. But this is just $x \in (A \backslash B) \cap (A \backslash C)$.

($\supset$) Suppose $x \in (A \backslash B) \cap (A \backslash C)$. Then $x \in (A \backslash B)$ and $x \in (A \backslash C)$. Thus $x \in A$ and $(x \notin B$ and $x \notin C)$. This implies $x \in A$ and $x \notin (B \cup C)$. But this is just $x \in A \backslash (B \cup C)$.   ∎

Exercise 2.2.7   Finish the proof of Theorem 2.2.6.

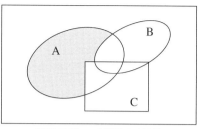

A\(B ∪ C) = (A\B) ∩ (A\C)

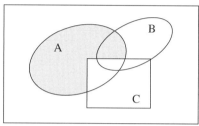

A\(B ∩ C) = (A\B) ∪ (A\C)

FIGURE 2.2.6

Definition 2.2.8 *For A a subset of X, the **power set of** A, denoted $\mathcal{P}(A)$, is the set of all subsets of A. A **collection** or **class of sets** is a subset of $\mathcal{P}(A)$, that is, a set of sets. A **family** is a set of collections.*

Example 2.2.9 *Let $X = \{a, b, c\}$. If $A = \{a, b\}$, $B = \{b\}$, $C = \{b, c\}$, then $\mathcal{P}(X)$, $\mathcal{C} = \{A\}$, $\mathcal{D} = \{A, B\}$, and $\mathcal{E} = \{A, C, \emptyset\}$ are collections, whereas $\mathbb{F} = \{\mathcal{D}, \mathcal{E}\}$ is a family.*

To a great extent, the distinction among sets, collections, and families depends on where one starts the analysis. For example, we define functions as sets of pairs $(x, f(x))$. We are often interested in the properties of different sets of functions. If the possible pairs are the "points," then a set of functions is a family. However, if X is the set of all functions, then a set of functions is just that, a set. We have the set/collection/family hierarchy in place for cases in which we have to distinguish among several levels in the same context. The following anticipates material on probability theory, where we assign probabilities to every set in a **field** of sets.

Example 2.2.10 *A **field** is a collection, $\mathcal{F}^\circ \subset \mathcal{P}(X)$, such that $\emptyset \in \mathcal{F}^\circ$, $[A \in \mathcal{F}^\circ] \Rightarrow [A^c \in \mathcal{F}^\circ]$ and $[A, B \in \mathcal{F}^\circ] \Rightarrow [(A \cup B \in \mathcal{F}^\circ) \wedge (A \cap B \in \mathcal{F}^\circ)]$. For any collection $\mathcal{E} \subset \mathcal{P}(X)$, let $\mathbb{F}^\circ(\mathcal{E})$ denote the family of all fields containing $\mathcal{E}$, that is, $\mathbb{F}^\circ(\mathcal{E}) = \{\mathcal{F}^\circ : \mathcal{E} \subset \mathcal{F}^\circ, \mathcal{F}^\circ \text{ a field}\}$. The **field generated by** $\mathcal{E}$ is defined as $\mathcal{F}^\circ(\mathcal{E}) = \bigcap\{\mathcal{F}^\circ : \mathcal{F}^\circ \in \mathbb{F}^\circ(\mathcal{E})\}$. This is a sensible definition because the intersection of any family of fields gives another field. In Example 2.2.9, $\mathcal{F}^\circ(\mathcal{C}) = \{\emptyset, X, A, \{c\}\}$ and $\mathcal{F}^\circ(\mathcal{D}) = \mathcal{F}^\circ(\mathcal{E}) = \mathcal{P}(X)$.*

The following are some of the most important sets we encounter in this book:

- $\mathbb{N} = \{1, 2, 3, \ldots\}$, the natural or "counting" numbers.
- $\mathbb{Z} = \{\ldots, -2, -1, 0, 1, 2, \ldots\}$, the integers.
- $\mathbb{Z}_+ = \{0, 1, 2, \ldots\}$, the nonnegative integers.
- $\mathbb{Q} = \{\frac{m}{n} : m, n \in \mathbb{Z}, n \neq 0\}$, the quotients, or rational numbers.
- $\mathbb{R}$, the set of "real numbers," that we construct in Chapter 3 by adding the so-called irrational numbers to $\mathbb{Q}$.

Note that $\mathbb{Q}$ contains all of the finite-length decimals, for example, $7.96518 = \frac{m}{n}$ for $m = 796{,}518$ and $n = 100{,}000$. This means that $\mathbb{Q}$ contains a representation for every physical measurement that we can make and every number we will ever see from a computer. The reason for introducing the extra numbers in $\mathbb{R}$ is not one of realism. Rather, we shall see that $\mathbb{Q}$ has "holes" in it, and even though the holes are infinitely small, they make analyzing some kinds of problems miserably difficult.

Even though we have not yet formally developed the set of numbers $\mathbb{R}$, the following example is worth seeing early and often.

Example 2.2.11 (The Field of Half-Closed Intervals) *Let $X = \mathbb{R}$ and for $a, b \in X$, $a < b$, define $(a, b] = \{x \in X : a < x \leq b\}$. Set $\mathcal{E} = \{(a, b] : a < b, \; a, b \in \mathbb{R}\}$ and let $\mathfrak{X} = \mathcal{F}^\circ(\mathcal{E})$. A set E belongs to $\mathfrak{X}$ iff it can be expressed as a finite union of disjoint intervals of one of the following three forms: $(a, b]$; $(-\infty, b] = \{x \in X : x \leq b\}$; or $(a, +\infty) = \{x \in X : a < x\}$.*

It is worth noting the style we used in this last example. When we write "define $(a, b] = \{x \in X : a < x \leq b\}$," we mean that whenever we use the symbols to the left of the equality, "$(a, b]$" in this case, we intend that you will understand these symbols to mean the symbols to the right of the equality, "$\{x \in X : a < x \leq b\}$" in this case. The word "let" is used in exactly the same way.

In a perfect world, we would take you through the construction of $\mathbb{N}$ starting from the idea of the empty set. Had we done this construction properly, the following would be a result.

Axiom 1 *Every nonempty $S \subset \mathbb{N}$ contains a smallest element, that is, $\leq$ is a well-ordering of $\mathbb{N}$.*

To be very explicit, we are assuming that if $S \in \mathcal{P}(\mathbb{N})$, $S \neq \emptyset$, then there exists $n \in S$ such that for all $m \in S$, $n \leq m$. There cannot be two such n, because $n \leq n'$ and $n' \leq n$ iff $n = n'$.

2.3 ◆ Products, Relations, Correspondences, and Functions

There is another way to construct new sets out of given ones, which involves the notion of an "ordered pair" of objects. In the set $\{a, b\}$, there is no preference

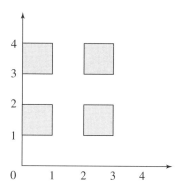

FIGURE 2.3.3   Cartesian product of $[0, 1] \cup [2, 3] \times [1, 2] \cup [3, 4]$.

given to a over b; that is, $\{a, b\} = \{b, a\}$, so that it is an unordered pair.[1] We can also consider ordered pairs (a, b), where we distinguish between the first and second elements.[2] Although what we mean when we use the phrase "we distinguish between the first and second elements" is very intuitive, it can be difficult to make it formal. One way is to say that "(a, b)" means the unordered pair of sets, "$\{\{a\}, \{a, b\}\} = \{\{a, b\}, \{a\}\}$," and we keep track of the order by noting that $\{a\} \subset \{a, b\}$, so that a comes before b. Throughout, A and B are any two sets, nonempty to avoid triviality.

Definition 2.3.1 *The **product** or **Cartesian product** of A **and** B, denoted $A \times B$, is the set of all ordered pairs $\{(a, b) : a \in A \text{ and } b \in B\}$. The sets A and B are the **axes** of the product $A \times B$.*

Example 2.3.2 $A = \{u, d\}$, $B = \{L, M, R\}$, $A \times B = \{(u, L), (u, M), (u, R), (d, L), (d, M), (d, R)\}$, *and* $A \times A = \{(u, u), (u, d), (d, u), (d, d)\}$.

Game theory is about the strategic interactions between people. If we analyze a two-person situation, we have to specify the options available to each. Suppose the first person's options are the set $A = \{u, d\}$, mnemonically, up and $d$own. Suppose the second person's options are the set $B = \{L, M, R\}$, mnemonically Left, Middle, and Right. An equilibrium in a game is a vector of choices, that is, an ordered pair, some element of $A \times B$. A continuous example is as follows.

Example 2.3.3 $A = [0, 1] \cup [2, 3]$, $B = [1, 2] \cup [3, 4]$, $A \times B$ *is the disjoint union of the four squares* $[0, 1] \times [1, 2]$, $[0, 1] \times [3, 4]$, $[2, 3] \times [1, 2]$, *and* $[2, 3] \times [3, 4]$. *[See Figure 2.3.3.]*

In game theory with three or more players, a vector of choices belongs to a larger product space. In general equilibrium theory, an allocation is a list of the consumptions of the people in the economy. The set of all allocations is a product

---

1. The curly brackets, "{" and "}" will always enclose a set so that "$\{a, b\}$" is the set containing elements a and b.

2. Hopefully, context will help you avoid confusing this notation with the interval consisting of all real numbers such that $a < x < b$.

space, just a product of n spaces. The following is an example of an inductive definition.

Definition 2.3.4   *Given a collection of sets,* $\{A_m : m \in \mathbb{N}\}$, *we define* $\times_{m=1}^{1} A_m = A_1$ *and inductively define* $\times_{m=1}^{n} A_m = \times_{m=1}^{n-1} A_m \times A_n$.

An ordered pair is called a 2-**tuple** and an n-**tuple** is an element of $\times_{m=1}^{n} A_m$. Sets of 2-tuples are called binary relations and sets of n-tuples are called n-ary relations. Relations are the mathematical objects of interest.

Definition 2.3.5   *Given two sets A and B, a **binary relation between A and B**, known simply as a **relation** if A and B can be inferred from context, is a subset $R \subset A \times B$. We use the notation $(a, b) \in R$ or $a R b$ to denote the relation R holding for the ordered pair (a, b) and read it "a is in the relation R to b." If $R \subset A \times A$, we say that R is a **relation on** A. The **range of a relation** R is the set of $b \in B$ for which there exists $a \in A$ with $(a, b) \in R$.*

Example 2.3.6   *$A = \{0, 1, 2, 3, 4\}$, so that $A \times A$ has twenty-five elements. With the usual convention that x is on the horizontal axis and y on the vertical, the relations $\leq$, $<$, $=$, and $\neq$ can be graphically represented by the $\otimes$'s in*

Note that $\leq$, $<$, $=$, and $\neq$ are sets. In terms of these sets, $\leq$ is the union of the disjoint sets, $<$ and $=$, and the complement of $=$ is $\neq$.

Relations can also be used in all kinds of cute ways.

Example 2.3.7   *Let $A = \{Austin, Des\ Moines, Harrisburg\}$ and $B = \{Texas, Iowa, Pennsylvania\}$. Then the relation $R = \{(Austin, Texas), (Des\ Moines, Iowa), (Harrisburg, Pennsylvania)\}$ expresses "is the state capital of."*

A relation between A and B is a subset of $A \times B$. A function from A to B is a special kind of relation, and a correspondence from A to B is a way to view a relation as a function; that is, it is an alternate definition of a relation.

Definition 2.3.8 *A **function (or mapping)** f, denoted $f : A \to B$, is a relation between A and B (i.e., $f \subset A \times B$) satisfying the following two properties:*

1. for all $a \in A$, there exists $b \in B$ such that $(a, b) \in f$, and

2. if $(a, b) \in f$ and $(a, b') \in f$, then $b = b'$.

*For each $a \in A$, the unique b such that $(a, b) \in f$ is denoted $f(a)$. A function may be written as $a \mapsto f(a)$ (read "a maps to $f(a)$"). The set A is called the **domain** of f, sometimes denoted $D(f)$. The **range** of f, denoted $Range(f)$ or $f(A)$, is $\{b \in B : \exists a \in A \text{ such that } (a, b) \in f\}$. The **graph** of f is $Gr(f) = \{(a, b) : (a, b) \in f\}$.*

Verbally, for each a in A, f associates a unique b, denoted $b = f(a)$. A function f and its graph are one and the same. It is odd to distinguish verbally between a function and its graph, but we (daringly) do it anyway.[3]

Example 2.3.9 *For $A = B = \{0, 1, 2, 3, 4\}$, the functions $f(x) = x$ and $g(x) = 4 - x$ can be represented by*

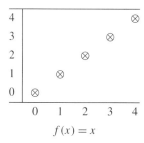

$f(x) = x$

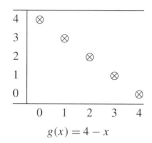

$g(x) = 4 - x$

Probabilities are an important example of functions.

Example 2.3.10 [↑*Example 2.2.10 (p. 20)*] *A **probability** is a function, $P : \mathcal{F}^\circ \to [0, 1]$, from a field of sets, $\mathcal{F}^\circ \subset \mathcal{P}(X)$, to the interval $[0, 1]$ with the properties that $P(\emptyset) = 0$, $P(X) = 1$, and for disjoint $A, B \in \mathcal{F}$, $P(A \cup B) = P(A) + P(B)$. From these properties, we see that $P(A^c) = 1 - P(A)$ (since A^c and A are disjoint and their union is X), that $[A \subset B] \Rightarrow [P(B) = P(A) + P(B \setminus A)]$ (for the same sort of reason), and that $P(A \cup B) = P(A) + P(B) - P(A \cap B)$ (since $A \cup B$ is the disjoint union of $A \setminus B$, $A \cap B$, and $B \setminus A$). Sometimes the number that a probability assigns to a set is called the **measure** of the set.*

Example 2.3.11 [↑*Example 2.2.11 (p. 21)*] *A **cumulative distribution function (cdf)** is a nondecreasing right-continuous function $F : \mathbb{R} \to [0, 1]$ that defines the probability, P_F, of an interval $(a, b]$ by $P_F((a, b]) = F(b) - F(a)$. We will see that P_F can be extended to the field of half-closed intervals and to the field of finite unions of disjoint intervals of all kinds: $(a, b]$, $(-\infty, b]$, $(a, +\infty)$, $[a, b) = \{x \in \mathbb{R} : a < x < b\}$, $[a, b] = \{x \in \mathbb{R} : a \leq X \leq b\}$; and (a, b), $(-\infty, b)$, $[a, +\infty)$ defined analogously. We will also see that P_F can be extended to a much larger collection of sets.*

---

3. Never let it be said that we lead dull lives of quiet desperation.

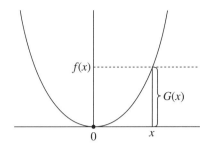

FIGURE 2.3.15

There are two equivalent ways to understand a correspondence from A to B: as a function from A to the subsets of B or as a subset of $A \times B$.

Definition 2.3.12 *A **correspondence** G, denoted $G : A \twoheadrightarrow B$, is a relation between A and B. For each $a \in A$, the set of b such that $(a, b) \in G$ is denoted $G(a)$. Equivalently, G is a function from A to $\mathcal{P}(B)$ assigning a set, $G(a)$, to each element $a \in A$.*

Exercise 2.3.13 For $A = B = [0, 1]$, draw three different correspondences from A to B that are not functions.

Exercise 2.3.14 Explicitly give the four relations in Example 2.3.6 (p. 23) as functions from A to $\mathcal{P}(A)$.

A correspondence G may have $G(a) = \emptyset$ or have $G(a)$ containing many elements. A function is a special kind of correspondence where for all a, $G(a)$ contains exactly one point.

Example 2.3.15 *In Figure 2.3.15, you can see the graph of the function $f(x) = x^2$ and the correspondence $G(x) = [0, x^2]$, $f : \mathbb{R} \to \mathbb{R}$, and $G : \mathbb{R} \twoheadrightarrow \mathbb{R}$. $G(0)$ consists of one point; for $x \neq 0$, $G(x)$ is an interval.*

Definition 2.3.16 *Given a function $f : A \to B$ and $E \subset A$, $E \neq \emptyset$, **the image of** E **under** f is written $f(E)$ and is defined by $f(E) = \{b \in B : (\exists e \in E)[f(e) = b]\}$. The **restriction of** f **to** E is written $f_{|E}$ and is defined as the function $f : E \to B$ having as a graph the set $Gr(f_{|E}) = Gr(f) \cap (E \times B)$.*

The image of a set E is the set of points to which it is mapped. The restriction of a function f to a set E ignores the behavior of the function outside of the set E.

Definition 2.3.17 *A set X is **finite** if it is empty, in which case it has 0 elements, or if there exists $n \in \mathbb{N}$ and a function $f : \{1, \ldots, n\} \to X$ such that $f(\{1, \ldots, n\}) = X$. The smallest n with this property is called the **cardinality of** X and is denoted $\#X$.*

This definition formalizes the idea that we would like to be able to count a finite set; that is, assign to each member of the set a number from $1, \ldots, n$ for some n. By Axiom 1, the cardinality of a finite set is well defined.[4]

Example 2.3.18 *If $X = \{a, b\}$, $a \neq b$, then the function $f : \{1, \ldots, 5\} \to X$ defined by $f(1) = f(2) = f(3) = a$ and $f(4) = f(5) = b$ shows that X is finite.*

The function f in this last example is a rather inefficient way to count the two-point set X. A function always takes a point to one single point. The given function f is **many-to-one**, that is, it takes many points to the same point. The following introduces the idea of a one-to-one function, and we return to it in more detail later.

Lemma 2.3.19 *If X is nonempty and finite and n is the cardinality of X, then there exists a function $f : \{1, \ldots, n\} \to X$ such that $f(\{1, \ldots, n\}) = X$, and for all $m \neq m'$, $m, m' \in \{1, \ldots, n\}$, we have $f(m) \neq f(m')$.*

Proof. By the definition of cardinality, we know that n is the smallest natural number with the property that there exists an $f : \{1, \ldots, n\} \to X$ with $f(\{1, \ldots, n\}) = X$. Suppose that for some $m \neq m'$ in $\{1, \ldots, n\}$, $f(m) = f(m')$. Consider the function $g : \{1, \ldots, n - 1\} \to X$ defined by $g(k) = f(k)$ for $k \in \{1, \ldots, m - 1\}$ and by $g(k) = f(k + 1)$ for $k \in \{m, \ldots, n - 1\}$. Since $g(\{1, \ldots, n - 1\}) = X$, n was not the cardinality of X, a contradiction that completes the proof. ∎

2.4 ◆ Equivalence Relations

As the name suggests, equivalence relations are relations of a special kind—a kind that appears frequently in the mathematics that economists use. The familiar equivalence classes from intermediate microeconomics are indifference curves, sets of consumption bundles that are all indifferent for the consumer, and isoprofit lines, sets of input-output vectors that yield the same profit for the producer. In game theory, one sees the strategic equivalence of strategies and the equivalence of games.

Definition 2.4.1 *An **equivalence relation** on a set A is a relation $\sim$ that is:*

1. ***reflexive**, $\forall a \in A$, $(a, a) \in \sim$,*
2. ***symmetric**, $\forall a, b \in A$, $[(a, b) \in \sim] \Leftrightarrow [(b, a) \in \sim]$, and*
3. ***transitive**, for all $a, b, c \in A$, $[[(a, b) \in \sim] \wedge [(b, c) \in \sim]] \Rightarrow [(a, c) \in \sim]$.*

This is perhaps more intuitive with the aRb notation: $\sim \subset (A \times A)$ is an equivalence relation iff for all $a, b, c \in A$, $a \sim a$, $[a \sim b] \Leftrightarrow [b \sim a]$, and $[a \sim b \wedge b \sim c] \Rightarrow [a \sim c]$.

Example 2.4.2 *Equality is an equivalence relation on $\mathbb{R}$. If $u : X \to \mathbb{R}$ is a utility function representing preferences on a set X, then defining $x \sim y$ by $u(x) = u(y)$ gives the indifference equivalence relation.*

---

4. This is a fancy way of saying that our definition makes sense—that if a set is finite, then the cardinality of that set exists.

Example 2.4.3 *Define the **congruence modulo 4** relation M_4 on $\mathbb{Z}$ by $\forall x, y \in \mathbb{Z}$, $x M_4 y$ if remainders obtained by dividing x and y by 4 are equal. For example, $13 M_4 65$ because dividing 13 and 65 by 4 gives a remainder of 1.*

Exercise 2.4.4 Show that congruence modulo 4 is an equivalence relation.

Definition 2.4.5 *Given an equivalence relation $\sim$ on a set A and an element $x \in A$, we define the **equivalence class determined by** x by $E_x = \{y \in A : y \sim x\}$. Note that $x \in E_x$ since $x \sim x$.*

Example 2.4.6 *The equivalence classes of $\mathbb{Z}$ for the relation M_4 are determined by $x \in \{0, 1, 2, 3\}$, where $E_x = \{z \in \mathbb{Z} : \exists k \in \mathbb{Z}, z = 4k + x\}$, that is, x is the remainder when z is divided by 4.*

Equivalence classes have the following property.

Theorem 2.4.7 *Two equivalence classes E and E' are either disjoint or equal.*

Proof. Let $E = \{y \in A : y \sim x\}$ and $E' = \{y \in A : y \sim x'\}$. If $E \cap E' = \emptyset$, then E and E' are disjoint. If $\exists z \in E \cap E'$, we show that $E = E'$. The first step is to demonstrate that $E \subset E'$, and the second step is to show that $E' \subset E$.

Let $w \in E$. We must show that $w \in E'$. Since $w \in E$, $w \sim x$. As $z \in E \cap E'$, we know that $z \sim x$ and $z \sim x'$. By transitivity $w \sim z$; hence $w \sim x'$, so that $w \in E'$.

Reversing the roles of E and E' in this argument demonstrates that $E' \subset E$. ∎

Looking at Example 2.4.6 in light of Theorem 2.4.7, we see that if two elements are in relation, they have the same equivalence class. So $E_1 = E_5 = E_9 = \ldots$ and $E_2 = E_6 = E_{10} = \ldots$. More generally, for all $n, k \in \mathbb{Z}$, $E_n = E_{4k+n}$.

Notation 2.4.8 *$A/\sim$ denotes the collection of all $\sim$-equivalence classes.*

Mnemonically, $\sim$ divides A into a collection of disjoint sets, so we write $A/\sim$. The union of all the sets in $A/\sim$ equals all of A because every element a of A belongs to exactly one of the equivalence classes. Another way to understand $A/\sim$ is as a partition of A.

Definition 2.4.9 *A **partition** of a set A is a collection of nonempty disjoint subsets of A whose union is all of A.*

We saw in Theorem 2.4.7 that equivalence relations give rise to partitions. The reverse is also true.

Exercise 2.4.10 For a partition, $\mathcal{C}$, of A, define $\sim_{\mathcal{C}}$ by $x \sim_{\mathcal{C}} y$ iff x and y belong to same element of $\mathcal{C}$. Show that $\sim_{\mathcal{C}}$ is an equivalence relation.

Example 2.4.11 *The equivalence classes of $\mathbb{Z}$ in Example 2.4.6 constitute a partition; $E_0 = \{\ldots, -8, -4, 0, 4, 8, \ldots\}$, $E_1 = \{\ldots, -7, -3, 1, 5, \ldots\}$, $E_2 = \{\ldots, -6, -2, 2, 6, \ldots\}$, and $E_3 = \{\ldots, -5, -1, 3, 7, \ldots\}$ are disjoint and their union is all of $\mathbb{Z}$. Generally, $\mathbb{Z}$ can be partitioned to n subsets via the equivalence relation $x \sim y$ iff x, y have the same remainder after division by n. The partitioning sets contain those subsets having remainders $0, 1, \ldots, n - 1$ to n. Another simple example is a coin toss experiment where the sample space $S = \{Heads, Tails\}$ has mutually exclusive events (i.e., $Heads \cap Tails = \emptyset$).*

Example 2.4.12 *Consider the relation $\sim$ on $\mathbb{R}$ given by $x \sim y$ iff $x - y \in \mathbb{Z}$. It can be easily checked that this is an equivalence relation. The equivalence of an arbitrary $x \in \mathbb{R}$ looks like $x + \mathbb{Z} = \{x + n : n \in \mathbb{Z}\}$. For all $n \in \mathbb{Z}$, x and $x + n$ are in the same equivalence class. Since for each x, there exists an $n \in \mathbb{Z}$ such that $n \leq x < n + 1$, x is in the same equivalence class as $x - n$, which we denote by (x), where $x - n \in [0, 1)$. Thus for each x, (x) is a representation of the equivalence class of x. Note that if $x, y \in [0, 1)$, then $x - y \notin \mathbb{Z}$, so $x \not\sim y$.*

What does the quotient space $\mathbb{R}/\sim$ looks like? This space consists of equivalence classes of $\sim$. By the above argument, we can make each member of $[0, 1)$ correspond to exactly one equivalence class of $\sim$. That is, we can think of $[0, 1)$ as $\mathbb{R}/\sim$.

Chapter 3 develops the real numbers, $\mathbb{R}$, as a collection of equivalence classes of sequences of elements of $\mathbb{Q}$.

2.5 ◆ Optimal Choice for Finite Sets

For all of this section, the set of options, X, is assumed to be finite.

Preference relations on a set of choices are at the core of economic theory. A decision maker's preferences are encoded in a preference relation, R, and "a is in the relation R to b" is interpreted as "a is at least as good as b." It is important to keep clear that the preference relation is assumed to be a property of the individual—your R is different than mine.

The two results in this section, Theorems 2.5.11 and 2.5.17, are the foundational results in the theory of *rational choice*:

- Theorem 2.5.11 shows that utility maximization is equivalent to preference maximization for complete and transitive preferences. This means that assuming that someone has a utility function and maximizes it is the same as assuming that the person can sensibly rank all of her options, perhaps allowing ties, and picks the option that she likes best or picks among the set of options that she likes best.

- Theorem 2.5.17 shows that preference maximizing behavior is equivalent to following a choice rule satisfying a minimal consistency condition called the weak axiom of revealed preference.

Theorem 2.5.14 shows that rational choice theory is not a mathematically empty one, and Theorem 2.5.15 gives the most basic comparative result for choice sets.

2.5.a The Basics

Let X be a finite set of options. We want to define the properties a relation $\succsim$ on X should have in order to represent preferences that are rational. Remember that we write "$x R y$" for "$(x, y) \in R$."

Definition 2.5.1 *A relation R on X is **complete** if for all $x, y \in X$, $x R y$ or $y R x$; it is **transitive** if for all $x, y, z \in X$, $[[x R y] \wedge [y R z]] \Rightarrow [x R z]$; and it is **rational** if it is both complete and transitive.*

Example 2.5.2 *One of the crucial order properties of the set of numbers, $\mathbb{R}$, is that $\leq$ and $\geq$ are complete and transitive.*

Completeness neither implies nor is implied by transitivity. To see this, the following exercise gives an example of a relation that satisfies both completeness and transitivity, gives other relations that satisfy one of the conditions but not the other, and gives a relation that satisfies neither. When you see a new concept, you should develop the two habits that this exercise exemplifies: finding examples in which the new concept does and does not hold and finding examples that demonstrate how the new concept interacts with other, possibly related concepts.

Exercise 2.5.3 In Example 2.3.6 (p. 23), show that $\leq$ is complete and transitive, that $<$ and $=$ are transitive but not complete, and that $\neq$ is neither transitive nor complete. Check that the relation $\succsim$ given later in Example 2.5.6 is complete but not transitive.

In thinking about preference relations, completeness is the requirement that any pair of choices can be compared for the purposes of making a choice. Given how much effort it is to make life decisions (jobs, marriage, kids), completeness is a strong requirement. When a relation is not complete, there are choices that cannot be compared and there may be two or more optimal choices in the set. For example, consider the relation $\subset$ on the set of all subsets of $A = \{1, \ldots, 10\}$ except A itself. Suppose we are looking for the largest subset. Then each of the subsets with nine elements is a largest element and they cannot be compared with each other. Transitivity is another rationality requirement. If violated, vicious cycles could arise among three or more options—any choice would have another that strictly beats it. To say "strictly beats" we need the following.

Definition 2.5.4 *Given a relation $\succsim$, define $x \succ y$ by $[x \succsim y] \wedge \neg[y \succsim x]$ and $x \sim y$ by $[x \succsim y] \wedge [y \succsim x]$.*

When talking about preference relations, "$x \succ y$" is read as "x is strictly preferred to y" and "$x \sim y$" is read as "x is indifferent to y." From the definitions, you can show that $[x \succsim y] \Leftrightarrow [[x \succ y] \vee [x \sim y]]$, and that the sets $\succ$ and $\sim$ are disjoint.

Exercise 2.5.5 Show that $x \sim y$ is an equivalence relation if $\succsim$ is rational.

Example 2.5.6 *Suppose you are at a restaurant and you have a choice among four meals, pork, beef, chicken, or fish, all costing the same. Suppose that your preferences, $\succsim$, and strict preferences, $\succ$, are given by*

	chic	fish	beef	pork			chic	fish	beef	pork
pork	⊗			⊗		pork	⊗			
beef			⊗	⊗		beef				⊗
fish		⊗	⊗	⊗		fish			⊗	⊗
chic	⊗	⊗	⊗			chic		⊗	⊗	

$$\succsim \qquad\qquad\qquad \succ$$

The basic behavioral assumption in economics is that you choose the option that you like best. Here $p \succ b \succ f \succ c \succ p$. Suppose you try to find your favorite *meal. Start by thinking about (say) c, discover you like f better so you switch your decision to f, but you like b better, so you switch again, but you like p better so you switch again, but you like c better so you switch again, coming back to where you started. You become confused and starve to death before you make up your mind.*

Exercise 2.5.7 Give the graphical representation $\succsim, \succ$, and $\sim$ for the complete transitive preferences satisfying $c \succ f \sim b \succ p$.

Exercise 2.5.8 Give the relation $\sim$ associated with the preferences given in Example 2.5.6. Is $\sim$ an equivalence relation? Can it reasonably be interpreted as indifference?

2.5.b Representing Preferences

Definition 2.5.9 *A utility function $u : X \to \mathbb{R}$ **represents** $\succsim$ if $[x \succ y] \Leftrightarrow [u(x) > u(y)]$ and $[x \sim y] \Leftrightarrow [u(x) = u(y)]$.*

Since u is a function, it assigns a numerical value to every point in X. Since we can compare any pair of numbers using $\geq$, any preference represented by a utility function is complete. As $\geq$ is transitive on $\mathbb{R}$, any preference represented by a utility function is transitive.

Exercise 2.5.10 Show that u represents $\succsim$ iff $[x \succsim y] \Leftrightarrow [u(x) \geq u(y)]$.

Theorem 2.5.11 *The relation $\succsim$ is rational iff there exists a utility function $u : X \to \mathbb{R}$ that represents $\succsim$.*

Since X is finite, we can replace $\mathbb{R}$ by $\mathbb{N}$ or by some set $\{1, \ldots, n\}$ in this result.

Proof. Suppose that $\succsim$ is rational. We must show that there exists a utility function $u : X \to \mathbb{N}$ that represents $\succsim$. Let $W(x) = \{y \in X : x \succsim y\}$; this is the set of options that are weakly worse than x. A candidate utility function is $u(x) = \#W(x)$. By transitivity, $[x \succsim y] \Rightarrow [W(y) \subset W(x)]$. By completeness, either $W(x) \subset W(y)$ or $W(y) \subset W(x)$, and $W(x) = W(y)$ if $x \sim y$. Also, $[x \succ y]$ implies that $W(y)$ is a proper subset of $W(x)$. When we combine, if $x \succ y$, then $u(x) > u(y)$, and if $x \sim y$, then $W(x) = W(y)$, so that $u(x) = u(y)$.

Now suppose that $u : X \to \mathbb{R}$ represents $\succsim$. We must show that $\succsim$ is complete and transitive. For $x, y \in X$, either $u(x) \geq u(y)$ or $u(y) \geq u(x)$ (or both). By the definition of representing, $x \succsim y$ or $y \succsim x$. Suppose now that $x, y, z \in X$, $x \succsim y$, and $y \succsim z$. We must show that $x \succsim z$. We know that $u(x) \geq u(y)$ and $u(y) \geq u(z)$. This implies that $u(x) \geq u(z)$, so that $x \succsim z$. ∎

The mapping $x \mapsto W(x)$ in the proof is yet another example of a correspondence, in this case from X to X. We now define the main correspondence used in rational choice theory.

Definition 2.5.12 *A **choice rule** is a function $C : \mathcal{P}(X) \to \mathcal{P}(X)$, equivalently a correspondence from $\mathcal{P}(X)$ to X, such that $C(B) \subset B$ for all $B \in \mathcal{P}(X)$, and $C(B) \neq \emptyset$ if $B \neq \emptyset$.*

The interpretation is that $C(B)$ is the set of options that might be chosen from the menu B of options. The best-known class of choice rules is made up of those of the form $C^*(B) = C^*(B, \succsim) = \{x \in B : \forall y \in B, \ x \succsim y\}$. In light of Theorem 2.5.11, $C^*(B) = \{x \in B : \forall y \in B, \ u(x) \geq u(y)\}$, that is, $C^*(B)$ is the set of utility maximizing elements of B.

The set of maximizers, the **argmax**, is a sufficiently important construct in economics that it has its own notation.

Definition 2.5.13 *For a nonempty set X and function $f : X \to \mathbb{R}$, $\arg\max_{x \in X} f(x)$ is the set $\{x^* \in X : (\forall x \in X)[f(x^*) \geq f(x)]\}$.*

The basic existence result tells us that the preference-maximizing choice rule yields a nonempty set of choices.

Theorem 2.5.14 *If B is a nonempty finite subset of X and $\succsim$ is a rational preference relation on X, then $C^*(B) \neq \emptyset$.*

Proof. Define $S^* = \cap_{x \in B}\{y \in B : y \succsim x\}$. It is clear that $S^* = C^*(B)$ (and you should check both directions of the inclusion if you are not used to writing proofs). All that is left is to show that $S^* \neq \emptyset$.

Let $n_B = \#B$ and pick a function $f : \{1, \ldots, n_B\} \to B$ such that $B = f(\{1, \ldots, n_B\})$. This means we order (or count) members of B as $f(1), \ldots, f(n_B)$. For $m \in \{1, \ldots, n_B\}$, let $S^*(m) = \{y \in B : \forall n \leq m, \ y \succsim f(n)\}$, so that $S^* = S^*(n_B)$. In other words, $S^*(m)$ contains the best elements between $f(1), \ldots, f(m)$ with respect to $\succsim$. Now using a function f^*, we inductively pick up the largest element among $f(1), \ldots, f(n)$ for all n. Define $f^*(1) = f(1)$. Given that $f^*(m-1)$ has been defined, define

$$f^*(m) = \begin{cases} f(m) & \text{if } f(m) \succsim f^*(m-1), \\ f^*(m-1) & \text{if } f^*(m-1) \succ f(m). \end{cases} \tag{2.1}$$

For each $m \in \{1, \ldots, n_B\}$, $S^*(m) \neq \emptyset$ because it contains $f^*(m)$, and by transitivity, $f^*(n_b) \in S^*$. ∎

The idea of the proof was simply to label the members of the finite set B and check its members step by step. We simply formalized this idea using logical tools and the definition of finiteness.

For $R, S \subset X$, we write $R \succsim S$ if $x \succsim y$ for all $x \in R$ and $y \in S$, and $R \succ S$ if $x \succ y$ for all $x \in R$ and $y \in S$. The basic comparison result for choice theory is that larger sets of options are at least weakly better.

Theorem 2.5.15 *If $A \subset B$ are nonempty finite subsets of X and $\succsim$ is a rational preference relation on X, then*

1. *$[x, y \in C^*(A)] \Rightarrow [x \sim y]$, optima are indifferent,*

2. *$C^*(B) \succsim C^*(A)$, larger sets are at least weakly better, and*

3. *$[C^*(B) \cap C^*(A) = \emptyset] \Rightarrow [C^*(B) \succ C^*(A)]$, a larger set is strictly better if it has a disjoint set of optima.*

Proof. The proof of (1) combines two proof strategies: contradiction and splitting into cases. Suppose that $[x, y \in C^*(A)]$ but $\neg[x \sim y]$. We split the statement $\neg[x \sim y]$ into two cases, $[\neg[x \sim y]] \Leftrightarrow [[x \succ y] \vee [y \succ x]]$. If $x \succ y$, then $y \notin C^*(A)$, a contradiction. If $y \succ x$, then $x \notin C^*(A)$, a contradiction.

To prove (2), we must show that $[[x \in C^*(B)] \wedge [y \in C^*(A)]] \Rightarrow [x \succsim y]$. We again give a proof by contradiction. Suppose that $[x \in C^*(B)] \wedge [y \in C^*(A)]$ but $\neg[x \succsim y]$. Since $\succsim$ is complete, $\neg[x \succsim y] \Rightarrow [y \succ x]$. As $y \in A$ and $A \subset B$, we know that $y \in B$. Therefore, $[y \succ x]$ contradicts $x \in C^*(A)$.

In what is becoming a pattern, we also prove (3) by contradiction. Suppose that $[C^*(B) \cap C^*(A) = \emptyset]$ but $\neg[C^*(B) \succ C^*(A)]$. By the definition of $R \succ S$ and the completeness of $\succsim$, $\neg[C^*(B) \succ C^*(A)]$ implies that there exists $y \in C^*(A)$ and $x \in C^*(B)$ such that $y \succsim x$. By (1), this implies that $y \in C^*(B, \succsim)$, which contradicts $[C^*(B) \cap C^*(A) = \emptyset]$. ∎

2.5.c Revealed Preference

We now approach the choice problem starting with a choice rule rather than with a preference relation. The question is whether there is anything new or different when we proceed in this direction. The answer is "No, provided the choice rule satisfies a minimal consistency requirement, and satisfying this minimal consistency requirement reveals a preference relation."

A choice rule C defines a relation, $\succsim^*$, "revealed preferred," defined by $x \succsim^* y$ if $(\exists B \in \mathcal{P}(X))[[x, y \in B] \wedge [x \in C(B)]]$. Note that $\neg[x \succsim^* y]$ is $(\forall B \in \mathcal{P}(X))[\neg[x, y \in B] \vee \neg[x \in C(B)]]$, equivalently $(\forall B \in \mathcal{P}(X))[[x \in C(B)] \Rightarrow [y \notin B]]$. In words, x is revealed preferred to y if there is a choice situation, B, in which both x and y are available, and x belongs to the choice set.

From the relation $\succsim^*$ we define "revealed strictly preferred," $\succ^*$, as in Definition 2.5.4 (p. 29). It is both a useful exercise in manipulating logic and a good way to understand a piece of choice theory to explicitly write out two versions of the meaning of $x \succ^* y$:

$$(\exists B_x \in \mathcal{P}(X))[[x, y \in B_x] \wedge [x \in C(B_x)]]$$
$$\wedge (\forall B \in \mathcal{P}(X))[[y \in C(B)] \Rightarrow [x \notin B]], \qquad (2.2)$$

equivalently

$$(\exists B_x \in \mathcal{P}(X))[[x, y \in B_x] \wedge [x \in C(B_x)] \wedge [y \notin C(B_x)]]$$
$$\wedge (\forall B \in \mathcal{P}(X))[[y \in C(B)] \Rightarrow [x \notin B]].$$

In words, the latter of these says that there is a choice situation where x and y are both available, x is chosen but y is not, and if y is ever chosen, then we know that x was not available.

A set $B \in \mathcal{P}(X)$ reveals a strict preference of y over x, written $y \succ_B x$, if $x, y \in B$ and $y \in C(B)$ but $x \notin C(B)$.

Definition 2.5.16 *A choice rule satisfies the* **weak axiom of revealed preference** *if* $[x \succsim^* y] \Rightarrow \neg(\exists B)[y \succ_B x]$.

This is the minimal consistency requirement. Satisfying this requirement means that choosing x when y is available in one situation is not consistent with choosing y but not x in some other situation where they are both available.

Theorem 2.5.17 *If C is a choice rule satisfying the weak axiom, then $\succsim^*$ is rational, and for all $B \in \mathcal{P}(X)$, $C(B) = C^*(B, \succsim^*)$. If $\succsim$ is rational, then $B \mapsto C^*(B, \succsim)$ satisfies the weak axiom, and $\succsim = \succsim^*$.*

Proof.   Suppose that C is a choice rule satisfying the weak axiom.

We must first show that $\succsim^*$ is complete and transitive.

Completeness: For all $x, y \in X$, $\{x, y\} \in \mathcal{P}(X)$ is a nonempty set. Therefore $C(\{x, y\}) \neq \varnothing$, so that $x \succsim^* y$ or $y \succsim^* x$.

Transitivity: Suppose that $x \succsim^* y$ and $y \succsim^* z$. We must show that $x \succsim^* z$. To do this, it is sufficient to demonstrate that $x \in C(\{x, y, z\})$. Since $C(\{x, y, z\})$ is a nonempty subset of $\{x, y, z\}$, we know that there are three cases: $x \in C(\{x, y, z\})$, $y \in C(\{x, y, z\})$, and $z \in C(\{x, y, z\})$. We must show that each of these cases leads to the conclusion that $x \in C(\{x, y, z\})$.

Case 1: This one is clear.

Case 2: $y \in C(\{x, y, z\})$, the weak axiom, and $x \succsim^* y$ implies that $x \in C(\{x, y, z\})$.

Case 3: $z \in C(\{x, y, z\})$, the weak axiom, and $y \succsim^* z$ implies that $y \in C(\{x, y, z\})$. As we just saw in Case 2, this means that $x \in C(\{x, y, z\})$.

We now show that for all $B \in \mathcal{P}(X)$, $C(B) = C^*(B, \succsim^*)$. Pick an arbitrary $B \in \mathcal{P}(X)$. It is sufficient to establish that $C(B) \subset C^*(B, \succsim^*)$ and $C^*(B, \succsim^*) \subset C(B)$.

Pick an arbitrary $x \in C(B)$. By the definition of $\succsim^*$, for all $y \in B$, $x \succsim^* y$. By the definition of $C^*(\cdot, \cdot)$, this means that $x \in C^*(B, \succsim^*)$.

Now pick an arbitrary $x \in C^*(B, \succsim^*)$. By the definition of $C^*(\cdot, \cdot)$, this means that $x \succsim^* y$ for all $y \in B$. By the definition of $\succsim^*$, for each $y \in B$, there is a set B_y such that $x, y \in B_y$ and $x \in C(B_y)$. As C satisfies the weak axiom, for all $y \in B$, there is no set B_y with the property that $y \succ_{B_y} x$. Since $C(B) \neq \varnothing$, if $x \notin C(B)$, then we would have $y \succ_B x$ for some $y \in B$, a contradiction.   ∎

Exercise 2.5.18   What is left to be proved in Theorem 2.5.17? Provide the missing step(s).

It is important to note the reach and the limitation of Theorem 2.5.17.

Reach: First, we did not use X being finite at any point in the proof, so it applies to infinite sets. Second, the proof would go through so long as C is defined on all two- and three-point sets. This means that we can replace $\mathcal{P}(X)$ with a family of sets $\mathcal{B}$ throughout, provided $\mathcal{B}$ contains all two- and three-point sets.

Limitation: In many of the economic situations of interest, the two- and three-point sets are not the ones that people are choosing from. For example, the leading case has $\mathcal{B}$ as the class of Walrasian budget sets.

2.6 ◆ Direct and Inverse Images, Compositions

Projections map products to their axes in a natural way: $\mathrm{proj}_A : A \times B \to A$ is defined by $\mathrm{proj}_A((a, b)) = a$; and $\mathrm{proj}_B : A \times B \to B$ is defined by $\mathrm{proj}_A((a, b)) = b$. The projections of a set $S \subset A \times B$ are defined by $\mathrm{proj}_A(S) = \{a : \exists b \in B, \ (a, b) \in$

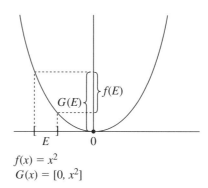

$f(x) = x^2$

$G(x) = [0, x^2]$

FIGURE 2.6.2

S} and $\text{proj}_B(S) = \{b : \exists a \in A, \ (a, b) \in S\}$. We use projections here to define direct and inverse images, we use them later in our study of sequence spaces and vector spaces.

2.6.a Direct Images

If $f : A \to B$ is a function and $E \subset A$, then $f(E) = \{f(a) \in B : a \in E\}$. We now extend this to correspondences/relations.

Definition 2.6.1 *Let R be a relation from A to B. If $E \subset A$, then the **(direct)** **image of E under the relation** R, denoted $R(E)$, is the set $\text{proj}_B(R \cap (E \times B))$.*

That is, for a relation R that consists of some ordered pairs, $R(E)$ contains the second component of all ordered pairs whose first component comes from E.

Exercise 2.6.2 Show that if $E \subset A$ and f is a function mapping A to B, then $f(E) = \cup_{a \in E}\{f(a)\}$, and if G is a correspondence mapping A to B, then $G(E) = \cup_{a \in E}G(a)$. [See Figure 2.6.2.]

Exercise 2.6.3 Consider functions $f, g : \mathbb{R} \to \mathbb{R}$, $f(x) = x^2$ and $g(x) = x^3$. Clearly $f(\mathbb{R})$ is contained in the positive real numbers. For every $r \geq 0$, $f(\sqrt{r}) = f(-\sqrt{r}) = r$, which implies that $f(\mathbb{R}) = f(\mathbb{R}_+) = \mathbb{R}_+$. Also $g(\sqrt[3]{r}) = r$ shows that all real numbers appear in the range of g, that is, $g(\mathbb{R}) = \mathbb{R}$.

Theorem 2.6.4 *Let f be a function from A to B and let $E, F \subset A$:*

1. If $E \subset F$, then $f(E) \subset f(F)$,

2. $f(E \cap F) \subset f(E) \cap f(F)$,

3. $f(E \cup F) = f(E) \cup f(F)$,

4. $f(E\backslash F) \subset f(E)$, and

5. $f(E \Delta F) \subset f(E) \Delta f(F)$.

Exercise 2.6.5 Prove Theorem 2.6.4 and give examples in which subset relations are proper.

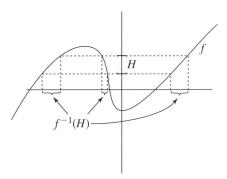

FIGURE 2.6.10

Exercise 2.6.6 Find and prove the analogue of Theorem 2.6.4 when the function f is replaced with a correspondence G, giving examples with the subset relations being proper.

2.6.b Inverse Relations

Inverse relations simply reverse the order in which we consider the axes.

Definition 2.6.7 *Given a relation R between A and B, the **inverse of** R is the relation R^{-1} between B and A defined by $R^{-1} = \{(b, a) : (a, b) \in R\}$. Images of sets under R^{-1} are called **inverse images**.*

The inverse of a function need not be a function, though it will always be a correspondence.

Example 2.6.8 *In general, functions are many-to-one. For example $f(x) = x^2$ from $\mathbb{R}$ to $\mathbb{R}$ maps both $+\sqrt{r}$ and $-\sqrt{r}$ to r when $r \geq 0$. In this case, the relation f^{-1}, viewed as a correspondence maps every nonnegative r to $\{-\sqrt{r}, +\sqrt{r}\}$, and maps every negative r to $\emptyset$.*

Example 2.6.9 *Let W be a finite set (of workers) and F a finite set (of firms). A function μ mapping W to $F \cup W$ is a **matching** if for all w, $[\mu(w) \in F] \vee [\mu(w) = w]$. We interpret $\mu(w) = w$ as the worker w being self-employed or unemployed. For $f \in F$, $\mu^{-1}(f) \subset W$ is the set of people who work at firm f.*

It is well worth the effort to be even more specific for functions.

Definition 2.6.10 *If f is a function from A to B and $H \subset B$, then the **inverse image of** H **under** f, denoted $f^{-1}(H)$, is the subset $\{a \in A | f(a) \in H\}$. [See Figure 2.6.10.]*

When $H = \{b\}$ is a one-point set, we write $f^{-1}(b)$ instead of $f^{-1}(\{b\})$.

Exercise 2.6.11 Just to be sure that the notation is clear, prove the following and illustrate the results with pictures: $f^{-1}(H) = \cup_{b \in H} f^{-1}(b)$, $\text{proj}_B^{-1}(H) = A \times H$, and $\text{proj}_A^{-1}(E) = E \times B$.

Exercise 2.6.12 (Level Sets of Functions) Let $f : A \to B$ be a function. Define $a \sim_f a'$ if $\exists b \in B$ such that $a, a' \in f^{-1}(b)$. These equivalence classes are called **level sets** of the function f.

1. Show that $\sim_f$ is an equivalence relation on A.
2. Show that $[a \sim_f a'] \Leftrightarrow [f(a) = f(a')]$.
3. Give an example with f, g being different functions from A to B but $\sim_f = \sim_g$.
4. Prove that the inverse images $f^{-1}(b)$ and $f^{-1}(b')$ are disjoint when $b \neq b'$. [This means that indifference curves never intersect.]

We return to inverse images under correspondences later. Since there are two ways to view them, as relations from A to B and as functions from A to $\mathcal{P}(B)$, there are two immediate possibilities for the definition of G inverse. It turns out that there is also a third possibility.

Inverse images under functions preserve the set operations, unions, intersections, and differences. As seen in Exercise 2.6.5, images need not have this property.

Theorem 2.6.13 *Let f be a function mapping A to B, and let $G, H \subset B$:*

1. *if $G \subset H$, then $f^{-1}(G) \subset f^{-1}(H)$,*
2. *$f^{-1}(G \cap H) = f^{-1}(G) \cap f^{-1}(H)$,*
3. *$f^{-1}(G \cup H) = f^{-1}(G) \cup f^{-1}(H)$, and*
4. *$f^{-1}(G \backslash H) = f^{-1}(G) \backslash f^{-1}(H)$.*

Proof. (1) If $a \in f^{-1}(G)$, then $f(a) \in G \subset H$ so $a \in f^{-1}(H)$. ∎

Exercise 2.6.14 Finish the proof of Theorem 2.6.13.

2.6.c Injections, Surjections, and Bijections

In general, functions are many-to-one. In Example 2.6.8 (p. 35), $f(x) = x^2$ maps both $+\sqrt{r}$ and $-\sqrt{r}$ to r. In Example 2.6.9 (p. 35), many workers may be matched to a single firm. When functions are not many-to-one but one-to-one, they have nice additional properties that flow from their inverses almost being functions.

Definition 2.6.15 $f : A \to B$ *is **one-to-one** or an **injection** if* $[f(a) = f(a')] \Rightarrow [a = a']$.

Since $[a = a'] \Rightarrow [f(a) = f(a')]$, being one-to-one is equivalent to $[f(a) = f(a')] \Leftrightarrow [a = a']$.

Recall the correspondence $G(b) = f^{-1}(b)$ from B to A introduced earlier. When f is many-to-one, then for some b, the correspondence $G(b)$ contains more than one point. When a correspondence always contains exactly one point, it is a function. Hence, the inverse of a one-to-one function is a function from the range of f to A. That is, the inverse of an injection $f : A \to B$ fails to be a function from B to A only in that it may not be defined for all of B.

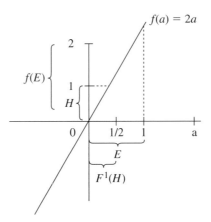

FIGURE 2.6.18 $-f : \mathbb{R} \to \mathbb{R}$ given by $f(a) = 2a$.

Example 2.6.16   *Let $E = \{2, 4, 6, \ldots\}$ be the set of even natural numbers and define $f(n) = 2n$. Then f is one-to-one, $f(\mathbb{N}) = E$, and f^{-1} is a function from E to $\mathbb{N}$.*

Definition 2.6.17   *If $Range(f) = B$, f maps A **onto** B and we call f a **surjection**; f is a **bijection** if it is one-to-one and onto, that is, if it is both an injection and a surjection, in which case we write $f : A \leftrightarrow B$.*

Note that surjectiveness of a map depends on the set into which the map is defined. For example, if we consider $f(x) = x^2$ as $f : \mathbb{R} \to \mathbb{R}_+$ then it is onto, whereas the same function viewed as $f : \mathbb{R} \to \mathbb{R}$ is not onto.

To summarize:

- injections are one-to-one and map A into B but may not cover all of B;
- surjections put A all over B but may not be one-to-one; and
- bijections from A to B are one-to-one onto functions, which means that their inverse correspondences are functions from B to A.

Example 2.6.18   *Let $E = [0, 1] \subset A = \mathbb{R}$, $H = [0, 1] \subset B = \mathbb{R}$, and $f(a) = 2a$. $Range(f) = \mathbb{R}$ so that f is a surjection; the image set is $f(E) = [0, 2]$; the inverse image set is $f^{-1}(H) = [0, \frac{1}{2}]$; and f is an injection, has inverse $f^{-1}(b) = \frac{1}{2}b$, and as a consequence of being one-to-one and onto is a bijection. [See Figure 2.6.18.]*

Exercise 2.6.19   Show that if $f : A \leftrightarrow B$ is a bijection between A and B, then the subset relations in Theorem 2.6.4 (p. 34) hold with equality.

2.6.d Compositions of Functions

If we first apply f to an $a \in A$ to get $b = f(a)$, and then apply g to b, we have a new, composite function, $h(a) = g(f(a))$.

Definition 2.6.20 *Let $f : A \to B$ and $g : B' \to C$, with $B' \subset B$ and $Range(f) \subset B'$. The* **composition** *$g \circ f$ is the function from A to C given by $g \circ f = \{(a, c) \in A \times C : (\exists b \in Range(f))[[(a, b) \in f] \wedge [(b, c) \in g]]\}$.*

The order matters greatly here. If $f : \mathbb{R}^2 \to \mathbb{R}^3$ and $g : \mathbb{R}^3 \to \mathbb{R}$, then $h(x) = g(f(x))$ is perfectly well defined for $x \in \mathbb{R}^2$, but "$f(g(y))$" is pure nonsense, since the domain of f is $\mathbb{R}^2$, not $\mathbb{R}$. In matrix algebra, this corresponds to matrices having to be comfortable in order for multiplication to be defined. However, even in comfortable cases, order matters.

Example 2.6.21 $\begin{bmatrix} 2 & 0 \\ 0 & 1 \end{bmatrix} \begin{bmatrix} 0 & 1 \\ 1 & 0 \end{bmatrix} \neq \begin{bmatrix} 0 & 1 \\ 1 & 0 \end{bmatrix} \begin{bmatrix} 2 & 0 \\ 0 & 1 \end{bmatrix}.$

Here is a nonlinear example of the order mattering.

Example 2.6.22 *Let $A \subset \mathbb{R}$, $f(a) = 2a$, and $g(a) = 3a^2 - 1$. Then $g \circ f = 3(2a)^2 - 1 = 12a^2 - 1$, whereas $f \circ g = 2(3a^2 - 1) = 6a^2 - 2$.*

Compositions preserve surjectiveness and injectiveness.

Theorem 2.6.23 *If $f : A \to B$ and $g : B \to C$ are surjections (injections), then their composition $g \circ f$ is a surjection (injection).*

Exercise 2.6.24 Prove Theorem 2.6.23.[5]

Example 2.6.25 *If $f(x) = x^2$ and $g(x) = x^3$, then g is a bijection between $\mathbb{R}$ and itself, whereas $g \circ f$ is not even a surjection.*

Theorem 2.6.26 *Suppose $f : A \to B$ and $g : B \to C$. Then*

1. if $g \circ f$ is onto, then g is onto, and

2. if $g \circ f$ is one-to-one, then f is one-to-one.

Proof. For the first part, we show that if g is not onto, then $g \circ f$ cannot be onto. The function g is not onto iff $g(B) \subsetneqq C$. Since $f(A) \subset B$, this implies that $g(f(A)) \subsetneqq C$.

In a similar fashion, f is not one-to-one iff $(\exists a \neq a')[f(a) = f(a')]$, which implies that $(g \circ f)(a) = (g \circ f)(a')$, so that $g \circ f$ is not one-to-one. ∎

To have $g \circ f$ be onto, one needs $g(Range(f)) = C$. To have $g \circ f$ be one-to-one, in addition to f being one-to-one, g should be injective on the $Range(f)$, but not necessarily on the whole B.

Exercise 2.6.27 Give an example of functions f, g where f is not onto but g and $g \circ f$ are onto. Also give an example where g is not injective, but $g \circ f$ is injective.

Definition 2.6.28 *The* **identity function on a set** *A is the function $f : A \to A$ defined by $f(a) = a$ for all $a \in A$.*

---

5. Hints: In the case of a surjection, we must show that for $(g \circ f)(a) := g(f(a))$, it is the case that $\forall c \in C$, there exists $a \in A$ such that $(g \circ f)(a) = c$. To see this, let $c \in C$. Since g is a surjection, $\exists b \in B$ such that $g(b) = c$. Similarly, since f is a surjection, $\exists a \in A$ such that $f(a) = b$. Then $(g \circ f)(a) = g(f(a)) = g(b) = c$.

Example 2.6.29 *Let M be a finite set (of men) and W a finite set (of women). A **matching** is a function μ from $M \cup W$ to itself such that for all $m \in M$, $[\mu(m) \in W] \vee [\mu(m) = m]$, for all $w \in W$, $[\mu(w) \in M] \vee [\mu(w) = w]$, and $\mu \circ \mu$ is the identity function.*

2.7 ◆ Weak and Partial Orders, Lattices

Rational choice theory for an individual requires a complete and transitive preference relation. Equilibria involve simultaneous rational choices by many individuals. While we think it reasonable to require an individual to choose among different options, requiring a group to be able to choose is a less reasonable assumption. One of the prominent examples of a partial ordering, that is, one that fails to be complete, arises when one asks for unanimous agreement among the individuals in a group.

Let X be a nonempty set and $\precsim$ a relation on X. We call $(X, \precsim)$ an ordered set. Orders are relations, but we use a different name because our emphasis is on orders with interpretations reminiscent of the usual order, $\leq$, on $\mathbb{R}$. Table 2.7 gives names to properties that an order $\precsim$ may or may not have.

Table 2.7

Property	Name
$(\forall x)[x \precsim x]$	Reflexivity
$(\forall x, y)[[x \precsim y] \Rightarrow [y \precsim x]]$	Symmetry
$(\forall x, y)[[x \precsim y] \wedge [y \precsim x]] \Rightarrow [x = y]]$	Antisymmetry
$(\forall x, y, z)[x \precsim y] \wedge [y \precsim z]] \Rightarrow [x \precsim z]]$	Transitivity
$(\forall x, y)[[x \precsim y] \vee [y \precsim x]$	Completeness
$(\forall x, y)(\exists u)[x, y \precsim u]$	Upper bound
$(\forall x, y)(\exists \ell)[\ell \precsim x, y]$	Lower bound
$(\forall x, y)(\exists u)[[x, y \precsim u] \wedge [[x, y \precsim u'] \Rightarrow [u \precsim u']]$	Least upper bound
$(\forall x, y)(\exists \ell)[[\ell \precsim x, y] \wedge [[\ell' \precsim x, y] \Rightarrow [\ell' \precsim \ell]]$	Greatest lower bound

To see that every complete relation is reflexive, take $x = y$ in the definition. The three main kinds of ordered sets we study are in the following.

Definition 2.7.1 $(X, \precsim)$ *is a **weakly ordered set** if $\precsim$ is complete and transitive.*[6] $(X, \precsim)$ *is a **partially ordered set (POSET)** if $\precsim$ is reflexive, antisymmetric, and transitive. $(X, \precsim)$ is a **lattice** if it is a POSET with the least upper bound and the greatest lower bound property.*

The following are, for economists, the most frequently used examples of these kinds of ordered sets.

---

6. Earlier, in §2.5, we used "rational" instead of "weak" for complete and transitive relations. It would have been too hard for us, as economists, to accept the idea that the foundation of our theory of choice was "weak"; hence the choice of a grander name, "rational."

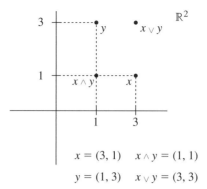

$$x = (3, 1) \quad x \wedge y = (1, 1)$$
$$y = (1, 3) \quad x \vee y = (3, 3)$$

FIGURE 2.7.2

Example 2.7.2 *Weak ordering: If $X = \mathbb{R}_+^2$ and $\precsim$ is defined by $x \precsim y$ iff $u(x) \leq u(y)$ for a utility function u, then $(X, \precsim)$ is a weakly ordered set. We often represent this weak ordering by drawing representative indifference curves.*

*   Partial ordering: If $X = \mathbb{R}_+^2$ and $\precsim$ is defined by the **usual vector order**, that is, by $\mathbf{x} \precsim \mathbf{y}$ iff $x_1 \leq y_1$ and $x_2 \leq y_2$, then $(X, \precsim)$ is a (POSET) that is not weakly ordered because it fails completeness, for example, the pair $(1, 3)$ and $(3, 1)$ are not comparable, as neither is greater in the vector order.*

*   Lattices: $\mathbb{R}^2$ with the vector order is the canonical example of a lattice. For all $\mathbf{x} = (x_1, x_2)$, $\mathbf{y} = (y_1, y_2) \in \mathbb{R}^2$, the greatest lower bound, denoted $\mathbf{x} \wedge \mathbf{y}$, is the point $(\min\{x_1, y_1\}, \min\{x_2, y_2\})$, and the least upper bound, denote $\mathbf{x} \vee \mathbf{y}$, is the point $(\max\{x_1, y_1\}, \max\{x_2, y_2\})$. [See Figure 2.7.2.] By the same logic, $\mathbb{R}^\ell$ is a lattice with the vector order.*

*   If $X = A \times B$, A, $B \subset \mathbb{R}$, is a rectangular set, then it is not only a POSET, but also it is a lattice. However, if X is the line $\{(x_1, x_2) \in \mathbb{R}^2 : x_1 + x_2 = 1\}$ with the vector ordering, then it is a POSET that is* not *a lattice.*

Notation 2.7.3 *If the greatest lower bound of a pair x, y in a POSET X exists, it is called the **infimum** and denoted $\inf_X(x, y)$, $\inf(x, y)$, or $x \wedge y$. If the least upper bound of a pair x, $y \in X$ exists, it is called the **supremum** and denoted $\sup_X(x, y)$, $\sup(x, y)$, or $x \vee y$.*

Example 2.7.4 *The relation $\leq$ on $\mathbb{R}$ is reflexive, antisymmetric, transitive, complete; $\sup(x, y)$ is the maximum of the two numbers x and y and $\inf(x, y)$ is the minimum of the two. Thus, $(\mathbb{R}, \leq)$ is a weakly ordered set, a POSET, and a lattice. On $\mathbb{R}$, the equality relation, $=$, is reflexive, symmetric, antisymmetric, and transitive, but fails the other properties.*

There is a tight connection with the logical connectives $\vee$ and $\wedge$ introduced earlier. Recall that the indicator of a set $A \subset X$ is that function $1_A : X \to \{0, 1\}$ defined by

$$1_A(x) := \begin{cases} 1 & \text{if } x \in A, \\ 0 & \text{if } x \notin A. \end{cases} \tag{2.3}$$

The set of indicator functions of subsets provides a classical example of lattices.

Example 2.7.5   *The relation $\leq$ on indicator functions is defined by $1_A \leq 1_B$ when $1_A(x) \leq 1_B(x)$ for every $x \in X$. Note that $1_A \leq 1_B$ iff $A \subset B$, so that $\leq$ is not complete. On the class of indicator functions, $\leq$ is reflexive, antisymmetric, transitive; $1_A \wedge 1_B = 1_{A \cap B}$ and $1_A \vee 1_B = 1_{A \cup B}$, so the class of indicators is a lattice with the relation $\leq$.*

Exercise 2.7.6   Even though the relation $\leq$ for vectors in $\mathbb{R}^2$ is not rational because it is not complete, defining $[(x_1, x_2) \sim (y_1, y_2)] \Leftrightarrow [[(x_1, x_2) \leq (y_1, y_2)] \wedge [(y_1, y_2) \leq (x_1, x_2)]]$ gives an equivalence relation. What is the equivalence relation?

You may be wondering how any reasonable relation with upper and lower bounds could fail to have the least upper or greatest lower bound property.

Example 2.7.7   *The relation $<$ on $\mathbb{R}$ (or on $\mathbb{Q}$) is not reflexive, is antisymmetric because $r < s$ and $s < r$ never happens, is transitive, and has the upper bound and the lower bound property but not the least upper bound nor the greatest lower bound property. This failure of $<$ to have the least upper and greatest lower bound has far-reaching consequences. To see why the failure occurs, let $r = s = 0$ and note that for any integer n, no matter how large, $1/10^n$ is an upper bound, but $1/10^{n+1}$ is a smaller upper bound.*

The theory of rational choice by individuals studies weakly ordered sets.

Example 2.7.8   *Suppose that $\precsim$ is a rational preference ordering on X, that is, it is complete and transitive so that $(X, \precsim)$ is a weakly ordered set. Completeness implies that $\precsim$ is also reflexive because we can take $x = y$ in the definition of completeness. Completeness also implies the existence of at least one upper bound for any pair $x, y \in X$ because we can take $u = x$ if $x \succsim y$ and $u = y$ if $y \succsim x$. Parallel logic implies that there exists at least one lower bound. If there are two or more indifferent points in X, then antisymmetry fails, so that $(X, \precsim)$ is not a lattice.*

The subset relation is not the only interesting one for sets.

Exercise 2.7.9   Let X be the nonempty subsets of Y, that is, $X = \mathcal{P}(Y) \setminus \{\emptyset\}$. Define the relation t on X by $A t B$ if $A \cap B \neq \emptyset$. Mnemonically, "$A t B$" if "A touches B." For $y \in Y$, let $X(y) = \{A \in X : y \in A\}$. Which, if any, of the properties given in Table 2.7 (p. 39) does the relation t have on X? On $X(y)$?

From Definition 2.5.4 (p. 29), given a relation (e.g., a partial ordering) $\succsim$ on a set X, we define $x \sim y$ if $x \succsim y$ and $y \succsim x$ and $x \succ x$ if $x \succsim y$ and $\neg(y \succsim x)$.

Exercise 2.7.10   Show that $\sim = (\succsim \cap \succsim^{-1})$ and that $\succ = (\succsim \setminus \sim)$.

Definition 2.7.11   *A point $x \in X$ is **undominated in** $\succsim$ if $\neg(\exists y \in X)[y \succ x]$.*

Example 2.7.12   *If $X = \{(x_1, x_2) \in \mathbb{R}^2 : x_1 + x_2 = 1\}$, then every point in X is undominated in $\leq$.*

Definition 2.7.13   *Let $\{\succsim_i : i \in I\}$ be a collection of rational preference orderings on a set X. Define the **unanimity order** on X by $x \succsim_U y$ if for all $i \in I$, $x \succsim_i y$. A point x is **Pareto optimal** or **Pareto efficient** if x is undominated in $\succsim_U$. If $x' \succ_U x$, then x' **Pareto improves on** x.*

In all but the simplest cases, the unanimity order will fail completeness.

Example 2.7.14 *Let $X = [0, 1]$ with $x \in X$ representing the proportion of the apple pie that person 1 will consume and $(1 - x)$ the proportion that person 2 will consume. Suppose that preferences are selfish, $[x > x'] \Leftrightarrow [x \succ_1 x']$ and $[x > x'] \Leftrightarrow [x \prec_2 x']$. No pair of points can be ranked by the unanimity order and every point is Pareto optimal.*

The definition of Pareto optimality in Definition 2.7.13 is the usual one. Note that $y \succ_U x$ iff $(\forall i \in I)[y \succsim_i x]$ and $(\exists j \in I)[y \succ_j x]$. Thus, x is Pareto efficient iff there is no y that everyone likes at least as well and someone strictly prefers. This differs from Definition 1.5.3 (p. 12), which is often known as *weak* Pareto efficiency: x is weakly Pareto efficient if there is no y such that $(\forall i \in I)[y \succ_i x]$.

Exercise 2.7.15 Show that if x is Pareto efficient, then it is weakly Pareto efficient. Modify the preferences of person 1 in Example 2.7.14 so that person 1 gains no extra utility from any pie beyond three-quarters of the pie; that is, three-quarters of the pie satiates person 1. With these preferences, give an example of an x that is weakly Pareto efficient but not Pareto efficient.

When preferences are nonsatiable, the allocation can be continuously adjusted, and preferences are continuous, then weak Pareto optimality and Pareto optimality are the same.

2.8 ◆ Monotonic Changes in Optima: Supermodularity and Lattices

Throughout economics, we are interested in how changes in one variable affect another variable. We answer such questions *assuming* that what we observe is the result of optimizing behavior. Part of learning to "think like an economist" involves internalizing this assumption. Given what we know about rational choice theory from §2.5, in many contexts optimizing behavior involves maximizing a utility function. In symbols, with $t \in T$ denoting a variable not determined by the individual, we let $x^*(t)$ denote the solution(s) to the problem $P(t)$,

$$\max_{x \in X} f(x, t), \tag{2.4}$$

and ask how $x^*(t)$ depends on t as t varies over T.

Since the problem $P(t)$ in (2.4) is meant as an approximation to, rather than a quantitative representation of, behavior, we are after "qualitative" results. These are results that are of the form "if t increases, then $x^*(t)$ will increase," and they should be "fairly immune to" details of the approximation.[7] If $x^*(\cdot)$ is differentiable, then we are after a statement of the form $dx^*/dt \geq 0$. If $x^*(\cdot)$ is not

---

7. Another part of learning to "think like an economist" involves developing an aesthetic sense of what "fairly immune" means. Aesthetics are complicated and subtle, best gained by immersion and indoctrination in and by the culture of economists, as typically happens during graduate school.

differentiable, then we are after a statement of the form that x^* is nondecreasing in t. We are going to go after such statements in three ways.

2.8.a. **The implicit function theorem**, using derivative assumptions on f when X and T are one-dimensional intervals.

2.8.b. **The simple univariate Topkis theorem**, using supermodularity assumptions on f when X and T are linearly ordered sets, for example, one-dimensional intervals.

2.8.c. **Monotone comparative statics**, using supermodularity when X is a lattice and T is partially ordered.

Unlike ranking by utility, supermodularity is not immune to monotonic rescaling, that is, supermodularity is a cardinal, not an ordinal concept. Quasi-supermodularity is the ordinal version of supermodularity, and we deal with it in the last part of this section.

Definition 2.8.1 *A partial order that also satisfies completeness is called a **total (or linear) ordering**, and $(X, \precsim)$ is called a **totally ordered set**. A **chain** in a partially ordered set is a subset, $X' \subset X$, such that $(X', \precsim)$ is totally ordered.*

The classical example of a linearly ordered set is the real line $(\mathbb{R}, \leq)$. In a total ordering, any two elements x and y in A can be compared, whereas in a partial ordering, there are noncomparable elements. For example, $(\mathcal{P}(\mathbb{N}), \subseteq)$ is a partially ordered set with many noncomparable elements. However, the set containing $\{1\}, \{1, 2\}, \ldots, \{1, 2, \ldots, n\}, \ldots$ is a chain in $\mathcal{P}(\mathbb{N})$.

Exercise 2.8.2 Show that if $A \subset B$ and B is totally ordered, then A is totally ordered.

2.8.a The Implicit Function Approach

Assume that X and T are interval subsets of $\mathbb{R}$ and that f is twice continuously differentiable.[8] The terms f_x, f_t, f_{xx}, and f_{xt} denote the corresponding partial derivatives of f. To have $f_x(x, t) = 0$ characterize $x^*(t)$, we must have $f_{xx} < 0$ (which is a standard result about concavity in microeconomics). From the implicit function theorem, we know that $f_{xx} \neq 0$ is what is needed for there to exist a function $x^*(t)$ such that

$$f_x(x^*(t), t) \equiv 0. \tag{2.5}$$

To find dx^*/dt, take the derivative on both sides with respect to t and find

$$f_{xx}\frac{dx^*}{dt} + f_{xt} = 0, \tag{2.6}$$

so that $dx^*/dt = -f_{xt}/f_{xx}$. Since $f_{xx} < 0$, this means that dx^*/dt and f_{xt} have the same sign.

───────────────

8. We study the continuity of functions in some detail later. For now, we are assuming that the reader has had a calculus-based microeconomics course.

This ought to be an intuitive result about the problem $P(t)$ in (2.4): if $f_{xt} > 0$, then increases in t increase f_x; increases in f_x are increases in the marginal reward of x; and as the marginal reward to x goes up, we expect the optimal level of x to go up. In a parallel fashion: if $f_{xt} < 0$, then increases in t decrease f_x; decreases in f_x are decreases in the marginal reward of x; and as the marginal reward to x goes down, we expect the optimal level of x to go down.

Exercise 2.8.3 Let $X = T = \mathbb{R}_+$ and $f(x,t) = x - \frac{1}{2}(x-t)^2$. Find $x^*(t)$ and verify directly that $dx^*/dt > 0$. Also find f_x, f_{xx}, and f_{xt}, and verify, using the sign test just given, that $dx^*/dt > 0$. If you can draw three-dimensional figures (and this is a skill worth developing), draw f and verify from your picture that $f_{xt} > 0$ and that it is this fact that makes $dx^*/dt > 0$. To practice with what goes wrong with derivative analysis when there are corner solutions, repeat this problem with $X = \mathbb{R}_+$, $T = \mathbb{R}$, and $g(x,t) = x - \frac{1}{2}(x+t)^2$.

Example 2.8.4 *The amount of a pollutant that can be emitted is regulated to be no more than $t \geq 0$. The cost function for a monopolist producing x is $c(x,t)$ with $c_t < 0$ and $c_{xt} < 0$. These derivative conditions mean that increases in the allowed emission level lower costs and lower marginal costs, so that the firm will always choose t. For a given t, the monopolist's maximization problem is therefore*

$$\max_{x \geq 0} f(x,t) = x p(x) - c(x,t), \qquad (2.7)$$

where $p(x)$ is the (inverse) demand function. Since $f_{xt} = -c_{xt}$, we know that increases in t lead the monopolist to produce more, provided $f_{xx} < 0$.

The catch in the previous analysis is that $f_{xx} = x p_{xx} + p_x - c_{xx}$, so that it seems we have to know that $p_{xx} < 0$, or concavity of the inverse demand, and $c_{xx} > 0$, or convexity of the cost function, before we can reliably conclude that $f_{xx} < 0$. The global concavity of $f(\cdot, t)$ seems to have little to do with the intuition that it is the lowering of marginal costs that makes x^* depend positively on t. However, global concavity of $f(\cdot, t)$ is *not* what we need for the implicit function theorem, rather only the concavity of $f(\cdot, t)$ in the region of $x^*(t)$. With differentiability, this local concavity is an *implication* of $x^*(t)$ being a strict local maximum for $f(\cdot, t)$. Supermodularity makes it clear that the local maximum property is all that is being assumed and allows us to work with optima that are nondifferentiable.

2.8.b The Simple Supermodularity Approach

The simplest case has X and T being linearly ordered sets. The most common example has X and T being intervals in $\mathbb{R}$ with the usual less-than-or-equal-to order. However, nothing rules out the sets X and T being discrete.

Definition 2.8.5 *For linearly ordered X and T, a function $f : X \times T \to \mathbb{R}$ is* **supermodular** *if for all $x' \succ x$ and all $t' \succ t$,*

$$f(x', t') - f(x, t') \geq f(x', t) - f(x, t), \qquad (2.8)$$

equivalently

$$f(x', t') - f(x', t) \geq f(x, t') - f(x, t). \tag{2.9}$$

It is **strictly supermodular** *if the inequalities are strict. For* **submodularity** *and* **strictly submodularity**, *reverse the inequalities.*

At t, the benefit of increasing from x to x' is $f(x', t) - f(x, t)$, at t', it is $f(x', t') - f(x, t')$. This assumption asks that the benefit of increasing x be increasing in t. A good verbal shorthand for this is that f *has increasing differences in x and t*. Three sufficient conditions in the differentiable case are: $\forall x$, $f_x(x, \cdot)$ is nondecreasing; $\forall t$, $f_t(\cdot, t)$ is nondecreasing; and $\forall x, t$, $f_{xt}(x, t) \geq 0$.

Theorem 2.8.6 (Topkis) *If X and T are linearly ordered, $f : X \times T \to \mathbb{R}$ is supermodular and $x^*(\tau)$ is the largest solution to $\max_{x \in X} f(x, \tau)$ for all τ, then $[t' \succ t] \Rightarrow [x^*(t') \gtrsim x^*(t)]$. Further, if there are unique, unequal maximizers at t' and t, then $x^*(t') \succ x^*(t)$.*

Proof.   The idea of the proof is that having $x^*(t') \prec x^*(t)$ can only arise if f has strictly decreasing differences. Suppose that $t' \succ t$ but that $x' := x^*(t') \prec x := x^*(t)$. As $x^*(t)$ and $x^*(t')$ are maximizers, $f(x', t') \geq f(x, t')$ and $f(x, t) \geq f(x', t)$. Since x' is the largest of the maximizers at t' and $x \succ x'$, that is, x is larger than the largest maximizer at t', we know a bit more—that $f(x', t') > f(x, t')$. Adding the inequalities, we get $f(x', t') + f(x, t) > f(x, t') + f(x', t)$, or

$$f(x, t) - f(x', t) > f(x, t') - f(x', t'),$$

that is, strictly decreasing differences in x and t.   ∎

Going back to the polluting monopolist of Example 2.8.4 (p. 44), we see that the supermodularity of f reduces to the supermodularity of $-c$. Thus assuming $-c$ (and hence f) is supermodular, we can use Theorem 2.8.6 to conclude that $x^*(t)$ is increasing. None of the second-derivative conditions except $c_{xt} < 0$ is necessary, and this can be replaced by the looser condition that $-c$ is supermodular.

Clever choices of T's and f's can make some analyses very easy.

Example 2.8.7 *Suppose that the one-to-one demand curve for a good produced by a monopolist is $x(p)$, so that $CS(p) = \int_p^\infty x(r)\, dr$ is the consumer surplus when the price p is charged. Let $p(\cdot)$ be $x^{-1}(\cdot)$, the inverse demand function. From intermediate microeconomics, you should know that the function $x \mapsto CS(p(x))$ is nondecreasing.*

The monopolist's profit when he produces x is $\pi(x) = x \cdot p(x) - c(x)$, where $c(x)$ is the cost of producing x. The maximization problems for the monopolist and for society are given by

$$\max_{x \geq 0} \pi(x) + 0 \cdot CS(p(x)), \quad \text{and} \tag{2.10}$$

$$\max_{x \geq 0} \pi(x) + 1 \cdot CS(p(x)). \tag{2.11}$$

Set $f(x, t) = \pi(x) + t\,CS(p(x))$, where $X = \mathbb{R}_+$ and $T = \{0, 1\}$. Since $CS(p(x))$ is nondecreasing, $f(x, t)$ is supermodular (and you should check this). Therefore $x^*(1) \geq x^*(0)$, so the monopolist always (weakly) restricts output relative to the social optimum.

Here is the externalities intuition: increases in x increase the welfare of people the monopolist does not care about, an effect external to the monopolist; the market gives the monopolist insufficient incentives to do the right thing. To fully appreciate how much simpler the supermodular analysis is, we have to see how complicated the differentiable analysis would be.

Example 2.8.8 [↑*Example 2.8.7 (p. 45)*] *Suppose that for every $t \in [0, 1]$, the problem*

$$\max_{x \geq 0} \pi(x) + t \cdot CS(p(x))$$

has a unique solution, $x^*(t)$, and that the mapping $t \mapsto x^*(t)$ is continuously differentiable. (This can be guaranteed if we make the right kinds of assumptions on $\pi(\cdot)$ and $CS(p(\cdot))$.) To find the sign of $dx^*(t)/dt$, we assume that the first-order conditions,

$$\pi'(x^*(t)) + t\,dCS(p(x^*(t)))/dx \equiv 0,$$

characterize the optimum. In general, this means that we have to assume that $x \mapsto \pi(x) + t\,CS(p(x))$ is a smooth, concave function. We then take the derivative of both sides with respect to t. This involves evaluating $d(\int_{p(x^*(t))}^{\infty} x(r)\,dr)/dt$. In general, $d(\int_{f(t)}^{\infty} x(r)\,dr)/dt = -f'(t)x(f(t))$, so that when we take derivatives on both sides, we have

$$\pi''(x^*(t))(dx^*/dt) + dCS(p(x^*(t)))/dx - p'(x^*)(dx^*/dt)x(p(x^*)) = 0.$$

Gathering terms yields

$$[\pi''(x^*) - p'(x^*)x(p(x^*))](dx^*/dt) + dCS(p(x^*(t)))/dx = 0. \quad (2.12)$$

Since we are assuming that we are at an optimum, we know that $\pi''(x^*) \leq 0$. By assumption, $p'(x^*) < 0$ and $x > 0$, so the term in the square brackets is negative. As argued earlier, $dCS(p(x^*(t)))/dx > 0$. Therefore, the only way that (2.12) can be satisfied is if $dx^*/dt > 0$. Finally, by the fundamental theorem of calculus (which says that the integral of a derivative is the function itself), $x^*(1) - x^*(0) = \int_0^1 \frac{dx^*(r)}{dt}\,dr$. The integral of a positive function is positive, so this yields $x^*(1) - x^*(0) > 0$.

2.8.c Monotone Comparative Statics

Suppose that $(X, \precsim_X)$ and $(T, \precsim_T)$ are lattices. Define the order $\precsim_{X \times T}$ on $X \times T$ by $(x', t') \succsim_{X \times T} (x, t)$ iff $x' \succsim_X x$ and $t' \succsim_T t$. (This is the unanimity order again.)

Lemma 2.8.9 $(X \times T, \precsim_{X \times T})$ *is a lattice.*

Proof. $(x', t') \vee (x, t) = (\max\{x', x\}, \max\{t', t\}) \in X \times T$, and $(x', t') \wedge (x, t)$
$= (\min\{x', x\}, \min\{t', t\}) \in X \times T$.   ■

Definition 2.8.10   *For a lattice* $(L, \precsim)$, $f : L \to \mathbb{R}$ *is **supermodular** if for all*
$\ell, \ell' \in L$,

$$f(\ell \wedge \ell') + f(\ell \vee \ell') \geq f(\ell) + f(\ell'), \tag{2.13}$$

equivalently

$$f(\ell \vee \ell') - f(\ell') \geq f(\ell) - f(\ell \wedge \ell'). \tag{2.14}$$

Taking $\ell' = (x', t)$ and $\ell = (x, t')$, we recover Definition 2.8.5 (and you should check this).

Exercise 2.8.11   Suppose that $(L, \precsim) = (\mathbb{R}^n, \leq), x \leq y$ iff $x_i \leq y_i, i = 1, \ldots, n$.
Show that L is a lattice. Show that $f : L \to \mathbb{R}$ is supermodular iff it has increasing differences in x_i and x_j for all $i \neq j$. Show that a twice continuously differentiable $f : L \to \mathbb{R}$ is supermodular iff $\partial^2 f / \partial x_i \partial x_j \geq 0$ for all $i \neq j$.

Example 2.8.12   *On the lattices* $(\mathbb{R}^n, \leq)$, *the function* $x \mapsto \min\{f_i(x_i) : i = 1, \ldots n\}$ *is supermodular if each* f_i *is nondecreasing, as is the function* $x \mapsto x_1 \cdot x_2 \cdots x_{n-1} x_n$. *For* $x, t \in \mathbb{R}^n$, *the function* $(x, t) \mapsto x \cdot t$ *is supermodular on* $\mathbb{R}^n \times \mathbb{R}^n$.

Exercise 2.8.13   Show the following.

1. If $f : \mathbb{R}^n \to \mathbb{R}$ is twice continuously differentiable, has a nonnegative gradient, and is supermodular, and $g : \mathbb{R} \to \mathbb{R}$ is twice continuously differentiable and convex, then $g(f(x))$ is increasing and supermodular.

2. If $f : [1, \infty) \times [1, \infty) \to \mathbb{R}$ is defined by $f(x_1, x_2) = x_1 + x_2$ and $g(r) = \log(r)$, then f is supermodular and strictly increasing, while g is strictly increasing and concave. However, $g(f(x))$ is strictly submodular.

Definition 2.8.14   *For* $A, B \subset L$, L *a lattice, the **strong set order** is defined by*
$A \precsim_{Strong} B$ *iff* $\forall (a, b) \in A \times B$, $a \wedge b \in A$ *and* $a \vee b \in B$.

Interval subsets of $\mathbb{R}$ are sets of the form $(-\infty, r), (-\infty, r], (r, s), (r, s],$
$[r, s], (r, \infty),$ or $[r, \infty)$.

Exercise 2.8.15   Show that for intervals $A, B \subset \mathbb{R}$, $A \precsim_{Strong} B$ iff every point in $A \setminus B$ is less than every point in $A \cap B$, and every point in $A \cap B$ is less than every point in $B \setminus A$. Also show that this is true when $\mathbb{R}$ is replaced with any linearly ordered set.

The strong set order is not, in general, reflexive.

Example 2.8.16   *If* $A = \{x \in \mathbb{R}_+^2 : x_1 + x_2 \leq 1\}$, *then* $\neg[A \succsim_{Strong} A]$. *However, if* $A \subset L$ *is itself a lattice, then* $[A \succsim_{Strong} A]$. *In particular, subsets of* $(\mathbb{R}, \leq)$ *are linearly ordered, hence they are lattices.*

Notation 2.8.17   *For* $S \subset L$ *and* $t \in T$, *let* $M(t, S) \subset S$ *be the set of solutions to the problem* $\max_{\ell \in S} f(\ell, t)$. *For* $t, t' \in T$, $S, S' \subset L$, *define* $(t', S') \succsim (t, S)$ *if* $t' \succsim_T t$ *and* $S' \succsim_{Strong} S$.

Theorem 2.8.18 *If $(L, \precsim_L)$ is a lattice, $(T, \precsim_T)$ is a partially ordered set, $f : L \times T \to \mathbb{R}$ is supermodular in ℓ for all t, and has increasing differences in ℓ and t, then $M(t, S)$ is nondecreasing in (t, S) for (t, S) with $M(t, S) \neq \emptyset$.*

Proof. Pick $(t', S') \succsim (t, S)$; we must show that $M(t', S') \succsim_{Strong} M(t, S)$.

Pick $\ell' \in M(t', S') \subset S'$ and $\ell \in M(t, S) \subset S$. By the definition of the strong set order, we have to show that $\ell \vee \ell' \in M(t', S')$ and $\ell \wedge \ell' \in M(t, S)$. We do the first one here; the second is an exercise.

Since $S' \succsim_{Strong} S$, $\ell' \wedge \ell \in S'$ and $\ell' \vee \ell \in S$. As ℓ is optimal in S and $\ell' \wedge \ell \in S$, we know that $f(\ell, t) - f(\ell' \wedge \ell) \geq 0$. Combining the supermodularity of $f(\cdot, t)$ and this last inequality, we have

$$f(\ell \vee \ell', t) - f(\ell', t) \geq f(\ell, t) - f(\ell \wedge \ell', t) \geq 0.$$

Increasing differences, $t' \succsim_T t$, $\ell \vee \ell' \succsim_L \ell'$, and this last inequality yield

$$f(\ell \vee \ell', t') - f(\ell', t') \geq f(\ell \vee \ell', t) - f(\ell', t) \geq 0.$$

Since ℓ' is optimal in S' and $\ell \vee \ell' \in S'$, we have just discovered that $\ell \vee \ell'$ is also optimal in S', that is, $\ell \vee \ell' \in M(t', S')$. ∎

Exercise 2.8.19 Complete the proof of Theorem 2.8.18.

The following is immediate from the last result, simply take $(t', S') = (t, S)$. However, a direct proof makes clearer how submodularity is working.

Corollary 2.8.20 *If $(L, \precsim)$ is a lattice, $f : L \to \mathbb{R}$ is supermodular, $S \subset L$ and $(S, \precsim)$ is a sublattice, that is, for $x, y \in S$, $x \vee y \in S$ and $x \wedge y \in S$, then $\arg \max_{x \in S} f(x)$ is a sublattice.*

2.8.d Quasi-Supermodularity

Sometimes people make the mistake of identifying supermodular utility functions as the ones for which there are complementarities. This is wrong.

Exercise 2.8.21 For $(x_1, x_2) \in \mathbb{R}^2_{++}$, define $u(x_1, x_2) = x_1 \cdot x_2$ and $v(x_1, x_2) = \log(u(x_1, x_2))$.

1. Show that $\partial^2 u / \partial x_1 \partial x_2 > 0$.
2. Show that $\partial^2 v / \partial x_1 \partial x_2 = 0$.
3. Find a monotonic transformation, f, of v such that $\partial^2 f(v)/\partial x_1 \partial x_2 < 0$ at some point $(x_1^\circ, x_2^\circ) \in \mathbb{R}^2_{++}$.

The problem is that supermodularity is not immune to monotonic transformations. That is, supermodularity, like expected utility theory, is a cardinal rather than an ordinal theory. Here is the ordinal version.

Definition 2.8.22 (Milgrom and Shannon) *A function $u : X \to \mathbb{R}$ is **quasi-supermodular** on the lattice X if, $\forall x, y \in X$,*

$$[u(x) \geq u(x \wedge y)] \Rightarrow [u(x \vee y) \geq u(y)], \quad \text{and}$$

$$[u(x) > u(x \wedge y)] \Rightarrow [u(x \vee y) > u(y)].$$

By way of contrast, $f : X \to \mathbb{R}$ is supermodular if $\forall x, y \in X$, and $f(x \vee y) + f(x \wedge y) \geq f(x) + f(y)$, which directly implies that it is quasi-supermodular. The reason for the adjective "quasi" comes from intermediate economics, where you should have learned that a monotonically increasing transformation of a concave utility function is quasi-concave.

Lemma 2.8.23 *A monotonic increasing transformation of a supermodular function is quasi-supermodular.*

Exercise 2.8.24 Prove Lemma 2.8.23.

Recall that a binary relation, $\precsim$, on a set X has a representation $u : X \to \mathbb{R}$ iff $[x \precsim y] \Leftrightarrow [u(x) \leq u(y)]$. For choice theory on *finite* lattices with monotonic preferences, quasi-supermodularity and supermodularity of preferences are indistinguishable.

Theorem 2.8.25 (Chambers and Echenique) *A binary relation on a finite lattice X has a weakly increasing and quasi-supermodular representation iff it has a weakly increasing and supermodular representation.*

Proof. Since supermodularity implies q-supermodularity, we need only show that a weakly increasing q-supermodular representation can be monotonically transformed to be supermodular. Let u be quasi-supermodular, set $u(X) = \{u_1 < u_2 < \cdots < u_N\}$, and define $g(u_n) = 2^{n-1}$, $n = 1, \ldots, N$. You can show that the function $v(x) = g(u(x))$ is supermodular. ■

Exercise 2.8.26 There are three results that you should know about the relation between monotonicity and quasi-supermodularity:

1. Strong monotonicity implies quasi-supermodularity, hence supermodularity: Show that if a binary relation $\precsim$ on a finite lattice X has a strictly increasing representation, then that representation is quasi-supermodular. [By the Chambers and Echenique result above, this implies that the binary relation has a supermodular representation.]

2. Weak monotonicity does not imply quasi-supermodularity: Let $X = \{(0, 0), (0, 1), (1, 0), \backslash(1, 1)\}$ with $(x, y) \precsim (x', y')$ iff $(x, y) \leq (x', y')$. Show that the utility function $u(x, y) = 0$ if $x = y = 0$ and $u(x, y) = 1$ otherwise is weakly monotonic, but no monotonic transformation of it is quasi-supermodular.

3. Supermodularity does not imply monotonicity: Let $X = \{(0, 0), (0, 1), (1, 0), (1, 1)\}$ with $(x, y) \precsim (x', y')$ iff $(x, y) \leq (x', y')$. Show that the utility function $u(0, 0) = 0$, $u(0, 1) = -1$, $u(1, 0) = 2$ and that $u(2, 2) = 1.5$ is strictly supermodular but not monotonic.

2.9 ◆ Tarski's Lattice Fixed–Point Theorem and Stable Matchings

In this section we give a lattice formulation of Gale-Shapley matching problems. Our aim is to create pairs by choosing one person from each side of a market, in such a fashion that everyone is happier with his or her part of the pairing than he or she would be alone, and there is no pair that could be rematched so as to make both better off.

2.9.a Matching Problems and Stability

A **matching problem** is a 4-tuple, $(M, W, (\succ_m)_{m \in M}, (\succ_w)_{w \in W})$. We assume that $M = \{m_1, \ldots, m_n\}$ and $W = \{w_1, \ldots, w_k\}$ are disjoint finite sets (mnemonically, *M*en and *W*omen). We also assume that each $m \in M$ has strict rational preferences, $\succ_m$, over the set $W \cup \{m\}$. By strict, we mean that no distinct pair is indifferent. The expression $w \succ_m w'$ means that m prefers to be matched with w rather than w', and $m \succ_m w$ means that m would prefer to be single, that is, matched with himself, rather than being matched to w. In exactly the same fashion, each $w \in W$ has strict rational preferences over $M \cup \{w\}$. Man m is *acceptable* to woman w if she likes him at least as much as staying single, and woman w is *acceptable* to man m if he likes her at least as much as staying single.

As in Example 2.6.29 (p. 39), a **matching** μ is a one-to-one onto function from $M \cup W$ to itself such that $\mu \circ \mu$ is the identity, $[\mu(m) \neq m] \Rightarrow [\mu(m) \in W]$, and $[\mu(w) \neq w] \Rightarrow [\mu(w) \in M]$. Interest focuses on properties of the set of matchings that are acceptable to everyone and that have the property that no one can find someone that he or she prefers to his or her own match and who prefers him or her to his or her match. Formally,

Definition 2.9.1 *A matching μ is **stable** if it satisfies:*

1. *individual rationality: $\forall m$, $\mu(m) \succeq_m m$ and $\forall w$, $\mu(w) \succeq_w w$, and*
2. *stability: $\neg \exists (m, w) \in M \times W$ such that $w \succ_m \mu(m)$ and $m \succ_w \mu(w)$.*

For the purposes of a later proof, the following is worth having as a separate result, even though it is an immediate implication of the strict preferences we have assumed.

Lemma 2.9.2 *If μ is individually rational, then it is stable iff $\neg \exists (m, w)$ such that $[w \succ_m \mu(m) \wedge m \succeq_w \mu(w)]$ or $[m \succ_w \mu(w) \wedge w \succeq_m \mu(m)]$.*

Exercise 2.9.3 A matching μ' is indicated by **boldface** and the *'s in following table, which shows preferences (Ex. 2.4, Roth and Sotomayor 1990).

$$w_2 \succ_{m_1} \boldsymbol{w_1^*} \succ_{m_1} w_3 \succ_{m_1} m_1 \qquad \boldsymbol{m_1^*} \succ_{w_1} m_3 \succ_{w_1} m_2 \succ_{w_1} w_1$$
$$w_1 \succ_{m_2} \boldsymbol{w_3^*} \succ_{m_2} w_2 \succ_{m_2} m_2 \qquad \boldsymbol{m_3^*} \succ_{w_2} m_1 \succ_{w_2} m_2 \succ_{w_2} w_2$$
$$w_1 \succ_{m_3} \boldsymbol{w_2^*} \succ_{m_3} w_3 \succ_{m_3} m_3 \qquad m_1 \succ_{w_3} m_3 \succ_{w_3} \boldsymbol{m_2^*} \succ_{w_3} w_3$$

1. Verify that μ' is stable.

2. Given a matching μ, for each m, define m's **possibility set** as $P_\mu(m) = \{w \in W : m \succeq_w \mu(w)\} \cup \{m\}$. Show that the matching μ' maximizes each m's preferences over $P_{\mu'}(m)$ and maximizes each w's preferences over $P_{\mu'}(w)$. [The simultaneous optimization of preferences over the available set of options is a pattern we see repeatedly.]

The possibility sets depend on who is matched with whom. More explicitly, if the other side of the market has higher quality matches, then my possibility set shrinks because there are fewer people on the other side who would like to switch to me.

2.9.b   Stable Matchings as Fixed Points

Another way to view functions is as elements of a product space. Let $\{A_i : i \in \Lambda\}$ be a collection of sets and define $A = \cup_i A_i$. A function $f : \Lambda \to A$ such that for all $i \in \Lambda$, $f(i) \in A_i$ is a point in $\times_{i \in \Lambda} A_i$, and vice versa.

Example 2.9.4   *Let $\Lambda = \{1, 2, 3\}$, $A_1 = \{a, b\}$, $A_2 = \{1, 2\}$, $A_3 = \{x, y\}$, and $A = \cup_i A_i = \{a, b, 1, 2, x, y\}$. The set of all functions from Λ to A contains 6^3 points. Of those 6^3 points, 8 have the property that for all $i \in \Lambda$, $f(i) \in A_i$.*

Definition 2.9.5   *A **fantasy** is a pair of functions $v = (v_M, v_W)$, $v_M \in \mathcal{V}_M := \times_{m \in M}(W \cup \{m\})$, $v_W \in \mathcal{V}_W := \times_{w \in W}(M \cup \{w\})$. The set of fantasies is $\mathcal{V} = \mathcal{V}_M \times \mathcal{V}_W$.*

Exercise 2.9.6   Show that any matching μ induces a fantasy, but fantasies may not induce matchings. Further, show that a fantasy induces a matching iff $[v_M(m) = w] \Leftrightarrow [m = v_W(w)]$.[9]

Equilibria are at the center of economic theorizing. The basic mathematical tools for the study of the existence of equilibria are fixed-point theorems.

Definition 2.9.7   *A point $x^* \in X$ is a **fixed point of the function** $f : X \to X$ if $f(x^*) = x^*$, and a **fixed point of the correspondence** $G : X \twoheadrightarrow X$ if $x^* \in G(x^*)$.*

Example 2.9.8   *The function $f(x) = ax + b$ from $\mathbb{R}$ to $\mathbb{R}$ has the unique fixed point $x^* = \frac{b}{1-a}$ if $a \neq 1$. The correspondence $F(x) = \{y : y \geq ax + b\}$ has a set of fixed points, $S = \{x^* : x^* \geq \frac{b}{1-a}\}$. The function $g(x) = x^2$ from $\mathbb{R}$ to $\mathbb{R}$ has two fixed points, $x^* = 0$ and $x^* = 1$. The correspondence $G(x) = \{y : y \geq x^2\}$ has a set of fixed points, $S = \{x^* : 0 \leq x^* \leq 1\}$.*

The arg max notation specifies the set of optimizers of a preference relation over a given set—for any finite set X and rational preference ordering $\succeq$, define arg max$_\succ X = \{x \in X : x \succeq X\}$. For any fantasy v, define a new fantasy $T(v) = (T_M(v), T_W(v))$ by

$$T_M(v)(m) = \text{arg max}_{\succ_m}(\{w \in W : m \succeq_w v_W(w)\} \cup \{m\}),$$

$$T_W(v)(w) = \text{arg max}_{\succ_w}(\{m \in M : w \succeq_m v_M(m)\} \cup \{w\}).$$

Here T is a function from $\mathcal{V}$ to itself because preferences are strict. With ties in the preferences, T would be a correspondence. T takes a fantasy, and for each person figures out who he or she could match to so as to make that other person better

---

9. We dedicate this exercise to Randall Wright.

off. It then picks everyone's favorite in these sets. This means that if a fantasy is Pareto worse for the men, then each woman is facing a (weakly) larger choice set, so is made happier.

The following indicates why we are interested in fixed points of T. It is the general version of what you checked in Exercise 2.9.3 (p. 50).

Lemma 2.9.9 *If a matching μ is stable, then the fantasy it induces is a fixed point of T. If a fantasy v is a fixed point of T, then it induces a stable matching.*

Proof. Check the definitions. ■

2.9.c Tarski's Fixed–Point Theorem

We now turn to the study of the properties of the set of fixed points of T, that is, to a study of the properties of stable matchings. Define the unanimity partial orders $\succeq_M$ on $\mathcal{V}_M$ and $\succeq_W$ on $\mathcal{V}_W$ by

$$v_M \succeq_M v'_M \quad \text{iff } \forall m \; v_M(m) \succeq_m v'_M(m),$$

$$v_W \succeq_W v'_W \quad \text{iff } \forall w \; v_W(w) \succeq_w v'_W(w).$$

Note carefully the *order reversal* in the following: define a partial order, $\succeq$, on $\mathcal{V} = \mathcal{V}_M \times \mathcal{V}_M$ by

$$(v_M, v_W) \succeq (v'_M, v'_W) \quad \text{iff } v_M \succeq_M v'_M \text{ and } v_W \preceq_W v'_W.$$

Recall that a nonempty POSET $(X, \precsim)$ is a **lattice** if every pair of elements has a greatest lower bound ($\inf_X$) and a least upper bound ($\sup_X$), both in X, in the order $\succsim$. Pairs of elements are not the only classes of sets that may have an infimum and a supremum.

Definition 2.9.10 *For $S \subset X$, $(X, \precsim)$ a lattice, ℓ is the **greatest lower bound for S** or the **infimum of S**, denoted $\ell = \inf(S)$, if $[\ell \precsim S] \land [[\ell' \precsim S] \Rightarrow [\ell' \precsim \ell]]$, and u is the **least upper bound for S** or the **supremum of S**, denoted $u = \sup(S)$, if $[u \succsim S] \land [[u' \succsim S] \Rightarrow [u' \succsim u]]$.*

Example 2.9.11 *$(\mathbb{R}, \leq)$ is a lattice; $(0, 1) = \{x \in \mathbb{R} : 0 < x < 1\}$ has a supremum, 1, and an infimum, 0, neither of which belongs to S. $\mathbb{Z}$ has neither a supremum nor an infimum. Any nonempty subset of $\mathbb{N}$ contains its infimum.*

Definition 2.9.12 *The **restriction of a relation $\succsim$ to a subset** $S \subset X$ is defined by $\succsim_{|S} = \succsim \cap (S \times S)$.*

Definition 2.9.13 *If $(X, \precsim)$ is a lattice, then a **sublattice** is a subset S of X such that $(S, \succsim_{|S})$ is a lattice, and a **lattice subset** is a subset T of X such that for every $x, y \in T$, there is a greatest lower bound and a least upper bound with respect to T.*

Every sublattice is a lattice subset, but the reverse is not true. The difference between sublattices and lattice subsets is whether or not one is defining $x \lor y$ and $x \land y$ in X or in the subset.

Example 2.9.14   *The set $S = \{(0, 0), (1, 0), (0, 1), (1, 1)\} \subset \mathbb{R}^2$ is the set of vertices of a square. It is a sublattice and a lattice subset of the lattice $(\mathbb{R}^2, \leq)$. The set $T = \{(0, 0), (2, 1), (1, 2), (3, 3)\} \subset \mathbb{R}^2$ is the set of vertices of a diamond, it is a not a sublattice because $(1, 1) = (2, 1) \vee (1, 2) \notin T$ and $(2, 2) = (2, 1) \wedge (1, 2) \notin T$. However, it is a lattice subset because, within T, every pair of elements has a unique greatest lower bound and a unique least upper bound. To see why, note that $(0, 0) \leq (2, 1)$, $(0, 0) \leq (1, 2)$ and there is nothing in T that is smaller than both $(2, 1)$ and $(1, 2)$ yet larger than $(0, 0)$. The other cases are similar.*

Exercise 2.9.15   Show the following.

1. $\{(x_1, x_2) \in \mathbb{R}^2 : x_1 \leq 0 \text{ or } x_2 \leq 0\} \cup \{(x_1, x_2) \in \mathbb{R}^2 : x_1 = x_2\}$ is not a sublattice but is a lattice subset.

2. Neither $T' = \{(0, 3), (1, 1), (2, 2), (3, 0)\}$ nor its convex hull are lattice subsets of $(\mathbb{R}^2, \leq)$.

3. The convex hull of $\{(0, 0), (2, 1), (1, 2), (3, 3)\} \subset \mathbb{R}^2$ is a sublattice.

4. If C is a convex lattice subset of $(\mathbb{R}^n, \leq)$, then it is a sublattice.

The upper bound of x, y in the set X is written $\sup_X(x, y)$. It may be different than its upper bound in S, written $\sup_S(x, y)$. In the definition of a sublattice, this is very important—x, $y \in S$ need not imply that $\sup_X(x, y) \in S$, only that $\sup_S(x, y) \in S$. This means that it is more than usually important to pay attention to subscripts, as you can see by trying to read the following example after having erased the subscripts.

Example 2.9.16   *$S := \{(0, 1), (0, 1), (2, 2)\}$ is a sublattice of $X := \{0, 1, 2\}^2 \subset \mathbb{R}^2$ with the usual vector ordering. $\sup_X((0, 1), (1, 0)) = (1, 1) \in X$ but $(1, 1) \notin S$, and $\sup_S((0, 1), (1, 0)) = (2, 2) \in S$.*

Definition 2.9.17   *A lattice is **complete** if every nonempty subset of X has an inf and a sup in X.*

Notation Alert 2.9.A   *In general, the meaning of the adjective "complete" depends critically on the noun that it modifies. The completeness of a lattice $(X, \precsim)$ neither implies nor is implied by the completeness of $\precsim$. Later, we will see complete metric spaces and complete probability spaces, and these spaces are not orders.*

Example 2.9.18   *$X := \{0, 1, 2\}^2 \subset \mathbb{R}^2$ with the usual vector ordering, $\leq$, is a complete lattice though $\leq$ is not a complete order. The order relation $\leq$ is complete on $\mathbb{R}$, but $(\mathbb{R}, \leq)$ is not a complete lattice because sets such as $(-\infty, a)$ have no lower bound, and $\mathbb{Z} \subset \mathbb{R}$ has neither an upper nor a lower bound.*

Lemma 2.9.19   *Every finite lattice is complete.*

Exercise 2.9.20   Prove Lemma 2.9.19 and show that $(\mathcal{V}, \succeq)$ is a finite lattice.

A function $R : X \to X$ from a lattice to itself is increasing if $[x \succsim y] \Rightarrow [R(x) \succsim R(y)]$. Repeated applications of the function R are denoted R^n, that is, $R^2 = R \circ R$, $R^3 = R \circ R^2$, and $R^n = R \circ R^{n-1}$. Note that x^* is a fixed point of R iff it is a fixed point of R^n for every $n \geq 1$, and that $R^n(x) = R^{n+1}(x)$ implies that $x^* = R^n(x)$ is a fixed point of R. Interest focuses on the properties of the set of fixed points of increasing mappings.

Example 2.9.21 *Consider the nine-point lattice, $X = \{0, 1, 2\}^2 \subset \mathbb{R}^2$ with the usual vector ordering. The function $R(i, j) = (i, j)$ for $(i, j) \notin \{(1, 1), (2, 1), (1, 2)\}$, and $R((1, 1)) = R((2, 1)) = R((1, 2)) = (2, 2)$ is increasing, $(0, 1)$ and $(1, 0)$ belong to the set, S, of fixed points, $(1, 1) = \sup_X((0, 1), (1, 0)) \notin S$, but $(2, 2) = \sup_S((0, 1), (1, 0)) \in S$.*

Theorem 2.9.22 (Tarski) *Every increasing map, R, from a complete lattice, $(X, \precsim)$, to itself has a largest fixed point, $x^*_{max} = \sup(\{x \in X : x \precsim R(x)\})$, and a least fixed point, $x^*_{min} = \inf(\{x \in X : R(x) \precsim x\})$.*

Proof. Suppose that $(X, \precsim)$ is a complete lattice, and that R is an increasing map from X to itself. We first show that $x^*_{max} = \sup(\{x \in X : x \precsim R(x)\})$ is a fixed point.

Define $U = \{x \in X : x \precsim R(x)\}$. Since X is complete, $\inf(X)$ exists. $U \neq \emptyset$ because $\inf(X) \in U$. As $U \neq \emptyset$ and $(X, \precsim)$ is complete, $x^*_{max} = \sup U$ exists. To show that x^*_{max} is a fixed point, it is sufficient to show that $x^*_{max} \precsim R(x^*_{max})$ and $R(x^*_{max}) \precsim x^*_{max}$ (by antisymmetry).

By the definition of supremum, for every $u \in U$, $u \precsim x^*_{max}$. As R is increasing, for all $u \in U$, $u \precsim R(u) \precsim R(x^*_{max})$, so that $R(x^*_{max})$ is an upper bound for U. Since x^*_{max} is the least upper bound, $x^*_{max} \precsim R(x^*_{max})$.

Since $x^*_{max} \precsim R(x^*_{max})$ and R is increasing, $R(x^*_{max}) \precsim R(R(x^*_{max}))$, so that $R(x^*_{max}) \in U$. As x^*_{max} is the least upper bound for U, $R(x^*_{max}) \precsim x^*_{max}$.

We now show that x^*_{max} is the largest fixed point of R. Let x^* be any fixed point, that is, satisfies $R(x^*) = x^*$. This implies that $x^* \precsim R(x^*)$, which in turn means that $x^* \in U$. Therefore, $x^* \precsim x^*_{max}$.

The proof that x^*_{min} is the least fixed point proceeds along very similar lines, and we leave it as an exercise. ∎

Exercise 2.9.23 Show that x^*_{min} is the least fixed point of R.

We can find fixed points by repeatedly applying an increasing R if we start in one of the right places.

Corollary 2.9.24 *If R is an increasing map from a finite, hence complete, lattice $(X, \precsim)$ to itself, if $x_0 = \sup X$ or $x_0 = \inf X$ and $x_{t+1} = R(x_t)$, then $x_{n+1} = x_n$ for all $n \geq N$ for some $N \leq \#X$. Further, the set S of fixed points is a complete lattice.*

Proof. Suppose first that $x_0 = \inf X$. This implies that $x_1 = R(x_0)$ satisfies $x_1 \succsim x_0$. Since R is increasing, $x_2 = R(x_1) \succsim R(x_0) = x_1$. More generally, $x_{n+1} = R(x_n) \succsim R(x_{n-1}) = x_n$. As X is finite, this process must end in at most $\#X$ steps. If $x_0 = \sup X$, then by the same logic $x_{n+1} \precsim x_n$ for all n.

Let S be the set of fixed points. Since X is finite, S is finite, so it is sufficient to show that S is a lattice as every finite lattice is complete.

Pick $x, y \in S$ and let $z = x \vee y$. Either $z \in S$ or $z \notin S$. We must show that $\sup_S(\{x, y\})$ exists in both cases.

Case 1: If $z \in S$, $z = \sup_S(\{x, y\})$.

Case 2: Suppose now that $z \notin S$. We must show that there is a fixed point of R that is an upper bound for both x and y. Let $z_0 = x \vee y$, so that $z_0 \succsim x$ and $z_0 \succsim y$. If $R(z_0) = z_0$, that is, if $z_0 \in S$, then we are done. Otherwise, define $z_1 = z_0 \vee R(z_0)$ so that $z_1 \succsim z_0$. Defining $z_{t+1} = R(z_t)$ for $t \geq 1$ gives an increasing sequence of points in X. Such a sequence must arrive at a fixed point: call it $z_T \in S$. By construction, $z_T \succsim z_{T-1} \succsim \cdots \succsim z_0$. By transitivity, $z_T \succsim x$ and $z_T \succsim y$, so that z_T is an upperbound for x and y that belongs to S. ■

This result is true without the finiteness, a fact which is useful in the study of supermodular games. As we have recently seen in Example 2.9.14, lattice subsets need not be sublattices.

Example 2.9.25 *Let X be the set $\{(n_1, n_2) \in \mathbb{Z}^2 : 0 \leq n_1, n_2 \leq 3\}$ so that $(X, \leq)$ is a sublattice of $\mathbb{Z}^2$. We are going to define an increasing function $f : X \to X$ that has as its fixed points the lattice subset $T = \{(0, 0), (2, 1), (1, 2), (3, 3)\} \subset X$. Since T is not a sublattice, this shows that the set of fixed points of an increasing function from a complete lattice to itself need not be a sublattice.*
For $(n_1, n_2) \in T$, define $f(n_1, n_2) = (n_1, n_2)$. For $(n_1, n_2) \notin T$, define $f(n_1, n_2) = (0, 0)$ if $n_1 + n_2 < 3$ and $f(n_1, n_2) = (3, 3)$ if $n_1 + n_2 \geq 3$.

Getting the starting point right is important in Corollary 2.9.24.

Exercise 2.9.26 Let $X = \{0, 1\}^2 \subset \mathbb{R}^2$ with the vector ordering. Define $R : X \to X$ by $R((0, 0)) = (0, 0)$, $R((1, 1)) = (1, 1)$, $R((1, 0)) = (0, 1)$, and $R((0, 1)) = (0, 1)$.

1. Verify that the set of fixed points is $S = \{(0, 0), (1, 1)\}$ and that S is a complete sublattice of X.
2. Find sup X and inf X.
3. Show that R is an increasing map from X to X.
4. If $x_0 = \sup X$ or $x_0 = \inf X$, show that for all $n \geq 1$, $x_{n+1} = x_n$.
5. If $x_0 = (0, 1)$ or $x_0 = (1, 0)$, show that $\neg(\exists N)(\forall n \geq N)[x_{n+1} = x_n]$.

Exercise 2.9.27 Show that the mapping $T : \mathcal{V} \to \mathcal{V}$ defined previously is increasing. Conclude that there is a nonempty set of stable matchings and that the stable matchings form a complete sublattice of $\mathcal{V}$.

The men's favorite fantasy is $v^{*M} = (\overline{v}_M, \underline{v}_W)$ defined by $\overline{v}_M(m) = \arg \max_{\succeq_m} W \cup \{m\}$ and $\underline{v}_W(w) = \{w\}$. The women's favorite fantasy, v^{*W}, is defined analogously.

Exercise 2.9.28 [↑Exercise 2.9.3 (p. 50)] Show that repeated applications of T to the men's favorite fantasy yields $((w_2, w_3, w_1), (m_3, m_1, m_2))$, whereas repeated applications of T to the women's favorite fantasy yields the matching μ' given earlier. Show that these two matchings are the largest and smallest fixed points of T. Give the set of all fixed points of T.

Exercise 2.9.29 Show that, for any matching problem, if v^* satisfies $\mu_M^* = T^n(v^{*M}) = T^{n+1}(v^{*M})$, then μ_M^* is the matching that is unanimously weakly preferred by the men to any stable matching. Formulate and prove the parallel result for μ_W^*.

2.10 ◆ Finite and Infinite Sets

The purpose of this section is to learn how to compare the sizes of sets by comparing the number of things they contain. This is easy when the sets are finite and difficult when they are not.

- Easy: If $A = \{1, 2, 3\}$, its cardinality, $\#A$, is 3, and if $B = \{1, 2, 3, 4\}$, $\#B = 4$, so that B is bigger than A. This might be captured by the observation that $A \subsetneq B$, but this does not help with $C = \{4, 5, 6, 7\}$. A better way to capture the difference in sizes is to note that any one-to-one function from A to B or from A to C cannot be onto.

- Difficult: Consider the set $C = \{2, 4, 6, \ldots\}$ of even numbers and the set $D = \{1, 2, 3, \ldots\}$ of natural numbers. $C \subsetneq D$ but the mapping $c \mapsto c/2$ is a bijection between C and D.

Definition 2.10.1 *Two sets A and B are **equivalent** or **have the same cardinality** if there is a bijection $f : A \leftrightarrow B$.*

Exercise 2.10.2 Show that Definition 2.10.1 really does define an equivalence relation.

Exercise 2.10.3 In each case, show that the sets have the same cardinality by giving a bijection between them:

1. $A = \{n^2 : n \in \mathbb{N}\}$ and $B = \{2n : n \in \mathbb{N}\}$.
2. The intervals $[0, 1]$ and $[a, b]$, where $a < b$.
3. The intervals $(0, 1)$ and (a, b), where $a < b$.
4. The set of real numbers and the interval $(-\pi/2, \pi/2)$. [Hint: Think about the tangent function.]

Bijections can be hard to find.

Example 2.10.4 *The function $F : \mathbb{R} \to (0, 1)$ given by $F(r) = \frac{e^r}{1+e^r}$ is a bijection, so $(0, 1)$ and $\mathbb{R}$ have the same cardinality. Since $(0, 1) \subset [0, 1] \subset \mathbb{R}$ and $\mathbb{R}$ and $(0, 1)$ have the same cardinality, we rather expect that $[0, 1]$ and $\mathbb{R}$ have the same cardinality, but while there are bijections, they may not leap to mind. However, there is a bijection between $(0, 1) \subset [0, 1]$ and $\mathbb{R}$ and a bijection between $[0, 1] \subset \mathbb{R}$ and $[0, 1]$. The main point of this section is the Cantor-Bernstein theorem, which tells us that this is sufficient to establish that $\mathbb{R}$ and $[0, 1]$ have the same cardinality.*

2.10.a Comparing the Sizes of Finite Sets

For finite sets, cardinality measures what we have always thought of as the size of a set. From Definition 2.3.17 (p. 25) and Lemma 2.3.19, a set X is finite if it is empty or if there exists $n \in \mathbb{N}$ and a bijection $f : \{1, \ldots, n\} \leftrightarrow X$.

The idea of **initial segments**, that is, the sets $\Psi_n := \{1, \ldots, n\}$, may appear quite early in one's childhood.[10] Taking a piece of candy away from a child who understands initial segments makes them unhappy. This is the observation that there is no bijection between Ψ_m and Ψ_n if $m < n$.

Lemma 2.10.5   *If B is a proper subset of a finite set A, then a bijection $f : A \leftrightarrow B$ does not exist.*

Exercise 2.10.6   Prove Lemma 2.10.5. [Hint: If you get stuck, take a look at the proof of Theorem 2.5.14.]

2.10.b   Comparing the Sizes of Infinite Sets

Lemma 2.10.5 says that a proper subset of a finite set cannot be equivalent to the whole set. This is quite clear. As the sets $C = \{2, 4, 6, 8, \ldots\}$ and $D = \{1, 2, 3, 4, \ldots\}$ above show, this is not true for $\mathbb{N}$. We must therefore conclude the following.

Theorem 2.10.7 $\mathbb{N}$ *is not finite.*

Definition 2.10.8   *A set A is **infinite** if it is not finite. It is **countably infinite** if there exists a bijection $f : \mathbb{N} \leftrightarrow A$. It is **countable** if it is finite or countably infinite. Otherwise it is **uncountable**.*

There are uncountable sets; we get to a specific one, $\mathcal{P}(\mathbb{N})$, later.

The function $n \mapsto n$ is a bijection showing that $\mathbb{N}$ is a countably infinite set. The list $\{0, +1, -1, +2, -2, \ldots\}$ gives the following bijection between $\mathbb{N}$ and $\mathbb{Z}$.

$$f(n) = \begin{cases} 0 & \text{if } n = 1, \\ n/2 & \text{if } n \geq 2 \text{ and } n \text{ is even,} \\ -(n-1)/2 & \text{if } n \geq 2 \text{ and } n \text{ is odd.} \end{cases} \tag{2.15}$$

A countably infinite set is the smallest kind of infinite set. To show this, we use Axiom 1 (p. 21) to prove the induction principle, which we use here and in the proof of the Cantor-Berstein theorem.

Theorem 2.10.9   *If $\mathbb{A}_n$ is a statement for each $n \in \mathbb{N}$, $\mathbb{A}_1$ is true, and $[\mathbb{A}_n \Rightarrow \mathbb{A}_{n+1}]$ for all $n \in \mathbb{N}$, then $\mathbb{A}_n$ is true for all $n \in \mathbb{N}$.*

Proof.   Let $S = \{n \in \mathbb{N} : \mathbb{A}_n \text{ is false }\}$. By Axiom 1, if $S \neq \emptyset$, then S has a least element that we call s. By assumption, $s \neq 1$. By the definition of S, $\mathbb{A}_{s-1}$ is true. By assumption, $[\mathbb{A}_{s-1} \Rightarrow \mathbb{A}_s]$, so that $s \notin S$.   ∎

Theorem 2.10.10   *Every infinite set contains a countably infinite set.*

---

10. One of the authors, in what may be a case of overzealous parenting, placed a set of three coins in front of his three-year-old daughter and asked her "Is that collection of coins countable?" She proceeded to pick up the first coin with her right hand, put it in her left hand, and say "one"; she picked up the second coin, put it in her left hand, and said "two" and picked up the final coin, put it in her left hand, and said "three." Thus, she put the set of coins into a one-to-one assignment with the first three natural numbers. We now make use of one-to-one assignments between elements of two sets.

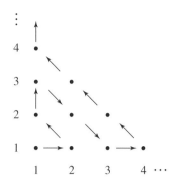

FIGURE 2.10.12

Proof. Let A be an infinite set. Since A is infinite, for all $n \in \mathbb{N}$ and all $f : \{1, \ldots, n\} \to A$, $(A \setminus f(\{1, \ldots, n\})) \neq \emptyset$. Define $f(1) = a_1$ for some $a_1 \in A$. Such an a_1 exists because A is infinite. Given a bijection $f : \{1, \ldots, n\} \leftrightarrow \{a_1, \ldots, a_n\}$ between $\{1, \ldots, n\}$ and a subset of A, define $f(n + 1)$ to be some point in the nonempty set $A \setminus f(\{1, \ldots, n\})$.

Let $\mathbb{A}_n$ be the statement "$f(n)$ exists." By induction, for all $n \in \mathbb{N}$, this statement is true. All that is left to show is that f is a bijection between $\mathbb{N}$ and a subset of A. Now, f is a bijection between $\mathbb{N}$ and a subset of A iff for all $n \in \mathbb{N}$, f is a bijection between $\{1, \ldots, n\}$ and $f(\{1, \ldots, n\})$. Let $\mathbb{A}_n$ be the statement "f is a bijection between $\{1, \ldots, n\}$ and $f(\{1, \ldots, n\})$". Again, by induction, for all $n \in \mathbb{N}$, this statement is true. ∎

Exercise 2.10.11 Prove that a finite union of countable sets is countable. Therefore, if A is uncountable and $B \subset A$ is countable, then $A \setminus B$ is uncountable. [Hint: Let $\mathbb{A}_n$ be the statement "a union of n countable sets is countable." The inductive proof now reduces to proving that the union of any two countable sets is countable.]

Example 2.10.12 $(\mathbb{N} \times \mathbb{N}) \sim \mathbb{N}$. *One can arrange $\mathbb{N} \times \mathbb{N}$ in an infinite matrix and follow any zigzag path that covers all the points.* [*See Figure 2.10.12.*]

Exercise 2.10.13 Show that if $A \sim A'$ and $B \sim B'$, then $A \times B \sim A' \times B'$. [Let $f : A \leftrightarrow A'$ and $g : B \leftrightarrow B'$ be bijections, and check properties of the mapping $(a, b) \mapsto (f(a), g(b))$.] Using this and previous results about $\mathbb{N}$ and $\mathbb{Z}$, show that $\mathbb{Z} \times \mathbb{Z} \sim \mathbb{N}$.

We can weaken the condition for proving the countability of a given set A.

Theorem 2.10.14 *Let A be a nonempty set. The following statements are equivalent:*

1. *There is a surjection $f : \mathbb{N} \to A$.*

2. *There is an injection $g : A \to \mathbb{N}$.*

3. *A is countable.*

Proof. A is countable iff it is either finite or countably infinite. It is finite iff there is a bijection $\varphi : \{1, \ldots, n\} \leftrightarrow A$ for some n. Define $f(m) = \varphi(m)$ if $m \leq n$, and define $f(m) = \varphi(1)$ if $m > n$. This gives a surjection $f : \mathbb{N} \to A$. Defining $g = \varphi^{-1}$ gives an injection $g : A \to \mathbb{N}$.

It is countably infinite iff there is a bijection, h, between A and $\mathbb{N}$. In this case, we take $f = g = h$.   ■

One way to understand the direction of maps in Theorem 2.10.14 is to think of a surjection $f : \mathbb{N} \to A$ as assigning natural numbers to all members of A. Since the map f can be many-to-one, A would have no more elements than $\mathbb{N}$, so it is countable. When there is an injection $g : A \to \mathbb{N}$, we embed A in $\mathbb{N}$, so again, informally, we can think of A as having no more elements than $\mathbb{N}$, and thus it is countable.

Theorem 2.10.15   *A countable union of countable sets is countable.*

Proof.   Let $\{A_i : i \in I\}$ be an indexed family of countable sets where I is countable. As each A_i is countable, for each i we can choose a surjection $f_i : \mathbb{N} \to A_i$. Similarly, we can choose a surjection $g : \mathbb{N} \to I$. Define $h : \mathbb{N} \times \mathbb{N} \to \cup_{i \in I} A_i$ by $h(n, m) = f_{g(n)}(m)$, which is a surjection.   ■

The following theorem is the basic tool that we use to show that uncountable infinite sets are equivalent.

Theorem 2.10.16 (Cantor–Bernstein)   *A and B are equivalent iff there exist $A_1 \subset A$ a bijection $f : A_1 \leftrightarrow B$, $B_1 \subset B$, and a bijection $g : B_1 \leftrightarrow A$.*

From Definition 2.9.12, the restriction of a function $f : A \to B$ to a subset $E \subset A$ is defined by $f_{|S} = Gr(f) \cap (S \times B)$. Equivalently, $f_{|S}$ is the function from S to B defined by $f_{|S}(s) = f(s)$ for each $s \in S \subset A$. Note that if $f : A \leftrightarrow B$ is a bijection, then its restriction to $A_1 \subseteq A$ is a bijection between A_1 and $f(A_1)$.

Proof of Cantor–Berstein.   If $A \sim B$, take $A_1 = A$ and $B_1 = B$.

Now suppose that $A_1 \subset A$, $A_1 \sim B$, and $B_1 \subset B$, $B_1 \sim A$. Let $f : A \leftrightarrow B_1$ and $g : B \leftrightarrow A_1$ be bijections. We must show that there is a bijection from all of A to all of B. For this, it is sufficient to show that there is a bijection from A to A_1 because g is a bijection between B and A_1. We do this by partitioning A and A_1 into a countably infinite collection of equivalent parts.

Set $A_0 = A$, $B_0 = B$ and inductively define

$$
\begin{aligned}
A_0 = A, \quad & A_1 = g(B_0), \quad A_2 = g(B_1), \quad A_3 = g(B_2), \quad \cdots \\
B_0 = B, \quad & B_1 = f(A_0), \quad B_2 = f(A_1), \quad B_3 = f(A_2), \quad \cdots
\end{aligned}
\tag{2.16}
$$

Check that $A_0 \supset A_1 \supset A_2 \supset A_3 \supset \cdots$. Define $A_\infty = \cap_n A_n$.

Check that the appropriate restrictions of $g \circ f$ are bijections between A_0 and A_2, between A_1 and A_3, and more generally between A_k and A_{k+2}. Since f, g are bijections, A_0 is equivalent to $f(A_0) = B_1$ and B_1 is equivalent to $A_2 = g(B_1)$, so by composing the bijections, A_0 is equivalent to A_2 via $g \circ f$. Similarly A_1 and A_3 are equivalent via $g \circ f_{|A_1}$. Generally A_k and A_{k+2} are equivalent through an appropriate restriction of $g \circ f$. This implies that (the appropriate restrictions of) $g \circ f$ are bijections between $(A_0 \setminus A_1)$ and $(A_2 \setminus A_3)$ and more generally between $(A_k \setminus A_{k+1})$ and $(A_{k+2} \setminus A_{k+3})$. Recall that the proof will be complete once we show that A_0 and A_1 are equivalent.

With these preliminaries, we can give partitions of A_0 and A_1, which, after rearrangement, shows that A_0 and A_1 are equivalent. The partitions are

$$A_0 = A_\infty \cup (A_0 \setminus A_1) \cup (A_1 \setminus A_2) \cup (A_2 \setminus A_3) \cup (A_3 \setminus A_4) \cup (A_4 \setminus A_5) \cup \cdots$$

$$A_1 = A_\infty \cup \quad \cup (A_1 \setminus A_2) \cup (A_2 \setminus A_3) \cup (A_3 \setminus A_4) \cup (A_4 \setminus A_5) \cup \cdots. \quad (2.17)$$

The rearrangement shifts all the sets $(A_{2k} \setminus A_{2k+1})$ in the bottom row two spots to the left, yielding

$$A_0 = A_\infty \cup (A_0 \setminus A_1) \cup (A_1 \setminus A_2) \cup (A_2 \setminus A_3) \cup (A_3 \setminus A_4) \cup (A_4 \setminus A_5) \cup \cdots$$

$$A_1 = A_\infty \cup (A_2 \setminus A_3) \cup (A_1 \setminus A_2) \cup (A_4 \setminus A_5) \cup (A_3 \setminus A_4) \cup (A_6 \setminus A_7) \cup \cdots.$$

$$(2.18)$$

On the right-hand side of the last two equalities, the odd terms are equal, and the even terms are equivalent by what we did above. ∎

Exercise 2.10.17   Show that any interval $[a, b]$ where $a < b$ has the same cardinality as $\mathbb{R}$.

Example 2.10.18   *The set of rationals, $\mathbb{Q}$, is countably infinite: $\mathbb{N} \subset \mathbb{Q}$ and the identity function is a bijection, and a subset of $\mathbb{Q}$ is equivalent to $\mathbb{N}$. Take a bijection $f : \mathbb{N} \leftrightarrow \mathbb{Z} \times \mathbb{N}$, and define $g : \mathbb{Z} \times \mathbb{N} \to \mathbb{Q}$ by $g(m, n) = \frac{m}{n}$; g is a surjection; hence $g \circ f : \mathbb{N} \to \mathbb{Q}$ is a surjection, removing duplicates (e.g., $\frac{18}{10}$ duplicates $\frac{9}{5}$), take a subset of $\mathbb{N}$ on which $g \circ f$ is a bijection.*
Define a surjection $g : \mathbb{N} \times \mathbb{N} \to \mathbb{Q}_{++}$ by $g(n, m) = \frac{m}{n}$. Since $\mathbb{N} \times \mathbb{N}$ is countable (Example 2.10.12), there is a surjection $h : \mathbb{N} \to \mathbb{N} \times \mathbb{N}$. Then $f = g \circ h : \mathbb{N} \to \mathbb{Q}_{++}$ is a surjection (Theorem 2.6.23), so by Theorem 2.10.14, $\mathbb{Q}_{++}$ is countable.

The intuition for the preceding example follows simply from Figure 2.10.12 if you replace the "," with "/." That is, replace $(1, 1)$ with the rational $\frac{1}{1}$, $(1, 2)$ with the rational $\frac{1}{2}$, $(3, 2)$ with the rational $\frac{3}{2}$, etc. The next theorem provides an alternative proof of Example 2.10.12.

Theorem 2.10.19   *A finite product of countable sets is countable.*

Proof.   Let A and B be two nonempty, countable sets. Choose surjective functions $g : \mathbb{N} \to A$ and $h : \mathbb{N} \to B$. Then the function $f : \mathbb{N} \times \mathbb{N} \to A \times B$ defined by $f(n, m) = (g(n), h(m))$ is surjective. By Theorem 2.10.14, $A \times B$ is countable. Proceed by induction for any finite product. ∎

While it is tempting to think that this result could be extended to show that a countable product of countable sets is countable, the next theorem shows that this is false and gives us our first example of an uncountable set.

Theorem 2.10.20   *Let $X = \{0, 1\}$. The set of all functions $\mathbf{x} : \mathbb{N} \to X$, denoted $X^{\mathbb{N}}$, is uncountable.*

Proof.   We show that any function $g : \mathbb{N} \to X^{\mathbb{N}}$ is not a surjection. The function g takes each integer n to a sequence $(x_{n,1}, x_{n,2}, \ldots, x_{n,m}, \ldots)$, where each $x_{i,j}$ is either 0 or 1. Define a point $\mathbf{y} = (y_1, y_2, \ldots, y_n, \ldots)$ of $X^{\mathbb{N}}$ by letting

$$y_n = \begin{cases} 0 & \text{if } x_{nn} = 1, \\ 1 & \text{if } x_{nn} = 0. \end{cases}$$

This can be visualized in the following table.

$n = 1$	$x_{1,1} \neq y_1$	$x_{1,1}$	$x_{1,3}$	$\cdots$	$x_{1,n}$	$\cdots$
$n = 2$	$x_{2,1}$	$x_{2,2} \neq y_2$	$x_{2,3}$	$\cdots$	$x_{1,n}$	$\cdots$
$n = 3$	$x_{3,1}$	$x_{3,2}$	$x_{3,3} \neq y_3$	$\cdots$	$x_{1,n}$	$\cdots$
$\vdots$	$\vdots$	$\vdots$	$\vdots$	$\ddots$	$\vdots$	
n	$x_{n,1}$	$x_{n,2}$	$x_{n,3}$	$\cdots$	$x_{n,n} \neq y_n$	$\cdots$
$\vdots$	$\vdots$	$\vdots$	$\vdots$		$\vdots$	$\ddots$

We note that $\mathbf{y} \in X^{\mathbb{N}}$ and $\mathbf{y}$ is not in the image of g. That is, given n, $g(n)$ and $\mathbf{y}$ differ in at least one coordinate, namely the nth. Thus g is not a surjection.   ■

The diagonal argument used above is useful for establishing the uncountability of the reals, which we save until Chapter 3. A function $f : \mathbb{N} \to \{0, 1\}$ is the indicator function of the set $A = \{n \in \mathbb{N} : f(n) = 1\}$. The set of all indicator functions is exactly the set of subsets of $\mathbb{N}$. Another way to express Theorem 2.10.20 is to say that there is no surjection of $\mathbb{N}$ onto its subsets, which is a special case of the next theorem. Its proof is another smart example of a diagonalization argument. Again we assume that the surjection exists and show that there is a point for which its definition requires it not to be in the range, leading to the desired contradiction.

Theorem 2.10.21   *If $X \neq \emptyset$, then there is no surjection from X to $\mathcal{P}(X)$.*

Proof.   Suppose that $f : X \to \mathcal{P}(X)$ is a surjection. Let $S = \{x \in X : x \notin f(x)\} \subset X$. Let $s \in f^{-1}(S)$, which is nonempty because by assumption, f is a surjection. Either $s \in S$ or $s \notin S$. If $s \in S$, then $s \notin S$, so we must conclude that $s \notin S$. However, if $s \notin S$, then $s \in S$.   ■

X and $\mathcal{P}(X)$ are not equivalent, as they have different cardinalities. When X is infinite, it is not clear if there are sets having cardinality strictly between the cardinality of X and the cardinality of $\mathcal{P}(X)$. The **continuum hypothesis** is the guess that there are no sets having cardinality strictly between the cardinality of $\mathbb{N}$ and $\mathcal{P}(\mathbb{N})$. We will see that $\#\mathbb{R} = \#\mathcal{P}(\mathbb{N})$, so the continuum hypothesis is that there is no set with cardinality strictly between that of $\mathbb{N}$ and that of $\mathbb{R}$.

The **generalized continuum hypothesis** is the guess that there is no set with cardinality between that of X and that of $\mathcal{P}(X)$ for any set X. It rarely matters for the mathematics that we do, but we accept the generalized continuum hypothesis just as we accept the axiom of choice, the topic of the next section.

Example 2.10.22   *Consider the following game known as "matching pennies." Two players each hold a penny and simultaneously reveal either "heads" (H) or "tails" (T) to each other. If both faces match (i.e., both heads or both tails) the first player wins a penny; otherwise the second player wins. Now suppose the players continue to play this game every day for the indefinite (infinite) future, which is being very optimistic about medical technology. The set of different sequences that* *could arise from play of this infinitely repeated game is, by what we just discussed, the same as that of $\mathbb{R}$.*

*In game theory, a strategy is a **complete contingent plan**, that is, a description of a choice of what to do at every point in time after every possible history of the game up till that point in time. One kind of strategy is to ignore what the other person does and to just play some sequence of heads and tails. This means that the set of strategies must have a cardinality of at least $\mathbb{R}$ since it contains an uncountable set. It turns out that the set of strategies for this game is what is called a **compact metric space**, and we show that such a space always has a cardinality of at most $\#\mathbb{R}$.*

2.11 ◆ The Axiom of Choice and Some Equivalent Results

The axiom of choice is, mostly, accepted by mathematicians. Without at least some version of it, we cannot use most of the mathematical models that have proved so useful in economics. For us, that is sufficient reason to accept it.

Axiom 2 (Axiom of Choice) *If A is a set and f is a function on A such that $f(a)$ is a nonempty set for each $a \in A$, then there exists a choice function, $a \mapsto c(a)$ such that for all $a \in A$, $c(a) \in f(a)$.*

Thus, if we have a class of nonempty sets, for example, the set of nonempty subsets of $\mathbb{R}$, we can choose a point from each of these sets. The argument that ensues arises because this axiom talks about the existence of a choice function without giving any clue as to how to find or construct such a function. This axiom has some equivalences such as Zorn's lemma, 2.11.1, and the well-ordering lemma, which are not as intuitive as the axiom of choice. To be fair, there are implications of the axiom of choice that border on the unbelievable, which is why some mathematicians reject it.

We use sequences, $(x_1, x_2, \ldots)$ with all kinds of properties relating $(x_1, x_2, \ldots, x_n)$ to x_{n+1}. The axiom of choice guarantees that these sequences exist. Now, to be fair, there is a weaker version of the axiom of choice, called the axiom of dependent choice, that delivers the sequences that we use and, indeed, everything that this book does with separable metric spaces. However, using it is a bit harder. In the later chapters, we will encounter models in which we cannot reasonably use metric spaces, and analyzing these requires the full axiom of choice.

The next result, Zorn's lemma, is equivalent to the axiom of choice. This equivalence is what mathematicians call a deep result, which means that the result is not at all obvious and that it is important for many parts of mathematics. We do not give a proof of the equivalence. We do, however, in later chapters use Zorn's lemma to prove the monotone class theorem, one of the most useful results in the measure theory that underpins probability theory, and the Hahn-Banach theorem, perhaps the key result in the study of convex optimization problems. In the next section, we use Zorn's lemma to show that a strengthening of the weak axiom of revealed preference is equivalent to rational choice.

Lemma 2.11.1 (Zorn)   *If A is a partially ordered set such that each chain (totally ordered subset) has an upper bound in A, then A has a maximal element.*

This is a very useful tool for proving the existence of maximal elements by checking the chain condition mentioned in the lemma. Here is an example of something that borders on unbelievable.

Theorem 2.11.2   *There is a probability μ on $\mathcal{P}(\mathbb{N})$ with $\mu(E) = 0$ for every finite $E \in \mathcal{P}(\mathbb{N})$.*

This result will turn out to be important in Chapter 9's study of probabilities on metric spaces and in Chapter 11's study of expanded spaces. Since $\mu(\mathbb{N}) = 1$, this seems a bit odd because $\mu(\{1, \ldots, n\}) \equiv 0$ and no finite subset of $\mathbb{N}$ does a good job of approximating the countably infinite set $\mathbb{N}$. Despite not needing the result for a long time, we give a proof here because the proof is a typical application of Zorn's lemma in that we prove the existence of what we want by showing that it is associated with the maximal element in some partially ordered class.[11]

Proof.   Let $\mathcal{F}^\circ \subset \mathcal{P}(\mathbb{N})$ denote the class of sets having finite complements. For our purposes, there are three important properties of $\mathcal{F}^\circ$.

1. $\emptyset \notin \mathcal{F}^\circ$,
2. $\mathcal{F}^\circ$ is closed under intersection, $[F, G \in \mathcal{F}]^\circ \Rightarrow [(F \cap G) \in \mathcal{F}^\circ]$, and
3. $\mathcal{F}^\circ$ contains the supersets of its elements, $[F \in \mathcal{F}^\circ, G \supset F] \Rightarrow [G \in \mathcal{F}^\circ]$.

Let $\mathfrak{F}$ denote all of the subsets of $\mathcal{P}(\mathbb{N})$ containing $\mathcal{F}^\circ$ and satisfying the three properties just named, not containing the empty set, being closed under intersection, and containing the supersets of all of its elements. $\mathfrak{F}$ is partially ordered by inclusion. By Zorn's lemma, to show that $\mathfrak{F}$ has a maximal element, it is sufficient to show that every chain in $\mathfrak{F}$ has an upper bound in $\mathfrak{F}$.

Let $\mathfrak{C} \subset \mathfrak{F}$ be a chain and define $\mathcal{F}^* = \cup\{\mathcal{F} : \mathcal{F} \in \mathfrak{C}\}$. To show that $\mathcal{F}^*$ is an upper bound for $\mathfrak{C}$ in $\mathfrak{F}$, we need to check that $\mathcal{F}^*$ has the requisite three properties.

1. If $\emptyset \in \mathcal{F}^*$, then, by the definition of unions, $\emptyset \in \mathcal{F}'$ for some $\mathcal{F}' \in \mathfrak{C}$, contradicting the definition of $\mathfrak{C}$.
2. If $F, G \in \mathcal{F}^*$, then $F, G \in \mathcal{F}'$ for some $\mathcal{F}' \in \mathfrak{C}$ because $\mathfrak{C}$ is a chain. Since $\mathcal{F}'$ is closed under intersections, $(F \cap G) \in \mathcal{F}' \subset \mathcal{F}^*$.
3. If $F \in \mathcal{F}^*$, then, by the definition of unions, $F, G \in \mathcal{F}'$ for some $\mathcal{F}' \in \mathfrak{C}$. Since $\mathcal{F}'$ contains the supersets of all of its elements, any superset of F belongs to $\mathcal{F}^*$.

By Zorn's lemma, this means that there is a maximal element, $\mathcal{F}^M$, of $\mathfrak{F}$. $\mathcal{F}^M$ is an element of $\mathfrak{F}$ that is not a proper subset of any other element of $\mathfrak{F}$. For our present purposes, the crucial aspect of this maximality is that for all $A \in \mathcal{P}(\mathbb{N})$, either $A \in \mathcal{F}^M$ or $A^c \in \mathcal{F}^M$, but not both.

---

11. The proof is fairly elaborate and could be skipped until the result is needed. The upcoming proof of Theorem 2.12.9 is another typical example of how we prove things using Zorn's lemma.

The last part of the assertion is easy, since $\mathcal{F}^M$ is closed under intersection and $A \cap A^c = \emptyset$, it cannot be that both belong to $\mathcal{F}^M$. Now, exactly one of the following must be true:

$$(\exists F \in \mathcal{F}^M)[A \cap F = \emptyset] \quad \text{or} \quad (\forall F \in \mathcal{F}^M)[A \cap F \neq \emptyset].$$

However, if $(\exists F \in \mathcal{F}^M)[A \cap F = \emptyset]$, then $A^c \supset F$ so that $A^c \in \mathcal{F}^M$. Therefore, either A has nonempty intersection with each $F \in \mathcal{F}^M$ or A^c does. Relabelling if necessary, we can suppose that it is A. Define $\mathcal{F}^A = \{G \in \mathcal{P}(\mathbb{N}) : G \supset (A \cap F) \text{ for some } F \in \mathcal{F}^M\}$. It is (fairly) immediate that $\mathcal{F}^A$ has the three properties, i.e. belongs to $\mathfrak{F}$. If $A \notin \mathcal{F}^M$, then $\mathcal{F}^A \supsetneq \mathcal{F}^*$, contradicting the maximality of $\mathcal{F}^*$.

We are now, at long last, in a position to define μ. Set $\mu(F) = 1$ if $F \in \mathcal{F}^M$ and $\mu(F) = 0$ otherwise. To show that μ is a probability, it is enough to show that for any disjoint pair of sets, A and B, $\mu(A \cup B) = \mu(A) + \mu(B)$.

Since A and B are disjoint, they cannot both belong to $\mathcal{F}^M$, if they did, so would their intersection, $\emptyset$, and we know that $\emptyset \notin \mathcal{F}^M$. If neither belongs to $\mathcal{F}^M$, then $(A^c \cap B^c)$ does, so that $\mu(A \cup B) = 0 = \mu(A) + \mu(B)$. If exactly one of the two sets belongs to $\mathcal{F}^M$, then $\mu(A \cup B) = 1 = 0 + 1$. ∎

A linearly ordered $(X, \precsim)$ is **well ordered** if every $S \subset X$ has a smallest element. Axiom 1 (p. 21) is the assumption that $(\mathbb{N}, \leq)$ is well ordered. The set $(\mathbb{Q}, \leq)$ is not well ordered since $S = \mathbb{Q}$ itself does not have a smallest element, nor does $\mathbb{Q}_{++}$. The following is also equivalent to the axiom of choice, a fact that is also deep.

Lemma 2.11.3 (Well Ordering) *For any set A, there exists a linear order on A, $\precsim$, such that $(A, \precsim)$ is well ordered.*

There is no guarantee that you will know what the order looks like. Recall that the usual $\leq$ is a linear order on the $\mathbb{R}$, but not a well ordering. No one has been able to construct a linear order, $\precsim$, on $\mathbb{R}$ such that $(\mathbb{R}, \precsim)$ is well ordered.

Example 2.11.4 *There are many ways of well ordering $\mathbb{Q}$, but they are all difficult to visualize and have very little to do with the usual order, $\leq$. Let $n \leftrightarrow q_n$ be any bijection from $\mathbb{N}$ to $\mathbb{Q}$. Define $q_n \precsim q_m$ iff $n \leq m$. What makes this difficult to visualize is that for any $\epsilon > 0$ and any $q \in \mathbb{Q}$, there exists $q' \in (0, \epsilon)$ such that $q \prec q'$. This is because $q = q_n$ for some $n \in \mathbb{N}$, and $\mathbb{Q} \cap (0, \epsilon)$ is an infinite set, hence contains q_m for some $m > n$.*

2.12 ◆ Revealed Preference and Rationalizability

If we choose x when y is also available, we have revealed that we prefer x to y. Further, we have done so in a very convincing fashion; we have opted for x when we could have had y.[12]

A *budget set* refers to a set of available choices. For the finite context, §2.5 showed that if we know the choices / revealed preferences for each set in a large

---

12. People *say* that they like to exercise and eat right, but there is a strong revealed preference for sloth and too-rich desserts.

collection of budget sets, then a very mild consistency requirement on choices is observationally equivalent to preference-maximizing behavior for a rational preference relation. The mild consistency requirement is called the weak axiom of revealed preference.

In this section we examine the equivalence of consistent revealed preferences when preferences are only revealed over smaller collections of budget sets and when the choice set may be infinite. Using Zorn's lemma, we show that choice behavior that does not demonstrate cycles is observationally equivalent to preference-maximizing behavior for a rational preference relation. The name for choice behavior not demonstrating cycles is the strong axiom of revealed preference.

Let X be an arbitrary set (of choices) and $\mathcal{B} \subset \mathcal{P}(X)$ a nonempty collection of nonempty budget sets. For this section, we suppose that the choice mapping is a singleton subset of B; that is, for all $B \in \mathcal{B}$, $C(B) \subset B$ and $\#C(B) = 1$. A relation, $\prec$, is irreflexive if for all x, $\neg[x \prec x]$. Considering the case $x = y$ in the definition of completeness in Table 2.7 (p. 39), we see that irreflexive relations cannot be complete.

Definition 2.12.1 *An irreflexive relation is **irreflexively complete** if $\forall x \neq y$, $[x \prec y] \vee [y \prec x]$.*

Though not complete on $\mathbb{R}$ because $\neg[x < x]$, $<$ is irreflexively complete.

Definition 2.12.2 *A single-valued choice function C on $\mathcal{B}$ is **rationalizable** if there exists an irreflexively complete transitive relation $\prec$ such that for all $B \in \mathcal{B}$, $C(B) = \{y : \forall x \in B, \; x \neq y, \; y \succ x\}$.*

Example 2.12.3 *Let $X = \{a, b, c\}$, $\mathcal{B} = \{\{a, b\}\}$, and $C(\{a, b\}) = \{a\}$. The following three transitive irreflexively complete preferences rationalize C: $a \succ_1 b \succ_1 c$, $a \succ_2 c \succ_2 b$, and $c \succ_3 a \succ_3 b$.*

Parallel to §2.5, we define the **irreflexive revealed preferred relation** by $x \succ^* y$ if $(\exists B \in \mathcal{B})[[x, y \in B] \wedge [C(B) = \{x\}]]$. The relation $\succ^*$ may fail to be complete for two kinds of reasons: $\mathcal{B}$ does not contain enough budget sets or $C(B)$ demonstrates intransitivities. Crudely, this section shows that if we rule out the intransitivities, the lack of a rich-enough collection of budgets sets is not a problem because Zorn's lemma guarantees that we can extend $\succ^*$.

An order, $\prec$, is a subset of $X \times X$. An **extension of** $\prec$ is an order, $\prec'$, that contains $\prec$. It is called an extension because $\prec'$ is **consistent with** $\prec$; that is, $[\prec \subset \prec'] \Leftrightarrow (\forall x, y \in X)[[x \prec y] \Rightarrow [x \prec' y]]$. It is easy to extend incomplete orders to complete orders, and the order $X \times X$ is an uninteresting example of such an extension. The following is a bit more interesting.

Example 2.12.4 *The relations $\leq$ and $<$ are not complete on $\mathbb{R}_+^2$, for example, $\neg[(2, 5) \leq (5, 2)]$, $\neg[(5, 2) \leq (2, 5)]$, $\neg[(2, 5) < (5, 2)]$, and $\neg[(5, 2) < (2, 5)]$. However, for any monotonic utility function $u : \mathbb{R}_+^2 \to \mathbb{R}$, the preference relation $\precsim_u = \{(x, y) : u(x) \leq u(y)\}$ extends $\leq$ and $\prec_u$ extends $<$. Since u is a utility function representing $\precsim_u$, $\precsim_u$ is a complete transitive extension of $\leq$.*

For our purposes, a more important extension of a relation is its transitive closure. The transitive closure of $\precsim$ is the smallest transitive relation consistent with $\precsim$. If R and R' are transitive relations, then $R'' = R \cap R'$ is also transitive.

More generally, if $\precsim_i$, $i \in \Lambda$, is a collection of transitive relations, then $\bigcap_i \precsim_i$ is transitive.

Definition 2.12.5 *For any relation $\precsim$, define the **transitive closure of** $\precsim$ as $\precsim_T = \bigcap\{\precsim' : \precsim \subset \precsim', \precsim' \text{ transitive}\}$. Equivalently, $\precsim_T$ is the intersection of all transitive relations that are consistent with $\precsim$.*

Note that in the above definition, $\{\precsim' : \precsim \subset \precsim', \precsim' \text{ transitive}\}$ is nonempty as it contains at least the full relation $X \times X$, so $\precsim_T$ is a nonempty transitive relation containing $\precsim$. One can easily check that if a relation is transitive, it is its own transitive closure.

Exercise 2.12.6 Suppose that $x \precsim y \precsim z \precsim x$. Show that $x \sim_T y \sim_T z$.

Definition x is **indirectly revealed preferred to** y if $x \succ_T^* y$.
The following shows why we call $\succ_T^*$ "indirectly revealed preferred."

Exercise 2.12.7 Show that $x \succ_T^* y$ iff there exists a finite collection $x = x_1 \succ^* \cdots \succ^* x_n = y$.

This exercise also suggests a way to construct $\precsim_T$. That is, for any finite sequence $x_1 \precsim x_2 \precsim \ldots \precsim x_n$, one adds $x_1 \precsim_T x_n$ to the set $\precsim$. You can easily check that the relation obtained this way is transitive.

Definition 2.12.8 *We say that the choice function C satisfies the **strong axiom of revealed preference** if $[x \succ_T^* y] \Rightarrow \neg[y \succ^* x]$.*

In words, if x is indirectly revealed preferred to y then it cannot be the case that y is revealed preferred to x. This is a stronger consistency requirement than the weak axiom because it concerns all possible finite sequences of choices.

Theorem 2.12.9 *Every transitive irreflexive order on X has a transitive irreflexively complete extension.*

The proof of this provides a template for our uses of Zorn's lemma. We construct a partially ordered class, in this case of preferences, in which any maximal element has the property we want, in this case irreflexive completeness. We then verify that every linearly ordered subset of the partially ordered class has a maximal element. By Zorn's lemma, this implies the existence of a maximal element.

Proof. Let S be the set of all transitive irreflexive extensions of $\succ$. For $A, B \in S$, we say that B is larger than A if B is an extension of A, that is, if $B \supset A$. There are two steps left in the argument: first, to show that any maximal element of $(S, \supset)$ is irreflexively complete; second, to show that any linearly ordered subset of $(S, \supset)$ has a maximal element.

Every maximal element of $(S, \supset)$ is irreflexively complete: Let M be a maximal element of $(S, \supset)$. We give a proof by contradiction. Suppose that M is not irreflexively complete; that is, suppose $\exists x \neq y$ such that $(x, y) \notin M$ and $(y, x) \notin M$. Let $M' = M \cup \{(x, y)\}$ and $M'' = M \cup \{(y, x)\}$. Since M' and M'' are both extensions of M and M is maximal in S, we know that neither M' nor M'' is a transitive irreflexive extension of M. As $x \neq y$, both M' and M'' are irreflexive, so they must fail to be transitive. This means that there exists an r and an s such that

$$x M' y, \quad y M' r, \quad \text{and} \quad r M' x, \quad \text{as well as} \quad y M'' x, \ x M'' s \text{ and } s M'' y. \quad (2.19)$$

Since M' and M'' were formed from M by adding only (x, y) or (y, x), we know that

$$y M r, \quad r M x, \quad x M s, \quad \text{and} \quad s M y. \quad \quad (2.20)$$

However, these and the transitivity of M imply that $y \succ y$ and $x \succ x$, contradicting the irreflexivity of M.

Any linearly ordered subset of $(S, \supset)$ *has a maximal element*: Let T be a linearly ordered subset of the partially ordered set $(S, \supset)$, that is, for all $A, B \in T$, either $B \supset A$ or $A \supset B$. We must show that S contains an upper bound, U, for T. Define $U = \cup\{A : A \in T\}$. U is clearly an extension of every element of T, so it is an upper bound. We must show that $U \in S$; that is, we must show that U is transitive and irreflexive.

Transitivity: Pick arbitrary $x, y, z \in X$. Suppose that $x U y$ and $y U z$. We must show that $x U z$. Since $x U y$ and U is defined as the union of the A's in T, we know there exists an $A \in T$ such that $x A y$ and an $A' \in T$ such that $y A' z$. Since T is a linearly ordered set, either A extends A' or A' extends A. Let $B = A \cup A'$. Note that $B \in T$, so B is transitive. As $x B y$ and $y B z$, we conclude that $x B z$. Since $B \subset U$, we conclude that $x U z$, as needed.

Irreflexivity: For all $x \in X$ and all $A \in S$, $(x, x) \notin A$. Therefore, since U is a union of sets not containing (x, x), we conclude that $(x, x) \notin U$.   ∎

Let us take some time and reflect on what we have achieved. We now know that if observable choice patterns satisfy the strong axiom of revealed preference, then we can explain choices as preference maximization for some transitive irreflexively complete preferences ordering. Furthermore, it is easy to see that preference maximization leads to choice patterns that satisfy the strong axiom. This is complete as far as it goes. However, in §2.5, we showed that with a rich set of budgets, choice behavior was not only consistent with preference maximization, but it completely identified the preferences involved. Here is an example of how far from identification of preferences Theorem 2.12.9 leaves us and how little it matters for observable behavior.

Example 2.12.10   *Let* $X = \mathbb{R}^2_{++}$; *define* $u(x_1, x_2) = x_1 \cdot x_2$, *and for each Walrasian budget set* $B = B((p_1, p_2), w) = \{x \in X : p_1 x_1 + p_2 x_2 \leq w\}$; *and let* $C(B)$ *be the unique solution to* $\max u(x_1, x_2)$ *subject to* $(x_1, x_2) \in B$. *[You should know, or be able to figure out, that* $C(B((p_1, p_2), w)) = \{(\frac{w}{2p_1}, \frac{w}{2p_2})\}.]$

Theorem 2.12.9 gives an extension of $\succ_u$ that is transitive and irreflexively complete. In particular, the extension has no indifference—all of the indifference curves of the utility u function disappear in the extension. Since behavior is the same for all budgets sets, this tells us that indifference curves, while a convenient tool for analysis, are a fiction, unrelated to observable behavior. We could have all the points on an indifference curve of the utility function u strictly ranked by preferences without changing neoclassical consumer theory.

2.13 ◆ Superstructures

We return now to where this chapter started—the notion of a set. Recall Bertrand Russell's paradox, the one that arises if we assume that to each property there corresponds a *set* of elements defined by the fact that they have that property. The paradox arises when we consider the property "is a set and does not belong to itself."

Suppose there is a *set*, A, consisting of elements, in this case sets, that has this property. If A is an element of A, that is, if it belongs to itself, then by the second part of the defining property of A, it does not belong to itself. On the other hand, if A is not an element of A, that is, it does not belong to itself, then by the second part of the defining property of A, it does belong to itself.

In other words, *if* we assume that every property corresponds to a set, then we are led to a contradiction. Arriving at a contradiction from our assumptions throws into doubt all that we thought we had derived. The solution to this problem is to limit the class of things that we are willing to call a set and to accept the concomitant limitations on the properties that can be identified with sets. Specifically, we are only willing to use the word "set" for elements of a **superstructure**. Corresponding to this limit is a limit on the properties that can be talked about. Fortunately for us, the limits are sufficiently large as to have no impact on any of the economic theorizing we are going to do.

We start with a set S (we have to start someplace primitive), and we agree that S is a set and that none of the elements of S contain any elements.

Definition 2.13.1 *Given a set S, define a sequence of sets $V_m(S)$ by $V_0(S) = S$ and $V_{m+1}(S) = V_m(S) \cup \mathcal{P}(V_m(S))$. The **superstructure over** S is the union $V(S) = \cup_{m \geq 0} V_n(S)$, and S is the **base set**. For any $x \in V(S)$, the **rank of** x is the smallest m such that $x \in V_m(S)$. S is a **set**, and the only other sets are elements of $V(S)$ with rank 1 or higher.*

This means that once we have decided on a set S, the elements of $V(S)$ that are not elements of S are the entirety of the class of sets that we are willing to call sets. One might worry about just what we have thrown out by this limitation on the use of the word "set." For example, according to this criterion $V(S)$ is not a set because it does not have a rank.

It takes a bit of work to see that the superstructure over S contains the objects we care about. To see why this should be true, let us examine some examples of what is contained in $V(S)$ when $S = \mathbb{R}$.

1. Elements of S have rank 0.

2. An ordered pair, (a, b), is defined as $\{\{a\}, \{a, b\}\}$, where $\{a\} \in V_1(S)$ and $\{a, b\} \in V_1(S)$. The set of all ordered pairs of elements of $\mathbb{R}$ belongs to $V_2(S)$.

3. A function $f : \mathbb{R} \to \mathbb{R}$ is a set of ordered pairs with the property that each first element occurs exactly once. Hence, f belongs to $V_3(S)$.

4. The set of continuous functions from $\mathbb{R}$ to $\mathbb{R}$ is a subset of the set of all functions from $\mathbb{R}$ to $\mathbb{R}$; hence it belongs to $V_4(S)$.

5. The set of continuously differentiable functions belongs to $V_4(S)$.

6. The set of ordered pairs of elements of $\mathbb{R}$ is $\mathbb{R}^2 \in V_2(S)$. The positive orthant, $\mathbb{R}_+^2$, is a subset of $\mathbb{R}^2$, hence an element of $V_3(S)$.

7. A preference relation on $\mathbb{R}_+^2$ is a subset of $\mathbb{R}_+^2 \times \mathbb{R}_+^2$, that is, a subset of the set of ordered pairs, where there first and the second element both belong to $\mathbb{R}_+^2$. Thus, any preference relation on $\mathbb{R}_+^2$ is an element of $V_5(S)$.

8. The set of rational preference relations on $\mathbb{R}_+^2$ is a subset of the preference relations on $\mathbb{R}_+^2$, hence an element of $V_6(S)$.

Exercise 2.13.2 Give the rank of the following sets in $V(S)$, $S = \mathbb{R}$, explaining your reasoning: $\mathbb{N}$; $\mathcal{P}(\mathbb{N})$; the set of all fields of subsets of $\mathbb{R}$; the set of all $\mathbb{R}$-valued sequences.

For our larger purpose, it is important to see some examples of what is *not* contained in $V(S)$. The examples all involve sequences of elements of $V(S)$ where the rank is unbounded along the sequence. Remember, a property of elements of a set is an indicator function that assigns to every element of the set an element of the two-point set, $\{0, 1\}$, that is, a subset of the set, and an element of the set has the property if it is sent to 1. Since $V(S)$ is not a set for us, we cannot define a function on $V(S)$. Remember what we are up to here: we are trying to see the implications of being careful about what we mean when we say "set," so we are going to, just for this one instance, allow ourselves to talk about properties of the nonset $V(S)$.

1. The sequence $(\mathbb{R}, \mathcal{P}(\mathbb{R}), \mathcal{P}(\mathcal{P}(\mathbb{R})), \ldots)$ is *not* an element of $V(S)$.

2. Any "property" on $V(S)$, that is, any indicator function on $V(S)$, that sends elements of unbounded rank to 1 defines something that is not a set.

Now, for the result we have been waiting for.

Lemma 2.13.3 *The Bertrand Russell property, "is a set and is not an element of itself," has elements of unbounded rank in $V(S)$.*

Exercise 2.13.4 Verify that if $S = \mathbb{R}$, then everything in the sequence $(\mathbb{R}, \mathcal{P}(\mathbb{R}), \mathcal{P}(\mathcal{P}(\mathbb{R})), \ldots)$ is a set and does not belong to itself. From this, prove Lemma 2.13.3.

Note that we have arrived at a place that might be slightly uncomfortable. We can conceive of things that we might be willing to call a set, for example, $V(S)$, or all those elements of $V(S)$ that do not belong to themselves. However, this is not a "set" for us because it is not an element of $V(S)$. The best advice is to learn to live with the discomfort and to, as used-car dealers say, trust us.

2.14 ◆ Bibliography

In making our notational choices, we tried to stick to what seem to us to be both modal among economists and internally consistent. We based our choices on our experiences with a number of textbooks and articles over many (too many?) years teaching graduate macro and micro courses.

The material on Russell's paradox and the "limit what we call sets" strategy for solving it, in §2.1 and §2.13, as well as a more extended treatment of the various equivalent formulations of the axiom of choice given in §2.11 is laid out in Chapter 1 of J. Dugundji's *Topology* (Boston: Allyn and Bacon, 1966).

Our particular limited class of sets, the superstructures introduced in §2.11, is taken from the branch of mathematics called nonstandard analysis; see T. Lindstrøm's "Introduction to nonstandard analysis," Chapter 1 in N. Cutland's *Nonstandard Analysis and Its Applications* (Cambridge: Cambridge University Press, 1988). The material on the optimal choice for finite sets, §2.5, and for infinite sets, §2.12, is better exposed as well as being tied into standard neoclassical choice theory in Chapter 1 of A. Mas-Colell, M. D. Whinston, and J. R. Green's *Microeconomic Theory* (New York: Oxford University Press, 1995).

Our coverage of lattices and monotone comparative statics, §2.7 and §2.8, deals almost exclusively with product lattices. A good monograph on the generalities of lattices and monotone comparative statics is the book by D. M. Topkis, *Supermodularity and Complementarity* (Princeton, N.J.: Princeton University Press, 1998). Stable matchings are wonderfully exposed in A. E. Roth and M. A. Sotomayor's *Two-Sided Matching: A Study in Game-Theoretic Modeling and Analysis* (Cambridge: Cambridge University Press, 1990). The approach taken here in §2.9, using Tarski's fixed-point theorem, is due to a short article by H. Adachi ("On a characterization of stable matchings," *Economic Letters* 68, 43–49).

Finally, our coverage of infinite sets in §2.10 owes much to Chapter 1 of A. N. Kolmogorov and S. V. Fomin's *Introductory Real Analysis* (translated and edited by R. A. Silverman; New York: Dover, 1975, 1970).

2.15 ◆ End-of-Chapter Problems

Exercise 2.15.1 The following refers to §2.5: If R_i, $i \in I$, is a collection of transitive relations, then $\cap_{i \in I} R_i$ is transitive. For any relation S, define the transitive closure of S as $\cap \{R : S \subset R, \ R \text{ transitive}\}$. If S is a relation that does not satisfy transitivity, then its transitive closure is the smallest transitive relation consistent with S that satisfies transitivity. Find the transitive closure of the intransitive relation in Example 2.5.6. Show (by example) that there is not a unique smallest complete relation containing a given relation S. This means that there is not a good definition of the complete closure of a relation.

Exercise 2.15.2 Let $f : A \to B$ be a function. Prove that the following statements are equivalent:

1. f is one-to-one on A.
2. $f(C \cap D) = f(C) \cap f(D)$ for all subsets C and D of A.
3. $f^{-1}[f(C)] = C$ for every subset C of A.
4. For all disjoint subsets C and D of A, the images $f(C)$ and $f(D)$ are disjoint.
5. For all subsets C and D of A with $D \subseteq C$, we have $f(C \backslash D) = f(C) \backslash f(D)$.

The monotone comparative statics results give sufficient but not necessary conditions for comparative statics results.

Exercise 2.15.3   Define $\Phi(r) = \frac{e^r}{1+e^r}$ and define $f(x, t) = \Phi(x - t)$ for $x, t \in \mathbb{R}$ ($\Phi(\cdot)$ is called the logistic cumulative distribution function). Show the following:

1. $f(\cdot, \cdot)$ is not supermodular.
2. If $S = [-10, -1]$ and $S' = [-10, 0]$, then $S \precsim_{Strong} S'$ and for all t, $M(t, S') > M(t, S)$.
3. $M(t, S)$ is nondecreasing in (t, S) for (t, S) with $M(t, S) \neq \emptyset$.
4. $f(\cdot, \cdot)$ is quasi-supermodular.
5. By modifying $f(\cdot, \cdot)$, one can make $f(\cdot, \cdot)$ fail to be quasi-supermodular without changing $M(t, S') > M(t, S)$ for many values of t.

The Space of Real Numbers

In this chapter we introduce the most common set that economists encounter in theoretical work—the "real" numbers, denoted $\mathbb{R}$. The set $\mathbb{R}$ provides far and away the most commonly used model of size. Quantities, prices, probabilities, rates of change, utility, and costs are all typically modeled using $\mathbb{R}$.

There are two main ways to arrive at $\mathbb{R}$, an axiomatic method and a completion method. We use the "complete the rationals" method, which adds exactly enough points to represent everything that we can approximate arbitrarily closely. Moreover the generalization of the completion method allows for the unified treatment of a number of later topics. The axiomatic method builds $\mathbb{R}$ using the algebraic operations and order relations introduced in Chapter 2. In our development, such axioms are results.

3.1 ◆ Why We Want More Than the Rationals

In elementary school, we learned about counting, a process that seems very natural. Counting uses what we call the **natural numbers**, $\mathbb{N} = \{1, 2, \ldots\}$. Soon after, we learned about 0 and the negative integers, giving us $\mathbb{Z} = \{\ldots, -2, -1, 0, 1, 2 \ldots\}$. From $\mathbb{Z}$, we form the set of ratios (or quotients) known as the rationals, $\mathbb{Q} := \{\frac{m}{n} : m, n \in \mathbb{Z}, \ n > 0\}$. Note the convention that the denominator, n, is positive, that is, belongs to the natural numbers, but the numerator, m, may be positive, negative, or 0.

We can do all conceivable physical measurements using $\mathbb{Q}$. As we live in a base 10 society, any measurement we take will belong to $\mathbb{Q}_{10} := \{\frac{m}{10^k} : m \in \mathbb{Z}, \ k \in \mathbb{N}\} \subsetneqq \mathbb{Q}$. This set of measurement values has a "fullness" property—for each k, no matter how large, between any pair $\frac{m}{10^k}$ and $\frac{m+1}{10^k}$ there are nine equally spaced points $\frac{10m+n}{10^{k+1}}$, $n = 1, 2, \ldots, 9$. This means that any desired degree of precision can be exceeded using $\mathbb{Q}_{10}$.

Though it is a sufficiently rich model of quantities for all conceivable empirical measurements, $\mathbb{Q}$ is a terrible model for constructing useful theories. The problem is that even though it has the fullness property and even though it is infinitely fine, it has "holes," and the holes matter. Two results make this point.

The length of the hypotenuse of a right triangle with sides 1 is $\sqrt{1^2 + 1^2} = \sqrt{2}$.

Lemma 3.1.1 $\sqrt{2} \notin \mathbb{Q}$.

Proof. Suppose that $\sqrt{2} = m/n$ for some $m, n \in \mathbb{N}$. From this we derive a contradiction.

Step 1: After canceling common factors, at most one of the integers m and n is even. Enumerate the prime numbers as $2 = p_1 < p_2 < \cdots$. The prime factorization of m is $m = p_1^{a_1} \cdot p_2^{a_2} \cdots$ and the prime factorization of n is $n = p_1^{b_1} \cdot p_2^{b_2} \cdots$, where the a_i, $b_i \in \mathbb{Z}_{\downarrow}$ and at most finitely many of them are nonzero. After cancellation of any common factors, we know that at most one of the integers m and n is even, depending on whether $a_1 > b_1$, $a_1 = b_1$, or $a_1 < b_1$.

Step 2: m must be even. Cross-multiplying and then squaring both sides of $\sqrt{2} = m/n$ yields $2n^2 = m^2$, so that m is even, that is, $a_1 > b_1$.

Step 3: If m is even, then n must be even. If m is even, it is of the form $2m'$ for some integer m', and $m^2 = 4(m')^2$, giving $2n^2 = 4(m')^2$, which is equivalent to $n^2 = 2(m')^2$, implying that n is even—a contradiction. ∎

If you believe that all geometric lengths must exist, that is, you believe in some kind of deep connection between numbers that we can imagine and idealized physical measurements, this observation might upset you. It might even make you want to add some new "numbers" to $\mathbb{Q}$. At the very least, we have to add numbers to make the geometry easier—Pythagoras's theorem should read that the length of the hypotenuse of a right triangle with adjacent sides of lengths a and b is $\sqrt{a^2 + b^2}$, without needing the proviso that a, b, and $\sqrt{a^2 + b^2}$ belong to $\mathbb{Q}$.

In statistics, one also wants to add numbers to $\mathbb{Q}$ to make the normal approximation work to all degrees of accuracy.

Example 3.1.2 *Suppose that $X_1, X_2, \ldots$ is a sequence of random variables that are independent and that all satisfy $P(X_n = +1) = P(X_n = -1) = \frac{1}{2}$. For each T, define the new random variable $S_T = \frac{1}{\sqrt{T^2}} \sum_{t=1}^{T^2} X_t$. It is clear that for all T, $P(S_T \in \mathbb{Q}) = 1$. From introductory probability and statistics, we know by the central limit theorem that for any interval (a, b), $\lim_T P(S_T \in (a, b)) = P(Z \in (a, b))$, where Z is a random variable with a standard normal (i.e., Gaussian) distribution. However, $P(Z \in \mathbb{Q}) = 0$ because Z is "smoothly" distributed and $\mathbb{Q}$ is countable.*

3.2 ◆ Basic Properties of Rationals

There is some redundancy in the definition we just gave for $\mathbb{Q}$—$\frac{6}{10} = \frac{3}{5}$ shows that there are (at least) two representations of the same number. Such bookkeeping details and some facts from elementary school are gathered in the following definition. We use "$A := B$" to mean that "A is defined by B," and we also use the convention that rationals have positive denominators.

Definition 3.2.1 *For $q = \frac{m}{n}, q' = \frac{m'}{n'} \in \mathbb{Q}$,*

 1. $q = q'$ iff $mn' = nm'$, otherwise $q \neq q'$,

 2. $q + q' := \frac{mn' + nm'}{nn'}$,

 3. $-q := \frac{-m}{n}$ and $q - q' := q + (-q')$,

 4. if $q \neq 0$, $q^{-1} := \frac{n}{m}$ (or $\frac{-n}{-m}$ if $m < 0$),

 5. $qq' := \frac{mm'}{nn'}$ and $q'/q = q'q^{-1}$ if $q \neq 0$,

 6. $q > 0$ if and only if $m > 0$, and

 7. $q > q'$ if and only if $q - q' > 0$.

Exercise 3.2.2 Show that $(\mathbb{Q}, \leq)$ is a linearly ordered set. In particular, for all $q \in \mathbb{Q}$, exactly one of the following three options holds: $q < 0$, $q = 0$, or $0 < q$.

Theorem 3.2.3 (Algebraic Properties of $\mathbb{Q}$) *For any x, y, $z \in \mathbb{Q}$,*

 A1. $x + y = y + x$,

 A2. $(x + y) + z = x + (y + z)$,

 A3. $x + y = x$ iff $y = 0$,

 A4. $x + y = 0$ iff $y = -x$,

 A5. $xy = yx$,

 A6. $(xy)z = x(yz)$,

 A7. $x(y + z) = (xy) + (xz)$, and

 A8. if $x \neq 0$, then $xy = 1$ iff $y = x^{-1}$.

Exercise 3.2.4 Convince yourself that the results in Theorem 3.2.3 follow from Definition 3.2.1 and properties of $\mathbb{Z}$. The exercise is mostly about naming the properties of $\mathbb{Z}$ that you need.

The property that $\mathbb{Q}$ does not have is called **completeness**. An example suggests the problem. We know that $\sqrt{2} \notin \mathbb{Q}$, but taking successively longer and longer decimal expansions, for example, 1.4, 1.41, 1.414, 1.4142, 1.41421, 1.414213, 1.4142135, etc., we come arbitrarily close to $\sqrt{2}$. Intuitively, a space is complete if anything that can be approximated to any degree of precision belongs to the space.[1] We define $\mathbb{R}$ so that it has this completeness property. As we will see, this in turn implies that many sets related to $\mathbb{R}$ have the same completeness property. Included in this list of related sets are: $\mathbb{R}^k$, sets of continuous functions on $\mathbb{R}^k$, and probability distributions on $\mathbb{R}$ or $\mathbb{R}^k$.

---

1. We use the adjective "completeness" here to follow a well-established convention, not because there is a paucity of serviceable alternative adjectives, nor because we have limited vocabularies. Note that completeness of an order is a very different concept, as is the completeness of a lattice.

3.3 ◆ Distance, Cauchy Sequences, and the Real Numbers

The absolute value of $q \in \mathbb{Q}$, is denoted/defined by

$$|q| = \begin{cases} q & \text{if } q \geq 0, \\ -q & \text{if } q < 0. \end{cases} \tag{3.1}$$

By considering cases, you should be able to show that for all $q, q' \in \mathbb{Q}$, $|q + q'| \leq |q| + |q'|$. The cases involve $q, q' \geq 0$, $q, q' \leq 0$, and q and q' having opposite signs.

3.3.a Distance and Sequences

The distance between rational numbers is the length of the line between them.

Definition 3.3.1   *The **distance between** $q, q' \in \mathbb{Q}$ is $d(q, q') = |q - q'|$.*

Exercise 3.3.2   Show that for all $x, y, z \in \mathbb{Q}$, $d : \mathbb{Q} \times \mathbb{Q} \to \mathbb{Q}_+$ satisfies:

1. $d(x, y) = d(y, x)$,
2. $d(x, y) = 0$ if and only if $x = y$, and
3. $d(x, y) + d(y, z) \geq d(x, z)$. [Hint: Set $q = x - y$, $q' = y - z$ and look at the cases above.]

The inequality in Exercise 3.3.2(3) is sufficiently important that it has a name: the **triangle inequality**. Imagine, as in Figure 3.3.2, that x, y, and z are the vertices of a triangle. Traveling along the edge from x to y covers a distance $d(x, y)$ and along the edge from y to z covers a distance $d(y, z)$. Along the edges of a triangle, this has to be at least as long as the distance covered traveling more directly, that is, along the edge from x to z, $d(x, z)$. Rewriting yields $d(x, y) + d(y, z) \geq d(x, z)$.

Definition 3.3.3   *A **sequence in a set** X is a mapping $n \mapsto x_n$ from $\mathbb{N}$ to X. The set of all sequences in X is $X^{\mathbb{N}}$.*

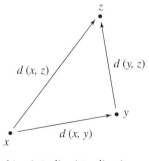

$$d(x, y) + d(y, z) \geq d(x, z)$$

FIGURE 3.3.2   The triangle inequality.

For notation, we will use $n \mapsto x_n$, $(x_n)_{n \in \mathbb{N}}$, $(x_n)_{n=1}^{\infty}$, (x_n), and sometimes simply x_n. This last notation appears to be an atrocious piece of laziness. After all, x_n is the nth element in the sequence, not the sequence itself. However, when we look at probabilities of complicated events, simplifying notation in this way will clarify the exposition.

Definition 3.3.4 *A map $k \mapsto n_k$ from $\mathbb{N}$ to $\mathbb{N}$ is **increasing** if for all $k \in \mathbb{N}$, $n_{k+1} > n_k$.*

The range of any increasing mapping must be infinite.

Definition 3.3.5 *Given a sequence $n \mapsto x_n$, any **increasing** mapping $k \mapsto n_k$ specifies another sequence $k \mapsto x_{n_k}$, called a **subsequence of** x_n.*

One can also define a subsequence of $n \mapsto x_n$ as the restriction of the sequence to an infinite subset of $\mathbb{N}$, namely the range of $k \mapsto n_k$.

Example 3.3.6 *Given the sequence $x_n = 1/n$, the increasing map $n_k = k^2$ gives rise to the subsequence $k \mapsto 1/k^2$, whereas the increasing map $n_k = 2^k$ gives rise to the subsequence $k \mapsto 1/2^k$. One could just as well understand these as the mapping $n \mapsto 1/n$ restricted to the sets $\{n^2 : n \in \mathbb{N}\}$ and $\{2^n : n \in \mathbb{N}\}$.*

3.3.b Cauchy Sequences

Intuitively, Cauchy sequences are the ones that "settle down," that stop "bouncing around." $\mathbb{Q}_{++}$ denotes the set of strictly positive rationals.

Definition 3.3.7 *A sequence q_n in $\mathbb{Q}$ is **Cauchy** if*

$$(\forall \epsilon \in \mathbb{Q}_{++})(\exists M \in \mathbb{N})(\forall n, m \geq M)[d(q_n, q_m) < \epsilon]. \tag{3.2}$$

In other words, for any $\epsilon > 0$, the elements in the tail of a Cauchy sequence remain at a distance less than ϵ from one another. As ϵ gets smaller, the later parts of the tail stay closer and closer together.

Example 3.3.8 *The following are Cauchy sequences.*

1. *$(r_n)_{n \in \mathbb{N}} = (7, 7, 7, 7, 7, 7, \ldots)$,*
2. *$(\frac{1}{n})_{n \in \mathbb{N}} = (1, \frac{1}{2}, \frac{1}{3}, \frac{1}{4}, \ldots)$,*
3. *$(z_n)_{n \in \mathbb{N}}$ given by $z_1 = 2$, $z_{n+1} = \frac{1}{2}(z_n + \frac{2}{z_n})$.*

(r_n) is a constant sequence, it does not bounce around at all. Intuitively this means it ought to be Cauchy, and it is. To see why, pick any rational $\epsilon > 0$ and note that for all n, m, $d(r_n, r_m) = |7 - 7| = 0 < \epsilon$.

$(\frac{1}{n})$ is also a Cauchy sequence. To see why, pick an arbitrary rational $\epsilon = \frac{m'}{n'} > 0$ so that m', $n' \geq 1$. Set $M = n' + 1$. For all n, $m \geq M$, $d(r_n, r_m) = |\frac{1}{n} - \frac{1}{m}| < \frac{1}{M} = \frac{1}{n'+1} < \frac{m'}{n'}$.

$(z_n) = (2, 1\frac{1}{2}, \frac{17}{12}, \frac{17}{24} + \frac{12}{17}, \ldots)$ is a Cauchy sequence, but proving this is a bit more difficult, and we will do it later. If we define $f(x) = \frac{1}{2}(x + \frac{2}{x})$, then $f(x) = x$ iff $x = \sqrt{2}$. We will see, also later, that the sequence z_n comes and stays arbitrarily close to $\sqrt{2}$, which is the notion of sequence converging to a number.

Example 3.3.9 *The following sequences are not Cauchy.*

1. $(x_n)_{n\in\mathbb{N}} = (+1, -1, +1, -1, +1, -1, \ldots)$,
2. $(y_n)_{n\in\mathbb{N}} = (\frac{1}{2}, \frac{1}{3}, \frac{2}{3}, \frac{1}{4}, \frac{2}{4}, \frac{3}{4}, \frac{1}{5}, \frac{2}{5}, \frac{3}{5}, \frac{4}{5}, \ldots)$, *and*
3. $(s_n)_{n\in\mathbb{N}}$ *given by* $s_n = \sum_{j=1}^{n} \frac{1}{j}$.

(x_n) is not a Cauchy sequence, for any $\epsilon < 1$, $d(x_n, x_{n+1}) = 1 > \epsilon$. However, if we look at the increasing mapping $k \mapsto 2k$, the subsequence $(x_{n_k}) = (x_{2n})$, we have $(x_{n_k})_{k\in\mathbb{N}} = (-1, -1, -1, \ldots)$, which is constant so that $d(x_n', x_m') = 0 < \epsilon$ for any $\epsilon > 0$.

(y_n) is also not a Cauchy sequence, indeed, it is quite badly behaved. Any rational $q = \frac{m'}{n'}$ which satisfies $0 < q < 1$ appears infinitely many times in the sequence. To see why, note that for a large enough n_1, $y_{n_1} = \frac{m'}{n'}$, for a larger n_2, $y_{n_2} = \frac{2m'}{2n'}$, for a yet larger $y_{n_3} = \frac{3m'}{3n'}$, and so on and so on. Thus, the subsequence y_{n_k} is the sequence constant at q, and for every q, there is a subsequence with $y_{n_k} = q$. It is this richness of the set of possible subsequences that means that (y_n) is not Cauchy. To see why, pick any $q \neq q'$ and set $\epsilon = d(q, q')$. For any M, there exist $n, m \geq M$ with $y_n = q$ and $y_m = q'$ so that $d(y_n, y_m) \geq \epsilon$.

$s_n - s_{n-1} = \frac{1}{n}$ and we have seen that the sequence $(\frac{1}{n})$ is Cauchy, so it is (perhaps) a bit surprising that (s_n) is not Cauchy. To show this, we will show that $s_{n+2^n} - s_n$ does not become small. $(s_{n+2^n} - s_n) = \sum_{j=n+1}^{2^n} \frac{1}{j} = \frac{1}{j} + \frac{1}{j+1} + \cdots + \frac{1}{n+2^n}$. This is a sum of 2^n terms, all of which are at least $\frac{1}{n+2^n}$. Thus, $(s_{n+2^n} - s_n) > \frac{2^n}{n+2^n} = \frac{1}{1+n/2^n}$. Now, since $n/2^n$ decreases as n increases, for all n, $\frac{1}{1+n/2^n} \geq \frac{2}{3}$.

If the later parts of the tail of a sequence are found at smaller and smaller intervals, then the same must be true of subsets of the tail, and this is what the next exercise asks you to prove.

Exercise 3.3.10 If q_n is a Cauchy sequence and q_{n_k} a subsequence, then $k \mapsto q_{n_k}$ is a Cauchy sequence. [The verbal shorthand for this result is: "Any subsequence of a Cauchy sequence is Cauchy."]

Any constant sequence is Cauchy, but there are less trivial examples.

Example 3.3.11 *Consider the sequence* $x_n = 1/2^n$. *We carefully show that it is a Cauchy sequence. For any given rational* $\epsilon > 0$, *we must show that by picking M sufficiently large, we can make* $|1/2^m - 1/2^n| < \epsilon$ *for* $n, m \geq M$. *Suppose that* $n \geq m$ *so that* $|1/2^m - 1/2^n| = 1/2^m - 1/2^n$. *[If* $m \geq n$, *reverse their roles in everything that follows.] Note that* $1/2^m - 1/2^n < 1/2^m \leq 1/2^M$, *so that we only have to require* $1/2^M < \epsilon$, *equivalently* $2^M \geq 1/\epsilon$. *Now,* $\epsilon = p/q$ *for positive integers p and q, so that* $1/\epsilon = q/p$. *Since* $p \geq 1$, $q \geq q/p = 1/\epsilon$. *As q is an integer greater than or equal to 1,* $2^q \geq q$, *so that setting* $M = q + 1$ *guarantees that* $2^M > 1/\epsilon$.

Definition 3.3.12 *A **sequence** q_n **is bounded** if there exists a $B \in \mathbb{Q}$ such that for all $n \in \mathbb{N}$, $|q_n| \leq B$.*

Lemma 3.3.13 *Any Cauchy sequence is bounded.*

Proof. As q_n is a Cauchy sequence,[2] we can pick $M \in \mathbb{N}$ such that $(\forall n, m \geq M)[d(q_n, q_m) < 1]$, set $B = \max\{|q_n| : n \leq M\} + 1$, and observe that for all $n \in \mathbb{N}$, $|q_n| \leq B$. ∎

The contrapositive is also useful: if a sequence is not bounded, then it cannot be Cauchy.

You showed that any subsequence of a Cauchy sequence is itself a Cauchy sequence in Exercise 3.3.10. Sometimes a sequence that is not Cauchy has a Cauchy subsequence.

Exercise 3.3.14 For each of the following sequences: if the sequence is Cauchy, prove it; if it is not Cauchy find a Cauchy subsequence or show that it is not possible to find a Cauchy subsequence.

1. $q_n = (-1)^n$,
2. $q_n = n$,
3. $q_n = \frac{n^2 - 5n + 3}{7n}$,
4. $q_n = 1/\sqrt{n}$, and
5. let $n \mapsto k_n$ be an arbitrary sequence in the ten-point set $\{0, 1, 2, \ldots, 8, 9\}$ and define $q_n = \sum_{i=1}^{n} \frac{k_i}{10^i}$ such that q_n is a succesively longer and longer decimal expansion.

Although we have not yet introduced the concept of convergence, since we can think of Cauchy sequences as those getting closer and closer to a point or a hole (a point that does not exist) in the space, we can appreciate the way in which it is possible to fill out those holes in the underlying space.

Let $\mathfrak{C}(\mathbb{Q}) \subset \mathbb{Q}^{\mathbb{N}}$ denote the set of all Cauchy sequences in $\mathbb{Q}$.

Definition 3.3.15 *Two Cauchy sequences, x_n and y_n, are **Cauchy equivalent**, written $x_n \sim_{\mathfrak{C}} y_n$, if*

$$(\forall \epsilon \in \mathbb{Q}_{++})(\exists N \in \mathbb{N})(\forall n \geq N)[d(x_n, y_n) < \epsilon]. \tag{3.3}$$

Exercise 3.3.16 Show that $\sim_{\mathfrak{C}}$ really is an equivalence relation.

Exercise 3.3.17 Show that the sequence $x_n = 1/n$ and the constant sequence $y_n = 0$ are equivalent, as are the sequences $x_n = r + 1/n$ and $y_n = r$, $r \in \mathbb{Q}$.

Return to Exercise 3.3.14(5). Consider the sequences $(k_1, k_2, k_3, \ldots) = (1, 0, 0, \ldots)$ and $(k'_1, k'_2, k'_3, \ldots) = (0, 9, 9, \ldots)$. These give rise to the Cauchy equivalent decimal expansions $0.10000\ldots$ and $0.09999\ldots$. For decimal expansions ending in unbroken strings of 9's, we substitute the Cauchy equivalent sequence ending in an unbroken string of 0's. This still leaves us with an uncountable collection of equivalence classes.

---

2. Note that we are starting to compress the proofs here. To be more complete, we would have written, "Let q_n be an arbitrary Cauchy sequence," and then continued. The idea is that you will ask yourself, before any proof starts, "What needs to be shown?" Moreover, you should presume that we have asked ourselves the same question.

Lemma 3.3.18 *The set of equivalence classes of Cauchy sequences is uncountable.*

Proof. Let $n \mapsto k_n$ and $n \mapsto k'_n$ be two sequences taking values in the two-point set $\{0, 5\}$ and let q_n and q'_n be the corresponding successively longer and longer decimal expansions. From Theorem 2.10.20, we know that this set is uncountable. Further, it is a proper subset of the set of sequences taking values in the ten-point set $\{0, 1, 2, \ldots, 8, 9\}$. Therefore, it is sufficient to show that $q_n \not\sim_{\mathfrak{C}} q'_n$. If $k_m \neq k'_m$, then for all $n \geq m$, $d(q_n, q'_n) > 4/10^m$. ∎

3.3.c The Real Numbers: Definition and Basic Properties

The previous lemma shows that $\mathbb{R}$ is uncountable.

Definition 3.3.19 *The set of **real numbers**, $\mathbb{R}$, is $\mathfrak{C}(\mathbb{Q})/\sim_{\mathfrak{C}}$, the set of equivalence classes of Cauchy sequences.*

Since Cauchy sequences of rationals tend to some points or holes that we would like to fill, we identify each of these holes with the class of equivalent sequences tending toward it. From our treatment of equivalence relations, $\mathbb{R}$ is a partition of $\mathfrak{C}(\mathbb{Q})$. In other words, $\sim_{\mathfrak{C}}$ partitions $\mathfrak{C}(\mathbb{Q})$, and each element of the partition represents one member of $\mathbb{R}$.

For any Cauchy sequence x_n, $[x_n]$ denotes the Cauchy equivalence class. For example,

$$\sqrt{2} = [1, 1.4, 1.41, 1.414, 1.4142, 1.41421, 1.414213, \ldots].$$

The constant sequences are important, and we identify any $q \in \mathbb{Q}$ with the constant sequence with the following abuse of notation:

$$q = [q, q, q, \ldots].$$

Note that we also have

$$q = [q, q, q, \ldots] = [q + 1, q + 1/2, q + 1/3, \ldots].$$

Looking at the constant sequences shows that we have embedded $\mathbb{Q}$ in $\mathbb{R}$. In particular, since $0 \in \mathbb{Q}$, we can write $[y_n] \neq 0$ without any ambiguity—$[y_n] \neq 0$ means that the equivalence class $[y_n]$ does not contain the sequence $(0, 0, 0, 0, 0, 0, 0, \ldots)$.

3.3.c.1 The Algebraic Operations
We understand addition, subtraction, multiplication, and division for $\mathbb{Q}$, and we wish to extend this understanding to $\mathbb{R}$. We do so in a fashion very close to the limit construction.

Definition 3.3.20 *For $r = [x_n]$ and $s = [y_n]$, $r, s \in \mathbb{R}$, we define the **sum of r and s** by $[x_n] + [y_n] = [x_n + y_n]$; the **product of r and s** by $[x_n] \cdot [y_n] = [x_n \cdot y_n]$; the **difference of r and s** by $[x_n] - [y_n] = [x_n - y_n]$; and, provided $[y_n] \neq [0, 0, 0, \ldots]$, the **quotient of r and s** by $[x_n]/[y_n] = [x_n/y_n]$.*

Although these definitions seem to make sense, we must actually prove that they do. The problem is that we could have $x_n \sim_{\mathfrak{C}} x_n'$ and $y_n \sim_{\mathfrak{C}} y_n'$ without having $[x_n \cdot y_n]$ existing or having it be the same as $[x_n' \cdot y_n']$. If either of these terrible things happens, then $[x_n] \cdot [y_n]$ is not defined.

Theorem 3.3.21 *If x_n, $y_n \in \mathfrak{C}(\mathbb{Q})$, then $x_n \cdot y_n$, $x_n + y_n$ and $x_n - y_n$ are Cauchy, as is x_n/y_n if $[y_n] \neq 0$. Further, if $x_n \sim_{\mathfrak{C}} x_n'$ and $y_n \sim_{\mathfrak{C}} x_n'$, then $x_n \cdot y_n \sim_{\mathfrak{C}} x_n' \cdot y_n'$ and $x_n + y_n \sim_{\mathfrak{C}} x_n' + y_n'$, $x_n - y_n \sim_{\mathfrak{C}} x_n' - y_n'$, and if $[y_n] \neq 0$, $x_n/y_n \sim_{\mathfrak{C}} x_n'/y_n'$.*

Proof.   Suppose that x_n, $y_n \in \mathfrak{C}(\mathbb{Q})$.

We begin by showing that $x_n \cdot y_n$ is Cauchy. Pick an arbitrary $\epsilon \in \mathbb{Q}_{++}$. We must show that $(\exists M \in \mathbb{N})(\forall n, m \geq M)[d(x_n \cdot y_n, x_m \cdot y_m) < \epsilon]$. By the triangle inequality and the definition of distance,

$$d(x_n \cdot y_n, x_m \cdot y_m) \leq d(x_n \cdot y_n, x_n \cdot y_m) + d(x_n \cdot y_m, x_m \cdot y_m)$$
$$= |x_n \cdot y_n - x_n \cdot y_m| + |x_n \cdot y_m - x_m \cdot y_m| \tag{3.4}$$
$$= |x_n| \cdot |y_n - y_m| + |y_m| \cdot |x_n - x_m|.$$

Since x_n and y_n are Cauchy, they are both bounded, say by $B \in \mathbb{Q}_{++}$, and we can choose M such that for all $n, m \geq M$, we have $|y_n - y_m| < \epsilon/2B$ and $|x_n - x_m| < \epsilon/2B$. Combining, we have for all $n, m \geq M$, $d(x_n \cdot y_n, x_m \cdot y_m) < B \cdot (\epsilon/2B + \epsilon/2B) = \epsilon$.

Now suppose that $x_n \sim_{\mathfrak{C}} x_n'$ and $y_n \sim_{\mathfrak{C}} x_n'$. We show that $x_n \cdot y_n \sim_{\mathfrak{C}} x_n' \cdot y_n'$. Pick an arbitrary $\epsilon \in \mathbb{Q}_{++}$. We must show that $(\exists N \in \mathbb{N})(\forall n \geq N)[d(x_n \cdot y_n, x_n' \cdot y_n') < \epsilon]$. Using the same logic as in (3.4), we have

$$d(x_n \cdot y_n, x_n' \cdot y_n') \leq |x_n| \cdot |y_n - y_n'| + |y_n'| \cdot |x_n - x_n'|. \tag{3.5}$$

Let $B \in \mathbb{Q}_{++}$ be a bound for the Cauchy sequences x_n and y_n'. Since $y_n \sim_{\mathfrak{C}} y_n'$ and $x_n \sim_{\mathfrak{C}} x_n'$, there exists an N such that for all $n \geq N$, $|y_n - y_n'| < \epsilon/2B$ and $|x_n - x_n'| < \epsilon/2B$. Combining yields $d(x_n \cdot y_n, x_n' \cdot y_n') < \epsilon$. ∎

Exercise 3.3.22   Complete the proof of Theorem 3.3.21. [Hints: For sums, you have to show that $(\exists N)(\forall n, m \geq N)[|(x_n + y_n) - (x_m + y_m)| < \epsilon]$. Now, $|(x_n + y_n) - (x_m + y_m)| = |(x_n - x_m) - (y_n - y_m)| \leq |x_n - x_m| + |y_n - y_m|$. For large N_x, the first of these is less than $\epsilon/2$; for large N_y, the second of these is less than $\epsilon/2$; and for $N = \max\{N_x, N_y\}$, you have what you need. For quotients, you can start by showing that $|\frac{1}{y_n} - \frac{1}{y_m}| \leq |y_n - y_m| \cdot \left(\frac{1}{\min\{|y_n|, |y_m|\}}\right)^2$.]

Theorem 3.3.23 (Algebraic Properties of $\mathbb{R}$)   *For any x, y, $z \in \mathbb{R}$,*

   A1. $x + y = y + x$,

   A2. $(x + y) + z = x + (y + z)$,

   A3. $x + y = x$ iff $y = 0$,

   A4. $x + y = 0$ iff $y = -x$,

   A5. $xy = yx$,

   A6. $(xy)z = x(yz)$,

A7. $x(y + z) = (xy) + (xz)$, *and*

A8. if $x \neq 0$, *then* $xy = 1$ *if and only if* $y = x^{-1}$.

Proof. A1, A2, A5, A6, and A7 follow directly from the corresponding statements for $\mathbb{Q}$. A3, A4, and A8 follow from Theorem 3.3.21. ∎

3.3.c.2 Order Properties

Since $\leq$ and $<$ are subsets of $\mathbb{Q} \times \mathbb{Q}$ and $\mathbb{Q} \subset \mathbb{R}$, it seems that we should be able to sensibly extend $\leq$ and $<$ to be a subsets of $\mathbb{R} \times \mathbb{R}$. Here is how it is done.

Definition 3.3.24 *A number* $r \in \mathbb{R}$ *is* **greater than** 0 *or* **strictly positive***, written* $r > 0$, *if for any* x_n *such that* $r - [x_n]$, $(\exists \delta \in \mathbb{Q}_{++})(\exists N \in \mathbb{N})(\forall n \geq N)[\delta \geq x_n]$. *It is* **strictly negative** *if* $[x_n \leq -\delta]$, *written* $r < 0$. *The set of strictly positive real numbers is denoted* $\mathbb{R}_{++}$, *and the set of strictly negative real numbers is denoted* $\mathbb{R}_{--}$.

Theorem 3.3.25 *For all* $r \in \mathbb{R}$, *exactly one of the following three options holds:* $r > 0$, $r = 0$, *or* $r < 0$.

Proof. We show first that if x_n is a Cauchy sequence of rationals, then only one of the following three can occur:

1. $(\exists \delta \in \mathbb{Q}_{++})(\exists N)(\forall n \geq N)[x_n \geq \delta]$,
2. $(\forall \delta \in \mathbb{Q}_{++})(\forall N)(\exists n \geq N)[|x_n| < \delta]$, or
3. $(\exists \delta \in \mathbb{Q}_{++})(\exists N)(\forall n \geq N)[x_n \leq -\delta]$.

For a given Cauchy sequence, x_n, either the second statement holds or it does not. If it does hold, then, because x_n is Cauchy, $(\forall \delta \in \mathbb{Q}_{++})(\exists N)(\forall n \geq N)[|x_n| < \delta]$, that is, x_n is Cauchy equivalent to 0. To see why this is true, pick an arbitrary $\delta > 0$. Since the second statement holds, $(\forall N)(\exists n \geq N)[|x_n| < \delta/2]$. As x_n is Cauchy, we can choose M such that for all n, $m \geq M$, $|x_n - x_m| < \delta/2$. Combining, we get that for all $n \geq \max\{N, M\}$, $|x_n| < \delta$.

If the second statement does not hold, we have

$$(\exists \delta \in \mathbb{Q}_{++})(\exists N)(\forall n \geq N)[|x_n| \geq \delta].$$

Since x_n is Cauchy, either the first or the third statement must hold, but not both. If the first statement holds, then $r > 0$, whereas if the third holds, then $r < 0$. ∎

Definition 3.3.26 *We say that* r *is* **nonnegative***, written* $r \geq 0$, *if* $[r > 0] \vee [r = 0]$, *and the set of nonnegative real numbers is* $\mathbb{R}_+$. *The* **nonpositive** *numbers are those with* $[r < 0] \vee [r = 0]$, *written* $r \leq 0$ *or* $r \in \mathbb{R}_-$.

$\leq$ is a linear order on $\mathbb{Q}$. Here is the extension of this order to $\mathbb{R}$.

Definition 3.3.27 *For* $r, s \in \mathbb{R}$, r *is* **less than or equal to** s, *written* $r \leq s$, *if* $r - s \leq 0$, r *is* **greater than or equal to** s, *written* $r \geq s$, *if* $r - s \geq 0$.

Theorem 3.3.28 $(\mathbb{R}, \leq)$ *is a linearly ordered set.*

Proof. We must show that for all $r, s \in \mathbb{R}$, $r \leq s$, or $s \leq r$, or both, in which case $r = s$. By Theorem 3.3.25, there are three mutually exclusive cases, $r - s \leq 0$, $r - s = 0$, or $r - s \geq 0$. ∎

Despite its simplicity, the following is very useful.

Lemma 3.3.29 (Basic Inequality Tool) *If $r = [x_n]$ and $(\forall \epsilon \in \mathbb{R}_{++})(\exists N)(\forall n \geq N)[x_n \leq s + \epsilon]$, then $r \leq s$.*

Proof. If it is not the case that $r \leq s$, then by Theorem 3.3.25, $r > s$, that is, $r - s > 0$. Set $\epsilon = (r - s)/2$, take y_n such that $[y_n] = s$, and derive a contradiction. ∎

3.3.c.3 Extending Functions

An easy application of Theorem 3.3.21 shows that if x is Cauchy so is x^n for each natural number n. By the linear properties in the theorem, any polynomial of x is Cauchy. If a function $f : \mathbb{Q} \to \mathbb{Q}$ has the property that $f(x_n)$ is a Cauchy sequence whenever x_n is a Cauchy sequence, then $f(\cdot)$ can be extended to a function $f : \mathbb{R} \to \mathbb{R}$ by $f([x_n]) = [f(x_n)]$. For example, Theorem 3.3.21 implies that $f(q) = P(q)$ satisfies this property for any polynomial $P(\cdot)$. Other examples of functions with this property are $f(q) = q^{-1}$ for $q \neq 0$ and $f(q) = |q|$. As distance is a crucial concept and is defined using the absolute value, we explicitly note and prove this last observation.

Lemma 3.3.30 *If q_n and r_n are Cauchy sequences of rationals, then $|q_n - r_n|$ is a Cauchy sequence of rationals.*

Proof. The basic properties of rationals given in Definition 3.2.1 show that each $|q_n - r_n|$ is rational. Theorem 3.3.21 shows that $s_n = q_n - r_n$ is a Cauchy sequence. Given that s_n is a Cauchy sequence, we must show that $|s_n|$ is a Cauchy sequence.

Pick $\epsilon \in \mathbb{Q}_{++}$. As s_n is Cauchy, there exists an M such that for all $n, m \geq M$, $|s_n - s_m| < \epsilon$. It is sufficient to show that $||s_n| - |s_m|| < \epsilon$. If s_n and s_m are both positive or both negative, then $|s_n - s_m| = ||s_n| - |s_m||$. In either of the other two cases, $|s_n - s_m| \geq ||s_n| - |s_m||$. ∎

Exercise 3.3.31 Suppose $f : \mathbb{Q} \to \mathbb{Q}$, $f(x_n)$ is Cauchy for every Cauchy sequence x_n. Show that the extension $f : \mathbb{R} \to \mathbb{R}$ by $f([x_n]) = [f(x_n)]$ is well-defined, that is, if for Cauchy sequences $x_n \sim_{\mathfrak{c}} y_n$, $f(x_n) \sim f(y_n)$. [Hint: Suppose that $x_n \sim_{\mathfrak{c}} y_n$, and show that both $x_1, y_1, \ldots, x_n, y_n, \ldots$ and $f(x_1), f(y_1), \ldots, f(x_n), f(y_n), \ldots$ are Cauchy.]

3.4 ◆ The Completeness of the Real Numbers

Completeness is the name given to the following property: every Cauchy sequence in a set converges to an element of the set. In this section we show that $\mathbb{R}$ has this completeness property. We will see later that this implies that the same is true for many of the sets that we use in economics. Intuitively, completeness is important because we can prove that the solutions to many problems are arbitrarily closely approximated by sequences that have the Cauchy property. If the set is complete, then we know that there is a solution.

Definition 3.4.1 *The **distance between** $x, y \in \mathbb{R}$ is $d(x, y) = |x - y|$.*

By Lemma 3.3.30, $|x - y| = [|x_n - y_n|]$, where $x = [x_n]$ and $y = [y_n]$, so everything is in place for this definition. We have defined our first metric; there will be many more in what follows.

Exercise 3.4.2   [↑Exercise 3.3.2] Show that d is a **metric on** $\mathbb{R}$; that is, show that for all x, y, $z \in \mathbb{R}$,

1. $d(x, y) = d(y, x)$,
2. $d(x, y) = 0$ iff $x = y$, and
3. $d(x, y) + d(y, z) \geq d(x, z)$.

Notation 3.4.3   *The last property, $d(x, y) + d(y, z) \geq d(x, z)$, is called the* **triangle inequality***.*

We show that $\mathbb{R}$ is a minimally expanded version of $\mathbb{Q}$ with the property called completeness. The minimality has to do with the fact that every real number is nearly a rational number.

Definition 3.4.4   *A set $A \subset \mathbb{R}$ is* **dense in** $\mathbb{R}$ *if for all $r \in \mathbb{R}$ and for all $\epsilon > 0$, there exists $a \in A$ such that $d(r, a) < \epsilon$.*

A subset is dense if it can arbitrarily closely approximate anything in the set that contains it. Trivially, $\mathbb{R}$ is dense in $\mathbb{R}$. Essentially by construction, we have the following.

Theorem 3.4.5 $\mathbb{Q}$ *is dense in* $\mathbb{R}$.

Proof.   Pick arbitrary $r \in \mathbb{R}$ and arbitrary $\epsilon > 0$. We must find a rational number q such that $d(q, r) < \epsilon$. By definition, $r = [r_n]$ for some Cauchy sequence of rationals r_n. Choose M such that for all m, $n \geq M$, $d(r_n, r_m) < \epsilon/2$. Let $q = [r_M, r_M, r_M, \ldots]$. We know that $d(q, r) = [q_n]$, where $q_n = |r_M - r_n|$. Since $q_n < \epsilon/2$ for all $n \geq M$, $[q_n] \leq \epsilon/2$ by Lemma 3.3.29, which is the basic inequality tool. Finally, $\epsilon/2 < \epsilon$, so that $d(q, r) < \epsilon$.   ∎

Here is another version of the rationals being everywhere: if $x < y$, x, $y \in \mathbb{R}$, then there must be a rational in the interval (x, y). This follows from the observation that there is a rational within $(y - x)/2$ of $(x + y)/2$. One implication of denseness is used so often in what follows that it has a special name.

Corollary 3.4.6 (The Archimedean Principle)   *If $x \in \mathbb{R}_{++}$, $\exists N \in \mathbb{N}$ such that $1/N < x < N$.*

Another way to put this is that $\mathbb{R}_{++} = \cup_{n \in \mathbb{N}} (\frac{1}{n}, n)$, where $(a, b) = \{x \in \mathbb{R} : a < x < b\}$.

Proof.   Since $\mathbb{Q}$ is dense, there is a rational, $q = \frac{m}{n}$, in the interval $(0, x)$ and another $q' = \frac{m'}{n'}$ in the interval $(x, x + 1)$. As $\frac{1}{n} \leq \frac{m}{n}$, $0 < \frac{1}{n} < x$. Since $\frac{m'}{n'} > x$, $n'm' > x$. Set $N = \max\{n, n'm'\}$.   ∎

The **irrational numbers** are the ones that are not rational, that is, $\mathbb{R} \setminus \mathbb{Q}$. As $\mathbb{R}$ is uncountable (Lemma 3.3.18) and $\mathbb{Q}$ is countable, the irrationals are uncountable.

Exercise 3.4.7 Prove that the irrationals are also dense in $\mathbb{R}$. [Hint: $\sqrt{2}$ is irrational and is less than 2. Consider irrational numbers of the form $q + \frac{\sqrt{2}}{2n}$ for $q \in \mathbb{Q}$ and large n.]

By definition, a sequence in $\mathbb{R}$ is a mapping from $\mathbb{N}$ to $\mathbb{R}$. If we keep in mind that $\mathbb{R}$ is a collection of equivalence classes of sequences, this is a mapping from $\mathbb{N}$ to a space of equivalence classes of sequences. In essence, we have a sequence of sequences.

Definition 3.4.8 *A sequence x_n in $\mathbb{R}$ is **Cauchy** if*

$$(\forall \epsilon \in \mathbb{R}_{++})(\exists M \in \mathbb{N})(\forall n, m \geq M)[d(x_n, x_m) < \epsilon]. \tag{3.6}$$

Again, Cauchy sequences are the ones that "settle down." The point is that Cauchy sequences in $\mathbb{R}$ settle down to something that belongs to $\mathbb{R}$.

Definition 3.4.9 *A sequence x_n in $\mathbb{R}$ **converges to** x **in** $\mathbb{R}$, if*

$$(\forall \epsilon \in \mathbb{R}_{++})(\exists N \in \mathbb{N})(\forall n \geq N)[d(x_n, x) < \epsilon]. \tag{3.7}$$

*A subsequence, x_{n_k}, **converges to** x **in** $\mathbb{R}$, if*

$$(\forall \epsilon \in \mathbb{R}_{++})(\exists K \in \mathbb{N})(\forall k \geq K)[d(x_{n_k}, x) < \epsilon]. \tag{3.8}$$

*If there exists an $x \in \mathbb{R}$ such that x_n converges to x, then x_n **is convergent**; otherwise it is **divergent**.*

Notation 3.4.10 *Convergence is written as $x_n \to x$, $d(x_n, x) \to 0$, $\lim_n x_n = x$, $\lim_{n \to \infty} x_n = x$, or $\lim x_n = x$. Convergence of subsequences is written as $x_{n_k} \to x$, $d(x_{n_k}, x) \to 0$, $\lim_k x_{n_k} = x$, $\lim_{k \to \infty} x_{n_k} = x$, or $\lim_k x_{n_k} = x$.*

Example 3.4.11 *In Example 3.3.9, we saw that, for every rational $q \in (0, 1)$, the sequence $(y_n)_{n \in \mathbb{N}} = (\frac{1}{2}, \frac{1}{3}, \frac{2}{3}, \frac{1}{4}, \frac{2}{4}, \frac{3}{4}, \frac{1}{5}, \frac{2}{5}, \frac{3}{5}, \frac{4}{5}, \ldots)$ had a subsequence y_{n_k} that was constant at q. From Theorem 3.4.5, this implies that for every $r \in [0, 1]$, there is a subsequence $y_{n_k} \to x$.*

We now show that if a sequence has a limit, that limit is unique.

Exercise 3.4.12 Show that if $x_n \to x$ and $x_n \to x'$, then $x = x'$. [Hint: By contradiction, suppose $x_n \to x$ and $x_n \to x'$ but $x \neq x'$, so that $d(x, x') = \epsilon > 0$. For large n, $d(x_n, x) < \epsilon/2$ and $d(x_n, x') < \epsilon/2$. What does this and the triangle inequality tell you about $d(x, x')$?]

Exercise 3.4.13 Show that if $x_n \to x$ and $y_n \to y$, then $x_n + y_n \to x + y$, $x_n - y_n \to x - y$ and $x_n y_n \to xy$. [Hint: This should look a great deal like the proof of Theorem 3.3.21.]

Notation Alert 3.4.A *The sequences $x_n = n$ or $x_n = 2^n$ satisfy $(\forall r \in \mathbb{R})(\exists N)(\forall n \geq N)[x_n \geq r]$. For any such sequence, we write $x_n \to \infty$. Even though we write $x_n \to \infty$, this does* not *mean that x_n is convergent or that ∞ is a number. Only elements of $\mathbb{R}$ are numbers, and they are equivalence classes of Cauchy, hence bounded, sequences. In a parallel fashion, we write $x_n \to -\infty$ if $(\forall r \in \mathbb{R})(\exists N)(\forall n \geq N)[x_n \leq r]$.*

Our first convergence result tells us that Cauchy sequences of rationals converge to their equivalence class.

Lemma 3.4.14 *If q_n is a Cauchy sequence of rationals and $x = [q_n]$, then $q_n \to x$.*

Note that we have x as the equivalence class of the sequence q_n of rationals, but when we write $q_n \to x$ we are talking about the convergence of the sequence of real numbers, $q_n = [q_n, q_n, q_n, \ldots]$ to the real number $x = [q_1, q_2, q_3, \ldots]$.

Proof. Pick an arbitrary $\epsilon > 0$. We must show that there exists an N such that for all $n \geq N$, $d(x, q_n) < \epsilon$. Since q_n is Cauchy, we can choose N such that for all $m, m' \geq N$, $d(q_m, q_{m'}) < \epsilon/2$. By definition, for any $n \geq N$,

$$d(x, q_n) = [|q_1 - q_n|, |q_2 - q_n|, \ldots, |q_{n-1} - q_n|, |0|, |q_{n+1} - q_n|, \ldots].$$

Note that the terms beyond the nth are of the form $|q_{n+\ell} - q_n|$, and by the choice of N, they are less than $\epsilon/2$. This implies that $d(x, q_n) \leq \epsilon/2$, so that by Lemma 3.3.29, $d(x, q_n) < \epsilon$. ■

Here is our second convergence result.

Lemma 3.4.15 *If x_n is a Cauchy sequence in $\mathbb{R}$ and any subsequence x_{n_k} converges to a limit, then x_n converges to the same limit.*

Proof. Let x_n be a Cauchy sequence and suppose that $x_{n_k} \to x$. Pick N such that for all $n, m \geq N$, $d(x_n, x_m) < \epsilon/2$. Pick K large enough that $n_K \geq N$ and for all $k \geq K$, $d(x_{n_k}, x) < \epsilon/2$. For all $n \geq n_K$, $d(x_n, x) \leq d(x_n, x_{n_K}) + d(x_{n_K}, x) < \epsilon/2 + \epsilon/2$.[3] ■

The following is essentially the last time that the real numbers appear as equivalence classes of rational numbers. With this next result in hand, we simply treat the elements of reals as the points and use properties of the distance. The proof is complicated in the sense that it has lots of details and subscripts, but it is simple in the sense that you should be able to intuit that the intermediate results are true and lead to the conclusion.

Theorem 3.4.16 (Completeness of $\mathbb{R}$) *If x_n is a Cauchy sequence in $\mathbb{R}$, then there exists a unique $x \in \mathbb{R}$ such that $x_n \to x$. Conversely, if $x_n \to x$ for some $x \in \mathbb{R}$, then x_n is a Cauchy sequence.*

Proof. Doing the easy part first, suppose that $x_n \to x$. Pick M such that for all $n \geq M$, $d(x_n, x) < \epsilon/2$. For all $n, m \geq M$, $d(x_n, x_m) \leq d(x_n, x) + d(x, x_m) < \epsilon/2 + \epsilon/2 = \epsilon$.

Now suppose that x_n is a Cauchy sequence in $\mathbb{R}$. By Lemma 3.4.15, it is sufficient to prove the existence of a convergent subsequence, x_{n_k}. Now, each x_n is of the form $[x_{n,m}]$ for some sequence $m \mapsto x_{n,m}$ of rationals. By Lemma

---

3. You have probably noticed that we are starting to leave more and more to you in these proofs. The parts that we have left out here include a reiteration of the definition of x_n being a Cauchy sequence, the definition of convergence, and citation of the triangle inequality.

3.4.14, $x_{n,m} \to x_n$. We choose a Cauchy sequence of rationals, x_{n_k,m_k}, and show that $x_{n_k} \to [x_{n_k,m_k}]$. To fix ideas, we arrange the sequence of sequences as follows:

$$
\begin{array}{c|ccccccccc}
x_1 = [x_{1,m}] & x_{1,1} & x_{1,2} & \cdots & x_{1,m-1} & x_{1,m} & x_{1,m+1} & \cdots & \to & x_1 \\
x_2 = [x_{2,m}] & x_{2,1} & x_{2,2} & \cdots & x_{2,m-1} & x_{2,m} & x_{2,m+1} & \cdots & \to & x_2 \\
\vdots & \vdots & \vdots & & \vdots & \vdots & \vdots & & & \vdots \\
x_n = [x_{n,m}] & x_{n,1} & x_{n,2} & \cdots & x_{n,m-1} & x_{n,m} & x_{n,m+1} & \cdots & \to & x_n \\
\vdots & \vdots & \vdots & & \vdots & \vdots & \vdots & & & \vdots
\end{array}
$$

Picking the subsequence x_{n_k}: Going down the far right column of the table, pick n_1 such that for all $n, m \geq n_1$, $d(x_n, x_m) < 1/2^1$, and given n_{k-1}, pick $n_k > n_{k-1}$ such that for all $n, m \geq n_k$, $d(x_n, x_m) < 1/2^k$.

Picking a Cauchy sequence of rationals, x_{n_k,m_k}: Going along the n_kth row of the table from left to right, pick m_k such that for all $m, m' \geq m_k$, $d(x_{n_k,m}, x_{n_k,m'}) < 1/2^k$. To see that $k \mapsto x_{n_k,m_k}$ is Cauchy, pick $\epsilon > 0$. We must show that for some K, and all $k, k' \geq K$, $d(x_{n_k,m_k}, x_{n_{k'},m_{k'}}) < \epsilon$. Pick K such that $1/2^K < \epsilon/3$ (which we can do by the Archimedean principle). By the triangle inequality and the choice of K,

$$d(x_{n_k,m_k}, x_{n_{k'},m_{k'}}) \leq d(x_{n_k,m_k}, x_{n_k}) + d(x_{n_k}, x_{n_{k'}}) + d(x_{n_{k'}}, x_{n_{k'},m_{k'}})$$

$$< \epsilon/3 + \epsilon/3 + \epsilon/3. \tag{3.9}$$

Let $x = [x_{n_k,m_k}]$ be the equivalence class of our chosen Cauchy sequence of rationals. The last step is to show that $x_{n_k} \to x$. This is again an implication of the triangle inequality,

$$d(x_{n_k}, x) \leq d(x_{n_k}, x_{n_k,m_k}) + d(x_{n_k,m_k}, x). \tag{3.10}$$

Choose $\epsilon > 0$. For K large enough that $1/2^K < \epsilon/2$, $d(x_{n_k}, x_{n_k,m_k}) < \epsilon/2$. By Lemma 3.4.14, for K large enough, $d(x_{n_k,m_k}, x) < \epsilon/2$. Combining these with (3.10) yields $d(x_{n_k}, x) < \epsilon/2 + \epsilon/2$. ∎

Since $\sqrt{2} \notin \mathbb{Q}$, Theorem 3.4.16 would not hold if we replaced $\mathbb{R}$ by $\mathbb{Q}$. We end this section with a useful extension of the basic inequality tool.

Lemma 3.4.17 *If $x_n \geq 0$ and $x_n \to x$, then $x \geq 0$; if x_n is an increasing sequence and $x_n \to x$, then for all n, $x_n \leq x$; and if x_n is a decreasing sequence and $x_n \to x$, then for all n, $x_n \geq x$.*

Proof. Since $x_n \to x$, x_n is Cauchy. If $x < 0$, pick $\epsilon < |x|/2$. For some N and all $n \geq N$, $d(x_n, x) < \epsilon$, implying that $x_n < 0$.

Suppose that for some N, $x_N > x$. Set $\epsilon = (x_N - x)$. For all $n \geq N$, $x_n \geq x_N > x + \epsilon/2$, so $d(x_n, x) \not\to 0$. The proof for decreasing sequences is similar. ∎

3.5 ◆ Examples Using Completeness

One of our original motivations for completing the rationals is that we wanted to have a representation of numbers x that are defined as roots of equations, that is, as the x solving $f(x) = 0$. This section examines properties of iterative procedures for constructing sequences that guarantee that the result is a Cauchy sequence. We begin with what are called geometric sums, then apply them to contraction mappings and to Newton's bisection method for finding roots, both of which generate Cauchy sequences with limits that solve equations. A classical example related to geometric sums is the exponential function, $e^x = \sum_{n=0}^{\infty} \frac{x^n}{n!}$ where $n!$, read "n factorial," is defined by $n! = n \cdot (n-1) \cdots 2 \cdot 1$. The infinite sum is defined as $\lim_N \sum_{n=0}^{N} \frac{x^n}{n!}$, and to show that the limit exists, we must show that this sequence of sums is Cauchy.[4]

3.5.a   Geometric Sums

The following preliminary result will allow us define our first class of infinite sums.

Lemma 3.5.1   *If $|\alpha| < 1$, then $\alpha^n \to 0$.*

Proof.   Pick $\epsilon > 0$. We must show that there exists an N such that for $n \geq N$, $d(\alpha^n, 0) = |\alpha^n - 0| = |\alpha^n| < \epsilon$, equivalently $(1/|\alpha|)^n > 1/\epsilon$. Since $|\alpha| < 1$, $1/|\alpha| = 1 + \eta$ for some $\eta > 0$. For $n \geq 2$, $(1 + \eta)^n > 1 + n\eta$ since $(1 + \eta)^n = 1 + n\eta + $ (strictly positive terms). Now, $1 + n\eta > 1/\epsilon$ iff $n > (1/\epsilon - 1)/\eta$. By the Archimedean principle, there exists an N such that $N > (1/\epsilon - 1)/\eta$. Therefore, for all $n \geq N$, $|\alpha^n| < \epsilon$.   ■

Here is a shorter version of the same proof (for practice reading compressed proofs).

Proof. $d(\alpha^n, 0) < \epsilon$ iff $(1/|\alpha|)^n > 1/\epsilon$. Since $|\alpha| < 1$, expanding $(1/|\alpha|)^n = (1 + \eta)^n$ shows that $(1/|\alpha|)^n > (1 + n\eta)$ for large n and $\eta = (1/|\alpha|) - 1 > 0$. For large n, $(1 + n\eta) > 1/\epsilon$.   ■

The sums in the following lemma are called **geometric sums**.

Lemma 3.5.2   *If $|\alpha| < 1$ and $x_n := \sum_{t=0}^{n} \alpha^t$, then $x_n \to \frac{1}{(1-\alpha)}$.*

We denote the limit of the Cauchy sequence x_n by $\sum_{t=0}^{\infty} \alpha^t$, or $\sum_{t \geq 0} \alpha^t$.

Proof. $x_n = \frac{x_n(1-\alpha)}{(1-\alpha)}$ and, by direct calculation, $x_n(1 - \alpha) = 1 - \alpha^{n+1}$. Therefore $x_n = \frac{(1-\alpha^{n+1})}{(1-\alpha)} = \frac{1}{(1-\alpha)} + \alpha^{n+1} \cdot \frac{1}{(1-\alpha)}$. Since $d(c + \epsilon_n, c) = d(\epsilon_n, 0)$, $\epsilon_n \to 0$ iff

---

4. Another even more classical example is π, the ratio of the circumference of a circle to its radius. This arises as the limit of a sequence if we inscribe larger and larger regular polygons on the inside of a circle of radius 1 and calculate their circumferences. As the number of sides increases, we have a bounded, increasing sequence that comes closer and closer to *pi*. We will study bounded increasing sequences in §3.7.e.

$c + \epsilon_n \to c$. Take $c = \frac{1}{(1-\alpha)}$ and $\epsilon_n = \alpha^{n+1} \cdot \frac{1}{(1-\alpha)}$. By Lemma 3.5.1, $\alpha^{n+1} \to 0$, so that $\epsilon_n \to 0$. ∎

As the x_n in Lemma 3.5.2 are convergent, we know that they are Cauchy. There is a direct proof that they are Cauchy that demonstrates a useful manipulation of geometric sums: For $n \geq m \geq M$, $x_n - x_m = \sum_{t=m}^{n} \alpha^t = \alpha^m \cdot \sum_{t=0}^{n-m} \alpha^t$. Since $m \geq M$, this is bounded above by $\alpha^M \cdot (\sum_{t=0}^{n-m} \alpha^t)$. Multiplying and dividing $(\sum_{t=0}^{n-m} \alpha^t)$ by $(1-\alpha)$ shows that this is equal to $\alpha^M \cdot (\frac{1-\alpha^{n-m+1}}{(1-\alpha)})$. Since $|\alpha| < 1$, this has an absolute value less than $|\alpha^M| \cdot (\frac{1}{(1-\alpha)})$. As $|\alpha|^M \to 0$, for M large, this value is less than ϵ.

Exercise 3.5.3 Define $s_N = \sum_{n=0}^{N} \frac{1}{n!}$, where, as above, $n!$ is defined by $n! = n \cdot (n-1) \cdots 2 \cdot 1$.

1. Give an M such that for all $m \geq M$, $\frac{s_{m+1}}{s_m} < \frac{1}{2}$.

2. Show that s_N is a Cauchy sequence. (We denote its limit by e.)

3. For $x \in$ define $r_N = \sum_{n=0}^{N} \frac{x^n}{n!}$. Show that for any x, there exists an M such that for all $m \geq M$, $\frac{|r_{m+1}|}{|r_m|} < \frac{1}{2}$.

4. Show that r_N is a Cauchy sequence. (We denote its limit by e^x.)

3.5.b Newton's Bisection Method

Suppose that $f : \mathbb{R} \to \mathbb{R}$ is a function and that we are interested in finding a solution (a.k.a. root) to the equation $f(x) = 0$. The leading examples in economics have $f(x) = g'(x)$ when we are interested in solving the problem $\max_x g(x)$ or $\min_x g(x)$. The granddaddy of numerical methods for solving an equation $f(x) = 0$ is due to Newton. In this section, we show that the method produces a Cauchy sequence of approximate solutions. Showing that the limit of the sequence of approximate solutions is a solution to $f(x) = 0$ requires continuity, and we study this formally in Chapter 4.

One finds an a_1 such that $f(a_1) < 0$ and a b_1 such that $f(b_1) > 0$. Let I_1 be the interval $[a_1, b_1]$ (where we allow $a_1 > b_1$ as a possibility, in which case we should formally write the interval as $[b_1, a_1]$). If f is well-enough behaved (this is where continuity comes in), we expect it to take value zero as it changes its sign in $[a_1, b_1]$. This root falls in one of the half intervals $[a_1, \frac{a_1+b_1}{2}]$ or $[\frac{a_1+b_1}{2}, b_1]$, and which half-interval is determined by the signs of f at $a_1, (a_1 + b_1)/2$ and b_1. By continuing the bisection of intervals, we track down the root in smaller and smaller intervals. Let us be more formal.

Given an interval $I_n = [a_n, b_n]$ with $f(a_n) < 0 < f(b_n)$, let $x_n = \frac{a_n+b_n}{2}$ bisect the interval I_n. There are three possible cases: $f(x_n) = 0$, $f(x_n) < 0$, and $f(x_n) > 0$.

1. If $f(x_n) = 0$, the numerical search terminates.

2. If $f(x_n) < 0$, define the new interval by $I_{n+1} = [a_{n+1}, b_{n+1}] = [x_n, b_n]$.

3. If $f(x_n) > 0$, define the new interval by $I_{n+1} = [a_{n+1}, b_{n+1}] = [a_n, x_n]$.

4. Let x_{n+1} be the bisector of the new interval, $x_{n+1} = \frac{a_{n+1}+b_{n+1}}{2}$, and return to the former three cases.

If the procedure terminates at the nth step, then $f(x_n) = 0$ and we have solved for a root of $f(x) = 0$. If the procedure never terminates, we show that the sequence x_n is Cauchy. By completeness, $x_n \to x$ for some x. As we will see in Chapter 4, if f is continuous, then we can guarantee that x is a root and that $f(x) = 0$.

Lemma 3.5.4 *If x_n is a nonterminating sequence produced by Newton's bisection method, then it is Cauchy and Cauchy equivalent to the sequences a_n and b_n of endpoints of the intervals I_n.*

Proof. Let $r_n = |a_n - b_n|$, where $I_n = [a_n, b_n]$ so that $r_n = \frac{1}{2}^{n-1}r_1$. For all $n, m \geq M$, all of the numbers $a_n, a_m, b_n, b_m, x_n, x_m$ belong to I_M, so that the distance between any two of them is less than or equal to $\frac{1}{2}^{M-1}r_1$. Since $|\frac{1}{2}| < 1$, Lemma 3.5.1 implies that $\lim_M \frac{1}{2}^{M-1}r_1 = 0$, so that all the sequences are Cauchy and Cauchy equivalent to each other. ∎

We do not yet have enough to guarantee that the limit of a sequence generated by Newton's bisection method converges to the root of an equation. For this we need the notion of the continuity of a function, which we study extensively in Chapter 4. For some intuition about what is needed, consider the function in the following example, it "jumps," that is, it is not "continuous."

Example 3.5.5 *Suppose we apply Newton's bisection method to the function $f(x) = 1$ if $x \geq 0$ and $f(x) = -1$ if $x < 0$. Note that f has no root; that is, for all x, $f(x) \neq 0$. Take $a_1 = -1$ and $b_1 = +1$. This gives $x_1 = 0$, $I_2 = [-1, 0]$, $x_2 = -\frac{1}{2}$, $I_3 = [-\frac{1}{2}, 0]$, $x_3 = -[(\frac{1}{2})^2]$, etc. Since $x_n = -[(\frac{1}{2})^{n-1}]$, $x_n \to 0$. However, $f(0) = 1$, and the limit of the Cauchy sequence is not a root for the equation.*

3.5.c Contraction Mappings

For $|\alpha| < 1$, the function $f(x) = \alpha x + r$ shrinks the distance between points, $|f(x) - f(y)| = |\alpha| \cdot |x - y|$. This is a special case of a function being **contractive**.

Definition 3.5.6 *For $M \subset \mathbb{R}$, $f : M \to M$ is a **contraction mapping on** M if for some $0 \leq \beta < 1$, for all $x, y \in M$, $d(f(x), f(y)) \leq \beta d(x, y)$.*

We often have sequences defined by a first value, x_1, and $x_{n+1} = f(x_n)$. Properties of f can determine whether or not the resultant sequence, $(x_n)_{n \in \mathbb{N}}$ is convergent. Being contractive is a useful sufficiently strong condition.

Lemma 3.5.7 *If $x_1 \in \mathbb{R}$, $x_n = f(x_{n-1})$, and f is contractive, then the sequence x_n is Cauchy.*

Proof. To see that the sequence $(x_n)_{n \in \mathbb{N}}$ is Cauchy, note that $d(x_3, x_2) \leq \beta d(x_2, x_1)$, so that $d(x_4, x_3) \leq \beta^2 d(x_2, x_1)$, and, continuing in this fashion, $d(x_{n+2}, x_{n+1}) \leq \beta^n d(x_2, x_1)$. Thus, for any M and any $n, m \geq M$, $d(x_n, x_m) \leq \sum_{j \geq M} d(x_j, x_{j+1}) \leq d(x_2, x_1)\beta^M \sum_{k=0}^{\infty} \beta^k$. Since $\beta^M \downarrow \infty$ as $M \uparrow$ and $\sum_{k=0}^{\infty} \beta^k = 1/(1-\beta)$, x_n is Cauchy. ∎

Exercise 3.5.8 For $x > 0$, define $f(x) = \frac{1}{2}(x + \frac{2}{x})$.

1. Show that if $M = [1, 2]$, then f is a contraction on M.
2. Show that if the starting point, x_1, satisfies $1 \le x_1 \le 2$, and $x_{n+1} = f(x_n)$, then $x_n \to \sqrt{2}$.
3. Show that if $M = \mathbb{R}_{++}$, then f is not a contraction on M.

3.5.d There Is Nothing So Useful as a Good Theory

If we are numerically solving for the root of an equation $f(x) = 0$ using an interative procedure in some applied setting, then after many, many steps, the difference between x_n and x_{n+1} and the difference between $f(x_n)$ and 0 will be less than any quantity we care about. If we are calculating the x_n and the $f(x_n)$ with a computer, they will be rational numbers, and we will therefore stop at a rational number when we have a sufficiently good approximation to a root of the equation $f(x) = 0$. In this sense, we could work with $\mathbb{Q}$ instead of $\mathbb{R}$. However, before starting the iterative procedure, we often try to convince ourselves that it should converge to some number, and a theorem to that effect is usually a pretty convincing argument. However, if f is the function $f(x) = x^2 - 2$, it is continuous and its two roots are $\pm\sqrt{2}$. A theorem based on $\mathbb{Q}$ cannot say that this f has a root.

To summarize this line of argument, we use $\mathbb{R}$ to prove theorems that guide us in our search for approximate numerical solutions. Often, the value of theorems is found in this guidance.

3.6 ◆ Supremum and Infimum

The supremum of a set of numbers is a generalized version of the maximum and the infimum a generalized version of the minimum. The sums of infinite discounted sequences of rewards are approached through the study of summability, which is based on properties of the supremum. Notions of patient preferences over infinite streams of rewards are studied using both the supremum and the infimum. Cost functions, expenditure functions, and profit functions are commonly used cases of functions defined as the suprema or infima of a set of numbers. They are functions (rather than empty sets) because the completeness of $\mathbb{R}$ implies that bounded sets have suprema and infima.

Definition 3.6.1 *A set $A \subset \mathbb{R}$ is **bounded above** if there exists an $x \in \mathbb{R}$ such that $a \le x$ for all $a \in A$, written $A \le x$. A is **bounded below** if $-A := \{-a : a \in A\}$ is bounded above, equivalently if $(\exists x \in \mathbb{R})[x \le A]$. A is **bounded** if it is bounded above and below.*

The following is a special case of Definition 2.9.10 applied to the lattice $(\mathbb{R}, \le)$.

Definition 3.6.2 *If it exists, the **supremum of a set** $A \subset \mathbb{R}$, denoted $\sup A$, is the least upper bound for A. More explicitly, $x = \sup A$ if $A \le x$, and for any $x' < x$, $A \not\le x'$. If it exists, the **infimum of a set** $A \subset \mathbb{R}$, denoted $\inf A$, is the greatest lower bound for A; $x = \inf A$ if $x \le A$ and for any $x' > x$, $x \not\le A$.*

Example 3.6.3 *There is an important difference between the minimum of a set and the infimum, and a parallel difference between the supremum of a set and the maximum.*

1. $\inf(-\infty, 1]$ *does not exist, which implies that* $\min(-\infty, 1]$ *does not exist.*
2. $\sup(-\infty, 1] = 1$ *and* $\max(-\infty, 1] = 1$.
3. $\min(0, 1]$ *does not exist, but* $\inf(0, 1] = 0$.
4. $\sup(0, \infty)$ *does not exist, which implies that* $\max(0, \infty)$ *does not exist.*

When $x = \sup A$, then for all $\epsilon > 0$, $(x - \epsilon, x) \cap A \neq \emptyset$.

Exercise 3.6.4 Show that if $\sup A$ exists, then it is unique.

The following is perhaps the most important consequence of the completeness of $\mathbb{R}$.

Theorem 3.6.5 *If a nonempty* $A \subset \mathbb{R}$ *is bounded above, then* $\sup A$ *exists; if it is bounded below, then* $\inf A$ *exists.*

The idea is to pick an upper bound for A, and then in each step to pick a closer upper bound to A. By appropriate choice, this forms a Cauchy sequence approaching the least upper bound of A.

Proof. Since A is bounded above, there exists x_1 with $A \leq x_1$. If x_n is the supremum of A, set $x_{n+1} = x_n$. If x_n is not the supremum of A, then there exists $\epsilon > 0$ such that $(x_n - \epsilon, x) \cap A \neq \emptyset$. By the Archimedean principle, there exists $n \in \mathbb{Z}$ such that $0 < 2^n < \epsilon$, hence $A \leq x_n - 2^n$. Define $z_n = \max\{z \in \mathbb{Z} : A \leq x_n - 2^z\}$ and define $x_{n+1} = x_n - 2^{z_n}$. Since $z_{n+1} < z_n$ and $|1/2| < 1$, x_n is a Cauchy sequence. Let $x = \lim_n x_n$ so that, by Lemma 3.4.17, for all $n \in \mathbb{N}$, $x \leq x_n$ and $A \leq x$.

If x is not $\sup A$, then $\exists m \in \mathbb{N}$ such that $(x - 1/2^m, x) \cap A = \emptyset$. As $x_n \to x$, there exists an M such that for all $n \geq M$, $|x_n - x| < 1/2^m$. However, for any such n, $x_{n+1} \leq x_n - (1/2^m) < x$, thus contradicting $x \leq x_n$ for all $n \in \mathbb{N}$. ∎

Notation Alert 3.6.A *We sometimes have unbounded sets. When A is not bounded above, we write $\sup A = \infty$, and when it is not bounded below, we write $\inf A = -\infty$. As with sequences that "converge" to $\pm\infty$, we are not suggesting that ∞ is a number. We are only giving ourselves a convenient notation for a circumstance that we run into (with distressing regularity).*

In neoclassical microeconomics, the supremum and infimum of linear functions over different sets play a leading role. The existence of suprema and infima in $\mathbb{R}$ is crucial in providing models of the quantities involved:

1. With $Y \subset \mathbb{R}^\ell$ denoting the set of technologically feasible points and $\mathbf{p} \in \mathbb{R}^\ell_{++}$, $\pi(\mathbf{p}) := \sup\{\mathbf{p} \cdot \mathbf{y} : \mathbf{y} \in Y\}$ is the profit function. From the profit function, we can recover all that neoclassical economics has to say about firm behavior as a supplier and as a demander of inputs. For many technologies, for example, constant returns to scale technologies, $\pi(\mathbf{p}) = \infty$ is quite possible.

2. Given a utility function $u : \mathbb{R}_+^\ell \to \mathbb{R}$ and a utility level $\bar{u}$, $C_u(\bar{u}) := \{\mathbf{x} \in \mathbb{R}_+^\ell : u(\mathbf{x}) \geq \bar{u}\}$ is an upper contour set of u. The expenditure function is $e(\mathbf{p}, \bar{u}) := \inf\{\mathbf{p} \cdot \mathbf{x} : \mathbf{x} \in C_u(\bar{u})\}$. From the expenditure function, we can recover all that neoclassical economics has to say about consumer behavior.

Exercise 3.6.6 Show the following:

1. Let D be nonempty and let $f : D \to \mathbb{R}$ have bounded range. If D_0 is a nonempty subset of D, prove that

$$\inf\{f(x) : x \in D\} \leq \inf\{f(x) : x \in D_0\} \leq \sup\{f(x) : x \in D_0\}$$

$$\leq \sup\{f(x) : x \in D\}.$$

2. Let X and Y be nonempty sets and let $f : X \times Y \to \mathbb{R}$ have bounded range in $\mathbb{R}$. Let

$$f_1(x) = \sup\{f(x, y) : y \in Y\}, \quad f_2(y) = \sup\{f(x, y) : x \in X\}.$$

Prove the **principle of iterated suprema**:

$$\sup\{f(x, y) : x \in X, \, y \in Y\} = \sup\{f_1(x) : x \in X\} = \sup\{f_2(y) : y \in Y\}. \tag{3.11}$$

(We sometimes express this as $\sup_{x,y} f(x, y) = \sup_x \sup_y f(x, y) = \sup_y \sup_x f(x, y)$.)

3. Let f and f_1 be as in the preceding exercise and let

$$g_2(y) = \inf\{f(x, y) : x \in X\}.$$

Prove that

$$\sup\{g_2(y) : y \in Y\} \leq \inf\{f_1(x) : x \in X\}. \tag{3.12}$$

(We sometimes express this as $\sup_y \inf_x f(x, y) \leq \inf_x \sup_y f(x, y)$.)

3.7 ◆ Summability

A dynamic optimization problem is one in which a decision must be made over time and in which early decisions affect later options. We model the times at which decisions have to be made as time periods $t \in T := \mathbb{Z}_+ = \{t \in \mathbb{Z} : t \geq 0\}$. This gives rise to a sequence of outcomes, one for each $t \in T$. Our first interest is assigning utility to such sequences, which is where the study of the summability of sequences comes in. Our second interest is in studying the properties of optimal choices. There are several notable aspects to the choice of $T = \mathbb{Z}_+$: (1) the time horizon over which decisions have to be made is infinite, $T = \mathbb{Z}_+$ rather than $T = \{0, 1, \ldots, \tau\}$; (2) decisions can only be made at discrete points in time, $T = \mathbb{Z}_+$ rather than $T = \mathbb{R}_+$; and (3) the time interval between decisions is not affected by the decisions themselves.

1. In principle, any "real world" time horizon is finite. However, not only is the mathematics often easier with an infinite horizon, but the results are

often more intuitive and sensible. This is an important point—a model that is less "realistic" in one of its basic assumptions is more "realistic" in the answers that it gives.

2. Whether time is discrete or continuous may eventually be known. At present there are at least two opinions. Many dynamic optimization problems are easier in discrete rather than in continuous time, while many have the reverse pattern. A discrete set of choice times is the easiest place to start learning about these models.

3. Often one spends resources to give oneself more options for when to move. In more advanced models of dynamic optimization, this can be handled satisfactorily, at least in principle.

3.7.a The Growth Model

We begin with a (rather long) canonical dynamic optimization problem, choosing how much to save or consume in a production economy. Special cases of this problem include the fishery problem or the cake-eating problem. As usual, we use first and second derivatives, f', f'', u', and u'', to make examples easier.

Notation 3.7.1 *The expression "$\lim_{x \downarrow 0} f'(x) > 1$" means "for any sequence $x_n \to 0$ such that $0 < x_{n+1} < x_n$ for all n, the limit of $f'(x_n)$ exists and is greater than 1." In a similar fashion, "$\lim_{x \to \infty} f'(x) < a$" means "for any sequence x_n such that $(\forall r \in \mathbb{R})(\exists N)(\forall n \geq N)[x_n \geq r]$, the limit of $f'(x_n)$ exists and is less a."*

Example 3.7.2 (Simple Growth Model) *There is a storable good that can be consumed or invested and used as a factor of production next period (known as one-period time-to-build). Production of the good is given by $y_t = F(x_{t-1})$, where $x_{t-1} \geq 0$ is chosen in period $t - 1$. The good depreciates at rate $\delta \in (0, 1]$. If $c_t \geq 0$ of the good is consumed in period t, resource feasibility requires $c_t + x_t \leq f(x_{t-1})$, where $f(x_{t-1}) = F(x_{t-1}) + (1 - \delta)x_{t-1}$. Starting with a stock x_{-1}, our objective is to find the most preferred feasible pattern c_t of consumptions.*

The assumptions on the production function are:

1. $f(0) = 0$,

2. $f'(x) > 1$ for all x in some interval $(0, a)$;

3. $f''(x) < 0$ for $x > 0$; and

4. $\lim_{x \to \infty} f'(x) < 1$.

Since $f'(x^\circ) < 1$ for some x°, the $f''(x) < 0$ assumption means that the graph of f crosses the 45° line, that is, there is a point x^C that satisfies $x^C = f(x^C)$. Toward the end of this section, we show that if there is no consumption, the amount of capital converges to x^C. Specifically, for $0 < x_t < x^C$, $f(x_t) > x_t$, so the stock increases, and for $x_t > x^C$, $f(x_t) < x_t$, so the stock decreases. This arises since for a small (large) enough capital stock, the additions to output from capital (i.e., marginal product of capital) are larger (smaller) than the depreciation of capital. The typical assumptions on the utility function are: $U(c_0, c_1, \ldots) = \sum_{t \in T} \beta^t u(c_t)$ for some $\beta \in (0, 1)$; and the "instantaneous" utility function $u : \mathbb{R} \to \mathbb{R}$ satisfies $u'(x) > 0$ for $x > 0$, and $u''(x) < 0$ for $x > 0$. Smaller β's *correspond to next period's utility being downweighted relative to the present period, which we interpret as more impatience. In a similar fashion, larger β's correspond to more patience.*

Let $r_t = \beta^t u(c_t)$ in the simple growth example. As a starting point in analyzing optima, we are interested in showing that the infinite sum that represents utility, $\sum_{t \in T} r_t$, is a number.[5] We revisit the properties of solutions to the growth model once we have developed a little bit of the theory of summable sequences.

3.7.b Sufficient Conditions for Summability

For finitely many terms, the order of summation does not matter. This is one of the basic algebraic properties of $\mathbb{R}$ that carries over from $\mathbb{Q}$. For infinitely many terms, one of the properties that summable sequences should have is that the order does not matter.

Example 3.7.3 *The sequence $1 - 1 + 1 - 1 + 1 - 1 \cdots$ can be arranged as $(1-1) + (1-1) + (1-1) \cdots = 0 + 0 + 0 + \cdots = 0$, and as $1 + (-1+1) + (-1+1) + \cdots = 1 + 0 + 0 + 0 = 1$.*

The order not mattering is not sufficient for a sequence to be summable. It is possible that regardless of the order in which we take the summations, the limit is ∞, rather than a number.

Definition 3.7.4 *Given a sequence $x = (x_t)_{t \in \mathbb{Z}_+}$, define the **partial sums** by $s_T = \sum_{t \leq T} x_t$. Bijections $\pi : \mathbb{Z}_+ \leftrightarrow \mathbb{Z}_+$ are called **permutations**. The **permutation of a sequence** x is the sequence x^π defined by $(x_{\pi(t)})_{t \in \mathbb{Z}_+}$, and the **permuted partial sums** are $s_T^\pi = \sum_{t \leq T} x_{\pi(t)}$.*

Example 3.7.5 *Consider the sequence $x = (1, 1/2, 1/3, 1/4, \ldots)$. There are several ways to show that $s_T \to \infty$. Here are three.*

1. *With $\int_a^b \frac{1}{x} dx = \ln b - \ln a$, an area-under-the-curve argument shows that $\sum_{t=2}^T 1/t \geq \ln(T-1) - \ln(2-1)$ and $\ln(T-1) \to \infty$.*

2.

$$\sum_{n=1}^{\infty} \frac{1}{n} = 1 + \frac{1}{2} + \frac{1}{3} + \frac{1}{4} + \frac{1}{5} + \frac{1}{6} + \frac{1}{7} + \frac{1}{8} + \cdots \qquad (3.13)$$

$$> 1 + \frac{1}{2} + \frac{1}{4} + \frac{1}{4} + \frac{1}{8} + \frac{1}{8} + \frac{1}{8} + \frac{1}{8} + \cdots \qquad (3.14)$$

$$= 1 + \frac{1}{2} + \frac{1}{2} + \frac{1}{2} + \cdots. \qquad (3.15)$$

3. *The first 3^0 terms in x sum to a number greater than or equal to 1, the next 3^1 terms in x sum to a number greater than 1, the next 3^2 terms sum to a number*

---

5. It is rather difficult to talk about maximization of a utility function if the function being maximized is meaningless.

greater than 1, *and this pattern continues; thus setting* $T_n = \sum_{i=0}^{n} 3^i$ *implies that* $\sum_{t=1}^{T_n} 1/t \geq n$.

Lemma 3.7.6 *If x is a sequence of nonnegative numbers, then for all permutations π, $s_T^\pi \to S$ where $S = \sup\{s_T : T \in \mathbb{Z}_+\}$.*

Proof. There are two cases, $\{s_T : T \in \mathbb{Z}_+\}$ is bounded and is not bounded.

Bounded: Let $S = \sup\{s_T : T \in \mathbb{Z}_+\} \in \mathbb{R}$ so that for all T, $s_T \leq S$. By the definition of sup, for any $\epsilon > 0$, there exists T such that $s_T > S - \epsilon$. Since $s_T \leq s_{T+1}$ for all T, this implies that for all $T' \geq T$, $s_{T'} > S - \epsilon$, so that $s_T \to S$.

Let $S^\pi = \sup\{s_T^\pi : T \in \mathbb{Z}_+\}$. By the previous step, we know that $s_T^\pi \to S^\pi$. This means that it is sufficient to show that $S^\pi = S$. We show that $S^\pi = S$ by demonstrating that (a) $S^\pi \geq S$, and (b) $S \geq S^\pi$.

(a) $S^\pi \geq S$: Pick an arbitrary $\epsilon > 0$. We show that $S^\pi > S - \epsilon$. For this it is sufficient to show that $S_t^\pi > S - \epsilon$ for large enough t. By the first step, there exists a T such that for all $T' \geq T$, $s_T > S - \epsilon$. Define $\tau(T) = \max\{\pi(t) : t \leq T\}$. Since all of the x_t are nonnegative, for any $t \geq \tau(T)$, $S_t^\pi \geq s_T$. Since $s_T > S - \epsilon$, for all large t, $S_t^\pi > S - \epsilon$.

(b) $S \geq S^\pi$: To show that $S \geq S^\pi$, repeat the above argument replacing x by x^π and π by π^{-1}.

Unbounded: Suppose that $\sup\{s_T : T \in \mathbb{Z}_+\} = \infty$. Since $s_T \leq s_{T+1}$, we must have $s_T \to \infty$. Let π be a permutation. To show that $s_T^\pi \to \infty$, it is sufficient to establish that for all T, there exists $\tau(T)$ such that for all $t \geq \tau(T)$, $s_t^\pi \geq s_T$. Define $\tau(T) = \max\{\pi(t) : t \leq T\}$. For $t \geq \tau(T)$, $s_t^\pi \geq s_T$ because each $x_t \geq 0$. To show that $[s_T^\pi \to \infty] \Rightarrow [s_T \to \infty]$, repeat the argument, replacing x by x^π and π by π^{-1}. ■

We are now in a position to define the sequences x_t (e.g., $r_t = \beta^t u(c_t)$) that can be legitimately summed. Note the use of the absolute value of the x_t in the following.

Definition 3.7.7 *A sequence x_t is **summable** if $\{\sum_{t=0}^{T} |x_t| : T \geq 0\}$ is bounded.*

Example 3.7.8 *The sequence $1, -1, 1, -1, \ldots$ is not summable as the partial sums of its absolute values are not bounded. The same is true of sequence $x_n = (-1)^n/n$, since the partial sums $\sum_{n=1}^{T} 1/n$ are unbounded. The sequence $x_n = (-1)^n/2^n$ is summable as its partial sums are bounded by Lemma 3.5.2.*

We have seen that when summability fails, all kinds of things can go wrong. Compare.

Theorem 3.7.9 *If x is summable, then there exists $S \in \mathbb{R}$ such that for all permutations π, $\lim_T s_T^\pi = S$.*

For summable x_t, the limit of the s_T is written $\sum_{t \in \mathbb{Z}_+} x_t$, $\sum_{t \geq 0} x_t$, $\sum_t x_t$, or even $\sum x_t$. The capital sigma symbol, $\sum$, is a mnemonic for sum.

Proof. Let $R = \sup\{\sum_{t=0}^{T} |x_t| : T \geq 0\}$. Define the remainder at T by $r_T = R - \sum_{t \leq T} |x_t|$. By Lemma 3.7.6, $r_T \to 0$. For $t, t' \geq T$, $|s_t - s_{t'}| \leq r_T$ so that s_T is a Cauchy sequence. Let $S = \lim_T s_T$. Pick a permutation π; we must show that

$s_T^\pi \to S$. Pick $\epsilon > 0$ and T such that for all $T' \geq T$, $r_{T'} < \epsilon$. For all $T' \geq \tau(T)$, $|s_{T'}^\pi - S| \leq r_{T'} < \epsilon$ where $\tau(T) := \max\{\pi(t) : t \leq T\}$. ■

To study the summability of $\beta^t u(c_t)$, it is simplest to have some easy-to-check sufficient conditions for a sequence to be summable. We use the verbal shorthand "for all large t" to mean "$(\exists T)(\forall t \geq T)$."

Definition 3.7.10 *A sequence x_t is **dominated by the sequence** y_t if for all t,* $|x_t| \leq |y_t|$.

Lemma 3.7.11 *The following imply that the sequence x_t is summable:*

1. *x_t is dominated by a summable sequence,*

2. *$x_t = \alpha y_t + \beta z_t$ for any α, $\beta \in \mathbb{R}$ and any summable sequences y_t and z_t,*

3. *the **ratio test**: there exists an $\eta < 1$ such that for all large t, $|x_{t+1}| \leq \eta|x_t|$, and*

4. ***Raabe's test**: for some $r > 1$ and $B > 0$, for all large t, $|x_t| \leq B/t^r$.*

Proof. If $\forall t$, $|x_t| \leq |y_t|$, then for all T, $\sum_{t \leq T} |x_t| \leq \sum_{t \leq T} |y_t|$. Therefore, the boundedness of $\{\sum_{t \leq T} |y_t| : T \geq 0\}$ implies the boundedness of $\{\sum_{t \leq T} |x_t| : T \geq 0\}$. In a similar fashion, the boundedness of $\{\sum_{t \leq T} |y_t|\}$ and $\{\sum_{t \leq T} |z_t|\}$ implies the boundedness of $\{\sum_{t \leq T} |\alpha y_t + \beta z_t|\}$.

Since geometric series are summable and what happens before any finite T cannot matter to the summability of a sequence, the previous result delivers the ratio test.

Finally, for Raabe's test, note that $\int_n^{n+1} \frac{1}{t^r} dt \geq \frac{1}{(n+1)^r}$. Therefore, the result is implied by the finiteness of $\lim_{T \to \infty} \int_1^T \frac{B}{t^r} dt$ (from your calculus class). ■

Exercise 3.7.12 Examine the convergence of the partial sums and the summability of the following sequences:

1. x defined by $x_t = a_t \delta^t$ where $0 < \delta < 1$ and $|a_t| < B$ for some B,

2. x defined by $x_t = a/(bt + c)$, and

3. x defined by $x_t = (-1)^t/(t + 1)$ [It can be shown, though not easily, that for all $a \in \mathbb{R}$, there exists a permutation function such that $s_T^\pi \to a$.]

3.7.c Summability in the Growth Model

Since any feasible sequence of consumptions, c_t, must have $c_t \in [0, f(x_{t-1})]$, if $f(x_{t-1})$ is bounded for all x_{t-1}, then the sequence $\beta^t u(c_t)$ is summable by the ratio test in Lemma 3.7.11. Many times in macroeconomics we use models in which there is no upper bound to productive capacities. In the growth model this means that $\lim_{x \to \infty} f(x) = \infty$. Despite this, we often have summability of the sequence $\beta^t u(c_t)$ in this case as well.

Lemma 3.7.13 *If $\lim_{x \to \infty} f'(x) < 1/\beta$, then for any feasible sequence of consumptions c_t, the sequence $\beta^t u(c_t)$ is summable.*

Proof. Consider the sequence of x_t's in which we never consume any of the good and instead invest all the output. This is $x_0 = f(x_{-1})$, $x_1 = f(x_0)$, and, continuing inductively, $x_{n+1} = f(x_n)$. Then, any feasible sequence of consumptions must satisfy $c_t \leq x_t$. Thus, it is sufficient to show that $\beta^t u(x_t)$ is summable.

Since $\lim_{x \to \infty} f'(x) < 1/\beta$, for all large t, $\frac{x_{t+1} - x_t}{x_t} \leq \beta - \eta$ for some $\eta > 0$. This directly implies that the sequence $\beta^t u(x_t)$ satisfies Raabe's test.   ∎

3.7.d Optimality, Euler Equations, and Steady States in the Growth Model

Suppose that $c^* = (c_0^*, c_1^*, \dots)$ is an optimal consumption path in the growth model. One possibility is that $c_0^* = y_0 + (1 - \delta)x_{-1}$, which means that the best thing to do is to eat all the goods right now and invest nothing for the future.

We concentrate on the case where $c_t^* > 0$ for all t. This would be guaranteed if, for example, $\lim_{x \downarrow 0} u'(x) = \infty$, so that the marginal utility of the first few units of goods is so large that it is worth waiting to consume, or $\lim_{x \downarrow 0} f'(x) = \infty$, so that productivity is so large that it overcomes any degree of impatience. To analyze this case, we develop the **Euler equation**, which is a condition on the derivatives of the u function at subsequent points in time.

Since c^* is optimal, it must be the case that, at all t, sacrificing an amount s of consumption today and eating all of the resulting increase in goods at $t + 1$ lowers utility. Define $g(s)$ to be the utility difference that is due to sacrificing s at t and eating the surplus at $t + 1$. There is a loss of utility at time t of $\text{loss}_t = \beta^t[u(c_t^*) - u(c_t^* - s)] > 0$ and a gain of utility at $t + 1$ $\text{gain}_{t+1} = \beta^{t+1}[u(c_{t+1}^* + [f(x_t^* + s) - f(x_t^*)]) - u(c_{t+1}^*)] > 0$. Combining, we get $g(s) = \text{gain}_{t+1} - \text{loss}_t$. Gathering the terms with s in them together yields

$$g(s) = \left(\beta^t u(c_t^*) + \beta^{t+1} u(c_{t+1}^*) \right)$$
$$- \left(\beta^t u(c_t^* - s) + \beta^{t+1} u(c_{t+1}^* + [f(x_t^* + s) - f(x_t^*)]) \right). \quad (3.16)$$

The optimality of c^* implies that $g(s) \leq 0$ and $g(0) = 0$. This in turn implies that $g'(0) = 0$. Calculating $g'(0)$ and dividing by β^t, we have $u'(c_t^*) - \beta u'(c_{t+1}^*) f'(x_t^*) = 0$. Rewriting gives the **Euler equation**,

$$u'(c_t^*) = \beta u'(c_{t+1}^*) f'(x_t^*). \quad (3.17)$$

The Euler equation says that along an optimal path with $c_t^* > 0$, the marginal utility of the last good not consumed at time t is equal to the discounted utility of consuming the output at $t + 1$ from having invested that good at t.

What remains is to examine the patterns of c_t^* that are implied by the Euler equation. Remember that derivatives being equal to zero is an implication of optimality. Therefore, implications of the Euler equation are implications of optimality. The easiest implications of the Euler equation arise in what are called steady states.

A *steady state solution* involves $c_t^* = c_{t+1}^* = c^s$ for all t. This happens when x^{ss} satisfies $c^{ss} = f(x^{ss}) - x^{ss}$. Using the Euler equation, we can answer the question of which steady states can be optimal.

If $c_t^* = c_{t+1}^* = c^{ss}$, then $u'(c^{ss}) = \beta u'(c^{ss}) f'(x^{ss})$, which implies that $\beta f'(x^{ss})$ = 1 (after dividing both sides by $u'(c^{ss})$). Lower β's, which correspond to impatience, must involve a higher f'. Since $f'' < 0$, this corresponds to a lower x^{ss} and lower c^{ss}. Thus, the steady state consistent with patient optimizers involves higher investments, x^{ss}, and higher consumption, c^{ss}. The highest possible level of steady state consumption occurs when we have maximized $f(x) - x$, that is, when $f'(x^{ss}) = 1$. Thus, for any $\beta < 1$, it is in principle possible to eventually have higher consumption in every period, but it is not worth lowering present consumption for long enough for that higher consumption to be realized.

To examine possible patterns of c_t^*'s away from steady state, it is easiest to develop some results on monotone sequences.

3.7.e Monotone Sequences

Recall our convention that sup $A = \infty$ if a set A is not bounded above and inf $A = -\infty$ when A is not bounded below.

Definition 3.7.14 A sequence x_t is **monotone increasing**, written $x_t \uparrow$, if $\forall t$, $x_{t+1} \geq x_t$, and it is **monotone decreasing**, written $x_t \downarrow$, if $\forall t$, $x_{t+1} \leq x_t$.

Theorem 3.7.15 If x_t is a monotone increasing sequence, then either $\sup\{x_t : t \geq 0\} = \infty$ or $x_t \to x$ for some $x \in \mathbb{R}$.

When a monotone increasing sequence converges to x, we may write $x_t \uparrow x$. Analogously, a convergent monotone decreasing sequence may be written $x_t \downarrow x$.

Exercise 3.7.16 Prove Theorem 3.7.15. [Hint: The sequences s_T in Lemma 3.7.6 were monotone, and that proof can be adapted to the present problem.]

Monotone decreasing sequences have the obvious analogous properties.

Exercise 3.7.17 For $x > 0$, define $f(x) = \frac{1}{2}(x + \frac{2}{x})$. We have seen that f is a contraction on $M = [1, 2]$, but is not a contraction on $\mathbb{R}_{++}$. Show the following:

1. If $0 < x_1 < \sqrt{2}$ and $x_{n+1} = f(x_n)$, then x_n is an increasing sequence.
2. If $x_1 > \sqrt{2}$ and $x_{n+1} = f(x_n)$, then x_n is a decreasing sequence.
3. If $x_1 \in \mathbb{R}_{++}$ and $x_{n+1} = f(x_n)$, then $x_n \to \sqrt{2}$.

3.7.f Optimality Away from Steady States

We can now return to the assertion that we made earlier that if there is no consumption, the stock of capital converges to the level x^C satisfying $x^C = f(x^C)$. This involves some assertions that use the continuity of the function f. Continuity is treated extensively in Chapter 4.

Let x_{-1} be the initial stock, $x_0 = f(x_{-1})$ the stock in period 0, and, more generally, $x_{t+1} = f(x_t)$ the stock in period $t + 1$. There are three cases, $x_0 < x^C$, $x_0 = x^C$, and $x_0 > x^C$. Throughout, we assume that $f'(x) > 0$ for all x.

1. If $x_0 = x^C$, then $\forall t$, $x_t = x^C$.

2. For $x < x^C$, $x^C > f(x) > x$. Therefore, if $x_0 < x^C$, then x_t is a monotone increasing sequence that is bounded above by x^C so that $x_t \uparrow x$ for some $x \leq x^C$. Suppose that $x < x^C$. Let $\epsilon = f(x) - x > 0$. Since f is continuous, it is possible to pick a $\delta > 0$ such that $[|x - x'| < \delta] \Rightarrow [|f(x) - f(x')| < \epsilon/2]$. For large t, $|x - x_t| < \delta$ because $x_t \to x$. Therefore, for large t, $x_{t+1} = f(x_t)$ satisfies $|x_{t+1} - f(x)| < \epsilon/2$. However, this implies that $x_{t+1} > x$, which contradicts $x_t < x$ for all t. That is, by continuity of f, if $x_n \to x$ then $f(x_n) \to f(x)$, but $x_{n+1} = f(x_n)$ so the sequence $f(x_n)$ is essentially the same as x_n, in which case by the uniqueness of a limit implies $f(x) = x$ (i.e. $x = x_C$).

3. In this case, for $x^C < x$, $x^C < f(x) < x$. Therefore, if $x^C < x_0$, x_t is a monotone decreasing sequence that is bounded below by x^C so that $x_t \downarrow x$ for some $x \geq x^C$. If $x > x^C$, a minor adaptation of the argument used in the previous case delivers $x_t \downarrow x^C$.

The same kind of monotone sequence arguments tell us about the behavior of optimal c_t^*'s away from steady states. Let $x(\beta)$ solve $\beta f'(x(\beta)) = 1$ or $f'(x(\beta)) = 1/\beta$, so that $x(\beta)$ is the steady state consistent with β. Since $f'' < 0$, higher β correspond to higher $x(\beta)$, that is, more patience leads to higher long-run consumption. The intuition is that with higher β, one is more willing to delay consumption and let the capital stock grow.

Suppose that $x_t^* < x(\beta)$. The Euler equation is $u'(x_t^*) = \beta u'(x_{t+1}^*) f'(x_t^*)$. Since $x_t^* < x(\beta)$, $f'(x_t^*) > 1/\beta$, which means that $u'(x_t^*) > u'(x_{t+1}^*)$. Since $u'' < 0$, this implies that $x_{t+1}^* > x_t^*$. Therefore, x_t^* is a monotone increasing sequence bounded above by $x(\beta)$. The same argument as used earlier shows that it can only converge to $x(\beta)$. When $x_t^* > x(\beta)$, $u'(x_t^*) < u'(x_{t+1}^*)$, so that $x_{t+1}^* < x_t^*$. In this case, x_t^* is a monotone decreasing sequence that can only converge to $x(\beta)$.

To summarize: If an optimal, positive solution exists for the growth model, then it is either monotonically increasing or monotonically decreasing toward the steady state $x(\beta)$, and $x(\beta)$ is an increasing function of β.

3.8 ◆ Products of Sequences and e^x

One of the most broadly useful mathematical functions is $E(x) = e^x$, defined by $e^x = \sum_{n=0}^{\infty} \frac{x^n}{n!}$, where $n!$, read "n factorial," is defined by $n! = n \cdot (n-1) \cdots 2 \cdot 1$.

Exercise 3.8.1   Show that for all x, the sequence $x_n = \frac{x^n}{n!}$ satisfies Raabe's test, hence is summable.

Properties that make e^x so useful include:

1. for all $x \in \mathbb{R}$, $e^x > 0$,

2. $\frac{d}{dx} e^x = e^x$, so that e^x is strictly increasing,

3. for all x, $y \in \mathbb{R}$, $e^x \cdot e^y = e^{x+y}$, so that $e^{-x} = \frac{1}{e^x}$,

4. $\lim_{x \to \infty} e^x = \infty$,

5. $\lim_{x \to -\infty} e^x = 0$, and

6. for any $n \in \mathbb{N}$, $\lim_{x \to \infty} x^n e^x = 0$.

The crucial property turns out to be the third one, that $e^x \cdot e^y = e^{x+y}$. Suppose that we show that $e^x \cdot e^y = e^{x+y}$. Since $e^{x+(-x)} = e^0 = 1$ and for $x \geq 0$, $e^x > 0$, we can conclude that $e^{-x} = 1/e^x > 0$. This in turn gives us $\frac{1}{h}(e^{x+h} - e^x) = e^x \frac{1}{h}(e^h - e^0)$. With a bit more work, this will yield $\frac{d}{dx} e^x = e^x$.

This means that we want to evaluate the product of two summable sequences, $(\sum_{n=0}^{\infty} \frac{x^n}{n!}) \cdot (\sum_{m=0}^{\infty} \frac{y^m}{m!})$ and show that it is equal to $\sum_{n=0}^{\infty} \frac{(x+y)^n}{n!}$). Here are some suggestive steps:

$$\sum_{n=0}^{\infty} \frac{x^n}{n!} \cdot \sum_{m=0}^{\infty} \frac{y^m}{m!} = \sum_{n=0}^{\infty} \sum_{k=0}^{n} \frac{n!}{n! \, k!(n-k)!} x^k y^{n-k} = \sum_{n=0}^{\infty} \frac{(x+y)^n}{n!} \qquad (3.18)$$

because $(x+y)^n = \frac{n!}{k!(n-k)!} x^k y^{n-k}$. The steps are suggestive, but we need to guarantee that they are legal; after all, we rearranged terms in an infinite product, and there is no guarantee that the result does not depend on how we rearrange the terms.

Definition 3.8.2 *Given sequences a_n and b_n, we define $c_n = \sum_{k=0}^{n} a_k b_{n-k}$ and define **the product of a_n and b_n** as $\sum_{n=0}^{\infty} c_n$ if the sum exists.*

Theorem 3.8.3 *If a_n and b_n are absolutely summable with $A = \sum_n a_n$ and $B = \sum_n b_n$, then the product of a_n and b_n is absolutely summable and $\sum_n c_n = AB$.*

Proof. Let A_n, B_n, and C_n be the partial sums of the sequences a_n, b_n, and c_n. Define $\beta_n = B_n - B$ so that $\beta_n \to 0$.

$$C_n = a_0 b_0 + (a_0 b_1 + a_1 b_0) + (a_0 b_2 + a_1 b_1 + a_2 b_0) + \cdots + (a_0 b_n + \cdots + a_n b_0)$$

$$= a_0 B_n + a_1 B_{n-1} + \cdots + a_n B_0$$

$$= a_0(B + \beta_n) + a_1(B + \beta_{n-1}) + \cdots + a_n(B + \beta_0)$$

$$= A_n B + \gamma_n, \qquad (3.19)$$

where $\gamma_n = (a_0 \beta_n + a_1 \beta_{n-1} + \cdots + a_n \beta_0)$. Since $\beta_n \to 0$ and a_n is absolutely summable, $\gamma_n \to 0$, so that $C_n \to AB$.

For the absolute summability of c_n, set $a'_n = |a_n|$ and $b'_n = |b_n|$, and note that the corresponding product terms, c'_n, satisfy $c'_n \geq |c_n|$. The previous argument applied to c'_n shows that c_n is absolutely summable. ∎

Exercise 3.8.4 Show the following:

1. $\frac{d}{dx} e^x = e^x$,

2. $\lim_{x \to \infty} e^x = \infty$,

3. $\lim_{x \to -\infty} e^x = 0$, and

4. for any $n \in \mathbb{N}$, $\lim_{x \to \infty} x^n e^x = 0$.

3.9 ◆ Patience, Lim inf, and Lim sup

This section examines notions of patience for dynamic programming. As seen above in the growth example, having a discount factor, β, close to 1 corresponds to being very patient. Here we examine other notions of patience.

In a finite (not infinite) horizon dynamic optimization problem, one receives $r = (r_0, r_1, \ldots, r_{T-1})$, a finite sequence of rewards. A good definition of patience is that for all permutations π of $\{0, \ldots, T-1\}$, r and r^π are indifferent. This captures the notion that the order in which rewards are received does not matter.

The function $\frac{1}{T} \sum_{t=0}^{T-1} r_t$ has the appropriate indifference to permutations. Judging a stream by the worst and best outcomes, that is, using $\min_t r_t$ and $\max_t r_t$, also gives utility functions that are indifferent to permutations. We briefly examine the infinite horizon versions of averages, minima, and maxima, and tie them, loosely, to discounting with a β close to 1. The main tools are the lim inf and lim sup of a sequence.[6]

3.9.a   Lim inf and Lim sup

For any sequence x_t and $T \in \mathbb{Z}_+$, define $\underline{x}_T = \inf\{x_t : t \geq T\}$ and $\overline{x}_T = \sup\{x_t : t \geq T\}$. Minimizing over smaller sets at least weakly increases the minimum, which implies that $T \mapsto \underline{x}_T$ is a monotone increasing sequence. By the same logic, $\overline{x}_T$ is a monotone decreasing sequence. Recall that Theorem 3.7.15 told us that monotone sequences are well behaved, meaning that the following is a sensible definition.

Definition 3.9.1   *For any sequence x_t in $\mathbb{R}$, the* lim inf *and the* lim sup *of x_t are defined by* $\liminf x_t := \lim_T \inf\{x_t : t \geq T\}$ *and* $\limsup x_t := \lim_T \sup\{x_t : t \geq T\}$.

We allow $\pm\infty$ as limits in these cases and agree that for all $r \in \mathbb{R}$, $-\infty < r < +\infty$. To see why we allow these limits, consider the sequence $x_t = -t$. For all T, $\underline{x}_T = -\infty$, and the only reasonable thing to do in this case is to define $\lim_T \underline{x}_T = -\infty$. In the same way, $\limsup t = \liminf t = \infty$. Note that we could just as easily define $\liminf x_t = \sup_T \inf\{x_t : t \geq T\}$ and $\limsup x_t = \inf_T \sup\{x_t : t \geq T\}$.

Example 3.9.2   *Consider the sequence $x_n = (-1)^n \frac{n}{1+n}$, so that $-1 < x_n < 1$ for all n. We know that $\frac{n}{n+1}$ converges to 1 monotonically from below, so the positive subsequence x_{2k} is increasing and tends to 1 from below, whereas the negative subsequence x_{2k+1} is decreasing and tends to -1 from above. Thus $\inf x_t : t \geq T = -1$, so $\liminf x_t = -1$ and similarly $\limsup x_t = 1$.*

Exercise 3.9.3   Find $\liminf x_t$ and $\limsup x_t$ for the following sequences:

1. $x_t = 1/(t+1)$,
2. $x_t = (-1)^t$,
3. $x_t = t$,

---

6. For a thorough analysis, see Massimo Marinacci's "An axiomatic approach to complete patience and time invariance," *Journal of Economic Theory* 83, 105–44 (1998).

4. $x_t = t \cdot (-1)^t$,

5. let π be a permutation of $\mathbb{Z}_+$ and let $x_t = \pi(t)$,

6. let $t \to x_t$ be a bijection between $\mathbb{Z}_+$ and $\mathbb{Q}$, and

7. let $t \to x_t$ be a bijection between $\mathbb{Z}_+$ and $\mathbb{Q}_+$.

[For the last two, you may want to prove that removing any finite set from $\mathbb{Q}$ leaves us with a dense set.]

A sequence is bounded if $\{x_t : t \in \mathbb{Z}_+\}$ is a bounded set. The basic result is the following theorem.

Theorem 3.9.4 *For any sequence x_t,*

1. $\inf x_t \leq \liminf x_t \leq \limsup x_t \leq \sup x_t$, *the inequalities may be strict,*

2. x_t *is bounded iff* $-\infty < \liminf x_t$ *and* $\limsup x_t < +\infty$, *and*

3. $\liminf x_t = \limsup x_t = x$ *iff* $x_t \to x$.

Proof. The inequalities about the inf and sup are immediate, and all four inequalities are strict for the sequence $(-2, +2, -1, +1, -1, +1, \ldots)$.

If x_t is bounded, then $\inf x_t$ and $\sup x_t$ exist and, by the previous step, provide bounds for $\liminf x_t$ and $\limsup x_t$. If $\liminf x_t$ and $\limsup x_t$ are finite, then with at most finitely many exceptions, $x_t \geq (\liminf x_t) - 1$ and $x_t \leq (\limsup x_t) + 1$, which implies that the sequence x_t is bounded.

If $x_t \to x$, then for all $\epsilon > 0$ and all large T, $\underline{x}_T > x - \epsilon$ and $\overline{x}_T < x + \epsilon$. Since ϵ is arbitrary, $\liminf x_t = \limsup x_t = x$.

Finally, suppose that $\liminf x_t = \limsup x_t = x$. Pick $\epsilon > 0$. For large enough T, $x - \epsilon < \underline{x}_T \leq \overline{x}_T < x + \epsilon$. For all $t \geq T$, $\underline{x}_T \leq x_t \leq \overline{x}_T$, so that $|x_t - x| < \epsilon$. ∎

Exercise 3.9.5 Completely examine whether the following can be satisfied:

1. $-\infty = \liminf x_n < \limsup x_n = 0$,

2. $0 = \liminf x_n < \limsup x_n = 2$,

3. $0 = \limsup(x_n + y_n) < \limsup x_n + \limsup y_n$, and

4. $\limsup(x_n + y_n) > \limsup x_n + \limsup y_n$.

3.9.b Patience

In the finite-horizon case, invariance to permutations is a convincing sign of patience. For example $\min_t r_t = \min_t r_t^\pi$ and $\max_t r_t = \max_t r_t^\pi$.

Lemma 3.9.6 *For any sequence x_t and permutation π, $\inf\{x_t : t \in \mathbb{Z}_+\} = \inf\{x_t^\pi : t \in \mathbb{Z}_+\}$ and $\sup\{x_t : t \in \mathbb{Z}_+\} = \sup\{x_t^\pi : t \in \mathbb{Z}_+\}$.*

Proof. Observe that the sets $\{x_t : t \in \mathbb{Z}_+\}$ and $\{x_t^\pi : t \in \mathbb{Z}_+\}$ are equal. ∎

Using inf to rank sequences means that $(0, 7, 7, 7, 7, 7, \ldots)$ and $(0, 3, 3, 3, 3, 3, \ldots)$ are indifferent. If "patience" is to mean anything sensible, it ought to be able to ignore the first period payoffs and look at the "long-run" behavior of the sequence. Thus, even though $\inf x_t$ and $\sup x_t$ are immune to permutations, they may not be sensible ways to get at patience in the infinite-horizon case. However, for both of the given sequences, the lim inf and lim sup give 7 and 3, respectively, and they ignore the early behavior of the sequence.

Lemma 3.9.7   *For any sequence x_t and any permutation π,* $\lim \inf x_t = \lim \inf x_t^\pi$, *and* $\lim \sup x_t = \lim \sup x_t^\pi$.

Proof.   We prove that $\lim \inf x_t = \lim \inf x_t^\pi$; the proof for lim sup is similar.

We show that $\lim \inf x_t \geq \lim \inf x_t^\pi$, arguing by contradiction. Suppose that $\lim \inf x_t < \lim \inf x_t^\pi$. If $\lim \inf x_t = \infty$, the inequality cannot hold. If $\lim \inf x_t = -\infty$ and the inequality holds, then $\lim \inf x_t^\pi = s \in \mathbb{R}$. This implies that for large T, $\inf\{x_t^\pi : t \geq T\} > s - \epsilon$. This in turn implies that for all but finitely many t, $x_t > s - \epsilon$, which contradicts $\lim \inf x_t = -\infty$. The last case has $\lim \inf x_t = r \in \mathbb{R}$ and $r < \lim \inf x_t^\pi$. Pick a number s strictly between r and $\lim \inf x_t^\pi$. Since $\underline{x}_T^\pi \uparrow \lim \inf x_t^\pi$, for large T, $\underline{x}_T^\pi = \inf\{x_t^\pi : t \geq T\} > s$. This means that, with the exception of at most finitely many numbers, $x_t > s$, which contradicts $\lim \inf x_t = r$.

To show that $\lim \inf x_t^\pi \geq \lim \inf x_t$, replace x by x^π and π by π^{-1}. ∎

One of the ways to judge whether utility functions are sensible is by examining their optima. With these criteria, inf, lim inf, sup, and lim sup seem peculiar.

Example 3.9.8   *Suppose that we try to maximize* $\inf u(c_t)$ *or* $\lim \inf u(c_t)$ *in the growth model. The first thing to observe about* $\inf u(c_t)$ *is that if we start with a small stock of capital, it does not matter if we invest enough to accumulate a large stock, which is the point of the* $(0, 7, 7, \ldots)$ *versus* $(0, 3, 3, \ldots)$ *example. If we try to maximize* $\lim \inf u(c_t)$, *then, for any T, what we do for the first T periods has no effect provided that it leaves us with a strictly positive stock of capital. Thus, over any finite time horizon, anything that does not completely destroy the capital stock is part of an optimal program.*

Suppose that we try to maximize $\sup u(c_t)$ *or* $\lim \sup u(c_t)$. *Again, optimality can imply anything except complete destruction of the capital stock over any finite time horizon. The many optimal patterns eventually have longer and longer delays between larger and larger feasts.*

Intuitively, the problem with the rankings so far is that they look at maximal or minimal payoffs rather than accounting for average behavior. In the finite-horizon case, the other sensible utility function that we examined was the average, $\frac{1}{T} \sum_{t=0}^{T-1} x_t$. The average is not invariant for infinite sequences.

Example 3.9.9   *Let $x = (0, 1, 0, 1, 0, 1, \ldots)$, so that $y = (\underbrace{11}_{2^1} \underbrace{0000}_{2^2} \underbrace{11111111}_{2^3}$*

$\cdots)$ is a permutation of x, and

$$0 = \inf x_t = \lim \inf x_t < \lim_T \frac{1}{T} \sum_{t=0}^{T-1} x_t = \tfrac{1}{2} < \lim \sup x_t = \sup x_t = 1, \quad \text{while}$$

$$\tfrac{1}{3} = \lim \inf_T \frac{1}{T} \sum_{t=0}^{T-1} y_t < \lim \sup_T \frac{1}{T} \sum_{t=0}^{T-1} y_t = \tfrac{2}{3}.$$

Since the lim inf *and* lim sup *differ, we know that the sequence $T \mapsto \frac{1}{T} \sum_{t=0}^{T-1} y_t$ is not convergent.*

3.9.c Discount Factors Close to One

In the growth model, discount factors, β, close to 1 corresponded to a great deal of patience. Since $\sum_{t=0}^{\infty} \beta^t = \frac{1}{(1-\beta)}$, it makes sense to consider ranking streams by $(1 - \beta) \sum_t \beta^t x_t$ and consider what happens as $\beta \uparrow 1$. This turns out to be intimately tied to $\lim_T \frac{1}{T} \sum_{t=0}^{T-1} x_t$. We offer the following without proof.

Theorem 3.9.10 (Froebenius, Littlewood) *If x_t is a bounded sequence, then $\lim_T \frac{1}{T} \sum_{t=0}^{T-1} x_t$ exists iff $\lim_{\beta \uparrow 1}(1 - \beta) \sum_t \beta^t x_t$ exists. In this case, the two limits are equal.*

If one is willing to drop completeness from the preference ordering, $\succsim$, over sequences, then there are a number of other patient orderings. For example, the **overtaking criterion** says that a sequence x_t is preferred to a sequence y_t iff $\liminf_n \sum_{t \leq n}(x_t - y_t) > 0$, and the **patient limit ordering** says that a sequence x_t is preferred to a sequence y_t iff there exists a $\beta_0 \in (0, 1)$ such that for all $\beta \in (\beta_0, 1)$, $\sum_t \beta^t x_t > \sum_t \beta^t y_t$.

Exercise 3.9.11 Give sequences x_t and y_t that cannot be ranked by the overtaking criterion, that is, for which $\liminf_n \sum_{t \leq n}(x_t - y_t) > 0$ and $\liminf_n \sum_{t \leq n}(y_t - x_t) > 0$.

Exercise 3.9.12 Give sequences x_t and y_t that cannot be ranked by the patient limit criterion, that is, sequences that have the property that for all $\beta_0 < 1$, there exist $\beta_x, \beta_y \in (\beta_0, 1)$ such that $\sum_t \beta_x^t x_t > \sum_t \beta_x^t y_t$ and $\sum_t \beta_y^t x_t < \sum_t \beta_y^t y_t$.

3.10 ◆ Some Perspective on Completing the Rationals

This chapter started with the rational numbers, $\mathbb{Q}$, which are more than sufficient for any physical measurement we can make. We then added new elements so that the limit of any Cauchy sequence in $\mathbb{Q}$ has a representation as a number $r \in \mathbb{R}$. This is a great deal of trouble, so it is worth spending a little time reflecting on what this accomplished.

If we use functions, $f : \mathbb{Q} \to \mathbb{Q}$, that are either polynomials with rational coefficients or ratios of such functions, then we have $f(q) \in \mathbb{Q}$ for all $q \in \mathbb{Q}$. This means that we can represent a rich class of functions and all of their values using only $\mathbb{Q}$. What we cannot do is represent all quantities that are defined implicitly, that is, defined by the fact that they satisfy an equation such as $f(x) = 0$.

By completing the rationals, we constructed a model of quantities that contains representations of the roots of all polynomials with rational coefficients. However, we did a great deal more, in good part because we can define and study infinite sums.

The function $f(x) = e^x$ is defined as an infinite sum of polynomials, as are the trigonometric functions $\sin(x)$ and $\cos(x)$. It can be shown that $e = e^1$ and π are not a root of any polynomial with rational coefficients. Indeed, it can be shown that, in a very strong sense, e and π are typical in this respect; "most" elements of $\mathbb{R}$ are not the roots of any polynomial with rational coefficients.

Whether or not these new numbers that we have introduced are "real" seems to be a complicated question. One should perhaps read the "real" in "real numbers" as the Spanish word *real*, meaning "royal." Real or not, almost all of the material in the rest of this book shows that the implications of the completeness property of $\mathbb{R}$ is tremendously useful.

3.11 ◆ Bibliography

For undergraduate textbooks on the material in this chapter, one of the most elementary and well written that we know of is K. G. Binmore's *The Foundations of Analysis: A Straightforward Introduction* (Cambridge: Cambridge University Press, 1980). W. Rudin's *Principles of Mathematical Analysis* (3rd Ed., New York: McGraw-Hill, 1976) and J. E. Marsden and M. J. Hoffman's *Elementary Classical Analysis* (2nd Ed., New York: W. H. Freeman, 1993) are both very good advanced undergraduate textbooks. Both follow the axiomatic approach to constructing the real numbers.

Real Analysis (3rd Ed., New York: Macmillan, 1988) by H. L. Royden is a classic graduate mathematics textbook that also begins with an axiomatic development of the real numbers. A more complete treatment of the completion approach to the real numbers, and of a huge number of mathematics topics relevant to economists, is available in the (also classic) graduate mathematics textbook *Real and Abstract Analysis: A Modern Treatment of the Theory of Functions of a Real Variable* (New York: Springer-Verlag, 1965) by E. Hewitt and K. Stromberg.

CHAPTER 4

The Finite-Dimensional Metric Space of Real Vectors

A metric space is a pair (M, d), where M is a nonempty set and $d(\cdot, \cdot)$ is a function that satisfies some properties (detailed later) that let us think of $d(x, y)$ as the distance between points x and y in M. Chapter 3 gave us our first metric space, $(\mathbb{R}, d)$, with the metric $d(x, y) = |x - y| = \sqrt{(x - y)^2}$.

The most fruitful immediate generalization of $(\mathbb{R}, d)$ is $(\mathbb{R}^\ell, d_E)$, the set of length-ℓ ordered lists of elements of $\mathbb{R}$. The distance between points $x = (x_1, \ldots, x_\ell)$ and $y = (y_1, \ldots, y_\ell)$ is defined as $d_E(x, y) = \sqrt{(x_1 - y_1)^2 + \cdots + (x_\ell - y_\ell)^2}$, more compactly written as $d_E(x, y) = \left(\sum_i (x_i - y_i)^2\right)^{1/2}$. The subscript "$E$" is for Euclid, and thinking about the Pythagorean theorem should tell you why we use this distance.

The metric space $(\mathbb{R}^\ell, d_E)$ appears throughout the study of optimization theory, game theory, exchange economies, probability, econometrics, and dynamic programming. As well as having a metric, it has vector space and algebraic structures. We start in §4.1 with most of the general definitions associated with metric spaces, but we give very few examples to go with these definitions. In §4.2 we go over all of these definitions in the context of what are called discrete metric spaces. These two sections are not enough, in our experience, to give the reader very much understanding of the concepts. Hopefully, seeing all of these concepts yet again, in the context of $\mathbb{R}^\ell$, where they are both more frequently used and more useful, will allow the reader to come back to §4.1 as a form of review.

In this chapter, we usually reserve "theorem" for results that are true in all metric spaces and use "lemma" for results that are specific to the metric space under study.

4.1 ◆ The Basic Definitions for Metric Spaces

This section gives most of the definitions and properties that are useful in the study of metric spaces: metrics; sequences and completeness; separability; open and closed sets; neighborhoods, topologies, and the topological equivalence of metrics; open covers, compactness, and connectedness; continuous functions; and homeomorphisms and isometries.

4.1.a A Word to the Reader

As we go through the various metric spaces, the reasons that these definitions and notations are useful and important should, hopefully, appear. Completeness and compactness are the most important properties of spaces. Generally, the proofs involving these two concepts are the most difficult. Except for $(\mathbb{Q}, d)$, we will not see many metric spaces that fail to be complete until we study metric spaces that consist of random variables that take on only finitely many values.

4.1.b Metrics

The following isolates the properties of the metric d on $\mathbb{R}$ that have proved the most useful.

Definition 4.1.1   *A **metric on** M, M nonempty, is a function $d : M \times M \to \mathbb{R}_+$ such that for all x, y, $z \in M$,*

1. *$d(x, y) = d(y, x)$,*
2. *$d(x, y) = 0$ iff $x = y$, and*
3. *$d(x, y) + d(y, z) \geq d(x, z)$, the **triangle inequality**.*

*A **metric space** is a pair (M, d), where M is nonempty and d is a metric on M.*

We not only have distances between points, but we also have distances between points and sets.

Definition 4.1.2   *For $E \subset M$, (M, d) a metric space, and $x \in M$, the **distance from** x **to** E is denoted $d(x, E)$ and defined by $d(x, E) = \inf\{d(x, e) : e \in E\}$.*

Note that $d(x, E) = 0$ need not imply that $x = E$.

Notation Alert 4.1.A   *Throughout this chapter, when we write "(M, d)" with no further explanation, we have assumed that M is nonempty and that d is a metric, and when we write "M" with no further explanation, it means that we got careless and forgot to write the metric, but we assume that it is there if we need it.*

If (M, d) is a metric space and E is a nonempty subset of M, then $(E, d_{|E \times E})$ is a metric space.

Notation Alert 4.1.B   *In yet another notational abuse for the purposes of clarity, we do not distinguish between d and $d_{|E \times E}$. Sometimes it is necessary to study several metrics simultaneously, and the other letter we use for metrics is ρ, or sometimes r or e.*

For any set, there exists at least one metric.

Example 4.1.3 *For any nonempty M, define the* **discrete metric** *by $e(x, y) = 1$ if $x \neq y$ and $e(x, y) = 0$ if $x = y$. To check that it is a metric, note that $e(x, y) = e(y, x)$, $e(x, y) = 0$ iff $x = y$, and $e(x, y) + e(y, z) \geq e(x, z)$ because the right-hand side of the inequality is either 0, making the inequality true, or 1, in which case the left-hand side is either 1 or 2.*

We will come to think of the discrete metric spaces as being pathologically badly behaved. In the next section, we study them extensively for two pedagogical reasons:

1. Discrete metric spaces are sufficiently simple that in one short section, it is possible to characterize the sets and spaces satisfying each of the definitions in this section.

2. Discrete metric spaces have a very interesting and useful class of continuous functions, which reappear in many guises throughout much of the rest of this book.

4.1.c Sequences and Completeness

Recall from Definition 3.3.3 (p. 75) that a sequence in M is a mapping $n \mapsto x_n$ from $\mathbb{N}$ to X, and from Definition 3.3.5 (p. 76), any increasing mapping $k \mapsto n_k$ specifies another sequence $k \mapsto x_{n_k}$, called a **subsequence of** x_n.

Definition 4.1.4 *A sequence x_n in (M, d)* **converges to** *x, written $x_n \to_d x$ or $x_n \to x$, if*

$$(\forall \epsilon \in \mathbb{R}_{++})(\exists N \in \mathbb{N})(\forall n \geq N)[d(x_n, x) < \epsilon],$$

in which case x_n is called **convergent** *and x is called its* **limit**. *A* **convergent subsequence** *is a subsequence that converges when regarded as a sequence. If x_n is not convergent, it is* **divergent**. *Finally, x_n is* **Cauchy** *if*

$$(\forall \epsilon \in \mathbb{R}_{++})(\exists M \in \mathbb{N})(\forall n, m \geq M)[d(x_n, x_m) < \epsilon].$$

The next theorem tells us that convergent sequences are Cauchy and that they can converge to at most one point.[1]

Theorem 4.1.5 *If $x_n \to x$, then x_n is Cauchy, and if $x' \neq x$, then $\neg[x_n \to x']$.*

Proof. Suppose that $x_n \to x$. Pick $\epsilon > 0$ and N such that for all $n \geq N$, $d(x_n, x) < \epsilon/2$. For all $n, m \geq N$, $d(x_n, x_m) \leq d(x_n, x) + d(x, x_m) < \epsilon/2 + \epsilon/2$.

Suppose now that $x_n \to x$, $x_n \to x'$, and $x \neq x'$. Let $\epsilon = d(x, x')/2$. Since $x \neq x'$, $\epsilon > 0$. For large n, $d(x, x_n) < \epsilon$ and $d(x_n, x') < \epsilon$. By the triangle inequality, $d(x, x') \leq d(x, x_n) + d(x_n, x') < 2\epsilon = d(x, x')$, so that $d(x, x') < d(x, x')$, a contradiction. ∎

---

1. As noted earlier, in this chapter, we usually reserve "theorem" for results that are true in all metric spaces, and use "lemma" for results specific to the metric space under study.

Convergent sequences are Cauchy sequences, but the reverse is not generally true—think of x_n being the n-length decimal expansion of $\sqrt{2}$ in the metric space $(\mathbb{Q}, d_E)$. The property that Cauchy sequences have limits is crucial and has a correspondingly crucial-sounding name: **completeness**.

Definition 4.1.6 (M, d) is **complete** if every Cauchy sequence is convergent.

We saw, in Chapter 3, that $(\mathbb{R}, d_E)$ is complete. In the next section we prove that discrete metric spaces are also complete.

4.1.d Separability

In $(\mathbb{R}, d_E)$, the countable set $\mathbb{Q}$ was dense. **Separability** names this property.

Definition 4.1.7 If (M, d) is a metric space and $E \subset M$, then E **is dense in** M if for all $x \in M$ and all $\epsilon > 0$, there exists an $e \in E$ such that $d(x, e) < \epsilon$. If there exists a countable dense E, then (M, d) is **separable**, and otherwise it is **nonseparable**.

Example 4.1.8 Separability is a property of both the set M and the metric d. For example, $(\mathbb{R}, d_E)$ is separable because $\mathbb{Q}$ is a countable dense set, but $(\mathbb{R}, e)$ is not separable with $e(\cdot, \cdot)$ being the discrete metric. To see why this last is true, let E be any countable subset of $\mathbb{R}$ so that $E^c \neq \emptyset$. For any $x \in E^c$ and any $e \in E$, $d(x, e) = 1$.

4.1.e Open and Closed Sets

Definition 4.1.9 The **open ball around a point** $x \in M$ **with radius** $r > 0$ **or the open r-ball around** x is the set $B_r^d(x) := \{y \in M : d(x, y) < r\}$. When the metric in question is clear from context, this is written as $B_r(x)$.

The reason for the word "ball" comes from using the Euclidean metric in $\mathbb{R}^2$ and $\mathbb{R}^3$, where $B_r(x)$ is, geometrically, a ball centered at x. The reason for the adjective "open" is contained in the next definition and theorem.

Definition 4.1.10 If (M, d) is a metric space, then $G \subset M$ is d-**open**, or simply **open** if $\forall x \in G$, $\exists \delta > 0$, such that $B_\delta(x) \subset G$. The class of d-open sets is denoted τ_d and is called the **metric topology**. $F \subset M$ is d-**closed** or **closed** if F^c is open.

The triangle inequality guarantees that open balls are in fact open.

Theorem 4.1.11 For any metric space, (M, d), each $B_r(x)$ is open.

The problem is that $x' \in B_r(x)$ can be close to the "edge" of the ball, which may dictate a very small δ to arrange that $B_\delta(x') \subset B_r(x)$.

Proof. Pick $x' \in B_\delta(x)$ so that $d(x, x') < r$. Let $\delta = r - d(x, x')$ so that $\delta > 0$, and pick any point $y \in B_\delta(x')$. We must show that $y \in B_r(x)$, that is, that $d(x, y) < r$. $d(x, y) \leq d(x, x') + d(x', y)$ by the triangle inequality, and $d(x, x') + d(x', y) < r + (r - d(x, x')) = r$. ■

Every open set can be expressed as a union of open balls. If G is open and $x \in G$, let $\delta(x) > 0$ satisfy $B_{\delta(x)}(x) \subset G$, so that $G = \cup_{x \in G} B_{\delta(x)}(x)$. In $(\mathbb{R}, d)$, this expresses every open set as an uncountable union of open balls.

4.1.f Neighborhoods, Topologies, and the Topological Equivalence of Metrics

Recall that $\tau_d \subset \mathcal{P}(M)$ is the class of d-open subsets of M.

Definition 4.1.12 *If $x \in G \in \tau_d$, then G **is an open neighborhood of** x, and if H contains an open neighborhood of x, then H **is a neighborhood of** x.*

The term "neighborhood" is meant to invoke a sense of "closeness." The invocation of "closeness" is somewhat misleading since M itself is a neighborhood of x. The idea is that a neighborhood G of x contains $B_\delta(x)$ for *very* small δ's.

Theorem 4.1.13 *For any metric space (M, d), $\emptyset$, $M \in \tau_d$, any union of sets in τ_d belongs to τ_d, as does any finite intersection of sets in τ_d.*

Proof. $\emptyset$ and M are open. Let $\{G_\alpha : \alpha \in A\} \subset \tau_d$ and pick $x \in G := \cup_{\alpha \in A} G_\alpha \in \tau$. Since $x \in G_{\alpha^\circ}$ for some $\alpha^\circ \in A$ and G_{α° is open, $\exists \delta > 0$ $B_\delta(x) \subset G_{\alpha^\circ} \subset G$. Finally, let $\{G_\alpha : \alpha \in A_F\} \subset \tau$, A_F finite, and choose $x \in \cap_{\alpha \in A_F} G_\alpha$. For each $\alpha \in A_F$, we can pick $\delta_\alpha > 0$ such that $B_{\delta_\alpha}(x) \subset G_\alpha$ because each G_α is open. Let $\delta = \min\{\delta_\alpha : \alpha \in A_F\}$. Since A_F is finite, $\delta > 0$. As $\delta \leq \delta_\alpha$ for each $\alpha \in A_F$, $B_\delta(x) \subset \cap_{\alpha \in A_F} G_\alpha$. ∎

These properties generalize.

Definition 4.1.14 *A class of sets $\tau \subset \mathcal{P}(M)$ is called a **topology** and (M, τ) is called a **topological space** if*

1. $\emptyset$, $M \in \tau$,

2. if $\{G_\alpha : \alpha \in A\} \subset \tau$, then $\cup_{\alpha \in A} G_\alpha \in \tau$, and

3. if $\{G_\alpha : \alpha \in A_F\} \subset \tau$, A_F finite, then $\cap_{\alpha \in A_F} G_\alpha \in \tau$.

*Sets in τ are τ-**open** or **open**. The complement of an open set is a **closed set**.*

Note how high the level of generality is becoming—metric spaces generalize $(\mathbb{R}, d)$ and topological spaces generalize metric spaces. There are topological spaces that are not metric spaces but are still useful to economists, but there are not that many of them.[2]

Since it is possible to have many metrics on the same space, we want to be able to identify metrics that are equivalent for topological purposes.

Definition 4.1.15 *If d and ρ are metrics on M, then we say that d and ρ are **topologically equivalent**, or simply **equivalent**, written $d \sim \rho$, if $\tau_d = \tau_\rho$.*

Example 4.1.16 *The metric $d(x, y) = |x - y|$ and the discrete metric $e(x, y) = 1$ if $x \neq y$ and $e(x, y) = 0$ if $x = y$ are not equivalent on $\mathbb{R}$, but they are equivalent*

---

2. This fact should, hopefully, help insulate you from any feelings of inferiority when/if people start throwing around phrases such as "topological spaces."

on $\mathbb{N}$. *To prove the first part, it suffices to show that there is one set in* τ_e *that is not in* τ_d. *The set* $\{0\} = B^e_\delta(0)$, $\delta < 1$, *is e-open because all open balls are open, but it is not d-open because for any* $\delta > 0$ *and any* $x \in \mathbb{R}$, $B^d_\delta(x) \not\subset \{0\}$ *(e.g., because* $B^d_\delta(x)$ *contains at least two points). For the second part, note that any subset of* $\mathbb{N}$ *is open with either metric.*

Theorem 4.1.17 *For any metric space* (M, d), *the metric d and the metric* $\rho(x, y) := \min\{d(x, y), 1\}$ *are equivalent.*

Proof. We must first check that ρ is a metric: $\rho(x, y) = \rho(y, x)$ immediately; $\rho(x, y) = 0$ iff $d(x, y) = 0$ iff $x = y$. A more involved argument is needed to show that $\rho(x, y) + \rho(y, z) \geq \rho(x, z)$. We know that $d(x, y) + d(y, z) \geq d(x, z)$ because d is a metric. Let $r, s, t \geq 0$ with $r + s \geq t$. It is sufficient to show that $\min\{r, 1\} + \min\{s, 1\} \geq \min\{t, 1\}$. We show that this inequality holds if $t > 1$ and if $t \leq 1$:

1. If $t > 1$, then $\min\{t, 1\} = 1$. If neither $r > 1$ nor $s > 1$, the inequality is satisfied because $r + s \geq t > 1$. If either $r > 1$ or $s > 1$, the inequality is satisfied because $1 + m \geq 1$ for any $m \geq 0$.

2. If $t \leq 1$, then $\min\{t, 1\} = t$. If neither $r > 1$ nor $s > 1$, the inequality is satisfied because $r + s \geq t$. If either $r > 1$ or $s > 1$, the inequality is satisfied because $1 + m \geq 1 \geq t$ for any $m \geq 0$.

Now let G be d-open. We must show that it is ρ-open. Pick a point $x \in G$. Since G is d-open, there exists $\delta > 0$ such that $B^d_\delta(x) \subset G$. If $\delta' = \min\{\delta, 1/2\}$, then $B^\rho_{\delta'}(x) \subset B^d_\delta(x) \subset G$ so that G is ρ-open. Now suppose that G is ρ-open. Pick a point $x \in G$ and $\delta > 0$ so that $B^\rho_\delta(x) \subset G$. Setting $\delta' = \min\{\delta, 1/2\}$ again yields $B^d_{\delta'}(x) \subset B^\rho_\delta(x) \subset G$. ∎

4.1.g Open Covers, Compactness, and Connectedness

If $B_\epsilon(x)$ is the open ϵ-ball around x, then it makes sense to think of $E^\epsilon := \cup_{x \in E} B_\epsilon(x)$ as the ϵ-ball around the set E. Being a union of open sets, E^ϵ is open. Note that $E^\epsilon = \{x \in M : d(x, E) < \epsilon\}$.

The collection $\{B_\epsilon(x) : x \in E\}$ is an example of what is called an **open cover of** E.

Definition 4.1.18 *An **open cover of a set** $E \subset M$ in a metric space (M, d) is a collection $\{G_\alpha : \alpha \in A\} \subset \tau_d$ of open sets such that $E \subset \cup_{\alpha \in A} G_\alpha$. A **finite subcover** is a subset, $\{G_\alpha : \alpha \in A_F\}$, $A_F \subset A$, A_F finite, such that $\{G_\alpha : \alpha \in A_F\}$ is an open cover of E.*

Definition 4.1.19 *If (M, d) is a metric space, then $K \subset M$ is **compact** if every open cover of K has a finite subcover. If M is compact, we say it is a **compact space**.*

It should be clear that every finite set is compact. We will see that compact sets are "approximately" finite.

Example 4.1.20 *We will see in §4.7.a that $[0, 1]$ is compact. The following sets are not compact.*

1. $(0, 1) = \cup_{n \in \mathbb{N}}(\frac{1}{n}, 1 - \frac{1}{n})$.

2. $\mathbb{R} = \cup_{n \in \mathbb{N}}(-n, +n)$.

3. $\mathbb{R}^{\ell} = \cup_{n \in \mathbb{N}}(-n, +n)^{\ell}$.

Connectedness is a property of metric spaces expressible in terms of open covers containing two elements.

Definition 4.1.21 *A metric space (M, d) is **disconnected** if it can be expressed as the union of two disjoint nonempty open sets, and otherwise it is **connected**.*

With a discrete metric, a space containing two or more points is disconnected. We show in §4.12 that $(\mathbb{R}, d_E)$, and $(\mathbb{R}^{\ell}, d_E)$, are connected.

4.1.h Continuous Functions

Let (M, d) and (M', d') be two metric spaces.

Definition 4.1.22 *A function $f : M \to M'$ is **continuous** if $f^{-1}(G)$ is an open subset of M for every open $G \subset M'$, equivalently if $f^{-1}(\tau_{d'}) \subset \tau_d$.*

This definition looks nothing like the intuitive idea of a discontinuity being a sudden jump in the value of a function. Have patience: it turns out that this captures the intuitive idea and is the best definition.[3]

The continuous functions are sufficiently important that they have their own notation.

Notation 4.1.23 *$C(M; M')$ denotes the set of continuous functions from M to M'. The set of continuous functions from M to $\mathbb{R}$ is written $C(M)$, that is, $C(M) = C(M; \mathbb{R})$. A function $f \in C(M)$ is **bounded** if there exists $B > 0$ such that $(\forall x \in M)[|f(x)| \leq B]$. The set of bounded continuous functions from M to $\mathbb{R}$ is written $C_b(M)$.*

For most of the metric spaces we consider, there are many functions that are not continuous. We spend a great deal of time studying a special class of these called the **measurable** functions in Chapter 7.

4.1.i Homeomorphisms and Isometries

The function $F(r) = \frac{e^r}{1+e^r}$ is a one-to-one onto function from $\mathbb{R}$ onto $(0, 1)$. Later we will see that $F(\cdot)$ and $F^{-1}(\cdot)$ are both continuous. For many mathematical purposes, the ones involving continuity and only continuity, this means that it is difficult to distinguish between $\mathbb{R}$ and its subset $(0, 1)$.

Definition 4.1.24 *A function $f : M \to M'$ is a **homeomorphism** if f is one-to-one onto continuous and has a continuous inverse, in which case we say that (M, d) and (M', d') are **homeomorphic**.*

---

3. One definition of mathematical maturity is the willingness to put up with reams of abstract nonsense with little-to-no evidence of usefulness. This part of the book does have a high ratio of definitions and theorems to applications. We try to improve that ratio soon.

In these terms, $\mathbb{R}$ and $(0, 1)$ are homeomorphic. They are not, however, **isometric**, and we are not trying to convince you that there is no distinction between $\mathbb{R}$ and $(0, 1)$.

Definition 4.1.25   *A function $f : M \to M'$ is an **isometry** if f is one-to-one onto and for every x, $y \in M$, $d(x, y) = d'(f(x), f(y))$, in which case we say that (M, d) and (M', d') are **isometric**.*

Isometry implies homeomorphism, but not the reverse. Given our usual (Euclidean) metric, $d(x, y) = |x - y|$, $(\mathbb{R}, d)$ and $((0, 1), d)$ are homeomorphic but not isometric, whereas $((0, 1), d)$ and $((r, r + 1), d)$ are isometric, and one isometry between them is $f(x) = x + r$.

4.2 ◆ Discrete Spaces

Let M be a nonempty set and define $e(x, y) = 1$ if $x \neq y$ and $e(x, y) = 0$ if $x = y$ as in Example 4.1.3. The metric space (M, d) is called a **discrete metric space**. In a discrete metric space, every point is an open set because $\{x\} = B_\epsilon(x)$ for any $\epsilon < 1$. This has many implications.

Lemma 4.2.1   *If x_n is a sequence in M, (M, e) a discrete metric space, then x_n is convergent iff it is constant for large n, that is, iff $(\exists x \in M)(\exists N \in \mathbb{N})(\forall n \geq N)[x_n = x]$.*

Proof.   If x_n is constant at, say, x, for large n, then $x_n \to x$. If $x_n \to x$, then for all large n, $x_n \in B_\epsilon(x) = \{x\}$ for $\epsilon < 1$.   ∎

Lemma 4.2.2   *Any discrete metric space is complete; it is separable if and only if it is countable.*

Proof.   Pick $\epsilon < 1$, so that $e(x, y) < \epsilon$ iff $x = y$. If x_n is a Cauchy sequence in a discrete metric space, then for large m, n, $e(x_n, x_m) < \epsilon$, so that $x_n = x_m$. If E is dense, then $E \cap B_\epsilon(x) \neq \emptyset$ for all x. Therefore E is dense iff $E = M$, so that M is separable iff it is countable.   ∎

Another implication of points being open sets is that every subset of M is both open and closed.

Lemma 4.2.3   *If (M, e) is a discrete metric space, then every $E \subset M$ is both open and closed.*

Another way to put this is that for discrete metric spaces (M, e), $\tau_e = \mathcal{P}(M)$.

Proof.   For $\epsilon < 1$, $E = \cup_{x \in E} B_\epsilon(x)$ expresses E as a union of open sets. In the same way, E^c is open, so that E is closed.   ∎

There are many metrics equivalent to the discrete metric.

Example 4.2.4   *If $M = \mathbb{N}$, e is the discrete metric on M, and $\rho(m, n) = |\frac{1}{m} - \frac{1}{n}|$, then $e \sim \rho$. To see why, note that for $\epsilon < \rho(m, m + 1)$, $B_\epsilon^\rho(m) = \{m\}$, so that $\tau_\rho = \tau_e = \mathcal{P}(\mathbb{N})$.*

Lemma 4.2.5   *If (M, e) is a discrete metric space containing two or more points, then M is disconnected.*

Proof. For any $x \in M$, let $G_1 = \{x\}$ and $G_2 = G_1^c$. This gives two, nonempty disjoint open sets whose union is all of M. ■

Since every subset of a discrete space is open, every function is continuous.

Lemma 4.2.6 *If (M, d) is a discrete metric space, (M', d') is a metric space, and $f : M \to M'$ is a function, then f is continuous.*

Proof. Pick an arbitrary open $G \subset M'$. We must show that $f^{-1}(G)$ is an open subset of M. Since every subset of M is open, $f^{-1}(G)$ is open. ■

Example 4.2.7 *If $M = \{1, 2, 3\}$ is a three-point discrete metric space, then any $f \in C(M)$, that is, any continuous $\mathbb{R}$-valued function f, is specified by three numbers $f(1)$, $f(2)$, and $f(3)$. If we put these together in a single list, we have the length-3 vector $(f(1), f(2), f(3))$. Thus, points (x, y, z) in $\mathbb{R}^3$ and functions in $C(M)$ can be identified, that is, treated as if they were one and the same thing.*

Example 4.2.8 *If $M = \mathbb{N}$, e is the discrete metric on M, and (M', d') is a metric space, then $C(\mathbb{N}; M')$ is the set of all sequences in M', $C(\mathbb{N})$ is the set of sequences in $\mathbb{R}$, and $C_b(\mathbb{N})$ is the set of all bounded sequences in $\mathbb{R}$. Sequences of rewards stretching into the indefinite future are often modeled as elements of $C(\mathbb{N})$ or $C_b(\mathbb{N})$.*

In general, compactness is intuitively similar to finiteness, and in discrete metric spaces, it is equivalent.

Lemma 4.2.9 (Compactness Criterion in Discrete Spaces) *If (M, d) is a discrete metric space, then $E \subset M$ is compact iff E is finite.*

Proof. Suppose that E is finite and let $\mathcal{G} = \{G_\alpha : \alpha \in A\}$ be an open cover of E. For each $x \in E$, pick $\alpha(x)$ such that $x \in G_{\alpha(x)}$. The collection $\{G_{\alpha(x)} : x \in E\}$ is a finite subcover of E.

Now suppose that E is compact, so that any open cover has a finite subcover. Pick $0 < \epsilon < 1$. Consider the open cover $\mathcal{G} = \{B_\epsilon(x) : x \in E\}$, a disjoint collection of sets. The only possible finite subcover of E is $\mathcal{G}$ itself. Therefore, $\mathcal{G}$ is finite, implying that E is finite. ■

Since all functions on discrete spaces are continuous, a function between two discrete spaces is a homeomorphism iff it is one-to-one and onto. This yields the following.

Lemma 4.2.10 *The discrete metric spaces (M, e) and (M', e') are homeomorphic iff $\#M = \#M'$.*

4.3 ◆ $\mathbb{R}^\ell$ as a Normed Vector Space

$\mathbb{R}^\ell$ is the ℓ-fold product of $\mathbb{R}$, equivalently the set of ordered lists of real numbers containing ℓ elements, or $\mathbb{R} \times \cdots \times \mathbb{R}$, where the product happens ℓ times. The spaces $\mathbb{R}^\ell$, $\ell = 1, 2, \ldots$ are the most heavily used metric spaces in economics. As well as being metric spaces, they are also normed vector spaces.

Elements of $\mathbb{R}^\ell$ are written **x** or **y**. We are used to thinking of $\mathbb{R}^1$ as a the line, of $\mathbb{R}^2$ as the plane, and $\mathbb{R}^3$ as three-space. The intuitions gained from that geometry are very valuable.

4.3.a Three Norms on $\mathbb{R}^\ell$

For **x**, **y** $\in \mathbb{R}^\ell$ and $\alpha, \beta \in \mathbb{R}$, we have

$$\mathbf{x} = \begin{bmatrix} x_1 \\ x_2 \\ \vdots \\ x_\ell \end{bmatrix}, \mathbf{y} = \begin{bmatrix} y_1 \\ y_2 \\ \vdots \\ y_\ell \end{bmatrix}, \quad \text{and} \quad \alpha\mathbf{x} + \beta\mathbf{y} := \begin{bmatrix} \alpha x_1 + \beta y_1 \\ \alpha x_2 + \beta y_2 \\ \vdots \\ \alpha x_\ell + \beta y_\ell \end{bmatrix}, \quad (4.1)$$

where each of the x_i and y_i are elements of $\mathbb{R}$, $i = 1, 2, \ldots, \ell$. Of special interest are the cases where $\beta = 0$, giving $\alpha\mathbf{x}$, and where the $x_i = 0$, giving the vector 0. In $\mathbb{R}^\ell$, 0 is the **origin**, $\alpha\mathbf{x}$ stretches the vector **x** by a factor of α, and one sums vectors by summing their components. $\mathbb{R}^\ell$ is a vector space.

Definition 4.3.1 *A set $\mathfrak{X}$ is a **vector space** (over $\mathbb{R}$) if the mappings $(x, y) \mapsto x + y$ from $\mathfrak{X} \times \mathfrak{X}$ to $\mathfrak{X}$ and $(r, x) \mapsto rx$ from $\mathbb{R} \times \mathfrak{X}$ to $\mathfrak{X}$ satisfy for all $x, y, z \in \mathfrak{X}$ and all $\alpha, \beta \in \mathbb{R}$:*

1. *$x + y = y + x$;*
2. *$(x + y) + z = x + (y + z)$;*
3. *the point $0 \in \mathfrak{X}$ defined as $0z$ for $0 \in \mathbb{R}$ and called the "origin," satisfies $y + (-1)y = 0$ and $0 + x = x$;*
4. *$(\alpha\beta)x = \alpha(\beta x)$; and $(\alpha + \beta)x = \alpha x + \beta x$.*

As defined above, $\mathbb{R}^\ell$ is a vector space but $\mathbb{N}^\ell$ is not. There are many vector spaces.

Example 4.3.2 *$C([a, b])$ denotes the set of continuous, $\mathbb{R}$-valued functions on the interval $[a, b]$. For $f, g \in C([a, b])$, we define $f + g$ to be the function taking on the value $f(x) + g(x)$ at $x \in [a, b]$, and we define αf to be the function taking on the value $\alpha f(x)$. The origin is the constant function 0.*

Example 4.3.3 *We have $c_0 \subset C(\mathbb{N})$ denoting the set of sequences $x = (x_1, x_2, x_3, \ldots)$ in $\mathbb{R}$ such that $x_n \to 0$. For $x, y \in c_0$, we define $x + y$ to be the sequence $(x_1 + y_1, x_2 + y_2, x_3 + y_3, \ldots)$, and we define αx to be the sequence $(\alpha x_1, \alpha x_2, \alpha x_3, \ldots)$. The origin is the constant sequence $(0, 0, 0, \ldots)$.*

We are going to define the three most commonly used norms for $\mathbb{R}^\ell$. There are many more, and we return to them later.[4]

---

4. They are for $p \in [1, \infty)$, $\|\mathbf{x}\|_p := \left[\sum_{i=1}^{\ell} |x_i|^p \right]^{1/p}$.

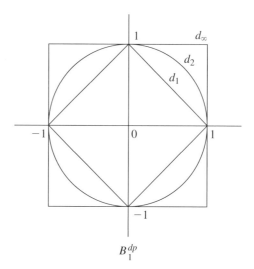

FIGURE 4.3.5

Definition 4.3.4 *For* $\mathbf{x} \in \mathbb{R}^{\ell}$, *define the* 1-*norm, the* 2-*norm, and the* ∞-*norm by*

$$\|\mathbf{x}\|_1 = \left[\sum_{i=1}^{\ell} |x_i| \right], \ \|\mathbf{x}\|_2 = \left[\sum_{i=1}^{\ell} |x_i|^2 \right]^{1/2}, \quad \text{and} \quad \|\mathbf{x}\|_{\infty} = \max_{i=1}^{\ell} |x_i|.$$

(4.2)

*The associated **norm distance** is defined by* $d_p(\mathbf{x}, \mathbf{y}) = \|\mathbf{x} - \mathbf{y}\|_p.$

We have $\|\mathbf{x}\|_2$ as the usual, Euclidean length of the vector $\mathbf{x}$, and $d_2(\mathbf{x}, \mathbf{y})$ is the Euclidean length of the line segment, $[\mathbf{x}, \mathbf{y}]$, joining $\mathbf{x}$ and $\mathbf{y}$.

Notation Alert 4.3.A *When we write* $\|\mathbf{x}\|$ *with no subscript, we usually mean* $\|\mathbf{x}\|_2$. *If we do not mean that, we mean that any of the norms work for what we are about.*

Example 4.3.5 *When working in the plane, that is,* $\mathbb{R}^2$, $d_1(\mathbf{x}, \mathbf{y}) = |x_1 - y_1| + |x_2 - y_2|$ *is sometimes called the "Manhattan distance" between* $\mathbf{x}$ *and* $\mathbf{y}$ *because it measures any of the shortest paths from* $\mathbf{x}$ *to* $\mathbf{y}$ *on the assumption that one can only travel vertically or horizontally on the map. The* d_1 *unit ball around* $(0, 0)$ *is the diamond with vertices* $(1, 0)$, $(0, 1)$, $(-1, 0)$, *and* $(0, -1)$. *[See Figure 4.3.5.] By contrast, the* d_{∞} *unit ball around* $(0, 0)$ *is the square with vertices* $(1, 1)$, $(-1, 1)$, $(-1, -1)$, *and* $(1, -1)$, *and the* d_2 *unit ball around* $(0, 0)$ *is the circle passing through the vertices of the* d_1 *unit ball diamond.*

Exercise 4.3.6 Let $\mathbf{e}_i \in \mathbb{R}^{\ell}$ be the unit vector in the ith direction, that is, the vector with all components equal to 0 except for a 1 in the ith position. Show that for $\mathbf{x}, \mathbf{y} \in \mathbb{R}^{\ell}$, the following are equivalent:

1. $\mathbf{y} = \mathbf{x} \pm r\mathbf{e}_i$ for some i,
2. $d_1(\mathbf{x}, \mathbf{y}) = d_2(\mathbf{x}, \mathbf{y}) = r$, and
3. $d_2(\mathbf{x}, \mathbf{y}) = d_\infty(\mathbf{x}, \mathbf{y}) = r$.

Definition 4.3.7   *If $\mathfrak{X}$ is a vector space, then a **norm** on $\mathfrak{X}$ is a function from $\mathfrak{X}$ to $\mathbb{R}_+$, denoted $\|\cdot\| : \mathfrak{X} \to \mathbb{R}_+$, which satisfies the following properties for $\forall x, y \in \mathfrak{X}$, and $\forall a \in \mathbb{R}$:*

 1. $\|y\| = 0$ iff $y = 0$,
 2. $\|ay\| = |a|\, \|y\|$, and
 3. $\|x + y\| \le \|x\| + \|y\|$.

*A vector space in which a norm has been defined has a **norm distance** between $x, y \in \mathfrak{X}$ defined by $d(x, y) = \|x - y\|$. A **normed vector space**, $(\mathfrak{X}, \|\cdot\|)$, is a metric vector space with a norm distance.*

If we have a norm, then the associated norm distance is really a distance function. To check the three properties that a metric must satisfy, note that: $d(x, y) = d(y, x)$ because $\|x - y\| = \|(-1)(y - x)\| = |-1|\|y - x\| = \|y - x\|$; $d(x, y) = 0$ iff $\|x - y\| = 0$ iff $x - y = 0$ iff $x = y$; and $d(x, z) \le d(x, y) + d(y, z)$ iff $\|x - z\| \le \|x - y\| + \|y - z\|$ iff $\|u + v\| \le \|u\| + \|y\|$, where $u = x - y$ and $v = y - z$.

Lemma 4.3.8 $\|\cdot\|_1$, $\|\cdot\|_2$, and $\|\cdot\|_\infty$ *are norms on $\mathbb{R}^\ell$.*

Proof.   (i) For each, it is easy to verify that $\|\mathbf{y}\|_p = 0$ iff $|y_i| = 0$ for $i = 1, \ldots, \ell$.

(ii)   For $p = 1, 2$, $\|a\mathbf{y}\|_p = \left[\sum_{i=1}^{\ell} |ay_i|^p\right]^{1/p} = \left[\sum_{i=1}^{\ell} |a|^p |y_i|^p\right]^{1/p} = (|a|^p)^{1/p}\|\mathbf{y}\|$. For $p = \infty$, observe that $\|a\mathbf{y}\|_\infty = \max_i |ay_i| = |a| \max_i |y_i|$.

(iii) For $p = 1$, this follows from the observation that $|x_i + y_i| \le |x_i| + |y_i|$, which implies that $\sum_i |x_i + y_i| \le \sum_i |x_i| + \sum_i |y_i|$. For $p = \infty$, this follows from two observations: $|x_i + y_i| \le |x_i| + |y_i|$, so that $\max_i |x_i + y_i| \le \max_i\{|x_i| + |y_i|\}$, and $\max_i\{|x_i| + |y_i|\} \le (\max_i |x_i|) + (\max_i |y_i|)$.

For $p = 2$, the proof is a bit more involved, and involves an inequality that is sufficiently important that it has a name—the Cauchy-Schwarz inequality, Lemma 4.3.10.   ∎

4.3.b   Inner (or Dot) Products and the Cauchy-Schwarz Inequality

We gather some basics about the inner (or dot) products of vectors.

Definition 4.3.9   *For $\mathbf{x}, \mathbf{y} \in \mathbb{R}^\ell$, the **inner (or dot) product of $\mathbf{x}$ and $\mathbf{y}$** is defined by $\mathbf{x} \cdot \mathbf{y} = \sum_i x_i y_i$.*

This means that $\|\mathbf{x}\|(= \|\mathbf{x}\|_2) = \sqrt{\mathbf{x} \cdot \mathbf{x}}$, equivalently $\mathbf{x} \cdot \mathbf{x} = \|\mathbf{x}\|^2$. When we think of vectors as $n \times 1$ matrixes, that is, as arrays of numbers into n rows and 1 column, we write $\mathbf{x} \cdot \mathbf{y}$ as $\mathbf{x}^T\mathbf{y}$. Here "$\mathbf{x}^T$" is read as "$\mathbf{x}$ transpose," and $\mathbf{x}^T\mathbf{y}$ is the multiplication of a $1 \times n$ matrix with an $n \times 1$ matrix. Sometimes, we are sloppier and just write $\mathbf{xy}$.

Lemma 4.3.10 (Cauchy–Schwarz Inequality) *For* $\mathbf{x}, \mathbf{y} \in \mathbb{R}^\ell$, $|\mathbf{x} \cdot \mathbf{y}| \leq \|\mathbf{x}\| \|\mathbf{y}\|$.

Proof. If either $\mathbf{x} = 0$, or $\mathbf{y} = 0$, the equality is satisfied automatically.

For any $\lambda \in \mathbb{R}$, $\|\mathbf{x} - \lambda \mathbf{y}\|^2 \geq 0$. $f(\lambda) = \|\mathbf{x} - \lambda \mathbf{y}\|^2 = (\mathbf{x} \cdot \mathbf{x}) - 2\lambda(\mathbf{x} \cdot \mathbf{y}) + \lambda^2(\mathbf{y} \cdot \mathbf{y})$. The function f is a quadratic in λ that opens upward. Therefore, by factorization, it is minimized by $\lambda^* = \frac{(\mathbf{x} \cdot \mathbf{y})}{(\mathbf{y} \cdot \mathbf{y})}$, so that $f(\lambda^*) = (\mathbf{x} \cdot \mathbf{x}) - 2\frac{(\mathbf{x} \cdot \mathbf{y})}{(\mathbf{y} \cdot \mathbf{y})}(\mathbf{x} \cdot \mathbf{y}) + \left(\frac{(\mathbf{x} \cdot \mathbf{y})}{(\mathbf{y} \cdot \mathbf{y})}\right)^2(\mathbf{y} \cdot \mathbf{y}) \geq 0$. Simplifying yields $(\mathbf{x} \cdot \mathbf{x}) - \frac{(\mathbf{x} \cdot \mathbf{y})^2}{(\mathbf{y} \cdot \mathbf{y})} \geq 0$. Rearranging, we get $(\mathbf{x} \cdot \mathbf{y})^2 \leq (\mathbf{x} \cdot \mathbf{x})(\mathbf{y} \cdot \mathbf{y})$ and taking square roots gives us $|\mathbf{x} \cdot \mathbf{y}| \leq \|\mathbf{x}\| \|\mathbf{y}\|$. ∎

To be fair, the way that we found the λ^* earlier was by solving $f'(\lambda) = -2(\mathbf{x} \cdot \mathbf{y}) + 2\lambda(\mathbf{y} \cdot \mathbf{y}) = 0$, that is, by taking the derivative and setting it equal to 0. Since we have not even developed the theory of continuous functions yet, that seems a bit like cheating, though the given proof still works. The following avoids even the appearance of cheating.

Exercise 4.3.11 (An Alternative Proof of Cauchy–Schwarz) For $\mathbf{x}, \mathbf{y} \neq 0$, expand $\left\| \frac{\mathbf{x}}{\|\mathbf{x}\|} + \frac{\mathbf{y}}{\|\mathbf{y}\|} \right\| \geq 0$ to find $\|\mathbf{x}\| \|\mathbf{y}\| \geq -\mathbf{x} \cdot \mathbf{y}$ and expand $\left\| \frac{\mathbf{x}}{\|\mathbf{x}\|} - \frac{\mathbf{y}}{\|\mathbf{y}\|} \right\| \geq 0$ to find $\|\mathbf{x}\| \|\mathbf{y}\| \geq \mathbf{x} \cdot \mathbf{y}$. Combine to find the Cauchy–Schwarz inequality.

The following finishes the proof of Lemma 4.3.8.

Corollary 4.3.12 *For* $\mathbf{x}, \mathbf{y} \in \mathbb{R}^\ell$, $\|\mathbf{x} + \mathbf{y}\| \leq \|\mathbf{x}\| + \|\mathbf{y}\|$.

Proof. $\|\mathbf{x} + \mathbf{y}\| \leq \|\mathbf{x}\| + \|\mathbf{y}\|$ iff $\|\mathbf{x} + \mathbf{y}\|^2 \leq (\|\mathbf{x}\| + \|\mathbf{y}\|)^2$ iff $((\mathbf{x} + \mathbf{y}) \cdot (\mathbf{x} + \mathbf{y})) \leq (\mathbf{x} \cdot \mathbf{x}) + 2\|\mathbf{x}\| \|\mathbf{y}\| + (\mathbf{y} \cdot \mathbf{y})$ iff $(\mathbf{x} \cdot \mathbf{x}) + 2(\mathbf{x} \cdot \mathbf{y}) + (\mathbf{y} \cdot \mathbf{y}) \leq (\mathbf{x} \cdot \mathbf{x}) + 2\|\mathbf{x}\| \|\mathbf{y}\| + (\mathbf{y} \cdot \mathbf{y})$ iff $2(\mathbf{x} \cdot \mathbf{y}) \leq 2\|\mathbf{x}\| \|\mathbf{y}\|$, which is Cauchy–Schwarz. ∎

There is an equality that is more precise than the Cauchy–Schwarz inequality. If $\mathbf{x} = r\mathbf{y}$ for some $r \geq 0$, we set θ, the angle between $\mathbf{x}$ and $\mathbf{y}$, equal to 0. If $r < 0$, $\theta = -180$. The origin and any two points, $\mathbf{x}, \mathbf{y}, \mathbf{x} \neq r\mathbf{y}$, in $\mathbb{R}^\ell$ determine a plane. In the plane, the angle between $\mathbf{x}$ and $\mathbf{y}$ is denoted θ.

Lemma 4.3.13 $\mathbf{xy} = \|\mathbf{x}\| \|\mathbf{y}\| \cos \theta$.

Since $|\cos \theta| \leq 1$, this equality implies the Cauchy–Schwarz inequality. The equality also implies that the Cauchy–Schwarz inequality is satisfied as an equality iff $\cos \theta = 1$. Now, $\cos \theta = 1$ iff $\theta \in \{0, \pm360, \pm720, \ldots\}$. Therefore, Cauchy–Schwarz holds exactly iff $\mathbf{x} = r\mathbf{y}$ for some $r \in \mathbb{R}$.

Proof. There are three cases: (1) $\mathbf{x} = 0$ or $\mathbf{y} = 0$; (2) $\mathbf{x} \neq 0$, $\mathbf{y} \neq 0$, and $\mathbf{x} = r\mathbf{y}$ for some $r \neq 0$; and (3) $\mathbf{x} \neq 0$, $\mathbf{y} \neq 0$, and $\mathbf{x} \neq r\mathbf{y}$ for any $r \in \mathbb{R}$. Cases (1) and (2) are immediate.

(3) In the plane determined by 0, $\mathbf{x}$, and $\mathbf{y}$, drop a perpendicular from $\mathbf{x}$ to $L = \{r\mathbf{y} : r \in \mathbb{R}\}$, the line spanned by $\mathbf{y}$. [See Figure 4.3.13.] Let $t\mathbf{y}$, $t \in \mathbb{R}$, be the point where the perpendicular intersects L, so that θ is the angle between the line segment $[0, \mathbf{x}]$ joining 0 and $\mathbf{x}$ and the line segment L. We first cover the case where $t \neq 0$.

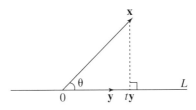

FIGURE 4.3.13

The triangle determined by 0, $\mathbf{x}$, and $t\mathbf{y}$ is a right triangle. From the Pythagorean theorem,[5] we know that $\|\mathbf{x}\|^2 = \|t\mathbf{y}\|^2 + \|\mathbf{x} - t\mathbf{y}\|^2$, equivalently $(\mathbf{xx}) = t^2(\mathbf{yy}) + (\mathbf{xx}) - 2t(\mathbf{xy}) + t^2(\mathbf{yy})$, so that $t^2(\mathbf{yy}) = t(\mathbf{xy})$. Since $t \neq 0$, this reduces to $\mathbf{xy} = t\|\mathbf{y}\|^2$. By the definition of $\cos\theta$, we know that $t = \cos\theta \frac{\|\mathbf{x}\|}{\|\mathbf{y}\|}$, so that $\mathbf{xy} = \|\mathbf{x}\|\|\mathbf{y}\| \cos\theta$, as claimed.

If $t = 0$, then $\mathbf{x}$ and $\mathbf{y}$ are perpendicular so that $\theta = 90$ or 270 and $\cos(\theta) = 0$. We must show that $\mathbf{xy} = 0$. If $\delta > 0$, then $(\mathbf{x} + \delta\mathbf{y})\mathbf{y} > 0$ because the angle between $(\mathbf{x} + \delta\mathbf{y})$ and $\mathbf{y}$ is less than 90 or greater than 270, and for similar reasons, $(\mathbf{x} - \delta\mathbf{y})\mathbf{y} < 0$. If we put these together, for every $\delta > 0$, $\mathbf{xy} + \delta\mathbf{yy} > 0 > \mathbf{xy} - \delta\mathbf{yy}$, which tells us that for all $\delta > 0$, $|\mathbf{xy}| < \delta(\mathbf{yy})$. Since this is true for all $\delta > 0$, $|\mathbf{xy}| = 0$. ∎

The inner product being 0 has very useful geometric content.

Definition 4.3.14 *We say that $\mathbf{x}$ and $\mathbf{y}$ are **perpendicular**, or **orthogonal**, written $\mathbf{x} \perp \mathbf{y}$, when $\mathbf{xy} = 0$. The set of all points orthogonal to $\mathbf{y} \neq 0$ is written as $\mathbf{y}^\perp$.*

4.3.c　The p-Norms, $p \in [1, \infty)$

The 1-norm, the 2-norm, and ∞-norm are in regular use in economics. There are others that are used much less frequently, but for completeness we give them here.[6]

For $p \in [1, \infty)$, the p-norms are $\|\mathbf{x}\|_p = \left[\sum_{i=1}^{\ell} |x_i|^p\right]^{1/p}$. Theorem 5.7.19 deals with the **Minkowski inequality**: for $\mathbf{x}, \mathbf{y} \in \mathbb{R}^\ell$,

$$\left[\sum_{i=1}^{\ell}(x_i + y_i)^p\right]^{1/p} \leq \left[\sum_{i=1}^{\ell}(x_i)^p\right]^{1/p} + \left[\sum_{i=1}^{\ell}(y_i)^p\right]^{1/p},$$

which is a direct consequence of the convexity and homogeneity[7] of degree 1 of the functions $f_p : \mathbb{R}_+^\ell \to \mathbb{R}$ defined $f_p(\mathbf{x}) = \left[\sum_{i=1}^{\ell}(x_i)^p\right]^{1/p}$. Just as the Cauchy-Schwarz inequality directly yields $\|\cdot\|_2$ being a norm, the Minkowski inequality directly yields $\|\cdot\|_p$ being a norm.

---

5. The square of the length of the hypotenuse of a right triangle is equal to the sum of the squares of the lengths of the adjacent edges of a right triangle.

6. In other words, feel free to ignore this material until you need it.

7. A function is homogeneous of degree r if $f(t\mathbf{x}) = t^r f(\mathbf{x})$.

Exercise 4.3.15 There are many more inner products and norms. Let M be an $\ell \times \ell$ symmetric matrix with the property that for all nonzero $\mathbf{x} \in \mathbb{R}^\ell$, $\mathbf{x}'M\mathbf{x} > 0$. Define the M-inner product of $\mathbf{x}$ and $\mathbf{y}$ as $\langle \mathbf{x}, \mathbf{y} \rangle_M = \mathbf{x}'M\mathbf{y}$ and the associated M-norm as $\|\mathbf{x}\|_M = \sqrt{\langle \mathbf{x}, \mathbf{x} \rangle_M}$. Show that:

1. taking $M = I$ (the identity matrix) gives the 2-norm above,
2. $\| \cdot \|_M$ is a norm,
3. for all $\mathbf{x}, \mathbf{y}$, $\langle \mathbf{x}, \mathbf{y} \rangle_M = \langle \mathbf{y}, \mathbf{x} \rangle_M$ (symmetry),
4. $\langle \alpha\mathbf{x} + \beta\mathbf{y}, \mathbf{z} \rangle_M = \alpha \langle \mathbf{x}, \mathbf{z} \rangle_M + \beta \langle \mathbf{y}, \mathbf{z} \rangle_M$ (linearity), and
5. $\langle \mathbf{x}, \mathbf{x} \rangle_M \geq 0$ with equality iff $\mathbf{x} = 0$.

Exercise 4.3.16 There are even more norms. Let $L : \mathbb{R}^\ell \to \mathbb{R}^\ell$ be a linear invertible mapping (i.e., one representable by an invertible matrix). Show that if $\| \cdot \|$ is a norm on $\mathbb{R}^\ell$, then so is $\| \cdot \|_L$ defined by $\|\mathbf{x}\|_L = \|L(\mathbf{x})\|$.

4.4 ◆ Completeness

Recall Definition 4.1.15 (p. 110), which says that two metrics d and ρ are equivalent iff $\tau_d = \tau_\rho$, that is, iff they give rise to the same class of open sets. We are now going to use the geometric insights we have just developed to show that the d_p-metrics are equivalent; that the spaces $(\mathbb{R}^\ell, d_1)$, $(\mathbb{R}^\ell, d_2)$, and $(\mathbb{R}^\ell, d_\infty)$ are homeomorphic to each other; and that each is a complete metric space. Lemma 4.4.2 is the crucial step in all of this proof.

4.4.a A Sufficient Condition for Topological Equivalence

The following says, roughly, that metrics that are everywhere approximately proportional to each other are equivalent.

Theorem 4.4.1 *If d and ρ are metrics on M and there exist $k_1, k_2 > 0$ such that $\forall x, y \in M$, $d(x, y) \leq k_1\rho(x, y) \leq k_2 d(x, y)$, then d and ρ are equivalent.*

Proof. We must show that $\tau_d = \tau_\rho$. Pick arbitrary $G \in \tau_d$ and arbitrary $x \in G$. Since G is d-open, $\exists \delta(x) > 0$ such that $B^d_{\delta(x)}(x) \subset G$. As $\rho(x, y) \leq k_2 d(x, y)$, $B^\rho_{k_2 \cdot \delta(x)}(x) \subset B^d_\delta(x)$. This means that $G = \cup_{x \in G} B^\rho_{k_2 \cdot \delta(x)}(x)$, which expresses G as a union of elements of τ_ρ. This in turn implies that $G \in \tau_\rho$. The argument for the reverse direction, $\tau_\rho \subset \tau_d$, is parallel. ■

Lemma 4.4.2 *For $\mathbf{x} \in \mathbb{R}^\ell$, $\|\mathbf{x}\|_1 \leq \sqrt{\ell}\|\mathbf{x}\|_2 \leq \ell\|\mathbf{x}\|_\infty \leq \ell^2\|\mathbf{x}\|_1$.*

Since $d_p(\mathbf{x}, \mathbf{y}) = \|\mathbf{x} - \mathbf{y}\|_p$, this implies that all three norm metrics are equivalent.

Exercise 4.4.3 [↑Exercise 4.3.6] Prove Lemma 4.4.2.

4.4.b The Completeness of $(\mathbb{R}^\ell, d_1)$, $(\mathbb{R}^\ell, d_2)$, and $(\mathbb{R}^\ell, d_\infty)$

The following is more useful than its obviousness might suggest.

Corollary 4.4.4 *For $p = 1, 2, \infty$, $d_p(\mathbf{x}_n, \mathbf{x}) \to 0$ iff for each i, $x_{i,n} \to x_i$.*

Lemma 4.4.5 *The spaces $(\mathbb{R}^\ell, d_1)$, $(\mathbb{R}^\ell, d_2)$, and $(\mathbb{R}^\ell, d_\infty)$ are complete.*

Proof. By Lemma 4.4.2, $\mathbf{x}_n$ is a d_1-Cauchy iff it is d_2-Cauchy iff it is d_∞-Cauchy. As $|x_{i,n} - x_{i,m}| \le d_\infty(\mathbf{x}_n, \mathbf{x}_m)$, for each $i \in \{1, \ldots, \ell\}$, $x_{i,n}$ is a Cauchy sequence in $\mathbb{R}$ if $\mathbf{x}_n$ is d_∞-Cauchy. As $\mathbb{R}$ is complete, for each $i \in \{1, \ldots, \ell\}$, $x_{i,n} \to x_i^*$ for some $x_i^* \subset \mathbb{R}$. Let $\mathbf{x}^* = (x_1^*, \ldots, x_\ell^*)^T$, so that $d_\infty(\mathbf{x}_n, \mathbf{x}^*) \to 0$, implying that $d_2(\mathbf{x}_n, \mathbf{x}^*) \to 0$ and $d_1(\mathbf{x}_n, \mathbf{x}^*) \to 0$. ∎

Exercise 4.4.6 Show that the mapping $f(\mathbf{x}) = \mathbf{x}$ from $\mathbb{R}^\ell$ to $\mathbb{R}^\ell$ is a homeomorphism between $(\mathbb{R}^\ell, d_1)$ and $(\mathbb{R}^\ell, d_2)$ and a homeomorphism between $(\mathbb{R}^\ell, d_1)$ and $(\mathbb{R}^\ell, d_\infty)$. Show also that f is not an isometry unless $\ell = 1$.

4.4.c Completeness and Contraction Mappings

One place where Cauchy sequences appear is as the result of the application of what is called a **contraction mapping**. The properties of contraction mappings on various complete metric spaces allow us to prove versions of the implicit function theorem, as well as some of the basic results in dynamical systems, Markov processes, and dynamic programming. It is also used, though not by us, to prove existence theorems for the solutions to many classes of differential equations.

4.4.c.1 Banach's Contraction Mapping Theorem

Definition 4.4.7 *A function $f : M \to M$ is a contraction mapping if there exists a contraction factor, $\beta \in (0, 1)$, such that $\forall x, y \in M$, $d(f(x), f(y)) \le \beta d(x, y)$.*

So, starting at any pair of points x and y, we measure the distance between them to find $d(x, y)$. Applying f to both of them, we arrive at the new pair of points $f(x)$ and $f(y)$. If f is a contraction mapping, the distance between the new pair of points is smaller than the original distance by some factor less than or equal to β.

Lemma 4.4.8 *If $f : M \to M$ is a contraction mapping, then for all $x \in M$, the sequence $(x, f(x), f(f(x)), \ldots)$ is Cauchy.*

Proof. Choose $x_0 \in M$ and define x_t by $x_{t+1} = f(x_t)$. Since f is a contraction,

$$d(x_2, x_1) = d(f(x_1), f(x_0)) \le \beta d(x_1, x_0). \tag{4.3}$$

Continuing inductively,

$$d(x_{t+1}, x_t) = d(f(x_t), f(x_{t-1})) \le \beta d(x_t, x_{t-1})$$
$$\le \beta^t d(x_1, x_0), \quad t = 1, 2, \ldots. \tag{4.4}$$

Therefore, for any $m > n \geq N$,

$$d(x_m, x_n) \leq d(x_m, x_{m-1}) + \cdots + d(x_{n+2}, x_{n+1}) + d(x_{n+1}, x_n) \quad (4.5)$$

$$\leq \left[\beta^{m-1} + \cdots + \beta^{n+1} + \beta^n \right] d(x_1, x_0)$$

$$= \beta^n [\beta^{m-n-1} + \cdots + \beta + 1] d(x_1, x_0)$$

$$\leq \frac{\beta^N}{1 - \beta} d(x_1, x_0).$$

Since $|\beta| < 1$, $\beta^N \to 0$, so that x_t is Cauchy. ∎

Definition 4.4.9 *If $f : M \to M$ and $f(x^*) = x^*$, we say that x^* is a **fixed point** of f.*

Theorem 4.4.10 (Banach) *If $f : M \to M$ is a contraction mapping and (M, d) is a complete metric space, then f has a unique fixed point, x^*, and from any x_0, the sequence $(x_0, f(x_0), f(f(x_0)), \ldots)$ converges to x^*.*

Proof. We first establish that if there is a fixed point, it is unique. Suppose that $f(x') = x'$ and $f(x°) = x°$. This implies that $d(x', x°) = d(f(x'), f(x°))$. However, since f is a contraction mapping, $d(x', x°) \leq \beta d(x', x°)$ where $\beta < 1$. This can only be satisfied by $d(x', x°) = 0$.

By the previous lemma, starting at any $x_0 \in M$ and iteratively defining $x_{t+1} = f(x_t)$ yields a Cauchy seqeunce. Since M is complete, $x_t \to x^*$ for some $x^* \in M$. The last step is to show that $f(x^*) = x^*$, i.e. $d(x^*, f(x^*)) = 0$.

Choose $\epsilon > 0$. Since $x_t \to x^*$, there exists a T such that for all $t \geq T$, $d(x^*, x_t) < \epsilon/3$ and $d(x_t, x_{t+1}) < \epsilon/3$. By the triangle inequality, for any t,

$$d(x^*, f(x^*)) \leq d(x^*, x_t) + d(x_t, f(x_t)) + d(f(x_t), f(x^*)). \quad (4.6)$$

Since $x_{t+1} = f(x_t)$, for $t \geq T$, the first and second terms on the right-hand side of the inequality (4.6) are less than $\epsilon/3$. Since $d(x^*, x_t) < \epsilon/3$ and f has a contraction factor $\beta < 1$, the third term, $d(f(x_t), f(x^*)) < \beta\epsilon/3$. When we combine, for any $\epsilon > 0$, $d(x^*, f(x^*)) < \epsilon$, that is, $d(x^*, f(x^*)) = 0$. ∎

4.4.c.2 Contractive Dynamical Systems

We give many applications of this result in §4.11. For now, we sketch some applications in **dynamical systems**. Dynamical systems on $\mathbb{R}^\ell$ are often defined as functions $f : \mathbb{R}^\ell \to \mathbb{R}^\ell$. Here x_0 is the starting point, $x_1 = f(x_0)$ is the first step, $x_t = f(x_{t-1})$ is the tth step, etc. Let f^t denote the t-fold application of the function f.

Example 4.4.11 (Linear Dynamical Systems in $\mathbb{R}^2$) *For a 2×2 matrix $M = \begin{bmatrix} m_{1,1} & m_{1,2} \\ m_{2,1} & m_{2,2} \end{bmatrix}$ and $x = \begin{bmatrix} x_1 \\ x_2 \end{bmatrix} \in \mathbb{R}^2$, $Mx = M(x) := \begin{bmatrix} m_{1,1}x_1 + m_{1,2}x_2 \\ m_{2,1}x_1 + m_{2,2}x_2 \end{bmatrix} \in \mathbb{R}^2$. For $\alpha, \beta \in \mathbb{R}$, $M(\alpha x + \beta y) = \alpha Mx + \beta My$, so that M is a **linear mapping** from $\mathbb{R}^2$ to $\mathbb{R}^2$.*

A linear dynamical system in $\mathbb{R}^2$ is specified by a vector $\mathbf{a} \in \mathbb{R}^2$ and a 2×2 matrix M. It starts at an initial point $\mathbf{x}_0 \in \mathbb{R}^2$, and the tth step is $\mathbf{x}_t = A^t(\mathbf{x}_0)$, where $A(\mathbf{x}) = \mathbf{a} + M\mathbf{x}$.

For example, if $M = \begin{bmatrix} 0 & \alpha \\ \beta & 0 \end{bmatrix}$ or $M = \begin{bmatrix} \alpha & 0 \\ 0 & \beta \end{bmatrix}$, then $A = \mathbf{a} + M$ is a contraction mapping iff $|\alpha| < 1$ and $|\beta| < 1$. More generally, the norm of M is defined by

$$\|M\| = \sup\{\|M\mathbf{x}\| : \|\mathbf{x}\| = 1\}.$$

One can use any of the norms for the vectors $\mathbf{x}$ and $M\mathbf{x}$; often the norms $\|\mathbf{x}\|_1$ and $\|M\mathbf{x}\|_1$ are the most useful. If $\beta = \|M\| < 1$, then $\mathbf{a} + M$ is a contraction mapping. To see why,

$$d(A(\mathbf{x}), A(\mathbf{y})) = \|M\mathbf{x} - M\mathbf{y}\| = \|M(\mathbf{x} - \mathbf{y})\| = \|bM\tfrac{(\mathbf{x}-\mathbf{y})}{b}\|,$$

where $b = \|\mathbf{x} - \mathbf{y}\| = d(\mathbf{x}, \mathbf{y})$. By the properties of norm, $\|bM\tfrac{(\mathbf{x}-\mathbf{y})}{b}\| = b\|M\tfrac{(\mathbf{x}-\mathbf{y})}{b}\|$ and $\|\tfrac{(\mathbf{x}-\mathbf{y})}{b}\| = 1$. Therefore, $d(A(\mathbf{x}), A(\mathbf{y})) \leq d(\mathbf{x}, \mathbf{y})\|M\| = \beta d(\mathbf{x}, \mathbf{y})$.

If $\|M\| < 1$, then the $\mathbf{x}_t$ are Cauchy. Since $\mathbb{R}^2$ is complete, $\mathbf{x}_t \to \mathbf{x}^*$ for some $\mathbf{x}^*$, and $\mathbf{x}^*$ is the unique solution to the equation $A(\mathbf{x}^*) = \mathbf{x}^*$.

Exercise 4.4.12 [↑Example 4.4.11] Setting $\mathbf{a} = 0$, sketch out several $\mathbf{x}_t$ paths for $\mathbf{x}_0$'s in the neighborhood of the origin for the cases $\alpha, \beta > 0$, $\alpha, \beta < 0$, $\alpha > 0 > \beta$, and $\alpha < 0 < \beta$.

Exercise 4.4.13 [↑Example 4.4.11] An example of a one-dimensional dynamical system with lag effects is $x_{t+1} = a + b_1 x_t + b_2 x_{t-1}$, where $x_t \in \mathbb{R}$. Show that this is a special case of linear dynamical systems in $\mathbb{R}^2$ by setting $\mathbf{x}_t = \begin{bmatrix} x_t \\ x_{t-1} \end{bmatrix}$, $\mathbf{a} = \begin{bmatrix} a \\ 0 \end{bmatrix}$, and $M = \begin{bmatrix} b_1 & b_2 \\ 1 & 0 \end{bmatrix}$ and examining the equation $\mathbf{x}_{t+1} = \mathbf{a} + M\mathbf{x}_t$. Give conditions on M guaranteeing that this is a contraction mapping for at least one of the metrics d_1, d_2, or d_∞.

Exercise 4.4.14 (Cournot Dynamics) Two firms, i and j, compete by producing quantities $q_i \geq 0$ and $q_j \geq 0$ of a homogeneous good and receiving profits of the form

$$\pi_i(q_i, q_j) = [p(q_i + q_j) - c]q_i,$$

where $p(\cdot)$ is the inverse market demand function for the good in question and $0 \leq c \ll 1$ is the marginal cost. Assume that $p(q) = 1 - q$. Firm i's best response to q_j, written $Br_i(q_j)$, is the solution to

$$\max_{q_i}[1 - (q_i + q_j) - c]q_i,$$

a quadratic in q_i that opens upward. Setting the derivative equal to 0 and solving shows that $Br_i(q_j) = \max\{\frac{1}{2}(1 - c) - \frac{1}{2}q_j, 0\}$. One version of the Cournot dynamics has

$$\begin{bmatrix} q_{1,t+1} \\ q_{2,t+1} \end{bmatrix} = \begin{bmatrix} Br_1(q_{2,t}) \\ Br_2(q_{1,t}) \end{bmatrix}.$$

Show that this is a contraction mapping on $\mathbb{R}_+^2$.

4.5 ◆ Closure, Convergence, and Completeness

In this section we give several characterizations of closed sets and define the closure of a set. Recall the following definition.

Definition 4.1.10.   *If (M, d) is a metric space, then $G \subset M$ is d-**open**, or simply **open** if $\forall x \in G$, $\exists \delta > 0$, such that $B_\delta(x) \subset G$. The class of d-open sets is denoted τ_d. $F \subset M$ is d-**closed** or **closed** if F^c is open.*

4.5.a Characterizing Closed Sets

Many make the mistake of assuming that sets must be either open or closed, perhaps because one thinks of doors as being either open or closed. Our coverage of discrete spaces (in which all sets are both open and closed) may not have dispelled this notion.

Example 4.5.1   *The set $(0, 1] \subset \mathbb{R}^1$ is neither open nor closed. It is not open because for all $\delta > 0$, $B_\delta(1) \not\subset (0, 1]$, and it is not closed because for all $\delta > 0$, $B_\delta(0) \not\subset (0, 1]^c$.*

Exercise 4.5.2   Show that the set $\mathbb{Q} \subset \mathbb{R}$ is neither open nor closed.

Theorem 4.1.13 showed that the class of open sets is closed arbitrary unions and finite intersections. This immediately implies that the class of closed sets is closed under arbitrary intersections and finite unions.

Corollary 4.5.3   *If F_α is closed for every $\alpha \in A$, then $\cap_{\alpha \in A} F_\alpha$ is closed. If A is finite, then $\cup_\alpha F_{\alpha \in A}$ is closed.*

Proof.   Observe that $(\cup_\alpha G_\alpha)^c = \cap_\alpha G_\alpha^c$, and that $(\cap_\alpha G_\alpha)^c = \cup_\alpha G_\alpha^c$.[8]   ■

Definition 4.5.4   *The **closure of a set** $E \subset M$ is denoted $\overline{E}$ or $\mathrm{cl}(E)$ and is defined as the smallest closed set containing E, $\mathrm{cl}(E) = \bigcap \{F : E \subset F, F \text{ closed}\}$.*

Since arbitrary intersections of closed sets are closed, $\mathrm{cl}(E)$ must be closed. There can be no smaller closed set containing E. If there were, it would be in the intersection on the right-hand side of the definition.

Definition 4.5.5   *The **boundary of** E is $\mathrm{cl}(E) \cap \mathrm{cl}(E^c)$, denoted ∂E.*

Recall that $B_\epsilon(x)$ is the open ϵ-ball around x and that $E^\epsilon = \cup_{e \in E} B_\epsilon(e)$ is the open ϵ-ball around the set E.

---

8. The proofs are becoming yet briefer. As you read, your job is to fill in the blanks. Here, you would "Suppose that F_α is a collection of closed sets," and "want to show that $\cap_\alpha F_\alpha$ is closed." You might continue, "Since $F_\alpha = G_\alpha^c$ for some open set G_α, ..."

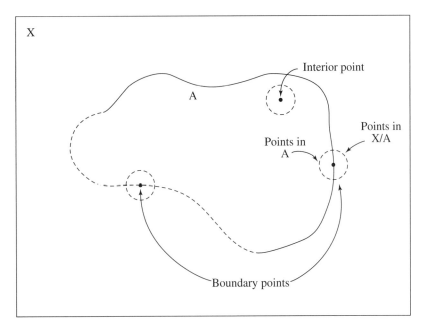

FIGURE 4.5.6

The following is the basic characterization result for closed sets.

Theorem 4.5.6   *For any $E \subset M$, the following are equivalent [see Figure 4.5.6]:*

1. *E is closed,*
2. *$E = \text{cl}(E)$,*
3. *$E = \bigcap \{E^\epsilon : \epsilon > 0\}$,*
4. *$E = \{x \in M : \forall \epsilon > 0, B_\epsilon(x) \cap E \neq \emptyset\}$,*
5. *$E = \{x \in M : \exists e_n \text{ in } E, e_n \to x\}$, and*
6. *$\partial E \subset E$.*

Condition (4) and the definition of ∂E immediately yields that $x \in \partial E$ iff $(\forall \epsilon > 0)[[B_\epsilon(x) \cap E \neq \emptyset] \wedge [B_\epsilon(x) \cap E^c \neq \emptyset]]$. Verbally, the boundary of a set is the set of points that are as close as you please to both the inside and the outside of the set.

Since these six conditions are equivalent, any one of them could have been used as the *definition* of a closed set. A suggestive name for condition (3) is the **shrink-wrapping definition of closure**. This is a term that is not in common use, but we like it. The names for points satisfying conditions (4) and (5) are **contact points** and **limit points**. These names are both suggestive and widely used.

Definition 4.5.7   *A point $x \in M$ is a **contact point** or **accumulation point** of a set $E \subset M$ if any open ball around x intersects E, that is, if for all $\epsilon > 0$,*

$B_\epsilon(x) \cap E \neq \emptyset$. x is a **limit point** of E if there exists a sequence x_n in E such that $x_n \to x$, and it is a **cluster point of** E if for all $\epsilon > 0$, $B_\epsilon(x) \cap E$ contains infinitely many points.

Verbally, conditions (4), (5), and (6) are that "a set is closed iff it contains all its contact points," "a set is closed iff it contains all its limit points," and "a set is closed iff it contains its boundary."

Exercise 4.5.8 Show the following, specifying which definitions or which pieces of Theorem 4.5.6 you are using. All of the indicated sets are subsets of $\mathbb{R}$:

1. If $A = (0, 1]$, then $[0, 1]$ is the set of cluster points of A, and $\partial A = \{0, 1\}$.

2. If $A = \{\frac{1}{n} : n \in \mathbb{N}\}$, then 0 is the only cluster point of A but $A \cup \{0\}$ is the closure of A, the set of contact points of A, and the boundary of A.

3. If $A = \{0\} \cup (1, 2)$, then $[1, 2]$ is the set of cluster points of A, while $\{0\} \cup [1, 2]$ is the set of contact points.

4. If $A = \mathbb{N}$, then A is closed and has no cluster points, but every point in A is a contact point of A, and $\partial A = A$.

5. If $A = \mathbb{Q}$, then the set of cluster points of A is $\mathbb{R}$, and $\partial A = \mathbb{R}$.

Proof of Theorem 4.5.6. (1) $\Leftrightarrow$ (2) is immediate.

(1) $\Rightarrow$ (3): Suppose that E is closed. We must show that $E = \bigcap\{E^\epsilon : \epsilon > 0\}$. For this it is sufficient to show that $x \notin E$ iff $x \notin \bigcap\{E^\epsilon : \epsilon > 0\}$.

Suppose that $x \in E^c$. Since E^c is open, $(\exists \epsilon > 0)[B_\epsilon(x) \subset E^c]$. This implies that $x \notin E^\epsilon$, so that $x \notin \bigcap\{E^\epsilon : \epsilon > 0\}$. Now suppose that $x \notin \bigcap\{E^\epsilon : \epsilon > 0\}$. This is equivalent to $(\exists \epsilon > 0)[x \notin E^\epsilon]$, in turn equivalent to $(\exists \epsilon > 0)[B_\epsilon(x) \cap E = \emptyset]$, which certainly implies that $x \notin E$.

(3) $\Leftrightarrow$ (4): We show that $\bigcap\{E^\epsilon : \epsilon > 0\} = \{x \in M : \forall \epsilon > 0, B_\epsilon(x) \cap E \neq \emptyset\}$. By the definition of union, $x \in E^\epsilon$ iff $(\exists e \in E)[x \in B_\epsilon(e)]$. Since $d(x, e) = d(e, x)$, $x \in B_\epsilon(e)$ iff $e \in B_\epsilon(x)$. Therefore, $x \in \bigcap\{E^\epsilon : \epsilon > 0\}$ iff $(\forall \epsilon > 0)[B_\epsilon(x) \cap E \neq \emptyset]$.

(4) $\Leftrightarrow$ (5): We show that $\{x \in M : \forall \epsilon > 0, B_\epsilon(x) \cap E \neq \emptyset\} = \{x \in M : \exists e_n \text{ ini } E, e_n \to x\}$. If $\forall \epsilon > 0$, $B_\epsilon(x) \cap E \neq \emptyset$, then for all $n \in \mathbb{N}$, $\exists e_n \in E$ such that $e_n \in B_{1/n}(x)$, so that $e_n \to x$. If $\exists e_n$ in E, $e_n \to x$, then for all $\epsilon > 0$, $\exists e_n \in B_\epsilon(x)$.

(3) $\Rightarrow$ (1): Suppose that $E = \bigcap\{E^\epsilon : \epsilon > 0\}$. We must show that E^c is open. Since $E = \bigcap\{E^\epsilon : \epsilon > 0\}$, $x \in E^c$ iff $(\exists \epsilon > 0)[x \notin E^\epsilon]$. This implies that $B_\epsilon(x) \cap E = \emptyset$, that is, that $B_\epsilon(x) \subset E^c$. ∎

Exercise 4.5.9 Complete the above proof.

Exercise 4.5.10 Provide a direct proof that (5) $\Rightarrow$ (1) and a direct proof that (1) $\Rightarrow$ (5).

Sometimes one wants to distinguish between those points x in $cl(E)$ that can only be reached by trivial sequences, $x_n = x$ for all large n, and those that can be reached by sequences x_n that are never equal to x. This is what motivated the definition of cluster points.

Exercise 4.5.11 Show that a set is closed iff it contains all of its cluster points.

4.5.b   Closure and Completeness

The Cournot dynamics in Exercise 4.4.14 led to a Cauchy sequence in $\mathbb{R}_+^2$. We know that $\mathbb{R}^2$ is complete, but we have not yet proved the same for $\mathbb{R}_+^2$. This follows from the next two results.

Lemma 4.5.12 $\mathbb{R}_+^\ell$ *is a closed subset of* $\mathbb{R}^\ell$.

Proof.   Let $\mathbf{x}_n \in \mathbb{R}_+^\ell$ converge to $\mathbf{x}$. This happens iff for each i, $x_{i,n} \to x_i$. Since each $x_{i,n} \geq 0$, $x_i \geq 0$. ■

Theorem 4.5.13   *Let* $E \subset M$. *If* (E, d) *is a complete metric space, then* E *is closed. If* E *is closed and* (M, d) *is complete, then* (E, d) *is a complete metric space.*

Proof.   Suppose that (E, d) is complete and $e_n \in E$, $e_n \to e$. We must show that $e \in E$ (the "limit points" part of Theorem 4.5.6). Since $e_n \to e$, e_n is Cauchy; thus it converges to a point in E, so that $e \in E$.

Suppose that E is closed. Let e_n be a Cauchy sequence in E. Since M is complete, $e_n \to x$ for some $x \in M$. Since E is closed, $x \in E$, so that (E, d) is complete. ■

Thus, if E is a closed subset of a complete metric space M and $f : E \to E$ is a contraction mapping, then the dynamic system associated with f gives rise to a Cauchy sequence that converges to a point in E if it starts in E.

A mapping is **nonexpansive** (rather than contractive) if for all $x \neq x'$, $d(f(x), f(x')) < d(x, x')$. The following example of a dynamic system on the closed set $E = \mathbb{R}_+$ shows that one needs the full power of a contraction in order to guarantee that the dynamical system converges. This provides a cautionary note about numerical procedures and a reason that one wants to **prove** that the system being investigated converges before trusting computer-based or other numerical procedures.

Example 4.5.14   *Define* $f : \mathbb{R}_+ \to \mathbb{R}_+$ *by* $f(x) = x + \frac{1}{x+1}$. *[See Figure 4.5.14.] To see that* f *is nonexpansive, pick* $0 \leq y < x$ *so that* $d(x, y) = x - y$. *By direct computation* $d(f(x), f(y)) = (x - y) + \left(\frac{1}{x+1} - \frac{1}{y+1}\right) < (x - y)$, *where the inequality follows from the observation that* $\left(\frac{1}{x+1} - \frac{1}{y+1}\right) < 0$ *because* $x > y$. *The associated dynamic system,* $x_{t+1} = f(x_t)$, *has two interesting properties:* $x_{t+1} > x_t$ *because* $f(r) > r$ *for all* $r \geq 0$ *and* $d(x_{t+1}, x_t) \to 0$ *because* $\lim_{r \to \infty}(f(r) - r) = 0$. *Note that this does* not *mean that* x_t *is a Cauchy sequence.*

We show that: (1) *the* x_t *do not converge, that is,* $\neg[x_t \uparrow x^*]$ *for any* $x^* \in \mathbb{R}_+$, *but* (2) *any numerical procedure will eventually report convergence.*

(1) *Suppose that* $x_t \uparrow x^*$ *for some* $x^* \in \mathbb{R}_+$. *By the definition of* f, $x_{t+1} = x_t + \frac{1}{x_t+1}$. *Since* $x_{t+1} < x^*$, $x_{t+1} > x_t + \frac{1}{x^*+1}$. *Let* $\eta = \frac{1}{x^*+1} > 0$. *Thus, if* x_t *converges to* x^*, *then for all* t, $x_{t+1} > x_t + \eta$, *contradiction the convergence of* x_t.

(2) *Numerical procedures specify "convergence" when the step size between subsequent values of the* x_t *(or the step size between values* x_t *and* x_{t+T} *for some* T*) are smaller than some predetermined* $\Delta > 0$. *Pick* r^* *such that* $f(r^*) - r^* < \Delta$ *(or* Δ / T*). Once* $x_t > r^*$, *the procedure reports convergence.*

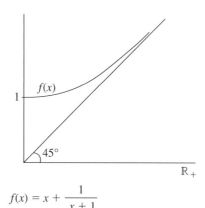

$$f(x) = x + \frac{1}{x + 1}$$

FIGURE 4.5.14

The point in this example can be quite serious. Numerically analyzing the limit points of a model—for example, a model such as the dynamical system given by f—allows one to announce, "The numerical procedure converged. Analyzing the numerical limit point, we see that. . . . " Everything that you fill in for the ". . ." is misleading nonsense about the model you are analyzing. By contrast, if we know that the f has a contraction factor β, then we know that once $d(x_{t+1}, x_t) < \Delta$, the limit point must be less than $\Delta/(1 - \beta)$ away, which not only implies that we are analyzing a point close to the true limit point but also gives guidance for the choice of Δ.

4.6 ◆ Separability

We use the separability of M or $\mathbb{R}^\ell$ in the next section which is on compactness. It will have extensive uses in probability theory. We repeat the following.

Definition 4.1.7. *If (M, d) is a metric space and $E \subset M$, then E **is dense in** M if for all $x \in M$ and all $\epsilon > 0$, there exists an $e \in E$ such that $d(x, e) < \epsilon$. If there exists a countable dense E, then (M, d) is **separable**, and otherwise it is **nonseparable**.*

We know that $\mathbb{Q}$ is a countable dense subset of $\mathbb{R}$ when we use the usual metric. However, with the discrete metric, $e(x, y) = 1$ if $x \neq y$ and $e(x, y) = 0$ if $x = y$, $(\mathbb{R}, e)$ is not separable.

Lemma 4.6.1 *$\mathbb{Q}^\ell$ is a countable dense subset of $\mathbb{R}^\ell$.*

Proof. The finite product of countable sets is countable. Pick $\mathbf{x} \in \mathbb{R}^\ell$ and $\epsilon > 0$. For each i, pick $q_i \in \mathbb{Q}$ such that $|x_i - q_i| < \epsilon$, and let $\mathbf{q} \in \mathbb{Q}^\ell$ be the vector with q_i in the ith spot, so that $d_\infty(\mathbf{q}, \mathbf{x}) < \epsilon$. A set is dense in one metric iff it is dense in every equivalent metric. ∎

Separability implies that all open sets can be represented as unions of sets in a relatively restricted class of sets.

Theorem 4.6.2 *If (M, d) is a separable metric space, then there exists a countable collection, $\mathcal{G}$, of open sets such that each open $G \subset M$ is a union of elements of $\mathcal{G}$.*

Proof. Set $\mathcal{G} = \{B_q(e) : e \in E, q \in \mathbb{Q}_{++}\}$, where E is a countable dense set. Pick an open $G \subset M$. For every $x \in G$, $(\exists \epsilon > 0)[B_\epsilon(x) \subset G]$. Pick $e_x \in E \cap B_\epsilon(x)$ and $q_x \in \mathbb{Q}_{++}$ such that $\epsilon - d(e_x, x) < q_x < \epsilon$, so that $x \in B_{q_x}(e_x) \subset B_\epsilon(x)$. Then $G = \cup_{x \in G} B_{q_x}(e_x)$ expresses G as a union of sets in $\mathcal{G}$. ∎

Exercise 4.6.3 Show that if $X \subset M$ and (M, d) is separable, then (X, d) is separable. [This may be a little bit trickier than it looks—E may be a countable dense subset of M with $X \cap E = \emptyset$.]

4.7 ♦ Compactness in $\mathbb{R}^\ell$

Compactness and continuity are the two concepts most used in guaranteeing the existence of optima. We give five characterizations of compactness applicable in any metric space and show how they apply to $\mathbb{R}^\ell$. The most important of the results in $\mathbb{R}^\ell$ is that a set is compact iff it is both closed and norm bounded.

We reproduce the following.

Definition 4.1.19. *If (M, d) is a metric space, then $K \subset M$ is **compact** if every open cover of K has a finite subcover.*

In order to apply this definition to show that a set E is compact we must examine *every* open covering of E, and this may be difficult. We have seen one easy case: Lemma 4.2.9 showed that a subset of a discrete metric space is compact iff it is finite. (In general, any finite subset of any metric space is necessarily compact.) On the other hand, to show that a set E is *not* compact, it is sufficient to show that only *one* open covering cannot be replaced by a finite subcollection that also covers E.

4.7.a The Compactness of $[0, 1]$

We have $[0, 1]$ as the canonical compact metric space. We introduce our five characterizations of compactness first in this space and give direct proofs that $[0, 1]$ has these properties. In all five characterizations, the essential idea is that compact sets are, to any degree of approximation, finite. Recall that $d(x, E) = \inf\{d(x, e) : e \in E\}$ is the distance between the point x and the set E.

Example 4.7.1 (Approximately Finite I) *Pick your favorite (tiny) $\epsilon > 0$. Let $K_\epsilon \subset [0, 1]$ be the finite set $\{0, \epsilon, 2\epsilon, 3\epsilon, \ldots, N\epsilon\}$, where N is the largest integer less than $1/\epsilon$. For any $x \in [0, 1]$, $d(x, K_\epsilon) < \epsilon$.*

A sequence in a finite set must take on at least one value infinitely often. Put another way, any sequence in a finite set must have a convergent, indeed eventually constant, subsequence.

Lemma 4.7.2 (Approximately Finite II) *Every sequence in $[0, 1]$ has a convergent subsequence.*

Proof. Infinitely many of the $n \in \mathbb{N}$ must have x_n in $[0, \frac{1}{2^1}]$ or have x_n in $[\frac{1}{2^1}, 1]$ (or both). Let $[a_1, b_1]$ be one of these two intervals containing infinitely many x_n, $n \in \mathbb{N}$, let $\mathbb{N}_1$ be the set of $n \in \mathbb{N}$ such that $x_n \in [a_1, b_1]$, and let n_1 be the first $n \in \mathbb{N}_1$ such that $x_{n_1} \in [a_1, b_1]$.

Infinitely many of the $n \in \mathbb{N}_1$ must have x_n either in $[a_1, a_1 + \frac{1}{2^2}]$ or in $[a_1 + \frac{1}{2^2}, b_1]$ (or both). Let $[a_2, b_2]$ be one of these intervals containing infinitely many x_n, $n \in \mathbb{N}_1$, let $\mathbb{N}_2$ be the set of $n \in \mathbb{N}_1$ such that $x_n \in [a_2, b_2]$, and let n_2 be the first $n \in \mathbb{N}_2$ greater than n_1 such that $x_{n_2} \in [a_2, b_2]$.

Continuing inductively, we find a Cauchy subsequence $x_{n_1}, x_{n_2}, x_{n_3}, \ldots$. Since $[0, 1]$ is complete (being a closed subset of the complete metric space $\mathbb{R}$), $x_{n_k} \to x$ for some $x \in [0, 1]$. ■

There is a more abstract version of being approximately finite, involving classes of open sets called **open covers**. An open cover of E is a collection $\{G_\alpha : \alpha \in A\}$ of open sets with $E \subset \cup_{\alpha \in A} G_\alpha$. A **subcover** is given by an $A' \subset A$ with $E \subset \cup_{\alpha \in A'} G_\alpha$. A **finite subcover** is given by a finite $A_F \subset A$ with $E \subset \cup_{\alpha \in A_F} G_\alpha$.

Exercise 4.7.3 Show that the collection $\{(-\infty, n) : n \in \mathbb{N}\}$ is an open cover of $[0, \infty)$ with no finite subcover. Give an open cover of $[0, 1)$ with no finite subcover.

If E is finite, then any open cover has a finite subcover. $[0, 1]$ also has this property.

Lemma 4.7.4 (Approximately Finite III) *Any open cover of $[0, 1]$ has a finite subcover.*

Proof. Suppose not. That is, suppose that there exists an open cover, $\mathcal{G} = \{G_\alpha : \alpha \in A\}$ with the property that every finite subcover misses some part of $[0, 1]$.

Any open set containing 0 must contain some interval $[0, \epsilon]$ for some $\epsilon > 0$. Let r be the largest element of $[0, 1]$ such that $[0, r]$ is covered by some finite subcover of $[0, 1]$; that is, define $r = \sup\{\epsilon > 0 : [0, \epsilon]$ is covered by some finite subcover of $\mathcal{G}\}$. By assumption, $r < 1$, and we have seen that $r > 0$.

Let A_F be a finite subset of A such that $[0, r] \subset \cup_{\alpha \in A_F} G_\alpha$. Since $\mathcal{G}$ is an open cover of $[0, 1]$, $r \in G_{\alpha^\circ}$ for some $\alpha^\circ \in A$. Therefore, there exists $\delta > 0$ such that $B_\delta(r) \subset G_{\alpha^\circ}$. Consider the finite subcover given by $A_F \cup \{\alpha^\circ\}$. It covers the interval $[0, r + \frac{1}{2}\delta]$, implying that r was not the supremum of the ends of the intervals covered by finite subcovers. ■

DeMorgan's duality rules are that $(\cup_\alpha E_\alpha)^c = \cap_\alpha E_\alpha^c$ and $(\cap_\alpha E_\alpha)^c = \cup_\alpha E_\alpha^c$. There is a DeMorgan dual to every open cover having a finite subcover.

Lemma 4.7.5 (Approximately Finite IV) *Let $\{F_\alpha : \alpha \in A\}$ be a collection of closed subsets of $[0, 1]$ with the property that for every finite $A_F \subset A$, $\cap_{\alpha \in A_F} F_\alpha \neq \emptyset$. Then $\cap_{\alpha \in A} F_\alpha \neq \emptyset$.*

Exercise 4.7.6 Show that $\{G_\alpha : \alpha \in A\}$ is an open cover of $[0, 1]$ iff $\cap_{\alpha \in A}(G_\alpha^c \cap [0, 1]) = \emptyset$. Using DeMorgan's rules, prove Lemma 4.7.5.

Consider the problem $\max_{x \in E} f(x)$. If E is finite, then this problem has a solution.

Lemma 4.7.7 (Approximately Finite V) *For any continuous function $f :$ $[0, 1] \to \mathbb{R}$, the problem $\max_{x \in [0,1]} f(x)$ has a solution.*

It is a little unfair that we have not yet really studied continuous functions. Still, recall the definition of a continuous function: $f^{-1}(G)$ is open for every open G.

Proof. The collection $\{f^{-1}((-n, +n)) : n \in \mathbb{N}\}$ is an open cover of $[0, 1]$. We have seen that it has a finite subcover. This in turn means that $f([0, 1])$ is bounded.

Let $r = \sup\{f(x) : x \in [0, 1]\}$. We must show that there exists an $x^* \in [0, 1]$ such that $f(x^*) = r$. Suppose not. Then the collection $\{f^{-1}((-\infty, r - \frac{1}{n})) : n \in \mathbb{N}\}$ is an open cover of $[0, 1]$. We have seen that it has a finite subcover. But this means that $f([0, 1]) \subset r - \frac{1}{N}$ for some integer N, which contradicts the definition of r as $\sup\{f(x) : x \in [0, 1]\}$. ∎

4.7.b Boundedness and Total Boundedness

In $\mathbb{R}^\ell$ we have the notion of a norm-bounded set.

Definition 4.7.8 *A subset E of $\mathbb{R}^\ell$ is **norm bounded**, or simply **bounded**, if $(\exists B)(\forall x \in E)[\|x\| \leq B]$.*

Example 4.7.9 *Let $(M, d) = (\mathbb{R}, d)$ and $E = \mathbb{R}_+$ and consider the open cover $\{(-1, n) : n \in \mathbb{N}\}$. No finite subcover can cover E (by the Archimedean principle), so E is not compact. More generally, if E is an unbounded subset of $\mathbb{R}^\ell$, then there is no finite subcover of the open cover $\{(-n, n)^\ell : n \in \mathbb{N}\}$, so compactness implies boundedness in $\mathbb{R}^\ell$.*

Boundedness captures a notion of limited size if we use norm distances. However, any metric d is equivalent to the metric $\rho(x, y) := \min\{d(x, y), 1\}$. For any $r > 1$ and any $x \in M$, $\{y \in M : \rho(x, y) < r\} = M$.

Definition 4.7.10 *$S \subset M$ is an **ϵ-net for a set** $E \subset M$ if for all $x \in E$, there exists an $s \in S$ such that $d(x, s) < \epsilon$. A set E is **totally bounded** if for every $\epsilon > 0$, it contains a **finite** ϵ-net.*

Intuitively, a finite ϵ-net for E is a finite subset that is within ϵ of "being" the set E. The set E is totally bounded if it is within ϵ of "being" a finite set for every ϵ, no matter how small. There are several reasons to use totally bounded instead of bounded. One of them is that it allows us to switch between equivalent metrics. The following is an example of this.

Exercise 4.7.11 *For $x, y \in \mathbb{R}$, define $\rho(x, y) = \min\{|x - y|, 1\}$. Show that $E \subset \mathbb{R}$ is a norm-bounded set iff E is totally bounded when we use the metric ρ.*

In $\mathbb{R}^\ell$ with the usual metric, there is no difference between boundedness and total boundedness.

Lemma 4.7.12 *$E \subset \mathbb{R}^\ell$ is norm bounded iff it is totally bounded.*

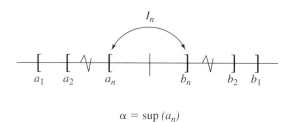

$$\alpha = \sup (a_n)$$

FIGURE 4.7.14 Common points of closed nested intervals.

Proof. If E is norm bounded by B, then for every $\epsilon > 0$, there is an ϵ-net containing at most $(2B/\epsilon)^{\ell}$ points. If E is not norm bounded, then there can be no finite 1-net. ■

4.7.c The Finite Intersection Property

DeMorgan's rules tightly relate the following definition to open covers.

Definition 4.7.13 *A collection* $\{E_\alpha : \alpha \in A\}$ *of subsets of M has the **finite intersection property (fip)** if for any finite $A_F \subset A$, $\cap_{\alpha \in A_F} E_\alpha \neq \emptyset$.*

The question we are going after is, "When does a collection having the fip have nonempty intersection?" This turns out to be intimately related to compactness. The collection of open sets $\{(0, 1/n) : n \in \mathbb{N}\}$ has the finite intersection property but has empty intersection, and the same is true of the collection of closed sets $\{[n, +\infty) : n \in \mathbb{N}\}$. The following gives an example of a sequence of closed sets having the fip and nonempty intersection. Compare it to Theorem 4.7.15(2) just below.

Exercise 4.7.14 Suppose that $\{[a_n, b_n] : n \in \mathbb{N}\}$ is nested, $[a_{n+1}, b_{n+1}] \subset [a_n, b_n]$. Show that the collection has the finite intersection property. Define $\overline{a} = \sup\{a_n : n \in \mathbb{N}\}$ and $\underline{b} = \inf\{b_n : n \in \mathbb{N}\}$. Show that $\overline{a} \leq \underline{b}$ and that $\cap_n [a_n, b_n] = [\overline{a}, \underline{b}]$. [See Figure 4.7.14.]

4.7.d Characterizations of Compactness

We now give four characterizations of compactness. Theorem 4.8.11 gives the fifth characterization, which is that every continuous function should achieve its optimum.

Theorem 4.7.15 *The following conditions on $K \subset M$ are equivalent:*

1. *every open cover of K has a finite subcover,*

2. *every collection of closed subsets of K with the finite intersection property has a nonempty intersection,*

3. *every sequence in K has a convergent subsequence converging to some point in K, and*

4. *K is totally bounded and complete.*

Using the sequence formulation of closure, we can easily to show that (3) implies that any compact set is closed, and any closed subset of a compact set is also compact. We record this as the following.

Corollary 4.7.16   *If $K \subset M$ is compact and F is a closed subset of K, then K is closed and F is compact.*

Proof of Theorem 4.7.15.   (1) $\Leftrightarrow$ (2) is an application of DeMorgan's laws because $\{G_\alpha : \alpha \in A\}$ is an open cover iff $\cap_\alpha G_\alpha^c = \emptyset$.

(2) $\Rightarrow$ (3): Let x_n be a sequence in K. Define $F_N = \mathrm{cl}\{x_n : n \geq N\} \cap K$, so that $\{F_N : N \in \mathbb{N}\}$ is a collection of closed subsets of K with the finite intersection property. By assumption, $\cap_N F_N \neq \emptyset$. Therefore, there exists $x \in K$ such that $x \in \cap_N F_N$. From Theorem 4.5.6, there exists $x_{m(1)} \in B(x, 1/1) \cap \{x_n : n \geq 1\}$. Given that $x_{m(N-1)}$ has been chosen, Theorem 4.5.6 implies that we can pick $m(N) > m(N-1)$ such that $x_{m(N)} \in B(x, 1/N) \cap \{x_n : n \geq N\}$. The subsequence $N \mapsto x_{m(N)}$ converges to x.

(3) $\Rightarrow$ (4): Suppose that K is not totally bounded. This means that there exists an $\epsilon > 0$ such that no finite set is an ϵ-net. Therefore, there is an infinite set of points at a distance ϵ or greater from each other. Every infinite set contains a countable set. Enumerating the countable set gives a sequence that can have no convergent subsequence because any convergent subsequence would have to be Cauchy.

Let x_n be a Cauchy sequence in K. By assumption, it has a convergent subsequence, $x_{n'}$, converging to some point $x \in K$. Since x_n is Cauchy, $x_n \to x$, so that K is complete.

(4) $\Rightarrow$ (1): This is the most involved part of the proof, so we begin with an outline. We assume that K is totally bounded and complete and show: (a) K is separable; (b) if K is not compact, there is a *countable* nested collection of closed sets, F_n, with the fip having empty intersection; (c) taking $x_n \in F_n$, we use total boundedness to extract a Cauchy sequence; (d) using completeness, we have a limit, x, and $x \in \cap_m F_m$, a contradiction.

(a) For each $n \in \mathbb{N}$, there is a finite $(1/n)$-net, S_n. $\cup_n S_n$ is a countable dense subset of K.

(b) Let $\{G_\alpha : \alpha \in A\}$ be an open cover of K with no finite subcover. Since K is separable, Theorem 4.6.2 tells us that there is a countable subcover $\{G_n : n \in \mathbb{N}\}$. Since $\{G_n : n \in \mathbb{N}\}$ is an open cover, $\{G_n^c : n \in \mathbb{N}\}$ is a collection of closed sets with empty intersection. Since no finite subcover of $\{G_n : n \in \mathbb{N}\}$ covers K, the collection $\{G_n^c : n \in \mathbb{N}\}$ has the fip. Defining $F_n = \cap_{m \leq n} G_m^c$ gives a nested collection of closed sets having the fip and empty intersection.

(c) For each $n \in \mathbb{N}$, pick $x_n \in F_n$. Since K is totally bounded, there is a finite cover of K by $1/k$-balls for every $k \in \mathbb{N}$.

At least one of the open $1/1$-balls, $B_1(s_1)$, contains infinitely many x_n. Let $\mathbb{N}_0 = \mathbb{N}$ and define $n_0 = 1 \in \mathbb{N}_0$. Define $\mathbb{N}_1 = \{n \in \mathbb{N}_0 : x_n \in B_1(s_1)\}$ so that $\mathbb{N}_1$ is an infinite subset of $\mathbb{N}_0$. Define n_1 to be the first element in $\mathbb{N}_1$ strictly greater than n_0.

Given an infinite set $\{x_n : n \in \mathbb{N}_{k-1}\} \subset K$ and an $n_{k-1} \in \mathbb{N}_{k-1}$, we know that at least one $1/k$-ball, $B_{1/k}(s_k)$, contains infinitely many of the x_n. Define $\mathbb{N}_k = \{n \in \mathbb{N}_{k-1} : x_n \in B_{1/k}(s_k)\}$ so that $\mathbb{N}_k$ is an infinite subset of $\mathbb{N}_{k-1}$. Define n_k to be the first element in $\mathbb{N}_k$ strictly greater than n_{k-1}.

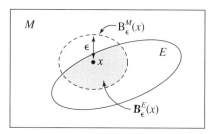

FIGURE 4.7.18

Since $1/k \to 0$, we know that x_{n_k} is a Cauchy sequence.

(d) Since K is complete, $x_{n_k} \to x$ for some $x \in K$. For all $k' \geq k$, $x_{n_{k'}} \in F_k$. Since F_k is closed, this implies that $x \in F_k$. Since k was arbitrary, $x \in \cap_k F_k$, a contradiction. ■

4.7.e Compactness Is Context Independent

We are dealing with subsets of a metric space (M, d). There is a hidden subtlety in the definition of open subsets of E. If $E \subset M$, then we regard (E, d) as a metric space. In the metric space (E, d), for $e \in E$, $B_\epsilon^E(e) = \{x \in E : d(x, e) < \epsilon\}$. This is a very different set than $B_\epsilon^M(e) = \{x \in M : d(x, e) < \epsilon\}$, though $B_\epsilon^E(e) = E \cap B_\epsilon^M(e)$. This means that open and closed subsets of E may look very different than open and closed subsets of M.

Example 4.7.17 (Context Matters) *Let $(M, d) = (\mathbb{R}, d)$, where $d(x, y) = |x - y|$, and let $E = (0, 1] \subset \mathbb{R}$. $G = (\frac{1}{2}, 1]$ is an open subset of the metric space (E, d), but is neither an open nor a closed subset of the metric space $(\mathbb{R}, d)$. $G^c = F = (0, \frac{1}{2}]$ is a closed subset of the metric space (E, d), but is neither an open nor a closed subset of the metric space $(\mathbb{R}, d)$.*

Theorem 4.7.18 *If $E \subset M$, then $G \subset E$ is an open subset of E iff $G = G' \cap E$ for some G' that is an open subset of M.*

Proof. G is an open subset of E iff $(\forall x \in G)(\exists \epsilon(x) > 0)[B_{\epsilon(x)}^E(x) \subset G]$. Set $G' = \cup_{x \in G} B_{\epsilon(x)}^M(x)$. [See Figure 4.7.18.] ■

Example 4.7.17 shows that whether or not a set is open may depend on the context. What is important is that compactness does *not* depend on the context.

Theorem 4.7.19 *If K is a compact subset of the metric space (M, d), and $K \subset E \subset M$, then K is a compact subset of the metric space (E, d).*

Proof. By Theorem 4.7.18, any open cover of K by sets that are open relative to K is equivalent to an open cover of K by sets that are open relative to E. ■

4.7.f Some Applications

The first application, Lemma 4.7.21, shows that continuous preferences on compact sets achieve their maximum. This uses the finite intersection property characterization of compactness to reach the result. The second tells us that compact intervals in $\mathbb{R}$ contain both their supremum and infimum. It uses the sequence characterization of compactness. The third application, Theorem 4.7.23, uses the totally bounded and complete characterization of compactness to show that a subset of $\mathbb{R}^\ell$ is compact iff it is closed and bounded. The fourth, Corollary 4.7.26, piggybacks on the second, using the sequence definition of closure to show that half-spaces are closed, which makes it easy to show that Walrasian budget sets are compact.

Definition 4.7.20 *Complete transitive preferences, $\succeq$, on a metric space (M, d) are* **continuous** *if for all $y \in M$, $\{x : x \succ y\}$ and $\{x : y \succ x\}$ are open, equivalently if $\{x : x \succeq y\}$ and $\{x : y \succeq x\}$ are closed.*

Lemma 4.7.21 *If K is a compact subset of M and preferences $\succeq$ are complete, transitive, and continuous, then the set of $x^* \in K$ such that for all $y \in K$, $x^* \succeq y$ is closed, hence compact and nonempty.*

Proof. Let $Wb(y, K) = \{x \in K : x \succeq y\}$; this is the set of points in K that are weakly better than y. $\{Wb(y, K) : y \in K\}$ is a collection of closed subsets of K. It has the finite intersection property because $\succeq$ is complete and transitive. Since K is compact, $\cap_{y \in K} Wb(y, K) \neq \emptyset$. Further $x^* \in \cap_{y \in K} Wb(y, K)$ iff for all $y \in K$, $x^* \succeq y$. Being an intersection of closed sets, it is closed. ∎

Many bounded subsets of $\mathbb{R}$ fail to contain their supremum or infimum.

Lemma 4.7.22 *If K is a compact subset of $\mathbb{R}$, then $\inf(K) \in K$ and $\sup(K) \in K$.*

Proof. Since K is compact, it must be bounded (since the open cover $\{(-n, +n) : n \in \mathbb{N}\}$ has a finite subcover). Let $s = \sup(K)$. For all $n \in \mathbb{N}$, $\exists k_n \in K$ such that $k_n > s - 1/n$. The only possible limit of a convergent subsequence of the k_n is s, so that $s \in K$. The argument for $\inf(K)$ is parallel. ∎

Theorem 4.7.23 (Compactness Criterion in $\mathbb{R}^\ell$) *$K \subset \mathbb{R}^\ell$ is compact iff it is closed and bounded.*

Proof. If K is totally bounded, it is bounded. If it is complete, then it is closed. Conversely, if K is closed and bounded, then it is complete because $\mathbb{R}^\ell$ is complete and it is totally bounded because it is bounded. ∎

For $\mathbf{p} > 0$ and $w > 0$, the associated Walrasian budget set is $B(\mathbf{p}, w) = \{\mathbf{x} \in \mathbb{R}^\ell_+ : \mathbf{p} \cdot \mathbf{x} \leq w\}$. This is an intersection of **closed half-spaces**, therefore closed.

Definition 4.7.24 *For $\mathbf{y} \in \mathbb{R}^\ell$, $\mathbf{y} \neq 0$, and $r \in \mathbb{R}$, the* **closed half-space determined by $\mathbf{y}$ and r** *is defined by $H^{\leq}_{\mathbf{y}}(r) = \{\mathbf{x} \in \mathbb{R}^\ell : \mathbf{y} \cdot \mathbf{x} \leq r\}$.*

In the coverage of convex analysis in Chapter 5, we examine the geometry behind the name "half-space" more thoroughly. Note that it makes sense to define $H^{\geq}_{\mathbf{y}}(r) = H^{\leq}_{-\mathbf{y}}(r)$.

Lemma 4.7.25 *The sets $H_{\mathbf{y}}^{\leq}(r)$ are closed.*

Proof. Let $\mathbf{x}_n$ be a sequence in $H_{\mathbf{y}}^{\leq}(r)$ converging to $\mathbf{x}$. We show that $\mathbf{y} \cdot \mathbf{x}_n \to \mathbf{y} \cdot \mathbf{x}$. Pick $\epsilon > 0$. For large n, $|y_i x_{i,n} - y_i x_i| = |y_i| \, |x_{i,n} - x_i| < \epsilon/\ell$, so that $|\mathbf{yx}_n - \mathbf{yx}| < \epsilon$. Since $\mathbf{yx}_n \leq r$, $\lim_n \mathbf{yx}_n = \mathbf{yx} \leq r$. ∎

Let $\mathbf{e}_i$ be the unit vector in the ith direction, that is, the vector that has 1's in the ith component and 0's in all other components. $B(\mathbf{p}, w) = H_{\mathbf{p}}^{\leq}(w) \cap \bigcap_i H_{\mathbf{e}_i}^{\geq}(0)$ expresses $B(\mathbf{p}, w)$ as an intersection of closed sets, so that it is closed. Since $\mathbf{p} > 0$, $B(\mathbf{p}, w)$ is bounded. As it is closed and bounded, it is compact.

Corollary 4.7.26 *Any continuous preference ordering on $\mathbb{R}^\ell$ has a preference maximal point in every $B(\mathbf{p}, w)$.*

Letting $x_{\preceq}(\mathbf{p}, w)$ be the set of preference optimal points gives the Walrasian **demand correspondence**. The correspondence maps $\mathbf{p}$, w, and $\succeq$ to closed subsets of $\mathbb{R}^\ell$.

4.8 ◆ Continuous Functions on $\mathbb{R}^\ell$

Throughout, (M, d) and (M', d') are metric spaces. Recall that $\tau_d \subset \mathcal{P}(M)$ is the collection of open subsets of M and $\tau_{d'} \subset \mathcal{P}(M')$ is the collection of open subsets of M'. For ease of reference, we repeat the following.

Definition 4.1.22. A function $f : M \to M'$ is **continuous** if $f^{-1}(G)$ is an open subset of M for every open $G \subset M'$, equivalently if $f^{-1}(\tau_{d'}) \subset \tau_d$.

The main result is Theorem 4.8.5, which gives four other formulations of continuity. In principle, we could have started with any one of them, and you should feel free to use whichever seems easiest in the context(s) you care about. We use this one because it facilitates the comparison with measurable functions, which can also be defined in terms of the properties of the inverse images of open sets.

4.8.a Jump and Oscillatory Discontinuities

At first it may not seem that Definition 4.1.22 captures the intuitive notion of continuity, which is an absence of jumps.

Example 4.8.1 (A Jump Discontinuity) *Define $f : \mathbb{R} \to \mathbb{R}$ by $f(x) = x \cdot 1_{[0,1)}(x)$; f has a jump discontinuity at $x = 1$, and for $0 < \epsilon < 1$, $f^{-1}(B_\epsilon(0))$ is not open. [See Figure 4.8.1.]*

The word "jump" is not appropriate for the next discontinuity.

Example 4.8.2 (An Oscillatory Discontinuity) *Define $g : \mathbb{R} \to \mathbb{R}$ by $g(0) = 0$ and $g(x) = \sin(\frac{1}{x})$ for $x \neq 0$; g has what is called an oscillatory discontinuity at $x = 0$, and for $0 < \epsilon < 1$, $g^{-1}(B_\epsilon(0))$ is not open because no δ-ball around 0 is a subset of $g^{-1}(B_\epsilon(0))$.*

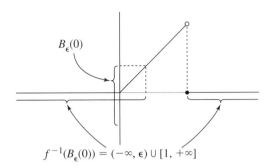

$$f^{-1}(B_\epsilon(0)) = (-\infty, \epsilon) \cup [1, +\infty]$$

FIGURE 4.8.1

4.8.b   Five Equivalent Formulations of Continuity

Since continuity is defined in terms of inverse images, it is worth recalling that set theoretic operations are preserved under inverse images, for $f : M \to M'$: $f^{-1}(\cup_a E_a) = \cup_a f^{-1}(E_a)$; $f^{-1}(\cap_a E_a) = \cap_a f^{-1}(E_a)$; $f^{-1}(E^c) = [f^{-1}(E)]^c$; and $f^{-1}(E \setminus F) = [f^{-1}(E)] \setminus [f^{-1}(F)]$. Inasmuch as set theoretic operations are preserved, the following turns out to be useful in finding easy proofs that functions are continuous.

Definition 4.8.3   *A class of open sets $S_{d'} \subset \tau_{d'}$ is a **subbasis** for $\tau_{d'}$ if every $G \in \tau_{d'}$ can be expressed as a union of sets, each of which is a finite intersection of elements of $S_{d'}$.*

Exercise 4.8.4   Show that each of the following classes of open sets is a subbasis for the usual τ_d on $\mathbb{R}$:

1. $\{(-\infty, r) : r \in \mathbb{R}\} \cup \{(s, \infty) : s \in \mathbb{R}\}$,
2. $\{(-\infty, p) : p \in \mathbb{Q}\} \cup \{(q, \infty) : q \in \mathbb{Q}\}$,
3. $\{B_\epsilon(r) : \epsilon \in \mathbb{R}_{++}, r \in \mathbb{R}\}$, and
4. $\{B_\epsilon(q) : \epsilon \in \mathbb{Q}_{++}, q \in \mathbb{Q}\}$.

Theorem 4.8.5   *The following are equivalent for a function $f : M \to M'$:*

1. *f is continuous;*
2. *for all closed $F \subset M'$, $f^{-1}(F)$ is a closed subset of M;*
3. *$f^{-1}(S_{d'}) \subset \tau_d$ for some subbasis $S_{d'}$ for $\tau_{d'}$;*
4. *for every $x \in M$ and $\epsilon > 0$, there exists $\delta > 0$ such that $[d(x, x') < \delta] \Rightarrow [d'(f(x), f(x')) < \epsilon]$; and*
5. *for every $x \in M$ and $x_n \to x$, $f(x_n) \to f(x)$.*

Before giving the proof of these equivalences, we use them.

Example 4.8.6   [↑*Exercise 4.8.4 and Theorem 4.8.5*]

1. *Constant functions are continuous because for every E, not merely the open ones, $f^{-1}(E)$ is either the open set $\emptyset$ or the open set M.*

2. *Contraction mappings are continuous because $d(f(x_n), f(x)) \leq \beta d(x_n, x)$, so that $[x_n \to x] \Rightarrow [f(x_n) \to f(x)]$.*

3. *For every $x' \in M$, the distance function $f(x) := d(x', x)$ is continuous. By the triangle inequality,*

$$d(x', x) \leq d(x', y) + d(x, y), \quad \text{and} \quad d(x', y) \leq d(x', x) + d(x, y),$$

so that $d(x', x) - d(x', y) \leq d(x, y)$ and $d(x', y) - d(x', x) \leq d(x, y)$. Combining, $|d(x', x) - d(x', y)| \leq d(x, y)$ so that $[d(x, y) < \epsilon] \Rightarrow [|f(x) - f(y)| < \epsilon]$.

4. *For every set $E \subset M$ and $x \in M$, we define the distance from x to E as $d(x, E) = \inf\{d(e, x) : e \in E\}$. The function $f(x) = d(x, E)$ is continuous because $|d(x, E) - d(y, E)| \leq d(x, y)$. To see why this inequality holds, note that*

$$d(y, E) \leq d(x, y) + d(x, E) \quad \text{and} \quad d(x, E) \leq d(x, y) + d(y, E).$$

Combining, $|d(x, E) - d(y, E)| \leq d(x, y)$.
   Note that because $\{0\}$ is closed, $f^{-1}(\{0\})$ is closed, and $f^{-1}(\{0\}) = \bigcap\{E^\epsilon : \epsilon > 0\} = \mathrm{cl}(E)$.

5. *If $f, g \in C(M)$, then $h(x) := \max\{f(x), g(x)\} \in C(M)$—$h(x) < r$ iff $f(x) < r$ and $g(x) < r$, so that $h^{-1}((-\infty, r))$ is the necessarily open intersection of the two open sets $f^{-1}((-\infty, r))$ and $g^{-1}((-\infty, r))$, and $h^{-1}((r, \infty))$ is the union of two open sets. More generally, the maximum or minimum of finitely many elements of $C(M)$ again belongs to $C(M)$.*

6. *The finiteness is needed in the previous part of this example. For $x' \in \mathbb{R}^\ell$, let $f_n(x) = \max\{0, 1 - n \cdot d(x', x)\}$. The function $f(x) = \sup\{f_n(x) : n \in \mathbb{N}\}$ is not continuous because $f^{-1}((1 - \epsilon, 1 + \epsilon)) = \{x'\}$ for any $\epsilon \in (0, 1)$, and $\{x'\}$ is not open.*

7. *For $x' \in M$ or $E \subset M$ and $\epsilon > 0$, let $f_\epsilon(x) = \max\{0, 1 - \frac{1}{\epsilon}d(x', x)\}$ or $g_\epsilon(x) = \max\{0, 1 - \frac{1}{\epsilon}d(x, E)\}$. Being the maxima of two continuous functions, each of these is continuous. For all x, $0 \leq f, g \leq 1$. $f(x') = 1$ while for $x \notin B_\epsilon(x')$, $f(x) = 0$ and $g(\mathrm{cl}(E)) = 1$ while for $x \notin E^\epsilon$, $g(x) = 0$.*

8. *For $f, g \in C(M)$ and $\alpha, \beta \in \mathbb{R}$, we define $f \cdot g$ by $(f \cdot g)(x) = f(x) \cdot g(x)$ and $\alpha f + \beta g$ by $(\alpha f + \beta g)(x) = \alpha f(x) + \beta g(x)$. This is called **pointwise multiplication and addition** because we do the multiplication or addition at each point x. If $f, g \in C(M)$, then $f \cdot g, \alpha f + \beta g \in C(\mathbb{R})$. Pick $x \in M$ and $\epsilon > 0$. We must show that there exists $\delta > 0$ such that $|x - y| < \delta$ implies that $|f(x)g(x) - f(y)g(y)| < \epsilon$. For any four numbers a_x, b_x, a_y, and b_y,*

$$|a_x b_x - a_y b_y| = |a_x b_x - a_x b_y + a_x b_y - a_y b_y|$$

$$\leq |a_x(b_x - b_y)| + |b_y(a_x - a_y)|$$

$$\leq |a_x| \cdot |b_x - b_y| + |b_y| \cdot |a_x - a_y|,$$

so $|f(x)g(x) - f(y)g(y)| \leq |f(x)| \cdot |g(x) - g(y)| + |g(y)| \cdot |f(x) - f(y)|$. By the continuity of g, we can choose δ_g such that $|x - y| < \delta_g$

implies $|g(x) - g(y)| < \epsilon/2(|f(x)| + 1)$. *By the continuity of f, we can choose δ_f such that $|x - y| < \delta_f$ means that $|f(x) - f(y)| < \epsilon/2(|g(x)| + 1)$. Setting $\delta = \min\{\delta_f, \delta_g\}$, $|x - y| < \delta$ implies that $|f(x)g(x) - f(y)g(y)| < \epsilon/2 + \epsilon/2$. Showing that $\alpha f + \beta g \in C(M)$ is similar.*

9. *The **(canonical) projection mappings** on $\mathbb{R}^\ell$ are defined by $\mathrm{proj}_i(\mathbf{x}) = x_i$, $i = 1, \ldots, \ell$. They are continuous because, as seen earlier, $[\mathbf{x}_n \to \mathbf{x}] \Leftrightarrow (\forall i)[x_{i,n} \to x_i]$. If we use (8) inductively, this implies that all multinomials on $\mathbb{R}^\ell$, that is, all finite sums of the form $a_0 + \sum_k a_k \Pi_i x_i^{n_{i,k}}$, are continuous. When $\ell = 1$, this is the set of polynomials.*

Exercise 4.8.7   Show that $\alpha f + \beta g \in C(M)$ if $f, g \in C(M)$ and $\alpha, \beta \in \mathbb{R}$.

Proof of Theorem 4.8.5. $(1) \Leftrightarrow (2)$ $f^{-1}(G)$ is open for every open G iff $f^{-1}(G^c) = [f^{-1}(G)]^c$ is closed for every closed G^c.

$(1) \Rightarrow (3)$ Suppose that $f^{-1}(\tau_{d'}) \subset \tau_d$. Since $S_{d'} \subset \tau_{d'}$, we know that $f^{-1}(S_{d'}) \subset \tau_d$.

$(3) \Rightarrow (1)$ Suppose that $S_{d'}$ is a subbasis for $\tau_{d'}$ and that $f^{-1}(S_{d'}) \subset \tau_d$. Let $T_{d'} = (S_{d'}, \cap^f)$ denote the class of finite intersections of elements of $S_{d'}$. Since $f^{-1}(\cap_a E_a) = \cap_a f^{-1}(E_a)$ and τ_d is closed under finite intersection, $f^{-1}(T_{d'}) \subset \tau_d$. Let $U_{d'} = (T_{d'}, \cup^a)$ denote the class of arbitrary unions of elements of $T_{d'}$. As $S_{d'}$ is a subbasis for $\tau_{d'}$, $U_{d'} = \tau_{d'}$. Since $f^{-1}(\cup_a E_a) = \cup_a f^{-1}(E_a)$ and τ_d is closed under arbitrary unions, $f^{-1}(\tau_{d'}) \subset \tau_d$.

$(1) \Rightarrow (4)$ Suppose that f is continuous. Pick $x \in M$ and $\epsilon > 0$. The ball $B_\epsilon(f(x))$ is an open subset of M', and x belongs to the open set $G = f^{-1}(B_\epsilon(f(x)))$. Therefore, there exists a $\delta > 0$ such that $B_\delta(x) \subset G$. Now, $d(x, x') < \delta$ iff $x' \in B_\delta(x)$, so that $[d(x, x') < \delta] \Rightarrow [f(x') \in f(G)]$. Since $f(G) = B_\epsilon(f(x))$, we know that $d'(f(x), f(x')) < \epsilon$.

$(4) \Rightarrow (1)$ Pick an arbitrary open $G \subset M'$. We must show that $f^{-1}(G)$ is an open subset of M. Pick $x \in f^{-1}(G)$ so that $f(x) \in G$. We must show there exists a $\delta > 0$ such that $B_\delta(x) \subset G$. Since G is open, there exists an $\epsilon > 0$ such that $B_\epsilon(f(x)) \subset G$. By assumption, there exists a $\delta > 0$ such that $[x' \in B_\delta(x)] \Rightarrow [f(x') \in B_\epsilon(f(x'))]$, so that $B_\delta(x) \subset G$.

$(4) \Rightarrow (5)$ Suppose that $x_n \to x$ and pick $\epsilon > 0$. We must show that $d(f(x_n), f(x)) < \epsilon$ for large n. Pick $\delta > 0$ such that $[d(x, x') < \delta] \Rightarrow [d'(f(x), f(x')) < \epsilon]$. For large n, $d(x, x_n) < \delta$.

$(5) \Rightarrow (4)$ Suppose, for the purposes of establishing a contradiction, that for some x in M and $\epsilon > 0$, for all $\delta > 0$, there is an x' such that $d(x, x') < \delta$ yet $d'(f(x), f(x')) \geq \epsilon$. This means that for each $n \in \mathbb{N}$, we can choose an x_n such that $d(x, x_n) < 1/n$ yet $d'(f(x), f(x')) \geq \epsilon$. This gives a sequence $x_n \to x$ with $f(x_n) \not\to f(x)$. ∎

Corollary 4.8.8   *If the graph of $f : M \to M'$ is closed and M' is compact, then f is continuous.*

Proof.   Let $x_n \to x$ in M. If $f(x_n) \not\to f(x)$, then for some $\epsilon > 0$, $d'(f(x_{n_k}), f(x)) \geq \epsilon$ for all k. Relabel x_{n_k} as x_k. Since M' is compact, x_k has a further subsequence, x_{k_m}, with $f(x_{k_m}) \to y$ for some $y \in M'$. As the graph of f is closed, $y = f(x)$, contradicting $d'(f(x_{n_k}), f(x)) \geq \epsilon$ for all k. ∎

Exercise 4.8.9 Show that the function $f(x) = 1/x$ for $x \neq 0$ and $f(0) = 0$ from $\mathbb{R}$ to $\mathbb{R}$ has a closed graph but is not continuous. [This shows why we need compactness for the previous result. It is informative to try to make the proof work for this function.]

Exercise 4.8.10 Theorem 4.8.5 gives five equivalent formulations of continuity. This means that it contains twenty statements of the form $(n) \Rightarrow (m)$, $n \neq m$, $n, m \in \{1, \ldots, 5\}$. The proof given above proves eight such statements. Pick another six of the possible statements and give 'direct' proofs of them.

4.8.c The Basic Existence Result

We are interested in continuous $\mathbb{R}$-valued functions for several reasons. One of the primary ones is that they achieve their maximum on compact sets; that is, for every compact K and $f \in C(M)$, there exists an $x^* \in K$ such that $f(x^*) = \sup f(K)$. In fact, one can go further and use the achievement of the maximum as a characterization of compactness.

Theorem 4.8.11 (Basic Existence) *$K \subset M$ is compact iff every $f \in C(M)$ achieves its maximum on K.*

The maximum of the function $-f$ is -1 times the minimum of the function f, so that the result could be formulated using continuous functions achieving their minima.

Compactness and continuity implying the existence of a maximum, the first part of Theorem 4.8.11, is the basic existence result in economic theory. It is sufficiently important that knowing several ways to prove it helps us understand the result, and Exercise 4.8.13 (directly below) sketches several alternate proofs. The "if" part of Theorem 4.8.11 tells us that to guarantee existence of an optimum for all preferences representable by a continuous utility function, we must assume that the choice sets are compact.

Proof. ($\Rightarrow$) Suppose that K is compact and $f \in C(M)$. Since f is continuous, the collection $\{f^{-1}((-n, n)) : n \in \mathbb{N}\}$ is an open cover of K. As K is compact, there is a finite subcover, so that $f(K)$ is bounded. Let $r = \sup f(K)$. Let $K_n = f^{-1}([r - 1/n, r]) \cap K$, so that $\{K_n : n \in \mathbb{N}\}$ is a class of closed subsets of K with the finite intersection property. Since K is compact, $\cap_n K_n \neq \emptyset$, and any $x^* \in \cap_n K_n$ satisfies $f(x^*) = r$, so that f achieves its maximum on K.

($\Leftarrow$) We give the easy proof available if $M = \mathbb{R}^\ell$. Exercise 6.11.29 sketches the general case. We must show that K is closed and bounded. The function $f(x) = \|x\| = d(0, x)$ is continuous, so it achieves its maximum, and thus K is bounded by that maximum. Let x_n be a sequence in K and suppose that $x_n \to x^\circ$. We must show that $x^\circ \in K$. The function $g(x) = -d(x, x^\circ)$ is continuous. Since $x_n \to x^\circ$, $\sup g(K) = 0$. The only way that g can achieve its maximum on K is if $x^\circ \in K$. ∎

Example 4.8.12 *The function $f(x) = x \cdot 1_{[0,1)}(x)$ maps the compact interval $[0, 1]$ into $\mathbb{R}$ and does not achieve its maximum. It is the jump of f at 1 that causes the problem with f. Ruling out jumps is not sufficient—the function $g(x) = 1/x$*

has no jumps and fails to achieve its maximum on the noncompact set (0, 1). *The unboundedness of g is at least part of the problem. Boundedness is not sufficient— the function h(x) = x has no jumps, is bounded, and fails to achieve its maximum on the noncompact set* (0, 1).

Exercise 4.8.13   Complete the following sketches of alternative proofs that every continuous function on a compact space achieves its maximum:

1. Sequences: If $f(K)$ is unbounded, then for every $n \in \mathbb{N}$, there exists x_n such that $f(x_n) \geq n$. Take a convergent subsequence of x_n to obtain a contradiction. Let $r = \sup\{f(x) : x \in K\}$. For every $n \in \mathbb{N}$, there exists x_n such that $f(x_n) \geq r - 1/n$. Take a convergent subsequence and show that its limit is an optimizer.

2. Finite intersection property: If $f(K)$ is unbounded, then for all $n \in \mathbb{N}$ the set $F_n = f^{-1}(-\infty, -n]) \cup f^{-1}([n, \infty))$ is nonempty and the class $\{F_n : n \in \mathbb{N}\}$ has the finite intersection property, hence an nonempty intersection, a contradiction. Let $r = \sup\{f(x) : x \in K\}$ and consider the collection $\{f^{-1}([r - 1/n, r] : n \in \mathbb{N}\}$.

3. Open cover property: $\{f^{-1}((-n, +n)) : n \in \mathbb{N}\}$ is an open cover of K. The existence of finite subcover implies that $f(K)$ is bounded. Let $r = \sup\{f(x) : x \in K\}$. Show that if f does not achieve its maximum, then $\{f^{-1}((-\infty, r - 1/n)) : n \in \mathbb{N}\}$ is an open cover with no finite subcover.

4. Totally bounded and complete: For each $n \in \mathbb{N}$, let S_n be a finite $1/n$-net for K. Define $T_n = \cup_{m \leq n} S_m$ and let $x_n \in T_n$ solve the problem $\max\{f(x) : x \in T_n\}$. Use total boundedness to show that we can find a Cauchy, hence convergent, subsequence from x_n, and let x be its limit. Since f is continuous and $T_n \subset T_{n+1}$, $f(x_n) \uparrow f(x)$. Suppose that $\sup f(K) - f(x) = \epsilon > 0$. There must be an open subset, G, of K on which f is greater than $f(x)$ by at least $\epsilon/2$. For large n, $T_n \cap G \neq \emptyset$, a contradiction.

4.8.d   Variations on the Existence Argument

Sometimes, one can get around the need for compactness by showing that the optimum, if one exists, must belong to a compact set. The following two examples use this idea.

Example 4.8.14   *Let $F \subset \mathbb{R}^\ell$ be a nonempty closed unbounded set, and let $\mathbf{x}$ be a point in $\mathbb{R}^\ell$. The problem $\min\{d(\mathbf{x}, \mathbf{y}) : \mathbf{y} \in F\}$ has a solution. To see why, pick an arbitrary $\mathbf{y}^\circ \in F$. If $\mathbf{y}^\circ = \mathbf{x}$, then we have solved the minimization problem. If not, let $r = d(\mathbf{x}, \mathbf{y}^\circ) > 0$. The set $F \cap \mathrm{cl}(B_r(\mathbf{x}))$ is compact because it is closed and bounded. Therefore, the problem $\min\{d(\mathbf{x}, \mathbf{y}) : \mathbf{y} \in (F \cap \mathrm{cl}(B_r(\mathbf{x})))\}$ has a solution, $\mathbf{y}^*$, and $d(\mathbf{x}, \mathbf{y}^*) \leq r$. For any omitted point, $\mathbf{y}' \in F \setminus \mathrm{cl}(B_r(\mathbf{x}))$, $d(\mathbf{x}, \mathbf{y}') > r$. Therefore, $\mathbf{y}^*$ is at least as close to $\mathbf{x}$ as any point in F.*

The following example also provides a review of monotone comparative statics.

Example 4.8.15   *A person of type θ, $\theta \in \Theta := [\underline{\theta}, \overline{\theta}] \subset (0, 1)$, chooses an effort, $e \in [0, \infty)$. An effort of e costs $c(\theta, e) := (1 - \theta)c(e)$ in utility terms, so that higher*

*types correspond to lower costs and to lower marginal costs. After the effort is
exerted, there are two possible outcomes: success, with a reward of R in utility
terms, and failure, with a reward of r in utility terms, R > r > 0. High efforts do
not guarantee success, but they do increase the probability of success. In particular,
there is a probability $P(e)$ of success and a probability $1 - P(e)$ of failure. Thus,
expected utility as a function of e, θ, R, and r is*

$$U(e, R, r, \theta) = R \cdot P(e) + r \cdot (1 - P(e)) - (1 - \theta)c(e),$$

and we define $V(R, r, \theta) = \sup_{e \in [0,\infty)} U(e, R, r, \theta)$.
  We assume that:

1. *$P : [0, \infty) \to [0, 1]$ is continuous and nondecreasing,*

2. *$c : [0, \infty) \to \mathbb{R}$ is continuous nondecreasing, $c(0) = 0$, and $\lim \sup_{e \to \infty}$
   $(1 - \bar{\theta})c(e) > R - r$.*

1. *To show that the problem $\max_{e \in [0,\infty)} U(e, R, r, \theta)$ always has a solution,
   pick arbitrary $\theta \in \Theta$. Choose $\bar{e}$ such that $(1 - \theta)c(\bar{e}) > R - r$. Since increases in effort can gain one at most $R - r$, we can replace the problem
   $\max_{e \in [0,\infty)} U(e, R, r, \theta)$ by the problem $\max_{e \in [0,\bar{e}]} U(e, R, r, \theta)$, which is
   compact and continuous.*

2. *The mappings $R \mapsto V(R, r, \theta)$, $r \mapsto V(R, r, \theta)$, and $\theta \mapsto V(R, r, \theta)$ are
   all at least weakly increasing because for any e, increases in R, r, or θ
   weakly increase $U(e, R, r, \theta)$.*

3. *Suppose that $R' > R$ and that $e' := e^*(R', r, \theta)$ and $e := e^*(R, r, \theta)$ contain
   only a single point. Intuitively, a higher reward for success ought to make
   someone work harder for that success. Checking for strict supermodularity
   with R as the parameter and e the variable that the decision maker is
   maximizing over and suppressing r and θ from the notation, we must check
   that if $e' > e$, $U(e', R') - U(e', R) > U(e, R') - U(e, R)$. The left-hand
   side is $(R' - R)P(e')$, while the right-hand side is $(R' - R)P(e)$. If $P(\cdot)$ is
   strictly increasing, the strict inequality holds so that $e' > e$, as we suspected.*

4. *Suppose that $\theta' > \theta$ and that $e' := e^*(R, r, \theta')$ and $e := e^*(R, r, \theta)$ contain only a single point. Intuitively, higher types have lower costs and
   lower marginal costs, so ought to be willing to work harder. Treating θ
   as the parameter and e as the decision variable and checking $U(e', \theta') -
   U(e', \theta) > U(e, \theta') - U(e, \theta)$ completes the problem. The left-hand side
   is $(\theta' - \theta)c(e')$, while the right-hand side is $(\theta' - \theta)c(e)$.*

4.8.e Continuous Images of Compact Sets

The following general result about $f(K)$ *also* implies that continuous functions
achieve their maxima. To see why, review Lemma 4.7.22 (p. 135).

Theorem 4.8.16 *If $f : M \to M'$ is continuous and $K \subset M$ is compact, then
$f(K)$ is a compact subset of M'.*

Proof. Let $\{G_\alpha^{M'} : \alpha \in A\}$ be an open cover of $f(K)$ so that $\{G_\alpha^M : \alpha \in A\}$ is an open cover of K when $G_\alpha^M := f^{-1}(G_\alpha^{M'})$. If A_F is a finite subset of A such that $\{G_\alpha^M : \alpha \in A_F\}$ covers K, then $\{G_\alpha^{M'} : \alpha \in A_F\}$ covers $f(K)$. ∎

Exercise 4.8.17 Reprove Theorem 4.8.16 using the sequence characterizations of compactness and continuity.

4.9 ◆ Lipschitz and Uniform Continuity

The technique in Exercise 4.8.13(4) shows that to any degree of approximation, we can replace the infinite problem max $f(x)$ subject to $x \in K$ by the conceptually far simpler, finite problem max $f(x)$ subject to $x \in T_n$. What is missing is guidance on how large n needs to be for the $1/n$-net to have a maximum within a given ϵ of the true optimum. For this we turn to a linked pair of concepts called **Lipschitz continuity** and **uniform continuity**.

It is easy to show that the function $g : (0, 1) \to \mathbb{R}$ defined by $g(r) = 1/r$ is continuous. However, it becomes arbitrarily steep as $r \downarrow 0$. This means that maximizing g over finite approximations to $(0, 1)$ gives very misleading answers as to the size of $\sup\{g(r) : r \in (0, 1)\}$. By contrast, the function $h(r) = 1 - r$ has a slope bounded in absolute value by $+1$, and maximizing h over any $1/n$-net for $(0, 1)$ misses the size of $\sup\{h(r) : r \in (0, 1)\}$ by at most $1/n$. Uniform continuity generalizes this notion of a bounded slope.

Definition 4.9.1 *A function $f : M \to M'$ is **Lipschitz continuous** if $\forall x, x'$, $d'(f(x), f(x')) \leq L d(x, x')$, where $L \in \mathbb{R}_+$ is called the **Lipschitz constant of** f and f is **uniformly continuous on** M if $\forall \epsilon > 0$, $\exists \delta > 0$ such that $\forall x \in M$, $[x' \in B_\delta(x)] \Rightarrow [f(x') \in B_\epsilon(f(x))]$.*

If f is Lipschitz continuous, then it is uniformly continuous; simply take $\delta = \epsilon/L$. The distinction between Lipschitz continuity and the $\epsilon - \delta$ formulation of continuity in Theorem 4.8.5(4) is the order of the quantifiers.

1. "$(\forall \epsilon > 0)(\exists \delta > 0)(\forall x \in M)[[x' \in B_\delta(x)] \Rightarrow [f(x') \in B_\epsilon(f(x))]]$" for uniform continuity, versus

2. "$(\forall \epsilon > 0)(\forall x \in M)(\exists \delta > 0)[[x' \in B_\delta(x)] \Rightarrow [f(x') \in B_\epsilon(f(x))]]$" for continuity.

For continuity, the δ can depend on both the x and the ϵ, whereas for uniform continuity, the δ can depend on the ϵ, but not on the x. The same δ must work at all x's. Being uniformly continuous implies being continuous, but not the reverse. As well as the example just given, $g(r) = 1/r$ on $(0, 1)$, we have the function $f(x) = e^x$ on $\mathbb{R}$.

Exercise 4.9.2 For any $x \in \mathbb{R}$, show that $e^x := \sum_{n=0}^{\infty} \frac{x^n}{n!}$ is well defined (summable). Show that the function $f(x) = e^x$ mapping $\mathbb{R}$ to $\mathbb{R}$ is continuous but not uniformly continuous.

Having a bounded derivative guarantees Lipschitz continuity, hence uniform continuity, but $h(x) = |x|$ has no derivative at 0 and yet has Lipschitz constant 1.

The function $f(x) = \sqrt{x}$ on $[0, 1]$ is not Lipschitz because its slope is unbounded, but as it is continuous on the compact set $[0, 1]$, it is uniformly continuous. This follows by direct calculation or from the following result.

Theorem 4.9.3 *If $f : M \to M'$ is continuous and M is compact, then f is uniformly continuous.*

Proof. Suppose that f is continuous but not uniformly continuous. For some $\epsilon > 0$ and all $n \in \mathbb{N}$, there exists $x_n, u_n \in M$ such that $d(x_n, u_n) < 1/n$, but $d'(f(x_n), f(u_n)) \geq \epsilon$. Since M is compact, $x_{n'} \to x$ for some $x \in M$ and some subsequence. As $d(x_n, u_n) < 1/n \to 0$, $u_{n'} \to x$. Since f is continuous, there exists $\delta > 0$ such that $[d(x', x) < \delta] \Rightarrow [d'(f(x'), f(x)) < \epsilon/2]$. For large n', $d(u_{n'}, x) < \delta$ and $d(x_{n'}, x) < \delta$, which implies that $d'(f(x_{n'}), f(u_{n'})) < \epsilon$, a contradiction. ∎

Returning to the theme of approximating maximization problems by (conceptually simpler) finite problems, we see that we have to know the size of the δ-balls across which f varies by at most ϵ. Then we know that optimizing over any δ-net gets us to within ϵ of the true maximum value of f.

4.10 ◆ Correspondences and the Theorem of the Maximum

Sets of optima may depend discontinuously on the set of choices that are offered, but the value of the set of choices is continuous. In this section we make these observations precise in a basic result called the theorem of the maximum, which says that under fairly general conditions, the set of optima depends **upper hemicontinuously** on the set of choices, whereas the value of the set of choices depends continuously. Upper hemicontinuity allows sets to "explode" when indifference is reached.

We begin with some basic examples from consumer demand theory. We then turn to what is called the **Hausdorff metric**, which is a way of measuring the distance between compact sets. With these in place, we state and prove the theorem of the maximum.

4.10.a Examples from Consumer Demand Theory

We begin with some examples of basic consumer demand theory from intermediate microeconomics. For $\mathbf{p} \in \mathbb{R}^{\ell}_{++}$ and $w \in \mathbb{R}^1_{++}$, the **Walrasian budget set** is

$$B_{\mathbf{p},w} = \{\mathbf{x} \in \mathbb{R}^{\ell}_+ : \mathbf{p} \cdot \mathbf{x} \leq w\}. \tag{4.7}$$

Facing prices $\mathbf{p}$ and having wealth w, $B_{\mathbf{p},w}$ is the set of affordable bundles of consumption for a consumer. If all consumers face the same price vector, then their market trade-offs among the goods must be equal.

Definition 4.10.1   *For a (continuous) utility function* $u : \mathbb{R}_+^\ell \to \mathbb{R}$, $\mathbf{x}^*(\mathbf{p}, w) :=$
$\{\mathbf{x} \in B_{\mathbf{p}, w} : \forall \mathbf{y} \in B_{\mathbf{p}, w}, u(\mathbf{x}) \geq u(\mathbf{y})\}$ *is the **demand set**, and* $v(\mathbf{p}, w) := u(\mathbf{x}^*(\mathbf{p}, w))$
$= \max\{u(\mathbf{x}) : \mathbf{x} \in B_{\mathbf{p}, w}\}$ *is the **indirect utility function**, a.k.a. the **value function**.*

There are two patterns to be noted in the following. First, the demand sets are discontinuous in the region where they "explode" and continuous otherwise. Second, the indirect utility function is continuous. This happens because, even though the demand set may explode, the points added in the explosion are all indifferent, so there is no jump in $v(\cdot, \cdot)$.

Example 4.10.2   *For* $\ell = 2$, *consider the following three utility functions:*

$$u_a(x_1, x_2) = x_1 \cdot x_2; \; u_b(x_1, x_2)$$

$$= x_1 + x_2; \; u_c(x_1, x_2) = \max\left(\min\{x_1, x_2\}, \tfrac{3}{2}\min\{\tfrac{1}{2}x_1, x_2\}\right),$$

where $u_c(\cdot, \cdot)$ *represents my nonconvex preferences for slices of bread,* x_1, *and slices of cheese of the same size,* x_2, *when I want a grilled cheese sandwich. I like open-faced sandwiches, where I have one slice of cheese per slice of bread. I also like "regular" sandwiches, where I have two slices of bread per slice of cheese. Other ratios of bread to cheese are useless to me. I also think that one open-faced sandwich is worth two-thirds of a regular sandwich.*

The demand sets and indirect utility functions are

$$x_a^*((p_1, p_2), w) = \left\{\left(\frac{w}{2p_1}, \frac{w}{2p_2}\right)\right\}, \; v_a((p_1, p_2), w) = \frac{w^2}{4p_1 p_2},$$

$$x_b^*((p_1, p_2), w) = \begin{cases} \{(\frac{w}{p_1}, 0)\} & \text{if } p_2 > p_1 \\ \{(0, \frac{w}{p_2})\} & \text{if } p_2 < p_1 \\ L_b & \text{if } p_2 = p_1, \end{cases} \; v_b((p_1, p_2), w) = \begin{cases} \frac{w}{p_1} & \text{if } p_2 > p_1 \\ \frac{w}{p_2} & \text{if } p_2 \leq p_1 \end{cases},$$

$$x_c^*((p_1, p_2), w) = \begin{cases} \{(\frac{w}{p_1 + p_2}, \frac{w}{p_1 + p_2})\} & \text{if } p_2 < p_1 \\ \{(\frac{w}{p_1 + \frac{1}{2}p_2}, \frac{1}{2}\frac{w}{p_1 + \frac{1}{2}p_2})\} & \text{if } p_2 > p_1 \\ S_c & \text{if } p_2 = p_1, \end{cases}$$

$$v_c((p_1, p_2), w) = \begin{cases} \frac{w}{p_1 + p_2} & \text{if } p_2 \leq p_1 \\ \frac{3}{4}\frac{w}{p_1 + \frac{1}{2}p_2} & \text{if } p_2 > p_1 \end{cases},$$

where L_b *is the line segment* $\{(\frac{\gamma w}{p_2}), \frac{(1-\gamma)w}{p_2}) : \gamma \in [0, 1]\}$ *and* S_c *is the two-point set* $\{(\frac{w}{2p_2}, \frac{w}{2p_2}), (\frac{2}{3}\frac{w}{p_2}, \frac{1}{3}\frac{w}{2p_2})\}$.

Exercise 4.10.3   Let $u_d(x_1, x_2) = \max\{x_1, x_2\}$ and $u_e(x_1, x_2) = \sqrt{x_1^2 + x_2^2}$. Find the demand sets and indirect utility functions for these utility functions and compare your answer to those for the utility function u_b from the previous exercise.

[You should find a striking similarity even though the three preferences are quite different.]

There are many situations in which peoples' purchase decisions appear to be motivated by liking a higher price. We can understand this in several fashions. In some situations, price might, arguably, indicate quality, say in the market for generic health supplements in an era of lax governmental oversight of corporate behavior. In others, consumers might want the people observing the purchase to be suitably impressed by the ability/willingness to pay the price, say for an expensive mechanical watch that works less well than modern electrical watches or for a holiday gift purchased for a boss. In such cases, it is as if the consumers are maximizing a utility function in which price enters positively. It turns out that we still have the same kinds of results: demand sets may explode around points of indifference, but the value function is continuous.

Exercise 4.10.4   Suppose that $u_f(x_1, x_2; p_1, p_2) = x_1 + p_2 x_2$. Find the demand set and the indirect utility function, and show that there is no utility function depending only on x_1 and x_2 giving these demand sets and indirect utility function. [Hint: What happens if w, p_1 and p_2 are all multiplied by a positive constant?]

Many goods are only sensibly purchased/consumed in integer portions rather than in continuously divisible portions. This kind of lumpiness also does not change the basic result that demand sets may explode, but since the opportunity set may explode as the price becomes low enough to afford the lumpy good, the value function may become discontinuous.

Exercise 4.10.5   Suppose that $u_b(x_1, x_2) = x_1 + x_2$, and the consumer is constrained to choose either 0 or 1 unit of good 2; that is, his budget set is $\{(x_1, x_2) \in B_{\mathbf{p},w} : x_2 \in \{0, 1\}\}$. Find the demand set and the discontinuous indirect utility function.

4.10.b The Hausdorff Metric

Recall that for $A \subset \mathbb{R}^\ell$ and $\epsilon > 0$, $A^\epsilon := \cup_{x \in A} B(x, \epsilon)$ is the open ϵ-ball around the set A.

Exercise 4.10.6   For $\epsilon = 1, \frac{1}{2}, \frac{1}{100}$, draw K^ϵ when $K = [0, 1]^2 \subset \mathbb{R}^2$, when $K = \{\mathbf{x} \in \mathbb{R}^2 : \|\mathbf{x}\| = 1\}$, and when $K = \{(x, x) \in \mathbb{R}^2 : 0 \le x \le 1\}$.

Definition 4.10.7   *For compact K_1, $K_2 \subset \mathbb{R}^\ell$, the **Hausdorff distance between** K_1 **and** K_2 is $d_H(K_1, K_2) = \inf\{\epsilon > 0 : [K_1 \subset K_2^\epsilon] \wedge [K_2 \subset K_1^\epsilon]\}$. [See Figure 4.10.7.]*

The Hausdorff metric, which applies to pairs of sets, is a direct generalization of the metric d, which applies to pairs of sets.

Example 4.10.8   *If $K_1 = \{x_1\}$ and $K_2 = \{x_2\}$, then $d_K(K_1, K_2) = d(x_1, x_2)$.*

Example 4.10.9   *If $K_1 \subset K_2$, then for all $\epsilon > 0$, $K_1 \subset K_2^\epsilon$. Thus, for $K_2 = [0, 1]$ and $K_1 = \{x\} \subset K_2$, $d_K(K_1, K_2) = \max\{x, 1 - x\}$.*

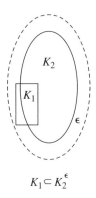

$$K_1 \subset K_2^\epsilon$$

FIGURE 4.10.7

The Hausdorff metric is only a metric when applied to compact sets.

Example 4.10.10 $E = \mathbb{Q} \cap [0, 1]$ *is bounded but not closed and* $d_H(E, [0, 1]) =$ 0. $\mathbb{R}$ *is closed but not bounded, and, using the convention that* inf $\emptyset = \infty$, $d_H([0, 1], \mathbb{R}) = \infty$.

Notation 4.10.11   *For* $E \subset \mathbb{R}^\ell$, $\mathcal{K}(E) := \{K \subset E : K \neq \emptyset$ *is compact*$\}$ *is the set of nonempty compact subsets of* E.

Theorem 4.10.12 $d_H(\cdot, \cdot)$ *is a metric on* $\mathcal{K}(\mathbb{R}^\ell)$, *that is, for all* K_1, K_2, $K_3 \in \mathcal{K}(\mathbb{R}^\ell)$, $d_H(K_1, K_2) \geq 0$,

    1. $d_H(K_1, K_2) = d_H(K_2, K_1)$,

    2. $d_H(K_1, K_2) = 0$ *iff* $K_1 = K_2$, *and*

    3. $d_H(K_1, K_3) \leq d_H(K_1, K_2) + d_H(K_2, K_3)$.

Proof. $d_H(K_1, K_2) \geq 0$ and $d_H(K_1, K_2) = d_H(K_2, K_1)$ are obvious.

If $K_1 = K_2$, then for all $\epsilon > 0$, $K_1 \subset K_2^\epsilon$ and $K_2 \subset K_1^\epsilon$, so that $d_H(K_1, K_2) =$ inf$\{\epsilon > 0\} = 0$. If $d_H(K_1, K_2) = 0$, then for all $\epsilon > 0$, $K_1 \subset K_2^\epsilon$ and $K_2 \subset K_1^\epsilon$, that is, $K_1 \subset \cap_{\epsilon > 0} K_2^\epsilon$ and $K_2 \subset \cap_{\epsilon > 0} K_1^\epsilon$. As compact sets are closed, $K_2 = \cap_{\epsilon > 0} K_2^\epsilon$ and $K_1 = \cap_{\epsilon > 0} K_1^\epsilon$. Combining yields $K_1 \subset K_2 \subset K_1$.

We must show that $d_H(K_1, K_3) \leq d_H(K_1, K_2) + d_H(K_2, K_3)$. Let $r = d_H(K_1, K_2)$, $s = d_H(K_2, K_3)$ and that $d_H(K_1, K_3) \leq (r + s)$. By the definition of the Hausdorff metric, for every $\epsilon > 0$,

$$(\forall e \in K_1)(\exists f(e) \in K_2)[d(e, f(e)) < r + \epsilon],$$

$$(\forall f \in K_2)(\exists e(f) \in K_1)[d(f, e(f)) < r + \epsilon],$$

$$(\forall f \in K_2)(\exists g(f) \in K_3)[d(f, g(f)) < s + \epsilon],$$

$$(\forall g \in K_3)(\exists f(g) \in K_2)[d(g, f(g)) < s + \epsilon].$$

Therefore, for all $e \in K_1$, there exists $g \in K_3$, namely $g = g(f(e))$, such that $d(e, g) < (r + s) + 2\epsilon$, and for all $g \in K_3$, there exists $e \in K_1$, namely

$e = e(f(g))$, such that $d(g, e) < (r + s) + 2\epsilon$. Therefore, for all $\epsilon > 0$, $d_H(K_1, K_3) < (r + s) + 2\epsilon$. Since ϵ was arbitrary, $d_H(K_1, K_3) \leq (r + s)$. ∎

For neoclassical demand theory, the important aspect of the Hausdorff metric is that the Walrasian budget set correspondence is continuous.

Lemma 4.10.13 *The function* $(\mathbf{p}, w) \mapsto B_{\mathbf{p}, w}$ *is continuous from* $\mathbb{R}_{++}^{\ell+1}$ *to* $\mathcal{K}(\mathbb{R}_+^\ell)$.

There is just a slight change in perspective on correspondences. Before, they were point-to-set mappings and now they are functions from the metric space $\mathbb{R}_{++}^{\ell+1}$ to the metric space $\mathcal{K}(\mathbb{R}_+^\ell)$.

Proof. The vertices of $B_{\mathbf{p}, w}$ are the points 0 and $(w/p_i)\mathbf{e}_i$, $i = 1, \ldots, \ell$, where $\mathbf{e}_i$ is the unit vector in the ith direction. If $(\mathbf{p}_n, w_n) \to (\mathbf{p}, w) \in \mathbb{R}_{++}^{\ell+1}$, then the vertices converge. If each vertex of $B_{\mathbf{p}_n, w_n}$ is within ϵ of the corresponding vertex of $B_{\mathbf{p}, w}$, then $d_H(B_{\mathbf{p}_n, w_n}, B_{\mathbf{p}, w}) < \epsilon$. ∎

Exercise 4.10.14 The last step in the preceding proof, "If each vertex of $B_{\mathbf{p}_n, w_n}$ is within ϵ of the corresponding vertex of $B_{\mathbf{p}, w}$, then $d_H(B_{\mathbf{p}_n, w_n}, B_{\mathbf{p}, w}) < \epsilon$," is certainly plausible, but it is a bit tricky. Fill in the details, at least when $\ell = 2$, but preferably for all ℓ.

A subset of D of $E \subset \mathbb{R}^\ell$ is dense in E if for all $\mathbf{x} \in E$ and all $\epsilon > 0$, there exists $\mathbf{y} \in D$ such that $d(\mathbf{x}, \mathbf{y}) < \epsilon$. There is a useful dense subset of $\mathcal{K}(E)$. For any $D \subset \mathbb{R}^\ell$, $\mathcal{P}_F(D)$ denotes the class of finite subsets of D. The following is (yet) another way of saying that compact sets are "approximately finite."

Lemma 4.10.15 *If D is a dense subset of E, then $\mathcal{P}_F(D)$ is a d_H-dense subset of $\mathcal{K}(E)$; that is, for all $K \in \mathcal{K}(E)$, there is a finite $F \subset D$ such that $d_H(K, F) < \epsilon$.*

Proof. Pick an arbitrary compact $K \subset E$ and $\epsilon > 0$. Let F' be a finite $\epsilon/2$ net for K, and for each $\mathbf{x}' \in F'$, choose $\mathbf{y}' = \mathbf{y}(\mathbf{x}') \in D$ such that $d(\mathbf{x}', \mathbf{y}') < \epsilon/2$. The finite set $F := \{\mathbf{y}(\mathbf{x}') : \mathbf{x}' \in F'\}$ satisfies $d_H(E, F) < \epsilon$. ∎

$(\mathcal{K}(E), d_H)$ inherits some of its properties from E.

Theorem 4.10.16 $E \subset \mathbb{R}^\ell$ *is complete iff* $(\mathcal{K}(E), d_H)$ *is complete.* $E \subset \mathbb{R}^\ell$ *is totally bounded iff* $(\mathcal{K}(E), d_H)$ *is totally bounded. Hence,* $E \subset \mathbb{R}^\ell$ *is compact iff* $(\mathcal{K}(E), d_H)$ *is compact.*

Proof. The compactness statement follows directly from the observation that a set is compact iff it is complete and totally bounded.

Total boundedness: If F_ϵ is a finite ϵ-net for E, then $\mathcal{P}(F_\epsilon) \setminus \{\emptyset\}$ is a finite ϵ-net for $\mathcal{K}(E)$. If $K_\epsilon = \{K_1, \ldots, K_M\}$ is a finite $\epsilon/2$-net for $\mathcal{K}(E)$ and F_m is a finite $\epsilon/2$-net for the compact set K_m, $m = 1, \ldots, M$, then $\cup_m F_m$ is a finite ϵ-net for E.

Completeness: Suppose that $(\mathcal{K}(E), d_H)$ is complete. It is easy to show that $\{\{x\} : x \in E\}$ is a closed subset of the complete metric space $\mathcal{K}(E)$, hence is itself complete. Since $d_H(\{x\}, \{y\}) = d(x, y)$, this means that E is complete.

Now suppose that (E, d) is complete, that is, that E is a closed subset of $\mathbb{R}^\ell$. Let K_n be a Cauchy sequence in $\mathcal{K}(E)$ and define $K = \{\mathbf{x} \in E : d(\mathbf{x}, K_n) \to 0\}$. We have to show that K is (1) nonempty, (2) closed, and (3) bounded.

(1) Nonemptiness: Recall that if $r_n := d(\mathbf{x}_n, \mathbf{x}_{n+1})$ is summable, then $\mathbf{x}_n$ is Cauchy. For each $k \in \mathbb{N}$, pick a strictly increasing subsequence n_k such that for all $n, m \geq n_k$, $d_H(K_n, K_m) < 1/2^k$. By the definition of the Hausdorff metric, for any $\mathbf{x}_k \in K_{n_k}$ and all $n \geq n_k$, $B(\mathbf{x}_k, 1/2^k) \cap K_n \neq \emptyset$. Choose a $\mathbf{x}_1 \in K_{n_1}$. Given $\mathbf{x}_{k-1}$, pick $\mathbf{x}_k \in [B(\mathbf{x}_{k-1}, 1/2^{k-1}) \cap K_{n_k}]$. Since $d(\mathbf{x}_k, \mathbf{x}_{k+1}) < 1/2^k$, $\mathbf{x}_k$ is a Cauchy sequence in E, and hence converges to a point $\mathbf{x} \in E$. For all $n \geq n_k$, $d(\mathbf{x}, K_n) < 1/2^k$, so $\mathbf{x} \in K$, showing that K is nonempty.

(2) Closedness: Suppose that $\mathbf{x}_m$ is a sequence in K and $\mathbf{x}_m \to \mathbf{x}$. We have to show that $\mathbf{x} \in K$. Pick $\epsilon > 0$. There exists an M such that for all $m \geq M$, $d(\mathbf{x}_m, \mathbf{x}) < \epsilon/2$. Since $\mathbf{x}_m \in K$, there exists an N such that for all $n \geq N$, $d(\mathbf{x}, K_n) < \epsilon/2$. By the triangle inequality, $d(\mathbf{x}, K_n) < \epsilon$ for all $n \geq N$, so that $\mathbf{x} \in K$.

(3) Boundedness: Pick N such that for all $n, m \geq N$, $d_H(K_n, K_m) < 1/2$. Then for all $n \geq N$, $K_n \subset K_N^{1/2}$, a bounded set, the closure of which must contain K. ■

4.10.c   The Theorem of the Maximum

The following formulation of the theorem of the maximum is a bit different than what one usually finds. Theorem 4.10.22 gives one version of the usual statement.

Theorem 4.10.17 (Theorem of the Maximum I)   *For $E \subset \mathbb{R}^\ell$, suppose that $u : E \times \mathcal{K}(E) \to \mathbb{R}$ is continuous and define the value function $K \mapsto v(K) := \max_{\mathbf{x} \in K} u(\mathbf{x}, K)$ and the arg max correspondence $K \mapsto \mathbf{x}^*(K) := \{\mathbf{x} \in K : \forall \mathbf{y} \in K, \; u(\mathbf{x}, K) \geq u(\mathbf{y}, K)\}$. Then*

> *1. $v(\cdot)$ is continuous, and*
>
> *2. $(\forall K \in \mathcal{K}(E))(\forall \epsilon > 0)(\exists \delta > 0)\big[[d_H(K, K') < \delta] \Rightarrow [\mathbf{x}^*(K') \subset (\mathbf{x}^*(K))^\epsilon]\big].$*

Further, if $\mathbf{x}^*(K)$ is always a singleton set, then the function $K \mapsto \mathbf{x}^*(K)$ is continuous.

The sequence version of the second statement looks a bit more like the "explosions" seen in our earlier examples. Specifically, it is true iff for all K_n and K with $d_H(K_n, K) \to 0$, for all $\epsilon > 0$, there exists an N such that for all $n \geq N$, $\mathbf{x}^*(K_n) \subset (\mathbf{x}^*(K))^\epsilon$. This allows $\mathbf{x}^*(K)$ to be much larger than $\mathbf{x}^*(K_n)$ for all large n, but does not allow $\mathbf{x}^*(K)$ to be much smaller than $\mathbf{x}^*(K_n)$ for large n. The proof requires a preliminary lemma related to this sequence formulation of the second condition.

Lemma 4.10.18   *If $d_H(K_n, K) \to 0$, then $F = K \cup \bigcup_{n \in \mathbb{N}} K_n$ is compact.*

Proof.   The set F is bounded; hence we need only show that it is closed, that is, that F^c is open. Pick $\mathbf{x} \notin F$ and consider the numbers $r_0 = d(\mathbf{x}, K) > 0$ and $r_n = d(\mathbf{x}, K_n) \to r_0$. If we set $\epsilon = r_0/2$, then for all but at most the first N of the n, $d(\mathbf{x}, K_n) > \epsilon$. Since N is finite, $\delta := \min\{r_n : n \leq N\} > 0$, and $B_\delta(\mathbf{x}) \cap F = \emptyset$. ■

Proof of Theorem of the Maximum I.   First, note that the continuity of $u(\cdot, K)$ implies that each problem $\max_{\mathbf{x} \in K} u(\mathbf{x}, K)$ has a nonempty closed set of solutions. Being a closed subset of the compact set K, the set of solutions, $\mathbf{x}^*(K)$, is itself compact. For both parts of the proof, let K_n, K be a sequence in $\mathcal{K}(E)$ with $d_H(K_n, K) \to 0$.

Continuity of $v(\cdot)$: We must show that $v(K_n) \to v(K)$. By the previous lemma, $F := K \cup \bigcup_{n \in \mathbb{N}} K_n$ is a compact subset of E. Since F is compact, so is $\mathcal{K}(F)$, and K_n is a convergent sequence in $\mathcal{K}(F)$. Since $F \times \mathcal{K}(F)$ is a compact subset of $E \times \mathcal{K}(E)$ and u is continuous on $E \times \mathcal{K}(E)$, it is uniformly continuous on $F \times \mathcal{K}(F)$.

Suppose first that for some $\epsilon > 0$, $\liminf_n v(K_n) = v(K) - \epsilon$. We show that this leads to a contradiction. Since u is uniformly continuous, we can pick $\delta > 0$ so that $d(\mathbf{x}, \mathbf{x}') < \delta$ and $d_H(K, K') < \delta$ imply that $|u(\mathbf{x}, K) - u(\mathbf{x}', K')| < \epsilon/2$. Pick N such that for $n \geq N$, $(\mathbf{x}^*(K))^\delta \cap K_n \neq \emptyset$ and $d_H(K, K_n) < \delta$. This implies that for all $n \geq N$ $v(K_n) > v(K) - \epsilon/2$, contradicting $\liminf_n v(K_n) = v(K) - \epsilon$.

Now suppose that for some $\epsilon > 0$, $\limsup_n v(K_n) = v(K) + \epsilon$. We show that this also leads to a contradiction. Pick a subsequence n_k such that $v(K_{n_k}) \to v(K) + \epsilon$. For each k, choose $\mathbf{x}_k \in \mathbf{x}^*(K_{n_k})$. Taking a subsequence if necessary, we get $\mathbf{x}_k \to \mathbf{x}$ for some $\mathbf{x} \in K$. By continuity, $u(\mathbf{x}_k, K_{n_k}) \to u(\mathbf{x}, K)$. Thus, $u(\mathbf{x}, K) = v(K) + \epsilon > v(K)$, contradicting the definition of $v(K)$ as the maximum.

Explosiveness: Pick $\epsilon > 0$. We must show that there exists an N such that for all $n \geq N$, $\mathbf{x}^*(K_n) \subset (\mathbf{x}^*(K))^\epsilon$. If not, then, taking a subsequence if necessary, we get $\mathbf{x}^*(K_n) \cap [E \setminus K^\epsilon] \neq \emptyset$. Let $\mathbf{x}_n$ be a sequence with $\mathbf{x}_n \in \mathbf{x}^*(K_n)$ for each n. $\mathbf{x}_n$ is a sequence in the compact set F; hence, taking a further subsequence if necessary, we have that it converges to some $\mathbf{x} \in F$. Since $[E \setminus K^\epsilon]$ is closed, it contains $\mathbf{x}$. However, since $\mathbf{x}_n$ is a sequence with each $\mathbf{x}_n$ belonging to K_n, any limit must belong to K.

Finally, if $\mathbf{x}^*(K)$ is always a singleton set, then $\mathbf{x}^*(K') \subset (\mathbf{x}(K))^\epsilon$ iff $d(\mathbf{x}^*(K'), \mathbf{x}^*(K)) < \epsilon$. ∎

To apply this to the continuity of the indirect utility function and the explosive behavior of the demand sets, we need the following.

Exercise 4.10.19 We know that if $d((\mathbf{p}_n, w_n), (\mathbf{p}, w)) \to 0$ for $(\mathbf{p}_n, w_n)$, $(\mathbf{p}, w) \in \mathbb{R}^{\ell+1}_{++}$, then $d_H(B_{\mathbf{p}_n, w_n}, B_{\mathbf{p}, w}) \to 0$. Show that the result may not hold if $(\mathbf{p}_n, w_n)$ converges to a vector with one or more 0's.

Since the mapping $(\mathbf{p}, w) \mapsto B_{\mathbf{p}, w}$ from $\mathbb{R}^{\ell+1}_{++}$ to $\mathcal{K}(\mathbb{R}^\ell_+)$ is continuous and the composition of continuous functions is continuous, with a small abuse of notation we have $(\mathbf{p}, w) \mapsto v(\mathbf{p}, w) = v(B_{\mathbf{p}, w})$ is continuous.

4.10.d Upper Hemicontinuity

The kind of explosiveness we have seen has a name.

Definition 4.10.20 *For a metric space (M, d), a correspondence $\Phi : M \to \mathcal{K}(E)$ is **upper hemicontinuous (uhc)** at x if for all $\epsilon > 0$, there exists $\delta > 0$ such that $[d(x, x') < \delta] \Rightarrow [\Phi(x') \subset (\Phi(x))^\epsilon]$. It is **upper hemicontinuous (uhc)** if it is uhc at all $x \in M$. [See Figure 4.10.20.]*

Just as there are equivalent sequence and ϵ-δ formulations of the continuity of functions, there are equivalent formulations of upper hemicontinuity.

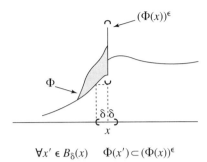

$$\forall x' \in B_\delta(x) \quad \Phi(x') \subset (\Phi(x))^\epsilon$$

FIGURE 4.10.20

Exercise 4.10.21 Show that Ψ is uhc iff for all $x \in M$ and all $x_n \to x$, $(\forall \epsilon > 0)(\exists N)(\forall n \geq N)[\Psi(x_n) \subset (\Psi(x))^\epsilon]$.

The following is an immediate corollary of theorem of the maximum I.

Theorem 4.10.22 (Theorem of the Maximum II) *For $\Theta \subset \mathbb{R}^k$, $E \subset \mathbb{R}^\ell$, $\Psi : \Theta \to \mathcal{K}(E)$ a correspondence, and $u : E \times \Theta \to \mathbb{R}$, we define the value function $\theta \mapsto v(\theta) := \max_{\mathbf{x} \in \Psi(\theta)} u(\mathbf{x}, \theta)$ and the arg max correspondence $\theta \mapsto \mathbf{x}^*(\theta) := \{\mathbf{x} \in \Psi(\theta) : \forall \mathbf{y} \in \Psi(\theta), \; u(\mathbf{x}, \theta) \geq u(\mathbf{y}, \theta))\}$. If u is (jointly) continuous and $\Psi(\cdot)$ is continuous, then*

1. *$v(\cdot)$ is continuous, and*
2. *$\mathbf{x}^*(\cdot)$ is upper hemicontinuous.*

Further, if $\mathbf{x}^*(\cdot)$ is always singleton valued, then $\theta \mapsto \mathbf{x}^*(\theta)$ is a continuous function.

4.10.e Some Complements

If there is an upper hemicontinuity, one might wonder what lower hemicontinuity might mean. The difference arises from the placement of x and x'. This allows "implosions."

Definition 4.10.23 *For a metric space (M, d), a correspondence $\Phi : M \to \mathcal{K}(E)$ is **lower hemicontinuous (lhc)** at x if for all $\epsilon > 0$, there exists $\delta > 0$ such that $[d(x, x') < \delta] \Rightarrow [\Phi(x) \subset (\Phi(x'))^\epsilon]$. It is **lower hemicontinuous (lhc)** if it is lhc at all $x \in M$. [See Figure 4.10.23.]*

Example 4.10.24 *The "implosive" correspondence $\Psi(x) = [0, x]$ if $x \in [0, 1)$ and $\Psi(1) = \{0\}$ from $[0, 1]$ to $\mathcal{K}([0, 1])$ is lhc, and is uhc at all x except 1.*

Exercise 4.10.25 Show that $\Psi : M \to \mathcal{K}(E)$ is continuous iff it is both uhc and lhc.

Exercise 4.10.26 Let x_n be a sequence in a compact $K \subset \mathbb{R}^\ell$ and set $T(m) = \text{cl}(\{x_n : m \geq n\})$. Give $\mathbb{N} \cup \{\infty\}$ the metric $d(n, m) = |\frac{1}{n} - \frac{1}{m}|$ with $\frac{1}{\infty} := 0$.

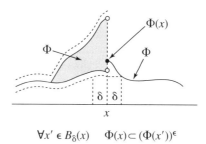

$$\forall x' \in B_\delta(x) \quad \Phi(x) \subset (\Phi(x'))^\epsilon$$

FIGURE 4.10.23

1. If $x_n \to x$ and we define $T(\infty) := \{x\}$, then the correspondence $T : \mathbb{N} \cup \{\infty\} \to K$ is continuous.

2. If we define $T(\infty)$ to be a closed subset of the accumulation points of the sequence x_n, then the correspondence $T : \mathbb{N} \cup \{\infty\} \to K$ is lhc.

3. If we define $T(\infty)$ to be a closed superset of the accumulation points of the sequence x_n, then the correspondence $T : \mathbb{N} \cup \{\infty\} \to K$ is uhc.

Exercise 4.10.27 Show that if $\Psi : M \to \mathcal{K}(E)$ is continuous, for all $x \in M$, and $\emptyset \neq \Phi(x) \subset \Psi(x)$ and the graph of Φ is a closed subset of $M \times E$, then Ψ is uhc. [The same is true if Ψ is merely uhc; see Lemma 6.1.32.]

The continuous image of a compact set is compact. There is a similar result here.

Exercise 4.10.28 If $\Psi : M \to \mathcal{K}(E)$ is uhc and $K \subset M$ is compact, show that $\Psi(K) := \cup_{x \in K} \Psi(x)$ is compact.

Correspondences are upper and lower *hemi*continuous, functions are upper and lower *semi*continuous. It turns out that the ideas are related.[9]

Definition 4.10.29 $f : M \to \mathbb{R}$ is **upper semicontinuous (usc)** *if for all* $x \in M$, $f(x) \geq \lim_{\epsilon \downarrow 0} \sup_{y \in B(x,\epsilon)} f(y)$.

Upper semicontinuity allows a function to jump, but only if it jumps upward.

Exercise 4.10.30 Let $f : M \to \mathbb{R}_+$ be usc. Show that the correspondence $F(x) := [0, f(x)]$ is uhc.

Exercise 4.10.31 Formulate lower semicontinuity and repeat the previous problem with "lower" replacing "upper."

Exercise 4.10.32 Show that any usc function on a compact set K achieves its maximum, but that $f(K)$ need not be compact.

Earlier, in Example 4.10.10, we saw that $d_H(\cdot, \cdot)$ is not a metric if applied to sets that are not closed, or are closed but not bounded. There is nothing to be done about the failure of closure, but boundedness can be fixed in a couple of ways. The first

---

9. There is a related notion of a demicontinuous correspondence. The mind boggles.

is to look at the distance between closed sets $F_1 \cap [-n, +n]^\ell$ and $F_2 \cap [-n, +n]^\ell$ for larger and large n, giving more weight to what happens for smaller n.

Exercise 4.10.33   For closed F_1, $F_2 \subset \mathbb{R}^\ell$, define

$$D_H(F_1, F_2) = \sum_{n \in \mathbb{N}} \frac{1}{2^n} d_H(F_1 \cap [-n, +n]^\ell, F_2 \cap [-n, +n]^\ell).$$

Let $\mathcal{C}(\mathbb{R}^\ell)$ denote the set of closed subsets of $\mathbb{R}^\ell$. Show the following:

1. D_H is a metric on $\mathcal{C}(\mathbb{R}^\ell)$.
2. $\mathcal{K}(\mathbb{R}^\ell)$ is dense in $\mathcal{C}(\mathbb{R}^\ell)$.
3. $(\mathcal{K}(\mathbb{R}^\ell), d_H)$ is topologically equivalent to $(\mathcal{K}(\mathbb{R}^\ell), D_H)$, that is, for compact K_n, K, $d_H(K_n, K) \to 0$ iff $D_H(K_n, K) \to 0$.
4. $(\mathcal{K}(\mathbb{R}^\ell), D_H)$ is not a complete metric space. [Consider the sequence of sets $K_n = \{n \cdot \mathbf{x}\}$ for some nonzero $\mathbf{x} \in \mathbb{R}^\ell$.]

Another way to extend $d_H(\cdot, \cdot)$ to unbounded closed subsets of $\mathbb{R}^\ell = (-\infty, +\infty)^\ell$ is to regard $(-\infty, +\infty)^\ell$ as a subset of the compact space $[-\infty, +\infty]^\ell$ and to use the associated metric on $[-\infty, +\infty]^\ell$ to (re-)define the Hausdorff metric. This general idea of adding ideal or fictional points, for example, $\pm\infty$, to a space so as to end up with a compact space is a very useful trick, which goes by the name compactification (see §6.6.f and §11.4.c for details).

Definition 4.10.34   *If (M, d) is a metric space; (M', d') is compact metric space; $\varphi : M \to M'$ is a homeomorphism.*

Example 4.10.35   *For $r \in \mathbb{R}$, define $\Phi(r) = \frac{1}{1+e^{-x}}$ and $G(r) = 2\Phi(r) - 1$. $\Phi(\cdot)$ is the **logistic** cumulative distribution function, and it is a homeomorphism between $(-\infty, +\infty)$ and $(0, 1)$. This means that G is a homeomorphism between $(-\infty, +\infty)$ and $(-1, +1)$. This means that $(G, ([-1, +1], d))$ is a compactification of $((-\infty, +\infty), d)$. If K is a compact subset of $\mathbb{R}$, then $G(K)$ is a compact subset of $(-1, +1)$, hence a compact subset of $[-1, +1]$. More generally, if F is a closed subset of $(-\infty, +\infty)$, then $K_F := \mathrm{cl}(G(F))$ is a compact subset of $[-1, +1]$, containing $+1$ if F is unbounded above and containing -1 if it is unbounded below. For closed subsets, F_1, F_2 of $\mathbb{R}$, we define the new Hausdorff metric by*

$$\rho_H(F_1, F_2) = \inf\{\epsilon > 0 : [K_{F_1} \subset K_{F_2}^\epsilon] \wedge [K_{F_2} \subset K_{F_1}^\epsilon]\}. \tag{4.8}$$

This procedure generalizes from $\mathbb{R}$ to $\mathbb{R}^\ell$: one embeds $(-\infty, +\infty)^\ell$ in $[-1, +1]^\ell$ one coordinate at a time.

Exercise 4.10.36   Show that ρ_H defines a metric on $\mathcal{C}(\mathbb{R})$, the closed subsets of $\mathbb{R}$. Further, show that d_H and ρ_H are topologically equivalent on $\mathcal{K}(\mathbb{R})$ and that $\mathcal{K}(\mathbb{R})$ is ρ_H-dense in $\mathcal{C}(\mathbb{R})$.

4.11 ◆ Banach's Contraction Mapping Theorem

Recall that a contraction mapping is a function $f : M \to M$ with the property that $\forall x, y \in M$, and $d(f(x), f(y)) \le \beta d(x, y)$ for some contraction factor $\beta < 1$. We saw in Lemma 4.4.8 and Theorem 4.4.10 (p. 122) that repeated applications of a contraction mapping yield a Cauchy sequence, and that if M is complete, then there is a unique fixed point, that is, a unique $x^* \in M$ such that $f(x^*) = x^*$. Fixed points of functions play a crucial role throughout economic theory.

As we saw, in dynamic systems, a fixed point is a rest point: if $x_t = x^*$, then for all k, $x_{t+k} = f(x_{t+k-1}) = x^*$. This means that if the system starts at or ends up at x^*, it will stay there forever. Some rest points have a stronger property called local stability or ϵ-stability, which means that if it starts in $B_\epsilon(x^*)$ the sytem will converge to x^*. Contraction mappings have the yet stronger property called global stability, which is ϵ-stability for all ϵ.

Example 4.11.1 (Linear Dynamical Systems in $\mathbb{R}^\ell$) [↑*Example 4.4.11*] *For an $\ell \times \ell$ matrix M and $\mathbf{x} \in \mathbb{R}^\ell$, $M\mathbf{x}$ is defined as the vector in $\mathbb{R}^\ell$ having the ith component equal to $M_{(i,\cdot)}\mathbf{x}$, where $M_{(i,\cdot)}$ is the ith row of the matrix M. A linear dynamical system starts at $\mathbf{x}_0$ and has $\mathbf{x}_{t+1} = \mathbf{a} + M\mathbf{x}_t$. If M is a contraction mapping for any of the metrics on $\mathbb{R}^\ell$ making it complete, then the unique rest point of the system is the solution to $\mathbf{x}^* = \mathbf{a} + M\mathbf{x}^*$; that is, $\mathbf{x}^* = (I - M)^{-1}\mathbf{a}$, where I is the $\ell \times \ell$ identity matrix. If $\mathbf{a} = 0$, then $\mathbf{x}^* = 0$.*

4.11.a Some Extensions

A function f might not be contractive even though f^n is for some $n > 1$. We will see examples of this property in our study of finite-state Markov chains (just below). The following is a mapping for which f^2 is contractive even though f is not.

Example 4.11.2 *Let $M = \mathbb{R}_+$, $d(x, y) = |x - y|$, and let $f(x) = 0$ if $x \le 1$, $f(x) = 2(x - 1)$ if $1 < x \le 1.5$, and $f(x) = 1$ for $1.5 < x$. For $x, y \in (1, 1.5)$, $d(f(x), f(y)) = 2d(x, y)$, but for all x, y, $d(f^2(x), f^2(y)) = 0$.*

The conclusions of Banach's contraction mapping theorem still hold.

Theorem 4.11.3 *If $f^n : M \to M$ is a contraction mapping for some n and (M, d) is complete, then f has a unique fixed point, x^*. Further, for any $x_0 \in M$, the sequence x_t defined by $x_{t+1} = f(x_t)$ converges to x^*.*

Proof. Since f^n is contractive, it has a unique fixed point x^*. We show that $f(x^*)$ is a fixed point of f. Note that $f(x^*) = f(f^n(x^*)) = f^{n+1}(x^*) = f^n(f(x^*))$. Since the fixed point of f^n is unique, this means that $f(x^*) = x^*$, so that x^* is a fixed point of f. If f has another fixed point, x', then $f^n(x') = x'$, a contradiction, so that f has a unique fixed point.

We now show that for any $x_0 \in M$, the sequence x_t defined by $x_{t+1} = f(x_t)$ converges to x^*. Any t is of the form $kn + i$ for some k and $i \in \{0, 1, \ldots, n-1\}$. Therefore, for any x_0, $x_t = f^t(x_0) = f^{kn}(f^i(x_0))$, so that $\lim_k f^{kn}(f^i(x_0)) = x^*$ since f^n is contractive. For each i, pick K_i such that for all $k \ge K_i$, $d(f^{kn}(f^i(x_0))$,

$x^*) < \epsilon$ and set $K = \max_i K_i$. For all $k \geq K$ and all i, $d(f^{kn}(f^i(x_0)), x^*) < \epsilon$. Therefore, for all large t, $d(f^t(x_0), x^*) < \epsilon$. ∎

Definition 4.11.4 *A mapping* $f : M \to M$ *is* **nonexpansive** *if for all* $x \neq x'$, $d(f(x), f(x')) < d(x, x')$.

As noted earlier in Example 4.5.14, on $\mathbb{R}_+$, the function $f(x) = x + \frac{1}{x+1}$ is nonexpansive, not on a compact domain, and has no fixed point. However, when M is compact, the conclusions of Banach's contraction mapping theorem still hold for nonexpansive mappings.

Theorem 4.11.5 *If* M *is compact and* $f : M \to M$ *is nonexpansive for all* $x \neq x'$, *then* f *has a unique fixed point,* x^*. *Further, for any* $x_0 \in M$, *the sequence* x_t *defined by* $x_{t+1} = f(x_t)$ *converges to* x^*.

Proof. If $x^* \neq x^\dagger$ are fixed points, then $d(f(x^*), f(x^\dagger)) = d(x^*, x^\dagger) < d(x^*, x^\dagger)$; thus if a fixed point exists, it is unique. Pick an arbitrary $x_0 \in M$ and define $x_{t+1} = f(x_t)$. Since M is compact, this has a convergent subsequence. As f is continuous, the limit is a fixed point. ∎

After we have shown that $C_b(M)$ is a complete metric space, we will use the contraction mapping theorem to prove the implicit function theorem, the basic result behind differentiable comparative statics, and to study deterministic and stochastic dynamic programming. At present, we apply contraction mapping results to finite-state Markov chains and to finite-state Markovian dynamic programming.

4.11.b Finite–State Markov Chains

At this point, we use only the most basic aspects of probabilities, leaning heavily on properties of simple conditional probabilities, $P(A \mid B) = P(A \cap B)/P(B)$ when $P(B) > 0$. For our purposes, a **stochastic process in** S is a sequence $(X_t)_{t=0}^\infty$ of random variables taking values in S. Stationary Markov chains are a special kind of stochastic process. We show later that the partial description of the distribution of the $(X_t)_{t=0}^\infty$ that we offer here determines the entire distribution.

We fix a finite set $S = \{1, \ldots, s, \ldots, S\}$ and define the set of probabilities on S as $\Delta(S) = \{p \in \mathbb{R}_+^S : \sum_s p_s = 1\}$. $\Delta(S)$ is closed and bounded, hence compact, hence complete. The distribution that puts mass 1 on a point s is called the **point mass on** s, written δ_s. In terms of vectors, δ_s is the unit vector in the sth direction.

A **probability transition matrix for** S is an $S \times S$ matrix $P = (P_{ij})_{i,j \in S}$, where $P_{ij} \geq 0$, and for all i, $\sum_j P_{ij} = 1$. Put another way, P is a matrix with the property that each row is an element of $\Delta(S)$.

Let $\mu_0 \in \Delta(S)$ and let P be a probability transition matrix. The pair (μ_0, P) determines a **stationary Markov chain**, which is a stochastic process with the properties that μ_0 is the distribution of X_0 and

$$(\forall t)[P(X_{t+1} = j | X_0 = i_0, \ldots, X_{t-1} = i_{t-1}, X_t = i)$$
$$= P(X_{t+1} = j | X_t = i) = P_{ij}. \tag{4.9}$$

Two aspects of (4.9) should be emphasized. First, the probability of moving from i to j does not depend on how the process arrived at i. Second, P_{ij} does not depend on t, and this is the reason for the adjective "stationary." Often μ_0 is taken to be δ_s, and one studies what happens for the different s.

The matrix P is called the **one-step transition matrix**, a name that comes from the following observation: if π is the (row) vector of probabilities describing the distribution of X_t, then $\pi'P$ is the (row) vector describing the distribution of X_{t+1}. This means that $\pi \mapsto \pi P$ is a mapping from the complete space $\Delta(S)$ to $\Delta(S)$. This motivates our interest in the question of whether the mapping $\pi \mapsto \pi P$ from $\Delta(S)$ to $\Delta(S)$ is a contraction, or is nonexpansive.

Example 4.11.6 *When the economy is in a good state, the distribution of returns is higher and in a bad state, it is lower. The transition matrix on $S = \{G, B\}$ is*

$$\begin{bmatrix} (1-\alpha) & \alpha \\ \beta & (1-\beta) \end{bmatrix},$$

where $P_{GG} = P(X_{t+1} = G | X_t = G) = 1 - \alpha$ and $P_{BG} = P(X_{t+1} = G | X_t = B) = \beta$. We now show that P is a contraction mapping if $\alpha, \beta \in (0, 1)$. Let $m_G = \min\{\alpha, 1 - \alpha\}$ and $m_B = \min\{\beta, 1 - \beta\}$, and set $m = m_G + m_B$. Note that $m_G, m_B \leq \frac{1}{2}$ and strict inequalities give us the interesting case. We now show that $p \mapsto pP$ is a contraction mapping with contraction factor $(1 - m)$. For $p, q \in \Delta(S)$,

$$\|pP - qP\|_1 = \sum_{j \in \{G,B\}} \left| \sum_{i \in \{G,B\}} (p_i - q_i)P_{i,j} \right|$$

$$= \sum_{j \in \{G,B\}} \left| \sum_{i \in \{G,B\}} (p_i - q_i)(P_{i,j} - m_j) + \sum_{i \in \{G,B\}} (p_i - q_i)m_j \right|$$

$$\leq \sum_{j \in \{G,B\}} \sum_{i \in \{G,B\}} |p_i - q_i|(P_{i,j} - m_j) + \sum_{j \in \{G,B\}} m_j \left| \sum_{i \in \{G,B\}} (p_i - q_i) \right|$$

$$= \sum_{i \in \{G,B\}} |p_i - q_i| \sum_{j \in \{G,B\}} (P_{i,j} - m_j) + \sum_{j \in \{G,B\}} m_j |0|$$

$$= \|p - q\|_1 \cdot (1 - m).$$

The unique fixed point in $\Delta(S)$ solves $\pi^*P = \pi^*$, and, solving two equations in two unknowns, this yields $\pi^* = (\frac{\beta}{\alpha+\beta}, \frac{\alpha}{\alpha+\beta})$. The contraction mapping theorem implies that defining $\mu_{t+1} = \mu_t P$, $\lim_t \mu_t = \pi^*$. Intuitively, in the long run, the influence of the initial state, μ_0, disappears, and the probability of being in the G state converges to π_G^*. The qualitative properties of π^* are fairly intuitive: since $\partial \pi_G^*/\partial \beta > 0$, as the moves from the B state to the G state become more likely, the process spends more time in the G state.

If $m = 1$, that is, $\alpha = \beta = 1/2$, then for all p, $pP = (1/2, 1/2)$. Therefore, for all p, q, $pP = qP$. This is finite convergence. If $\alpha = \beta = 0.99$, then $m = 0.98$, and we know that $d_1(\mu_t, \pi^*) \leq C(0.98)^t$ for some constant C.

Definition 4.11.7 $\pi \in \Delta(S)$ *is an **ergodic distribution for** P if $\pi P = \pi$.*

Exercise 4.11.8 Solve for the set of ergodic distributions and the contraction factor for each of the following P, where $\alpha, \beta \in (0, 1)$:

$$\begin{bmatrix} 1 & 0 \\ 0 & 1 \end{bmatrix} \begin{bmatrix} 0 & 1 \\ 1 & 0 \end{bmatrix} \begin{bmatrix} \alpha & (1-\alpha) \\ (1-\alpha) & \alpha \end{bmatrix} \begin{bmatrix} \alpha & (1-\alpha) \\ (1-\beta) & \beta \end{bmatrix} \begin{bmatrix} 1 & 0 \\ (1-\beta) & \beta \end{bmatrix}.$$

We write $M \gg 0$ for a matrix M if each entry is strictly positive.

Theorem 4.11.9 *If $P^n \gg 0$ for some n, then P has a unique ergodic distribution.*

Proof. From Theorem 4.11.3, it is sufficient to show that if $P \gg 0$, then $p \mapsto pP$ is contractive on $\Delta(S)$.

Suppose that $P \gg 0$. For each $j \in S$, let $m_j = \min_i P_{i,j}$. Since $P \gg 0$, we know that for all j, $m_j > 0$. Define $m = \sum_j m_j$. We show that for $p, q \in \Delta(S)$, $\|pP - qP\|_1 \le (1-m)\|p - q\|_1$:

$$\|pP - qP\|_1 = \sum_{j \in S} \left| \sum_{i \in S} (p_i - q_i) P_{i,j} \right|$$

$$= \sum_{j \in S} \left| \sum_{i \in S} (p_i - q_i)(P_{i,j} - m_j) + \sum_{i \in S} (p_i - q_i) m_j \right|$$

$$\le \sum_{j \in S} \sum_{i \in S} |p_i - q_i|(P_{i,j} - m_j) + \sum_{j \in S} m_j \left| \sum_{i \in S} (p_i - q_i) \right|$$

$$= \sum_{i \in S} |p_i - q_i| \sum_{j \in S} (P_{i,j} - m_j) + 0$$

$$= \|p - q\|_1 \cdot (1 - m),$$

where the next-to-last equality follows from the observation that $p, q \in \Delta(S)$, and the last equality follows from the observation that for all $i \in S$, $\sum_{j \in S} P_{i,j} = 1$ and $\sum_{j \in S} m_j = m$. ∎

The condition that $P^n \gg 0$ for some n is sufficient for a unique ergodic distribution, but it is not necessary. Consider the contractive matrix

$$P = \begin{bmatrix} 1 & 0 \\ (1-\beta) & \beta \end{bmatrix}.$$

Note that $(1, 0)P^n = (1, 0)$, $(0, 1)P^n = ((1-\beta^n), \beta^n)$, so that for all $a \in [0, 1]$, $(a, 1-a)P^n = a(1, 0) + (1-a)((1-\beta^n), \beta^n) \to (1, 0)$. Another instructive example is the noncontractive matrix

$$P = \begin{bmatrix} 0 & 1 \\ 1 & 0 \end{bmatrix},$$

which has $(\frac{1}{2}, \frac{1}{2})$ as its unique ergodic distribution even though $P^n \not\gg 0$ for all n.

Exercise 4.11.10 If $P^n \gg 0$ for some n, show that the matrix P^t converges and characterize the limit. [Note that $\delta_s P^t \to \pi^*$, where π^* is the unique ergodic distribution.]

4.11.c Finite-State Markovian Dynamic Programming

Dynamic programming is the study of interlinked optimization problems that occur over time. Most often, any uncertainty involved has a Markovian structure. The crucial tool for solving these dynamic problems is the notion of a value function.

The value function maps times and states to payoffs. At a given node, that is, at a given point in time and state, the **value** is the best that can be done thereafter. If you know the value of every possible consequence of your present actions, you face a simple one-period problem. Solving these one-period problems one after the other eventually solves the whole problem.

We show that the value function exists and how it works in finite problems. We use Banach's contraction mapping theorem to show that, given enough special structures on the problem, a value function exists in the infinite-horizon Markovian case.

4.11.c.1 The Stagecoach Problem

To see how the value function works, we begin with a finite-horizon problem with no uncertainty. A traveler, call him Mark Twain, has to travel by stagecoach from node O, call it St. Joseph, Missouri, to node z, call it Carson City, Nevada, in a fashion that yields the largest number of incidents that he can write about. This is formulated using a set of nodes, a set of directed paths, and a set of expected numbers of incidents.

The nodes are the stagecoach stops, unimaginatively named

$$\mathcal{N} = \{O, a, b, c, d, e, f, g, h, z\}.$$

The set of directed paths is a set of ordered pairs representing routes that one can travel between nodes using a stagecoach,

$$\mathcal{P} = \{(O, a), (O, b), (a, c), (a, d), (b, c), (b, d), (b, e), (c, f), (c, g),$$
$$(d, f), (d, h), (e, g), (e, h), (f, z), (g, z), (h, z)\}.$$

This means that from O, one can take a stagecoach to either a or b, from a, one can go to either c or d, from b one can go to c or d or e, and so forth. The set of expected rewards is given in the following table:

(s, t)	$u(s, t)$	(s, t)	$u(s, t)$	(s, t)	$u(s, t)$	(s, t)	$u(s, t)$
(O, a)	19	(O, b)	23	(a, c)	11	(a, d)	6
(b, c)	6	(b, d)	19	(b, e)	3	(c, f)	16
(c, g)	23	(d, f)	23	(d, h)	29	(e, g)	3
(e, h)	22	(f, z)	12	(g, z)	21	(h, z)	2

This means that taking the path $(O, b), (b, e), (e, g)$, and (g, z), that is, traveling from O to b to e to g to z, gives rewards $23 + 3 + 3 + 21 = 50$.

The dynamic programming problem is to find the feasible path(s) that maximizes the sum of the rewards accrued along the path.

In Twain's problem: there are ten possible paths (check that for yourself); calculating the reward associated with each path involves adding four numbers, and maximizing among the ten numbers requires making ten comparisons. This is not a particularly difficult problem, and we do not need the machinery we are developing here to solve it.

Example 4.11.11 *Suppose we were looking at a travel problem with one initial node,* 12 *stages, and one terminal node. Suppose further that at each stage there are* 12 *nodes, all of them possible choices from any node in the previous stage: the table giving the rewards will contain* $12 + 11 \cdot 144 + 12 = 1608$ *numbers; there are* $12^{12} = 8{,}916{,}100{,}448{,}256$, *or roughly* 9 *trillion, possible paths; finding the reward associated with each one requires adding* 13 *numbers, over* 100 *trillion calculations; and then there are the roughly* 9 *trillion comparisons to be done. By contrast, the machinery developed here requires* $(11 \cdot 144) + 12$ *additions and the same number of comparisons. While tedious, this can be organized onto a single sheet of an artist's sketch pad.*

Here are the steps in solving Twain's problem:

1. There are three nodes, f, g, and h, that have the single terminal node z as a successor. If we end up at f, the reward from there to z is 12. If we end up at g, the reward from there to z is 21. If we end up at h, the reward from there to z is 2. We record these numbers as the values of being at f, g, and h, that is, $V(f) = 12$, $V(g) = 21$, and $V(h) = 2$.

2. There are three nodes, c, d, and e, that have all of their direct successors among the nodes for which we have values.

   a. At c, one chooses between going to f, which gives rewards $16 + V(f) = 28$, and going to g, which gives rewards $23 + V(g) = 44$. The best response to being at c is g, which we record as $Br(c) = g$, and the value to being at c is 44, which we record as $V(c) = 44$.

   b. At d, one chooses between going to f, which gives rewards $23 + V(f) = 35$, and going to h, which gives rewards $29 + V(h) = 31$. We record this as $Br(d) = f$ and $V(d) = 35$.

   c. At e, solve $\max\{3 + V(g), 22 + V(h)\}$ and record this as $Br(e) = \{g, h\}$ and $V(e) = 24$.

3. There are two nodes, a and b, that have all of their direct successors among the nodes for which we have values. $V(a) = \max\{11 + V(c), 6 + V(d)\} = 55$ and $Br(a) = c$, while $V(b) = \max\{6 + V(c), 19 + V(d), 3 + V(e)\} = 54$ and $Br(b) = d$.

4. Finally, $V(O) = \max\{19 + V(a), 23 + V(b)\} = 77$ and $Br(O) = b$.

5. There is only one path that follows the Br choices, O, b, d, f, z. As a check, the payoffs to this path are $23 + 19 + 23 + 12 = 77$.

Exercise 4.11.12   Solve Twain's problem using the steps just described if the objective is to minimize the rewards; that is, view the numbers in the table as costs rather than rewards and minimize costs.

Exercise 4.11.13   Verify the $(11 \cdot 144) + 12 = 1596$ figure for additions and comparisons in Example 4.11.11.

4.11.c.2 Generalizing the Stagecoach Problem

We say that node s is a **parent**, respectively **child**, of node t if (s, t), respectively (t, s), is one of the ordered pairs in $\mathcal{P}$. The set of parents of a node t is denoted $P(t)$, and the set of children of t is denoted $C(t)$. A node can have any number of parents and any number of children. For example, the original node, O, has 0 parents, $P(O) = \emptyset$. A terminal node, z, is one with no children, $C(z) = \emptyset$, and the set of terminal nodes is written Z. Consistent with this terminology, we assume that no node is its own parent, grandparent, great-grandparent, and so on, and that no node is its own child, grandchild, great-grandchild, and so on. Let $S(t)$ denote the set of all successors (or descendants) of node t. We assume that each terminal node, z, is a successor of O, the original node.

For dynamic programming, payoffs u depend on the branch taken from O to the terminal nodes. For each pair $(s, t) \in \mathcal{P}$, there is a payoff $u(s, t)$, and the payoff to a branch $O \rightsquigarrow z = (O = t_0, t_1, \ldots, t_n = z)$ is $\sum_{i=0}^{n-1} u(t_i, t_{i+1})$. More generally, for nonterminal node s and terminal node z, define $u(s \rightsquigarrow z)$ to be the part of the payoffs accrued on the $s \rightarrow z$ branch. Then the value function is defined by

$$V(t) := \max_{z \in S(t) \cap Z}[u(t \rightsquigarrow z)].$$

This is the mathematical formulation of the idea that the value at a node is the best that can be starting from that node. The dynamic programming problem is to find $V(O)$, that is, to solve

$$V(O) := \max_{z \in Z}[u(O \rightsquigarrow z)].$$

The basic recursion relation that the value function satisfies, known as **Bellman's equation** or **Pontryagin's formula**, is

$$V(s) = \max_{t \in C(s)}[u(s, t) + V(t)].$$

When the stage structure is an important part of the problem, it becomes

$$V_n(s) = \max_{t \in C(s)}[u(s, t) + V_{n+1}(t)],$$

where s is in stage n and t is in stage $n + 1$.

To summarize: The basic device is to start at the terminal nodes z, figure out the value for all parents of terminal nodes that have nothing but terminal children, calling these nodes T_1, then determine the value for all parents of $T_1 \cup Z$ nodes that have nothing but $T_1 \cup Z$ children, and so on, until you have worked all the way back through the tree.

Repeating the description of this process with more notation, we set $T_0 = Z$, set $T_1 = \{s \in \mathcal{N} : C(s) \subset T_0\}$ and inductively set $T_{n+1} = \{s \in \mathcal{N} : C(s) \subset \cup_{0 \leq m \leq n} T_m\}$.

Exercise 4.11.14 Show that there exists an n^* such that $T_{n^*} = \{O\}$ and that for all $n < n^*$, $T_n \neq \emptyset$. By "opt" we mean either max or min, depending on the application. For $s \in T_1 \setminus T_0$, define $V(s) = \text{opt}_{z \in C(s)}\{u(s, z)\}$, for $s \in T_n \setminus T_{n-1}$, $n \geq 2$, define $V(s) = \text{opt}_{t \in C(s)}\{u(s, t) + V(t)\}$. Let $Br(s)$ be the set of solutions to $\text{opt}_{t \in T_{n-1}}\{u(s, t) + V(t)\}$. Show that any element of $Br(Br(\cdots (Br(O)) \cdots))$ is an optimal path.

4.11.c.3 Adding Stochastic Structure and an Infinite Horizon

The stagecoach problem and its generalizations were finite problems and there was no randomness. It is easy to add randomness if one supposes that the choice one makes at a node leads to a distribution over the places one might end up.[10] Formally, this has the effect of adding expectations to the Bellman-Pontryagin equation, and the solution optimizes expected rewards.

The growth problem given in Example 3.7.2 was an example of a stationary problem with an infinite set of nodes at each point in time and an infinite set of choices available at most nodes. Here we add randomness to an infinite horizon problem, but to make the complications tractable, we assume finiteness of the nodes at any point in time and finiteness of the set of choices available at any node, and we also assume that the problem is stationary. Stationarity means that the set of nodes and the set of choices and the set of rewards associated with these choices do not change over time. Utility becomes the summed, discounted stream of rewards.

One problem with the "start at the end and work backward" solution method described for stagecoach problems is that when the time horizon is infinite, there is no longer a set of terminal nodes. The essential idea of the method we use for these stationary problems is to pretend that some stage is the last possible and to keep working backward until our solution converges.

The maximizer faces a sequence of interlinked decision at times $t \in \{0, 1, \ldots\}$. At each t, he or she learns the value of the random state, $X_t = s$, with s in a finite-state space S. For each $s \in S$, the maximizing person has a finite set of available actions $A(s)$. The choice of $a \in A(s)$ when the state is s gives utility $u(a, s)$. When the choice is made at $t \in \{0, 1, \ldots\}$, it leads to a random state, X_{t+1}, at time $t + 1$, according to a transition probability $P_{i,j}(a) = P(X_{t+1} = j | X_t = i, a_t = a)$, at which point the whole process starts again. If the sequence $(a_t, s_t)_{t \in \{0,1,\ldots\}}$ is the outcome, the utility is $U = \sum_t \beta^t u(a_t, s_t)$ for some $0 < \beta < 1$.

The **stationary Markovian dynamic programming problem** is to find a sequence(s) $a_t^*(s_t) \in A(s_t)$ that maximizes the expected value of U conditional on the various possible values of X_0.

One of the methods for solving the infinite-horizon discounted dynamic programming problems just described is called the **method of successive approximation**: one pretends that the problem has only one decision period left and that

---

10. This rather stretches the interpretation of the stagecoach problem as a travel story.

if one ends up in state s after this last decision, one will receive $\beta V_0(s)$, often $V_0(s) \equiv 0$. The function V_0 can be represented as a vector on $\mathbb{R}^S$. Given V_0, define a new vector, V_1, in $\mathbb{R}^S$ by

$$V_1(s) = \max_{a \in A(s)} \left[u(a, s) + \beta \sum_{j \in S} V_0(j) P_{s,j}(a) \right].$$

Given the finiteness of $A(s)$, this maximization problem has a solution. More generally, once $V_t \in \mathbb{R}^S$ has been defined, define

$$V_{t+1}(s) = \max_{a \in A(s)} \left[u(a, s) + \beta \sum_{j \in S} V_t(j) P_{s,j}(a) \right].$$

Again, the maximization problem just specified has a solution.

We have just given a mapping from possible value functions to other possible value functions. Since it is an equation that gives a mapping from one function to another, we call it a **functional equation**. We define the mapping $v \mapsto T_v$, from $\mathbb{R}^S$ to itself, by defining the sth component of T_v. Specifically, define $T_v(s)$ by

$$T_v(s) = \max_{a \in A(s)} \left[u(a, s) + \beta \sum_{j \in S} v_j P_{s,j}(a) \right]. \tag{4.10}$$

Exercise 4.11.15 Demand, D_t, in period t is iid with $P(D_t = 0) = P(D_t = 1) = \frac{1}{4}$ and $P(D_t = 2) = P(D_t = 3) = P(D_t = 4) = \frac{1}{6}$. One keeps inventory on hand to meet potential demand. If you do not have the good on hand, you lose the sale and lose a reputation for reliability valued at r. If you do have the good on hand, your revenue is p per sale. Storing the good for one period costs s and ordering $a \geq 1$ of the good costs $C + c \cdot a$. At the end of a period, after there have been some sales, you check your inventory, S_t, and decide how many to order, a_t. S_{t+1} is the random variable $\max\{S_t + a_t - D_t, 0\}$. The objective function is $\sum_t \beta^t \Pi_t$, where Π_t is profits in period t. Write out several of the $P_{s,j}(a)$, the one- and the two-stage value functions.

Theorem 4.11.16 *The function $v \mapsto T_v$ is a contraction mapping.*

Proof. We begin with a useful result that we refer to later. ∎

Lemma 4.11.17 *If $f : A \to \mathbb{R}$ and $g : A \to \mathbb{R}$ are both bounded, then $\sup_{a \in A} f(a) - \sup_{a \in A} g(a) \leq \sup_{a \in A}[f(a) - g(a)]$. If f and g achieve their maxima, then this implies that $\max_{a \in A} f(a) - \max_{a \in A} g(a) \leq \max_{a \in A}[f(a) - g(a)]$.*

Proof. Suppose first that f achieves its maximum at a point $a_f \in A$. In this case, $\sup_{a \in A}[f(a) - g(a)] \geq [f(a_f) - g(a_f)] \geq \sup_{a \in A} f(a) - \sup_{a \in A} g(a)$.

Suppose now that f does not achieve its maximum. Pick $\epsilon > 0$. Let $a_\epsilon \in A$ satisfy $f(a_\epsilon) \geq \sup_{a \in A} f(a) - \epsilon$. In this case, $\sup_{a \in A}[f(a) - g(a)] \geq [f(a_\epsilon) - g(a_f)] \geq (\sup_{a \in A} f(a) + \epsilon) - \sup_{a \in A} g(a)$. ∎

Returning now to the proof of the Theorem 4.11.16, we choose arbitrary $v, w \in \mathbb{R}^S$ and arbitrary $s \in S$. We show that $\|T_v - T_w\|_\infty \leq \beta \|v - w\|_\infty$:

$$T_v(s) - T_w(s) = \max_{a \in A(s)} f(a) - \max_{a \in A(s)} g(a),$$

where $f(a) = u(a, s) + \beta \sum_{j \in S} v_j P_{s,j}(a)$ and $g(a) = u(a, s) + \beta \sum_{j \in S} w_j P_{s,j}(a)$. By the just-proved Lemma 4.11.17, $T_v(s) - T_w(s) \leq \max_{a \in A(s)}[f(a) - g(a)]$. Canceling the $u(a, s)$ terms and rearranging yields

$$\max_{a \in A(s)} [f(a) - g(a)] = \max_{a \in A(s)} \beta\Big[\sum_{j \in S}(v_j - w_j)P_{s,j}(a)\Big].$$

Since the $P_{s,j}$ are nonnegative and sum to 1, for all $a \in A(s)$,

$$\beta\Big[\sum_{j \in S}(v_j - w_j)P_{s,j}(a)\Big] \leq \beta \max_{j \in S}|v_j - w_j| = \beta\|v - w\|_\infty.$$

Combining, for all $s \in S$, we have $T_v(s) - T_w(s) \leq \beta\|v - w\|_\infty$. Reversing the roles of v and w yields $T_w(s) - T_v(s) \leq \beta\|w - v\|_\infty$. Combining these two gives us $\forall s \in S, |T_v(s) - T_w(s)| \leq \beta\|v - w\|_\infty$, or $\|T_v - T_w\|_\infty \leq \|v - w\|_\infty$. ∎

Let v^* denote the unique fixed point of T. We show that $s \mapsto v^*(s)$ is the **value function** for the optimization problem, and in the process, we give the Bellman-Pontryagin equation. Let $a^*(s)$ belong to the solution set to the problem

$$\max_{a \in A(s)} \Big[u(a, s) + \beta \sum_{j \in S} v_j^* P_{s,j}(a)\Big].$$

The function (or correspondence) $s \mapsto a^*(s)$ is called the **policy function**. We find the policy function by solving a relatively simple problem. We maximize the sum of two terms, the immediate payoff, $u(a, s)$, plus the expected discounted value of the next period's value, $\beta \sum_{j \in S} v_j^* P_{s,j}(a)$. The point of all of this is that by solving this relatively simple problem, we have solved the infinite-horizon problem.

Lemma 4.11.18   *Using the policy $a^*(\cdot)$ at all points in time gives the expected payoff $v^*(s)$ if started from state s at time 0.*

Proof.   If you follow a^* for one period and receive $v^*(j)$ for ending up in j, then for all s, you receive

$$v_1(s) = u(a^*(s), s) + \beta \sum_j v^*(j)P_{s,j}(a^*(s)).$$

Since $a^*(s)$ solves $\max_{a \in A(s)} u(a, s) + \beta \sum_j v^*(j)P_{s,j}(a)$, this means that for all s, $v_1(s) = v^*(s)$.

If you follow a^* for one period and then receive $v_1(j)$ for ending up in j, you find the payoff for following a^* for two periods and then receiving $v^*(j)$ for ending up in j. Thus,

$$v_2(s) = u(a^*(s), s) + \beta \sum_j v_1(j)P_{s,j}(a^*(s)).$$

Since $v_1 = v^*$, a^* is the policy function and T has the unique fixed point v^*, $v_2 = v^*$.

If we continue in this fashion, for all t, the payoff to following a^* for t periods and then receiving $v^*(j)$ for ending up in j is v^*. Since the payoffs are summable, this means that following a^* for all t gives v^*. ∎

Define $\widehat{v}(s)$ to be the supremum of the expected payoffs achievable starting at s, the supremum being taken over all possible feasible policies, $\alpha = (a_t(\cdot, \cdot))_{t \in \mathbb{N}}$,

$$\widehat{v}(s) = \sup_{(a_t(\cdot, \cdot))_{t \in \mathbb{N}}} E\left(\sum_t \beta^t u(a_t, X_t) | X_0 = s\right).$$

Exercise 4.11.19 Show that for all s, $v^*(s) = \widehat{v}(s)$. To answer this, first suppose that for every $j \in S$, there is a feasible policy, a_j, that achieves the supremum of possible payoffs starting at j. Then for all $s \in S$,

$$\widehat{v}(s) = \max_{a \in A(s)} u(a, s) + \beta \sum_j \widehat{v}(j) P_{s,j}(a)$$

because the optimal policy starting from s does something optimal in the first period on the assumption that you will use a_j if you end up in j. Since v^* is the unique fixed point of this mapping, we know that $\widehat{v} = v^*$.

Now suppose that the supremum is not achievable. For every $\epsilon > 0$ and each $j \in S$, let α_j^ϵ be a policy that gives expected payoffs $V(\alpha_j^\epsilon) \geq \widehat{v}(j) - \epsilon$ starting in j. Show that

$$\widehat{v}(s) \geq \max_{a \in A(s)} \left[u(a, s) + \beta \sum_j V(\alpha_j^\epsilon) P_{s,j}(a)\right],$$

which implies that

$$\widehat{v}(s) \geq \max_{a \in A(s)} \left[u(a, s) + \beta \sum_j \widehat{v}(j) P_{s,j}(a)\right] - \epsilon.$$

Going the other way, show that

$$\widehat{v}(s) \leq \max_{a \in A(s)} \left[u(a, s) + \beta \sum_j [V(\alpha_j^\epsilon) + 2\epsilon] P_{s,j}(a)\right],$$

which implies that

$$\widehat{v}(s) \leq \max_{a \in A(s)} \left[u(a, s) + \beta \sum_j \widehat{v}(j) P_{s,j}(a)\right] + 2\epsilon.$$

Combining yields $\widehat{v}(s) = \max_{a \in A(s)} \left[u(a, s) + \beta \sum_j \widehat{v}(j) P_{s,j}(a)\right]$.

To summarize: First, you can find the value function by applying a contraction mapping; second, once you have found the value function, you are one step away from finding an optimal policy by solving the Bellman-Pontryagin equation; and third, the optimal policy is stationary.

Exercise 4.11.20 This problem asks you to find the optimal inventory policy for a small operation, one that has space enough to store only three units, so that S, the state space, is $S = \{0, 1, 2, 3\}$. Thus, for each $s \in S$, the set of actions is the possible number to order, $A(s) = \{0, \dots, 3 - s\}$.

Time line: First, s_t is observed and action a_t taken, which leads to a total of $a_t + s_t$ on hand; second, random demand, D_t, happens, leading to a sale of $X_t =$

$\min\{D_t, a_t + s_t\}$, revenues of $300 \cdot X_t$, and to $s_{t+1} = \max\{(a_t + s_t) - D_{t+1}, 0\}$. The discount factor per period is $\beta = 0.9$.

Assume that D_t is a sequence of independent random variables satisfying $P(D_t = 1) = 0.8$ and $P(D_t = 0) = 0.2$.

The costs of ordering $a > 0$ are $40 + 100a$, and the costs of ordering 0 are 0. Per unit, per period, storage costs are 20, so that if $s_{t+1} = s$, then a cost of $20 \cdot s$ is incurred.

1. Fill in the rest of the $u(a, s)$ payoff matrix, a "*" entry reflects an action not being available.

$a_\downarrow \setminus s_\rightarrow$	0	1	2	3
0	0	236		
1	100			*
2	−24		*	*
3	−144	*	*	*

Note that the 100 in the $a = 1$, $s = 0$ box corresponds to $0.8 \cdot [300 - 140 - 0] + 0.2 \cdot [0 - 140 - 0] = 100$, where the first terms in the brackets are revenues, the second terms are ordering costs, and the last terms are storage costs.

2. Fill in the rest of the $P_{ij}(a) = P(S_{t+1} = j | S_t = j, a_t = a)$ tables:

$a = 0$

$i_\downarrow \setminus j_\rightarrow$	0	1	2	3
0	1	0	0	0
1	0.8	0.2	0	0
2				
3				

$a = 1$

$i_\downarrow \setminus j_\rightarrow$	0	1	2	3
0	0.8	0.2	0	0
1				
2				

$a = 2$

$i_\downarrow \setminus j_\rightarrow$	0	1	2	3
0	0	0.8	0.2	0
1	0	0	0.8	0.2

$a = 3$

$i_\downarrow \setminus j_\rightarrow$	0	1	2	3
0	0	0	0.8	0.2

3. For $v_0 = 0$, $v_0 \in \mathbb{R}^S$, show that $v_1 = (100, 236, 216, 196)$ using the recursion relation

$$v_1(s) = \max_{a \in A(s)} \left[u(a, s) + \beta \sum_j P_{sj}(a) v_0(j) \right].$$

4. Find $v_2 \in \mathbb{R}^S$ using the recursion relation

$$v_2(s) = \max_{a \in A(s)} \left[u(a, s) + \beta \sum_j P_{sj}(a) v_1(j) \right].$$

5. Show that $v = (v_0, v_1, v_2, v_3) = (1205.61, 1346.39, 1445.61, 1508.34)$ satisfies the Bellman-Pontryagin equation (to within a couple of digits) and give the optimal policy.

The following extended example shows other uses of these kinds of dynamic optimization logic and problems; in particular, it shows how an equilibrium price may be determined by a contraction mapping.

Example 4.11.21 *This dynamic general equilibrium model is drawn from Lucas (1978), "Asset prices in an exchange economy,"* Econometrica, *pp. 1429–45. There is a representative agent, that is, a "stand in" for a large number of identical consumers, with preferences given by*

$$E\left[\sum_{t=0}^{\infty} \beta^t u(c_t)\right],$$

where $u : \mathbb{R}_+ \to \mathbb{R}$ is assumed to be continuously differentiable and strictly increasing and $0 < \beta < 1$.

The agent is endowed with a project that yields uncertain exogenous random output $y_t \in S$, where S is a finite subset of $\mathbb{R}_{++}$. The output is perishable, so output in period t cannot be stored to period $t + 1$ for later consumption. The randomness is modeled as a finite-state Markov chain, where $Prob(y_{t+1} = y' | y_t = y) = P_{yy'}$.

Ownership of the project is determined each period in a competitive stock market. The project has one outstanding perfectly divisible equity share. Owning $z_t \in [0, 1]$ of the equity share entitles the owner to z_t of the project's output at the beginning of period t. Shares are traded after payment of dividends at price p_t. An agent's budget set in any period t is given by

$$c_t + p_t z_{t+1} = z_t(y_t + p_t), \quad \text{equivalently } c_t = z_t y_t + p_t(z_t - z_{t+1}). \ (4.11)$$

Let primed variables denote period $t + 1$ variables and unprimed variables denote period t values. An equilibrium for this economy is a pair of functions:

1. *$p(y)$, giving prices as a function of the random output, that is, the random state of the economy, and*

2. *an agent's value function, $v(y, z)$, which depends on y, the exogenous stochastic dividends, and z, her holdings of stocks.*

If she knows the price function and is optimizing, then $v(y, z)$ must be given by

$$v(y, z) = \max_{z'}\left[u(zy + p(y)(z - z')) + \beta \sum_{y'} P_{yy'} v(y', z')\right].$$

Remember, the agent's demand stands in for the demand of the large number of identical consumers. For the agent's demand in the stock market, $z'(y, z)$, to clear the stock market, $z'(y, z)$ must be equal to the supply of shares, that is, $z'(y, z) = 1$. Now, since $z' = 1$, going back to the agent's budget set in (4.11), we have $c = y + p(1 - 1)$, or $c = y$. In equilibrium, the stock market and the goods market clear.

Along the lines of Theorem 4.11.16, one can show that v is a fixed point of a contraction mapping $v \mapsto T_v$, specifically the mapping defined by

$$T_v(v)(y, z) = \max_{z'}\left[u(zy + p(y)(z - z')) + \beta \sum_j P_{yy'}v(y', z')\right].$$

To see that $p(y)$ is well defined, the agent's problem must satisfy, if $u''(\cdot) < 0$, the necessary condition

$$p(y)u'(y) = \beta \sum_{y' \in S} P_{yy'}\left[u'(y')(y' + p(y'))\right]. \tag{4.12}$$

Let $f(y) \equiv p(y)u'(y)$ and $g(y) \equiv \beta \sum_{y' \in S} P_{yy'}u'(y')y'$. Then we can define an operator $T_p : \mathbb{R}_+^S \to \mathbb{R}_+^S$, the set of possible nonnegative price functions. For any element $f \in \mathbb{R}_+^S$, we can express the right-hand side of the Euler equation (4.12) as

$$T_p(f)(y) = g(y) + \beta \sum_{y' \in S} P_{yy'}f(y').$$

Note that since $y \in S \subset \mathbb{R}_{++}$, $u'(y)$ is finite. Thus, $g(y)$ is bounded and we can use a contraction-mapping argument to establish the existence of the fixed point $f = T_p f$.

4.12 ◆ Connectedness

Chapter 5 deals with convex sets. A set is convex if it is possible to travel between any two points in the set along a straight line. Here we study what happens when we allow curvy "lines." We begin with a reprise.

Definition 4.1.21.   A metric space (M, d) is **disconnected** if it can be expressed as the union of two disjoint nonempty open sets; otherwise it is **connected**.

Interval subsets of $\mathbb{R}$ are, intuitively, connected sets.

Definition 4.12.1 *An **interval in** $\mathbb{R}$ is a subset of $\mathbb{R}$ taking on one of the following nine forms: $(-\infty, b]$; $(-\infty, b)$; $(a, b]$; $[a, b]$; $[a, b)$; (a, b); $[a, \infty)$; (a, ∞); or $(-\infty, +\infty)$. When we write "$|a, b|$," we mean one of these nine.*[11]

Theorem 4.12.2 *A subset of $\mathbb{R}^1$ is connected iff it is an interval.*

Note that $[a, a]$ is an interval; it is not a very long interval, but it still meets the definition. It is also connected, for the trivial reason that a one-point set cannot be expressed as the disjoint union of two nonempty sets.

Proof.   Suppose that $E \subset \mathbb{R}$ is an interval containing more than one point. If E is disconnected, then $E = G_1 \cup G_2$, where G_1 and G_2 are disjoint open sets.

---

11. The "$|a, b|$" notation comes from the French tradition. They write "$]a, b[$" for the set we write as "(a, b)," and "$]a, b]$" for the set we write as "$(a, b]$," and so on. The "$|a, b|$" notation is meant to suggest that at either end of the interval, we have either "$]$" or "$[$."

Relabeling the G's if necessary, there exists $r \in G_1$ and $t \in G_2$ such that $r < t$. Let $s = \sup\{x : [r, x) \subset G_1\}$. We know that $s \leq t$ because $t \notin G_1$. Since E is an interval, $s \in E$. If $s \in G_1$, there exists $\delta > 0$ such that $[s, s + \delta) \subset G_1$, contradicting s being the supremum. If $s \in G_2$, there exists $\delta > 0$ such that $(s - \delta, s] \subset G_2$, again contradicting s being the supremum.

Suppose that $E \subset \mathbb{R}$ is not an interval. This implies that there exists $r, t \in E$ and $s \notin E$ such that $r < s < t$. $E \cap (-\infty, s)$ and $E \cap (s, \infty)$ are nonempty disjoint open subsets of E, so that E is disconnected. ■

Theorem 4.12.3 *If (M, d) is a metric space, the following are equivalent:*

1. *M is connected;*

2. *$\emptyset$ and M are the only subsets of M that are both open and closed; and*

3. *If $f : M \to \{0, 1\}$ is a continuous mapping from M to the two-point set $\{0, 1\}$, then f is constant.*

Proof. Suppose that M is connected. If there exists a $G \neq \emptyset$, M that is both open and closed, then $M = G \cup G^c$ expresses M as the disjoint union of two nonempty open sets.

Suppose that $\emptyset$ and M are the only subsets of M that are both open and closed. If f is a nonconstant continuous mapping from M to the two-point set $\{0, 1\}$, then $M = f^{-1}(0) \cup f^{-1}(1)$ expresses M as the disjoint union of two nonempty open sets.

Suppose that every continuous $f : M \to \{0, 1\}$ is constant. If M is not connected, then $M = G_1 \cup G_2$ for a pair of disjoint nonempty open sets. The function $f = 1_{G_1}$ is then continuous, but nonconstant. ■

The following yields the intermediate value theorem, one of the main results in calculus.

Lemma 4.12.4 *If $f : M \to M'$ is continuous and M is connected, then $f(M)$ is a connected subset of M'.*

Proof. If $E = f(M)$ is disconnected, there exists a continuous onto $g : E \to \{0, 1\}$, and $g \circ f$ is a continuous onto function from M to $\{0, 1\}$. ■

Theorem 4.12.5 (Intermediate Value) *If $f : |a, b| \to \mathbb{R}$ is continuous, $y, y' \in f(|a, b|)$, and $y < y'' < y'$, then there exists $x \in |a, b|$ such that $f(x) = y''$.*

Proof. The interval $|a, b|$ is connected so that $f(|a, b|)$ is a connected subset of $\mathbb{R}$. The only connected subsets of $\mathbb{R}$ are intervals. ■

A verbal shorthand is: a continuous function on an interval takes on all of its intermediate values.

Example 4.12.6 [↑*Newton's Bisection Method, §3.5*] *Starting with a continuous $f : \mathbb{R} \to \mathbb{R}$ and a pair $a_1, b_1 \in \mathbb{R}$ such that $f(a_1) < 0 < f(b_1)$, Newton's bisection method provides an iterative procedure converging to a solution to the equation $f(x) = 0$. We have seen that the method provides a Cauchy sequence whether or not f is continuous. The intermediate value theorem guarantees the existence of a solution because 0 is intermediate between $f(a_1)$ and $f(b_1)$. Since*

a_n and b_n are Cauchy and Cauchy equivalent to x_n by Lemma 3.5.4 and f is continuous, $\lim_n f(a_n) = \lim_n f(b_n) = \lim_n f(x_n)$. As $f(a_n) < 0 < f(b_n)$, the only possibility is that $\lim_n f(x_n) = 0$.

The main use of Theorem 4.12.5 in calculus is in proving Taylor's theorem with remainder. We use it instead to provide a proof of the existence of general equilibrium when there are two goods. The following recapitulates material at the end of Chapter 1.

Definition 4.12.7 *An **exchange economy model with I consumers and ℓ goods** is a collection $\mathcal{E} = \{X_i, \succsim_i, \mathbf{y}_i\}_{i \in I}$, where:*

1. *I is a finite set (of consumers),*
2. *for each $i \in I$, $X_i \subset \mathbb{R}^\ell$ is a nonempty consumption set $X_i \subset \mathbb{R}^\ell$ (usually $X_i = \mathbb{R}_+^\ell$),*
3. *for each $i \in I$, $\succsim_i$ is a rational preference ordering on X_i, and*
4. *for each $i \in I$, $\mathbf{y}_i \in X_i$ is an initial endowment or allocation.*

*An **allocation for** $\mathcal{E}$ is a list of consumption bundles, $\mathbf{x} = (\mathbf{x}_i)_{i \in I}$ in $\times_{i \in I} X_i$. An allocation is **feasible** if $\sum_i \mathbf{x}_i \leq \sum_i \mathbf{y}_i$. A feasible allocation is **weakly Pareto efficient** if there is no other feasible allocation $(\mathbf{x}_i')_{i \in I}$ such that $\mathbf{x}_i' \succ \mathbf{x}_i$ for all $i \in I$.*

*A price-allocation pair $(\mathbf{p}, \mathbf{x})$ in $\mathbb{R}_+^\ell \times \times_{i \in I} X_i$ is an **equilibrium for** $\mathcal{E}$ if $\mathbf{x}$ is feasible, and if $\forall i \in I$, $\mathbf{p}\mathbf{x}_i \leq \mathbf{p}\mathbf{y}_i$ and $\mathbf{x}_i' \in X_i$ satisfies $\mathbf{x}_i' \succ \mathbf{x}_i$, then $\mathbf{p}\mathbf{x}_i' > \mathbf{p}\mathbf{y}_i$.*

It is worth repeating, several times, "an equilibrium is a price and an allocation."

What we need for such an existence result appears in the following, rather long, example of a general equilibrium model with two people and two goods. Despite the small number of people, both act as if they were price takers.

Example 4.12.8 (Edgeworth Box Economy) *Let $I = \{1, 2\}$, $\ell = 2$, $X_i = \mathbb{R}_+^2$, with $\succeq_1$ being represented by the utility function $u_1(x_{1,1}, x_{1,2}) = x_{1,1}^2 x_{1,2}$ and $\succeq_2$ by $u_1(x_{1,1}, x_{1,2}) = \sqrt{x_{1,1}} + \sqrt{x_{1,2}}$; also $y_1 = (3, 1)$ and $y_2 = (1, 5)$.*

Taking prices as given, consumer 1 solves $\max_{(x_{1,1}, x_{1,2}) \in X_1} x_{1,1}^2 x_{1,2}$ subject to $p_1 x_{1,1} + p_2 x_{1,2} \leq 3p_1 + 1p_2$. The Lagrangian is

$$L(x_{1,1}, x_{1,2}, \lambda) = x_{1,1}^2 x_{1,2} + \lambda(3p_1 + 1p_2 - [p_1 x_{1,1} + p_2 x_{1,2}]).$$

*Setting $\partial L/\partial x_{1,1} = \partial L/\partial x_{1,2} = \partial L/\partial \lambda = 0$ and solving for the optimal $x_{1,1}^*$ and $x_{1,2}^*$ yields $x_1^*(\mathbf{p}) = x_1^*(p_1, p_2) = (x_{1,1}^*, x_{1,2}^*) = (\frac{2}{3} \frac{(3p_1 + p_2)}{p_1}, \frac{1}{3} \frac{(3p_1 + p_2)}{p_2})$. In terms of $\mathbf{p} = (p_1, p_2)$, 1's **excess demand** is $z_1(\mathbf{p}) = x_1^*(\mathbf{p}) - (3, 1)$. Since preferences are strictly monotonic, the consumers spends all of their income, which means that $\mathbf{p} \cdot z_1(\mathbf{p}) \equiv 0$.*

As should happen, 1's demand functions are homogeneous of degree 0, that is, for all $t > 0$, $x_1^*(tp_1, tp_2) = t^0 x_1^*(p_1, p_2)$. This means that there is no loss in dividing both prices by p_1. Letting $\rho = p_2/p_1$ be the relative price of good 2, we have $x_1^*(\rho) := x_1^*(1, \rho) = (\frac{2}{3}(3 + \rho), \frac{1}{3} \frac{(3+\rho)}{\rho})$. In terms of ρ, 1's excess demand function is $z_1(\rho) = x_1^*(\rho) - (3, 1) = (\frac{2}{3}(3 + \rho) - 3, \frac{1}{3} \frac{(3+\rho)}{\rho} - 1)$.

For consumer 2, $z_2(\mathbf{p}) = (\frac{p_1+5p_2}{\sqrt{p_1}(\sqrt{p_1}+\sqrt{p_2})} - 1, \frac{p_1+5p_2}{\sqrt{p_2}(\sqrt{p_1}+\sqrt{p_2})} - 5)$, and it too satisfies $\mathbf{p} \cdot z_2(\mathbf{p}) \equiv 0$. In terms of ρ, it is $z_2^*(\rho) := z_2^*(1, \rho) = (\frac{1+5\rho}{1+\sqrt{\rho}} - 1, \frac{1+5\rho}{\sqrt{\rho}(1+\sqrt{\rho})} - 5)$.

*For both consumers, for large ρ, that is, when good 2 is much more expensive than good 1, the excess demand for good 1 is positive. For very small ρ, it is negative. This means that **aggregate excess demand**, $z(\mathbf{p}) := \sum_i z_i(\mathbf{p})$, has the same properties. Since $\mathbf{p}z_i(\mathbf{p}) \equiv 0$, we have $\mathbf{p}z(\mathbf{p}) \equiv 0$, which is known as **Walras's law**. Combining these observations, if we can find a ρ such that $z_{1,1}(\rho) + z_{2,1}(\rho) = 0$, that is, such that the excess demand for good 1 is equal to 0, then we can conclude that the excess demand for good 2 is also equal to 0 by Walras's law, and determine that $((1, \rho), (x_1^*(1, \rho), x_2^*(1, \rho))$ is an equilibrium.*

Theorem 4.12.9 *Suppose that $\mathcal{E}$ is an exchange economy with two goods. Suppose further that preferences are such that the excess demand functions, $\mathbf{p} \mapsto z_i(\mathbf{p})$, are continuous, and that $\sum_i z_{i,1}(\rho) > 0$ for small ρ and $\sum_i z_{i,1}(\rho) < 0$ for large ρ. Then $\mathcal{E}$ has an equilibrium.*

Proof. By the intermediate value theorem, there exists ρ such that $\sum_i z_{i,1}(\rho) = 0$. Observe that $\mathbf{p}z_i(\mathbf{p}) \leq 0$ by the definition of $x_i^*(\mathbf{p})$. Since $\sum_i z_{i,1}(\rho) = 0$, we know that $\rho \sum_i z_{i,2}(1, \rho) \leq 0$. Combining, we find that $((1, \rho), (x_i^*(1, \rho))_{i \in I})$ is a price-allocation pair that satisfies the definition of an equilibrium.   ∎

We will see that the equilibrium existence conclusion is true under much weaker assumptions, but that the argument requires a much deeper result than the intermediate value theorem.

For many purposes, a weaker notion than connectedness is useful.

Definition 4.12.10 *A metric space (M, d) is **path connected** if for every $x, x' \in M$, there is a continuous function $f : [0, 1] \to M$ such that $f(0) = x$ and $f(1) = x'$. The set $\{f(t) : t \in [0, 1]\}$ is then a **path joining x and x'**.*

The notion of a "path" comes from imagining moving continuously from 0 to 1. At each point $t \in [0, 1]$, imagine being located at $f(t) \in M$. The path is the set of points in M that are passed through on the way from x to x'. If f is a path from x to x' and g a path from x' to x'', then defining $h(t) = f(2t)$ for $t \leq \frac{1}{2}$ and $h(t) = g(2(t - \frac{1}{2}))$ for $\frac{1}{2} \leq t$ gives a path joining x and x''. We can formalize this as follows.

Lemma 4.12.11 *Being connected by a path is an equivalence relation.*

Theorem 4.12.12 *If (M, d) is path connected, then it is connected.*

Proof. Suppose that M is path connected but that $M = G_1 \cup G_2$ expresses M as the disjoint union of two nonempty open sets. Pick $x \in G_1$ and $x' \in G_2$ and let f be a path joining x and x'. Then $f^{-1}(G_1) \cup f^{-1}(G_2)$ expresses the interval $[0, 1]$ as the disjoint union of two nonempty open sets, a contradiction.   ∎

Take C to be the countable dense set $\mathbb{Q}^n$ in the following result. It is rather surprising.

Theorem 4.12.13 *For every n, $\mathbb{R}^n$ is path connected. If $C \neq \emptyset$, $\mathbb{R}^1 \setminus C$ is disconnected. For every $n \geq 2$ and every countable set C, $\mathbb{R}^n \setminus C$ is path connected.*

Proof.   The first statement is an immediate consequence of the third. It is also a consequence of the observation that $f(t) = t\mathbf{x} + (1-t)\mathbf{x}'$, $t \in [0,1]$, is a (straight-line) path joining $\mathbf{x}$ and $\mathbf{x}'$.

For the second statement, let $C \neq \emptyset$ and let $D = C^c = \mathbb{R}^1 \setminus C$. $(D \cap (-\infty, c)) \cup (D \cap (c, \infty))$ expresses D as the disjoint union of nonempty open sets.

Let C be a countable subset of $\mathbb{R}^n$, $n \geq 2$. Pick distinct $\mathbf{x}, \mathbf{x}' \notin C$. Let ℓ be a line segment of nonzero length, perpendicular to the segment $[\mathbf{x}, \mathbf{x}']$ and intersecting it at $\frac{1}{2}\mathbf{x} + \frac{1}{2}\mathbf{x}'$. For each of the uncountably many $\mathbf{z} \in \ell$, the path $\ell_{\mathbf{z}} := [\mathbf{x}, \mathbf{z}] \cup [\mathbf{z}, \mathbf{x}']$ connects $\mathbf{x}$ and $\mathbf{x}'$. To show that $\mathbb{R}^n \setminus C$ is path connected, it is sufficient to show that at least one of the paths $\ell_{\mathbf{z}}$ stays in $\mathbb{R}^n \setminus C$. If not, then each $\ell_{\mathbf{z}}$ must intersect C. However, if $\mathbf{z} \neq \mathbf{z}'$, then the only two points that $\ell_{\mathbf{z}}$ and $\ell_{\mathbf{z}'}$ have in common are $\mathbf{x}$ and $\mathbf{x}'$, which are not in C. Therefore, if each $\ell_{\mathbf{z}}$ intersects C, then C must be uncountable. ■

4.13 ◆ Bibliography

The material in this chapter was drawn mostly from undergraduate real analysis courses. There are many excellent relevant textbooks. J. E. Marsden and M. J. Hoffman's *Elementary Classical Analysis* (2nd Ed., New York: W. H. Freeman, 1993), W. Rudin's *Principles of Mathematical Analysis* (3rd Ed., New York: McGraw-Hill, 1976), and J. R. Munkres's *Topology* (2nd Ed., Upper Saddle River, N.J.: Prentice-Hall, 2000) are three that we are familiar with and like very much.

Finite-Dimensional Convex Analysis

A set C in $\mathbb{R}^\ell$ is convex if the line joining any two points in C stays inside C. Perfectly round balls are convex subsets of $\mathbb{R}^3$, as are lines, planes, and triangles. As good as they are otherwise, donuts are not convex.

This simple geometric notion interacts with the existence, characterization, and behavior of optima. This happens because in many of the models of optimizing behavior when facing constraints used by economists, in a region of an optimum, x^*, it looks as though x^* is the solution to the maximization of a continuous linear function, L, mapping a convex set C to $\mathbb{R}$. This makes the study of both convex sets and the set of continuous linear functions important.

The set of continuous linear functions on $\mathbb{R}^\ell$ is called the **dual space** of $\mathbb{R}^\ell$, which turns out to be the same as (i.e., isometric to) $\mathbb{R}^\ell$. The dual space is intimately tied to the separation of convex sets, say $C, D \subset \mathbb{R}^\ell$, from one another. This involves finding a continuous linear $L : \mathbb{R}^\ell \to \mathbb{R}$ such that $L(x) \le L(y)$ for all $x \in C$ and $y \in D$. The tie to optimization theory comes from the observation that $(\forall x \in C, y \in D)[L(x) \le L(y)] \Leftrightarrow [\sup_{x \in C} L(x) \le \inf_{y \in D} L(y)]$.

Our coverage begins with the basic geometry of convexity and the dual space. With this in hand, we turn to the three degrees of separation of convex sets and their associated theorems. The first, and strongest, kind of separation gives us neoclassical duality theory. The second strongest kind of separation gives us concave and convex functions and their derivative conditions, which characterize their optima. The weakest kind of separation leads directly to the workhorse of optimization theory, the Kuhn-Tucker theorem.

These finite-dimensional results on convexity are part of the background one needs to understand much of the material in the first semester of a graduate microeconomics course. This part of the chapter ends with a tour through the relation between differentiability and convexity.

By the second semester in graduate microeconomics, one also needs another crucial property of convex sets. This is the fixed-point property, and it comes in two forms: Brouwer's fixed-point theorem states that a continuous function from a compact convex set to itself has a fixed point; Kakutani's fixed-point theorem states that an upper hemicontinuous nonempty-valued convex-valued correspondence from a compact convex set to itself has a fixed point. We give these results and show how they are used in the foundational parts of general equilibrium theory and game theory.

5.1 ◆ The Basic Geometry of Convexity

Our starting point is the weighted average of two numbers, r and s. Weighted averages are given by $\gamma r + (1 - \gamma)s$, $\gamma \in [0, 1]$. The larger γ is, the closer the weighted average is to r, and the smaller it is the closer the weighted average is to s.

5.1.a Convex Combinations in $\mathbb{R}^\ell$

In $\mathbb{R}^2$ and $\mathbb{R}^3$, one can (and should) draw the following constructions. For $\mathbf{x}, \mathbf{y} \in \mathbb{R}^\ell$ and $\gamma \in [0, 1]$, we have

$$\mathbf{x} = \begin{bmatrix} x_1 \\ x_2 \\ \vdots \\ x_\ell \end{bmatrix}, \quad \mathbf{y} = \begin{bmatrix} y_1 \\ y_2 \\ \vdots \\ y_\ell \end{bmatrix}, \quad \text{and} \quad \gamma\mathbf{x} + (1 - \gamma)\mathbf{y} := \begin{bmatrix} \gamma x_1 + (1 - \gamma)y_1 \\ \gamma x_2 + (1 - \gamma)y_2 \\ \vdots \\ \gamma x_\ell + (1 - \gamma)y_\ell \end{bmatrix}.$$

(5.1)

Example 5.1.1 *If $\mathbf{x} = (1, 0)' \in \mathbb{R}^2$ and $\mathbf{y} = (0, 1)'$, then $\gamma\mathbf{x} + (1 - \gamma)\mathbf{y} = (\gamma, (1 - \gamma))'$. As γ varies from 0 to 1, the point $\gamma\mathbf{x} + (1 - \gamma)\mathbf{y}$ moves down and to the right from $\mathbf{y}$ to $\mathbf{x}$ in the plane.*

Definition 5.1.2 *The weighted average, $\gamma\mathbf{x} + (1 - \gamma)\mathbf{y}$, $\gamma \in [0, 1]$, of two points, $\mathbf{x}$ and $\mathbf{y}$, in $\mathbb{R}^\ell$ is a called a **convex combination of $\mathbf{x}$ and $\mathbf{y}$**. A set $C \subset \mathbb{R}^\ell$ is **convex** if for all $\mathbf{x}, \mathbf{y} \in C$, every convex combination of $\mathbf{x}$ and $\mathbf{y}$ belongs to C (Figure 5.1.2).*

Notation 5.1.3 *We often write "$\gamma\mathbf{x} + (1 - \gamma)\mathbf{y}$" as "$\mathbf{x}\gamma\mathbf{y}$." In this notation, C is convex iff for every $\mathbf{x}, \mathbf{y} \in C$ and every $\gamma \in [0, 1]$, $\mathbf{x}\gamma\mathbf{y} \in C$. When we write $\|\mathbf{x}\|$ for $\mathbf{x} \in \mathbb{R}^\ell$, we mean $\|\mathbf{x}\|_2 = \sqrt{\mathbf{x} \cdot \mathbf{x}}$ unless we explicitly note otherwise.*

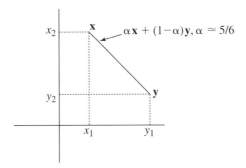

FIGURE 5.1.2   Weighted or convex combinations.

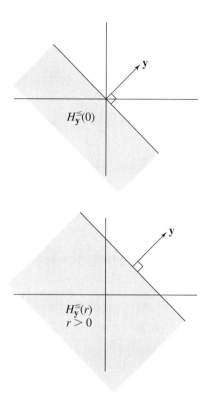

FIGURE 5.1.5

Theorem 5.1.4 *The intersection of any class of convex sets is convex.*

Proof. Pick $\mathbf{x}$, $\mathbf{x}' \in \cap_{\alpha \in A} C_\alpha$ and $\gamma \in [0, 1]$. Since each C_α is convex, $\mathbf{x}\gamma\mathbf{x}' \in C_\alpha$, which implies that $\mathbf{x}\gamma\mathbf{x}' \in \cap_{\alpha \in A} C_\alpha$. ∎

5.1.b Half-Spaces

An important class of convex sets is comprised of the **half-spaces**. These are defined by inner products, and we use the result that $\mathbf{xy} = \|\mathbf{x}\|\,\|\mathbf{y}\|\cos\theta$, from Lemma 4.3.13, to understand their geometry.

Definition 5.1.5 *For every nonzero* $\mathbf{y} \in \mathbb{R}^\ell$ *and* $r \in \mathbb{R}$, *we have the four half-spaces*, $H_{\mathbf{y}}^{\leq}(r) = \{\mathbf{x} \in \mathbb{R}^\ell : \mathbf{y}\cdot\mathbf{x} \leq r\}$, $H_{\mathbf{y}}^{<}(r) = \{\mathbf{x} \in \mathbb{R}^\ell : \mathbf{y}\cdot\mathbf{x} < r\}$, $H_{\mathbf{y}}^{\geq}(r) = \{\mathbf{x} \in \mathbb{R}^\ell : \mathbf{y}\cdot\mathbf{x} \geq r\}$, *and* $H_{\mathbf{y}}^{>}(r) = \{\mathbf{x} \in \mathbb{R}^\ell : \mathbf{y}\cdot\mathbf{x} > r\}$ *(Figure 5.1.5)*.

For $\mathbf{y} \neq 0$, $H_{\mathbf{y}}^{\leq}(0)$ is the set of $\mathbf{x}$ such that the angle between $\mathbf{x}$ and $\mathbf{y}$ is between 90 and 270, and $\mathbf{a} \in H_{\mathbf{y}}^{\leq}(0)$ iff $\mathbf{a} = \mathbf{x} - s\mathbf{y}$, where $\mathbf{x} \in \mathbf{y}^\perp$ and $s \geq 0$. This is the geometry behind the name "half-space": to go from the origin to a point in $H_{\mathbf{y}}^{\leq}(0)$, one goes to any $\mathbf{x}$ perpendicular to $\mathbf{y}$, and then some nonnegative distance in the direction directly opposite to $\mathbf{y}$.

For $\mathbf{y} \in \mathbb{R}^\ell$, define $L_\mathbf{y} : \mathbb{R}^\ell \to \mathbb{R}$ by $L_\mathbf{y}(\mathbf{x}) = \mathbf{x} \cdot \mathbf{y}$. The mappings $L_\mathbf{y}$ are continuous and linear. Further, $H_\mathbf{y}^\leq(r) = L_\mathbf{y}^{-1}((-\infty, r])$, $H_\mathbf{y}^<(r) = L_\mathbf{y}^{-1}((-\infty, r))$, $H_\mathbf{y}^\geq(r) = L_\mathbf{y}^{-1}([r, \infty))$, and $H_\mathbf{y}^>(r) = L_\mathbf{y}^{-1}((r, \infty))$. This means that $H_\mathbf{y}^\leq(r)$ and $H_\mathbf{y}^\geq(r)$ are closed, whereas $H_\mathbf{y}^<(r)$ and $H_\mathbf{y}^>(r)$ are open.

Example 5.1.6 *The half-spaces are convex. Consider, for example, $H_\mathbf{y}^\leq(r) = \{\mathbf{x} \in \mathbb{R}^\ell : \mathbf{y} \cdot \mathbf{x} \leq r\}$. If $\mathbf{y}\mathbf{x} \leq r$ and $\mathbf{y}\mathbf{x}' \leq r$, then $\mathbf{y} \cdot (\mathbf{x}\gamma\mathbf{x}') = (\mathbf{y}\mathbf{x})\gamma(\mathbf{y}\mathbf{x}') \leq r$ because the weighted average of two numbers weakly less than r is weakly less than r. The arguments for the other half-spaces, $H_\mathbf{y}^<(r)$, $H_\mathbf{y}^\geq(r)$, and $H_\mathbf{y}^>(r)$, are completely analogous.*

Example 5.1.7 *For S a finite set, $\Delta(S) = \{p \in \mathbb{R}_+^S : \sum_s p_s = 1\}$ is the set of probabilities on S. It is convex because it is the intersection of half-spaces. For $p, q \in \Delta(S)$ and $\gamma \in (0, 1)$, the point $p\gamma q$ can be interpreted as a **compound gamble** or **compound lottery**: in the first stage, p is chosen with probability γ and q with probability $1 - \gamma$; independently, in the second stage, a point $s \in S$ is chosen according to either the distribution p or the distribution q.*

5.1.c Convex Preferences and Convex Technologies

Walrasian budget sets, $\{\mathbf{x} \in \mathbb{R}_+^\ell : \mathbf{p} \cdot \mathbf{x} \leq w\}$, are convex because they are the intersection of the half-spaces $H_{\mathbf{e}_i}^\geq(0)$, $i = 1, \ldots, \ell$, $\mathbf{e}_i$ the unit vector in the ith direction, and the half-space $H_\mathbf{p}^\leq(w)$.

In consumer theory, we use convex preferences, which we interpret as reflecting a taste for variety.

Example 5.1.8 *A rational preference relation $\succeq$ in $\mathbb{R}_+^\ell$ is **convex** if for all $\mathbf{y}$, the set $\{\mathbf{x} \in \mathbb{R}_+^\ell : \mathbf{x} \succeq \mathbf{y}\}$ is a convex set, and it is **strictly convex** if for all $\mathbf{x} \succsim \mathbf{y}$ and all $\gamma \in (0, 1)$, $\mathbf{x}\gamma\mathbf{y} \succ \mathbf{y}$. If we think of consuming the bundle $\mathbf{x}\gamma\mathbf{y}$ as having the bundle $\mathbf{x}$ on γ of the days and the bundle $\mathbf{y}$ the rest of the time, then convexity of preferences captures the notion that variety is valued—if $\mathbf{x} \sim \mathbf{y}$, then $\mathbf{x}\gamma\mathbf{y} \succsim \mathbf{x}$ for convex preferences and $\mathbf{x}\gamma\mathbf{y} \succ \mathbf{x}$ for strictly convex preferences.*

Later in Lemma 5.6.16, we use a supporting hyperplane argument to show that convexity and monotonicity yield decreasing marginal rates of substitution.

Lemma 5.1.9 *If preferences $\succeq$ or $\mathbb{R}_+^\ell$ are continuous and convex (respectively, strictly convex), then the Walrasian demand set is nonempty, compact, and convex (respectively, contains exactly one point).*

Proof. Preferences are continuous if for all $\mathbf{y} \in \mathbb{R}_+^\ell$, $\{\mathbf{x} : \mathbf{x} \succeq \mathbf{y}\}$ and $\{\mathbf{x} : \mathbf{y} \succeq \mathbf{x}\}$ are closed. Let $K = B(\mathbf{p}, w)$ be a compact convex Walrasian budget set. As in Lemma 4.7.21, let $Wb(\mathbf{y}, K) = \{\mathbf{x} \in K : \mathbf{x} \succeq \mathbf{y}\}$. The demand set is the set of preference maximal points, that is, $x(\mathbf{p}, w) = \cap_{\mathbf{y} \in K} Wb(\mathbf{y}, K)$. This is a collection of closed convex subsets of a compact set with the finite intersection property, so the intersection is closed, hence compact, nonempty, and convex.

Finally, if the preferences are strictly convex and $x(\mathbf{p}, w)$ contains two points, $\mathbf{x} \neq \mathbf{y}$, then $\mathbf{x} \sim \mathbf{y}$. Since $x(\mathbf{p}, w)$ is convex, it contains $\mathbf{x}\frac{1}{2}\mathbf{y}$. By strict convexity, $\mathbf{x}\frac{1}{2}\mathbf{y} \succ \mathbf{y}$, which contradicts $\mathbf{y} \in x(\mathbf{p}, w)$. ∎

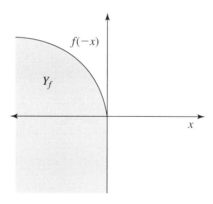

FIGURE 5.1.10 Y_f, a convex technology.

We also often assume that technologies are convex.

Example 5.1.10 *From intermediate microeconomics, the simplest kind of production theory involves one input, $x \geq 0$, and one output, $y \geq 0$. These are connected by a **production function**, $x \mapsto f(x)$, with the interpretation that $f(x)$ is the most y that can be produced using input level y. Hence, we always assume that $f(0) = 0$, that is, that there is no free lunch. The set of technologically feasible points is known as a **technology**. In this case, it is $\{(x, y) : x \geq 0, y \leq f(x)\}$, known, for fairly obvious reasons, as the **subgraph of** f.*

*The convention in graduate economics is to measure inputs as negatives and outputs as positives. This makes the technology associated with f into the set $Y_f = \{(x_1, x_2) : x_1 \leq 0, x_2 \leq f(-x_1)\}$. A point $(x_1, x_2) \in Y_f$ is called a **netput vector**, and a technology, Y_f, is a set of feasible netput vectors (Figure 5.1.10).*

We always assume that $0 \in Y$, that is, that shutdown is possible.

Definition 5.1.11 *A technology $Y \subset \mathbb{R}^\ell$ is **convex** if the set Y is convex.*

The convexity of a technology has a time interpretation.

Example 5.1.12 *Let $Y \subset \mathbb{R}^\ell$ denote the set of technologically feasible netput vectors. The netput combination $\mathbf{y}\gamma\mathbf{y}'$ can be interpreted as running the technology at $\mathbf{y}$ for γ of the time and running it at $\mathbf{y}'$ for the rest of the time. The convexity of a technology Y captures the notion that this kind of shifting of resources over time is possible. If Y represents the set of all possible factory configurations for producing, say, cars, one might well believe that it is convex. However, if one already has a factory set up to produce large cars with powerful, inefficient engines, being able to run it half of each year to produce small cars with efficient engines seems unlikely. This is the difference between the short run, in which the technology is not convex, and the (perhaps very) long run, in which it is convex.*

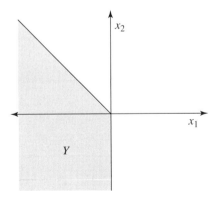

FIGURE 5.1.16   The technology Y is a convex cone; it has constant returns to scale.

The convexity of a technology also has a "returns to scale" interpretation. The idea of constant returns to scale is that if one multiplies all inputs by some constant factor, then outputs will be multiplied by the same factor. More generally we have the following.

Definition 5.1.13   *A technology $Y \subset \mathbb{R}^\ell$ has*

1. ***nonincreasing returns to scale*** *if for all $\mathbf{y} \in Y$ and all $\alpha \in [0, 1]$, $\alpha\mathbf{y} \in Y$,*

2. ***nondecreasing returns to scale*** *if for all $\mathbf{y} \in Y$ and all $\alpha \in [1, \infty)$, $\alpha\mathbf{y} \in Y$, and*

3. ***constant returns to scale*** *if for all $\mathbf{y} \in Y$ and all $\alpha \in \mathbb{R}_+$, $\alpha\mathbf{y} \in Y$.*

Exercise 5.1.14   Show that a one-input/one-output technology with production function $y = f(x) = x^r$, $r > 0$, has nonincreasing (respectively nondecreasing, respectively constant) returns to scale iff $r \leq 1$ (respectively $r \geq 1$, respectively $r = 1$).

Exercise 5.1.15   Show that if Y is a convex technology, then it has nonincreasing returns to scale.

Having constant returns to scale has a geometric name.

Definition 5.1.16   *A nonempty $X \subset \mathbb{R}^\ell$ is a **cone** if $[\mathbf{x} \in X] \Rightarrow (\forall \alpha \geq 0)[\alpha\mathbf{x} \in X]$; it is a **convex cone** if it is a cone and it is convex, that is, if $(\forall \alpha, \beta \geq 0)(\forall \mathbf{x}, \mathbf{y} \in X)[(\alpha\mathbf{x} + \beta\mathbf{y}) \in X]$ (Figure 5.1.16).*

Some people take constant returns to scale as a self-obvious statement, as an axiom if you will. Perhaps this is right. If one had the powers of a major deity, then one could create a second, alternate universe containing an exact duplicate of a given technology. Thus, $\mathbf{y} \in Y$ would lead to $2\mathbf{y} \in Y$. If one ran the second universe only α of the time, then, in principle, the netput would be $(1 + \alpha)\mathbf{y} \in Y$. If one duplicate universe is possible, why not many?

One way to understand this argument is that it is vacuously true, a statement of the form "if one can do an impossible thing, then for all $\mathbf{y} \in Y$ and for all $\alpha \geq 0$, $\alpha\mathbf{y} \in Y$." While logically true, the argument does not seem to have much more going for it. There are at least two counters to constant returns to scale, one involving physics and the other involving the empirical observation that decreasing returns to scale always set in.

1. Physics: If I "increase an input pipe by a factor of 2," what have I done? Have I increased the diameter of the pipe by 2, in which case I have increased its capacity by 4, or have I increased its capacity by 2, in which case I have increased the volume of, say, metal in the pipe by $(\sqrt{2})^3$? Or do I increase the metal in the pipe by 2, in which case the capacity of the valve goes up by $(2^{1/3})^2$? Now, these calculations are based on capacity at a given pressure, and real pipes deliver a whole schedule of flows. In more sophisticated terms, a pipe is a function from pressure to flow. The simple-minded calculations just done are based on what are called smooth or laminar flows, and when there is turbulence in the pipe, nobody knows the exact formula relating cross section to flow at various pressures. All of this means that increasing $\mathbf{y}$ to $\alpha\mathbf{y}$ is rather trickier than it might appear, even when $\alpha = 2$.

2. Empirics: In situation after situation, we have found that there are decreasing returns to scale. One reason might be that there is some unmeasured input and that observing decreasing returns to scale in the measured $\mathbf{y}$ is consistent with constant returns to scale with the unmeasured input fixed. Sometimes the unmeasured input is thought of as "entrepreneurial talent," a rather hard concept to define, but very intuitive.[1]

5.1.d Vector Subspaces, Spans, and Dimension

From Definition 4.3.1, a vector space $\mathfrak{X}$ over $\mathbb{R}$ must contain $\alpha\mathbf{x} + \beta\mathbf{y}$ for every $\alpha, \beta \in \mathbb{R}$ and $\mathbf{x}, \mathbf{y} \in \mathfrak{X}$.

Definition 5.1.17 *A subset of $\mathbb{R}^\ell$ that is itself a vector space is called a **vector subspace** or a **linear subspace**.*

In $\mathbb{R}^\ell$, the smallest vector subspace is $V = \{0\}$, known as the trivial subspace. The next largest kind of vector subspace is a line through the origin. Since we can always take $\alpha \in (0, 1)$ and $\beta = (1 - \alpha)$, vector subspaces are convex.

Exercise 5.1.18 Show that the intersection of any collection of vector subspaces is another vector subspace.

Exercise 5.1.19 Show that the intersection of any collection of cones is a cone. A **wedge** is a convex cone that does not contain any nontrivial linear subspaces. Show that the intersection of any collection of wedges is a wedge.

Recall that $\mathbf{x}^\perp := \{\mathbf{y} : \mathbf{xy} = 0\}$.

Example 5.1.20   *In* $\mathbb{R}^\ell$, $\mathbf{x}^\perp$ *is a vector subspace of* $\mathbb{R}^\ell$. *To see why, pick* $\mathbf{y}, \mathbf{z} \in \mathbf{y}^\perp$ *and* $\alpha, \beta \in \mathbb{R}$. *Then* $\alpha \mathbf{y} + \beta \mathbf{z} \in \mathbf{x}^\perp$ *because* $(\alpha \mathbf{y} + \beta \mathbf{z}) \cdot \mathbf{y} = \alpha \mathbf{y} \cdot \mathbf{x} + \beta \mathbf{z} \cdot \mathbf{x} = 0$.

The easiest way to find vector subspaces is to span them.

Definition 5.1.21   *For any collection* $S \subset \mathbb{R}^\ell$, *the* **span of** S *is defined as* $\mathbf{span}(S) = \{\sum_h \beta_h s_h : \beta_h \in \mathbb{R},\ s_h \in S\}$, *where only finite sums are considered.*

Lemma 5.1.22   *For any* $S \subset \mathbb{R}^\ell$, $\mathbf{span}(S)$ *is a vector subspace of* $\mathbb{R}^\ell$, *and for any vector subspace* V, $\mathbf{span}(V) = V$.

Proof.   The weighted sum of two finite sums is another finite sum.   ∎

Linear dependence and independence appear here and reappear in the study of spaces of functions.

Definition 5.1.23   *A set* $\{\mathbf{y}_i : i \in F\} \subset \mathbb{R}^\ell$ *is* **linearly independent** *if* $[\sum_i \beta_i \mathbf{y}_i = 0] \Rightarrow (\forall i)[\beta_i = 0]$; *otherwise it is* **linearly dependent**.

If $\mathbf{e}_i$ is the unit vector in the ith direction, $\{\mathbf{e}_i : i \in \{1, \ldots, \ell\}\}$ is a linearly independent set that spans $\mathbb{R}^\ell$.

Definition 5.1.24   *If* $V = \{0\}$, *then by convention its* **dimension** *is equal to* 0. *If* $V \neq \{0\}$ *is a vector subspace of* $\mathbb{R}^\ell$, *then its* **dimension**, *denoted* $\dim(V)$, *is the smallest cardinality of any set containing* V *in its span.*

Theorem 5.1.25   *For any vector subspace* $V \neq \{0\}$ *of* $\mathbb{R}^\ell$, *there is a linearly independent set* $\{\mathbf{x}_1, \ldots, \mathbf{x}_m\}$ *with* $V = \mathbf{span}(\{\mathbf{x}_1, \ldots, \mathbf{x}_m\})$ *and* $m = \dim(V)$.

Proof.   Since $V \subset \mathbb{R}^\ell$ and $\mathbb{R}^\ell$ is the span of the set $\{\mathbf{e}_i : i \in \{1, \ldots, \ell\}\}$, V is spanned by some m-element set, $\{\mathbf{x}_1, \ldots, \mathbf{x}_m\}$. Further, since $\mathbb{N}$ is well ordered, there is a smallest m with this property. If $\{\mathbf{x}_1, \ldots, \mathbf{x}_m\}$ is not linearly independent, then some strict subset has the same span so that $\dim(V) \leq (m - 1)$.   ∎

This also allows us to establish that any ℓ-dimensional vector subspace of $\mathbb{R}^n$, $n \geq \ell$, is isometric with $\mathbb{R}^\ell$.

Definition 5.1.26   *An* **orthonormal subset of** $\mathbb{R}^\ell$ *is a collection of vectors* $\{x_i : i \in I\}$ *such that* $\|x_i\| = 1$ *for all* $i \in I$ *and* $x_i \perp x_j$ *for* $i \neq j$.

We know that there exist orthonormal sets for any $I \subset \ell$; consider the set $\{\mathbf{e}_i : i \in I\}$, where $\mathbf{e}_i$ is the unit vector in the ith direction.

Theorem 5.1.27   *Suppose that* $\mathcal{X} = \{\mathbf{x}_1, \ldots, \mathbf{x}_\ell\}$ *is an orthonormal subset of* $\mathbb{R}^n$. *If* $\mathbf{x} \in \mathbb{R}^\ell$ *and* $\beta_i := \mathbf{x} \cdot \mathbf{x}_i$, *then* $\sum_i |\beta_i|^2 \leq \|\mathbf{x}\|$ *(***Bessel's inequality***),* $\mathbf{x}' = \mathbf{x} - \sum_i \beta_i \mathbf{x}_i$ *is orthogonal to* $\mathbf{span}(\mathcal{X})$, *and* $\mathbf{x} \in \mathbf{span}(\mathcal{X})$ *iff* $\mathbf{x} = \sum_i \beta_i \mathbf{x}_i$.

Proof.   Observe that $\|\mathbf{x}'\|^2 = (\mathbf{x} - \sum_i \beta_i \mathbf{x}_i) \cdot (\mathbf{x} - \sum_i \beta_i \mathbf{x}_i) \geq 0$. Expanding, we get $(\mathbf{x} \cdot \mathbf{x}) - 2 \sum_i \beta_i (\mathbf{x}, \mathbf{x}_i) + \sum_i \sum_j \beta_i \beta_i (\mathbf{x}_i \cdot \mathbf{x}_j) \geq 0$. Since $(\mathbf{x}_i \cdot \mathbf{x}_j) = 0$ if $i \neq j$ and 1 if $i = j$, this gives $\|\mathbf{x}\|^2 - \sum_i |\beta_i|^2 \geq 0$.

Any $\mathbf{x} \in \mathbf{span}(\mathcal{X})$ is the finite sum of points of the form $r_i \mathbf{x}_i$, $r_i \in \mathbb{R}$. For such a $\mathbf{y}$, $(\mathbf{x}' \cdot \mathbf{y}) = \sum_i r_i (\mathbf{x}' \cdot \mathbf{x}_i)$. However, $(\mathbf{x}' \cdot \mathbf{x}_i) = (\mathbf{x} - \sum_j \beta_j \mathbf{x}_j) \cdot \mathbf{x}_i) = (\mathbf{x} \cdot \mathbf{x}_i) - \sum_j \beta_j (\mathbf{x}_j \cdot \mathbf{x}_i) = \beta_j - \beta_j = 0$.

If $\mathbf{x} = \sum_i \beta_i \mathbf{x}_i$, then $\mathbf{x} \in \mathbf{span}(\mathfrak{X})$ by the definition of span. If $\mathbf{x} \in \mathbf{span}(\mathfrak{X})$, then $\mathbf{x} = \sum_i r_i \mathbf{x}_i$, and $\beta_i = (\mathbf{x} \cdot \mathbf{x}_i) = r_i$ by direct calculation. ■

Exercise 5.1.28   Use Bessel's inequality to give (yet another) proof of the Cauchy-Schwarz inequality. [Hint: The singleton set $\{\mathbf{y}/\|\mathbf{y}\|\}$ is orthonormal for any $\mathbf{y} \neq 0$.]

Using Theorem 5.1.27, we can generate an orthonormal set from any linearly independent set. Further, the orthonormal set will have exactly the same span. If we start from a set spanning a vector subspace V, there is no loss in assuming that our spanning set is orthonormal. The algorithm is called the **Gram-Schmidt procedure**.

Let $\mathfrak{X} = \{\mathbf{y}_i : i \in I\}$ be a linearly independent subset of $\mathbb{R}^n$. Since $\mathbf{y}_1 \neq 0$, we can set $\mathbf{x}_1 = \mathbf{y}_1/\|\mathbf{y}_1\|$. Observe that $\mathbf{span}(\mathbf{x}_1) = \mathbf{span}(\mathbf{y}_1)$.

Given that $\mathbf{x}_n$ has been defined, that $\{\mathbf{x}_1, \ldots, \mathbf{x}_n\}$ is orthonormal, and that $\mathbf{span}(\{\mathbf{x}_1, \ldots, \mathbf{x}_n\}) = \mathbf{span}(\{\mathbf{y}_1, \ldots, \mathbf{y}_n\})$, set $\mathbf{y}'_{n+1} = \mathbf{y}_{n+1} - \sum_{i \leq n} \beta_i \mathbf{x}_i$, where $\beta_i = \langle \mathbf{y}_{n+1}, \mathbf{x}_i \rangle$. Since $\mathfrak{X}$ is a linearly independent set, $\mathbf{y}'_{n+1} \neq 0$. By Theorem 5.1.27, $\mathbf{y}'_{n+1}$ is orthogonal to $\mathbf{span}(\{\mathbf{x}_1, \ldots, \mathbf{x}_n\})$. Define $\mathbf{x}_{n+1} = \mathbf{y}'_{n+1}/\|\mathbf{y}'_{n+1}\|$. The set $\{\mathbf{x}_1, \ldots, \mathbf{x}_{n+1}\}$ is orthonormal and $\mathbf{span}(\{\mathbf{x}_1, \ldots, \mathbf{x}_{n+1}\}) = \mathbf{span}(\{\mathbf{y}_1, \ldots, \mathbf{y}_{n+1}\})$.

Exercise 5.1.29   The last step in the Gram-Schmidt procedure made the claim that $\mathbf{span}(\{\mathbf{x}_1, \ldots, \mathbf{x}_{n+1}\}) = \mathbf{span}(\{\mathbf{y}_1, \ldots, \mathbf{y}_{n+1}\})$. Prove this claim.

Corollary 5.1.30   *If W is an ℓ-dimensional subspace of $\mathbb{R}^n$, then W is isometric to $\mathbb{R}^\ell$.*

Proof. $W = \mathbf{span}(\{\mathbf{y}_1, \ldots, \mathbf{y}_\ell\})$. Using Gram-Schmidt, produce an orthonormal set $\{\mathbf{x}_1, \ldots, \mathbf{x}_\ell\}$ with $W = \mathbf{span}(\{\mathbf{x}_1, \ldots, \mathbf{x}_\ell\})$. By Theorem 5.1.27, each $\mathbf{w} \in W$ is of the form $\sum_i \beta_i (\mathbf{w} \cdot \mathbf{x}_i)$ and $\|\mathbf{w}\|^2 = \sum_i |\beta_i|^2$. The isometry is $\mathbf{w} \leftrightarrow (\beta_1, \ldots, \beta_\ell)$. ■

Starting with different linearly independent sets gives us different isometries. However, the composition of isometries is another isometry, so we might as well treat all of the ℓ-dimensional vector spaces as being the same.

Exercise 5.1.31   Show that any m-dimensional vector subspace of $\mathbb{R}^\ell$ can be expressed as an intersection of $\ell - m$ sets of the form $\mathbf{x}^\perp$. Show that $\mathbf{x}^\perp$ is closed. Thus, any vector subspace of $\mathbb{R}^\ell$ is closed.

5.1.e Sums of Sets

The sum and the difference of sets are very different from the set theoretic sum, that is, the union, and the set theoretic difference.

Definition 5.1.32   *For A, $B \subset \mathfrak{X}$, the **(Minkowski) sum** is $A + B = \{\mathbf{a} + \mathbf{b} : \mathbf{a} \in A, \mathbf{b} \in B\}$, and the **Minkowski difference** is $A - B = \{\mathbf{a} - \mathbf{b} : \mathbf{a} \in A, \mathbf{b} \in B\}$. When $B = \{b\}$, $A + B$ is the **translate of A by** b and $A - B$ is the **translate of A by** $-b$.*

Example 5.1.33   *For $E \subset \mathbb{R}^\ell$, $E^\epsilon = E + B_\epsilon(0)$. The "+" and the "−" notation can be a bit deceptive. If $B = \{0\}$, then $A + B = A$. However, unless A contains exactly one point, $A - A \neq \{0\}$.*

Exercise 5.1.34   Give the set $A - A$ when $A = \{\mathbf{x} \in \mathbb{R}^2 : \|\mathbf{x}\| \leq 1\}$ and when $A = \mathbb{R}^2_+$.

Example 5.1.35   *For $\mathbf{y} \neq 0$ and $r \in \mathbb{R}$, let $t = \frac{r}{\mathbf{yy}}$. $H_\mathbf{y}^{\leq}(r)$ is the translate of $H_\mathbf{y}^{\leq}(0)$ by the vector $t\mathbf{y}$, that is, $H_\mathbf{y}^{\leq}(r) = H_\mathbf{y}^{\leq}(0) + \{t\mathbf{y}\}$. Thus, to go from $t\mathbf{y}$ to a point in $H_\mathbf{y}^{\leq}(r)$, one first goes in a direct $\mathbf{x}$ perpendicular to $\mathbf{y}$ and then some nonnegative distance in the direction directly opposite to $\mathbf{y}$.*

Example 5.1.36   *Preferences on $\mathbb{R}^\ell_+$ are **monotonic** if for all $\mathbf{x}$, $\{\mathbf{x}\} + \mathbb{R}^\ell_{++} \subset \{\mathbf{y} : \mathbf{y} \succ \mathbf{x}\}$. A technology Y satisfies **free disposal** if for all $\mathbf{y} \in Y$, $\{\mathbf{y}\} + \mathbb{R}^\ell_- \subset Y$.*

The translate of a convex set is convex. One can say further that if one uses a convex set of translations, one has a convex set.

Lemma 5.1.37   *If A and B are convex, then $A + B$ and $A - B$ are convex.*

Proof.   Let $\mathbf{x} = \mathbf{a} + \mathbf{b} \in A + B$ and $\mathbf{x}' = \mathbf{a}' + \mathbf{b}' \in A + B$, and let $\gamma \in (0, 1)$. Since A and B are convex, $\mathbf{a}\gamma\mathbf{a}' \in A$ and $\mathbf{b}\gamma\mathbf{b}' \in B$. Since $\mathbf{x}\gamma\mathbf{x}' = (\mathbf{a}\gamma\mathbf{a}') + (\mathbf{b}\gamma\mathbf{b}')$, we know that $\mathbf{x}\gamma\mathbf{x}' \in A + B$.   ■

Example 5.1.38   *If $A = \{(x_1, x_2) : x_2 \geq e^{x_1}\}$ and $B = \{(x_1, x_2) : x_2 \geq e^{-x_1}\}$, then A and B are closed and convex and $A + B = \{(x_1, x_2) : x_2 > 0\}$ is convex and not closed.*

Lemma 5.1.39   *If A and B are compact subsets of $\mathbb{R}^\ell$, then $A + B$ is compact; if A is closed and B is compact, then $A + B$ is closed; and if either A or B is open, then $A + B$ is open.*

Proof.   Suppose that A and B are compact. The mapping $(\mathbf{x}, \mathbf{y}) \mapsto (\mathbf{x} + \mathbf{y})$ from $\mathbb{R}^{2\ell}$ to $\mathbb{R}^\ell$ is continuous, the set $A \times B \subset \mathbb{R}^{2\ell}$ is compact, and $A + B$ is the continuous image of a compact set.

Suppose that A is closed and B is compact. Let $\mathbf{x}_n = \mathbf{a}_n + \mathbf{b}_n \in (A + B)$, $\mathbf{a}_n \in A$, $\mathbf{b}_n \in B$, and suppose that $\mathbf{x}_n \to \mathbf{x}$. We must show that $\mathbf{x} \in (A + B)$. Taking a convergent subsequence if necessary, we set $\mathbf{b}_n \to \mathbf{b}$ for some $\mathbf{b} \in B$. As $\mathbf{x}_n \to \mathbf{x}$ and $\mathbf{b}_n \to \mathbf{b}$ and thus $\mathbf{a}_n \to \mathbf{a} := (\mathbf{x} - \mathbf{b})$. As A is closed, $\mathbf{a} \in A$, and thus $\mathbf{x} = \mathbf{a} + \mathbf{b}$, so that $\mathbf{x} \in A + B$.

Suppose that A is open and pick $\mathbf{x} = \mathbf{a} + \mathbf{b} \in (A + B)$. Since $\mathbf{a} \in A$, $\exists \epsilon > 0$ such that $B_\epsilon(\mathbf{a}) \subset A$. This implies that $B_\epsilon(\mathbf{a} + \mathbf{a}) \subset (A + B)$.   ■

Exercise 5.1.40   [↑Example 5.1.38 and Lemma 5.1.39] Which part of the proof of Lemma 5.1.39 goes wrong in Example 5.1.38?

5.2 ◆ The Dual Space of $\mathbb{R}^\ell$

Separation theorems in convex analysis talk about ways of separating disjoint convex sets, C and D, from each other using continuous linear functions. They assert the existence of a nonzero continuous linear function $L : \mathbb{R}^\ell \to \mathbb{R}$ and an

$r \in \mathbb{R}$ such that $C \subset L^{-1}((-\infty, r])$ and $D \subset L^{-1}([r, \infty))$ for disjoint convex C and D. Since continuous, linear functions play such a crucial role, we study them first.

5.2.a $\mathbb{R}^\ell$ is Self-Dual

The **dual space of** $\mathbb{R}^\ell$ is the set of all continuous linear functions. In this finite-dimensional case, it has a very simple structure.

Definition 5.2.1   *A function* $L : \mathbb{R}^\ell \to \mathbb{R}$ *is **linear** if for all* $\mathbf{x}, \mathbf{y} \in \mathbb{R}^\ell$ *and all* $\alpha, \beta \in \mathbb{R}$, $L(\alpha\mathbf{x} + \beta\mathbf{y}) = \alpha L(\mathbf{x}) + \beta L(\mathbf{y})$.

Theorem 5.2.2 $L : \mathbb{R}^\ell \to \mathbb{R}$ *is a continuous, linear function iff there exists a* $\mathbf{y} \in \mathbb{R}^\ell$ *such that for all* $\mathbf{x} \in \mathbb{R}^\ell$, $L(\mathbf{x}) = \mathbf{y} \cdot \mathbf{x}$.

Let $\mathfrak{X}^*$ denote the set of continuous linear functions on a vector space $\mathfrak{X}$. $\mathfrak{X}^*$ is called the **dual space of** $\mathfrak{X}$. Theorem 5.2.2 shows that $(\mathbb{R}^\ell)^*$ can be identified with $\mathbb{R}^\ell$ and that $\mathbb{R}^\ell$ is its own dual space, that is, $\mathbb{R}^\ell$ is **self-dual**.

Proof.   If $L(\mathbf{x}) = \mathbf{y} \cdot \mathbf{x}$, then L is continuous and linear.

Now suppose that L is linear. Let $\mathbf{e}_i$ be the unit vector in the ith direction. Every $\mathbf{x} \in \mathbb{R}^\ell$ has a unique representation as $\mathbf{x} = \sum_{i=1}^\ell x_i \mathbf{e}_i$, $x_i \in \mathbb{R}$. The linearity of L implies that $L(\mathbf{x}) = \sum_{i=1}^\ell x_i L(\mathbf{e}_i)$, which suggests in turn that L is continuous.[2] If $\mathbf{y}$ is the vector having an ith component $y_i = L(\mathbf{e}_i)$, $L(\mathbf{x}) = \sum_{i=1}^\ell x_i y_i = \mathbf{y} \cdot \mathbf{x}$.   ■

Note that $H_{\mathbf{y}}^\leq(r) = L^{-1}((-\infty, r])$ when $L(\mathbf{x}) = \mathbf{y} \cdot \mathbf{x}$, which expresses $H_{\mathbf{y}}^\leq(r)$ as the inverse image of a closed set under a continuous linear function. Since we can identify nonzero continuous L's on $\mathbb{R}^\ell$ with nonzero $\mathbf{y}$'s, separation of C and D is equivalent to the existence of a nonzero $\mathbf{y}$ and $r \in \mathbb{R}$ such that $C \subset H_{\mathbf{y}}^\leq(r)$ and $D \subset H_{\mathbf{y}}^\geq(r)$. Thus, the separation theorems that follow are also geometric existence results.

The dual space of a normed vector space has its own norm. In $\mathbb{R}^\ell$, this norm is equal to the usual norm.

Definition 5.2.3   *When* $\mathfrak{X}$ *is a normed vector space, and* $L \in \mathfrak{X}^*$, *we define the norm of* L *by* $\|L\|^* = \sup\{|Lx| : x \in \mathfrak{X}, \|x\| \leq 1\}$.

Recall that $xy = \|x\|\|y\| \cos\theta$ in $\mathbb{R}^\ell$ and that $\cos(\theta)$ is maximized at $\theta = 0$. This means that the supremum is achieved—if $L \in (\mathbb{R}^\ell)^*$ is represented by $\mathbf{y}$, then $\|L\|^* = \|\mathbf{y}\|$.

5.2.b   A Brief Infinite–Dimensional Detour

Infinite-dimensional vector spaces have vector subspaces that are dense and not closed. Related to this, there exist discontinuous linear functions. We give exam-

---

2. In the infinite-dimensional vector spaces that we study later, there are discontinuous linear functions.

ples of this first in $C([0, 1])$, the space of continuous $\mathbb{R}$-valued functions on $[0, 1]$, and then in $\mathcal{D}$, the set of cumulative distribution functions on $\mathbb{R}$.

5.2.b.1 Continuous Functions on $[0, 1]$

$C([0, 1])$ denotes the set of continuous functions on $[0, 1]$. The distance between continuous functions is $d_{\infty}(f, g) := \max_{x \in [0,1]} |f(x) - g(x)|$. We will see in §6.4 that $(C([0, 1]), d_{\infty})$ is a complete separable metric space.

Example 5.2.4 *The set of polynomials is a dense, infinite-dimensional vector subspace of $C([0, 1])$ that is not closed.*

  1. *It is a vector space because a linear combination of polynomials is another polynomial.*

  2. *It is infinite-dimensional because, for every n, the set of functions $\{1, x^1, \ldots, x^n\}$ is linearly independent. To see why the linear independence claim is true, suppose that $f(x) = \sum_{i=0}^{n} \beta_i x^i \equiv 0$ with at least one $\beta_i \neq 0$. Let j be the largest index of a nonzero β_i and take the jth derivative of both sides of the "$\equiv$"; observe that $j! \cdot \beta_j = 0$.*

  3. *The Stone-Weierstrass theorem, to be covered in §6.4.b, shows that the polynomials are dense, hence, if the set of polynomials is closed, then it is all of $C([0, 1])$.*

  4. *The set of polynomials is not closed. To see why, consider the exponential function defined in Chapter 3 by $f(x) = e^x = \sum_{n=0}^{\infty} \frac{x^n}{n!}$. The sequence of polynomials $f_N(x) = \sum_{n=0}^{N} \frac{x^n}{n!}$ has $d_{\infty}(f_N, f) = |f(1) - f_N(1)| = \sum_{n=N+1}^{\infty} \frac{x^n}{n!} \downarrow 0$. However, $f(x)$ is not a polynomial because $f'(x) = f(x)$, and no polynomial has this property.*

As soon as one has an infinite set of linearly independent points in a vector space, one can construct a discontinuous linear function.

Example 5.2.5 *We can define a linear function, L, from $\mathcal{P} \subset C([0, 1])$, the set of polynomials, to $\mathbb{R}$. For each $j = 0, 1, \ldots$, define $L(x^j) = j^2$. For any polynomial, $p(x) = \sum_{j=0}^{N} \beta_j x^j$, $L(p) = \sum_j j^2 \cdot \beta_j$. The function L is clearly linear, however it is also discontinuous. The sequence of functions $f_j(x) = x^j / j$ has the property that $d_{\infty}(f_j, 0) = 1/j \to 0$, but $L(f_j) = j \to \infty$. Thus we have a sequence converging to 0 while $L(\cdot)$ is strictly increasing and unbounded.*

5.2.b.2 The Cumulative Distribution Functions

Probability distributions are determined by their cumulative distribution functions (cdf's). In your probability and statistics class, you may have seen the definition of $F_n \to_{w^*} F$ for sequences of cdf's. The subscripted "w" refers to "weak" convergence. The reason for the name "weak" will, hopefully, be cleared up later. Weak convergence requires that $F_n(x) \to F(x)$ for every x at which $F(\cdot)$ is continuous.

Example 5.2.6 *$F(x) = 1_{[0,\infty)}(x)$, then $F(\cdot)$ is the cdf associated with point mass at 0, that is, the cdf of the degenerate random variable taking on only the value 0. Let $F_n(x) = 1_{[1/n,\infty)}(x)$. This is the cdf of the degenerate random variable taking on only the value $1/n$. For every $x \neq 0$, $F_n(x) \to F(x)$, so that $F_n \to_{w^*} F$.*

We are now ready for an example of a discontinuous linear function.

Example 5.2.7 *The set $\mathcal{D}$ of distribution functions is convex, and later we will see it as a subset of an infinite-dimensional vector space. We now give a linear function on $\mathcal{D}$ that is not continuous, $L(F) = F(0)$. Linearity follows from the observation that $L(\alpha F + \beta G) = \alpha F(0) + \beta G(0) = \alpha L(F) + \beta L(G)$. L is discontinuous because $F_n(x) = 1_{[1/n,\infty)}(x)$ converges to $F(x) = 1_{[0,\infty)}$, but $L(F_n) \equiv 0$ while $L(F) = 1$.*

One of the reasons that the foregoing convergence is called "weak" is because it is weaker than, that is, implied by, the following notion of "strong" or "norm" convergence of cdf's.

Recall that for any cdf F, we define the probability P_F on $\mathcal{B}^\circ$, the field of finite disjoint unions of intervals (Definition 4.12.1 *et seq.*) We can define the **norm distance** between cdfs by $\|F - G\| = \sup_{E \in \mathcal{B}^\circ} |F(E) - G(E)|$. In terms of linear functions, this is $\|F - G\| = \sup_{E \in \mathcal{B}^\circ} |\int 1_E(x)\, dF(x) - \int 1_E(x)\, dG(x)|$.

The norm distance between probabilities is extensively used in studying the distance between information structures, but it is so "tight" a measure of distance that it is useless for many, even most, other purposes.

Exercise 5.2.8 [↑Example 5.2.7] We have seen that $F_n \to_{w^*} F$ when $F_n(x) = 1_{[1/n,\infty)}(x)$ and $F(x) = 1_{[0,\infty)}$. Show that $\|F_n - F\| \equiv 1$.

If G_n is the cdf of the uniform distribution on the points $k/n, k = 1, \ldots, n$ and G is the cdf of the uniform distribution on $[0, 1]$, show that $G_n \to_{w^*} G$, but that $\|G_n - G\| \equiv 1$.

Show that $L(F) = F(0)$ is norm continuous, that is, $[\|F_n - F\| \to 0] \Rightarrow [L(F_n) \to L(F)]$.

5.3 ◆ The Three Degrees of Convex Separation

Being disjoint is one way to say that two sets are apart from each other. Under additional conditions on disjoint convex sets C and D, one can reach conclusions about stronger kinds of apartness. Throughout, $L : \mathbb{R}^\ell \to \mathbb{R}$ is a nonzero continuous linear function.

Definition 5.3.1 *C and D can be **separated** if $\exists L \neq 0$ and $r \in \mathbb{R}$ such that $C \subset L^{-1}((-\infty, r])$, while $D \subset L^{-1}([r, -\infty))$; they can be **strictly separated** if $C \subset L^{-1}((-\infty, r])$, while $D \subset L^{-1}((r, -\infty))$; and they can be **strongly separated** if, in addition, $\exists \epsilon > 0, C \subset L^{-1}((-\infty, r - \epsilon])$, while $D \subset L^{-1}([r + \epsilon, -\infty))$.*

Separation allows both sets to intersect, or even belong to, the boundary of the half-space $L^{-1}((-\infty, r])$. Strict separation allows no intersection of the sets at the boundary, though one of the sets can belong to the boundary. Strong separation imposes a uniform and strictly positive distance from the boundary.[3]

---

3. There is a degree of separation between strict and strong, which involves $C \subset L^{-1}((-\infty, r))$ and $D \subset L^{-1}((r, -\infty))$, but it is rarely used. We can formulate strong separation equivalently as $\exists \epsilon > 0, C \subset L^{-1}((-\infty, r - \epsilon))$ and $D \subset L^{-1}((r + \epsilon, -\infty))$.

We have results for each degree of separation:

1. For strong separation, Theorem 5.4.1 shows that a closed convex set can be strongly separated from a point not in the set, and Exercise 5.4.5 shows that this is easily extended to a strong separation result for disjoint closed convex sets, one of which is compact.

2. For strict separation, the supporting hyperplane theorem, 5.5.16, shows that a point on the relative boundary of a convex set C has a supporting hyperplane that strictly separates it from the relative interior of C.

3. Finally, for separation, Theorem 5.7.1 shows that arbitrary disjoint convex sets can be separated.

The separation of points from closed convex sets leads to characterizations of the convex hull and the closed convex hull of a set. These in turn lead to neoclassical duality theory. The supporting hyperplane theorem leads to the second fundamental theorem of welfare economics. It also leads to the separation of convex sets, Theorem 5.7.1, and to the Kuhn-Tucker theorem, which is the basic result in constrained optimization.

Without convexity, no variety of separation by continuous linear functions is possible, even for sets with disjoint closures.

Example 5.3.2   Let $C = \{\mathbf{x} \in \mathbb{R}^\ell : \|\mathbf{x}\| = 1\}$ and $D = \{0\}$. For any $\epsilon < \frac{1}{2}$, $C^\epsilon \cap D^\epsilon = \emptyset$ For any nonzero linear $L : \mathbb{R}^\ell \to \mathbb{R}$, there exists $\mathbf{x}, -\mathbf{x} \in C$ such that $L(-\mathbf{x}) < L(0) < L(\mathbf{x})$. For example, if L is represented by $\mathbf{y} \neq 0$, take $\mathbf{x} = \mathbf{y}/\|\mathbf{y}\|$. In this example, it is impossible for C to belong to one side of a half-space and D to the other.

For convex sets, different degrees of separation are achievable. We suggest that you draw each of the situations in the following.

Example 5.3.3   (See Figure 5.3.3.) (a) Let $C = \{(x_1, x_2) : [x_2 < 0] \vee [[x_2 = 0] \wedge [x_1 \leq 0]]\}$ and $D = C^c$. C and D are disjoint convex sets separated by $\mathbf{y} = (0, 1)$ and $r = 0$, but C and D cannot be strictly separated.

(b) Let $C = \{(x_1, x_2) \in \mathbb{R}^2_+ : x_1 x_2 \geq 1\} \subset \mathbb{R}^2$ and let $D = \{(x_1, x_2) \in \mathbb{R}^2 : (-x_1, x_2) \in C\}$. C and D are disjoint closed convex sets that can be strictly separated by $\mathbf{y} = (-1, 0)$ and $r = 0$, but C and D cannot be strongly separated.

(c) Let $C = B_1((0, 0))$ and let $D = \{(1, 0)\}$ so that $D \subset \partial C$. C and D can be strictly separated by $\mathbf{y} = (1, 0)$ and $r = 1$, but cannot be strongly separated.

(d) Let $C = B_1((-2, -2))$ and $D = \mathbb{R}^2_{++}$. C and D have disjoint closures, one of which is compact, and they can be strongly separated by many L's.

There is a fact about translation invariance that makes some proofs a bit easier: there is no loss in assuming that one of the sets contains the origin. Suppose that one chooses $\mathbf{v} \in D$ and translates C to $C - \mathbf{v}$ and D to $D - \mathbf{v}$. If L is such that $C \subset L^{-1}((-\infty, r])$ and $D \subset L^{-1}([r, \infty))$, then $(C - \mathbf{v}) \subset L^{-1}((-\infty, r'])$ and $(D - \mathbf{v}) \subset L^{-1}([r', \infty))$, where $r' = r - L(\mathbf{v})$. In other words, there is no loss in assuming, for our separation results, that one of the convex sets, here, $D - \mathbf{v}$, contains the origin.

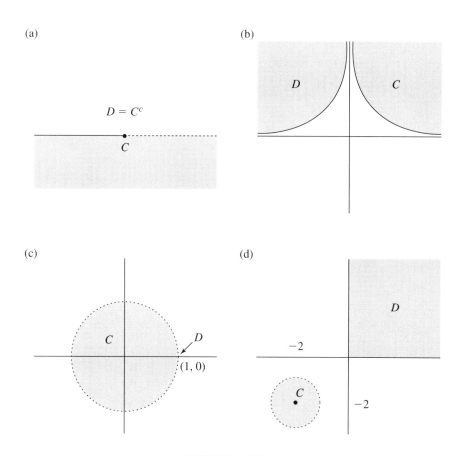

FIGURE 5.3.3

5.4 ◆ Strong Separation and Neoclassical Duality

Neoclassical duality theory gives us several ways to characterize preferences and technology sets. The strong separation theorem is the basic mathematics behind neoclassical duality results, though we need two additional constructs, the convex hull and the affine hull of a set.

5.4.a Strongly Separating Points from Close Convex Sets

Theorem 5.4.1 (Strong Separation) *If $C \subset \mathbb{R}^\ell$ is closed and convex and $D = \{\mathbf{v}\} \subset C^c$, then C and D can be strongly separated.*

By translation, we can take $\mathbf{v} = 0$, and Theorem 5.4.1 becomes the following statement: If $C \subset \mathbb{R}^\ell$ is convex, closed, and does not contain 0, then there exists a

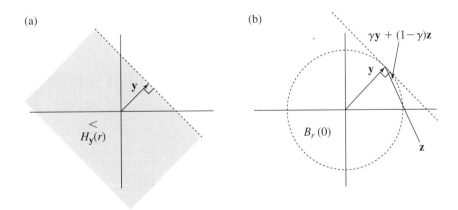

FIGURE 5.4.2

nonzero L and a strictly positive r such that $L(C) \subset (r, \infty)$. The geometry behind Theorem 5.4.1 reduces to the following.

Lemma 5.4.2 *For any* $\mathbf{y} \neq 0$ *and any* $\mathbf{z} \in H_{\mathbf{y}}^<(r)$, $r = \mathbf{yy}$, *the line segment joining* $\mathbf{y}$ *and* $\mathbf{z}$ *contains points with lengths strictly less than* $\mathbf{y}$ *(Figure 5.4.2).*

Proof. Since $\mathbf{z} \in H_{\mathbf{y}}^<(r)$, it is of the form $\mathbf{z} = \mathbf{x} - s\mathbf{y}$ for some $\mathbf{x} \perp \mathbf{y}$ and $s > 0$. We show that for γ close to 1, $\|\gamma\mathbf{y} + (1-\gamma)\mathbf{z}\|^2 < \|\mathbf{y}\|^2$. Use $\mathbf{z} = \mathbf{x} - s\mathbf{y}$ and expand

$$f(\gamma) = \|\gamma\mathbf{y} + (1-\gamma)\mathbf{z}\|^2 = (\gamma\mathbf{y} + (1-\gamma)[\mathbf{x} - s\mathbf{y}]) \cdot (\gamma\mathbf{y} + (1-\gamma)[\mathbf{x} - s\mathbf{y}]),$$

$$f(\gamma) = \gamma^2\mathbf{yy} + 2\gamma(1-\gamma)\mathbf{y}(\mathbf{x} - s\mathbf{y}) + (1-\gamma)^2(\mathbf{x} - s\mathbf{y})(\mathbf{x} - s\mathbf{y}) \quad (5.2)$$

$$= \mathbf{yy}[\gamma^2 - 2s\gamma(1-\gamma) + s^2(1-\gamma)^2] + (1-\gamma)^2\mathbf{xy}$$

$$= \mathbf{yy}[\gamma - s(1-\gamma)]^2 + (1-\gamma)^2\mathbf{xy}.$$

Since $f(1) = \mathbf{yy}$ and $f'(1) = 2(1+s) > 0$, the length of $\mathbf{y}\gamma\mathbf{z}$ is strictly increasing at $\gamma = 1$, which means that $f(\gamma) < f(1)$ when $(1 - \epsilon) < \gamma < 1$. ■

Proof of Theorem 5.4.1. Translating by $-\mathbf{v}$ if need be, there is no loss in assuming that $\mathbf{v} = 0$. Pick $n > 0$ such that $K = \mathrm{cl}(B_n(\mathbf{v})) \cap C \neq \emptyset$. K is closed and bounded, hence compact, so the continuous function $f(\mathbf{x}) = \|\mathbf{x}\|$ achieves its minimum, say at $\mathbf{y} \in C$. Let L be the linear function determined by $\mathbf{y}$. We claim that $C \subset L^{-1}([r, \infty))$, where $r = \mathbf{yy}$. If not, there exists $\mathbf{z} \in C \cap H_{\mathbf{y}}^<(r)$. By the previous lemma, for γ close to 1, $\|\mathbf{y}\gamma\mathbf{z}\| < \|\mathbf{y}\|$. Since C is convex, $\mathbf{y}\gamma\mathbf{z} \in C$, contradicting the definition of $\mathbf{y}$ as the minimizer of $\|\cdot\|$ in C. For strong separation, note that $C \subset L^{-1}([r/2 + \epsilon, \infty))$ and $\mathbf{v} \subset L^{-1}((-\infty, r/2 - \epsilon])$ for any $\epsilon < r/2$. ■

Exercise 5.4.3 Show that if $C \subset \mathbb{R}^\ell$ is convex, $D = \{\mathbf{v}\}$, and there exists $\epsilon > 0$ such that $B_\epsilon(\mathbf{v}) \cap C = \emptyset$, then C and D can be strongly separated.

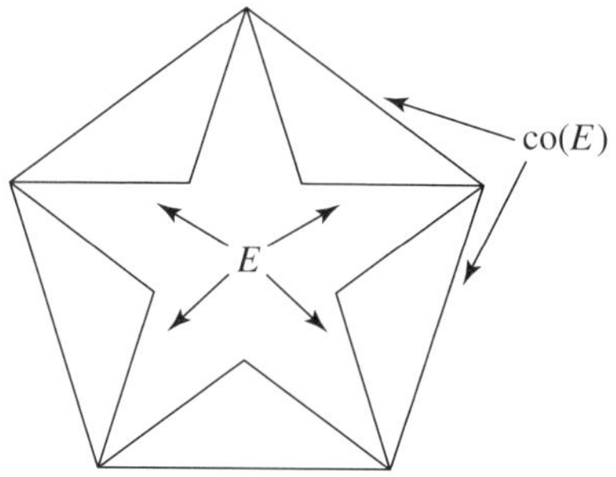

FIGURE 5.4.6 Convex hull $co(E)$ of a lone star E.

Exercise 5.4.4 The closed convex set $B = \{\mathbf{x} : \|\mathbf{x}\| \le 1\}$ can be strongly separated from every point in $B^c = \{\mathbf{x} : \|\mathbf{x}\| > 1\}$. Using this, express B as the intersection of a set of closed half-spaces, of a set of open half-spaces, and as the countable intersection of a set of closed half-spaces.

We saw that convex closed sets can be strongly separated from points not in them. They can also be strongly separated from disjoint compact convex sets.

Exercise 5.4.5 Show that if C and D are disjoint, C is closed and convex and D is compact and convex, then C and D can be strongly separated. [Show that there exist $\mathbf{c}^* \in C$ and $\mathbf{d}^* \in D$ that solve the problem $\min\{\|\mathbf{c} - \mathbf{d}\| : \mathbf{c} \in C, \mathbf{d} \in D\}$. Translate so that $\mathbf{v} = \frac{1}{2}\mathbf{c}^* + \frac{1}{2}\mathbf{d}^*$ becomes the origin. There is a vector $\mathbf{y}$ and $r > 0$ such that $C \subset L^{-1}((r, \infty))$ and $D \subset L^{-1}((-\infty, -r))$.]

5.4.b Convex Hulls and Affine Hulls

There is an "outside" and an "inside" way to get at the convex hull.

5.4.b.1 Shrink Wrapping from the Outside

Definition 5.4.6 *The **convex hull of a set** E is defined as the smallest convex set containing E, $\operatorname{co}(E) = \bigcap\{C : E \subset C, C \text{ convex}\}$ (Figure 5.4.6). The **closed convex hull of a set** E is defined as the smallest closed convex set containing E, $\overline{\operatorname{co}}(E) = \bigcap\{C : E \subset C, C \text{ convex and closed}\}$.*

This is a "shrink-wrapping" kind of definition. It means that $\operatorname{co}(E)$ is the smallest convex set containing E and that $\overline{\operatorname{co}}(E)$ is the smallest closed convex set containing E.

Corollary 5.4.7 *The closure of a convex set is closed, and if $E \subset \mathbb{R}^\ell$ is convex, then $\operatorname{cl}(E) = \overline{\operatorname{co}}(E)$.*

Proof. Suppose that E is convex, let $\mathbf{x}, \mathbf{y} \in \operatorname{cl}(E)$, and pick $\gamma \in (0, 1)$. We must show that $\mathbf{x}\gamma\mathbf{y} \in \operatorname{cl}(E)$. Since $\mathbf{x}, \mathbf{y} \in \operatorname{cl}(E)$, there are sequences $\mathbf{x}_n$ and $\mathbf{y}_n$ in E converging to $\mathbf{x}$ and $\mathbf{y}$, respectively. As E is convex, for all n, $\mathbf{x}_n\gamma\mathbf{y}_n \in E$. Since multiplication and addition are continuous, $\mathbf{x}_n\gamma\mathbf{y}_n \to \mathbf{x}\gamma\mathbf{y}$, so that $\mathbf{x}\gamma\mathbf{y} \in \operatorname{cl}(E)$.

If E is convex, then, by what was just shown, $\mathrm{cl}(E)$ is a convex set containing E, so $\overline{\mathrm{co}}(E) \subset \mathrm{cl}(E)$. If $\mathbf{x} \in \mathrm{cl}(E)$ but $\mathbf{x} \notin \overline{\mathrm{co}}(E)$, then $\mathbf{x}$ can be strongly separated from $\overline{\mathrm{co}}(E)$, hence strongly separated from E. This contradicts $\mathbf{x} \in \mathrm{cl}(E)$. ∎

Lemma 5.4.8 *A closed convex set is the intersection of the closed half-spaces that contain it.*

Exercise 5.4.9 Prove Lemma 5.4.8; that is, for any $E \subset \mathbb{R}^\ell$, show that $\overline{\mathrm{co}}(E) = \bigcap \{H_{\mathbf{y}}^{\leq}(r) : E \subset H_{\mathbf{y}}^{\leq}(r)\}$. Compare this with Exercise 5.4.4.

5.4.b.2 Weighted Averaging from the Inside
The given definition of the convex hull of E works from the "outside," taking the intersection of convex sets containing E. Another way to arrive at the convex hull of E works from the "inside."

Definition 5.4.10 *For $E \subset \mathbb{R}^\ell$, $\mathbf{x}$ is a **convex combination of points in** E if there exists a finite collection of $\gamma_i \geq 0$ and $\mathbf{y}_i \in E$ with $\sum_i \gamma_i = 1$ and $\sum_i \gamma_i \mathbf{y}_i = \mathbf{x}$.*

The set of all convex combinations of points in E is convex, and $\mathrm{co}(E)$ must contain all convex combinations of points in E. Therefore, $\mathrm{co}(E)$ is exactly the set of all convex combinations of points in E. This is the inside approach.

We know more. Carathéodory's theorem (just below) says that we need only take convex combinations of at most $\ell + 1$ points in E for the inside approach. We need some preparation.

Definition 5.4.11 *A set $\{\mathbf{y}_i : i \in F\} \subset \mathbb{R}^\ell$ is **linearly independent** if $[\sum_i \beta_i \mathbf{y}_i = 0] \Rightarrow (\forall i)[\beta_i = 0]$; otherwise they are **linearly dependent**. A set $\{\mathbf{y}_i : i \in F\} \subset \mathbb{R}^\ell$ is **affinely independent** if for all $j \in F$, the collection $\{(\mathbf{y}_i - \mathbf{y}_j) : i \in F, i \neq j\}$ is linearly independent.*

Adding the origin to a set of linearly independent vectors gives an affinely independent set of vectors. The vectors $(2, 0)$, $(0, 2)$, and $(1, 1)$ are linearly dependent since $\frac{1}{2}(2, 0) + \frac{1}{2}(0, 2) + (-1)(1, 1) = (0, 0)$. They are affinely independent because the sets $\{(1, -1), (-1, 1)\}$, $\{(2, -2), (1, -2)\}$, and $\{(-2, 2), (-1, 1)\}$ are linearly independent. Geometrically, any two of the edges of the triangle determined by $(2, 0)$, $(0, 2)$, and $(1, 1)$ are linearly independent.

Exercise 5.4.12 Show that a set $\{\mathbf{y}_i : i \in F\} \subset \mathbb{R}^\ell$ is affinely dependent iff there exist $\{\alpha_i : i \in F\}$, not all equal to 0, such that $\sum_i \alpha_i \mathbf{y}_i = 0$ and $\sum_i \alpha_i = 0$.

As most ℓ-vectors can be linearly independent in $\mathbb{R}^\ell$, which means at most $(\ell + 1)$-vectors in $\mathbb{R}^\ell$ can be affinely independent.

Theorem 5.4.13 (Carathéodory) *If $E \subset \mathbb{R}^\ell$ and $\mathbf{x} \in \mathrm{co}(E)$, then $\mathbf{x}$ is a convex combination of affinely independent points of E. In particular, $\mathbf{x}$ is a convex combination of at most $\ell + 1$ points of E.*

Proof. Since $\mathbf{x} \in \mathrm{co}(E)$, it can be expressed as a finite sum of k elements of E, $\sum_{i=1}^{k} \gamma_i \mathbf{y}_i = \mathbf{x}$ with $\gamma_i > 0$ and $\mathbf{y}_i \in E$. Let $K \subset \mathbb{N}$ be the set of all k for which such a representation is available. We know that $K \neq \emptyset$. As $\mathbb{N}$ is well ordered, there is a least such k, and we work with that k.

Suppose that the set $\{\mathbf{y}_i : i \leq k\}$ is affinely dependent. This means there are numbers $\{\alpha_i : i \leq k\}$, not all equal to 0, such that $\sum_i \alpha_i \mathbf{y}_i = 0$ and $\sum_i \alpha_i = 0$. All of the γ_i are greater than 0 and at least one of the α_i is positive. Among the positive pairs, pick the one with γ_j/α_j as small as possible. Note that

$$\mathbf{x} = \left(\sum_{i=1}^{k} \gamma_i \mathbf{y}_i\right) + 0 = \left(\sum_{i=1}^{k} \gamma_i \mathbf{y}_i\right) + \left(\frac{\gamma_j}{\alpha_j} \sum_i \alpha_i \mathbf{y}_i\right).$$

Rearranging gives

$$\mathbf{x} = \left(\sum_{i=1}^{k} \left(\gamma_i - \frac{\gamma_j}{\alpha_j}\alpha_i\right) \mathbf{y}_i\right),$$

where all of the coefficients of the $\mathbf{y}_i$, the $\left(\gamma_i - \frac{\gamma_j}{\alpha_j}\alpha_i\right)$, are nonnegative and at least one of them, the jth, is equal to 0. This means that we have expressed $\mathbf{x}$ as a linear combination of fewer than k of the $\mathbf{y}_i$, contradicting the minimality of k. ∎

Corollary 5.4.14 *If E is compact, then $\mathrm{co}(E)$ is compact.*

Proof. Let Δ be the compact set $\{\gamma \in \mathbb{R}_+^{\ell+1} : \sum_i \gamma_i = 1\}$. The mapping $(\gamma, e) \mapsto \sum_i \gamma_i e_i$ from $\Delta \times E^{\ell+1}$ to $\mathbb{R}^\ell$ is continuous so that its image is compact. By Carathéodory's theorem, its range is $\mathrm{co}(E)$. ∎

We will see infinite-dimensional vector spaces in which the convex hull of a compact set is not compact.

5.4.b.3 Affine Hulls

If we take the line $L = \{(x_1, x_2) : x_2 = x_1\}$ in $\mathbb{R}^2$ and shift it, for example, to $L + (0, 1) = \{(x_1, x_2) : x_2 = 1 + x_1\}$, the new set is not a linear subspace of $\mathbb{R}^2$. Though on its own it looks like one, it is what we call an **affine** subspace.

Definition 5.4.15 $W \subset \mathbb{R}^\ell$ *is an **affine subspace** if $W = \{\mathbf{z}\} + V$ for some vector subspace $V \subset \mathbb{R}^\ell$ and point $\mathbf{z} \in \mathbb{R}^\ell$. The **dimension of** W is defined as the dimension of V.*

Contrast convex combinations, $\alpha\mathbf{x} + (1 - \alpha)\mathbf{y}$ for $\alpha \in [0, 1]$, with the following.

Definition 5.4.16 *An **affine combination** of $\mathbf{x}$, $\mathbf{y}$ is a point of the form $r\mathbf{x} + (1 - r)\mathbf{y}$ for $r \in \mathbb{R}$.*

Example 5.4.17 *If $\mathbf{x}$ is not a multiple of $\mathbf{y}$ in $\mathbb{R}^\ell$, then the set of affine combinations of $\mathbf{x}$ and $\mathbf{y}$, $\{r\mathbf{x} + (1 - r)\mathbf{y} : r \in \mathbb{R}\}$, is a one-dimensional affine subspace of $\mathbb{R}^\ell$. If $\mathbf{x}$ is a multiple of $\mathbf{y}$, then it is a one-dimensional vector subspace (Figure 5.4.17).*

Lemma 5.4.18 W *is an affine subspace of $\mathbb{R}^\ell$ iff for all $\mathbf{x}, \mathbf{y} \in W$ and all $r \in \mathbb{R}$, $r\mathbf{x} + (1 - r)\mathbf{y} \in W$, that is, iff W is closed under affine combinations.*

Proof. Suppose that $W = \{\mathbf{z}\} + V$, V a vector subspace. For any $\mathbf{x}, \mathbf{y} \in W = \{\mathbf{z}\} + V$, $r\mathbf{x} + (1 - r)\mathbf{y} = r(\mathbf{x}' + \mathbf{z}) + (1 - r)(\mathbf{y}' + \mathbf{z}) = (r\mathbf{x}' + (1 - r)\mathbf{y}') + \mathbf{z}$ for some $\mathbf{x}', \mathbf{y}' \in V$. Since V is a vector subspace, $(r\mathbf{x}' + (1 - r)\mathbf{y}') \in V$.

Now suppose that for all $\mathbf{x}, \mathbf{y} \in W$ and all $r \in \mathbb{R}$, $r\mathbf{x} + (1 - r)\mathbf{y} \in W$. We proceed in two steps: first, we show that for any $\mathbf{w} \in W$, $W = \{\mathbf{w}\} + (W - W)$, and second, we show that $(W - W)$ is a vector subspace of $\mathbb{R}^\ell$.

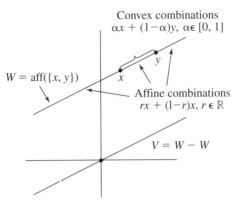

FIGURE 5.4.17

Step 1: Pick $\mathbf{w} \in W$. We show that $W \subset \{\mathbf{w}\} + (W - W) \subset W$. For any $\mathbf{x}, \mathbf{w} \in W$, $\mathbf{x} = \mathbf{w} + (\mathbf{x} - \mathbf{w})$, so that $\mathbf{x} \in \{\mathbf{w}\} + (W - W)$, that is, $W \subset \{\mathbf{w}\} + (W - W)$. Any $\mathbf{z} \in \{\mathbf{w}\} + (W - W)$ is of the form $\mathbf{z} = \mathbf{w} + (\mathbf{x} - \mathbf{y})$ for some $\mathbf{x}, \mathbf{y} \in W$. We wish to show that $\mathbf{z} \in W$. Now, $\mathbf{v} := (2\mathbf{w} + (1 - 2)\mathbf{y}) \in W$ and $\mathbf{v}' := (2\mathbf{x} + (1 - 2)\mathbf{y}) \in W$ by assumption (taking $r = 2$). Further, $\mathbf{z} = \frac{1}{2}\mathbf{v} + \frac{1}{2}\mathbf{v}'$ (taking $r = \frac{1}{2}$); hence it belongs to W as well, since $\frac{1}{2}\mathbf{v} + \frac{1}{2}\mathbf{v}' = \mathbf{w} + (\mathbf{x} - \mathbf{y})$, showing that $\{\mathbf{w}\} + (W - W) \subset W$.

Step 2: For $\mathbf{u}, \mathbf{v} \in (W - W)$, $\alpha, \beta \in \mathbb{R}$, we show that $\alpha\mathbf{u} + \beta\mathbf{v} \in (W - W)$. We do this in two steps: (a) for all $r \in \mathbb{R}$, $r(W - W) \subset (W - W)$, and (b) $(W - W) + (W - W) \subset (W - W)$. Since $\alpha\mathbf{x} \in (W - W)$ and $\beta\mathbf{y} \in (W - W)$, this completes the proof.

(a) For $\mathbf{u} \in (W - W)$, there exist $\mathbf{x}, \mathbf{y} \in W$ such that $\mathbf{u} = (\mathbf{x} - \mathbf{y})$. Pick $r \in \mathbb{R}$, and $r(\mathbf{x} - \mathbf{y}) = (r\mathbf{x} + (1 - r)\mathbf{y}) - \mathbf{y} \in (W - W)$.

(b) For $\mathbf{x}, \mathbf{x}', \mathbf{y}, \mathbf{y}' \in W$, $\mathbf{v} := (\frac{1}{2}\mathbf{x} + \frac{1}{2}\mathbf{x}') \in W$ and $\mathbf{v}' := (\frac{1}{2}\mathbf{y} + \frac{1}{2}\mathbf{y}') \in W$. $(\mathbf{x} - \mathbf{y}) + (\mathbf{x}' - \mathbf{y}') = 2(\mathbf{v} + -\mathbf{v}') \in 2 \cdot (W - W)$. By (a), $2 \cdot (W - W) \subset (W - W)$, so $(\mathbf{x} - \mathbf{y}) + (\mathbf{x}' - \mathbf{y}') \in (W - W)$. ■

A striking aspect of the previous proof is that for any $\mathbf{w} \in W$, W an affine subspace, $W = \{\mathbf{w}\} + V$, where $V = (W - W)$. This implies that the class of affine subspaces is closed under intersection: let $\{W_i = \{\mathbf{w}_i\} + V_i : i \in I\}$ be a collection of affine spaces, pick $\mathbf{w} \in \cap_i W_i$, note that $W_i = \{\mathbf{w}\} + V_i$, and conclude that $\cap_i W_i = \{\mathbf{w}\} + \cap_i V_i$. Since arbitrary intersections of vector subspaces are vector subspaces, $\cap_i W_i = \{\mathbf{w}\} + V$ for the vector subspace $V = \cap_i V_i$, which means that the following shrink-wrapping definition is sensible.

Definition 5.4.19   *For $E \subset \mathbb{R}^\ell$, $\mathrm{aff}(E) = \cap\{W : E \subset W, W$ is an affine subspace\} is the **affine hull** of E.*

Later we define the dimension of a convex set as the dimension of its affine hull. This gives us a "natural" vector space in which to think about the boundary and interior of a convex set. We also use affine hulls in proving the Krein-Milman theorem for $\mathbb{R}^\ell$, which says that every compact convex subset of $\mathbb{R}^\ell$ is the convex hull of its extreme points.

5.4.c Neoclassical Duality

The neoclassical theory of the firm assumes that firms are characterized by a technology, Y, which specifies which input-output vectors are possible, that they take prices, $\mathbf{p}$, as given, and that they follow the behavioral rule, $\max_{\mathbf{y} \in Y} \mathbf{p}\mathbf{y}$.

In intermediate microeconomics, you learned about production functions, $f : \mathbb{R}^N_+ \to \mathbb{R}$, where $y = f(\mathbf{x})$ is the maximal technologically feasible amount of output produced by an input vector $\mathbf{x}$. Setting $\ell = (N + 1)$, in graduate microeconomics, we express this as a set of technologically feasible netput vectors: $Y = \{(y, \mathbf{x}) \in \mathbb{R} \times \mathbb{R}^N_- : y \le f(-\mathbf{x})\}$. A further difference between production theory in intermediate microeconomics and in graduate microeconomics is that graduate microeconomics allows firms to produce more than one good, that is, $\mathbf{y} \in Y$ may have two or more positive components.

Example 5.4.20 *Most firms produce multiple outputs. This is easiest to visualize when there is one input and two outputs. For example, consider the technologies* $Y_\rho = \{(y_1, y_2, y_3) : y_1 \le 0, y_2, y_3 \ge 0, (y_2^2 + y_3^2)^\rho \le -y_1\}$. *For* $0 < \rho < \frac{1}{2}$, *this is a convex technology, for* $\rho = \frac{1}{2}$, *it is a convex cone.*

Definition 5.4.21 *With* $\mathbf{p} \in \mathbb{R}^\ell_{++}$, *the neoclassical **profit function** for a firm with technology characterized by Y is*

$$\pi(\mathbf{p}) = \pi_Y(\mathbf{p}) = \sup\{\mathbf{p}\mathbf{y} : \mathbf{y} \in Y\}. \tag{5.3}$$

If $y(\mathbf{p})$ *solves the maximization problem* $\max\{\mathbf{p}\mathbf{y} : \mathbf{y} \in Y\}$, *then it is the profit maximizing netput behavior for the firm; the positive component(s) of $y(\mathbf{p})$ are the **supply function(s)/correspondence** and the negative component(s) are the **input demand function(s)/correspondence**.*

Exercise 5.4.22 [↑Lemma 5.1.39] The technologically feasible input-output pairs for a profit-maximizing price-taking firm is given by

$$Y = \{(0, 0), (-4, 12), (-9, 17), (-7, 13)\} \subset \mathbb{R}^2.$$

1. Find the firm's supply and demand function $y(\mathbf{p})$.
2. Find the firm's profit function, $\pi(\mathbf{p}) = \pi_Y(\mathbf{p})$.
3. Show that $Y' = \{\mathbf{y} \in \mathbb{R}^2 : (\forall \mathbf{p} \gg 0)[\mathbf{p}\mathbf{y} \le \pi(\mathbf{p})]\}$ is equal to the necessarily closed $\overline{\mathrm{co}}(Y) + \mathbb{R}^2_-$. Further, show that $Y' = \bigcap\{H^{\le}_{\mathbf{p}}(\pi(\mathbf{p})) : \mathbf{p} \gg 0\}$, which also implies that Y' is closed and convex. [Some hints for finding Y': You know that $Y \subset Y'$. You have shown that Y' is closed and convex. You can also show that it satisfies free disposal as defined in Example 5.1.36.]
4. Show that $\pi_Y(\mathbf{p}) = \pi_{Y'}(\mathbf{p})$ for all $\mathbf{p} \gg 0$ and characterize the difference(s) between the supply and input demand functions for Y and for Y'.

If we think that profits go to shareholders/owners of the firm and observe that this profit enters their budget sets as consumers, then we do not care if we use Y or Y' in thinking about their consumption behavior. At least for these purposes, there is no loss in assuming that a technology is convex and satisfies free disposal.

The neoclassical theory of consumer demand assumes that people have fixed preferences and maximize them over their budget sets, taking prices as given.

Assuming the existence of a utility function representing a complete preference ordering rules out nothing of particular interest to us. Duality theory shows that assuming convexity also matters very little.

Definition 5.4.23   *Given a utility function $u : \mathbb{R}_+^\ell \to \mathbb{R}$ and a utility level $\overline{u}$, $C_u(\overline{u}) := \{\mathbf{x} \in \mathbb{R}_+^\ell : u(\mathbf{x}) \geq \overline{u}\}$ is an **upper contour set of** u. Preferences are **convex** if all upper contour sets are convex.*

*For $\mathbf{p} \gg 0$ and wealth $w > 0$, the indirect utility function is $v(\mathbf{p}, w) = \sup\{u(\mathbf{x}) : \mathbf{x} \geq 0, \mathbf{px} \leq w\}$. If $x(\mathbf{p}, w)$ solves the problem $\max\{u(\mathbf{x}) : \mathbf{x} \geq 0, \mathbf{px} \leq w\}$, we call $x(\cdot, \cdot)$ the **demand function**. For $\mathbf{p} \gg 0$, the **expenditure function** is given by $e_u(\mathbf{p}, \overline{u}) = e(\mathbf{p}, \overline{u}) = \inf\{\mathbf{px} : \mathbf{x} \in C_u(\overline{u})\}$. If $h_u(\mathbf{p}, \overline{u}) = h(\mathbf{p}, \overline{u})$ solves the problem $\inf\{\mathbf{px} : \mathbf{x} \in C_u(\overline{u})\}$, we call $h(\cdot, \cdot)$ the **Hicksian demand functions**.[4]*

Exercise 5.4.24   For $(x_1, x_2) \in \mathbb{R}_+^2$, let $f(x_1, x_2) = \min\{x_1, 2x_2\}$, $g(x_1, x_2) = \min\{2x_1, x_2\}$ and define

$$u(x_1, x_2) = \max\{f(x_1, x_2), g(x_1, x_2)\}.$$

1. Show that preferences represented by f, g, or u are monotonic (as defined in Example 5.1.36).
2. Find the sets $C_f(6) = \{(x_1, x_2) : f(x_1, x_2) \geq 6\}$, $C_g(6) = \{(x_1, x_2) : g(x_1, x_2) \geq 6\}$, and $C_u(6) = \{(x_1, x_2) : u(x_1, x_2) \geq 6\}$.
3. Show that $u(\cdot)$ does not represent convex preferences.
4. Find the expenditure functions, $e_f(\mathbf{p}, \overline{u})$, $e_g(\mathbf{p}, \overline{u})$, and $e_u(\mathbf{p}, \overline{u})$. Also find the associated demand functions and the associated Hicksian demand functions.
5. Let $C_f^{exp}(6) = \{\mathbf{x} \in \mathbb{R}_+^2 : (\forall \mathbf{p} \gg 0)[\mathbf{px} \geq e_f(\mathbf{p}, 6)]\}$ with similar definitions for C_g^{exp} and C_u^{exp}. Show that $C_f^{exp}(6) = \bigcap\{H_{\mathbf{p}}^{\geq}(e_f(\mathbf{p}, 6)) : \mathbf{p} \gg 0\}$ with similar results for C_g^{exp} and C_u^{exp}.
6. Give a continuous concave utility function $v(\cdot)$ having $C_v(r) = C_u^{exp}(r)$ for all $r \in \mathbb{R}$ and $e_v(\mathbf{p}, r) = e_u(\mathbf{p}, r)$. What is the difference between the Hicksian demand functions for v and those for u? What is the difference between the demand functions for v and those for u?
7. Show that $C_f^{exp}(\overline{u}) = C_f(\overline{u})$, that $C_g^{exp}(\overline{u}) = C_g(\overline{u})$, and that $C_u^{exp}(\overline{u}) = \overline{co}(C_u(\overline{u})) + \mathbb{R}_+^2$. Further, show that

$$\inf\{\mathbf{px} : \mathbf{x} \in C_u(\overline{u})\} = \inf\{\mathbf{px} : \mathbf{x} \in C_u^{exp}(\overline{u})\}.$$

[In doing welfare comparisons, we often try to calculate the least money needed to achieve a given level of utility. For such calculations, it does not matter if we use $\overline{co}(C_u) + \mathbb{R}_+^\ell$ or C_u.]

Exercises 5.4.22 and 5.4.24 showed that we can often, for economic purposes, replace sets by the intersection of a class of closed convex half-spaces that contain

---

4. And we use the word "functions" even when there is more than one solution to the minimization problem, that is, even when we should say "Hicksian demand correspondence."

them. For these analyses, convexity is an irrelevant assumption, as are monotonicity and free disposal.

5.5 ◆ Boundary Issues

The boundary of a general set can be pretty weird, but the pathologies are essentially absent when we look at the boundaries of convex sets. In this section we first explicate the previous sentence and then look at the separation of points on the boundaries of convex sets from the interior of the set. The main result is the supporting hyperplane theorem, and along the way to proving it, we investigate the main properties of support functions, which include the profit functions, the cost functions, and the expenditure functions from neoclassical economics.

5.5.a Interiors and Boundaries

Intuitively, the "interior" of a half-space $H_{\mathbf{y}}^{\leq}(r)$ is the set $H_{\mathbf{y}}^{<}(r)$ because any point in $H_{\mathbf{y}}^{<}(r)$ can be surrounded by a (perhaps tiny) ϵ-ball that stays entirely within $H_{\mathbf{y}}^{\leq}(r)$. The "boundary" of $H_{\mathbf{y}}^{\leq}(r)$ ought to be $\{\mathbf{x} : \mathbf{x} \cdot \mathbf{y} = r\}$, because this is the set of points right at the "edge" of the set $H_{\mathbf{y}}^{\leq}(r)$ and its complement. These sets have intuitive interiors and boundaries because they are convex. In general, the boundary and the interior of a set can seem rather odd. The following general treatment of interiors and boundaries could as well have appeared in Chapter 4 on metric spaces, but, until now, we have had little use for it.

Definition 5.5.1 *If E is a subset of M, (M, d) a metric space, the **interior of** E is defined as the largest open subset of E, $\mathrm{int}(E) = \cup\{G : G \subset E, G \text{ open}\}$. The **boundary of** E is defined as $\partial E = \mathrm{cl}(E) \cap \mathrm{cl}(E^c)$.*

Notation Alert 5.5.A *The symbol "∂" is also used for partial derivatives, for example, $\partial f(x, y)/\partial x$.*

Boundaries and interiors can have odd properties, $\mathrm{int}(\mathbb{Q}) = \emptyset$ and $\mathrm{cl}(\mathbb{Q}) = \mathbb{R} = \mathrm{cl}(\mathbb{Q}^c)$, so that $\partial \mathbb{Q} = \mathbb{R}$. Further, $\mathrm{int}(\mathbb{Q}) \cup \mathrm{int}(\mathbb{Q}^c) = \emptyset \cup \emptyset \subsetneqq \mathrm{int}(\mathbb{Q} \cup \mathbb{Q}^c) = \mathbb{R}$. As we will see, the boundaries and interiors of convex sets are better behaved. Meanwhile, here is some information on the general situation.

Theorem 5.5.2 *For all subsets E of the metric space (M, d),*

1. $\mathrm{int}(E) = (\mathrm{cl}(E^c))^c$,
2. *E is open iff $E = \mathrm{int}(E)$,*
3. $\partial E = \mathrm{cl}(E) \setminus \mathrm{int}(E) = \mathrm{cl}(E) \cap \mathrm{cl}(E^c)$,
4. $\partial E \cap \mathrm{int}(E) = \emptyset$,
5. $\mathrm{cl}(E) = \partial E \cup \mathrm{int}(E)$,
6. *$M = \mathrm{int}(E) \cup \partial E \cup \mathrm{int}(E^c)$ is a partition of M, and*
7. *$x \in \partial E$ iff $(\forall \epsilon > 0)[[B_\epsilon(x) \cap E \neq \emptyset] \wedge [B_\epsilon(x) \cap E^c \neq \emptyset]]$.*

Proof. For (1), note that an open $G \subset E$ iff the closed $G^c \supset E^c$. Therefore $\operatorname{int}(E) = \cup\{F^c : F \text{ closed}, F \supset E^c\} = (\cap\{F : F \text{ closed}, F \supset E^c\})^c = (\operatorname{cl}(E^c))^c$. Since E is closed iff $E = \operatorname{cl}(E)$, (1) implies the (2).

For (3), by definition $\partial E = \operatorname{cl}(E) \cap \operatorname{cl}(E^c)$. From the (1), $(\operatorname{cl}(E^c))^c = \operatorname{int}(E)$, so that $\partial E = \operatorname{cl}(E) \setminus \operatorname{int}(E)$ (because $A \cap B = A \setminus (B^c)$ for every A and B). ∎

Exercise 5.5.3 Complete the proof of Theorem 5.5.2.

Exercise 5.5.4 Show that $\operatorname{int}(H_{\mathbf{y}}^{\leq}(r)) = H_{\mathbf{y}}^{<}(r)$ and that $\partial H_{\mathbf{y}}^{\leq}(r) = \{\mathbf{x} : \mathbf{x} \cdot \mathbf{y} = r\}$. [Therefore, by Theorem 5.5.2, $H_{\mathbf{y}}^{\leq}(r)$ is not only convex, but is also closed.

Often, context matters. For example, the interval (a, b) is an open subset of $\mathbb{R}$, but if $\mathbb{R}$ is embedded in $\mathbb{R}^2$, say as the horizontal axis, the interval is no longer open because it has no interior. However, it does have an interior relative to $\mathbb{R} \subset \mathbb{R}^2$.

Definition 5.5.5 *If E is a subset of $M' \subset M$, (M, d) a metric space, the **interior of E relative to M'** or the **relative interior of E** is defined as the largest subset of E that is open relative to M', $\operatorname{rel int}(E) = \cup\{G \cap M' : (G \cap M') \subset E, G \text{ open}\}$.*

Exercise 5.5.6 Formulate the definition of the relative boundary and relative closure of $E \subset M'$. Check that you have the correct formulation by verifying that all of the results of Theorem 5.5.2 hold with the adjective "relative" inserted in the appropriate places.

The following gives a sense of how very weird closed sets and boundaries can be.

Exercise 5.5.7 Let $\{q_n : n \in \mathbb{N}\}$ be an enumeration of the rationals in $\mathbb{R}$. Let G be the open set $\cup_n B_{\epsilon/2^n}(q_n)$, and let F be the closed set G^c. The total "length" of the set G, spread out over the whole real line, must be less than $\sum_n \epsilon/2^n = \epsilon$.

1. Show that G is dense.
2. Show that $\operatorname{int} F = \emptyset$.
3. Show that $\partial F = \partial G = F$.

5.5.b Support Functions

Replacing a set by the intersection of all closed half-spaces containing them gives the closed convex hull.

Corollary 5.5.8 *For any $E \subset \mathbb{R}^{\ell}$, $\overline{\operatorname{co}}(E) = \cap\{H_{\mathbf{y}}^{\leq}(r) : E \subset H_{\mathbf{y}}^{\leq}(r)\}$.*

Proof. If $\mathbf{x} \in \overline{\operatorname{co}}(E)$, then $\mathbf{x}$ belongs to every closed half-space containing E because half-spaces are convex.

If $\mathbf{x} \notin \overline{\operatorname{co}}(E)$, then by Theorem 5.4.1, $\mathbf{x}$ and $\overline{\operatorname{co}}(E)$ can be strongly separated, which implies that for some $H_{\mathbf{y}}^{\leq}(r)$ containing E, $\mathbf{x} \notin H_{\mathbf{y}}^{\leq}(r)$. ∎

Note that $E \subset H_{\mathbf{y}}^{\leq}(r)$ iff $r \geq \sup\{\mathbf{ye} : \mathbf{e} \in E\}$. Since $H_{\mathbf{y}}^{\leq}(r') \subset H_{\mathbf{y}}^{\leq}(r)$ if $r' < r$, this means that the only half-spaces that we have to consider are the ones where $r = \sup\{\mathbf{ye} : \mathbf{e} \in E\}$.

Definition 5.5.9 *The **support function** or* sup-***based support function** of a set E is $\mu_E^{\sup}(\mathbf{y}) = \sup\{\mathbf{ye} : \mathbf{e} \in E\}$. The inf-**based support function** of E is $\mu_E^{\inf}(\mathbf{y}) = \inf\{\mathbf{ye} : \mathbf{e} \in E\}$.*

Notation Alert 5.5.B $\mu_E^{\sup}(\mathbf{y}) = -\mu_E^{\inf}(-\mathbf{y})$, *so the* inf-*based and the* sup-*based support functions are essentially the same. We are not as careful as we might be about specifying which one we mean when we write "μ_E is a support function." As an example of why we have this carefree attitude, we note that support functions are not affected by the convexity of a set E; that is, $\mu_E(\mathbf{y}) = \mu_{\mathrm{co}(E)}(\mathbf{y})$ for all $\mathbf{y}$, and this is true for either version of the support function.*

As a further example of this carefree attitude, we summarize the definitions of the profit function and the expenditure function by saying they are "support functions evaluated at positive vectors."

Corollary 5.5.10 *For any $E \subset \mathbb{R}^\ell$, [(1)]*

1. $\overline{\mathrm{co}}(E) = \bigcap\{H_{\mathbf{y}}^{\leq}(\mu_E^{\sup}(\mathbf{y})) : \mathbf{y} \in \mathbb{R}^\ell\}$, *and*

2. $\mathrm{cl}(\mathrm{co}(E) + \mathbb{R}_-^\ell) = \bigcap\{H_{\mathbf{y}}^{\leq}(\mu_E^{\sup}(\mathbf{y})) : \mathbf{y} \gg 0\}$.

Exercise 5.5.11 Prove Corollary 5.5.10.

Given a technology Y, we can find $\pi(\mathbf{p}) = \mu_Y^{\sup}(\mathbf{p})$. Given the profit function $\mathbf{p} \mapsto \pi(\mathbf{p})$, we have defined a new technology, Y', by the set $Y' = \bigcap\{H_{\mathbf{p}}^{\leq}(\mu^{\sup}(\mathbf{p})) : \mathbf{p} \gg 0\}$. Corollary 5.5.10 tells us that Y' is $\mathrm{cl}(\mathrm{co}(Y) + \mathbb{R}_-^\ell)$. Therefore, if Y is convex, closed, and satisfies free disposal, $Y = Y'$.

This is the essence of the results in neoclassical duality theory, which indicate that knowing one representation of the technology is the same as knowing another. More specifically, under some assumptions meant to deal with some pesky details, it makes no difference if we specify the technology, Y, and derive the profit function, $\pi(\mathbf{p}) = \mu_Y(\mathbf{p})$, or if we specify the profit function, $\pi(\cdot)$, and derive the technology, $Y = \{\mathbf{y} : \mathbf{py} \leq \pi(\mathbf{p})$ for all $\mathbf{p} \gg 0\}$. In the same fashion, knowing the utility function of a consumer is the same as knowing the upper contour sets, and knowing the upper contour sets is the same as knowing the expenditure function.

5.5.c Supporting Hyperplanes

Definition 5.5.12 *A **closed hyperplane** is defined as a set of the form $L^{-1}(r)$ for some nonzero continuous linear function $L : \mathbb{R}^\ell \to \mathbb{R}$. For $C \subset \mathbb{R}^\ell$, a **supporting hyperplane** at a point $\mathbf{x} \in \partial C$ is a closed hyperplane containing $\mathbf{x}$ such that $C \subset L^{-1}((\infty, r])$ or $C \subset L^{-1}([r, +\infty))$.*

Geometrically, if $C \subset H_{\mathbf{y}}^{\leq}(r)$ and $\mathbf{x} \in (C \cap \partial H_{\mathbf{y}}^{\leq}(r))$, then $\partial H_{\mathbf{y}}^{\leq}(r)$ is the support hyperplane at $\mathbf{x}$ (Figure 5.5.12). The first part of the following exemplifies the general relation between support hyperplanes and support functions.

Example 5.5.13 *Let $C = \mathrm{cl}(B_1(0)) \subset \mathbb{R}^2$ so that $\partial C = \{\mathbf{x} \in \mathbb{R}^2 : \|\mathbf{x}\| = 1\}$. For every $\mathbf{y} \neq 0$, $\mu_C(\mathbf{y}) = \|\mathbf{y}\|$, the point in ∂C that solves the problem $\sup\{\mathbf{yc} : \mathbf{c} \in C\}$ is $\mathbf{c}^*(\mathbf{y}) = \frac{\mathbf{y}}{\|\mathbf{y}\|}$. Let $L_{\mathbf{y}}(\mathbf{x}) = \mathbf{yx}$. The support hyperplane at $\mathbf{c}^*(\mathbf{y})$ is the set $L_{\mathbf{y}}^{-1}(r)$, where $r = \|\mathbf{y}\| = \mu_C(\mathbf{y})$.*

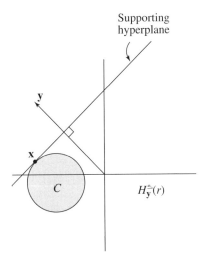

FIGURE 5.5.12

Let D be the nonconvex *set* $\{(x_1, x_2) : x_2 \leq (x_1)^2\}$. *D has no support hyperplanes, though the convex set* $\mathrm{cl}(D^c)$ *does have them.*

Lemma 5.5.14   *If* $\mathbf{y} \neq 0$, $\mu_C(\mathbf{y}) < \infty$, *and* $\mathbf{x} \in \partial C$ *satisfies* $\mathbf{yx} = \mu_C(\mathbf{y})$, *then* $L_{\mathbf{y}}^{-1}(r)$ *is a support hyperplane for C at* $\mathbf{x}$ *when* $r = \mu_C(\mathbf{y})$.

Proof.   This follows directly from the definitions.   ■

For the boundary and interior of a convex set, context matters. The set $\{(x_1, x_2) : x_1^2 + x_2^2 \leq 1\}$ is the unit disk in the $\mathbb{R}^2$, the plane. It can also be viewed as a subset of $\mathbb{R}^3$, that is, $\{(x_1, x_2, x_3) : x_1^2 + x_2^2 \leq 1, x_3 = 0\}$. In the plane, the boundary is $\{(x_1, x_2) : x_1^2 + x_2^2 = 1\}$ and the interior is $\{(x_1, x_2) : x_1^2 + x_2^2 < 1\}$. In $\mathbb{R}^3$, the boundary is the whole set and the interior is empty.

We are after context-independent definitions of the boundary and interior of a convex set. To this end, we define the relative interior and the relative boundary of a convex set as the interior relative to the smallest affine set containing it. That is, for the purposes of defining the relative interior, relative closure, and relative boundary of a convex $C \subset \mathbb{R}^\ell$, we regard C as a subset of the metric space $(\mathrm{aff}(C), d)$.

Definition 5.5.15   *The **relative interior of a convex** $C \subset \mathbb{R}^\ell$ is defined as* $\mathrm{rel\ int}(C) = \cup\{G : G \subset C, G$ *an open subset of* $\mathrm{aff}(C)\}$; *the **relative closure** is* $\cap\{F : C \subset F, F$ *a closed subset of* $\mathrm{aff}(C)\}$; *and the **relative boundary** is the set theoretic difference of relative closure and the relative interior.*

Theorem 5.5.16 (Supporting Hyperplane)   *If* $C \subset \mathbb{R}^\ell$ *is convex and* $\mathbf{x}$ *is in the relative boundary of C, then* $\mathbf{x}$ *can be strictly separated from* $\mathrm{rel\ int}(C)$ *by a support hyperplane at* $\mathbf{x}$.

Proof. Let C be a convex set containing more than one point,[5] and let $\mathbf{x}$ be a point in the relative boundary of C. Since C contains $\mathbf{x}' \neq \mathbf{x}''$, it contains all $\mathbf{x}'\gamma\mathbf{x}''$, $\gamma \in (0, 1)$, so that $\dim(\mathrm{aff}(C)) \geq 1$. Therefore rel $\mathrm{int}(C) \neq \emptyset$. Pick $\mathbf{c} \in$ rel $\mathrm{int}(C)$ and let $D = C - \mathbf{c}$ so that $0 \in$ rel $\mathrm{int}(D)$. As D contains the origin, $V = \mathrm{aff}(D)$ is a vector subspace of $\mathbb{R}^{\ell}$. There are two steps left: (1) find a linear function that defines a support hyperplane for $\mathbf{z} := \mathbf{x} - \mathbf{c}$, and (2) show how to translate it so that it is a support plane for C at $\mathbf{x}$.

(1) For this paragraph, we work in the finite-dimensional vector space V so that $\mathrm{int}(D) \neq \emptyset$. Since $\mathbf{z}$ is in ∂D, there exists a sequence $\mathbf{z}_n \in \mathrm{cl}(D)^c$, $\mathbf{z}_n \to \mathbf{z}$. For each $n \in \mathbb{N}$, pick $\mathbf{y}'_n \neq 0$ such that $D \subset H^{<}_{\mathbf{y}'_n}(r_n)$, where $r_n = \mathbf{y}'_n\mathbf{z}_n$. We can do this by strong separation. The sequence $\mathbf{y}_n := \mathbf{y}'_n/\|\mathbf{y}'_n\|$ belongs to the compact set $\{\mathbf{x} \in V : \|\mathbf{x}\| = 1\}$. Taking a convergent subsequence if necessary gives us $\mathbf{y}_n \to \mathbf{y}$. For all $n \in \mathbb{N}$ and all $\mathbf{d} \in \mathrm{int}(D)$, $\mathbf{y}_n\mathbf{d} < \mathbf{y}_n\mathbf{z}$, so that $\mathbf{y}\mathbf{d} \leq \mathbf{y}\mathbf{z}$. Suppose that for some $\mathbf{d} \in \mathrm{int}(D)$, $\mathbf{y}\mathbf{d} = \mathbf{y}\mathbf{z}$. Since $\mathbf{d} \in \mathrm{int}(D)$, there exists an $\epsilon > 0$ such that $\mathbf{d} + \epsilon\mathbf{y} \in \mathrm{int}(D)$. However, this implies that $\mathbf{y}(\mathbf{d} + \epsilon\mathbf{y}) = \mathbf{y}\mathbf{d} + \epsilon\mathbf{y}\mathbf{y} = \mathbf{y}\mathbf{z} + \epsilon$, a contradiction.

(2) Set $r = \mathbf{y}(\mathbf{z} + \mathbf{c}) = \mathbf{y}\mathbf{x}$. ∎

5.5.d The Two Fundamental Theorems of Welfare Economics

At the end of Chapter 1 we discussed the first fundamental theorem (FFT) of welfare economics and we gave the basics of the neoclassical model of an exchange economy and a Walrasian equilibrium in Definition 4.12.7. The FFT said that if $(\mathbf{x}, \mathbf{p})$ is a Walrasian equilibrium, then $\mathbf{x}$ is Pareto efficient. The second fundamental theorem of welfare economics (SFT) is a partial converse to the FFT.

We interpret the FFT as meaning that an equilibrium in price-mediated, voluntary trade, where prices are taken as given, results in an efficient allocation—all possible mutually improving trades have been made. There are essentially no conditions on the model of the economy needed in order for this to be true. The SFT says that if the economy meets a number of conditions, then any efficient allocation is an equilibrium (in price-mediated, voluntary trade, where prices are taken as given), provided that the initial allocation is correct. The proviso is important— a grotesquely unfair efficient allocation gives me all of the pie and you none of it. Such an allocation could not result from voluntary trade if you started with all the pie and all of everything else as well. It could, however, result from involuntary "trade."

Let $M = \mathbb{R}^{\ell}_{+} \setminus \{0\}$. We say that preferences $\succeq_i$ are **strictly monotonic on** X_i if for all allocations $x_i \in X_i$, $(x_i + M) \subset B_i(x_i)$, where $B_i(x_i) = \{y_i \in X_i : y_i \succ_i x_i\}$. Strict monotonicity means that an increase in the consumption of any good is strictly preferred.

Theorem 5.5.17 (Second Fundamental Theorem of Welfare Economics) *Suppose that preferences are continuous, convex, and strictly monotonic. If*

---

5. If C contains just one point, then $C = $ rel $\mathrm{int}(C)$ and $\partial C = \emptyset$, so the result is vacuously true. We are just being fussy here. We know you will never worry about the one-point case, but just in case you do, we have it covered.

$\mathbf{x} = (\mathbf{x}_i)_{i \in I}$ *is Pareto efficient, then there is an initial allocation* $\mathbf{y} = (\mathbf{y}_i)_{i \in I}$ *and a price vector* $\mathbf{p}$ *such that* $(\mathbf{p}, \mathbf{x})$ *is an equilibrium when the initial endowments are* $\mathbf{y}$.

Proof.   Since preferences are continuous and monotonic, $\mathbf{x}_i \in \partial B_i(\mathbf{x}_i)$.

We first show that $\mathbf{z} := \sum_i \mathbf{x}_i \in \partial B$, where $B := \sum_i B_i(\mathbf{x}_i)$. Pick $\epsilon > 0$. We must show that $B_\epsilon(\mathbf{z}) \cap B \neq \emptyset$ and $B_\epsilon(\mathbf{z}) \cap B^c \neq \emptyset$. For the first, note that since $\mathbf{x}_i \in \partial B_i(\mathbf{x}_i)$ and $B_{\epsilon/I}(\mathbf{x}_i) \cap B_i \neq \emptyset$, so that $B_\epsilon(\sum_i) \cap \sum_i B_i(\mathbf{x}_i) \neq \emptyset$. For the second note that $\mathbf{z} \notin B$.

Each $B_i(\mathbf{x}_i)$ is open because preferences are continuous. The set B is open because it is the sum of open sets (Lemma 5.1.39). We now show that B is convex. For this, it is sufficient to show that each B_i is convex. Suppose that $\mathbf{x}_i', \mathbf{x}_i'' \succ_i \mathbf{x}_i$ and $\gamma \in (0, 1)$. We must show that $\mathbf{x}_i' \gamma \mathbf{x}_i'' \succ_i \mathbf{x}_i$. There are two possible cases, $\mathbf{x}_i' \succeq \mathbf{x}_i''$ and $\mathbf{x}_i'' \succeq \mathbf{x}_i'$. If $\mathbf{x}_i' \succeq \mathbf{x}_i''$, then $\mathbf{x}_i' \gamma \mathbf{x}_i'' \succeq_i \mathbf{x}_i'' \succ_i \mathbf{x}_i$. If $\mathbf{x}_i'' \succeq \mathbf{x}_i'$, then $\mathbf{x}_i' \gamma \mathbf{x}_i'' \succeq_i \mathbf{x}_i' \succ_i \mathbf{x}_i$.

By the supporting hyperplane theorem, there exists a nonzero vector $\mathbf{p}$ such that for all $\mathbf{y} \in B$, $\mathbf{p}\mathbf{z} < \mathbf{p}\mathbf{y}$. We now show that $\mathbf{p} \gg 0$. Suppose that $p_k \leq 0$ for good k. Let $\mathbf{e}_k$ be the unit vector in the kth direction. By monotonicity, the vector $\mathbf{y} = \mathbf{z} + \mathbf{e}_k \in B$, but $\mathbf{p}(\mathbf{z} + \mathbf{e}_k) \leq \mathbf{p}\mathbf{z}$, a contradiction.

Finally, let $(\mathbf{y}_i)_{i \in I}$ be any allocation such that for all $i \in I$, $\mathbf{p}\mathbf{x}_i = \mathbf{p}\mathbf{y}_i$.   ∎

5.6 ◆ Concave and Convex Functions

Concave and convex functions are defined on convex subsets of $\mathbb{R}^\ell$. There is no such thing as a concave set. Throughout this section, C is a nonempty convex subset of $\mathbb{R}^\ell$ that serves as the domain of the functions we are interested in. Rather than worry about relative interiors and boundaries, we assume that $\text{int}(C) \neq \emptyset$ so that $\text{aff}(C) = \mathbb{R}^\ell$.

5.6.a   Basic Definitions and Examples

The essential geometric picture is that if rain falls on the graph of a concave function, it will run off. This is the same as saying that everything below the function is a convex set.

Definition 5.6.1   *For* $f : C \to \mathbb{R}$, *the* **subgraph of** f *is* $\text{sub}(f) = \{(\mathbf{x}, y) \in C \times \mathbb{R} : y \leq f(\mathbf{x})\}$ *and the* **epigraph of** f *is* $\text{epi}(f) = \{(\mathbf{x}, y) \in C \times \mathbb{R} : f(\mathbf{x}) \leq y\}$.

We have seen subgraphs before; the subgraph of a production function is the technology that it defines.

Definition 5.6.2   *A function* $f : C \to \mathbb{R}$ *is* **concave** *if* $\forall \mathbf{x}, \mathbf{x}' \in C$, $\forall \gamma \in (0, 1)$, *and* $f(\mathbf{x}\gamma\mathbf{x}') \geq f(\mathbf{x})\gamma f(\mathbf{x}')$. *More explicitly,* $f(\gamma\mathbf{x} + (1 - \gamma)\mathbf{x}') \geq \gamma f(\mathbf{x}) + (1 - \gamma)f(\mathbf{x}')$. *A function is* **convex** *if the function* $(-f)$ *is concave, equivalently if for all* $\mathbf{x}, \mathbf{x}' \in C$ *and all* $\gamma \in (0, 1)$, $f(\mathbf{x}\gamma\mathbf{x}') \leq f(\mathbf{x})\gamma f(\mathbf{x}')$ *(Figure 5.6.2).*

Theorem 5.6.3 $f : C \to \mathbb{R}$ *is concave iff* $\text{sub}(f)$ *is a convex set.*

Proof.   Suppose that f is concave. If $(\mathbf{x}, y)$, $(\mathbf{x}', y') \in \text{sub}(f)$ and $\gamma \in (0, 1)$, then $(y\gamma y') \leq f(\mathbf{x})\gamma f(\mathbf{x}')$ because $y \leq f(\mathbf{x})$ and $y' \leq f(\mathbf{x}')$.

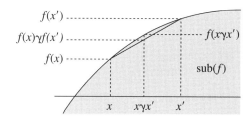

FIGURE 5.6.2 sub(f) is a convex set iff f is concave.

Suppose that sub(f) is a convex set. Since $(\mathbf{x}, f(\mathbf{x}))$ and $(\mathbf{x}', f(\mathbf{x}')) \in$ sub(f), so is $(\mathbf{x}, f(\mathbf{x}))\gamma(\mathbf{x}', f(\mathbf{x}'))$, that is, $f(\mathbf{x})\gamma f(\mathbf{x}') \leq f(\mathbf{x}\gamma\mathbf{x}')$. ∎

A line connecting $(\mathbf{x}, f(\mathbf{x}))$ and $(\mathbf{y}, f(\mathbf{y}))$ is called a **chord** of f. Concave functions are always above their chords, whereas convex functions are always below them.

Example 5.6.4 *Linear functions are both concave and convex. If $f : (a, b) \to \mathbb{R}$ and f' is decreasing, f is concave (this is proved in Theorem 5.10.6). Thus, $f(r) = \sqrt{r}$ is concave on $\mathbb{R}_+$ and $g(r) = |r|^p$ is convex for $p \geq 1$. Since $\mathrm{epi}(\min\{f, g\}) = \mathrm{epi}(f) \cap \mathrm{epi}(g)$, the minimum of two concave functions is concave. By induction, the minimum of any finite number of concave functions is concave (Figure 5.6.4).*

Concave functions are well behaved on the interior of their domain, but may behave badly at the boundaries.

Example 5.6.5 *The function $f(r) = 1$ if $r \in (0, 1)$ and $f(0) = f(1) = 0$ is concave but not continuous on the convex set $C = [0, 1]$ and continuous on $\mathrm{int}(C) = (0, 1)$. For a more elaborate example, let $C = \{\mathbf{x} \in \mathbb{R}^2 : \|\mathbf{x}\| \leq 1\}$ and $f : C \to \mathbb{R}$. For $\|\mathbf{x}\| < 1$, set $f(\mathbf{x}) = 0$ and for $\|\mathbf{x}\| = 1$, $f(\mathbf{x})$ can be anything less than or equal to 0. This means that $f(\cdot)$ can be arbitrarily badly behaved on the boundary of its domain (Figure 5.6.5).*

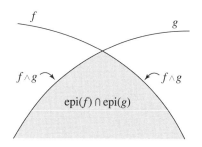

FIGURE 5.6.4 $\mathrm{epi}(f \wedge g) = \mathrm{epi}(f) \cap \mathrm{epi}(g)$.

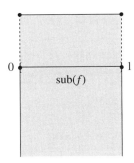

FIGURE 5.6.5

Definition 5.6.6 *A function $h : \mathbb{R}^\ell \to \mathbb{R}$ is **affine** if it is the sum of a linear function and a constant, that is, if $h(\mathbf{x}) = a + \mathbf{xy}$ for some $a \in \mathbb{R}$ and $\mathbf{y} \in \mathbb{R}^\ell$.*

The graph of an affine function is an affine subspace of $\mathbb{R}^\ell \times \mathbb{R}$, and the graph of a linear function is a vector subspace.

Definition 5.6.7 *An affine function $h : \mathbb{R}^\ell \to \mathbb{R}$ **majorizes** a concave function $f : C \to \mathbb{R}$ if for all $\mathbf{x} \in C$, $h(\mathbf{x}) \geq f(\mathbf{x})$.*

Theorem 5.6.8 *On the interior of its domain, any concave function is the lower envelope of the set of affine functions that majorize it; that is, for all $\mathbf{x} \in \mathrm{int}(C)$, $f(\mathbf{x}) = \inf\{h(\mathbf{x}) : h \text{ is affine and majorizes } f\}$.*

Lemma 5.4.8 tells us that this should be true, $\mathrm{sub}(f)$ is the intersection of the closed half-spaces that contain it, and the graph of the boundary of the half-space is the graph of an affine function. The proof fills in the details of this intuition.

Proof. It is clear that for all $\mathbf{x} \in C$, $f(\mathbf{x}) \leq \inf\{h(\mathbf{x}) : h \text{ is affine and majorizes } f\}$. To prove the result, it is sufficient to show that for any $\mathbf{x} \in \mathrm{int}(C)$, there is at least one affine h that majorizes f and satisfies $f(\mathbf{x}) = h(\mathbf{x})$.

Pick $\mathbf{x}^\circ \in \mathrm{int}(C)$, and let $y^\circ = f(\mathbf{x}^\circ)$; $(\mathbf{x}^\circ, y^\circ)$ belongs to the boundary of $\mathrm{sub}(f)$ (because $(\mathbf{x}^\circ, y^\circ \pm \epsilon)$ belong to $\mathrm{sub}(f)$ and its complement). By the supporting hyperplane theorem, $(\mathbf{x}^\circ, y^\circ)$ can be strictly separated from $\mathrm{int}(\mathrm{sub}(f))$ by a nonzero linear function $L : \mathbb{R}^\ell \times \mathbb{R} \to \mathbb{R}$, $L(\mathbf{x}, y) = \mathbf{v}_x \mathbf{x} + v_y y$, with $L((\mathbf{x}^\circ, y^\circ)) > L(\mathrm{int}(\mathrm{sub}(f)))$. Since $(\mathbf{x}, f(\mathbf{x}) - r) \in \mathrm{int}(\mathrm{sub}(f))$ for all $r > 0$, we know that $v_y > 0$.

Consider the affine function $h(\mathbf{x}) = \frac{1}{v_y}(\mathbf{x}^\circ - \mathbf{x}) + y^\circ$; $h(\mathbf{x}^\circ) = 0 + y^\circ = f(\mathbf{x}^\circ)$, so all that is left to show is that h majorizes f, that is, for any $\mathbf{x} \in C$,

$$\frac{1}{v_y}(\mathbf{x}^\circ - \mathbf{x}) + y^\circ \geq f(\mathbf{x}). \tag{5.4}$$

By the choice of L,

$$\mathbf{v}_x \mathbf{x}^\circ + v_y y^\circ \geq \mathbf{v}_x \mathbf{x} + v_y f(\mathbf{x}). \tag{5.5}$$

Subtracting $\mathbf{v}_x\mathbf{x}$ from both sides and dividing by the strictly positive v_y yields (5.4). ■

Example 5.6.9 *The support function* $\mu_E(\mathbf{y}) = \sup\{\mathbf{xy} : \mathbf{x} \in E\}$ *is finite if* E *is bounded. Let* $L_\mathbf{x}(\mathbf{y}) = \mathbf{xy}$. *Being the supremum of a collection of linear functions,* $\{L_\mathbf{x}(\cdot) : \mathbf{x} \in E\}$, *the epigraph of* μ_E *is a convex set, so that* μ_E *is a convex function.*

5.6.b Convex Preferences and Quasi-Concave Utility Functions

Recall that preferences on C are convex iff for all $\mathbf{y} \in C$, $\{\mathbf{x} \in C : \mathbf{x} \succsim \mathbf{y}\}$ is a convex set. Convexity of preferences is interpreted as a desire for variation in consumption. Concave utility functions represent convex preferences.

Lemma 5.6.10 *A concave utility function,* $u : C \to \mathbb{R}$, *represents convex preferences.*

Proof. For all r, $\{\mathbf{x} \in C : u(\mathbf{x}) \geq r\}$ is a convex set because, if $u(\mathbf{x}), u(\mathbf{x}') \geq r$, then $u(\mathbf{x}\gamma\mathbf{x}') \geq u(\mathbf{x})\gamma u(\mathbf{x}') \geq \min\{u(\mathbf{x}), u(\mathbf{x}')\} \geq r$. ■

Exercise 5.6.11 Show that the lower contour sets of a convex function are convex sets.

If u represents preferences $\succsim$, then $f \circ u$ represents the same preferences if $f : \mathbb{R} \to \mathbb{R}$ is monotonic. Monotonic transformations of concave functions are **quasi-concave**.

Definition 5.6.12 *A function* $u : C \to \mathbb{R}$ *is* **quasi-concave** *if for all* $\mathbf{y} \in C$, *the upper contour set* $\{\mathbf{x} \in C : u(\mathbf{x}) \geq u(\mathbf{y})\}$ *is a convex set, and* **strictly quasi-concave** *if the upper contour set is strictly convex.*

Exercise 5.6.13 Show that u is quasi-concave iff $\mathbf{x}, \mathbf{x}' \in C$ and all $\gamma \in (0, 1)$, $u(\mathbf{x}\gamma\mathbf{x}') \geq \min\{u(\mathbf{x}), u(\mathbf{x}')\}$, and that it is strictly quasi-concave iff the inequality is strict for all $\mathbf{x} \neq \mathbf{x}'$.

Convex subsets of $\mathbb{R}$ are intervals, which makes quasi-concave functions on $\mathbb{R}$ easier to visualize.

Definition 5.6.14 *A function* $u : \mathbb{R} \to \mathbb{R}$ *is* **single peaked** *if there exists* $x \in \mathbb{R}$ *such that*

1. $[r < r' \leq x] \Rightarrow [u(r) \leq u(r') \leq u(x)]$, *and*
2. $[x \leq s' < s] \Rightarrow [u(x) \geq u(s') \geq u(s)]$.

Exercise 5.6.15 If $u : \mathbb{R} \to \mathbb{R}$, then u is quasi-concave iff it is either monotonic or single peaked.

The supporting hyperplane theorem gives another interpretation to convex preferences, one that involves decreasing marginal rates of substitution (MRS).

Lemma 5.6.16 *If* $C_u(\overline{u}) = \{\mathbf{x} \in \mathbb{R}^2 : u(\mathbf{x}) \geq \overline{u}\}$ *is an upper contour set for a continuous, quasi-concave monotonic utility function* $u : \mathbb{R}^2 \to \mathbb{R}$, *then for* $\mathbf{x} \neq \mathbf{x}'$ *in* $\partial C_u(\overline{u})$ *with* $x_1 < x_1'$ *and* $x_2 > x_2'$, *any supporting hyperplane at* $\mathbf{x}'$ *must have a shallower (lower in absolute value) slope than any supporting hyperplane at* $\mathbf{x}$.

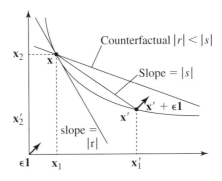

Counterfactual $|r| < |s|$

Slope $= |s|$

$\mathbf{x}' + \epsilon\mathbf{1}$

slope $= |r|$

FIGURE 5.6.16

Proof. Monotonicity implies that the supporting hyperplanes at $\mathbf{x}$ and $\mathbf{x}'$ have negative slopes. Let $|r|$ be the absolute value of the slope of a supporting hyperplane at $\mathbf{x}$, $|t|$ be the absolute value of the slope of a supporting hyperplane at $\mathbf{x}'$, and $|r|$ the slope of the line segment $[\mathbf{x}, \mathbf{x}']$. We show that $|r| \geq |s| \geq |t|$ (Figure 5.6.16).

Let $\mathbf{1} = (1, 1)$ be the vector of 1's. By monotonicity, for any $\epsilon > 0$, $\mathbf{x} + \epsilon\mathbf{1}$ and $\mathbf{x}' + \epsilon\mathbf{1}$ belong to int(C). If $|r| < |s|$, then for some small $\epsilon > 0$, $\mathbf{x}' + \epsilon\mathbf{1}$ is not separated from $\mathbf{x}$ by the hyperplane at $\mathbf{x}$. By parallel reasoning, $|s| < |t|$ leads to a contradiction. ∎

We interpret the (absolute value of the) slope of the support hyperplanes as the **marginal rate of substitution (MRS) between goods** 1 **and** 2. What we have shown is that convexity and monotonicity imply that the MRS weakly decreases as more of good 1 is consumed. To see that decreasing MRS implies the convexity of $C_u(\overline{u})$, note that decreasing MRS implies that $C_u(\overline{u})$ is the intersection of the half-spaces above the support hyperplanes, so it must be convex.

5.6.c Maximization and Derivative Conditions in the Interior

Even if the boundaries of the upper contour sets just considered are not differentiable, we find marginal rates of substitution, a concept usually based on calculus considerations. This is part of a larger pattern—concave and convex functions may not be differentiable, but they are nearly differentiable.

Definition 5.6.17 *For a concave function* $f : C \to \mathbb{R}$ *and* $\mathbf{x}° \in$ int(C), *the* **supdifferential** *of* f *at* $\mathbf{x}°$ *is the set of linear functions,* $h : \mathbb{R}^\ell \to \mathbb{R}$ *such that* $(\forall \mathbf{x} \in C)[f(\mathbf{x}) \leq f(\mathbf{x}°) + h(\mathbf{x} - \mathbf{x}°)]$. *The* **subdifferential** *of a convex function reverses the inequality (Figure 5.6.17).*

The supdifferential of f at $\mathbf{x}° \in$ int(C) is written $D_{\mathbf{x}}^{\text{sup}} f(\mathbf{x}°)$ and the subdifferential is written $D_{\mathbf{x}}^{\text{sub}} f(\mathbf{x}°)$ to mimic the differential notation for the derivative of f at $\mathbf{x}°$, $D_{\mathbf{x}} f(\mathbf{x}°)$. If $D_{\mathbf{x}}^{\text{sup}} f(\mathbf{x}°)$ contains just one linear function, that linear function is the derivative of f at $\mathbf{x}°$.

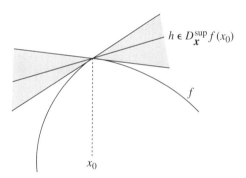

FIGURE 5.6.17　Any h with graph in the shaded region belongs to $D_{\mathbf{x}}^{\sup} f(x_0)$.

Theorem 5.6.18 *The supdifferential of a concave $f : C \to \mathbb{R}$ is nonempty at any $\mathbf{x}^\circ \in \operatorname{int}(C)$.*

Multiplying by -1 shows that the same is true for the subdifferential of convex functions.

Proof. Let $y^\circ = f(\mathbf{x}^\circ)$. The point $(\mathbf{x}^\circ, y^\circ)$ belongs to $\partial \operatorname{sub}(f)$ because $(\mathbf{x}^\circ, y^\circ + 1/n)$ converges from the outside of $\operatorname{sub}(f)$, whereas $(\mathbf{x}^\circ, y^\circ - 1/n)$ converges from inside $\operatorname{sub}(f)$. Therefore, the support hyperplane theorem can be applied to give a nonzero linear $L : \mathbb{R}^\ell \times \mathbb{R} \to \mathbb{R}$ such that $L((\mathbf{x}^\circ, y^\circ)) = r$ and $L(\operatorname{sub}(f)) \leq r$.

The function L can be written $L((\mathbf{x}, y)) = \mathbf{v}_x \mathbf{x} + v_y y$ for $\mathbf{v}_x \in \mathbb{R}^\ell$ and $v_y \in \mathbb{R}$. We now show that for $\mathbf{x}^\circ \in \operatorname{int}(C)$, any support hyperplane to $\operatorname{sub}(f)$ must have the property that $v_y > 0$. Geometrically, this follows from the observation that the vector normal to the hyperplane separating $(\mathbf{x}^\circ, f(\mathbf{x}^\circ))$ from $\operatorname{sub}(f)$ must point upward. Algebraically, there are two cases to consider, $v_y = 0$ and $v_y < 0$.

If $v_y = 0$, then $L((\mathbf{x}^\circ, y^\circ)) = \mathbf{v}_x \mathbf{x}^\circ + 0 y^\circ = r$. Pick $\epsilon > 0$ but small enough that $(\mathbf{x}^\circ + \epsilon \mathbf{v}_x) \in \operatorname{int}(C)$. For any y, including those for which $y < f(\mathbf{x}^\circ + \epsilon \mathbf{v}_x)$, $L((\mathbf{x}^\circ + \epsilon \mathbf{v}_x), y) = \mathbf{v}_x \mathbf{x}^\circ + \epsilon \mathbf{v}_x \mathbf{v}_x > r$, contradicting $L(\operatorname{sub}(f)) \leq r$.

If $v_y < 0$, then the points $\mathbf{x}_n = (\mathbf{x}^\circ, y^\circ - n)$ all belong to $\operatorname{sub}(f)$ but $L(\mathbf{x}_n) \uparrow +\infty$, which contradicts $L(\operatorname{sub}(f)) \leq r$.

Finally, rearranging $\mathbf{v}_x \mathbf{x}^\circ + v_y f(\mathbf{x}^\circ) \geq \mathbf{v}_x \mathbf{x} + v_y f(\mathbf{x})$ yields $\frac{1}{v_y} \mathbf{v}_x (\mathbf{x}^\circ - \mathbf{x}) \geq f(\mathbf{x}^\circ) - f(\mathbf{x})$, that is, that $h(\mathbf{x}) = \frac{1}{v_y} \mathbf{v}_x \mathbf{x}$ is in the supdifferential of f at $\mathbf{x}^\circ$. ∎

Definition 5.6.19 *Let $f : C \to \mathbb{R}$. A point $\mathbf{x}^* \in C$ is a **local maximizer** if*

$$(\exists \epsilon > 0)[[\|\mathbf{x} - \mathbf{x}^*\| < \epsilon] \Rightarrow [f(\mathbf{x}^*) \geq f(\mathbf{x})]],$$

*and $\mathbf{x}^* \in C$ is a **global maximizer** if*

$$(\forall \mathbf{x} \in C)[f(\mathbf{x}^*) \geq f(\mathbf{x})].$$

Theorem 5.6.20 *If $f : C \to \mathbb{R}$ is concave, then $\mathbf{x}^*$ is a local maximizer iff $\mathbf{x}^*$ is a global maximizer, and if $\mathbf{x}^* \in \operatorname{int}(C)$, then $\mathbf{x}^*$ is a global maximizer iff $0 \in D_{\mathbf{x}}^{\sup} f(\mathbf{x}^*)$.*

Proof.   If $\mathbf{x}^*$ is a global maximizer, then it must be a local maximizer. Suppose that $\mathbf{x}^*$ is a local maximizer but $(\exists \mathbf{x}' \in C)[f(\mathbf{x}') > f(\mathbf{x}^*)]$. For all $n \in \mathbb{N}$, let $\gamma_n = 1 - 1/n$. By concavity, $f(\mathbf{x}^* \gamma_n \mathbf{x}') \geq f(\mathbf{x}^*) \gamma_n f(\mathbf{x}') > f(\mathbf{x}^*)$. However, for large n, $d(\mathbf{x}^* \gamma_n \mathbf{x}') < \epsilon$, contradicting the assumption that $\mathbf{x}^*$ is a local maximizer.

Now suppose that $\mathbf{x}^* \in \mathrm{int}(C)$. If $0 \in D_{\mathbf{x}}^{\sup} f(\mathbf{x}^*)$, then by the definition of supdifferentials, for all $\mathbf{x} \in C$, $f(\mathbf{x}) \leq f(\mathbf{x}^*) + 0 \cdot (\mathbf{x} - \mathbf{x}^*) = f(\mathbf{x}^*)$, so that $\mathbf{x}^*$ is a global maximizer. Now suppose that $\mathbf{x}^*$ is a global maximizer, so that for all $\mathbf{x} \in C$, $f(\mathbf{x}) \leq f(\mathbf{x}^*) + 0 \cdot (\mathbf{x} - \mathbf{x}^*)$. This implies that $0 \in D_{\mathbf{x}}^{\sup} f(\mathbf{x}^*)$.   ∎

The need for the condition that $\mathbf{x}^* \in \mathrm{int}(C)$ in Theorem 5.6.20 appears in the following example.

Example 5.6.21 $f(x) = x$ *for* $x \in [0, 1]$ *is concave, and it is maximized by* $x^* = 1$, *but* $f'(x) \equiv 1$.

Recall that a function is quasi-concave iff its upper contour sets are convex and that monotonic transformations of concave functions are quasi-concave.

Corollary 5.6.22   *If* $f : C \to \mathbb{R}$ *is quasi-concave, then* $\mathbf{x}^*$ *is a local maximizer iff* $\mathbf{x}^*$ *is a global maximizer.*

Exercise 5.6.23   Show that the proof of the first part of Theorem 5.6.20 works for quasi-concave functions.

Exercise 5.6.24   Show that the function $f(x) = x^3$ is quasi-concave, but that $f'(x^*) = 0$ does not characterize the optimum of f over the convex set $\mathbb{R}$.

Corollary 5.6.25   *When* $f : C \to \mathbb{R}$ *is concave and* $h : \mathbb{R} \to \mathbb{R}$ *is continuously differentiable with* $h'(r) > 0$ *for all* $r \in \mathbb{R}$, *if* $D_{\mathbf{x}}(h \circ f)(\mathbf{x}^*) = 0$, *then* $\mathbf{x}^*$ *is a global maximizer.*

The statement "if $D_{\mathbf{x}}(h \circ f)(\mathbf{x}^*) = 0$" requires that the derivative exist.

Proof. $D_{\mathbf{x}}(h \circ f)(\mathbf{x}^*) = h'(f(\mathbf{x}^*)) D_{\mathbf{x}} f(\mathbf{x}^*)$ and $h'(f(\mathbf{x}^*)) > 0$.   ∎

5.6.d   Maximization and Derivative Conditions at the Boundary

Suppose that we are maximizing a linear function, $f(\mathbf{x}) = \mathbf{x} \cdot \mathbf{y}$, over a compact convex $C \subset \mathbb{R}^l$. Now, the gradient of f is $\mathbf{y}$, so solving for a point at which the gradient is 0 is not going to be useful. Despite this, there are useful derivative conditions.

Suppose that we can identify a set of extreme points, $\mathrm{extr}(C) \subset \partial C$ with the property that every point $\mathbf{x} \in C$ is a convex combination of points, $\mathbf{x}_i$, in $\mathrm{extr}(C)$, $\mathbf{x} = \sum_i \alpha_i \mathbf{x}_i$. By linearity,

$$\max_{\mathbf{x} \in C} f(\mathbf{x}) = \max_{\alpha} \sum_i \alpha_i f(\mathbf{x}_i). \tag{5.6}$$

To solve this maximization problem, we just pick the $\mathbf{x}_i \in \mathrm{extr}(C)$ that maximizes f, set the corresponding $\alpha_i = 1$, and we are done. At the optimum, $\mathbf{x}^*$, the gradient of f will not point into the interior of C, because if it did, moving in that direction would increase f.

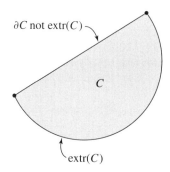

∂C not extr(C)

C

extr(C)

FIGURE 5.6.27

This means that if we are at the boundary of C and the gradient "points outward," we expect that we have maximized the function. This "pointing outward" is the substitute for the condition that $0 \in D_{\mathbf{x}}^{\sup} f(x^*)$ in Theorem 5.6.20. To make all of this precise requires the notion of an **extreme point** and the Krein-Milman theorem.

Notation 5.6.26 $[\mathbf{x}, \mathbf{y}] := \{\alpha \mathbf{x} + (1 - \alpha)\mathbf{y} : \alpha \in [0, 1]\}$ *denotes the line segment joining* $\mathbf{x}$ *and* $\mathbf{y}$.

Definition 5.6.27 $\mathbf{v} \in S$ *is an* **extreme point of** $S \subset \mathbb{R}^\ell$ *if for all* $[\mathbf{x}, \mathbf{y}] \subset S$, $\mathbf{v} \in [\mathbf{x}, \mathbf{y}]$ *implies that* $\mathbf{v} = \mathbf{x}$ *or* $\mathbf{v} = \mathbf{y}$. extr(S) *denotes the set of extreme points of* S *(Figure 5.6.27).*

Example 5.6.28   *Here* $(0, 1) \subset \mathbb{R}^1$ *has no extreme points;* $[0, 1] \subset \mathbb{R}^1$ *has the extreme points* 0 *and* 1; $\mathbb{R}_+^\ell$ *has the single extreme point* 0; *and for* $w > 0$, $\mathbf{p} \gg 0$, $\{\mathbf{x} \in \mathbb{R}_+^\ell : \mathbf{p} \cdot \mathbf{x} \leq w\}$ *has the extreme points* $\{0, \{\frac{w}{\mathbf{p}_i} \mathbf{e}_i\} : i = 1, \dots, \ell\}\}$.

Theorem 5.6.29 (Krein–Milman for $\mathbb{R}^\ell$)   *A compact convex* $C \subset \mathbb{R}^\ell$ *is equal to the convex hull of its extreme points. In particular,* extr$(C) \neq \emptyset$ *for compact convex* C.

Proof.   The result is trivial if $C = \emptyset$ or if $\dim(\mathrm{aff}(C)) = 0$.

The first part of the proof inductively argues that extr$(C) \neq \emptyset$ when $\dim(\mathrm{aff}(C)) = k - 1$ implies that extr$(C) \neq \emptyset$ when $\dim(\mathrm{aff}(C)) = k$. The second shows that $C \subset \mathbb{R}^\ell$ is equal to the convex hull of extr(C).

For extr$(C) \neq \emptyset$: Suppose that extr$(C) \neq \emptyset$ if $\dim(\mathrm{aff}(C)) \leq k - 1$. We inductively show that extr$(C) \neq \emptyset$ if $\dim(\mathrm{aff}(C)) = k$. Suppose that $\dim(\mathrm{aff}(C)) = k$ and pick $\mathbf{x}$ in the boundary of C relative to $\mathrm{aff}(C)$. By the supporting hyperplane theorem, $\mathbf{x}$ can be strictly separated from rel int(C) by a support hyperplane at $\mathbf{x}$. Let H be the support hyperplane and consider the set $C \cap H$. This is a compact convex set with dimension $(k - 1)$ or smaller. By the inductive hypothesis, $C \cap H$ has an extreme point, $\mathbf{v}$. We must show that $\mathbf{v}$ is an extreme point of C; that is, if $\mathbf{v} \in [\mathbf{x}, \mathbf{y}] \subset C$, we must show that $\mathbf{v} = \mathbf{x}$ or $\mathbf{v} = \mathbf{y}$.

If $[\mathbf{x}, \mathbf{y}] \subset (C \cap H)$, then $\mathbf{v} = \mathbf{x}$ or $\mathbf{v} = \mathbf{y}$ because $\mathbf{v}$ is an extreme point of $(C \cap H)$.

Suppose now that $[\mathbf{x}, \mathbf{y}] \subset C$ but $[\mathbf{x}, \mathbf{y}] \not\subset H$. Since H is a hyperplane, either $\mathbf{x} \notin H$ or $\mathbf{y} \notin H$, but not both. This means that H strongly separates $\mathbf{x}$ from $\mathbf{y}$, which contradicts H supporting C and $\mathbf{x}, \mathbf{y} \in C$.

For $C = \mathrm{co}(\mathrm{extr}(C))$: Since $\mathrm{extr}(C) \subset C$, it is clear that $\overline{\mathrm{co}}(\mathrm{extr}(C)) \subset C$. Suppose that there exists $\mathbf{v} \in C$ that is not in $\overline{\mathrm{co}}(\mathrm{extr}(C))$. Since $\overline{\mathrm{co}}(\mathrm{extr}(C))$ is compact and convex, it can be strongly separated from $\mathbf{v}$, that is, there exists a nonzero $\mathbf{y}$ such that $\mathbf{y} \cdot \mathbf{v} > \max\{\mathbf{y} \cdot \mathbf{x} : \mathbf{x} \in \overline{\mathrm{co}}(\mathrm{extr}(C))\}$. Let $r = \max\{\mathbf{y} \cdot \mathbf{x} : \mathbf{x} \in C\}$, so that

$$r > \max\{\mathbf{y} \cdot \mathbf{x} : \mathbf{x} \in \overline{\mathrm{co}}(\mathrm{extr}(C))\}. \tag{5.7}$$

Set $H = \{\mathbf{x} : \mathbf{y} \cdot \mathbf{x} = r\}$. By construction, $H \cap C \neq \emptyset$. Further, by the first step, $H \cap C$ has an extreme point. By the same argument as used in the first step, an extreme point of $H \cap C$ is an extreme point of C. But this contradicts (5.7).   ∎

Exercise 5.6.30   Show that any compact $S \subset \mathbb{R}^\ell$ has an extreme point (whether or not S is convex).

Exercise 5.6.31   We usually maximize concave functions and minimize convex functions over convex domains. Suppose we reverse this pattern. Show that if $f : C \to \mathbb{R}$ is a continuous convex function on the compact convex set C, then the problem $\max_{\mathbf{x} \in C} f(\mathbf{x})$ has at least one solution $\mathbf{x}^e \in \mathrm{extr}(C)$.

Lemma 5.6.32   *If $f : \mathbb{R}^\ell \to \mathbb{R}$ is linear and $C \subset \mathbb{R}^\ell$ is compact and convex, then at least one solution to $\max_{\mathbf{x} \in C} f(\mathbf{x})$ belongs to $\mathrm{extr}(C)$. Further, at any extreme solution, the gradient of f points outward; that is, if $\mathbf{x}^e$ is an extreme solution, then for all $\mathbf{x} \in C$, $D_\mathbf{x} f(\mathbf{x}^e) \cdot (\mathbf{x}^e - \mathbf{x}) \geq 0$.*

Recall that $\mathbf{xy} \geq 0$ iff $\cos(\theta) \geq 0$, where θ is the angle between $\mathbf{x}$ and $\mathbf{y}$. This only happens if $-90 \leq \theta \leq 90$, that is, only if $\mathbf{x}$ and $\mathbf{y}$ are pointing in more or less the same direction. A vector $(\mathbf{x}^e - \mathbf{x})$ is pointing from $\mathbf{x}$ toward the extreme point $\mathbf{x}^e$, that is, pointing outward. Thus, $D_\mathbf{x} f(\mathbf{x}^e)(\mathbf{x}^e - \mathbf{x}) \geq 0$ for all $\mathbf{x} \in C$ has the geometric interpretation that the gradient is also pointing outward.

Proof.   There exists a solution, $\mathbf{x}^*$, because f is continuous and C is compact. By Carathéodory's theorem and the Krein-Milman theorem, $\mathbf{x}^* = \sum_{i=1}^{\ell+1} \alpha_i \mathbf{x}_i$ for some collection of $\mathbf{x}_i$ in $\mathrm{extr}(C)$. Since $f(\mathbf{x}^*) = \sum_{i=1}^{\ell+1} \alpha_i f(\mathbf{x}_i)$ and $f(\mathbf{x}^*) \geq f(\mathbf{x}_i)$, if $\alpha_i > 0$, then $f(\mathbf{x}_i) = f(\mathbf{x}^*)$, and at least one of the α_i must be strictly positive. The last part follows from the observation that $D_\mathbf{x} f = \mathbf{y}$ for some $\mathbf{y}$, so that $f(\mathbf{x}^e) \geq f(\mathbf{x})$ is equivalent to $D_\mathbf{x} f(\mathbf{x}^e)(\mathbf{x}^e - \mathbf{x}) \geq 0$.   ∎

In view of this result, if a concave function has a derivative at an extreme point and that derivative points outward, then we expect that we have found a global maximizer. The following theorem shows that this is true, even if the concave function is only supdifferentiable.

Theorem 5.6.33   *If f is a concave function defined on an open neighborhood of a compact convex $C \subset \mathbb{R}^\ell$ and $\mathbf{x}^e$ is an extreme point of C with $h(\mathbf{x}^e - \mathbf{x}) \geq 0$ for all $h \in D_\mathbf{x}^{sup} f(\mathbf{x}^e)$ and $\mathbf{x} \in C$, then $\mathbf{x}^e$ solves the problem $\max_{\mathbf{x} \in C} f(\mathbf{x})$.*

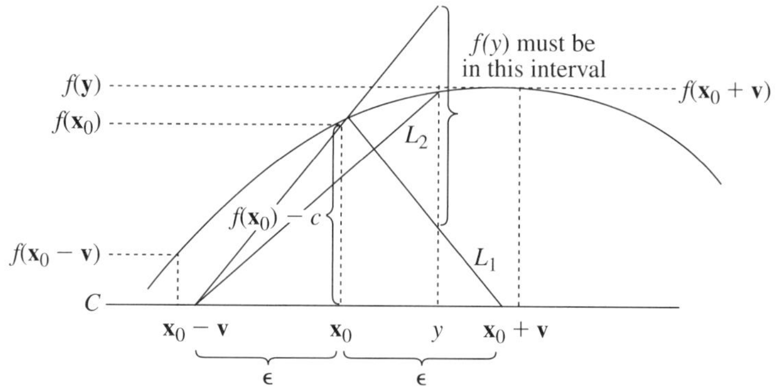

FIGURE 5.6.34

Proof. By the definition of h being in the supdifferential, for any $\mathbf{x} \in C$, $f(\mathbf{x}) \leq f(\mathbf{x}^e) + h(\mathbf{x} - \mathbf{x}^e)$, equivalently $f(\mathbf{x}) + h(\mathbf{x}^e - \mathbf{x}) \leq f(\mathbf{x}^e)$. If $h(\mathbf{x}^e - \mathbf{x}) \geq 0$ for all $\mathbf{x} \in C$, then $f(\mathbf{x}) \leq f(\mathbf{x}^e)$; that is, $\mathbf{x}^e$ solves the problem $\max_{\mathbf{x} \in C} f(\mathbf{x})$. ∎

5.6.e The Continuity of Concave Functions

We have seen that concave functions are subdifferentiable at all interior points of their domain. They are also continuous on the interior of their domain.

Theorem 5.6.34 *If f is concave (or convex) on C, then f is continuous on* int(C).

Proof. Pick $\mathbf{x}_0 \in \text{int}(C)$ and choose $\delta > 0$ such that $\text{cl}(B_\delta(\mathbf{x}_0)) \subset \text{int}(C)$. Let $S = \{\mathbf{x}_0 \pm \delta\mathbf{e}_i : i \leq \ell\}$, where $\mathbf{e}_i$ is the unit vector in the ith direction. For all $\mathbf{x} \in \text{co}(S)$, $f(\mathbf{x}) \geq c := \min\{f(s) : s \in S\}$. Pick $\epsilon \in (0, \delta)$ such that $B_\epsilon(\mathbf{x}_0) \subset \text{co}(S)$. We now show that for all $\mathbf{y} \in B_\epsilon(\mathbf{x}_0)$, $|f(\mathbf{y}) - f(\mathbf{x}_0)| \leq \frac{1}{\epsilon}(f(\mathbf{x}_0) - c)\|\mathbf{y} - \mathbf{x}_0\|$, which completes the proof.

Choose an arbitrary $\mathbf{y} \in B_\epsilon(\mathbf{x}_0)$. If $\mathbf{y} = \mathbf{x}_0$, the inequality holds because $0 \leq 0$. Suppose that $\mathbf{y} \neq \mathbf{x}_0$.

Let $\mathbf{v} = \epsilon(\mathbf{y} - \mathbf{x}_0)/\|\mathbf{y} - \mathbf{x}_0\|$, so that $\|\mathbf{v}\| = \epsilon$. Note that $\mathbf{x}_0 - \mathbf{v}$ and $\mathbf{x}_0 + \mathbf{v}$ belong to co(S), so that $f(\mathbf{x}_0 - \mathbf{v}) \geq c$ and $f(\mathbf{x}_0 + \mathbf{v}) \geq c$. Restrict attention to the line in C determined by the three points $\mathbf{x}_0 - \mathbf{v}$, $\mathbf{x}_0$, and $\mathbf{x}_0 + \mathbf{v}$. [See Figure 5.6.34.]

Since $\|\mathbf{y} - \mathbf{x}_0\| < \epsilon$, there exists $\alpha \in (0, 1)$ such that $\mathbf{y} = \alpha\mathbf{x}_0 + (1 - \alpha)(\mathbf{x}_0 + \mathbf{v})$ and $\beta \in (0, 1)$ such that $\mathbf{x}_0 = \beta(\mathbf{x}_0 - \mathbf{v}) + (1 - \beta)\mathbf{y}$. Therefore, the concavity of f implies that:

1. the point $(\mathbf{y}, f(\mathbf{y}))$ is above the line segment $[(\mathbf{x}_0, f(\mathbf{x}_0)); (\mathbf{x}_0 + \mathbf{v}, f(\mathbf{x}_0 + \mathbf{v}))]$, hence above the line segment $L_1 = [(\mathbf{x}_0, f(\mathbf{x}_0)), (\mathbf{x}_0 + \mathbf{v}, c)]$, which has slope equal to $\frac{1}{\epsilon}(f(\mathbf{x}_0) - c)$ in absolute value, and

2. the point $(\mathbf{x}_0, f(\mathbf{x}_0))$ is above the line segment $[((\mathbf{x}_0 - \mathbf{v}), f(\mathbf{x}_0 - \mathbf{v})), (\mathbf{y}, f(\mathbf{y}))]$, hence above the line segment $L_2 = [((\mathbf{y}_0 - \mathbf{v}), c), (\mathbf{v}, f(\mathbf{v}))]$.

From these relations, comparing the absolute value of the slopes of the two line segments shows that the first one, $[(\mathbf{x}_0, f(\mathbf{x}_0)), (\mathbf{x}_0 + \mathbf{v}, c)]$, has the larger of the two. Finally, $(\mathbf{x}_0, f(\mathbf{x}_0))$ is above the line segment $[(\mathbf{x}_0, f(\mathbf{x}_0)), (\mathbf{x}_0 + \mathbf{v}, c)]$, so that $|f(\mathbf{y}) - f(\mathbf{x}_0)| \leq \frac{1}{\epsilon}((\mathbf{x}_0) - c)\|\mathbf{y} - \mathbf{x}_0\|$. ■

5.7 ◆ Separation and the Hahn–Banach Theorem

The Hahn-Banach theorem is, essentially, an algebraic restatement of the geometry of the separation results. Sometimes algebra is easier than geometry. This is especially true in Chapter 8, where we give the Hahn-Banach theorem for the infinite-dimensional normed spaces that generalize $\mathbb{R}^\ell$, and in Chapter 10, where we use it for infinite-dimensional locally convex spaces, which generalize the infinite-dimensional normed spaces.

Once we have proved the separation theorem, we study convex cones and wedges. The epigraphs of what are called sublinear functions are convex cones, and they are often wedges. Sublinear functions are the Minkowski functions (or Minkowski gauges) of convex sets containing the origin. The Hahn-Banach theorem relates sublinear functions to linear functions; hence it relates linear functions to convex sets, just as the separation results do.

5.7.a Separating Disjoint Convex Sets

Theorem 5.7.1 (Separation) *Disjoint convex sets can be separated.*

Proof. If $C \cap D = \emptyset$, then $0 \notin C - D$. If C and D are convex, then $C - D$ is convex and 0 is either on the relative boundary of $C - D$ or else disjoint from the relative closure of $C - D$. In either case, by Theorem 5.5.16 there exists a nonzero, linear L such that $L(C - D) \geq L(0)$. This implies that $L(c) \geq L(d)$ for every $c \in C$ and $d \in D$. Since $L(C)$ is bounded below (by any $L(d)$), we can set $r = \inf L(C)$. ■

Recall that $C^\epsilon = C + B_\epsilon(0)$.

Theorem 5.7.2 (Strict Separation) *Convex sets C and D can be strictly separated iff for some $\epsilon > 0$, $C^\epsilon \cap D^\epsilon = \emptyset$.*

Exercise 5.7.3 Prove Theorem 5.7.2, and show, by example, that having C and D contained in disjoint open convex sets is **not** enough to guarantee strict separation.

5.7.b Cones and Wedges

The following repeats part of Definition 5.1.16 (p. 177).

Definition 5.7.4 *A nonempty $X \subset \mathbb{R}^\ell$ is a **convex cone** if for all $\mathbf{x}, \mathbf{y} \in X$ and all $\alpha, \beta \geq 0$, $\alpha\mathbf{x} + \beta\mathbf{y} \in X$. A convex cone is a **wedge** if it contains no nontrivial vector subspaces.*

The convexity follows from the observation that $\beta = (1 - \alpha)$ is allowed when $\alpha \in [0, 1]$.

Example 5.7.5   *Intuitively, a wedge is a convex cone with a pointy end. $\mathbb{R}^3_+$ is a wedge, but the cone $C = \{(x_1, x_2, x_3) \in \mathbb{R}^3 : x_1 \geq 0, x_2 \geq 0\}$ is not. A line through the origin in $\mathbb{R}^\ell$ is a cone, but it is not a wedge. The trivial wedge is one that contains just the pointy part, $X = \{0\}$.*

Example 5.7.6   *We anticipate later developments and state that cones also appear in spaces of functions. The set of concave functions mapping a convex set C to $\mathbb{R}$ is a cone—if f and g are concave and α, $\beta \geq 0$, then $\alpha f + \beta g$ is concave (as is easily verified).*

Theorem 5.7.7   *A nonempty $X \subset \mathbb{R}^\ell$ is a cone iff it is closed under addition and nonnegative scalar multiplication, that is, iff $X = X + X$ and $X = \mathbb{R}_+ \cdot X$.*

Proof.   Suppose that X is a convex cone. Then, taking $\alpha = \beta = 1$ in Definition 5.7.4, $X + X \subset X$. Any $x \in X$ is equal to $\frac{1}{2}x + \frac{1}{2}x$, so $X \subset X + X$. Since $1 \in \mathbb{R}_+$, $X \subset \mathbb{R}_+ \cdot X$ for any set X. Taking $\alpha \in \mathbb{R}_+$ and $\beta = 0$ delivers $\mathbb{R}_+ \cdot X \subset X$.

Suppose that $X = X + X$ and $X = \mathbb{R}_+ \cdot X$. For $\mathbf{x}, \mathbf{y} \in X$ and α, $\beta \in \mathbb{R}_+$, $\alpha\mathbf{x} \in X$, $\beta\mathbf{y} \in X$ because $X = \mathbb{R}_+ \cdot X$. Therefore, $\alpha\mathbf{x} + \beta\mathbf{y} \in X$ because $X = X + X$, so that X is a convex cone. ■

Since $\mathbb{R}^\ell$ is a cone and the intersection of cones is a cone, there is always a smallest cone and a smallest closed cone containing any given set. As the intersection of wedges is a wedge, there is often, but not always, a smallest wedge containing any given set.

Lemma 5.7.8   *If $0 \notin \overline{\text{co}}(E)$ and $\overline{\text{co}}(E)$ is compact, then there is a smallest closed wedge containing E.*

Exercise 5.7.9   If $E = \{(x_1, x_2) \in \mathbb{R}^2 : x_1 = 1\}$, give the smallest wedge containing E and show that no closed wedge contains E.

Exercise 5.7.10   If K is a compact convex subset of $\mathbb{R}^\ell$ and $0 \notin K$, show that $\{r \cdot K : r \in \mathbb{R}_+\}$ is closed and is a wedge. Show that this means that it is the smallest closed wedge containing K. From this, prove Lemma 5.7.8.

Wedges give useful lattice orders on $\mathbb{R}^\ell$.

Exercise 5.7.11   For $A \subset \mathbb{R}^\ell$, define $\mathbf{x} \leq_A \mathbf{y}$ if $\mathbf{x} \cdot \mathbf{a} \leq \mathbf{y} \cdot \mathbf{a}$ for all $\mathbf{a} \in A$.

1. With $A = \{\mathbf{e}_i : i \in \{1, \ldots, \ell\}\}$ being the set of unit vectors, or $A = \mathbb{R}^\ell_+$, show that this recovers the usual $\leq$ ordering.

2. Show that if A contains ℓ elements and spans $\mathbb{R}^\ell$, then $(\mathbb{R}^\ell, \leq_A)$ is a lattice.

3. Give conditions on A such that the ordering $\leq_A$ is trivial, that is, $x \leq_A y$ iff $x = y$.

4. Show that for any $A \subset \mathbb{R}^\ell$, $C_A := \{\mathbf{x} : \mathbf{x} \cdot A \leq 0\}$ is a cone.

5. Let A be a closed cone, let $C = \{\mathbf{x} : \mathbf{x} \cdot A \leq 0\}$, and let $B = \{\mathbf{y} : \mathbf{y} \cdot C \leq 0\}$. Show that $A = B$. (For this reason, A and C are called **dual cones** [Figure 5.7.11].)

6. Let $\mathbf{x} \neq 0$ be a point in $\mathbb{R}^\ell$ and $r > 0$ such that $\|\mathbf{x}\| > r$. Show that there is a smallest wedge containing $\{\mathbf{y} : \|\mathbf{y} - \mathbf{x}\| \leq r\}$ and that it spans $\mathbb{R}^\ell$. [For $n \geq 3$, this is a cone that is *not* defined by finitely many linear inequalities.]

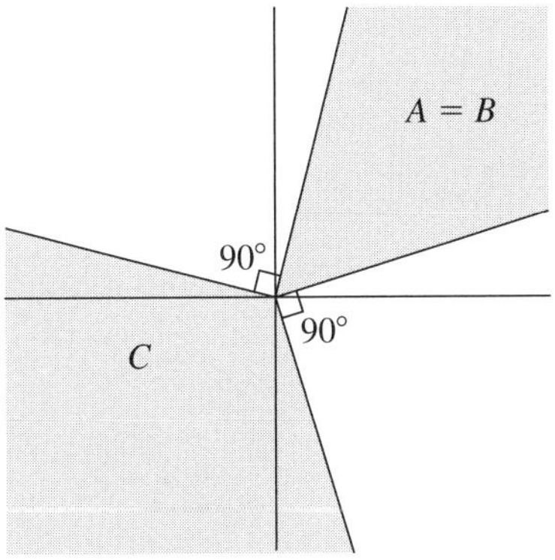

FIGURE 5.7.11

7. Let A be a finite spanning linearly independent set such that C_A is a wedge, and let $f : \mathbb{R}^\ell \to \mathbb{R}$ be a twice continuously differentiable function. Give a condition on the derivatives of f that is equivalent to f being supermodular on the lattice $(\mathbb{R}^\ell, \leq_A)$. [Doing this in $\mathbb{R}^2$ is enough, and you already know the answer if $A = \{\mathbf{e}_1, \mathbf{e}_2\}$.]

Exercise 5.7.12   Let $ND \subset \mathbb{R}^\ell$, $\ell \geq 2$, be the set of vectors $\mathbf{x}' = (x_1, \ldots, x_n)$ with the property that $x_1 \leq x_2 \leq \cdots \leq x_n$, that is, the set of vectors with non-decreasing components. Define $\mathbf{x} \leq_{ND} \mathbf{y}$ if $(\mathbf{x} - \mathbf{y}) \in ND$.

1. Show that ND is a closed convex cone in $\mathbb{R}^\ell$, but that ND is not a wedge.

2. For $\mathbf{x}, \mathbf{y} \in \mathbb{R}^2$, draw ND and show that $\min\{\mathbf{x}, \mathbf{y}\}$ and $\max\{\mathbf{x}, \mathbf{y}\}$ are not well defined. From this, you know that $(\mathbb{R}^\ell, \leq_{ND})$ is a *not* a lattice.

3. In $\mathbb{R}^2$, $\mathbf{y} \leq_{ND} \mathbf{y}$ iff $\mathbf{x} \leq_A \mathbf{y}$ for some set of vectors A. Give the set A. Generalize to $\mathbb{R}^\ell$. [Here you should see that $\leq_{ND}$ is not a lattice ordering because the set A does not span.]

5.7.c Sublinear Functions and Minkowski Gauges

Much of the geometric insight in this section comes from the next example and the exercise.

Example 5.7.13   *The set $K = [-1, 2] \subset \mathbb{R}$ is a convex norm-bounded subset of $\mathbb{R}$ with $0 \in \text{rel int}(K)$. Let $K' = \{(x, 1) \in \mathbb{R}^2 : x \in K\}$. The smallest closed convex cone in $\mathbb{R}^2$ containing K' is the epigraph of the convex function*

$$m_K(x) = \begin{cases} \frac{1}{2}x & \text{if } x \geq 0, \\ -x & \text{if } x < 0. \end{cases}$$

*Further, $m_K(x) \geq 0$; m_K is **homogeneous of degree** 1, that is, for any $\alpha \geq 0$, $m_K(\alpha x) = \alpha m_K(x)$; m_K **goes up sublinearly**, that is, for any $x, y \in \mathbb{R}$, $m_K(x + y) \leq m_K(x) + m_K(y)$; and $m_K(x) > 1$ iff $x \notin K$. Finally, $m_K(x) = \inf\{\alpha \geq 0 : x \in \alpha K\}$ (Figure 5.7.13).*

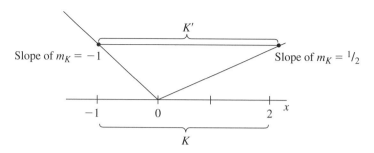

FIGURE 5.7.13

Exercise 5.7.14 Let $K = (-\infty, 2)$ and $K' = \{(x, 1) \in \mathbb{R}^2 : x \in K\}$.

1. Find the smallest closed convex cone in $\mathbb{R}^2$ containing K', and find the function, $m_K(\cdot)$, for which it is the epigraph.

2. Verify that m_K is homogeneous of degree 1 and goes up sublinearly.

3. Show that $m_K(x) > 1$ iff $x \notin \mathrm{cl}(K)$.

4. Show that $m_K(x) = \inf\{\alpha \geq 0 : x \in \alpha K\}$.

Definition 5.7.15 *A function* $m : \mathbb{R}^\ell \to \mathbb{R}_+$ *is called* **sublinear** *if:*

1. it is **homogeneous of degree** *1, that is, for any* $\alpha \geq 0$, $m(\alpha\mathbf{x}) = \alpha m(\mathbf{x})$, *and*

2. it **goes up sublinearly**, *for any* $\mathbf{x}, \mathbf{y} \in \mathbb{R}^\ell$, $m(\mathbf{x} + \mathbf{y}) \leq m(\mathbf{x}) + m(\mathbf{y})$.

Sublinear functions are necessarily convex, $m(\alpha\mathbf{x} + (1 - \alpha)\mathbf{y}) \leq m(\alpha\mathbf{x}) + m((1 - \alpha)\mathbf{y}) = \alpha m(\mathbf{x}) + (1 - \alpha)m(\mathbf{x})$. The trivial sublinear function is $m(\mathbf{x}) \equiv 0$.

Definition 5.7.16 *For K a convex subset of $\mathbb{R}^\ell$ with $0 \in \mathrm{int}(K)$, the* **Minkowski gauge** *or* **Minkowski function** *of K is $m_K(\mathbf{x}) := \inf\{\alpha \geq 0 : \mathbf{x} \in \alpha K\}$. We say that $m : \mathbb{R}^\ell \to \mathbb{R}_+$ is a Minkowski gauge if it is the gauge for some convex K with $0 \in \mathrm{int}(K)$.*

The reason for the restriction that $0 \in \mathrm{int}(K)$ is that if $K = [-1, 0]$, then for any $x > 0$, $m_K(x) = \infty$.

Theorem 5.7.17 *The following are equivalent:*

1. m is sublinear.

2. m is a Minkowski gauge.

3. $\mathrm{epi}(m) \subset \mathbb{R}^\ell \times \mathbb{R}_+$ is a closed convex cone spanning $\mathbb{R}^\ell \times \mathbb{R}$.

Further, if m is sublinear, then it is the Minkowski gauge of the convex set $\{(\mathbf{x}, 1) : m(\mathbf{x}) \leq 1\}$.

Proof. $(3) \Rightarrow (1)$ and $(3) \Rightarrow (2)$. If $\mathrm{epi}(m) \subset \mathbb{R}^\ell \times \mathbb{R}_+$ is a closed convex cone spanning $\mathbb{R}^\ell \times \mathbb{R}$, then Theorem 5.7.7 shows that m is sublinear, and simple checking shows that m is the Minkowski gauge of the set $K = \{\mathbf{x} : (\mathbf{x}, 1) \in \mathrm{epi}(m)\}$.

$(1) \Rightarrow (3)$. If m is sublinear, then Theorem 5.7.7 shows that $\mathrm{epi}(m)$ is a convex cone. Since m goes up sublinearly, some ϵ-neighborhood of the point $(0, 1)$ belongs to epi(m), so that it spans $\mathbb{R}^\ell \times \mathbb{R}$. Since epi($m$) is convex, m is convex, so that Theorem 5.6.34 shows that it is continuous, which implies that epi(m) is closed.

(2) $\Rightarrow$ (3). If m is a Minkowski gauge for some convex set with $0 \in \text{int}(K)$, then it is nonnegative, homogeneous of degree 1, cl(K) = $\{\mathbf{x} : m(\mathbf{x}) \leq 1\}$, and epi($m$) is the smallest closed, necessarily convex cone containing $\{(\mathbf{x}, 1) : \mathbf{x} \in \text{cl}(K)\}$. Since $0 \in \text{int}(K)$, epi(m) spans $\mathbb{R}^\ell \times \mathbb{R}$. ∎

5.7.d Convexity, Minkowski's Inequality, and p-Norms

Before turning to the Hahn-Banach theorem, let us go back and finish a long-delayed piece of work. We defined the 1-norm, the 2-norm, and the ∞-norm for $\mathbb{R}^\ell$. The only hard work in showing that they were norms involved the Cauchy-Schwarz inequality, which was needed to show that the 2-norm satisfies the triangle inequality. We now turn to the Minkowski inequality, which implies that the p-norms satisfy the triangle inequality.

Lemma 5.7.18 *If f and g are concave (convex) functions and α, $\beta \geq 0$, then $\alpha f + \beta g$ is concave (convex), that is, the set of concave (convex) functions is a convex cone in the space of functions.*

Proof. For all f, g, α, and β, $(\alpha f + \beta g)(x \gamma x') = \alpha f(x \gamma x') + \beta g(x \gamma x')$. If f and g are concave, this is weakly greater than $\alpha[f(x) \gamma f(x')] + \beta[g(x) \gamma g(x')]$. Rearranging yields $[(\alpha f + \beta g)(x)] \gamma [(\alpha f + \beta g)(x')]$. Reverse the inequality for convexity. ∎

Theorem 5.7.19 (Minkowski) *For any $\mathbf{u}$, $\mathbf{v} \in \mathbb{R}^\ell$, we have*

$$\left[\sum_{i=1}^{\ell}(u_i + v_i)^p\right]^{1/p} \leq \left[\sum_{i=1}^{\ell}(u_i)^p\right]^{1/p} + \left[\sum_{i=1}^{\ell}(v_i)^p\right]^{1/p}.$$

Proof. For $p \in [1, \infty)$, the function $\mathbf{x} \mapsto (x_i)^p$ from $\mathbb{R}_+^\ell$ to $\mathbb{R}$ is convex since the second derivative of r^p with respect to r is greater than or equal to 0. Being the sum of convex functions, $g_p(\mathbf{x}) = \sum_i (x_i)^p$ is convex. Being a monotonic transformation af g_p and homogeneous of degree 1, the function $f_p(\mathbf{x}) = \left[\sum_{i=1}^{\ell}(x_i)^p\right]^{1/p}$ is the Minkowski gauge of the set $\{\mathbf{x} : g_p(\mathbf{x}) \leq 1\}$. ∎

Exercise 5.7.20 Show that for all $p \in [1, \infty)$, $d_p(x, y) \leq (\ell)^{1/p} d_\infty(x, y)$ and $d_\infty(x, y) \leq d_p(x, y)$, so that the p-norm distances are all equivalent metrics.

Exercise 5.7.21 Show that every monotonic quasi-convex homogeneous of degree 1 function on $\mathbb{R}_+^\ell$ is convex and that every monotonic quasi-concave homogeneous of degree 1 function on $\mathbb{R}_+^\ell$ is concave.

5.7.e The Hahn-Banach Theorem for $\mathbb{R}^\ell$

Theorem 5.7.22 (Hahn-Banach) *If V is a vector subspace of $\mathbb{R}^\ell$, $h : V \to \mathbb{R}$ is linear, $m : \mathbb{R}^\ell \to \mathbb{R}$ is sublinear, and for all $\mathbf{x} \in V$, $h(\mathbf{x}) \leq m(\mathbf{x})$, then there exists a linear function $H : \mathbb{R}^\ell \to \mathbb{R}$ such that $H_{|V} = h$ and for all $\mathbf{x} \in \mathbb{R}^\ell$, $H(\mathbf{x}) \leq m(\mathbf{x})$.*

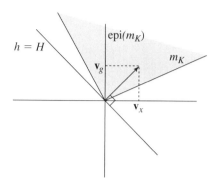

FIGURE 5.7.22   Hahn-Banach for $\mathbb{R}^l$, $l = 1$, $\mathbf{v}_y = 0$.

To see why this delivers the separation results, let K be a convex set with $0 \in \text{int}(K)$ and $\mathbf{x}^\circ$ a point not in $\text{cl}(K)$. Let m_K be the Minkowski gauge of K, so that $r := m_K(\mathbf{x}^\circ) > 1$. For $\mathbf{y} \in V = \text{span}(\{\mathbf{x}^\circ\})$, define $h(\mathbf{y}) = r\mathbf{y}$. On this linear subspace, $h(\mathbf{y}) = m_K(\mathbf{y})$. Extending to H, we have $H(\mathbf{x}^\circ) = r > 1$ and for all $\mathbf{x} \in K$, $H(\mathbf{x}) \leq m_K(\mathbf{x}) \leq 1$. Thus, H is a linear function separating $\mathbf{x}^\circ$ from K.

Recall that one identifies functions with their graphs.

Proof.   In $\mathbb{R}^\ell \times \mathbb{R}$, we can separate (the graph of) h from the convex set $\text{int}(\text{epi}(m))$ by a nonzero linear function $L(\mathbf{x}, y) = \mathbf{v}_x \mathbf{x} + v_y y$ with $L(\text{epi}(m)) \geq L(h)$. Since h is a linear subspace of $\mathbb{R}^\ell \times \mathbb{R}$, $L(h) = \{0\}$. $L(\text{int}(\text{epi}(m)) > 0$ follows from the observation that $(\mathbf{0}, 1) + \epsilon(\mathbf{v}_x, \mathbf{y}) \in \text{int}(\text{epi}(m)) > 0$, and this in turn implies that $v_y > 0$ because $(\mathbf{0}, 1) \in \text{int}(\text{epi}(m))$ (Figure 5.7.22). Define $H(\mathbf{z}) = -\frac{1}{v_y}\mathbf{v}_x \cdot \mathbf{z}$.   ∎

Exercise 5.7.23   Show that the Hahn-Banach theorem is true with the weaker assumption that m is convex. Relate this to Theorem 5.6.8 and to the subdifferentiability of convex functions.

5.8 ◆ Separation and the Kuhn–Tucker Theorem

Constrained optimization problems pervade economic theory. The Kuhn-Tucker theorem gives information about the solutions to these problems. In particular, through the separation theorem, it yields Lagrange multipliers, which measure the sensitivity of the solution to changes in the constraints.

5.8.a Examples of Lagrangian Functions

We begin with many examples of how Lagrangian functions and multipliers help solve constrained maximization problems. These examples all have the same underlying geometry, and after understanding the examples, we turn to what are called Kuhn-Tucker conditions. These are first be presented as an extension of the reasoning used in Lagrangian multipliers. To really understand why, rather than how, the Kuhn-Tucker conditions work requires saddle points.

Consider the following version of the neoclassical consumer demand problem: A consumer has preferences over the nonnegative levels of consumption of two goods. Consumption levels of the two goods are represented by $x = (x_1, x_2) \in \mathbb{R}^2_+$. We assume that this consumer's preferences can be represented by the concave[6] utility function

$$u(x_1, x_2) = \sqrt{x_1 x_2}.$$

The consumer has an income of 50 and faces prices $p = (p_1, p_2) = (5, 10)$. The standard behavioral assumption is that the consumer chooses among her affordable levels of consumption so as to make herself as happy as possible. This can be formalized as solving the constrained optimization problem:

$$\max_{(x_1,x_2)} \sqrt{x_1 x_2} \text{ subject to } 5x_1 + 10x_2 \leq 50, \quad x_1, x_2 \geq 0,$$

which in turn can be rewritten as

$$\max_{\mathbf{x}} u(\mathbf{x}) \text{ subject to } \mathbf{p} \cdot \mathbf{x} \leq w, \quad \mathbf{x} \geq 0.$$

The function being maximized is called the **objective function**.[7]

Exercise 5.8.1 The problem asks you to solve the previous maximization problem using a particular sequence of steps:

1. Draw the set of affordable points (i.e., the points in $\mathbb{R}^2_+$ that satisfy $\mathbf{p} \cdot \mathbf{x} \leq 50$).
2. Find the slope and equation of the budget line.
3. Find the equations for the indifference curves (i.e., solve $\sqrt{x_1 x_2} = c$ for $x_2(x_1, c)$).
4. Find the slope of the indifference curves.
5. Algebraically set the slope of the indifference curve equal to the slope of the budget line. This gives one equation in the two unknowns. Solve the equation for x_2 in terms of x_1.
6. Solve the two-equation system consisting of the previous equation and the budget line.
7. Explain geometrically why the solution to the two-equation system is in fact the solution to the constrained optimation problem.

---

6. It is monotonic homogeneous of degree 1, and as it is monotonic transformation of the concave $v = \log x_1 + \log x_2$, it is quasi-concave.

7. From Webster, the second definition of the word "objective" is "2a: something toward which effort is directed: an aim, goal, or end of action."

8. Explain economically why the solution you found is in fact the consumer's demand. Phrases that should come to your mind from intermediate microeconomics are "marginal rate of substitution," and "market rate of substitution."

Exercise 5.8.2 Construct the **Lagrangian function** for the optimization problem just given,

$$L(x_1, x_2, \lambda) = \sqrt{x_1 x_2} + \lambda(50 - [5x_1 + 10x_2]), \quad \text{or}$$

$$L(\mathbf{x}, \lambda) = u(\mathbf{x}) + \lambda(w - \mathbf{p} \cdot \mathbf{x}),$$

and show that the solution to the three-equation system,

$$\frac{\partial L(x_1, x_2, \lambda)}{\partial x_1} = 0, \quad \frac{\partial L(x_1, x_2, \lambda)}{\partial x_2} = 0, \quad \frac{\partial L(x_1, x_2, \lambda)}{\partial \lambda} = 0,$$

is the same as the solution you found in the previous problem. Be sure to solve for the extra variable, λ.

Exercise 5.8.3 Solve the previous problem for general w, that is, find the demands $\mathbf{x}^*(w)$ as a function of w, so that $\mathbf{x}^*(50)$ gives you the previous answer. Define $v(w) = u(\mathbf{x}^*(w))$ and find $(\partial v / \partial w)_{|w=50}$. Your answer should be the solution for λ that you found earlier. Interpret the derivative you just found economically. A phrase that should come to your mind from intermediate microeconomics is "marginal utility."

Note that the gradient of the function defining the budget line is

$$D_{\mathbf{x}}\mathbf{p} \cdot \mathbf{x} = \mathbf{p} = \begin{bmatrix} 5 \\ 10 \end{bmatrix}.$$

Exercise 5.8.4 Let $\mathbf{x}^*$ denote the solution in the previous two problems. Show that

$$D_{\mathbf{x}}u(\mathbf{x}^*) = \lambda \begin{bmatrix} 5 \\ 10 \end{bmatrix} = \lambda \mathbf{p}$$

for the same $\lambda > 0$ that you found in those problems. Interpret this geometrically.

Exercise 5.8.5 Draw the set of $\mathbf{x}$ such that $u(\mathbf{x}) > u(\mathbf{x}^*)$ in the previous problem. Show that it is disjoint from the set of affordable bundles. This is another way of saying that $\mathbf{x}^*$ does in fact solve the problem. Since these are disjoint convex sets, they can be separated. Find the half-space that separates them and the supporting hyperplane at $\mathbf{x}^*$.

Exercise 5.8.6 Suppose that $u(x_1, x_2) = x_2 \sqrt{x_1}$. Set up the Lagrangian function and use it to solve the problem

$$\max_{\mathbf{x}} u(\mathbf{x}) \text{ s.t. } \mathbf{x} \geq 0, \quad \mathbf{p} \cdot \mathbf{x} \leq w$$

for $\mathbf{x}^*(p, w)$ and $v(\mathbf{p}, w) = u(\mathbf{x}^*(\mathbf{p}, w))$, where $\mathbf{p} \gg 0$, $w > 0$. Check (for yourself) that the geometry and algebra of the previous several problems holds, especially the separation of the "strictly better than" set and the feasible set, and $\lambda = (\partial v/\partial w)$.

The following problem has a nonlinear constraint. This is different than the linear budget constraints, but the same technique works.

Exercise 5.8.7   Suppose that $u(x_1, x_2) = 7x_1 + 3x_2$. Set up the Lagrangian function and use it to solve the problem

$$\max_{\mathbf{x}} u(\mathbf{x}) \text{ s.t. } \mathbf{x} \cdot \mathbf{x} \leq c$$

as a function of $c > 0$. Check (for yourself) that the geometry and algebra of the previous several problems holds, especially the separation of the "strictly better than" set and the feasible set, and that $\lambda = (\partial v/\partial c)$.

In the next problem, the geometry is a bit trickier because the solution happens at a corner of the feasible set.

Exercise 5.8.8   Consider the problem

$$\max_{(x_1, x_2)} \sqrt{x_1 x_2} \text{ subject to } 2x_1 + 1x_2 \leq 12, \quad 1x_1 + 2x_2 \leq 12, x_1, x_2 \geq 0.$$

Find the optimum x^*, and note that

$$D_x u(x^*) = \lambda_1 \begin{bmatrix} 2 \\ 1 \end{bmatrix} + \lambda_2 \begin{bmatrix} 1 \\ 2 \end{bmatrix},$$

where $\lambda_1, \lambda_2 > 0$. Interpret this geometrically (think cones). Construct the Lagrangian function

$$L(x_1, x_2, \lambda_1, \lambda_2) = \sqrt{x_1 x_2} + \lambda_1(12 - [2x_1 + x_2]) + \lambda_2(12 - [x_1 + 2x_2]),$$

and look at the four-equation system

$$\frac{\partial L(x_1, x_2, \lambda_1, \lambda_2)}{\partial x_1} = 0, \qquad \frac{\partial L(x_1, x_2, \lambda_1, \lambda_2)}{\partial x_2} = 0,$$

$$\frac{\partial L(x_1, x_2, \lambda_1, \lambda_2)}{\partial \lambda_1} = 0, \qquad \frac{\partial L(x_1, x_2, \lambda_1, \lambda_2)}{\partial \lambda_2} = 0.$$

Show that solving this set of equations gives x^*.

So far all of the questions have been consumer maximizations. There are also constrained optimization questions in producer theory.

Exercise 5.8.9   Here is an example of the simplest kind of production theory—one input, with level $x \geq 0$, and one output, with level, $y \geq 0$. One formalization of this runs as follows: an (*input,output*) vector $(x, y) \in \mathbb{R}^2$ is **feasible** if $y \leq f(x)$, $x, y \geq 0$, where $f : \mathbb{R}_+ \rightarrow \mathbb{R}_+$ (this function is called the production function). The behavioral assumption is that the owner of the right to use the production technology represented by the feasible set takes the prices, p for output and w for input, as given and chooses the profit-maximizing feasible point. Letting Π denote profits, we come to the Π-maximization problem,

$$\max_{x,y} \Pi(x, y) = py - wx \text{ subject to } g(x, y) := y - f(x) \leq 0, \quad x, y \geq 0.$$

For this problem, assume that $f(x) = \sqrt{x}$:

1. The Lagrangian function for this optimization problem is

   $$L(x, y, \lambda) = \Pi(x, y) + \lambda(0 - g(x, y)) = py - wx + \lambda(f(x) - y).$$

   Solve the equations $\partial L/\partial x = 0$, $\partial L/\partial y = 0$, and $\partial L/\partial \lambda = 0$ for $y^*(p, w)$ (the supply function), $x^*(p, w)$ (the input demand function), and for $\lambda^*(p, w)$.

2. Find the gradient of the objective function, $\Pi(\cdot, \cdot)$ and the gradient of the constraint function, $g(\cdot, \cdot)$, and draw the geometric relationship between the two gradients and λ^* at the solution. Also show the separation of the "strictly better than" set from the feasible set.

3. Define the profit function $\Pi(p, w) = py^*(p, w) - wx^*(p, w)$. Show that $\partial \Pi/\partial p = y^*(p, w)$ and $\partial \Pi/\partial w = x^*(p, w)$. [The cancellation of all of the extra terms is a general phenomenon, and we will see why when we get to the envelope theorem.]

Exercise 5.8.10 Another optimization problem with implications for producer theory involves producing some amount of output, y^0, using inputs, $\mathbf{x} \in \mathbb{R}^\ell$. Suppose that $L = 2$ and that the set of feasible $(\mathbf{x}, y)$ combinations satisfies $y \leq f(x_1, x_2)$, $\mathbf{x}, y \geq 0$, where (for this problem)

$$f(x_1, x_2) = x_1^{0.2} x_2^{0.6}.$$

If we assume that the producer takes the prices $\mathbf{w} = (w_1, w_2)$ of inputs as given, the cost minimization problem is

$$\min_{x_1, x_2} w_1 x_1 + w_2 x_2 \text{ subject to } f(x_1, x_2) \geq y^0, \quad x_1, x_2 \geq 0.$$

To set this up as a standard maximization problem, multiply by -1:

$$\max_{x_1, x_2} -(w_1 x_1 + w_2 x_2) \text{ subject to } -f(x_1, x_2) \leq -y^0, \quad x_1, x_2 \geq 0.$$

1. The Lagrangian function for this optimization problem is

   $$L(x_1, x_2, \lambda) = -(w_1 x_1 + w_2 x_2) + \lambda(-y^0 - (-f(x_1, x_2)))$$

   $$= -\mathbf{wx} + \lambda(f(\mathbf{x}) - y^0).$$

   Solve the equations $\partial L/\partial x = 0$, $\partial L/\partial y = 0$, and $\partial L/\partial \lambda = 0$ for $x_1^*(\mathbf{w}, y^0)$ and $x_2^*(\mathbf{w}, y^0)$, the conditional factor demands, and for $\lambda^*(\mathbf{w}, y^0)$.

2. Find the gradient of the objective function and the gradient of the constraint function, and draw the geometric relationship between the two gradients

and λ^* at the solution. Also show the separation of the "strictly better than" set from the feasible set.

3. Define the cost function $c(w, y^0) = w\mathbf{x}^*(p, y^0)$. Show that $\lambda^* = \partial c/\partial y^0$.

5.8.b Complementary Slackness

The time has come to talk of many things,[8] and to expand on them. The pattern has been that we have one nonzero multiplier for each constraint that is relevant to or **binding** on the problem. Further, the value of that multiplier tells us how the objective function changes for a small change in the constraint.

5.8.b.1 Looking Back and Binding and Nonbinding Constraints
For example, our first problem,

$$\max_{(x_1, x_2)} \sqrt{x_1 x_2} \text{ subject to } 5x_1 + 10x_2 \le 50, \quad x_1, x_2 \ge 0,$$

has three constraints: first, $5x_1 + 10x_2 \le 50$, second, $-x_1 \le 0$, and third, $-x_2 \le 0$. However, at any solution, the second and third constraints are not binding (i.e., not relevant), and we use a Lagrangian with only one multiplier.

For another example, the problem in Exercise 5.8.8,

$$\max_{(x_1, x_2)} \sqrt{x_1 x_2} \text{ subject to } 2x_1 + 1x_2 \le 12, \quad 1x_1 + 2x_2 \le 12, \quad x_1, x_2 \ge 0,$$

has four constraints: first, $2x_1 + 1x_2 \le 12$, second, $1x_1 + 2x_2 \le 12$, third, $-x_1 \le 0$, and fourth, $-x_2 \le 0$. However, only the first two are binding at the solution, and we could solve the problem using only two multipliers.

5.8.b.2 The General Pattern
The general pattern is that we can use a Lagrangian with multipliers for each of the binding constraints to solve our constrained optimization problems. We include all of the multipliers, but set the irrelevant ones equal to 0. The general form of the problem is

$$\max_{\mathbf{x} \in X} f(\mathbf{x}) \text{ subject to } g(\mathbf{x}) \le b,$$

where $X \subset \mathbb{R}^\ell$, $f : X \to \mathbb{R}$, $g : X \to \mathbb{R}^m$.

For example, with $L = 2$ and $m = 4$, the previous four-constraint problem had

$$g(x_1, x_2) = \begin{bmatrix} 2x_1 + 1x_2 \\ 1x_1 + 2x_2 \\ -x_1 \\ -x_2 \end{bmatrix}, \qquad \mathbf{b} = \begin{bmatrix} 12 \\ 12 \\ 0 \\ 0 \end{bmatrix}. \tag{5.8}$$

---

8. "Of shoes—and ships—and sealing-wax—Of cabbages—and kings—And why the sea is boiling hot—And whether pigs have wings."

The general form of the Lagrangian function, $L : \mathbb{R}^\ell \times \mathbb{R}^m_+ \to \mathbb{R}$, is

$$L(\mathbf{x}, \boldsymbol{\lambda}) = f(\mathbf{x}) + \boldsymbol{\lambda}(\mathbf{b} - g(\mathbf{x})). \tag{5.9}$$

With x_k's and g_m's, this is

$$L(x_1, \ldots, x_\ell, \lambda_1, \ldots, \lambda_m) = f(x_1, \ldots, x_\ell) + \sum_{m=1}^{m} \lambda_m (b_m - g_m(x_1, \ldots, x_\ell)).$$

$$\tag{5.10}$$

Definition 5.8.11 *The **Lagrangian function** for the problem without non-negativity constraints on* $\mathbf{x}$*,*

$$\max_{\mathbf{x} \in X} f(\mathbf{x}) \text{ subject to } g(\mathbf{x}) \leq \mathbf{b}, \tag{5.11}$$

is $L(\mathbf{x}, \boldsymbol{\lambda}) = f(\mathbf{x}) + \boldsymbol{\lambda}(\mathbf{b} - g(\mathbf{x}))$. *The components of the vector* $\boldsymbol{\lambda}$ *are called **multipliers** or **Lagrange multipliers**. The associated **Kuhn-Tucker (K-T) conditions** are*

$$\text{(a)} \ \frac{\partial L}{\partial \mathbf{x}} = 0, \quad \text{(b)} \ \frac{\partial L}{\partial \boldsymbol{\lambda}} \geq 0, \ \boldsymbol{\lambda} \geq 0, \quad and \quad \boldsymbol{\lambda} \cdot \frac{\partial L}{\partial \boldsymbol{\lambda}} = 0. \tag{5.12}$$

Taken together, the last three parts of (5.12), $\frac{\partial L}{\partial \boldsymbol{\lambda}} \geq 0$, $\boldsymbol{\lambda} \geq 0$, and $\boldsymbol{\lambda} \cdot \frac{\partial L}{\partial \boldsymbol{\lambda}} = 0$, are equivalent to

$$\mathbf{b} - g(\mathbf{x}) \geq 0, \ \boldsymbol{\lambda} \geq 0, \quad \text{and} \quad \boldsymbol{\lambda} \cdot (\mathbf{b} - g(\mathbf{x})) = 0. \tag{5.13}$$

If $\mathbf{x}, \mathbf{y} \geq 0$, then $[\mathbf{x} \cdot \mathbf{y} = 0] \Rightarrow [[[x_i > 0] \Rightarrow [y_i = 0]] \wedge [[y_i > 0] \Rightarrow [x_i = 0]]$. Applying this to the two vectors $(\mathbf{b} - g(\mathbf{x}))$ and $\boldsymbol{\lambda}$ yields complementary slackness.

Lemma 5.8.12 (Complementary Slackness) *If the K-T conditions hold, then* $[\lambda_m > 0] \Rightarrow [b_m - g_m(\mathbf{x}) = 0]$ *and* $[b_m - g_m(\mathbf{x}) > 0] \Rightarrow [\lambda_m = 0]$.

In words, only the binding constraints can have strictly positive multipliers, and nonbinding constraints have 0 multipliers. When we think about earlier the $\lambda = \partial v / \partial w$ results, this is good, and changing a nonbinding constraint should have no effect on the optimum.

Exercise 5.8.13 Write out the Kuhn-Tucker conditions in (5.12) with x_k's and g_m's.

5.8.b.3 Special Treatment of Nonnegativity Constraints

One can always include nonnegativity constraints, $\mathbf{x} \geq 0$, into the $g(\mathbf{x}) \leq \mathbf{b}$ constraint as in (5.8) above. However, because nonnegativity constraints are so common, they often have their own separate notation for multipliers, $\boldsymbol{\mu}$, as well as having a formulation without multipliers.

With separate multipliers, the four-constraint problem,

$$\max_{(x_1, x_2)} \sqrt{x_1 x_2} \text{ subject to } 2x_1 + 1x_2 \leq 12, \ 1x_1 + 2x_2 \leq 12, \ -x_1 \leq 0, \ -x_2 \leq 0,$$

has the Lagrangian

$$L(x_1, x_2, \lambda_1, \lambda_2, \mu_1, \mu_2) = \sqrt{x_1 x_2} + \lambda_1(12 - [2x_1 + 1x_2]) + \lambda_2(12 - [1x_1 + 2x_2])$$
$$+ \mu_1(0 - (-x_1)) + \mu_2(0 - (-x_2)).$$

This can be written as

$$L(\mathbf{x}, \lambda, \mu) = f(\mathbf{x}) + \lambda(\mathbf{b} - g(\mathbf{x})) + \mu\mathbf{x}, \tag{5.14}$$

with the understanding that g does not contain the nonnegativity constraints.

Definition 5.8.14   *The **Lagrangian function** for the problem with nonnegativity constraints,*

$$\max_{\mathbf{x} \in X} f(\mathbf{x}) \quad \text{subject to } g(\mathbf{x}) \le \mathbf{b} \text{ and } \mathbf{x} \ge 0, \tag{5.15}$$

*is $L(\mathbf{x}, \lambda, \mu) = f(\mathbf{x}) + \lambda(\mathbf{b} - g(\mathbf{x})) + \mu\mathbf{x}$. The associated **Kuhn-Tucker (K-T) conditions** are*

$$\text{(a)} \ \frac{\partial L}{\partial \mathbf{x}} = 0, \quad \text{(b)} \ \frac{\partial L}{\partial \lambda} \ge 0, \ \lambda \ge 0, \ \lambda \cdot \frac{\partial L}{\partial \lambda} = 0,$$
$$\text{(c)} \frac{\partial L}{\partial \mu} \ge 0, \ \mu \ge 0, \ \text{and } \mu \cdot \frac{\partial L}{\partial \mu} = 0. \tag{5.16}$$

There is an equivalent formulation of (5.16) that drops the explicit mention of μ even though we mean to keep the nonnegativity constraints. If we define $\mathcal{L}(\mathbf{x}, \lambda) = f(\mathbf{x}) + \lambda(\mathbf{b} - g(\mathbf{x}))$, then

$$\frac{\partial L}{\partial \mathbf{x}} = 0 \Leftrightarrow \frac{\partial \mathcal{L}}{\partial \mathbf{x}} = -\mu. \tag{5.17}$$

Since $\partial L/\partial \mu = \mathbf{x}$, we can rewrite (5.16) avoiding all mention of μ as

$$\text{(a)} \ \frac{\partial \mathcal{L}}{\partial \mathbf{x}} \le 0, \ \mathbf{x} \ge 0, \ \mathbf{x} \cdot \frac{\partial \mathcal{L}}{\partial \mathbf{x}} = 0,$$
$$\text{(b)} \ \frac{\partial \mathcal{L}}{\partial \lambda} \ge 0, \ \lambda \ge 0, \ \text{and } \lambda \cdot \frac{\partial \mathcal{L}}{\partial \lambda} = 0. \tag{5.18}$$

Notation Alert 5.8.A   *When seeing these variants on the Kuhn-Tucker conditions for the first time, it makes sense to have different symbols for the Lagrangian function L and the different Lagrangian function $\mathcal{L}$, which is what makes (5.16) and (5.18) different from each other. In the future, we will not distinguish by symbols, and just use whichever variant of the Lagrangian function is appropriate to the problem being analyzed.*

Exercise 5.8.15   Write out these last two versions of the Lagrangian conditions with x_k's and g_m's.

5.8.b.4 Corner Solutions and Complementary Slackness
The next problem demonstrates complementary slackness in the K-T conditions.

Exercise 5.8.16 Let $c = (2, 12)$, $p = (1, 5)$, $w = 5$, and $u(\mathbf{x}) = -(\mathbf{x} - \mathbf{c})(\mathbf{x} - \mathbf{c})$ and consider the problem

$$\max_{\mathbf{x}} u(\mathbf{x}) \text{ subject to } \mathbf{px} \le w, \quad -\mathbf{x} \le 0.$$

We ask you to solve this problem by making many wrong guesses about which constraints are binding and which are not. The idea is not that we want you to practice getting things wrong—if you are anything like us, that will happen often enough without any outside help. Rather, we want you to have practice in seeing what happens when you guess wrong, so that you learn to recognize it, and can then get better at guessing right.

1. Write out the Lagrangian both in vector and in x_k and g_m notation using $\boldsymbol{\mu} = (\mu_1, \mu_2)$ for the multipliers of the nonnegativity constraints.

2. Write out the K-T conditions.

3. Try to solve the K-T conditions on the assumption that only the first non-negativity constraint is binding, that is, on the assumption that $\mu_1 > 0$ and $\lambda_1 = \mu_2 = 0$. Interpret.

4. Try to solve the K-T conditions on the assumption that only the second nonnegativity constraint is binding, that is, on the assumption that $\mu_2 > 0$ and $\lambda_1 = \mu_1 = 0$. Interpret.

5. Try to solve the K-T conditions on the assumption that only the budget constraint is binding, that is, on the assumption that $\lambda_1 > 0$ and $\mu_1 = \mu_2 = 0$. Interpret.

6. Try to solve the K-T conditions on the assumption that the budget constraint and the second nonnegativity constraint are both binding, that is, on the assumption that $\lambda_1 > 0$, $\mu_2 > 0$, and $\mu_1 = 0$. Interpret.

7. Try to solve the K-T conditions on the assumption that the budget constraint and the first nonnegativity constraint are both binding, that is, on the assumption that $\lambda_1 > 0$, $\mu_1 > 0$, and $\mu_2 = 0$. Interpret.

The previous problem had a solution at a corner. This is, for obvious reasons, called a **corner solution**. Corner solutions are essentially always what happens in consumer demand. After all, how many gas centrifuges or cart horses do you own? However, we rarely draw or analyze them in intermediate microeconomics. The next two problems have corner solutions (at least some of the time). You can proceed in one of three ways: put the nonnegativity constraints into the $g(\cdot)$ function; use the $\boldsymbol{\mu}$ multipliers for the nonnegativity constraint; or use the Kuhn-Tucker conditions in (5.18) that avoid all mention of $\boldsymbol{\mu}$. We chose the notation in the next questions as if you were going to proceed in the second fashion, but you should not feel constrained by our choice.

Exercise 5.8.17 Using the K-T conditions, completely solve the problem

$$\max_{\mathbf{x}} u(\mathbf{x}) \text{ s.t. } \mathbf{px} \le w, \quad -\mathbf{x} \le 0,$$

where $u(x_1, x_2) = x_1 + 2x_2$, $\mathbf{p} \gg 0$, and $w > 0$. Letting $(\mathbf{x}^*(\mathbf{p}, w), \boldsymbol{\lambda}^*(\mathbf{p}, w))$ denote the solution to the K-T conditions, define $v(\mathbf{p}, w) = u(\mathbf{x}^*(\mathbf{p}, w))$ and show that $\boldsymbol{\lambda}^* = \partial v / \partial w$ at all points where $v(\cdot, \cdot)$ is differentiable.

Exercise 5.8.18   Using the K-T conditions, completely solve the problem

$$\max_{\mathbf{x}} u(\mathbf{x}) \text{ s.t. } \mathbf{p}\mathbf{x} \leq w, \quad -\mathbf{x} \leq 0,$$

where $u(x_1, x_2) = x_1 + 2\sqrt{x_2}$, $p \gg 0$, and $w > 0$. Letting $(\mathbf{x}^*(p, w), \boldsymbol{\lambda}^*(\mathbf{p}, w))$ denote the solution to the K-T conditions, define $v(\mathbf{p}, w) = u(\mathbf{x}^*(\mathbf{p}, w))$ and show that $\boldsymbol{\lambda}^* = \partial v / \partial w$.

Exercise 5.8.19   Using the K-T conditions, completely solve the problem

$$\max_{\mathbf{x}} u(\mathbf{x}) \text{ s.t. } \mathbf{p}\mathbf{x} \leq w, \quad -\mathbf{x} \leq 0,$$

where $u(x_1, x_2) = (\frac{1}{x_1} + \frac{1}{x_2})^{-1}$, $\mathbf{p} \gg 0$, and $w > 0$. Letting $(\mathbf{x}^*(\mathbf{p}, w), \boldsymbol{\lambda}^*(\mathbf{p}, w))$ denote the solution, define $v(\mathbf{p}, w) = u(\mathbf{x}^*(\mathbf{p}, w))$ and show that $\boldsymbol{\lambda}^* = \partial v / \partial w$.

Exercise 5.8.20   Return to the simplest kind of production theory—one input, with level $x \geq 0$, and one output, with level, $y \geq 0$. If we let Π denote profits, the Π-maximization problem is

$$\max_{x,y} \Pi(x, y) = py - wx \text{ subject to } g(x, y) := y - f(x) \leq 0, \quad x, y \geq 0.$$

For this problem, assume that $f(x) = \sqrt{x + 1} - 1$.

1. Write out and solve the Lagrangian function for $y^*(p, w)$ (the supply function), $x^*(p, w)$ (the input demand function), and for $\lambda^*(p, w)$.

2. Find the gradient of the objective function, $\Pi(\cdot, \cdot)$ and show how to express it as a positive linear combination of the binding constraint(s).

3. Define the profit function $\Pi(p, w) = py^*(p, w) - wx^*(p, w)$. Show that $\partial \Pi / \partial p = y^*(p, w)$ and $\partial \Pi / \partial w = x^*(p, w)$. [The cancellation of all of the extra terms is, as before, an implication of the envelope theorem.]

Exercise 5.8.21   Some amount of output, y^0, is produced using inputs $\mathbf{x} \in \mathbb{R}_+^2$. The set of feasible $(\mathbf{x}, y)$ combinations satisfies $y \leq f(x_1, x_2)$ and $\mathbf{x}, y \geq 0$, where (for this problem)

$$f(x_1, x_2) = (x_1 + \sqrt{x_2})^{0.75}.$$

If we assume that the producer takes the prices $w = (w_1, w_2)$ of inputs as given, the cost-minimization problem is

$$\min_{x_1, x_2} w_1 x_1 + w_2 x_2 \text{ subject to } f(x_1, x_2) \geq y^0, \quad x_1, x_2 \geq 0.$$

To set this up as a standard maximization problem, multiply by -1:

$$\max_{x_1, x_2} -(w_1 x_1 + w_2 x_2) \text{ subject to } -f(x_1, x_2) \leq -y^0, \quad x_1, x_2 \geq 0.$$

1. Write out and solve the Lagrangian function for $x_1^*(w, y^0)$ and $x_2^*(w, y^0)$, the conditional factor demands, and for $\lambda^*(w, y^0)$.

2. Find the gradient of the objective function and the gradient of the constraint function, and draw the geometric relationship between the two gradients and λ^* at the solution. Also show the separation of the "strictly better than" set from the feasible set.

3. Define the cost function $c(w, y^0) = wx^*(p, y^0)$. Show that $\lambda^* = \partial c / \partial y^0$.

5.8.b.5 Pitfalls and Problems

So far, Lagrangian functions and the Kuhn-Tucker conditions have worked quite well. Further, they give some extra information in the form of the vector or scalar λ^*'s. Would that life were always so simple.

- Our first exercise below has a strictly quasi-concave utility function that we maximize over the standard convex Walrasian budget set. There is a solution to the Kuhn-Tucker conditions that is not an optimum. At the nonoptimal solution to the Kuhn-Tucker conditions, a constraint is binding but the associated multiplier is equal to 0.

- Our second exercise below has a strictly convex utility function being maximized over a convex Walrasian budget set. By the Krein-Milman theorem, at least one solution must happen at an extreme point of the constraint set. There are two solutions to the Kuhn-Tucker conditions, one at an extreme point of the set. Neither solution to the K-T conditions is optimal.

- Our third exercise below has a concave utility function being maximized over a nonconvex constraint set. There is a solution to the Kuhn-Tucker conditions that is a local minimum.

Exercise 5.8.22 (Strict Quasi–Concavity Is Not Enough) Let $u(x_1, x_2) = \left(\sqrt{x_1 x_2} - 3\right)^3$, and consider the problem

$$\max_{x_1, x_2} u(x_1, x_2) \text{ s.t. } 3x_1 + x_2 \leq 12, \quad -x_1 \leq 0, \quad -x_2 \leq 0.$$

Show that u is strictly quasi-concave on $\mathbb{R}_{++}^2$ and that the point $(x_1^*, x_2^*, \lambda) = (3, 3, 0)$ satisfies the Kuhn-Tucker conditions but is not optimal.

Exercise 5.8.23 Let $f(x_1, x_2) = (x_1^2 + x_2^2)/2$, $g(x_1, x_2) = x_1 + x_2$ and consider the problem

$$\max_{x_1, x_2} f(x_1, x_2) \text{ s.t. } g(x_1, x_2) \leq 10, \quad -x_1 \leq 0, \quad -x_2 \leq 0.$$

Show that $(x_1^*, x_2^*, \lambda^*) = (5, 5, 5)$ satisfies the K-T conditions but does not solve the maximization problem.

Exercise 5.8.24 Let $f(x_1, x_2) = x_1 + x_2$, $g(x_1, x_2) = \sqrt{x_1} + \sqrt{x_2}$ and consider the problem

$$\max_{x_1, x_2} f(x_1, x_2) \text{ s.t. } g(x_1, x_2) \leq 4, \quad -x_1 \leq 0, \quad -x_2 \leq 0.$$

Show that $(x_1^*, x_2^*, \lambda^*) = (4, 4, \frac{1}{2})$ satisfies the K-T conditions but does not solve the maximization problem.

The point of the previous problems is that for the K-T conditions to characterize an optimum, we need more.

5.8.c Saddle Points, or Why Does It Work?

The more that we are looking for is that the problems be **concave-convex**, that the function $f(\cdot)$ be concave and the functions $g_m(\cdot)$ be convex. In the process of seeing how this works, we use a fairly weird-looking construct called a **saddle point**, and the derivative conditions in §5.6.c and §5.6.d reappear.

Put the nonnegativity constraints into $g(\cdot)$, let X be the domain of the functions, and construct the Lagrangian,

$$L(\mathbf{x}, \boldsymbol{\lambda}) = f(\mathbf{x}) + \boldsymbol{\lambda}(\mathbf{b} - g(\mathbf{x})).$$

Definition 5.8.25 $(\mathbf{x}^*, \boldsymbol{\lambda}^*) \in X \times \mathbb{R}_+^m$ *is a **saddle point** for $L(\cdot, \cdot)$ if*

$$(\forall \mathbf{x} \in X)(\forall \boldsymbol{\lambda} \geq 0)[L(\mathbf{x}, \boldsymbol{\lambda}^*) \leq L(\mathbf{x}^*, \boldsymbol{\lambda}^*) \leq L(\mathbf{x}^*, \boldsymbol{\lambda})].$$

In words, $\mathbf{x}^*$ is a maximum with respect to $\mathbf{x}$ for $\boldsymbol{\lambda}$ fixed at $\boldsymbol{\lambda}^*$ and $\boldsymbol{\lambda}^*$ is a minimum with respect to $\boldsymbol{\lambda}$ for $\mathbf{x}$ fixed at $\mathbf{x}^*$. To see why these are called saddle points, imagine looking at the saddle on a horse from the side: there is a minimum in the direction from your left to your right at the same point that there is a minimum in the direction from your side of the horse to the other.

Example 5.8.26 *The function $f(x_1, x_2) = x_2^2 - x_1^2$ has a saddle point at $(0, 0)$, the function $x_1 \mapsto f(x_1, 0)$ has a maximum at $x_1^* = 0$, and the function $x_2 \mapsto f(0, x_2)$ has a minimum at $x_2^* = 0$ (Figure 5.8.26).*

Any saddle point for a Lagrangian function contains an optimum for the constrained maximization problem. If the maximization problem is well behaved, that is, concave-convex, then every optimum is associated with a saddle point of the Lagrangian function.

The next result, K-T Part I, says that if $(\mathbf{x}^*, \boldsymbol{\lambda}^*)$ is a saddle point, then $\mathbf{x}^*$ is a solution to the constrained optimization problem. Its proof is quite easy and requires no conditions on the functions f and g, nor on the set X. We do not even assume that X is a subset of $\mathbb{R}^\ell$.

The following result, K-T Part II, says that if $\mathbf{x}^*$ is a solution to the constrained optimization problem and the problem is concave-convex, then there exists a $\boldsymbol{\lambda}^*$ such that $(\mathbf{x}^*, \boldsymbol{\lambda}^*)$ is a saddle point. The proof of Part II is quite a bit harder, and the vector $\boldsymbol{\lambda}^*$ arises from the separation of two convex sets.

Theorem 5.8.27 (Kuhn–Tucker, Part I) *If $(\mathbf{x}^*, \boldsymbol{\lambda}^*)$ is a saddle point for $L(\mathbf{x}, \boldsymbol{\lambda}) = f(\mathbf{x}) + \boldsymbol{\lambda}(\mathbf{b} - g(\mathbf{x}))$, $X \neq \emptyset$, $f : X \to \mathbb{R}$, and $g : X \to \mathbb{R}^m$, then $\mathbf{x}^*$ solves the problem*

$$\max_{\mathbf{x} \in X} f(\mathbf{x}) \text{ subject to } g(\mathbf{x}) \leq \mathbf{b}.$$

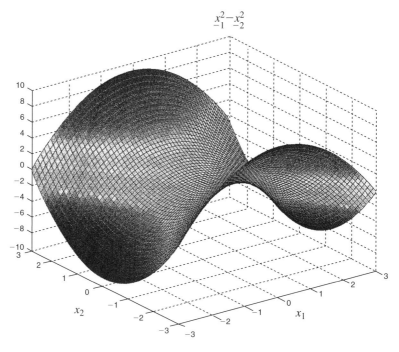

$$\frac{x_1^2 - x_2^2}{}$$

FIGURE 5.8.26

Proof. Suppose that $(\mathbf{x}^*, \lambda^*)$ is a saddle point for $L(\cdot, \cdot)$. There are three steps:

1. feasibility, $g(\mathbf{x}^*) \leq \mathbf{b}$,

2. complementary slackness, $\lambda^*(\mathbf{b} - g(\mathbf{x}^*)) = 0$, and

3. optimality, for all $\mathbf{x}'$ such that $g(\mathbf{x}') \leq \mathbf{b}$, $f(\mathbf{x}') \leq f(\mathbf{x}^*)$.

(1) If $\neg[g(\mathbf{x}^*) \leq \mathbf{b}]$, then for some k, $b_k - g_k(\mathbf{x}^*) < 0$. Letting $\lambda_k = n\mathbf{e}_k$, $\mathbf{e}_k$ the unit vector in the kth direction, shows that $\inf\{L(\mathbf{x}^*, \lambda) : \lambda \geq 0\} = -\infty$, so that $(\mathbf{x}^*, \lambda^*)$ is not a saddle point.

(2) Since $\mathbf{b} - g(\mathbf{x}^*) \geq 0$, the minimum of $\lambda(\mathbf{b} - g(\mathbf{x}^*))$ subject to $\lambda \geq 0$ is 0, since $\lambda = 0$ is feasible.

(3) Now suppose that there exists an $\mathbf{x}'$ such that $g(\mathbf{x}') \leq \mathbf{b}$ and $f(\mathbf{x}') > f(\mathbf{x}^*)$. We show that $L(\mathbf{x}', \lambda^*) = f(\mathbf{x}') + \lambda^*(\mathbf{b} - g(\mathbf{x}')) > f(\mathbf{x}^*) + \lambda^*(\mathbf{b} - g(\mathbf{x}^*))$, which contradicts the saddle-point assumption and completes the proof.

By assumption, $f(\mathbf{x}') > f(\mathbf{x}^*)$. Since $\mathbf{b} - g(\mathbf{x}') \geq 0$, $\lambda^*(\mathbf{b} - g(\mathbf{x}')) \geq 0$ because $\lambda^* \geq 0$. By step (2), $\lambda^*(\mathbf{b} - g(\mathbf{x}^*)) = 0$ so that $\lambda^*(\mathbf{b} - g(\mathbf{x}')) \geq \lambda^*(\mathbf{b} - g(\mathbf{x}^*))$. Combining yields $L(\mathbf{x}', \lambda^*) > L(\mathbf{x}^*, \lambda^*)$. ∎

We replace the general domain X with an open convex subset of $\mathbb{R}^\ell$ for the next result.

Theorem 5.8.28 (Kuhn–Tucker, Part II) *Suppose that $X \subset \mathbb{R}^\ell$ is an open convex set, that $f : X \to \mathbb{R}$ is concave, and that each $g_k : X \to \mathbb{R}$, $k = 1, \ldots m$ is* *convex. If* $\mathbf{x}^*$ *solves the problem* $\max_{\mathbf{x} \in X} f(\mathbf{x})$ subject to $g(\mathbf{x}) \leq \mathbf{b}$, *then there exists a nonzero vector* $\boldsymbol{\lambda}^* \in \mathbb{R}_+^m$ *such that* $(\mathbf{x}^*, \boldsymbol{\lambda}^*)$ *is a saddle point for the Lagrangian function* $L(\mathbf{x}, \boldsymbol{\lambda}) = f(\mathbf{x}) + \boldsymbol{\lambda}(\mathbf{b} - g(\mathbf{x}))$.

A Brief Detour: This result guarantees the existence of multipliers that lead to a saddle point. It is often the case that the multipliers have important information about the sensitivity of the optimum to changes in the value of b. For the multipliers to have this information, some variant on a condition called the **constraint qualification (CQ)** must hold. We have never encountered in our own work an optimization problem in which the multipliers were not unique and informative. However, such examples can be constructed if you are careful and have a perverse turn of mind. Having the requisite perverse turn of mind, we give an example in §5.9.d.

Proof. Suppose that $\mathbf{x}^*$ solves the maximization problem. We use the separation theorem to find the existence of $\boldsymbol{\lambda}^*$ in the following two cases: (a) $g(\mathbf{x}^*) \ll \mathbf{b}$, that is, no constraints are binding at the optimum, and (b) $g(\mathbf{x}^*) \leq \mathbf{b}$ but not $g(\mathbf{x}^*) \ll \mathbf{b}$, that is, at least one constraint is satisfied with equality.

(a) Suppose that $g(\mathbf{x}^*) \ll \mathbf{b}$. Set $\boldsymbol{\lambda}^* = 0$. We must show that for all $\mathbf{x} \in X$ and all $\boldsymbol{\lambda} \geq 0$,

$$\underbrace{f(\mathbf{x}) + 0 \cdot (\mathbf{b} - g(\mathbf{x})) \leq f(\mathbf{x}^*) + 0 \cdot (\mathbf{b} - g(\mathbf{x}^*))}_{\text{first}} \underbrace{\leq f(\mathbf{x}^*) + \boldsymbol{\lambda} \cdot (\mathbf{b} - g(\mathbf{x}^*))}_{\text{second}}.$$

(5.19)

The second inequality follows from $(\mathbf{b} - g(\mathbf{x}^*)) \geq 0$. Suppose that the first inequality is violated; that is, there exists an $\mathbf{x}' \in X$ such that $f(\mathbf{x}') > f(\mathbf{x}^*)$. By the concavity of $f(\cdot)$, for all $\gamma \in (0, 1)$, $f(\mathbf{x}'\gamma\mathbf{x}^*) > f(\mathbf{x}^*)$. Theorem 5.6.34 shows that convex functions are continuous on the interior of their domains. By the continuity of $g(\cdot)$ and $g(\mathbf{x}^*) \ll \mathbf{b}$, for γ close to 0, $g(\mathbf{x}'\gamma\mathbf{x}^*) \ll \mathbf{b}$. This means that for small γ, $(\mathbf{x}'\gamma\mathbf{x}^*)$ is feasible, contradicting the optimality of $\mathbf{x}^*$.

(b) Suppose that $g(\mathbf{x}^*) \leq \mathbf{b}$ but not $g(\mathbf{x}^*) \ll \mathbf{b}$, that is, at least one constraint is satisfied with equality. The proof shows that the following sets are convex and disjoint:

$$A = \{(a_0, \mathbf{a}) \in \mathbb{R} \times \mathbb{R}^m : (\exists \mathbf{x} \in X)[a_0 \leq f(\mathbf{x}) \wedge \mathbf{a} \leq \mathbf{b} - g(\mathbf{x})]\}, \quad \text{and}$$

$$C = \{(c_0, \mathbf{c}) \in \mathbb{R} \times \mathbb{R}^m : c_0 > f(\mathbf{x}^*) \wedge \mathbf{c} \gg 0\}.$$

Being a (hyper-)rectangle opening "up and to the right," the set C is convex. A is also convex. Pick $(a_0, \mathbf{a})$, $(a_0', \mathbf{a}') \in A$, and $\gamma \in (0, 1)$. Choose $\mathbf{x}$ such that $a_0 \leq f(\mathbf{x})$ and $\mathbf{a} \leq \mathbf{b} - g(\mathbf{x})$, $\mathbf{x}'$ such that $a_0' \leq f(\mathbf{x}')$ and $\mathbf{a}' \leq \mathbf{b} - g(\mathbf{x}')$, By the concavity of $f(\cdot)$ and each $b - g_k(\cdot)$, $(a_0 \gamma a_0') \leq f(\mathbf{x}\gamma\mathbf{x}')$ and $(\mathbf{a} + \mathbf{b})\gamma(\mathbf{a}' + \mathbf{b}) \leq g(\mathbf{x}\gamma\mathbf{x}')$, so that $(a_0, \mathbf{a})\gamma(a_0', \mathbf{a}') \in A$.

A and C are disjoint. If not, then there exists $c_0 > f(\mathbf{x}^*)$ and $\mathbf{c} \gg 0$ with $(c_0, \mathbf{c}) \in A$, contradicting the definition of $\mathbf{x}^*$. By the separation of convex sets, there exists a nonzero, continuous linear $L : \mathbb{R} \times \mathbb{R}^m \to \mathbb{R}$ with $L(A) \leq \inf L(C)$. Let $(y_1, \mathbf{y})$ represent L. Since C opens "up and to the right," the $(y_1, \mathbf{y}) \geq 0$, because, if not, $L(C)$ is unbounded below, contradicting $L(A) \leq \inf L(C)$. Moreover, since $(y_1, \mathbf{y}) \geq 0$ and the condition on $(c_0, \mathbf{c}) \in C$ is only that $\mathbf{c} \gg 0$, $\inf L(C) = y_1 f(\mathbf{x}^*)$.

There are two cases: (1) $y_1 = 0$, and (2) $y_1 > 0$. We show that if $y_1 = 0$, then $(\mathbf{x}^*, 0)$ is a saddle point for the Lagrangian and that if $y_1 > 0$, then $(\mathbf{x}^*, \boldsymbol{\lambda}^*)$ is a saddle point with $\boldsymbol{\lambda}^* := \frac{1}{y_1} \mathbf{y}$.

(1) Suppose that $y_1 = 0$, so that $\mathbf{y} > 0$ and $\inf L(C) = 0$. We must show that for all $\mathbf{x} \in X$ and all $\boldsymbol{\lambda} \geq 0$,

$$f(\mathbf{x}) \underbrace{\leq}_{\text{first}} f(\mathbf{x}^*) + 0 \cdot (\mathbf{b} - g(\mathbf{x}^*)) \underbrace{\leq}_{\text{second}} f(\mathbf{x}^*) + \boldsymbol{\lambda} \cdot (\mathbf{b} - g(\mathbf{x}^*)).$$

The second inequality follows from $(\mathbf{b} - g(\mathbf{x}^*)) \geq 0$. Suppose that the first inequality is violated, that is, for some $\mathbf{x}' \in X$, $f(\mathbf{x}') > f(\mathbf{x}^*)$. Pick any $\mathbf{x}' \gg 0$ with $g(\mathbf{x}') - \mathbf{b} \leq \mathbf{a}'$ so that the point $(f(\mathbf{x}'), \mathbf{a}') \in A$. Since $\mathbf{y} > 0$ and $\mathbf{a}' \gg 0$, $L((f(\mathbf{x}'), \mathbf{a}') > 0$, which contradicts $L(A) \leq \inf L(C) = 0$.

(2) Suppose now that $y_1 > 0$ and define $\boldsymbol{\lambda}^* := \frac{1}{y_1} \mathbf{y}$. We must show that for all $\mathbf{x} \in X$ and all $\boldsymbol{\lambda} \geq 0$,

$$f(\mathbf{x}) + \boldsymbol{\lambda}^*(\mathbf{b} - g(\mathbf{x})) \underbrace{\leq}_{\text{first}} f(\mathbf{x}^*) + \boldsymbol{\lambda}^* \cdot (\mathbf{b} - g(\mathbf{x}^*)) \underbrace{\leq}_{\text{second}} f(\mathbf{x}^*) + \boldsymbol{\lambda} \cdot (\mathbf{b} - g(\mathbf{x}^*)).$$

The second inequality is easy, as usual.

For the first inequality, we begin by noting that the vector $(1, \boldsymbol{\lambda}^*)$ also separates the sets A and C. Thus, for all $(a_0, \mathbf{a}) \in A$ and $(c_0, \mathbf{c}) \in C$, we have

$$1 \cdot a_0 + \boldsymbol{\lambda}^* \cdot \mathbf{a} \leq 1 \cdot c_0 + \boldsymbol{\lambda}^* \mathbf{c}.$$

From $\inf L(C) = y_1 f(\mathbf{x}^*)$, the infimum of the right-hand side of this inequality takes the value $f(\mathbf{x}^*)$, yielding, for all $(a_0, \mathbf{a}) \in A$,

$$a_0 + \boldsymbol{\lambda}^* \cdot \mathbf{a} \leq f(\mathbf{x}^*).$$

If the first inequality is violated, say at $\mathbf{x}'$, then $f(\mathbf{x}') + \boldsymbol{\lambda}^*(\mathbf{b} - g(\mathbf{x}')) > f(\mathbf{x}^*) + 0$. Taking $(a_0, \mathbf{a})$ arbitrarily close to $((f(\mathbf{x}'), (\mathbf{b} - g(\mathbf{x}'))$ yields a contradiction. ∎

Corollary 5.8.29 *The same result is true if f is a continuous strictly increasing transformation of a concave function and each g_k is a continuous strictly decreasing transformation of a convex function.*

5.9 ◆ Interpreting Lagrange Multipliers

We now turn to the question, "What use are these multipliers anyway?" In the first place, they are sometimes solutions to the derivative conditions for an optimum. But more usefully than this, they often contain information about how sensitive an optimal value is to a change in the parameters.

5.9.a Kuhn-Tucker Conditions as Derivative Conditions for a Saddle Point

We saw that derivative conditions are necessary and sufficient for $\mathbf{x}^*$ maximizing a concave function or minimizing a convex function (Theorem 5.6.20). We now show that the K-T conditions *are* the derivative conditions for a saddle point. First, for $\mathbf{x}$ fixed at $\mathbf{x}^*$, the function $L(\mathbf{x}^*, \cdot)$ is linear in $\boldsymbol{\lambda}$.

Lemma 5.9.1   *If* $\mathbf{b} - g(\mathbf{x}^*) \geq 0$, *then* $\boldsymbol{\lambda}^*$ *solves the derivative conditions*

$$\frac{\partial L(\mathbf{x}^*, \boldsymbol{\lambda}^*)}{\partial \boldsymbol{\lambda}} \geq 0, \boldsymbol{\lambda}^* \geq 0, \quad \text{and} \quad \boldsymbol{\lambda}^* \cdot \frac{\partial L(\mathbf{x}^*, \boldsymbol{\lambda}^*)}{\partial \boldsymbol{\lambda}} = 0. \qquad (5.20)$$

if and only if $\boldsymbol{\lambda}^*$ *solves the problem* $\min_{\boldsymbol{\lambda} \geq 0} L(\mathbf{x}^*, \boldsymbol{\lambda})$.

The conditions in (5.20) are part of the K-T conditions.

Exercise 5.9.2   Prove the previous lemma.

This means that one way to understand the problems you worked on earlier is that you wrote down the first-order derivative conditions for a saddle point and solved them. The following is a direct implication of Theorem 5.6.20.

Lemma 5.9.3   *If* $h : C \to \mathbb{R}$ *is a concave continuously differentiable function and C is open and convex, then* $\mathbf{x}^*$ *solves* $\max_{\mathbf{x} \in C} h(\mathbf{x})$ *if and only if* $D_{\mathbf{x}} h(\mathbf{x}^*) = 0$.

The first-order conditions for $\max_{\mathbf{x} \in C} f(\mathbf{x}) + \boldsymbol{\lambda}^*(\mathbf{b} - g(\mathbf{x}))$ are the remaining part of the K-T conditions. Combining what we have so far yields the following.

Theorem 5.9.4   *Suppose that $X \subset \mathbb{R}^\ell$ is an open convex set, that $f : X \to \mathbb{R}$ is concave and continuously differentiable, and that the $g_k : X \to \mathbb{R}$ are convex and continuously differentiable. The pair* $(\mathbf{x}^*, \boldsymbol{\lambda}^*)$ *satisfies the derivative conditions*

$$D_{\mathbf{x}} L(\mathbf{x}^*, \boldsymbol{\lambda}^*) = 0, D_{\boldsymbol{\lambda}} L(\mathbf{x}^*, \boldsymbol{\lambda}^*) \geq 0, \quad \boldsymbol{\lambda}^* \geq 0, \text{ and } \boldsymbol{\lambda}^* \cdot D_{\boldsymbol{\lambda}} L(\mathbf{x}^*, \boldsymbol{\lambda}^*) = 0 \quad (5.21)$$

iff $(\mathbf{x}^*, \boldsymbol{\lambda}^*)$ *is a saddle point.*

5.9.b The Implicit Function Theorem

The Lagrangian equations characterizing a solution to a constrained optimization problem are of the form

$$D_{\mathbf{x}} f(\mathbf{x}) + \boldsymbol{\lambda}(\mathbf{b} - g(\mathbf{x})) = 0.$$

Often, they characterize an optimum, and we are interested in the dependence of the optimal $\mathbf{x}^*$ on the parameters $\mathbf{b}$. Letting $h(\mathbf{x}, \mathbf{b})$ denote $D_{\mathbf{x}}(f(\mathbf{x}) + \boldsymbol{\lambda}(\mathbf{b} - g(\mathbf{x})))$, we are interested in a function $\mathbf{x}(\mathbf{b})$ that makes

$$h(\mathbf{x}(\mathbf{b}), \mathbf{b}) \equiv 0, \qquad (5.22)$$

where "$\equiv$" is read as "is equivalent to" or "is identically equal to." In words, the equation $h(\mathbf{x}, \mathbf{b}) = 0$ implicitly defines $\mathbf{x}$ as a function of $\mathbf{b}$.

Exercise 5.9.5 Explain when you can and when you cannot solve $h(\mathbf{x}, \mathbf{b}) = 0$ for $\mathbf{x}$ as a function of $\mathbf{b}$ when

1. $x, b \in \mathbb{R}^1$, $h(x, b) = rx + sb + t$, where $r, s, t \in \mathbb{R}^1$. Express your answer in terms of the Jacobean matrix for $h(x, b)$.

2. $x, b \in \mathbb{R}^1$, $h(x, b) = r(x - b)^n + t$, where $r, t \in \mathbb{R}^1$, with $n \geq 1$ an integer. Express your answer in terms of the Jacobean matrix for $h(x, b)$.

3. $\mathbf{x} \in \mathbb{R}^N$, $\mathbf{b} \in \mathbb{R}^m$, $h(\mathbf{x}, \mathbf{b}) = R\mathbf{x} + S\mathbf{b} + t$, R is an $N \times N$ matrix, S is an $N \times m$ matrix, and $t \in \mathbb{R}^N$. Express your answer in terms of the Jacobean matrix for $h(\mathbf{x}, \mathbf{b})$.

Sometimes all we need is information about $D_{\mathbf{b}}\mathbf{x}$ rather than the whole function $\mathbf{x}(\mathbf{b})$. Suppose that $h(\mathbf{x}, \mathbf{b}) = 0$ defines $\mathbf{x}(\mathbf{b})$ implicitly, and that $\mathbf{x}(\cdot)$ is differentiable. Then, setting $\psi(\mathbf{b}) = h(\mathbf{x}(\mathbf{b}), \mathbf{b})$,

$$\psi(\mathbf{b}) \equiv 0 \text{ implies } D_{\mathbf{b}}\psi(\mathbf{b}) = 0.$$

What we did in the last part of the above was to *totally differentiate* $h(\mathbf{x}(\mathbf{b}), \mathbf{b}) \equiv 0$ *with respect to* $\mathbf{b}$. $D_{\mathbf{b}}\psi(\mathbf{b}) = 0$ is

$$D_{\mathbf{x}}h(\mathbf{x}(\mathbf{b}), \mathbf{b})D_{\mathbf{b}}\mathbf{x}(\mathbf{b}) + D_{\mathbf{b}}h(\mathbf{x}(\mathbf{b}), \mathbf{b}) = 0.$$

Provided $D_{\mathbf{x}}h(\mathbf{x}(\mathbf{b}), \mathbf{b})$ is invertible, we find that

$$D_{\mathbf{b}}\mathbf{x}(\mathbf{b}) = -[D_{\mathbf{x}}h(\mathbf{x}(\mathbf{b}), \mathbf{b})]^{-1}D_{\mathbf{b}}h(\mathbf{x}(\mathbf{b}), \mathbf{b}),$$

which looks like more of a mess than it really is.

Exercise 5.9.6 Suppose that $x, b \in \mathbb{R}$ and $h(x, b) = (x - b)^3 - 1$.

1. Solve for $x(b)$ implicitly defined by $h(x(b), b) \equiv 0$ and find dx/db.

2. Totally differentiate $h(x(b), b) \equiv 0$ and find dx/db.

Exercise 5.9.7 Suppose that $h(x, b) = \ln(x + 1) + x + b$ and that $x(b)$ is implicitly defined by $h(x(b), b) \equiv 0$. Find dx/db.

5.9.c The Envelope Theorem

Here is our favorite version of the envelope theorem. Suppose we have a differentiable function $f : \mathbb{R}^K \times \Theta \to \mathbb{R}$, $\Theta \subset \mathbb{R}^N$, int $\Theta \neq \emptyset$. Consider the function

$$v(\theta) = \max_{\mathbf{x} \in \mathbb{R}^K} f(\mathbf{x}, \theta). \tag{5.23}$$

We are interested in $\partial v/\partial \theta$. In words, we want to know how the maximized value depends on θ, we think of the vector θ as being parameters, and we think of the vector $\mathbf{x}$ as being under the control of a maximizer.

Technical stuff: For each θ in some neighborhood of $\theta° \in$ int Θ suppose that the unique maximizer $\mathbf{x}^*(\theta)$ is locally characterized by the first-order conditions (FOC)

$$\frac{\partial f(\mathbf{x}^*(\theta), \theta)}{\partial \mathbf{x}} = 0.$$

Suppose that the conditions of the implicit function hold so that locally $\mathbf{x}^*(\cdot)$ is a differentiable function of θ. Note that $v(\theta) = f(\mathbf{x}^*(\theta), \theta)$, so that θ has two effects on $v(\cdot)$, the direct one and the indirect one that operates through $\mathbf{x}^*(\cdot)$. The envelope theorem says that the indirect effect does not matter.

Theorem 5.9.8 (Envelope)   *Under the conditions just given,*

$$\frac{\partial v(\theta°)}{\partial \theta} = \frac{\partial f(\mathbf{x}^*(\theta°), \theta°)}{\partial \theta}.$$

Proof.   To see why, start taking derivatives and apply the FOC,

$$\frac{\partial v(\theta°)}{\partial \theta} = \underbrace{\frac{\partial f(\mathbf{x}^*(\theta°), \theta°)}{\partial \mathbf{x}}}_{= 0 \text{ by the FOC}} \frac{\partial \mathbf{x}^*(\theta°)}{\partial \theta} + \frac{\partial f(\mathbf{x}^*(\theta°), \theta°)}{\partial \theta}. \quad ■$$

This is particularly useful if we think about Lagrangians and the fact that they turn constrained optimization problems into unconstrained optimization problems. For example, in the utility-maximization problems you had before,

$$v(\mathbf{p}, w) = \max_{\mathbf{x} \in \mathbb{R}^\ell} \left[u(\mathbf{x}) + \lambda^*(w - p \cdot \mathbf{x}) \right],$$

equivalently

$$v(\mathbf{p}, w) = u(\mathbf{x}(\mathbf{p}, w)) + \lambda^*(w - \mathbf{p} \cdot \mathbf{x}(\mathbf{p}, w)),$$

where λ^* was part of the saddle point. Note that this is really the same as setting $v(\mathbf{p}, w) = u(\mathbf{x}(\mathbf{p}, w))$ because $\lambda^*(w - \mathbf{p} \cdot \mathbf{x}(\mathbf{p}, w)) \equiv 0$. Directly by the envelope theorem, we do not have to consider how $\mathbf{p}$ or w affects v through the optimal $\mathbf{x}(\cdot, \cdot)$, but we can directly conclude that $\partial v(\mathbf{p}, w)/\partial w = \lambda^*$ and $\partial v(\mathbf{p}, w)/\partial p_k = \lambda^* x_k(p, w)$.

Exercise 5.9.9   Directly taking the derivative of $v(\mathbf{p}, w) = u(\mathbf{x}(\mathbf{p}, w)) + \lambda^*(w - \mathbf{p} \cdot \mathbf{x}(\mathbf{p}, w))$ with respect to $\mathbf{p}$ and w and using the FOC, check that the results just given are true.

Exercise 5.9.10   Using the envelope theorem, explain why $\partial c/\partial y^0 = \lambda$ in the cost-minimization problems you solved earlier.

5.9.d   Pesky Details about Constraint Qualification

Consider the following, rather artificial, maximization problem.

Example 5.9.11   *Maximize $f(x_1, x_2) = x_1 + x_2$ subject to $x_1, x_2 \geq 0$, $x_1 \leq 1$, $x_2 \leq 1$, $x_1 + 2x_2 \leq 3$, and $2x_1 + x_2 \leq 3$. The solution is clearly $(x_1^*, x_2^*) = (1, 1)$.*

However, no single constraint is binding in the sense that relaxing it improves the achievable maximum. There are many ways of expressing the gradient of f as a positive linear combination of the constraints satisfied with inequality at the optimum. This means that the Lagrange multipliers that solve the K-T conditions are not unique.

The easiest way to get around such problems is to assume them away, that is, to assume that at the optimum, the gradients of all the constraints satisfied with equality are linearly independent. When we care about the answer, we check whether or not the assumption is true. It has never failed us except when we work at constructing silly examples. Any such assumption is generally known as a **constraint qualification**, which is a reasonable name.

5.10 ◆ Differentiability and Concavity

This section develops the negative semidefiniteness of the matrix of second derivatives as being equivalent to the concavity of a twice continuously differentiable function. It also develops the determinant test for negative semidefiniteness. As we need them, we mention, but not prove, some basic facts about matrix multiplication and determinants.[9]

5.10.a The Two Results

Before giving the results, we need some terminology.

Definition 5.10.1 *An $n \times n$ matrix $\mathbf{A} = (a_{ij})_{i,j=1,\ldots,n}$ is **symmetric** if $a_{ij} = a_{ji}$ for all i and j. A symmetric $\mathbf{A}$ is **negative semidefinite** if for all vectors $\mathbf{z} \in \mathbb{R}^n$, $\mathbf{z}^T \mathbf{A} \mathbf{z} \leq 0$, and it is **negative definite** if for all $\mathbf{z} \neq 0$, $\mathbf{z}^T \mathbf{A} \mathbf{z} < 0$.*

Notation Alert 5.10.A *Previously, we denoted the transpose of a vector $\mathbf{z}$ by $\mathbf{z}'$. For this section, we will instead use $\mathbf{z}^T$. This is because it is convenient to have f' and f'' denoting the first and second derivatives of a function.*

This section is devoted to proving two results. Here is the first one.

Theorem 5.10.2 *A twice continuously differentiable $f : \mathbb{R}^n \to \mathbb{R}$ defined on an open convex set C is concave iff for all $\mathbf{x}^\circ \in C$, $D_{\mathbf{x}}^2 f(\mathbf{x}^\circ)$ is negative semidefinite if for all $\mathbf{x}^\circ \in C$, $D_{\mathbf{x}}^2 f(\mathbf{x}^\circ)$ is negative definite, then f is strictly concave.*

The **principal submatrices** of a symmetric $n \times n$ matrix $\mathbf{A} = (a_{ij})_{i,j=1,\ldots,n}$ are the $m \times m$ matrices $(a_{ij})_{i,j=1,\ldots,m}$, $m \leq n$. Thus, the three principal submatrices of the 3×3 matrix

$$\mathbf{A} = \begin{bmatrix} 3 & 0 & 0 \\ 0 & 4 & \sqrt{3} \\ 0 & \sqrt{3} & 6 \end{bmatrix}$$

---

9. Multiplying by -1 makes this a section on convexity, something that we mention only here.

are

$$[3], \quad \begin{bmatrix} 3 & 0 \\ 0 & 4 \end{bmatrix}, \quad \text{and} \quad \begin{bmatrix} 3 & 0 & 0 \\ 0 & 4 & \sqrt{3} \\ 0 & \sqrt{3} & 6 \end{bmatrix}.$$

Here is the second result that we prove in this section.

Theorem 5.10.3 *A matrix* **A** *is negative semidefinite (respectively, negative definite) iff the sign of mth principal submatrix is either* 0 *or* -1^m *(respectively, the sign of the mth principal submatrix is* -1^m*). It is positive semidefinite (respectively, positive definite) if "*-1^m*" is replaced with "*$+1^m$*" throughout.*

In the following two problems, use Theorem 5.10.2 and 5.10.3, even though we have not yet proved them.

Exercise 5.10.4   The function $f : \mathbb{R}_+^2 \to \mathbb{R}$ defined by $f(x, y) = x^\alpha y^\beta$, α, $\beta > 0$, is strictly concave on $\mathbb{R}_{++}^2$ if $\alpha + \beta < 1$ and concave on $\mathbb{R}_{++}^2$ if $\alpha + \beta = 1$.

Exercise 5.10.5   The function $f : \mathbb{R}_+^2 \to \mathbb{R}$ defined by $f(x, y) = (x^p + y^p)^{1/p}$ is convex on $\mathbb{R}_{++}^2$ if $p \geq 1$ and concave if $p \leq 1$.

5.10.b The One-Dimensional Case, $f : \mathbb{R}^1 \to \mathbb{R}$

Theorem 5.10.6   *If* $f : (a, b) \to \mathbb{R}$ *is twice continuously differentiable, then* f *is concave iff* $f''(x) \leq 0$ *for all* $x \in (a, b)$.

Exercise 5.10.7   Fill in the details in the following sketch of a proof of Theorem 5.10.6.

1. Show that if $f''(x) \leq 0$ for all $x \in (a, b)$, then f is concave. [Hint: We know that f' is nonincreasing. Pick x, y with $a < x < y < b$, choose $\alpha \in (0, 1)$, and define $z = \alpha x + (1 - \alpha)y$. Note that $(z - x) = (1 - \alpha)(y - x)$ and $(y - z) = \alpha(y - x)$. Show that

$$f(z) - f(x) = \int_x^z f'(t)\, dt \geq f'(z)(z - x) = f'(z)(1 - \alpha)(y - x),$$

$$f(y) - f(z) = \int_z^y f'(t)\, dt \leq f'(z)(y - z) = f'(z)\alpha(y - x).$$

Therefore,

$$f(z) \geq f(x) + f'(z)(1 - \alpha)(y - x), \quad f(z) \geq f(y) - f'(z)\alpha(y - x).$$

Multiply the left side by α and the right side by $(1 - \alpha)$, and . . . .]

2. Show that if f is concave, then $f''(x) \leq 0$ for all $x \in (a, b)$. [If not, then $f''(x^\circ) > 0$ for some $x^\circ \in (a, b)$, which implies that f'' is strictly positive on some interval $(a', b') \subset (a, b)$. Reverse the above argument.]

The proof you just gave immediately yields the following.

Corollary 5.10.8 *If $f : (a, b) \to \mathbb{R}$ is twice continuously differentiable and $f''(x) < 0$ for all $x \in (a, b)$, then f is strictly concave.*

The exercise that follows shows that one can have $f''(x) = 0$ for a small set of $x \in (a, b)$ and still maintain strict concavity.

Exercise 5.10.9 Suppose that $f : (-1, +1) \to \mathbb{R}$ is twice continuously differentiable and $f''(x) = -|x|$, so that $f''(0) = 0$. Show that f is strictly concave.

5.10.c The Multidimensional Case, $f : \mathbb{R}^n \to \mathbb{R}$

Theorem 5.10.10 *If $f : C \to \mathbb{R}$ is twice continuously differentiable and C an open convex subset of $\mathbb{R}^n$, then f is concave iff $D_{\mathbf{x}}^2 f(\mathbf{x})$ is negative semidefinite for all $\mathbf{x} \in C$.*

Exercise 5.10.11 For each $\mathbf{y}, \mathbf{z} \in \mathbb{R}^n$, define $g_{\mathbf{y}, \mathbf{z}}(\lambda) = f(\mathbf{y} + \lambda \mathbf{z})$ for those λ in the interval $\{\lambda : \mathbf{y} + \lambda \mathbf{z} \in C\}$. Fill in the details in the following sketch of a proof of Theorem 5.10.10.

1. Show that f is concave iff each $g_{\mathbf{y}, \mathbf{z}}$ is concave.

2. Show that $g''(\lambda) = \mathbf{z}^T D_{\mathbf{x}}^2 f(\mathbf{x}^\circ) \mathbf{x}$, where $\mathbf{x}^\circ = \mathbf{y} + \lambda \mathbf{z}$.

3. Conclude that f is concave iff for all $\mathbf{x}^\circ \in C$, $D^2 f(\mathbf{x}^\circ)$ is negative semidefinite.

The proof you just gave immediately yields the following.

Corollary 5.10.12 *If $f : C \to \mathbb{R}$ is twice continuously differentiable, C an open convex subset of $\mathbb{R}^n$, and $D_{\mathbf{x}}^2 f(\mathbf{x})$ is negative definite for all $\mathbf{x} \in C$, then f is strictly concave.*

5.10.d The Theory of the Firm and Differentiable Support Functions

We have seen that for a set $Y \subset \mathbb{R}^\ell$, $\mu_Y(\mathbf{p}) := \sup\{\mathbf{p}\mathbf{y} : \mathbf{y} \in Y\}$ is a convex function because it is the supremum of a collection of affine functions. When Y is the set of technologically feasible points for a firm and $\mathbf{p} \gg 0$, then $\mathbf{p} \mapsto \mu_Y(\mathbf{p})$ is the **profit function** of a neoclassical price-taking firm. Let us suppose that Y is a bounded set in order to avoid the problems that arise when $\mu_Y(\mathbf{p}) = \infty$, and let us assume that it is closed in order to avoid the problems of the existence of nonempty supply and demand correspondences.

Recall that with $\mathbf{p} \in \mathbb{R}^\ell_{++}$, the neoclassical profit function for a firm with technology characterized by Y is

$$\pi_Y(\mathbf{p}) = \mu_Y(\mathbf{p}) = \sup\{\mathbf{p}\mathbf{y} : \mathbf{y} \in Y\}, \tag{5.24}$$

and if $y(\mathbf{p})$ solves the maximization problem $\max\{\mathbf{p}\mathbf{y} : \mathbf{y} \in Y\}$, then it is the profit-maximizing netput behavior for the firm. A convex function is differentiable at a point $\mathbf{p}^\circ$ iff $D_{\mathbf{p}}^{\mathrm{sub}} \mu_Y(\mathbf{p}^\circ)$ contains only one point.

Theorem 5.10.13 *If* $Y \subset \mathbb{R}^\ell$ *is compact, then* $\mu_Y(\cdot)$ *is differentiable at* $\mathbf{p}^\circ$ *iff* $y(\mathbf{y}^\circ)$ *contains one point, in which case* $D_{\mathbf{p}}\mu_Y(\mathbf{p}^\circ) = y(\mathbf{p}^\circ)$.

Proof. For any $\mathbf{z} \in y(\mathbf{p}^\circ)$ and any $\mathbf{p}$, $\mu_Y(\mathbf{p}) \geq \mathbf{z}\mathbf{p}$. (Since $\mu_Y(\mathbf{p})$ is the best one can do facing prices $\mathbf{p}$, it must be at least as good as choosing to do $\mathbf{z}$.) As $\mu_Y(\mathbf{p}^\circ) = \mathbf{p}^\circ \mathbf{z}$, adding and subtracting $\mu_Y(\mathbf{p}^\circ)$ yields $\mu_Y(\mathbf{p}) \geq \mu_Y(\mathbf{p}^\circ) + \mathbf{z}(\mathbf{p} - \mathbf{p}^\circ)$, which is the definition of $\mathbf{z}$ being in the subdifferential of μ_Y at $\mathbf{p}^\circ$. This means that $y(\mathbf{p}^\circ) \subset D_{\mathbf{p}}^{\mathrm{sub}}\mu_Y(\mathbf{p}^\circ)$. Hence, if $D_{\mathbf{p}}^{\mathrm{sub}}\mu_Y(\mathbf{p}^\circ)$ contains only one point, that point must be $y(\mathbf{p}^\circ)$. ∎

Exercise 5.10.14 Complete the proof of Theorem 5.10.13 by showing that if $y(\mathbf{p}^\circ)$ contains only one point, then $\mu_Y(\cdot)$ is differentiable at $\mathbf{p}^\circ$.

The first part of Theorem 5.10.13 gives us Hotelling's lemma, which says that profits move down if any input price goes up or any output price goes down.

Lemma 5.10.15 (Hotelling) *If the profit function is differentiable at* $\mathbf{p}^\circ$, *then* $D_{\mathbf{p}}\pi(\mathbf{p}^\circ) = y(\mathbf{p}^\circ)$.

Suppose now that $y(\cdot)$ is continuously differentiable on some neighborhood of $\mathbf{p}^\circ$. Then Hotelling's lemma holds on the entire neighborhood, that is, $D_{\mathbf{p}}\pi(\mathbf{p}^\circ) \equiv y(\mathbf{p}^\circ)$ on the neighborhood. Taking derivatives on both sides yields the following.

Theorem 5.10.16 *If* $y(\cdot)$ *is continuously differentiable on some neighborhood of* $\mathbf{p}^\circ$, *then* $D_{\mathbf{p}}y(\mathbf{p}^\circ) = D_{\mathbf{p}}^2\pi(\mathbf{p}^\circ)$, $D_{\mathbf{p}}^2\pi(\mathbf{p}^\circ)$ *is negative semidefinite and* $D_{\mathbf{p}}y(\mathbf{p}^\circ)\mathbf{p}^\circ = 0$.

Proof. The negative semidefiniteness comes from the convexity of $\pi(\cdot)$ and the second condition comes from the observation that $\pi(\cdot)$ is homogeneous of degree 1, that is, that for all $r > 0$ and all $\mathbf{p}$, $\pi(r\mathbf{p}) = r\pi(\mathbf{p})$. ∎

$D_{\mathbf{p}}y(\mathbf{p}^\circ)$ is the supply substitution matrix, and negative semidefiniteness gives us the law of supply, that $\partial y_k / \partial p_k \geq 0$, that if the price of an output rises, then the supply increases (at least weakly), and if the price of an input rises, then the demand for the input decreases.

5.10.e A Review of Some Matrix Algebra

The foregoing demonstrated that we sometimes want to know conditions on $n \times n$ symmetric matrices $\mathbf{A}$ such that $\mathbf{z}^T \mathbf{A} \mathbf{z} \leq 0$ for all $\mathbf{z}$ or $\mathbf{z}^T \mathbf{A} \mathbf{z} < 0$ for all $\mathbf{z} \neq 0$. We are trying to prove that $\mathbf{A}$ is negative semidefinite (respectively, negative definite) iff the sign of mth principal submatrix is either 0 or -1^m (respectively, the sign of the mth principal submatrix is -1^m). This will take a longish detour through eigenvalues and eigenvectors, but as the detour is useful for the study of linear regression as well, this section is also background for first-year econometrics.

Throughout, all matrices have only real number entries.

$|\mathbf{A}|$ denotes the determinant of the square $\mathbf{A}$. Recall that $\mathbf{A}$ is invertible, as a linear mapping, iff $|\mathbf{A}| \neq 0$. (If these statements do not make sense to you, you missed linear algebra and/or have to do some review.)

Exercise 5.10.17 Remember, or look up, how to find determinants for 2×2 and 3×3 matrices.

A vector $\mathbf{x} \neq 0$ is an **eigenvector** and the number $\lambda \neq 0$ is an **eigenvalue**[10] for $\mathbf{A}$ if $\mathbf{Ax} = \lambda \mathbf{x}$. Note that $\mathbf{Ax} = \lambda \mathbf{x}$ iff $\mathbf{A}(r\mathbf{x}) = \lambda(r\mathbf{x})$ for all $r \neq 0$. Therefore, we can, and do, normalize eigenvectors by $\|\mathbf{x}\| = 1$, which corresponds to setting $r = 1/\|\mathbf{x}\|$. There is still some ambiguity, since we could just as well set $r = -1/\|\mathbf{x}\|$.

In general, one might have to consider λ's and $\mathbf{x}$'s that are imaginary numbers, that is $\lambda = a + bi$ with $i = \sqrt{-1}$. This means that $\mathbf{x}$ has to be imaginary as well. To see why, read on.

Lemma 5.10.18 $\mathbf{A} = \lambda \mathbf{x}$, $\mathbf{x} \neq 0$, *iff* $(\mathbf{A} - \lambda \mathbf{I})\mathbf{x} = 0$ *iff* $|\mathbf{A} - \lambda \mathbf{I}| = 0$.

Proof. You should know why this is true. If not, you need some more review. ∎

Define $g(\lambda) = |\mathbf{A} - \lambda \mathbf{I}|$ so that g is an nth-degree polynomial in λ. The fundamental theorem of algebra tells us that any nth-degree polynomial has n roots, counting multiplicities, in the complex plane. To be a bit more concrete, this means that there are complex numbers λ_i, $i = 1, \ldots, n$ such that

$$g(y) = (\lambda_1 - y)(\lambda_2 - y) \cdots (\lambda_n - y).$$

The "counting multiplicities" phrase means that the λ_i need not be distinct.

Exercise 5.10.19 Using the quadratic formula, show that if $\mathbf{A}$ is a symmetric 2×2 matrix, then both of the eigenvalues of $\mathbf{A}$ are real numbers. Give a 2×2 nonsymmetric matrix with real entries having two imaginary eigenvalues. [This can be done with a matrix having only 0's and 1's as entries.]

The conclusion about real eigenvalues in the previous problem is true for general $n \times n$ matrices, and we turn to this result.

From your trigonometry class (or from someplace else), $(a + bi)(c + di) = (ac - bd) + (ad + bc)i$ defines multiplication of complex numbers and $(a + bi)^* := a - bi$ defines the complex conjugate of the number $(a + bi)$. Note that $rs = sr$ and $r = r^*$ iff r is a real number for complex r, s. By direct calculation, $(rs)^* = r^* s^*$ for any pair of complex numbers r, s. Complex vectors are vectors with complex numbers as their entries. Their dot product is defined in the usual way, $\mathbf{x} \cdot \mathbf{y} := \sum_i x_i y_i$. Notationally, $\mathbf{x} \cdot \mathbf{y}$ is written $\mathbf{x}^T \mathbf{y}$. The next proof uses

Exercise 5.10.20 If r is a complex number, show that $rr^* = 0$ iff $r = 0$. If $\mathbf{x}$ is a complex vector, show that $\mathbf{x}^T \mathbf{x}^* = 0$ iff $\mathbf{x} = 0$.

Lemma 5.10.21 *Every eigenvalue of a symmetric $\mathbf{A}$ is real and distinct eigenvectors are orthogonal to each other.*

Proof. The eigenvalue part: Suppose that λ is an eigenvalue and $\mathbf{x}$ an associated eigenvector, so that

$$\mathbf{Ax} = \lambda \mathbf{x}. \tag{5.25}$$

---

10. "Eigen" is a German word meaning "own." Sometimes eigenvalues are called characteristic roots. The idea that we are building to is that the eigenvalues and eigenvectors tell us everything there is to know about the matrix $\mathbf{A}$.

Taking the complex conjugate of both sides yields

$$\mathbf{A}\mathbf{x}^* = \lambda^* \mathbf{x}^* \tag{5.26}$$

because $\mathbf{A}$ has only real entries:

$$[\mathbf{A}\mathbf{x} = \lambda\mathbf{x}] \Rightarrow [(\mathbf{x}^*)^T \mathbf{A}\mathbf{x} = (\mathbf{x}^*)^T \lambda\mathbf{x} = \lambda\mathbf{x}^T\mathbf{x}^*],$$

$$[\mathbf{A}\mathbf{x}^* = \lambda^*\mathbf{x}^*] \Rightarrow [\mathbf{x}^T \mathbf{A}\mathbf{x}^* = \mathbf{x}^T \lambda^*\mathbf{x}^* = \lambda^*\mathbf{x}^T\mathbf{x}^*].$$

Subtracting, we get

$$(\mathbf{x}^*)^T \mathbf{A}\mathbf{x} - \mathbf{x}^T \mathbf{A}\mathbf{x}^* = (\lambda - \lambda^*)\mathbf{x}^T\mathbf{x}^*.$$

Since the matrix $\mathbf{A}$ is symmetric,

$$(\mathbf{x}^*)^T \mathbf{A}\mathbf{x} - \mathbf{x}^T \mathbf{A}\mathbf{x}^* = 0.$$

Since $\mathbf{x} \neq 0$, $\mathbf{x}^T\mathbf{x}^* \neq 0$. Therefore,

$$[(\lambda - \lambda^*)\mathbf{x}^T\mathbf{x}^* = 0] \Rightarrow [(\lambda - \lambda^*) = 0],$$

which can only happen if λ is a real number.

The eigenvector part: From the previous part, all eigenvalues are real. Since $\mathbf{A}$ is real, this implies that all eigenvectors are also real.

Let $\lambda_i \neq \lambda_j$ be distinct eigenvalues and $\mathbf{x}_i$, $\mathbf{x}_j$ their associated eigenvectors so that

$$\mathbf{A}\mathbf{x}_i = \lambda_i\mathbf{x}_i, \quad \mathbf{A}\mathbf{x}_j = \lambda_j\mathbf{x}_j.$$

Premultiplying by the appropriate vectors gives us

$$\mathbf{x}_j^T \mathbf{A}\mathbf{x}_i = \lambda_i\mathbf{x}_j^T\mathbf{x}_i, \quad \mathbf{x}_i^T \mathbf{A}\mathbf{x}_j = \lambda_j\mathbf{x}_i^T\mathbf{x}_j.$$

We know that $\mathbf{x}_i^T\mathbf{x}_j = \mathbf{x}_j^T\mathbf{x}_i$ (by properties of dot products). Because $\mathbf{A}$ is symmetric,

$$\mathbf{x}_j^T \mathbf{A}\mathbf{x}_i = \mathbf{x}_i^T \mathbf{A}\mathbf{x}_j.$$

Combining yields

$$(\lambda_i - \lambda_j)\mathbf{x}_j^T\mathbf{x}_i = 0.$$

Since $(\lambda_i - \lambda_j) \neq 0$, we conclude that $\mathbf{x}_i \cdot \mathbf{x}_j = 0$, the orthogonality we were looking for. ∎

The following uses basic linear algebra definitions.

Exercise 5.10.22   If the $n \times n$ $\mathbf{A}$ has n distinct eigenvalues, then its eigenvectors form an orthonormal basis for $\mathbb{R}^n$.

A careful proof shows that if $\mathbf{A}$ has an eigenvalue λ_i with multiplicity $k \geq 2$, then we can pick k orthogonal eigenvectors spanning the k-dimensional set of all x

such that $\mathbf{A}x = \lambda_i x$. There are infinitely many different ways of selecting such an orthogonal set. You either accept this on faith or go review a good matrix algebra textbook.

Exercise 5.10.23　Find eigenvalues and eigenvectors for

$$\begin{bmatrix} 4 & \sqrt{3} \\ \sqrt{3} & 6 \end{bmatrix} \quad \text{and} \quad \begin{bmatrix} 3 & 0 & 0 \\ 0 & 4 & \sqrt{3} \\ 0 & \sqrt{3} & 6 \end{bmatrix}.$$

Let $\lambda_1, \ldots, \lambda_n$ be the eigenvalues of $\mathbf{A}$ (repeating any multiplicities) and let $\mathbf{u}_1, \ldots, \mathbf{u}_n$ be a corresponding set of orthonormal eigenvectors. Let $\mathbf{Q} = (\mathbf{u}_1, \ldots, \mathbf{u}_n)$ be the matrix with the eigenvectors as columns. Note that $\mathbf{Q}^T \mathbf{Q} = \mathbf{I}$, so that $\mathbf{Q}^{-1} = \mathbf{Q}^T$. A matrix with its transpose as its inverse is an **orthogonal matrix**. Let Λ be the $n \times n$ matrix with $\Lambda_{ii} = \lambda_i$, with 0's in the off-diagonal.

Exercise 5.10.24　Show that $\mathbf{Q}^T \mathbf{A} \mathbf{Q} = \Lambda$, equivalently $\mathbf{A} = \mathbf{Q} \Lambda \mathbf{Q}^T$.

Expressing a symmetric matrix $\mathbf{A}$ in this form is called **diagonalizing the matrix**. We have shown that any symmetric matrix can be diagonalized so as to have its eigenvalues along the diagonal, and the matrix that achieves this is the matrix of eigenvectors.

Theorem 5.10.25　$\mathbf{A}$ *is negative (semi-)definite iff all of its eigenvalues are less than (or equal to)* 0.

Proof.　$\mathbf{z}^T \mathbf{A} \mathbf{z} = \mathbf{z}^T \mathbf{Q}^T \Lambda \mathbf{Q} \mathbf{z} = \mathbf{V}^T \Lambda \mathbf{V}$ and the matrix $\mathbf{Q}$ is invertible.　■

5.10.f The Alternating Signs Determinant Test for Concavity

Now we have enough matrix algebra background to prove what we set out prove: $\mathbf{A}$ is negative semidefinite (respectively, negative definite) iff the sign of mth principal submatrix is either 0 or -1^m (respectively, the sign of the mth principal submatrix is -1^m).

We defined $g(y) = |\mathbf{A} - y\mathbf{I}|$ so that g is an nth-degree polynomial in λ and used the fundamental theorem of algebra (and some calculation) to tell us that

$$g(y) = (\lambda_1 - y)(\lambda_2 - y) \cdots (\lambda_n - y),$$

where the λ_i are the eigenvalues of $\mathbf{A}$. Note that $g(0) = |\mathbf{A} - 0\mathbf{I}| = |\mathbf{A}| = \lambda_1 \cdot \lambda_2 \cdots \lambda_n$, that is, as follows.

Lemma 5.10.26　*The determinant of a matrix is the product of its eigenvalues.*

We did not use symmetry for this result.

Recall that the principal submatrices of a symmetric $n \times n$ matrix $\mathbf{A} = (a_{ij})_{i,j=1,\ldots,n}$ are the $m \times m$ matrices $(a_{ij})_{i,j=1,\ldots,m}$, $m \leq n$. The following is pretty obvious, but it is nevertheless useful.

Exercise 5.10.27　$\mathbf{A}$ is negative definite iff for all $m \leq n$ and all nonzero x having only the first m components not equal to 0, $x^T \mathbf{A} x < 0$.

Looking at $m = 1$, we must check whether

$$(x_1, 0, 0, \ldots, 0) \begin{bmatrix} a_{11} & a_{12} & \cdots & a_{1n} \\ a_{21} & a_{22} & \cdots & a_{2n} \\ \vdots & \vdots & & \vdots \\ a_{n1} & a_{n2} & \cdots & a_{nn} \end{bmatrix} \begin{pmatrix} x_1 \\ 0 \\ 0 \\ \vdots \\ 0 \end{pmatrix} = a_{11} x_1^2 < 0.$$

This is true iff the first principal submatrix of $\mathbf{A}$ has the same sign as $-1^m = -1^1 = -1$.

Looking at $m = 2$, we must check whether

$$(x_1, x_2, 0, \ldots, 0) \begin{bmatrix} a_{11} & a_{12} & \cdots & a_{1n} \\ a_{21} & a_{22} & \cdots & a_{2n} \\ \vdots & \vdots & & \vdots \\ a_{n1} & a_{n2} & \cdots & a_{nn} \end{bmatrix} \begin{pmatrix} x_1 \\ x_2 \\ 0 \\ \vdots \\ 0 \end{pmatrix} < 0.$$

This is true iff the matrix

$$\begin{bmatrix} a_{11} & a_{12} \\ a_{21} & a_{22} \end{bmatrix}$$

is negative definite, which is true iff all of its eigenvalues are negative. There are two eigenvalues, and as the product of two negative numbers is positive, the $m = 2$ case is handled by having the sign of the determinant of the 2×2 principal submatrix be -1^2.

Looking at $m = 3$, we must check whether

$$\begin{bmatrix} a_{11} & a_{12} & a_{13} \\ a_{21} & a_{22} & a_{23} \\ a_{31} & a_{32} & a_{33} \end{bmatrix}$$

is negative definite, which is true iff all of its eigenvalues are negative. There are three eigenvalues, and as the product of three negative numbers is negative, the $m = 3$ case is handled by having the sign of the determinant of the 3×3 principal submatrix as -1^3.

Continue in this fashion, and you have a proof of Theorem 5.10.3. Your job is to fill in the details for the negative semidefinite, the positive definite, and the positive semidefinite cases as well.

Exercise 5.10.28 Prove Theorem 5.10.3.

5.11 ◆ Fixed-Point Theorems and General Equilibrium Theory

We have already seen two fixed-point theorems: Tarski's for monotone functions from a lattice to itself and Banach's for contraction mappings from a complete metric space to itself. The first gave us the existence of stable matchings and the second conditions under which dynamical systems converge to an equilibrium.[11] There are two more fixed-point theorems that economists use regularly to prove that equilibria exist: Brouwer's fixed-point theorem for functions and Kakutani's fixed-point theorem for correspondences. The general pattern is that using Kakutani's theorem yields an easier proof of a more general result.

Here, we state these two theorems and show how they fail if any one of their assumptions fails individually. We do not prove them.[12] We use these theorems to give two different proofs of equilibrium existence in neoclassical models of exchange economies. In the two proofs we see the general pattern mentioned earlier, an easier proof of a more general result with Kakutani's theorem.

Economists think that prices are tremendously useful coordination devices, and these existence results tell us that our models of this kind of coordination are not entirely empty. In §5.12, we use these fixed-point theorems to prove the existence of Nash equilibria and perfect equilibria. There are also proper equilibria and stable sets of equilibria for finite games, and we also use Kakutani's theorem for the proofs of their existence.

5.11.a Brouwer's Fixed-Point Theorem

Recall that a fixed point of a function $f : X \to X$ is a point x^* with $f(x^*) = x^*$.

Theorem 5.11.1 (Brouwer) *If $K \subset \mathbb{R}^\ell$ is compact and convex and $f : K \to K$ is continuous, then f has a fixed point.*

There are three assumptions in Brouwer's theorem, continuity, compactness, and convexity. The following example shows that none can be dispensed with.

Example 5.11.2 *Let $K = [0, 1]$, the dis*continuous $f(x) = 1_{[0, \frac{1}{2})}(x)$ *has no fixed point. The continuous function $f(x) = x + 1$ has no fixed point in the* non*compact, convex set $\mathbb{R}_+$. The continuous function $f(x) = 3 - x$ has no fixed point in the compact,* non*convex set $[0, 1] \cup [2, 3]$.*

Example 5.11.3 *When K is a one-dimensional compact convex set, Brouwer's theorem is equivalent to the intermediate value Theorem 4.12.5 that tells us that a continuous function on an interval takes on an interval of values. Suppose that $f : [0, 1] \to [0, 1]$ is continuous. Define $g(x) = f(x) - x$ so that $g(x^*) = 0$ iff $f(x^*) = x^*$, that is, iff x^* is a fixed point. As $f(0) \geq 0$, $g(0) \geq 0$. As $f(1) \leq 1$, $g(1) \leq 1$, and since g is continuous, it must take on all values between $g(1)$ and $g(0)$ (Figure 5.11.3).*

Let f^n be the n-fold application of the function f. We have seen examples of functions that are not contractions but that have f^n as a contraction and thus have a fixed point for some $n \geq 1$.

---

11. In Chapter 6, it gives us the implicit function theorem, the existence and uniqueness of the value function in stochastic dynamic programming, Mom, and apple pie too.

12. There are many excellent proofs available in the literature, and we give many pointers at the end of this chapter.

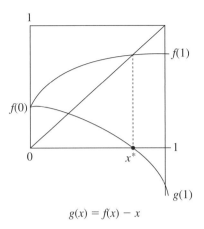

$$g(x) = f(x) - x$$

FIGURE 5.11.3

Exercise 5.11.4 Let K be the compact nonconvex set $\{x \in \mathbb{R}^2 : \|x\| = 1\}$. Let $f_\theta : K \to K$ defined by rotation by the angle θ.

1. Give the fixed points of f if θ is rational.

2. Show that f_θ has no fixed point, and neither does any f^n, $n \geq 1$, if θ is irrational.

When P is the $S \times S$ transition matrix for a Markov chain, one solves for **ergodic distributions** by solving the equations $\pi = \pi P$ for some $\pi \in \Delta(S)$. [See §4.11.b]. The existence of a solution to these equations follows from Brouwer's theorem.

Example 5.11.5 *If the mapping $\pi \mapsto \pi P$ or one of its iterates is a contraction, then there is a unique π^* such that $\pi^* = \pi^* P$. Any such π^* is called an ergodic distribution. Many P's are not contraction mappings nor have they any iterate that is a contraction mapping. For such P, note that the mapping $\pi \mapsto \pi P$ is continuous from $\Delta(S)$ to itself. By Brouwer's theorem, it has a fixed point; that is, ergodic distributions always exist.*

Theorem 5.11.6 *Let $\mathcal{E} = \{X_i, \succsim_i, \mathbf{y}_i\}_{i \in I}$ be an exchange economy model. Assume that for every $i \in I$,*

1. *X_i is a compact convex subset of $\mathbb{R}^\ell_+$ with $\mathbf{y}_i \in \text{int}(X_i)$, and*

2. *$\succsim_i$ can be represented by a $u_i : X_i \to \mathbb{R}$ that is monotonic, $[\mathbf{x} > \mathbf{y}] \Rightarrow [u_i(\mathbf{x}) > u_i(\mathbf{y})]$, and strictly quasi-concave, $[u(\mathbf{x}) = u(\mathbf{y})] \Rightarrow (\forall \alpha \in (0, 1))[u(\mathbf{x}\alpha\mathbf{y}) > u(\mathbf{x})]$.*

Then there exists an equilibrium $(\mathbf{p}, \mathbf{x})$ for $\mathcal{E}$.

Proof. Let $\Delta^\ell \equiv \{\mathbf{p} \in \mathbb{R}^\ell_+ : \sum_{i=1}^\ell p_i = 1\}$ be the simplex of normalized prices. For $\mathbf{p} \in \Delta^\ell$ and $i \in I$, define $\mathbf{x}_i(\mathbf{p}) = \arg\max\{u_i(\mathbf{x}) : \mathbf{x} \in X_i, \mathbf{px} \leq \mathbf{py}_i\}$. Since u_i is continuous and $\{\mathbf{x} \in X_i : \mathbf{px} \leq \mathbf{py}_i\} \neq \emptyset$, $\mathbf{x}_i(\mathbf{p}) \neq \emptyset$. As u_i is strictly quasi-concave, $\#\mathbf{x}_i(\mathbf{p}) = 1$; that is, the demand correspondence is single-valued. By the theorem of the maximum, $x_i(\cdot)$ is upper hemicontinuous. Since it is single valued, it is a continuous function.

Define i's **excess demand function** by $z_i(\mathbf{p}) = x_i(\mathbf{p}) - y_i$ and define the **excess demand function** by $z(\mathbf{p}) = \sum_{i \in I} z_i(\mathbf{p})$. Note that $z(\cdot)$ is a continuous function. We define a function $f : \Delta^\ell \to \mathbb{R}^\ell$ by defining each component, $k \in \{1, \ldots, \ell\}$, by

$$f_k(\mathbf{p}) = \frac{p_k + \max\{0, z_k(\mathbf{p})\}}{1 + \sum_{j=1}^\ell z_j(\mathbf{p})}.$$

For each $\mathbf{p} \in \Delta^\ell$, $\sum_{k=1}^\ell p_k = 1$, so that $\mathbf{p} \mapsto f(\mathbf{p})$ is a continuous function from the compact convex set Δ^ℓ to itself. By Brouwer's theorem, there exists $\mathbf{p}^*$ such that $f(\mathbf{p}^*) = \mathbf{p}^*$. Let $\mathbf{x}^*$ be the allocation $(x_i(\mathbf{p}^*))_{i \in I}$. We claim that $(\mathbf{p}^*, \mathbf{x}^*)$ is a Walrasian equilibrium for $\mathcal{E}$.

By the definition of $x_i(\mathbf{p}^*)$, we know that if $x_i' \succ x_i(\mathbf{p}^*)$, then $\mathbf{p}^* x_i' > \mathbf{p}^* y_i$, so all that is left to show is feasibility.

Since $f(\mathbf{p}^*) = \mathbf{p}^*$, for each $k \in \{1, \ldots, \ell\}$, $f_k(\mathbf{p}^*) = p_k^*$; that is,

$$p_k^* = \frac{p_k^* + \max\{0, z_k(\mathbf{p}^*)\}}{1 + \sum_{j=1}^\ell z_j(\mathbf{p}^*)}.$$

Cross-multiplying, simplifying, and then multiplying each equation of $z_k(\mathbf{p}^*)$ gives

$$\left[\sum_j \max\{0, z_j(\mathbf{p}^*)\} \right] \cdot \left[\sum_k p_k^* z_k(\mathbf{p}^*) \right] = \sum_k z_k(\mathbf{p}^*) \max\{0, z_k(\mathbf{p}^*)\}. \quad (5.27)$$

We now use the monotonicity of the u_i. Since each u_i is monotonic, for all $\mathbf{p} \in \Delta^\ell$, $\mathbf{p} x_i(\mathbf{p}) = \mathbf{p} y_i$. This gives *Walras's law*: for all $\mathbf{p} \in \Delta^\ell$, $\mathbf{p} z(\mathbf{p}) = 0$. Applying this to the left-hand side of (5.27) yields that the right-hand side is equal to 0, that is, $\sum_k z_k(\mathbf{p}^*) \max\{0, z_k(\mathbf{p}^*)\} = 0$. Each of the k terms in this expression is either 0 or strictly positive. Since the sum is 0, each term must equal 0. Thus, $z(\mathbf{p}^*) = \sum_i (x_i^* - y_i) \leq 0$, that is, $\mathbf{x}^*$ is feasible. ∎

The compactness assumption on the X_i in the existence Theorem 5.11.6 is not perfectly satisfactory. On the one hand, by replacing $\mathbb{R}_+^\ell$ by $X_i(B) := \{\mathbf{x} \in \mathbb{R}_+^\ell : \|\mathbf{x}\| \leq B\}$ where B is huge hardly seems like much of a restriction. However, we have no guarantee that it is not some subtle aspect of compactness that is driving the existence result. For example, we could only apply the theorem of the maximum in the middle of the proof because X_i was compact.

Exercise 5.11.7 Suppose that the assumptions of Theorem 5.11.6 hold and that all goods are *desirable* in the sense that if $\mathbf{p}_n \to \partial \Delta^\ell$, then $\|z(\mathbf{p}_n)\| \to \infty$ (where $\partial \Delta^\ell$ is the relative boundary in $\mathrm{aff}(\Delta^\ell)$). Prove that Theorem 5.11.6 holds with each X_i being the noncompact set $\mathbb{R}_+^\ell$. [Let $B^n \uparrow \infty$ and apply the theorem to the model economy with $X_i(B_n)$. This gives a sequence of equilibria $(\mathbf{p}_n^*, \mathbf{x}_n^*)$. Show that they have a convergent subsequence. Show that the limit of the subsequence must have $\mathbf{p}^* \gg 0$ and must be an equilibrium.]

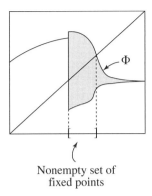

Nonempty set of
fixed points

FIGURE 5.11.8

5.11.b Kakutani's Fixed-Point Theorem

Recall that x^* is a fixed point of a correspondence $\Phi : X \twoheadrightarrow X$ if $x^* \in \Phi(x^*)$. Brouwer's theorem is about continuous functions from a compact convex set to itself. Kakutani's theorem is about upper hemicontinuous correspondences from a compact convex set to itself. Recall in reading the following that we have assumed that correspondences are nonempty valued. If $\Phi(x) \equiv \varnothing$, then it can have no fixed point.

Theorem 5.11.8 (Kakutani) *If $K \subset \mathbb{R}^\ell$ is compact and convex and $\Phi : K \twoheadrightarrow K$ is upper hemicontinuous and convex valued, then Φ has a fixed point.*

There are four assumptions in Kakutani's theorem: the two on the set K are compactness and convexity and the two on the correspondence are upper hemicontinuity convex valuedness. None can be dispensed with. Figure 5.11.8 shows how the convexity comes in at explosions of the correspondence. The following is essentially a repeat of Example 5.11.2 (p. 240). The reason the repeat works is that if the graph if Φ is the graph of a continuous function from a compact set to itself, then Φ is upper hemicontinuous and convex valued.

Example 5.11.9 *Let $K = [0, 1]$ and let Φ be the graph of the discontinuous function $f(x) = 1_{[0,\frac{1}{2})}(x)$; Φ fails to be upper hemicontinuous and has no fixed point. If $\widehat{\Phi}$ has as its graph the closure of Φ, then it is not convex valued at $x = \frac{1}{2}$ and has no fixed point. If Φ has as its graph the continuous function $f(x) = x + 1$, it is not only upper hemicontinuous but continuous, but has no fixed point in the noncompact convex set $\mathbb{R}_+$. If Φ has as its graph the continuous function $f(x) = 3 - x$, then it is continuous but has no fixed point in the compact nonconvex set $[0, 1] \cup [2, 3]$.*

Exercise 5.11.10 Suppose that $K \subset \mathbb{R}^\ell$ is nonempty and compact and that the graph of $\Phi : K \twoheadrightarrow K$ is the graph of a continuous function. Show that Φ is upper hemicontinuous and convex valued. From this, show that Kakutani's theorem implies Brouwer's theorem.

A closed graph is often easier to verify than upper hemicontinuity. Fortunately, in this context they are the same.

Lemma 5.11.11 *A nonempty-valued compact-valued correspondence from a compact set K to itself is upper hemicontinuous iff it has a closed graph.*

Proof. Suppose that Φ is upper hemicontinuous. Let $y_n \in \Phi(x_n)$ and $(x_n, y_n) \to (x, y)$. We must show that $y \in \Phi(x)$. If not, $y \notin (\Phi(x))^\epsilon$ for some $\epsilon > 0$. By upper hemicontinuity, for large n, $\Phi(x_n) \subset (\Phi(x))^\epsilon$, a contradiction. The reverse implication is present in Exercise 4.10.27. ∎

Notation 5.11.12 *For $x \in \times_{i \in I} X_i$ and $y_j \in X_j$, $j \in I$, $x \backslash y_j$ is the vector in $\times_{i \in I} X_i$ with $\mathrm{proj}_k(x \backslash y_j) = x_k$ if $k \neq j$ and $\mathrm{proj}_j(x \backslash y_j) = y_j$.*

Example 5.11.13 *Let $X_1 = \{a, b, c\}$, $X_2 = \{\alpha, \beta, \gamma\}$, and $X_3 = \{1, 2, 3\}$. Let $x = (a, \alpha, 1)$, $y_1 = b$, $y_2 = \gamma$, and $y_3 = 2$; $(x \backslash y_1) = (b, \alpha, 1)$, $(x \backslash y_2) = (a, \gamma, 1)$, and $(x \backslash y_3) = (b, \alpha, 2)$.*

Definition 5.11.14 *A **finite-dimensional quasi-game** is a collection $(C_i, \Gamma_i, v_i)_{i \in I}$, where each C_i is a compact subset of $\mathbb{R}^\ell$, each Γ_i is a continuous correspondence from $C = \times_{j \in I} C_j$ to C_i, and each $v_i : C \to \mathbb{R}$ is a continuous utility function. A **quasi-equilibrium** is a vector c^* such that for each $i \in I$, c_i^* solves the problem $\max\{v_i(c^* \backslash b_i) : b_i \in \Gamma_i(c^*)\}$.*

Theorem 5.11.15 *If the C_i are convex, the Γ_i convex valued, and for all $i \in I$ and all c, the mappings $b_i \mapsto v_i(c \backslash b_i)$ are quasi-concave, then the quasi-game $(C_i, \Gamma_i, v_i)_{i \in I}$ has a quasi-equilibrium.*

Proof. Let $Br_i(c) = \arg\max\{v_i(c \backslash b_i) : b_i \in \Gamma_i(c)\}$. (Mnemonically, "$Br_i$" is "$i$'s best responses.") Since the conditions of the theorem of the maximum are satisfied, $c \mapsto Br_i(c)$ is an upper-hemicontinuous correspondence. Since v_i is quasi-concave, for all c, $Br_i(c)$ is a convex set. Define $Br(c) = \times_{i \in I} Br_i(c)$. By Kakutani's theorem, there exists an $c^* \in Br(c^*)$. ∎

The following drops the monotonicity and strict quasi-concavity from Theorem 5.11.6.

Theorem 5.11.16 *Let $\mathcal{E} = \{X_i, \succsim_i, \mathbf{y}_i\}_{i \in I}$ be an exchange economy model. Assume that for every $i \in I$,*

1. *X_i is a compact convex subset of $\mathbb{R}_+^\ell$ with $\mathbf{y}_i \in \mathrm{int}(X_i)$, and*
2. *$\succsim_i$ represented by a $u_i : X_i \to \mathbb{R}$ that is quasi-concave, $[u(\mathbf{x}) = u(\mathbf{y})] \Rightarrow (\forall \alpha \in (0, 1))[u(\mathbf{x} \alpha \mathbf{y}) > u(\mathbf{x})]$.*

Then there exists an equilibrium $(\mathbf{p}, \mathbf{x})$ for $\mathcal{E}$.

Proof. We construct a quasi-game from $\mathcal{E}$ by adding an extra person, $i = 0$, called the "auctioneer." For $i \in I = \{1, \ldots, I\}$, let $C_i = X_i$ and set $C_0 = \Delta^\ell$.

For $i \in I$, define $\Gamma_i(\mathbf{p}) = \{\mathbf{x} \in X_i : \mathbf{px} \leq \mathbf{py}_i\}$. Since $\mathbf{y}_i \in \mathrm{int}(X_i)$, Γ_i is a continuous correspondence from $C := \times_{j=1}^{I+1} C_j$ to C_i. Define $\Gamma_{+1}(c) \equiv \Delta^\ell$ for each $c \in C$. As it is constant, the auctioneer's correspondence is continuous.

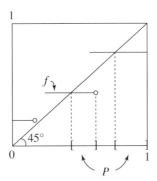

FIGURE 5.11.17

For $i \in I$, define $v_i(c) = u_i(\mathbf{x}_i)$ and define $v_{I+1}(c) = \mathbf{p} \cdot \sum_{i \in I} (\mathbf{x}_i - \mathbf{y}_i)$. Thus, the auctioneer tries to choose prices to maximize the value of excess demand.

Since the v_i are quasi-concave in b_i and v_0 is linear in c_0, hence quasi-concave, Theorem 5.11.15 implies that this quasi-game has a quasi-equilibrium, $(\mathbf{p}^*, \mathbf{x}^*)$. We claim that $(\mathbf{p}^*, \mathbf{x}^*)$ is a Walrasian equilibrium.

It is clear that each $\mathbf{x}_i^*$ is optimal in $\{\mathbf{x} \in X_i : \mathbf{px} \le \mathbf{py}_i\}$. All that is left to show is that $\mathbf{x}^*$ is feasible. Since $\mathbf{px}_i^* \le \mathbf{py}_i$ for each i, we know that $v_0(\mathbf{p}^*, \mathbf{x}^*) = \mathbf{p}^* \sum_{i \in I}(\mathbf{x}_i^* - \mathbf{y}_i) \le 0$. Since $\mathbf{p}^*$ maximizes the auctioneers utility, for all $\mathbf{p} \in \Delta$, $\mathbf{p} \sum_{i \in I}(\mathbf{x}_i^* - \mathbf{y}_i) \le 0$. Taking $\mathbf{p} = \mathbf{e}_k$ for $k \in \{1, \ldots, \ell\}$, $\sum_{i \in I}(\mathbf{x}_i^* - \mathbf{y}_i) \le 0$, so that $\mathbf{x}^*$ is feasible. ∎

A comparison of the proofs of Theorem 5.11.6 and Theorem 5.11.16 shows a typical pattern—using Kakutani's theorem allows an easier proof of a more general result.

Sometimes we need neither continuity nor convex valuedness for a fixed-point result.

Exercise 5.11.17 Tarski's fixed-point theorem implies that any nondecreasing function from [0, 1] to itself has a fixed point. Using the special properties of [0, 1], prove this directly. [Let $P = \{x \in [0, 1] : x \le f(x)\}$. Show that $0 \in P$, so that $P \ne \emptyset$. Let $\bar{x} = \sup P$. Show that $f(\bar{x}) = \bar{x}$ by showing that $f(\bar{x}) \le \bar{x}$ and $f(\bar{x}) \ge \bar{x}$. See Figure 5.11.17.]

5.12 ◆ Fixed-Point Theorems for Nash Equilibria and Perfect Equilibria

A "game" is a model of strategic interactions. One specifies a "game" by specifying the players, their possible actions, and their payoffs. An equilibrium is a vector of actions, one for each person, that are mutual best responses.

5.12.a Some Informational Considerations

A subtlety in the concept of "possible actions" is that it includes informational considerations.

Example 5.12.1 (Agatha Christie) *One possible action for a murderer is to leave a false piece of evidence that implicates him. This may have the effect of leading the detective, who does not know the murderer's identity, to first conclude that the murderer is guilty, then to become convinced of his innocence after the evidence turns out to be false.*

Example 5.12.2 (Spence's Labor Market) *If it is easier for high-ability people to get diplomas and difficult for potential employers to distinguish among peoples' abilities. Then those with more ability may choose to get a diploma (even if it adds nothing to their eventual productivity).*

Example 5.12.3 *One of the possible actions for a chief executive officer (CEO) of a firm is to engage in creative bookkeeping that results in understatements of the size of losses that a company has had or is facing. This affects the information that is available to other actors, say an investor, a regulator, or a corporate buyout specialist.*

These kinds of examples become even more interesting when we start asking about mutual best responses. What is the detective's best response to the possibility that a murderer may leave false clues? How much education should one obtain to give a better signal depends on what people with higher and with lower ability are doing, which in turn depends on how employers interpret educational attainments, which in turn depends on how much education people of different abilities choose to get. What are employee and potential investor best responses to a CEO who may be choosing to disguise the state of the firm? And what is a CEO's best response to those choices?

5.12.b Examples of Pure Strategy Equilibria for Finite Games

Here is the basic formalism of game theory.

Definition 5.12.4 *A **game** Γ is a collection $(A_i, u_i)_{i \in I}$, where*

1. *I is a nonempty set (of people, sometimes called players or agents),*
2. *for each $i \in I$, A_i is i's set of possible actions, and*
3. *for each $i \in I$, $u_i : A \to \mathbb{R}$ is i's utility function, where $A := \times_{i \in I} A_i$ is the product of the actions.*

*A point in A is called an **action profile**. If A, the set of action profiles, is finite, that is, if I is finite and each A_i is finite, then Γ is a **finite game**.*

Note that the utility functions of the agents can be represented as points in $(\mathbb{R}^A)^I$ because $u_i \in \mathbb{R}^A$. Thus, the set of games having a given set of agents and action sets can be represented as a finite-dimensional vector. The following is repetitive of previous notation.

Notation 5.12.5   *For* $a \in \times_{i \in I} A_i$ *and* $b_j \in A_j$, $j \in I$, $a \backslash b_j$ *is the vector in* $\times_{i \in I} A_i$ *with* $\mathrm{proj}_k(a \backslash b_j) = a_k$ *if* $k \neq j$ *and* $\mathrm{proj}_j(a \backslash b_j) = b_j$.

Here is the "mutual best response" condition.

Definition 5.12.6   *A **pure strategy Nash equilibrium for** Γ is a vector of actions, $a^* \in A$, such that for all $i \in I$, a_i^* solves* $\max_{b_i \in A_i} u_i(a^* \backslash b_i)$.

5.12.b.1   The Advantage of Being Small

Here is a story known as **Rational Pigs**. There are two pigs, one Little and one Big. They are in a long room. There is a lever at one end of the room, which, when pushed, gives an unpleasant electric shock, $-c$ in utility terms, and gives food at the trough across the room, worth utility b. The Big pig can shove the Little pig out of the way and take all the food if they are at the food output together, and the two pigs are equally fast getting across the room. However, during the time that it takes the Big pig to cross the room, the Little pig can eat α of the food. With $0 < \alpha < 1$ and $(1 - \alpha)b > c > 0$, Little soon figures out that, no matter what Big is doing, pushing gets nothing but a shock on the (sensitive) snout, and so will wait by the trough. In other words, the strategy Push is **dominated** by the strategy Wait.

There is a standard representation of this kind of game. Here $I = \{\text{Little,Big}\}$, $A_1 = A_2 = \{\text{Push, Wait}\}$. This is called a 2×2 game because there are two players with two strategies apiece. The utilities are given in the following table.

Rational Pigs

	Push	Wait
Push	$(-c, b - c)$	$(-c, b)$
Wait	$(\alpha b, (1 - \alpha)b - c)$	$(0, 0)$

where $0 < \alpha < 1$ and $(1 - \alpha)b > c > 0$.

Some conventions: The representation of the choices has player 1, listed first, in this case Little, choosing which row occurs and player 2, in this case Big, choosing which column; each entry in the matrix is uniquely identified by the actions a_1 and a_2 of the two players; each entry has two numbers, (x, y)—$(u_1(a_1, a_2), u_2(a_1, a_2))$—so that x is the utility of player 1 and y the utility of player 2 when the vector $a = (a_1, a_2)$ is chosen.

If the utility cost of the shock to the snout is $c = 2$, the utility benefit of the food is $b = 10$, and Litte can eat $\alpha = \frac{1}{2}$ of the food in the time it takes Big to cross the room, then the payoffs are

Rational Pigs

	Push	Wait
Push	$(-2, 8)$	$(-2, 8)$
Wait	$(5, 3)$	$(0, 0)$

We claim that the unique equilibrium for this game is (Wait, Push), Big does all the work, and Little gets food for free. This is the advantage of being small. One can see this by inspection: if Big is pushing, Little is better off waiting, and if Little is waiting, then Big is better off pushing; for each of the other three action profiles, one or the other would be better off by changing his/her action. Thus, (Wait, Push) is the unique action profile that involves mutual best responses.

5.12.b.2 Dominated Actions
The Rational Pigs game is one having what are called dominated actions.

Definition 5.12.7 *In a game* $\Gamma = (A_i, u_i)_{i \in I}$, *an action* b_i *is **dominated for** i **by a strategy** c_i if for all $a \in A$, $u_i(a \backslash c_i) > u_i(a \backslash b_i)$. b_i is **dominated** if it is dominated by some c_i.*

If we look at the payoff matrix, since $5 > -2$ and $0 > -2$, the action Wait dominates the action Push for Little. No one in their right mind should play a dominated strategy. This has a more measured tone in the following.

Lemma 5.12.8 *If b_i is dominated and a^* is an equilibrium, then $a_i^* \neq b_i$.*

Proof. Suppose that $a_i^* = b_i$. Since $u_i(a^* \backslash c_i) > u_i(a^*)$ for some $C_i \in A_i$, a^* is not an equilibrium. ∎

The classic game with dominant strategies is called the Prisoners' Dilemma. The story is as follows: two criminals have been caught, but it is after they have destroyed the evidence of serious wrongdoing. Without further evidence, the prosecuting attorney can charge them both for an offense carrying a term of $b > 0$ years. However, if the prosecuting attorney gets either prisoner to give evidence on the other (Squeal), the latter will get a term of $B > b$ years. The prosecuting attorney makes a deal with the judge to reduce any term given to a prisoner who squeals by an amount r, $b \geq r > 0$, $B - b > r$ (equivalent to $-b > -B + r$). With the utilities of the years in prison being given by $B = 15$, $b = r = 1$, this gives

	Squeal	Silent
Squeal	$(-14, -14)$	$(0, -15)$
Silent	$(-15, 0)$	$(-1, -1)$

In the Prisoners' Dilemma, Squeal dominates Silent for both players. Another way to put this is that the only possible rational action for either player is Squeal. This does not depend on i knowing what $-i$ is doing—whatever $-i$ does, Squeal is best. We might as well solve the optimization problems independently of each other.

What makes it interesting is that when you put the two individual's solutions together, you have a common disaster. They are both spending 14 years in prison, and by cooperating with each other and being Silent, they could both spend only 1 year in prison. Their individualistic choice of an action imposes costs on someone else. The individual calculation and the joint calculation are very very different.

One useful way to view many economists is as apologists for the inequities of a moderately classist version of the political system called laissez-faire capitalism. Perhaps this is the driving force behind the large body of literature that tries to explain why we should expect cooperation in this situation. After all, if economists' models come to the conclusion that equilibria without outside intervention can be quite bad for all involved, they become an attack on the justifications for laissez-faire capitalism. Another way to understand this literature is that we are, in many ways, a cooperative species, so a model predicting extremely harmful noncooperation is very counterintuitive.

5.12.b.3 Multiple Equilibria and the Need for Coordination

Here is another story: Two hunters who live in villages at some distance from each other in the era before telephones have to decide whether to hunt for Stag or for Rabbit; hunting a stag requires that both hunters have their stag equipment with them, as one hunter with stag equipment will not catch anything; hunting for rabbits requires only one hunter with rabbit-hunting equipment. The payoffs have $S > R > 0$, for example, $S = 20$, $R = 8$, which gives

	Stag	Rabbit
Stag	(20, 20)	(0, 8)
Rabbit	(8, 0)	(8, 8)

This is a **coordination game**. If the players' coordinate their actions they can both achieve higher payoffs. There are two obvious Nash equilibria for this game. (Stag, Stag) and (Rabbit, Rabbit). There is a role, then, for some agent to act as a coordinator.

It is tempting to look for social roles and institutions that coordinate actions, such as matchmakers, advertisers, and publishers of schedules, for example, for trains and planes. Sometimes we might imagine a tradition that serves as a coordinator—something like we hunt stags on days following full moons except during the springtime.

Many macroeconomists tell stories like this but use the word "sunspots" to talk about coordination. This may be because overt reference to our intellectual debt to Keynes is out of fashion. In any case, any signals that are correlated and observed by the agents can serve to coordinate the peoples' actions. We will not go down this road, which leads us to what are called **correlated equilibria**.

5.12.c Mixed or Random Strategies

If you have played hide and seek with very young children, you may have noticed that they always hide in the same place, and that you have to search, while loudly explaining your actions, "Is he hiding in the closet? No. What about behind the couch," while they giggle helplessly. Once children actually understand hiding, they begin to *vary* where they hide; they *mix* it up; they *randomize*. Randomizing where one hides is the only sensible strategy in games of hide and seek.

If you have played or watched a game such as tennis, squash, ping pong, or volleyball, you will have noticed that the servers, if they are any good at all, purposefully randomize where they serve. If one, for example, always serves the tennis ball to the same spot, the opponent will move so as to be able to hit that ball in the strongest possible fashion.

5.12.c.1 Notation and Definitions

Notation 5.12.9 *For a finite set, S, $\Delta(S) := \{p \in \mathbb{R}^S : p \geq 0,\ \sum_s p_s = 1\}$ denotes the set of probability distributions over a set S.*

Definition 5.12.10 *A **mixed strategy for** i is an element $\sigma_i \in \Delta_i := \Delta(A_i)$. A **mixed strategy for a game** Γ is an element $\sigma \in \times_{i \in I} \Delta_i$.*

We assume that when people pick their strategies at random, they do so *independently*.

Inasmuch as independent probabilities are multiplied, the expected utility of j to a strategy $\sigma = (\sigma_i)_{i \in I}$ is

$$U_j(\sigma) = \sum_a u_j(a) \Pi_{i \in I} \sigma_i(a_i). \tag{5.28}$$

Example 5.12.11 *Let $I = \{1, 2\}$, $A_1 = \{Up, Down\}$, $A_2 = \{Left, Right\}$, $\sigma_1 = (\frac{1}{3}, \frac{2}{3})$ and $\sigma_2 = (\frac{1}{3}, \frac{2}{3})$. The following three distributions over A all have (σ_1, σ_2) as marginal distributions, but only the first one has the choices of the two players being independent:*

	Left	Right		Left	Right		Left	Right
Up	$\frac{1}{9}$	$\frac{2}{9}$	Up	$\frac{1}{3}$	0	Up	$\frac{1}{6}$	$\frac{1}{6}$
Down	$\frac{2}{9}$	$\frac{4}{9}$	Down	0	$\frac{2}{3}$	Down	$\frac{1}{6}$	$\frac{1}{2}$

If 1's payoffs are given by

	Left	Right
Up	9	4
Down	17	-3

then 1's payoffs to independent randomization with the marginals $\sigma_1 = (\frac{1}{3}, \frac{2}{3})$ and $\sigma_2 = (\frac{1}{4}, \frac{3}{4})$ are $U_1(\sigma_1, \sigma_2) = 9 \cdot (\frac{1}{3} \cdot \frac{1}{4}) + 4 \cdot (\frac{1}{3} \cdot \frac{3}{4}) + 17 \cdot (\frac{2}{3} \cdot \frac{1}{4}) - 3 \cdot (\frac{2}{3} \cdot \frac{3}{4})$.

If player 2 is playing $Left$ with probability β and $Right$ with probability $(1 - \beta)$, then 1's payoff to Up is $U_1((1, 0), (\beta, (1 - \beta))) = 9\beta + 4(1 - \beta)$, to $Down$ is $U_1((0, 1), (\beta, (1 - \beta))) = 17\beta - 3(1 - \beta)$, and to playing Up with probability α and $Down$ with probability $(1 - \alpha)$ is $U_1((\alpha, (1 - \alpha)), (\beta, (1 - \beta))) = \alpha U_1((1, 0), (\beta, (1 - \beta))) + (1 - \alpha) U_1((0, 1), (\beta, (1 - \beta)))$.

For any $\sigma_{-i} \in \times_{j \neq i} \Delta_j$, a player's payoffs are linear in their own probabilities.

Lemma 5.12.12 (Own Probability Linearity) *For all $\sigma \in \Delta$ and all $i \in I$, the mapping $\mu_i \mapsto U_i(\sigma \backslash \mu_i)$ is linear, that is, $U_i(\sigma \backslash \gamma \mu_i + (1 - \gamma) v_i) = \gamma U_i(\sigma \backslash \mu_i) v_i) + (1 - \gamma) U_i(\sigma \backslash v_i)$.*

Proof. For i's action $b \in A_i$, let $v_i(b) = \sum_{a_{-i}} u_i(b, a_{-i}) \cdot \Pi_{j \neq i} \sigma_j(a_j)$. Fon any μ_i, $U_i(\sigma \backslash \mu_i) = \sum_b v_i(b) \mu_i(b)$. ∎

Here is the "mutual best response" condition when we consider mixed strategies.

Definition 5.12.13 *$\sigma^* \in \Delta$ is a **Nash equilibrium for** Γ if for all $i \in I$ and all $\mu_i \in \Delta_i$, $U_i(\sigma^*) \geq U_i(\sigma^* \backslash \mu_i)$. The set of Nash equilibria for Γ is denoted $Eq(\Gamma)$.*

For $b \in A_i$, we use $b_i \in \Delta_i$ to denote the probability distribution on A_i that puts mass 1 on b_i. If we wanted to be more careful and make this harder to read, we would use δ_{b_i}, the probability defined by $\delta_{b_i}(E) = 1_E(b_i)$.

Exercise 5.12.14   Using the linearity lemma, show that σ^* is a Nash equilibrium iff for all $i \in I$ and all $b_i \in A_i$, $U_i(\sigma^*) \geq U_i(\sigma^* \backslash b_i)$.

Definition 5.12.15   *A mixed strategy $\mu_i \in \Delta_i$ **dominates** a strategy $a_i \in A_i$ and a_i is a **dominated strategy** for agent i if for all $\sigma \in \Delta$, $U_i(\sigma \backslash \mu_i) > U_i(\sigma \backslash a_i)$.*

Example 5.12.16   *It is perhaps surprising that a strategy a_i can be dominated by a mixed strategy even though it is not dominated by a pure strategy $b_i \in A_i$. With the following payoffs, the mixed strategy $(\frac{1}{2}, 0, \frac{1}{2})$ on (T, M, B) dominates the pure strategy M, also known as $(0, 1, 0)$, for player 1, but no pure strategy dominates it (only player 1's utilities are given):*

	L	R
T	9	0
M	2	2
B	0	9

5.12.c.2 Inspection Games

In the following game, there is a worker who can either Shirk or put in an Effort. The boss can either Inspect or Not. Inspecting someone who is working has an opportunity cost, $c > 0$, finding a shirker has a benefit $b > c$. The worker receives w if he shirks and is not found out, 0 if she shirks and is inspected, and $w - e > 0$ if he puts in the effort, whether or not he is inspected. In matrix form, the game is

	Inspect	Don't inspect
Shirk	$(0, b - c)$	$(w, 0)$
Effort	$(w - e, -c)$	$(w - e, 0)$

Just as in the childrens' game of hide and seek, there cannot be an equilibrium in which the two players always choose one strategy. For there to be an equilibrium, there must be randomization. What is interesting is that one player's randomization equilibrium probability depends on the other player's payoffs. This is a result of the own linearity lemma, 5.12.12, and the following result.

Theorem 5.12.17   *If σ^* is an equilibrium with $\sigma_i^*(a_i) > 0$ and $\sigma_i^*(b_i) > 0$ for $a_i, b_i \in A_i$, then $U_i(\sigma^* \backslash a_i) = U_i(\sigma^* \backslash b_i)$.*

In words, if someone is randomizing his actions in equilibrium, he must be indifferent between the actions he is randomizing over.

Proof.   Suppose that $U_i(\sigma^* \backslash a_i) > U_i(\sigma^* \backslash b_i)$. Consider the mixed strategy μ_i with $\mu_i'(a_i) = \sigma_i^*(a_i) + \sigma_i^*(b_i)$ and $\mu'(c_i) = \sigma_i^*(c_i)$ for $c_i \notin \{a_i, b_i\}$. $U_i(\sigma^* \backslash \mu_i') > U_i(\sigma^*)$ because it has shifted probability toward something with higher payoffs. ∎

With $w = 10$, $e = 2$, $c = 1$, and $b = 2$, the inspection game above is represented by

	Inspect	Don't inspect
Shirk	(0, 1)	(10, 0)
Effort	(8, −1)	(8, 0)

Exercise 5.12.18 Referring to the inspection game just given, let α be the probability that 1 shirks and β the probability that 2 inspects.

1. As a function of β, 2's probability of inspecting, find 1's best response. In particular, find the critical value β^* at which 1 is indifferent between shirking and putting in an effort.

2. As a function of α, 1's probability of shirking, find 2's best response. In particular, find the critical value α^* at which 2 is indifferent between inspecting and not inspecting.

3. Show that $((\alpha^*, (1 - \alpha^*)), (\beta^*, (1 - \beta^*)))$ is the unique Nash equilibrium for this game. [The point is that you find the equilibrium by finding a mixed strategy for i that makes j indifferent and vice versa. This means that i's equilibrium-mixed strategy is determined by j's payoffs and vice versa.]

4. In the general version of this inspection game, show that the equilibrium probability of shirking goes down as the cost of monitoring, c, goes down, and that the probability of monitoring is independent of the monitor's costs and benefits.

5.12.c.3 Trust and Electronic Commerce

The eBay auction for a doggie-shaped vase of a particularly vile shade of green has just ended. Now the winner should send the seller the money and the seller should send the winner the vase. If both act honorably, the utilities are $(u_b, u_s) = (1, 1)$; if the buyer acts honorably and the seller dishonorably, the utilities are $(u_b, u_s) = (-2, 2)$; if the reverse, the utilities are $(u_b, u_s) = (2, -2)$; and if both act dishonorably, the utilities are $(u_b, u_s) = (-1, -1)$. In matrix form the game is:

		Seller	
		Honorable	Dishonorable
Buyer	Honorable	(1, 1)	(−2, 2)
	Dishonorable	(2, −2)	(−1, −1)

In many ways, this is a depressing game to think about—no matter what the other player is doing, acting dishonorably is a best response for both players. This is, again, a case of dominance. Returning to the eBay example, let us suppose that for a (utility) cost s, $0 < s < 1$, the buyer and the seller can mail their obligations to a third party intermediary who will hold the payment until the vase arrives or hold the vase until the payment arrives, forward each on to the correct party if both arrive, and return the vase or the money to the correct party if one side acts dishonorably. Thus, each person has three choices, send to the intermediary, honorable, dishonorable. The payoff matrix for the symmetric 3×3 game just described is:

		Seller		
		Intermediary	Honorable	Dishonorable
	Intermediary	1−s,1−s	1−s,1	−s,0
Buyer	Honorable	1,1−s	1,1	−2,2
	Dishonorable	0,−s	2,−2	−1,−1

Exercise 5.12.19   The first three questions are about finding the unique equilibrium and the last two lead to interpretations.

1. Show that there is no pure-strategy equilibrium for this game.

2. Show that there is no mixed-strategy equilibrium involving the seller playing exactly two strategies with strictly positive probability.

3. Find the unique mixed-strategy equilibrium and its expected utility as a function of s.

4. For what values of s is the probability of dishonorable behavior lowest? Highest?

5. If the intermediary is a monopolist, what will she charge for her services?

One of the really perverse aspects of this situation is that the availability of an intermediary is what makes trade possible, but people are willing to incur the cost of the intermediary only because there continues to be cheating. This is much like a monitoring game.

It is the SEC (Securities and Exchange Commission) and honest accounting that has, historically, made the vigor of the U.S. stock market possible. We expect to have a positive, though hopefully low, frequency of cheating: Ponzi schemes; insider trading; backdating of options; accounting fraud or other breaches of fiduciary duty; or ratings agencies giving specially favorable treatment to their larger clients. The complicated question is how much cheating is too much?

5.12.d   Nash Equilibrium Existence

In the 3×3 game of Exercise 5.12.19, it was a bit of work to find the equilibrium. It could have been worse: there could have been five players, each with more than twenty strategies, for a matrix having 20^5 utility entries and no pure-strategy equilibrium. It is useful, or at least comforting, to know that equilibria always exist, even if they might be difficult to find.

As before, $Br_i(\sigma) := \{\mu_i \in \Delta_i : \forall b \in A_i, \ U_i(\sigma \backslash \mu_i) \geq U_i(\sigma \backslash b)\}$ and $Br(\sigma) := \times_{i \in I} Br_i(\sigma)$; $\sigma^* \in Eq(\Gamma)$ iff $\sigma^* \in Br(\sigma^*)$, that is, iff σ^* is a fixed point of the best response correspondence.

Definition 5.12.20   *A correspondence Ψ from a nonempty compact convex subset, K, of $\mathbb{R}^n$ to $\mathbb{R}^n$ is a **game theoretic correspondence (gtc)** if*

1. for all $x \in K$, $\Psi(x) \neq \emptyset$,

2. for all $x \in K$, $\Psi(x)$ is convex,

3. Ψ is upper hemicontinuous.

Since the theorem of the maximum and the own probability linearity lemma, 5.12.12 (p. 250), tell us that $Br(\cdot)$ is a gtc, the following is an immediate implication of Kakutani's theorem.

Theorem 5.12.21 (Nash) *Every finite game has an equilibrium.*

Proof. For each $i \in I$, the (graph of the) correspondence $\sigma \mapsto Br_i(\sigma)$ is defined by finitely many weak inequalities, $\{(\sigma, \mu_i) : \forall b \in A_i, U_i(\sigma \backslash \mu_i) \geq U_i(\sigma \backslash b)\}$. Since U_i is continuous (even polynomial), the graph of the best response correspondence is closed. Since at least one of the pure strategies must be a best response, it is nonempty valued. By the own linearity lemma, it is convex valued. Hence $\sigma \mapsto \times_{i \in I} Br_i(\sigma)$ is a gtc and so has a fixed point, σ^*. ∎

Corollary 5.12.22 *For any finite game Γ, the set of equilibria, $Eq(\Gamma)$, is a nonempty compact set.*

Proof. We just showed that the set of equilibria is nonempty. It is closed because it is defined as the set of mixed strategies satisfying polynomial inequalities, and polynomials are continuous. ∎

5.12.e Idle Threats and Perfection

In Puccini's *Gianni Schicchi*, the wealthy Buoso Donati has died and left his large estate to a monastery. Before the will is read by anyone else, the relatives call in a noted mimic, Gianni Schicchi, to play Buoso on his deathbed, rewrite the will, and then convincingly die. The relatives explain, very carefully, to Gianni Schicchi just how severe are the penalties for anyone caught tampering with a will (at the time, the penalities included having one's hand cut off). The plan is put into effect, but on the deathbed, Gianni Schicchi, as Buoso Donati, rewrites the will leaving the entire estate to the noted mimic and great artist, Gianni Schicchi. The relatives can expose him *and thereby expose themselves as well*, or they can remain silent.

Let player 1 be Gianni Schicchi, who has two actions while playing Donati on his deathbed, L being leave the money to the relatives and R being will it to the noted artist and mimic Gianni Schicchi. Let player 2 represent the relatives, who have two actions, l being reporting Gianni to the authorities and r letting him get away with the counterswindle. Inventing some utility numbers with $y \gg 200$ to make the point, we have

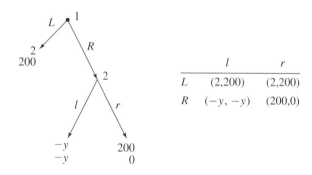

	l	r
L	$(2,200)$	$(2,200)$
R	$(-y,-y)$	$(200,0)$

The box gives the 2×2 representation of the extensive form game. There are two kinds of equilibria for this game.

Exercise 5.12.23 Show the following:

1. For all $\beta \geq 198/(200 + y)$, the strategy $(L, (\beta, (1 - \beta)))$ is an equilibrium giving payoffs of $(2, 200)$.

2. (R, r) is an equilibrium giving payoffs $(200, 0)$.

3. There are no other equilibria.

The equilibria with $\beta > 0$ can be interpreted as the relatives making the threat, "If you double-cross us, we go to the authorities." Your analysis just showed Gianni believing that threat is an equilibrium. However, since following through on the threat means that the relatives also suffer the loss of their right hands, you might believe that the threat is an "idle" one; that is, it is a threat that the relatives would not follow through on if Gianni "called their bluff."

One way to get at this logic is to note that $\beta > 0$ is only a best response because Gianni's choice of L means that there is no possibility that the relatives will be called on to follow through on their threat. Reinhart Selten's notion of a perfect equilibrium forces people to pay attention to all possibilities.

5.12.f Perfect Equilibrium Existence

For each $i \in I$, let $\eta^i \gg 0$ be a vector in $\mathbb{R}^{A_i}$ with $\sum_{b \in A_i} \eta_b^i \ll 1$. Define $\Delta_i(\eta^i) = \{\sigma_i \in \Delta_i : \sigma_i \geq \eta^i\}$. This is a compact convex subset of Δ_i, and as $\eta^i \gg 0$, every element of $\Delta_i(\eta^i)$ is strictly positive. The small number η_b^i represents the minimum possible probability that i plays the action b.

Example 5.12.24 In the 2×2 version of Gianni Schicchi, let $\eta^1 = (\frac{1}{205}, \frac{3}{791})$ and $\eta^2 = (\frac{2}{519}, \frac{1}{836})$. $\Delta_1(\eta^1)$ is the set of $(\alpha, (1 - \alpha))$ with $\alpha \in [\frac{1}{205}, \frac{788}{791}]$ and $\Delta_2(\eta^2)$ is the set of $(\beta, (1 - \beta))$ with $\beta \in [\frac{2}{519}, \frac{835}{836}]$.

Definition 5.12.25 The sets $\Delta_i(\eta^i)$ are called η^i-**perturbations** of Δ_i and the game $\Gamma(\eta) = (\Delta_i(\eta^i), U_i)_{i \in I}$ is an η-**perturbation** of Γ.

Given $\eta^i \gg 0$ and $\sigma \in \Delta$, define $Br_i(\sigma : \eta^i) = \{\mu_i \in \Delta_i(\eta^i) : U_i(\sigma \backslash \mu_i) \geq U_i(\sigma \backslash \Delta_i(\eta^i))\}$. This gives i's best response when his minimum probability of playing strategy b is η_b^i. For $\eta = (\eta^i)_{i \in I} \in \mathbb{R}_{++}^{\times_{i \in I} A_i}$, define $Br(\sigma : \eta) = \times_{i \in I} Br_i(\sigma : \eta^i)$. The basic idea is that we find fixed points of $\sigma \mapsto Br(\sigma : \eta)$ and then take limits as $\eta \to 0$.

Definition 5.12.26 For $\epsilon \in \mathbb{R}_{++}$, $\sigma^\circ = \sigma(\epsilon)$ is an ϵ-**perfect equilibrium** if for all $i \in I$,

$$[u_i(\sigma^\circ \backslash a_i) < u_i(\sigma^\circ \backslash b_i)] \Rightarrow [\sigma_i^\circ(a_i) \leq \epsilon].$$

A fixed point of $\sigma \mapsto Br(\sigma : \eta)$ is an ϵ-perfect equilibrium if ϵ is greater than every component of the vector η. Note that any fixed point of $\sigma \mapsto Br(\sigma : \eta)$ is strictly positive. At work here is the device of having the correspondence take on only a subset of the values of Δ.

Exercise 5.12.27 Show that for small ϵ, the only ϵ-perfect equilibria for the Gianni Schicchi game are $((\epsilon_L^1, (1 - \epsilon_L^1)), ((1 - \epsilon_r^2), \epsilon_r^2)$.

Lemma 5.12.28 $\sigma \mapsto Br(\sigma : \epsilon)$ is a gtc.

Exercise 5.12.29 Prove Lemma 5.12.28.

Definition 5.12.30 σ^* is a **perfect equilibrium** if there is a sequence $\epsilon_n \to 0$ and a sequence σ_n of ϵ_n-perfect equilibria such that $\sigma_n \to \sigma^*$. The set of perfect equilibria for Γ is denoted $Per(\Gamma)$.

Intuitively, this requires that for some tiny perturbation of the game Γ, there must exist an equilibrium of the perturbed game that is arbitrarily close to σ^*. Since the strategy sets are perturbed, an ϵ_n-perfect equilibrium is typically not an equilibrium. However, it is very nearly an equilibrium for large n.

Lemma 5.12.31 If $\sigma_n \to \sigma^*$ is a sequence of ϵ_n-perfect equilibria, $\epsilon_n \to 0$, then for all $r > 0$, all $i \in I$, and all $b \in A_i$, there exists N such that for all $n \geq N$, $U_i(\sigma_n \backslash b) \leq U_i(\sigma_n) + r$.

Proof. Every element of $Br_i(\sigma : \epsilon^i)$ puts as much mass as possible on the set $B_n := \{b \in A_i : U_i(\sigma \backslash b) \geq U_i(\sigma \backslash A_i)\}$. Therefore $|U_i(\sigma_n) - U_i(\sigma \backslash B_n)| \to 0$. For all n such that $|U_i(\sigma_n) - U_i(\sigma \backslash B_n)| < r$, for all $b \in A_i$, $U_i(\sigma_n \backslash b) \leq U_i(\sigma_n) + r$. ∎

Using this result, we have the following.

Theorem 5.12.32 (Selten) For finite Γ, $Per(\Gamma)$ is a closed nonempty subset of $Eq(\Gamma)$.

Proof. Since $\sigma \mapsto Br(\sigma : \epsilon)$ is a gtc, ϵ-perfect equilibria exist. For $\delta > 0$, let

$$E(\delta) = cl(\{\sigma : \sigma \text{ is an } \epsilon\text{-perfect equilibrium for some } \epsilon \text{ with } \epsilon_b^i < \delta\}). \quad (5.29)$$

The collection of closed sets $\{E(\delta) : \delta > 0\}$ has the finite intersection property in the compact set Δ, hence has nonempty intersection, and $Per(\Gamma) = \cap_{\delta > 0} E(\delta)$.

All that is left to show is that any perfect equilibrium, σ^*, is a Nash equilibrium. If it is not, then for some $i \in I$ and some $b \in A_i$, $U_i(\sigma^* \backslash b) > U_i(\sigma^*) + 2r$ for some $r > 0$. Let $\sigma_n \to \sigma^*$ be a sequence of ϵ_n-perfect equilibria, $\epsilon_n \to 0$. By continuity, for all large n, $U_i(\sigma_n \backslash b) > U_i(\sigma_n) + r$, contradicting the previous lemma. ∎

Exercise 5.12.33 Carefully verify that $Per(\Gamma) = \cap_{\delta > 0} E(\delta)$.

Exercise 5.12.34 Show that in Gianni Schicchi, there is only one perfect equilibrium.

5.12.g A Short Overview of Equilibrium Refinements

In the following game, there is only one perfect equilibrium.

	Left	Right
Up	(1, 1)	(0, 0)
Down	(0, 0)	(0, 0)

To see why, fix a vector of perturbations (ϵ^1, ϵ^2). If 1 is playing an action σ_1 with $\sigma_1(\text{Up}) \geq \epsilon_1(\text{Up}) > 0$ and $\sigma_1(\text{Down}) \geq \epsilon^1(\text{Down}) > 0$, then two payoffs satisfy

$$U_2(\sigma_1, \text{Left}) \geq \epsilon^1(\text{Up}) > 0, \quad \text{and} \quad U_2(\sigma_1, \text{Right}) = 0.$$

This means that in any perturbed game, Left strictly dominates Right for 2. Therefore, in any equilibrium of the perturbed game, 2 puts as much mass as possible on Left; that is, 2 plays the action $\sigma_2(\text{Left}) = 1 - \epsilon^2(\text{Right})$, $\sigma_2(\text{Right}) = \epsilon^2(\text{Right})$. (This is Lemma 5.12.31 at work.) By symmetry, 1 puts as much mass as possible on Up. If we take the limits as $(\epsilon^1, \epsilon^2) \to 0$, the unique perfect equilibrium is (Up,Left).

Perfect equilibria do not survive deletion of strictly dominated actions, an observation due to Roger Myerson:

	L	R	A_2
T	$(1, 1)$	$(0, 0)$	$(-1, -2)$
B	$(0, 0)$	$(0, 0)$	$(0, -2)$
A_2	$(-2, -1)$	$(-2, 0)$	$(-2, -2)$

Delete the two strictly dominated actions, A_1 and A_2, and you are back at the previous 2×2 game. However, in the present 3×3 game, (B, R) is a perfect equilibrium. To see why, suppose that 2's perturbations satisfy $\epsilon^2(A_2) > \epsilon^2(L)$ and that 2 is playing the perturbed action $\sigma_2 = (\epsilon^2(L), 1 - (\epsilon^2(L) + \epsilon^2(A_2)), \epsilon^2(A_2))$. In this case, the payoff to T is strictly less than 0, while the payoff to B is 0. Therefore, if 1's perturbations satisfy the parallel pattern, there is only one equilibrium, (B, R) played with as much probability as possible. If we take limits as the perturbations go to 0, (B, R) is perfect. After deletion of the strongly dominated actions, (B, R) is not perfect. Oooops.

This is a serious problem. When we write down game-theoretic models of strategic situations, we do not include all of the dominated actions that players could take but which we think are irrelevant. For example, in Gianni Schicchi, we did not include the option for Gianni to drop his disguise to the lawyers while imitating Donato on his deathbed. That was certainly an option, but not one he would ever want to take. Whether or not we include this crazy option ought not to affect our analysis.

One could argue, as Myerson did, that the perturbations we just gave are not reasonable. They required that the agents play really bad strategies, A_2, with higher probability than they play strategies that are only moderately bad. This leads to proper equilibria, a closed nonempty subset of the perfect equilibria.

Definition 5.12.35 *An ϵ-perfect equilibrium, $\sigma^\circ = \sigma(\epsilon)$, is ϵ-**proper** if for all $i \in I$ and all $a_i, b_i \in A_i$,*

$$[u_i(\sigma^\circ \backslash a_i) < u_i(\sigma^\circ \backslash b_i)] \Rightarrow [\sigma_i^\circ(a_i) \leq \epsilon \sigma_i^\circ(b_i)].$$

*σ^* is a **proper equilibrium** if there is a sequence $\epsilon_n \to 0$ and a sequence σ_n of ϵ_n-proper equilibria such that $\sigma_n \to \sigma^*$. The set of proper equilibria for Γ is denoted $Pro(\Gamma)$.*

A clever fixed-point argument shows that the set of ϵ-proper equilibria is nonempty, and taking a convergent subsequence in the compact set Δ as $\epsilon_n \to 0$ shows that proper equilibria exist.

Shrinking the set of Nash equilibria to the set of perfect equilibria is just the beginning of what is called **equilibrium refinement**. There are three basic approaches: a perfection approach, due to Selten, a properness approach, due to Myerson, and the stable-sets approach due to Kohlberg and Mertens. Throughout the following, Γ_ϵ denotes an ϵ-perturbation of a game Γ. One example, but not the only one, of the kinds of perturbations under consideration are the restrictions that players play in $\Delta_i(\epsilon^i)$. Roughly, the three approaches are:

1. (perfection) accept σ if there exists any Γ_ϵ, $\epsilon \to 0$, $\sigma^\epsilon \in Eq(\Gamma_\epsilon)$, $\sigma^\epsilon \to \sigma$,

2. (properness) accept σ if there exists any *reasonable* Γ_ϵ, $\epsilon \to 0$, $\sigma^\epsilon \in Eq(\Gamma_\epsilon)$, $\sigma^\epsilon \to \sigma$, and

3. (stable sets) accept a set of equilibria, E, only if for all Γ_ϵ, $\epsilon \to 0$, there exists $\sigma^\epsilon \in Eq(\Gamma_\epsilon)$, $\sigma^\epsilon \to E$.

Imagine a ball resting on a wavy surface. Any rest point is an equilibrium. The rest point is perfect if there exists some kind of tiny tap that we can give to the ball and have it stay in the same area. The rest point is proper if it stays in the same area after reasonable kinds of taps (e.g., we might agree that taps from directly above are not a reasonable test of the equilibrium). A set of points (e.g., a valley with a flat bottom) is stable if no matter what kind of tiny tap it receives, the ball stays in the set after being tapped.

To learn more about these issues, you should take a course in game theory.

5.13 ◆ Bibliography

For an authoritative source on finite-dimensional convex analysis, the classic monograph is R. T. Rockafellar's *Convex Analysis* (Princeton, N.J.: Princeton University Press, 1970), but we would not recommend it as a starting point. T. Ichiishi's *Game Theory for Economic Analysis* (New York: Academic Press, 1983) provides a great deal of extremely well-exposited convex analysis as well as a complete proof of both Brouwer's and Kakutani's fixed-point theorems.

There are many other easily available proofs of the two main fixed-point theorems in this chapter. *Topological Methods in Euclidean Spaces* (Cambridge: Cambridge University Press, 1980) by G. Naber has a complete proof, at an undergraduate level, of Brouwer's theorem. *Linear Operators,* Vol. 1 (New York: Interscience, 1958) by N. Dunford and J. T. Schwartz has a proof of Brouwer's theorem that examines the behavior of determinants of smooth functions from a compact convex set to itself, then appeals to the Stone-Weierstrass theorem to show that the smooth functions are dense. K. C. Border's *Fixed Point Theorems with Applications to Economics and Game Theory* (Cambridge: Cambridge University Press, 1985) is an excellent book, well described by its title.

CHAPTER 6

Metric Spaces

We have examined the metric spaces $(\mathbb{R}^{\ell}, d_E)$, $\ell = 1, 2, \ldots$ in great detail. More general metric spaces also appear in the study of optimization theory, game theory, exchange economies, probability, dynamic programming, and more or less anything else that is a concern of economic theory. Throughout, completeness and compactness are the crucial properties, guaranteeing that many of the sequences we care about have limits or have subsequences that have limits. Many of the metric spaces we study have additional vector space and algebraic structure.

The subsequent sections are devoted to specific metric spaces, covering the list of topics mentioned. Often interest focuses on particularly useful subsets of these spaces. The spaces to be covered include:

1. $\mathcal{K}(M)$—the space of compact subsets M, especially when $M = \mathbb{R}^{\ell}$;

2. $C(M)$ and $C_b(M)$—the space of continuous and the space of continuous bounded functions on a metric space M;

3. $\mathcal{D}(\mathbb{R})$—the space of cumulative distribution functions on $\mathbb{R}$;

4. $\mathbb{R}^{\mathbb{N}}$—the space of $\mathbb{R}$-valued sequences; and

5. the Lebesgue measure space of integrable random variables.

Toward the end of this chapter, we discuss the metric completion theorem, which tells us how to add elements to any metric space so as to make it complete. This is the general version of what we did to construct $\mathbb{R}$ from $\mathbb{Q}$. Two classes of important spaces, the measure spaces and the L^p spaces, can be developed through the metric completion theorem. We introduce these briefly here, but do not develop them in detail until we get to Chapters 7 and 8.

As before, we develop results that are useful for all metric spaces, but only as we need them for specific spaces. This means that the early sections of this chapter tend to be longer than the later ones, in which we can apply the results that we have proved earlier.

6.1 ◆ The Space of Compact Sets and the Theorem of the Maximum

Much of economic theory is concerned with the effects of exogenous parameters on behavior. Much of the study of these effects can take place in the following framework: there is a parameter, x, in a set X, that is not under the control of the decision maker; the parameter x determines $\Gamma(x)$, the set of options available to the decision maker; $\Gamma(x)$ is a subset of Y, the set of all possibly relevant options; knowing x, the decision maker picks $y \in \Gamma(x)$ to maximize a utility function $f : X \times Y \to \mathbb{R}$. We define the **value function**, $v(x) = \sup\{f(x, y) : y \in \Gamma(x)\}$ and the **arg max correspondence**, $G(x) = \{y \in \Gamma(x) : f(x, y) = v(x)\}$.

After a number of examples, we introduce the **Hausdorff metric** on the space of compact subsets of a metric space. This metric space plays a crucial role in the theorem of the maximum, which tells us how v and G behave.

Throughout, (X, d_X) and (Y, d_Y) are metric spaces.

6.1.a Examples

We present a number of examples to demonstrate both the reach of the approach just sketched and to demonstrate the kinds of continuity results we might expect. The typical pattern has $x \mapsto \Gamma(x)$ being a continuous correspondence, $x \mapsto v(x)$ is a continuous function, and $x \mapsto G(x)$ is a correspondence that fails to be continuous by "exploding," but does not fail to be continuous by "imploding." For those who prefer to skip intuition and go straight to the math of the matter, upper and lower hemicontinuity are defined for compact-valued correspondences in §6.1.d and for general correspondences in §6.1.g.

Example 6.1.1 *In consumer choice theory, the parameters are $(\mathbf{p}, w)$, the set of affordable options is the Walrasian budget set, $B_{(\mathbf{p}, w)} \subset \mathbb{R}_+^\ell$, and the consumer maximizes $u : \mathbb{R}_+^\ell \to \mathbb{R}$.*

*Here $x = (\mathbf{p}, w)$, $\Gamma(x)$ is the Walrasian budget set, $Y = \mathbb{R}_+^\ell$, and the function f depends only on consumption, $y \in Y$, and not on the parameters, $x = (\mathbf{p}, w)$. In this context, the value function is known as the **indirect utility function**, and the arg max correspondence is known as the Walrasian demands.*

Example 6.1.2 *[↑Example 2.8.7] The downward-sloping demand curve for a good produced by a monopolist is $q(p) \geq 0$, so that the consumer surplus when p is charged is $CS(p) = \int_p^\infty q(r)\, dr$. Let $p(\cdot) = q^{-1}(\cdot)$ be the inverse demand function. Since the demand curve is downward sloping, the function $q \mapsto CS(p(q))$ is nondecreasing. If we let $c(q)$ be the cost of producing q, the monopolist's problem and society's problem are, respectively, $\max_{q \in \mathbb{R}_+} \pi(q) + 0 \cdot CS(p(q))$ and $\max_{q \in \mathbb{R}_+} \pi(q) + x \cdot CS(p(q))$. Set $f(q, x) = \pi(q) + x\,CS(p(q))$, where $x \in X = \{0, 1\}$.*

Here X is the two-point set $\{0, 1\}$, $\Gamma(x)$ is identically equal to $Y = \mathbb{R}_+$, and f is as given. The welfare loss stemming from monopoly power is $v(1) - v(0)$. We are also interested in the relation between $G(1)$ and $G(0)$.

The best response correspondence in game theory provides a leading class of examples of correspondences that "explode."

Example 6.1.3 *Consider the following example of a best response correspondence from game theory. The game is played between two individuals who can choose between two actions, say go to the museum (M) or go to the art show (A). If both choose M or both choose A, they meet. If one chooses M and the other chooses A, they do not meet. Meetings are pleasurable and yield each player payoff 1, whereas if they do not meet they receive payoff 0. This is known as a coordination game. The players choose probability distributions over the two actions: say player 1 chooses M with probability p and A with probability $1 - p$, while player 2 chooses M with probability q and A with probability $1 - q$. As above, the game is represented by the left-hand matrix. The right-hand matrix gives the probabilities of the four outcomes:*

$1\downarrow \backslash 2 \rightarrow$	M	A		M (q)	A $(1-q)$
M	1, 1	0, 0	M (p)	pq	$p(1-q)$
A	0, 0	1, 1	A $(1-p)$	$(1-p)q$	$(1-p)(1-q)$

Agent 1's expected payoff from playing M with probability p while his opponent is playing M with probability q is denoted $\pi_1(p, q)$ or $U_1(p, q)$, and given by

$$\pi_1(p, q) = p \cdot \big[q \cdot 1 + (1 - q) \cdot 0 \big] + (1 - p) \cdot \big[q \cdot 0 + (1 - q) \cdot 1 \big]$$
$$= p(2q - 1) + (1 - q). \tag{6.1}$$

*Agent 1 chooses $p \in [0, 1]$ to maximize $\pi_1(p, q)$, giving the best response correspondence $Br_1(q)$. For any $q < \frac{1}{2}$, payoffs are decreasing in p so that $Br_1(q) = 0$; for any $q > \frac{1}{2}$, payoffs are increasing in p so that $Br_1(q) = 1$; and at $q = \frac{1}{2}$, payoffs are independent of p so that $Br_1(\frac{1}{2}) = [0, 1]$. The correspondence $Br_1(\cdot)$ is not **lower hemicontinuous** at $q = \frac{1}{2}$ since it "explodes." It is rather **upper hemicontinuous** since upper hemicontinuity allows explosions. This game is symmetric so that agent 2's payoffs and best response correspondence are identical to those of agent 1.*

Example 6.1.4 *A household has preferences represented by the utility function $u : \mathbb{R}_+^2 \rightarrow \mathbb{R}$ given by $U(x_1, x_2) = x_1 + x_2$. The household has a positive endowment of good 2 denoted $\omega \in \mathbb{R}_+$. The household can trade its endowment on a competitive market to obtain good 1, where the price of good 1 in terms of good 2 is given by $p \in \mathbb{R}_+$. The household's purchases are constrained by its income; its budget set is given by*

$$B(p, \omega) = \{(x_1, x_2) \in \mathbb{R}_+^2 : px_1 + x_2 \leq \omega\}.$$

If we take prices and ω as given, the household's problem is

$$v(p, \omega) = \max_{(x_1, x_2) \in B(p, \omega)} u(x_1, x_2), \tag{6.2}$$

and the arg max correspondence is

$$(x_1^*, x_2^*) = \begin{cases} (0, \omega) & \text{if } p > 1 \\ (x, \omega - x) \text{ with } x \in [0, \omega] & \text{if } p = 1 \\ (\frac{\omega}{p}, 0) & \text{if } p < 1 \end{cases}.$$

Goods 1 and 2 are perfect substitutes for each other, so that if good 1 is more expensive (inexpensive), the household consumes none of (only) it, whereas if the two goods are the same price, there is an entire interval of indifferent consumptions. The continuous value function is increasing in w and decreasing in p,

$$v(p, \omega) = \begin{cases} \omega & \text{if } p \geq 1 \\ \frac{\omega}{p} & \text{if } 1 > p > 0 \end{cases}.$$

Intuitively, $B(\cdot, \cdot)$ is a compact set that depends continuously, even smoothly, on p and ω as long as p > 0. However, the arg max correspondence is not continuous at p = 1—it "blows up."

6.1.b　The Metric Space of Compact Sets

Recall that $E^\epsilon := \cup_{e \in E} B_\epsilon(e)$ is the ϵ-ball around the set E.

Definition 6.1.5 $\mathcal{K}(Y)$ *denotes the space of nonempty compact subsets of* Y. *For* $E, F \in \mathcal{K}(Y)$, $e_+(E, F) := \inf\{\epsilon > 0 : E \subset F^\epsilon\}$. *The* **Hausdorff distance** *between* $E, F \in \mathcal{K}(Y)$ *is* $d_H(E, F) = \max\{e_+(E, F), e_+(F, E)\}$.

Intuitively, $d_H(E, F) < \epsilon$ iff the ϵ-ball around F contains E, $e_+(E, F) < \epsilon$ and the ϵ-ball around E contains F, $e_+(F, E) < \epsilon$. If E is the noncompact set $\mathbb{Q} \cap [0, 1]$ and $F = [0, 1]$, then $d_H(E, F) = 0$, even though $E \neq F$. If E is the noncompact set $\mathbb{N}$ and $F = [0, 1]$, then $d_H(E, F) \notin \mathbb{R}$. For these reasons, we apply d_H only to compact sets.

Exercise 6.1.6 If $E \subset F$, then $e_+(E, F) = 0$. If $E = \{y\}$ and $F = \{y'\}$, then $d_H(E, F) = e_+(E, F) = e_+(F, E) = d_Y(y, y')$.

The mapping $y \leftrightarrow \{y\}$ is an isometry, $d(x, y) = d_H(\{x\}, \{y\})$.

Lemma 6.1.7 $(\mathcal{K}(Y), d_H)$ *is a metric space.*

Proof. $d_H(E, F) = d_H(F, E)$. If $E = F$, then $d_H(E, F) = 0$. If $d_H(E, F) = 0$, then $E \subset \text{cl}(F)$ and $F \subset \text{cl}(E)$ (using the shrink-wrapping characterization of closures). Since compact sets are closed, $E \subset F$ and $F \subset E$, so that $E = F$.

Pick $E, F, G \in \mathcal{K}(Y)$; we must show that $d_H(E, F) + d_H(F, G) \geq d_H(E, G)$. Let $r = d_H(E, F)$, $s = d_H(F, G)$. We have to show that $d_H(E, G) \leq (r + s)$. By the definition of the Hausdorff metric, for every $\epsilon > 0$,

$$(\forall e \in E)(\exists f(e) \in F)[d(e, f(e)) < r + \epsilon],$$

$$(\forall f \in F)(\exists e(f) \in E)[d(f, e(f)) < r + \epsilon],$$

$$(\forall f \in F)(\exists g(f) \in G)[d(f, g(f)) < s + \epsilon],$$

$$(\forall g \in G)(\exists f(g) \in F)[d(g, f(g)) < s + \epsilon].$$

Therefore, for all $e \in E$, there exists $g \in G$, namely $g = g(f(e))$, such that $d(e, g) < (r + s) + 2\epsilon$, and for all $g \in G$, there exists $e \in E$, namely $e = e(f(g))$, such that $d(g, e) < (r + s) + 2\epsilon$. Therefore, for all $\epsilon > 0$, $d_H(E, G) < (r + s) + 2\epsilon$.   ∎

The rest of the arguments use a pair of limit sets that are of independent interest because they provide a second exposure to ideas of "infinitely often" and "almost always," which are crucial in limit arguments, especially in probability theory.

Given a sequence of statements $\mathbb{A}_n$, we say "$\mathbb{A}_n$ infinitely often" if for all N there exists an $n \geq N$ such that $\mathbb{A}_n$ is true, and we say "$\mathbb{A}_n$ almost always" if there exists an N such that for all $n \geq N$, $\mathbb{A}_n$ is true. We often write these as "$\mathbb{A}_n$ i.o.," and "$\mathbb{A}_n$ a.a."

Example 6.1.8 *If $\mathbb{A}_n$ is the statement "$n \geq 10^{17}$," then $\mathbb{T} = \{n \in \mathbb{N} : \mathbb{A}_n$ is true$\}$ is the set $\{10^{17}, 10^{17} + 1, 10^{17} + 2, \ldots\}$, so that $[\mathbb{A}_n$ a.a.$]$ and $[\mathbb{A}_n$ i.o.$]$. If $\mathbb{A}_n$ is the statement "n is odd," the corresponding $\mathbb{T}$ is the set $\{1, 3, 5, 7, 9, \ldots\}$, so that $\neg[\mathbb{A}_n$ a.a.$]$ but $[\mathbb{A}_n$ i.o.$]$. If $\mathbb{A}_n$ is the statement "$n \leq 3$," then $\mathbb{T} = \{1, 2, 3\}$, so that $\neg[\mathbb{A}_n$ a.a.$]$ and $\neg[\mathbb{A}_n$ i.o.$]$.*

We have already used these concepts in studying the limit behaviors of sequences.

Example 6.1.9 *$x_n \to x$ iff $\forall \epsilon > 0$, $x_n \in B_\epsilon(x)$ a.a., and x is a contact point of the sequence x_n iff $\forall \epsilon > 0$, $x_n \in B_\epsilon(x)$ i.o.*

We record the following direct implications of these definitions.

Theorem 6.1.10 *For any sequence of statements $\mathbb{A}_n$,*

   1. $[\mathbb{A}_n$ a.a.$] \Rightarrow [\mathbb{A}_n$ i.o.$]$,

   2. $[(\neg\mathbb{A}_n)$ a.a.$] \Leftrightarrow \neg[\mathbb{A}_n$ i.o.$]$, and

   3. $[(\neg\mathbb{A}_n)$ i.o.$] \Leftrightarrow \neg[\mathbb{A}_n$ a.a.$]$.

Verbally, the second of these statements should be read as "$\mathbb{A}_n$ is false almost always iff it is not the case that $\mathbb{A}_n$ is true infinitely often."

Proof of Theorem 6.1.10.   Given a sequence of statements $\mathbb{A}_n$, let $\mathbb{T} = \{n \in \mathbb{N} : \mathbb{A}_n$ is true$\}$, so that $\mathbb{T}^c = \{n \in \mathbb{N} : \mathbb{A}_n$ is false$\}$.

(1) $[\mathbb{A}_n$ a.a.$]$ is equivalent to $\mathbb{T}$ containing a tail set (also known as a cofinal set), that is, a set of the form $\{N, N + 1, N + 2, \ldots\}$. $[\mathbb{A}_n$ i.o.$]$ is equivalent to $\mathbb{T}$ being infinite. Since every tail set is infinite, $[\mathbb{A}_n$ a.a.$] \Rightarrow [\mathbb{A}_n$ i.o.$]$.

(2) $[(\neg\mathbb{A}_n)$ a.a.$]$ is equivalent to the statement that $\mathbb{T}^c$ contains a cofinal set. The complement of a cofinal set is necessarily finite. Observe that $\neg[\mathbb{A}_n$ i.o.$]$ is equivalent to the statement that $\mathbb{T}$ is finite.   ∎

Exercise 6.1.11   Complete the proof of Theorem 6.1.10.

If we have a sequence of sets A_n and a sequence of points $x_n \in A_n$, the sequence may converge, have a contact point, or have no limits along any subsequence. We are interested in the set of points to which some sequence x_n in A_n can converge and in the set of points that are contact points of some such sequence.

Definition 6.1.12 *The **closed lim sup** of a sequence of sets A_n in $\mathcal{P}(Y)$ is defined as* $\operatorname{clim\,sup}_n A_n = \{y \in Y : (\forall \epsilon > 0)[B_\epsilon(y) \cap A_n \neq \emptyset \text{ i.o.}\}$, *and the **closed lim inf** of is defined as* $\operatorname{clim\,inf}_n A_n = \{y \in Y : (\forall \epsilon > 0)[B_\epsilon(y) \cap A_n \neq \emptyset \text{ a.a.}\}$.

Exercise 6.1.13 For $A_n = [0, n]$ find $\operatorname{clim\,sup}_n A_n$ and $\operatorname{clim\,inf}_n A_n$ in $\mathbb{R}$.

1. For $B_{2n} = [0, n]$ and $B_{2n-1} = [-n, 0]$, $n = 1, 2, \ldots$, find $\operatorname{clim\,sup}_n B_n$ and $\operatorname{clim\,inf}_n B_n$ in $\mathbb{R}$.

2. For $C_n = \{1/n, n\}$ find $\operatorname{clim\,sup}_n C_n$ and $\operatorname{clim\,inf}_n C_n$ in $\mathbb{R}$.

3. For $D_{2n} = \{1/n, n\}$ and $D_{2n-1} = \{n\}$, $n = 1, 2, \ldots$, find $\operatorname{clim\,sup}_n D_n$ and $\operatorname{clim\,inf}_n D_n$ in $\mathbb{R}$.

4. For $E_n = \{n\}$, find $\operatorname{clim\,sup}_n E_n$ and $\operatorname{clim\,inf}_n E_n$ in $\mathbb{R}$.

Lemma 6.1.14 *For any sequence of sets, A_n, $\operatorname{clim\,inf}_n A_n \subset \operatorname{clim\,sup}_n A_n$, and both sets are closed.*

Proof. $\operatorname{clim\,inf}_n A_n \subset \operatorname{clim\,sup}_n A_n$ because almost always implies infinitely often.

We show that $(\operatorname{clim\,inf}_n A_n)^c$ is open. Note that $y \notin \operatorname{clim\,inf}_n A_n$ iff $(\exists \epsilon > 0)[B_\epsilon(y) \cap A_n = \emptyset \text{ a.a.}]$. Since $B_\epsilon(y)$ is open, for all $y' \in B_\epsilon(y)$, $\exists \delta > 0$ such that $B_\delta(y') \subset B_\epsilon(y)$. Therefore, for all $y' \in B_\epsilon(y)$, $[B_\delta(y) \cap A_n = \emptyset \text{ a.a.}\}$, so that $B_\epsilon(y) \subset (\operatorname{clim\,inf}_n A_n)^c$. The proof for $\operatorname{clim\,sup}_n A_n$ is very similar. ■

If we have a sequence of sets, A_n, for which there exists a set A with $A = \operatorname{clim\,sup} A_n = \operatorname{clim\,inf} A_n$, then A is called the **Painlevé-Kuratowski** limit of the sequence. This kind of convergence can be metrized for closed subsets of separable metric spaces. We now show that, restricted to $\mathcal{K}(Y)$, Painlevé-Kuratowski convergence and Hausdorff convergence are equivalent.

Exercise 6.1.15 Let A_n be a sequence of compact sets and A a compact set. $A = \operatorname{clim\,sup} A_n = \operatorname{clim\,inf} A_n$ iff $d_H(A_n, A) \to 0$.

6.1.c Completeness, Separability, and Compactness

$(\mathcal{K}(Y), d_H)$ inherits separability, completeness, total boundedness, and compactness from (Y, d).[1]

Theorem 6.1.16 *We have the following:*

1. *$(\mathcal{K}(Y), d_H)$ is separable iff (Y, d) is separable,*

2. *$(\mathcal{K}(Y), d_H)$ is totally bounded iff (Y, d) is totally bounded,*

3. *$(\mathcal{K}(Y), d_H)$ is complete iff (Y, d) is complete, and*

4. *$(\mathcal{K}(Y), d_H)$ is compact iff (Y, d) is compact.*

Proof. Observe that $\{\{y\} : y \in Y\}$ is a closed subset of $(\mathcal{K}(Y), d_H)$ and that $d(x, y) = d_H(\{x\}, \{y\})$. Since any subset of a separable (totally bounded) metric space is separable (totally bounded), the separability (total boundedness) of

---

1. These results are not directly relevant to our study of the continuity properties of correspondences.

$(\mathcal{K}(Y), d_H)$ implies the separability (total boundedness) of (Y, d). The completeness (compactness) of $(\mathcal{K}(Y), d_H)$ implies the completeness (compactness) of any closed set.

Suppose that Y is separable and let C be a countable dense subset of Y. Let $C' \subset \mathcal{K}(Y)$ be the class of finite subsets of C. C' is a countable dense subset of $\mathcal{K}(Y)$. In a similar fashion, if C is an ϵ-net for Y, then C' is an ϵ-net for $\mathcal{K}(Y)$.

Suppose that Y is complete and let E_n be a Cauchy sequence in $\mathcal{K}(Y)$. We show: (a) that $\operatorname{clim\,inf}_n E_n \neq \emptyset$; (b) that $E = \operatorname{clim\,inf}_n E_n = \operatorname{clim\,sup}_n E_n$; (c) that $d_H(E_n, E) \to 0$; and (d) that E is compact.

(a) Pick N_1 such that for all $n, m \geq N_1, d_H(E_n, E_m) < 1/2^1$. Pick $y_1 \in E_{N_1}$. For all $n \geq N_1, B_{1/2^1}(y_1) \cap E_n \neq \emptyset$.

Given N_{k-1} and y_{k-1}, pick $N_k > N_{k-1}$ such that for all $n, m \geq N_k, d_H(E_n, E_m) < 1/2^k$. Choose $y_k \in B_{1/2^{k-1}}(y_{k-1}) \cap E_{N_1}$. For all $n \geq N_k, B_{1/2^k}(y_k) \cap E_n \neq \emptyset$.

This inductive procedure gives a Cauchy sequence y_k. Since Y is complete, $y_k \to y$ for some $y \in Y$. By construction, $y \in \operatorname{clim\,inf}_n E_n$.

(b) From Lemma 6.1.14, we know that $\operatorname{clim\,inf}_n E_n \subset \operatorname{clim\,sup}_n E_n$, and both are closed. Suppose that the subset relation is strict. Pick $y \in \operatorname{clim\,sup}_n E_n$ that does not belong to $\operatorname{clim\,inf}_n E_n$. Since $\operatorname{clim\,inf}_n E_n$ is closed, there exists $\epsilon > 0$, $B_\epsilon(y) \cap \operatorname{clim\,inf}_n E_n = \emptyset$. This means that $d_H(E_n, E_m) > \epsilon$ for infinitely many n, m pairs, contradicting E_n being a Cauchy sequence.

(c) Pick $\epsilon > 0$ and N so that $d_H(E_n, E_m) < \epsilon/2$ for all $n, m \geq N$. It is sufficient to show that $d_H(E_N, E) \leq \epsilon/2$. By the definition of $\operatorname{clim\,sup}_n E_n$, every $y \in \operatorname{clim\,sup}_n E_n$ must be a limit of points of that are within $\epsilon/2$ of E_N so that $e_+(E_N, \operatorname{clim\,sup}_n E_n) \leq \epsilon/2$. In a similar fashion, $e_+(\operatorname{clim\,inf}_n E_n, E_N) \leq \epsilon/2$.

(d) We know that E is a closed nonempty (by step (a)) subset of the complete space Y; hence it is complete. We now show that it is totally bounded. Pick $\epsilon > 0$; we must show that there is a finite ϵ-net for E. Pick n large so that $d_H(E_n, E) < \epsilon/3$. Since E_n is compact, it has a finite $\epsilon/3$-net, S. For each point in S, choose $e(s) \in E$ such that $d(e(s), s) < \epsilon/3$. We finish the proof by showing that the collection $S_E = \{e(s) : s \in S\}$ is a finite ϵ-net for E. Pick $x \in E$. There exists $e_n(x) \in E_n$ such that $d(x, e_n(x)) < \epsilon/3$. Because S is an $\epsilon/3$-net, there exists $s(x) \in S$ such that $d(e_n(x), s(x)) < \epsilon/3$. The point $e(s(x)) \in S_E$ satisfies $d(e(s(x)), s(x)) < \epsilon/3$. Combining yields $d(x, e(s(x))) < \epsilon$.

Finally, suppose (Y, d) is compact. Then it is totally bounded and complete, which implies that $(\mathcal{K}(Y), d_H)$ is totally bounded and complete, hence compact. ∎

6.1.d Compact-Valued Upper and Lower Hemicontinuity

Correspondences are relations.

Definition 6.1.17 [↑*Definition 2.3.12*]*A set $S \subset (X \times Y)$ defines a **correspondence** $\Gamma : X \to \mathcal{P}(Y)$ by $\Gamma(x) = \{y : (x, y) \in S\}$. The set S is the **graph of** Γ. Γ is **compact valued** if $\Gamma(x)$ is a compact subset of Y for all $x \in X$.*

Correspondences that neither explode nor implode are called continuous; those that only fail to be continuous by blowing up are called upper hemicontinuous, and those that only fail to be continuous by imploding are called lower hemicontinuous.

Definition 6.1.18 *A compact-valued* $\Gamma : X \to \mathcal{K}(Y)$ *is*

1. ***upper hemicontinuous (uhc)*** *if* $(\forall x)[[x_n \to x] \Rightarrow [e_+(\Gamma(x_n), \Gamma(x)) \to 0]]$,

2. ***lower hemicontinuous (lhc)*** *if* $(\forall x)[[x_n \to x] \Rightarrow [e_+(\Gamma(x), \Gamma(x_n)) \to 0]]$, *and*

3. ***continuous*** *if* $(\forall x)[[x_n \to x] \Rightarrow [d_H(\Gamma(x), \Gamma(x_n)) \to 0]]$.

Exercise 6.1.19 Which of the following correspondences from $[0, 1]$ to itself are uhc, which are lhc, and which are continuous? Give both the "exploding/imploding" intuition and a proof of your answers:

$$(a)\ \Gamma(x) = \begin{cases} 1 & \text{if } x < \frac{1}{2} \\ \{0, 1\} & \text{if } x = \frac{1}{2} \\ 0 & \text{if } x > \frac{1}{2} \end{cases}, \quad (b)\ \Gamma(x) = \begin{cases} 1 & \text{if } x < \frac{1}{2} \\ [0, 1] & \text{if } x = \frac{1}{2} \\ 0 & \text{if } x > \frac{1}{2} \end{cases},$$

$$(c)\ \Gamma(x) = \begin{cases} \left[0, \frac{1}{2}\right] & \text{if } x \neq \frac{1}{2} \\ [0, 1] & \text{if } x = \frac{1}{2} \end{cases}, \quad (d)\ \Gamma(x) = \begin{cases} [0, 1] & \text{if } x \neq \frac{1}{2} \\ \left[0, \frac{1}{2}\right] & \text{if } x = \frac{1}{2} \end{cases},$$

$(e)\ \Gamma(x) = [x, 1]$.

Singleton-valued correspondences are functions. For functions, there is no distinction between being continuous, being lhc, and being uhc.

Exercise 6.1.20 Suppose that for all $x \in X$, $\Gamma(x)$ contains exactly one point, that is, $\Gamma(x) = \{f(x)\}$. Show that the function $x \mapsto f(x)$ is continuous iff Γ is lhc iff Γ is uhc iff Γ is continuous (as a correspondence). [Hint: If $E = \{y\}$ and $F = \{y'\}$, then $d_H(E, F) = e_+(E, F) = e_+(F, E) = d_Y(y, y')$.]

Just as it was useful to develop several characterizations of continuity, it is useful to develop some alternate characterizations of uhc and lhc. The starting point is the observation that there are three possible definitions of the inverse of a correspondence. The nearly useless one defines $\Gamma^{-1}(E) = \{x : \Gamma(x) = E\}$, which is clearly "correct" if we regard Γ as a function from X to $\mathcal{P}(Y)$.

Definition 6.1.21 *Let* Γ *be a correspondence from* X *to* Y. *The* ***upper inverse of*** Γ *at* E *is* $\Gamma^+(E) = \{x : \Gamma(x) \subset E\}$ *and the* ***lower inverse of*** Γ *at* E *is* $\Gamma^-(E) = \{x : \Gamma(x) \cap E \neq \emptyset\}$ *(Figure 6.1.21).*

If $\Gamma(x)$ is always a singleton set, then its graph is the graph of a function, $x \mapsto f(x)$, and $\Gamma^+(E) = \Gamma^-(E) = f^{-1}(E)$. Thus, the following shows that upper hemicontinuity of correspondences is a generalization of continuity for functions.

Exercise 6.1.22 If Γ is a nonempty-valued correspondence from X to Y, then for all $E \subset Y$, $\Gamma^{-1}(E) \subset \Gamma^+(E) \subset \Gamma^-(E)$. Give examples for which the subset relations are proper.

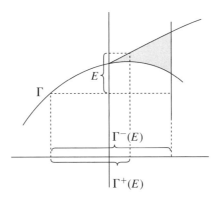

FIGURE 6.1.21

Theorem 6.1.23 *For $\Gamma : X \to \mathcal{K}(Y)$, the following conditions are equivalent:*

1. Γ *is uhc;*

2. *for all open $G \subset Y$, $\Gamma^+(G)$ is open; and*

3. *for all closed $F \subset Y$, $\Gamma^-(F)$ is closed.*

Proof. Note that $(\Gamma^+(E))^c = \{x : \neg[\Gamma(x) \subset E]\} = \{x : \Gamma(x) \cap E^c \neq \emptyset\} = \Gamma^-(E^c)$. Therefore, $(2)\Leftrightarrow(3)$.

$(1)\Rightarrow(3)$: We suppose that $[x_n \to x] \Rightarrow [e_+(\Gamma(x_n), \Gamma(x)) \to 0]$. Let F be a closed subset of Y; we must show that $\Gamma^-(F)$ is closed. Let x_n be a sequence in $\Gamma^-(F)$ and suppose that $x_n \to x$; we have to show that $x \in \Gamma^-(F)$, that is, that there exists $y \in \Gamma(x) \cap F$.

For all n, $\Gamma(x_n) \cap F \neq \emptyset$. For all $m \in \mathbb{N}$, for large n, $\Gamma(x_n) \subset (\Gamma(x))^{1/m}$. Therefore, for all m, there exists $y_m \in \Gamma(x)$ such that $d(y_m, F) < 1/m$. Since $\Gamma(x)$ is compact, taking a subsequence if necessary, we get that $y_m \to y$ for some $y \in \Gamma(x)$. Since $d(\cdot, F)$ is a continuous function, $d(y, F) = 0$. Since F is closed, $y \in F$, so that $y \in \Gamma(x) \cap F$.

$(2)\Rightarrow(1)$: Now suppose that for all open $G \subset Y$, $\Gamma^+(G)$ is open. Let $x_n \to x$; we must show that for all $\epsilon > 0$, for large n, $\Gamma(x_n) \subset (\Gamma(x))^\epsilon$. Let G be the open set $(\Gamma(x))^\epsilon$, so that x belongs to the open set $\Gamma^+(G)$. For large n, $x_n \in \Gamma^+(G)$ so that $\Gamma(x_n) \subset (\Gamma(x))^\epsilon$. ■

The following shows that lower hemicontinuity of correspondences is also a generalization of continuity for functions.

Theorem 6.1.24 *For $\Gamma : X \to \mathcal{K}(Y)$, the following conditions are equivalent:*

1. Γ *is lhc;*

2. *for all open $G \subset Y$, $\Gamma^-(G)$ is open; and*

3. *for all closed $F \subset Y$, $\Gamma^+(F)$ is closed.*

Exercise 6.1.25 Prove Theorem 6.1.24.

For $A \subset X$, we define $\Gamma(A) = \cup_{x \in A} \Gamma(x)$ as the **image of** A **under** Γ. We know that $f(K)$ is compact if K is compact and f is a continuous function, that is, the continuous image of a compact set is compact. We now show that the upper hemicontinuous image of a compact set is compact. This is another of the ways in which upper hemicontinuous correspondences are generalizations of continuous functions.

Lemma 6.1.26 *The upper hemicontinuous image of a compact set is compact.*

Proof. Let $\Psi : X \to \mathcal{K}(Y)$ be uhc and let K be a compact subset of X. We show that $\Psi(K)$ is compact.

Let y_n be a sequence in $\Psi(K)$; we must show that some subsequence converges to some y in $\Psi(K)$. By the definition of union, we can choose $x_n \in K$ such that $y_n \in \Psi(x_n)$. Taking a convergent subsequence in K if necessary, $x_n \to x$ for some $x \in K$. For each $m \in \mathbb{N}$, pick $n(m)$ such that for all $n \geq n(m)$, $d_Y(y_n, \Psi(x)) < 1/m$. This means that we can pick $y'_m \in \Psi(x)$ such that $d_Y(y_{n(m)}, y'_m) < 1/m$. Taking a convergent subsequence of the y'_m if necessary yields $y'_m \to y$ for some y in the compact set $\Psi(x)$. Since $d(y_{n(m)}, y'_m) \to 0$, $y_{n(m)} \to y$ as well. Since $x \in K$, $y \in \Psi(K)$. ∎

Just as with functions, it is sometimes useful to take the composition of correspondences. Let Γ be a correspondence from X to Y and Ψ a correspondence from Y to Z. We define the composition by $(\Psi \circ \Gamma)(x) = \Psi(\Gamma(x))$.

Corollary 6.1.27 *The composition of uhc correspondences is uhc, the composition of lhc correspondences is lhc, and the composition of continuous correspondences is continuous.*

Proof. If Γ and Ψ are uhc, then $(\Psi \circ \Gamma)^+(G)$ is open for every open $G \subset Z$. ∎

Exercise 6.1.28 Complete the proof of Corollary 6.1.27.

6.1.e The Theorem of the Maximum

We are interested in correspondences from the metric space (X, d_X) to (Y, d_Y) and utility functions $f : X \times Y \to \mathbb{R}$. To talk of closed subsets of $X \times Y$ and continuous functions on $X \times Y$, we need the product metric, which is defined so as to have the property that $[(x_n, y_n) \to (x, y)] \Leftrightarrow [[x_n \to x] \wedge [y_n \to y]]$.

Definition 6.1.29 *For metric space (X, d_X) to (Y, d_Y), the **product metric on** $X \times Y$ is $d_{X \times Y}((x, y), (x', y')) = \max\{d_X(x, x'), d_Y(y, y')\}$.*

Exercise 6.1.30 Show that the product metric is equivalent to $\rho((x, y), (x', y')) = d_X(x, x') + d_Y(y, y')$ and to $e((x, y), (x', y')) = \sqrt{d_X(x, x')^2 + d_Y(y, y')^2}$.

Theorem 6.1.31 (Theorem of the Maximum) *If $x \mapsto \Gamma(x)$ is a continuous compact-valued nonempty-valued correspondence from X to Y and $f : X \times Y \to \mathbb{R}$ is continuous, then $v(x) := \sup\{f(x, y) : y \in \Gamma(x)\}$ is a continuous function and $G(x) := \{y \in \Gamma(x) : f(x, y) = v(x)\}$ is a nonempty-valued compact-valued upper hemicontinuous correspondence.*

Proof.   We first show that for all x, $G(x)$ is a nonempty compact set. For all x, the function $f_x(\cdot) := f(x, \cdot) : Y \to \mathbb{R}$ is continuous because $d_Y(y_n, y) \to 0$ iff $d_{X \times Y}((x, y_n), (x, y)) \to 0$. Therefore, $G(x) \neq \emptyset$ because the continuous function f_x achieves its maximum on the compact set $\Gamma(x)$. Further, $G(x) = f_x^{-1}(\{v(x)\}) \cap \Gamma(x)$ is a closed subset of $\Gamma(x)$ because $f_x^{-1}(\{v(x)\})$ is closed (because it is the inverse image of a point), and $\Gamma(x)$ is closed. Since the closed subset of a compact set is compact, this implies that $G(x)$ is compact.

We now show that the graph of G is closed. The next lemma, 6.1.32, implies that any closed graph subset of a continuous correspondence is uhc, which completes the proof. Let (x_n, y_n) be a sequence in the graph such that $(x_n, y_n) \to (x, y)$. Suppose that $y \notin G(x)$; that is, suppose that y is not an optimal choice at x. Pick an optimal $y' \in G(x)$ and $\epsilon > 0$ such that $f(x, y') > f(x, y) + \epsilon$. Since Γ is continuous, there exists $y'_n \in \Gamma(x_n)$ such that $y'_n \to y'$. By continuity, $f(x_n, y_n) \to f(x, y)$ and $f(x_n, y'_n) \to f(x, y')$. For large n, this means that $f(x_n, y'_n) > f(x_n, y_n)$, implying $y_n \notin G(x_n)$, contradicting the choice of (x_n, y_n) in the graph of G.

Finally, v is the composition of the uhc correspondences $x \mapsto (\{x\} \times G(x))$ and $(x, y) \mapsto \{f(x, y)\}$. As it is the composition of uhc correspondences, v is uhc, and because it is uhc and singleton valued, Exercise 6.1.20 implies that v is a continuous function.   ∎

The tactic of proof to show that the graph of G is closed is used frequently—suppose that the graph is not closed, so that the limit of a sequence of optimal responses, y_n, to a sequence of parameters, x_n, converges to something, (x, y), such that y is not a best response to x. Since something, y', is a best response to x, continuity of the set of options, Γ, implies that for large n there is an option, y'_n, very close to y'. For large n, y'_n must be better than y_n, by the continuity of the utility function, f.

For uhc, it is frequently sufficient to show that the graph is closed. We say that a correspondence Φ is a subset of the correspondence Ψ if the graph of Φ is a subset of the graph of Ψ, equivalently if $\Phi(x) \subset \Psi(x)$ for all $x \in X$.

Lemma 6.1.32   *If $\Psi : X \to \mathcal{K}(Y)$ is uhc and $\Phi : X \to \mathcal{K}(Y)$ is a closed graph subset of Ψ, then Φ is uhc.*

Proof.   Let F be a closed subset of Y, so that $(X \times F)$ is a closed subset of $X \times Y$. Since S, the graph of Φ, is closed, this means that $T = S \cap (X \times F)$ is closed. We show that $\Phi^-(F) = \{x : (\exists y \in Y)[(x, y) \in T]\}$ is closed.

Let x_n be a sequence in $\Phi^-(F)$, $x_n \to x$. We must show that $x \in \Phi^-(F)$, that is, that there exists $y \in F \cap \Phi(x)$. Since $x_n \in \Phi^-(F)$, there exists $y_n \in (F \cap \Phi(x_n))$. Since Ψ is uhc, for all $m \in \mathbb{N}$, there exists an $n(m)$ such that for all $n \geq n(m)$, $y_n \in (\Psi(x))^{1/m}$. Since $\Psi(x)$ is compact, taking a subsequence if necessary yields $y_{n(m)} \to y \in \Psi(x)$. Since F is closed, $y \in F$. As the graph of Φ is closed, $(x_{n(m)}, y_{n(m)}) \to (x, y)$ implies that $y \in \Phi(x)$.   ∎

Recall that a closed-graph function $f : X \to Y$ with Y compact is continuous, which is a special case of the following.

Corollary 6.1.33   *If $\Phi : X \to \mathcal{K}(Y)$ has a closed graph and Y is compact, then Φ is uhc.*

Proof.   Take $\Psi(x) \equiv Y$ and apply the foregoing.   ∎

The following shows why the proof of Lemma 6.1.32 had to be as tricky as it was.

Exercise 6.1.34 Show that the correspondence $\Gamma : \mathbb{R}_+ \to \mathcal{K}(\mathbb{R}_+)$ defined by $\Gamma(x) = \{\frac{1}{x}\}$ if $x > 0$ and $\Gamma(0) = \{0\}$ has a closed graph but is not upper hemi-continuous.

6.1.f Applications

For any finite product, $\times_{i \in I} X_i$, of metric spaces (X_i, d_i), the product metric is $d(x, x') := \max_i \{d_i(x_i, x_i')\}$. This generalizes the product metric on the product of two spaces.

Example 6.1.35 *Let $u : \mathbb{R}_+^\ell \to \mathbb{R}$ be continuous. For any $(\mathbf{p}, w) \in \mathbb{R}_{++}^\ell \times \mathbb{R}_{++}$, let $B(\mathbf{p}, w) = \{\mathbf{x} \in \mathbb{R}_+^\ell : \mathbf{px} \leq w\}$, let $v(\mathbf{p}, w) = \max\{u(\mathbf{x}) : \mathbf{x} \in B(\mathbf{p}, w)\}$ and let $x(\mathbf{p}, w)$ be the arg max correspondence. By the theorem of the maximum, if we show that B is continuous, then we know that v, the indirect utility function, is continuous and that $x(\cdot, \cdot)$, the Walrasian demand correspondence, is uhc. As singleton-valued uhc correspondences are continuous, we also know that if x is single valued, then it is continuous.*

Exercise 6.1.36 Show that $(\mathbf{p}, w) \mapsto B(\mathbf{p}, w)$ is continuous so that the indirect utility function is continuous and the demand correspondence is uhc when u is continuous. [Hint: $B(\mathbf{p}, w)$ is the convex hull of the points 0 and $\mathbf{e}_i(w/p_i)$, $i = 1, \ldots, \ell$, where $\mathbf{e}_i$ is the unit vector in the ith direction. These points are continuous functions of $(\mathbf{p}, w)$. Appeal to Carathéodory's theorem.]

Exercise 6.1.37 Preferences, $\succeq$, on $\mathbb{R}_+^\ell$ are **strictly convex** if $\mathbf{x} \sim \mathbf{x}'$, $\mathbf{x} \neq \mathbf{x}'$, and $\gamma \in (0, 1)$ imply that $\mathbf{x}\gamma\mathbf{x}' \succ \mathbf{x}$. If u is a continuous utility function representing strictly convex preferences, then the Walrasian demands are continuous.

Example 6.1.38 *In general equilibrium theory models of exchange economies, people do not have monetary income, w, but rather have endowments, $\omega_i \in \mathbb{R}_+^\ell$. Their budget sets are $B(\mathbf{p}, \omega_i) = \{\mathbf{x} \in \mathbb{R}_+^\ell : \mathbf{px} \leq \mathbf{p}\omega_i\}$. The mapping $\omega_i \mapsto \mathbf{p}\omega_i$ is continuous; therefore the correspondence $(\mathbf{p}, \omega_i) \mapsto B(\mathbf{p}, \omega_i)$ is continuous, being the composition of continuous correspondences.*

Part of specifying a game involves specifying the vector of utilities, u, associated with the choices of the players. Let $Eq(u)$ be the set of equilibria when u is the vector of utilities. Showing the uhc of the correspondence $u \mapsto Eq(u)$ means that if we mistakenly specify the utilities as being u_n instead of u and use $\sigma_n \in Eq(u_n)$ as the equilibrium to analyze, then we are close to analyzing an equilibrium of the game at u as $u_n \to u$. This is the best that one could possibly do.

Example 6.1.39 *For the following game: when $r > 0$, the unique element of $Eq(r)$ is $\{Up, Left\}$; when $r < 0$, the unique element of $Eq(r)$ is $\{Down, Right\}$; when $r = 0$, any mixed strategy is an equilibrium:*

	Left	Right
Up	(r, r)	$(0, 0)$
Down	$(0, 0)$	$(-r, -r)$

We now show that the Nash equilibrium correspondence is uhc.

Example 6.1.40   *Let* $\mathcal{G}(u) = (A_i, u_i)_{i \in I}$ *be a finite game parameterized by the vector* $u \in \mathbb{R}^{A \cdot I}$. *For each* u, *let* $Eq(u)$ *denote the set of Nash equilibria for* $\mathcal{G}(u)$, *so that* Eq *is a correspondence from* $\mathbb{R}^{A \cdot I}$ *to the compact spaces* $\times_{i \in I} \Delta_i$. *The graph of the correspondence* $u \mapsto Eq(u)$ *is*

$$S = \{(u, \sigma) \in \mathbb{R}^{A \cdot I} \times \times_{i \in I} \Delta_i : (\forall i \in I)(\forall a_i \in A_i)[U_i(\sigma) \geq U_i(\sigma \backslash a_i)].$$

The set S *is closed since the* U_i *are continuous functions of* u *and* σ. *As* S *is a subset of the constant correspondence* $\Gamma(u) \equiv \times_{i \in I} \Delta_i$, Eq *is uhc.*

6.1.g   General Upper and Lower Hemicontinuity

We have only studied compact-valued nonempty-valued correspondences from one metric space to another, which covers every case that is of concern in this book. Most authors deal with more general classes of correspondences. The theory is trickier. The starting point for the general treatments uses the first conditions in Theorems 6.1.23 and 6.1.24 as definitions of uhc and lhc. To be more explicit, the following definition, which applies to arbitrary nonempty-valued correspondences, agrees with Definition 6.1.18 when applied to compact-valued correspondences.

Definition 6.1.41   *A nonempty-valued correspondence* $\Gamma : X \to \mathcal{P}(Y)$ *is* **upper hemicontinuous** *if for all open* $G \subset Y$, $\Gamma^+(G)$ *is open, it is* **lower hemicontinuous** *if for all open* $G \subset Y$, $\Gamma^-(G)$ *is open, and it is* **continuous** *if it is both upper and lower hemicontinuous.*

   The following two examples demonstrate the tricky aspects of the general theory and more or less exhaust what we will say on this topic. The first example demonstrates why one wants only closed-valued correspondences. It gives what seems to be a Lipschitz continuous, open-valued correspondence that fails to be upper hemicontinuous according to Definition 6.1.41. The second shows that requiring closure alone is not enough to make Definition 6.1.41 sensible. It gives a closed-valued correspondence that is, intuitively, continuous, but that fails to be upper hemicontinuous.

Example 6.1.42   *Define* Γ *from* $(0, 1]$ *to* $(0, 1]$ *by* $\Gamma(r) = (0, r)$. *Let* G *be the open set* $(0, s)$. $\Gamma^+(G) = (0, s]$, *which is not open, so* Γ *is not uhc according to Definition* 6.1.41.

Example 6.1.43   *Let* $F = \{(x_1, x_2) \in \mathbb{R}_+^2 : x_1 \cdot x_2 \geq 1\}$, *and for* $r \in [1, \infty)$, *define* $F(r) = \{(x_1, x_2) \in F : x_1 \leq r, \ x_2 \leq r\} \cup \{(x_1, x_2) : 0 \leq x_1 \leq \frac{1}{r}, \ x_2 = r\}$. *Define the correspondence* $\Gamma : [0, 1] \to \mathcal{P}(\mathbb{R}_+^2)$ *by* $\Gamma(0) = F$ *and* $\Gamma(x) = F(1/x)$ *if* $x > 0$.
   For every $\epsilon > 0$, $\Gamma^+(F^\epsilon)$ *is an open subset of* $[0, 1]$ *containing* 0. *However, for* $0 < \delta < 1$, *define an open set containing* F *by* $G(\delta) = \{(x_1, x_2) \in \mathbb{R}_+^2 : x_1 \cdot x_2 > 1 - \delta\}$. *Note that for all* $x > 0$, $\Gamma(x) \not\subset G(\delta)$ *for any* $\delta \in (0, 1)$, *so that* $\Gamma^+(G) = \{0\}$, *which is not an open subset of* $[0, 1]$, *so* Γ *fails Definition* 6.1.41.

If $x_n \downarrow 0$ and $F_n = \Gamma(x_n)$ in Example 6.1.43, then clim sup F_n = clim inf F_n = F. This is called **Painlevé-Kuratowski** convergence, for which there are metrics, but they take us rather far afield.

6.2 ◆ Spaces of Continuous Functions

We now look at some of the basic properties of the set of continuous real-valued functions and the set of bounded continuous real-valued functions on a metric space M.

Definition 6.2.1 *A function $f : M \to \mathbb{R}$ is **bounded** if $\exists B$ such that $|f(x)| \le B$ for all $x \in M$. For any metric space (M, d), $C(M)$ denotes the set of continuous, $\mathbb{R}$-valued functions on M and $C_b(M)$ denotes the set of bounded continuous functions on M. The **sup norm** of $f \in C_b(M)$ is defined by $\|f\|_\infty = \sup\{|f(x)| : x \in M\}$, and the **sup norm distance** between $f, g \in C_b(M)$ is defined as $d_\infty(f, g) = \|f - g\|_\infty = \sup\{|f(x) - g(x)| : x \in M\}$. If $d_\infty(f_n, f) \to 0$, we say that f_n **converges uniformly to** f or that f_n **is uniformly convergent**.*

$C(M)$ is a vector space and $C_b(M)$ a normed vector space.

Theorem 6.2.2 *$(C_b(M), \|\cdot\|_\infty)$ is a normed vector space, so that d_∞ is a metric.*

Proof. As the other properties of a norm are clear, proving the triangle inequality for d_∞ is sufficient.[2] Pick $f, g, h \in C_b(M)$; we must show that $d_\infty(f, g) + d_\infty(g, h) \ge d_\infty(f, h) = 0$.

$$d_\infty(f, g) + d_\infty(g, h) = \sup_x (|f(x) - g(x)|) + \sup_x (|g(x) - h(x)|) \ge$$
$$\sup_x (|f(x) - g(x)| + |g(x) - h(x)|) \ge \sup_x |f(x) - h(x)| = d_\infty(f, h).$$

The first inequality comes from the observation that for any two bounded real-valued functions, $r(\cdot), s(\cdot)$, $[\sup_x r(x) + \sup_y s(s)] \ge \sup_z [r(z) + s(z)]$ because on the left-hand side of the inequality, one can choose two different arguments for the two functions, whereas on the right, one must choose the same argument for both. The last inequality comes from the triangle inequality in $\mathbb{R}$, $|f(x) - g(x)| + |g(x) - h(x)| \ge |f(x) - h(x)|$ for all $x \in M$. ∎

Notation 6.2.3 *Recall that $C(M; M')$ denotes the set of continuous functions from M to M' and $C(M) = C(M; \mathbb{R})$. After $M' = \mathbb{R}$, the most frequent choice is $M' = \mathbb{R}^\ell$. In this case $f(x) = (f_1(x), \ldots, f_\ell(x))$, where each $f_i \in C(M)$. This means that there is no useful distinction between $C(M; \mathbb{R}^\ell)$ and $(C(M))^\ell$.*

---

2. In case they do not look clear: If $|f(x)| \le B_f$ and $|g(x)| \le B_g$, then $d_\infty(f, g) \le B_f + B_g$, so that $d_\infty(f, g) \in \mathbb{R}_+$; $d_\infty(f, g) = d_\infty(g, f)$ because $|f(x) - g(x)| = |g(x) - f(x)|$. $d_\infty(f, g) = 0$ iff for all x, $|f(x) - g(x)| = 0$ iff $f = g$.

6.2.a Some Examples

We begin by examining how properties of (M, d) affect properties of $C_b(M)$ and $C(M)$.

6.2.a.1 $C(M) = C_b(M)$ iff (M, d) Is Compact

When M is compact, every continuous function is bounded so that $C_b(M) = C(M)$. Further, for compact M, $d_\infty(f, g) = \max_{x \in M} |f(x) - g(x)|$ since the continuous function $|f(x) - g(x)|$ achieves its maximum on the compact set M. One can go further.

We know that (M, d) is compact iff every $f \in C(M)$ achieves its maximum and minimum on M. The following is a strengthening of this result—M is compact iff every $f \in C(M)$ is bounded.

Theorem 6.2.4 *A metric space (M, d) is compact iff $C_b(M) = C(M)$.*

Proof. All that is left to show is that the boundedness of every continuous function implies the compactness of M. We show that M is totally bounded and complete.

If M is not totally bounded, then there exists an $\epsilon > 0$ and an infinite set of points, F, at distance ϵ or more from each other. Let $\{x_n : n \in \mathbb{N}\}$ be a countable infinite subset of F and let f_n be the continuous function $[1 - \frac{2}{\epsilon} d(x, x_n)] \vee 0$. Note that f_n is continuous and that for $n \neq m$ and all $x \in M$, $f_n(x) \cdot f_m(x) = 0$. Therefore, $f(x) := \sum_{n \in \mathbb{N}} n \cdot f_n(x)$ is continuous, but it is clearly not bounded.

If M is not complete, then there exists a Cauchy sequence x_n that has no limit. Taking a subsequence if necessary, we get that for all $n \neq m$, $x_n \neq x_m$. Since the sequence has no limit, we know in particular that no x_n is the limit of the sequence. Therefore, for every n, there exists an $\epsilon_n > 0$ such that $d(x_n, x_m) > \epsilon_n$ for all $m \neq n$. Define $g_n(x) = [1 - \frac{2}{\epsilon_n} d(x, x_n)] \vee 0$ and $g(x) = \sum_{n \in \mathbb{N}} n \cdot g_n(x)$. This is a continuous unbounded function. ∎

6.2.a.2 $C(M)$ and $C_b(M)$ when (M, d) Is Discrete

The simplest discrete spaces are the finite ones, in which case $C(M) = \mathbb{R}^M$. If we order M as $M = \{m_1, \ldots, m_\ell\}$, then any function $f \in C(M)$ is specified by the ℓ numbers $(f(m_1), \ldots, f(m_\ell))$, and this is a vector in $\mathbb{R}^\ell$. In other words, one can view all of our study of $\mathbb{R}^\ell$ as the study of a space of continuous functions.

More generally, whenever M is discrete, any function is continuous, so that $C(M)$ is the set of all functions $f : M \to \mathbb{R}$ and $C_b(M)$ is the set of all bounded functions. A very useful example is $M = \mathbb{N}$.

6.2.a.3 $C(\mathbb{N})$ and Its Vector Subspaces

$C(\mathbb{N})$ is the set of sequences in $\mathbb{R}$ and $C_b(\mathbb{N})$ is the set of bounded sequences in $\mathbb{R}$. The following generalizes to $C(\mathbb{N}; \mathbb{R}^\ell)$ and $C_b(\mathbb{N}; \mathbb{R}^\ell)$ in obvious fashions.

Example 6.2.5 (Nonseparability) *$C_b(\mathbb{N})$ is not separable because $1_A \in C_b(\mathbb{N})$ for every $A \in \mathcal{P}(\mathbb{N})$, and $d_\infty(1_A, 1_B) = 1$ if $A \neq B$. Since $\mathcal{P}(\mathbb{N})$ is uncountable, this means that $C_b(\mathbb{N})$ contains uncountably many points at distance 1 from each other. Therefore, no countable set can be dense.*

Exercise 6.2.6 Show that $C_b(\mathbb{R})$ is not separable.

Example 6.2.7 (c_0 and c) *The set $c_0 \subset \mathbb{R}^{\mathbb{N}}$ is the set of $\mathbb{R}$-valued sequences converging to 0, $c = \{c_0 + r\mathbf{1} : r \in \mathbb{R}\}$, where $\mathbf{1}$ is the function constant at 1. A nonnegative point in c_0 might represent the stream of benefit from a nonrenewable resource that is being exhausted, and a point in c the stream of benefits from a renewable resource that is stable, that is, has a limit in the long run.*

Lemma 6.2.8 *$(c, \|\cdot\|_{\infty})$ and $(c_0, \|\cdot\|_{\infty})$ are complete separable normed vector spaces.*

Proof. Verifying the properties of a vector space is easy. For completeness it is sufficient to observe that both are closed subsets of $C_b(\mathbb{N})$. For the separability of c_0, consider the countable set of points of the form $x = (q_1, \ldots, q_n, 0, 0, \ldots)$, where $q_i \in \mathbb{Q}, i = 1, \ldots, n$, and $n \in \mathbb{N}$. For c, consider the countable set of points of the form $y = (q_0 + q_1, \ldots, q_0 + q_n, q_0, q_0, \ldots)$, where $q_i \in \mathbb{Q}, i = 0, 1, \ldots, n$, and $n \in \mathbb{N}$. ■

c_0 has further vector subspaces that have different norms and often these are used as models of benefit streams. It is convenient to introduce the **projection map** and the closely related **evaluation map**, which are applicable to all spaces of functions, not just the continuous ones on $\mathbb{N}$.

Definition 6.2.9 *For $f : M \to M'$ and $x \in M$, the **projection map at x** is defined by $\text{proj}_x(f) := f(x)$ and the **evaluation map** is the mapping $(x, f) \mapsto (x, f(x)) = (x, \text{proj}_x(f))$. For any $S \subset M$, $\text{proj}_S(f) = \{f(x) : x \in S\} \subset M'$ and for any finite ordered $(x_1, \ldots, x_n) \in M^n$, $\text{proj}_{(x_1,\ldots,x_n)} f = (f(x_1), \ldots, f(x_n)) \in (M')^n$.*

The name "projection" comes from $\mathbb{R}^{\ell}$, where we project a point in $\mathbb{R}^{\ell}$ onto one of the coordinate axes. In the case of $C(\mathbb{N})$, for $x = (x_1, x_2, \ldots)$ and $k \in \mathbb{N}$, $\text{proj}_k(x) = x_k \in \mathbb{R}$. Thus, $\text{proj}_x : C(M) \to \mathbb{R}$, and the evaluation map is from $M \times C(M)$ to $M \times \mathbb{R}$. More generally, $\text{proj}_x : C(M : \mathbb{R}^{\ell}) \to \mathbb{R}^{\ell}$ and $\text{proj}_x : C(M : M') \to M'$, while the corresponding evaluation maps are from $M \times C(M : \mathbb{R}^{\ell})$ to $M \times \mathbb{R}^{\ell}$ and from $M \times C(M : M')$ to $M \times M'$.

Example 6.2.10 (ℓ^p, $p \in [1, \infty)$) *For $p \in [1, \infty)$, $\ell^p \subset c_0$ is defined as the set of all points, $x \in c_0$, for which the norm $\|x\|_p := \left(\sum_{n \in \mathbb{N}} |\text{proj}_n(x)|^p\right)^{1/p} < \infty$. To see some of what is involved, $x = (1, 1/2, 1/3, \ldots) \in \ell^p$ for all $p \in (1, \infty)$, but $x \notin \ell_1$. If $x \in \ell^p$, $p \in [1, \infty)$, then $\|\text{proj}_{1,\ldots,n}(x)\|_p \uparrow \|x\|_p$. Using this and the facts about p-norms for $\mathbb{R}^n$ shows that $\|\cdot\|_p$ is indeed a norm.*

Exercise 6.2.11 For all $p \in [1, \infty)$, give a point $x(p) \in c_0$ such that $\|x\|_p = \infty$, but for all $\epsilon > 0$, $\|x\|_{p+\epsilon} < \infty$. [Look at variants of the observation that $\sum_n (1/n)^p < \infty$ for $p > 1$ but $\sum_n (1/n)^1 = \infty$.]

Exercise 6.2.12 Show that for $p \in [1, \infty)$, $(\ell^p, \|\cdot\|_p)$ is a separable normed vector space. [This space is also complete, but this more difficult fact is treated later. For separability, either consider the class of points of the form $x = (q_1, \ldots, q_n, 0, 0, \ldots)$, where $n \in \mathbb{N}$ and $q_i \in \mathbb{Q}$, or use the next example and observe that ℓ^p is $\|\cdot\|_{\infty}$ separable and that this implies the result.]

Example 6.2.13 (ℓ^{∞}) *When $p = \infty$, $\ell^{\infty} := C_b(\mathbb{N})$ with the sup norm, $\|x\|_{\infty} = \sup\{|\text{proj}_k(x)| : k \in \mathbb{N}\}$. Note that c and c_0 are closed vector subspaces of ℓ^{∞} and that they are strict subsets of ℓ^{∞} because, for example, $x = (+1, -1, +1, -1, \ldots)$*

$\in \ell^\infty \setminus c$. *A nonnegative point in* ℓ^∞ *might represent all possible bounded sequences of consumption over time.*

The following is about the convergence of $\text{proj}_k(x_n)$ and the relation between this and the convergence of the x_n.

Exercise 6.2.14 Show the following:

1. If $\|x_n - x\|_p \to 0$ in ℓ^p, $p \in [1, \infty]$, then for all $k \in \mathbb{N}$, $\text{proj}_k(x_n) \to \text{proj}_k(x)$.

2. Let $x_n \in \ell^p$ be the point that is 1 up to the nth coordinate and 0 thereafter and let $\mathbf{1} \in \ell^\infty$ be the vector that is equal to 1 in each coordinate; $\text{proj}_k(x_n) \to \text{proj}_k(\mathbf{1})$ for all k, but $\mathbf{1} \notin \ell^p$ for $p < \infty$.

3. Let $x_n \in C(\mathbb{N})$ be k in the kth coordinate for $k \leq n$, and 0 in all other coordinates. Show that each $x_n \in \ell^\infty$, give the vector $x \in C(\mathbb{N})$ with the property that $\text{proj}_k(x_n) \to \text{proj}_k(x)$ for all k, and show that $x \notin \ell^\infty$.

6.2.a.4 Cumulative Distribution Functions

We now embed a vitally important class of possibly *discontinuous* functions on $\mathbb{R}$ in $C_b(\mathbb{R})$. It may be worth reviewing Examples 2.2.11 and 2.3.11 (pp. 21 and 24) and Lipschitz continuity, Definition 4.9.1 (p. 143).

Definition 6.2.15 *A **cumulative distribution function (cdf)** F is a nondecreasing function mapping $\mathbb{R}$ to $[0, 1]$ such that $\lim_{n \to \infty} F(-n) = 1 - \lim_{n \to \infty} F(n) = 0$, and F is **right continuous**, that is, for all $x \in \mathbb{R}$, $\lim_{\epsilon \downarrow 0} F(x + \epsilon) = F(x)$. $\mathcal{D} = \mathcal{D}(\mathbb{R})$ denotes the set of cumulative distribution functions.*

We interpret $F(x)$ as the probability that a random variable takes a value less than or equal to x, $F(x) = P(X \leq x)$. This means that $P(X \in (a, b]) = F(b) - F(a)$ is the probability that X takes a value in the interval $(a, b]$.

Example 6.2.16 *If $F(x) = \frac{e^x}{1+e^x}$ so that X takes a value less than or equal to x with probability $\frac{e^x}{1+e^x}$, then X follows what is called a **logistic distribution**. If X takes the values ± 1 with probability $\frac{1}{2}$ each, then the associated cdf is $G(x) = 0$ if $x < -1$, $G(x) = \frac{1}{2}$ if $-1 \leq x < +1$, and $G(x) = 1$ if $+1 \leq x$. If Y is uniformly distributed on the interval $[0, 1]$, the cdf is $H(x) = 0$ if $x < 0$, $H(x) = x$ if $0 \leq x < 1$, and $H(x) = 1$ if $1 \leq x$.*

Jumps, or discontinuities, in a cdf occur at x's such that $F^-(x) := \lim_{\epsilon \downarrow 0} F(x - \epsilon) < F(x)$. Jumps correspond to $P(\{x\}) > 0$. Recall that interval subsets on $\mathbb{R}$ are denoted $|a, b|$ and may take on any of the following nine forms: $(-\infty, b]$, $(-\infty, b)$, $(a, b]$, $[a, b]$, $[a, b)$, (a, b), $[a, \infty)$, (a, ∞), or $(-\infty, +\infty)$. Let $\mathcal{B}^\circ \subset \mathcal{P}(\mathbb{R})$ denote the collection of finite disjoint unions of intervals.

Exercise 6.2.17 Show that $\mathcal{B}^\circ$ is a field of subsets of $\mathbb{R}$.

Any cdf F defines a function $P_F : \mathcal{B}^\circ \to [0, 1]$. To be explicit, define $F(-\infty) = 0$, $F(\infty) = F^-(\infty) = 1$ and have $P_F((a, b]) = F(b) - F(a)$; $P_F((a, b)) = F^-(b) - F(a)$; $P_F([a, b)) = F^-(b) - F^-(a)$; and $P_F([a, b]) = F(b) - F^-(a)$. For any finite disjoint union $\cup_i |a_i, b_i|$, we define $P(\cup_i |a_i, b_i|) = \sum_i P_F(|a_i, b_i|)$. By checking cases, we see that if $A = \cup_i |a_i, b_i| = \cup_j |c_j, d_j|$ are two ways of expressing a disjoint collection of intervals, then $\sum_i P_F(|a_i, b_i|) = \sum_j P_F(|c_j, d_j|)$.

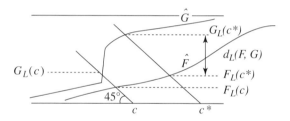

FIGURE 6.2.18 c^* solves $\max_c[F_L(c) - G_L(c)]$, giving $d_L(F, G) = [F_L(c^*) - G_L(c^*)]$.

Example 6.2.18 *Given a cdf F with graph $gr(F)$, let $\tilde{F}$ be the correspondence with $gr(\tilde{F}) = \mathrm{cl}(gr(F))$ and let $\hat{F}$ be the correspondence defined by $\hat{F}(x) = \mathrm{co}(\tilde{F}(x))$. When F is continuous, $\hat{F} = F$. If F is not continuous at x°, then $\hat{F}(x^\circ)$ is the line segment joining $(x^\circ, F^-(x^\circ))$ and $(x^\circ, F(x^\circ))$.*

To each $\hat{F}$, we now associate a function $F_L \in C_b(\mathbb{R})$. For each $c \in \mathbb{R}$, the line $\{(x, y) : x + y = c\}$ intersects the graph of $\hat{F}$ exactly once, at a point (x_c, y_c). Define $F_L(c) = y_c$. The function F_L is nondecreasing, $\lim_{c \downarrow -\infty} F_L(c) = 0$, $\lim_{c \uparrow \infty} F_L(c) = 1$ and has Lipschitz constant 1, and every $g \in C_b(\mathbb{R})$ with these properties arises from some cdf (Figure 6.2.18).

Example 6.2.19 *[↑Example 6.2.16] The cdf's for X taking the values ± 1 with probability $\frac{1}{2}$ each and Y uniformly distributed on $[0, 1]$ were given earlier as F and G. The associated F_L and G_L are given by*

$$F_L(c) = \begin{cases} 0 & \text{if } c < -1 \\ 1 + c & \text{if } -1 \leq c < -0.5 \\ 0.5 & \text{if } -0.5 \leq c \leq 1.5 \\ c - 1 & \text{if } 1.5 < c \leq 2 \\ 1 & \text{if } 2 < c \end{cases} \quad \text{and} \quad G_L(c) = \begin{cases} 0 & \text{if } c < 0 \\ c/2 & \text{if } 0 \leq c < 2 \\ 1 & \text{if } 2 \leq c \end{cases}.$$

Definition 6.2.20 *For cdf's F and G, we define the **Levy distance** by $d_L(F, G) = d_\infty(F_L, G_L)$. If $d_L(F_n, F) \to 0$ we say that F_n **converges weakly** or **converges weak*** to F.[3]*

Every F_L is bounded, so d_L is a metric. For every pair, F_L and G_L, the maximum of $|F_L(c) - G_L(c)|$ is achieved since both go to 0 and to 1 at $-\infty$ and ∞.

There are two equivalent formulations of d_L:

- $\rho(F, G) = \inf\{h > 0 : (\forall x)[F(x - h) - h \leq G(x) \leq F(x + h) + h]\}$, and
- $\rho(F, G) = \inf\{\epsilon > 0 : (\forall x) \left[[P_F((-\infty, x]) \leq P_G((-\infty, x]^\epsilon)] \wedge [P_G((-\infty, x]) \leq P_F((-\infty, x]^\epsilon)]\right]$.

---

3. The reason for the names "weak" and "weak*" may seem opaque right now, and we are not explicit about them until §9.3.a. One way to remember why the convergence is not strong is that $\sup_x |F_n(x) - F(x)| \not\to 0$. Rather, the supremum of the distance between the cdf's looked at from a slanted perspective goes to 0.

Lemma 6.2.21 *For all F, G, $\rho(F, G) = d_L(F, G)$. [Start at any c^* solving $\max_{c \in \mathbb{R}} |F_L(c) - G_L(c)|$.]*

In particular, $d_L(F_n, F) \to 0$ if and only if $\forall h > 0$, $P_{F_n}((-\infty, x - h]) - h \leq P_F((-\infty, x]) \leq P_{F_n}((-\infty, x + h]) + h$ for all large n.

Exercise 6.2.22 Prove Lemma 6.2.21.

We return to the study of the set of cdf's with the Levy metric once we have gathered the requisite facts about completeness and compactness in $C_b(M)$.

6.2.a.5 Continuous Time Stochastic Processes

As our last example, we introduce probabilities on the space $C([0, 1])$ as a way of talking about **stochastic processes**.[4] Suppose that a random $f \in C([0, 1])$ is chosen according to some probability distribution. Then, as time, t, moves continuously from 0 to 1, we observe $f(t)$. What is traced out is random, hence the term "stochastic," and happens over time, hence the term "process."

We are very explicit later about what we mean by the terms "random variables" and "independent." For now, since this is by way of an example, we presume only a rudimentary acquaintance with these ideas. Suppose that we have a sequence of random variables, $X_0, X_1, X_2, \ldots$, that are mutually independent and that satisfy $Prob(X_0 = 0) = 1$, $Prob(X_n = 1) = Prob(X_n = -1) = \frac{1}{2}$ for $n \geq 1$. We define a sequence of probability distributions on $C([0, 1])$. The major device we use is **linear interpolation**.

In drawing the graph, linear interpolation connects the points, $(t_{k-1}, f(t_{k-1}))$ and $(t_k, f(t_k))$ with a line segment. The formal starting point is a set of time points, $0 = t_0 < t_1 < \cdots < t_n = 1$, and a set of values, $f(0), f(t_1), \ldots, f(t_{n-1}), f(1)$. Any $t \in (t_{k-1}, t_k)$ is a convex combination of t_{k-1} and t_k, specifically $t = t_k \gamma t_{k-1}$, where $\gamma = \frac{t - t_{k-1}}{t_k - t_{k-1}}$. For any such t, linear interpolation defines $f(t) = f(t_k) \gamma f(t_{k-1})$.

Example 6.2.23 *To define the random f_0, start with the time points $t_0 = 0$, $t_1 = 1$ and the values $f_0(t_0) = X_0$ and $f_0(t_1) = X_1$. Define f_0 to be the associated linear interpolation. This gives a random function in $C([0, 1])$, which starts at 0 and with probability $\frac{1}{2}$ goes upward with slope 1 and downward with slope 1.*

To define the random f_1, start with the time points $t_0 = 0$, $t_1 = 1/2^1$ and $t_2 = 2/2^1$ and values $f_1(t_0) = \frac{1}{\sqrt{2^1}} X_0$, $f_1(t_1) = \frac{1}{\sqrt{2^1}} X_1$, and $f_1(t_2) = \frac{1}{\sqrt{2^1}}(X_1 + X_2)$. Define f_1 to be the associated linear interpolation. This gives one of four possible random functions in $C([0, 1])$.

*To define the random f_n, start with the time points $t_k = k/2^n$, $k = 0, \ldots, 2^n$ and values $f_n(t_k) = \frac{1}{\sqrt{2^n}} \sum_{j \leq k} X_j$. This gives one of 2^{n+1} possible random functions in $C([0, 1])$. Let μ_n be the probability distribution for f_n just described. **Brownian motion** is the stochastic process described by $\mu = \lim_n \mu_n$.*

The Levy distance gives a metric on the probability distributions on $\mathbb{R}$. Later we give the Prokhorov metric on the set of probability distributions on a metric space M. It is equivalent to the Levy distance on $M = \mathbb{R}$ and, applied to probabilities on the metric space $M = C([0, 1])$, gives the metric in which the μ_n converge to μ.

---

4. And this is the way we regard them since Prokhorov's work in the mid-1950s.

Exercise 6.2.24 [↑Example 6.2.23] This problem assumes that you have seen the undergraduate prob/stats version of the central limit theorem. An increment of a function f over an interval $[r, s]$ is the number $f(s) - f(r)$, where $r < s$. The idea that you are after in this problem is that Brownian motion involves the independent normally distributed increments. Throughout this problem, N is a fixed large integer, say, larger than 7.

1. For $r = i/2^N$ and $s = j/2^N$, $i < j$, show that the distribution of the random variables $f_m(s) - f_m(r)$ converges to $N(0, s - r)$, that is, a normal (or Gaussian) random variable with variance $s - r$ as $m \uparrow \infty$.

2. For $r = i/2^N$, $s = j/2^N$, $r' = i'/2^N$, $s' = j'/2^N$, $i < j \leq i' < j'$, show that for all $m \geq N$, the two random variables $f_m(s) - f_m(r)$ and $f_m(s') - f_m(r')$ are functions of independent sets of random variables (which means that they are themselves independent).

3. The previous two problems consider increments between numbers of the form $k/2^N$, that is, at dyadic rationals. We know that the set of all numbers of the form $k/2^N$, $N \in \mathbb{N}$, $0 \leq k \leq 2^N$, is dense in $[0, 1]$. The following is a step toward showing that knowing the behavior of increments between dyadic rationals tells you about the behavior of all increments.

   Show that for all $s \in (0, 1]$, $\lim_{r \uparrow s} \lim_m \text{Var}(f_m(r) - f_m(s)) = 0$, where $\lim_{r \uparrow s} \lim_m \text{Var}(f_m(r) - f_m(s)) = 0$, $\text{Var}(X)$ is the variance of a random variable.

A summary of the previous problem is that for any collection of times, $0 = t_0 < t_1 < \cdots < t_n \leq 1$, the random variables $((f_n(t_1) - f_n(t_0), (f_n(t_2) - f_n(t_1), \ldots, (f_n(t_n) - f_n(t_{n-1}))$ converge to a $N(0, \Sigma)$, where Σ is the $n \times n$ matrix with 0's off the diagonal and with (k, k)th entry $(t_k - t_{k-1})$. This gives a stochastic process over the time interval $[0, 1]$. We can extend it to the same kind of process over the time interval $[0, 2]$ in several ways. The most direct is to independently pick f_1 and f_2 in $C([0, 1])$ according to μ and to define $g(t) = f_1(t)$ if $t \in [0, 1]$ and $g(t) = f_1(1) + f_2(t - 1)$ if $t \in [1, 2]$. One can continue gluing functions together in this fashion for as long as one pleases, leading to distributions on $C(\mathbb{R}_+)$.

6.2.b Pointwise and Uniform Convergence

We say that a sequence f_n in $C(M)$ **converges uniformly** to f if $d_\infty(f_n, f) \to 0$. This is because the distance between $f_n(x)$ and $f(x)$ becomes uniformly small as $n \to \infty$. Sometimes we do not have uniformity.

Example 6.2.25 *In $C([0, 1])$, the continuous functions t^n converge pointwise, but not uniformly, to the discontinuous function $1_{\{0\}}(t)$. In $C(\mathbb{R})$, the functions t/n converge pointwise, but not uniformly, to the function 0.*

Definition 6.2.26 *A sequence f_n **converges pointwise** to f if for all $x \in M$, $\text{proj}_x(f_n) \to \text{proj}_x(f)$, that is, if for all $x \in M$, $f_n(x) \to f(x)$.*

Since $|f_n(x) - f(x)| \leq d_\infty(f_n, f)$, if f_n converges uniformly to f, it converges pointwise. That is, uniform convergence is stronger than pointwise convergence.

Sometimes we have convergence of a kind of intermediate between uniform and pointwise.

Example 6.2.27 *Consider the sequence of functions $f_n(x) := \frac{e^{x/n}}{1+e^{x/n}}$ in $C_b(\mathbb{R})$. For any r, $f_n(r) \to \frac{1}{2}$, but $\sup_{x \in \mathbb{R}} |f_n(x) - \frac{1}{2}| = \frac{1}{2}$ for all n. Thus, the f_n converge pointwise, but not uniformly to the function identically equal to $\frac{1}{2}$. We can say more. For any compact $K \subset \mathbb{R}$, f_n converges uniformly over the compact set K; that is, for all compact $K \subset \mathbb{R}$, $\max_{x \in K} |f_n(x) - \frac{1}{2}| \to 0$.*

Definition 6.2.28 *A sequence f_n in $C(M)$ converges uniformly on compact sets or uniformly on compacta to f if for all compact $K \subset M$, $\max_{x \in K} |f_n(x) - f(x)| \to 0$.*

Uniform convergence clearly implies uniform convergence on compact sets. Considering the special compact sets $K = \{x\}$ shows that uniform convergence on compact sets implies pointwise convergence.

Exercise 6.2.29 For any closed $F \subset \mathbb{R}^\ell$, we define the following metric on $C(F)$:

$$\rho(f, g) = \sum_{n \in \mathbb{N}} \frac{1}{2^n} \min\{1, \max_{x \in F, \; \|x\| \le n} |f(x) - g(x)|\}. \qquad (6.3)$$

Show that $\rho(f_n, f) \to 0$ iff f_n converges to f uniformly on compact sets.

Exercise 6.2.30 For a utility function $u \in C(\mathbb{R}_+^\ell)$, $\mathbf{p} \in \mathbb{R}_{++}^\ell$, and $w > 0$, let $x(\mathbf{p}, w : u)$ denote the Walrasian demand set for the utility function u and $v(\mathbf{p}, w : u)$ the corresponding indirect utility function. Show that $x(\cdot, \cdot : \cdot)$ is upper hemi-continuous and that if $\rho(u_n, u) \to 0$, then the sequence of functions $v(\cdot, \cdot : u_n)$ converges pointwise to the function $v(\cdot, \cdot : u)$.

Example 6.2.25 gave a sequence of continuous functions converging pointwise but not uniformly to a discontinuous function. We give an example of a sequence of continuous functions converging pointwise but not uniformly on compact sets to a continuous function. Thus, pointwise convergence is strictly weaker than uniform convergence on compact sets, even when the limit functions are continuous.

Example 6.2.31 *Define $f_n \in C_b(\mathbb{R}_+)$ by $f_n(x) = [1 - n|x - \frac{1}{n}|] \vee 0$, so that $f_n(0) = f_n([\frac{2}{n}, \infty)) = 0$ and $f_n(\frac{1}{n}) = 1$. The sequence f_n converges pointwise to the function $f(x) \equiv 0$, but not uniformly on any compact set of the form $[0, b]$, $b > 0$.*

6.2.c Completeness

We now show that for any metric space M, $(C_b(M), \|\cdot\|_\infty)$ is a complete normed vector space. We use this result in several places. A short list includes:

- §6.2.d, where we prove that stationary deterministic dynamic programming problems have continuous value functions.
- §6.2.e, where we give conditions on stationary deterministic dynamic programming problems that imply the concavity of the value function.

- §6.3, where we identify the set of cdf's with a closed, hence complete, subset of $C_b(\mathbb{R})$.

- §6.7, where we prove the implicit function theorem by applying Banach's contraction mapping theorem to spaces of continuous bounded functions.

Theorem 6.2.32 $(C_b(M), \|\cdot\|_\infty)$ *is a complete normed vector space. In particular,* $(C_b(M), d_\infty)$ *is a complete metric space.*

Proof. In view of Theorem 6.2.2, all that is left to show is that $(C_b(M), d_\infty)$ is complete. Suppose that f_n is a Cauchy sequence in $(C_b(M), d_\infty)$. For all $x \in M$, $f_n(x)$ is Cauchy because $|f_n(x) - f_m(x)| \leq d_\infty(f_n, f_m)$. Let $f(\cdot)$ be the pointwise limit of the sequence f_n; that is, define $f : M \to \mathbb{R}$ by $f(x) = \lim_n f_n(x)$. We must show that f is continuous and that $d_\infty(f_n, f) \to 0$.

Continuity: Pick an arbitrary $x \in M$ and arbitrary $\epsilon > 0$. To show that f is continuous at x, we show that there exists $\delta > 0$ such that $d(x, x') < \delta$ implies $|f(x) - f(x')| < \epsilon$.

Pick N such that $\forall n, m \geq N$, $d_\infty(f_n, f_m) < \epsilon/3$. This implies that $|f(x) - f_N(x)| \leq \epsilon/3$. Choose a $\delta > 0$ such that $d(x, x') < \delta$ implies $|f_N(x) - f_N(x')| < \epsilon/3$. Combining, we get $\forall x'$ such that $d(x, x') < \delta$,

$$|f(x) - f(x')| = |f(x) - f_N(x) + f_N(x) - f_N(x') + f_N(x') - f(x')|$$

$$\leq |f(x) - f_N(x)| + |f_N(x) - f_N(x')| + |f_N(x') - f(x')| < \epsilon.$$

Uniform convergence: To see that $d_\infty(f_n, f) \to 0$, pick $\epsilon > 0$ and N such that for all $n, m \geq N$, $d_\infty(f_n, f_m) < \epsilon$. Then for all $x \in M$ and all $n \geq N$, $|f_n(x) - f(x)| \leq \epsilon$. ∎

6.2.d Continuous Value Functions in Dynamic Programming

We are going to revisit the growth model in §3.7.a with slightly different notation. The problem has the value function

$$v(x_0) = \sup_{x_1, x_2, x_3, \dots} \sum_{t=0}^{\infty} \beta^t r(x_t, x_{t+1}) \quad \text{subject to } \forall t, \, x_{t+1} \in \Gamma(x_t) = [0, f(x_t)].$$

(6.4)

The idea is that we start with x_0 units of capital and we can consume, at period $t = 0$, any amount $c_0 \in [0, f(x_0)]$, receiving a utility of $u(c_0)$, while the rest of the good, $x_1 = f(x_0) - c_0$, can be invested in order to produce tomorrow, when we face the same problem. To make this fit, note that we can take $r(x_t, x_{t+1}) = u(f(x_t) - x_{t+1})$.

This is an infinite-horizon problem, and it is stationary in the sense that the function r and the correspondence Γ do not depend on time. We approximate the problem in (6.4) by the following sequence of finite-horizon problems, indexed by $\tau = 0, 1, \dots$, where τ counts the number of periods into the future that matter. The problems are

$$v_\tau(x_0) = \max_{x_1, x_2, x_3, \ldots} \sum_{t=0}^{\tau} \beta^t r(x_t, x_{t+1}) \quad \text{subject to } \forall t, \, x_{t+1} \in \Gamma(x_t) = [0, \, f(x_t)].$$

(6.5)

If $\tau = 0$, the universe is about to end, so we should use all the capital right now, $c_0^*(x) = f(x)$, which leads to the value function $v_0(x) = u(f(x))$.

Now suppose that $\tau = 1$. We can reformulate the problem when we have one period left as

$$v_1(x_0) = \max_{y \in \Gamma(x_0)} [r(x_0, y) + \beta v_0(y)],$$

(6.6)

because we know that when we have zero periods left and y capital available, we will use all the proceeds of the capital and have utility $v_0(y)$. $x_1^*(x_0)$ denotes the solution, so that we use $c_0^* = f(x_0) - x_1^*(x_0)$ in this period and all the proceeds of the capital in the second period.

Now, $v_1(x)$ is the best we can do if there is one period after today and we have x capital right now. We now solve the problem for $\tau = 2$, assuming that, in one period, we do the best we can for ourselves. This gives

$$v_2(x_0) = \max_{y \in \Gamma(x_0)} [r(x_0, y) + \beta v_1(y)].$$

(6.7)

You should be able to give a complete argument for this: if there were a better policy for the $\tau = 2$ problem, then. . . . Induction has been started, and

$$v_{\tau+1}(x_0) = \max_{y \in \Gamma(x_0)} [r(x_0, y) + \beta v_\tau(y)].$$

(6.8)

This process takes a value function, v_τ, and delivers a new value function $v_{\tau+1}$. That is, it is a mapping from value function to value function. From the theorem of the maximum, since r and Γ are continuous, v_0 is continuous. Since r, Γ, and v_0 are continuous, the theorem of the maximum implies that v_1 is continuous. Continuing, we get that all of the v_τ are continuous. If we assume that there is a maximal capital stock, $\overline{x}$, then the relevant range of x_0 is $[0, \overline{x}]$, so each v_τ belongs to the complete space $C([0, \overline{x}])$.

By Theorem 4.4.10 (p. 122), (i.e., by Banach's contraction mapping theorem), if we know that the mapping $T : C([0, \overline{x}]) \to C([0, \overline{x}])$ defined by

$$T_v(x_0) = \max_{y \in \Gamma(x_0)} [r(x_0, y) + \beta v(y)]$$

(6.9)

is a contraction mapping, then v_τ converges to the unique fixed point of the mapping T.

Lemma 4.11.17 (p. 162) implies that for all $v, w \in C([0, \overline{x}])$ and all $x_0 \in [0, \overline{x}]$,

$$T_v(x_0) - T_w(x_0) \leq \max_{y \in \Gamma(x_0)} \left([r(x_0, y) + \beta v(y)] - [r(x_0, y) + \beta w(y)] \right)$$

$$= \max_{y \in \Gamma(x_0)} \beta[v(y) - w(y)].$$

(6.10)

Thus, $T_v(x_0) - T_w(x_0) \le \beta \|v - w\|_\infty$. Reversing v and w yields $T_w(x_0) - T_v(x_0)$ $\le \beta \|v - w\|_\infty$. When we combine, for all x_0, $|T_v(x_0) - T_w(x_0)| \le \beta \|v - w\|_\infty$, so that $v \mapsto T_v$ is a contraction mapping.

Finally, note that following the policy $x_{t+1}^*(x_t) = \arg\max_{y \in \Gamma(x_t)} [r(x_t, y) + \beta v(y)]$ must give an optimum for the problem in (6.4). If there is a better policy, say, $(x_0, x_1', x_2', \ldots)$ it gives value $v' = v(x_0) + \epsilon$ for some $\epsilon > 0$. As $T \uparrow \infty$, $(x_0, x_1', x_2', \ldots, x_T)$ must give utility converging to v', which contradicts the observation that the optimal $v_T(x_0)$ converge to $v(x_0)$.

Blackwell gave us an alternative way to show that the mapping $v \mapsto T_v$ is a contraction, one that works even in stochastic dynamic programming contexts.

Lemma 6.2.33 (Blackwell) *Suppose that $(B(X), \|\cdot\|_\infty)$ is the set of bounded functions on a nonempty set X with the sup norm and the usual lattice order. If $T : B(X) \to B(X)$ is monotonic, $[f \ge g] \Rightarrow [T(f) \ge T(g)]$, and for some $\beta \in (0, 1)$, $T(f + c) \le T(f) + \beta c$ for all $c \ge 0$, then T is a contraction mapping.*

Since $(B(X), \|\cdot\|_\infty)$ is a complete metric space, Banach's contraction mapping can be used here.

Exercise 6.2.34 Prove Blackwell's lemma. [Consider $f \le \|f - g\|_\infty$ and $g \le \|f - g\|_\infty$ with $c = \|f - g\|_\infty$.]

6.2.e Concave Value Functions in Dynamic Programming

We begin with the following simple observation.

Theorem 6.2.35 *For any convex $S \subset \mathbb{R}$, if $f_n \in C(S)$ is a sequence of concave functions and f_n converges pointwise to f, then f is concave.*

Proof. Pick $\mathbf{x}, \mathbf{y} \in S$ and $\alpha \in [0, 1]$. We must show that $f(\alpha\mathbf{x} + (1 - \alpha)\mathbf{y}) \ge \alpha f(\mathbf{x}) + (1 - \alpha)f(\mathbf{y})$. We know that for all $n \in \mathbb{N}$, $f_n(\alpha\mathbf{x} + (1 - \alpha)\mathbf{y}) \ge \alpha f_n(\mathbf{x}) + (1 - \alpha)f_n(\mathbf{y})$. By pointwise convergence, $f_n(\alpha\mathbf{x} + (1 - \alpha)\mathbf{y}) \to f(\alpha\mathbf{x} + (1 - \alpha)\mathbf{y})$, $f_n(\mathbf{x}) \to f(\mathbf{x})$ and $f_n(\mathbf{y}) \to f(\mathbf{y})$. ∎

In particular, this means that the set of concave functions in $C([0, \overline{x}])$ is closed, hence complete. The relevance for dynamic programming can be seen in the following (long) example and exercises.

Example 6.2.36 [↑§3.7.a] *A planner chooses how much people should consume and how much to invest in a productive technology. Preferences are given by $u(c_t) = \log(c_t)$ and what is not consumed out of production today, say k_{t+1}, becomes productive tomorrow, $y_{t+1} = F(k_{t+1}) = k_{t+1}^\alpha$, where $\alpha \le 1$. Here we assume that capital depreciates fully in the period or $\delta = 1$. If people discount the future at rate $\beta < 1$ and there is a general time horizon represented by T (which could be ∞), the sequence version of the programming problem can be written as*

$$v_T(k_0) = \max_{\{c_t, k_{t+1}\}_{t=0}^T} \sum_{t=0}^T \beta^t u(c_t)$$

$$s.t. c_t + k_{t+1} = f(k_t), \forall t$$

$$c_t \ge 0, k_{t+1} \ge 0, k_0 \text{ given,}$$

where $v_\tau(x)$ denotes the maximized present discounted value of the problem with τ future periods remaining starting in state x.

One way to solve this problem is to write down the system of first-order conditions for each $t \in \{0, \ldots, T\}$. That was the approach in §3.7.a. An alternative approach is to use dynamic programming. The basic solution concept for dynamic programming is backward induction. Beginning at the end (so that there are $\tau = 0$ periods remaining), consider what happens if a person enters the last period with k_T units of capital and acts optimally. In that case he solves:

$$v_0(k_T) = \max_{c_T \geq 0, k_{T+1} \geq 0} \log(c_T)$$

$$s.t. c_T + k_{T+1} \leq k_T^\alpha.$$

The solution is simple, $k_{T+1} = 0$ and $c_T = k_T^\alpha$. In that case,

$$v_0(k_T) = \phi(0) \log(k_T^\alpha) + \gamma(0),$$

where $\phi(0) = 1$ and $\gamma(0) = 0$.[5]

Now consider the problem if a person enters the next to last period $t = T - 1$ with k_{T-1} units of capital and acts optimally. The problem can be stated as

$$v_1(k_{T-1}) = \max_{c_{T-1} \geq 0, k_T \geq 0} \log(c_{T-1}) + \beta v_0(k_T)$$

$$s.t. c_{T-1} + k_T \leq k_{T-1}^\alpha.$$

Since utility is strictly increasing in consumption, the constraint holds with equality and we can substitute out consumption. Further, we substitute $v_0(k_T) = \alpha \log(k_T)$ into the objective to yield the problem

$$v_1(k_{T-1}) = \max_{k_T \geq 0} \log(k_{T-1}^\alpha - k_T) + \beta \alpha \log(k_T) \tag{6.11}$$

Figure 6.2.36 provides some intuition as to why the right-hand side of (6.11) may in fact attain a maximum.

The first-order condition for (6.11) is given by

$$\frac{1}{k_{T-1}^\alpha - k_T} = \frac{\alpha\beta}{k_T}$$

$$\Longleftrightarrow k_T = \frac{\alpha\beta k_{T-1}^\alpha}{(1 + \alpha\beta)}. \tag{6.12}$$

---

5. The argument in the functions ϕ and γ just denotes the number of periods remaining, that is, τ, which in this case is zero.

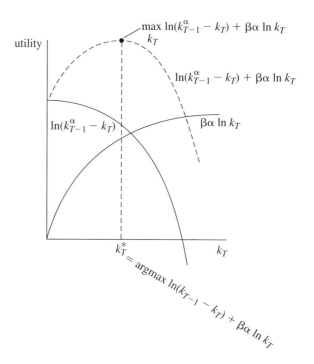

FIGURE 6.2.36

Substituting the decision rule (6.12) back into the objective (6.11) yields

$$v_1(k_{T-1}) = \log\left(k_{T-1}^\alpha - \frac{\alpha\beta k_{T-1}^\alpha}{(1+\alpha\beta)}\right) + \alpha\beta \log\left(\frac{\alpha\beta k_{T-1}^\alpha}{1+\alpha\beta}\right)$$

$$\Longleftrightarrow v_1(k_{T-1}) = (1+\alpha\beta)\log\left(k_{T-1}^\alpha\right)$$

$$- (1+\alpha\beta)\log(1+\alpha\beta) + \alpha\beta\log(\alpha\beta)$$

$$\Longleftrightarrow v_1(k_{T-1}) = \phi(1)\alpha\log\left(k_{T-1}\right) + \gamma(1), \qquad (6.13)$$

where the coefficients are obviously defined as $\phi(1) = (1+\alpha\beta)$ and $\gamma(1) = \alpha\beta\log(\alpha\beta) - (1+\alpha\beta)\log(1+\alpha\beta)$.

By induction, let there be τ future periods remaining :

$$v_\tau(k_{T-\tau}) = \max_{k_{T-\tau+1}\geq 0} \log(k_{T-\tau}^\alpha - k_{T-\tau+1}) + \beta v_{\tau-1}(k_{T-\tau+1}),$$

which upon substituting the $\tau - 1$ period analogue of (6.13) given by $v_{\tau-1}(k_{T-\tau+1})$ $=\phi(\tau-1)\alpha\log\left(k_{T-\tau+1}\right) + \gamma(\tau-1)$ into the above yields

$$v_\tau(k_{T-\tau}) = \max_{k_{T-\tau+1}\geq 0} \log(k_{T-\tau}^\alpha - k_{T-\tau+1})$$

$$+ \beta\left[\phi(\tau-1)\alpha\log\left(k_{T-\tau+1}\right) + \gamma(\tau-1)\right] \qquad (6.14)$$

The first-order condition is

$$\frac{1}{k_{T-\tau}^{\alpha} - k_{T-\tau+1}} = \frac{\alpha\beta\phi(\tau - 1)}{k_{T-\tau+1}}$$

$$\Longleftrightarrow k_{T-\tau+1} = \frac{\alpha\beta\phi(\tau - 1)k_{T-\tau}^{\alpha}}{(1 + \alpha\beta\phi(\tau - 1))} \qquad (6.15)$$

Substituting the decision rule (6.15) back into the objective (6.14) yields

$$v_{\tau}(k_{T-\tau}) = \log\left(k_{T-\tau}^{\alpha} - \frac{\alpha\beta\phi(\tau - 1)k_{T-\tau}^{\alpha}}{(1 + \alpha\beta\phi(\tau - 1))}\right)$$

$$+ \alpha\beta\phi_{T-\tau+1}\log\left(\frac{\alpha\beta\phi(\tau - 1)k_{T-\tau}^{\alpha}}{(1 + \alpha\beta\phi(\tau - 1))}\right) + \beta\gamma(\tau - 1)$$

$$\Longleftrightarrow v_{\tau}(k_{T-\tau}) = \phi(\tau)\alpha\log\left(k_{T-\tau}\right) + \gamma(\tau), \qquad (6.16)$$

where $\phi(\tau) = (1 + \alpha\beta\phi(\tau - 1))$ *and* $\gamma(\tau) = \alpha\beta\phi(\tau - 1)\log(\alpha\beta\phi(\tau - 1)) + \beta\gamma(\tau - 1) - (1 + \alpha\beta\phi(\tau - 1))\log(1 + \alpha\beta\phi(\tau - 1))$.
Note that the term that matters for decisions is $\phi(\tau) = (1 + \alpha\beta(1 + \alpha\beta (1 + \alpha\beta (... ((1 + \alpha\beta))))) = 1 + \alpha\beta + (\alpha\beta)^2 + ... + (\alpha\beta)^{\tau}$. *But this expression can be written as*

$$\phi(\tau) = \frac{1 - (\alpha\beta)^{\tau+1}}{1 - \alpha\beta}.$$

Since $\alpha\beta < 1$, *note further that as* $\tau \to \infty$, $\phi(\infty)$ *is independent of time and given simply by* $1/(1 - \alpha\beta)$. *Furthermore, it can be shown that* $\gamma(\infty)$ *is a constant* γ *independent of time. In that case, the value function analogue of (6.16) for the infinite-horizon case for any given capital stock k is given by*

$$v(k) = \frac{\alpha\log(k)}{1 - \alpha\beta} + \gamma.$$

Note that the value function, like the preferences and production technology, is continuous, strictly increasing, strictly concave, and differentiable. Furthermore, the decision rule (or policy function) analogue of (6.15) for the infinite-horizon case for any given capital stock k is given by

$$k' = \alpha\beta k^{\alpha},$$

which again is an increasing continuous function.
    An alternative (simpler) way to construct a solution for the value function in the infinite-horizon case can be found using the method of undetermined coefficients. In particular, conjecture $v(k) = A + B\log k$. *Then for the infinite-horizon case,*

$$v(k) = \max_{k' \geq 0} \log(k^{\alpha} - k') + \beta v(k'),$$

or under the conjecture

$$v(k) = \max_{k' \geq 0} \log(k^\alpha - k') + \beta \left[A + B \log k' \right]. \tag{6.17}$$

The first-order condition is

$$\frac{1}{k^\alpha - k'} = \frac{\beta B}{k'}$$

$$\Longleftrightarrow k' = \frac{\beta B k^\alpha}{(1 + \beta B)}.$$

Substituting k' back into (6.17)

$$v(k) = \log \left(k^\alpha - \frac{\beta B k^\alpha}{(1 + \beta B)} \right) + \beta \left[A + B \log \frac{\beta B k^\alpha}{(1 + \beta B)} \right]$$

$$\Leftrightarrow v(k) = (1 + \beta B)\alpha \log(k) - (1 + \beta B) \log(1 + \beta B)$$

$$+ \beta A + \beta B \log(\beta B) - \beta B \log(1 + \beta B).$$

Grouping terms this implies

$$A = -(1 + 2\beta B) \log(1 + \beta B) + \beta A + \beta B \log(\beta B)$$

$$B = (1 + \beta B)\alpha$$

or

$$B = \frac{\alpha}{(1 - \alpha\beta)} > 0$$

$$A(1 - \beta) = \frac{\beta\alpha}{(1 - \alpha\beta)} \log \left(\frac{\alpha\beta}{(1 - \alpha\beta)} \right) - \left(1 + \frac{2\beta\alpha}{(1 - \alpha\beta)} \right) \log \left(1 + \frac{\alpha\beta}{(1 - \alpha\beta)} \right)$$

$$\Longleftrightarrow A = \frac{\beta\alpha \log(\alpha\beta) + \log(1 - \alpha\beta)}{(1 - \alpha\beta)(1 - \beta)}.$$

Given this, the conjectured value function satisfies

$$v(k) = A + \frac{\alpha}{(1 - \alpha\beta)} \log k$$

(so that we can see $\gamma(\infty) = A$ independent of time, etc.) and the decision rule is given by

$$k' = \frac{\beta \left(\frac{\alpha}{(1 - \alpha\beta)} \right) k^\alpha}{(1 + \beta \left(\frac{\alpha}{(1 - \alpha\beta)} \right))}$$

$$\Longleftrightarrow k' = \alpha\beta k^\alpha.$$

But these expressions are identical to the inductive construction above.

Exercise 6.2.37 Suppose that (M, d) is a complete metric space, that $T : M \to M$ is a contraction mapping, that S is a closed subset of M, and that $T(S) \subset S$. Show that the unique fixed point of T belongs to S.

Exercise 6.2.38 In the growth problem, show that if $u(\cdot)$ is concave and the graph of Γ is convex, then for any concave v, $T(v)$ is concave. [Taking S to be the set of concave functions in the previous exercise shows that the value function is concave.]

6.2.f Dimensionality, Finite and Infinite

Linear dependence and independence in $\mathbb{R}^\ell$ provide the starting point for the study of dimensionality in a vector space $\mathfrak{X}$.

Definition 6.2.39 *Fix a vector space $\mathfrak{X}$.*

1. *A finite set $F = \{f_1, \ldots, f_n\} \subset \mathfrak{X}$ is **linearly dependent** if there exist $\alpha_1, \ldots, \alpha_n$, not all equal to 0, such that $\sum \alpha_k f_k = 0$. Otherwise it is **linearly independent**.*

2. *An infinite set F is **linearly dependent** if there exists a linearly dependent finite subset. Otherwise it is **linearly independent**.*

3. *If there exists a finite subset of $\mathfrak{X}$ set containing ℓ linearly independent elements but no set containing $\ell + 1$ linearly independent elements, then $\mathfrak{X}$ is ℓ-**dimensional**, in particular **finite dimensional**.*

4. *If for every $n \in \mathbb{N}$ there is a set containing n linearly independent elements, then $\mathfrak{X}$ is **infinite dimensional**.*

Spanning sets are crucial to understanding the structure of vector spaces. Note that the sums in the following are finite.

Definition 6.2.40 *The **span of a set** $E \subset \mathfrak{X}$ is the set of finite sums*

$$\mathbf{span}(E) := \left\{ \sum_{n \leq N} \beta_n e_n : N \in \mathbb{N}, \ \beta_n \in \mathbb{R}, \ e_n \in E \right\}.$$

No finite set can span an infinite-dimensional $\mathfrak{X}$ because it contains an infinite linearly independent set. The notion of a basis in $\mathbb{R}^k$ has two possible generalizations for the infinite-dimensional $\mathfrak{X}$'s that we care about. The first turns out to be useful for many theoretical concerns, whereas the second is more broadly useful.

Definition 6.2.41 *A linearly independent $H \subset \mathfrak{X}$ is a **Hamel basis** if $\mathbf{span}(H) = \mathfrak{X}$. When $\mathfrak{X}$ is a complete separable normed vector space, a countable $S = \{s_n : n \in \mathbb{N}\} \subset \mathfrak{X}$ is a **Schauder basis** if every $v \in \mathfrak{X}$ has a unique sequence r_n in $\mathbb{R}$ such that $\|v - \sum_{i \leq n} r_i s_i\| \to 0$.*

Given a Hamel basis containing ℓ points, every point in an ℓ-dimensional vector space is completely specified by ℓ real numbers. For all practical purposes, this means that any ℓ-dimensional vector space is $\mathbb{R}^\ell$.

Lemma 6.2.42 $C([0, 1])$ *is infinite dimensional.*

Proof. Let $t_n = 1 - \frac{1}{2^n} \uparrow 1$ and define $s_n = (t_{n+1} + t_n)/2$ (the midpoint of the interval $[t_n, t_{n+1}]$). For each n, define $f_n(x) = (1 - 4|x - s_n|/(t_{n+1} - t_n)) \vee 0$. For $n \neq m$, $f_n \cdot f_m = 0$, so that any finite subset of $\{f_n : n \in \mathbb{N}\}$ is linearly independent. ∎

Exercise 6.2.43 The set $U = \{f \in C([0, 1]) : \|f\| \leq 1\}$ is closed and bounded, but not compact.

Exercise 6.2.44 If (M, d) is a metric space and M is infinite, then $C_b(M)$ is infinite dimensional. Further, the set $U = \{f \in C_b(M) : \|f\| \leq 1\}$ is closed and bounded, but not compact. [The proof of Lemma 6.2.42 uses a sequence of points x_n and radii ϵ_n such that the $B_{\epsilon_n}(x_n)$ are disjoint.]

Example 6.2.45 [↑*Example 6.2.10*] *Any ℓ^p space, $p \in [1, \infty]$, is infinite dimensional. To see why, let e_n be the element of ℓ^p having $\text{proj}_n(e_n) = 1$ and $\text{proj}_m(e_n) = 0$ for $m \neq n$. The collection $\{e_n : n \in \mathbb{N}\}$ is linearly independent.*

Definition 6.2.46 $W \subset \mathfrak{X}$ *is a **linear subspace of** $\mathfrak{X}$ if W is a vector space.*

It turns out that compact subsets of infinite-dimensional vector spaces are "nearly" finite dimensional.

Definition 6.2.47 *A subset F of a metric vector space $(\mathfrak{X}, d)$ is **approximately flat** if for all $\epsilon > 0$ there exists a finite-dimensional linear subspace, $W \subset \mathfrak{X}$, such that $F \subset W^\epsilon$.*

Lemma 6.2.48 *Any compact $K \subset \mathfrak{X}$, $\mathfrak{X}$ a metric vector space, is approximately flat.*

Proof. For $\epsilon > 0$, take a finite subcover, $\{B_\epsilon(f_n) : n \leq N\}$, of $\{B_\epsilon(f) : f \in K\}$. Let $W = \{\sum_{n \leq N} \beta_n f_n : \beta_n \in \mathbb{R}\}$. ∎

The approximate flatness of compact sets has implications for the estimation of nonlinear functions by the regression methods discussed later in §6.5. The span of a set $E \subset \mathfrak{X}$ is

$$\mathbf{span}(E) := \left\{ \sum_{n \leq N} \beta_n e_n : N \in \mathbb{N}, \ \beta_n \in \mathbb{R}, \ e_n \in E \right\}.$$

Much of the theory of nonparametric estimation is based on compact sets $K \subset C([0, 1])$ such that $\mathbf{span}(K)$ is dense in $C([0, 1])$. For example, $K = \{x \mapsto G(ax + b) : -1 \leq a, b \leq 1\}$ is compact when G is continuous, and $\mathbf{span}(K)$ is dense when $G(r) = e^r/(1 + e^r)$. (This is not obvious and we prove it later.) The functions in $\mathbf{span}(K)$ are examples of single hidden layer feedforward artificial neural networks.

In order for $\mathbf{span}(K)$ to be dense, it must have elements that point in every possible direction. However, since K is compact, it is approximately flat, which means that it does not point "very far" into very many directions. An estimator of a function f is $\widehat{f} = \sum_{n \leq N} \widehat{\beta}_n \widehat{e}_n$, where the $\widehat{e}_n \in K$. When K does not point very far in the direction of f, there is a precision problem; small errors in the estimation of the $\widehat{e}_n$ can have very large effects on the estimator.

6.2.g The Levy Distance on cdf's

Let $\mathcal{D}(\mathbb{R})$ be the set of cumulative distribution functions. As in Example 6.2.18, there is a one-to-one onto function between the F's in $\mathcal{D}$ and the set $\mathcal{D}_L$ of F_L's in $C_b(\mathbb{R})$ with the following properties: (1) $[c < c'] \Rightarrow [F_L(c) \leq F_L(c')]$; (2) for all c, c', $|F(c) - F(c')| \leq |c - c'|$; (3) $\lim_{c \downarrow -\infty} F_L(c) = 0$; and (4) $\lim_{c \uparrow \infty} F_L(c) = 1$.

Theorem 6.2.49 $\mathcal{D}_L$ *is a closed subset of* $C_b(\mathbb{R})$.

Proof. Suppose that g_n is a sequence in $\mathcal{D}_L$ converging to g. We must show that g has properties (1)–(4).

(1) If $c < c'$, then for all n, $g_n(c) \leq g_n(c')$. Since $g_n(c) \to g(c)$ and $g_n(c') \to g(c')$, $g(c) \leq g(c')$.

(2) Pick c, c'; since $|g_n(c) - g_n(c')| \leq |c - c'|$ for all n, $|g(c) - g(c')| \leq |c - c'|$.

(3) Pick $\epsilon > 0$; we must show that there exists a c such that for all $c' < c$, $g(c') \leq \epsilon$. For this it is sufficient to show that for large n, $g_n(c') < \epsilon$. Choose N such that for $n, m \geq N$, $d_\infty(g_n, g_m) < \epsilon/2$. Pick c such that for all $c' < c$, $g_M(c') < \epsilon/2$. For all $n \geq N$, $g_n(c') < \epsilon$. The proof of (4) is nearly identical to the proof of (3). ∎

Since $C_b(\mathbb{R})$ is complete, any closed subset is complete. Therefore, $(\mathcal{D}_L, d_L)$ is a complete metric space.

For classical reasons, the subsets of $\mathcal{D}$ that have compact closure are called **tight** subsets. If we know that a sequence of cdf's is itself tight or belongs to a tight set, then we know that it must have a convergent subsequence. The proof of the central limit theorem[6] requires a tightness argument.

Definition 6.2.50 *A set* S *of cdf's is* **tight** *if for all* $\epsilon > 0$, *there exists* $x > 0$ *such that for all* $F \in S$, $F(-x) < \epsilon$ *and* $F(+x) > 1 - \epsilon$; *equivalently there exists* $c > 0$ *such that for all* F_L *with* $F \in S$, $F_L(-c) < \epsilon$ *and* $F_L(+c) > 1 - \epsilon$.

For any cdf, F, the set $\{F\}$ is tight because $\lim_{n \to \infty} F(-n) = 1 - \lim_{n \to \infty} F(n) = 0$. The tightness of a set S is a uniform version of this condition. That is, S is tight iff $\lim_{n \to \infty} \sup\{F(-n) : F \in S\} = 0$ and $\lim_{n \to \infty} \inf\{1 - F(n) : F \in S\} = 0$.

Example 6.2.51 $F_r(x) = 1_{[r, \infty)}(x)$ *corresponds to the degenerate random variable that takes the value* r *with probability* 1. *The set* $\{F_r : r \in \mathbb{R}\}$ *is* **not** *tight. Defining* $G_n = \frac{1}{2}F_{-n} + \frac{1}{2}F_n$ *and* $S = \{G_n : n \in \mathbb{N}\}$, S *is* **not** *tight, this time because half of the mass goes to* $+\infty$ *and half to* $-\infty$. $H_r = \frac{x - (-r)}{3r}1_{(-r, +2r]}(x) + 1_{(2r, \infty)}(x)$ *is the cdf of the uniform distribution on the interval* $(-r, +2r]$. *The collection* $S = \{H_r : r \in E\}$ *is tight iff* E *is a bounded subset of* $\mathbb{R}$.

The following is another formulation of tightness. Recall that P_F is defined for any finite union of intervals, and the compact set that appears in the following is assumed to be in the domain of P_F.

---

6. The central limit theorem is possibly the most amazing result about statistical regularity in the known universe.

Lemma 6.2.52 *A set S of cdf's is tight iff for every $\epsilon > 0$, there is a compact set K such that for all $F \in S$, $P_F(K) > 1 - \epsilon$.*

Proof. Suppose that $P_F(K) > 1 - \epsilon$ for some compact finite union of intervals. For any x such that $K \subset [-x, x]$, $F(-x) < \epsilon$ and $F(+x) > 1 - \epsilon$. On the other hand, if $F(-x) < \epsilon/2$ and $F(+x) > 1 - \epsilon/2$, then $P_F([-x, x]) > 1 - \epsilon$. ∎

Theorem 6.2.53 *A set $S \subset \mathcal{D}$ has compact closure iff it is tight.*

There is a fairly easy, direct proof that constructs a finite ϵ-net of functions in $C_b(\mathbb{R})$ for every tight set. It utilizes the fact that the F_L are Lipschitz continuous and looks at linear interpolations with slopes less than or equal to 1 with kinks only at points of the form $-c + kc/2^n$, $0 \le k \le 2^{n+1}$, for large n. Rather than give that proof, we use a very powerful result called the **Arzela-Ascoli theorem**. This is rather in the nature of using a sledgehammer to crack a walnut, but this theorem gives the basic compactness criterion for subsets of $C(M)$ when M is compact, and we need this for later developments.

6.2.h Compactness in $C(M)$ when M Is Compact

Suppose that M is a compact metric space. We now present a condition on $S \subset C(M)$ equivalent to S having a compact closure. The compactness criterion is the content of the Arzela-Ascoli theorem. The proof that we give uses the diagonalization theorem, a result that we reuse several times. Examples demonstrate what is involved.

Example 6.2.54 (Unbounded Range Problems in $C(M)$) *If M is finite, then $C(M) = \mathbb{R}^M$ and $S \subset C(M)$ has compact closure iff S is bounded. If $M = [0, 1]$, the sequence $f_n(t) = n$ is unbounded and has no convergent subsequence.*

Example 6.2.55 (Steepness Problems in $C(M)$) *We have seen two examples of sequences of functions in $C([0, 1])$ that are uniformly bounded and become arbitrarily steep, $g_n(t) = t^n$ and $h_n(t) = [1 - n \cdot |x - \frac{1}{n}|] \vee 0$. The first sequence converges pointwise to a discontinuous function and the second to 0, but neither sequence converges uniformly nor has a uniformly convergent subsequence.*

We can see the steepness problem interacting with optimization in the following example, which makes more sense if you remember that a subset of a metric space is compact iff every continuous function achieves its maximum on the set.

Example 6.2.56 (Steepness Revisited) *Define $u(f) = \int_0^{0.5} f(t)\,dt - \int_{0.5}^1 f(t)\,dt$ for $f \in C([0, 1])$. If $d_\infty(f_n, f) < \epsilon$, then $|u(f_n) - u(f)| < \epsilon$, so that u is continuous. We have $\sup\{u(f) : f \in B_1(0)\} = 1$, but there is no $f \in B_1(0)$ such that $u(f) = 1$. To see why the first claim is true, consider the function f_ϵ which linearly interpolates the points $0 = t_0$, $t_1 = \frac{1}{2} - \epsilon$, $t_2 = \frac{1}{2} + \epsilon$, and $t_3 = 1$ with the associated values 1, 1, -1, and -1. Calculation shows that $u(f_\epsilon) = 1 - \epsilon$. As $\epsilon \to 0$, the f_ϵ become steeper and steeper and there is no convergent subsequence.*

These examples teach us that we need controls on the range and the "steepness" of the functions in a set in order for a subset of $C(M)$ to be compact.

Definition 6.2.57   *Let S be a subset of $C(M)$. The set S is*

1. **equicontinuous at** $x \in M$ *if* $(\forall \epsilon > 0)(\exists \delta > 0)(\forall x' \in M)[[d(x', x) < \delta] \Rightarrow (\forall f \in S)[|f(x') - f(x))| < \epsilon]]$,

2. **equicontinuous** *if S is equicontinuous at all $x \in M$, and*

3. **uniformly equicontinuous** *if* $(\forall \epsilon > 0)(\exists \delta > 0)(\forall x, x' \in M)[[d(x', x) < \delta] \Rightarrow (\forall f \in S)[|f(x') - f(x))| < \epsilon]]$,

Equicontinuity at x controls the maximal "variability" of a function in the set S at the point x. The set of functions $S = \{t^n : n \in \mathbb{N}\}$ in $C([0, 1])$ is not equicontinuous at 1, though it is equicontinuous at every $t < 1$. Since this class S fails to be equicontinuous at a single point, $t = 1$, it is not equicontinuous.

The distinction between equicontinuity and uniform equicontinuity is the same as that between continuity and uniform continuity, Definition 4.9.1 (p. 143). To be uniformly equicontinuous, for every $\epsilon > 0$, there must be a $\delta > 0$ that "works" at every x and every $f \in S$, where "works" means $[[d(x', x) < \delta] \Rightarrow (\forall f \in S)[|f(x') - f(x))| < \epsilon]]$. A function is uniformly continuous iff the set $S = \{f\}$ is uniformly equicontinuous.

Exercise 6.2.58   If M is finite, then $C(M) = \mathbb{R}^M$. Show that any subset of $C(M)$ is equicontinuous in this case.

Every continuous function on a compact set is uniformly continuous, Theorem 4.9.3. That result is but a special case of the following, which has almost the same proof.

Theorem 6.2.59   *If $S \subset C(M)$ is equicontinuous and M is compact, then S is uniformly equicontinuous.*

Proof.   Suppose that S is equicontinuous but not uniformly equicontinuous. For some $\epsilon > 0$ and all $n \in \mathbb{N}$, there exist $x_n, u_n \in M$ and $f_n \in S$ such that $d(x_n, u_n) < 1/n$ but $|f_n(x_n) - f_n(u_n)| \geq \epsilon$. Since M is compact, $x_{n'} \to x$ for some $x \in M$ and some subsequence. Since $d(x_n, u_n) < 1/n \to 0$, $u_{n'} \to x$. Since S is equicontinuous, there exists $\delta > 0$ such that $[d(x', x) < \delta] \Rightarrow [|f_n(x') - f_n(x)| < \epsilon/2]$. For large n', $d(u_{n'}, x) < \delta$ and $d(x_{n'}, x) < \delta$, which implies that $|f_{n'}(x_{n'}) - f_{n'}(u_{n'})| < \epsilon$, a contradiction.   ■

Definition 6.2.60   *A set $S \subset C(M)$ is **norm bounded**, or simply **bounded**, if $\exists B > 0$ such that for all $f \in S$, $\|f\|_\infty < B$.*

The following is the basic compactness criterion for subsets of $C(M)$, M compact.

Theorem 6.2.61 (Arzela–Ascoli)   *Let M be a compact space. $S \subset C(M)$ has compact closure iff S is bounded and equicontinuous.*

If M is finite so that $C(M) = \mathbb{R}^M$, the equicontinuity is automatically satisfied. In this case, the Arzela-Ascoli theorem reduces to the result that a subset of $\mathbb{R}^\ell$ has compact closure iff it is bounded.

Proof.   Suppose that $S \subset C(M)$ has compact closure. The continuous function $f \mapsto d_\infty(0, f)$ achieves its maximum on the closure, so S is bounded.

Pick $\epsilon > 0$. To show that S is equicontinuous, we must show that there exists $\delta > 0$ such that for all $x, x' \in M$, $[d(x, x') < \delta] \Rightarrow [|f(x) - f(x')| < \epsilon]$. Let $\{f_n : k = 1, \ldots, N\}$ be a finite $\epsilon/3$ net for S; that is, for each $f \in S$, there exists f_n such that $d_\infty(f, f_n) = \max_{x \in M} |f(x) - f_n(x)| < \epsilon/3$. Since each f_n is continuous and M is compact, each f_n is uniformly continuous. We can therefore choose $\delta > 0$ such that for each f_n, $[d(x, x') < \delta] \Rightarrow [|f_n(x) - f_n(x')| < \epsilon/3]$. Pick $f \in S$ and $x, x' \in M$ such that $d(x, x') < \delta$. For f_n such that $d_\infty(f, f_n) < \epsilon/3$,

$$|f(x) - f(x')| = |f(x) - f_n(x) + f_n(x) - f_n(x') + f_n(x') - f(x')| \leq$$

$$|f(x) - f_n(x)| + |f_n(x) - f_n(x')| + |f_n(x') - f(x')| < \epsilon/3 + \epsilon/3 + \epsilon/3.$$

Now suppose that S is bounded and equicontinuous. Let f_n be a sequence in S. We must show that a convergent subsequence exists. Since $C(M)$ is complete, it is sufficient to show the existence of a Cauchy subsequence.

Let E be a countable dense subset of M (e.g., $E = \cup_q E_q$, where E_q is a finite q-net for M, with q a positive rational). Enumerate E as $E = \{e_i : i \in \mathbb{N}\}$. For each i, let $n \mapsto r_{i,n}$ be the sequence of real numbers $r_{i,n} = f_n(x_i)$. Since S is bounded, each sequence $r_{i,n}$ is bounded. By the diagonalization theorem (given directly below), there exists an increasing sequence of integers $n_1, n_2, \ldots$ such that for all i, $r_{i,n_k} = f_{n_k}(x_i)$ converges to a limit r_i. We now show that $g_k := f_{n_k}$ is a Cauchy subsequence of f_n.

Pick $\epsilon > 0$; we must show that there exists N such that for $k, k' \geq N$, $d_\infty(g_k, g_{k'}) = \max_{x \in M} |g_k(x) - g_{k'}(x)| \leq \epsilon$. Since S is an equicontinuous set and M is compact, S is uniformly equicontinuous. This means that we can choose $\delta > 0$ such that for all $f \in S$, $[d(x, x') < \delta] \Rightarrow |f(x) - f(x')| < \epsilon/3]$. Since E is dense in M, there exists an I such that $\{e_i : i \leq I\}$ is a finite δ-net for M.

Pick N large enough that for any $k \geq N$ and all e_i, $i \leq I$, $|f_{n_k}(e_i) - r_i| = |g_k(e_i) - r_i| < \epsilon/6$. This yields, for all $k, k' \geq N$, $|g_k(e_i) - g_{k'}(e_i)| < \epsilon/3$. Choose arbitrary $x \in M, k, k' \geq N, x \in M$, and $e_i, i \leq I$ such that $d(x, e_i) < \delta$. Combining, we get

$$|g_k(x) - g_{k'}(x)| = |g_k(x) - g_k(e_i) + g_k(e_i) - g_{k'}(e_i) + g_{k'}(e_i) - g_{k'}(x)|$$

$$= |g_k(x) - g_k(e_i)| + |g_k(e_i) - g_{k'}(e_i)| + |g_{k'}(e_i) - g_{k'}(x)| < \epsilon/3 + \epsilon/3 + \epsilon/3.$$

∎

Theorem 6.2.62 (Diagonalization) *Suppose that for each $i \in \mathbb{N}$, $n \mapsto r_{i,n}$ is a bounded sequence in $\mathbb{R}$. Then there exists an increasing sequence $k \mapsto n_k$ of integers such that for each i, $\lim_k r_{i,n_k}$ exists.*

It may help to visualize the $r_{i,n}$ sequences as an array,

$$\begin{matrix} r_{1,1} & r_{1,2} & r_{1,3} & \cdots \\ r_{2,1} & r_{2,2} & r_{2,3} & \cdots \\ r_{3,1} & r_{3,2} & r_{3,3} & \cdots \\ \vdots & \vdots & \vdots & \end{matrix},$$

with $n \mapsto r_{i,n}$ being the ith row in the array.

Proof.   From the sequence $n \mapsto r_{1,n}$, select a convergent subsequence,

$$r_{1,n_{1,1}} \quad r_{1,n_{1,2}} \quad r_{1,n_{1,3}} \quad \cdots .$$

This can be done because $n \mapsto r_{1,n}$ is bounded. We take $k \mapsto n_{1,k}$ to be an increasing sequence, and $\lim_k r_{1,n_{1,k}}$ exists.

Now consider the sequence $n \mapsto r_{2,n}$, but along the subsequence $n_{1,1}, n_{1,2}, \ldots,$

$$r_{2,n_{1,1}} \quad r_{2,n_{1,2}} \quad r_{2,n_{1,3}} \quad \cdots$$

Since this is a subsequence of a bounded sequence, it is bounded, which means that we can pick a convergent subsequence

$$r_{2,n_{2,1}} \quad r_{2,n_{2,2}} \quad r_{2,n_{2,3}} \quad \cdots ,$$

where $k \mapsto n_{2,k}$ is increasing and is a subsequence of $k \mapsto n_{1,k}$, and $\lim_k r_{2,n_{2,k}}$ exists.

Continuing in this fashion gives an array of integers,

$$
\begin{array}{cccc}
n_{1,1} & n_{1,2} & n_{1,3} & \cdots \\
n_{2,1} & n_{2,2} & n_{2,3} & \cdots \\
n_{3,1} & n_{3,2} & n_{3,3} & \cdots \\
\vdots & \vdots & \vdots &
\end{array}
, \tag{6.18}
$$

with the property that each row is increasing, that the $(i + 1)$th row is a subset of the ith row, and that for all i, $\lim_k r_{i,n_{i,k}}$ exists.

We now go "down the diagonal" in the array (6.18) by setting $n_k = n_{k,k}$. Since each row in (6.18) is increasing and is a subset of the previous row, n_k is increasing. For each i, the collection $\{n_k : k \geq i\}$ is a subset of the ith row in (6.18). Therefore, for each i, $\lim_k r_{i,n_k}$ exists.   ∎

6.3   ◆ $\mathcal{D}(\mathbb{R})$, the Space of Cumulative Distribution Functions

In this section we show that the space of cdf's with the Levy distance is complete and separable, that $d_L(F_n, F) \to 0$ iff $F_n(x) \to F(x)$ at all x that are continuity points of F characterizes compact sets of cdf's, and that the set of cdf's with finitely many jumps is dense.

Example 6.3.1   *The sequence of cdf's $F_n(x) = 1_{[1/n,\infty)}(x)$ corresponds to the sequence of probabilities putting probability 1 on the set $1/n$. The associated $g_n = (F_n)_L$ are*

$$
g_n(c) = \begin{cases} 0 & \text{if } c < 1/n \\ c - 1/n & \text{if } 1/n \leq c < 1 + 1/n \ . \\ 1 & \text{if } 1 + 1/n \leq c \end{cases}
$$

Point masses at $1/n$ should converge to point mass at 0. The sequence of cdf's does not converge either pointwise or uniformly to $F(x) = 1_{[0,\infty)}(x)$ because $F_n(0) = 0$ for all $n \in \mathbb{N}$. However, the g_n converge uniformly to

$$g(c) = \begin{cases} 0 & \text{if } c < 0n \\ c & \text{if } 0 \leq c < 1 \\ 1 & \text{if } 1 \leq c \end{cases}.$$

6.3.a Tightness in $\mathcal{D}(\mathbb{R})$

We now prove Theorem 6.2.53, which says that a set $S \subset \mathcal{D}$ has compact closure iff it is tight, that is, iff for every $\epsilon > 0$, there exists an x such that for all $F \in S$, $F(-x) < \epsilon$ and $F(x) > 1 - \epsilon$. The set $\mathcal{D}_L$ is uniformly equicontinuous because every $g \in \mathcal{D}_L$ satisfies $|g(c) - g(c')| \leq |c - c'|$. It is also norm bounded because every $g \in \mathcal{D}_L$ satisfies $\|g\|_\infty = 1$. However, $\mathbb{R}$ is not compact, nor is $\mathcal{D}_L \subset C_b(\mathbb{R})$.

Exercise 6.3.2 Let $F_n(x) = 1_{[n,\infty)}(x)$. Show that F_n is a sequence of cdf's that converges pointwise to 0 and has no d_L-convergent subsequence. Let $G_n(x) = \frac{1}{2}1_{[-n,n)}(x) + 1_{[n,\infty)}(x)$. Show that G_n is a sequence of cdf's that converges pointwise to $\frac{1}{2}$ and has no d_L-convergent subsequence. [Intuitively, the probability mass in the F_n sequence "escapes to $+\infty$," while half of the mass in the G_n sequence "escapes to $+\infty$" and the other half "escapes to $-\infty$." Tightness is a condition that prevents the mass from escaping by requiring that it stay on compact sets.]

Proof of Theorem 6.2.53. Suppose that S is tight. Pick $\epsilon > 0$; we show that there exists a compact subset, K_ϵ, such that $S \subset (K_\epsilon)^\epsilon$. This is sufficient since a finite $\epsilon/2$-net for $K_{\epsilon/2}$ contains a finite ϵ-net for S.

Pick $c > 0$ such that for all $g \in S$, $g(-c) < \epsilon$ and $g(c) > 1 - \epsilon$. Let $K_\epsilon = \{g \in \mathcal{D}_L : g(-c) = 0, \ g(c) = 1\}$. Since any $g \in K_\epsilon$ is constant outside of the interval $[-c, c]$ and d_∞ agrees with the sup norm metric on the space $C([-c, c])$, then elements of K_ϵ restricted to $[-c, c]$ form a compact set because they are uniformly bounded (by 1) and equicontinuous (because they have a Lipschitz constant of 1). Finally, $S \subset (K_\epsilon)^\epsilon$ because for all $g \in S$, $d(g_{\|[-c,c]}, K_\epsilon) < \epsilon$.

Now suppose that S is not tight; that is, for some $\epsilon > 0$ and for all $c > 0$, there exists $g \in S$ such that $g(-c) \geq \epsilon$ or $g(c) \leq 1 - \epsilon$. For each $n \in \mathbb{N}$, choose $g_n \in S$ such that $g_n(-n) \geq \epsilon$ or $g_n(n) \leq 1 - \epsilon$. No subsequence of g_n can converge to any $h \in \mathcal{D}_L$ because $\lim_n h(-n) = 1 - \lim_n h(n) = 0$, so that $d_\infty(h, g_n) > \epsilon/2$ a.a. ∎

Corollary 6.3.3 $\mathcal{D}_L$ *is separable.*

Proof. For $n \in \mathbb{N}$, let $K_n = \{g \in \mathcal{D}_L : g(-n) = 0, \ g(n) = 1\}$ and let E_n be a finite $1/n$-net for K_n. The set $E = \cup_n E_n$ is countable and dense. ∎

6.3.b Weak Convergence of Cumulative Distribution Functions

We saw in Example 6.3.1 that cdf's representing point mass at $1/n$ fail to converge, even pointwise, to the cdf representing point mass at 0. Specifically, we saw that

$F_n(x) \equiv 0$ but $F(x) = 1$. The point 0 is special, as it is a discontinuity point of the limit cdf.

Definition 6.3.4 *A sequence F_n of cdf's **converges weakly to** F, written $F_n \to_{w^*} F$, if $F_n(x) \to F(x)$ at every x for which F is continuous.*

We are often interested in sequences of discontinuous cdf's that converge to a continuous cdf.

Example 6.3.5 *Suppose that the random variables $X_1, X_2, \ldots$ are independent and identically distributed with $P(\{X_i = 1\}) = r$ and $P(\{X_i = 0\}) = 1 - r$ for some $r \in (0, 1)$ so that $\sigma^2 = \mathrm{Var}(X_i) = r(1 - r)$. Consider the sequence of random variables $S_n = \frac{1}{\sigma \sqrt{n}} \sum_{i \le n} (X_i - r)$ and let F_n be the cdf of S_n. For each n, F_n has discontinuities at the $n + 1$ points of the form $\frac{k - nr}{\sigma \sqrt{n}}$, $k = 0, 1, \ldots, n$. The central limit theorem tells us that F_n converges weakly to Φ, to the (uniformly) continuous cdf of the standard normal, defined by $\Phi(x) = \frac{1}{\sqrt{2\pi}} \int_{-\infty}^{x} e^{-t^2/2} \, dt$. In particular, for all $x \in \mathbb{R}$, $\mathrm{Prob}(S_n \le x) \to \Phi(x)$, since every x is a continuity point of Φ.*

Suppose that x is a discontinuity of F so that $P_F(\{x\}) := F(x) - F^-(x)$. Since F is nondecreasing and its range is $[0, 1]$, there are at most n points x such that $P(x) \ge 1/n$. Therefore, F has at most countably many discontinuity points.

Lemma 6.3.6 *The complement of a countable set of points in $\mathbb{R}$ is dense.*

Proof. Let E be any countable subset of $\mathbb{R}$. It is sufficient to show that (a, b) intersects E^c for any nonempty open interval, (a, b). It is sufficient to observe that (a, b) is uncountable—$E \cap (a, b)$ is at most countable so that $E^c \cap (a, b)$ is uncountable, hence not empty. ∎

Corollary 6.3.7 *If $F, G \in \mathcal{D}$, $F \ne G$, then $F(x) \ne G(x)$ for some x that is a continuity point of both F and G.*

Proof. Let E_F and E_G be the set of discontinuity points of F and G. The set $E = E_F \cup E_G$ is countable, so that E^c is dense.

Pick x such that $|F(x) - G(x)| = \delta > 0$. Since both F and G are right continuous, there exists an $\epsilon > 0$ such that for all $x' \in [x, x + \epsilon)$, $|F(x) - F(x')| < \delta/2$ and $|G(x) - G(x')| < \delta/2$. $E^c \cap [x, x + \epsilon) \ne \emptyset$. ∎

It is surprising that the following extremely obvious fact is so useful. We have already utilized it in the proof of Corollary 4.8.8, which showed that a closed graph function taking values in a compact metric space is continuous.

Theorem 6.3.8 *In a metric space (M, d), $x_n \to x$ iff for every subsequence $x_{n'}$, there is a further subsequence, $x_{n''} \to x$.*

Proof. If $x_n \to x$, then for any subsequence $x_{n'}$, take $x_{n''} = x_{n'}$. If $x_n \not\to x$, there exists $\epsilon > 0$ such that $d(x, x_n) \ge \epsilon$ infinitely often. Let n' enumerate $\{n : d(x, x_n) \ge \epsilon\}$. No subsequence of $x_{n'}$ can converge to x. ∎

Theorem 6.3.9 $F_n \to_{w^*} F$ *if and only $d_L(F_n, F) \to 0$.*

Proof. Suppose first that $F_n \to_{w^*} F$. We first show that $S = \{F\} \cup \{F_n : n \in \mathbb{N}\}$ is tight and then that any subsequence, $F_{n'}$, has a further subsequence, $F_{n''}$, with $d_L(F_{n''}, G) \to 0$.

Tightness: We claim that there exists a dense set of points, x_0, such that $\pm x_0$ are both continuity points of F, namely the complement of the countable set $E \cup -E$, where E is the set of discontinuity points of F.

As it is dense, we can pick arbitrarily large positive x_0 with $\pm x_0$ a continuity point of F. Pick such an x_0 such that $F(-x_0) < \epsilon/2$ and $F(x_0) > 1 - \epsilon/2$. Since $\pm x_0$ are continuity points, $\exists N$ such that for all $n \geq N$, $|F_n(-x_0) - F(-x_0)| < \epsilon/2$ and $|F_n(x_0) - F(x_0)| < \epsilon/2$. For each $n < N$, pick positive $x_n \in E \cup -E$ such that $F_n(-x_n) < \epsilon$ and $F_n(x_n) > 1 - \epsilon$. Let $x = \max\{x_i : 0 \leq i \leq N\}$. For all $G \in S$, $G(-x) < \epsilon$ and $G(x) > 1 - \epsilon$.

Every sequence has a further, convergent subsequence: Since S is tight, every subsequence $F_{n'}$ has a further subsequence $F_{n''}$ with $d_L(F_{n''}, G) \to 0$ for some cdf G. By Theorem 6.3.8, it is sufficient to show that $G = F$. If not, then F and G disagree at some point x_0 that is a continuity point for both cdf's (Corollary 6.3.7). Suppose that $F(x_0) > G(x_0)$. (The reverse inequality is essentially identical.) Let $4r = F(x_0) - G(x_0)$. For large n'', $|F_{n''}(x_0) - F(x_0)| < r$. Pick $\delta > 0$ such that for all $x' \in [x_0, x_0 + 2\delta]$, $|F(x_0) - F(x')| < r$ and $|G(x_0) - G(x')| < r$. Let $\epsilon = \min\{\delta, r\}$. For $c = x_0 + \epsilon + G(x_0)$, $|(F_n)_L(c) - G_L(c)| > \epsilon$ for all large n'', contradicting $d_L(F_{n''}, G) \to 0$.

Now suppose that $d_L(F_n, F) \to 0$. We must show that $F_n \to_{w*} F$. Let x_0 be a continuity point of F and pick $\epsilon > 0$. As F is continuous at x_0, there exists $\eta \in (0, \epsilon)$ such $F(x_0) \leq F(x_0 + \eta) < F(x_0) + \epsilon$. Pick $\delta < \min\{\eta, \epsilon - \eta\}$. Moving no more than δ along the line with slope -1 through the point $(x_0 + \eta, F(x_0 + \eta))$, the height stays within the interval $(F(x_0) - \epsilon, F(x_0) + \epsilon)$. For large enough n that $d_L(F_n, F) < \delta$, the correspondence $\hat{F}_n$, defined in Example 6.2.18, must intersect this line segment. By parallel arguments, within ϵ to the left of x_0, the value of $\hat{F}_n$ is also within ϵ of $F(x_0)$. Since each F_n is nondecreasing, $|F_n(x_0) - F(x_0)| < \epsilon$. ∎

6.3.c The Denseness of Finitely Supported Probabilities

A probability with $P(E) = 1$ for a finite set E is called **finitely supported**. In this case, the smallest such finite set is called the **support** of P. The associated cumulative distribution function has finitely many jumps at the points in E.

Theorem 6.3.10 *The finitely supported cdf's are dense in $\mathcal{D}(\mathbb{R})$.*

Remember that real numbers are nearly rational numbers. This result is saying that probabilities are nearly finitely supported. Example 6.3.5 shows that it is often very useful to have more probabilities than the finitely supported ones. Exercises 6.11.10 and 6.11.11 show that cdf's can be very strange indeed when we get outside of the set of those that are finitely supported.

The idea of the proof is simple. Suppose that G is the cdf of a probability supported on n points, $x_1 < x_2 < \cdots < x_n$. The associated G_L has slope 0 before some point y_1, slope 1 until the point $z_1 = y_1 + P(x_1) > y_1$, slope 0 until $y_2 = z_1 + (x_2 - x_1) > z_1$, slope 1 until $z_2 = y_2 + P(x_2) > y_2$, and so on, By adjusting n and the $y_i < z_i < y_{i+1}$ pairs, $i \leq n - 1$, we can get a function arbitrarily close to any F_L, because F_L is everywhere increasing at a rate between 0 and 1.

Exercise 6.3.11 Give the details of the proof of Theorem 6.3.10.

6.4 ◆ Approximation in $C(M)$ when M Is Compact

In this section, we maintain the assumption that M is compact and, to avoid trivialities, that it contains at least two points. We prove a lattice version and an algebraic version of the **Stone-Weierstrass theorem**, both of which give conditions under which subsets of $C(M)$ are dense in $C(M)$.

The Stone-Weierstrass theorem is one of the most powerful and useful results in the study of spaces of functions. We use it to demonstrate that three main classes of nonparametric regressions—Fourier series, polynomial expansions, and feedforward artificial neural networks with monotonic activation functions—are all dense. This is the crucial property leading to consistency of nonparametric estimators of functions in the statistical and econometrics literature. We also use the Stone-Weierstrass theorem to give yet another characterization of a compact metric space—M is compact iff $C(M)$ is separable.

When $M = [0, 1]$, the separability of $C(M)$ can be proved fairly directly. One interpolates linearly at rational time points $0 = t_0 < t_1 < \cdots < t_n = 1, n \in \mathbb{N}$, for all possible rational values, $q_0, q_1, \ldots, q_n$, which gives a countable set of functions. Since every continuous function on $[0, 1]$ is uniformly continuous, the set is dense, which means that every continuous function on $[0, 1]$ can be approximated by a piecewise linear function. How one generalizes this construction to an arbitrary compact M is not at all clear. Instead, the Stone-Weierstrass theorem goes after this problem from another direction, looking for conditions on subsets, E, of $C(M)$ that are sufficient to guarantee that E is dense.

6.4.a   The Lattice Version of Stone-Weierstrass

Recall that a **lattice** is a partially ordered set in which every pair of elements has a supremum and an infimum. We define the partial order, $\leq$, on $C(M)$ by $f \leq g$ if $f(x) \leq g(x)$ for all $x \in M$. For $f, g \in C(M)$, $(f \wedge g)(x) := \min[f(x), g(x)] \in C(M)$ and $(f \vee g)(x) := \max[f(x), g(x)] \in C(M)$ as shown in Example 4.8.6.[7]

Definition 6.4.1 $\mathcal{H} \subset C(M)$ is a **lattice subset** of $C(M)$ if it contains $f \wedge g$ and $f \vee g$ for every $f, g \in \mathcal{H}$. For $S \subset C(M)$, the **lattice subset generated by** S is $\mathcal{H}(S) = \bigcap\{L \subset C(M) : S \subset L, \ \ L$ a lattice subset of $C(M)\}$.

Since the intersection of any collection of lattice subsets is another lattice subset, $\mathcal{H}(S)$ is well defined. There is a distinction between a lattice subset and a sublattice (see Definition 2.9.13 and Example 2.9.16). Every lattice subset is a sublattice, but the reverse need not be true.

Example 6.4.2   *In* $C([0, 1])$: *the three-point set* $\{0, t, 1 - t\}$ *is a sublattice but not a lattice subset; the set of linear interpolations is a lattice subset and a sublattice.*

Definition 6.4.3   *A subset* $\mathcal{H}$ *of* $C(M)$ ***separates points*** *if for any two distinct points* $x, x' \in M$, $\exists h \in \mathcal{H}$ *with* $h(x) \neq h(x')$. *It* ***separates points to arbitrary values***

---

7. Another way to prove that $f \wedge g$ and $f \vee g$ are continuous is to note that $(f \wedge g)(x) = \frac{1}{2}(f + g)(x) - \frac{1}{2}|f - g|(x)$ and $(f \vee g)(x) = \frac{1}{2}(f + g)(x) + \frac{1}{2}|f - g|(x)$.

if for all distinct $x, x' \in M$ *and all* $r, r' \in \mathbb{R}$, $\exists h \in \mathcal{H}$ *such that* $h(x) = r$ *and* $h(x') = r'$.

Separating points to arbitrary values implies separating points.

Example 6.4.4 *In* $C([0, 1])$: *the set of constant functions is a lattice subset, but is not separating; the set* $\{t, 1-t\}$ *is separating, but is neither a lattice subset nor a sublattice; the set* $\{at + b : a, b \in \mathbb{R}\}$ *separates points to arbitrary values, is not a lattice subset, but is a sublattice; the set of linear interpolations in* $C([0, 1])$ *is* $\mathcal{H}(S)$, *where* $S = \{at + b : a, b \in \mathbb{R}\}$; *it is a lattice subset of* $C([0, 1])$ *that separates points to arbitrary values, and we know that it is dense.*

The lattice version of the Stone-Weierstrass theorem is a consequence of the following lemma. Recall that we have assumed that M contains at least two points.

Theorem 6.4.5 (Kakutani–Krein) *If* $\mathcal{H} \subset C(M)$ *is a lattice subset that separates points to arbitrary values, then* $\mathcal{H}$ *is dense in* $C(M)$.

Proof. Take $f \in \mathcal{C}(M)$ and $\epsilon > 0$. We want to find an element η of $\mathcal{H}$ that is within ϵ of f. We do this in two steps. First, for every $y \in M$, we find a function $f_y \in \mathcal{H}$ with two properties: $\forall z \in M$ and $f_y(z) < f(z) + \epsilon$, and for all z in an open neighborhood $V(y)$ of y, we also have $f(z) - \epsilon < f_y(z)$. Second, we take a finite subcover of M by neighborhoods $V(y_j)$, $j = 1, \ldots, J$, and define $\eta = \max_j f_j$.

First step: Since $\mathcal{H}$ separates points to arbitrary values, for every $x, y \in M$, there exists a function $\eta_{x,y} \in \mathcal{H}$ that agrees with f at the points x and y, that is, with $\eta_{x,y}(x) = f(x)$ and $\eta_{x,y}(y) = f(y)$. Note that $\eta_{x,y}$ cannot differ from f by more than ϵ in small open sets around x and y.

Let $U(x, y)$ be the open set $\{z \in M : \eta_{x,y}(z) < f(z) + \epsilon\}$. Both x and y belong to $U(x, y)$. In a similar fashion, let $V(x, y)$ be the open set $\{z \in M : f(z) - \epsilon < \eta_{x,y}(z)\}$. Again, both x and y belong to $V(x, y)$. Thus, x, y are in the open set $U(x, y) \cap V(x, y)$ and $\eta_{x,y}$ is within ϵ of f everywhere in the open set $U(x, y) \cap V(x, y)$.

For any y, consider the open cover of M given by $\{U(x, y) : x \in M\}$. Since M is compact, there is a finite open subcover, $\{U(x_i, y) : i = 1, \ldots, I\}$. As $\mathcal{H}$ is a lattice subset, $f_y := \min\{\eta_{x_i,y} : i = 1, \ldots, I\}$ belongs to $\mathcal{H}$. Since each x belongs to some $U(x_i, y)$ and for every $z \in U(x_i, y)$, $\eta_{x_i,y}(z) < f(z) + \epsilon$, we have constructed, for every y, a function $f_y \in \mathcal{H}$ with the properties that

$$f_y(z) \leq \eta_{x_i,y}(z) < f(z) + \epsilon \quad \forall z \in M, \quad \text{and}$$

$$f(z) - \epsilon < f_y(z) \forall z \in V(y) := \cap_{i \leq I} V(x_i, y). \tag{6.19}$$

Second step: Since $\{V(y) : y \in M\}$ is an open cover of M, it has a finite subcover, $\{V(y_j) : j = 1, \ldots, J\}$. Define $\eta = \max\{f_{y_j} : j = 1, \ldots, J\}$. As each f_{y_j} belongs to $\mathcal{H}$, η belongs to $\mathcal{H}$. As each $z \in M$ belongs to some $V(y_j)$, (6.19) yields $f(z) - \epsilon < \eta(z) < f(z) + \epsilon$ for all $z \in M$. ∎

Separating points to arbitrary values looks like a difficult condition, but that is not so.

Theorem 6.4.6 (Stone–Weierstrass L)   *If $\mathcal{H} \subset C(M)$ is a lattice subset, a vector subspace, contains the constant functions, and separates points, then $\mathcal{H}$ is dense in $C(M)$.*

Proof.   We show that $\mathcal{H}$ separates points to arbitrary values.

Let $x, x' \in M$ with $x \neq x'$. Since $\mathcal{H}$ is separating, $\exists h \in \mathcal{H}$ such that $h(x) \neq h(x')$. Let $r, r' \in \mathbb{R}$; then the system of linear equations $r = \mu + \lambda h(x)$ and $r' = \mu + \lambda h(x')$ has a unique solution $(\mu, \lambda) \in \mathbb{R}^2$ because $h(x) \neq h(x')$. Set $g(x) = \mu + \lambda h(x)$. As $\mathcal{H}$ is a vector subspace containing constant functions, $g \in \mathcal{H}$. Moreover, we see that $g(x) = r$ and $g(x') = r'$.   ∎

6.4.b   The Algebraic Version of Stone-Weierstrass

We now turn to the algebraic case of the Stone-Weierstrass theorem. Following this, we apply whichever version is more suitable to some concrete examples.

Definition 6.4.7   *A vector subspace $\mathcal{H} \subset C(M)$ is an **algebra** if it is closed under multiplication.*

$C(M)$ is itself an algebra. The classical algebra of functions is the set of polynomials in $C([0, 1])$. Multinomials are most easily defined using **multi-index** notation: a multi-index $\alpha = (\alpha(1), \ldots, \alpha(k))$ is a point in $\{0, 1, 2, \ldots\}^k$; for $\mathbf{x} = (x_1, \ldots, x_k) \in \mathbb{R}^k$, $\mathbf{x}^\alpha := x_1^{\alpha(1)} \cdot x_2^{\alpha(2)} \cdots x_k^{\alpha(k)}$. A **multinomial** on $\mathbb{R}^k$ is a function of the form $f(\mathbf{x}) = \sum_{m \leq M} \beta_m \mathbf{x}^{\alpha_m}$ for $M \in \mathbb{N}$, each $\beta_m \in \mathbb{R}$, and α_m a multi-index. In $C([0, 1]^k)$, the set of multinomials is the classical algebra of functions. Multinomials play a larger role.

Lemma 6.4.8 $\mathcal{H} \subset C(M)$ *is an algebra of functions iff for all finite $\{f_1, \ldots, f_k\} \subset \mathcal{H}$, $k \in \mathbb{N}$ and for all multinomials P on $\mathbb{R}^k$, $P(f_1, \ldots, f_k) \in \mathcal{H}$.*

Proof.   The terms $\alpha x_1 + \beta x_2$ and $x_1 \cdot x_2$ are multinomials; a multinomial in multinomials is another multinomial.   ∎

The intersection of algebras is another algebra and $C(M)$ is an algebra, so that there is always a smallest algebra of continuous functions containing any given class of functions.

Definition 6.4.9   *For $S \subset C(M)$, let $\mathcal{A}(S)$ be the **algebra generated by** S, which is defined as the intersection of all algebras in $C(M)$ containing S.*

Exercise 6.4.10   For any $S \subset C(M)$, show that the set of multinomials in finite subsets of S is equal to $\mathcal{A}(S)$.

Example 6.4.11   *In $C([0, 1])$, $\mathcal{A}(S)$ is the vector space of all polynomials when $S = \{t, 1\}$. When $S' = \{\sqrt{t}, 1\}$, $\mathcal{A}(S')$ is the set of multinomials in the two functions. $\mathcal{A}(S')$ is also a vector space containing the constants and separating points.*

Theorem 6.4.12 (Stone–Weierstrass A)   *If $\mathcal{H} \subset C(M)$ is an algebra, contains the constant functions, and separates points, then $\mathcal{H}$ is dense in $C(M)$.*

Since dense subsets of dense sets are dense, we have the following.

Corollary 6.4.13 *If $\overline{\mathcal{H}}$ contains an algebra, the constants, and separates points, then $\mathcal{H}$ is dense in $C(M)$.*

Example 6.4.14 *The following sets satisfy the conditions of Stone-Weierstrass A: the set of polynomials on $[a, b] \subset \mathbb{R}$; the set of multinomials in $\{\sqrt{t}, 1\}$ on $[a, b] \subset \mathbb{R}_+$; the set of multinomials in $\{h(t), 1\}$ on $[a, b]$, where $h : [a, b] \to \mathbb{R}$ is continuous and invertible; for $h_i : [a_i, b_i] \to \mathbb{R}$ continuous and invertible, $i = 1, \dots, \ell$, the set of multinomials in $\{1, h_1(x_1), \dots, h_\ell(x_\ell)\}$ on $\times_{i=1}^{\ell}[a_i, b_i]$; the set of functions of the form $\{\cos(A(\mathbf{x})) : A$, an affine function from $\mathbb{R}^\ell$ to $\mathbb{R}\}$ on any compact $K \subset \mathbb{R}^\ell$.*

The essential idea in the proof is to show that containing all multinomials implies coming arbitrarily close to being a lattice and that coming arbitrarily close is enough for denseness. There are two preliminary lemmas.

Lemma 6.4.15 *There exists a sequence of polynomials $P_n \in C([-1, 1])$ that converges uniformly to $f(x) = |x|$.*

Recall that $C(M; M')$ is the set of continuous functions from M to M'.

Proof. We define a monotone mapping, T, from a compact, lattice subset $K \subset C := C([-1, 1]; [0, 1])$ to itself, and we show that the mapping T takes polynomials to polynomials and that iterative applications of T converge to the unique fixed point, $g(x) = |x|$.

Let $K = \{f \in C : f(0) = 0, \ f(x) = f(-x), \ |f(x) - f(x')| \leq |x - x'|\}$. If $f \in K$, the combination of the $f(0) = 0$ and the Lipschitz constant being 1 implies that $f(x) \leq |x|$. K is closed, bounded, and equicontinuous, hence compact. Define $\varphi : [0, 1] \to [0, 1]$ by $\varphi(r) = r - \frac{1}{2}r^2$ and $T : K \to C$ by $T(f)(x) = f(x) + \frac{1}{2}(x^2 - [f(x)]^2) = \varphi(f(x)) + \frac{1}{2}x^2$. By inspection, T takes polynomials to polynomials, and if $g(x) = |x|$, then $T(g) = g$. We now show that $g(x) = |x|$ is the only fixed point of T in K.

We first show that $T : K \to K$. Since $d\varphi(r)/dr = 1 - r$ and $d(\frac{1}{2}x^2)/dx = x$, if $f \in K$, then $f(x) \leq |x|$ implies that $T(f)$ also has 1 as its Lipschitz constant. We also have: $[f(0) = 0] \Rightarrow [T(f)(0) = 0]$; $[f(x) = f(-x)] \Rightarrow [T(f)(x) = T(f)(-x)]$; $[f \in C] \Rightarrow [T(f) \in C]$. Combining, we get $T : K \to K$. Further, for $g \in K$, $T(g) = g$ iff $g(x) = |x|$, since $g(x) = g(x) + \frac{1}{2}(x^2 - [g(x)]^2)$ iff $x^2 = [g(x)]^2$.

We now show that T is monotone; that is, for all $x \in [-1, 1]$, $T(f)(x) \geq f(x)$. By inspection, $T(f)(x) - f(x) = \frac{1}{2}(x^2 - [f(x)]^2)$. Since $f(x) \leq |x|$, $\frac{1}{2}(x^2 - [f(x)]^2) \geq 0$ and is strictly positive if $f(x) < |x|$.

We now show that iterative applications of T converge to the unique fixed point of T. Pick a function P_0 in K (e.g., the polynomial $P_0 = 0$), and define $P_{t+1} = T(P_t)$. Since K is compact, taking a subsequence if necessary, we find that P_t converges to some function $h \in K$. If for some $x \in [-1, 1]$, $h(x) < |x|$, set $\epsilon = \frac{1}{2}(x^2 - [h(x)]^2)$. For large t, $P_t(x) > h(x) - \epsilon$, which implies that $P_{t+1}(x) > h(x)$, a contradiction. ∎

Lemma 6.4.16 *A vector subspace $\mathcal{H} \subset C(M)$ containing the constants is a lattice subset iff for every element $h \in \mathcal{H}$, the function $|h| \in \mathcal{H}$ as well.*

Proof.   ($\Rightarrow$) If $h \in \mathcal{H}$, then $\max(h, 0), \min(h, 0) \in \mathcal{H}$, so that $|h| = \max(h, 0) - \min(h, 0) \in \mathcal{H}$.

($\Leftarrow$) If $f, g \in \mathcal{H}$, then $|f - g| \in \mathcal{H}$, so that $\max(f, g) = \frac{1}{2}\left[(f + g) + \frac{1}{2}|f - g|\right]$ $\in \mathcal{H}$ and $\min(f, g) = \frac{1}{2}\left[(f + g) - \frac{1}{2}|f - g|\right] \in \mathcal{H}$.   ∎

Proof of Theorem 6.4.12, Stone–Weierstrass A.   If $\mathcal{H}$ is a separating subalgebra of $C(M)$ containing constant functions, then so is its closure $\overline{\mathcal{H}}$. Therefore it suffices to show that $\overline{\mathcal{H}}$ is a lattice and apply the lattice version of Stone-Weierstrass, Theorem 6.4.6.

Let $f \in \overline{\mathcal{H}}$ be nonzero. By Lemma 6.4.15, there exists a sequence P_n of polynomials that converges uniformly on $[-1, 1]$ to $f(x) = |x|$. Since $-1 \leq \frac{f}{\|f\|} \leq 1$, the sequence of fuctions $P_n\left(\frac{f}{\|f\|}\right)$ converges uniformly to $\left|\frac{f}{\|f\|}\right| = \frac{|f|}{\|f\|}$, so that $\|f\| \, P_n\left(\frac{f}{\|f\|}\right) \to |f|$. As $\overline{\mathcal{H}}$ is an algebra, all terms in this sequence are in $\overline{\mathcal{H}}$. Since $\overline{\mathcal{H}}$ is closed, $|f| \in \overline{\mathcal{H}}$. By Lemma 6.4.16, $\overline{\mathcal{H}}$ is a lattice.   ∎

Exercise 6.4.17   Prove that if $\mathcal{H}$ is a separating subalgebra of $C(M)$, then $\overline{\mathcal{H}}$ is as well.

6.4.c   Applications of Stone–Weierstrass

Both versions of the Stone-Weierstrass theorem are very general statements about the density of a subset $\mathcal{H}$ in $C(M)$ or, equivalently, about approximation in $C(M)$. The next examples show that they cover many of the known approximation theorems.

Example 6.4.18 (Piecewise Linear Functions)   *As seen in Example 6.4.4, the set of piecewise linear functions in $C([a, b])$ is a vector subspace, is a lattice subset, contains the constants, and separates points, so that it is dense.*

Example 6.4.19 (Lipschitz Functions)   *Let* $\text{Lip} \subset C(M)$ *be the set of Lipschitz functions,* $\text{Lip} = \{h \in C(M) : (\exists L \in \mathbb{R})[|h(x) - h(y)| \leq L d(x, y)]\}$. *To show that* Lip *is dense in $C(M)$, it is sufficient to establish either that it is an algebra containing the constants and separating points or that it is a vector subspace, lattice subset containing the constants and separating points. We do the latter.*

Since $f(x) = d(x, y)$ is Lipschitz (with constant 1), Lip *separates points. If $f, g \in \text{Lip}$ with Lipschitz constants L_f and L_g and $\alpha, \beta \in \mathbb{R}$, then*

$$|(\alpha f + \beta g)(x) - (\alpha f + \beta g)(y)| = |\alpha(f(x) - f(y)) + \beta(g(x) - g(y))|$$

$$\leq |\alpha||f(x) - f(y)| + |\beta||g(x) - g(y)|$$

$$\leq (|\alpha|L_f + |\beta|L_g)d(x, y),$$

so that Lip *is a vector subspace.*

Since $f(x) = 1$ is Lipschitz (with constant 0), Lip *being a vector subspace means that* Lip *contains the constant functions.*

Finally, if $f \in \text{Lip}$, then $|f| \in \text{Lip}$ because $||f|(x) - |f|(y)| \leq |f(x) - f(y)| \leq L_f d(x, y)$, so that L is a lattice (by Lemma 6.4.16).

The last example shows that any continuous function, no matter how irregular, is within ϵ of a Lipschitz function. We expect that as $\epsilon \downarrow 0$, the minimal Lipschitz constant of the function that ϵ-approximates will increase.

Example 6.4.20 *Let* Mult *be the set of all multinomials* $h : M \to \mathbb{R}$, *where* M *is a compact subset of* $\mathbb{R}^n$. *It is easy to show that* Mult *is a subalgebra of* $C(M)$ *containing the constants and is separating. Thus* Mult *is dense in* $C(M)$; *that is, for every continuous* $f \in C(M)$, *no matter how strange, and for every* $\epsilon > 0$, *no matter how small, there is a multinomial,* p_ϵ, *such that* $\max_{x \in M} |f(x) - p_\epsilon(x)| < \epsilon$.

A special case of Example 6.4.20 is $M = [a, b]$, known as the Weierstrass approximation theorem. Note that all the previous examples guarantee the existence of a dense set in $C(M)$ but do not present a constructive method of approximating a continuous function. Example 7.2.20 (p. 374) gives an explicit construction using Theorem 7.2.19, which is a result from probability theory.

Example 6.4.21 *For compact* $M \subset \mathbb{R}^\ell$, $C^r(M)$ *denotes the set of functions with* r *continuous derivatives on some open set containing* M. $C^r(M)$ *is a separating algebra containing the constant functions, so is dense in* $C(M)$. *Alternately,* $C^r(M)$ *contains the multinomials, which are dense, implying that* $C^r(M)$ *is dense in* $C(M)$.

One way to approximate continuous functions on a noncompact set M is to approximate them on larger and larger compact subsets. This works best when M is the countable union of compact sets, for example, $\mathbb{R}_+ = \cup_n [0, n]$ or $\mathbb{R}^\ell = \cup_n [-n, n]^\ell$.

Example 6.4.22 *Let* $\mathcal{H} \subset C_b(\mathbb{R}_+)$ *denote the set of functions of the form* $f(x) = \sum_{n=0}^{n_k} a_n e^{-nx}$, $a_n \in \mathbb{R}$, $n_k \in \{0, 1, \ldots\}$. *If* $n_k = 0$, *then* $\mathcal{H}$ *contains the constant functions. If* $n_k = 1$ *and* $a_1 \neq 0$, *then* $\mathcal{H}$ *separates points. For* $f, g \in \mathcal{H}$, $fg \in \mathcal{H}$ *since* $e^{-nx} \cdot e^{-mx} = e^{-(n+m)x}$. *Therefore, for any compact* $M \subset \mathbb{R}_+$, $\mathcal{H}_{|M} = \{f_{|M} : f \in \mathcal{H}\}$ *is dense in* $C(M)$. *Here* $\mathcal{H} = \mathcal{A}(\{1, e^{-x}\})$.

This means that to within any $\epsilon > 0$, any continuous function is approximated by a sum of exponentials with integer weights. To get a sense of how surprising one might find that result, imagine trying to construct such a sum to get within (say) $1/10^8$ of a piecewise linear function in $C([0, 1])$ that varies between ± 200 and has $2,117$ kinks. While the Stone-Weierstrass theorem says that it can be done, it may take a great deal of additional ingenuity and patience to actually do it.

Exercise 6.4.23 Show that for any $r \neq 0$, the set of functions of the form $f(x) = \sum_{n=0}^N a_n e^{-n(rx)}$, $a_n \in \mathbb{R}$, $N \in \{0, 1, \ldots\}$ is dense in $C(M)$ when M is a compact subset of $\mathbb{R}$ (not merely $\mathbb{R}_+$). Further, this implies that the set of functions of the form $f(x) = \sum_{n=0}^N a_n e^{b_n x}$, $a_n, b_n \in \mathbb{R}$, $N \in \{0, 1, \ldots\}$ is dense in $C(M)$ when M is a compact subset of $\mathbb{R}$.

The product metric, $d_{M \times M'}$, on $M \times M'$ was given in Definition 6.1.29, and equivalent forms are in Exercise 6.1.30. Its defining characteristic is that $(x_n, y_n) \to (x, y)$ iff $x_n \to x$ and $y_n \to y$.

Lemma 6.4.24 *If* (M, d) *and* (M', d') *are compact, then* $(M \times M', d_{M \times M'})$ *is compact.*

Proof.   Let (x_n, y_n) be a sequence in $M \times M'$. Take a subsequence $x_{n'}$ of x_n such that $x_n \to x$. Take a subsequence $y_{n''}$ of $y_{n'}$ such that $y_n \to y$. The subsequence $(x_{n''}, y_{n''}) \to (x, y)$.   ∎

For $u \in C(M)$ and $v \in C(M')$, M and M' compact, the function $u \cdot v : M \times M' \to \mathbb{R}$ is defined by $(u \cdot v)(x, y) = u(x)v(y)$. Since multiplication is continuous, $u \cdot v \in C(M \times M')$.

Exercise 6.4.25   With M and M' compact, show that the set of functions of the form $f = \sum_{n=1}^{N} u_n \cdot v_n$, $u_n \in C(M)$ and $v_n \in C(M')$, is dense in $C(M \times M')$.

Exercise 6.4.26   [↑Exercise 6.4.23] Show that for any compact $M \subset \mathbb{R}^2$, the set, $\mathcal{H}$, of functions of the form $f(x_1, x_2) = \sum_{n_1, n_2=0}^{N} a_{n_1, n_2} e^{-(n_1 x_1 + n_2 x_2)}$, $a_{n_1, n_2} \in \mathbb{R}$, $N \in \{0, 1, \ldots\}$, is dense in $C(M)$. Give a set S such that $\mathcal{H} = \mathcal{A}(S)$.

To examine Fourier series, we need some reminders from your trigonometry class from many years ago: $\cos(x) = \cos(-x)$; $\sin(x) = -\sin(-x)$; $\cos(x)\cos(y) = (\cos(x + y) + \cos(x - y))/2$; $\sin(x)\sin(y) = (\cos(x - y) - \cos(x + y))/2$; and $\sin(x)\cos(y) = (\sin(x + y) + \sin(x - y))/2$.

Example 6.4.27 (Fourier Series)   *Let $\mathcal{H} \subset C_b(\mathbb{R}_+)$ denote the set of functions of the form*

$$h(x) = a_0 + \sum_{n=1}^{N}[a_n \cos(nx) + b_n \sin(nx)], \qquad (6.20)$$

where $N \in \mathbb{N}$ and $a_n, b_n \in \mathbb{R}$. From the trigonometry rules just noted, $\mathcal{H}$ is an algebra. Taking $N = 1$, $a_n = b_n = 0$, and $a_0 \in \mathbb{R}$ shows that $\mathcal{H}$ contains the constants. Finally, for $x \neq y$, $[\cos(x) = \cos(y)] \Rightarrow [\sin(x) \neq \sin(y)]$, so that $\mathcal{H}$ separates points. Here, $\mathcal{H} = \mathcal{A}(\{1, \cos(x), \sin(x)\})$.

Exercise 6.4.28   Show that one can replace the "nx" by "$n/2^k$" for any fixed k in (6.20). [The choice above was $k = 0$.]

Exercise 6.4.29   Show that for any compact $M \subset \mathbb{R}^2$, the set $\mathcal{H}$ of functions of the form $f(x_1, x_2) = a_0 + \sum_{n_1, n_2=0}^{N}[a_{n_1, n_2} \cos(n_1 x_1 + n_2 x_2) + b_{n_1, n_2} \sin(n_1 x_1 + n_2 x_2)]$, $N \in \mathbb{N}$, $a_{n_1, n_2}, b_{n_1, n_2} \in \mathbb{R}$, is dense in $C(M)$. Give a set S such that $\mathcal{H} = \mathcal{A}(S)$.

The proof of the following uses the Stone-Weierstrass theorem, but in a rather indirect fashion.

Theorem 6.4.30 (Single Hidden Layer Feedforward Networks)   *If $G : \mathbb{R} \to [0, 1]$ is a continuous cdf and M is a compact subset of $\mathbb{R}^\ell$, then* **span**$(\{G(\mathbf{a}x + b) : (\mathbf{a}, b) \in \mathbb{R}^\ell \times \mathbb{R}\})$ *is dense in $C(M)$.*

Exercise 6.4.31   Fill in the details of the following sketch of a proof of Theorem 6.4.30.

1. Let Aff^ℓ denote the set of affine functions from $\mathbb{R}^\ell$ to $\mathbb{R}$. Show that if $\mathcal{G} \subset C(\mathbb{R})$ is uniformly dense on compacta in $C(\mathbb{R})$, then $\mathcal{G} \circ Aff^\ell$ is uniformly dense on compacta in $C(\mathbb{R}^\ell)$. [Using the trigonometry rules review above, show that the span of the functions $\{\sin(A(\mathbf{x})) : A \in Aff^\ell\}$

is an algebra of functions in $C_b(\mathbb{R}^\ell)$ that contain the constants and separate points. Show that this implies that for any affine $A : \mathbb{R}^\ell \to \mathbb{R}$ and any $\epsilon > 0$, there is a function in $\mathcal{G}$ that is within ϵ of sin on the interval $A(M)$, M a compact subset of $\mathbb{R}^\ell$.]

2. Show that $G(n(x - x_0))$ converges weakly to the cdf $1_{[x_0,\infty)}(x)$.

3. Let F_{x_0} be the cdf giving point mass at x_0. Show that over any compact interval $[a, b] \subset \mathbb{R}$, for any $\epsilon > 0$, $\mathbf{span}\{F_{x_0} : x_0 \in [a, b]\}$ contains a function that is within ϵ of sin on the interval $[a, b]$.

4. Combining the previous steps, show that $\mathbf{span}(\{G(\mathbf{a}x + b) : (\mathbf{a}, b) \in \mathbb{R}^\ell \times \mathbb{R}\})$ is uniformly dense on compacta in $C(\mathbb{R}^\ell)$.

6.4.d Separability of $C(M)$

In a more abstract vein, we show the following.

Theorem 6.4.32 *M is compact iff $C(M)$ is separable.*

Proof. If M is not compact, then it either fails to be totally bounded or it fails to be complete. In either case, there exists a sequence of distinct points, x_n, and $\epsilon_n > 0$ such that for all $m \neq n$, $B_{\epsilon_m}(x_m) \cap B_{\epsilon_n}(x_n) = \emptyset$. The functions $f_n(x) := [1 - \frac{1}{\epsilon_n} d(x, x_n)] \vee 0$ are all at d_∞-distance 1 from each other and are only nonzero on the sets $B_{\epsilon_n}(x_n)$. Thus, for any nonempty $A \subset \mathbb{N}$, $f_A(x) := \sum_{n \in A} f_n(x)$ is a continuous function. This gives uncountably many continuous functions at distance 1 from each other.

Suppose now that M is compact. Compactness implies the existence of a countable dense $E \subset M$. Define $f_e(x) = d(e, x)$. The class $S_E = \{f_e : e \in E\}$ separates points because E is dense. Therefore $\mathcal{A}(\{1\} \cup S_E)$ is dense in $C(M)$. All that remains is to show that there is a countable subset, F, of $\mathcal{A}(\{1\} \cup S_E)$ that is dense in $\mathcal{A}(S_E)$. Take F to be the set of multinomials in finite collections of f_e's with rational coefficients. ∎

6.5 ◆ Regression Analysis as Approximation Theory

The starting point is a compact $K \subset \mathbb{R}^\ell$, a finite set $(\mathbf{x}_n)_{n \leq N}$ of points in K, and a finite set $(y_n)_{n \leq N}$ of points in $\mathbb{R}$. The aim is to find a function $f \in C(K)$ that approximates the *data* $(\mathbf{x}_n, y_n)_{n \leq N} \subset (\mathbb{R}^{\ell+1})^N$. This may arise purely as a curve-fitting exercise in which one wants a convenient summary of the data. More often, it arises from a situation in which the $\mathbf{x}_n$ are chosen (by some process) and the y_n are of the form $f(\mathbf{x}_n) + \varepsilon_n$. Here the unknown function f is the object of our interest,[8] and the ε_n are random variables with mean 0.

---

8. If the $\mathbf{x}_n$ are the demographic descriptions of person n and y_n is her yearly income, then the unknown f summarizes the relation between average yearly income and demographic variables. If the $\mathbf{x}_n$ are the costs, including time costs, of the various legal ways to dispose of trash in

One specifies a set $S \subset C(K)$ and looks for a function $\widehat{f} \in S$ that comes "closest" to the data. There are several important observations about this process, all of which flow from there being only finitely many of the $\mathbf{x}_n$ and point to limitations on the set S.

1. Suppose that K is infinite and the $\mathbf{x}_n$ are distinct. Let $(\mathbf{x}'_m, y'_m)_{m \le M}$ be any arbitrary collection of distinct points in $(\mathbb{R}^{\ell+1})^M$ with the set of $\mathbf{x}'_m$ disjoint from the set of $\mathbf{x}_n$. There will always be functions $f \in C(K)$ such that $y_n = f(\mathbf{x}_n)$ for $n \le N$ and $y'_m = f(\mathbf{x}'_m)$ for $m \le M$. This means that if $S = C(K)$, then there are perfect summaries of the data, $y_n = f(\mathbf{x}_n)$ for $n \le N$, that are arbitrary outside of the data, $y'_m = f(\mathbf{x}'_m)$ for $m \le M$.

2. Suppose that S_r is the set of functions with a Lipschitz constant less than or equal to $r \ge 0$. If it were known that $f \in S_r$, then matching the data puts restrictions on the possible values of $f(\mathbf{x}'_m)$, specifically $|f(\mathbf{x}'_m) - f(\mathbf{x}_n)| \le r|\mathbf{x}'_m - \mathbf{x}_n|$ for all n and m. As we gather more data, the compact set $(\mathbf{x}_n)_{n \le N}$ becomes a better and better approximation to the compact set K. This means that for all of the $\mathbf{x}'_m$, the minimal value of $|\mathbf{x}'_m - \mathbf{x}_n|$ becomes small, so that larger values of r are possible with the same degree of restrictiveness. From Example 6.4.19, we know that $\cup_r S_r$ is dense in $C(K)$, so that slowly increasing the r as we gather more data makes it feasible to approximate any unknown f. This pattern contains the essence of nonparametric regression, allowing richer sets S as more data are gathered.

We need some background before we examine the curve-fitting problem.

6.5.a   Orthogonal Projections in $\mathbb{R}^\ell$

The geometry and the algebra of orthogonal projections in $\mathbb{R}^N$ form the essential background for linear and nonlinear regression.[9] For this section, the convention that all vectors are column vectors matters; $\mathbf{x}'$ denotes the (row vector) transpose of the vector $\mathbf{x}$; $\mathbf{x}'\mathbf{y}$ denotes the dot product of $\mathbf{x}$ and $\mathbf{y}$; and if $L : \mathbb{R}^N \to \mathbb{R}^k$ is a linear map, we use "$L\mathbf{x}$" instead of "$L(\mathbf{x})$."

Let A be an m-dimensional linear subspace of $\mathbb{R}^N$ spanned by vectors $\mathbf{x}_1, \ldots, \mathbf{x}_m$, $m < k$. This means that every $a \in A$ has a unique representation as $a = \sum_{i \le m} \beta_i \mathbf{x}_i$, where the vector $\beta = (\beta_1, \ldots, \beta_m)' \in \mathbb{R}^m$. In matrix notation, $A = \{\mathbf{X}\beta : \beta \in \mathbb{R}^m\}$, where $\mathbf{X}$ is the $N \times m$ matrix having $\mathbf{x}_i$ as its ith column and A is the span of the columns of $\mathbf{X}$.

Recall that two vectors are **orthogonal**, written $\mathbf{x} \perp \mathbf{y}$, if $\mathbf{x}'\mathbf{y} = 0$. We write $\mathbf{x} \perp A$ if $\mathbf{x} \perp \mathbf{a}$ for all $\mathbf{a} \in A$. Let $B = A^\perp = \{\mathbf{x} \in \mathbb{R}^N : \mathbf{x} \perp A\}$ denote the **orthogonal complement** of A. Since A is m-dimensional, B is the $(N - m)$-dimensional

---

neighborhood n and y_n is the amount of illegal trash dumping that neighborhood n generates, then the unknown f summarizes the propensity to dump as a function of the price of legal alternatives.

9. Orthogonal projections in Hilbert spaces, the infinite-dimensional vector spaces that most resemble $\mathbb{R}^N$, are the essential background for nonparametric regression. We return to nonparametric regression in §8.4.

orthogonal complement of A. One proves the following fact from linear algebra by working with a basis for A and a basis for B.

Lemma 6.5.1 *If A is a linear subspace of $\mathbb{R}^N$ and $B = A^\perp$, then every $\mathbf{y} \in \mathbb{R}^N$ has a unique representation as $\mathbf{y} = \mathbf{a} + \mathbf{b}$, $\mathbf{a} \in A$, $\mathbf{b} \in B$.*

Definition 6.5.2 $\text{proj}_A : \mathbb{R}^N \to A$ is the **orthogonal projection onto** A if $\text{proj}_A \, \mathbf{y} = \mathbf{a}$, where $\mathbf{y} = \mathbf{a} + \mathbf{b}$, $\mathbf{b} \perp A$. A mapping is an **orthogonal projection** if it is an orthogonal projection onto some vector subspace of $\mathbb{R}^N$.

Note that an orthogonal projection proj_A is linear, that $\text{proj}_A(\mathbb{R}^N) = A$, $\text{proj}_A^2 = \text{proj}_A$, and that $\text{proj}_A : A \to A$ is the identity map from $\mathbb{R}^N$ to itself (defined by $I(\mathbf{x}) = \mathbf{x}$). These are the defining properties of a possibly nonorthogonal projection. What makes the projection orthogonal is the requirement that $(\mathbf{y} - \text{proj}_A \, \mathbf{y}) \perp \text{proj}_A \, \mathbf{y}$.

Theorem 6.5.3 (Geometry of Orthogonal Projection) proj_A *is the orthogonal projection onto A iff for all $\mathbf{y} \in \mathbb{R}^N$, $\text{proj}_A \, \mathbf{y}$ solves the problem $\min_{\mathbf{a} \in A} \|\mathbf{y} - \mathbf{a}\|_2$.*

In particular, the solution to $\min_{\mathbf{a} \in A} \|\mathbf{y} - \mathbf{a}\|_2$ is unique. We must use the 2-norm in order for this result to be true. If we take $A = \{(x_1, x_2)' \in \mathbb{R}^2 : x_1 = x_2\}$ to be the diagonal in the plane and let $\mathbf{y} = (0, 1)'$, then the set of $a \in A$ that solves $\min \|\mathbf{y} - \mathbf{a}\|_1$ is the line segment joining $(0, 0)'$ and $(1, 1)'$.

Exercise 6.5.4 Give a vector subspace A and $\mathbf{y} \notin A$ such that $\min_{\mathbf{a} \in A} \|\mathbf{y} - \mathbf{a}\|_\infty$ has a (nontrivial) segment of solutions.

Theorem 6.5.1 (Proof of Theorem 6.5.3) *Suppose that proj_A is the orthogonal projection onto A. Pick arbitrary $\mathbf{y} \in \mathbb{R}^N$. There are two cases, $\mathbf{y} \in A$ and $\mathbf{y} \notin A$. If $\mathbf{y} \in A$, then $\text{proj}_A \, \mathbf{y} = \mathbf{y}$, so that $\|\mathbf{y} - \text{proj}_A \, \mathbf{y}\|_2 = 0$, and this solves the problem $\min_{\mathbf{a} \in A} \|\mathbf{y} - \mathbf{a}\|_2$. Suppose that $\mathbf{y} \notin A$. By definition, $\text{proj}_A \, \mathbf{y} = \mathbf{a}$, where $\mathbf{y} = \mathbf{a} + \mathbf{b}$ and $\mathbf{b} \perp A$. Pick $\mathbf{a}^\circ \in A$, $\mathbf{a}^\circ \neq \mathbf{a}$. Since $\mathbf{b} \perp A$, $\mathbf{b}'(\mathbf{a} - \mathbf{a}^\circ) = 0$, so that the triangle determined by the points $\mathbf{y}$, $\mathbf{a}$, and $\mathbf{a}^\circ$ is a right triangle with the line segment $[\mathbf{y}, \mathbf{a}^\circ]$ being the hypotenuse. This means that $\|\mathbf{y} - \mathbf{a}^\circ\|_2 > \|\mathbf{y} - \mathbf{a}\|_2$.*

Now let $Q(\mathbf{y})$ be the solutions to $\min_{\mathbf{a} \in A} \|\mathbf{y} - \mathbf{a}\|_2$. A solution exists because $f(\mathbf{a}) = d(\mathbf{y}, \mathbf{a})$ is continuous, hence achieves its minimum on the closed and bounded set $\{\mathbf{a} \in A : d(\mathbf{y}, \mathbf{a}) \leq d(\mathbf{y}, \mathbf{a}^\circ)\}$, where $\mathbf{a}^\circ$ is an arbitrary point in A. We have to show that $\mathbf{y} \mapsto Q(\mathbf{y})$ is the orthogonal projection onto A. By the previous part of the proof, if $Q(\mathbf{y}) \neq \text{proj}_A \, \mathbf{y}$, then $Q(\mathbf{y})$ does not solve $\min_{\mathbf{a} \in A} \|\mathbf{y} - \mathbf{a}\|_2$.

Crucial to linear regression is a particular algebraic expression of the solution to $\min_{\mathbf{a} \in A} \|\mathbf{y} - \mathbf{a}\|_2$. Since each $\mathbf{a} \in A$ has a unique representation of the form $\mathbf{X}\beta$ for some $\beta \in \mathbb{R}^m$, the minimization problem is equivalent to

$$\min_\beta h(\beta) := (\mathbf{y} - \mathbf{X}\beta)'(\mathbf{y} - \mathbf{X}\beta). \tag{6.21}$$

Rewriting gives us $h(\beta) = \mathbf{y}'\mathbf{y} - 2\mathbf{y}'\mathbf{X}\beta + \beta'\mathbf{X}'\mathbf{X}\beta$. Since the m columns of $\mathbf{X}$ are linearly independent, $(\mathbf{X}'\mathbf{X})$ is an $m \times m$ invertible matrix. Further, it is symmetric, $(\mathbf{X}'\mathbf{X})' = (\mathbf{X}'\mathbf{X})$, and positive definite, $\mathbf{z}'(\mathbf{X}'\mathbf{X})\mathbf{z} = (\mathbf{z}'\mathbf{X}')(\mathbf{X}\mathbf{z}) > 0$ if $\mathbf{z} \neq 0$. Since $D_\beta^2 h$, the matrix of second derivatives of f with respect to β is $\mathbf{X}'\mathbf{X}$, which means that f is convex. This in turn implies that solving $D_\beta h = 0$ yields the solution to the minimization problem.[10]

Setting the derivative of $D_\beta h = 0$ yields the equation $-2\mathbf{y}'\mathbf{X} + 2\beta'(\mathbf{X}'\mathbf{X}) = 0$, equivalently $(\mathbf{X}'\mathbf{X})\beta = \mathbf{X}'\mathbf{y}$. Solving gives us $\beta^* = (\mathbf{X}'\mathbf{X})^{-1}\mathbf{X}'\mathbf{y}$, so that $\mathbf{a}^* = \mathbf{X}\beta^* = \mathbf{X}(\mathbf{X}'\mathbf{X})^{-1}\mathbf{X}'\mathbf{y}$. This means that $\mathbf{b}^* = \mathbf{y} - \mathbf{a}^* = \mathbf{I}\mathbf{y} - \mathbf{X}(\mathbf{X}'\mathbf{X})^{-1}\mathbf{X}'\mathbf{y}$, where $\mathbf{I}$ is the $N \times N$ identity matrix. This can be rewritten as $\mathbf{b}^* = (\mathbf{I} - \mathbf{X}(\mathbf{X}'\mathbf{X})^{-1}\mathbf{X}')\mathbf{y}$. Summarizing this algebra, we have the following.

Theorem 6.5.5 (Algebra of Orthogonal Projection) *If $A = \{\mathbf{X}\beta : \beta \in \mathbb{R}^m\}$ is an m-dimensional subspace of $\mathbb{R}^N$, then the mapping $\mathbf{y} \mapsto \mathbf{X}(\mathbf{X}'\mathbf{X})^{-1}\mathbf{X}'\mathbf{y}$ is the orthogonal projection onto A and $\mathbf{y} \mapsto (\mathbf{I} - \mathbf{X}(\mathbf{X}'\mathbf{X})^{-1}\mathbf{X}')\mathbf{y}$ is the orthogonal projection onto $A^\perp$.*

Exercise 6.5.6   Explicitly calculate $(\mathbf{a}^*)'(\mathbf{b}^*)$ and show that it is equal to 0.

Exercise 6.5.7   Let $A = \left\{ \begin{bmatrix} 1 \\ 1 \end{bmatrix} \beta : \beta \in \mathbb{R}^1 \right\}$. For $\mathbf{y} \in \mathbb{R}^2$, algebraically find the orthogonal projection onto A and onto $A^\perp$ and explain the geometry. More generally, let $\mathbf{X}$ be the $N \times 1$ matrix of 1's and $A = \{\mathbf{X}\beta : \beta \in \mathbb{R}^1\}$. Find the orthogonal projection of $\mathbf{y} \in \mathbb{R}^N$ onto A and onto $A^\perp$. [Note that if we let $\mathbf{1}$ be the vector of 1's, then $\mathbf{z} \perp \mathbf{1}$ iff $\mathbf{z} \cdot \mathbf{1} = 0$ iff $\sum_i z_i = 0$.]

Exercise 6.5.8   Let $\mathbf{X}$ be the 3×2 matrix with the first column being all 1's and the second column being $(5, 2, 10)'$. Set $A = \{\mathbf{X}\beta : \beta \in \mathbb{R}^1\}$ and $\mathbf{y} = (3, 9, 0)' \in \mathbb{R}^3$. Find the orthogonal projection of $\mathbf{y}$ onto A and onto $A^\perp$.

6.5.b Linear Regression

Returning to curve fitting, recall that we have a compact $K \subset \mathbb{R}^\ell$, a finite set $(\mathbf{x}_n)_{n \leq N}$ of points in K, and a finite set $(y_n)_{n \leq N}$ of points in $\mathbb{R}$. The aim is to find a function $f \in S \subset C(K)$ that approximates these data. The first class $S \subset C(K)$ to be considered is the set of affine functions, $S = \{f(\mathbf{x}; \beta) = \beta_0 + \mathbf{x}'\beta_1 : \beta = (\beta_0, \beta_1) \in \mathbb{R}^{1+\ell}\}$.

We pick $\widehat{f} \in S$ to solve

$$\min_{f \in S} \sum_n (y_n - f(\mathbf{x}_n))^2. \tag{6.22}$$

If we let

$$\mathbf{y} = \begin{bmatrix} y_1 \\ y_2 \\ \vdots \\ \vdots \\ y_N \end{bmatrix}, \quad \mathbf{X} = \begin{bmatrix} 1 & x_{1,1} & x_{1,2} & \cdots & x_{1,\ell} \\ 1 & x_{2,1} & x_{2,2} & \cdots & x_{2,\ell} \\ \vdots & \vdots & \vdots & \vdots & \vdots \\ \vdots & \vdots & \vdots & \vdots & \vdots \\ 1 & x_{N,1} & x_{N,2} & \cdots & x_{N,\ell} \end{bmatrix}, \quad \text{and} \quad \beta = \begin{bmatrix} \beta_0 \\ \beta_1 \\ \vdots \\ \beta_\ell \end{bmatrix},$$

---

10. The basic derivative properties of the minima of convex functions can be found in §5.6.c and §5.10.

with $m = \ell + 1$, the problem in (6.22) is equivalent to

$$\min_{\beta \in \mathbb{R}^m} h(\beta) = (\mathbf{y} - \mathbf{X}\beta)'(\mathbf{y} - \mathbf{X}\beta), \qquad (6.23)$$

which repeats (6.21). The solution is $\widehat{\beta} = (\mathbf{X}'\mathbf{X})^{-1}\mathbf{X}'\mathbf{y}$. For $\mathbf{x} \in \mathbb{R}^\ell$, let $\widetilde{\mathbf{x}} = (1, \mathbf{x}')' \in \mathbb{R}^{1+\ell}$. The affine function that approximates the data is $\widehat{f}(\mathbf{x}) = f(\mathbf{x}; \widehat{\beta}) = \widetilde{\mathbf{x}}'\widehat{\beta}$.

The **fitted errors** are the vector $\epsilon := \mathbf{y} - \widehat{f}(\mathbf{x}) = \mathbf{y} - \widetilde{\mathbf{x}}'\widehat{\beta}$. Therefore, another way to understand the minimization problem in (6.23) is that we are trying to find the β that minimizes the size of the fitted errors, $\min_\beta \epsilon'\epsilon$.

Exercise 6.5.9 [↑Exercise 6.5.8] Suppose that $K = [2, 10] \subset \mathbb{R}^1$ and that the data are $(\mathbf{x}_1, \mathbf{x}_2, \mathbf{x}_3) = (5, 2, 10)$ and $(y_1, y_2, y_3) = (3, 9, 0)$. Graph the data and $\widehat{f}$. Identify the components of the projection onto $A^\perp$ on your graph.

Though $\widehat{f}$ is estimated on the basis of data points $(\mathbf{x}_n)_{n=1}^N$ that belong to $K \subset \mathbb{R}^\ell$, it maps all of $\mathbb{R}^\ell$ to $\mathbb{R}$. Since it is approximates the data as well as an affine function can, $\widehat{f}$ captures the affine structure of the data points. If the affine structure of the data points is similar to the structure of the true relation between points in K and $\mathbb{R}$, then $\widehat{f}$ also resembles the true relation on K. If the true relation on K generalizes to $\mathbb{R}^\ell$, then the extension of $\widehat{f}$ from K to $\mathbb{R}^\ell$ also captures that generalization. In particular, $\partial \widehat{f}/\partial x_i = \widehat{\beta}_i, i = 1, \dots, \ell$, captures the independent effect of x_i on y.

Being able to declare "the effect of x_i and y is $\widehat{\beta}_i$" is so tempting that one wants to assume both that the true relation is affine and that it generalizes. There are two problem areas that one should address as well as possible before making such a declaration—functional form assumptions, in the form of specification testing, and endogeneity/omitted variable biases.

1. The existence some nonlinear relation between $\mathbf{X}$ and ϵ, the fitted errors, provides evidence that the relation is not in fact affine.

2. If some omitted variable is correlated with x_i and y, then $\widehat{\beta}_i$ is not, on average, correct even if the relation is affine. For example, if one estimates the slope of the relation between fertilizer per acre and yield per acre without taking account of the fact that higher-quality farmland responds more strongly to fertilizer and that farmers put more fertilizer on higher quality land, one confounds the effects of fertilizer and land quality on yield.

6.5.c Nonlinear Regression

Nonlinear regression takes S to be the span of a compact, often finite, $F \subset C(K)$, and may have the set F larger for larger n. The finite case is easiest and covers the classical series regressions cases. We start there.

6.5.c.1 The Span-of-a-Finite-Set Case

Let $F = \{f_h : h = 0, 1, \dots, H\}$ and set $S = \mathbf{span}(F) := \left\{ \sum_{h \leq H} \beta_h f_h : \beta_h \in \mathbb{R} \right\}$ as the **span** of F. The estimation problem is to pick the $\widehat{f} \in S$ that best approximates the data. As before, this involves solving the problem

$$\min_{f \in S} \sum_n (y_n - f(\mathbf{x}_n))^2. \qquad (6.24)$$

When S is the span of a finite set and f_0 is the constant function 1, one sets

$$\mathbf{y} = \begin{bmatrix} y_1 \\ y_2 \\ \vdots \\ \vdots \\ y_N \end{bmatrix}, \mathbf{X} = \begin{bmatrix} 1 & f_1(\mathbf{x}_1) & f_2(\mathbf{x}_1) & \cdots & f_H(\mathbf{x}_1) \\ 1 & f_1(\mathbf{x}_2) & f_2(\mathbf{x}_2) & \cdots & f_H(\mathbf{x}_2) \\ \vdots & \vdots & \vdots & \vdots & \vdots \\ \vdots & \vdots & \vdots & \vdots & \vdots \\ 1 & f_1(\mathbf{x}_N) & f_2(\mathbf{x}_N) & \cdots & f_H(\mathbf{x}_N) \end{bmatrix}, \text{ sets } \beta = \begin{bmatrix} \beta_0 \\ \beta_1 \\ \vdots \\ \beta_H \end{bmatrix}$$

and solves $\min_{\beta \in \mathbb{R}^m} h(\beta) = (\mathbf{y} - \mathbf{X}\beta)'(\mathbf{y} - \mathbf{X}\beta)$. This is linear regression projecting $\mathbf{y}$ onto the span of vectors of nonlinear functions of the observed $\mathbf{x}$'s.

1. If $S = \mathbf{span}(\{1, (\text{proj}_i)_{i=1}^{\ell}\})$, where proj_i is the canonical projection mapping, $\text{proj}_i(\mathbf{x}) = x_i, i = 1, \ldots, \ell$, we have **linear regression**.

2. If $S = \mathbf{span}(\{1, (\log(x_i))_{i=1}^{\ell}\})$ and the y_n are replaced with $\log(y_n)$, we have **log-linear regression**. Here, if we are careful, or lucky, the β_i measure the marginal effects of $\log(x_i)$ on $\log(y)$; that is, they measure the elasticities of y with respect to x_i.

3. If $S_H = \mathbf{span}(\{1, (\cos(hx))_{h=1}^{H}, (\sin(hx))_{h=1}^{H}\}) \subset C(\mathbb{R})$, we have one-dimensional **Fourier series regression**.

4. If $S_H = \mathbf{span}(\{1, (\cos(\sum_{i \leq \ell} h_i x_i)_{h_i \leq H}), (\sin(\sum_{i \leq \ell} h_i x_i)_{h_i \leq H})\}) \subset C_b(\mathbb{R}^{\ell})$, we have ℓ-dimensional **Fourier series regression**.

5. If $S_H = \mathbf{span}(\{1, x^1, x^2, \ldots, x^H\}) \subset C(\mathbb{R})$, we have one-dimensional H-**order polynomial regression**.

6. If $S_H = \mathbf{span}(\{1, \{\Pi_{i \leq \ell} x_i^{n_i} : \sum n_i \leq H\}\}) \subset C(\mathbb{R}^{\ell})$, we have ℓ-dimensional H-**order multinomial regression**.

6.5.c.2 Degrees of Complexity

The first two regression techniques, linear and log-linear, are fairly straightforward. The others leave us facing the problem of how large we should choose H, that is, how complex a set of functions we should use. There is a technique known as **cross-validation**, which gives an estimate based only on the data of how well $\widehat{f}$ generalizes, and this contains information about how large we should choose H.[11]

Fix a finite set $F_H = \{f_h : h = 0, 1, \ldots, H\}$ and let $S_H = \mathbf{span}(F_H)$. For each $m \leq N$, let $\widehat{f}_{H,m}$ solve the minimization problem

$$\min_{f \in S_H} \sum_{n \neq m} (y_n - f(\mathbf{x}_n))^2 \tag{6.25}$$

for $\widehat{f}_{H,m}$. In words, $\widehat{f}_{H,m}$ does the best that S_H can do in capturing the pattern of all of the data except the one left out, the mth. Define the mth **generalization error** as $e_m = (y_m - \widehat{f}_{H,m}(\mathbf{x}_m))^2$ and define $\text{err}_H = \frac{1}{N} \sum_m e_m$.

---

11. The full name for the procedure about to be described is "leave-one-out cross-validation." The reason for this name will become clear. Replacing "one" with some other integer or some fraction of N gives other variants of the procedure.

As far as $\widehat{f}_m$ is concerned, the data point $(\mathbf{x}_m, y_m)$ is new. This means that e_m gives a measure of how $\widehat{f}_{H,m}$ generalizes to a new data point and that, insofar as the data can tell, err_H gives an estimate of the average generalization error of S_H.

As H grows, one expects that the minimum in (6.25) goes down. Indeed, if $H \geq N$, one often finds the minimum of 0 to be attained. On the other hand, if $\widehat{f}_m$ fits all the points except the mth "too well," then we expect its generalization error to the mth point to be large.

Definition 6.5.10 *The **cross-validated choice of** H is the solution to*

$$\min_{H} \text{err}_H. \tag{6.26}$$

6.5.c.3 The Span-of-a-Compact-Set Case

The class of functions called **artificial neural networks (ANNs)** has many exemplars in which the set S is compact and $\mathbf{span}(S)$ is dense in $C(K)$. The set $\mathbf{span}(S)$ is a vector subspace, but is typically not an algebra.

1. Let G be a function in $C(\mathbb{R})$ and let $F = \{G(\alpha_0 + \alpha_1 x) : \|\alpha\| \leq B\} \subset C(\mathbb{R})$. The H-span of 1 and S is defined as $\mathbf{span}_H(S) = \{\beta_0 + \sum_{h \leq H} \beta_h f_h : f_h \in F\}$. This gives an example of what are known as **artificial neural networks (ANN)**. More specifically, this class of functions is known as a **single hidden layer feedforward (slff) artificial neural network** (with H hidden nodes based on the activation function G). Theorem 6.4.30 gives sufficient conditions for $\cup_H \mathbf{span}_H$ to be uniformly dense on compact sets in $C(\mathbb{R})$. These conditions can be relaxed; essentially, one need only assume that G is not a polynomial.

2. As above, set $\widetilde{\mathbf{x}} = (1, \mathbf{x}')' \in \mathbb{R}^{1+\ell}$. Let G be a nonpolynomial function in $C(\mathbb{R})$ and let $F = \{G(\alpha'\widetilde{\mathbf{x}}) : \|\alpha\| \leq B\}$. Here $F \subset C(\mathbb{R}^\ell)$. With $\mathbf{span}_H(S)$ defined as above, we have a multidimensional slff ANN. Again, G not being a polynomial is essentially the only condition needed to guarantee that $\cup_H \mathbf{span}_H$ is uniformly dense on compact sets in $C(\mathbb{R}^\ell)$.

Once again, the problem of choosing an H appears and, once again, cross-validation can be used.[12] Let $\theta_H = (\alpha, \beta) \in \mathbb{R}^{H \cdot 1 + \ell} \times \mathbb{R}^{1+H}$ and let $f_H(\mathbf{x}; \theta_H) = \beta_0 + \sum_{h \leq H} \beta_h f_h$, where $f_h(\mathbf{x}) = G(\alpha'_h \widetilde{\mathbf{x}})$. The computational problem of solving

$$\min_{\theta_H} \sum_{n \neq m} (y_n - f_H(\mathbf{x}_n; \theta_H))^2$$

can be quite difficult since the α_h enter nonlinearly. Doing it N times is (roughly) N times harder, though that is what is needed for cross-validation.

Especially when ℓ, the dimensionality of the data, is high, this class of ANNs has often done exceedingly well at fitting data. Comparing them to Fourier series provides a good intuition as to why this should be so: Suppose that the nonpolynomial G that we choose for our ANN is $G(r) = \cos(r)$. When we run a Fourier series regression, one of the terms, a $\cos(\sum_{i \leq \ell} h_i x_i)_{h_i \leq H}$ or a $\sin(\sum_{i \leq \ell} h_i x_i)_{h_i \leq H}$ has the largest β. An ANN with a single hidden unit and $B \geq H$ has at least that

---

12. In principle, one must also choose the bound B, and cross-validation can be used for that too, though it seems not to matter much in practice.

large a β. In other words, ANNs can pick the nonlinear functions having the high-est explanatory power first. When the data are high dimensional, this turns out to be a huge savings.

6.5.d Testing for Neglected Nonlinearity

As noted earlier, one of the basic assumptions in linear regression is that the true relation between $\mathbf{x}$ and y is affine. Nonlinear regression techniques let us check, to the extent that the data allow, whether or not the assumption is valid.

Setting $F = \{1, (\mathrm{proj}_i)_{i=1}^{\ell}\}$ and $S = \mathbf{span}(F)$ and solving

$$\min_{f \in S} \sum_n (y_n - f(\mathbf{x}_n))^2 \tag{6.27}$$

gives us linear regression. Suppose that we expand F to $F \cup \{g\}$ for some nonlinear g, set $S^g = \mathbf{span}(F \cup \{g\})$, and solve

$$\min_{f \in S^g} \sum_n (y_n - f(\mathbf{x}_n))^2. \tag{6.28}$$

If the β associated with the new function g in (6.28) is "too large," that is, both statistically and economically different from 0, or if the difference between the minimum in (6.27) and the minimum in (6.28) is "too large," then we have evidence that we have, at the very least, neglected the nonlinearity captured by the function g.

For a long time, the functions g that were chosen were squares of one of the x_i and sometimes the products of some pair x_i and x_j. The advantage of such a choice is that it is relatively easy to understand what finding "too large" an associated β implies. The downside is that such a test has little power to find nonlinearities that are almost orthogonal to the chosen function g. The ANN variant of this kind of neglected nonlinearity test does not have this downside, though it may be less easy to understand what finding "too large" a β implies.

Let $T = \{G(\alpha'\widetilde{x}) : \|\alpha\| \le B\}$, and solve the problem

$$\min_{f \in S, \beta \in \mathbb{R}, g \in T} \sum_n (y_n - [f(\mathbf{x}_n) + \beta g(\mathbf{x}_n)])^2. \tag{6.29}$$

Here the optimization is over β and α, and the ability to vary α means that there are fewer nonlinearities that are almost orthogonal to T. Again, if the β associated with the new function g is "too large" or if the difference between this minimum and the linear regression minimum is "too large," we have evidence that we have neglected a nonlinearity in the data.

6.6 ◆ Countable Product Spaces and Sequence Spaces

We have previously considered finite products of spaces, for example, $\mathbb{R}^{\ell}$ and $M \times M'$, and given them product metrics. In $M \times M'$, these are metrics with the property that $(x_n, y_n) \to (x, y)$ iff $x_n \to x$ and $y_n \to y$. There are many circumstances in which we need the products of countably many spaces.

For each $a \in A$, when A is a finite or countably infinite set, let (M_a, d_a) be a metric space.

Definition 6.6.1 *The **product space** $M = \times_{a \in A} M_a$ is the set of choice functions, $a \mapsto c(a)$, from A to $\cup_{a \in A} M_a$, such that for all $a \in A$, $c(a) \in M_a$.*

In other words and symbols, M is the set of all possible lists $(x_a)_{a \in A}$ with the property that for each a, x_a is a point in M_a.

Definition 6.6.2 *For $x \in M$ and $a \in A$, the **coordinate function**, **canonical projection**, or **projection** is the function $\mathrm{proj}_a : M \to M_a$ defined by $\mathrm{proj}_a(x) = x_a$. For $B \subset A$, $\mathrm{proj}_B : M \to \times_{b \in B} M_b$ is defined by $\mathrm{proj}_B(x) = (x_b)_{b \in B}$.*

With this notation, a point $x \in M$ is represented by $x = (\mathrm{proj}_a(x))_{a \in A}$. When $A = \mathbb{N}$, this can also be written as $x = (\mathrm{proj}_1(x), \mathrm{proj}_2(x), \ldots)$.

Definition 6.6.3 *$\times_{a \in A} M_a$ is a **sequence space** if $A = \mathbb{N}$ and there exists a single set S such that for each $n \in \mathbb{N}$, $M_n = S$. Such a sequence space is written $S^{\mathbb{N}}$.*

$S^{\mathbb{N}}$ is the set of all S-valued sequences, whence the name.

6.6.a Some Examples

The set $C(\mathbb{N})$ of continuous functions on the discrete set $\mathbb{N}$ is a different notation for $\mathbb{R}^{\mathbb{N}}$. Several useful subsets of $\mathbb{R}^{\mathbb{N}}$ are included in §6.2.a.3.

There is another class of uses of countable product spaces in learning theory and in the theory of infinitely repeated games. In order for there to be something to learn about in learning theory, situations must be repeated many times. Rather than try to figure out exactly what we mean by "many times," we send the number of times to infinity and look at where this process leads.

Example 6.6.4 *If $S = \{H, T\}$, then $S^{\mathbb{N}}$ is the set of all possible results of flipping a coin once a day from now into eternity. Theorem 2.10.20 shows that $S^{\mathbb{N}}$ is uncountable. A leading example in statistical learning theory draws a random number $\theta \in [0, 1]$ according to some cdf, and then has P_θ a distribution on $\{H, T\}^{\mathbb{N}}$ with the property that $(\mathrm{proj}_t(x))_{t \in \mathbb{N}}$ is a sequence of independent random variables, all having the probability that $\mathrm{proj}_t(x) = H$ equal to θ. At each point in time t, the problem is to estimate θ from the history $\mathrm{proj}_{\{1,\ldots,t\}}(x) = (\mathrm{proj}_1(x), \mathrm{proj}_2(x), \ldots \mathrm{proj}_t(x))$.*

Example 6.6.5 *$S = \mathbb{R}_+^2$ or $S = [0, M]^2$ when we are considering an infinitely repeated quantity-setting game with two players, or $S = \times_{i \in I} A_i$ when we are considering repeating a game with player set I and each $i \in I$ has action set A_i. In both of these cases, the space of possible outcomes is $S^{\mathbb{N}}$. A strategy for a player must specify an initial choice and a sequence of mappings $\mathrm{proj}_{\{1,\ldots,t\}}(x) \mapsto a_i \in A_i$. This is a complicated object, and we no longer have the indentically distributed properties that made learning relatively easy in the previous example.*

Example 6.6.6 (Strategy Sets for Infinitely Repeated Games) *Fix a finite set of players I and for each $i \in I$, a finite set of actions A_i. Let $A = \times_{i \in I} A_i$. Let $S^0 = \{h^0\}$ be a singleton set. Given S^{t-1}, define $S^t = S^{t-1} \times A$ so that $S^t = S^0 \times A^t$. A **pure strategy** for player i at time t is a function $\sigma_{i,t} : S^{t-1} \to A_i$. The* *set of all pure strategies for player i at time t is therefore* $\Sigma_{i,t} := A_i^{S^{t-1}}$, *the set of all such functions. The set of all pure strategies for player i is the countable product space* $\Sigma_i := \times_{t \in \mathbb{N}} \Sigma_{i,t}$. *The set of all pure strategies in the game is* $\Sigma := \times_{i \in I} \Sigma_i$. *The set of all mixed, or randomized, strategies for i is the set of all probability distributions on the uncountable space* Σ_i. *We will see that when* A_i *is finite,* Σ_i *is a compact metric space. Chapter 9 presents a systematic study of probabilities on metric spaces.*

Exercise 6.6.7 [↑Example 6.6.6] Let $I = \{1, 2\}$ and suppose that each A_i contains two elements. As a function of t, find $\#\Sigma_{i,t}$, the cardinality of i's set of pure strategies at time t. [This gives superexponential growth; you should find that $\log(\#\Sigma_{i,t})$ grows exponentially fast.]

Strategies for infinitely repeated games are an example of a dynamical system.

Example 6.6.8 (Cournot and Other Dynamical Systems) [↑*Exercise 4.4.14*] *Two firms selling to a market decide on their quantities,* $a_i \in A_i = [0, M]$, $i = 1, 2$. *Set* $A = A_1 \times A_2$. *There is an initial state* $\theta_0 = (\theta_{i,0})_{i \in I} = (\theta_{1,0}, \theta_{2,0}) \in \Theta := A$ *at time* $t = 0$. *When* t *is an odd period, player 1 changes* $\theta_{1,t-1}$ *to* $\theta_{1,t} = Br_1(\theta_{2,t-1})$ *and when* t *is an even period, player 2 changes* $\theta_{2,t-1}$ *to* $\theta_{2,t} = Br_2(\theta_{1,t-1})$. *Or, if you want to combine the periods,*

$$(\theta_{1,t-1}, \theta_{2,t-1}) \mapsto (Br_1(\theta_{2,t-1}), Br_2(Br_1(\theta_{2,t-1}))).$$

In either case, we have specified a dynamical system. For $t = 1$, H^{t-1}, *the history up to and including time* $t - 1$ *is defined as* Θ *and for* $t \geq 2$, H^{t-1} *is defined as* $\Theta \times A^{t-1}$. *A **dynamical system on** A is a class of functions* $f_t : H^{t-1} \to A$, $t \in \mathbb{N}$. *A pure strategy for an infinitely repeated game gives a dynamical system on* A.

Whatever dynamical system we study, for each θ_0, *the result is the **outcome**,*

$$\mathbb{O}(\theta_0) := (\theta_0, f_1(\theta_0), f_2(\theta_0, f_1(\theta_0)), \ldots),$$

a point in $\Theta \times A^{\mathbb{N}}$.

If each f_t *depends only on what happened in period* $t - 1$, *we have a **Markovian system**. A Markovian system is **stationary** if there exists an* $f : A \to A$ *such that for all* t, $f_t = f$. *For a stationary Markovian system, the outcome is*

$$\mathbb{O}(\theta_0) = (\theta_0, f(\theta_0), f(f(\theta_0)), f(f(f(\theta_0))), \ldots).$$

Definition 6.6.9 *A point* $\hat{a} \in A$ *is **stable** for the dynamical system* $(f_t)_{t \in \mathbb{N}}$ *on* A *if* $\exists \theta_0$ *such that* $\mathbb{O}(\theta_0) = (\theta_0, \hat{a}, \hat{a}, \hat{a}, \ldots)$.

Thus, $\hat{a}$ is stable for a stationary Markovian dynamical system iff $\hat{a}$ is a fixed point of f.

With the best response dynamics specified in Example 6.6.8, the stable points are exactly the Nash equilibria. Stability is a weak notion. A point is stable if, starting at that point, one never moves from there if there are no perturbations. Think of a marble balanced on the point of a pin. Stronger notions of stability require convergence.

Suppose that S is a metric space. Points in $S^{\mathbb{N}}$ are sequences, so that it makes sense to say that a point $s(\in S^{\mathbb{N}})$ converges to $\hat{a}(\in S)$, written $s \to \hat{a}$, if $\forall \epsilon > 0$, $\text{proj}_n(s) \in B_\epsilon(\hat{a})$ a.a.

Definition 6.6.10 *A point $\hat{s} \in S$ is **asymptotically stable** or **locally stable** for a dynamical system $(f_t)_{t \in \mathbb{N}}$ on S if it is stable and $\exists \epsilon > 0$ such that for all $\theta_0 \in B(\hat{\theta}, \epsilon)$, $\mathbb{O}(\theta_0) \to \hat{s}$.*

Example 6.6.11 *If $\Theta = S = \mathbb{R}_+$ and the stationary dynamical system is Markovian with $f(r) = r^2$, then 1 is a stable point that is not asymptotically stable, whereas 0 is asymptotically stable, and for all $\theta_0 < 1$, $\mathbb{O}(\theta_0) \to 0$.*

Exercise 6.6.12 If $\Theta = S = \mathbb{R}$ (not just $\mathbb{R}_+$) and the stationary dynamical system is Markovian with $f(r) = r^2$, find the stable and the asymptotically stable points. Find the stable and the asymptotically stable points if $f(r) = r^3$.

Definition 6.6.13 *A point $\hat{s} \in S$ is **globally stable** if it is stable and $\forall \theta_0$, $\mathbb{O}(\theta_0) \to \hat{s}$.*

If more than one point is stable, then there cannot be a globally stable point. If a Markovian dynamical system, $f : S \to S$, is a contraction mapping and S is a complete metric space, then the unique fixed point is globally stable. Many dynamical systems of interest have no stable points but still have reasonable "limit" behavior.

Definition 6.6.14 $s \in S^{\mathbb{N}}$ ***accumulates at*** u *if $\forall \epsilon > 0$, $\text{proj}_n(s) \in B_\epsilon(u)$ infinitely often. The set of accumulation points of a sequence s is denoted $\rightsquigarrow (s)$.*

Accumulation points of sequences give us (yet) another way to phrase closure and compactness.

Exercise 6.6.15 Let (M, d) be a metric space. Show that:

1. $E \subset M$ is closed iff for all $x \in E^{\mathbb{N}}$, $\rightsquigarrow (x) \subset E$;
2. a nonempty $K \subset M$ is compact iff $\forall x \in K^{\mathbb{N}}$, $\rightsquigarrow (x)$ subset of K; and
3. for all $x \in M^{\mathbb{N}}$, $\rightsquigarrow (x)$ is closed.

Definition 6.6.16 *For a dynamic system $(f_t)_{t \in \mathbb{N}}$ on S, the set $\bigcup_{\theta \in \Theta} \rightsquigarrow (\mathbb{O}(\theta))$ is the set of ω-**limit points**.*

If S is compact, for all θ, $\rightsquigarrow (\mathbb{O}(\theta)) \neq \emptyset$.

Exercise 6.6.17 (A Discontinuous Markovian Dynamical System) Let $\Theta = S = [0, 8]$, and define $f : S \to S$ by

$$f = \begin{cases} 5 + x/4 & \text{if } 0 \le x < 4, \\ x/4 & \text{if } 4 \le x \le 8 . \end{cases}$$

Graph f and give the set of ω-limit points of the associated Markovian dynamical system, proving your answer.

Exercise 6.6.18 (Another Discontinuous Markovian Dynamical System) Let $\Theta = S = [0, 8]$ and define $f : S \to S$ by

$$f = \begin{cases} 5 + x/4 & \text{if } 0 \leq x < 4, \\ -7 + 7x/4 & \text{if } 4 \leq x \leq 8 . \end{cases}$$

Graph f and give the set of ω-limit points of the associated Markovian dynamical system, proving your answer.

When S is compact, there is a useful metric on $C(S; S)$, the continuous functions from S to S, which is given by $d(f, g) = \max_{x \in S} d(f(x), g(x))$. The dependence of the set of accumulation points of a continuous Markovian dynamical system on the $f \in C(S; S)$ can be amazingly sensitive.

Example 6.6.19 *Let $S = \{x \in \mathbb{R}^2 : \|x\|_2 = 1\}$ and define $f_r(x)$ to be clockwise rotation of x by r degrees. If r is rational, then $\leadsto (\mathbb{O}((1, 0)))$ contains finitely many equally spaced points. By contrast, if r is irrational, then it can be shown (by some rather difficult arguments) that $\leadsto (\mathbb{O}((1, 0))) = S$.*

If a dynamical system cycles, it will have ω-limit points. Cycles are invariant sets and are often special cases of attractors.

Definition 6.6.20 *A set $S' \subset S$ is **invariant** under the dynamical system $(f_t)_{t \in \mathbb{N}}$ if $\theta \in S'$ implies $\forall k$, $\text{proj}_k(\mathbb{O}(\theta)) \in S'$. An invariant S' is an **attractor** if $\exists \epsilon > 0$ such that for all θ with $d(\theta, S') < \epsilon$, $\leadsto (\mathbb{O}(\theta)) \subset S'$.*

A final class of uses is found in discrete time stochastic process theory.

Example 6.6.21 *Suppose that a point $x \in \mathbb{R}^{\mathbb{N}}$ is chosen according to some probability distribution and that as time, t, moves from 1 to 2 to 3 and onward, we observe $\text{proj}_t(x)$. What we observe is random and happens over time. Of special interest is the joint distribution of the random variables $\text{proj}_{\{1,\ldots,t\}}(x) \in \mathbb{R}^t$ and $\text{proj}_{t+1}(x)$. Knowing these joint distributions allows us to give, at each t, the distribution of the next realization/observation, $\text{proj}_{t+1}(x)$, conditional on what we have seen so far—the random variables in the vector $\text{proj}_{\{1,\ldots,t\}}(x)$. The same considerations apply to probability distributions on $(\mathbb{R}^\ell)^{\mathbb{N}}$, which give the set of vector-valued discrete time stochastic processes.*

6.6.b   The Product Metric on $\times_{a \in A} M_a$

Parallel to the case of finite products, we want a metric on $M = \times_{a \in A} M_a$ with the property that $x_n \to x$ iff $\forall a \in A$, $\text{proj}_a(x_n) \to \text{proj}_a(x)$ in the metric space (M_a, d_a). Since A is finite or countable, summing over $a \in A$ is sensible.

Definition 6.6.22 *The **product metric on** M is defined as*

$$d_{Prod}(x, x') = \sum_{a \in A} 2^{-a} \min\{d_a(\text{proj}_a(x), \text{proj}_a(x')), 1\}. \tag{6.30}$$

Lemma 6.6.23 *$d_{Prod}(x_n, x) \to 0$ in M iff for all $a \in A$, $\text{proj}_a(x_n) \to \text{proj}_a(x)$.*

Proof. Suppose that for some $a \in A$, $\text{proj}_a(x_n) \nrightarrow \text{proj}_a(x)$. This means that there exists $\epsilon > 0$ and a subsequence $x_{n'}$ such that $d_a(\text{proj}_a(x_{n'}), \text{proj}_a(x)) \geq \epsilon$ almost always. Along the subsequence $x_{n'}$, $d_{Prod}(x_{n'}, x) \geq \min\{\epsilon, 1\}/2^a$, so that $x_n \nrightarrow x$.

Suppose that for all $a \in A$, $\text{proj}_a(x_n) \to \text{proj}_a(x)$. Pick $\epsilon > 0$. Choose $\overline{a}$ large so that $\sum_{a > \overline{a}} 2^{-a} < \epsilon/2$. For $a \leq \overline{a}$, pick N_a such that for $n \geq N_a$, $d_a(\text{proj}_a(x_n), \text{proj}_a(x)) < \epsilon/2\overline{a}$. For $n \geq \max\{N_a : a \leq \overline{a}\}$, $d_{Prod}(x_n, x) < \overline{a} \cdot (\epsilon/2\overline{a}) + \epsilon/2$. ∎

Exercise 6.6.24 [↑Example 6.6.6] If each (M_a, d_a) is discrete, then $x_n \to x$ in $\times_{a \in A} M_a$ iff for all $a \in A$, $\text{proj}_a(x_n) = \text{proj}_a(x)$ almost always. [This implies that the outcome function, $\mathbb{O}$, mapping strategies in infinitely repeated games to $(\times_{i \in I} A_i)^{\mathbb{N}}$ is continuous.]

As we have seen, $C(\mathbb{N}) = \mathbb{R}^{\mathbb{N}}$. Keep in mind that a point x in $\mathbb{R}^{\mathbb{N}}$ is a sequence of real numbers and that a sequence in $\mathbb{R}^{\mathbb{N}}$ is therefore a sequence of sequences. Rereading Lemma 6.6.23, we see that convergence in the product metric on $\mathbb{R}^{\mathbb{N}}$ is the same as pointwise convergence in $C(\mathbb{N})$.

Exercise 6.6.25 [↑Example 6.2.7] Let $c_F \subset c_0$ be the set of sequences in $\mathbb{R}^{\mathbb{N}}$ that are nonzero at most finitely often, that is, almost always equal to 0. Let $c_F(\mathbb{Q})$ denote the elements of c_F having only rational coordinates. Show that $c_F(\mathbb{Q})$ is a countable dense subset of $\mathbb{R}^{\mathbb{N}}$.

Exercise 6.6.26 Show that $\mathbb{R}^{\mathbb{N}}$ is complete.

Exercise 6.6.27 [↑Diagonalization, Theorem 6.2.62] For each $n \in \mathbb{N}$, let $[a_n, b_n]$ be a nonempty compact interval in $\mathbb{R}$. Show that $\times_{n \in \mathbb{N}}[a_n, b_n]$ is compact.

6.6.c Cylinder Sets

For $b \in A$ and $E_b \subset M_b$, $\text{proj}_b^{-1}(E_b)$ is the subset of $\times_{a \in A} M_a$ having the bth coordinate in the set E_b. Drawing the picture in $\mathbb{R}^2$ suggests why such a set is called a cylinder. Finiteness is crucial in the following.

Definition 6.6.28 *Let B be a finite subset of A, and for each $b \in B$, let $E_b \subset M_b$. A **cylinder set with base** $\times_{b \in B} E_b$ is a set of the form $\text{proj}_B^{-1}(\times_{b \in B} E_b)$. A base set is **open** if each E_b is an open subset of M_b. $\mathcal{C} \subset \mathcal{P}(M)$ denotes the class of cylinder sets.*

Exercise 6.6.29 Show that $\mathcal{C}$ is closed under finite intersection; that is, if $\{C_1, \ldots, C_M\} \subset \mathcal{C}$, then $\cap_{m \leq M} C_m \in \mathcal{C}$. Show that the class of cylinder sets with open base sets is also closed under finite intersection.

Recall that a topology on a space is the class of open sets and that a class of open sets, S, is a subbasis for a topology if every open G can be expressed as a union of sets, each of which is a finite intersection of elements of S (Definition 4.8.3). Proving continuity using subbases is often easier, for instance, Example 4.8.6 (p. 137).

Lemma 6.6.30 *The class of cylinder sets with open base sets is a subbasis for the product metric topology.*

Proof. Let G be an open subset of $M = \times_{a \in A} M_a$ and pick $x \in G$. We must show that there is a cylinder set with an open base set. Since G is open, $\exists \epsilon > 0$, so that $B_\epsilon(x) \subset G$. Choose $\bar{a}$ such that $\sum_{a \geq \bar{a}} 2^{-a} < \epsilon/2$. For each $a \leq \bar{a}$, let E_a be the open ball around $\text{proj}_a(x)$ with radius $\epsilon/2\bar{a}$ in M_a. The cylinder set $C = \text{proj}_{a \leq \bar{a}}^{-1}(\times_{a \leq \bar{a}} E_a)$ has an open base, and $x \in C \subset B_\epsilon(x) \subset G$. ■

Exercise 6.6.31 [↑Exercise 6.6.25] Show directly that every nonempty open cylinder set in $\mathbb{R}^{\mathbb{N}}$ contains a point in $c_F(\mathbb{Q})$.

6.6.d Compactness, Completeness, and Separability in $\times_{a \in A} M_a$

We have already seen that $\mathbb{R}^{\mathbb{N}}$ is complete and separable and that $\times_{n \in \mathbb{N}}[a_n, b_n]$ is a compact subset of $\mathbb{R}^{\mathbb{N}}$. These results are implications of the following.

Theorem 6.6.32 $(\times_{a \in A} M_a, d_{Prod})$ *is complete iff each* (M_a, d_a) *is complete, is separable iff each* (M_a, d_a) *is separable, and is compact iff each* (M_a, d_a) *is compact.*

Proof. Completeness: Suppose that each M_a is complete and let x_n be a Cauchy sequence in $\times_{a \in A} M_a$. For each $a \in A$, $\text{proj}_a(x_n)$ is Cauchy and so converges to some $x_a^\circ \in M_a$. By Lemma 6.6.23, $x_n \to x^\circ$.

Suppose that (M_b, d_b) is not complete for some $b \in A$. Let $x_{b,n}$ be a Cauchy sequence in M_b that does not converge to any point in M_b. Let x° be a point in M and define the sequence x_n by $\text{proj}_b(x_n) = x_{b,n}$ and $\text{proj}_a(x_n) = \text{proj}_a(x^\circ)$ for all $a \neq b$. Then x_n is a Cauchy sequence that does not converge to any point in M.

Separability: Suppose that E_a is a countable dense subset of M_a for each $a \in A$. Let B_n be a sequence of finite subsets of A such that $A = \cup_n B_n$. Pick $x^\circ \in \times_{a \in A} M_a$. For each $n \in \mathbb{N}$, let $E_n = \{x \in M : (\forall b \in B_n)[\text{proj}_b(x) \in E_b] \land (\forall a \notin B_n)[\text{proj}_a(x) = \text{proj}_a(x^\circ)]\}$. E_n is countable so that $E = \cup_n E_n$ is countable. Every open cylinder set contains a point in E.

Suppose now that M_b is not separable for some $b \in A$. If E is dense in M, then $\text{proj}_b(E)$ is dense in M_b, so that no countable subset of M can be dense.

Compactness: The argument here is closely related to proof of diagonalization, Theorem 6.2.62 (p. 292). Let $x_n = (x_{a,n})_{a \in A}$ be a sequence in M. Enumerate A using $\mathbb{N}$. From the sequence $n \mapsto x_{1,n}$ we can select a convergent subsequence, $x_{1,n_{1,1}}, x_{1,n_{1,2}}, x_{1,n_{1,3}}, \ldots$ because M_1 is compact. Now consider the sequence $n \mapsto x_{2,n}$ along the subsequence $n_{1,1}, n_{1,2}, \ldots$. We can choose a convergent subsequence (of the subsequence) $x_{1,n_{2,1}}, x_{1,n_{2,2}}, x_{1,n_{2,3}}, \ldots$, where $k \mapsto n_{2,k}$ is a strictly increasing subsequence of $k \mapsto n_{1,k}$. Continuing in this fashion gives a sequence of sequences, $(k \mapsto n_{a,k})_{a \in A}$, with the property that each row is increasing, the $(a + 1)$st row is a subset of the ith row, and for all a, $\lim_k x_{n_{a,k}}$ exists. Let $n_k = n_{k,k}$ and the subsequence x_{n_k} of x_n converges.

Finally, suppose that for some b, M_b contains a sequence, $x_{b,n}$, with no convergent subsequence. Let x° be a point in M and define the sequence x_n by $\text{proj}_b(x_n) = x_{b,n}$ and $\text{proj}_a(x_n) = \text{proj}_a(x^\circ)$ for all $a \neq b$. Then x_n is a sequence with no convergent subsequence. ■

We know that for optima to exist for all continuous preferences, we must have a compact domain. The structure of continuous preferences on compact $M = \times_{a \in A} M_a$ is informative.

Definition 6.6.33 *A function $f : \times_{a \in A} M_a \to \mathbb{R}$ is **finitely determined** if there is a finite $B \subset A$ and function $g : \times_{b \in B} M_b \to \mathbb{R}$ such that $f(x) = g(\text{proj}_B(x))$.*

Lemma 6.6.34 *If $M = \times_{a \in A} M_a$ is compact, then the finitely determined functions in $C(M)$ are dense.*

Proof. The class of continuous, finitely determined functions is an algebra that separates points and contains the constants. ■

This means that continuous functions on compact sequence spaces are **tail insensitive**, that is, to within any $\epsilon > 0$, they do not depend on any but the first N_ϵ-coordinates.

Example 6.6.35 [↑*Growth Model Example 3.7.2*] *Let $u : \mathbb{R}_+ \to \mathbb{R}_+$ be a continuous increasing instantaneous utility function. For any sequence $b \in \mathbb{R}_+^{\mathbb{N}}$, let $K = K(b) = \times_{t \in \mathbb{N}}[0, b_t]$. Consider the class of functions $U : K \to \mathbb{R}_+ \cup \{\infty\}$ defined by $U(x) = \sum_t \beta^t u(x_t)$ for some $\beta \in (0, 1)$. The function U belongs to $C(K)$ iff $\sum_t \beta^t u(b_t) < \infty$. To see why, note that if $\sum_t \beta^t u(b_t) < \infty$, then $U_T(x) := \sum_{t \leq T} \beta^t u(x_t)$ gives a Cauchy sequence of (finitely determined) continuous functions converging pointwise to U. Conversely, if $\sum_t \beta^t u(b_t) = \infty$, then the function U fails to achieve its maximum on the compact set K, so cannot be continuous.*

Exercise 6.6.36 [↑*Growth example*] Show directly, using the theorem of the maximum, that the value function for the infinite-horizon dynamic problem in (6.4) (p. 280) is continuous. [Remember in checking the compactness of the set of possible consumption paths that the upper hemicontinuous image of a compact set is compact.]

There are many utility functions on product spaces that are not discounted sums. For games that last into the indefinite future and players that choose among finitely many actions at each point in time, tail insensitivity of the payoffs is equivalent to continuity.

Lemma 6.6.37 *If each (M_a, d_a) is a finite metric space, then $U : \times_{a \in A} M_a$ is continuous in the product metric iff U is tail insensitive.*

Exercise 6.6.38 Prove Lemma 6.6.37.

Exercise 6.6.39 Which, if any, of the following functions mapping $[0, 1]^{\mathbb{N}} \to \mathbb{R}$ are continuous? Explain.

1. $U_1(x) = \liminf_T \frac{1}{T} \sum_{t=1}^{T} x_t$.
2. $U_2(x) = \limsup_T \frac{1}{T} \sum_{t=1}^{T} x_t$.
3. $U_3(x) = U_2(x) - U_1(x)$.
4. $V_\beta(x) = (1 - \beta) \sum_t \beta^{t-1} x_t, \ |\beta| < 1$.

5. $V(x) = \liminf_{\beta \uparrow 1} V_\beta(x)$.

6. $V_p(x) = \sup\{V_\beta(x^\pi) : \pi$ is a finite permutation of $\mathbb{N}\}$. [A finite permutation is a bijection of $\mathbb{N}$ and itself with $\pi(t) \neq t$ for at most finitely many t and, as before, x^π is defined by $\text{proj}_t(x^\pi) = x_{\pi(t)}$ for all $t \in \mathbb{N}$.]

6.6.e   The Universal Separable Metric Space

In principle, we can learn everything that there is to learn about every possible separable metric space by studying subsets of the compact metric space $[0, 1]^{\mathbb{N}}$. This turns out to be very convenient when we examine the properties of probabilities on metric spaces, a study that includes much of the material on stochastic processes.

Theorem 6.6.40   *A metric space (M, d_M) is separable iff it is homeomorphic to a subset of $[0, 1]^{\mathbb{N}}$ with the product metric.*

Proof.   Let $E = \{e_a : a \in \mathbb{N}\}$ be a countable dense subset of M. Define $\varphi : M \to [0, 1]^{\mathbb{N}}$ by $\text{proj}_a(\varphi(x)) = \min\{1, d_M(x, e_a)\}$. We now show that M and $\varphi(M)$ are homeomorphic. The φ is one-to-one, hence invertible on $\varphi(M)$, because $x \neq x'$ iff $(\exists a \in \mathbb{N})[\min\{1, d(x, e_a)\} \neq \min\{1, d(x', e_a)\}]$, and it is a homeomorphism since $[x_n \to x] \Leftrightarrow (\forall a \in \mathbb{R})[\min\{1, d(x_n, e_a)\} \to \min\{1, d(x, e_a)\}]$.   ∎

6.6.f   Compactifications

The mapping $x \mapsto x$ from the noncompact metric space $(0, 1)$ to the compact metric space $[0, 1]$. embeds $(0, 1)$ as a dense subset of $[0, 1]$. This addition of extra point(s) so as to make a space compact is called **compactification**. The following repeats Definition 4.10.34.

Definition 6.6.41   *A **compactification of a metric space** (M, d) is a pair $(\varphi, (M', d'))$, where (M', d') is compact, φ is a homeomorphism between M and $\varphi(M)$, and $\varphi(M)$ is a dense subset of M'.*

Theorem 6.6.42   *Every separable metric space can be compactified.*

Proof. M is homeomorphic to some $\varphi(M) \subset [0, 1]^{\mathbb{N}}$, and the compactification is $(\varphi, (\text{cl}(\varphi(M)), d_{Prod}))$.   ∎

One problem with compactifications is that some functions in $C_b(M)$ may not have continuous extensions to M', for example, the function $\sin(1/x)$ is a continuous bounded function on $(0, 1)$, but has no continuous extension to $[0, 1]$. This example is an instance of a general result.

Lemma 6.6.43   *If $(\varphi, (M', d'))$ is a compactification of (M, d), then $(C(M') \circ \varphi) = \{f \circ \varphi : f \in C(M')\}$ is a strict subset of $C_b(M)$ iff M is not compact.*

Proof.   Recall that M is compact iff $C_b(M) = C(M)$ iff $C_b(M)$ is separable. Since M' is compact, $(C(M') \circ \varphi)$ is a separable subset of $C_b(M)$. If M is not compact, then $(C(M') \circ \varphi)$ is a separable subset of a nonseparable space, hence a strict subset. If M is compact, then φ must be a homeomorphism and the mapping $f \leftrightarrow (f \circ \varphi)$ an isometry between $C(M')$ and $C(M)$.   ∎

In studying probabilities on metric spaces, we use the separability of $C([0, 1]^{\mathbb{N}})$ to prove some of the properties of what is called weak* convergence of probabilities.

6.6.g Homeomorphisms and Isometries

The function $F(r) = \frac{e^r}{1+e^r}$ is a homeomorphism of $\mathbb{R}$ and $(0, 1)$ (Definition 4.1.24). A homeomorphism between the domains $(0, 1)$ and $\mathbb{R}$ allows us to relabel the functions in $C_b((0, 1))$ as functions in $C_b(\mathbb{R})$. This relabeling is what we call an isometry, and it captures the notion of "everything but the name is the same." The following repeats and relabels Definition 4.1.25.

Definition 6.6.44 *An **isometry** between two metric spaces (M, d) and (M', d') is a one-to-one onto mapping $T : M \leftrightarrow M'$ such that for all $x, y \in M$, $d'(T(x), T(y)) = d(x, y)$. If any isometry between (M, d) and (M', d') exists, we say that they are **isometric**.*

Exercise 6.6.45   Show that being isometric is an equivalence relation for metric spaces and that T is an isometry between M and M' iff T^{-1} is an isometry between M' and M.

Example 6.6.46 (Isometries as Relabeling)   *If $g \in C_b((0, 1))$, then $(g \circ F) \in C_b(\mathbb{R})$, where $(F(r) = e^r/(1 + e^r)$. Let us have some notation for this mapping: define $T_F : C_b((0, 1)) \to C_b(\mathbb{R})$ by $T_F(g) = (g \circ F)$. Note that T_F is one-to-one, onto and satisfies $d_\infty(T_F(g), T_F(g')) = d_\infty(g, g')$; that is, T_F is an **isometry** between $C_b((0, 1))$ and $C_b(\mathbb{R})$. Intuitively, the existence of an isometry shows us that up to relabeling using T_F, $C_b((0, 1))$ and $C_b(\mathbb{R})$ are the same space.*

Exercise 6.6.47   [↑Example 6.6.46] Define $T_F^{-1} : C_b(\mathbb{R}) \to C_b((0, 1))$ by $T_F^{-1}(h) = (h \circ F^{-1})$. Show that T_F^{-1} is an isometry and that it really is the inverse of T_F.

Metric spaces can be isometric to proper subsets of themselves.

Exercise 6.6.48   Show that $[0, \infty)$ and $[1, \infty)$ are isometric. Let $F \subset C([0, 1])$ be the set of continuous functions on $[0, 1]$ that are constant on the interval $[\frac{1}{2}, 1]$. Show that F and $C([0, 1])$ are isometric.

The homeomorphism of $\mathbb{R}$ and $(0, 1)$ has an implication for preferences.

Example 6.6.49   *The preferences representable by utility functions in $C(\mathbb{R}_+^\ell)$ are the same as the preferences representable by utility functions in $C_b(\mathbb{R}_+^\ell)$. To see why, let $F(r) = \frac{e^r}{1+e^r}$, and for $u \in C(\mathbb{R}_+^\ell)$, define $v = F \circ u \in C_b(\mathbb{R}_+^\ell)$. The u and v represent the same preferences.*

Since every separable metric space (M, d) is homeomorphic to a subset of $[0, 1]^{\mathbb{N}}$, continuous functions on subsets of $[0, 1]^{\mathbb{N}}$ become useful objects to study. For any $E \subset [0, 1]^{\mathbb{N}}$ and $f \in C([0, 1]^{\mathbb{N}})$, $f_{|E}$, the restrictions of f to E must belong to $C(E)$, hence must be isometric to an element of $C(M)$ if E and M are homeomorphic. This does not mean that any $g \in C(E)$ is the restriction of some function in $C([0, 1]^{\mathbb{N}})$, though it is for compact E. This leads us to our next topic.

6.7 ◆ Defining Functions Implicitly and by Extension

There are two main contexts in which we want to know that a continuous function satisfying infinitely many conditions exists. First, we would like there to be a function $y^*(x)$ such that for each $x \in M$, $y^*(x)$ solves the maximization problem $\max_{y \in \Gamma(x)} f(x, y)$. If there is such a function, we would like to know about its properties. Second, given a continuous (or continuous bounded) function $f : E \to M'$, $E \subset M$, we would like there to be a continuous (or continuous bounded) $F : M \to M'$ such that $F_{|E} = f$. If it exists, the function F is a **continuous extension** of f. The existence of continuous extensions tells us that the set of continuous functions is very rich. Here are sketches of some applications.

6.7.a Applications of Extension Ideas

The theorem of the maximum, 6.1.31, shows that, given a continuous compact-valued correspondence Γ from M to M' and a continuous $f : M \times M' \to \mathbb{R}$, $x \mapsto y^*(x)$ is an upper hemicontinuous correspondence. If $\#y^*(x) = 1$, then y^* is a function. Since uhc functions are continuous, this tells us that y^* is continuous. When $M = \mathbb{R}^n$ and $M' = \mathbb{R}^m$, the implicit function theorem, which we prove using extension ideas, gives conditions under which the derivative, $D_x y^*$, exists and provides information about the sign of the derivative.

We have seen that any cdf, F, defines a probability, P_F, on the field, $\mathcal{F}^\circ$, of interval subsets of $\mathbb{R}$. We show that the function P_F is uniformly continuous on $\mathcal{F}^\circ$ and that $\mathcal{F}^\circ$ is dense in the class of subsets of $\mathbb{R}$ that we care about—the Borel σ-field, denoted $\mathcal{F}$ or $\sigma(\mathcal{F}^\circ)$. The extension theorem for dense sets implies that any uniformly continuous function on a dense set can be extended to a continuous function on the whole set.

The basic existence result, Theorem 4.8.11, shows that $K \subset M$ is compact iff every $F \in C(M)$ achieves its maximum on K. This is a bit puzzling. Compactness is an intrinsic property of K and should have nothing to do with the space M that contains K. The Tietze extension theorem implies that a function f belongs to $C(K)$ iff it is of the form $F_{|K}$ for some $F \in C(M)$.

Finally, we will see that the countably additive probability, μ, on metric spaces satisfies a sandwiching property—for every (measurable) set E and every $\epsilon > 0$, there exists a closed set F and an open set G with $F \subset E \subset G$ and $\mu(G \setminus F) < \epsilon$. This "squeezes" E between a closed and an open set. The last result in this section shows that there are continuous functions, $f : M \to [0, 1]$ such that for all $x \in F$, $f(x) = 1$ and for all $y \in G^c$, $f(y) = 0$. This means that the continuous functions are very close to the indicator functions—$\mu(F) = \int 1_F(x) \, d\mu(x) \simeq \int f(x) \, d\mu(x)$. Since μ is determined by its values on closed sets, this shows that μ is also determined by its integrals against continuous functions. In other words, we can view μ as a function $L_\mu : C_b(M) \to \mathbb{R}$ defined by $L_\mu(f) = \int f \, d\mu$, and this function tells us, at least in principle, everything there is to know about μ. Further, the function L_μ is linear, $L_\mu(\alpha f + \beta g) = \alpha L_\mu(f) + \beta L_\mu(g)$. Sometimes, probabilities are easier to study when we look at them as linear functions.

6.7.b The Implicit Function Theorem

There is one basic technique in the differentiable comparative statics that economists use.

Example 6.7.1 *Suppose that one solves the class of optimization problems* $\max_y f(x, y)$ *by solving the first-order condition,* $\partial f(x, y^*)/\partial y = 0$, *for* y^* *for each* x. *This implicitly defines* y^* *as a function of* x. *If* $y^* = y^*(x)$ *is differentiable, then one calculates the sign of* dy^*/dx *by noting that the first-order condition,* $\partial f(x, y^*(x))/\partial x$, *is identically equal to* 0. *Taking derivatives on both sides of* $\partial f(x, y^*(x))/\partial x \equiv 0$ *with respect to* x, *we have*

$$\frac{\partial^2 f(x, y^*(x))}{\partial y \partial x} + \frac{\partial^2 f(x, y^*(x))}{\partial y^2} \frac{dy^*(x)}{dx} = 0. \tag{6.31}$$

Mostly, when $y^*(x)$ *is a strict optimum,* $\frac{\partial^2 f(x, y^*(x))}{\partial y^2} < 0$. *In this case,* $dy^*(x)/dx > 0$ *if* $\partial^2 f/\partial y \partial x > 0$ *and* $dy^*(x)/dx < 0$ *if* $\partial^2 f/\partial y \partial x < 0$.

The intuition for this should be clear (after enough exposure to economics). One picks the optimal y^* to make the marginal value of increasing y equal to 0. If increases in x make the marginal value of y higher, $\partial^2 f/\partial y \partial x > 0$, then y^* must move up as x moves up, or if increases in x make the marginal value of y lower, $\partial^2 f/\partial y \partial x < 0$, then y^* must move down as x moves up.

One can go far as an economist by assuming that the equation $\partial f(x, y^*)/\partial y = 0$ really does implicitly define y^* as a differentiable function of x, provided one stays away from the cases where y^* is not a function of x, or not a differentiable function of x. Thus, one reason to study the conditions under which y^* is a differentiable function of x is to avoid embarassment. A more important reason is that the intuition for why y^* is a well-behaved function of x contains real insight into the structure of the problems that economists care about.

We give several results that are all called the implicit function theorem. The first is a continuously parameterized version of Banach's contraction mapping theorem, one that may, initially, seem a bit removed from the differentiable y^*. The starting point is a pair of compact intervals, $I_a(x_0) = \{x \in \mathbb{R} : |x - x_0| \le a\}$, $I_b(y_0) = \{y \in \mathbb{R} : |y - y_0| \le b\}$, $a, b > 0$, and a continuous function $g : I_a \times I_b \to \mathbb{R}$ such that $g(x_0, y_0) = 0$. (The leading case has g related to $\partial f/\partial y$.) Suppose that each $g(x, \cdot) : \mathbb{R} \to \mathbb{R}$ is a contraction mapping with the same contraction factor β, that is, $\forall x \in I_a$ and $\forall y, y' \in I_b$, $|g(x, y) - g(x, y')| \le \beta |y - y'|$.

For $x \in I_a$, $y \mapsto y_0 + g(x, y)$ is a contraction mapping from I_b to $\mathbb{R}$. Provided that $y \mapsto y_0 + g(x, y)$ maps I_b to itself, there is a unique fixed point we call $h(x)$. If there is an interval of x's on which there is a fixed point, then we have $h(x) = y_0 + g(x, h(x))$ for that interval. There is a fixed point at x_0 because if we get $h(x_0) = y_0$, $y_0 = y_0 + g(x_0, y_0) = y_0 + 0$. Since g is continuous it seems plausible that the fixed point of $y \mapsto y_0 + g(x, y)$ should depend continuously on x in a small area of the form $I_s(x_0) = \{x \in \mathbb{R} : |x - x_0| \le s\}$, $s \in (0, a]$.

Theorem 6.7.2 (Implicit Function Theorem)   *If $g : I_a(x_0) \times I_b(y_0) \to \mathbb{R}$ is continuous, $g(x_0, y_0) = 0$ and $\exists \beta < 1$ such that $\forall x \in I_a(x_0)$ and all $y, y' \in I_b(y_0)$, $|g(x, y) - g(x, y')| \le \beta |y - y'|$, then there exists $s \in (0, a]$ and a unique continuous function $x \mapsto h(x)$ on $I_s(x_0)$ such that for all $x \in I_s(x_0)$, $h(x) = y_0 + g(x, h(x))$.*

Proof.   For $s \in (0, a]$, let $C_s = C(I_s(x_0); I_b(y_0))$. For $\psi \in C_s$, define the function $x \mapsto T(\psi)(x)$ by $T(\psi)(x) = y_0 + g(x, \psi(x))$. As it is the composition of continuous functions, $T(\psi)$ is a continuous $\mathbb{R}$-valued function. We want to find s small enough that $T : C_s \to C_s$ and show that T is a contraction. If $\psi \in C_s$, then

$$|y_0 - T(\psi)(x)| = |g(x, \psi(x))|$$
$$= |g(x, \psi(x)) - g(x, y_0) + g(x, y_0)|$$
$$\le |g(x, \psi(x)) - g(x, y_0)| + |g(x, y_0)|$$
$$\le \beta |\psi(x) - y_0| + |g(x, y_0)|.$$

If $\psi \in C_s$, then $\beta |\psi(x) - y_0| \le \beta b$. Since $g(x_0, y_0) = 0$ and g is continuous, we can pick $s \in (0, a]$ such that $|g(x, y_0)| < (1 - \beta)b$ for all $x \in I_s(x_0)$. For any such s, any $\psi \in C_s$, and any $x \in I_s(x_0)$, $|y_0 - T(\psi)(x)| \le \beta b + (1 - \beta)b = b$, implying that $T(\psi) \in C_s$. For $\psi, \varphi \in C_s$ and all $x \in I_s(x_0)$,

$$|T(\psi)(x) - T(\varphi)(x)| = |g(x, \psi(x)) - g(x, \varphi(x))| \le \beta |\psi(x) - \varphi(x)|;$$

that is, $\|T(\psi) - T(\varphi)\|_\infty \le \beta \|\psi - \varphi\|_\infty$, and $T : C_s \to C_s$ is a contraction mapping.   ■

Exercise 6.7.3   Verify that $C_s = C(I_s(x_0); I_b(y_0))$ is complete by showing that it is a closed subset of $C(I_s(x_0))$.

If $G(x, y) = a + bx + cy$, $G(x_0, y_0) = 0$, and $c \ne 0$, then we can solve $G(x, h(x)) \equiv 0$ for $y = h(x)$. Specifically, $h(x) = -\frac{b}{c}x - \frac{a}{c}$ and $h(x_0) = y_0$. The smaller $c = D_y G(x, y)$, the more y must move in order to compensate for any change in G that arises from changes in x. If $c = D_y G(x, y) = 0$, then there is, in general, no hope of determining y as a function of x so as to guarantee that $G(x, h(x)) = 0$.

Suppose now that F is not affine, but is continuously differentiable and satisfies $F(x_0, y_0) = 0$. Then for $a = 0$, $b = \partial F(x_0, y_0)/\partial x$, and $c = \partial F(x_0, y_0)/\partial y$, the function $G(x, y) = a + bx + cy$ provides a very good approximation to F at (x_0, y_0),

$$\lim_{\|(x, y) - (x_0, y_0)\| \to 0} \frac{F(x, y) - G(x, y)}{\|(x, y) - (x_0, y_0)\|} = 0.$$

It is reasonable to hope that for some interval around x_0, there is a differentiable function, $y = h(x)$ such that $F(x, h(x)) \equiv 0$ and that $\frac{dh(x_0)}{dx} = -\frac{b}{c} = -\frac{\partial F(x_0, y_0)/\partial x}{\partial F(x_0, y_0)/\partial y}$.

Corollary 6.7.4 (Implicit Function Theorem) *If $g : I_a(x_0) \times I_b(y_0) \to \mathbb{R}$ is continuously differentiable, $g(x_0, y_0) = 0$, and $\partial g(x_0, y_0)/\partial y \neq 0$, then there exists $s \in (0, a]$ and a unique continuous function $x \mapsto h(x)$ on $I_s(x_0)$ such that $h(x_0) = y_0$ and for all $x \in I_s(x_0)$, $g(x, h(x)) = 0$.*

Proof. Let $c = \partial g(x_0, y_0)/\partial y$ and define $G(x, y) = (y - y_0) - \frac{1}{c}g(x, y)$. We check that G satisfies the conditions of Theorem 6.7.2. G is continuously differentiable (hence continuous), and $G(x_0, y_0) = 0$. Since $\partial G(x_0, y_0)/\partial y = 1 - 1 = 0$ and G is continuously differentiable, for any $\beta \in (0, 1)$, there is an $a' \in (0, a]$ and $b' \in (0, b]$ such that for all $(x, y) \in (I_{a'}(x_0) \times I_{b'}(y_0))$, $|\partial G(x, y)/\partial y| \leq \beta$. This means that for all $x \in I_{a'}(x_0)$ and $y, y' \in I_{b'}(y_0)$, $|G(x, y) - G(x, y')| \leq \beta|y - y'|$.

From Theorem 6.7.2, we conclude that there exists $h(x)$ satisfying $h(x) = y_0 + G(x, h(x))$. The definition of $G(x, y)$ implies that $\forall x \in I_s(x_0)$, $h(x) = y_0 + [h(x) - y_0] + \frac{1}{c}g(x, h(x))$, that is, that $\forall x \in I_s(x_0)$, $g(x, h(x)) = 0$. ∎

Example 6.7.5 [↑*Example 6.7.1*] *For each value of x, suppose that we solve the problem $\max_y f(x, y)$ by solving $\partial f(x, y)/\partial y = 0$. If we take $g(x, y) = \partial f(x, y)/\partial y = 0$, the condition $\partial g(x_0, y_0)/\partial y \neq 0$ comes from $\partial^2 f(x_0, y_0)/\partial y^2 < 0$, which is what guarantees that y_0 is a strict maximum at x_0. If y_0 is not a strict maximum, then it is one of many maxima and y^* is a correspondence, not a function.*

Example 6.7.6 *Suppose that $u : \mathbb{R}^2_{++} \to \mathbb{R}$ is continuously differentiable and that $\partial u/\partial x_i > 0$, $i = 1, 2$. Let $u_0 = u(x_{1,0}, x_{2,0})$ and define $g(x_1, x_2) = u(x_1, x_2) - u_0$ so that $g(x_{1,0}, x_{2,0}) = 0$. Since $\partial g/\partial x_1 > 0$ and $\partial g/\partial x_2 > 0$, on some interval around $x_{2,0}$ we can find a continuous $x_2 \mapsto x_1(x_2)$ such that $u(x_1(x_2), x_2) \equiv u_0$ and on some interval around $x_{1,0}$ we can find a continuous $x_1 \mapsto x_2(x_1)$ such that $u(x_1, x_2(x_1)) \equiv u_0$. This means that in a region around $(x_{1,0}, x_{2,0})$ the indifference set is a continuous, invertible function.*

Missing so far is the derivative of $h(x)$.

Corollary 6.7.7 (Implicit Function Theorem) *Under the conditions of Corollary 6.7.4, the function $h(x)$ is continuously differentiable on a neighborhood of x_0, and at all points x_1 in that neighborhood, $dh(x_1)/dx = -\frac{\partial g(x_1, y_1)/\partial x}{\partial g(x_1, y_1)/\partial y}$, where $y_1 = h(x_1)$.*

Applying this to Examples 6.7.1 and 6.7.5 yields

$$dy^*(x_0)/dx = -\frac{\partial^2 f(x, y^*(x))/\partial y \partial x}{\partial^2 f(x, y^*(x))/\partial y^2}, \tag{6.32}$$

which comes from (6.31). Since we expect $\partial^2 f(x, y^*(x))/\partial y^2 < 0$, we now have the mathematics to say that the sign of $dy^*(x_0)/dx$ is equal to the sign of $\partial^2 f(x, y^*(x))/\partial y \partial x$. Applying this to Example 6.7.6 yields

$$dx_2(x_{1,0})/dx_1 = -\frac{\partial u(x_{1,0}, x_{2,0})/\partial x_1}{\partial u(x_{1,0}, x_{2,0})/\partial x_2}, \tag{6.33}$$

so that the slope of an indifference curve is negative when the utility function is monotonic.

Proof of Corollary 6.7.7.   Fix x_1 in $(x_0 - s, x_0 + s)$ and set $y_1 = h(x_1)$. Let Δx be a small nonzero number and define $\Delta y = h(x_1 + \Delta x) - y_1$. By the definition of h, $g(x_1 + \Delta x, y_1 + \Delta y) = 0$. By the mean value theorem from calculus, there exists $\alpha \in [0, 1]$ such that

$$\frac{\partial g(x_1 + \alpha \Delta x, y_1 + \alpha \Delta y)}{\partial x} \Delta x + \frac{\partial g(x_1 + \alpha \Delta x, y_1 + \alpha \Delta y)}{\partial y} \Delta y = 0. \tag{6.34}$$

Solving for $\Delta y / \Delta x$ yields

$$\frac{\Delta y}{\Delta x} = -\frac{\partial g(x_1 + \alpha \Delta x, y_1 + \alpha \Delta y)/\partial x}{\partial g(x_1 + \alpha \Delta x, y_1 + \alpha \Delta y)/\partial y}. \tag{6.35}$$

As g is continuously differentiable, the right-hand side converges to $-\frac{\partial g(x_1, y_1)/\partial x}{\partial g(x_1, y_1)/\partial y}$ as $\Delta x \to 0$. Since $dh(x_1)/dx = \lim_{\Delta x \to 0} \frac{\Delta y}{\Delta x}$, $dh(x_1)/dx = -\frac{\partial g(x_1, y_1)/\partial x}{\partial g(x_1, y_1)/\partial y}$. Since the right-hand side is continuous and x_1 is arbitrary, h is continuously differentiable. ∎

Essentially the same proof, but with more notation, gives us the following.

Corollary 6.7.8 (Implicit Function Theorem)   *If $g : \mathbb{R}^\ell \times \mathbb{R} \to \mathbb{R}$, $g(\mathbf{x}_0, y_0) = 0$, g is continuously differentiable on a neighborhood of $(\mathbf{x}_0, y_0)$, and $\partial g(\mathbf{x}_0, y_0)/\partial y \neq 0$, then there exists a continuous differentiable $\mathbf{x} \mapsto h(\mathbf{x})$ from a neighborhood of $\mathbf{x}_0$ in $\mathbb{R}^\ell$ to $\mathbb{R}$ such that $g(\mathbf{x}, h(\mathbf{x})) \equiv 0$ and the partial derivatives of h are $\partial h(\mathbf{x})/\partial x_i = -\frac{\partial g(\mathbf{x}, h(\mathbf{x}))/\partial x_i}{\partial g(\mathbf{x}, h(\mathbf{x}))/\partial y}$.*

Example 6.7.9 (Cobb–Douglas Demand Theory)   *One can solve the problem* max $x_1 \cdot x_2$ *subject to* $p_1 x_1 + p_2 x_2 \leq w$, $x_1, x_2 \geq 0$ *explicitly for* $x^*(p_1, p_2, w)$ *and then explicitly calculate the derivatives* $\partial x_1/\partial \theta$, $\theta = p_1, p_2, w$. *Here we use the implicit function theorem. At the solution,* $p_1 x_1 + p_2 x_2 = w$, *so that* $x_2(x_1) = \frac{1}{p_2}(w - p_1 x_1)$. *This means that we solve can the problem* max$_{x_1}$ $x_1 \cdot x_2(x_1) = \frac{w x_1}{p_2} - p_1(x_1)^2$. *The derivative condition for the solution is* $g(x_1; (p_1, p_2, w)) = \frac{w}{p_2} - 2 p_1 x_1 = 0$. *Since* $\partial g/\partial x_1 \neq 0$, *there exists* $x_1(p_1, p_2, w)$ *such that* $g(x_1(p_1, p_2, w); (p_1, p_2, w)) \equiv 0$ *and* $\partial x_1/\partial \theta = -\frac{\partial g/\partial \theta}{\partial g/\partial x_1}$.

Exercise 6.7.10   Find the demand function $x_1(p_1, p_2, w)$ in Example 6.7.9; take its derivatives with respect to p_1, p_2, and w; and verify that the answer is the same as one finds using the implicit function theorem.

Exercise 6.7.11   Consider the problem $\max (x_1)^{2/3} \cdot (x_2)^{1/3} + \log(x_1 + x_2) \cdot \sqrt{x_2}$ subject to $p_1 x_1 + p_2 x_2 \leq w$, $x_1, x_2 \geq 0$. Find the partial derivatives of $x_1(p_1, p_2, w)$ and their signs.

Exercise 6.7.12   Consider the problem $\max x_1 + \sqrt{x_2}$ subject to $p_1 x_1 + p_2 x_2 \leq w$, $x_1, x_2 \geq 0$. Where they exist, find the partial derivatives of $x_1(p_1, p_2, w)$. [The point of this problem in demand theory is to make sure that you know how to handle corner solutions and that you can figure out when the conditions of the implicit function theorem apply.]

It is rare that maximizers control only a single variable. When they control many variables, there are many derivative conditions. To handle such situations, we use the following.

Corollary 6.7.13 (Implicit Function Theorem) *If* $g : \mathbb{R}^n \times \mathbb{R}^\ell \to \mathbb{R}^\ell$, $g(\mathbf{x}_0, \mathbf{y}_0) = 0$, g *is continuously differentiable on a neighborhood of* $(\mathbf{x}_0, \mathbf{y}_0)$, *and* $D_{\mathbf{y}}g(\mathbf{x}_0, \mathbf{y}_0)$ *is invertible (as an* $\ell \times \ell$ *matrix), then there exists a continuous differentiable* $\mathbf{x} \mapsto h(\mathbf{x})$ *from a neighborhood of* $\mathbf{x}_0$ *in* $\mathbb{R}^n$ *to* $\mathbb{R}^\ell$ *such that* $g(\mathbf{x}, h(\mathbf{x})) \equiv 0$ *and the derivative of* h *is* $D_{\mathbf{x}}h(\mathbf{x}) = -(D_{\mathbf{y}}g(\mathbf{x}, h(\mathbf{x})))^{-1} \cdot D_{\mathbf{x}}g(\mathbf{x}, h(\mathbf{x}))$.

If $\ell = 1$, then $(D_{\mathbf{y}}g(\mathbf{x}, h(\mathbf{x})))^{-1} = \frac{1}{\partial g(\mathbf{x}, h(\mathbf{x}))/\partial y}$, and this result gives the previous one. The following encapsulates the differentiable comparative statics in the theory of the firm.

Exercise 6.7.14 Suppose that $f : \mathbb{R}^N_{++} \to \mathbb{R}^1$ is strictly concave, twice continuously differentiable, and satisfies $D_{\mathbf{x}}f \gg 0$. Further suppose that for some $p^\circ > 0$ and $\mathbf{w}^\circ \gg 0$, the (profit) maximization problem

$$\max\nolimits_{\mathbf{x} \in \mathbb{R}^N_{++}} \Pi(\mathbf{x}) = pf(\mathbf{x}) - \mathbf{w} \cdot \mathbf{x}$$

has a strictly positive solution.

1. Show that the derivative conditions for the optimum, $\mathbf{x}^*$, are $pD_{\mathbf{x}}f(\mathbf{x}^*) = \mathbf{w}$ or $D_{\mathbf{x}}f(\mathbf{x}^*) = \frac{1}{p}\mathbf{w}$, The $\mathbf{x}^*$ are called the factor demands.

2. Show that the solution to the above problem, $\mathbf{x}^*(p, \mathbf{w})$, is a differentiable function of p and $\mathbf{w}$ on a neighborhood of $(p^\circ, \mathbf{w}^\circ)$.

3. The supply function is defined by $y(p, \mathbf{w}) = f(\mathbf{x}^*(p, \mathbf{w}))$, so that $D_p y(p, \mathbf{w}) = D_{\mathbf{x}}f(\mathbf{x}^*)D_p \mathbf{x}^*$. Take the derivative with respect to p on both sides of the equivalence $D_{\mathbf{x}}f(\mathbf{x}^*(p, \mathbf{w})) \equiv \frac{1}{p}\mathbf{w}$. Show that this implies that $D_p \mathbf{x}^* = -\frac{1}{p^2}(D^2_{\mathbf{x}}f(\mathbf{x}^*))^{-1}\mathbf{w}$.

4. Using the negative definiteness of $D^2_{\mathbf{x}}f$, show that $D_p y(p, \mathbf{w}) > 0$.

5. Using the previous part of this problem, show that it is not the case that $D_p \mathbf{x}^* \leq 0$.

6. Let e_n be the unit vector in the nth direction. Taking the derivative with respect to w_n on both sides of the equivalence $D_{\mathbf{x}}f(\mathbf{x}^*(p, \mathbf{w})) \equiv \frac{1}{p}\mathbf{w}$ gives the equation $D^2_{\mathbf{x}}f(\mathbf{x}^*)D_{w_n}\mathbf{x}^* = \frac{1}{p}e_n$. Premultiply both sides of $D_{w_n}\mathbf{x}^* = \frac{1}{p}(D^2_{\mathbf{x}}f(\mathbf{x}^*))^{-1}e_n$ by e_n. Conclude that $\partial x^*_n/\partial w_n < 0$.

In §2.8, we saw lattice theory at work in finding comparative statics results. In Chapter 5, we saw that convex analysis allows us to obtain these differentiable comparative statics results with much less work. It is conceivable that many, or even most, of the uses of the implicit function theorem in economics will be supplanted by more sophisticated tools, as was the case in the theory of the firm, but it certainly has not happened yet.

6.7.c The Extension Theorem for Dense Sets

Let E be a dense subset of M and $f : E \to M'$ be a continuous function. A couple of examples show the difficulties in finding an $F \in C(M; M')$ such that $F_{|E} = f$.

Example 6.7.15 *Both $E = \mathbb{R}\backslash\{0\}$ and $E' = \mathbb{Q}\backslash\{0\}$ are dense in $\mathbb{R}$. The unbounded function $f(r) = \frac{1}{r}$ is continuous on both, as is the bounded function $g(r) = \sin(1/r)$. However, there is no $F \in C(\mathbb{R})$ that agrees with f and no $G \in C(\mathbb{R})$ that agrees with g on E' or on E.*

What goes wrong in the example is that sequences $e_n \to 0$ can have $f(e_n)$ either not converging or going to different limits.

Theorem 6.7.16 (Extension of Continuous Functions on Dense Sets)
If E is dense in M, then $f \in C(E; M')$ has a continuous extension $F \in C(M; M')$ iff for each $x \in M$ and all sequences e_n in E converging to x, the sequences $n \mapsto f(e_n)$ converge to the same limit in M'. Further, if an extension of f exists, it is unique.

Proof. If $F \in C(M; M')$ and $F_{|E} = f$, then $\forall x \in M$ and $e_n \to x$, we have $f(e_n) \to F(x)$.

Now suppose that $\forall x \in M$ and sequences e_n in E converging to x, the sequences $n \mapsto f(e_n)$ converge to the same limit in M'. For $x \in M$, define $F(x) = \lim f(e_n)$ for any $e_n \to x$. We must show that F is continuous. Let W be an open subset of M' containing $F(x)$. Pick $\epsilon > 0$ such that $\text{cl}(B_\epsilon(F(x))) \subset W$. Since f is continuous, there exists $\delta > 0$ such that $f(B_\delta(x) \cap E) \subset B_\epsilon(F(x))$. Taking $e_n \to x'$ for $x' \in B_\delta(x)$ shows that $F(x') \in \text{cl}(B_\epsilon(F(x)))$. This yields $[d(x', x) < \delta] \Rightarrow [F(x') \in W]$. ∎

Exercise 6.7.17 Complete the proof of Theorem 6.7.16.

Corollary 6.7.18 *If E is dense in M, $f \in C(E; M')$ is uniformly continuous, and (M', d') is complete, then there exists a unique $F \in C(M; M')$ such that $F_{|E} = f$ and F is uniformly continuous.*

Proof. Let e_n be a sequence in E converging to $x \in M$. We claim that $f(e_n)$ is Cauchy. Pick $\epsilon > 0$. Because f is uniformly continuous on E, $\exists \delta > 0$ such that $\forall e, e' \in E, [d_M(e, e') < \delta] \Rightarrow [d'_M(f(e), f(e')) < \epsilon]$. Pick N such that for all $n \geq N, d_M(e_n, x) < \delta/2$. By the triangle inequality, for all $n, m \geq N, d_M(e_n, e_m) < \delta$, so that $d'(f(e_n), f(e_m)) < \epsilon$. Since M' is complete, $f(e_n)$ converges to some $y \in M'$. Let e'_n be another sequence in E converging to x. By uniform continuity, for any $\epsilon > 0$, for large n, $d'(f(e'_n), y) < \epsilon$ because $d_M(e'_n, e_n) < \delta$. By Theorem 6.7.16, f has a unique continuous extension, $F \in C(M; M')$. We show that F is uniformly continuous.

Pick $\epsilon > 0$. Since f is uniformly continuous on E, there exists δ such that for all $e, e' \in E, [d_M(e, e') < \delta] \Rightarrow [d'(f(e), f(e')) < \epsilon/2]$. Since F is a continuous extension of f, $[d_M(x, x') < \delta] \Rightarrow [d'(f(x), f(x')) \leq \epsilon/2]$, and $\epsilon/2 < \epsilon$. ∎

The completeness of M' and the denseness of E are crucial to Corollary 6.7.18, but the uniform continuity is not crucial for the existence of a continuous extension.

Exercise 6.7.19 Show the following:

1. The mapping $f(e) = e$ from $E = \mathbb{Q}$ to $M' = \mathbb{Q}$ is uniformly continuous, but has no continuous extension to $M = \mathbb{R}$.

2. The mapping $f(e) = e$ from $E = \{0, 1\}$ to $M' = \{0, 1\}$ is uniformly continuous, but has no continuous extension to $M = [0, 1]$.

3. The mapping $f(e) = e^2$ from $E = \mathbb{Q}$ to $M' = \mathbb{R}$ is not uniformly continuous, but has a continuous extension to $M = \mathbb{R}$.

6.7.d Limits of Approximate Optima as Exact Optima

All too often economic models have the property that, to any degree of approximation, an optimum exists, but there is no exact optimum. At least for single-person optimization problems, such problems are readily soluble, and a representation of the approximate optimum is readily available. The device is a compact embedding, which is a compactification without the homeomorphism requirement, and the extension of uniformly continuous functions from a dense set.

Example 6.7.20 *Consider the bounded discontinuous function $u : [0, 1] \to \mathbb{R}$ defined by $u(x) = x1_{[0,1)}(x)$. For any $\epsilon > 0$, $x_\epsilon := 1 - \epsilon$ has the property that $u(x_\epsilon) \geq \sup_{x \in [0,1]} u(x) - \epsilon$. In this sense, u nearly achieves its maximum on $[0, 1]$, but does not achieve it because there is no way to represent "the point just to the left of 1." The compactness of $[0, 1]$ does not help because the objective function is not continuous.*

The easy solution is to add such a point, call it "$1 - dt$" or "$1-$," and define an extension of u, $\widetilde{u}$, to have the property that $\widetilde{u}(1-) = 1$. In order to make this construction applicable to a wider range of problems, we have to isolate some of the structures in what we just did with this addition of a point. The extended function $\widetilde{u} : [0, 1-] \cup \{1\} \to \mathbb{R}$ is defined by $\widetilde{u}(x') = x'$ if $0 \leq x' < 1$, $\widetilde{u}(x') = 1$ if $x' = 1-$, and $\widetilde{u}(1) = 0$. If we give $[0, 1-] \cup \{1\}$ the appropriate metric, we can make $[0, 1'] \cup \{1\}$ isometric to the set $[0, 1] \cup \{2\}$, the set $[0, 1) \cup \{1\}$ dense in $[0, 1'] \cup \{1\}$, and the function $\widetilde{u}$ continuous on this compact set.

After a bit of work, the following provides us with a more systematic way of representing the limits of approximate solutions to optimization problems.

Definition 6.7.21 *For a nonempty set X, $(f, (X', d'))$ is a **compact embedding** if $f : X \to X'$ is one-to-one, (X', d') is compact, and $f(X)$ is dense in X'.*

Since f is a bijection of X and $f(X)$, then any $u : X \to \mathbb{R}$ is automatically identified with the function $x' \mapsto u(f^{-1}(x'))$ on $f(X)$, also denoted u. Let $\mathcal{U}$ be a sup norm separable set of functions, for example, any countable set, separating points in X. It turns out that one can choose f and X' so as to guarantee that all $u \in \mathcal{U}$ have a unique continuous extension from the dense set $f(X')$ to all of X'. This means that the failure of existence of optima can be viewed as the failure of our models of the choice set to contain representations of the correct limits of approximate solutions.

Theorem 6.7.22 (Compact Embeddings) *If $\mathcal{U}$ is a sup norm separable set of bounded functions that separates points in X, then there exists a compact embedding $(f, (X', d'))$ with the property that each $(u \circ f^{-1})$ has a unique continuous extension to X'.*

Proof. Let $\mathcal{V}$ be a countable sup norm dense subset of $\mathcal{U}$. For each $u \in \mathcal{V}$, let I_v be the compact interval $[\inf_{x \in X} v(x), \sup_{x \in X} v(x)]$. Define $M' = \times_{v \in \mathcal{V}} I_v$ and give M' the product metric. Since $\mathcal{V}$ separates points (being sup norm dense in a set that separates points), the mapping $f(x)$ defined by $\text{proj}_v(f(x)) = v(x)$ for $v \in \mathcal{V}$ gives a bijection between X and a subset $f(X)$ in the compact space M'. Let X' be the closure of $f(X)$ and d' the restriction of the product metric to X'. ∎

The last step in the proof and a number of implications are given in the following.

Exercise 6.7.23 For the compact embedding construction just given, show the following.

1. Each $(v \circ f^{-1})$ has a unique continuous extension, $\widetilde{v} : X' \to \mathbb{R}$. [Consider the uniformly continuous function $\text{proj}_v : X' \to \mathbb{R}$.]
2. For all $v_1, v_2 \in \mathcal{V}$, $\sup_{x \in X} |v_1(x) - v_2(x)| = \max_{x' \in X'} |\widetilde{v}_1(x') - \widetilde{v}_2(x')|$.
3. If v_n is a Cauchy sequence of functions in $\mathcal{V}$, then $\widetilde{v}_n$ is a Cauchy sequence in $C(X')$, hence has a continuous limit.
4. Each $(u \circ f^{-1})$, $u \in \mathcal{U}$, has a unique extension, $\widetilde{u}$, in $C(X')$. [The set $\mathcal{V}$ is dense in $\mathcal{U}$, and $\|v_n - u\|_\infty \to 0$ implies that v_n is a Cauchy sequence.]
5. The closure of $\mathcal{A}(\widetilde{\mathcal{U}})$, that is, the smallest algebra containing $\{\widetilde{u} : u \in \mathcal{U}\}$, is equal to $C(X')$.

With the previous results in mind, compare, for any u in the closure of $\mathcal{A}(\mathcal{U})$, the maximization problems

$$\max_{x \in X} u(x) \text{ and } \max_{x' \in X'} \widetilde{u}(x') \qquad (6.36)$$

In particular, show that if x is a solution to the left-hand problem, then $f(x)$ is a solution to the right-hand problem; that if x_n is a sequence in X with $u(x_n) \to \sup_{x \in X} u(x)$, then $x'_n := f(x_n)$ has a subsequence that converges to a solution in the right-hand problem; and that if x' is a solution to the right-hand problem, then there exists a sequence x_n in X such that $f(x_n) \to x'$ and $u(x_n) \to \sup_{x \in X} u(x)$.

Compact embedding allows us to represent solutions to the problem of maximizing any bounded function on any set.[13] However, one cannot simultaneously perform compact embeddings on each A_i in a game $\Gamma = (A_i, u_i)_{i \in I}$ in such a fashion that the u_i have a jointly continuous extension to the product of the compact spaces A'_i.

Example 6.7.24 *Let $I = \{1, 2\}$, $A_1 = A_2 = \mathbb{N}$, and let $u_i(a_i, a_j) = G(a_i - a_j)$, where $r \mapsto G(r)$ is the bounded strictly monotonic function $e^r/(1 + e^r)$. For each $i \in I$, each a_i is strictly dominated by, say, $a_i + 17$, so that there is no equilibrium for this game. Further, the set of slices of u_i given by $\{u_i(\cdot, a_j) : a_j \in A_j\}$ cannot have compact closure because there are countably many points at distance $1/2$ or greater from each other.*

---

13. Here we are using embeddings in compact metric spaces, which have at most the cardinality of the continuum. For large X, one can work with larger compact spaces having nonmetric topologies, as in §11.4.

Let $(f_i, (A'_i, d'))$ be any compact embedding of A_i, $i \in I$. $u_i(f_i^{-1}(a'_i), f_j^{-1}(a'_j))$ can have no continuous extension $\widetilde{u}_i$ from $(f_i, f_j)(A_i \times A_j)$ to the compact space $A'_i \times A'_j$. If it did, its slices would be a sup norm compact set (being the continuous image of the compact space A'_j). However, each slice of $\widetilde{u}_i$ is the extension of a slice of u_i, so that the slices of $\widetilde{u}_i$ cannot have compact closure.

6.7.e Two Extension Theorems for Closed Sets

The proof of the following result is much easier for metric spaces than it is for what are called "normal topological spaces" because we can give an extension more explicitly. §6.11.e introduces this class of nonmetric topological spaces and sketches the proof of the extension result.

Theorem 6.7.25 (Tietze Extension) *If F is a closed subset of M and $g \in C_b(F)$, then there is a continuous extension $G \in C_b(M)$ such that $\|G\|_\infty = \|g\|_\infty$.*

The proof uses the calculation in the following exercise.

Exercise 6.7.26 Show that the function $f(s) = 1/\left[(1+s^2)^{1/s}\right]$

1. maps $(0, \infty)$ into the interval $(0, 1)$,

2. is strictly decreasing,

3. $\lim_{s \downarrow 0} f'(s) = -\infty$, and

4. $\lim_{s \downarrow 0} f(s) = 1$. [Apply l'Hôpital's rule to $g(s) := \log(f(s))$.]

Proof of Theorem 6.7.25. If $h \in C_b(F)$ is constant or $F = M$, existence of an extension is immediate, so we suppose that $h \geq 0$ is not constant and $F \subsetneqq M$. Let $a = \inf_{t \in F} h(t)$ and $b = \sup_{t \in F} h(t)$, $a < b$. We extend the function $g = (h - a)$ to a function G. The function $G + a$ extends h.

For $t \in F$ and $x \in F^c$, define $\varphi(t, x) = g(t)/\left[(1 + d(t, x)^2)^{1/d(x,F)}\right]$. Since division and exponentiation are continuous, φ is continuous on $F \times F^c$. Define G by

$$G(x) = \begin{cases} g(x) & \text{if } x \in F, \\ \sup_{t \in F} \varphi(t, x) & \text{if } x \notin F. \end{cases} \tag{6.37}$$

$G \geq 0$, $G_{|F} = g$, and G is bounded above by $\sup_{t \in F} g(t)$, hence $a \leq G(x) \leq b$. To show that G is continuous, it suffices to show that $\lim_{x_n \to x} G(x_n) = G(x)$ for $x \in \text{int } F$, $x \in F^c$, and $x \in \partial F$, and the first two cases are immediate.

If $x \in \partial F$ and the sequence x_n is a.a. in F, then $G(x_n) = g(x_n) \to g(x) = G(x)$. All that is left is the case where x_n is i.o. in F^c, and by which was just argued, there is no loss in assuming that x_n is a.a. in F^c. For all $t \in F$, $t \neq x$, as $x_n \to x$, $1/[(1 + d(t, x_n)^2]^{1/d(x_n, F)} \to 0$, while $1/[(1 + d(x, x_n)^2]^{1/d(x_n, F)} \to 1$ (by the previous exercise). Thus, $G(x_n) = \sup_{t \in F} g(t)/[(1 + d(t, x_n)^2]^{1/d(x_n, F)} \to G(x)$. ∎

Corollary 6.7.27 *If K is a compact subset of M, then $f \in C(K)$ iff there exists an $F \in C_b(M)$ that extends f.*

Proof. Compact sets are closed; therefore any $f \in C(K)$ has a bounded continuous extension. Restricting any $F \in C_b(M)$ to K yields an $f \in C(K)$. ∎

Compare the following to the basic existence result, Theorem 4.8.11.

Corollary 6.7.28 *K is compact iff every $f \in C(K)$ achieves its maximum on K.*

Another useful extension result has an approximation feel: continuous functions come arbitrarily close to the indicators of closed sets. This is useful in studying probabilities on metric spaces.

Lemma 6.7.29 (Urysohn for Metric Spaces) *If $F \subset G \subset M$, (M, d) a metric space, F closed and G open, there is a continuous function $f : M \to [0, 1]$ such that $f(F) = 1$ and $f(G^c) = 0$.*

Proof. Use either the Tietze extension theorem or the function $f(x) = d(x, B)/(d(x, B) + d(x, A))$, where $A = F$ and $B = G^c$. ∎

There is yet another an easy direct proof if F is compact.

Exercise 6.7.30 Let K be a compact subset of an open set G. Complete the following steps to show that there is a continuous function $f : M \to [0, 1]$ such that $f(K) = 1$ and $f(G^c) = 0$.

1. Show that there exists an $\epsilon > 0$ such that $K \subset K^\epsilon \subset G$.

2. Show that the function $f(x) = \max\{0, 1 - \frac{1}{\epsilon}d(x, K)\}$ has the desired properties.

6.8 ◆ The Metric Completion Theorem

For all $x, y \in \mathbb{R}^1$, $d(x, y) = |x - y|$ and $(\mathbb{R}, d)$ is a complete metric space. There are many metrics on $\mathbb{R}^1$ with the property that $\rho(x_n, x) \to 0$ iff $d(x_n, x) \to 0$, but without the property that $(\mathbb{R}, \rho)$ is complete. When $\rho(x_n, x) \to 0$ iff $d(x_n, x) \to 0$, we say that ρ and d are equivalent, written $\rho \sim d$.

We begin with a number of examples of what must be added to $(\mathbb{R}, \rho)$ to make it ρ-complete. The main result in this section is the metric completion theorem, which proves that one can always add elements to a metric space so as to make it complete. It is useful that the way one must add the new points is, up to isometry, unique.

6.8.a Equivalent Metrics and Completeness

Define $\rho(x, y) = |G(x) - G(y)|$, where $G(r) = e^r/(1 + e^r)$. The relevant properties of $G(\cdot)$ are that it is continuous, strictly increasing, that $\lim_{x \to -\infty} G(x) = 0$, and $\lim_{x \to +\infty} G(x) = 1$.

Exercise 6.8.1 Show that ρ is a metric equivalent to d. By considering sequences $x_n \to +\infty$ and $x_n \to -\infty$, show that $(\mathbb{R}, \rho)$ is not complete.

Let $\mathfrak{C}(\mathbb{R}, \rho)$ denote the set of Cauchy sequences in $(\mathbb{R}, \rho)$; define two Cauchy sequences, x_n, x'_n, to be equivalent if $\lim_n \rho(x_n, x'_n) \to 0$; and define $\overline{\mathbb{R}}$ to be the set of equivalence classes of Cauchy sequences in $\mathfrak{C}(\mathbb{R}, \rho)$, identifying each $r \in \mathbb{R}$ with the equivalence class of the constant sequence $(r, r, r, \ldots)$. This gives us $\mathbb{R} \subset \overline{\mathbb{R}}$. To define the metric $\overline{\rho}$ on $\overline{\mathbb{R}}$, define $\overline{\rho}(x, x')$ as $\lim_n \rho(x_n, x'_n)$ for any x_n in the equivalence class of x and x'_n in the equivalence class of x'. You should see the following.

Lemma 6.8.2 $\mathbb{R}$ *is dense in the complete, separable metric space* $(\overline{\mathbb{R}}, \overline{\rho})$.

One can visualize $\overline{\mathbb{R}}$ as $[-\infty, +\infty]$ with $x_n \to_\rho x$ iff (a) $x \in \mathbb{R}$ and $x_n \to_d x$, or (b) $x = -\infty$ and for all $r > 0$, $x_n < -r$ for large n, or (c) $x = +\infty$ and for all $r > 0$, $x_n > +r$ for large n.

Exercise 6.8.3 Show that $(\overline{\mathbb{R}}, \overline{\rho})$ is compact, but that $\mathbb{R}$ is not a compact subset of $\overline{\mathbb{R}}$ and that $\overline{\mathbb{R}}$ is homeomorphic to $[0, 1]$.

We adopt the following conventions for additions and subtractions involving $\pm\infty$.

Definition 6.8.4 *For all* $x \in \mathbb{R}$, $x + \infty := \infty$; *and for all* $x \in \mathbb{R}$, $x - \infty := -\infty$.

If $x_n \to x$ and $y_n \to \infty$, then the sequence $x_n + y_n \to \infty$. Hence, if we want to define addition in $\overline{\mathbb{R}}$ by extending its behavior on sequences, we must set $x + \infty = \infty$ for $x \in \mathbb{R}$. In just the same way, we must define $x - \infty = -\infty$. To the extent possible, these conventions make it possible to treat $\pm\infty$ as if they were numbers. Note very carefully that we do not define $\infty - \infty$. This is because we can have $x_n \to -\infty$ and $y_n \to \infty$ without having $x_n + y_n$ be convergent to anything.

The space $\overline{\mathbb{R}} = [-\infty, \infty]$ is the second most frequently used completion of the rationals. There are other completions, and we introduce some of them now, not because they are intrinsically important, but because they help us understand the scope and the power of the metric completion theorem.

Exercise 6.8.5 Show that $v(x, y) = |e^x - e^y|$ is a metric on $\mathbb{R}$, show that $v \sim d$, characterize the Cauchy sequences in $(\mathbb{R}, v)$, and describe a way to visualize the set of equivalence classes of Cauchy sequences in $(\mathbb{R}, v)$.

Exercise 6.8.6 Define $g(x) = 1/x$ if $x \neq 0$ and define $g(0) = 0$. Show that $w(x, y) = |g(x) - g(y)|$ is a metric on $\mathbb{R}$ and that $w \not\sim d$, characterize the Cauchy sequences in $(\mathbb{R}, w)$, and describe a way to visualize the set of equivalence classes of Cauchy sequences in $(\mathbb{R}, w)$.

6.8.b The Completion Theorem

We originally defined $\mathbb{R}$ as the set of equivalence classes of Cauchy sequences of $\mathbb{Q}$, with $\mathbb{Q}$ being dense in the resulting complete metric space. We have just seen this pattern three more times, and it is time to cover the general version of the construction.

Definition 6.8.7 *For a metric space* (M, d_M), $\mathfrak{C}(M, d_M)$ *denotes the set of Cauchy sequences. For any* $x_n, y_n \in \mathfrak{C}(M, d_M)$, *we define* $\widehat{d}_M(x_n, y_n) = \lim_n d_M(x_n, y_n)$, *which exists because* $\mathbb{R}_+$ *is complete. We define sequences*

$x_n, y_n \in \mathfrak{C}(M, d_M)$ to be equivalent, written $x_n \sim y_n$, if $\widehat{d_M}(x_n, y_n) = 0$. The **canonical completion** of (M, d_M) is $(\widehat{M}, \widehat{d_M})$, where $\widehat{M} = \mathfrak{C}(M, d_M)/\sim$ is the set of equivalence classes of d_M-Cauchy sequences. We embed each $x \in M$ as $\varphi(x) \in \widehat{M}$, defined as the equivalence class of the constant sequence $(x, x, x, \ldots)$.

The following guarantees that this definition makes sense.

Theorem 6.8.8 (Canonical Metric Completion) $(\widehat{M}, \widehat{d_M})$ is a complete metric space, $\varphi(M)$ is dense in $\widehat{M}$, and φ is an isometry between M and $\varphi(M)$.

Proof.   That $\widehat{d_M}$ is a metric and φ an isometry are immediate.

Denseness of $\varphi(M)$: Pick $\widehat{x} \in \widehat{M}$ and $\epsilon > 0$. We must show that there exists $\widehat{y} \in \varphi(M)$ such that $\widehat{d_M}(\widehat{y}, \widehat{x}) < \epsilon$. To this end, let x_n be a Cauchy sequence in the equivalence class $\widehat{x}$. Pick $A \in \mathbb{N}$ such that for all $a, b \geq A$, $d_M(x_a, x_b) < \epsilon/2$ and let $\widehat{y} = \varphi(x_A) \in \varphi(M)$; then $\widehat{d_M}(\widehat{y}, \widehat{x}) = \lim_n d_M(x_A, x_n) \leq \epsilon/2 < \epsilon$.

Completeness of $(\widehat{M}, \widehat{d_M})$: Let $\widehat{x}_n$ be a Cauchy sequence in $\widehat{M}$. We must show that there exists an $\widehat{x} \in \widehat{M}$ such that $\widehat{d_M}(\widehat{x}_n, \widehat{x}) \to 0$. This is where things get a little tricky.

Any subsequence of a Cauchy sequence is Cauchy, and if the subsequence has a limit, the original sequence has the same limit. Therefore, we can assume, by relabeling if necessary, that for all n, $\sum_{m \geq n} \widehat{d}(\widehat{x}_n, \widehat{x}_m) < 1/2^n$. For each $\widehat{x}_n$, pick a Cauchy sequence, $k \mapsto x_{n,k}$, in the equivalence class of $\widehat{x}_n$. Arrange the sequence of sequences in an array,

$$\begin{array}{ccccc}
x_{1,1} & x_{1,2} & x_{1,3} & x_{1,4} & x_{1,5} \cdots \\
x_{2,1} & x_{2,2} & x_{2,3} & x_{2,4} & x_{2,5} \cdots \\
x_{3,1} & x_{3,2} & x_{3,3} & x_{3,4} & x_{3,5} \cdots \\
x_{4,1} & x_{4,2} & x_{4,3} & x_{4,4} & x_{4,5} \cdots \\
\vdots & \vdots & \vdots & \vdots &
\end{array}$$

The idea is to choose a sequence $x_{n,k(n)}$ that travels down and to the right in the array. To show that it is Cauchy, let $\widehat{x}$ denote its equivalence class, and show that $\widehat{d_M}(\widehat{x}_n, \widehat{x}) \to 0$.

Picking the sequence: Since each $x_{n,k}$ is a Cauchy sequence we can inductively find $k(n) > k(n-1)$ such that for all $j, j' \geq k(n)$, $d_M(x_{n,j}, x_{n,j'}) < 1/2^n$.

$x_{n,k(n)}$ is Cauchy: For any $x_{n,k}$, let $[x_{n,k}] = \varphi(x_{n,k})$ be the equivalence class of $(x_{n,k}, x_{n,k}, \ldots)$. Note that $\widehat{d_M}([x_{n,k}], [x_{n',k'}]) = d_M(x_{n,k}, x_{n',k'})$. By choice of $k(n)$, for each n, $\widehat{d_M}([x_{n,k(n)}], \widehat{x}_n) \leq 1/2^n$. By the triangle inequality, for $n' > n$,

$$\widehat{d_M}([x_{n,k(n)}], [x_{n',k(n')}]) \leq \widehat{d_M}([x_{n,k(n)}], \widehat{x}_n) + \widehat{d_M}(\widehat{x}_n, \widehat{x}_{n'}) + \widehat{d_M}(\widehat{x}_{n'}, [x_{n',k(n')}]).$$

Each of the three terms is less than or equal to $1/2^n$ so that $\sum_{n' > n} d_M(x_{n,k(n)}, x_{n',k(n')}) \leq 6/2^n < \infty$, implying that $x_{n,k(n)}$ is Cauchy. Let $\widehat{x}$ denote the equivalence class of the Cauchy sequence $x_{n,k(n)}$.

Finally, $\widehat{d_M}(\widehat{x}_n, \widehat{x}) \to 0$ because $\widehat{d_M}([x_{n,k(n)}], \widehat{x}_n) \to 0$.   ∎

Isometries are a form of relabeling. It is the use of isometries with dense subsets of complete spaces rather than homeomorphisms with dense subsets of compact spaces that distinguishes completions from compactifications (Definition 6.6.41 on

p. 319). What is similar between the two is that they are both methods of adding new points to spaces.

Definition 6.8.9 *If f is an isometry between M and a dense subset of M' and (M', d') is complete, $(f, (M', d'))$ is called a* **completion of** *(M, d).*

The following is a direct result of the extension theorem for dense sets.

Lemma 6.8.10 *Any completion of (M, d) is isometric to the canonical completion of (M, d).*

Proof. Suppose that f is an isometry of M and a dense $f(M) \subset M'$, and g is an isometry of M and a dense $g(M) \subset M''$, where (M', d') and (M'', d'') are complete. The function $g \circ f^{-1}$ is an isometry between a dense subset of M' and M'', and as isometries are uniformly continuous, it has a unique continuous extension. The extension is also an isometry, so that M' and M'' are isometric. ∎

There is no logical reason to distinguish between isometric sets, as an isometry changes the names of points without changing anything else. Therefore we call any completion of (M, d) *the* completion of (M, d). The following summarizes.

Theorem 6.8.11 (Metric Completion) *Any metric space has a completion, and all completions are isometric.*

Exercise 6.8.12 Show that if (M, d) is separable, then so is its completion.

Lemma 6.8.13 *For any subset E of a metric space (M, d), if $(f, (M', d'))$ is the completion of (M, d), then $(f_{|E}, (\mathrm{cl}(f(E)), d'))$ is the completion of (E, d). In particular, the completion of any dense subset of M is the completion of M.*

Proof. $f_{|E}$ is a homeomorphism of E and $f(E)$, which is dense in its closure, $\mathrm{cl}(f(E))$. As it is a closed subset of a complete metric space, $\mathrm{cl}(f(E))$ is itself complete. If E is dense in M, then $\mathrm{cl}(f(E)) = M'$. ∎

6.8.c Representation and "Reality"

We have seen that the set of finitely supported cdf's is dense in $\mathcal{D}(\mathbb{R})$ and that $\mathcal{D}(\mathbb{R})$ is complete. In light of the last lemma, 6.8.13, this means that we could have *defined* $\mathcal{D}(\mathbb{R})$ as the completion of the finitely supported cdf's.

The advantage of defining $\mathcal{D}(\mathbb{R})$ as the completion of the finitely supported cdf's is that it makes it clear that we are nearly working with finitely supported probabilities since they are dense. However, if we had proceeded in that fashion, the question might well have been, "What are these new cdf's that completion has given us in reality?" The advantage of defining cdf's directly is that we have representations of the new cdf's that the completion process adds, and we can answer the question with, "The new cdf's are, in 'reality,' the right-continuous nondecreasing functions that go to 0 at $-\infty$ and to 1 at $+\infty$."

We devote Chapter 7 to developing the theory of measurable functions and their integrals. We begin with the class of simple functions. These are functions for which the definition of the integral is clear, easy, and a uniformly continuous function on the set of simple functions. When we complete the class of functions, we arrive back at the measurable functions. In this way, the measurable functions are nearly simple functions, since they are the completion of the simple functions, and we have a "real" representation of them.

6.9 ◆ The Lebesgue Measure Space

In this section we develop measurable functions as the completion of simple functions, and Chapter 7 develops the measurable functions more directly. The most widely used model of randomness is Kolmogorov's model of probability space, $(\Omega, \mathcal{F}, P)$. A probability, P, is a function from sets, $E \in \mathcal{F} \subset \mathcal{P}(\Omega)$, to $[0, 1]$. The probability that a point randomly drawn according to P lies in the set E is then $P(E)$. A random variable is a function $X : \Omega \to \mathbb{R}$ with the interpretation that when $\omega \in \Omega$ is drawn, $X(\omega)$ is the observed random number, and the probability that we observe, for example, $X(w_i) \leq r$ is $P(X^{-1}((-\infty, r])))$.

Definition 6.9.1 *A **probability space** is a triple, $(\Omega, \mathcal{F}, P)$, where Ω is a nonempty set, $\mathcal{F}$ is a collection of subsets to which we assign probability, and $P : \mathcal{F} \to [0, 1]$ is the probability. An $E \in \mathcal{F}$ is an **event**. For $E \in \mathcal{F}$, $P(E) \in [0, 1]$ is the **probability of the event** E. A function $X : \Omega \to \mathbb{R}$ is a **random variable**, or an $\mathbb{R}$-**valued measurable function**, if for all $r \in \mathbb{R}$, $X^{-1}((-\infty, r]) \in \mathcal{F}$; that is, it is an event to which we can assign probability.*

This section is an extended example demonstrating some of the properties that we would like our model of randomness, $(\Omega, \mathcal{F}, P)$, to have.

6.9.a Simple Measurable Functions and Their Integrals

Let $\Omega = (0, 1]$ and let $\mathcal{B}^\circ$ denote the empty set and the class of finite disjoint unions of interval subsets of $(0, 1]$. Recall that all the possibilities are allowed, (a, b), $[a, b)$, $[a, b]$, and $(a, b]$, and that the notation "$|a, b|$" means any one of these possibilities. When $a = b$, $[a, b]$ is a singleton set. Classes of sets having the properties that $\mathcal{B}^\circ$ has are called fields.

Definition 6.9.2 *$\mathcal{F}^\circ \subset \mathcal{P}(\Omega)$ is a **field** (or **algebra**) if $\emptyset \in \mathcal{F}^\circ$, $[A \in \mathcal{F}^\circ] \Rightarrow [A^c \in \mathcal{F}^\circ]$, and $[A, B \in \mathcal{F}^\circ] \Rightarrow [(A \cup B \in \mathcal{F}^\circ) \wedge (A \cap B \in \mathcal{F}^\circ)]$. $\mathcal{F} \subset \mathcal{P}(\Omega)$ is a σ-**field** (or σ-**algebra**) if it is a field and also satisfies the condition that $[\{A_n : n \in \mathbb{N}\} \subset \mathcal{F}] \Rightarrow [(\cup_n A_n \in \mathcal{F}^\circ) \wedge (\cap_n A_n \in \mathcal{F}^\circ)]$.*

Notation 6.9.3 *Fields may or may not be σ-fields. When we mean to emphasize that a field may fail to satisfy the extra countable union and countable intersection condition that makes it a σ-field, we usually have a superscript $^\circ$.*

We always always always assume that the domain of a probability is a field, and in the vast majority of contexts, we assume that it is a σ-field.

Exercise 6.9.4 Show that $\mathcal{B}^\circ$ is a field of subsets of Ω, that $\{1_E : E \in \mathcal{B}^\circ\}$ is a lattice of functions on Ω, and that **span**($\{1_E : E \in \mathcal{B}^\circ\}$) is an algebra of functions that separates points.[14] Show that it is not a σ-field by showing that for every rational $q_n \in (0, 1]$, $\{q_n\} \in \mathcal{B}^\circ$, but $\mathbb{Q} \cap (0, 1] \notin \mathcal{B}^\circ$.

Definition 6.9.5 *A function $P : \mathcal{G} \to [0, 1]$, $\mathcal{G}$ either a field or a σ-field, is a **finitely additive probability** if $P(\emptyset) = 1 - P(\Omega) = 0$, and for all disjoint A, $B \in \mathcal{G}$, $P(A \cup B) = P(A) + P(B)$. A finitely additive probability is **countably additive** if for all disjoint sequences A_n in $\mathcal{G}$, if $\cup_n A_n \in \mathcal{G}$, then $P(\cup_n A_n) = \sum_n P(A_n)$.*

We hardly ever bring up probabilities that are finitely but not countably additive unless we want to make a point about how badly things work out if countable additivity is violated.

Definition 6.9.6 (Uniform Distribution) *For $E = \cup_{n=1}^N |a_n, b_n| \in \mathcal{B}^\circ$ a disjoint union of intervals, define $\lambda : \mathcal{B}^\circ \to [0, 1]$ by $\lambda(E) = \sum_{n=1}^N (b_n - a_n)$. The probability λ is called the **uniform distribution**.*

It is clear that λ is a finitely additive probability; we show later that it is also countably additive.

Definition 6.9.7 *A function $f : \Omega \to \mathbb{R}$ is **simple** if its range is finite. It is a **simple $\mathcal{B}^\circ$-measurable function or random variable** if it is simple and for all $r \in \mathbb{R}$, $f^{-1}(r) \in \mathcal{B}^\circ$. The set of simple measurable functions is written M_s or $M_s(\mathcal{B}^\circ)$.*

Any simple random variable is of the form $f(x) = \sum_{k \leq K} \beta_k 1_{E_k}(x)$ for $E_k \in \mathcal{B}^\circ$ and $\beta_k \in \mathbb{R}$. The representation of a simple random variable is not unique because $\beta_1 1_{E_1} + \beta_2 1_{E_2} = \beta_1 1_{E_1 \setminus E_2} + (\beta_1 + \beta_2) 1_{E_1 \cap E_2} + \beta_2 1_{E_2 \setminus E_1}$. However, there is a unique such representation of f when the E_k form a partition of Ω and the β_k are distinct. When uniqueness is needed, that is the representation we work with. You should check that if we use different representations of f, it does not change the integral we now define.

Definition 6.9.8 *For $f = \sum_k \beta_k 1_{E_k} \in M_s$, the **integral, or expectation, of f with respect to the uniform distribution** λ is $\int f \, d\lambda := \sum_k \beta_k \lambda(E_k)$.*

Notation 6.9.9 *$\int_\Omega f \, d\lambda$, $\int_\Omega f(\omega) \, d\lambda(\omega)$, $\int_\Omega f(\omega) \, \lambda(d\omega)$, $E \, f$, and $E^\lambda f$ are other notations in common use for the integral.*

The basic connection between probabilities and expectations is $\int 1_E \, d\lambda = \lambda(E)$. Since the mapping $f \mapsto \int f \, d\lambda$ from M_s to $\mathbb{R}$ is linear, knowing $\int f \, d\lambda$ for all $f \in M_S$ gives us λ, we can identify λ with a linear function from M_s to $\mathbb{R}$.

Notation Alert 6.9.A *There is a tradition of using the word "functional" for functions on spaces of functions. Thus, "the integral is a linear functional" is the same as "the integral is a linear function (on a space of functions)."*

---

14. **span**($\{1_E : E \in \mathcal{B}^\circ\}$) is a sup norm separable set of bounded functions separating points, hence eminently suitable for use in constructing a compact embedding (see Definition 6.7.21). The resulting compact metric space is an example of something called a Stone space. It has many peculiar properties, for example, every function 1_E, which takes on only the two values 0 and 1, has a unique continuous extension. This means that the Stone space is highly disconnected; see §6.11.a.

For the most part, we develop the theory of probability as a theory of functions mapping a class of sets, $\mathcal{F}$, to $[0, 1]$. There is an alternate development that defines probabilities as linear functionals, and we study this approach extensively in Chapter 9.

6.9.b Norms and Metrics on the Space of Simple Functions

The first metric on $M_s(\mathcal{B}°)$ asks that two random variables be close iff it is unlikely that they are very far apart.

Definition 6.9.10 $\alpha(f, g) := \inf\{\epsilon \geq 0 : \lambda(\{\omega : |f(\omega) - g(\omega)| > \epsilon\}) \leq \epsilon$ *is the* **Ky Fan pseudometric** *on* $M_s(\mathcal{B}°)$. *If* $\alpha(f_n, f) \to 0$, *we say that* f_n **converges to** f **in probability**.

This is called a pseudometric rather than a metric because of the following problem: if $E = [a, a]$, then $\alpha(1_E, 0) = 0$. We circumvent this issue by defining $f \sim g$ if $\alpha(f, g) = 0$ and regarding α as a metric on the equivalence classes. From this point of view, the function 0 and the function $1_{[a,a]}$ are identical because $\lambda([a, a]) = a - a = 0$, and the probability of picking exactly the point $a \in (0, 1]$ is 0.

Definition 6.9.11   *A* **pseudometric on** M, M *nonempty, is a function* $d : M \times M \to \mathbb{R}_+$ *such that for all* $x, y, z \in M$,

1. $d(x, y) = d(y, x)$,
2. $d(x, y) = 0$ *if* $x = y$, *but* $d(x, y) = 0$ *and* $x \neq y$ *can happen, and*
3. $d(x, y) + d(y, z) \geq d(x, z)$.

A **pseudometric space** *is a pair* (M, d) *where* M *is nonempty and* d *is a pseudometric on* M.

Any pseudometric is a metric on $M/\sim$, where $\sim$ is defined by $[x \sim y] \Leftrightarrow [d(x, y) = 0]$.

Example 6.9.12   *You have seen several pseudometrics used in microeconomics, though perhaps not with this name:*

1. *The utility distance between bundles* x *and* y *is defined as* $d_u(x, y) = |u(x) - u(y)|$. *The* d_u-*equivalence classes are indifference curves.*

2. *Suppose a consumer faces initial prices* $\mathbf{p}^0$ *with an income* w, *derives a utility* $u^0 = v(\mathbf{p}^0, w)$, *and prices change to* $\mathbf{p}^1$. *The compensating variation,* $CV(\mathbf{p}^1, \mathbf{p}^0) = w - e(\mathbf{p}^1, u^0)$, *is the amount that the consumer would need to be compensated for the price changes in order to be left indifferent, which can easily be 0 for* $\mathbf{p}^1 \neq \mathbf{p}^0$.

In the following, to be absolutely correct, we would have to talk of pseudonorms and norm pseudometrics.

Definition 6.9.13   *For* $p \in [1, \infty)$ *and* $f \in M_s$, *the* L^p-**norm of** f *is* $\|f\|_p = \left[\int |f|^p \, d\lambda\right]^{1/p}$, *and the associated* **norm metric** *is* $d_p(f, g) = \|f - g\|_p$. *The* L^∞-**norm of** f *is* $\|f\|_\infty = \inf\{B \geq 0 : \lambda(|f| > B) = 0\}$, *and the associated* **norm metric** *is* $d_\infty(f, g) = \|f - g\|_\infty$.

It is worth reviewing the properties of norms on vector spaces, Definition 4.3.7 (p. 117). Except for the triangle inequality, it is immediate that $\|\cdot\|_p$ satisfies the properties of a norm. For $p \in (1, \infty)$, we prove the triangle inequality in Theorem 8.3.4 (p. 461). For $p = 1$, we prove it here.

Lemma 6.9.14 *For any* $f, g \in M_s$, $\|f + g\|_1 \leq \|f\|_1 + \|g\|_1$.

Proof. For all ω, $|f + g|(\omega) \leq |f|(\omega) + |g|(\omega)$ and the integral is a monotonic functional; that is, $r \leq s$ in M_s implies $\int r \, d\lambda \leq \int s \, d\lambda$. ∎

Exercise 6.9.15 Show that $\|f + g\|_\infty \leq \|f\|_\infty + \|g\|_\infty$.

Definition 6.9.16 *The **Lebesgue space** (L^0, α) is defined as the completion of (M_s, α), and for $p \in [1, \infty)$, the **Lebesgue space** (L^p, d_p) is defined as the completion of the metric space (M_s, d_p), with the norm $\|f\|_p := d_p(f, 0)$.*

Exercise 6.9.17 Show that (M_s, d_∞) is not separable. [This is related to the reason that we have to give a separate definition of (L^∞, d_∞).]

We will see that the L^p spaces, $p \in [1, \infty]$ are complete normed vector spaces; that is, they are **Banach spaces**, which we cover in Chapter 8.

When we completed $\mathbb{Q}$, we named the new space $\mathbb{R}$. Given that we have $\mathbb{R}$, it is possible to understand the points that are added to M_s by giving them representations in terms of $\mathbb{R}$-valued functions on $(0, 1]$. These functions are known as "measurable functions," and we cover their properties extensively in Chapter 7.

6.9.c The Basic Properties of (M_s, d_1)

First we have the following.

Lemma 6.9.18 *The mapping* $E : M_s \to \mathbb{R}$ *is* d_1-*uniformly continuous, linear, and nondecreasing.*

This means that $E : M_s \to \mathbb{R}$ has a unique continuous extension to L^1.

Proof. Linearity is the property that for all $\alpha, \beta \in \mathbb{R}$ and all $f, g \in M_s$, $E(\alpha f + \beta g) = \alpha E f + \beta E g$. This is immediate from the definition of the integral. The function $E : M_s \to \mathbb{R}$ is nondecreasing if $[f \leq g] \Rightarrow [\int f \leq \int g]$, which is also immediate.

If $f = \sum_k \beta_k 1_{E_k}$, then $E f = \sum_k \beta_k \lambda(E_k) \leq \sum_k |\beta_k| 1_{E_k} = E |f| = \|f\|_1$. In the same way, $-E f \leq \|f\|_1$, so that $|E f| \leq \|f\|_1$.

Suppose that $\|f_n - f\|_1 \to 0$. By linearity, $|E f_n - E f| = |E(f_n - f)|$. Combining, we have $|E f_n - E f| = |E(f_n - f)| \leq E|f_n - f| = \|f_n - f\|_1 \to 0$. Rewriting, if $d_1(f_n, f) < \epsilon$, then $|E f_n - E f| < \epsilon$, so that we have uniform continuity. ∎

Lemma 6.9.19 (M_s, d_1) *is an incomplete separable infinite-dimensional metric vector space.*

Proof.   Since $\| \cdot \|_1$ is a norm, d_1 is a metric.

For $n \in \mathbb{N}$ and $0 \le k \le 2^n - 1$, let $I(k, n) = (k/2^n, (k + 1)/2^n] \in \mathcal{B}^\circ$ be the kth dyadic interval of order n. The sequence $f_n = \sum_k \frac{k}{2^n} 1_{I(k,n)}$ is d_1-Cauchy, but does not converge to any simple function.

The $\mathcal{B}^\circ_\mathbb{Q} \subset \mathcal{B}^\circ$ denote the set of finite unions of intervals with rational endpoints. The set of simple measurable functions taking on only rational values and having each $f^{-1}(r) \in \mathcal{B}^\circ_\mathbb{Q}$ is a countable dense set.

Suppose that $d_1(f_n, f) \to 0$ and $d_1(g_n, g) \to 0$ in M_s. The metric vector space properties follow from two observations: first, for all ω, $|(f_n + g_n) - (f + g)|(\omega) \le |f_n - f|(\omega) + |g_n - g|(\omega)$ so that, second, $d_1(f_n + g_n, f + g) = \int |(f_n + g_n) - (f + g)| \, d\lambda \le \int |f_n - f| \, d\lambda + \int |g_n - g| \, d\lambda$.

Finally, we construct a sequence of increasingly "jagged" functions in M_s that are linearly independent. Note that $I(k, n) = I(2k, n + 1) \cup I(2k + 1, n + 1)$ expresses the kth dyadic interval of order n as the disjoint union of two dyadic intervals of order $n + 1$. Let $E_n = \cup \{I(k, n) : k \text{ is even}\}$ so that $E_n^c = \cup \{I(k, n) : k \text{ is odd}\}$. The function $f_n = 1_{E_n} - 1_{E_n^c}$ bounces between ± 1 over each dyadic interval of order $n - 1$, so that the collection $\{f_n : n \le N\}$ is linearly independent for all N.   ∎

We will see the dyadic intervals, $I(k, n)$, and the sequence of functions, $f_n = 1_{E_n} - 1_{E_n^c}$, used in the previous proof on several occasions in what follows.

We now show that every $f \in C([0, 1])$ can be "sandwiched" between two equivalent Cauchy sequences in L^1 and that $C([0, 1])$ is d_1-dense in M_s. This means that we could, in principle, develop (L_1, d_1) as the completion of $(C([0, 1]), d_1)$.

Lemma 6.9.20   *If $f \in C([0, 1])$, then there is a pair of equivalent Cauchy sequences, f_n^- and f_n^+, in (M_s, d_1), such that for all $\omega \in (0, 1]$, $f_n^-(\omega) \le f(\omega) \le f_n^+(\omega)$. Further, the continuous functions are d_1-dense.*

Proof.   Pick arbitrary $f \in C([0, 1])$ and B such that $\|f\|_\infty < B$. Let $I(k, n) = (k/2^n, (k + 1)/2^n]$ be the kth dyadic interval of order n, $k \in \mathbb{Z}$. Define $r_{k,n}^+ = \sup\{f(I(k, n))\}$, $r_{k,n}^- = \inf\{f(I(k, n))\}$, and $s_{k,n} = r_{k,n}^+ - r_{k,n}^-$.

Define

$$f_n^+(\omega) = \sum_{k=-B2^n}^{k=B2^n} r_{k,n}^+ 1_{I(k,n)}(\omega) \quad \text{and} \quad f_n^-(\omega) = \sum_{k=-B2^n}^{k=B2^n} r_{k,n}^- 1_{I(k,n)}(\omega).$$

By construction, for all $\omega \in (0, 1]$, $f_n^-(\omega) \le f(\omega) \le f_n^+$. We now show that f_n^+ and f_n^- are equivalent Cauchy sequences. As f is uniformly continuous, $s_N := \max_k s_{k,N}$ converges to 0. By construction, for all $m, n \ge N$, $d_1(f_n^+, f_m^+) \le s_N$, $d_1(f_n^-, f_m^-) \le s_N$, and $d_1(f_n^+, f_m^-) \le s_N$.

We now show that the (equivalence classes of the) $C([0, 1])$ in L^1 are d_1-dense. Since M_s is dense, it is sufficient to show that for all $g \in M_s$ and all $\epsilon > 0$, there is a continuous function f such that $d_1(f, g) < \epsilon$.

Every $g \in M_s$ is at d_1-distance 0 from a function $\sum_{k \le K} \beta_k 1_{|a_k, b_k|}$, where $a_k < b_k, b_k \le a_{k+1}, \beta_k \ne 0$, and the intervals are disjoint. Pick $\epsilon > 0$. For m large, define

$$f_{k,m}(\omega) = (1 - m \cdot d(\omega, [a_k + \tfrac{1}{m}, b_k - \tfrac{1}{m}])) \vee 0.$$

For m large, $d_1(f_{k,m}, 1_{|a_k,b_k|}) < \epsilon/K \cdot |\beta_k|$, so that $d_1(\sum \beta_k 1_{|a_k,b_k|}, \sum \beta_k f_{k,m})$ $< \epsilon$. ∎

6.9.d The Probability Metric Space

The **symmetric difference of** E, $F \subset \Omega$ is defined by $E \triangle F = (E \setminus F) \cup (F \setminus E)$.

Definition 6.9.21 *For E, $F \in \mathcal{B}^\circ$, we define the **probability pseudometric** by* $\rho_\lambda(E, F) = \lambda(E \triangle F)$.

Exercise 6.9.22 Show that the mapping $E \mapsto 1_E$ from $(\mathcal{B}^\circ, d_\lambda)$ to (M_s, d_1) is an isometry.

This implies that the metric completion of (M_s, d_1) contains the metric completion of $(\mathcal{B}^\circ, \rho_\lambda)$. Since M_s is the span of the set $\{1_E : E \in \mathcal{B}^\circ\}$, it is possible to study the structure of L^1 after first studying the completion of the metric space $(\mathcal{B}^\circ, \rho_\lambda)$. Limits of ρ_λ-Cauchy sequences appear in the constructions of limits in the study of the two laws of large numbers.

6.9.e Stochastic Independence

Stochastic independence is the key concept underlying an enormous amount of probability and statistics. To make our notation more consonant with the fact that we are starting to deal with probability theory, we substitute "P" for "λ" and call $P(E)$ the probability of the set E.

Recall that $I(k, n) = (k/2^n, (k + 1)/2^n]$. For each $n \in \mathbb{N}$, define the simple random variable

$$X_n(\omega) = \begin{cases} 1 & \text{if } s \in I(k, n), \, k \text{ odd,} \\ 0 & \text{if } s \in I(k, n), \, k \text{ even.} \end{cases} \tag{6.38}$$

Intuitively, the X_n are independent, so knowing the value of $X_n(\omega)$ for all n in a finite set N tells you nothing new about the value of $X_m(\omega)$ for any m not in N.

Definition 6.9.23 *A pair of events, E, $F \in \mathcal{B}^\circ$, is **stochastically independent**, or simply **independent**, written $E \perp\!\!\!\perp F$, if $P(E \cap F) = P(E) \cdot P(F)$.*

The multiplication of probabilities is a very intuitive expression of independence. We think that the probability of rolling two sixes in a row with a fair die is $(1/6) \cdot (1/6)$ because we think the outcomes of the two rolls have nothing to do with each other. More generally, we have the following.

Definition 6.9.24 *A collection $\{E_\alpha : \alpha \in A\}$ of events is **independent** if for any finite $A' \subset A$, $P(\cap_{\alpha \in A'} E_\alpha) = \Pi_{\alpha \in A'} P(E_\alpha)$. A collection $\{\mathcal{C}_\alpha : \alpha \in A\}$ of subsets of $\mathcal{B}$ is **independent** if for all collections $\{E_\alpha \in \mathcal{C}_\alpha : \alpha \in A\}$ are independent. $\{X_\alpha : \alpha \in A\}$ is an **independent collection of simple random variables** if for every collection $\{r_\alpha : \alpha \in A\} \subset \mathbb{R}$, the collection of events, $\{X_\alpha^{-1}(r_\alpha) : \alpha \in A\}$, is independent.*

Some of the interpretation of independence is contained in the following.

Exercise 6.9.25   Let E, $F \in \mathcal{B}^\circ$. If $P(F) > 0$, the conditional probability of E given F, or "P of E given F," is $P(E|F) := \frac{P(E \cap F)}{P(F)}$. The idea is that $P(E)$ is the likelihood we assign to E before we learn anything and $P(E|F)$ is the likelihood after we have learned that F has occurred.

Prove the following:

(a) If $E \cap F = \emptyset$ and $P(E)$, $P(F) > 0$, then $\neg E \perp\!\!\!\perp F$.

(b) $\emptyset \perp\!\!\!\perp E$ and $(0, 1] \perp\!\!\!\perp E$.

(c) If $E \perp\!\!\!\perp F$ then $E \perp\!\!\!\perp F^c$.

(d) If $P(F) > 0$, then $E \perp\!\!\!\perp F$ iff $P(E|F) = P(E)$.

Independence is not transitive, and pairwise independence does not imply independence.

Exercise 6.9.26   Let $E_1 = (0, \frac{1}{2}]$, $E_2 = (\frac{1}{4}, \frac{3}{4}]$, and $E_3 = (\frac{1}{2}, 1]$. Show that $E_1 \perp\!\!\!\perp E_2$ and $E_2 \perp\!\!\!\perp E_3$, but that $\neg E_1 \perp\!\!\!\perp E_3$. Let $A_1 = (0, \frac{4}{16}]$, $A_2 = (\frac{3}{16}, \frac{7}{16}]$, and $A_3 = (0, \frac{1}{16}] \cup (\frac{6}{16}, \frac{9}{16}]$. Show that every pair A_i, A_j, $i \neq j$, is independent, but that the collection $\{A_1, A_2, A_3\}$ is *not* independent.

There is a weak connection between stochastic independence and linear independence and a stronger connection between independence and orthogonality.

Definition 6.9.27   *The **inner product** of two simple random variables f, $g \in M_s$ is defined as $E\, fg$, equivalently $\int fg \, dP$, and written as $\langle f, g \rangle$. Two simple random variables are **orthogonal**, written $f \perp g$, if $\langle f, g \rangle = 0$.*

The inner product of functions in M_s is a generalization of the inner (or dot) product of vectors in $\mathbb{R}^\ell$. In particular, it is **bilinear** for every f, g, $h \in M_s$ and α, $\beta \in \mathbb{R}$, $\langle (\alpha f + \beta g), h \rangle = \alpha \langle f, h \rangle + \beta \langle g, h \rangle$, and, since $\langle f, g \rangle = \langle g, f \rangle$, $\langle h, (\alpha f + \beta g) \rangle = \langle \alpha h, f \rangle + \beta \langle h, g \rangle$, and $\langle f, f \rangle = 0$ iff $f = 0$.

Exercise 6.9.28   Show the following:

1. If $P(E)$, $P(F) > 0$ and $E \perp\!\!\!\perp F$, then 1_E and 1_F are linearly independent.

2. For any E with $P(E)$, $P(E^c) > 0$, 1_E and 1_{E^c} are linearly independent, but $\neg(E \perp\!\!\!\perp F)$.

3. Show that if $E \perp\!\!\!\perp F$, then $\forall f : \mathbb{R} \to \mathbb{R}$ and $\forall g : \mathbb{R} \to \mathbb{R}$, $[f(1_E) - E\, f(1_E)] \perp [g(1_F) - E\, g(1_F)]$. [The converse of this statement is in Lemma 6.9.29.]

For any $\mathcal{S} \subset \mathcal{P}(\Omega)$, let $\mathcal{F}^\circ(\mathcal{S})$ be the smallest field containing $\mathcal{S}$ and $M_s(\mathcal{S})$ the set of simple $\mathcal{F}^\circ(\mathcal{S})$-measurable random variables. $\mathcal{F}^\circ(\emptyset) = \{\emptyset, \Omega\}$ is the smallest field of all.

Lemma 6.9.29   *A collection $\{E_\alpha : \alpha \in A\}$ of events is independent iff for any finite, disjoint A_F, $B_F \subset A$, any $f \in M_s(\{E_\alpha : \alpha \in A_F\})$, any $g \in M_s(\{E_\alpha : \alpha \in B_F\})$, $(f - E\, f) \perp (g - E\, g)$.*

Proof. Suppose that $\{E_\alpha : \alpha \in A\}$ is an independent collection of events. Pick finite $A_F, B_F \subset A$. For each $\omega \in \Omega$, let $A_F(\omega) = \bigcap\{E_\alpha : \alpha \in A_F, \ \omega \in E_\alpha\}$. The partition of Ω generated by $\{E_\alpha : \alpha \in A_F\}$ is defined as $\mathbb{P}(A_F) = \{A_F(\omega) : \omega \in \Omega\}$. Any $f \in M_s(\{E_\alpha : \alpha \in A_F\})$ is[15] of the form $\sum_k r_k 1_{F_k}$, where the finite sum is over F_k in $\mathbb{P}(A_F)$, and any $g \in M_s(\{E_\alpha : \alpha \in B_F\})$ is of the form $\sum_\ell s_\ell 1_{G_\ell}$, where the sum is over G_ℓ in $\mathbb{P}(B_F)$. We must show that $\langle f - E\,f, g - E\,g \rangle = 0$. By direct calculation, $\langle f - E\,f, g - E\,g \rangle = \langle f, g \rangle - E\,f \cdot E\,g$, so it is sufficient to show that $E\,fg = E\,f \cdot E\,g$. Again by direct calculation, $fg = \sum_k r_k(\sum_\ell s_\ell 1_{G_\ell \cap F_k})$, so that $\langle f, g \rangle = E\,fg = \sum_k r_k(\sum_\ell s_\ell P(G_\ell \cap F_k))$. Independence implies that $P(G_\ell \cap F_k) = P(G_\ell) \cdot P(F_k)$. Using this to rearrange the sum yields $\sum_k r_k P(F_k) \cdot (\sum_\ell s_\ell P(G_\ell)) = E\,f \cdot E\,g$.

Now suppose that for finite disjoint $A_F, B_F \subset A$, any $f \in M_s(A_F)$ and any $g \in M_s(B_F)$, we have $(f - E\,f) \perp (g - E\,g)$. For all $A' \subset A$ of size 2, we see that $P(\cap_{\alpha \in A'} A_\alpha) = \Pi_{\alpha \in A'} P(E_\alpha)$ by taking A' as the disjoint union of nonempty A_F and B_F. Now suppose that for all $A' \subset A$ of size n, $P(\cap_{\alpha \in A'} A_\alpha) = \Pi_{\alpha \in A'} P(E_\alpha)$. The same holds for all $A'' \subset A$ of size $n + 1$ because $1_{\cap_{\alpha \in A'} E_\alpha} \in M_s(A')$, so that $\langle 1_{\cap_{\alpha \in A'} E_\alpha} - \Pi_{\alpha \in A'} P(E_\alpha), 1_{E_\gamma} - P(E_\gamma) \rangle = 0$ for $\gamma \notin A'$. ∎

The following is not conceptually difficult, but the notation can get out of hand if you are not careful.

Exercise 6.9.30   The collection $\{X_n : n \in \mathbb{N}\}$ defined in (6.38) is independent.

6.9.f The Weak and the Strong Law of Large Numbers

There is a simple inequality that gives the weak law of large numbers.

Lemma 6.9.31 (Chebyschev)   *For any* $X \in M_s$, $X \geq 0$, *and* $r > 0$, $P(\{X \geq r\}) \leq (E\,X)/r$.

Proof. $X \leq r \cdot 1_{\{X \geq r\}}$, so that $E\,X \leq r \cdot E1_{\{X \geq r\}} = r \cdot P(\{X \geq r\})$. ∎

In particular, for any simple random variable, $P(|X| \geq r) \leq (E\,|X|)/r$, and $P((X - E\,X)^2 \geq r^2) = P(|X - E\,X| \geq r) \leq \text{Var}(X)/r^2$, where $\text{Var}(X) := E\,(X - E\,X)^2$ is the variance of X.

Exercise 6.9.32   Show the following:

1. $\text{Var}(r\,X) = r^2\,\text{Var}(X)$.
2. $\text{Var}(X) = E\,X^2 - (E\,X)^2$.
3. If $X \perp\!\!\!\perp Y$, then $\text{Var}(X + Y) = \text{Var}(X) + \text{Var}(Y)$.

The following is the weak law of large numbers for the X_n defined in (6.38).

Theorem 6.9.33   *For any* $\epsilon > 0$, $\lim_n P\left\{\omega : \left|\frac{1}{n} \sum_{t=1}^n (X_t(\omega) - \frac{1}{2})\right| \geq \epsilon \right\} = 0$.

Proof.   For all n, $E(\frac{1}{n} \sum_{t=1}^n (X_t - \frac{1}{2})) = 0$ and $\text{Var}(\frac{1}{n} \sum_{t=1}^n (X_t - \frac{1}{2})) = \frac{n}{n^2}$. By Chebyschev, $P(\{|S_n - 0| \geq \epsilon\}) \leq \frac{1/n}{\epsilon} \to 0$. ∎

---

15. Formally, "is" should be "has a member in its equivalence class."

The previous result says that if n is large, it is unlikely that the sample average, $\frac{1}{n} \sum_{t=1}^{n} X_t$, is very far from the true average, $\frac{1}{2}$. The strong law of large numbers is the statement that, outside of a set of ω having probability 0, $\lim_n \frac{1}{n} \sum_{t=1}^{n} X_t(\omega) = \frac{1}{2}$. This is a very different kind of statement, and much stronger. It rules out some set E, with positive probability, with the property that for each $\omega \in E$, there exists an infinite sequence of times, $T_n(\omega)$, with $\left| \frac{1}{T_n(\omega)} \sum_{t=1}^{T_n(\omega)} X_t(\omega) - \frac{1}{2} \right| > \epsilon$. If the $T_n(\omega)$ were arranged to become sparser and sparser as n grows larger, this could still be consistent with Theorem 6.9.33.

Before going any further, let us look carefully at the set of ω we are talking about. For any ω and any T, $\lim \inf_n \frac{1}{n} \sum_{t=1}^{n} X_t(\omega) = \lim \inf_n \frac{1}{n} \sum_{t=T+1}^{n} X_t(\omega)$, and the same is true with "lim sup" replacing "lim inf." In words, for all T, the first T values of X_t do not affect limit behavior. This indicates that we need more sets than are contained in $\mathcal{B}^\circ$.

For any sequence y_n in $\mathbb{R}$, $\lim \inf_n y_n \leq r$ iff $\forall \epsilon > 0$, $\forall N \in \mathbb{N}$, $\exists n \geq N$ $y_n \leq r + \epsilon$. Also, $\lim \sup_n y_n \leq r$ iff $\forall \epsilon > 0$, $\exists N \in \mathbb{N}$, $\forall n \geq N$ $y_n \leq r + \epsilon$. Let $\omega \mapsto Y_n(\omega)$ be a sequence of random variables, so that $\{\omega : Y_n \leq s\} \in \mathcal{B}^\circ$ for all $s \in \mathbb{R}$. We are interested in $\omega \mapsto \lim \inf_n Y_n(\omega)$ and $\omega \mapsto \lim \sup_n Y_n(\omega)$ being random variables. This requires that the following two sets belong to the domain of our probability,

$$\{\omega : \lim \inf_n Y_n(\omega) \leq r\} = \bigcap_{\epsilon \in \mathbb{Q}_{++}} \bigcap_{N \in \mathbb{N}} \bigcup_{n \geq N} \{\omega : Y_n(\omega) \leq r + \epsilon\}, \quad \text{and}$$

$$\{\omega : \lim \sup_n Y_n(\omega) \leq r\} = \bigcap_{\epsilon \in \mathbb{Q}_{++}} \bigcup_{N \in \mathbb{N}} \bigcap_{n \geq N} \{\omega : Y_n(\omega) \leq r + \epsilon\}.$$

We have $\bigcap_{N \in \mathbb{N}} \bigcup_{n \geq N} \{\omega : Y_n(\omega) \leq r + \epsilon\} = \{\omega : Y_n(\omega) \leq r + \epsilon \text{ i.o.}\}$, and $\bigcup_{N \in \mathbb{N}} \bigcap_{n \geq N} \{\omega : Y_n(\omega) \leq r + \epsilon\} = \{\omega : Y_n(\omega) \leq r + \epsilon \text{ a.a.}\}$. These are written, with no loss of precision but considerable gain in clarity, as $\{Y_n \leq r + \epsilon\}$ a.a. and $\{Y_n \leq r + \epsilon\}$ i.o.

To talk about the probability of different kinds of limit behavior, we must have a class of sets that is closed under countable unions and intersections. As noted earlier, such classes are called σ-fields (or σ-algebras), and we must extend P from its domain, $\mathcal{B}^\circ$, to a σ-field containing $\mathcal{B}^\circ$. As we will see, this is the same as completing the metric space $(\mathcal{B}^\circ, \rho_P)$, where $\rho_P(E, F) := P(E \Delta F)$. That is, the completion of the metric space is isometric to a σ-field containing $\mathcal{B}^\circ$. At the center of making the extension work well are the notions of countably additive measures and measurable functions. We cover that material extensively in Chapter 7.

6.10 ◆ Bibliography

J. R. Munkres's *Topology* (2nd Ed., Upper Saddle River, N.J.: Prentice-Hall, 2000) and A. N. Kolmogorov and S. V. Fomin's *Introductory Real Analysis* (translated and edited by R. A. Silverman; New York: Dover, 1975, 1970) both develop metric spaces in great detail, and almost all of the analysis material covered in this chapter can be found in these two books. However, our applications to the space of cdf's, the space of compact sets, regression analysis, and the Lebesgue measure space are not found in either of those volumes. Our coverage of the implicit function theorem owes a great deal to E. Ward Cheney's *Analysis for Applied Mathematics* (New York: Springer-Verlag, 2001), and this is a good book to own for several other reasons as well.

6.11 ◆ End-of-Chapter Problems

These problems are loosely grouped around the following themes.

(a) Isolated points and connected spaces.

(b) More on the types of convergence of continuous functions.

(c) The convergence of closed sets.

(d) Upper semicontinuity and the existence of maxima.

(e) Urysohn's lemma as an introduction to nonmetric topologies.

(f) Finitely determined functions on compact product spaces.

(g) Changing metrics.

(h) Small of subsets of a metric space.

(i) Another proof of the metric completion theorem.

6.11.a Isolated Points and Connected Sets

You should review Definition 4.1.21 (p. 112) and Theorem 4.12.12 (p. 170).

Exercise 6.11.1 Let (X, d) be a metric space. Show that for all $x \in X$, $\{x\}$ is a closed subset of X, as is any finite subset of X.

Definition 6.11.2 *A point $x \in M$ is **isolated** if $B_\epsilon(x) = \{x\}$ for some $\epsilon > 0$.*

Thus, x is isolated iff $\{x\}$ is both open and closed, equivalently iff $M \setminus \{x\}$ is both open and closed. Recall that a space (M, d) is **connected** iff it cannot be expressed as the union of two nonempty sets, both of which are open and closed.

Exercise 6.11.3 Show the following.

1. x is isolated iff $\text{int}(\{x\}) \neq \emptyset$.

2. No point in $\mathbb{R}^k$ is isolated with any of the usual metrics.

3. No point in $C_b(\mathbb{R})$ is isolated.

4. If (M, d) is a discrete metric space, then every point is isolated.

5. Let $M = \mathbb{N} \cup \{\infty\}$ with the metric $d(x, y) = |\frac{1}{x} - \frac{1}{y}|$, where we define $\frac{1}{\infty} = 0$. Every point in M except $\{\infty\}$ is isolated.

Isolated points are not continuously connected to anything.

Lemma 6.11.4 *If x is an isolated point in the metric space (M, d), then there is no continuous $f : [0, 1] \to M$ with $f(0) = x$ and $f(1) = y$ for $y \neq x$.*

Exercise 6.11.5 Fill in the details of the following sketch of a proof. Suppose there is such a continuous function. Show that $f^{-1}(x)$ is both open and closed and has a nonempty complement that is also open and closed. We know $0 \in f^{-1}(x)$. Let $r = \sup\{t : f(t) = 1\}$ and show that $r < 1$. Show that $r \in f^{-1}(x)$ because $f^{-1}(x)$ is closed. Show that $r \notin f^{-1}(x)$ because $f^{-1}(x)$ is open.

Theorem 6.11.6 *If E is a connected set and $E \subset F \subset \mathrm{cl}(E)$, then F is connected.*

Exercise 6.11.7 Using Lemma 6.11.4, prove Theorem 6.11.6.

Theorem 4.12.12 showed that path connectedness implies connectedness. The following exercise gives a connected set that is not path connected.

Exercise 6.11.8 Show the following:

1. The set $E = \{(x_1, x_2) \in \mathbb{R}^2 : x_2 = \sin(1/x_1), \ x_1 \in (0, 1]\}$ is connected. [The interval $(0, 1]$ is connected and $\sin(1/r)$ is continuous on the interval.]

2. The set $\mathrm{cl}(E)$ is equal to $E \cup \{(0, x_2) : x_2 \in [-1, +1]\}$.

3. The set $\mathrm{cl}(E)$ is connected, but there is no path connecting $(0, 0) \in \mathrm{cl}(E)$ to $(1/\pi, 0) \in E$. [Let f be a path from $(0, 0)$ to $(1/\pi, 0)$ with $f(0) = (0, 0)$ and $f(1) = (1/\pi, 0)$. Since proj_1 and proj_2 are continuous functions from $\mathbb{R}^2$ to its axes, the functions $\mathrm{proj}_1 \circ f$ and $\mathrm{proj}_2 \circ f$ are continuous. $\mathrm{proj}_1 \circ f(0) = 0$, $\mathrm{proj}_1 \circ f(1) = 1/\pi$, and the continuous image of a connected set is connected. This means that $[0, 1/\pi] \subset \mathrm{proj}_1 \circ f([0, 1])$. In particular, this implies that for every $n \in \mathbb{N}$, there exists $t_n \in [0, 1]$ such that $\mathrm{proj}_1(f(t_n)) = 1/n\pi$. All but finitely many of the t_n must be in any interval of the form $[0, \delta), \delta > 0$. However, $\mathrm{proj}_2 \circ f(t_n) = \pm 1$ as n is even or odd, so that $t_n \to 0$. However, $+1 = \limsup_n \mathrm{proj}_2 \circ f(t_n) > \liminf_n \mathrm{proj}_2 \circ f(t_n) = -1$, implying that $\mathrm{proj}_2 \circ f$ is not continuous.]

6.11.b More on Types of Convergence of Functions

The following shows (again) the stark distinction between pointwise and uniform convergence of continuous functions.

Exercise 6.11.9 [↑Example 6.6.46] Consider the sequence $g_n(x) = \max\{0, 1 - |x - n|\}$ in $C_b(\mathbb{R})$. Show that:

1. for all $x \in \mathbb{R}$, $g_n(x) \to 0$, that is, g_n converges pointwise to 0,

2. $\neg[\|g_n\|_\infty \to 0]$,

3. $T_F^{-1}(g_n)$ converges pointwise to 0 in $C_b((0, 1))$, and

4. there is a sequence h_n in $C([0, 1])$ converging pointwise to 0 and having $\|h_n\|_\infty \equiv 1$.

Recall that a sequence of cdf's, F_n, converges weakly to a cdf F iff F_n converges to F pointwise, but only at the continuity points of F. A side benefit of the next two exercises is that they show that cumulative distribution functions can be hard to visualize.

Exercise 6.11.10 Let $\{q_m : m \in \mathbb{N}\}$ be an enumeration of $\mathbb{Q} \subset \mathbb{R}$. Define $F(x) = \sum_{\{m : q_m \leq x\}} 2^{-m}$ and $F_n(x) = \frac{1}{1-(1/2^n)} \sum_{\{m \leq n, \, q_m \leq x\}} 2^{-m}$. Show the following:

1. F_n and F are cdf's.
2. $0 < F(r) < 1$ for all r.
3. F is strictly increasing, $[x < x'] \Rightarrow [F(x) < F(x')]$.
4. if $q \in \mathbb{Q}$, then F is discontinuous at q.
5. if $x \notin \mathbb{Q}$, then x is a continuity point of F.
6. $F_n \to_{w^*} F$.

The previous limit cdf was strictly increasing over every interval and discontinuous at a dense set of points. The limit cdf in the following is continuous, but fractal.

Exercise 6.11.11 Parallel to (6.38) on p. 340, show that you can define a sequence Y_n of independent simple, $\{0, 1\}$-valued random variables in M_s with the property that $\lambda(Y_n = 1) = \frac{1}{3}$. Let $S_n = \sum_{i \leq n} Y_i/2^i$ and let F_n be the cdf of S_n. Show that F_n is a tight sequence of cdf's. Show that $F_n \to_{w^*} F$ for some continuous strictly increasing F. [For the continuity and strictly increasing properties, let $S = \lim_n S_n$, and for every interval $I_{k,n} = [k/2^n, (k+1)/2^n]$, $(1/3)^n \leq \lambda(Y \in I_{k,n}) \leq (2/3)^n$.]

The sequence of functions $f_n(t) = [1 - n|x - \frac{1}{n}|] \vee 0$ in $C([0, 1])$ demonstrates that pointwise convergence to a continuous function does not imply uniform convergence. However, if we ask that the convergence be monotone, we have a positive result.

Theorem 6.11.12 (Dini's Theorem) *If f_n is a sequence of functions in $C(M)$, M compact, $f_{n+1} \leq f_n$, and the function $f_\infty(x) := \lim_n f_n(x)$ is continuous, then f_n is Cauchy, that is, it converges uniformly to f_∞.*

Exercise 6.11.13 Fill in the details in the following sketch of a proof of Dini's theorem. First, defining $g_n = f_n - f$, we can reduce the proof to the case that the g_n converge pointwise downward to the function $g = 0$. Second, for $\epsilon > 0$ and $x \in M$, define $N(\epsilon, x)$ as the minimum N such that for all $n \geq N$, $g_n(x) < \epsilon$. Let $G(\epsilon, x) = \{y : g_{N(\epsilon,x)}(y) < \epsilon\}$. Show that $\{G(\epsilon, x) : x \in M\}$ is an open cover of M. Take a finite subcover and complete the proof.

Often the spaces $C(K)$ with K compact are enough for applications. But sometimes one wants more generality, for example, $C(\mathbb{R})$ rather than $C([a, b])$. The easiest cases involve $C(M)$ when M fails compactness, but satisfies a countable version of compactness.

Definition 6.11.14 *A metric space (M, d) is σ-compact if it can be expressed as a countable union of compact sets; it is strongly σ-compact if it can be expressed as a countable union of open sets with compact closure.*

Clearly every strongly σ-compact space is compact, but the reverse is not true.

Exercise 6.11.15 With d being the usual metric on $\mathbb{R}$, $(\mathbb{Q}, d)$ is σ-compact but not strongly σ-compact.

Exercise 6.11.16   Show that $\mathbb{R}^k$ is strongly σ-compact and that $C([0, 1])$ is not σ-compact.

Exercise 6.11.17   Show that if $M = \cup_n K_n$ with each K_n compact, then $M = \cup_n K_n'$, where each K_n' is compact and $K_n' \subset K_{n+1}'$. In other words, there is no loss in assuming that a σ-compact space or a strongly σ-compact space is the countable union of an increasing sequence of compact sets.

Recall that a sequence of functions f_n in $C(M)$ **converges to f uniformly on compact sets** if $\forall K$ compact, $\sup_{x \in K} |f_n(x) - f(x)| \to 0$.

Exercise 6.11.18   In $C(\mathbb{R})$, the sequence of functions $f_n(r) = \max\{0, |r| - n\}$ converges uniformly on compact sets to $f = 0$, but $\sup_{x \in \mathbb{R}} |f_n(x) - f(x)| \equiv \infty$.

When M is strongly σ-compact, there are many metrics on $C(M)$ with the property that $d(f_n, f) \to 0$ iff f_n converges uniformly to f on compact sets. Let $M = \cup_n G_n$ with $G_n \subset G_{n+1}$ and each G_n open and having compact closure K_n. Here is one of the metrics:

$$d_\sigma(f, g) = \sum_n \frac{1}{2^n} \min\{1, \max_{x \in K_n} |f(x) - g(x)|\}. \qquad (6.39)$$

Theorem 6.11.19   *If M is strongly σ-compact, then $d_\sigma(f_n, f) \to 0$ iff f_n converges uniformly to f on compact sets.*

Exercise 6.11.20   Under the conditions of Theorem 6.11.19, show that if K is a compact subset of M, then there exists an N such that for all $n \geq N$, $K \subset K_N$. With this result in place, give a proof of Theorem 6.11.19.

Recall from Definition 6.2.9 (p. 274) that the mapping $\varphi : M \times C(M) \to \mathbb{R}$ defined by $\varphi(x, f) = f(x)$ is called the **evaluation mapping**. We give $M \times C(M)$ the product metric that comes from d and d_σ, that is, $(x_n, f_n) \to (x, f)$ iff $d(x_n, x) + d_\sigma(f_n, f) \to 0$.

Exercise 6.11.21   Show that if M is strongly σ-compact, then φ is jointly continuous. [If $x_n \to x$, then $K = \{x\} \cup \{x_n : n \in \mathbb{N}\}$ is compact.]

Joint continuity of the evaluation mapping has information about what distance means in $C(M)$.

Exercise 6.11.22   Show that if M is strongly σ-compact and ρ is a metric on $C(M)$ such that φ is jointly continuous, then $[\rho(f_n, f) \to 0] \Rightarrow [d_\sigma(f_n, f) \to 0]$.

6.11.c   Convergence of Closed Sets

The following type of convergence for closed sets has proved to be important in the theory of random sets and in dynamic programming. Let $\mathcal{C}(M)$ denote the set of nonempty closed subsets of the metric space M.

Definition 6.11.23   *A sequence F_n in $\mathcal{C}(M)$ converges to $F \in \mathcal{C}(M)$ in the **Painlevé-Kuratowski** sense, written $F_n \to_{PK} F$, if $F = \text{clim sup } F_n = \text{clim inf } F_n$.*

Exercise 6.11.24 Show that $F_n \to_{PK} F$ iff the sequence of functions $f_n(x) := d(x, F_n)$ converges pointwise to the function $f(x) := d(x, F)$ iff the uniformly bounded sequence of functions $g_n(x) := [d(x, F_n) \wedge 1]$ converges pointwise to the function $g(x) := [d(x, F) \wedge 1]$.

Exercise 6.11.25 Let $\{x_a : a \in \mathbb{N}\}$ be a countable dense subset of the separable metric space (M, d). Show that the mapping $F \mapsto \varphi(F) := (\min\{d(x_a, F), 1\})_{a \in \mathbb{N}}$ from $\mathcal{C}(M)$ to $[0, 1]^{\mathbb{N}}$ is one-to-one.

Definition 6.11.26 *The metric space (M, d) is **locally compact** if for all $x \in M$ there is an $\epsilon > 0$ such that $\mathrm{cl}(B_\epsilon(x))$ is compact.*

$\mathbb{R}^\ell$ is locally compact, but $C([0, 1])$ and the other infinite-dimensional spaces that we have seen are not. Metrizing Painlevé-Kuratowski convergence is fairly easy in locally compact spaces.

Exercise 6.11.27 [↑Exercise 6.11.25] Suppose that (M, d) is a locally compact separable metric space. Define a metric $\rho(F_1, F_2)$ on $\mathcal{C}(M)$ by $\rho(F_1, F_2) = d_{Prod}(\varphi(F_1), \varphi(F_2))$. Show that $\rho(F_n, F) \to 0$ iff $F_n \to_{PK} F$.

Not being locally compact makes a difference.

Exercise 6.11.28 Show the following:

1. $C([0, 1])$ is not locally compact.

2. For each $n \in \mathbb{N}$, let $K_n = \{f \in C([0, 1]) : \|f\|_\infty \leq 1, \ f(1) = -1$, and $f([0, 1 - \frac{1}{n}]) = 1\}$. Show that K_n is a decreasing set of bounded closed convex sets with $\cap_n K_n = \emptyset$.

3. This ruins the applicability of the metric ρ of the previous problem to closed subsets of $C([0, 1])$.

4. Show that if K_n is a decreasing set of bounded closed convex sets in $\mathbb{R}^\ell$, then $\cap_n K_n \neq \emptyset$.

6.11.d Upper Semicontinuity and the Existence of Maxima

Theorem 4.8.11 shows that every continuous function on a compact set achieves its maximum and proves the converse for any $K \subset \mathbb{R}^\ell$. The following is a sketch of the general proof of the converse.

Exercise 6.11.29 Fill in the following sketch of a proof that if every $f \in C(M)$ achieves its maximum on $K \subset M$, then K is compact.

Suppose that every $f \in C(M)$ achieves its maximum on K. We show that K is totally bounded and complete.

Step 1. K is totally bounded. Suppose not.

1. Show that there exists $\epsilon > 0$ such that there is no finite ϵ-net for K.

2. Show that there is a countable set of points $x_n, n \in \mathbb{N}$, such that for all $n \neq m$, $d(x_n, x_m) \geq \epsilon$. [Let x_1 be a point in K, and show that there exists an x_2 such that $d(x_1, x_2) \geq \epsilon$. Let $F = \{x_1, \ldots, x_N\}$ be a finite set of points such that

for all $n \neq m, n, m \leq N, d(x_n, x_m) \geq \epsilon$. Show that the there exists a point x_{N+1} at a distance at least ϵ from F.]

3. Define $g_n(x)$ to be the continuous function $\max\{0, 1 - \frac{2}{\epsilon}d(x_n, x)\}$. Show that $g(x) := \sum_n n \cdot g_n(x)$ is continuous.

4. Show that $\sup_{x \in K} g(x) = \infty$. Therefore, by assumption, there exists $x^* \in K$ such that $g(x^*) = \infty$, a contradiction.

Step 2. K is complete. If not, there exists a Cauchy sequence x_n in K that does not converge to any point in K.

1. Show that we can find a subsequence with the property that no point appears twice in the subsequence. Continue to use x_n to denote the subsequence.

2. Show that for each x_n in the (sub)sequence, there exists an $\epsilon_n > 0$ such that for all $m \neq n$, $x_m \notin B_{x_n}(\epsilon_n)$.

3. Show that the function $f_n(x) = \max\{0, 1 - \frac{2}{\epsilon_n}d(x_n, x)\}$ is continuous on M, nonnegative, and strictly positive only on $B_{x_n}(\epsilon_n/2)$, and that for all $m \neq n$ and all $x' \in B_{x_m}(\epsilon_m)$, $f_n(x') = 0$.

4. Show that the function $f(x) = \sum_n n \cdot f_n(x)$ is continuous.

5. Show that $\sup_{x \in K} f(x) = \infty$ so that, by assumption, there exists $x^* \in K$ such that $f(x^*) = \infty$, a contradiction.

Continuity of a function on a compact set implies the existence of both a maximum and a minimum. To achieve only the existence of a maximum requires something strictly weaker—upper semicontinuity as given in Definition 4.10.29 (p. 152). Here is an alternative definition.

Definition 6.11.30 *A function $f : M \to \mathbb{R}$ is **upper semicontinuous** if $f^{-1}([a, \infty))$ is closed for all $a \in \mathbb{R}$. It is **lower semicontinuous** if $-f$ is upper semicontinuous, that is, if $f^{-1}((-\infty, a])$ is closed for all $a \in \mathbb{R}$.*

cdf's are usc, but unless they are continuous, they fail to be lsc.

Exercise 6.11.31 Show that Definitions 4.10.29 and 6.11.30 are equivalent.

Exercise 6.11.32 $f(x) := 1_F(x)$ is usc for all closed sets F. In particular, $f(r) = 1$ if $r = 0$ and $f(r) = 0$ otherwise is usc.

Lemma 6.11.33 *Any usc function on a compact set achieves its maximum.*

Exercise 6.11.34 Prove Lemma 6.11.33. [If f is unbounded, then $f^{-1}([n, \infty))$ is a collection of closed sets with the finite intersection property. Establish a contradiction. Now let $r = \sup f(x)$ and consider the sets $f^{-1}([r - \frac{1}{n}, \infty))$.]

Exercise 6.11.35 Show that a usc function on a compact set need not achieve its minimum.

6.11.e Urysohn's Lemma and Nonmetric Topologies

Recall Definition 4.1.14: a topology on a set X is a collection of subsets that contains $\emptyset$ and X and is closed under finite intersections and arbitrary unions.

The following definitions and facts for topological spaces are the same as their counterparts in metric spaces:

1. $f : X \to \mathbb{R}$ is continous if $f^{-1}(G) \in \tau$ for all open $G \subset \mathbb{R}$.
2. $C(X)$ denotes the set of continuous $\mathbb{R}$-valued functions.
3. The closure of a set is the intersection of all closed sets containing it.

The general study of topologies begins with their separation properties. We focus on the following two:

1. (X, τ) is **Hausdorff, or** T_2 if for all $x \neq y$, there exists $U, V \in \tau$ such that $x \in U$, $y \in V$ and $U \cap V = \emptyset$.
2. (X, τ) is **normal, or** T_4 if it is Hausdorff and for all closed sets F, F' with $F \cap F' = \emptyset$, there exists $U, V \tau$ such that $F \subset U$, $F' \subset V$ and $U \cap V = \emptyset$.

Exercise 6.11.36   Show that every normal space is Hausdorff and that every metric space is normal.

Topological spaces that fail the separation axioms often do not have very useful sets of continuous functions, as in the following, wonderfully perverse case.

Exercise 6.11.37 (Sierpinski Space)   Consider the topological space (X, τ) with $X = \{0, 1\}$, $\tau = \{\emptyset, \{0\}, X\}$. Show that (X, τ) is not Hausdorff and that $C(X)$ contains only the constant functions.

For metric spaces, compactness is equivalent to every continuous function achieving its maximum, Theorem 4.8.11. The proofs depend on there being a rich class of continuous functions.

Exercise 6.11.38   Consider the topological space (X, τ) with $X = \mathbb{N}$, $\tau = \{\emptyset, X, \{1\}, \{1, 2\}, \{1, 2, 3\}, \ldots\}$. Show that τ is a topology, give an open cover with no finite subcover, and show that every continuous function on (X, τ) is constant, hence achieves its maximum.

Urysohn's lemma shows that normal spaces have a rich set of continuous functions, in particular that they approximate the indicators of closed sets.

Lemma 6.11.39 (Urysohn)   *If (X, τ) is normal, $F \subset G$, F closed and G open, then there exists an $f \in C(X)$ such that $f : X \to [0, 1]$ and $\forall x \in F$, $f(x) = 1$ and for all $y \in G^c$, $f(y) = 0$.*

Exercise 6.11.40   Fill in the details in the following sketch of a proof of Urysohn's lemma: For the first time through, you might want to use the metric structure to construct open sets having the requisite properties.

1. In any Hausdorff space, points are closed, that is, $\{x\}$ is a closed set for all $x \in X$.
2. For each $x \in F$, show that we can pick a disjoint pair of open sets U_x, V_x such that $x \in U_x$ and $G^c \subset V_x$.
3. Define $G_0 = \cup_{x \in F} U_x$. Show that $F \subset G_0 \subset G$.

For dyadic rationals $q = k/2^n$ strictly between 0 and 1; that is, with $k \in \{1, \ldots, 2^n - 1\}$, we are going to inductively define a collection of closed and open sets $G_q \subset F_q$ with the property that $q < q'$ implies that

$$G_0 \supset F_q \supset G_q \supset F_{q'} \supset G_{q'} \supset F_1 := F. \qquad (6.40)$$

For any $x \in G_0$, the function f is defined as the supremum of the dyadic rationals, $k/2^n$, such that $x \in F_{k/2^n}$, and for $x \in G_0^c$, f is set equal to 0. This yields $f(x) = 1$ for $x \in F_1 = F$ and $f(x) = 0$ for all $x \in G^c$, and all that is left is to show that f is in fact continuous.

4. The initial step: Show that there exists an open set $G_{1/2}$ with closure $F_{1/2}$ such that $F_1 \subset G_{1/2} \subset F_{1/2} \subset G_0$.

5. The inductive step: Suppose that for all dyadic rationals, $1 > q_1 > \cdots > q_{2^n - 1} > 0$ of order n, we have open and closed sets with $F_1 \subset G_{q_1} \subset F_{q_1} \subset \cdots \subset G_{q_{2^n-1}} \subset F_{q_{2^n-1}} \subset G_0$. Show that we can define open and closed G's and F's for the dyadic rationals of order $n + 1$ that are not of order n so that (6.40) holds for all dyadic rationals of order $n + 1$.

6. Define $f(x) = \sup\{q : x \in F_q\}$. Show that $\{x : f(x) > r\} = \cup_{q > r} G_q$ is open and that $\{x : f(x) < r\} = \cup_{q < r} F_q^c$ is also open, so that f is continuous.

6.11.f Finitely Determined Functions on Compact Product Spaces

The Stone-Weierstrass theorem is a bit of sledgehammer, and one can prove things with it without having a good sense of why they *should* be true. Here is an example.

Exercise 6.11.41   This problem asks you to prove that the continuous finitely determined functions on a compact product space are dense *without* using the Stone-Weierstrass theorem.

Let $M = \times_{a \in A} M_a$ be compact. Pick $f \in C(M)$ and $\epsilon > 0$. Pick an arbitrary $x^\circ \in M$. Let B_n be a nested sequence of finite subsets of A such that $\cup_n B_n = A$. For each $n \in \mathbb{N}$ and $x \in M$, define $y_n(x) = (\mathrm{proj}_{B_n}(x), \mathrm{proj}_{A \setminus B_n}(x^\circ)) \in M$, and define $f_n(x) = f(y_n(x))$. [Note that $y_n(x)$ is the vector in $\times_{a \in A} M_a$ that is equal to x in the B_n coordinates and equal to x° in the others.]

1. Show that for all $x \in M$, $y_n(x) \to x$.

2. Show that each f_n is continuous and finitely determined.

3. Show that $f_n \to f$ uniformly. [Hint: f is uniformly continuous and the δ-balls in the product metric have a useful structure.]

4. Conclude that the continuous finitely determined functions are dense in $C(M)$.

6.11.g Changing Metrics

Switching between equivalent metrics can be useful.

Exercise 6.11.42   If (M', d') is a metric space and $f : M \to M'$ is one-to-one, then we can define a metric on M by $d_f(x, x') = d'(f(x), f(x'))$. Show that if f is a homeomorphism between (M, d) and (M', d'), then d_f and d are equivalent metrics.

There is also a way to operate more directly on the distance function itself to give a new metric—note that $d(x_n, x) \to 0$ iff $\min\{d(x_n, x), 1\} \to 0$.

Exercise 6.11.43   Let (M, d) be a metric space. Let $f : \mathbb{R}_+ \to \mathbb{R}_+$, be f non-decreasing and subadditive (i.e., $f(a + b) \le f(a) + f(b)$) and satisfy $f(r) = 0$ iff $r = 0$. Define $\rho(x, x') = f(d(x, x'))$, that is, $\rho = f \circ d$. Show that:

1. (M, ρ) a metric space;
2. ρ is equivalent to d;
3. $f(r) = \min\{r, 1\}$ satisfies these criteria, as does $g(r) = r/(1 + r)$;
4. $(M, \min\{d, 1\}) = (M, f \circ d)$ is complete iff (M, d) is complete; and
5. $(M, g \circ d)$ need not be complete even if (M, d) is complete.

The following characterizes compactness as a strong form of completeness.

Exercise 6.11.44   Show that:

1. if (M, d) is compact, then (M, ρ) is complete for every $\rho \sim d$; and
2. if (M, ρ) is complete for every $\rho \sim d$, then (M, d) is compact.

6.11.h Small Subsets of Complete Metric Spaces

The next problems examine a definition of subset A of a metric space M being "small" relative to M.

Definition 6.11.45   *A subset A of a metric space is **nowhere dense** if its closure has no interior.*

Exercise 6.11.46   Prove that the following are equivalent:

1. A is nowhere dense.
2. For all open balls $B_r(x)$, $A \cap B_r(x)$ is not dense in $B_r(x)$.
3. Every open ball $B_r(x)$ contains another open ball $B_s(y)$ such that $B_s(y) \cap A = \emptyset$.

A set that is nowhere dense certainly seems small. The following shows that a countable union of nowhere dense sets must be a strict subset of a complete metric space.

Theorem 6.11.47 (Baire)   *If M is a complete metric space, then there is no countable collection, A_n, of nowhere dense sets such that $M = \cup_{n\in\mathbb{N}}A_n$.*

Exercise 6.11.48   Prove Baire's theorem. [Suppose that $M = \cup_{n\in\mathbb{N}}A_n$, where each A_n is nowhere dense. Pick an arbitrary open ball with radius 2, $B_2(x_0)$. Since A_1 is nowhere dense, there is a ball $B_1(x_1) \subset B_2(x_0)$ such that $B_1(x_1) \cap A_1 = \emptyset$. Since A_2 is nowhere dense, there is a ball $B_{1/2}(x_2) \subset B_1(x_1)$ such that $A_2 \cap B_{1/2}(x_2) = \emptyset$. Continuing, we find a nested sequence of open balls $B_{1/n}(x_n) \subset B_{1/(n+1)}(x_{n+1})$ with $B_{1/n}(x_n) \cap A_n = \emptyset$. Show that $\cap_n B_{1/n}(x_n)$ is nonempty because M is complete. Since $1/n \to 0$, $\cap_n B_{1/n}(x_n) = \{x\}$ for some $x \in M$. Show that $x \notin \cup_{n\in\mathbb{N}}A_n$.]

Note that the class of countable unions of nowhere dense sets is closed under countable union. Let us agree to call any countable union of nowhere dense sets a **Baire small** set. Baire's theorem can be rephrased as no complete metric space is Baire small.

The following contains (yet) another proof of the uncountability of $\mathbb{R}^k$.

Exercise 6.11.49 If M is complete and contains no isolated points, then M is uncountable. [Show that $\{x\}$ is nowhere dense. Consider countable unions of the nowhere dense sets of the form $\{x\}$.]

The following shows that there are Baire small subsets of $\mathbb{R}^k$ that are quite "large."

Exercise 6.11.50 Enumerate the point in $\mathbb{R}^k$ with rational coordinates as $\{q_n : n \in \mathbb{N}\}$. For $\epsilon > 0$, let $A_n = B_{\epsilon/2^n}(q_n)$. The volume of A_n is some constant, κ, involving π, times $\left(\frac{\epsilon}{2^n}\right)^k$, which means that $C(\epsilon) := \cup_n A_n$ has volume strictly smaller than $\kappa \cdot \epsilon$ (assuming that the volume of a countable union is smaller than the sum of the volumes, a "countable additivity" condition, which reappears in our study of probability theory).

1. Show that $C(\epsilon) = \cup_n A_n$ is an open dense set and that its complement is a closed nowhere dense set. In particular, its complement is a closed Baire small set.
2. Let $\epsilon_n \downarrow 0$. Show that the complement of $\cap_{n \in \mathbb{N}} C(\epsilon_n)$ is Baire small and that $\cap_{n \in \mathbb{N}} C(\epsilon_n)$ has a volume smaller than any $\delta > 0$.
3. Using Exercise 6.11.49, show that $\cap_{n \in \mathbb{N}} C(\epsilon_n)$ is uncountable.

Baire's theorem also delivers dense infinite-dimensional vector subspaces of $C([0, 1])$.

Exercise 6.11.51 Let $\{f_n : n \in \mathbb{N}\}$ be a dense subset of $C([0, 1])$, and define $A_n = \mathbf{span}(\{f_m : m \le n\})$. Show that A_n has an empty interior, that $A := \cup_n A_n$ is a dense linear subspace of $C([0, 1])$, and that $A \subsetneq C([0, 1])$.

6.11.i Another Proof of the Metric Completion Theorem

The construction of the completion of a metric space given in Theorem 6.8.11 was a bit painful, not because it was particularly complicated, but because you had to be finicky about so many details. The following gives another construction of the completion of a metric space (M, d) in which there are fewer details that you have to be finicky about.

Exercise 6.11.52 Let (M, d_M) be a metric space and let d_M be bounded. For an arbitrary point $u \in M$, define $f_u(x) = d_M(u, x)$ and $F = F(M, d_M) = \{f_u + h : h \in C_b(M)\}$.

1. Show that $F \subset C_b(M)$, so that $e(f, g) = \sup_x |f(x) - g(x)|$ is a metric on $F(M, d_M)$.
2. Show that for any $u, v \in M$, $f_u - f_v \in C_b(M)$ and $e(f_u, f_v) = d_M(u, v)$.

3. Show that for any $x \in M$, $f_x \in F$. [Note that $f_x = f_u + (f_x - f_u)$.]

4. Show that $x \mapsto f_x$ is an isometry from M to F.

5. Let T be the closure of the range of the isometry $x \mapsto f_x$. Show that (T, e) is the completion of (M, d_M).

6. Show that the boundedness of f_u can be dispensed with in this construction. [There are at least two ways to do this. First, repeat the steps being careful that e is a metric on F even when f_u is not bounded. Second, replace an unbounded d_M with $d'_M(x, x') := \min\{1, d_M(x, x')\}$, construct the completion, (T, e), using F as above, and check that the isometric embedding works for d_M as well.]

CHAPTER 7

Measure Spaces and Probability

The most fruitful generalizations of continuous functions on metric spaces are the measurable functions on measure spaces. Measure spaces and measurable functions appear throughout all parts of economics: general equilibrium models, econometrics, stochastic dynamic programming, and more or less everything else that is dealt with in economic theory and econometrics.

This chapter begins in §7.1 with the definitions of the basic concepts we study in measure spaces: σ-fields of measurable sets; measurable functions, also known as random variables; countably additive probability measures; integrals of random variables; and continuity of the integral. The continuity result is the most involved of these, and we delay its proof for quite some time, until §7.5, being careful not to create any logical circularities. These make up the minimum background needed to understand and to prove the first three of the four basic limit results presented in §7.2: the weak law of large numbers (WLLN), the Borel-Cantelli lemma, and the strong law of large numbers (SLLN). The fourth limit result is Kolmogorov's 0-1 law, and its proof requires yet more background on measurability, which is provided in §7.3 with other related material; §7.4 proves Kolmogorov's 0-1 law and another 0-1 law due to Hewitt and Savage. After this, we have two sections of fairly serious measure theory, §7.5 on the continuity of the integral and §7.6, which shows that interesting countably additive probabilities exist. The approach we take to the existence of interesting probabilities is directly based on continuity and approximation, viewing σ-fields as metric spaces and probabilities as uniformly continuous functions on them. The appendix to this chapter covers a more traditional approach to measure and integral with respect to the Lebesgue measure on $\mathbb{R}$.

The σ-fields of measurable sets are the collections of sets to which we can meaningfully assign probabilities. The Hewitt-Savage 0-1 law allows us, in §7.8, to demonstrate, in the context of studying intergenerational equity, the existence of seriously nonmeasurable sets. Serious nonmeasurability involves sets to which one cannot sensibly assign probability.

There are nonmeasurable sets to which one can sensibly assign the probability 0. These are called null sets and are covered in §7.9. Including null sets in our σ-fields is called "completing" them, yet another use of the word "complete." Combined with information about analytic sets, this allows us to solve a great many measurability problems simultaneously.

In Chapter 8 we turn to the classical function spaces of random variables with pth moments, the so-called L^p spaces. One in particular, L^2, is the essential mathematical model used in regression analysis, covered in §8.4. This material leads, in §8.5, back to measure theory, to the Radon-Nikodym theorem and the Hahn-Jordan decomposition, both of which are crucial measure theory tools.

7.1 ◆ The Basics of Measure Theory

The underlying construct in probability theory is a **measure space** or **measurable space**, which is a pair $(\Omega, \mathcal{F})$. Ω is a nonempty set and $\mathcal{F} \subset \mathcal{P}(\Omega)$ is a σ-field (or σ-algebra) of subsets of Ω. Elements $E \in \mathcal{F}$ are called **measurable subsets of Ω**, or **measurable sets**, or, in probabilistic contexts, **events**.

A **probability** on $(\Omega, \mathcal{F})$ is a function $P : \mathcal{F} \to [0, 1]$ satisfying rules given later that guarantee that it behaves as a "probability" should. The "should" also imposes conditions on $\mathcal{F}$, and the requisite conditions on P and $\mathcal{F}$ are closely related. A **probability space** is a triple, $(\Omega, \mathcal{F}, P)$.

The main use of probability spaces is as the domain for **random variables** or **measurable functions**. A random variable is a function $X : \Omega \to M$, M a metric space, such that $X^{-1}(G) \in \mathcal{F}$ for every open $G \subset M$. Mostly $M = \mathbb{R}$ or $M = \mathbb{R}^k$.

The idea is that a point $\omega \in \Omega$ is randomly drawn according to P. We interpret this to mean that for each $E \in \mathcal{F}$, $P(E)$ is the probability that the point ω belongs to E. When ω is drawn, we observe $X(\omega)$, not ω itself. For example, if $X(\omega) = 1_E(\omega)$, then we observe 1 when ω is in E and 0 otherwise. When we observe a sequence of random variables over time, we observe the sequence $X_t(\omega)$.

7.1.a Measurable Sets

A topology is a collection of sets that contains the whole space and the empty set, as well as being closed under arbitrary union and finite intersection. The following repeats and renumbers Definition 6.9.2 (p. 335).

Definition 7.1.1 *A class $\mathcal{F} \subset \mathcal{P}^{\Omega}$ is an **field** of subsets if it satisfies 1, 2, and 3 below. If it also satisfies 4, it is a σ-**field**. Elements of $\mathcal{F}$ are called **measurable subsets of Ω**, or the **measurable sets**, or **events**.*

1. *$\emptyset, \Omega \in \mathcal{F}$,*

2. *$[E \in \mathcal{F}] \Rightarrow [E^c \in \mathcal{F}]$, closure under complementation,*

3. *$[(E_k)_{k=1}^K \subset \mathcal{F}] \Rightarrow [[\cup_{k=1}^K E_k \in \mathcal{F}] \wedge [\cap_{k=1}^K E_k \in \mathcal{F}]]$, closure under finite intersections and unions, and*

4. *$[(E_n)_{n \in \mathbb{N}} \subset \mathcal{F}] \Rightarrow [[\cup_{n \in \mathbb{N}} E_n \in \mathcal{F}] \wedge [\cap_{n \in \mathbb{N}} E_n \in \mathcal{F}]]$, closure under countable intersections and unions.*

Fields and σ-fields are also closed under other useful set theoretic operations: if $E, F \in \mathcal{F}$, then $E \setminus F = E \cap (F^c) \in \mathcal{F}$ and $[E \Delta F] = [E \setminus F] \cup [F \setminus E] \in \mathcal{F}$. The following are used in many, many limit constructions.

Definition 7.1.2   *Given* $\{E_n : n \in \mathbb{N}\} \subset \mathcal{F}$, *the event* E_n **infinitely often (i.o.)** *is defined by* $[E_n \text{ i.o.}] = \cap_N \cup_{n \geq N} E_n$ *and the event* E_n **almost always (a.a.)** *is defined by* $[E_n \text{ a.a.}] := \cup_N \cap_{n \geq N} E_n$.

Exercise 7.1.3   Let S_n be a sequence of random variables and define $E_n(\epsilon) = \{\omega : |S_n - r| < \epsilon\}$. Show that $\cap_m [E_n(1/m) \text{ i.o.}]$ is the set of ω for which r is an accumulation point of the sequence $S_n(\omega)$ and that $\cap_m [E_n(1/m) \text{ a.a.}]$ is the set of ω for which $S_n(\omega) \; \textgreater \; r$.

Since $A_N := \cup_{n \geq N} E_n \in \mathcal{F}$ for all N, $[E_n \text{ i.o.}] = \cap_N A_N \in \mathcal{F}$. Similarly, since $B_N := \cap_{n \geq N} E_n \in \mathcal{F}$ for all N, $[E_n \text{ a.a.}] = \cup_N B_N \in \mathcal{F}$. Sometimes $[E_n \text{ i.o.}]$ is written $\limsup_n E_n$ and $[E_n \text{ a.a.}]$ is written $\liminf_n E_n$. The reason for this is that $\omega \in [E_n \text{ i.o.}]$ iff $\limsup_n 1_{E_n}(\omega) = 1$ and $\omega \in [E_n \text{ a.a.}]$ iff $\liminf_n 1_{E_n}(\omega) = 1$. From these, one easily proves the intuitive result that $[E_n \text{ a.a.}] \subset [E_n \text{ i.o.}]$.

Notation 7.1.4   *When we have a sequence* E_n *in* $\mathcal{F}$, $E_n \subset E_{n+1}$ *for all* n *and* $E = \cup_n E_n$, *we write* $E_n \uparrow E$; *when* $E_n \supset E_{n+1}$ *for all* n *and* $E = \cap_n E_n$, *we write* $E_n \downarrow E$.

Exercise 7.1.5   If $E_n \uparrow E$ or $E_n \downarrow E$, then $E = [E_n \text{ a.a.}] = [E_n \text{ i.o.}]$.

7.1.b   Basic Examples of Operations with Measurable Sets

For any Ω, $\mathcal{P}(\Omega)$ is a σ-field. When Ω is finite or countable, this is a useful collection of measurable sets, but when Ω is uncountable, it is rarely useful.

Notation 7.1.6   *When a collection of sets is a field, but may fail to be a σ-field, it will have "$\circ$" beside it, for example,* $\mathcal{F}^\circ$ *rather than* $\mathcal{F}$.

Example 7.1.7   *Let* $\Omega = (0, 1]$, *let* $\mathcal{B}^\circ$ *denote the empty set and the class of finite disjoint unions of interval subsets of* $(0, 1]$. *(Recall that all the possibilities are allowed,* (a, b), $[a, b)$, $[a, b]$, *and* $(a, b]$, *and that the notation "$|a, b|$" means any one of the four possibilities.) The class* $\mathcal{B}^\circ \subset \mathcal{P}((0, 1])$ *is a field, but not a σ-field. To see why, note that for every* $q \in \mathbb{Q} \cap (0, 1]$, $\{q\} = [q, q] \in \mathcal{B}^\circ$, *so that* $\mathbb{Q}$ *is a countable union of elements of* $\mathcal{B}^\circ$. *However, the countable set* $\mathbb{Q}$ *cannot be expressed as a finite union of intervals: if one of the intervals has strictly positive length, the union is an uncountable set; if none of the intervals has strictly positive length, the union is a finite set.*

Another way to understand σ-fields is to think of them as fields that contain $[E_n \text{ i.o.}]$ and $[E_n \text{ a.a.}]$ for any sequence of events E_n.

Theorem 7.1.8   *A class of sets* $\mathcal{G}$ *is a σ-field iff it is closed under complementation and contains* $\emptyset$, Ω *and* $[(E_n)_{n \in \mathbb{N}} \subset \mathcal{F}] \Rightarrow [[E_n \text{ i.o.}] \in \mathcal{G} \wedge [E_n \text{ a.a.}] \in \mathcal{G}]$.

Proof.   That σ-fields contain $[E_n \text{ i.o.}]$ and $[E_n \text{ a.a.}]$ was established earlier. Suppose that $\mathcal{G}$ is closed under complementation and contains $\emptyset$ and $[(E_n)_{n \in \mathbb{N}} \subset \mathcal{F}] \Rightarrow [[E_n \text{ i.o.}] \in \mathcal{G} \wedge [E_n \text{ a.a.}] \in \mathcal{G}]$. To show that $\mathcal{G}$ is a field requires only that we show that it is closed under finite unions and intersections. Let $(E_n)_{n=1}^N \subset \mathcal{G}$. Consider the infinite sequence $(A_n)_{n \in \mathbb{N}}$ defined by

$$A_1, \ldots, A_N, A_{N+1} \cdots$$

$$= E_1, E_2, \ldots, E_{N-1}, E_N, E_1, E_2, \ldots, E_{N-1}, E_N, E_1, \ldots.$$

$[A_n \text{ i.o.}] = \cup_{n=1}^N E_n$ and $[A_n \text{ a.a.}] = \cap_{n=1}^N E_n$, so that $\mathcal{G}$ is a field.

We now show that $\mathcal{G}$ is a σ-field. Since $\mathcal{G}$ is a field, it turns out that it is sufficient to show that $\mathcal{G}$ is closed under monotonic unions and intersections. Here is why. Let $(E_n)_{n \in \mathbb{N}} \subset \mathcal{G}$. For each $m \in \mathbb{N}$, $D_m := \cup_{n=1}^m E_n \in \mathcal{G}$ and $F_m := \cap_{n=1}^m E_n \in \mathcal{G}$ because $\mathcal{G}$ is a field. $[D_m \text{ a.a.}] = [D_m \text{ i.o.}] = \cup_{n \in \mathbb{N}} E_n \in \mathcal{G}$, and $[F_m \text{ a.a.}] = [F_m \text{ i.o.}] = \cap_{n \in \mathbb{N}} E_n \in \mathcal{G}$. ∎

In a metric space, the class of closed sets is closed under arbitrary intersections. In this sense, σ-fields, which are collections of sets rather than collections of points, resemble closed sets, which are collections of points.

Lemma 7.1.9 *If $\{\mathcal{G}_\alpha : \alpha \in A\}$ is a collection of σ-fields of subsets of Ω, then $\mathcal{G} = \cap_{\alpha \in A} \mathcal{G}_\alpha$ is a σ-field.*

Proof. $\emptyset, \Omega \in \mathcal{G}$ because $\emptyset, \Omega \in \mathcal{G}_\alpha$ for all α. If $E \in \mathcal{G}$ then $E \in \mathcal{G}_\alpha$ for all α; hence $E^c \in \mathcal{G}$ because $E^c \in \mathcal{G}_\alpha$ for all α. Similarly, if $\{E_n : n \in \mathbb{N}\} \subset \mathcal{G}$, then for all α, $\cup_n E_n \in \mathcal{G}_\alpha$ and $\cap_n E_n \in \mathcal{G}_\alpha$, so that $\cup_n E_n \in \mathcal{G}$ and $\cap_n E_n \in \mathcal{G}$. ∎

The closure of a set is the smallest closed set containing it. Here is the parallel statement for σ-fields.

Definition 7.1.10 *Given a class $\mathcal{E} \subset \mathcal{P}(\Omega)$, the σ-**field generated by** $\mathcal{E}$ is denoted $\sigma(\mathcal{E})$ and defined by $\sigma(\mathcal{E}) = \cap \{\mathcal{G} : \mathcal{E} \subset \mathcal{G}, \ \mathcal{G} \text{ a } \sigma\text{-field}\}$. If $\mathcal{G} = \sigma(\mathcal{E})$, then we say that $\mathcal{E}$ is a **generating class of sets for** $\mathcal{G}$.*

Exercise 7.1.11 Show that arbitrary intersections of fields is a field and give the appropriate definition of $\mathcal{F}^\circ(\mathcal{E})$, the field generated by $\mathcal{E}$.

With the field $\mathcal{B}^\circ$ of finite disjoint unions of interval subsets of $(0, 1]$, $\mathcal{B} := \sigma(\mathcal{B}^\circ)$ is called the **Borel σ-field** for $(0, 1]$ in honor of Emile Borel.

Notation 7.1.12 *The **Borel σ-field** for a metric space (M, d) is the σ-field $\mathcal{B}_M := \sigma(\tau_d)$, where τ_d is the class of open subsets of M. When the metric space is $\mathbb{R}$, it is denoted $\mathcal{B}$ (rather than $\mathcal{B}_\mathbb{R}$), and when the metric space is $\mathbb{R}^k$, it is denoted $\mathcal{B}^k$ (rather than $\mathcal{B}_{\mathbb{R}^k}$).*

Exercise 7.1.13 It seems that we have given two definitions for $\mathcal{B}_{(0,1]}$—as the smallest σ-field containing the finite disjoint unions of intervals and as the smallest σ-field containing the open subsets of $(0, 1]$. Show that the definitions are equivalent. [Show that $\tau_d \subset \mathcal{B}^\circ$, $\sigma(\tau_d) \subset \sigma(\mathcal{B}^\circ)$, and $\mathcal{B}^\circ \subset \sigma(\tau_d)$.]

7.1.c Probabilities

For a number of good reasons, we limit our study almost exclusively to countably additive probabilities on σ-fields. We will see that these are the extensions, by continuity, of countably additive probabilities on fields.

Definition 7.1.14   *A function* $P : \mathcal{G} \to [0, 1]$, $\mathcal{G}$ *a field or a* σ-*field, is a **finitely additive probability** if* $P(\emptyset) = 0$, $P(\Omega) = 1$, *and for disjoint* E, $F \in \mathcal{G}$, $P(E \cup F) = P(E) + P(F)$. *A finitely additive probability is a **countably additive probability** if for disjoint* $\{E_n : n \in \mathbb{N}\} \subset \mathcal{G}$ *with* $\cup_n E_n \in \mathcal{G}$ *(which is guaranteed if* $\mathcal{G}$ *is a* σ-*field)*, $P(\cup_n E_n) = \sum_n P(E_n)$.

Countable additivity is a form of continuity.

Theorem 7.1.15   *For a finitely additive probability* $P : \mathcal{G} \to [0, 1]$, $\mathcal{G}$ *a field, the following are equivalent:*

1. *P is countably additive on* $\mathcal{G}$,

2. *P is continuous from above,* $[[E_n \downarrow E] \wedge [E_n, E \in \mathcal{G}]] \Rightarrow [P(E_n) \downarrow P(E)]$,

3. *P is continuous from below,* $[[E_n \uparrow E] \wedge [E_n, E \in \mathcal{G}]] \Rightarrow [P(E_n) \uparrow P(E)]$, *and*

4. *P is continuous from above at* $\emptyset$, $[[E_n \downarrow \emptyset] \wedge [E_n \in \mathcal{G}]] \Rightarrow [P(E_n) \downarrow 0]$.

Proof.   (2) $\Leftrightarrow$ (3) because $E_n \uparrow E$ iff $E_n^c \downarrow E^c$ and fields are closed under complementation.

(2) $\Leftrightarrow$ (4) because $E_n \downarrow \emptyset$ iff $\forall E \in \mathcal{G}$, $(E_n \cup E) \downarrow E$.

(1) $\Rightarrow$ (3): Suppose that P is countably additive. Let $E_n \uparrow E$ with E_n, $E \in \mathcal{G}$. We must show that $P(E_n) \uparrow P(E)$. Define $E_0 = \emptyset$ and $D_n = E_n \setminus E_{n-1}$. Since $\mathcal{G}$ is a field, each $D_n \in \mathcal{G}$. As P is finitely additive on $\mathcal{G}$, the D_n are disjoint, $E_n = \cup_{k \le n} D_k$, and $P(E_n) = \sum_{k \le n} P(D_k)$. Since P is countably additive, $P(E_n) \uparrow P(E)$.

(3) $\Rightarrow$ (1): Let $\{E_n : n \in \mathbb{N}\} \subset \mathcal{G}$ be a collection of disjoint sets with $\cup_n E_n \in \mathcal{G}$. Define $F_n = \cup_{i \le n} E_i$. Since P is finitely additive, $P(F_n) = P(\cup_{i \le n} E_i) = \sum_{i \le n} P(E_i)$. As $F_n \uparrow E$, $P(E) = \lim_n \sum_{i \le n} P(E_i) = \sum_n P(E_n)$.   ∎

The right continuity of a cumulative distribution function, F (see Definition 6.2.15, p. 275) if your memory is hazy) is a necessary condition for the countable additivity of the associated probability P_F on $\mathcal{B}^\circ$.[1]

Example 7.1.16   *Let* $E_n = (a, a + \frac{1}{n}]$, *so that* $E_n \downarrow \emptyset$. *The countable additivity of* P_F *requires that* $P_F(E_n) \downarrow 0$. *Since* $P_F(E_n) = F(a + \frac{1}{n}) - F(a)$, *this requires that* $F(a + \frac{1}{n}) \downarrow F(a)$.

Example 7.1.17 (Lebesgue's uniform distribution)   *In Example 7.1.7, we introduced* $\mathcal{B}^\circ$, *the field of finite disjoint unions of interval subsets of* $(0, 1]$. *We define Lebesgue's uniform distribution by* $\lambda : \mathcal{B}^\circ$ *by* $\lambda(|a, b|) = (b - a)$, *and by additivity for finite disjoint unions of intervals (we use "lambda" because it begins with the letter "L," for Lebesgue). In 7.6, we will show that* λ *has a unique continuous extension to a countably additive probability, also denoted by* λ, *on* $\mathcal{B} = \sigma(\mathcal{B}^\circ)$. *As well as being countably additive,* λ *is **nonatomic**. In this context, being nonatomic is the same as assigning probability 0 to every point,* $\lambda(\{x\}) = (x - x) = 0$. *Since* λ *is countably additive, this means that* $\lambda(\mathbb{Q}) = \lambda(\cup_n \{q_n\}) = 0$ *where* $n \mapsto q_n$ *is an enumeration of the rationals in* $(0, 1]$.

---

1. It is also sufficient; see Theorem 7.6.4.

7.1.d Measurable Functions Taking Values in Metric Spaces

A function, f, from a metric space (M, d) to a metric space (M', d') is continuous if $f^{-1}(G)$ is open for every open $G \subset M'$. Put another way, f is continuous iff $f^{-1}(\tau_{d'}) \subset \tau_d$.

Definition 7.1.18 (Measurable Functions Taking Values in Metric Spaces) *A function, X, from a measure space $(\Omega, \mathcal{F})$ to a metric space (M, d) is **measurable** or $\mathcal{F}$-**measurable** if $X^{-1}(G) \in \mathcal{F}$ for every open set G, that is, iff $X^{-1}(\tau_d) \subset \mathcal{F}$.*

Since $X^{-1}(A^c) = (X^{-1}(A))^c$, $X^{-1}(\cup_n A_n) = \cup_n X^{-1}(A_n)$, and $X^{-1}(\cap_n A_n) = \cap_n X^{-1}(A_n)$, $X^{-1}(\mathcal{B}_M)$ is a σ-field and we have the following, which could have been used as the definition of measurable functions taking values in metric spaces.

Lemma 7.1.19 $X : \Omega \to M$ *is measurable iff* $X^{-1}(\mathcal{B}_M) \subset \mathcal{F}$.

Notation 7.1.20 $X^{-1}(\mathcal{B}_M)$ *is the σ-**field generated by** X. It is denoted $\sigma(X)$ and, as it is a subset of the σ-field $\mathcal{F}$, it is called a **sub-σ-field**.*

We study sub-σ-fields systematically in §7.2.e.1 (p. 381).
The leading case has $M = \mathbb{R}$ and has its own notation.

Notation 7.1.21 $L^0 = L^0(\Omega, \mathcal{F}, P)$ *is the set of $\mathbb{R}$-valued measurable functions, and elements of L^0 are also called **random variables**.*

Theorem 7.1.22 *For a function $X : \Omega \to \mathbb{R}$, the following are equivalent:*

1. X is measurable, that is, X is a random variable,

2. for all $r \in \mathbb{R}$, $X^{-1}((-\infty, r)) \in \mathcal{F}$,

3. for all $r \in \mathbb{R}$, $X^{-1}((-\infty, r]) \in \mathcal{F}$, and

4. for all $r < s$, $r, s \in \mathbb{R}$, $X^{-1}((r, s]) \in \mathcal{F}$.

Proof. $(1) \Rightarrow (2)$ because $(-\infty, r)$ is open.
$(2) \Rightarrow (3)$ because $X^{-1}((-\infty, r]) = \cap_n X^{-1}((-\infty, r + \frac{1}{n}))$ and $\mathcal{F}$ is closed under countable intersection.
$(3) \Rightarrow (4)$ because $X^{-1}((r, s]) = X^{-1}((-\infty, s]) \setminus X^{-1}((-\infty, r])$ and $\mathcal{F}$ is closed under set theoretic differencing.
$(4) \Rightarrow (1)$: Any open $G \subset \mathbb{R}$ is of the form $G = \cup_n (a_n, b_n)$ for some set of open intervals. Further, $X^{-1}(G) = \cup_n X^{-1}((a_n, b_n))$, so it is sufficient to show that each $X^{-1}((a_n, b_n)) \in \mathcal{F}$. Let $r_{n,m}$ be a strictly increasing sequence in (a_n, b_n) with $\lim_m r_{n,m} = b_n$. $X^{-1}((a_n, b_n)) = \cup_m X^{-1}((a_n, r_{n,m}])$. ∎

Notation 7.1.23 *For $X \in L^0(\Omega, \mathcal{F}, P)$ and $B \subset \mathbb{R}$, we write $\{X \in B\}$ as shorthand for $\{\omega : X(\omega) \in B\}$, that is, for $X^{-1}(B)$. There are some special cases: when $B = (-\infty, r]$, we write $\{X \leq r\}$; when $B = (-\infty, r)$, we write $\{X < r\}$; when $B = [r, +\infty)$, we write $\{X \geq r\}$; and when $B = (r, +\infty)$, we write $\{X > r\}$.*

Exercise 7.1.24 Using the notation $\{X \in B\}$ for the various sets B that appear in the argument, show that one can replace "$\mathbb{R}$" by "$\mathbb{Q}$" in conditions (2), (3), and (4) in Theorem 7.1.22.

The space $C_b(M)$ of continuous bounded functions is an algebra (a vector space that is closed under multiplication), a lattice (closed under the operations $\wedge$ and $\vee$), and closed under uniform convergence and composition with countinuous bounded functions on $\mathbb{R}$. For the following two results, keep in mind that L^0 may contain unbounded functions and that uniform convergence implies pointwise convergence.

Theorem 7.1.25 $L^0(\Omega, \mathcal{F}, P)$ *is an algebra and a lattice.*

Proof. Suppose that $X, Y \in L^0$ and that $\alpha, \beta \in \mathbb{R}$. We show that $\alpha X + \beta Y \in L^0$ and that $X \cdot Y \in L^0$.

$\alpha X, \beta Y \in L^0$: For $\alpha > 0$, $\{\alpha X \in (-\infty, r]\} = \{X \in (-\infty, r/\alpha]\} \in \mathcal{F}$; for $\alpha < 0$, $\{\alpha X \in (-\infty, r]\} = \{X \in [r/\alpha, \infty)\} \in \mathcal{F}$; and for $\alpha = 0$, $\{\alpha X \in (-\infty, r]\} = \emptyset \in \mathcal{F}$ or $\{\alpha X \in (-\infty, r]\} = \Omega \in \mathcal{F}$. The same argument shows that $\beta Y \in L^0$.

$\alpha X + \beta Y \in L^0$: By the previous step, it is sufficient to show that $X + Y \in L^0$. $\{X + Y < r\} = \cup_{q \in \mathbb{Q}} [\{X < q\} \cap \{Y < r - q\}]$. Since $\{X < q\}$, $\{Y < r - q\} \in \mathcal{F}$, $[\{X < q\} \cap \{Y < r - q\}] \in \mathcal{F}$. As $\mathbb{Q}$ is countable and $\mathcal{F}$ is closed under a countable union, $\{X + Y < r\} \in \mathcal{F}$.

$X \cdot Y \in L^0$: It is sufficient to show that $\{X \cdot Y > r\} \in \mathcal{F}$ for all $r \in \mathbb{R}$. First fix $r \geq 0$. For each pair of strictly positive rationals, (q_1, q_2), let $D(q_1, q_2) = [\{X > q_1\} \cap \{Y > q_2\}]$ and $E(q_1, q_2) = [\{X < -q_1\} \cap \{Y < -q_2\}]$. Since $X, Y \in L^0$, $F(q_1, q_2) := D(q_1, q_2) \cup E(q_1, q_2) \in \mathcal{F}$. Because $\mathcal{F}$ is closed under countable unions, $\{X \cdot Y > r\} = \cup \{F(q_1, q_2) : q_1 \cdot q_2 > r\} \in \mathcal{F}$. For $r < 0$, the arguments are parallel. ∎

Theorem 7.1.26 $L^0(\Omega, \mathcal{F}, P)$ *is closed under pointwise convergence and under composition with continuous functions.*

Proof. Suppose that $\{X_n : n \in \mathbb{N}\} \subset L^0$ and for all $\omega \in \Omega$, $X_n(\omega) \to X(\omega)$. It is sufficient to show that $\{X < r\} \in \mathcal{F}$. But $\{X < r\} = \cup_{q \in \mathbb{Q}_{++}} [\{X_n < r - q\} \text{ a.a.}]$.

Now suppose that $f : \mathbb{R} \to \mathbb{R}$ is continuous and that $X \in L^0$. To show that $\omega \mapsto f(X(\omega))$ is measurable, it is sufficient to show that $\{f(X) > r\} \in \mathcal{F}$. But $\{f(X) > r\} = \{X \in G\}$, where G is the open set $f^{-1}((r, +\infty))$. ∎

7.1.e Approximating Measurable Functions

Definition 7.1.27 *For $X, Y \in L^0$, we say that $X = Y$ almost everywhere (a.e.) if $P(\{|X - Y| = 0\}) = 1$.*

Definition 7.1.28 *The Ky Fan pseudometric on $L^0(\Omega, \mathcal{F}, P)$ is $\alpha(X, Y) = \inf\{\epsilon \geq 0 : P(\{|X - Y| > \epsilon\}) \leq \epsilon\}$. If $\alpha(X_n, X) \to 0$, we say that X_n converges to X in probability.*

Being almost everywhere equal is an equivalence relation, and $X = Y$ a.e. iff $\alpha(X, Y) = 0$, so the Ky Fan pseudometric is a metric on the almost everywhere equivalence classes. With very few exceptions, we work with the equivalence classes of random variables. Here are some examples of how "almost everywhere" is used:

1. X is *a.e. constant* if X is a.e. equal to $c \cdot 1_\Omega$ for some constant c, that is, if $P(\{\omega : X(\omega) = c\}) = 1$;

2. X is *a.e. nonnegative*, or $X \geq 0$ a.e., if $P(\{X \geq 0\}) = 1$; equivalently if $\{X < 0\}$ is a **null set**, that is, a set having probability 0; and

3. $X \geq Y$ a.e. if $X - Y$ is a.e. nonnegative.

Every continuous function on $[0, 1]$ is nearly piecewise linear in the sense that there is a piecewise linear function uniformly within any ϵ of any continuous function. For some approximation arguments, this is useful. For measurable functions, we have to have a way of saying "nearly"; that is, we need a metric. Loosely, we say that X and Y are **close in probability** if it is unlikely that they are very far apart, that is, if they are probably close. The Ky Fan metric gets at this intuition, and we now verify that it is a metric.

Lemma 7.1.29 $\alpha(\cdot, \cdot)$ *is a metric on a.e. equivalence classes of elements of* L^0.

Proof. It is immediate that $\alpha(X, Y) = \alpha(Y, X)$. If $\alpha(X, Y) = 0$, then $X = Y$ a.e. If $\alpha(X, Y) > 0$, then $P(\{|X - Y| > 0\}) > 0$, so it is not the case that $X = Y$ a.e. For the triangle inequality, let $X, Y, Z \in L^0$, for all ω in a set $E_{X,Y}$ with probability at most $\alpha(X, Y)$, $|X(\omega) - Y(\omega)| \leq \alpha(X, Y)$, and for all ω in a set $E_{Y,Z}$ with probability at most $\alpha(Y, Z)$, $|Y(\omega) - Z(\omega)| \leq \alpha(Y, Z)$. For all $\omega \in E_{X,Y} \cup E_{Y,Z}$,

$$|X(\omega) - Z(\omega)| \leq |X(\omega) - Y(\omega)| + |Y(\omega) - Z(\omega)| \leq \alpha(X, Y) + \alpha(Y, Z).$$

(7.1)

Since the set $E_{X,Y} \cup E_{Y,Z}$ has probability at most $\alpha(X, Y) + \alpha(Y, Z)$, $\alpha(X, Z) \leq \alpha(X, Y) + \alpha(Y, Z)$. ∎

Exercise 7.1.30 $\alpha(X_n, X) \to 0$ iff for all $\epsilon > 0$, $P(\{|X_n - X| > \epsilon\}) \to 0$.

The simpler subclass of $L^0(\Omega, \mathcal{F}, P)$ that we often use to approximate measurable functions is the class of **simple functions**.

Definition 7.1.31 *A function is* **simple** *if its range is finite, and* $M_s = M_s(\Omega, \mathcal{F}, P) \subset L^0(\Omega, \mathcal{F}, P)$ *denotes the set of simple random variables.*

Exercise 7.1.32 A simple function $X : \Omega \to \mathbb{R}$ is $\mathcal{F}$-measurable iff $X^{-1}(r) \in \mathcal{F}$ for all $r \in \mathbb{R}$.

Note that $X \in M_s$ iff $X = \sum_{k \leq K} \beta_k 1_{E_k}$ for $\beta_k \in \mathbb{R}$, $E_k \in \mathcal{F}$, and $K \in \mathbb{N}$. If we allow the E_k to overlap, the representation is not unique, for example, $1_{(0, \frac{3}{4}]} + 1_{(\frac{1}{2}, 1]} = 1_{(0, \frac{1}{2}] \cup (\frac{3}{4}, 1]} + 2 \cdot 1_{(\frac{1}{2}, \frac{3}{4}]}$.

Exercise 7.1.33 Show that every $X \in M_s$ has a unique representation $\sum_{k \leq K} \beta_k 1_{E_k}$ in which the $\{E_k : k = 1, \ldots, K\}$ are a partition of Ω and for all $k \neq k'$, $\beta_k \neq \beta_{k'}$. [Hint: Let $B = \{\beta_1, \ldots, \beta_K\}$ denote the range of X, being sure to include a $\beta_k = 0$ if 0 is in the range of X.]

When we need a unique representation of a simple function, we use the one you just proved existed—the one with the E_k forming a partition of Ω, and the β_k's distinct.

The basic approximation result is that all measurable functions are nearly simple.

Theorem 7.1.34   *M_s is a dense subset of the metric space (L^0, α), and $X : \Omega \to \mathbb{R}$ is measurable iff there exists a sequence, X_n, of simple measurable functions with the property that for all ω, $X_n(\omega) \to X(\omega)$.*

Proof.   If $X \in L^0$, consider the sequence of simple random variables that is due to H. Lebesgue:

$$X_n(\omega) = \sum_{k=-n \cdot 2^n}^{k=+n \cdot 2^n} k/2^n \cdot 1_{X^{-1}([k/2^n, (k+1)/2^n))}(\omega). \tag{7.2}$$

For all $n > |X(\omega)|$, $X(\omega) \geq X_n(\omega) > X(\omega) - 1/2^n$, so that $X_n(\omega) \to X(\omega)$. The converse holds because L^0 is closed under pointwise convergence.

To see that M_s is dense in L^0, note that for all ϵ, $\{|X - X_n| > \epsilon\} \downarrow \emptyset$. By countable additivity of P, $P(\{|X - X_n| > \epsilon\}) \downarrow 0$. Therefore, by Exercise 7.1.30, $\alpha(X_n, X) \to 0$.   ∎

Detour. M_s is an algebra of bounded functions, and it separates points if for all $\omega \neq \omega'$, there exists $E \in \mathcal{F}$ such that $1_E(\omega) \neq 1_E(\omega')$. These properties make it look like a set of continuous functions on a compact space. Indeed, if there is a countable sup norm dense subset of M_s, one can identify the compact space using the embedding techniques of §6.7.d. For more general spaces, see §11.4.c. This leads to very interesting mathematics, which is sometimes even useful to economists.

7.1.f   Integrals of Random Variables

We define the integral of $\mathbb{R}$-valued or measurable functions in three steps: simple random variables, nonnegative random variables, and general random variables.

7.1.f.1   Integrals of Simple Random Variables
For a simple random variable, the integral is the expected value from elementary probability theory.

Definition 7.1.35 (The Integral of Simple Random Variables)   *For $X(\omega) = \sum \beta_k 1_{E_k}(\omega)$ a simple random variable, the **integral of X (with respect to P) or expectation of X (with respect to P)** is defined by $\int X \, dP = \sum \beta_k P(E_k)$.*

Notation 7.1.36   *$E \, X$, $E^P X$, $\int_\Omega X(\omega) \, P(d\omega)$, $\int_\Omega X(\omega) \, dP(\omega)$, and $\int_\Omega X \, dP$ are other notations in common use for the integral.*

If $P(E_k) = 0$, then $E \, X$ does not depend on the value of the associated β_k, that is, if $X, Y \in M_s$ and $X = Y$ a.e., then $E \, X = E \, Y$.

Some further comments:

1. It is worth checking that different representations of a simple X all have the same integral.

2. Every mapping $\omega \mapsto X(\omega) \in \mathbb{R}^\ell$ can be written as $\omega \mapsto (X_1(\omega), \ldots, X_\ell(\omega))$. We could simultaneously develop the integrals of random vectors by defining $E^P X = (E^P X_1, \ldots, E^P X_\ell)$.

3. When $\omega \mapsto X(\omega) \in V$, V an infinite-dimensional vector space, integration theory is more complicated: definitions multiply, counterexamples to sensible statements multiply, and the results that are true become harder to prove.

The integral with respect to P is a mapping from the space of simple functions to $\mathbb{R}$. For historical reasons, functions defined on spaces of functions are often called **functionals**. The following contains the most basic properties of the functional we call the integral.

Exercise 7.1.37 Show that for simple random variables, the mapping $X \mapsto E^P X$ is

1. **positive**, if $0 \leq X$ a.e., then $0 \leq E^P X$,
2. **monotonic**, if $Y \leq X$ a.e., then $0 \leq E^P Y \leq E^P X$, and
3. **linear**, $E^P(\alpha X + \beta Y) = \alpha E^P X + \beta E^P Y$.

The indicator functions, 1_E, $E \in \mathcal{F}$, are particularly important because $P(E) = E^P 1_E$. Since the span of the set of indicator functions is the set of simple functions and the integral is a linear functional, the integral of all simple functions is determined by its values on the indicator functions.

7.1.f.2 Integrals of Nonnegative Random Variables
The following extends the integral from simple functions to nonnegative measurable functions.

Definition 7.1.38 (The Integral of Nonnegative Random Variables) *For $X \in L^0$, $X \geq 0$ a.e., $L_{\leq}(X)$ denotes the set $Y \in M_s$ such that $0 \leq Y \leq X$ a.e. The **integral of** X (**with respect to** P) or **expectation of** X (**with respect to** P) is defined by $\int X \, dP = \sup\{\int Y \, dP : Y \in L_{\leq}(X)\}$ when this supremum is finite. When the supremum is infinite, X is **nonintegrable**.*

Notation 7.1.39 *When X is nonintegrable, we write $\int X \, dP = \infty$, as if this defined an integral.*

There are surprisingly pedestrian situations in which nonintegrability occurs, for example, when Y_1 and Y_2 are independent normal random variables with mean 0 and variance 1 and $X := |Y_1/Y_2|$, where we define $Y_1/0$ to be any number we please.

7.1.f.3 Properties of the Integral, Part I
The following provides our first serious use of the countable additivity of P.

Theorem 7.1.40 (Properties of the Integral, I) *For integrable nonnegative X, Y and $\{X_n : n \in \mathbb{N}\}$,*

1. *positivity, if $0 \leq X$ a.e., then $0 \leq E^P X$,*
2. *monotonicity, if $X \leq Y$ a.e., then $0 \leq E^P X \leq E^P Y$,*
3. *positive linearity, for $X, Y \geq 0$ a.e. and $\alpha, \beta \in \mathbb{R}_+$, $E^P(\alpha X + \beta Y) = \alpha E^P X + \beta E^P Y$, and*

4. *monotone convergence, or continuity from below*, if $0 \leq X_n \uparrow X$ a.e., then $E^P X_n \uparrow E^P X$.

Example 7.1.41 *Suppose that P is Lebesgue's uniform distribution, λ, from Example 7.1.17 and that $n \mapsto q_n$ is an enumeration of the rationals in $(0, 1]$. Let $E_m = \{q_n : n \leq m\}$ so that $E_m \uparrow \mathbb{Q}$. Because λ is nonatomic, $\lambda(E_m) = 0$ so that $\lambda(E_m) = E^\lambda 1_{E_m} = 0$. Because (we will show that) λ is countably additive, $E^\lambda 1_{E_m} \uparrow E^\lambda 1_{\mathbb{Q}} = 0$.*

Note that $E_n \uparrow E$ iff $1_{E_n}(\omega) \uparrow 1_E(\omega)$ for all ω. Monotone convergence fails if P is not countably additive.

Lemma 7.1.42 *If X_n is a sequence in L^0, then the set of ω for which $X_n(\omega)$ converges is measurable.*

Proof. Let X_n be a sequence in L^0. We want to show that $C = \{\omega : \exists r \ X_n(\omega) \to r\} \in \mathcal{F}$. Since $\mathbb{R}$ is complete, $C = \{\omega : X_n(\omega) \text{ is Cauchy}\}$. But $C = \cap_{\epsilon \in \mathbb{Q}_{++}} \cup_{N \in \mathbb{N}} \cap_{n,m \geq N} \{|X_n - X_m| < \epsilon\}$, which belongs to $\mathcal{F}$ because each $\{|X_n - X_m| < \epsilon\}$ belongs to $\mathcal{F}$. ∎

Proof of Theorem 7.1.40. We begin with positivity: Since the constant function, 0, belongs to $L_{\leq}(X)$, and the integral is positive on simple functions, $0 \leq E^P X$.

Monotonicity: If $P(\{0 \leq Y \leq X\}) = 1$, then $L_{\leq}(Y) \subset L_{\leq}(X)$, so $E^P Y \leq E^P X$.

Monotone convergence: From the monotonicity property, we know that $E^P X_n$ is a nondecreasing sequence and $E^P X_n \leq E^P X$, so that $\lim_n E^P X_n \leq E^P X$. Therefore, it is sufficient to show that for all $\epsilon > 0$, $\lim_n E^P X_n \geq E^P X - \epsilon$. The idea is to show that $E^P X_n$ is eventually greater than the integral of any simple function that is below X.

Pick a simple $Y = \sum_k \beta_k 1_{E_k} \leq X$ such that $E^P Y \geq E^P X - \epsilon/3$. Let $\beta'_k = \max\{\beta_k - \epsilon/3, 0\}$ and define $Y' = \sum_k \beta'_k 1_{E_k}$, so that $E^P Y' \geq E^P Y - \epsilon/3$.

For each k, let $E_{k,n} = \{\omega \in E_k : X_n(\omega) \geq \beta'_k\}$. Since $X_n(\omega) \uparrow X(\omega)$ for all ω, $E_{k,n} \uparrow E_k$. By the continuity from below of P, $P(E_{k,n}) \uparrow P(E_k)$. As $X_n \geq \sum_k \beta'_k 1_{E_{k,n}}$,

$$E^P X_n \geq E^P \sum_k \beta'_k 1_{E_{k,n}} = \sum_k \beta'_k P(E_{k,n}) \uparrow \sum_k \beta'_k P(E_k) = E^P Y'.$$

For large n, $E^P X_n > E^P Y' - \epsilon/3 \geq E^P Y - 2\epsilon/3 \geq E^P X - \epsilon$.

Positive linearity: We know that there are sequences of simple functions $X_n \uparrow X$ and $Y_n \uparrow Y$. For $\alpha, \beta \in \mathbb{R}_+$, $\alpha X_n(\omega) \uparrow \alpha X$, $\alpha Y_n(\omega) \uparrow \alpha Y$, and $(\alpha X_n + \beta Y_n)(\omega) \uparrow (\alpha X + \beta Y)(\omega)$. By continuity from below, $E^P \alpha X_n(\omega) \uparrow E^P \alpha X$, $E^P \alpha Y_n(\omega) \uparrow E^P \alpha Y$, and $E^P(\alpha X_n + \beta Y_n) \uparrow E^P(\alpha X + \beta Y)$. From linearity on simple functions, $E^P(\alpha X_n + \beta Y_n) = \alpha E^P X_n + \beta E^P Y_n$. ∎

The situation in Theorem 7.1.40(2) happens often enough that it has its own name.

Definition 7.1.43 $X \in L^0$ *is **dominated** by Y if $0 \leq |X| \leq Y$ a.e.*

Corollary 7.1.44 *If $X \geq 0$ is dominated by an integrable Y, then X is integrable.*

Exercise 7.1.45 For $X \in L^0$ and $\{X_n : n \in \mathbb{N}\} \subset L^0$, if $0 \leq X_n \uparrow X$ a.e. and $E^P X = \infty$, then $E^P X_n \uparrow E^P X$.

7.1.f.4 Integrals of Random Variables

For a measurable X, we define the **positive part of X** as $X^+ = X \vee 0$ and **negative part of X** as $X^- = -(X \wedge 0)$. More explicitly,

$$X^+(\omega) = \begin{cases} X(\omega) & \text{if } X(\omega) \geq 0 \\ 0 & \text{if } X(\omega) \leq 0 \end{cases} \quad \text{and} \quad X^-(\omega) = \begin{cases} -X(\omega) & \text{if } X(\omega) \leq 0 \\ 0 & \text{if } X(\omega) \geq 0 \end{cases},$$

so that $X = X^+ - X^-$, $|X| = X^+ + X^-$, and $X^+ \cdot X^- = 0$. Since $f(r) = r 1_{[0,\infty)}(r)$ is continuous, $X^+ = f(X)$ is measurable. Similarly, $g(r) = -r 1_{(-\infty,0]}(r)$ is continuous and $X^- = g(X)$ is measurable.

Exercise 7.1.46 For $X, Y \in L^0$, give three proofs that $X \wedge Y \in L^0$ and $X \vee Y \in L^0$ using the following lines of argument:

1. Show that there is a sequence of polynomials, $f_n(r)$, with $f_n(r) \to |r|$ for all r, and use the fact that L^0 is an algebra that is closed under pointwise convergence.

2. Give $\{X \wedge Y < r\}$ and $\{X \vee Y < r\}$ in terms of $\{X < r\}$ and $\{Y < r\}$.

3. Show that for $r, s \in \mathbb{R}$, $r \wedge s = \frac{1}{2}(r+s) + \frac{1}{2}|r - s|$ and $r \vee s = \frac{1}{2}(r+s) - \frac{1}{2}|r - s|$; use the fact that L^0 is closed under composition with continuous functions.

Definition 7.1.47 (The Integral of Random Variables) *An $X \in L^0(\Omega, \mathcal{F}, P)$ is **integrable** if the nonnegative random variable $|X|$ is integrable, and its integral is defined as $\int X \, dP = \int X^+ \, dP - \int X^- \, dP$. A measurable X is **nonintegrable** if the nonnegative random variable $|X|$ is nonintegrable, that is, if either $\int X^+ \, dP = \infty$ or $\int X^- \, dP = \infty$. $L^1 = L^1(\Omega, \mathcal{F}, P) \subset L^0(\Omega, \mathcal{F}, P)$ denotes the set of integrable random variables on $(\Omega, \mathcal{F}, P)$.*

L^0 is a vector space and a lattice, that is, a **vector lattice**, and an algebra.

Theorem 7.1.48 *$L^1(\Omega, \mathcal{F}, P)$ is a vector lattice but not, in general, an algebra.*

Proof. Suppose that $X, Y \in L^1$. We have to show that $|X \vee Y| \in L^1$. For all ω, $|X|(\omega) \vee |Y|(\omega) \geq |X \vee Y|(\omega)$, so it is sufficient to show that $|X| \vee |Y| \in L^1$. $|X| \vee |Y| = |X| \cdot 1_{\{|X| \geq |Y|\}} + |Y| \cdot 1_{\{|Y| > |X|\}}$. Since $(|X| \cdot 1_{\{|X| \geq |Y|\}}) \leq |X|$, $E^P(|X| \cdot 1_{\{|X| \geq |Y|\}}) \leq E^P|X|$, and as $(|Y| \cdot 1_{\{|Y| > |X|\}}) \leq |Y|$, $E^P(|Y| \cdot 1_{\{|Y| > |X|\}}) \leq E^P|Y|$. Therefore, $E^P(|X| \vee |Y|) \leq E^P|X| + E^P Y$, so that $|X| \vee Y$ is integrable. To finish the proof, we give an example of $X, Y \in L^1$ with $X \cdot Y \notin L^1$. Take $(\Omega, \mathcal{F}, P) = ((0, 1], \mathcal{B}, \lambda)$ and $X = Y = \sum_n \beta_n 1_{(1/2^{n+1}, 1/2^n]}$, $\beta_n = \sqrt{n \cdot 2^n}$. ∎

7.1.f.5 Properties of the Integral, Part II

The continuity part of the next result is called **Lebesgue's dominated convergence theorem**.

Theorem 7.1.49 (Properties of the Integral, II) *For* X_n, X, Y *in* $L^1(\Omega, \mathcal{F}, P)$,

1. *positivity, if* $0 \leq X$ *a.e., then* $0 \leq E^P X$,
2. *monotonicity, if* $Y \leq X$ *a.e., then* $E^P Y \leq E^P X$,
3. *linearity, for* $\alpha, \beta \in \mathbb{R}$, $E^P(\alpha X + \beta Y) = \alpha E^P X + \beta E^P Y$, *and*
4. *continuity, or dominated convergence, if* $|X_n| \leq |Y|$ *and* $X_n \to X$ *a.e., then* $E^P X_n \to E^P X$.

The dominated convergence theorem is a crucial piece of continuity, yielding an enormous number of results in probability and statistics. In §7.5, we prove the dominated convergence theorem, replacing "$X_n \to X$ a.e." with the weaker "$\alpha(X_n, X) \to 0$." Until then, we use these properties of the integral without proof, being careful not to create any logical circularities.

7.1.g $L^1(\Omega, \mathcal{F}, P)$ Is a Normed Vector Space

The relevant norm is very similar to the 1-norm on $\mathbb{R}^\ell$, $\|\mathbf{x}\|_1 := \sum_{i=1}^\ell |x_i|$. It even has the same notation.

Definition 7.1.50 *For* $X \in L^1(\Omega, \mathcal{F}, P)$, *the* L^1-*norm is* $\|X\|_1 = \int |X| \, dP$ *and the associated* **norm distance** *is* $d_1(X, Y) = \|X - Y\|_1$.

It is not for the purpose of confusing the reader that the 1-norm on $\mathbb{R}^\ell$ and the L^1-norm on $L^1(\Omega, \mathcal{F}, P)$ have the same notation. Take $\Omega = \{1, \ldots, \ell\}$, $\mathcal{F} = \mathcal{P}(\Omega)$, and $P(\{\omega\}) = 1/\ell$. Any $X \in L^1(\Omega, \mathcal{F}, P)$ is of the form $X = (X(1), \ldots, X(\ell))$, that is, it is a point in $\mathbb{R}^\ell$, and $\|X\|_1 = \int_\Omega |X(\omega)| \, dP(\omega) = \frac{1}{n} \sum_{\omega=1}^\ell |X(\omega)|$.

The following comes directly from the properties of the integral.

Corollary 7.1.51 $(L^1(\Omega, \mathcal{F}, P), \|\cdot\|_1)$ *is a normed vector space, and for* $X \in L^1$, $|\int X \, dP| \leq \int |X| \, dP$,

Proof. By Definition 4.3.7 (p. 117), in order to show that $\|\cdot\|_1$ is a norm on the equivalence classes of integrable functions, we must verify that $\|X\|_1 = 0$ iff $X = 0$ a.e., $\|\alpha X\|_1 = |\alpha| \|X\|_1$ for all $\alpha \in \mathbb{R}$ and that $\|X + Y\|_1 \leq \|X\|_1 + \|Y\|_1$.

$X = 0$ a.e. iff $0 \leq X$ a.e. and $0 \leq -X$ a.e. By linearity and positivity, these yield $0 \leq \int X \, dP$ and $0 \leq -\int X \, dP$, so that $\|X\|_1 = 0$. If it is not the case that $X = 0$ a.e., then $P(\{|X| > 0\}) > 0$. Define $A_n = \{|X| > 1/n\}$ so that $A_n \uparrow \{|X| > 0\}$. By countable additivity, there exists n such that $P(A_n) > 0$. The simple function $\frac{1}{n} 1_{A_n} \in L_{\leq}(|X|)$ and has a strictly positive integral, so that $\|X\|_1 > 0$.

$\|\alpha X\|_1 = \int |\alpha X| \, dP$ by defininition and $\int |\alpha X| \, dP = \int |\alpha| \, |X| \, dP$ by linearity and the previous step because $(|\alpha| \, |X| - |\alpha X|) = 0$ a.e. By linearity again, $\int |\alpha| \, |X| \, dP = |\alpha| \int |X| \, dP$.

$\|X + Y\|_1 = \int |X + Y| \, dP$ by definition, $\int |X + Y| \, dP \leq \int (|X| + |Y|) \, dP$ by monotonicity, and $\int (|X| + |Y|) \, dP = \int |X| \, dP + \int |Y| \, dP$ by linearity.

$X \leq |X|$ a.e. and $-X \leq |X|$ a.e. By monotonicity, $\int X \, dP \leq \int |X| \, dP$ and $-\int X \, dP \leq \int |X| \, dP$. ∎

7.1.h The Probability Metric Space

Recall that the **symmetric difference of** A, $B \subset \Omega$ is defined by $A \triangle B = (A \setminus B) \cup (B \setminus A)$.

Definition 7.1.52 *For A, $B \in \mathcal{F}$, we define the **probability metric** by $\rho_P(A, B) = P(A \triangle B)$.*

Remember, $\rho_P(\cdot, \cdot)$ is a metric on the a.e. equivalence classes of $\mathcal{F}$.

Lemma 7.1.53 *For any A, $B \in \mathcal{F}$, $\rho_P(A, B) = E\,|1_A - 1_B| = d_1(1_A, 1_B) = \alpha(1_A, 1_B)$.*

Proof. The second equality is the definition of d_1. $|1_A - 1_B|(\omega) = 1$ iff $\omega \in A \triangle B$; otherwise $|1_A - 1_B|(\omega) = 0$, giving the first and the third equality. ∎

Lemma 7.1.54 *For A, B, $C \in \mathcal{F}$,*

1. $\rho_P(\emptyset, A) = P(A)$,
2. $\rho_P(A, B) = \rho_P(A^c, B^c)$,
3. $A \mapsto P(A)$ *is uniformly continuous, and*
4. $\rho_P(A, B) + \rho_P(B, C) \geq \rho_P(A, C)$ *(the triangle inequality).*

Proof. (1) follows from $\emptyset \triangle A = A$. If we denote 1_Ω by 1, (2) follows from $1_{A \triangle B} = |1_A - 1_B| = |(1 - 1_A) - (1 - 1_B)| = |1_{A^c} - 1_{B^c}|$. (3) follows from the observation that $|P(A) - P(B)| = |E\,(1_A - 1_B)| \leq E\,|1_A - 1_B| = \rho_P(A, B)$. (4) follows from $|1_A - 1_B| + |1_B + 1_C| \geq |1_A - 1_C|$ and the monotonicity of the integral. ∎

7.1.i Comparison with the Riemann Integral; $\int_a^b f(x)\,dx$

The integral $\int_a^b f(x)\,dx$ from your calculus class is called the **Riemann integral**, when $(b - a) = 1$, the one defined here is called the **Lebesgue integral**. We sketch the argument that after the obvious rescaling, the Riemann integral is a special case of the Lebesgue integral. This means that you should not be shy about using any of your knowledge about integration from calculus; indeed, that knowledge is sometimes sufficient for what we do here. We are building on what you know, not replacing it.

The difference between the two integrals is the class of simple functions that are used to approximate functions. The Lebesgue approximations to a nonnegative X in Definition 7.1.38 (p. 364) use $L_{\leq}(X)$, the set of $Y \in M_s(\Omega, \mathcal{F}, P)$ for which $0 \leq Y \leq X$ a.e., and for general X one approximates X^+ and X^- and adds the results.

By contrast, the Riemann approximations to a function $f : [a, b] \to \mathbb{R}_+$ use two approximating sets, $R_{\leq}(f)$, which approximates from below, and $R_{\geq}(f)$, which approximates from above.

Definition 7.1.55 *The **lower Riemann approximation set** for $f : [a, b] \to \mathbb{R}$ is*

$$R_{\leq}(f) = \{g(x) = \textstyle\sum_{k \leq K} \beta_k 1_{|a_k, b_k|}(x) : (\forall x \in [a, b])[g(x) \leq f(x)], \quad K \in \mathbb{N}\},$$

*and the **upper Riemann approximation set** is*

$$R_\geq(f) = \{h(x) = \textstyle\sum_{k \leq K} \beta_k 1_{|a_k, b_k|}(x) : (\forall x \in [a, b])[f(x) \leq h(x)], \quad K \in \mathbb{N}\}.$$

In both the Lebesgue and the Riemann case, the approximating functions are simple. The difference is between interval sets $|a_k, b_k|$ and measurable sets E_k, the comparison between

$$\textstyle\sum_{k \leq K} \beta_k 1_{|a_k, b_k|}(x) \text{ and } \sum_{k \leq K} \beta_k 1_{E_k}(x).$$

The E_k *might be* intervals, but they *need not be* intervals. This small change in the richness of the approximating set solves a large number of difficulties very neatly.

Definition 7.1.56 *The **Riemann integral** of $g = \sum_k \beta_k 1_{|a_k, b_k|}$ is $\int_a^b g(x)\,dx :=$ $\sum_k \beta_k (b_k - a_k)$, and a bounded f is **Riemann integrable** if*

$$\sup\left\{\int_a^b g(x)\,dx : g \in R_\leq(f)\right\} = \inf\left\{\int_a^b h(x)\,dx : h \in R_\geq(f)\right\}, \quad (7.3)$$

*in which case $\int_a^b f(x)\,dx$ is defined as the common value. To extend this to unbounded functions, one takes all sequences $N, M \to \infty$ and defines the functions $f_{N,M} = (f \wedge N) \vee -M$ (which bounds them above by N and below by $-M$). Then f is **Riemann integrable** if $\lim_{N,M \to \infty} \int_a^b f_{N,M}(x)\,dx$ exists and does not depend on the chosen sequences N, M.*

Example 7.1.57 *If $f : [0, 1] \to \mathbb{R}$ is $f(x) = 1_\mathbb{Q}(x)$, then $R_\leq(f)$ contains only the function identically equal to 0 and $R_\geq(f)$ contains only the simple $\mathcal{B}^\circ$-measurable functions that satisfy $h(x) \geq 1$. Thus, f is not Riemann integrable, but it is Lebesgue integrable, with respect to the uniform distribution, λ, for which we have $\lambda(\mathbb{Q} \cap [0, 1]) = 0$ and $\int f\,d\lambda = 0$. [Both $R_\leq(f)$ and $R_\geq(f)$ contain functions that violate the given inequalities at finitely many points when one uses the degenerate intervals $[q, q]$.]*

For the Lebesgue integral, the integrability of $|X|$ gives us the integrability of X. The same is not true for the Riemann integral.

Example 7.1.58 *Suppose that $f(x) = 2 \cdot 1_\mathbb{Q}(x) - 1$ for $x \in [0, 1]$. In this case, f^+ is the indicator of the rationals in $[0, 1]$ and f^- is the indicator of the irrationals. In this case, $|f|(x) = 1$ for all $x \in [0, 1]$, but neither f^+ nor f^- is Riemann integrable.*

Some functions with infinitely many discontinuities are Riemann integrable.

Example 7.1.59 *If $E = \{\frac{1}{n} : n \in \mathbb{N}\} \subset [0, 1]$ and $f = 1_E$, then $\int_0^1 f(x)\,dx = \int f\,d\lambda = 0$. To see why $\int_0^1 f(x)\,dx = 0$, note that $R_\leq$ again contains only the function identically equal to 0. Pick $\epsilon > 0$, set $\delta = \epsilon/3$, and consider the intervals $I_n := |\frac{1}{n} - \frac{\delta}{2^n}, \frac{1}{n} + \frac{\delta}{2^n}|$. The function*

$$h_\delta(x) = 1_{[0, \delta)}(x) + \textstyle\sum_{\{n : n > 1/\delta\}} 1_{I_n}(x)$$

belongs to $R_\geq$, and $\int_0^1 h_\delta(x)\,dx < \delta + 2\delta = \epsilon$.

It should be clear that the Riemann approximations to a Riemann integrable function are measurable with respect to $\mathcal{B}^\circ$ and that any upper envelope of such a set is measurable with respect to $\sigma(\mathcal{B}^\circ)$. It is both plausible and true that for a Riemann integrable f, $\int_a^b f(x)\,dx = (b-a)\int_{[a,b]} f(\omega)\,dP(\omega)$, where P is the uniform distribution on $[a, b]$, that is, the one with the cdf $F(x) = (x-a)/(b-a)$ for $x \in [a, b]$.

Thus, to find $X, Y \in L^1((0, 1], \mathcal{B}, \lambda)$ such that $XY \notin L^1$, we could have leaned on what you learned in your calculus class—that $\int_0^1 \frac{1}{x}\,dx = \infty$, that is, it is not Riemann integrable, and taken $X(\omega) = Y(\omega) = 1/\sqrt{\omega}$ in the proof of Theorem 7.1.48 (p. 366).

Finally, for integrals of the form $\int_{-\infty}^\infty f(x)\,dx$, we "paste together" countably many copies of the uniform probability distribution on the intervals ..., $(-2, -1]$, $(-1, 0]$, $(0, 1]$, $(1, 2]$, ..., and then add integrals over the intervals. That is, $\int_{-\infty}^\infty f(x)\,dx := \sum_{n=-\infty}^{+\infty} \int_n^{n+1} f(x)\,dx$. Now there are two ways that an integral can be undefined: it can be undefined over some interval(s) $(n, n+1]$ or it can be finite over each interval but have the countable sum undefined.

7.2 ◆ Four Limit Results

In this section, we give and prove the weak law of large numbers (WLLN) and the Borel-Cantelli lemma. We give the strong law of large numbers (SLLN) but only prove it under more restrictive conditions than are needed. All three results tell us that the probabilities of some limit events, for example, the long-run average equaling the theoretical average, are equal either to 0 or to 1. Kolmogorov's 0-1 law is the fourth result. It tells us that this is exactly what we should expect because the limit events are all what are called "tail events," that is, events that depend on the tail behavior of sequences of independent random variables.

The measurability results of the previous section cover what we need for the WLLN, the SLLN, and the Borel-Cantelli lemma, but they are not enough to give a complete proof of Kolmogorov's result. For the presentation of Kolmogorov's result, we need the basic ideas and results on sub-σ-fields, and we cover these here. We return to measurability considerations and complete the proof in a subsequent section.

7.2.a Independence, Variance, and Covariance

The following expands Definition 6.9.24 (p. 340) to include all random variables and gives the expanded version a new number.

Definition 7.2.1 *A collection $\{E_\alpha : \alpha \in A\} \subset \mathcal{F}$ is **independent** if for finite $A_F \subset A$, $P(\cap_{\alpha \in A_F} E_\alpha) = \Pi_{\alpha \in A_F} P(E_\alpha)$. A collection $\{\mathcal{C}_\alpha : \alpha \in A\}$ of subsets of $\mathcal{F}$ is **independent** if for all collections $\{E_\alpha \in \mathcal{C}_\alpha : \alpha \in A\}$ is independent. A collection of random variables $\{X_\alpha : \alpha \in A\}$ is **independent** if for every collection $\{G_\alpha : \alpha \in A\}$ of open subsets of the ranges of the X_α, the collection of events, $\{X_\alpha^{-1}(G_\alpha) : \alpha \in A\}$, is independent.*

For the sake of concreteness, we use a particular probability space in many examples and exercises, $(\Omega, \mathcal{F}, P) = ((0, 1], \mathcal{B}, \lambda)$. For intervals $|a, b|, 0 < a \le b \le 1, \lambda(|a, b|) = b - a$, and for $E = \cup_k |a_k, b_k|$ a finite disjoint union of intervals, $\lambda(E) = \sum_k (b_k - a_k)$. This defines λ on the field $\mathcal{B}^\circ$, of finite disjoint unions of intervals. We have not yet shown that λ can be extended to a countably additive probability on $\mathcal{B} = \sigma(\mathcal{B}^\circ)$, the smallest σ-field containing $\mathcal{B}^\circ$, but we will in §7.6. In the meantime, it is a good source of examples.

Exercise 7.2.2 For $(\Omega, \mathcal{F}, P) = ((0, 1], \mathcal{B}, \lambda)$, $n \in \mathbb{N}$ and $0 \le k \le 2^n - 1$, let $I(k, n) = (k/2^n, (k + 1)/2^n]$. Let $\mathcal{C}_n = \{I(k, n) : k = 0, 1, \ldots, 2^n - 1\}$. Show the following:

1. For all n and $k \ne k'$, $I(k, n)$ and $I(k', n)$ are not independent.

2. The collection $\{\mathcal{C}_n : n \in \mathbb{N}\}$ is not independent.

3. For each $n \in \mathbb{N}$, let $X_n(\omega) = \sum_{k \text{ odd}} 1_{I(k,n)}(\omega)$. The collection $\{X_n : n \in \mathbb{N}\}$ is independent.

Definition 7.2.3 For $p \in [1, \infty)$, $L^p = L^p(\Omega, \mathcal{F}, P) := \{X \in L^0 : \int |X|^p \, dP < \infty\}$, and $L^\infty = L^\infty(\Omega, \mathcal{F}, P) := \{X \in L^0 : \exists B \in \mathbb{R}, P(|X| \le B) = 1\}$. For $X \in L^p$, $p \in [1, \infty)$, we define the p-norm as $\|X\|_p = \left[\int |X|^p \, dP\right]^{1/p}$, and we define $\|X\|_\infty = \inf\{B : P(|X| \le B) = 1\}$.

For $1 \le p < p' \le \infty$, $L^{p'} \subset L^p$, and you should see how to prove this (think about integrating over the disjoint sets $|X| \le 1$ and $|X| > 1$). We will see that for $1 \le p < p' \le \infty$, $L^{p'} \subsetneq L^p$ iff $(\Omega, \mathcal{F}, P)$ is not composed of a finite set of atoms.

Definition 7.2.4 For a probability space $(\Omega, \mathcal{F}, P)$, $E \in \mathcal{F}$ is an **atom** of P if $P(E) > 0$ and $[E' \subset E] \Rightarrow [P(E') = 0$ or $P(E') = P(E)]$. P is **nonatomic** or **nonatomic on** $\mathcal{F}$ if it has no atoms and **purely atomic** if it has only atoms.

The function $\lambda : \mathcal{B} \to [0, 1]$ is nonatomic in the probability space $((0, 1], \mathcal{B}, \lambda)$, λ. However, it is purely atomic in the probability space $((0, 1], \mathcal{G}, \lambda)$, where $\mathcal{G} = \{\emptyset, (0, \frac{1}{2}], (\frac{1}{2}, 1], (0, 1]\} \subset \mathcal{B}$.

Here are three examples to show part of what is involved.

Example 7.2.5 In the probability space $((0, 1], \mathcal{G}, \lambda)$ with $\mathcal{G} = \{\emptyset, (0, \frac{1}{2}], (\frac{1}{2}, 1], (0, 1]\} \subset \mathcal{B}$, $L^0 = \{\beta_1 1_{(0, \frac{1}{2}]} + \beta_2 1_{(\frac{1}{2}, 1]} : \beta_1, \beta_2 \in \mathbb{R}\}$, and for all $p \in [1, \infty]$, $L^0 = L^p$.

Example 7.2.6 For $(\Omega, \mathcal{F}, P) = ((0, 1], \mathcal{B}, \lambda)$ and $0 < r < 1$, define $X_r(\omega) = 1/\omega^r$ so that $E X_r = r/(1 - r)$. For any $p \in [1, \infty)$, $X_r \in L^p((0, 1], \mathcal{B}, \lambda)$ iff $r < \frac{1}{p}$. Thus, for $\frac{1}{p'} < r < \frac{1}{p}$, we have a random variable that is in $L^{p'}$ but not in L^p when $1 \le p < p' < \infty$.

Example 7.2.7 For the countable probability space $(\Omega, \mathcal{F}, P) = (\mathbb{N}, \mathcal{P}(\mathbb{N}), P)$, set $P(\{n\}) = 1/2^n$ and $P(E) = \sum_{n \in E} P(\{n\})$. This gives rise to a purely atomic countably additive probability with countably many atoms. For $n \in \Omega$ and $0 \le r \le 2$, define $X_r(n) = r^n$. This gives $E X_r = \sum_n \left(\frac{r}{2}\right)^n$ and $E |X_r|^p = \sum_n \left(\frac{r^p}{2}\right)^n$, *which in turn yield* $E |X_r|^p < \infty$ *iff* $r^p < 2$ *iff* $r < 2^{1/p}$. *Picking* r *such that* $2^{1/p'} < r < 2^{1/p}$ *gives a random variable in* $L^{p'}$ *but not in* L^p, $1 \le p < p' < \infty$.

The most commonly used L^p space is L^2, and it has its own terminology.

Definition 7.2.8 *For* $X \in L^2(\Omega, \mathcal{F}, P)$, *the **variance of** X is written either as* σ_X^2 *or as* $\mathrm{Var}(X)$ *and defined as* $E (X - E X)^2$. *The **standard deviation of** X is* $\sigma_X = (E (X - E X)^2)^{1/2} = \|X - E X\|_2$.

The following shows that $L^2(\Omega, \mathcal{F}, P)$ is exactly the set of random variables having finite variance.

Exercise 7.2.9 $(X - r)^2$ is integrable for all $r \in \mathbb{R}$ iff X^2 is integrable.

Definition 7.2.10 *The **covariance of** $X, Y \in L^2$ is defined as* $\sigma_{X,Y} = E (X - E X)(Y - E Y)$, *and the **correlation** is defined as* $\rho_{X,Y} = \sigma_{X,Y}/\sigma_X\sigma_Y$.

The next result shows that the covariance is a number. However, the product of elements of L^2 is an element of L^1, so that L^2 is not, in general, an algebra.

Lemma 7.2.11 *If* $X, Y \in L^2(\Omega, \mathcal{F}, P)$, *then* $XY \in L^1(\Omega, \mathcal{F}, P)$.

Proof. For any $x, y \in \mathbb{R}$, $(x + y)^2 \ge 0$ implies $x^2 + y^2 \ge -2xy$, whereas $(x - y)^2 \ge 0$ implies $x^2 + y^2 \ge +2xy$. Combining yields $x^2 + y^2 \ge 2|xy|$. Applying this at each ω gives

$$|X(\omega)Y(\omega)| \le \tfrac{1}{2}|X(\omega)|^2 + \tfrac{1}{2}|Y(\omega)|^2. \tag{7.4}$$

Since $|X|^2 + |Y|^2$ is integrable, $|XY|$ is integrable as well. ∎

Lemma 7.2.12 $L^2(\Omega, \mathcal{F}, P)$ *is a vector lattice.*

Proof. L^2 is a clearly a vector space. To show that L^2 is a lattice, observe that $|X \vee Y| \le |X| \vee |Y|$ a.e. and $|X \wedge Y| \le |X| \vee |Y|$. Since squaring is monotonic on the nonnegative numbers and the integral is monotonic, to show that $E |X \vee Y|^2 < \infty$ and $E |X \wedge Y|^2 < \infty$, it is sufficient to show that $E (|X| \vee |Y|)^2 < \infty$. But $(|X| \vee |Y|)^2 = |X|^2 \vee |Y|^2$, $|X|^2, |Y|^2 \in L^1$ and L^1 is a vector lattice. ∎

Exercise 7.2.13 Give two examples of random variables with $X \in L^2$ but $X \cdot X \notin L^2$. [Look at Examples 7.2.6 and 7.2.7 if you get stuck.]

Proving the basic properties of variance and covariance is a good exercise.

Exercise 7.2.14 Show the following basic properties:

1. For all $a \in \mathbb{R}$, $\mathrm{Var}(aX) = a^2 \mathrm{Var}(X)$ and $\mathrm{Var}(X - a) = \mathrm{Var}(X)$;

2. $\mathrm{Cov}(X, Y) = E^P XY - (E^P X)(E^P Y)$;

3. $\mathrm{Var}(X) = E^P X^2 - (E^P X)^2$;

4. For all $a, b \in \mathbb{R}$, $\mathrm{Cov}(aX, bY) = ab\,\mathrm{Cov}(X, Y)$ and $\mathrm{Cov}(X - a, Y - b) = \mathrm{Cov}(X, Y)$;

5. $\mathrm{Var}(X + Y) = \mathrm{Var}(X) + \mathrm{Var}(Y) + 2\,\mathrm{Cov}(X, Y)$;

6. $\mathrm{Var}(aX + bY) = a^2 \mathrm{Var}(X) + b^2 \mathrm{Var}(Y) + 2ab\,\mathrm{Cov}(X, Y)$;

7. $\rho(X, Y) = \frac{a}{|a|}\rho(aX, Y)$ for $a \ne 0$ (which means that correlation is a **unitless** measure of the relation between two random variables, changing from

measuring X in, say, gallons to, say, liters does not change X's correlation with Y; however, changing from gallons to negative gallons reverses the sign of the correlation.); and

8. $-1 \leq \rho(X, Y) \leq 1$. [Hint: Prove this for simple X and Y using the Cauchy-Schwarz inequality for $\mathbb{R}^n$, and then pass to the limit using dominated convergence.]

Lemma 7.2.15 *If $X \perp\!\!\!\perp Y \in L^2$, then $E\ XY = E\ X \cdot E\ Y$ and $\mathrm{Cov}(X, Y) = 0$.*

Proof. If $X \perp\!\!\!\perp Y$, then $(X - E\ X) \perp\!\!\!\perp (Y - E\ Y)$ from the definition of independence. Since $\mathrm{Cov}(X, Y) = E\ (X - E\ X)(Y - E\ Y)$, if we prove the first part, the second part is immediate. If X and Y are simple, the result is easy.[2] Define

$$X_n(\omega) = \sum_{j=0}^{n2^n} \frac{j}{2^n} 1_{X \in [j/2^n, (j+1)/2^n)}(\omega) + \sum_{k=0}^{n2^n} (-\frac{k}{2^n}) 1_{X \in (-(k+1)/2^n, k/2^n]}(\omega),$$

(7.5)

with a similar definition for Y_n. For all ω, $|X_n(\omega)| \leq |X(\omega)|$, $|Y_n(\omega)| \leq |Y(\omega)|$, and $|X_n(\omega) Y_n(\omega)| \leq |X(\omega) Y(\omega)|$. Also, $X_n(\omega) \to X(\omega)$ and $Y_n(\omega) \to Y(\omega)$, so that $X_n(\omega) Y_n(\omega) \to X(\omega) Y(\omega)$. Thus, X_n is dominated by X and converges to X for all ω; Y_n is dominated by Y and converges to Y for all ω; and $X_n Y_n$ is dominated by XY and converges to XY for all ω. By dominated convergence, $E\ X_n \to E\ X$, $E\ Y_n \to E\ Y$, and $E\ X_n Y_n \to E\ XY$. By independence, $X_n \perp\!\!\!\perp Y_n$. Since X_n and Y_n are simple, $E\ X_n Y_n = E\ X_n \cdot E\ Y_n$. ∎

Vectors $\mathbf{x}, \mathbf{y} \in \mathbb{R}^\ell$ are orthogonal if $\mathbf{x} \cdot \mathbf{y} = 0$, that is, if $\sum_{i \in I} x_i y_i = 0$. For $X, Y \in L^2$, $E\ X \cdot Y = \int_\Omega X(\omega) Y(\omega)\, dP(\omega)$. The summation and the integral have very similar forms, which is the initial observation in studying the geometry of L^2.

Definition 7.2.16 *$X, Y \in L^2$ are **orthogonal**, written $X \perp Y$, if $E\ XY = 0$.*

In these terms, $\mathrm{Cov}(X, Y) = 0$ iff $(X - E\ X) \perp (Y - E\ Y)$.

7.2.b Result 1: The Weak Law of Large Numbers

The weak law of large numbers says that it is unlikely that the average of independent random variables with the same distribution is very far from their (common) expectation. We prove it and some of its generalizations under the assumption that the random variables have finite variance.

Theorem 7.2.17 (Weak Law of Large Numbers [WLLN]) *If $\{X_n : n \in \mathbb{N}\} \subset L^2(\Omega, \mathcal{F}, P)$ is an independent collection of random variables and*

---

2. Well, maybe not. If $X = \sum_i \beta_i 1_{E_i}$ and $Y = \sum_j \gamma_j 1_{E_j}$, then for all i and j, $E_i \perp\!\!\!\perp E_j$, and you should check the definition with small open neighborhoods of β_i and γ_j. Therefore, since $X \cdot Y = \sum_i \sum_j \beta_i \gamma_j 1_{E_i \cap E_j}$, we have $E\ (X \cdot Y) = \sum_i \sum_j \beta_i \gamma_j P(E_i \cap E_j)$. By independence, this is equal to $\sum_i \sum_j \beta_i \gamma_j P(E_i) \cdot P(E_j)$. Rearranging the sums yields $[\sum_i \beta_i P(E_i)] \cdot [\sum_j \gamma_j P(E_j)]$.

$\exists \mu, \sigma^2 \in \mathbb{R}$ *such that for all* $n, m \in \mathbb{N}$, $E\, X_n = E\, X_m = \mu$ *and* $\mathrm{Var}(X_n) = \mathrm{Var}(X_m)$ $= \sigma^2$, *then for all* $\epsilon > 0$,

$$\lim_n P(\{\omega : |\tfrac{1}{n} \sum_{i \leq n} X_n(\omega) - \mu| > \epsilon\}) = 0. \tag{7.6}$$

The random variable $S_n(\omega) = \frac{1}{n} \sum_{i \leq n} X_n(\omega)$ is the **sample average** when we have observed the values of $X_1(\omega)$ through $X_n(\omega)$. The WLLN says that when n is large, the probability that the sample average is very far from the true average of the random variables is very small. Put another way, $\alpha(S_n, \mu) \to 0$, where μ is the constant random variable taking the value μ and α is the Ky Fan metric for convergence in probability.

The proof uses an easy inequality.

Theorem 7.2.18 (Basic Inequality)   *For integrable* $X \geq 0$ *and* $r > 0$,

$$P(\{X \geq r\}) \leq \tfrac{E\, X}{r}. \tag{7.7}$$

Proof.   Since $0 \leq r \cdot 1_{\{X \geq r\}}(\omega) \leq X(\omega)$, $E\,(r \cdot 1_{\{X \geq r\}}) \leq E\, X$ and $E\,(r \cdot 1_{\{X \geq r\}})$ $= r\, P(\{X \geq r\})$.   ∎

Here are some implications:

1. For any integrable X, $|X| \geq 0$ is integrable, so that $P(\{|X| \geq r\}) \leq \frac{E\,|X|}{r}$, a special case of the following.

2. **Markov's inequality**, for $X \in L^p$ and $p \in [1, \infty)$, $P(\{|X| \geq r\}) \leq \frac{E\,|X|^p}{r^p}$, which holds because $|X| \geq r$ iff $|X|^p \geq r^p$.

3. For any X with finite variance, subtract $E\, X$ from X and apply Markov's inequality with $p = 2$ to get $P(\{|X - E\, X| \geq r\}) \leq \frac{\mathrm{Var}(X)}{r^2}$, which is **Chebyshev's inequality**.

Proof of Theorem 7.2.17 (WLLN).   Letting $S_n(\omega) = \frac{1}{n} \sum_{i \leq n} X_n$, we have $E\, S_n = \mu$ and $\mathrm{Var}(S_n) = \frac{\sigma^2}{n}$. The Chebyshev inequality yields $P(\{|S_n - \mu| > \epsilon\}) \leq \frac{\mathrm{Var}(S_n)}{\epsilon^2} = \frac{\sigma^2}{n \cdot \epsilon^2} \to 0$.   ∎

One way to understand the result is that S_n is very close to μ in the sense that the probability that S_n falls within an arbitrarily small δ-ball around μ is very close to 1 for large n. This means that if $g : \mathbb{R} \to \mathbb{R}$ is continuous and bounded, then one would guess that $g(S_n)$ should be very close to $g(\mu)$. In the following proof that this guess is correct, note that σ^2 is also a function of μ.

Theorem 7.2.19   *Suppose that* $S_n = S_n(\mu)$ *is a sequence of random variables with mean* μ *and variance* $\sigma_n^2(\mu)$, $\mu \in [a, b] \subset \mathbb{R}$; *that* $g : \mathbb{R} \to \mathbb{R}$ *is continuous and bounded; and that for each* μ, $\sigma_n^2(\mu) \to 0$. *Then for all* μ, $E\, g(S_n) \to g(\mu)$, *and if* $\sigma_n^2(\mu) \to 0$ *uniformly in an interval* $[a, b]$, *then* $\sup_{\mu \in [a,b]} |E\, g(S_n) - g(\mu)| \to 0$.

Before proving this, let us give an application.

Example 7.2.20 (Bernstein Polynomials)   *Let* $\{X_n(\mu) : n \in \mathbb{N}\}$ *be an independent collection of random variables with* $P(X_n = 1) = \mu = 1 - P(X_n = 0)$ *for some* $\mu \in [0, 1]$, *and let* $S_n(\mu) = \frac{1}{n} \sum_{i \leq n} X_n$, *so that* $\sigma_n^2(\mu) = \frac{\mu(1-\mu)}{n}$, *which* *converges uniformly to 0 for all* $\mu \in [0, 1]$. *Note that* $P(S_n(\mu) = k) = {}_nC_k\mu^k(1 - \mu)^{n-k}$.

Let g be a continuous function from $[0, 1]$ *to* $\mathbb{R}$. *Then*

$$f_n(\mu) := E\ g(S_n(\mu)) = \sum_{k=0}^{n} g(k/n)P(S_n(\mu) = k)$$

$$= \sum_{k=0}^{n} g(k/n)\,{}_nC_k\mu^k(1 - \mu)^{n-k} \qquad (7.8)$$

is a polynomial of order n. By Theorem 7.2.19, $f_n(\cdot)$ *converges uniformly to* $g(\cdot)$ *on* $[0, 1]$.

The Stone-Weierstrass approximation theorem tells us that some sequence of polynomials approaches g uniformly on $[0, 1]$; the Bernstein polynomials are an explicit sequence of polynomials that do the job.

Proof of Theorem 7.2.19.   For any random variable Y and $r \in \mathbb{R}$, $|Y(\omega) - r| \geq (Y(\omega) - r)$ and $|Y(\omega) - r| \geq (r - Y(\omega))$ for all ω. If Y is integrable, the monotonicity of the integral implies that

$$E\ |Y - r| \geq \max\{E\ (Y - r),\ E\ (r - Y)\} = |E\ (Y - r)|. \qquad (7.9)$$

We apply this with $Y = g(S_n)$ and $r = g(\mu)$.

Fix an arbitrary μ and an $\epsilon > 0$. We must show that $|E\ g(S_n) - g(\mu)| < 2\epsilon$ for large n. From (7.9),

$$|E\ g(S_n) - g(\mu)| \leq \int |g(S_n) - g(\mu)|\,dP. \qquad (7.10)$$

Pick $\delta > 0$ so that $|x - \mu| < \delta$ implies that $|g(x) - g(\mu)| < \epsilon$. We can break $\int |g(S_n) - g(\mu)|\,dP$ into two parts,

$$\int |g(S_n) - g(\mu)|\,dP = \int_{|S_n - \mu| < \delta} |g(S_n) - g(\mu)|\,dP$$

$$+ \int_{|S_n - \mu| \geq \delta} |g(S_n) - g(\mu)|\,dP. \qquad (7.11)$$

The first part is less than or equal to ϵ for all n. Since g is bounded, there exists an $M > 0$ such that $|g(S_n) - g(\mu)| \leq M$ for all ω. Therefore the second part is less than or equal to $\int_{|S_n - \mu| \geq \delta} M\,dP = M \cdot P(|S_n - \mu| \geq \delta)$. By Chebyshev's inequality, $P(|S_n - \mu| \geq \delta) \leq \frac{\sigma_n^2(\mu)}{\delta^2}$. For all n such that $\sigma_n^2(\mu) < M \cdot \epsilon\delta^2$, the second part is also less than ϵ.

For the uniform convergence part, note that g is uniformly continuous on a compact interval $[a, b]$. Therefore, δ can be chosen to work as above for all $\mu \in [a, b]$. Further, if $\sigma_n^2(\mu)$ goes to 0 uniformly, we can choose n as above. ∎

7.2.c Result 2: The Borel-Cantelli Lemma

Both directions of the proof of the SLLN rely on the Borel-Cantelli lemma, which is one of the most useful results for studying limits in probability theory. We use it to prove the completeness of $(L^0(\Omega, \mathcal{F}, P), \alpha)$ and to relate α-convergence (i.e., convergence in probability) to almost everywhere convergence, which in turn yields a version of the basic continuity result for expected utility theory.

7.2.c.1 Statement and Proof

Lemma 7.2.21 (Borel-Cantelli) *If* $\{A_n : n \in \mathbb{N}\} \subset \mathcal{F}$ *and* $\sum_n P(A_n) < \infty$, *then* $P([A_n \text{ i.o.}]) = 0$. *If* $\{A_n : n \in \mathbb{N}\} \subset \mathcal{F}$ *is independent and* $\sum_n P(A_n) = \infty$, *then* $P([A_n \text{ i.o.}]) = 1$.

Proof. $P([A_n \text{ i.o.}]) = P(\cap_m \cup_{n \geq m} A_n) \leq \sum_{n \geq m} P(A_n) \downarrow 0$ as $m \to \infty$.

Now suppose that $\{A_n : n \in \mathbb{N}\}$ is a sequence of independent events with $\sum_n P(A_n) = \infty$. We want to show that $P([A_n \text{ i.o.}]^c) = 0$, that is, that $P([A_n^c \text{ a.a.}]) = 0$. Since $[A_n^c \text{ a.a.}] \subset \cap_{n \geq m} A_n^c$ for all m, it is sufficient to show that $P(\cap_{n \geq m} A_n^c) = 0$ for all m. Since $(1 - x) \leq e^{-x}$, for all j,

$$P(\cap_{m \leq n \leq m+j} A_n^c) = \Pi_{n=m}^{m+j}(1 - P(A_n)) \leq \Pi_{n=m}^{m+j} e^{-P(A_n)} = e^{-\sum_{n=m}^{m+j} P(A_n)} \downarrow 0$$

as $j \to \infty$. From this, $P(\cap_{n \geq m} A_n^c) = \lim_j P(\cap_{m \leq n \leq m+j} A_n^c) = 0$. ∎

7.2.c.2 The Completeness of (L^0, α) and $(\mathcal{F}, \rho_P)$

Theorem 7.2.22 *The pseudometric space* (L^0, α) *is complete.*

Proof. Let X_n be a Cauchy sequence. Taking a subsequence if necessary, there is no loss in assuming that $\sum_n \alpha(X_n, X_{n+1}) < \infty$. Pick $\epsilon > 0$. There exists N such that $\sum_{n \geq N} \alpha(X_n, X_{n+1}) < \epsilon$. The Borel-Cantelli lemma implies that $P([\{|X_N - X_{N+n}| > \epsilon\} \text{ i.o.}]) = 0$. This in turn implies that $P(C) = 1$, where C is the convergence set for the sequence X_n. Define $X(\omega) = 1_C(\omega) \cdot \lim_n X_n(\omega)$. Since a subsequence of the Cauchy sequence converges, the whole sequence converges. ∎

Corollary 7.2.23 $(\mathcal{F}, \rho_P)$ *is complete.*

Proof. $\{1_A : A \in \mathcal{F}\}$ is a closed subset of (L^0, α), and $\rho_P(A, B) = \alpha(1_A, 1_B)$. ∎

Exercise 7.2.24 Prove Corollary 7.2.23 directly after showing the following.

1. $[[A_n \text{ i.o.}] \setminus [A_n \text{ a.a.}]] = [(A_n \triangle A_{n+1}) \text{ i.o.}]$.
2. If $\sum_n \rho_P(A_n, A_{n+1}) < \infty$, then $\rho_P([A_n \text{ i.o.}], [A_n \text{ a.a.}]) = 0$.
3. If $\sum_n \rho_P(A_n, A_{n+1}) < \infty$, then $\rho_P(A_n, A) \to 0$ for any A with $\rho_P(A, [A_n \text{ i.o.}]) = 0$.

7.2.c.3 Convergence in Probability and Convergence Almost Everywhere
In the proof of the completeness of L^0, the sequence X_n had the property that for all $\omega \in C$, $X_n(\omega) \to X(\omega)$ and $P(C) = 1$. This kind of convergence has the obvious name.

Definition 7.2.25 *A sequence X_n in $L^0(\Omega, \mathcal{F}, P)$ **converges almost everywhere (a.e.) to** $X \in L^0$ if $P(\{\omega : X_n(\omega) \to X(\omega)\}) = 1$, written $X_n \to_{a.e.} X$ or $X_n \to X$ a.e.*

Notation Alert 7.2.A *There is an immense difference between "almost always," which refers to the behavior of all elements of a sequence that have a sufficiently large index, and "almost everywhere," which refers to a set of ω. Since we are often interested in the convergence of sequences of random variables, the two phrases can occur in the same sentence.*

Theorem 7.2.26 *If $X_n \to X$ a.e., then $\alpha(X_n, X) \to 0$.*

Proof. Since $X_n \to X$ a.e., for any $\epsilon > 0$, $P(|X_n - X| \geq \epsilon) \to 0$. Therefore, $P(|X_n - X| \geq \epsilon) < \epsilon$ for almost all n, that is, $\alpha(X_n, X) < \epsilon$ for almost all n. ∎

The converse is not true; converging in probability does not mean converging almost everywhere.

Example 7.2.27 *Take $X_n = 1_{E_n}$, where the sequence E_n is given by*

$$\{E_1, E_2, E_3, \ldots\} = \{(0, 1], (0, \tfrac{1}{2}], (\tfrac{1}{2}, 1], (0, \tfrac{1}{3}], (\tfrac{1}{3}, \tfrac{2}{3}], (\tfrac{2}{3}, 1], (0, \tfrac{1}{4}], \ldots\}$$

in our favorite probability space, $((0, 1], \mathcal{B}, \lambda)$. It is clear that $\alpha(X_n, 0) \to 0$. However, for each and every $\omega \in \Omega$, $\liminf X_n(\omega) = 0 < \limsup X_n(\omega) = 1$— the set on which $X_n(\omega) \to 0$ is the empty set. However, one could consider the subsequence $1_{(0,1]}, 1_{(0,\frac{1}{2}]}, 1_{(0,\frac{1}{3}]}, \ldots$, which is converging a.e. to 0.

The following is a general version of the device just seen.

Theorem 7.2.28 *If $\alpha(X_n, X) \to 0$, then there exists a subsequence $X_{k(n)}$ such that $X_{k(n)}$ converges to X a.e.*

Proof. Take a subsequence $k(n)$ such that $\sum_n \alpha(X_{k(n)}, X) < \infty$. For any $\epsilon > 0$, Borel-Cantelli tells us that $P([\{|X_{k(n)} - X| > \epsilon\} \text{ a.a.}]) = 0$. ∎

Almost everywhere convergence only has a metric if P is purely atomic.

Exercise 7.2.29 If P is purely atomic in the probability space $(\Omega, \mathcal{F}, P)$, then $X_n \to X$ a.e. iff $\alpha(X_n, X) \to 0$. [Once $\alpha(X_n, X) < 1/k$, X_n must be within $1/k$ of X on all of the atoms with $P(E) \geq 1/k$.]

Recall that $r_n \to 0$ iff for every subsequence, $r_{n'}$, there is a further subsequence, $r_{n''}$, such that $r_{n''} \to 0$.

Lemma 7.2.30 *There is no metric d on any $L^p((0, 1], \mathcal{B}, \lambda)$, $p = 0$ or $p \in [1, \infty]$ with the property that $d(X_n, X) \to 0$ iff $X_n \to X$ a.e.*

Proof. Suppose there were such a metric. Example 7.2.27 gives a sequence X_n in $L^\infty \subset L^p$, $p = 0$ or $p \in [1, \infty]$ with $r_n := d(X_n, X) \not\to 0$. However, each subsequence, $X_{n'}$, has a further subsequence, $X_{n''}$, with the property that $d(X_{n_k}, X) \to 0$. ∎

7.2.c.4 Expected Utility Theory

The basic model in the theory of choice under uncertainty associates a random reward, modeled as a random variable $X_a \in L^0$ with every action, $a \in A$, that an individual might take. A consumer's preferences over L^0 are assumed to be representable by a continuous utility function $U : L^0 \to \mathbb{R}$, which means that the individual is modeled as solving the problem

$$\max_{a \in A} U(X_a). \tag{7.12}$$

If the set $\{X_a : a \in A\}$ is compact, there is always an optimum.[3]

The most commonly used type of preference is representable by what is called an **expected utility function** and the associated choice theory is known as **expected utility theory**. In this theory, the utility of X_a is $U(X_a) = E u(X_a)$, where $u \in C_b(\mathbb{R})$. Borel-Cantelli yields the basic continuity result in expected utility theory.

Theorem 7.2.31 *For $u \in C_b(\mathbb{R})$, $X \mapsto E u(X)$ is continuous.*

Proof. Let $B = \|u\|_\infty$ and let X_n be a sequence in L^0 converging in probability to X. We must show that $E u(X_n) \to E u(X)$. Take any subsequence, $X_{n'}$. By what we have just shown (Theorem 7.2.28), there is a further subsequence, $X_{n''}$, such that $X_{n''} \to X$ a.e. Define $Y_{n''} = u(X_{n''})$ and $Y = u(X)$. Each $Y_{n''}$ satisfies $|Y_{n''}| \leq B$ and $Y_{n''} \to Y$ a.e. Therefore, by dominated convergence, $E Y_{n''} \to E Y$. ∎

7.2.d Result 3: The Strong Law of Large Numbers

Knowing that $\lim_n P(\{\omega : |S_n(\omega) - \mu| > \epsilon\}) = 0$ does not tell us that $P(\{\omega : S_n(\omega) \to \mu\}) = 1$. In principle, it could be the case that there is a set $E \in \mathcal{F}$, $P(E) > 0$ and an $\epsilon > 0$ such that for all $\omega \in E$, $|S_n(\omega) - \mu| > \epsilon$ for infinitely many n. Let $N(\omega, \epsilon) \subset \mathbb{N}$ be the set of n such that $|S_n(\omega) - \mu| > \epsilon$. The observation is that one could have $N(\omega, \epsilon)$ being infinite for all $\omega \in E$ by arranging for $N(\omega, \epsilon)$ and $N(\omega', \epsilon)$ to be eventually disjoint. Note that $P(\{\omega : S_n(\omega) \to \mu\}) = 1$ iff for all $\epsilon > 0$, $P(\{\omega : N(\omega, \epsilon) \text{ is finite }\}) = 1$.

Saying the following phrase as "eye-eye-dee" should help you feel like a statistician.

Notation 7.2.32 *We say that $\{X_n : n \in \mathbb{N}\} \subset L^0(\Omega, \mathcal{F}, P)$ is an **independent and identically distributed** sequence, or simply **iid**, if the collection $\{X_n : n \in \mathbb{N}\}$ is independent and if for all $A \in \mathcal{B}$ and for all n, m, $P(X_n \in A) = P(X_m \in A)$.*

Theorem 7.2.33 (Strong Law of Large Numbers [SLLN]) *If $\{X_n : n \in \mathbb{N}\} \subset L^1(\Omega, \mathcal{F}, P)$ is an iid sequence with $E X_n = \mu$, then $P(\{\omega : S_n(\omega) \to \mu\}) = 1$. If $\{X_n : n \in \mathbb{N}\}$ is an iid sequence with $E |X_n| = \infty$, then $P(\{\omega : S_n(\omega) \text{ converges }\}) = 0$.*

We prove a weaker version of the first part, substituting $L^4(\Omega, \mathcal{F}, P)$ for $L^1(\Omega, \mathcal{F}, P)$ in the first, and more important, part.[4] It is important to understand that it is *not* the case that for all ω, $S_n(\omega) \to \mu$.

---

3. Theorem 8.12.14 (p. 545) gives the compactness criterion for subsets of L^0.

4. For a sketch of a complete proof see Exercise 7.2.60. For other good proofs see, for example, Dudley's theorem 8.3.5, or Feller's theorem VII.8.1 and 2.

Example 7.2.34 *For* $(\Omega, \mathcal{F}, P) = ((0, 1], \mathcal{B}, \lambda)$, $k = 0, \ldots, 2^n - 1$, *let* $Y_{k,n}(\omega) = 1_{(k/2^n, (k+1)/2^n]}(\omega)$ *and* $X_n(\omega) = \sum_k$ *even* $Y_{k,n}(\omega) - \sum_k$ *odd* $Y_{k,n}(\omega)$. $\{X_n : n \in \mathbb{N}\}$ *is an iid sequence with* $\lambda(\{X_n = -1\}) = \lambda(\{X_n = +1\}) = \frac{1}{2}$, *so that* $E\, X_n = 0$. *Theorem 7.2.33 implies that* $\lambda(\{\omega : S_n(\omega) \to 0\}) = 1$. *However, if* $\omega = k/2^n$, *then for all* $m \ge n$, $X_m(\omega) = +1$, *so that* $S_n(\omega) \to 1$. *If we look at* $(0, 1]$ *as a metric space, there is a dense set of exceptions to* $S_n(\omega)$ *converging. There is more: it can be shown that there are uncountably many* ω *such that* $S_n(\omega) \not\to 0$. *[To see that there are uncountably many exceptions, take a typical* $\omega = 0.\omega_1\omega_2\omega_3 \ldots$, *where the* $\omega_n \in \{0, 1\}$ *is the nth element of the binary (or dyadic) expansion and* $\lim_n \frac{1}{n}\#\{i \le n : \omega_i = 1\} = \frac{1}{2}$. *Consider* $\omega' = 0.0\omega_1 0\omega_2 0\omega_3 \ldots$. *for which* $\lim_n \frac{1}{n}\#\{i \le n : \omega_i' = 1\} = \frac{1}{4}$. *For denseness, start the process of adding 0's into every other spot after* ω_N *and consider the union over N of these sets.]*

Here is the version of the strong law that we prove here.

Theorem 7.2.35 (Strong Law of Large Numbers [SLLN]) *If* $\{X_n : n \in \mathbb{N}\} \subset L^4(\Omega, \mathcal{F}, P)$ *is an independent sequence with* $E\, X_n = \mu$ *and* $E\, |X_n|^4$ *bounded, then* $P(\{\omega : S_n(\omega) \to \mu\}) = 1$. *If* $\{X_n : n \in \mathbb{N}\}$ *is iid and* $E\, |X_n| = \infty$, *then* $P(\{\omega : S_n(\omega) \text{ converges }\}) = 0$.

Exercise 7.2.36 For $r > 0$, let $E_r = [\{|S_n - \mu| < r\} \text{ a.a.}]$. Show the following:

1. For $0 < r < s$, $E_r \subset E_s$.

2. The *uncountable* intersection, $\cap_{r>0} E_r$, is equal to the countable intersection $\cap_{n \in \mathbb{N}} E_{1/n}$.

3. $\cap_{r>0} E_r = \{S_n \to \mu\}$.

4. If $P([\{|S_n - \mu| \ge r\} \text{ i.o.}]) = 0$, then $P(\{\omega : S_n(\omega) \to \mu\}) = 1$.

Proof of Theorem 7.2.35. Let B satisfy $E\, |X_n - \mu|^4 \le B$ for all n and let $r > 0$. From Markov's inequality,

$$P(|S_n - \mu| \ge r) \le p_n := \frac{E\, |S_n - \mu|^4}{r^4}. \tag{7.13}$$

We show that $\sum_n p_n < \infty$, which implies that $P([\{|S_n - \mu| \ge r\} \text{ i.o.}]) = 0$. By the previous exercise, this completes the first half of the proof.

Since $S_n - \mu = \frac{1}{n}\sum_{i \le n}(X_i - \mu)$, $E\,(S_n - \mu)^4 = \frac{1}{n^4} E\,(\sum_{i \le n}(X_i - \mu))^4$. Let $Y_i = X_i - \mu$. We have to bound $E\, \sum_{i,j,k,m \le n} Y_i Y_j Y_k Y_m$. Since the expectation of the product of independent random variables is the product of the expectations, Lemma 7.2.15, $E\, Y_i Y_j Y_k Y_m = 0$ unless all of the indexes are equal or two of the indexes are equal to one number and the other two to another. There are n of the first type of term and $n(n-1)$ of the second type, so that $E\,(\sum_{i \le n}(X_i - \mu))^4 \le n^2 \cdot B$. Combining yields $p_n \le \frac{B}{r^4}\frac{1}{n^2}$ and $\sum_n \frac{B}{r^4}\frac{1}{n^2} < \infty$. By Borel-Cantelli, $P([\{|S_n - \mu| \ge r\} \text{ i.o.}]) = 0$, completing the first part of the proof.

Now suppose that X_n is iid with $E\, |X_n| = \infty$ for all n. We

1. show that $\frac{1}{n}\sum_{i \le n} X_i(\omega) \to r$ implies that $\frac{1}{n}X_n(\omega) \to 0$,

2. show that $P([|X_n| > n \text{ i.o.}]) = 1$, and

3. observe that if $|X_n|(\omega) > n$ for infinitely many n, then $\frac{1}{n}X_n(\omega)$ is not a Cauchy sequence.

(1) Note that $\frac{1}{n} \sum_{i \leq n} X_i(\omega) \to r$ iff $\frac{1}{n+1} \sum_{i \leq n} X_i(\omega) \to r$ iff $\frac{1}{n} \sum_{i \leq (n-1)} X_i(\omega)$ $\to r$. Therefore, if $\frac{1}{n} \sum_{i \leq n} X_i(\omega)$ is convergent, then

$$\frac{1}{n}[(\sum_{i \leq n} X_i(\omega)) - (\sum_{i \leq (n-1)} X_i(\omega))] = \frac{1}{n} X_n(\omega) \to 0.$$

(2) Let $A_n = \{|X_n| > n\}$. This is a sequence of independent events, and we show that $\sum_n P(A_n) = \infty$, so that $P[A_n \text{ i.o.}] = 1$. We first show that for any $Y \geq 0$ with $E\,Y = \infty$, $\sum_n P(\{Y > n\}) = \infty$. Let $Y_M = Y \wedge M$, that is, $Y_M(\omega) = \min\{Y(\omega), M\}$, so that $E\,Y_M \uparrow \infty$. For all ω, $Y_M(\omega) \leq \sum_{n=0}^{M} 1_{\{Y>n\}}(\omega)$, so $E\,Y_M \leq E \sum_{n=0}^{M} 1_{\{Y>n\}}(\omega) = \sum_{n=0}^{M} P(\{Y > n\})$.
Taking $Y = |X_n|$ completes the proof.   ■

Some Commentary: This result is sufficiently central to probability, statistics, and economics that it is worth looking more carefully at the proof.

1. The proof could have picked a sequence r_n going to 0 sufficiently slowly that $\sum_n \frac{B}{r_n^4} \frac{1}{n^2} < \infty$. For example, if we had used $r_n = \frac{1}{n^{1/4}}$ or $r_n = \frac{1}{\sqrt{n}}$, the proof would not have worked because $\sum_n 1 = \sum_n \frac{1}{n} = \infty$. From the Gaussian part of the central limit theorem (from your introductory course in probability and statistics), the distribution of $\frac{1}{\sqrt{n}} \sum_{i \leq n}(X_i - \mu)$ should be approximately Gaussian. The law of the iterated logarithm shows that for the sequence $r_n = 1/\sqrt{2n \log(\log(n))}$, $P(\limsup_n |\frac{1}{n} \sum_{i \leq n}(X_i - \mu)|/r_n = 1) = 1$ if the X_i are iid and have variance 1. The function $\sqrt{2n \log(\log(n))}$ is only defined for n for which $\log(\log(n)) > 0$; to fix this, we could redefine log as $\max\{1, \log\}$ or else not start paying attention until n is large enough. Either way, the function $\log(\log(n))$ does go to infinity, but at a very slow crawl.

2. We used the assumption that the X_n belong to L^4. Somewhat finicky truncation arguments are needed to make the argument with iid X_n belonging to L^1. One replaces X_n with $Y_n = n1_{\{X_n \geq n\}} - n1_{\{X_n \leq -n\}} + (X_n \vee -n) \wedge n$ and shows that the result is true for the Y_n (which is not easy) and that, in the limit, there is no difference between using Y_n or X_n (which is not too hard). In Exercise 7.2.60 (p. 386), we give a sketch of the details.

3. The X_n need not have exactly the same mean. This can be seen by a process called "centering." Suppose that $E\,X_n = \mu_n$ and consider the centered random variables $Y_n = X_n - \mu_n$. If $E\,|Y_n|^4$ is bounded, then the given proof shows that $|\frac{1}{n} \sum_{i \leq n} X_n - \frac{1}{n} \sum_{i \leq n} \mu_i| \to 0$ with probability 1. This is mostly useful if $\frac{1}{n} \sum_{i \leq n} \mu_i$ is itself convergent.

4. In the converse to the main result, we showed that when the iid X_n satisfy $E\,|X_n| = \infty$, the probability that the sample average is convergent is equal to 0. A bounded sequence can fail to be convergent, but this does not fully capture how badly behaved the sample average is in this case. With a fairly simple modification of the argument, one can go further and show that the increments between S_n and S_{n-1} are, with probability 1, arbitrarily large infinitely often. The argument runs as follows: for any $a > 0$ and $Y \geq 0$ with $E\,Y = \infty$, $E\,Y/a = \infty$. Therefore, using the argument given, for any

$a > 0$, $P([\{|X_n|/a > n\} \text{ i.o.}]) = 1$, equivalently $P([\{|X_n| > an\} \text{ i.o.}]) = 1$. A sequence of numbers, y_n, with arbitrarily large increments must satisfy $\lim\sup_n |y_n| = \infty$. Hence, $P(\lim\sup_n |\frac{1}{n}\sum_{i\leq n} X_n| = \infty) = 1$; that is, with probability 1, the absolute value of the average becomes arbitrarily large i.o. rather than converging.

7.2.e Result 4: Kolmogorov's 0-1 Law

The SLLN concerns $\lim_n \frac{1}{n}\sum_{i\leq n} X_n$. The limit behavior of a sequence of numbers depends only on "tail" behavior. In this context, the word "tail" comes from the observation that for any M,

$$\lim_n \frac{1}{n}\sum_{i\leq n} X_n = \lim_n \frac{1}{n}\sum_{M\leq i\leq n} X_n. \tag{7.14}$$

The same is true with $\lim\inf_n$ and $\lim\sup_n$ replacing $\lim_n$. If we take $X_n = 1_{A_n}$, $[A_n \text{ i.o.}]$ and $[A_n \text{ a.a.}]$ are tail events in the same sense.

We often use limit theorems to tell us about the behavior of "really big" things. In this context, it would be good to know what we can tell about the behavior of "really big" numbers of independent random variables. Kolmogorov's 0-1 law tells us that the answer is that limit behavior must be deterministic. Here are some examples where we have already seen a 0-1 law at work.

Example 7.2.37 *If $A = \{\lim\inf_n \frac{1}{n} X_n = \lim\sup_n \frac{1}{n} X_n\}$ for an iid sequence of random variables X_n, then the SLLN tells us that $P(A)$ is equal to 1 or 0 depending on whether the X_n have finite expectation.*

If A_n is a sequence of independent events, Borel-Cantelli tells us that whether $P[A_n \text{ i.o.}] = 0$ or $P[A_n \text{ i.o.}] = 1$ depends on whether $\sum_n P(A_n) < \infty$ or $\sum_n P(A_n) = \infty$.

To say what tail events are, we have to understand the basics of sub-σ-fields.

7.2.e.1 Sub-σ-Fields

If $(\Omega, \mathcal{F})$ is a measure space and $\mathcal{G} \subset \mathcal{F}$ is a σ-field of sets, then $\mathcal{G}$ is a **sub-σ-field** **(of $\mathcal{F}$)**. The important differences between sub-σ-fields $\mathcal{G}$ and $\mathcal{H}$ come from the differences between the $\mathcal{G}$-measurable functions and the $\mathcal{H}$-measurable functions, that is, between $L^0(\Omega, \mathcal{G}, P)$ and $L^0(\Omega, \mathcal{H}, P)$.

Here is a sketch of how sub-σ-fields are used to model information: utility depends on $\omega \in \Omega$ and on the action $a \in A$ that a person chooses, $u = u(\omega, a)$; we interpret a person's information being contained in $\mathcal{G}$ as the statement that he can choose $a \in A$ as a $\mathcal{G}$-measurable function, $f(\omega)$; her expected utility if she chooses $f(\cdot)$ is then $\int_\Omega u(\omega, f(\omega)) \, dP(\omega)$; and he chooses $f(\cdot)$ so as to maximize expected utility. The larger $\mathcal{G}$, the more measurable functions the person can choose between, hence the (weakly at least) higher her expected utility.

Definition 7.2.38 *$\mathcal{G} \subset \mathcal{F}$ is a **sub-σ-field** (of $\mathcal{F}$) if $\mathcal{G}$ is itself a σ-field. A random variable $X : \Omega \to M$, (M, d) a metric space, is $\mathcal{G}$-**measurable** if $X^{-1}(\mathcal{B}_M) \subset \mathcal{G}$. The smallest σ-field making X measurable is denoted $\sigma(X)$ and defined by $X^{-1}(\mathcal{B}_M)$.*

As is so often true, the finite case is the easiest to understand.

Example 7.2.39 *Let $\Omega = \{(\omega_1, \omega_2) : \omega_i \in \{1, 2, 3, 4, 5, 6\}\}$. This is, for example, the set of possible outcomes of rolling two dice. Let $\mathcal{F} = \mathcal{P}(\Omega)$ and let $P(\omega) = 1/36$ for each $\omega \in \Omega$.*

Let $\mathcal{G}_1 = \{A \times \{1, 2, 3, 4, 5, 6\} : A \subset \{1, 2, 3, 4, 5, 6\}\}$. $\mathcal{G}_1$ is a sub-σ-field. A random variable $X : \Omega \to M$, M a metric space, depends only on ω_1 if for all $\omega_1, \omega_2, \omega_2'$, $X(\omega_1, \omega_2) = X(\omega_1, \omega_2')$. For any such X and any measurable $B \subset M$, $X^{-1}(G) \in \mathcal{G}_1$. $\mathcal{G}_2$ is defined analogously.

For $k = 2, 3, \ldots, 12$, let $E_k = \{(\omega_1, \omega_2) \in \Omega : \omega_1 + \omega_2 = k\}$. E_k is the event that the sum of the roll of two dice is equal to k. Let $\mathcal{H} = \{\cup_{k \in S} E_k : S \subset \{2, 3, \ldots, 12\}\}$. $\mathcal{H}$ is another sub-σ-field, and in this case, it is the one that "depends" only on the sum of the two dice. More specifically, suppose that for some k and for all $\omega, \omega' \in E_k$, $X(\omega) = X(\omega')$. For any such X and any open $G \subset M$, $X^{-1}(G) \in \mathcal{H}$.

Exercise 7.2.40 Draw several of the smallest, nonempty elements of the sub-σ-fields in Example 7.2.39. Also, define $S(\omega_1, \omega_2) = \omega_1 + \omega_2$ and $T(\omega_1, \omega_2) = \omega_1 + r \cdot \omega_2$, r irrational. Give $\sigma(S)$ and $\sigma(T)$.

Exercise 7.2.41 Let $\mathcal{F}$ be a (σ-)field of subsets of a **finite** Ω and let $\{E_s : s \in S\}$ be the partition of Ω into the smallest nonempty elements of $\mathcal{F}$. Show that $\mathcal{F} = \sigma(\{E_s : s \in S\})$.

In the case of an infinite Ω, partitions no longer characterize σ-fields. This is a shame.

Example 7.2.42 *In $((0, 1], \mathcal{B}, \lambda)$, the sub-$\sigma$-field $\mathcal{G} := \sigma\{\{\omega\} : \omega \in (0, 1]\}$ is generated by the finest possible partition of $(0, 1]$, which consists of the class of sets that are either countable or have a countable complement. It is fairly useless since P has one atom in $((0, 1], \mathcal{G}, \lambda)$, that is, every $X \in L^0((0, 1], \mathcal{G}, \lambda)$ is almost everywhere constant. To see why, note that since $\{X \leq r\} \in \mathcal{G}$, it is either a countable set, so that it has probability 0, or its complement is countable, so that it has probability 1. Thus, for all r, the cdf of X, $F_X(r) = P(\{X \leq r\})$ is equal to either 0 or 1.*

The set of random variables that are functions of X is the central concept in regression analysis. The key result, due to Doob, is that Y is $\sigma(X)$-measurable iff $Y = f(X)$ for some measurable function f. This means that regression analysis is about finding a function that is $\sigma(X)$-measurable. The proofs of the following two results use the finiteness of Ω in crucial ways. Doob's theorem (shown later) is about the infinite case.

Lemma 7.2.43 *If Ω is **finite**, then for random variables X, Y, Y is $\sigma(X)$-measurable iff $[X(\omega) = X(\omega')] \Rightarrow [Y(\omega) = Y(\omega')]$.*

Proof. $X(\omega) = X(\omega')$ iff $\omega, \omega' \in E_s$ for some E_s in the partition generated by $\sigma(X)$. Y is measurable iff it is constant on the E_s. ∎

Lemma 7.2.44 *If Ω is **finite** and both $X : \Omega \to \mathbb{R}$ and $Y : \Omega \to \mathbb{R}$ are random variables, then Y is $\sigma(X)$-measurable iff there exists a continuous function $f : \mathbb{R} \to \mathbb{R}$ such that $Y(\omega) = f(X(\omega))$ for all $\omega \in \Omega$.*

Proof. If $Y(\omega) = f(X(\omega))$, then Y is constant when X is constant.

If Y is $\sigma(X)$ measurable, then for every x in the range of X, $X^{-1}(x)$ is an element of the partition generated by $\sigma(X)$. For each such x, pick $\omega \in X^{-1}(s)$ and define $f(x) = Y(\omega)$. This determines the value of f at finitely many points, and it can be set arbitrarily, for example, by linear interpolation, at all other points. ∎

The case of an infinite Ω requires a much more complicated proof.

Theorem 7.2.45 (Doob) *If $X : \Omega \to M$, then $Y : \Omega \to \mathbb{R}$ is $\sigma(X)$-measurable iff there exists a measurable function $f : M \to \mathbb{R}$ such that $Y(\omega) = f(X(\omega))$ for all $\omega \in \Omega$.*

Proof. First suppose that $Y(\omega) = f(X(\omega))$ for all $\omega \in \Omega$ and some measurable $f : M \to \mathbb{R}$. By measurability, $f^{-1}(\mathcal{B}) \subset \mathcal{B}_M$ and $X^{-1}(\mathcal{B}_M) \subset \mathcal{F}$, so that Y is $\sigma(X)$-measurable.

Now suppose that Y is $\sigma(X)$-measurable and define $E_{k,n} = Y^{-1}([\frac{k}{2^n}, \frac{k+1}{2^n}))$ and $Y_n = \sum_{|k| \leq n \cdot 2^n} \frac{k}{2^n} 1_{E_{k,n}}$. Since Y is $\sigma(X)$-measurable, $E_{k,n} \in \sigma(X)$, so Y_n is $\sigma(X)$-measurable. For each n, the collection $\{E_{k,n} : k \in \mathbb{Z}\}$ is disjoint, so that each ω lies in at most one of the $E_{k,n}$.

Since $\sigma(X) = X^{-1}(\mathcal{B}_M)$, there is at least one $B_{k,n} \in \mathcal{B}_M$ such that $E_{k,n} = X^{-1}(B_{k,n})$. Note that the $B_{k,n}$ need not be disjoint.[5] Define $f_n : M \to \mathbb{R}$ by $f_n(x) = \sum_{|k| \leq n \cdot 2^n} \frac{k}{2^n} 1_{B_{k,n}}(x)$. Since the $E_{k,n}$ are disjoint, for each ω, $f_n(X(\omega)) := \sum_{k \in \mathbb{Z}} \frac{k}{2^n} 1_{B_{k,n}}(X(\omega)) = Y_n(\omega)$. This means that we have established that for each ω,

$$Y(\omega) = \lim_n Y_n(\omega) = \lim_n f_n(X(\omega)). \tag{7.15}$$

All that is left to show is that we can find a limit function f on the range of $X(\omega)$. The convergence set $C = \{x \in M : f_n(x) \text{ converges}\}$ belongs to $\mathcal{B}_M$ because

$$C = \cap_m \cup_M \cap_{n,m \geq M} \{d(f_n, f_m) < 1/m\}. \tag{7.16}$$

Therefore, the function $f(x) = 1_C(x) \cdot \lim_n f_n(x)$ is measurable. From (7.15), for all ω, $X(\omega) \in C$, so that for all ω, $Y(\omega) = f(X(\omega))$. ∎

7.2.e.2 Measurability of Vector–Valued Random Variables
We often care about $\sigma(X)$ when X takes values in $\mathbb{R}^\ell$, that is, $X(\omega) = (X_1(\omega), \ldots, X_\ell(\omega))$, or takes values in $\mathbb{R}^\mathbb{N}$, $X(\omega) = (X_1(\omega), X_2(\omega), \ldots)$. To interpret the following, recall that every metric space, by assumption, is given its Borel σ-field.

Theorem 7.2.46 *The $\mathbb{R}$-valued mappings $\{X_i : i = 1, \ldots, \ell\}$ are measurable iff the mapping $(X_1, \ldots, X_\ell) : \Omega \to \mathbb{R}^\ell$ is measurable.*

Proof. Suppose that each of the $\mathbb{R}$-valued mappings $\{X_i : i = 1, \ldots, \ell\}$ is measurable. For any collection of open sets, G_i, $i = 1, \ldots, \ell$, $E = \cap_{i=1}^\ell \{X_i \in G_i\}$ is measurable. Now, $E = \{(X_1, \ldots, X_\ell) \in \times_{i=1}^\ell G_i\}$. Since every open subset of $\mathbb{R}^n$ is a countable union of open rectangles, the mapping $X(\omega) = (X_1(\omega), \ldots, X_\ell(\omega))$

---

5. For example, take $X(\omega) = 1_{(\frac{1}{2}, 1]}(\omega)$, $X^{-1}(\{0, 2\}) = (0, \frac{1}{2}]$ and $X^{-1}(\{1, 2\}) = (\frac{1}{2}, 1]$, and the sets $\{0, 2\}$ and $\{1, 2\}$ are not disjoint.

is measurable. Suppose now that $(X_1, \ldots, X_\ell) : \Omega \to \mathbb{R}$ is measurable. Each X_i is the composition of the continuous mapping proj_i and $(X_1, \ldots, X_\ell)$, hence is measurable. ∎

From Doob's theorem, this means that Y is $\sigma(X_1, \ldots, X_\ell)$-measurable iff there is a measurable $f : \mathbb{R}^\ell \to \mathbb{R}$ such that $Y = f(X_1, \ldots, X_\ell)$.

Corollary 7.2.47 *The $\mathbb{R}$-valued mappings $\{X_n : n \in \mathbb{N}\}$ are measurable iff $X(\omega) := (X_1(\omega), X_2(\omega), \ldots)$ from Ω to $\times_{n \in \mathbb{N}} \mathbb{R}$ is measurable.*

Proof. Suppose that the $\mathbb{R}$-valued mappings $\{X_n : n \in \mathbb{N}\}$ are measurable. A subbasis for the topology on $\times_{n \in \mathbb{N}} \mathbb{R}$ is given by $\text{proj}_{1,\ldots,\ell}^{-1}(G)$ for G an open subset of $\mathbb{R}^\ell$. By the previous result, this is measurable. Since $\times_{n \in \mathbb{N}} \mathbb{R}$ is separable, this is sufficient. If X is measurable, then $X_i = \text{proj}_i(X)$ is measurable. ∎

From Doob's theorem, this means that Y is $\sigma(X_1, X_2, \ldots)$-measurable iff there is a measurable $f : \mathbb{R}^\mathbb{N} \to \mathbb{R}$ such that $Y = f(X_1, X_2, \ldots)$. The next set of results uses this for sequences of random variables of the form $X_M, X_{M+1}, \ldots$.

7.2.e.3 Tail σ-Fields

Let X_n be a sequence of random variables. Define $\mathcal{H}_n = \sigma(X_1, \ldots, X_n) \subset \mathcal{F}$ as the smallest σ-field making all of the random variables $X_1, \ldots, X_n$ measurable. In other words, $\mathcal{H}_n$ is the small σ-field containing sets of form $X_i^{-1}((-\infty, r])$, $r \in \mathbb{R}$, $1 \le i \le n$. Doob's theorem shows that events in $\mathcal{H}_n$ are the ones that can be expressed in terms of measurable functions of the random vector $(X_1, \ldots, X_n)$.

In a similar fashion, define $\mathcal{H}_{M+} = \sigma(X_M, X_{M+1}, \ldots) \subset \mathcal{F}$ as the smallest σ-field making all of the random variables $X_M, X_{M+1}, \ldots$ measurable.

Definition 7.2.48 *The **tail σ-field** for a sequence of random variables X_n is $\mathcal{H}_{tail} = \cap_M \mathcal{H}_{M+}$.*

Lemma 7.2.49 *If A_n is a sequence of events and $X_n = 1_{A_n}$, then $[A_n \text{ i.o.}] \in \mathcal{H}_{tail}$, $[A_n \text{ a.a.}] \in \mathcal{H}_{tail}$, and the event $A = \{\liminf_n \frac{1}{n} X_n = \limsup_n \frac{1}{n} X_n\} \in \mathcal{H}_{tail}$.*

Proof. $[A_n \text{ i.o.}] \in \mathcal{H}_{tail}$ iff for all M, $[A_n \text{ i.o.}] \in \mathcal{H}_{M+}$. Since $\cup_{n \ge N} A_n \downarrow [A_n \text{ i.o.}]$, for all M, $[A_n \text{ i.o.}] = \cap_N \cup_{n \ge N} A_n = \cap_{N \ge M} \cup_{n \ge N} A_n \in \mathcal{H}_{M+}$. A similar argument covers $[A_n \text{ a.a.}]$.

For any $r \in \mathbb{R}$ and $\omega \in \Omega$, $\liminf_n \frac{1}{n} X_n(\omega) < r$ iff for all M, $\inf_{n \ge M} \frac{1}{n} X_n(\omega) < r$. Since $\{\inf_{n \ge M} \frac{1}{n} X_n(\omega) < r\} \in \mathcal{H}_{M+}$, this means that $\{\liminf_n \frac{1}{n} X_n(\omega) < r\} \in \mathcal{H}_{M+}$ for all M, which implies that $R := \liminf_n \frac{1}{n} X_n$ is $\mathcal{H}_{tail}$-measurable. By similar arguments, so is $S = \limsup_n \frac{1}{n} X_n$. A is the measurable event $\{R - S = 0\}$. ∎

When the X_n are independent, the tail σ-field is pretty trivial.

Theorem 7.2.50 (Kolmogorov) *If X_n is an independent sequence and $A \in \mathcal{H}_{tail}$, then $P(A) = 0$ or $P(A) = 1$.*

We give a proof of this later (p. 397) after we have proved the requisite measurability results.

7.2.e.4 Two Examples

Perhaps a more intuitive way to express what is involved with sequences of independent random variables is that the influence of early events disappears from limit calculations. The following shows just how much this result depends on independence.

Definition 7.2.51 *X is a **Bernoulli random variable with parameter** r, written $X \sim Bern(r)$, if $P(X = 1) = r$ and $P(X = 0) = 1 - r$.*

Example 7.2.52 *For $1 > p > q > 0$, define a sequence of Bernoulli random variables with the following properties: $X_0 \sim Bern(\frac{1}{2})$; if $X_0 = 1$, then $X_1, X_2, \ldots$ are iid $Bern(p)$; and if $X_0 = 0$, then $X_1, X_2, \ldots$ are iid $Bern(q)$. We do not observe X_0, but we do observe the X_n, $n \geq 1$. The SLLN implies that $\frac{1}{n} \sum_{i \leq n} X_i \to p$ if $X_0 = 1$ and $\to q$ if $X_0 = 0$. This means that $\lim_n \frac{1}{n} \sum_{i \leq n} X_i$ is a random variable, being half of the time equal to p and half of the time equal to q.*

In Example 7.2.52, P has two atoms in $\mathcal{H}_{tail}$, the sets $\{X_0 = 0\}$ and $\{X_0 = 1\}$, and the influence of these early events never fades.

In §4.11.b (p. 155 *et seq.*), we briefly introduced finite Markov chains. We showed that if the transition matrix P has the property that $P^n \gg 0$ for some n, then the influence of early events washes out, and the distribution of X_t converges to the unique ergodic distribution. We will see that such Markov chains also have a trivial tail σ-field. An interesting class of Markov chains with nontrivial tail σ-fields is the following.

Example 7.2.53 (Gambler's Ruin) *The state space is the set $S = \{0, 1, \ldots, M\}$. The transition matrix is given by $P_{i,j} = 1$ if $i = j = 0$ or $i = j = M$; $P_{i,j} = p$, $p \in (0, 1)$, if $j = (i + 1)$; $P_{i,j} = (1 - p)$ if $j = (i - 1)$; and $P_{i,j} = 0$ otherwise. With an initial distribution μ_0 being point mass on m, the interpretation is that a gambler starts with a stake of m and in each period gains 1 with probability p and loses 1 with probability $(1 - p)$. The gambler quits if his stake reaches M and is ruined if the stake reaches 0.*

The tail σ-field contains two events: $\lim_t X_t = 0$ and $\lim_t X_t = M$. The probability of these two events depends on the starting point.

7.2.f Complements and Extensions

Here is an explicit construction of a sequence of random variables on the probability space $((0, 1], \mathcal{B}, \lambda)$ with the properties of Example 7.2.52.

Example 7.2.54 *Let $(\Omega, \mathcal{F}, P) = ((0, 1], \mathcal{B}, \lambda)$. For any $1 > p > 0$ and any $(a, b] \subset (0, 1]$, the **p-split of** $(a, b]$ is the partition of $(a, b]$ into a first interval containing p of $(a, b]$ and a second interval containing the remaining $(1 - p)$ of $(a, b]$. Algebraically, the first interval is $(a, pb + (1 - p)a]$ and the second is $(pb + (1 - p)a, b]$. Define $X_0(\omega) = 1_{(\frac{1}{2}, 1]}(\omega)$. We define $X_n(\omega)$ inductively on $(0, \frac{1}{2}]$ and $(\frac{1}{2}, 1]$.*

For $\omega \in (0, \frac{1}{2}]$, define $X_1(\omega) = 1$ if ω is in the first part of a q-split of $(0, \frac{1}{2}]$ and $X_1 = 0$ otherwise. Given X_n as an indicator of interval subsets, $(a_k, b_k]$ of $(0, \frac{1}{2}]$, *define $X_{n+1}(\omega)$ to be equal to 1 on the first part of a q-split of every interval $(a_k, b_k]$ and equal to 1 otherwise.*

In an exactly parallel fashion, for $\omega \in (\frac{1}{2}, 1]$, define $X_1(\omega) = 1$ if ω is in the first part of a p-split of $(0, \frac{1}{2}]$ and $X_1 = 0$ otherwise. Given X_n as an indicator of interval subsets, $(a_k, b_k]$ of $(0, \frac{1}{2}]$, define $X_{n+1}(\omega)$ to be equal to 1 on the first part of a p-split of every interval $(a_k, b_k]$ and equal to 1 otherwise.

If we define $Q(E) = 2 \cdot \lambda(E \cap (0, \frac{1}{2}])$, then using Q gives the sequence X_1, X_n, . . . is iid Bern(q); if we define $Q'(E) = 2 \cdot \lambda(E \cap (\frac{1}{2}, 1])$, then using Q' yields the sequence $X_1, X_n, \ldots$ is iid Bern(p). By the SLLN, $Q(\{\frac{1}{n} \sum_{i \le n} X_n \to q\}) = 1$ and $Q'(\{\frac{1}{n} \sum_{i \le n} X_n \to p\}) = 1$. Combining and letting $L(\omega) = \liminf_n \frac{1}{n} \sum_{i \le n} X_n$ gives us $\lambda(L = q) = \lambda(L = p) = \frac{1}{2}$.

Here are two more generalizations of the WLLN.

Theorem 7.2.55 (WLLN with Orthogonality) *If $\{X_n : n \in \mathbb{N}\} \subset L^2(\Omega, \mathcal{F}, P)$ is an orthogonal collection of random variables and $\exists \mu, \sigma^2 \in \mathbb{R}$ such that for all $n, m \in \mathbb{N}$, $E\, X_n = E\, X_m = \mu$ and $\mathrm{Var}(X_n) = \mathrm{Var}(X_m) = \sigma^2$, then for all $\epsilon > 0$,*

$$\lim_n P(\{\omega : |S_n(\omega) - \mu| > \epsilon\}) = 0. \qquad (7.17)$$

Exercise 7.2.56 Verify that the same proof works with orthogonality replacing independence.

Theorem 7.2.57 (WLLN with a Bounded Higher Moment) *If $\{X_n : n \in \mathbb{N}\}$ is an independent collection of random variables with $\|X_n\|_{1+\delta} \le B$ for all n, then $\alpha(S_n - \mu_n, 0) \to 0$, where $\mu_n := \frac{1}{n} \sum_{i \le n} E\, X_n$.*

Exercise 7.2.58 Modify the given proof of the WLLN (by use of Markov's inequality) to prove this stronger version of the result.

Here is a version of the SLLN that does not assume that higher moments exist, slightly weakens the independence condition, and assumes that the random variables are identically distributed.

Theorem 7.2.59 *If $\{X_n : n \in \mathbb{N}\} \subset L^1(\Omega, \mathcal{F}, P)$ is a sequence of pairwise independent identically distributed random variables with mean μ, then $P(\{\omega : S_n(\omega) \to \mu\}) = 1$.*

Exercise 7.2.60 (Finicky Truncation Arguments) Fill in the following sketch of Etemadi's proof of Theorem 7.2.59.

1. Show that X_n^+ and X_n^- both satisfy the assumptions of the theorem, so that if we prove it for $X_n \ge 0$, we are done. For the rest of the proof, assume that $X_n \ge 0$.

2. Define $S_n = \frac{1}{n} \sum_{i \le n} X_i$, $Y_n = X_n \cdot 1_{\{X_n \le n\}}$ and $S_n^Y = \frac{1}{n} \sum_{i \le n} Y_i$. Let X be a random variable with the same distribution as the X_n. Show that $P(\{X_n = Y_n\} \text{ a.a.}) = 1$ by Borel-Cantelli and the following:

$$\sum_n P(\{Y_n \neq X_n\}) = \sum_n P(\{X_n > n\}) = \sum_n \sum_{i \geq n} P(\{X \in (i, i+1]\}). \quad (7.18)$$

$$= \sum_{i \geq 1} i P(\{X \in (i, i+1]\}) \leq \sum_{i \geq 1} E\, X \cdot 1_{\{X \in (i, i+1]\}}. \quad (7.19)$$

$$\leq E\, X < \infty. \quad (7.20)$$

3. For $\alpha > 1$, let k_n be the integer part of α^n. Show that if $P(\{S_{k_n}^Y \to \mu\}) = 1$, then $P(\{S_{k_n} \to \mu\}) = 1$.

4. Show that $E\, X = \lim_n E\, X \cdot 1_{\{X < n\}} = \lim_n E\, Y_n = \lim_n E\, S_{k_n}^Y$.

5. Pick arbitrary $\epsilon > 0$. Using Chebyshev, show that there are positive constants c for each of the following inequalities (remembering that we allow c to vary across the inequalities):

$$\sum_n P(\{|S_{k_n}^Y - E\, S_{k_n}^Y| > \epsilon\}) \leq c \sum_n \mathrm{Var}(S_{k_n}^Y)$$

$$= c \sum_n \frac{1}{k_n^2} \sum_{i \leq k_n} \mathrm{Var}(Y_i) \leq c \sum_i \frac{1}{i^2} E\, Y_i^2$$

$$= c \sum_i \frac{1}{i^2} E\, X^2 1_{\{X \in (0, i]\}} = c \sum_i \frac{1}{i^2} \left[\sum_{k=0}^{i-1} E\, X^2 \cdot 1_{\{X \in (k, k+1]\}} \right]$$

$$\leq c \sum_{k \geq 0} \frac{1}{k+1} E\, X^2 \cdot 1_{\{X \in (k, k+1]\}} \leq c \sum_{k \geq 0} E\, X \cdot 1_{\{X \in (k, k+1]\}} = cE\, X < \infty.$$

6. Using the previous and Borel-Cantelli, show that $P(\{S_{k_n}^Y \to \mu\}) = 1$. From above, this means that $P(\{S_{k_n} \to \mu\}) = 1$.

7. The problem is that we want to show that $P(\{S_n \to \mu\}) = 1$ and we have only shown that it is true if we skip from k_n to k_{n+1}. Using the monotonicity of $\sum_{i \leq n} X_i$, show that

$$\frac{1}{\alpha} \mu \leq \liminf_n S_n \leq \limsup_n S_n \leq \alpha \mu. \quad (7.21)$$

8. Show that, since α is an arbitrary number greater than 1, this implies that $P(\{S_n \to \mu\}) = 1$.

Exercise 7.2.61 (↑Herds, Based on Example 7.2.54) Suppose that $0 < q < \frac{1}{2} < p < 1$, and that no one observes X_0. Suppose that person 1 observes X_1 and then picks an optimal $a_1^* \in \{0, 1\}$. Person n, $n > 1$, observes X_n and $h_n := (a_1^*, \ldots, a_{n-1}^*)$ and then picks an optimal $a_1^* \in \{0, 1\}$. Suppose that everyone's utility function is $u(X_0, a_n) = 1_{\{a_n = X_0\}}$.

1. Show that if $h_3 = (0, 0)$ or $h_3 = (1, 1)$, then person 3 ignores X_3 and matches the actions of the first two people.

2. Show further that all the people thereafter will ignore their own information.

3. (A good deal harder) Show that with probability 1, there comes a time after which all the people ignore their own information.

[The story behind this exercise is that all people want to guess the true value of X_0 and match it; for example, they want to go to the better restaurant/show/movie. They each receive an independent, noisy signal, X_n. If $X_0 = 1$, then X_n is more likely to equal 1 than 0, and if $X_0 = 0$, then X_n is more likely to equal 0 than 1. This is called a herd model because everyone follows some finite set of early movers. If all the people were to share their signals, the SLLN tells us that there would be perfect information. Since only finitely many signals are used, there is a positive probability of error. This has an externalities intuition: me acting on my signal has a positive externality for those who move after me, but since I do not take account of this effect, I am more likely to ignore my signal.]

7.3 ◆ Good Sets Arguments and Measurability

We begin with the measurability of mappings from one measure space to another. When the range of the mapping is a metric space and its Borel σ-field, the definition is equivalent to the "inverse image of open sets is measurable" definition given earlier. The proof uses a *good sets* argument. In particular, we call the class of sets having the property we want the good ones, and then show that the good sets contain a field or other class of sets generating the σ-field we care about and satisfy some property guaranteeing that they contain the generated σ-field. We present two results that help make good sets arguments: the monotone class theorem and Dynkin's π-λ theorem.

To see that measurability is worth studying, we look at the change of variables theorem, at countable products of measure spaces, and at the way that fields approximate σ-fields. The countable product of a space S can serve as the space of realizations of a discrete-time stochastic process taking values in S, and we take a detour through Markov processes to illustrate the use of the definitions of measurable subsets of product spaces.

7.3.a Measurable Functions Taking Values in Measure Spaces

Compare the following with Definition 7.1.18 (p. 360).

Definition 7.3.1 (Measurable Functions Taking Values in Measure Spaces) *A function, X, from a measure space $(\Omega, \mathcal{F})$ to a measure space $(\Omega', \mathcal{F}')$ is **measurable** or $\mathcal{F}'\backslash\mathcal{F}$-**measurable** if $X^{-1}(E') \in \mathcal{F}$ for every $E' \in \mathcal{F}'$, that is, iff $X^{-1}(\mathcal{F}') \subset \mathcal{F}$.*

Compositions of measurable functions are measurable.

Lemma 7.3.2 *If $X : \Omega \to \Omega'$ and $Y : \Omega' \to \Omega''$ are both measurable, then the composition $\omega \mapsto Y(X(\omega))$ from Ω to Ω'' is measurable.*

Proof. $Y^{-1}(\mathcal{F}'') \subset \mathcal{F}'$ and $X^{-1}(\mathcal{F}') \subset \mathcal{F}$, so that $X^{-1}(Y^{-1}(\mathcal{F}'')) \subset \mathcal{F}$. ∎

Measurability involves $X^{-1}(\mathcal{F}') \subset \mathcal{F}$. Since the set theoretic operations of unions, intersections, and complements pass directly through inverses of functions, $X^{-1}(\mathcal{F}')$ is itself a σ-field, which we denote as $\sigma(X)$.

Since the Borel σ-field is generated by the open sets, the following implies that a function from a measure space $(\Omega, \mathcal{F})$ to a metric space (M, d) is measurable iff it is a measurable function from the measure space $(\Omega, \mathcal{F})$ to the measure space $(M, \mathcal{B}_M)$.

Theorem 7.3.3 *$X : \Omega \to \Omega'$ is measurable iff $X^{-1}(\mathcal{E}') \subset \mathcal{F}$ for some class of sets $\mathcal{E}'$ that generates $\mathcal{F}'$.*

The proof is our first use of the **good sets principle**, which is a very useful way of proving measurability, and we use it often. You name a class of sets that has the property that you want and then show that it contains or is equal to the σ-field you are interested in. As this is our first good sets argument, we give the details.

Proof. If X is measurable, $X^{-1}(\mathcal{F}') \subset \mathcal{F}$. Since $\mathcal{E}' \subset \mathcal{F}'$, we know that $X^{-1}(\mathcal{E}') \subset \mathcal{F}$.

Now suppose that $\mathcal{E}'$ generates $\mathcal{F}'$ and that $X^{-1}(\mathcal{E}') \subset \mathcal{F}$. We now define a class of **good sets**. They are "good" because they have the property we want. Let $\mathcal{G}' = \{E' \in \mathcal{F}' : X^{-1}(E) \in \mathcal{F}\}$. We now show that $\mathcal{G}' = \mathcal{F}'$. We do this by showing that $\mathcal{G}'$ is a σ-field containing $\mathcal{E}'$. Here are the steps:

1. We check that $\mathcal{G}'$ satisfies the properties of a σ-field and that it contains $\mathcal{E}'$.

2. We know that $\mathcal{G}' \subset \mathcal{F}'$ because we defined it that way.

3. Since $\mathcal{F}'$ is the smallest σ-field containing $\mathcal{E}'$, by the definition of generating, this means that $\mathcal{F}' \subset \mathcal{G}'$.

First check that $\mathcal{E}' \subset \mathcal{G}'$. This is just a rewrite of the assumption that $X^{-1}(\mathcal{E}') \subset \mathcal{F}$.

Now check that $\mathcal{G}'$ satisfies the three properties of being a σ-field: (1) since $X^{-1}(\emptyset) = \emptyset \in \mathcal{F}$ and $X^{-1}(\Omega') = \Omega \in \mathcal{F}$, we know that $\emptyset, \Omega' \in \mathcal{G}'$; (2) if $E' \in \mathcal{G}'$, then $(E')^c \in \mathcal{G}'$ because $X^{-1}((E')^c) = (X^{-1}(E'))^c$; and (3) if E'_n is a sequence in $\mathcal{G}'$, then $\cup_n E'_n \in \mathcal{G}'$ and $\cap_n E'_n \in \mathcal{G}'$ because $X^{-1}(\cup_n E'_n) = \cup_n X^{-1}(E'_n)$ and $X^{-1}(\cap_n E'_n) = \cap_n X^{-1}(E'_n)$.

Therefore $\mathcal{G}' \subset \mathcal{F}'$, the class of good sets, is a σ-field containing $\mathcal{E}'$ so that $\mathcal{G}' \subset \mathcal{F}' \subset \mathcal{G}'$. ∎

Corollary 7.3.4 *For (M, d) a metric space with topology τ_d and Borel σ-field $\mathcal{B}_M$ and a function $X : \Omega \to M$, $X^{-1}(\tau_d) \subset \mathcal{F}$ iff $X^{-1}(\mathcal{B}_M) \subset \mathcal{F}$.*

Exercise 7.3.5 Show that $\mathcal{E} = \{(-\infty, r] : r \in \mathbb{R}\}$ generates $\mathcal{B}$, as does $\mathcal{E} = \{(r, s] : r < s, \ r, s \in \mathbb{R}\}$. Compare with Theorem 7.1.22. [Since $\mathcal{E} \subset \mathcal{B}$, argue that it is sufficient to show that $\tau_d \subset \sigma(\mathcal{E})$. Now show that every open subset of $\mathbb{R}$ belongs to $\sigma(\mathcal{E})$.]

7.3.b The Monotone Class Theorem

Let $\mathcal{G} \subset \mathcal{F}$ be the class of sets with the property under question—the "good" sets. It is often quite easy to show that $\mathcal{G}$ contains a field $\mathcal{F}^\circ$ that generates $\mathcal{F}$. Extra considerations, sometimes quite subtle, show that the good sets are closed under monotone unions and monotone intersections. This turns out to be sufficient to show that $\mathcal{G} = \mathcal{F}$.

Definition 7.3.6 *A class of sets $\mathcal{G}$ is **a monotone class** if for any monotone sequences, $E_n \downarrow E$, $F_n \uparrow F$, E_n, $F_n \in \mathcal{G}$, E, $F \in \mathcal{G}$.*

Any σ-field is a monotone class, as is any chain of sets.

Exercise 7.3.7 Show that if $\mathcal{M}$ is a field and a monotone class, then it is a σ-field. [Hint: Let E_n be a sequence in $\mathcal{M}$ and define $F_n = \cup_{k \leq n} E_k$ and $G_n = \cap_{k \leq n} E_k$. As $\mathcal{M}$ is a field, F_n, $G_n \in \mathcal{M}$. Finally, $F_n \uparrow \cup_k E_k$ and $G_n \downarrow \cap_k E_k$.]

Theorem 7.3.8 (Set Theoretic Monotone Class) *If $\mathcal{G}$ is monotone class containing a field, $\mathcal{F}^\circ$, then $\mathcal{F} = \sigma(\mathcal{F}^\circ) \subset \mathcal{G}$.*

Proof. Let $\mathcal{M}$ be the smallest monotone class containing $\mathcal{F}^\circ$, that is, the intersection of all monotone classes containing $\mathcal{F}^\circ$. Since $\mathcal{M} \subset \mathcal{G}$, it is (more than) sufficient to show that $\mathcal{F} = \mathcal{M}$.

We first show that $\mathcal{M}$ is a field. Since it is a monotone class, we know that it is a σ-field containing $\mathcal{F}^\circ$, that is, $\mathcal{F} \subset \mathcal{M}$. Since $\mathcal{F}$ is a monotone class containing $\mathcal{F}^\circ$ and $\mathcal{M}$ is the smallest such monotone class, we conclude that $\mathcal{F} = \mathcal{M}$.

Fix an arbitrary $E \in \mathcal{M}$ and define $\mathcal{M}_E = \{F \in \mathcal{M} : E \cap F, \ E \cap F^c, \ E^c \cap F \in \mathcal{M}\}$. We now show that:

1. $\mathcal{M}_E$ is a monotone class—let F_n be a monotone sequence in $\mathcal{M}$, which makes $E \cap F_n$, $E \cap F_n^c$, and $E^c \cap F_n$ into monotone sequences in $\mathcal{M}$.

2. If $E \in \mathcal{F}^\circ$, then $\mathcal{F}^\circ \subset \mathcal{M}_E$—$[E \in \mathcal{F}^\circ] \Rightarrow (\forall F \in \mathcal{F}^\circ)[E \cap F, \ E \cap F^c, \ E^c \cap F \in \mathcal{F}^\circ \subset \mathcal{M}]$.

3. If $E \in \mathcal{F}^\circ$, then $\mathcal{M}_E = \mathcal{M}$—from the previous two steps, $\mathcal{M}_E \subset \mathcal{M}$ is a monotone class containing $\mathcal{F}^\circ$, and $\mathcal{M}$ is the smallest monotone class containing $\mathcal{F}^\circ$.

4. If $F \in \mathcal{M}$, then $\mathcal{F}^\circ \subset \mathcal{M}_F$—because $\mathcal{M}_E = \mathcal{M}$ for any $E \in \mathcal{F}^\circ$, $E \cap F$, $E \cap F^c$, and $E^c \cap F$ belong to $\mathcal{M}_F$.

5. If $F \in \mathcal{M}$, then $\mathcal{M}_F = \mathcal{M}$—$\mathcal{M}_F \subset \mathcal{M}$ is a monotone class containing $\mathcal{F}^\circ$, and $\mathcal{M}$ is the smallest monotone class containing $\mathcal{F}^\circ$.

From the previous steps, we conclude that $\mathcal{M}$ is a field—for E, $F \in \mathcal{M} = \mathcal{M}_E$, we know that $E \cap F$, $E \cap F^c$, and $E^c \cap F$ belong to $\mathcal{M}$. ∎

It is worth repeating what we have shown in a pithier fashion: the smallest monotone class containing a field is the σ-field it generates. Here is an easy corollary.

Corollary 7.3.9 *If P and Q are two countably additive probabilities on $\mathcal{F} = \sigma(\mathcal{F}^\circ)$ that agree on a field $\mathcal{F}^\circ$, then they agree on $\mathcal{F}$.*

Proof. Let $\mathcal{G}$ denote good sets, that is, $\mathcal{G} = \{E \in \mathcal{F} : P(E) = Q(E)\}$. By assumption, $\mathcal{F}^\circ \subset \mathcal{G}$. Therefore, it is sufficient to show that $\mathcal{G}$ is a monotone class. Let $E_n \uparrow E$, each $E_n \in \mathcal{G}$. Since $P(E_n) = Q(E_n)$ and both P and Q are countably additive, $P(E) = \lim_n P(E_n) = \lim_n Q(E_n) = Q(E)$, so that $E \in \mathcal{G}$. The argument for $E_n \downarrow E$ is the same. ∎

Exercise 7.3.10 $P(X \in A) = P(Y \in A)$ for all $A \in \mathcal{B}$ iff $P(X \le r) = P(Y \le r)$ for all $r \in \mathbb{R}$. [This follows directly from Corollary 7.3.9, but giving a direct proof is a chance to use a good sets argument.]

Exercise 7.3.11   If X, Y are measurable functions taking values in a separable metric space M, then $\{X = Y\}, \{X \ne Y\} \in \mathcal{F}$. [Proving one of these proves the other. Let $\{x_m : m \in \mathbb{N}\}$ be a countable dense subset of M. For each m, define $f_m(x)$ to be the closest point in $\{x_1, \ldots, x_m\}$ breaking ties in favor of the lower index. Show that for all $x \in M$, $f_m(x) \to x$. Show that $\omega \mapsto d(f_m(X(\omega)), f_m(Y(\omega))$ is measurable. Consider the set ω, where this function converges to 0.]

7.3.c   Change of Variables

It is often the case that we have a probability space $(\Omega, \mathcal{F}, P)$, a measure space $(\Omega', \mathcal{F}')$, and a measurable function $X : \Omega \to \Omega'$. The probability that $X \in E' \in \mathcal{F}'$ is denoted P_X. The basic change of variables result is that if $h : \Omega' \to \mathbb{R}$, one can calculate either $\int_\Omega h(X(\omega)) \, dP(\omega)$ or $\int_{\Omega'} h(\omega') \, dP_X(\omega')$ and get the same answer.

Definition 7.3.12   *For a probability space $(\Omega, \mathcal{F}, P)$, a measure space $(\Omega', \mathcal{F}')$, and a measurable $X : \Omega \to \Omega'$, the countably additive probability P_X defined by $P_X(E') = P(\{X \in E'\})$, $E' \in \mathcal{F}'$, is called the **distribution induced by** X or the **image law of** X.*

Theorem 7.3.13   *If X is a measurable function on $(\Omega, \mathcal{F}, P)$ taking values in the measure space $(\Omega', \mathcal{F}')$ and $h : \Omega' \to \mathbb{R}$ is measurable, then $\int_{\Omega'} h(\omega') \, dP_X(\omega')$ is defined iff $\int_\Omega h(X(\omega)) \, dP(\omega)$ is defined, and whenever they are defined, the two integrals are equal, $\int h \, dP_X = \int (h \circ X) \, dP$.*

Proof.   If $h = 1_{E'}$ for   some $E' \in \mathcal{F}'$,   then $\int h(X(\omega)) \, dP(\omega)$ $= \int 1_{E'}(X(\omega)) \, dP(\omega) = P(\{X \in E'\}) = P_X(E') = P(\{X \in E'\})$. By linearity of the integral, this implies that $\int h \, dP_X = \int (h \circ X) \, dP$ for all simple measurable h. Finally, $P_X(\{h_n \to h\}) = 1$ iff $P(\{(h_n \circ X) \to (h \circ X)\}) = 1$, so by dominated convergence, $\int h \, dP_X$ is well defined iff $\int (h \circ X) \, dP$ is well defined, and in this case, $\int h \, dP_X = \int (h \circ X) \, dP$.   ∎

7.3.d   Countable Products of Measure Spaces

Given a finite or countable collection $\{(M_n, d_n) : n \in A\}$, $A \subset \mathbb{N}$, of metric spaces, we define the product metric space as $M = \times_{n \in A} M_n$ with the product metric $d_{Prod}(x, y) = \sum_{n \in A} \frac{1}{2^n} \min\{1, d_n(x_n, y_y)\}$. Given a finite or countable collection $\{(\Omega_n, \mathcal{F}_n) : n \in A\}$, $A \subset \mathbb{N}$, of measure spaces, the product space is $\times_{n \in A} \Omega_n$, and we define the product σ-field, denoted $\mathcal{F} = \otimes_{n \in A} \mathcal{F}_n$. For our purposes, it is important that the Borel σ-field on M be $\otimes_{n \in A} \mathcal{B}_{M_n}$, the product of the Borel σ-fields.

7.3.d.1   Cylinder Sets, Product Fields, and Product σ-Fields
Recall that a **cylinder set with base set** $E_i \subset \Omega_i$ in a product space $\times_{n \in \mathbb{N}} \Omega_n$ is the set of $\omega = (\omega_1, \omega_2, \ldots)$ having $\omega_i \in E_i$, that is, $\text{proj}_i^{-1}(E_i)$. A cylinder set

puts restrictions on the value of only one component of a vector ω. Measurable rectangles put restrictions on only finitely many.

Definition 7.3.14 *For a finite or countable collection* $\{(\Omega_n, \mathcal{F}_n) : n \in A\}$, *a **measurable rectangle** in* $\times_{n \in A} \Omega_n$ *is a finite intersection of cylinder sets having measurable base sets. The **product measure space** or **product** is the measure space* $(\Omega, \mathcal{F}) = (\times_{n \in A} \Omega_n, \otimes_{n \in A} \mathcal{F}_n)$, *where the **product field**, $\mathcal{F}^\circ$ denoted $\times_{n \in A} \mathcal{F}_n$, is the smallest field containing all cylinder sets with base sets in $\mathcal{F}_n$, $n \in A$ and the **product σ-field**, $\mathcal{F}$ denoted $\otimes_{n \in A} \mathcal{F}_n$, is the smallest σ-field containing $\mathcal{F}^\circ$, that is, $\mathcal{F} = \sigma(\mathcal{F}^\circ)$.*

Exercise 7.3.15 For $i, j \in \mathbb{N}$, $E_i \subset \Omega_i$, $E_j \subset \Omega_j$,

1. show that $[\text{proj}_i^{-1}(E_i) \cup \text{proj}_j^{-1}(E_j)]^c = [\text{proj}_i^{-1}(E_i^c) \cap \text{proj}_j^{-1}(E_j^c)]$,
2. draw the pictures representing this equality when $\Omega_i = \Omega_j = (0, 1]$, and
   a. $E_i = E_j = (0, \frac{1}{2}]$;
   b. $E_i = E_j = (0.3, 0.7]$; and
   c. $E_i = (0.1, 0.3] \cup (0.5, 0.7]$ and $E_j = (0.3, 0.8]$.

Exercise 7.3.16 Show that the product σ-field is the smallest σ-field making each of the canonical projections measurable, that is, $\otimes_{n \in A} \mathcal{F}_n = \sigma(\{\text{proj}_n^{-1}(\mathcal{F}_n) : n \in A\})$. [This is mostly an exercise in writing things down carefully.]

Lemma 7.3.17 *Let $\Omega = \times_{n \in A} \Omega_n$, $A \subset \mathbb{N}$. For every finite A, the product field, $\times_{n \in A} \mathcal{F}_n$, is the class of finite unions of measurable rectangles. Further, for $A = \mathbb{N}$, $\mathcal{F}^\circ = \cup_N \{\times_{n \le N} \mathcal{F}_n\}$.*

Proof. The class of measurable rectangles contains $\emptyset$ and Ω. Generalizing Exercise 7.3.15 shows that it is closed under complementation, and it is obviously closed under finite unions, hence, taking complements, under finite intersections. The last follows from the observation that any measurable rectangle puts restrictions on only finitely many components of an $\omega \in \Omega$. ∎

7.3.d.2 Measurable Slices
For $(x, y) \in \mathbb{R}^2$ and a continuous $f \in C(\mathbb{R}^2)$, every "slice" of f is continuous. That is to say, for every x°, the function **slice of f taken at x°**, that is, the function $y \mapsto f(x^\circ, y)$, is continuous (with parallel statements for slices at y°). A very easy way to prove this is to note that the function $m(y) := (x^\circ, y)$ is continuous and $f(x^\circ, \cdot)$ is the composition of the continuous functions f and m. At this point, a piece of notation from game theory is quite useful.

Notation 7.3.18 *For $m \in A$, $\omega_{-m} \in \Omega_{-m} := \times_{n \in A \setminus \{m\}} \Omega_n$, and $\omega_m^\circ \in \Omega_m$, the point $(\omega_m^\circ, \omega_{-m}) \in \Omega$ is defined by $\text{proj}_n(\omega) = \omega_n$ if $n \neq m$, and $\text{proj}_m(\omega) = \omega_m^\circ$.*

Lemma 7.3.19 *For $(\Omega, \mathcal{F}) = (\times_{n \in A} \Omega_n, \otimes_{n \in A} \mathcal{F}_n)$, $A \subset \mathbb{N}$, any measurable $X : \Omega \to \Omega'$, $(\Omega', \mathcal{F}')$ a measure space, any $n \in A$, and any $\omega_n^\circ \in \Omega_n$, the slice of X taken at ω_n° is measurable; that is, the function $\omega_{-m} \mapsto X(\omega_m^\circ, \omega_{-m})$ is measurable.*

Proof. If $\{\omega^\circ\} \in \mathcal{F}_n$, then the mapping $m(\omega_{-m}) = (\omega_m^\circ, \omega_{-m})$ is measurable and the composition of measurable functions is measurable. More generally, the class of X for which the result is true contains all simple functions measurable with respect to the product field, and is closed under pointwise limits. ∎

7.3.d.3 A Markovian Detour

A stationary discrete-time Markov chain with finite state space $(S, \mathcal{S})$, $\mathcal{S} = \mathcal{P}(S)$, is a special kind of stochastic process, that is, a special kind of probability distribution on $\times_{n=0}^{\infty} S$. In this context, it is interesting to note that these Markov chains are defined by the value of the probability on the product field, $\cup_N \{\times_{n=0}^{N} \mathcal{S}\}$. From §4.11.b, specifying a finite-state Markov chain requires specifying an initial distribution and an $S \times S$ transition matrix.

Definition 7.3.20 *The **history space for a Markov chain with state space** S is the product measure space $\Omega := \times_{n \in \{0\} \cup \mathbb{N}} S$ with σ-field $\mathcal{F} := \otimes_{n \in \{0\} \cup \mathbb{N}} \mathcal{S}$. For $t = 0, 1, \ldots$, we define the random variable $X_t(\omega) = \text{proj}_t(\omega)$.*

A stationary Markov chain is specified by an initial distribution, $\mu_0 \in \Delta(S)$, and a transition matrix, $(P_{i,j})_{i,j \in S}$, which satisfies $\sum_j P_{i,j} = 1$ for all $i \in S$. The interpretation is that $P(X_0 = i) = \mu_0(i)$ and

$$P(X_{t+1} = j \mid X_0 = i_0, \ldots, X_{t-1} = i_{t-1}, X_t = i) = P_{i,j}. \qquad (7.22)$$

For any $\omega = (s_0, s_1, \ldots) \in \Omega$ and $N \in \mathbb{N}$, the measurable rectangle with base set $\{\text{proj}_{0,\ldots,N}(\omega)\}$ is the set of all histories that begin with $h := (s_0, \ldots, s_N)$. The probability of this rectangle is defined by $\nu_N(h) = \mu_0(s_0) \cdot P_{s_0, s_1} \cdot \ldots \cdot P_{s_{N-1}, s_N}$. For any $A \in \cup_N \times_{n=0}^{N} \mathcal{S}$, $P(\text{proj}_{0,\ldots,N}^{-1}(A))$ is defined by $\sum_{h \in A} \nu_N(h)$.

The P just defined is a finitely additive probability on the product field. We will see in Exercise 7.6.28 (p. 422), that every finitely additive probability on this particular type of field is automatically countably additive, hence, by Carathéodory's extension theorem (p. 412), has a unique countably additive extension to the product σ-field.

For finite S, Theorem 4.11.9 (p. 157) shows that if the transition matrix P has the property that $P^n \gg 0$ for some n, then $\pi \mapsto \pi P$ is a contraction mapping from $\Delta(S)$ to itself, so there exists a unique $\pi \in \Delta(S)$ such that for any μ_0, $\mu_0 P^n \to \pi$. Many interesting transition matrices do not have this property, but a great deal of the behavior of the associated Markov chain can be deduced by other methods.

The set of ergodic distributions for the Gambler's Ruin, Example 7.2.53 (p. 385), is the set of π of the form $\pi = (\alpha, 0, \ldots, 0, (1 - \alpha))$, $0 \le \alpha \le 1$. We now show that starting with any μ_0, the process eventually ends up staying either at 0 or at M, with the probability of the outcome depending on μ_0 and p.

Let $\pi_{m,M}$ be the probability that the gambler eventually quits with M given that he starts with m, $\pi_{m,0}$ being the probability that she is eventually ruined. We wish to show that for all m, $\pi_{m,0} + \pi_{m,M} = 1$ and to indicate how to calculate the $\pi_{m,0}$.

The first observation is that for $m \neq 0$, $\pi_{m,M} \ge p^{M-m} > 0$ because p^{M-m} is the probability that the gambler starting at m never loses on the way from m to M. [For instance, if $M - m = 2$, it is possible that the gambler loses a round and then wins three in a row, and p^{M-m} does not count this possibility.] By the same logic, for $m \neq M$, $\pi_{m,0} > 0$. Let $A_t = \{\omega : X_t(\omega) = m\}$. We wish to show that $P([A_t \text{ i.o.}]) = 0$.

Let $\tau_1(\omega) = \min\{t \geq 1 : X_t(\omega) = m\}$ be the random time till the process first returns to m, with $\min \emptyset := \infty$. Since $\pi_{m,0} + \pi_{m,M} > 0$, $q^* := P(\tau_1 < \infty) < 1$. Given τ_k, the random time until the kth return to m, define τ_{k+1} on $\tau_k < \infty$ by $\tau_{k+1}(\omega) = \min\{t > \tau_k : X_t(\omega) = m\}$, again with $\min \emptyset := \infty$, and with $\tau_{k+1}(\omega) := \infty$ for $\omega \in \{\tau_k = \infty\}$. As we are working with a Markov process, for all t, $P(\tau_{k+1} < \infty \mid \tau_k = t) = q^*$. Therefore, $P([\tau_1 < \infty] \cap [\tau_2 < \infty]) = (q^*)^2$, and more generally, $P(\cap_{k \leq K}[\tau_k < \infty]) = (q^*)^K$. Since $[A_t \text{ i.o.}] \subset \cap_{k \leq K}[\tau_k < \infty])$ for all K, $P([A_t \text{ i.o.}]) = 0$.

The argument just given does not depend on which m is chosen; hence, starting at any $m \neq 0$, M, the Markov chain will eventually leave m never to return, so that $\pi_{m,0} + \pi_{m,M} = 1$.

To calculate the $\pi_{m,0}$, consider that for each $m \neq 0$, M, $\pi_{m,0} = p\pi_{m+1,0} + (1 - p)\pi_{m-1,0}$. Since $\pi_{M,0} = 0$ and $\pi_{0,0} = 1$, this gives $M - 2$ equations in $M - 2$ unknowns, and these can be readily solved. When $p = \frac{1}{2}$, we can skip most of even this small bit of algebra.

Exercise 7.3.21   If $p = \frac{1}{2}$ in the Gambler's Ruin, show that for all $i \in \{0, \ldots, M\}$, $E(X_{t+1} \mid X_t = i) = i$. From this show that $E X_t \equiv m$ if the gambler starts with m. We already know that $X_t \to X$ a.e. for some random variable taking values in the two-point set $\{0, M\}$. Using dominated convergence, show that $E X = m$. Solve for $\pi_{m,0}$ and $\pi_{m,M}$.

7.3.d.4 The Borel σ-Field on the Product of Metric Spaces
The relation between the product of metric spaces and the product of the associated measure spaces is direct.

Theorem 7.3.22   *For $\{(M_n, d_n) : n \in A\}$ a finite or countably infinite collection of metric spaces with Borel σ-fields $\mathcal{B}_n$ and product space (M, d_{Prod}), $\otimes_{n \in A} \mathcal{B}_n$ is the Borel σ-field on M.*

Proof.   The class of cylinder sets having open base sets generates the Borel σ-field. ∎

This result allows us to look further into the 0-1 law for finite recurrent Markov chains and the structure of tail σ-fields.

Definition 7.3.23   *A set of S-valued random variables $\{X_t : t = 0, \ldots\}$ is called a **stochastic process in** S because it is stochastic, that is, random, and because it is a process, that is, it unfolds over time. If S is a finite-state space with σ-field $\mathcal{S} = \mathcal{P}(S)$, and for all t,*

$$P(X_{t+1} = j \mid (X_0, \ldots, X_t) = (i_0, \ldots, i_t)) = P(X_{t+1} = j \mid X_t = i_t), \quad (7.23)$$

*we have a **finite-state space Markov process**. The Markov process is **stationary** if there exists an $S \times S$ **transition matrix** $P = (P_{ij})_{i,j \in S}$ such that for all t and all $i, j \in S$, $P(X_{t+1} = j \mid X_t = i) = P_{ij}$. A stationary Markov process with transition matrix P that starts in i, that is, has $P(X_0 = i) = 1$, is called an (i, P)-**chain**.*

We can take $(\Omega, \mathcal{F})$ to be $(\times_{t=0}^{\infty} S, \otimes_{t=0}^{\infty} \mathcal{S})$ and $X_t(\omega) = \text{proj}_t(\omega)$. Define $\mathcal{F}_t = \sigma(\{X_\tau : \tau \leq t\})$, $\mathcal{F}_{t+} = \sigma(\{X_\tau : \tau \geq t\})$, $\mathcal{F}_{tail} = \cap_t \mathcal{F}_{t+}$, and $\mathcal{F} = \sigma\{\mathcal{F}_t : t = 0, \ldots\}$.

Lemma 7.3.24   *For an (i, P)-chain with $P^n \gg 0$ for some n, $\mathcal{F}_{tail}$ is trivial.*

Proof.   Starting with $X_0 = i$, $P^n \gg 0$ means that we spend a random amount of time, T_i, until we first visit i, and that random amount of time has finite expectation.

The **pre-T_i fragment** is the sequence of points in S, $(X_0, \ldots, X_{T_i-1})$. The post-T_i process, $(X_{T_i}, X_{T_i+1}, \ldots$ is another (i, P)-chain. From the condition in (7.23), we can deduce what is called the **strong Markov property**—that the pre-T_i fragment and the post-T_i chain are independent.

A pre-T_i fragment is called an **i-block**. (For instance, if $S = \{1, 2, 3\}$, then, e.g., $(1, 2)$, $(1, 3)$ are 1-blocks representing returning to 1 after a 2 and after a 3, respectively. Longer examples of 1-blocks are $(1, 2, 2, 3, 2)$ and $(1, 3, 3, 2)$.) The strong Markov property and induction tell us that an (i, P)-chain is a sequence of iid i-blocks. Let B_k be the kth i-block in the sequence, $\mathcal{G}_k = \sigma(\{B_j : j \leq k\})$, $\mathcal{G}_{k+} = \sigma(\{B_j : j \geq k\})$, and $\mathcal{G}_{tail} = \cap_k \mathcal{G}_{k+}$. Kolmogorov's 0-1 law implies that for any $A \in \mathcal{G}_{tail}$, $P(A) = 0$ or $P(A) = 1$. Finally, for $k \geq t$, $\mathcal{F}_{t+} \subset \mathcal{G}_{k+}$ because i-blocks last at least one period. This implies that $\mathcal{F}_{tail}$ is a subset of the trivial σ-field $\mathcal{G}_{tail}$.   ∎

7.3.e Fields Approximate σ-Fields

Measurable sets can be hard to visualize, but they are very close, in a metric sense, to being sets that are relatively easy to visualize.

7.3.e.1 Difficult Visualizations
Drawing the set in following example is not easy.

Example 7.3.25   *Enumerate the rationals in $(0, 1]$ as $\{q_n : n \in \mathbb{N}\}$. For $\delta > 0$, let $E(\delta) = \cup_n B_{\delta/2^n}(q_n)$. This gives an open set that is dense because it contains the rationals, yet has "length" strictly less than δ. Its complement therefore "ought to" have "length" greater than $1 - \delta$, yet it contains no open interval because every open interval contains one of the q_n. Exercise 6.11.50 (p. 353) showed that $E = \cap_m E(1/m)$ is dense and uncountable, and its "length" is less than $1/m$ for every m.*

Each $E(\delta)$ is, to any degree of approximation, a finite union of the form $\cup_{n \leq N} B_{\delta/2^n}(q_n)$, which is a set that is very easy to visualize. The set E is well approximated by the set $\varnothing$, and if this is not easy to visualize, then at least we can visualize its complement easily. We show that this kind of pattern is more general—from a probabilistic point of view, all of the measurable sets nearly belong to a smaller class. The sense of "nearly belong" is metric.

7.3.e.2 Approximating σ-Fields by Fields
Recall that the probability metric distance between sets A and B is $\rho_P(A, B) = P(A \triangle B)$. The following shows that a generating field, $\mathcal{F}^\circ$, is ρ_P-dense in the metric space $\mathcal{F}$. The proof is another good sets argument.

Theorem 7.3.26 (Approximating σ-Fields by Fields)   *If $\mathcal{F} = \sigma(\mathcal{F}^\circ)$, $\mathcal{F}^\circ$ a field, then for every $E \in \mathcal{F}$ and every $\epsilon > 0$, there exists an $E^\circ \in \mathcal{F}^\circ$ such that $\rho_P(E, E^\circ) < \epsilon$.*

Proof. Let $\mathcal{G} = \{E \in \mathcal{F} : \forall \epsilon > 0, \exists E^\circ \in \mathcal{F}^\circ, \ \rho_P(E, E^\circ) < \epsilon\}$. $\mathcal{G}$ is a field, so let $E_n \uparrow E$, $E_n \in \mathcal{G}$. For each n, pick $E_n^\circ \in \mathcal{F}^\circ$ such that $\rho_P(E_n, E_n^\circ) < \epsilon/2^{n+1}$. Pick N large so that $\rho_P(E, \cup_{n \leq N} E_n) < \epsilon/2$. Then $\rho_P(E, \cup_{n \leq N} E_n^\circ) < \epsilon$. ∎

Thus, if $\mathcal{B}^\circ$ is the field of disjoint unions of intervals with rational endpoints, then every set in the Borel σ-field, $\mathcal{B}$, is nearly an element of $\mathcal{B}^\circ$. This makes the Borel σ-field more manageable—any measurable set is nearly a finite disjoint union of intervals with rational endpoints.

Detour: Anticipating future developments, suppose that we have a countably additive P on a σ-field $\mathcal{F}$. From Corollary 7.2.23 (p. 376), we know that $(\mathcal{F}, \rho_P)$ is complete. Therefore, $\mathcal{F}$ is the completion of $\mathcal{F}^\circ$ for any field generating $\mathcal{F}$. In §7.6, we start with a P that is countably additive on a field $\mathcal{F}^\circ$ and from there go backward through this logic, showing that $\mathcal{F}$ is, up to isometry, the completion of the metric space $(\mathcal{F}^\circ, \rho_P)$. Since $P : \mathcal{F}^\circ \to [0, 1]$ is uniformly continuous, it has a unique continuous extension from $\mathcal{F}^\circ$ to the completion, a result known as Carathéodory's extension theorem.

7.3.f Good Sets and Dynkin's π-λ Theorem

Dynkin's π-λ theorem is a systematic way to make good sets arguments. Those we have used so far have shown that the good sets contain a class of sets that generate the σ-field and that the good sets are a σ-field. If there have been any difficulties, they have involved this last step, showing that the good sets are a σ-field. Showing that something is a λ-system is easier because there are fewer properties to check, which is what makes Dynkin's result so useful.

Definition 7.3.27 *A class* $\mathfrak{P}$ *of sets is called a* π**-system** *if it closed under finite intersection, that is, if* $[A, B \in \mathfrak{P}] \Rightarrow [(A \cap B) \in \mathfrak{P}]$.

A class $\mathfrak{L}$ *of sets is called a* λ**-system** *if it contains* Ω *and is closed under proper differencing and monotone unions, that is, if* $\Omega \in \mathfrak{L}$, $[[A, B \in \mathfrak{L}] \wedge [A \subset B]] \Rightarrow [(B \setminus A) \in \mathfrak{L}]$, *and* $[[(A_n)_{n \in \mathbb{N}} \subset \mathfrak{L}] \wedge [A_n \uparrow A]] \Rightarrow [A \in \mathfrak{L}]$.

Theorem 7.3.28 (Dynkin) *If* $\mathfrak{P}$ *is a* π-system, $\mathfrak{L}$ *is a* λ-system, *and* $\mathfrak{P} \subset \mathfrak{L}$, *then* $\sigma(\mathfrak{P}) \subset \mathfrak{L}$.

Before we give the proof, here is a typical application.

Corollary 7.3.29 *If P and Q are two countably additive probabilities on $\mathcal{B}$ that have the same cdf, that is, for all x, $F_P(x) = P(-\infty, x] = F_Q(x) = Q(-\infty, x]$, then P and Q agree on $\mathcal{B}$.*

Proof. Observe that $\mathfrak{P} = \{(-\infty, x] : x \in \mathbb{R}\}$ is a π-system because it is closed under finite intersection. The candidate for the class of good sets is $\mathfrak{L} = \{E \in \mathcal{B} : P(E) = Q(E)\}$. This clearly contains Ω, is closed under proper differences, and, by the countable additivity of both P and Q, is closed under monotone unions. Therefore $\sigma(\mathfrak{P}) = \mathcal{B} \subset \mathfrak{L}$. ∎

The proof of Dynkin's π-λ theorem uses the following, which shows that λ-systems can only fail to be σ-fields by failing to be closed under finite intersection.

Exercise 7.3.30 If $\mathcal{H}$ is a π-system and a λ-system, then it is a σ-field. [Prove the following steps: since $\Omega \in \mathcal{H}$ and $\mathcal{H}$ is closed under proper differences, it is closed under complementation; since it is a π-system, it is closed under finite intersection, hence under finite union, so it is a field; for fields, closure under monotone countable unions and intersections implies closure under arbitrary countable unions and intersections.]

Proof of Dynkin's π-λ Theorem. Let $\mathfrak{P}$ be a π-system and $\mathfrak{L}$ a λ-system containing $\mathfrak{P}$. We must show that $\sigma(\mathfrak{P}) \subset \mathfrak{L}$. It is sufficient to show that $\ell(\mathfrak{P})$, the smallest λ-system containing $\mathfrak{P}$, is a π-system (because this implies that it is a σ-field by the exercise that you just finished).

For sets A and B, let $\mathcal{G}_A = \{B : A \cap B \in \ell(\mathfrak{P})\}$ and $\mathcal{G}_B = \{C : B \cap C \in \ell(\mathfrak{P})\}$. We show that $[B \in \ell(\mathfrak{P})] \Rightarrow [\ell(\mathfrak{P}) \subset \mathcal{G}_B]$, which implies that if B and C belong to $\ell(\mathfrak{P})$, then so does $B \cap C$.

Step 1: If $A \in \ell(\mathfrak{P})$, then $\mathcal{G}_A$ is a λ-system. (To see this, simply check the properties.)

Step 2: If $A \in \mathfrak{P}$, then $\mathfrak{P} \subset \mathcal{G}_A$, so that $\ell(\mathfrak{P}) \subset \mathcal{G}_A$. (To see this, note that $\mathfrak{P} \subset \ell(\mathfrak{P})$ so that $\mathcal{G}_A$ is a λ-system by Step 1. Since $\mathfrak{P}$ is a π-system, this means that $\mathfrak{P} \subset \mathcal{G}_A$. Finally, by the minimality of $\ell(\mathfrak{P})$, we get $\ell(\mathfrak{P}) \subset \mathcal{G}_A$.)

Step 3: If $A \in \mathfrak{P}$ and $B \in \ell(\mathfrak{P})$, then $A \cap B \in \ell(\mathfrak{P})$. (To see this, by definition, $\mathcal{G}_A$ is the set of all B whose intersection with A belongs to $\ell(\mathfrak{P})$ and by the previous step, $\mathcal{G}_A \supset \ell(\mathfrak{P})$.) ∎

7.4 ◆ Two 0-1 Laws

Kolmogorov's 0-1 law explains why our limit theorems for sequences of independent random variables keep giving us only probabilities of either 0 or 1. The Hewitt-Savage 0-1 law expands the set of events that must have probability either 0 or 1, but requires that the random variables not only be independent but also be identically distributed. This second 0-1 law is also a key ingredient in delivering a rather surprising *non*measurability result in what seems like a pedestrian kind of question in social choice theory.

7.4.a Proving Kolmogorov's 0-1 Law

The ingredients in the proof are the approximation of σ-fields by fields and Dynkin's theorem.

Proof of Theorem 7.2.50, Kolmogorov's 0-1 Law The field $\mathcal{H}^\circ = \cup_n \mathcal{H}_n$ generates $\mathcal{H} := \sigma(\{\mathcal{H}_n : n \in \mathbb{N}\})$. Since $A \in \mathcal{H}_{tail} \subset \mathcal{H}$, for every $k > 0$, there exists $A_k \in \mathcal{H}^\circ$ such that $P(A_k \triangle A) < 1/k$, so that $|P(A_k) - P(A)| < 1/k$.

We now state something that seems entirely plausible, but which needs a separate proof, given below: Since $A_k \in \mathcal{H}_n$ for some n and $A \in \mathcal{H}_{(n+1)+}$, $A_k \perp\!\!\!\perp A$.

This means that $P(A_k \triangle A) = P(A_k)(1 - P(A)) + P(A)(1 - P(A_k)) < 1/k$. Since $|P(A_k) - P(A)| \to 0$, $[P(A_k)(1 - P(A)) + P(A)(1 - P(A_k))] \to 2P(A)(1 - P(A))$, so that $P(A)(1 - P(A)) = 0$.

Here is the proof of the plausible step.

Lemma 7.4.1 *If* $\{\mathcal{H}_\tau : \tau \in T\}$ *is an independent collection of π-systems and* U, V *are disjoint subsets of* T, *then* $\sigma(\{\mathcal{H}_\tau : \tau \in U\}) \perp\!\!\!\perp \sigma(\{\mathcal{H}_\tau : \tau \in V\})$.

Proof. Let $\mathcal{H}(U) = \sigma(\{\mathcal{H}_\tau : \tau \in U\})$ and $\mathcal{H}(V) = \sigma(\{\mathcal{H}_\tau : \tau \in V\})$. We must show that for any $E \in \mathcal{H}(U)$ and any $F \in \mathcal{H}(V)$, $E \perp\!\!\!\perp F$, that is, $P(E \cap F) = P(E) \cdot P(F)$. The proof has two steps: first we use the π-λ theorem to show that every $E \in \mathcal{H}(U)$ is independent of any F in any $\mathcal{H}_\tau$, $\tau \in V$; then we use the π-λ theorem again to show that every $F \in \mathcal{H}(V)$ is independent of every $E \in \mathcal{H}(U)$.

Pick arbitrary $\tau \in V$ and $F \in \mathcal{H}_\tau$. We first show that any $E \in \mathcal{H}(U)$, $E \perp\!\!\!\perp F$. Let $\mathfrak{P}(U)$ be the class of finite intersections of sets in $\cup_{\tau \in U} \mathcal{H}_\tau$. Let $\mathfrak{L}(U) = \{E \in \mathcal{H}(U) : E \perp\!\!\!\perp F\}$. By the definition of independence, $\mathfrak{P}(U) \subset \mathfrak{L}(U)$. To check that $\mathfrak{L}(U)$ is a λ-system, note that $\Omega \in \mathfrak{L}(U)$, and that countable additivity implies that $\mathfrak{L}(U)$ is closed under monotonic unions. Now suppose that $A \subset B$, A, $B \in \mathfrak{L}(U)$. We must show that $(B \setminus A) \in \mathfrak{L}(U)$. Since $A \perp\!\!\!\perp F$ and $B \perp\!\!\!\perp F$, this follows from $P((B \setminus A) \cap F) = P((B \cap F) \setminus (A \cap F)) = P(B \cap F) - P(A \cap F) = P(B) \cdot P(F) - P(A) \cdot P(F) = (P(B) - P(A)) \cdot P(F) = P(B \setminus A) \cdot P(F)$.

Thus, for every $E \in \mathcal{H}(U)$ and every F in any $\mathcal{H}_\tau$, $\tau \in V$, $E \perp\!\!\!\perp F$. Now let $\mathfrak{P}(V)$ be the class of finite intersections of sets in $\cup_{\tau \in V} \mathcal{H}_\tau$ and let $\mathfrak{L}(V) = \{E \in \mathcal{H}(V) : \forall E \in \mathcal{H}(U),\ E \perp\!\!\!\perp F\}$. By the definition of independence, $\mathfrak{P}(V) \subset \mathfrak{L}(V)$. Checking that $\mathfrak{L}(V)$ is a λ-system proceeds just as in the first step. ∎

Recall that when defining independence in Definition 7.2.1, we used collections of events $\{X_\alpha^{-1}(G_\alpha) : \alpha \in A\}$, where the G_α were open. Since the class of open sets is a π-system, Lemma 7.4.1 means that we could have just as well have used the collections $\{X_\alpha^{-1}(B_\alpha) : \alpha \in A\}$, where the B_α are Borel measurable.

7.4.b The Hewitt–Savage 0-1 Law

Kolmogorov's 0-1 law applies to the tail σ-field in situations where the random variables are independent. When the variables are not only independent but identically distributed, we involve the Hewitt-Savage 0-1 law, which applies to the larger σ-field that contains symmetric sets, that is, finite-permutation invariant sets.

Definition 7.4.2 *A **finite permutation** of* $\mathbb{N}$ *is a bijection* $\pi : \mathbb{N} \leftrightarrow \mathbb{N}$ *with* $\pi(t) \neq t$ *for at most finitely many t;* x^π *is defined by* $\text{proj}_t(x^\pi) = x_{\pi(t)}$ *for all* $t \in \mathbb{N}$.

Definition 7.4.3 *For $M \in \mathbb{N}$, a measurable subset, A, of the product space $\mathbb{R}^\mathbb{N}$ is* *M-symmetric* *if* $[x \in A] \Leftrightarrow [x^\pi \in A]$ *for all finite permutations π. It is **symmetric** if it is M-symmetric for all $M \in \mathbb{N}$. The class of M-symmetric sets is denoted* $\mathcal{F}_{M\text{-symmetric}}$, *the class of symmetric sets is denoted* $\mathcal{F}_{\text{symmetric}}$.

The Borel σ-field for $\mathbb{R}^\mathbb{N}$ is denoted $\mathcal{B}^\mathbb{N}$ and is equivalently defined as $\otimes_{n \in \mathbb{N}} \mathcal{B}$. The tail σ-field for $\mathbb{R}^\mathbb{N}$ is defined in two steps, $\mathcal{F}_{M+} = \sigma\{\text{proj}_m^{-1}(\mathcal{B}) : m \geq M + 1\}$ and $\mathcal{F}_{tail} = \cap_M \mathcal{F}_{M+}$.

Lemma 7.4.4 $\mathcal{F}_{\text{symmetric}}$ *is a σ-field strictly containing* $\mathcal{F}_{\text{tail}}$.

Proof. Let $\mathfrak{P}_M$ denote the set of measurable rectangles in $\mathcal{F}_{M+}$ and $\mathfrak{L}_M$ the set $\{E \in \mathcal{F}_{M+} : E \in \mathcal{F}_{M\text{-symmetric}}\}$. Any element of $\mathfrak{P}_M$ is M-symmetric so that $\mathfrak{P}_M \subset \mathfrak{L}_M$. It is easy to check that $\mathfrak{L}_M$ contains Ω and is closed under proper differencing and under monotone unions. Therefore, by the π-λ theorem, $\mathcal{F}_{M+} \subset \mathcal{F}_{M\text{-symmetric}}$. If we take intersections over M, $\mathcal{F}_{\text{tail}} \subset \mathcal{F}_{\text{symmetric}}$.

To see that the subset relation is strict, fix a sequence c_n and set $A = [\{\sum_{i \le n} x_i > c_n\}$ i.o.]. For any permutation π, the sums defining A are almost always equal; hence $[x \in A]$ iff $[x^\pi \in A]$, so $A \in \mathcal{F}_{\text{symmetric}}$. However, if $A \in \mathcal{F}_{2+}$, then A is of the form $\mathbb{R} \times A'$, where $A' \subset \times_{n=2}^\infty \mathbb{R}$. For any bounded sequence of c_n, changing the value of the first component changes whether or not $x \in A$, so that $A \notin \mathcal{F}_{\text{tail}}$. ∎

If X_n is a sequence of random variables, then for any $A \in \mathcal{B}^\mathbb{N}$, $\{(X_1, X_2, \ldots) \in A\}$ is measurable. To see why, note that $\mathcal{B}^\mathbb{N}$ is generated by sets of the form $A = \text{proj}_n^{-1}(E_n)$, $E_n \in \mathcal{B}$, and that for such a set A, $\{(X_1, X_2, \ldots) \in A\}$ is $X_n^{-1}(E_i)$.

Definition 7.4.5 *Given a sequence of random variables X_n, the **associated symmetric σ-field** is defined as $\mathcal{S} := \{B \in \mathcal{F} : (X_1, X_2, \ldots) \in A\}$ for some symmetric measurable $A \subset \mathbb{R}^\mathbb{N}$.*

Theorem 7.4.6 (Hewitt–Savage) *If X_n is an iid sequence of random variables and A is in the associated symmetric σ-field, then $P(A) = 0$ or $P(A) = 1$.*

Proof. Pick $A \in \mathcal{S}$ and $\epsilon > 0$. We show that $|P^2(A) - P(A)| < 4\epsilon$. Since the only nonnegative numbers $r = 0$ and 1 satisfy $|r^2 - r| = 0$, this completes the proof. Define the field $\mathcal{F}^\circ = \cup_{n \in \mathbb{N}} \sigma(X_1, \ldots, X_n)$ and let $\mathcal{F} = \sigma(\mathcal{F}^\circ)$.

By the definition of $\mathcal{S}$, there exists some $E \subset \mathbb{R}^\mathbb{N}$ with

$$A = \{(X_1, \ldots, X_n, X_{n+1}, \ldots, X_{2n}, X_{2n+1}, \ldots) \in E\}.$$

Since A is symmetric, it is equal to the event

$$\{(X_{n+1}, \ldots, X_{2n}, X_1, \ldots, X_n, X_{2n+1}, \ldots) \in E\},$$

as this involves permuting only the first $2n$ random variables.

Since $\mathcal{S} \subset \mathcal{F}$, $\exists B \in \mathcal{F}^\circ$ such that $P(A \triangle B) < \epsilon$. By the definition of $\mathcal{F}^\circ$ and $\sigma(X_1, \ldots, X_n)$, for some n and some $F \subset \mathbb{R}^n$, $B = \{(X_1, \ldots, X_n) \in F\}$. Let

$$C = \{(X_{n+1}, \ldots, X_{2n}) \in F\} = \{(X_1, \ldots, X_n, X_{n+1}, \ldots, X_{2n}) \in \mathbb{R}^n \times F\}.$$

$$(7.24)$$

Since the X_n are iid, $P(A \triangle B) = P(A \triangle C) < \epsilon$.

Now, $|P(A) - P(B)| < \epsilon$, so that $|P^2(A) - P^2(B)| < 2\epsilon$ (since x^2 has a slope less than or equal to 2 on the interval $[0, 1]$). Further, $P^2(B) = P(B \cap C)$ because B and C are independent and have the same probability. As $A \triangle (B \cap C) \subset (A \triangle B) \cup (A \triangle C)$, we have $|P(A) - P(B \cap C)| \le P(A \triangle (B \cap C)) < 2\epsilon$. Combining, we see that $P(A)$ is within 2ϵ of $P^2(B)$ and $P^2(A)$ is also within 2ϵ. Hence, $|P^2(A) - P(A)| < 4\epsilon$. ∎

We use the Hewitt-Savage 0-1 law in §7.8 to prove the existence of sets that are seriously nonmeasurable. The other ingredients in the proof are product measures and Fubini's theorem. As with many results in probability theory, Fubini's theorem depends on Lebesgue's dominated convergence theorem. We have put off proving it for long enough.

7.5 ◆ Dominated Convergence, Uniform Integrability, and Continuity of the Integral

It is now time to prove the basic properties of the integral listed in Theorem 7.1.49 (p. 367): positivity, linearity, monotonicity, and continuity. The continuity property goes by the name Lebesgue's dominated convergence theorem. In Theorem 7.1.40, we proved monotone convergence, that is, if $0 \leq X_n \leq X$ a.e. and $E\,|X| < \infty$, then $E\,X_n \uparrow E\,X$.

7.5.a Perspective and Examples

$L^1 = L^1(\Omega, \mathcal{F}, P)$ is a vector subspace of the metric vector space (L^0, α). We are looking for useful conditions on sets $S \subset L^1$ on which the integral is an α-continuous function. The first example shows that being an L^1-bounded set is not sufficient.

Example 7.5.1 *The integral is not α-continuous on the set $\overline{U} := \{X \in L^1 : \|X\|_1 \leq 1\}$. To see why, consider a sequence of the form $X_n = \beta_n 1_{(0,1/n]}$ in $L^1((0, 1], \mathcal{B}, \lambda)$. For all sequences β_n, $\alpha(X_n, 0) \to 0$. If $\beta_n = n$, then $E\,X_n \equiv 1 \not\to E\,0$; if $\beta_n = (-1)^n n$, then $E\,X_n = (-1)^n$, which is not convergent; whereas if $\beta_n = n/\log(n)$, then $E\,X_n \to E\,0$.*

The integral is a linear function on the metric vector space (L^1, α), but we have just seen that it is a discontinuous linear function. The discontinuity is more extreme: arbitrarily small amounts of distance in the metric α can lead to arbitrarily large changes in the value of the integral.

Exercise 7.5.2 (↑Example 7.5.1) Show that for every $r > 0$, no matter how large, and every $\epsilon > 0$, no matter how small, there exists an $X \in L^1((0, 1], \mathcal{B}, \lambda)$ with $\alpha(X, 0) < \epsilon$ and $E\,X > r$.

Another way of phrasing the result that you just proved is that within L^1, every α-ball is a norm unbounded set. Put yet another way, the L^1-diameter of $L^1 \cap B_\epsilon^\alpha(0)$ is infinite.

7.5.b Three Conditions for a Continuous Integral

The three conditions for a continuous integral are that S is L^∞ dominated, that S is L^1 dominated, and that S is uniformly integrable. Being L^∞-dominated implies being L^1-dominated, and being L^1-dominated implies being uniformly integrable. None of the implications reverse. All three conditions guarantee a "small-event continuity" condition.

Definition 7.5.3 $S \subset L^1(\Omega, \mathcal{F}, P)$ is L^∞-***dominated*** if for some $Y \geq 0$, $Y \in L^\infty$, $|X| \leq Y$ a.e. for all $X \in S$.

Theorem 7.5.4   *The integral is α-continuous on any L^∞-dominated set.*

Proof.   Let $B = \|Y\|_\infty$, where Y dominates S. For n such that $\alpha(X_n, X) < \delta/(2B + 1)$,

$$E\,|X_n - X| = \int_{|X_n - X| \leq \delta} |X_n - X|\,dP + \int_{|X_n - X| > \delta} |X_n - X|\,dP \leq \delta$$

$$+ 2B\delta < \delta(2B + 1) = \epsilon, \tag{7.25}$$

so that $E\,|X_n - X| \to 0$. Since $|E\,X_n - E\,X| \leq E\,|X_n - X|$, the integral is α-continuous. ∎

The "small event" in the proof is the set $\{|X_n - X| > \delta\}$. It is small in the sense that it has probability less than δ for large n, but more importantly for our purposes, integrating $|X_n - X|$ over the small set contributes at most $2B\delta$ to the integral. As $\delta \to 0$, the contribution to the integral over the small set goes to 0, and this is the small-event continuity condition.

Definition 7.5.5 $S \subset L^1(\Omega, \mathcal{F}, P)$ is L^1-***dominated***, *or simply* ***dominated***, *if for some $Y \geq 0$, $Y \in L^1$, $|X| \leq Y$ a.e. for all $X \in S$.*

Theorem 7.5.6 (Dominated Convergence)   *The integral is α-continuous on any L^1-dominated set.*

Proof.   Let Y dominate $S \subset L^1$. By monotone convergence, if $0 \leq Y_n \uparrow Y$ a.e., then $E\,Y_n \uparrow E\,Y$. Thus, $E\,Y \cdot 1_{\{Y \leq B\}} \uparrow E\,Y$ and $E\,Y \cdot 1_{\{Y > B\}} \downarrow 0$ as $B \uparrow \infty$.

Let X_n be a sequence dominated by Y and $\alpha(X_n, X) \to 0$. Then X is dominated by Y, hence integrable. Pick $\epsilon > 0$. We must show that for large n, $\int |X_n - X|\,dP < \epsilon$. For n large enough that $\alpha(X_n, X) < \delta$,

$$\int |X_n - X|\,dP = \int_{\{|X_n - X| \leq \delta\}} |X_n - X|\,dP + \int_{\{|X_n - X| > \delta\} \cap \{Y \leq B\}} |X_n - X|\,dP$$

$$+ \int_{\{|X_n - X| > \delta\} \cap \{Y > B\}} |X_n - X|\,dP$$

$$\leq \delta(1 - \delta) + 2B\delta + 2E\,Y \cdot 1_{\{Y > B\}}.$$

Pick B large enough that $2E\,Y \cdot 1_{\{Y > B\}} < \epsilon/3$, which can be done because $E\,Y \cdot 1_{\{Y > B\}} \downarrow 0$ as $B \to \infty$. Then pick δ small enough that $\delta < \epsilon/3$ and $2B\delta < \epsilon/3$. ∎

The "small event" in the proof is the set $\{|X_n - X| > \delta\}$, and we have to split it into two parts, the part on which $Y \leq B$ and the part on which $Y > B$. The small-event continuity shows up directly in the following, which allows us to skip the splitting.

Definition 7.5.7 $S \subset L^1(\Omega, \mathcal{F}, P)$ is ***uniformly integrable (u.i.)*** if

1. $\sup_{X \in S} E\,|X| < \infty$, *and*
2. $(\forall \epsilon > 0)(\exists \delta > 0)\left[\, [P(A) < \delta] \Rightarrow (\forall X \in S)[\int_A |X|\,dP < \epsilon]\,\right]$.

The last condition can be written $\lim_{P(A)\to 0} \sup_{X\in S} E\,|X| \cdot 1_A = 0$.

Lemma 7.5.8 *If S is dominated, then it is uniformly integrable.*

Proof. For S dominated by Y, $\sup_{X\in S} E\,|X| \le E\,Y < \infty$ and $\forall X \in S$, $\int_A |X|\, dP$ $\le \int_A Y\, dP$, so it is sufficient to show that if $P(A_n) \to 0$, $\int_{A_n} Y\, dP \to 0$. Since $Y \cdot 1_{A_n}$ is also dominated by Y and $\alpha(Y \cdot 1_{A_n}, 0) \to 0$, the result follows from dominated convergence, that is, from Theorem 7.5.6. ■

An immediate implication is that if S is u.i. and $X \in L^1$, then $S \cup \{X\}$ is u.i.

Theorem 7.5.9 *The integral is α-continuous on any uniformly integrable S.*

Proof. Let X_n be a sequence in a uniformly integrable S with $\alpha(X_n, X) \to 0$. We first show that X is integrable, which is a more subtle matter than it might appear, and then show that the small-event continuity makes the integral continuous.

Integrability of X: Taking a subsequence if necessary gives us $|X| = \lim_n |X_n|$ $= \liminf_n |X_n|$ a.e. We claim that $E\,|X| \le \liminf_n E\,|X_n|$, and since S is u.i., this means that $E\,|X| < \infty$, that is, that X is integrable. To see why $E\,|X| \le \liminf_n E\,|X_n|$, define $Y_n(\omega) = \inf\{|X_m|(\omega) : m \ge n\}$ so that Y_n is an increasing sequence of random variables with $Y_n \uparrow |X|$ and $Y_n \le |X_n|$. By monotone convergence, $E\,Y_n \uparrow E\,|X|$, and by the definition of $\liminf$, $E\,|X| \le \liminf_n E\,|X_n|$.

Continuity of the integral: Pick $\epsilon > 0$ and $\delta < \epsilon/3$ sufficiently small that $[P(A) < \delta] \Rightarrow (\forall R \in S \cup \{X\})[E\,|R| \cdot 1_A < \epsilon]$. This can be done because $S \cup \{X\}$ is u.i. For n large enough that $\alpha(X_n, X) < \delta$,

$$
\int |X_n - X|\, dP = \int_{\{|X_n-X|\le\delta\}} |X_n - X|\, dP + \int_{\{|X_n-X|>\delta\}} |X_n - X|\, dP
$$

$$
\le \delta(1-\delta) + \int_{\{|X_n-X|>\delta\}} (|X_n| + |X|)\, dP
$$

$$
\le \epsilon/3 + \epsilon/3 + \epsilon/3,
$$

where the last two inequalities follow from monotonicity and $P(\{|X_n - X| \ge \delta\})$ $< \delta$, respectively. ■

We have seen, either from monotone convergence or dominated convergence, that every integrable $Y \ge 0$ has the property that $\lim_{B\uparrow\infty} E\,Y \cdot 1_{\{Y>B\}} \to 0$. Having this property uniformly is equivalent to uniform integrability.

Lemma 7.5.10 $S \subset L^1(\Omega, \mathcal{F}, P)$ *is uniformly integrable iff* $\lim_{B\uparrow\infty} \sup_{X\in S} E\,|X| \cdot 1_{\{|X|>B\}} = 0$.

Proof. Suppose that $\lim_{B\uparrow\infty} \sup_{X\in S} E\,|X| \cdot 1_{\{|X|>B\}} = 0$. If $\sup_{X\in S} E\,|X| = \infty$, then for every n we can pick $X_n \in S$ with $E\,|X_n| > 2n$. Now, $E\,|X_n| = \int_{|X_n|\le n} |X_n|\, dP + \int_{|X_n|>n} |X_n|\, dP$. Since the first term on the right-hand side is less than or equal to n, the second term must be greater than n. Thus, for every $n \in \mathbb{N}$, $E\,|X_n| \cdot 1_{\{|X_n|>n\}} > n$, a contradiction. Now pick $\epsilon > 0$; we must find a $\delta > 0$ such that $[P(A) < \delta] \Rightarrow (\forall X \in S)[E\,|X| \cdot 1_A < \epsilon]$. Pick B such that $\sup_{X\in S} E\,|X| \cdot 1_{\{|X|>B\}} < \epsilon/2$ and choose $\delta < \epsilon/2$ sufficiently small that $B\delta < \epsilon/2$. For any A with $P(A) < \delta$, $\int_A |X|\, dP = \int_{A\cap\{|X|\le B\}} |X|\, dP + \int_{A\cap\{|X|>B\}} |X|\, dP < B\delta + \epsilon/2 < \epsilon$.

Suppose that S is u.i. and let $\sup_{X \in S} E\,|X| < R$. Pick $\epsilon > 0$. We must show that for large B, $\sup_{X \in S} E\,|X| \cdot 1_{\{|X| > B\}} < \epsilon$. By u.i. there is a δ such that $P(A) < \delta$ implies that $\sup_{X \in S} \int_A |X|\,dP < \epsilon$. By Chebyshev, $\sup_{X \in S} P(|X| > B) \leq \frac{R}{B}$. Choose B large enough that $\frac{R}{B} < \delta$. ∎

The classical Lebesgue's dominated convergence theorem seems slightly weaker than Theorem 7.5.6 as it uses a.e. convergence rather than α-convergence. However, since every α-convergent sequence has a further a.e. convergent sequence, the difference is illusory.

Theorem 7.5.11 (Lebesgue's Dominated Convergence) *If $Y \in L^1$, X_n is a sequence of random variables dominated by Y, and $X_n \to X$ a.e., then X is integrable and $E\,X_n \to E\,X$.*

Theorem 7.5.12 (Levi) *If $X_n \geq 0$, $X_n \uparrow X$ a.e., and $\sup_n E\,X_n = B < \infty$, then $E\,X_n \uparrow E\,X = B$.*

Proof. An easy argument shows that $E\,X = B$, so the sequence X_n is dominated by X. ∎

Exercise 7.5.13 Give the "easy argument" just mentioned.

We can also state a "series version" of Levi's theorem, which shows that under certain conditions, integration and infinite summation are interchangeable.

Corollary 7.5.14 *If $Y_k \geq 0$ is a sequence in L^1 with $\sum_k E\,Y_k < \infty$, then $\sum_k E\,Y_k = E\,\sum_k Y_k$.*

Proof. Apply Levi's theorem to the functions $X_n := \sum_{k \leq n} Y_k$. ∎

7.5.c Positivity, Linearity, and Monotonicity

The remaining conditions left to prove from Theorem 7.1.49 (p. 367) are positivity, linearity, and monotonicity.

Positivity is easy—if $0 \leq X$ a.e., then the integral of every element of $L_\leq(X)$ is nonnegative.

Pick integrable X, Y, and $\alpha, \beta \in \mathbb{R}$. We want to show that $E\,(\alpha X + \beta Y) = \alpha E\,X + \beta E\,Y$.

Exercise 7.5.15 For integrable X and $r \in \mathbb{R}$, $E\,rX = rE\,X$.

From the result in the exercise, to prove linearity, it is sufficient to show that $E\,(X + Y) = E\,X + E\,Y$.

Lemma 7.5.16 *For integrable X, Y, $E\,(X + Y) = E\,X + E\,Y$.*

Proof. Let $X_n^+ \uparrow X^+$, $X_n^- \uparrow X^-$, $Y_n^+ \uparrow Y^+$, and $Y_n^- \uparrow Y^-$ be sequences of simple functions. Define $X_n = X_n^+ - X_n^-$ and $Y_n = Y_n^+ - Y_n^-$. Since X_n and Y_n are simple, $E\,(X_n + Y_n) = E\,X_n + E\,Y_n$. For all ω, $(X_n + Y_n)(\omega) \to (X + Y)(\omega)$ and $|(X_n + Y_n)(\omega)| \leq 2 \cdot \max\{|X(\omega)|, |Y(\omega)|\}$. Dominated convergence finishes the argument. ∎

7.5.d Interchange of Integration and Differentiation

We are interested in the existence and calculation of $\frac{d}{dt} \int_a^b f(x, t) \, dx$ and of $\frac{d}{dt} \int_{a(t)}^{b(t)} f(x, t) \, dx$. For the first, define $\varphi(t) = \int_a^b f(x, t) \, dx$. By definition $\frac{d}{dt} \varphi(t_0) = \lim_{h_n \to 0} \frac{1}{h_n} [\varphi(t_0 + h_n) - \varphi(t_0)]$. The essential idea is that

$$\frac{1}{h_n} [\varphi(t_0 + h_n) - \varphi(t_0)] = \int_a^b \frac{[f(x, t_0 + h_n) - f(x, t_0)]}{h_n} \, dx, \quad (7.26)$$

and that, under reasonable kinds of conditions, $\eta_n(x) := \frac{[f(x, t_0 + h_n) - f(x, t_0)]}{h_n} \to \frac{\partial f(x, t_0)}{\partial t}$ a.e. in $[a, b]$. If the set $\{\eta_n : n \in \mathbb{N}\}$ is uniformly integrable, we can conclude that

$$\lim_n \int_a^b \eta_n(x) \, dx = \lim_n \int_a^b \frac{[f(x, t_0 + h_n) - f(x, t_0)]}{h_n} \, dx = \int_a^b \frac{\partial f(x, t_0)}{\partial t} \, dx.$$

$$(7.27)$$

Theorem 7.5.17 *Suppose that $f(\cdot, \cdot)$ is measurable on $[a, b] \times B_\delta(t_0)$ for some $\delta > 0$, that $f(\cdot, t)$ is integrable for each $t \in B_\delta(t_0)$, and that for almost all x in $[a, b]$ and for all $t \in B_\delta(t_0)$, $\left| \frac{\partial f(x, t)}{\partial t} \right| \leq g(x)$ for some integrable $g(x)$. Then*

$$\frac{d}{dt} \int_a^b f(x, t_0) \, dx = \int_a^b \frac{\partial f(x, t_0)}{\partial t} \, dx, \quad (7.28)$$

that is, we can interchange the derivative and the integral.

Proof. Dominated convergence. ∎

A subtle issue arises in evaluating $\frac{d}{dt} \int_{a(t)}^{b(t)} f(x, t) \, dx$. Suppose for simplicity, that $f(x, t) = f(x)$, that is, f has no dependence on t, that $a(t) \equiv a$, and that $b(t) = b_0 + \gamma t$ for $\gamma > 0$. Letting $\varphi(t) = \int_a^{b(t)} f(x) \, dx$ gives the derivative that we are interested in:

$$\lim_{h_n \to 0} \frac{1}{h_n} [\varphi(t_0 + h_n) - \varphi(t_0)] = \lim_{h_n \to 0} \frac{1}{h_n} \left[\int_a^{b_0 + \gamma h_n} f(x) \, dx - \int_a^{b_0} f(x) \, dx \right].$$

$$(7.29)$$

This is equal to $\lim_{h_n \to 0} \frac{1}{h_n} [\int_{b_0}^{b_0 + \gamma h_n} f(x) \, dx]$, and one's best guess as to the limit is $\gamma f(b_0)$. The subtlety is that in our entire theory of integration, functions can differ on null sets without it making any difference, and $\{b_0\}$ is a null set. We avoid this issue by assuming that f is continuous.[6]

Theorem 7.5.18 (Leibniz's Rule) *If $a(\cdot)$, $b(\cdot)$ are continuous differentiable on $B_\delta(t_0)$ for some $\delta > 0$, $f(\cdot, \cdot)$ is jointly continuous and bounded on $(a(t_0) -$*

---

6. If you ever need more generality, consider the measurable functions $\overline{f}(x) := \lim_{h \downarrow 0} \frac{1}{h} \int_x^{x+h} f(y) \, dy$ and $\underline{f}(x) := \lim_{h \downarrow 0} \frac{1}{h} \int_{x-h}^x f(y) \, dy$.

ϵ, $b(t_0) - \epsilon) \times B_\delta(t_0)$ for some $\epsilon > 0$, and for all $t \in B_\delta(t_0)$, $\left| \frac{\partial f(x,t)}{\partial t} \right| \leq g(x)$ for some integrable $g(x)$, then for $\varphi(t) := \int_{a(t)}^{b(t)} f(x, t) \, dx$,

$$\frac{d\varphi(t_0)}{dt} = \int_{a(t_0)}^{b(t_0)} \frac{\partial f(x, t_0)}{\partial t} \, dx - a'(t_0) f(a(t_0), t_0) + b'(t_0) f(b(t_0), t_0). \quad (7.30)$$

Proof.   Dominated convergence.   ∎

The domain of f can be $\Omega \times B_\delta(t_0)$.

Theorem 7.5.19   *Suppose that $(\Omega, \mathcal{F}, P)$ is a probability space, $f : \Omega \times B_\delta(t_0) \to \mathbb{R}$ for some $\delta > 0$, and [(a)]*

(a) *there exists B such that for all $t \in B_\delta(t_0)$, $\int |f(\omega, t)| \, dP(\omega) \leq B$,*

(b) *$\partial f(\cdot, t_0)/\partial t$ exists a.e.,*

(c) *$\partial f(\cdot, t_0)/\partial t$ belongs to L^1, and*

(d) *for some $\delta_1 > 0$, the collection $\{ \frac{f(\omega, t_0 + h) - f(\omega, t_0)}{h} : h \neq 0, \ h \in B_{\delta_1}(0) \}$ is uniformly integrable,*

then

1. *$\frac{d}{dt} \int_\Omega f(\omega, t_0) \, dP(\omega) = \int_\Omega \frac{\partial f(\omega, t_0)}{\partial t} \, dP(\omega)$, and*

2. *$\lim_{h \to 0} \int_\Omega | \frac{\partial f(\omega, t_0)}{\partial t} - \frac{f(\omega, t_0 + h) - f(\omega, t_0)}{h} | \, dP(\omega) = 0$.*

Proof.   Dominated convergence.   ∎

7.5.e   Loss Functions and Quantile Estimation

Suppose that we must choose a number, r, that makes $E(X - r)^2$ as small as possible, where $X \in L^2(\Omega, \mathcal{F}, P)$; that is, we want to solve

$$\min_{r \in \mathbb{R}} E(X - r)^2. \quad (7.31)$$

Let $\mu = EX$. Since $E(X - r)^2 = E([X - \mu] + [\mu - r])^2 = E[X - \mu]^2 + 2E[X - \mu] \cdot [\mu - r] + [\mu - r]^2$ and $2E[X - \mu] \cdot [\mu - r] = 0$ (because $[\mu - r]$ is a constant), the solution to (7.31) is $r^* = \mu$.

In statistics, the function $g(x, r) = (x - r)^2$ is an example of a **loss function**. The idea is that one must estimate, ahead of time, the best value for X knowing that if $X = x$ and your estimate was r, you suffer a loss $g(x, r)$. One's expected loss is $E g(X, r)$ and we solve the problem $\min_r E g(X, r)$.

Definition 7.5.20   *If F is a cdf and $F(r) = \alpha$, then r is an α-quantile of F.*

Lemma 7.5.21 (Quantile Estimation)   *Suppose that $X \in L^1$ has a continuously differentiable cumulative distribution function $F(x) = P(\{X \leq x\})$, and that for some $\alpha \in (0, 1)$ the loss function is*

$$g_\alpha(x, r) = \begin{cases} \alpha(x - r) & \text{if } r < x, \\ (1 - \alpha)(r - x) & \text{if } r \geq x. \end{cases}$$

The solution to $\min_r E g(X, r)$ is any r such that $F(r) = \alpha$.

The idea is that if we overestimate by z, that is, if $r = X + z$ for some $z \geq 0$, we suffer a loss of $(1 - \alpha)z = (1 - \alpha)(r - X)$, whereas if we underestimate, $r = X - z$, we suffer a loss of αz. If α goes down, that is, if underestimates become less costly relative to overestimates, then we expect the optimal r to move down. Since the cdf is nondecreasing, α-quantiles move down as α moves down, and this is something that the lemma delivers. We need a quick detour.

7.5.f Detour: Integrals of the Form $\int_{-\infty}^{\infty} f(x)\, dx$

We are interested in distributions with densities on all of $\mathbb{R}$, a topic that requires integrals with respect to measures that are not finite. Basically, we paste together countably many copies of the uniform probability distribution on the intervals

$$\ldots, (-2, -1], (-1, 0], (0, 1], (1, 2], \ldots,$$

and then add integrals over the intervals. After we paste this together, there are two ways that an integral can be undefined/infinite: it can be infinite over some interval $(n, n + 1]$ or it can be finite over each interval but have the countable sum nonconvergent.[7]

Definition 7.5.22 *For measurable $g : \mathbb{R} \to \mathbb{R}_+$, $\int g(x)\, dx = \int_{-\infty}^{\infty} g(x)\, dx :=$ $\sum_{n=-\infty}^{\infty} \int_{(n,n+1]} g(x)\, dx$. If g is integrable on each interval $(n, n + 1]$ and the sequence of numbers is absolutely summable, then $g \geq 0$ is **Lebesgue integrable** or **integrable with respect to the Lebesgue measure**; otherwise it is not and we write $\int g\, dx = \infty$. When $g = 1_E$ for some measurable E, the **Lebesgue measure of E** is $\int 1_E\, dx \in [0, \infty]$.*

For functions that may take on both positive and negative values, we proceed as we did for the integral above, by considering $g^+ = g \vee 0$ and $g^- = -(g \wedge 0)$.

Definition 7.5.23 *If $g : \mathbb{R} \to \mathbb{R}$ is measurable, then g is **Lebesgue integrable** if g^+ and g^- are integrable and $\int g\, dx := \int g^+\, dx - \int g^-\, dx$.*

If $g \geq 0$ and $\int g(x)\, dx = 1$, then g is the **density function** or **probability density function (pdf)** of a random variable with the distribution defined by $P_g(X \in A) = \int [g(x) \cdot 1_A(x)]\, dx = \int_A g(x)\, dx$.

Exercise 7.5.24 Show that if g is a density function, then $P_g(\cdot)$ is a countably additive probability on $\mathcal{B}$. [You have to prove a version of the monotone convergence theorem for Lebesgue integrable functions.]

If g is a pdf, then the cdf is $G(r) = \int_{(-\infty,r]} g(x)\, dx$, written $\int_{-\infty}^{r} g(x)\, dx$.

Exercise 7.5.25 Show that if g is a density function, then the associated cdf is continuous. [This again uses a monotone convergence theorem for Lebesgue integrable functions.]

---

7. If it has been a while since you have seen summability, we recommend a review of §3.7.b.

Exercise 7.5.26   Show that if $h : \mathbb{R} \to \mathbb{R}$ is measurable, X is a random variable with density g, and $h(X)$ is integrable, then $E\, h(X) = \int h(x)g(x)\, dx$. [Show first that it is true for simple h and then take limits.]

7.5.g   Scheffé's Theorem and the Norm Convergence of Probabilities

When the densities of a sequence of probabilities converge, the corresponding probabilities converge in the strongest sense that one ever uses.

Definition 7.5.27   *For probabilities P, Q on a σ-field $\mathcal{F}$, the **variation norm distance** is $\|P - Q\|_{vn} = \sup_{A \in \mathcal{F}} |P(A) - Q(A)|$.*

Example 7.5.28   *Let P_n be the uniform distribution on $\{\frac{k}{n} : k = 1, 2, \ldots, n\}$ and let $P = \lambda$ be the uniform distribution on $[0, 1]$. $\|P_n - P\|_{vn} \equiv 1$ because $P_n(\mathbb{Q}) \equiv 1$ and $P(\mathbb{Q}) = 0$.*

Theorem 7.5.29 (Scheffé)   *If each P_n has a density function g_n, P has a density function g, and $g_n \to g$ a.e. in each of the intervals $(n, n+1]$, then $\|P_n - P\|_{vn} \to 0$.*

Proof.   First suppose that g and the g_n are 0 outside of a bounded set $[-M, +M]$. For all $A \in \mathcal{B}_\mathbb{R}$, $A \subset [-M, +M]$, $|\int_A g(x)\, dx - \int_A g_n(x)\, dx| \leq \int |g(x) - g_n(x)|\, dx$. We wish to apply dominated convergence to conclude that $\int |g(x) - g_n(x)|\, dx \to 0$.

To this end, define $h_n = g - g_n$ such that the sequence h_n^+ is dominated by the integrable g and $h_n^+ \to 0$ a.e. By dominated convergence, $\int h_n^+\, dx \to 0$.

Now, $\int (g - g_n)\, dx = 0$, so that $\int_{h_n \geq 0} h_n\, dx = -\int_{h_n < 0} h_n\, dx$. From this, $\int |g - g_n|\, dx = \int |h_n|\, dx = 2 \int h_n^+\, dx \to 0$.

To finish the proof, for any $\epsilon > 0$, there is an M such that $\int_{[-M, +M]} g(x)\, dx > 1 - \epsilon/4$. For large enough n, the previous step shows that $\int_{[-M, +M]} |g(x) - g_n(x)|\, dx < \epsilon/4$. Note that

$$\int_{[-M, +M]^c} |g(x) - g_n(x)|\, dx \leq \int_{[-M, +M]^c} |g(x)|\, dx$$

$$+ \int_{[-M, +M]^c} |g_n(x)|\, dx < \epsilon/4 + 2\epsilon/4.$$

Putting these together yields $\int |g_n(x) - g(x)|\, dx < \epsilon$.   ■

7.5.h   Loss Functions and Quantile Estimation, Again

We now give the delayed proof.

Proof of Lemma 7.5.21.   Let $f(x) = F'(x)$ so that $\varphi(r) := E\, g_\alpha(X, r) = \int g_\alpha(x, r) f(x)\, dx$. $\varphi(\cdot)$ is convex because for each x, $g_\alpha(x, \beta r + (1 - \beta) r') f(x) \leq \beta g_\alpha(x, r) f(x) + (1 - \beta) g_\alpha(x, r') f(x)$ since $f(x) \geq 0$. Therefore, to find the minimum, it is sufficient to take the derivative and set it equal to 0.

Rewriting gives us $\varphi(r) = \int_{-\infty}^{r} (1-\alpha)(x-r)f(x)\,dx + \int_{r}^{\infty} \alpha(r-x)f(x)\,dx$. By Leibniz's rule,

$$\varphi'(r) = \frac{d}{dr} \int_{-\infty}^{r} (1-\alpha)(x-r)f(x)\,dx + \frac{d}{dr} \int_{r}^{\infty} \alpha(r-x)f(x)\,dx$$

$$= (1-\alpha)(r-r)f(r) + \int_{-\infty}^{r} -(1-\alpha)f(x)\,dx -$$

$$\alpha(r-r)f(r) + \int_{r}^{\infty} \alpha f(x)\,dx$$

$$= 0 - (1-\alpha)F(r) - 0 + \alpha(1-F(r)),$$

and this last term is equal to 0 iff $F(r) = \alpha$. ∎

Exercise 7.5.30 In the previous proof, we assumed that the cdf had a density so that we could use Leibniz's rule. Prove the same theorem with a continuous F that need not have a density. Then prove it for any cdf.

In the simplest linear regression theory, one has data $(X_n, Y_n)_{n=1}^{N}$ and one solves the problem

$$\min_{\beta_0, \beta_1} \sum_n (Y_n - (\beta_0 + \beta_1 X_n))^2. \tag{7.32}$$

If the distribution of the data is close to the distribution of the random vector (X, Y), then we have, at least approximately, found the (β_0^*, β_1^*) that solve the problem

$$\min_{\beta_0, \beta_1} E\,(Y - (\beta_0 + \beta_1 X))^2. \tag{7.33}$$

Once we have developed the theory of conditional expectations, we examine the conditions under which the conditional expectation of Y given X, $E\,(Y \mid X)$, is equal to $\beta_0^* + \beta_1^* X$. Returning to the loss functions g_α, we consider the problem

$$\min_{\beta_0, \beta_1} E\, g_\alpha(Y, (\beta_0 + \beta_1 X)). \tag{7.34}$$

Under the hypothesis just given, the solutions $(\beta_0^\alpha, \beta_1^\alpha)$ to this problem have the property that $\beta_0^\alpha + \beta_1^\alpha$ is the α-quantile of $F_Y(\cdot \mid X)$, the conditional distribution of Y given X.

Exercise 7.5.31 (Bencivenga) The monsoons will come at some random time T in the next month, $T \in [0, 1]$. A farmer must precommit to a planting time, a. As a function of the action, a, and the realized value of T, t, the harvest will be

$$h(a, t) = \begin{cases} K - r|a - t| & \text{if } a \le t \\ K - s|a - t| & \text{if } a > t \end{cases},$$

where $K > r, s > 0$. The random arrival time of the monsoon has the cumulative distribution function $F(\cdot)$, and $f(x) = F'(x)$ is its strictly positive probability density function.

1. Verify that for any $t \in [0, 1]$, $h(\cdot, t)$ is concave.

2. Suppose that the farmer is risk neutral and let $V(a) = E\, h(a, T)$. Show that the solution to the problem $\max_{a \in [0,1]} V(a)$ is the interior point $a^* = a^*(r, s)$ satisfying $F(a^*) = r/(r + s)$. Find $\partial a^*/\partial r$ and $\partial a^*/\partial s$ and give their signs.

3. Suppose now that the farmer is risk averse; that is, he chooses a to maximize $\Psi(a) = E\, u(h(a, T))$, where u is twice continuously differentiable, strictly increasing, and strictly concave.[8] Show that $\Psi(\cdot)$ is strictly concave and that the solution to the problem $\max_{a \in [0,1]} \Psi(a)$ is an interior point $a^\dagger(r, s)$. Find expressions for $\partial a^\dagger/\partial r$ and $\partial a^\dagger/\partial s$ and give their signs.

4. Return to the case of a risk-neutral farmer, but suppose now that her pre-commitment is not total. Specifically, at any point in time $\tau \in [-\frac{1}{10}, 1]$, the farmer, if she has not already decided to plant, can decide to plant at $a = \tau + \frac{1}{10}$; that is, there is a (roughly) 3-day lag between deciding to plant and actually having the job done. Supposing that $F(r) = r$, that is, that the distribution of monsoon times is uniform, find the farmer's optimal policy as a function of r and s.

7.5.i   Complements and Extensions

We begin with a sufficient condition for uniform integrability.

Exercise 7.5.32   If $\sup_{X \in S} \|X\|_p < \infty$ for some $p > 1$, then S is uniformly integrable. [Hint: Markov's inequality.]

We already know that if $0 \le X_n \uparrow X$, then $E\, X_n \uparrow E\, X$. We can replace 0 by any integrable X_0

Theorem   7.5.33   (Monotone   Convergence)   *If a sequence X_n in $L^0(\Omega, \mathcal{F}, P)$ satisfies $X_0 \le X_n \uparrow X$ a.e. and both X_0 and X are integrable, then $E\, X_n \uparrow E\, X$.*

A better, but longer, name for this result is "monotone convergence for sequences with an integrable lower bound." The need for the integrable lower bound appears in the following.

Exercise 7.5.34   On $((0, 1], \mathcal{B}, \lambda)$, define $X_0(\omega) = -1/\omega$, $X_n = 1_{(0,1/n]} \cdot X_0$, and $X(\omega) \equiv 0$. Show that $X_n(\omega) \to X(\omega)$ for all ω, but that for all n, $E\, X_n = -\infty < E\, X = 0$.

Exercise 7.5.35   Under the assumptions of Theorem 7.5.33, show that $0 \le (X_n - X_0) \uparrow (X - X_0)$, that $E\, (X_n - X_0) \uparrow E\, (X - X_0)$, and that $E\, X_n \uparrow E\, X$. Also formulate and prove the monotone convergence theorem for decreasing sequences of random variables.

---

8. See §7.11.c for more on risk aversion.

The following is, basically, what we use to show the integrability of the limit random variable, X, in our result about the integral being continuous on uniformly integrable sets. Note that we are allowing infinite integrals in the following, which causes no problems since all the random variables are nonnegative.

Lemma 7.5.36 (Fatou) *For any sequence $X_n \geq 0$ of measurable functions,*

$$E \liminf_n X_n \leq \liminf_n E X_n \leq \limsup_n E X_n \leq E \limsup_n X_n.$$

One can have strict inequalities.

Example 7.5.37 [↑*Example 7.2.34*] *For $(\Omega, \mathcal{F}, P) = ((0, 1], \mathcal{B}, \lambda)$ and $k = 0, \ldots, 2^n - 1$, let $Y_{k,n}(\omega) = 1_{(k/2^n, (k+1)/2^n]}(\omega)$ and $X'_n(\omega) = \sum_{k \text{ even}} Y_{k,n}(\omega)$. $\{X'_n : n \in \mathbb{N}\}$ is an iid sequence with $\lambda(\{X'_n = 1\}) = \lambda(\{X'_n = 0\}) = \frac{1}{2}$, so that $E X'_n \equiv \frac{1}{2}$, while $E \liminf_n X'_n = 0$.*

Exercise 7.5.38 Give a sequence $X_n \geq 0$ with $\limsup_n E X_n < E \limsup_n X_n$.

Exercise 7.5.39 Fill in the following sketch of a proof of Fatou's lemma:

1. To show that $\liminf_n X_n \leq \liminf_n E X_n$, it is sufficient to consider the case with $\liminf_n E X_n = r < \infty$.

2. Define $Y_n(\omega) = \inf\{X_m(\omega) : m \geq n\}$ so that $Y_n \leq X_n$ and $Y_n \uparrow \liminf_n X_n$. Show that $E \liminf_n X_n \leq \liminf_n E X_n$.

3. Finish the rest of the proof by multiplying each X_n by -1 and applying the monotone convergence theorem for decreasing sequences of random variables.

Exercise 7.5.40 For $(\Omega, \mathcal{F}, P) = ((0, 1], \mathcal{B}, \lambda)$ and all $n \geq 3$, let $X_n(\omega) = \frac{n}{\log n} \cdot 1_{(0, 1/n]}(\omega)$ and $X = 0$. Show that $\sup_{n \geq 3} E |X_n| < \infty$, that the set $S = \{X_n : n \geq 3\}$ is not dominated, that it is uniformly integrable, and that $E X_n \to E X$.

Given the following, the u.i. result in the previous exercise is not surprising.

Theorem 7.5.41 *For X and X_n in L^1, $E |X_n - X| \to 0$ iff $\alpha(X_n, X) \to 0$ and $S = \{X_n : n \in \mathbb{N}\}$ is uniformly integrable.*

Proof. ($\Leftarrow$): This is directly from the proof of the α-continuity of the integral on u.i. sets.

($\Rightarrow$): If $E |X_n - X| < \epsilon^2$, then Chebyshev delivers $P(\{|X_n - X| > \epsilon\}) < \epsilon$, that is, $\alpha(X_n, X) < \epsilon$.

We first show that $\sup_n E |X_n| < \infty$: since $|X_n| \leq |X| + |X_n - X|$ for all ω, $E |X_n| \leq E |X| + E |X_n - X|$. As $E |X_n - X| \to 0$, $\sup_n E |X_n| < \infty$.

Pick $\epsilon > 0$. We now show that there exists $\delta > 0$ such that $[P(A) < \delta] \Rightarrow (\forall X_n)[E |X_n| \cdot 1_A < \epsilon]$. Choose δ_0 so that $P(A) < \delta_0$ implies $E |X| \cdot 1_A < \epsilon/2$ (which we can do by dominated convergence). For some N and all $n \geq N$, $E |X_n - X| < \epsilon/2$. Pick $\delta \in (0, \delta_0)$ such that for each $n < N$, $P(A) < \delta$ implies that $E |X_n| \cdot 1_A < \epsilon/2$. For $n \geq N$, $E |X_n| \cdot 1_A \leq E (|X| + |X_n - X|) \cdot 1_A < \epsilon/2 + \epsilon/2$. ∎

7.6 ◆ The Existence of Nonatomic Countably Additive Probabilities

If $(\Omega, \mathcal{F}, P)$ is a finite probability space, then P must be countably additive. If $\Omega = \{\omega_n : n \in \mathbb{N}\}$ is countable and each $\{\omega_n\} \in \mathcal{F}$, then any sequence $p_n \geq 0$ with $\sum_n p_n = 1$ gives rise to a countably additive probability defined by $P(E) = \sum_{\{n:\omega_n \in E\}} p_n$. If those were the only examples of countably additive probabilities, the analyses in this chapter would be far less interesting and the proofs we give much more complicated than need be. At issue is whether or not there exist countably additive **nonatomic probabilities** on σ-fields. This turns out to be a question about the existence of extensions of probabilities from fields to the σ-fields they generate.

7.6.a The Importance of Nonatomic Probabilities on σ-Fields

Recall from Definition 7.2.4 (p. 371) that the idea of an "atom" is something that is irreducibly small. The existence question for finitely additive nonatomic probabilities on fields has an easy answer. For an interval $E = |a, b| \subset (0, 1]$, we defined $\lambda(E) = b - a$ and extended λ to $\mathcal{B}^\circ$, the set of finite disjoint unions of intervals additively. λ is clearly nonatomic and finitely additive on the field $\mathcal{B}^\circ$. We will see in Theorem 7.6.4 that it is also countably additive on $\mathcal{B}^\circ$. The question is whether or not λ can be extended to a countably additive probability on $\mathcal{B} = \sigma(\mathcal{B}^\circ)$. This is not "just a technical issue." If we do not have countably additive nonatomic probabilities on σ-fields, then our limit results so far are true only because the "if" part of the statements cannot be satisfied in any interesting case.

Lemma 7.6.1 *If X_n is an iid sequence of random variables in $L^0(\Omega, \mathcal{F}, P)$ and the X_n are not a.e. constant, then P is nonatomic.*

Proof. Since the X_n are not a.e. constant, there exists $r \in \mathbb{R}$ and $p \in (0, 1)$ such that $P(X_n \leq r) = p$. Let $q = \min\{p, 1 - p\}$, let $S = \sum_n Y_n/2^n$, and let $F_S(x) = P(S \leq x)$, where $Y_n = 1_{\{X_n \leq r\}}$.

F_S is continuous; if it is not, there is a sequence of intervals $A_n = (x - \epsilon_n, x + \epsilon_n)$, $\epsilon_n \downarrow 0$ such that $\lim_n P(S \in A_n) = \delta \geq 0$. But the probability that S is in any interval $[k/2^n, (k + 1)/2^n]$ is strictly between $q^n \downarrow 0$ and $(1 - q)^n \downarrow 0$.

F_S is also strictly increasing on $[0, 1]$ because there is a dyadic rational interval subset, $[k/2^n, (k + 1)/2^n]$, of every interval, and $P(S \in [k/2^n, (k + 1)/2^n]) > 0$.

Since F_S is continuous, it takes every value in the interval $[0, 1]$. Let $E_r = \{S \leq F_S^{-1}(r)\}$, so that the class of events $\{E_r : 0 \leq r \leq 1\}$ is nested and $P(E_r) = r$. Choose any E with $P(E) > 0$. The function $r \mapsto P(E \cap E_r)$ is continuous (by dominated convergence); hence it takes on every value in the interval $[0, P(E)]$. ∎

To reiterate, if we do not have a countably additive nonatomic probability, then we do not have sequences of nondegenerate iid random variables.

7.6.b Extension and Countable Additivity

The time has come to show that there are many more interesting countably additive probabilities. In this section we prove three results in this direction and a subsidiary result about the relation between countable additive probabilities and compact sets.

Theorem 7.6.2 (Carathéodory's Extension) *If P is a countably additive probability on a field $\mathcal{F}^\circ$, then it has a unique countably additive extension to $\mathcal{F} = \sigma(\mathcal{F}^\circ)$.*

Let us quickly dispense with the uniqueness part of the assertion.

Lemma 7.6.3 (Uniqueness) *If Q, Q' are countably additive probabilities that agree on $\mathcal{F}^\circ$, then they agree on $\mathcal{F} = \sigma(\mathcal{F}^\circ)$.*

Proof. The class $\mathcal{G} = \{E \in \mathcal{F} : Q(E) = Q'(E)\}$ is a monotone class containing $\mathcal{F}^\circ$. ∎

Recall that for a cdf F, $F^-(x) := \lim_{x_n \uparrow x} F(x_n)$. For any cdf F, remember that we defined P_F on $\mathcal{B}^\circ$, the class of finite disjoint unions of interval subsets of $\mathbb{R}$, by $P_F(-\infty, b] = F(b)$, $P_F(\infty, b) = F^-(b)$, $P_F[b, \infty) = 1 - P_f(-\infty, b)$, $P_F(b, \infty) = 1 - P_F(-\infty, b]$, $P_F(a, b] = F(b) - F(a)$, $P_F(a, b) = F(b) - F^-(a)$, $P_F[a, b] = F(b) - F^-(a)$, $P_F[a, b) = F^-(b) - F^-(a)$, and by additivity for finite unions of sets of the forms just given.

Theorem 7.6.4 (Countable Additivity from cdf's) *Every P_F is countably additive on $\mathcal{B}^\circ$.*

Combined with Carathéodory's extension theorem, this theorem says that every cdf F gives rise to a unique countably additive probability P_F on $\mathcal{B}$. Since every countably additive P on $\mathcal{B}$ gives rise to a unique cdf F_P, this establishes a bijection between cdf's and countably additive probabilities on $\mathcal{B}$.

Countably additive probabilities on $\mathbb{R}$ give rise to countably additive probabilities on metric spaces M. Given a countably additive probability P on $\mathbb{R}$, every random variable $X : \mathbb{R} \to M$ defines a countably additive probability on M, namely $\mu(A) = P_F(X^{-1}(A))$. If M is a complete, separable metric space (csm), we show that every countably additive probability on $\mathcal{B}_M$ arises this way. Let λ denote the uniform distribution on $(0, 1]$, that is, the countably additive extension of P_F, where $F(x) = x 1_{(0,1]}(x) + 1_{(1,\infty]}(x)$.

Theorem 7.6.5 (Countably Additive Probabilities on csm's) *If M is a csm, then μ is a countably additive probability on $\mathcal{B}_M$ iff there is a measurable function $X : (0, 1] \to M$ such that $\mu(A) = \lambda(X^{-1}(A))$.*

An essential property of countably additive probabilities on csm's is their **tightness**.

Definition 7.6.6 *A probability μ on a metric space $(M, \mathcal{B}_M)$ is **tight** if for every $\epsilon > 0$, there exists a compact set K_ϵ such that $\mu(K_\epsilon) > 1 - \epsilon$.*

We also prove that probabilities on csm's are tight.

Theorem 7.6.7 (Tight Probabilities on csm's) *If (M, d) is a complete separable metric space and μ is a countably additive probability on $(M, \mathcal{B}_M)$, then μ is tight.*

7.6.c Proving Carathéodory's Extension Theorem

Throughout this subsection, $\mathcal{F}^\circ$ is a field of sets generating $\mathcal{F}$. For $E, F \in \mathcal{F}^\circ$ and a probability $P : \mathcal{F}^\circ \to [0, 1]$ that is countably additive on $\mathcal{F}^\circ$, $\rho_P(E, F) :=$ $P(E \triangle F)$. Recall that $\rho_P(E, F) = \rho_P(E^c, F^c)$ because $(E \triangle F) = (E^c \triangle F^c)$. Also recall that it is quite possible that $E \neq F$ even though $\rho_P(E, F) = 0$. In this case, we say that $E \sim_P F$ and do not distinguish between E and F.

Our aim is to show that the completion of the metric space $(\mathcal{F}^\circ, \rho_P)$ is, up to isometry, $(\mathcal{F}, d_{\widehat{P}})$, where $d_{\widehat{P}}(E, F) := \widehat{P}(E \triangle F)$ and $\widehat{P}$ is the unique continuous extension of the function $E \mapsto P(E) = \rho_P(\emptyset, P)$.[9]

Observe that $|P(A) - P(B)| = |E\, 1_A - E\, 1_B| \leq E\, |1_A - 1_B| = \rho_P(A, B)$, so that $E \mapsto P(E)$ is uniformly continuous. The following recapitulates two earlier results, Theorem 7.3.26 (p. 395) and Corollary 7.2.23 (p. 7.2.23), and gives us further reason to believe that we can achieve our aim.

Theorem 7.6.8 *If P is countably additive on $\mathcal{F} = \sigma(\mathcal{F}^\circ)$ and $\mathcal{F}^\circ$ is a field, then $(\mathcal{F}, \rho_P)$ is complete and $\mathcal{F}^\circ$ is dense in $\mathcal{F}$.*

So, *if* we have a countably additive probability *already* defined on $\sigma(\mathcal{F}^\circ)$, then the completion of $(\mathcal{F}^\circ, \rho_P)$ is indeed $(\mathcal{F}, \rho_P)$ (up to renaming by an isometry). Countable additivity is crucial to this representation. Suppose that $E_n \downarrow \emptyset$ but that $\lim_n P(E_n) = \epsilon > 0$. The sequence E_n is Cauchy in $\mathcal{F}^\circ$, so in any completion it must have a representation as a nonempty set, but there is none in $\mathcal{F}$. This arises because $E_n \downarrow \emptyset$, which means that the representation ought to be $\emptyset$ itself.[10] For clarity we repeat the first basic result of interest.

Theorem 7.6.1 (Theorem 7.6.2 [Carathéodory's Extension]) *If P is a countably additive probability on a field $\mathcal{F}^\circ$, then it has a unique countably additive extension to $\mathcal{F} = \sigma(\mathcal{F}^\circ)$.*

Proof. The space $(\mathcal{F}^\circ, \rho_P)$ has a completion, $(\widehat{\mathcal{F}^\circ}, \widehat{d}_P)$. The proof will identify the elements of $\widehat{\mathcal{F}^\circ}$ with the sets in $\mathcal{F}$ and thereby extend the function $E \mapsto P(E) = \rho_P(\emptyset, E)$ from $\mathcal{F}^\circ$ to $\mathcal{F}$.

---

9. As a reminder, Theorem 6.8.11 (p. 334), the metric completion theorem, says that every metric space has a completion and that up to isometry, the completion is unique. Also, Corollary 6.7.18 tells us that every uniformly continuous function from a metric space to a complete metric space, for example [0, 1], has a unique continuous extension to the completion.

10. It is possible to add points to Ω so as to represent the limit, in other words, to "complete" the space Ω, so that the P that fails countable additivity has a countably additive representation on the expansion of Ω, see, for example, §11.4.c.

Step 1 adds to $\mathcal{F}^\circ$ all of the countable unions of elements of $\mathcal{F}^\circ$ and checks that the extension of P to this class of sets is countably additive. Step 2 completes the extension.

Step 1. $(\mathcal{F}^\circ, \bigcup^c) \subset \mathcal{F}$ is defined as the class of all countable unions of sets in $\mathcal{F}^\circ$. As a countable union of countable unions is a countable union, $(\mathcal{F}^\circ, \bigcup^c)$ is itself closed under countable unions.

Since $\mathcal{F}^\circ$ is a field, $E \in (\mathcal{F}^\circ, \bigcup^c)$ iff there exists $A_n \in \mathcal{F}^\circ$ such that $A_n \uparrow E$. We now show that $A_n \uparrow E$ implies that A_n is a Cauchy sequence in $(\mathcal{F}^\circ, \rho_P)$. Let $\widehat{A}$ be its limit and define $\widehat{P}(E) = \widehat{d}_P(\emptyset, \widehat{A})$. The following lemma shows that this leads to a unique definition of $\widehat{P}(E)$ on $(\mathcal{F}^\circ, \bigcup^c)$, and the subsequent lemma shows that $\widehat{P}$ is finitely additive, and continuous from below on $(\mathcal{F}^\circ, \bigcup^c)$. ∎

Lemma 7.6.9 *If A_n, $B_n \in \mathcal{F}^\circ$, A_n, $B_n \uparrow E$, then A_n and B_n are Cauchy equivalent.*

Proof. We must show that $\lim_n P(A_n \triangle B_n) = 0$. Let $p_A = \lim_n P(A_n)$ and $p_B = \lim_n P(B_n)$. For all fixed k, $(B_k \setminus A_n) \downarrow \emptyset$. Countable additivity implies that $P(B_k \setminus A_n) \downarrow 0$, so that $P(B_k) \leq p_A$; hence $p_B \leq p_A$. Reversing roles gives us $p_A \leq p_B$, so $p = p_A = p_B$.

Since $(A_n \cap B_n) \uparrow E$ and $(A_n \cup B_n) \uparrow E$, $\lim_n P(A_n \cap B_n) = \lim_n P(A_n \cup B_n) = p$. Finally, $\lim_n P(A_n \triangle B_n) = \lim_n \left[P(A_n \cup B_n) - P(A_n \cap B_n) \right] = 0$. ∎

Lemma 7.6.10 *$\widehat{P}$ is finitely additive, and continuous from below on $(\mathcal{F}^\circ, \bigcup^c)$.*

Proof. We begin with finite additivity. If $E \cap F = \emptyset$, E, $F \in (\mathcal{F}^\circ, \bigcup^c)$, take sequences A_n, B_n in $\mathcal{F}^\circ$ such that $A_n \uparrow E$ and $B_n \uparrow F$. Since $A_n \cap B_n = \emptyset$, $P(A_n \cup B_n) = P(A_n) + P(B_n)$. The left-hand side converges up to $\widehat{P}(E \cup F)$, while the right-hand side converges up to $\widehat{P}(E) + \widehat{P}(F)$.

Now suppose that $F_n \in (\mathcal{F}^\circ, \bigcup^c)$ and $F_n \uparrow F$. We must show that $\widehat{P}(F_n) \uparrow \widehat{P}(F)$. For each n, let $A_{n,k} \uparrow F_n$ be a sequence in $\mathcal{F}^\circ$. Define $B_m = \cup_{n,k \leq m} A_{n,k}$ such that $B_m \subset F_m$ and $B_m \uparrow F$. By Lemma 7.6.9, $P(B_m) \uparrow \widehat{P}(F)$ and $P(B_m) \leq \widehat{P}(F_m)$, so that $\widehat{P}(F_m) \uparrow \widehat{P}(F)$. ∎

Step 2. Let $\mathcal{N}(\epsilon) = \left\{ F \in \mathcal{F} : \left(\exists F^\epsilon \in (\mathcal{F}^\circ, \bigcup^c) \right) \left\{ \left[F \subset F^\epsilon \wedge \widehat{P}(F^\epsilon) < \epsilon \right] \right\} \right\}$. If P had already been extended to $\mathcal{F}$, $\mathcal{N}(\epsilon)$ would be the ϵ-ball around $\emptyset$.

The good sets are

$$\mathcal{G} = \left\{ E \in \mathcal{F} : (\forall \epsilon > 0) \left(\exists A \in \mathcal{F}^\circ \right) \left[E \triangle A \in \mathcal{N}(\epsilon) \right] \right\}, \tag{7.35}$$

and they represent the limits of sequences of sets in $\mathcal{F}^\circ$. For $E \in \mathcal{G}$, we define $\widehat{P}(E) = \cap_{\epsilon > 0} \mathrm{cl}\{P(A) : A \in \mathcal{F}^\circ, (E \triangle A) \in \mathcal{N}(\epsilon)\}$.

Step 2a shows that $\mathcal{G} = \mathcal{F}$, and Step 2b shows that for each $E \in \mathcal{F}$, $\widehat{P}(E)$ is a single number (rather than a set of numbers) and that $\widehat{P}$ is a countably additive extension of P to $\mathcal{F}$.

Step 2a. To show that $\mathcal{G} = \mathcal{F}$, we begin with the following.

Lemma 7.6.11 *If $E_k \in \mathcal{N}(\epsilon_k)$, then $\cup_k E_k \in \mathcal{N}(\sum_k \epsilon_k)$.*

Proof. For each k, pick $A_k \in \mathcal{F}^\circ$ with $(E_k \triangle A_k) \subset F_k$, $F_k \in (\mathcal{F}^\circ, \bigcup^c)$, and $\widehat{P}(F_k) < \epsilon/2^k$. Observe that $(\cup_k E_k) \triangle (\cup_k A_k) \subset \cup_k (E_k \triangle A_k) \subset \cup_k F_k$. By the countable additivity of $\widehat{P}$ on $(\mathcal{F}^\circ, \bigcup^c)$, $\widehat{P}(\cup_k F_k) \leq \sum_k \epsilon_k$. ∎

Lemma 7.6.12 $\mathcal{G} = \mathcal{F}$.

Proof. $\mathcal{G}$ contains $\mathcal{F}°$; thus it contains $\emptyset$ and Ω and is closed under complementation and finite unions. Hence it is sufficient to show that $\mathcal{G}$ is closed under countable monotonic unions.

Let E_n be a sequence in $\mathcal{G}$, $E_n \uparrow E$. For each n, pick $A_n \in \mathcal{F}°$ with $(E_n \Delta A_n) \in \mathcal{N}(\epsilon/2^n)$. Since $E \Delta (\cup_n A_n) \subset \cup_n (E_n \Delta A_n)$, $E \in \mathcal{N}(\sum_n \epsilon/2^n)$ (by Lemma 7.6.11) and ϵ is arbitrary.   ∎

Step 2b. $\widehat{P}$ is a countably additive extension of P to $\mathcal{F}$.

We first show that our definition of $\widehat{P}$, $\widehat{P}(E) = \cap_{\epsilon>0} \mathrm{cl}\{P(A) : A \in \mathcal{F}°,$ $(E \Delta A) \in \mathcal{N}(\epsilon)\}$, really does give a number. Define $R_\epsilon := \mathrm{cl}\{P(A) : A \in \mathcal{F}°, \ (E \Delta A) \in \mathcal{N}(\epsilon)\}$. The set $\mathcal{G}_2 \subset \mathcal{F}$ for which $R(\epsilon) \neq \emptyset$ for all $\epsilon > 0$ is a field containing $\mathcal{F}°$, so it is equal to $\mathcal{F}$. For any $E \in \mathcal{F}$, $\{R(\epsilon) : \epsilon > 0\}$ is a collection of compact sets with the finite-intersection property, so $\widehat{P}(E) := \cap_{\epsilon>0} R(\epsilon) \neq \emptyset$. The following shows that $\widehat{P}(E)$ is a single number.

Lemma 7.6.13   *If A, $B \in \mathcal{F}°$, $(E \Delta A) \in \mathcal{N}(\epsilon)$, and $(E \Delta B) \in \mathcal{N}(\epsilon)$, then $|P(A) - P(B)| < 2\epsilon$.*

Proof.   We know that $|P(A) - P(B)| \leq P(A \Delta B)$. By assumption,

$$\exists F_A^\epsilon \in (\mathcal{F}°, \bigcup{}^c) \text{such that } E \Delta A \subset F_A^\epsilon, \ \widehat{P}(F_A^\epsilon) < \epsilon, \tag{7.36}$$

and

$$\exists F_B^\epsilon \in (\mathcal{F}°, \bigcup{}^c) \text{such that } E \Delta B \subset F_B^\epsilon, \ \widehat{P}(F_B^\epsilon) < \epsilon. \tag{7.37}$$

Define $F^\epsilon = F_A^\epsilon \cup F_B^\epsilon$ such that $F^\epsilon \in (\mathcal{F}°, \bigcup{}^c)$ and $\widehat{P}(F^\epsilon) < 2\epsilon$. For any E, $(A \Delta B) \subset (E \Delta A) \cup (E \Delta B)$ (because $|1_A - 1_B| \leq \max\{|1_E - 1_A|, |1_E - 1_B|\}$), so that $(A \Delta B) \subset F_A^\epsilon \cup F_B^\epsilon$, implying that $P(A \Delta B) \leq \widehat{P}(F^\epsilon) < 2\epsilon$.   ∎

To summarize what we have so far, $\widehat{P} : \mathcal{F} \to [0, 1]$ agrees with P on $\mathcal{F}°$. All that is left to complete the proof of Carathéodory's extension theorem is to show that $\widehat{P}$ is a countably additive probability on $\mathcal{F}$. This follows from the continuity of $\widehat{P}$ on $\widehat{\mathcal{F}}°$.

Lemma 7.6.14   *$\widehat{P}$ is a countably additive probability on $\mathcal{F}$.*

Proof.   Finite additivity is immediate. Let $E_n \uparrow E$ and pick arbitrary $\epsilon > 0$. For each n, pick $A_n \in \mathcal{F}°$ such that $(E_n \Delta A_n) \in \mathcal{N}(\epsilon/2^{n+1})$. This yields $|\sum_n P(A_n) - \sum_n \widehat{P}(E_n)| < \epsilon/2$. We know that $E \Delta (\cup_m A_m) \subset \cup_n (E_n \Delta A_n)$, so that $E \Delta (\cup_m A_m) \in \mathcal{N}(\epsilon/2)$. For M large, $(\cup_m A_m) \Delta (\cup_{m \leq M} A_m) \in \mathcal{N}(\epsilon/2)$. Therefore, $E \Delta (\cup_{m \leq M} A_m) \in \mathcal{N}(\epsilon)$. Combining yields $|\widehat{P}(E) - \sum_n \widehat{P}(E_n)| < \epsilon$.   ∎

7.6.d   Countable Additivity from cdf's

Before showing that every P_F is countably additive on $\mathcal{B}°$, we need a definition and a lemma.

Definition 7.6.15   *A subset $\mathcal{C}$ of a field $\mathcal{F}°$ is a **compact class** if it is closed under finite unions and for any sequence C_n in $\mathcal{C}$, $[\cap_n C_n = \emptyset] \Rightarrow (\exists N)[\cap_{n \leq N} C_n = \emptyset]$.*

It is clear that the nonempty compact subsets of a metric space are a compact class, and in our first applications, this is the only kind of compact class we use. Later, in proving Theorem 7.7.3 (p. 426) we use a compact class that is related to, but not the same as, this class of compact sets.

Lemma 7.6.16 *If P is finitely additive on a field $\mathcal{F}^\circ$, $\mathcal{C}$ is a compact class, and for all $E \in \mathcal{F}^\circ$, $P(E) = \sup\{P(C) : C \subset E, \ C \in \mathcal{C}\}$, then P is countably additive.*

Proof. Let $E_n \downarrow \emptyset$, $E_n \in \mathcal{F}^\circ$ and pick $\epsilon > 0$. We must show that $P(E_n) < \epsilon$ for large n. Pick $C_n \subset E_n$ such that $P(E_n \setminus C_n) < \epsilon/2^n$. Since $\cap_n C_n \subset \cap_n E_n = \emptyset$ and $\mathcal{C}$ is a compact class, there exists N such that $\cap_{n \le N} C_n = \emptyset$. We show that for all $n \ge N$, $E_n \subset \cup_{m=1}^n [E_m \setminus C_m]$. This completes the proof because it implies that $P(E_n) \le \sum_{m=1}^n \epsilon/2^m < \epsilon$.

If $\omega \in E_n$, then $\omega \in E_m$ for all $m \le n$ because E_n is a decreasing sequence of sets. Since $\cap_{m=1}^n C_m = \emptyset$, we know that $\exists m' \le n$ such that $\omega \notin C_{m'}$. Therefore $\omega \in E_{m'} \setminus C_{m'}$. ■

We now repeat and prove the second basic result of interest.

Theorem 7.6.4 *Every P_F is countably additive on $\mathcal{B}^\circ$.*

Proof. Let $\mathcal{C} \subset \mathcal{B}^\circ$ be the compact class consisting of finite disjoint unions of compact intervals $[a, b]$. Checking the cases shows that $P_F(E) = \sup\{P(C) : C \subset E, \ C \in \mathcal{C}\}$ for all $E \in \mathcal{B}^\circ$. Apply the previous lemma. ■

It is worth reiterating what has been done here: every countably additive P gives rise to a unique cdf, F_P, every cdf F gives rise to a unique probability P_F, and F is the cdf of P_F. Thus, we have a bijection between cdfs and countably additive probabilities on $\mathcal{B}$.

Recall that a metric space is complete if all Cauchy sequences converge and separable if it has a countable dense set. The inner approximability by compact sets that we used in the proof of Theorem 7.6.4 is at the heart of countable additivity for probabilities on complete separable metric spaces (csm's).

7.6.e Countably Additive Probabilities on csm's

Given a countably additive P on $(\mathbb{R}, \mathcal{B})$, it is easy to produce countably additive probabilities on any metric space, M. For a measurable $X : \mathbb{R} \to M$, define $\mu(A) = P(X^{-1}(A))$ for $A \in \mathcal{B}_M$. We have μ as a countably additive probability on $(M, \mathcal{B}_M)$ because inverse images preserve set operations. The third of the four results we said we would prove in this section tells us that this is the only way to generate countably additive probabilities on csm's. We now repeat the statement of the third basic result of interest.

Theorem 7.6.5 *If M is a csm, then μ is a countably additive probability on $\mathcal{B}_M$ iff there is a measurable function $X : (0, 1] \to M$ such that $\mu(A) = \lambda(X^{-1}(A))$.*

We saw that $M = [0, 1]^{\mathbb{N}}$ with the product metric was the universal separable metric space in the sense that there is a continuous bijection with a continuous inverse between any separable metric space and some subset of M. The Borel isomorphism theorem shows that there is a measurable bijection with a measurable inverse between any pair of complete separable metric spaces having the same cardinality. In this sense, the measure space $((0, 1], \mathcal{B})$ is the universal measure space for all of the uncountable measure spaces $(M, \mathcal{B}_M)$, where $\mathcal{B}_M$ is the Borel σ-field when (M, d) is a csm.

Theorem 7.6.5 is a direct implication of the Borel isomorphism theorem. However, proving the theorem would take us further afield than we wish to go. Thus, we state it, give the short proof of Theorem 7.6.5 that uses it, and give a longer, more elementary proof.

There is a related result, the measure algebra isomorphism theorem, which tells us that there is an isometry between the separable metric space $(\mathcal{B}, \rho_\lambda)$ and any separable $(\mathcal{F}, \rho_P)$ when P is nonatomic. This makes $(\mathcal{B}, \rho_\lambda)$ into a universal measure algebra.

7.6.e.1 Borel Isomorphisms

Definition 7.6.17   *Two measurable spaces* $(A, \mathcal{A})$ *and* $(A', \mathcal{A}')$ *are* **measurably isomorphic** *if there exists a bijection* $f : A \leftrightarrow A'$ *that is measurable and has a measurable inverse.*

When two measurable spaces are measurably isomorphic, probabilities on one space correspond directly to probabilities on the other. If μ is a probability on A, then $f(\mu)$ is a probability on A', and all probabilities on A' are of the form $f(\mu)$ for some probability μ on A.

Theorem 7.6.18 (Borel Isomorphism)   *If* (M, d) *and* (M', d') *are complete separable metric spaces and* B *and* B' *are Borel measurable subsets of* M *and* M', *respectively, then they are measurably isomorphic iff they have the same cardinality.*

In particular, $(0, 1]$, $\mathbb{R}$, $C(X)$, X a compact metric space, and the separable $L^p(\Omega, \mathcal{F}, P)$ spaces, $p \in [1, \infty)$, are all measurably isomorphic. We do not prove the Borel isomorphism theorem.[11] However, we offer a short proof of Theorem 7.6.5 based on it, as well as a longer, clunkier, more direct proof.

Proof of Theorem 7.6.5 based on Borel isomorphisms.   If $M = \{x_n : n \in \mathbb{N}\}$ is countable, the requisite X is easy to construct: define $p_0 = 0$, $p_n = \sum_{k \leq n} \mu(x_k)$ and define $X(\omega) = x_k$ for all $\omega \in (p_{k-1}, p_k]$. If M is uncountable, let $f : \mathbb{R} \leftrightarrow M$ be a Borel isomorphism and let G be the cdf of $f^{-1}(\mu)$ and P_G the associated probability. Define $X_G : (0, 1) \to \mathbb{R}$ by $X_G(\omega) = \inf\{x : G(x) \geq \omega\}$. It is clear that $P_G = X_G(\lambda)$, so that $\mu = f(X_G(\lambda))$.   ∎

The function X_G used in this proof is called the **probability integral transform of** G. It directly shows that any probability distribution on $\mathbb{R}$ is of the form $X(\lambda)$ for a nondecreasing function.

A Clunkier Proof of Theorem 7.6.5.   The requisite X is the limit of a sequence of measurable functions. To prepare for the construction of the sequence of measurable functions, let $M' = \{y_n : n \in \mathbb{N}\}$ be a countable dense subset of

---

11. See Dudley, Ch. 13; Dellacherie and Meyer, Ch. III; and Cohn, *Measure Theory* (Boston: Birkhäuser, 1980), Ch. 8, for proofs.

M and let $\mathcal{E} = \{G_n : n \in \mathbb{N}\}$ be the set of open balls around points in M' with a rational radius. Let $\mathcal{B}_M^\circ = \cup_n \sigma(\{G_k : k \leq n\})$. $\mathcal{B}_M^\circ$ is a field, and a simple good sets argument verifies that $\mathcal{B}_M = \sigma(\mathcal{B}_M^\circ)$ (because every open $G \subset M$ is a union of elements of $\mathcal{E}$).

For each n, let $\mathcal{P}_n = \{E_1^n, \ldots, E_{K(n)}^n\}$ be the partition generated by $\{G_k : k \leq n\}$. For each $E_k^n \in \mathcal{P}_n$, let x_k^n be a point in E_k^n. This sequence of partitions has three properties relevant for the present construction:

1. The partitions become finer, and for all n, the partition $\mathcal{P}_{n+1}$ is a refinement of the partition $\mathcal{P}_n$.

2. The diameter of the elements shrinks to 0, if $E_{k(n+1)}^{n+1} \subset E_{k(n)}^n$ for all n, then $\sup_{x,y \in E_{k(n)}^n} d(x,y) \to 0$.

3. For all $x \in M$, if $x \in E_{k(n)}^n$ for all n, then $x_{k(n)}^n \to x$.

For each n, define $p_0^n = 0$, and for $k \leq K(n)$, define $p_k^n = \sum_{\kappa \leq k} \mu(E_\kappa^n)$. For each $\omega \in (p_{k-1}^n, p_k^n]$, define $X_n(\omega) = x_k^n$. Each X_n is measurable, for each ω, $X_n(\omega)$ is Cauchy, and we define $X(\omega) = \lim_n X_n(\omega)$. We must show that $\lambda(X \in E) = \mu(E)$ for each $E \in \mathcal{B}$.

Note that for all $n' \leq n \leq m$, $\lambda(X_m \in G_{n'}) = \mu(G_{n'})$, so that $\lambda(X_m \in G_{n'})$ gets to and stays at the value $\mu(G_{n'})$. It would be tempting to directly conclude that $\lambda(\lim_m X_m \in G_{n'}) = \mu(G_{n'})$. However, a sequence $X_m(\omega)$ that belongs to an open set, $G_{n'}$, can converge to something outside of the open set, so a more delicate argument is needed.[12]

For closed $F \in \sigma(\{G_k : k \leq n\})$, $\lambda(X \in F) \geq \mu(F)$, where the "$\geq$" comes from the observation that the $X_n(\omega) \in F^c$ might converge to points in F. To show that $\lambda(X \in F) = \mu(F)$, it is sufficient to show that $\lambda(X \in F^c) \geq \mu(F^c)$.

Recall the dense set $M' = \{y_n : n \in \mathbb{N}\}$ with which we started the construction. For each $y_n \in F^c$, let $\epsilon_n = d(y_n, F) > 0$ and pick a rational $q_n \in (\frac{1}{2}\epsilon_n, \epsilon_n)$. For $\delta > 0$, $F^\delta = \cup_{x \in F} B_\delta(x)$ is the δ-ball around F. For any sequence $\delta_n \downarrow 0$, we can express the open set F^c as the countable union of the closed sets $(F^{\epsilon_m})^c$. Therefore, $F^c = \cup_n \mathrm{cl}\, B_{q_n}(y_n)$, which implies that $\cup_{n \leq N} \mathrm{cl}\, B_{q_n}(y_n) \uparrow F^c$ and $\mu(\cup_{n \leq N} \mathrm{cl}\, B_{q_n}(y_n)) \uparrow \mu(F^c)$. Since $B_{q_n}(y_n) \in \mathcal{E}$, $\lambda(\lim_m X_m \in \mathrm{cl}\, B_{q_n}(y_n)) \geq \mu(\mathrm{cl}\, B_{q_n}(y_n))$. ∎

7.6.f Nonatomic Probabilities and Measure Algebra Isomorphisms

Our first result, the measure algebra isomorphism theorem, tells us that if $(\Omega, \mathcal{F}, P)$ is nonatomic and countably generated, then it very much "looks like" the probability space $((0, 1], \mathcal{B}, \lambda)$. Our second result gives six different characterizations of nonatomic probabilities.

7.6.f.1 The Measure Algebra Isomorphism Theorem

Definition 7.6.19 *Let* $(\mathcal{A}, \mu)$ *and* $(\mathcal{A}', \mu')$ *be two σ-fields and countably additive probabilities. A **measure algebra isomorphism** is a bijection* $T : \mathcal{A} \leftrightarrow$

---

12. We did warn you that this is a clunkier proof.

A' such that for all E, F and sequences E_n in A', $T(E \setminus F) = T(E) \setminus T(F)$, $T(\cup_n E_n) = \cup_n T(E_n)$ and $\mu(E) = \mu'(T(E))$.

Measure algebra isomorphisms simply rename a set $E \in A$ as the set $T(E) \in A'$. We have seen renaming before for metric spaces. This is the same idea.

Exercise 7.6.20 T is a measure algebra isomorphism between (A, μ) and (A', μ') iff T is an isometry between the metric spaces (A, ρ_μ) and $(A', \rho_{\mu'})$.

A measure algebra isomorphism is a bijection between σ-fields, whereas a Borel isomorphism is a bijection between sets. Every measurable isomorphism, $f : A \leftrightarrow A'$, gives rise to a measure algebra isomorphism, defined by $T(E) = \{f(\omega) : \omega \in E\}$, but the reverse is not true.

Reprising Definition 7.2.4 for measure algebras, an **atom** of a measure algebra (A, μ) is a set $E \in A$ such that $\mu(E) > 0$, and for all $F \subset E$, $F \in A$, $\mu(F) = 0$ or $\mu(F) = \mu(E)$. Both the measure μ and the measure algebra (A, μ) are called **nonatomic** if the measure algebra has no atoms. A measure algebra is **separable** if the metric space (A, d_μ) is separable. To be specific here, let $\mathcal{B}_{(0,1]}$ denote the Borel subsets of $(0, 1]$ and let λ be the uniform distribution on $\mathcal{B}_{(0,1]}$. The measure algebra $(\mathcal{B}_{(0,1]}, \lambda)$ is separable and nonatomic. The following theorem says that any separable, nonatomic measure algebra, (A, μ), is, to within a measure algebra isomorphism, the same as $(\mathcal{B}_{(0,1]}, \lambda)$.

Theorem 7.6.21 (Measure Algebra Isomorphism) (A, ρ_μ) is separable and nonatomic iff there is an isometry $T : A \leftrightarrow \mathcal{B}_{(0,1]}$ between (A, ρ_μ) and $(\mathcal{B}_{(0,1]}, \rho_\lambda)$.

The following lemma is the crucial piece of the proof of this theorem.

Lemma 7.6.22 $(\Omega, \mathcal{F}, P)$ is nonatomic iff for all $\epsilon > 0$, there is a finite, measurable partition, $\{E_k : k \leq K\}$ of Ω such that for all k, $P(E_k) < \epsilon$.

Proof.   If P has an atom of size ϵ, then for all finite partitions, some E_k has $P(E_k) \geq \epsilon$.

Now suppose that P is nonatomic. For any finite partition $\mathcal{E} = \{E_1, \ldots, E_K\}$, define $|\mathcal{E}| = \max_k P(E_k)$. We wish to show that there is a sequence of finite partitions, $\mathcal{E}_n$, with $|\mathcal{E}_n| \downarrow 0$.

Suppose that $\inf |\mathcal{E}| = \epsilon > 0$, where the infimum is taken over all finite partitions. Let $\mathcal{E}'_n$ be a sequence with $|\mathcal{E}'_n| \downarrow \epsilon$. For each $n \in \mathbb{N}$, define $\mathcal{E}_n$ as the partition generated by $\{\mathcal{E}'_i : i \leq n\}$, so that $\mathcal{E}_n$ is a sequence of increasingly fine partitions. There must be at least one element $E_k \in \mathcal{E}_1 = \{E_1, \ldots, E_K\}$ such that $|P_n \cap E_k| \downarrow \epsilon$. Relabel each such E_k as $F_{1,k}$, $k \in K'$.

For each $F_{1,k}$, the sequence of partitions given by $\mathcal{E}_n \cap F_{1,k}$ satisfies $|\mathcal{E}_n \cap F_{1,k}| \downarrow \epsilon$ for all $n \in \mathbb{N}$. Therefore, we can find an $F_{2,k} \in \mathcal{E}_2 \cap F_{1,k}$ such that $|P_n \cap F_{2,k}| \downarrow \epsilon$ for all $n \in \mathbb{N}$. Continuing inductively gives a sequence $F_{n,k} \downarrow F_k := \cap_n F_{n,k}$ with $P(F_{n,k}) \downarrow \epsilon$, so that $P(F_k) = \epsilon$. Since P is nonatomic, there exists an $F_{0,k} \subset F_k$ such that $0 < P(F_{0,k}) < P(F_k)$. Let $\mathcal{E}''_n$ be the partition generated by $\mathcal{E}_n$ and $\{F_{0,k} : k \in K'\}$, so that $\lim_n |\mathcal{E}''| < \epsilon$, which contradicts $\inf |\mathcal{E}| = \epsilon$.   ∎

Proof of Theorem 7.6.21.   The metric space (A, ρ_P) is separable iff A is countably generated. Let $\{A_n : n \in \mathbb{N}\}$ generate A and let $\mathcal{B}_n$ be an increasingly fine

sequence of partitions with $|\mathcal{B}_n| \downarrow 0$. For each $n \in \mathbb{N}$, let $\mathcal{E}_n$ be the partition gener-
ated by $\mathcal{B}_n$ and $\{A_1, \ldots, A_n\}$, so that $|\mathcal{E}_n| \leq |\mathcal{B}_n|$. Since $\mathcal{A}^\circ := \cup_n \sigma(\mathcal{E}_n)$ is a field
that generates $\mathcal{A}$, it is ρ_P-dense.

For $\mathcal{E}_1 = \{E_1, \ldots, E_{K_1}\}$, define $T(E_1) = (0, \lambda(E_1)] \subset (0, 1]$, $T(E_2) = (\lambda(E_1), \lambda(E_1) + \lambda(E_2)] \subset (0, 1]$, or more generally,

$$T(E_k) = (\textstyle\sum_{j \leq k-1} \lambda(E_j), \sum_{j \leq k} \lambda(E_j)] \subset (0, 1]$$

for $k = 1, \ldots, K_1$. For the inductive part, T has been defined for $\mathcal{E}_n = \{E_1, \ldots, E_K\}$ and maps each E_k to the interval $T(E_k) = (\sum_{i<k} P(E_i), \sum_{i \leq k} P(E_i)]$. For
each $E_k \in \mathcal{E}_n$, $\mathcal{E}_{n+1} \cap E_k$ partitions each E_k. Subdivide the interval $T(E_k) \subset (0, 1]$
in the same fashion.

This defines T for all elements of the partitions in the sequence $\mathcal{E}_n$. For finite
unions of such elements, define T to be the corresponding finite union in $\mathcal{B}$.
This defines T on $\mathcal{A}^\circ$. $T(\mathcal{A}^\circ)$ is dense in $\mathcal{B}_{(0,1]}$ because $T(\mathcal{A}^\circ)$ is a field and
$\mathcal{B}^\circ := \sigma(T(\mathcal{A}^\circ)) = \mathcal{B}_{(0,1]}$ because $|\mathcal{E}_n| \downarrow 0$. Combining these observations, we see
that T is an isometry from the dense set $\mathcal{A}^\circ$ to the dense set $\mathcal{B}^\circ$. Hence it has a
unique uniformly continuous extension, $\overline{T}$, which is also an isometry. ∎

7.6.f.2 Six Ways to Be Nonatomic

In principle, any one of the following could have been used as the definition of
P being nonatomic. However, the weaker conditions are easier to verify than the
others, and the chosen definition of nonatomic probabilities, 7.2.4, is the easiest
of all of them.

Theorem 7.6.23 *The following are equivalent:*

1. *P is nonatomic.*

2. *For all $\epsilon > 0$, there is a finite measurable partition, $\{E_k : k \leq K\}$, of Ω such that for all k, $P(E_k) < \epsilon$.*

3. *For all $E \in \mathcal{F}$, there exists an $E' \subset E$ such that $P(E') = \frac{1}{2}P(E)$.*

4. *For all $E \in \mathcal{F}$, there is a subfamily of subsets of E, $\{E_r : r \in [0, 1]\}$, that is nested for all $r < s$, $E_r \subset E_s$, and for all r, $P(E_r) = r \cdot P(E)$.*

5. *There exists $X : \Omega \to [0, 1]$ such that $\forall r \in [0, 1]$, $P(X \leq r) = r$.*

6. *There exists an iid sequence of nondegenerate random variables on Ω.*

Proof. $(1) \Leftrightarrow (2)$ is Lemma 7.6.22.

$(3) \Rightarrow (4)$. Define E_1 so that $P(E_1) = \frac{1}{2}P(\Omega) = \frac{1}{2}$, and define $E_2 = E_1^c$. This
gives an ordered partition of Ω, $\mathcal{E}_1 = (E_1, E_2)$ into 2^1 elements each having
probability $1/2^1$. Split each element of $\mathcal{E}_1$ into a first half and a second half,
each having probability $1/2^2$, and order them as $\mathcal{E}_2 = (E_1, \ldots, E_4)$. Continuing
inductively gives a sequence of ordered partitions $\mathcal{E}_n = (E_1, \ldots, E_{2^n})$. For $r \in
(0, 1]$, let $k_n(r)$ be the largest integer with $k/2^n \leq r$. Let E_n be the union of the
first $k_n(r)$ elements of $\mathcal{E}_n$. Then $E_n \uparrow E := \cup_n E_n$, and $P(E_n) = k_n(r)/2^n \uparrow r$.

$(4) \Rightarrow (3)$ is immediate.

$(2) \Rightarrow (4)$. Suppose first that there is a measure algebra isomorphism between
$(\Omega, \mathcal{F}, P)$ and $((0, 1], \mathcal{B}, \lambda)$, as would happen if $(\Omega, \mathcal{F}, P)$ is countably generated
(Theorem 7.6.21). Define $E_x = E \cap (0, x]$ such that $g(0) = 0$ and $g(1) = P(E)$.

By dominated convergence, $g(\cdot)$ is continuous; hence for some $x \in [0, 1]$, $g(x) = r P(E)$. By the measure algebra isomorphism, this guarantees that the same is true for $(\Omega, \mathcal{F}, P)$.

In order to apply the measure algebra isomorphism, it is sufficient to find a countably generated sub-σ-field of $\mathcal{F}$ on which P is nonatomic. Let $\mathcal{E}_n$ be a sequence of finite partitions with $|\mathcal{E}_n| \downarrow 0$. $\mathcal{G} := \sigma(\{\mathcal{E}_n : n \in \mathbb{N}\})$ is countably generated, and because (2) $\Rightarrow$ (1), P is nonatomic on $\mathcal{G}$.

(4) $\Leftrightarrow$ (5) is immediate. (5) $\Rightarrow$ (6) because we know how to define countably many independent random variables on $((0, 1], \mathcal{B}, \lambda)$, and (6) $\Rightarrow$ (1) is Lemma 7.6.1.   ■

7.6.g   The Tightness of Probabilities on csm's

We begin with the following "sandwiching" theorem, valid for any metric space, not only the complete separable ones.

Theorem 7.6.24   *If μ is a countably additive probability on $\mathcal{B}_M$, the Borel σ-field of a metric space M, then for all $E \in \mathcal{B}_M$ and all $\epsilon > 0$, $\exists F$ closed, $\exists G$ open, such that*

$$F \subset E \subset G, \quad \text{and} \quad \mu(G \setminus F) < \epsilon.$$

Thus any set E can be sandwiched between a closed set F and an open set G, and the sandwich can be arbitrarily thin.

Exercise 7.6.25   Prove Theorem 7.6.24 using a good sets argument.

Adding completeness to the assumptions gives us sandwiching between compact and open sets.

Theorem 7.6.26   *If μ is a countably additive probability on $\mathcal{B}_M$, the Borel σ-field of a complete separable metric space M, then for all $E \in \mathcal{B}_M$ and all $\epsilon > 0$, $\exists K$ compact, $\exists G$ open, such that*

$$K \subset E \subset G, \quad \text{and} \quad \mu(G \setminus K) < \epsilon.$$

If we let $E = M$, this means that for all $\epsilon > 0$, there exists a compact set having probability at least $1 - \epsilon$. If this is satisfied, it implies that compact sets are almost all that there is, from a probabilistic point of view. This property has its own name, and we reproduce the definition here.

Definition 7.6.6.   A probability μ on a metric space $(M, \mathcal{B}_M)$ is **tight** if for every $\epsilon > 0$, there exists a compact set K_ϵ such that $\mu(K_\epsilon) > 1 - \epsilon$.

To prove Theorem 7.6.26, it is sufficient to show that μ is tight—if $\mu(K) > 1 - \epsilon/2$ and F-G form a closed-open sandwich of E with $\mu(G \setminus F) < \epsilon/2$, then $(K \cap F) \subset E \subset G$, $K \cap F$ is compact and $\mu(G \setminus (K \cap F)) < \epsilon$. The easiest proof of the tightness of μ uses the fact that a set K is compact iff it is totally bounded (for all $\epsilon > 0$, there is a finite ϵ-net) and complete. Since any closed subset of a csm is complete, the closure of a totally bounded subset of a csm is compact.

Proof of Theorem 7.6.26. By the arguments above, it is sufficient to show that μ is tight. Pick arbitrary $\epsilon > 0$. We must show the existence of a compact K such that $\mu(K) > 1 - \epsilon$.

Since X is separable, for each n there is a sequence $(A_{n,k})_{k=1}^{\infty}$ of $1/n$-balls covering X. As μ is countably additive, there exists k_n such that $\mu(\cup_{k \leq k_n} A_{n,k}) > 1 - \epsilon/2^n$. The set $C = \cap_n \cup_{k \leq k_n} A_{n,k}$ is totally bounded. Using countable additivity again yields $\mu(C) > 1 - \epsilon$. Since X is complete, $K = \text{cl}(C)$ is compact. Finally, $C \subset K$, so that $\mu(K) > 1 - \epsilon$. ∎

Exercise 7.6.27 There is an easy proof that a countably additive probability on the Borel subsets of $\mathbb{R}^k$ is tight—each $E_n = \text{cl}(B_n(0))$ is compact and $P(E_n) \uparrow P(\mathbb{R}^k) = 1$. Why does this proof not work for $C([0, 1])$?

7.6.h Complements and Problems

Sometimes all finitely additive probabilities are countably additive, and Carathéodory's extension theorem is not needed.

Exercise 7.6.28 Let S be a finite set, $M = S^{\mathbb{N}}$, and for $x = (x_1, x_2, \ldots)$ and $y = (y_1, y_2, \ldots)$ in M. The product metric, $d(x, y) = \sum_n \frac{1}{2^n} 1_{x_n \neq y_n}$, makes M into an uncountable compact metric space.

1. Show that $x_n = (x_{n,1}, x_{n,2}, \ldots, x_{n,k}, \ldots)$ converges to $y = (y_1, y_2, \ldots, y_k, \ldots)$ in M iff for all $K \in \mathbb{N}$, there exists an N such that for all $n \geq N$ and all $k \leq K$, $x_{n,k} = y_k$.

2. Show that for all K, $\mathcal{B}_M^\circ(K) = \{\text{proj}_{n:n \leq K}^{-1}(A), \ A \subset S^K\}$ is a field of compact subsets of M.

3. Show that $\mathcal{B}_M = \sigma(\cup_K \mathcal{B}_M^\circ(K) : K \in \mathbb{N})$.

4. Show that any finitely additive probability on $\mathcal{B}_M$ is countably additive. [Hint: Compact classes.]

Every countably additive probability on $(\Omega, \mathcal{F}) = (\mathbb{N}, \mathcal{P}(\mathbb{N}))$ is purely atomic; that is, Ω can be partitioned into countably many atoms. The same is not true if we allow countable additivity to fail. One can measure the degree of failure of countable additivity of a probability μ as $\delta(\mu) = \sup\{\lim_n \mu(B_n) : B_n \downarrow \emptyset\}$. If $\delta(\mu) = 1$, then μ is **purely finitely additive**. Sometimes, probabilities on fields are hopelessly finitely additive.

Exercise 7.6.29 (An Atomic Purely Finitely Additive Probability) Consider the left-continuous function $F(x) = \sum_n \frac{1}{2^n} 1_{(n,+\infty)}(x)$. For any left-open, right-closed interval $(a, b]$, define $P_F((a, b]) = F(b) - F(a)$. Let $\mathcal{I}^\circ$ denote the smallest field (not σ-field) containing the intervals $(a, b]$. Characterize $\mathcal{I}^\circ$, show that $\sigma(\mathcal{I}^\circ) = \mathcal{B}$, and show that there is a sequence of sets $B_n \downarrow \emptyset$ in $\mathcal{I}^\circ$ with $P_F(B_n) = 1$ for all n.

Example 8.8.21 uses the Hahn-Banach theorem to construct a purely finitely additive probability on any nonatomic probability space. Here is another way to get at purely finitely additive probabilities.

Example 7.6.30 (A Nonatomic Purely Finitely Additive Probability)
Let $\mathcal{A} \subset \mathcal{P}(\mathbb{N})$ be the set of A such that $\lim_n \frac{1}{n}\#\{A \cap \{1, \ldots, n\}\}$ exists, and define $\mu : \mathcal{A} \to [0, 1]$ by $\mu(A) = \lim_n \frac{1}{n}\#\{A \cap \{1, \ldots, n\}\}$. For any finite $A \subset \mathbb{N}$, $\mu(A) = 0$; hence $B_n = \{n + 1, n + 2, \ldots\} \downarrow \emptyset$ but $\mu(B_n) \equiv 1$. We shall see in §10.6.d that μ can be extended to all of $\mathcal{P}(\mathbb{N})$, essentially by taking limits in the compact space $[0, 1]^{\mathcal{P}(\mathbb{N})}$, which means that μ is purely finitely additive. To see that μ is nonatomic, pick $\epsilon > 0$, $n > 1/\epsilon$, and consider the sets of the form $B_k = \{(m \cdot n) + k : m \in \mathbb{N}\}$, $k = 0, 1, \ldots, n - 1$. The B_k partition $\mathbb{N}$ into n sets, each with $\mu(B_k) = 1/n < \epsilon$.

Definition 7.6.31   *The **support of a probability** μ on a metric space (M, d) is denoted $F(\mu)$ and defined as the smallest closed set having probability 1, $F(\mu) = \cap\{F : \mu(F) = 1 \ F \text{ closed}\}$.*

Exercise 7.6.32   Show that $\mu(F(\mu)) = 1$. Show that $F(\mu)^c = \cup\{G : \mu(G) = 0, \ G \text{ open}\}$. If (M, d) is a csm, show that $F(\mu)$ is σ-compact.

7.7 ◆ Transition Probabilities, Product Measures, and Fubini's Theorem

Transition probabilities give rise to probabilities on product spaces, and integration can be done iteratively. When the transition probability is constant, the probability is called a product measure, and the iterated integration result is called Fubini's theorem.

7.7.a Perspective and Preview

Transition probabilities arise in two kinds of models of interest to economists: stochastic process models and sequential decision models.

1. Suppose that the history of a process up till time t is a point in a measure space $(\Omega_1, \mathcal{F}_1)$ distributed according to P_1 and that the distribution over what happens next, which is a point in the measure space $(\Omega_2, \mathcal{F}_2)$, depends on ω_1.

2. Suppose that someone sees a signal, ω_1, in a measure space $(\Omega_1, \mathcal{F}_1)$ that is distributed according to P_1 and chooses, as a function of ω_1, a random action in a measure space $(\Omega_2, \mathcal{F}_2)$.

There are three ways to model this kind of random process or random strategy, and in most of the contexts we care about, they are equivalent. The first one is used in stochastic process theory and the second and third in game theory.

1. A **transition probability** is a function $P : \Omega_1 \times \mathcal{F}_2 \to [0, 1]$ with the following properties:
   a. for each ω_1, $P(\omega_1, \cdot) \in \Delta(\mathcal{F}_2)$, and
   b. for each $E_2 \in \mathcal{F}_2$, $P(\cdot, E_2)$ is a measurable function.

2. Define on $\Delta(\mathcal{F}_2)$, the set of countably additive probabilities on $\mathcal{F}_2$, $\mathcal{D}_2$, the smallest σ-field containing the sets $\{\mu : \mu(E) \leq r, \; r \in [0, 1], \; E_2 \in \mathcal{F}_2\}$. The random strategy is a measurable function $\omega_1 \mapsto s(\omega_1)$ in $\Delta(\mathcal{F}_2)$. This is the direct analogue of a behavioral strategy in a finite extensive form game.

3. Let X be a continuously distributed random variable independent of ω_1, and consider the random transition rule, or random strategy, $f(\omega_1, X)$.

For the equivalence of (1) and (2), we need $s(\omega_1)(E_2) = P(\omega_1, E_2)$ for all ω_1 and E_2. For the equivalence of (2) and (3), $s(\omega_1)$ has to be the distribution of $f(\omega_1, X)$ for all ω_1.

For each E in the product σ-field $\mathcal{F}_1 \otimes \mathcal{F}_2$ and $\omega_1 \in \Omega_1$, let $E_{\omega_1} = \{\omega_2 \in \Omega_2 : (\omega_1, \omega_2) \in E\}$. That is, E_{ω_1} is the slice of E at ω_1. The first result in this section shows that there is a unique probability Q on $\mathcal{F}_1 \otimes \mathcal{F}_2$ with the property that for all $E \in \mathcal{F}_1 \otimes \mathcal{F}_2$,

$$Q(E) = \int_{\Omega_1} P(\omega_1, E_{\omega_1}) \, P_1(d\omega_1), \tag{7.38}$$

and that for every integrable $u : \Omega_1 \times \Omega_2 \to \mathbb{R}$,

$$\int_{\Omega_1 \times \Omega_2} u \, dQ = \int_{\Omega_1} \left[\int_{\Omega_2} u(\omega_1, x) \, P(\omega_1, dx) \right] P_1(d\omega_1). \tag{7.39}$$

This kind of iterated integration is known as Fubini's theorem when $P(\omega_1, E_2)$ does not depend on ω_1, that is, when $P(\omega_1, E_2) = P_2(E_2)$ for some fixed P_2. Intuitively and formally, this is independence of ω_1 and ω_2, and Q is what is called a product measure. Product measures over countable products of csm's require just a bit more work involving compact classes.

7.7.b Transition Probabilities Induce Probabilities on Product Spaces

Recall that slices of measurable functions are measurable. We begin.

Theorem 7.7.1 *If $(\Omega_1, \mathcal{F}_1, P_1)$ is a probability space, $(\Omega_2, \mathcal{F}_2)$ is a measure space, and $P : \Omega_1 \times \mathcal{F}_2 \to [0, 1]$ is a transition probability, then*

1. *there is a unique probability Q on $\mathcal{F}_1 \otimes \mathcal{F}_2$ satisfying (7.38), defined by*

$$Q(A_1 \times A_2) = \int_{A_1} P(\omega_1, A_2) \, dP_1(\omega_1), \tag{7.40}$$

2. *for every $u \in L^1(\Omega_1 \times \Omega_2, \mathcal{F}_1 \otimes \mathcal{F}_2, Q)$, the function $Y(\omega_1) := \int_{\Omega_2} u(\omega_1, x) \, P(\omega_1, dx)$ is measurable, and*

3. *$\int_{\Omega_1} Y(\omega_1) \, dP(\omega_1) = \int_{\Omega_1 \times \Omega_2} u(\omega_1, \omega_2) \, dQ(\omega_1, \omega_2)$.*

Proof. A good sets argument shows that (7.38) is satisfied if it is satisfied for measurable rectangles. By Carathéodory's extension theorem, to prove the first claim, it is sufficient to show that $\int_{A_1} P(\omega_1, A_2)\, dP_1(\omega_1)$ defines a countably additive probability on the field of measurable rectangles.

Suppose that $B_1 \times B_2$ is a measurable rectangle that is a countable disjoint union of measurable rectangles, that is, $B_1 \times B_2 = \cup_{n \in \mathbb{N}}(A_{1,n} \times A_{2,n})$ with the $A_{1,n} \times A_{2,n}$ being disjoint. We must show that $\int_{B_1} P(\omega_1, B_2)\, dP_1(\omega_1)$ is equal to $\sum_{n \in \mathbb{N}} \int_{A_{1,n}} P(\omega_1, A_{2,n})\, dP_1(\omega_1)$.

Now, the disjointness delivers $1_{B_1 \times B_2}(\omega_1, \omega_2) = 1_{B_1}(\omega_1) \cdot 1_{B_2}(\omega_2) = \sum_{n \in \mathbb{N}} 1_{A_{1,n}}(\omega_1) \cdot 1_{A_{2,n}}(\omega_2)$. By countable additivity of each $P(\omega_1, \cdot)$, for every ω_1, $1_{B_1}(\omega_1) P(\omega_1, B_2) = \sum_{n \in \mathbb{N}} 1_{A_{1,n}}(\omega_1) P(\omega_1, A_{2,n})$. Integrating with respect to P_1 delivers $P(B_1 \times B_2) = \sum_{n \in \mathbb{N}} P(A_{1,n} \times P_{2,n})$.

Let $\mathcal{G}$ be the class of good *functions*, that is, the class of u for which the second claim is true. $\mathcal{G}$ includes all the simple functions measurable with respect to the field of measurable rectangles. If u_n is a sequence of bounded measurable functions in $\mathcal{G}$ and converges pointwise to a bounded u, then by dominated convergence, the associated Y_n converge pointwise and the pointwise limit of measurable functions is measurable, so that $u \in \mathcal{G}$. Finally, if u is integrable it is the pointwise limit of bounded measurable functions that are dominated by $|u|$, so dominated convergence again delivers the result.

The proof of the third claim directly parallels the proof of the second. ∎

7.7.c Fubini's Theorem

In the special case that there exists a single probability P_2 such that $P(\omega_1, A_2) = P_2(A_2)$ for all $A_2 \in \mathcal{F}_2$, the distribution of ω_2 in no way depends on ω_1. This in turn means that ω_1 and ω_2 are independent.

Theorem 7.7.2 (Fubini) *If $(\Omega_1, \mathcal{F}_1, P_1)$ and $(\Omega_2, \mathcal{F}_2, P_2)$ are probability spaces, then*

1. there is a unique probability Q on $\mathcal{F}_1 \otimes \mathcal{F}_2$ with the property that

$$Q(A_1 \times A_2) = P_1(A_1) \cdot P_2(A_2), \quad \text{and} \qquad (7.41)$$

2. for every $u \in L^1(\Omega_1 \times \Omega_2, \mathcal{F}_1 \otimes \mathcal{F}_2, Q)$,

$$\int_{\Omega_1 \times \Omega_2} u\, dQ = \int_{\Omega_1} \left[\int_{\Omega_2} u(\omega_1, \omega_2)\, dP_2(\omega_2) \right] dP_1(\omega_1) = \qquad (7.42)$$

$$\int_{\Omega_2} \left[\int_{\Omega_1} u(\omega_1, \omega_2)\, dP_1(\omega_1) \right] dP_2(\omega_2).$$

Proof. Apply the previous theorem to the transition probabilities $P_{1,2}(\omega_1, A_2) \equiv P_2(A_2)$ and $P_{2,1}(\omega_2, A_1) \equiv P_1(A_1)$. The two probabilities that arise are equal. ∎

7.7.d Product Measures on Countable Products of csm's

Let $(M_n, \mathcal{B}_n, P_n)_{n \in \mathbb{N}}$ be a collection of probability spaces where, for each n, there is a metric, d_n, making (M_n, d_n) a complete separable metric space and for which $\mathcal{B}_n$ is the Borel σ-field. The question is whether Fubini's theorem can be extended to give a probability Q on $\mathcal{B} := \otimes_{n \in \mathbb{N}} \mathcal{B}_n$. If we use Fubini's theorem inductively, for each N, there is a probability Q_N on $(\times_{n \leq N} M_n, \otimes_{n \leq N} \mathcal{B}_n)$. We identify any $E_N \in \otimes_{n \leq N} \mathcal{B}_n$ with the set $\text{proj}_{n \leq N}^{-1}(E_N) \subset \times_{n \in \mathbb{N}} M_n$. With this identification, it is clear that $\mathcal{B}° := \cup_N \otimes_{n \leq N} \mathcal{B}_n$ is a field and that there is a finitely additive probability, $Q : \mathcal{B}° \to [0, 1]$. The question is whether or not it can be extended to $\mathcal{B} = \sigma(\mathcal{B}°)$. Lemma 7.6.16 (p. 416) and the tightness of measures on csm's deliver the following.

Theorem 7.7.3 *Under the conditions just given, there is a unique probability Q on $\mathcal{B}$ determined by its values on $\cup_N \otimes_{n \leq N} \mathcal{B}_n$.*

Proof. Let $\mathcal{C}_N$ be the collection of sets of the form $\text{proj}_{n \leq N}^{-1}(K_N)$, where K_N is a compact subset of $\times_{n \leq N} M_n$, and let $\mathcal{C} = \cup_N \mathcal{C}_N$. Since $Q = Q_N$ on $\otimes_{n \leq N} \mathcal{B}_n$, the tightness of probabilities on csm's implies that Q satisfies the assumptions of Lemma 7.6.16. ∎

7.8 ◆ Seriously Nonmeasurable Sets and Intergenerational Equity

We begin with a short discussion of what serious nonmeasurability entails and then show how it arises in the study of intergenerational equity.

7.8.a Thick Sandwiches

In metric spaces, we saw that any measurable set, E, can be thinly sandwiched between a closed set, F, and an open set, G, that is, for all $\epsilon > 0$, there is a closed F and an open G with $F \subset E \subset G$ and $P(G \setminus F) < \epsilon$.

Definition 7.8.1 *A set $E \subset \Omega$ is **seriously nonmeasurable** if for some probability,*

$$\sup\{P(A) : A \subset E, \quad A \text{ measurable }\} < \inf\{P(B) : E \subset B, \quad B \text{ measurable }\}. \tag{7.43}$$

The left-hand side of the inequality gives an inner approximation to the set E and the right-hand side gives an outer approximation. The next section gives us a systematic, unambiguous way to handle $E \notin \mathcal{F}$ for which the inner and outer approximations are almost everywhere equal, that is, when there is equality in (7.43). It involves what are called **null sets**. There is no systematic way to handle serious nonmeasurability, and we are about to see that it arises in studying models of the basic issues of intergenerational equity.

7.8.b   Intergenerational Equity

As a piece of background, we note that the sets we care about are measurable when preferences can be represented by a measurable $\mathbb{R}$-valued utility function.

Exercise 7.8.2   If we have a preference relation, $\succsim$, on a separable metric space, (X, d), represented by a measurable utility function, $u : X \to \mathbb{R}$, then the sets $R = \{(x, y) \in X \times X : y \succ x\}$, $L = \{(x, y) \in X \times X : x \succ y\}$, and $I = \{(x, y) \in X \times X : x \sim y\}$ are measurable. [Look at the function $v(x, y) := u(x) - u(y)$.]

We now turn to social preferences on infinite streams of utilities, one utility level, $x_t \in [0, 1]$, for the generation living at time $t \in \mathbb{N}$. We study preferences on $X := [0, 1]^{\mathbb{N}}$ that display intergenerational equity. Recall that a finite permutation of $\mathbb{N}$ is a bijection of $\mathbb{N}$ and itself with $\pi(t) \neq t$ for at most finitely many t and that for $x \in X$, x^{π} is defined by $\text{proj}_t(x^{\pi}) = x_{\pi(t)}$ for all $t \in \mathbb{N}$.

Definition 7.8.3   *A preference relation on X **displays intergenerational equity** if finite permutations do not affect the preference ordering, that is, if $x \succ y$ iff $x^{\pi} \succ y$ for all finite permutations π.*

We write that $x > y$ in X iff $x_t \geq y_t$ for all t and $x \neq y$, and $x \gg y$ iff $x_t > y_t$ for all t.

Exercise 7.8.4   This problem concerns the following four orderings on X: better by permutation, $x \succ y$ iff there exists a finite permutation π such that $x^{\pi} > y$; long-run average differences, $x \succ y$ iff $\liminf_n \frac{1}{n} \sum_{t \leq n}(x_t - y_t) > 0$; overtaking, $x \succ y$ iff $\liminf \sum_{t \leq n}(x_t - y_t) > 0$; and patience limit, $x \succ y$ iff $(\exists \beta_0 \in (0, 1))(\forall \beta \in (\beta_0, 1)[\sum_t \beta^t x_t > \sum_t \beta^t y_t]$.

1. Show that all of the orderings are irreflexive, that is, $\neg[x \sim x]$; that two are transitive and two are not; and that all of them display intergenerational equity.

2. Show that none of the orders is complete.

3. Which of the four orderings respect the weak Pareto ordering, that is, satisfy $[x \gg y] \Rightarrow [x \succ y]$? Which satisfy the strong Pareto ordering, that is, $[x > y] \Rightarrow [x \succ y]$?

Theorem 7.8.5   *There exists a complete transitive preference relation $\succeq$ on X that respects intergenerational equity and the weak Pareto ordering.*

Proof.   Let $(E, \succ)$ be a nonempty subset of X and $\succ$ a complete ordering on E respecting intergenerational equity and the weak Pareto ordering. (Any of the incomplete orderings you looked at earlier with these properties will do; simply take E to be a subset of X with every $(x, y) \in E \times E$, $x \neq y$, comparable.) Let Ψ denote the set of all such pairs and partially order Ψ by $(E_1, \succ_1) \succeq (E_2, \succ_2)$ if E_1 is at least as large as E_2 and $\succ_1$ is an extension of $\succ_2$, that is, if $E_1 \supset E_2$ and $\succ_1 \cap (E_2 \times E_2) = \succ_2$. Let $\{(E_a, \succ_a) : a \in A\}$ be a chain of pairs in Ψ. The chain has an upper bound, namely $(E, \succ)$, where $E = \cup_a E_a$ and $\succ = \cup_a \succ_a$. By Zorn's lemma, Ψ has a maximal element, call it $(E^*, \succ^*)$. We claim that $E^* = X$. If not, there exists $z \notin E^*$ that is not comparable to any $x \in E^*$. Since $\succ^*$ respects intergenerational equity, no permutation of z can give rise to any $x \in E^*$ and as $\succ^*$ respects the weak Pareto order, z cannot be Pareto comparable to any

$x \in E^*$. Therefore, $\succ^*$ can be extended arbitrarily to $E^* \cup \{z\}$ and still respect intergenerational equity and the Pareto order, contradicting the three-maximality of $(E^*, \succ^*)$. ■

We are now going to define the "uniform" distribution, λ, on $X = [0, 1]^{\mathbb{N}}$ by requiring that $\{\text{proj}_n(x) : n \in \mathbb{N}\}$ be an iid collection of $U[0, 1]$ random variables. In a corresponding fashion, we define the "uniform" distribution, Λ, on $X \times X$, as the product of λ with itself, that is, two independent draws according to λ. $\mathcal{X}$ and $\mathcal{X} \otimes \mathcal{X}$ denote the two Borel σ-fields. By Theorem 7.7.3 (p. 426), λ and Λ exist.

You just showed that each of four orderings displaying intergenerational equity is incomplete. There are two results in the next theorem: first, they are really really incomplete, and second, if they also respect the Pareto ordering, they are nonmeasurable.

As before, define $R = \{(x, y) : y \succ x\}$, $L = \{(x, y) : x \succ y\}$, and $I = X \setminus (R \cup L)$, that is, $I = \{(x, y) : \neg[x \succ y] \cap \neg[y \succ x]\}$.

Theorem 7.8.6 (Zame) *If $\succ$ is an irreflexive ordering on X that displays intergenerational equity and F is a measurable set containing I, then $\Lambda(F) = 1$. If $\succ$ also respects the weak Pareto ordering, then I is seriously nonmeasurable.*

Three comments:

1. If for some $\epsilon > 0$, x° satisfies $(1 - \epsilon) \geq x_t^\circ$ for infinitely many t and $x_t^\circ > \epsilon$ for infinitely many t, then $\lambda(\{y : x^\circ \succ y\}) \leq \lim_T \epsilon^T = 0$ and $\lambda(\{y : y \succ x^\circ\}) \leq \lim_T (1 - \epsilon)^T = 0$. This shows that the relation "$\succ$" already has the property that the set of y comparable to any x° is pretty small according to λ.

2. We might interpret $\Lambda(F) = 1$ for all measurable $F \supset I$ as telling us that incomparable or indifferent pairs are the most likely to arise when we require intergenerational equity; that is, we cannot compare most pairs of streams of utility. This interpretation becomes a bit strained when we also require a respect for the Pareto ordering—our model says we cannot determine the likeliness of the set.

3. The interaction of this result with Theorem 7.8.5 is instructive. Take $\succsim$ to be a complete transitive relation on X respecting intergenerational equity. We say that $\succsim$ is measurable if $G = \{(x, y) : y \succsim x\}$ is measurable. If G and I are measurable, then, since $G = R \cup I$, R is also measurable. The first part of Zame's theorem implies that $\Lambda(I) = 1$, so that $\Lambda(R) = \Lambda(L) = 0$. The second part says that if we require that the relation $\succsim$ also satisfy the Pareto principle, then I is seriously nonmeasurable, which in turn implies that R and L are also seriously nonmeasurable.

Exercise 7.8.7 Fill in the details of the following sketch of a proof of the first part of Theorem 7.8.6. Let $R = \{(x, y) \in X \times X : y \succ x\}$, $L = \{(x, y) \in X \times X : x \succ y\}$, and $I = \{(x, y) \in X \times X : \neg[x \succ y] \wedge \neg[y \succ x]\}$. Define the mapping $\iota(x, y) = (y, x)$ from $X \times X$ to itself.

1. $\iota(x, y) = (y, x)$ is measurable and measure preserving, that is, $\Lambda(\iota(E)) = \Lambda(E)$.

2. R, L, I are a partition of $X \times X$. Further, to show that any F containing I satisfies $\Lambda(F) = 1$, it is sufficient to show that $\Lambda(E) = 0$ for any measurable $E \subset R$.

3. There is a measurable $E^\circ \subset R$ with $\Lambda(E^\circ) = \sup\{\Lambda E' : E' \subset R, \ E' \in \mathcal{X} \otimes \mathcal{X}\}$.

4. For finite permutations π, π', define $E^{\pi,\pi'} = \{(x^\pi, y^{\pi'}) : (x, y) \in E^\circ\}$. Show that $E^{\pi,\pi'} \subset E^\circ$, that $A := \cup_{\pi,\pi'} E^{\pi,\pi'}$ is measurable and that $E^\circ \subset A \subset R$.

5. For $x \in X$, let $A_x = \{y : (x, y) \in A\}$. Show that $x \mapsto \lambda(A_x)$ is measurable and that each A_x is symmetric (in the Hewitt-Savage sense).

6. Show that $\Lambda(A) = \int_X \lambda(A_x) \, d\lambda(x)$, and that the Hewitt-Savage 0-1 law implies that $\lambda(X_1) = \Lambda(A)$, where $X_1 = \{x : \lambda(A_x) = 1\}$. [Fubini's theorem.]

7. Show that X_1 is symmetric, so has probability either 0 or 1.

8. Conclude that if $\Lambda(A) > 0$, then $\Lambda(X \times X) \geq 2$, so that $\Lambda(A) = \Lambda(E) = 0$.

Exercise 7.8.8   Fill in the details of the following sketch of a proof of the second part of Theorem 7.8.6.

1. Let $Gr = \{(x, y) : x \succsim y\}$. Show that Gr is measurable iff R, L, and I are measurable.

2. Suppose that I is not seriously nonmeasurable. By the first part, any measurable superset of I has probability 1. Now show that there is a Borel measurable $J \subset I$ such that $\Lambda(J) = 1$ and for all finite permutations, π, π', and all $(x, y) \in J$, $(x^\pi, y^{\pi'}) \in J$.

3. As before, let J_x be the x section of J and show that $\lambda(J_x) = 1$ for λ almost all $x \in X$. Pick x^* such that $\lambda(J_{x^*}) = 1$. [Fubini's theorem.]

4. Let $0 < b_n \uparrow 1$ so that $\Pi_n b_n > \frac{1}{2}$ and set $D = \times_{n \in \mathbb{N}}[0, b_n]$. Let $b = (b_1, b_2, \ldots) \in X$ and $(1 - b) = ((1 - b_1), \ldots)$. Show that the mapping $f(x) := x + \frac{1}{2}(1 - b)$ from D to X is one-to-one, measurable, and measure preserving.

5. Show that $\lambda(D \cap J_{x^*}) = \lambda(f(D \cap J_{x^*})) > \frac{1}{2}$.

6. Show that $D \cap J_{x^*}$ and $f(D \cap J_{x^*})$ are disjoint because $f(y) \gg y$ for any $y \in (D \cap J_{x^*})$.

7. Since the assumption of a measurable I leads us to conclude that there exist disjoint measurable sets each having probability greater than $\frac{1}{2}$, we must conclude that I is nonmeasurable.

Exercise 7.8.9   Using the relation $>$ defined above, show directly that $\Lambda(\{(x, y) : y > x\}) = 0$. [The first part of the theorem is saying that the relation ">" must have this property.]

7.8.c   Seriously Nonmeasurable Subsets of $[0, 1]$

The sets $[0, 1]$ and $[0, 1]^\mathbb{N} \times [0, 1]^\mathbb{N}$ are Borel isomorphic. Indeed, one can show that the bijection can be chosen to have the property that $\Lambda = f(\mu)$, where μ is the uniform distribution on [0, 1]. This means that there exists an $E \subset [0, 1]$ with the property that $\sup\{\mu(A) : A \subset E\} < \inf\{\mu(B) : E \subset B\}$, where the sup and inf are taken over the class of measurable sets. More generally, every uncountable complete separable metric space has seriously nonmeasurable subsets for any nonatomic probability.

7.9 ◆ Null Sets, Completions of σ-Fields, and Measurable Optima

Null sets are possibly nonmeasurable subsets of measurable sets having 0 probability. Adding them to a σ-field makes the σ-field bigger in a fashion that makes some of the sets and functions we care about measurable. Without this addition, they are not measurable, only nearly measurable.

7.9.a The Null Sets

The name "null set" suggests something that does not matter. Often they do not, and one way to understand this section is to realize that we are developing the tools that make sure that they do not matter.

Definition 7.9.1 *$A' \subset \Omega$ is **P-null** if there exists $A \in \mathcal{F}$ with $A' \subset A$ and $P(A) = 0$.*

Here is a partial explanation of why we say they often do not matter.

Theorem 7.9.2 *If $X \in L^1(\Omega, \mathcal{F}, P)$ and A is a measurable P-null set, then $\int 1_A |X| \, dP = 0$.*

Note that we could just as well write $\int_A |X| \, dP$ for $\int 1_A |X| \, dP = 0$.

Proof. If Y is any element of $L_{\leq}(1_A|X|)$, then $\int Y \, dP = 0$. ∎

What is missing from the result that null sets do not affect integrals is any explanation of what we do when $A' \subset A$, but $A' \notin \mathcal{F}$. That is, how do we go about defining the integral $\int 1_{A'} |X| \, dP$ when A' is a nonmeasurable null set? The answer is that we just expand $\mathcal{F}$ to include A' and define the extension of P to A' by $P(A') = 0$. This has the effect of solving a raft of messy technical problems that will show up later.[13]

7.9.b The Completions

$\mathcal{N}_P = \{N : \exists A \in \mathcal{F}, \ N \subset A, \ P(A) = 0\}$ is the class of P-null sets.[14]

Definition 7.9.3 *The **completion of a σ-field** $\mathcal{F}$ **with respect to** P is $\mathcal{F}^P = \{F : F = E \cup N, \ E \in \mathcal{F}, \ N \in \mathcal{N}_P\}$. The **universal completion** of $\mathcal{F}$ is $\mathcal{F}^U = \bigcap\{\mathcal{F}^P :$*

---

13. If the collective noun for technical problems is not "raft," then it ought to be.
14. This set is the set of subsets of $\bigcap_{\epsilon>0} \mathcal{N}(\epsilon)$ in the proof of Carathéodory's extension theorem.

P is a probability on $\mathcal{F}$ } and elements of $\mathcal{F}^U$ are called **universally measurable sets***. We define the* **extension of P to** $\mathcal{F}^P$ *by* $\widehat{P}(E') = P(E)$ *whenever* $E' = E \cup N$ *with* $N \in \mathcal{N}_P$.

We have just added nonmeasurable sets to the σ-fields, but they are null sets rather than seriously nonmeasurable sets. We have to check that this definition makes sense.

Lemma 7.9.4 $\mathcal{F}^P$ *and* $\mathcal{F}^U$ *are* σ*-fields, and for each P,* $\widehat{P}$ *is a probability on* $\mathcal{F}^P$.

Proof. $\emptyset, \Omega \in \mathcal{F}^P$. Let $F = E \cup N \in \mathcal{F}^P$ with $E \in \mathcal{F}$ and $N \subset A$, $A \in \mathcal{F}$ and $P(A) = 0$. $F^c = (E \cup A)^c \cup (A \setminus N)$ expresses F^c as the union of a set in $\mathcal{F}$ and a null set, showing that $\mathcal{F}^P$ is closed under complementation. Since P is countably additive, $\mathcal{F}^P$ is closed under countable unions, hence is a σ-field. To check that $\widehat{P}$ is a probability, we must show that it is countably additive. Let $F_n \uparrow F$ in $\mathcal{F}^P$. Since $F_n = E_n \cup N_n$, $N_n \subset A_n$, and $P(A_n) = 0$, $F = (\cup_n E_n) \cup (\cup_n N_n) = E \cup N$. $E \in \mathcal{F}$ and $N \subset \cup_n A_n \in \mathcal{F}$. Finally, by the countable additivity of P, $P(\cup_n A_n) = 0$.   ∎

There are two corollaries worth noting. The first tells us that $\mathcal{F}^P$-measurable functions can be sandwiched between $\mathcal{F}$-measurable functions that are equal with probability 1. The second tells us that the universally measurable σ-field is the "natural" domain for probabilities.

Corollary 7.9.5 $X \in L^0(\Omega, \mathcal{F}^P, \widehat{P})$ *iff there exist* $R, S \in L^0(\Omega, \mathcal{F}, P)$ *and* $E \in \mathcal{F}$ *such that* $P(E) = 1$*, for all* $\omega \in E$*,* $R(\omega) \leq X(\omega) \leq S(\omega)$*,* $P(R = S) = 1$.

Proof.   The result is true if X is a simple $\mathcal{F}^P$-measurable function, and continues to be true if we take pointwise limits.   ∎

Corollary 7.9.6   *The mapping* $P^U \mapsto P$ *from probabilities on* $\mathcal{F}^U$ *to their restrictions to* $\mathcal{F}$ *is a bijection between the set of probabilities on* $\mathcal{F}^U$ *and the probabilities on* $\mathcal{F}$*. In particular, every probability P on* $\mathcal{F}$ *has a unique extension to* $\mathcal{F}^U$.

Proof.   We must show that $P^U \mapsto P^U_{|\mathcal{F}}$ is one-to-one and onto.

One-to-one: Suppose that P^U, Q^U are two probabilities on $\mathcal{F}^U$ that agree on $\mathcal{F}$, and let P be their common restriction to $\mathcal{F}$. The relevant class of good sets is $\mathcal{G} = \{F \in \mathcal{F}^U : P^U(F) = Q^U(F)\}$. By assumption, $\mathcal{G}$ contains $\mathcal{F}$. By definition, any $F \in \mathcal{F}^U$ belongs to $\mathcal{F}^P$. This means that there exist $E, A \in \mathcal{F}$ such that $E \subset F \subset E \cup A$ and $P(A) = 0$. Since $P(A) = Q(A)$ and $P(E) = Q(E)$, both $P^U(F)$ and $Q^U(F)$ are between $P(E)$ and $P(E \cup A)$. Since $P(E) = P(E \cup A)$, $\mathcal{G} = \mathcal{F}^U$.

Onto: Every probability P on $\mathcal{F}$ has an extension to $\mathcal{F}^P \supset \mathcal{F}^U$.   ∎

The following is Theorem 7.9.2 with a different σ-field and probability; hence it has exactly the same proof.

Theorem 7.9.7   *If* $X \in L^1(\Omega, \mathcal{F}^P, \widehat{P})$ *and* $E \in \mathcal{F}^P$ *is a null set, then* $\int 1_E |X| \, dP = 0$.

Notation Alert 7.9.A   *In view of the previous results, we do not distinguish between* $\widehat{P}$ *and P unless we really really have to. Theorem 7.9.7 allows us to say* *that "the integral over a null set is 0" and not have to worry about whether the null set belongs to $\mathcal{F}$ or to $\mathcal{F}^P$. This may seem a bit sloppy, but it is actually clearer in the long run—one just ignores null sets whenever possible, acknowledging that one is doing so by putting in the "almost everywhere" qualification.*

Though the completion adds only null sets, it may add many of them. The next example gives a famous set, due to Cantor. It is an uncountable subset of $[0, 1]$ with λ probability 0 (λ being, as usual, the uniform distribution). The completion of the Borel σ-field contains all of its subsets, and the set of all subsets of an uncountable set has cardinality larger than the cardinality of $[0, 1]$. By contrast, the set of λ-equivalence classes of measurable subsets of $[0, 1]$ has the same cardinality as $[0, 1]$ because $(\mathcal{B}, \rho_\lambda)$ is a complete separable metric space; hence it is homeomorphic to a subset of $[0, 1]^\mathbb{N}$, and $[0, 1]^\mathbb{N}$ has the cardinality of the continuum.

Example 7.9.8 (The Cantor Set) *For each x in the uncountable set $\{0, 2\}^\mathbb{N}$, $x = (x_1, x_2, x_3, \ldots)$, define $r(x) \in [0, 1]$ by*

$$r(x) = \sum_n \frac{x_n}{3^n}.$$

*$C := r(\{0, 2\}^\mathbb{N})$ is called the **Cantor set** and is the set of all numbers with trinary expansions containing only 0's and 2. A good name for C is "the omitted middle-thirds set."*

Theorem 7.9.9 *The Cantor set*

1. *is compact,*

2. *is uncountable, and*

3. *satisfies $\lambda(C) = 0$, so that C has empty interior.*

Proof. For (1), note that r is continuous when $\{0, 2\}^\mathbb{N}$ has the product topology. This means that C is the continuous image of a compact set.

For (2), since $\{0, 2\}^\mathbb{N}$ is uncountable, it is sufficient to show that r is one-to-one. The intuition is pretty easy: the mapping $r : \{0, 1, 2\}^\mathbb{N} \to [0, 1]$ is onto, but because $0.1000000 \ldots = 0.02222222 \ldots$ for trinary expansions, it is many-to-one. However, 0.1000000 cannot be in C since it has a 1 in one of its components.

More formally, pick $x \neq y$ in $\{0, 2\}^\mathbb{N}$ and let $n = n_{x,y}$ be the smallest natural number with $x_n \neq y_n$. Without loss, $x_n = 0$ and $y_n = 1$, so that $r(y) \geq r(x)$. The difference $r(y) - r(x)$ is at least $(\frac{2}{3})^n - \sum_{m \geq n+1}(\frac{2}{3})^m$, which is $(\frac{2}{3})^n[\frac{2}{3} - \frac{1}{3}] > 0$, so that $r(y) > r(x)$.

For (3), for each $n \in \mathbb{N}$, let $A_n \subset \{0, 1, 2\}^\mathbb{N}$ be set $\text{proj}_n^{-1}(\{0, 2\})$ and let $B_n = \cap_{m \leq n} A_n$. Note that $r(B_n) \downarrow C$. Further, $\lambda(r(B_1)) = \frac{2}{3}$ and for all $n \in \mathbb{N}$, $\lambda(r(B_{n+1})) = \frac{2}{3}\lambda(r(B_n))$. Thus, $\lambda(C) = \lim_n (\frac{2}{3})^n = 0$. Finally, if $\text{int}(C) \neq \emptyset$, then $\lambda(C) > 0$. ∎

Exercise 7.9.10 The Cantor set is hard to visualize. Give $\Omega = \{0, 2\}^\mathbb{N}$ the uniform distribution, P, that is, the distribution with $\{\text{proj}_n : n \in \mathbb{N}\}$ being an iid collection with $P(\{\text{proj}_n = 0\}) = P(\{\text{proj}_n = 2\}) = \frac{1}{2}$. Let F_r be the cdf of the random variable r used in defining the Cantor set. Show that $F_r(0) = 0$, $F_r(1) = 1$, and that F_r is continuous. [This is a continuous cdf with a derivative that is almost everywhere equal to 0. For all x not in the compact set C having empty interior, F_r is constant on an interval around x. The set of x with such an interval around it is C^c, which has Lebesgue measure 1.]

7.9.c   Projections of Measurable Sets Are Analytic Sets

Projections of measurable sets give rise to a class of sets called **analytic sets**, and these are universally measurable, which means that they differ from a measurable set by at most a null set.

Definition 7.9.11   *For $(\Omega, \mathcal{F})$ a measurable space, $A \subset \Omega$ is $\mathcal{F}$-analytic if there exists a complete separable metric space (M, d) and a measurable $B \in \mathcal{F} \otimes \mathcal{B}_M$ such that $A = \mathrm{proj}_\Omega(B)$. The class of $\mathcal{F}$-analytic sets is denoted $\mathbf{A}(\mathcal{F})$.*

It is, perhaps, surprising that analytic sets need not be measurable.[15] However, they are nearly measurable. The following result gathers together a number of the results about analytic sets and σ-fields in Dellacherie and Meyer (1978, Ch. III). We do not offer a proof.

Theorem 7.9.12   *For all measure spaces $(\Omega, \mathcal{F})$ and all probabilities P on $\mathcal{F}$,*

$$\mathcal{F} \subset \mathbf{A}(\mathcal{F}) \subset \sigma(\mathbf{A}(\mathcal{F})) \subset \mathcal{F}^U = \mathbf{A}(\mathcal{F}^U) \subset \mathcal{F}^P = \mathbf{A}(\mathcal{F}^P). \qquad (7.44)$$

It turns out that $\mathbf{A}(\mathcal{F})$ fails to be a σ-field by failing to be closed under complementation.

Exercise 7.9.13 $\mathbf{A}(\mathcal{F})$ is closed under countable union and countable intersection. [Show that the projection of a union is a union of projections and that the projection of an intersection is the intersection of projections.]

This is a result that allows us to prove that a number of objects of interest are "nearly" measurable. If we prove that a set belongs to any of the given classes between $\mathcal{F}$ and $\mathcal{F}^U$ in (7.44), then we know that for any probability, the set has a uniquely determined probability. The last equality in (7.44) tells us that if we start with a completed σ-field, $\mathcal{F}^P$, then projection of measurable sets gives us another measurable set.

7.9.d   Two Examples

The following examples demonstrate the utility of projection arguments.

---

15. Indeed, the whole study of analytic sets arose from the work of Henri Lebesgue in 1905, in which he offered a mistaken argument that projections of measurable sets are measurable. Souslin, a student of Lusin's, found the error and started the systematic study of these sets in a short 1917 note. Souslin died soon after, and in 1930 Lusin published a book on analytic sets with a preface by Lebesgue.

Example 7.9.14 (Hitting Times in Stochastic Processes) *For a probability space* $(\Omega, \mathcal{F}, P)$ *and the time set* $[0, \infty)$, *one defines a measurable continuous-time stochastic process as a function* $X : \Omega \times [0, \infty) \to \mathbb{R}$ *that is* $\mathcal{F} \otimes \mathcal{B}_{[0,\infty)}$-*measurable. The interpretation is that the random draw of an* ω *gives rise to the random time path* $X(\omega, \cdot)$; *that previous to time t, one has observed the process* $\{X(\omega, s) : s \in [0, t)\}$; *and that at time t, one has observed the process* $\{X(\omega, s) : s \in [0, t]\}$. *For measurable* $B \subset \mathbb{R}$, *we are interested in the time until the process first enters, or hits, the set* B, *which is the number* $\tau_B(\omega) := \inf\{t \in [0, \infty) : X(\omega, t) \in B\}$.

The set $\{\omega : \tau_B(\omega) < r\}$ *can be expressed as the uncountable union of measurable sets,* $\cup_{t \in [0,r)} A(t)$, *where* $A(t) = \{\omega : X(\omega, t) \in B\}$. *In general, uncountable unions of measurable sets are not measurable, and it can be shown that there are measurable* B *for which* $\{\tau_B < r\}$ *is not a measurable set. However, projection arguments allow us to show that it is nearly measurable.*

As X is a measurable function, its graph, gr X, is a $\mathcal{F} \otimes \mathcal{B}_{[0,\infty)} \otimes \mathcal{B}$-*measurable subset of* $\Omega \times [0, \infty) \times \mathbb{R}$. *The event* $\{\tau_B < r\}$ *belongs to* $\mathbf{A}(\mathcal{F})$ *because*

$$\{\tau_B < r\} = \mathrm{proj}_\Omega(gr\ X \cap [\Omega \times [0, r) \times B]).$$

Thus, $\{\tau_B < r\}$ *is an element of* $\mathcal{F}^U$, *and if* $\mathcal{F} = \mathcal{F}^P$, *then* $\{\tau_B < r\}$ *is measurable.*

In probability theory and stochastic process arguments, one often finds the assumption that the probability space is complete, that is, that $\mathcal{F} = \mathcal{F}^P$. This allows projection arguments to be used when one has to show measurability.

The next example involves the question of whether or not a signal is invertible. The situation is that if $h \in H$ occurs, then $f(h)$ is observed in some metric space (M', d'). If P is a probability on (a σ-field of subsets of) H, then one might want to answer the question, "What is the probability that $f(h)$ perfectly reveals the value of h?" For this to have a sensible answer, the event that f is one-to-one must be nearly measurable.

Definition 7.9.15 *For any measure space* $(\Omega, \mathcal{F})$ *and any nonempty* $H \subset \Omega$, *measurable or not, the **trace of** $\mathcal{F}$ **on** H is the σ-field* $\mathcal{H} := \{E \cap H : E \in \mathcal{F}\}$.

Example 7.9.16 (The Event That a Function Is One-to-One) *Let H be a measurable subset of a complete separable metric space* (M, d), $\mathcal{H}$ *the trace of the Borel* σ-field, $\mathcal{B}_M$, *on* H, *and* $f : H \to M'$ *be* $\mathcal{H}$-*measurable, where* (M', d') *is a metric space. The event that f is 1-1 is* $\{h \in H : \forall h' \neq h, \ f(h) \neq f(h')\}$. *To see why, define* $\varphi : H \times H \to \mathbb{R}_+$ *by* $\varphi(h, h') = d'(f(h), f(h'))$. φ *is measurable because it is the composition of the continuous function* $d : M' \times M' \to \mathbb{R}$ *and the measurable function* $(h, h') \mapsto (f(h), f(h'))$. $\varphi^{-1}(0)$ *is the set of pairs* (h, h') *that are mapped to the same value in* M'. *Included in* $\varphi^{-1}(0)$ *is the diagonal, that is, the set of pairs with* $h = h'$.

Let $D \in \mathcal{H} \otimes \mathcal{H}$ *be the diagonal in* $H \times H$, *that is,* $D = \{(h, h') \in H \times H : h = h'\}$. *Since* $H \subset M$ *and* M *is a separable metric space,* $D \in \mathcal{H} \otimes \mathcal{H}$. *Thus,* $B := \varphi^{-1}(0) \setminus D \in \mathcal{H} \otimes \mathcal{H}$. $A := \mathrm{proj}_H(B)$ *belongs to* $\mathbf{A}(\mathcal{H})$ *and consists of the set of h for which there exists an* $h' \neq h$ *satisfying* $f(h) = f(h')$. *Being analytic, this set, hence its complement, is universally measurable.*

7.9.e Projections of Analytic Sets Are Analytic Sets

Analytic sets require complete separable metric space (M, d) and a $B \in \mathcal{F} \otimes \mathcal{B}_M$ to be projected onto Ω. One can substitute any $H \in \mathbf{A}(\mathcal{B}_M)$ for the csm space (M, d).

Lemma 7.9.17 *If $(\Omega, \mathcal{F})$ is a measure space, (M, d) is a csm, and $B \in \mathbf{A}(\mathcal{F} \otimes \mathcal{B}_M)$, then $\mathrm{proj}_\Omega(B)$ is analytic. In particular, if $H \in \mathbf{A}(\mathcal{B}_M)$, $\mathcal{H}$ is the trace of $\mathcal{B}_M$ on H, and $B \in \mathcal{F} \otimes \mathcal{H}$, then $\mathrm{proj}_\Omega(B)$ is analytic.*

Proof. Suppose that $B \in \mathbf{A}(\mathcal{F} \otimes \mathcal{B}_M)$ and $A = \mathrm{proj}_\Omega(B)$. We wish to show that A is analytic.

Let (M', d') be a csm space and $B' \in (\mathcal{F} \otimes \mathcal{B}_M) \otimes \mathcal{B}_{M'}$ such that $\mathrm{proj}_{\Omega \times M}(B')$ $= B$. Consider the product csm space $M \times M'$ with the corresponding Borel σ-field $\mathcal{B}_M \otimes \mathcal{B}_{M'}$. $B' \in \mathcal{F} \otimes (\mathcal{B}_M \otimes \mathcal{B}_{M'})$ and $A = \mathrm{proj}_\Omega(B')$. ■

Exercise 7.9.18 Complete the proof of Lemma 7.9.17. Using this, show that the result in Example 7.9.16 continues to be true if H is assumed to be analytic rather than measurable.

We now turn to the measurability of value functions and sets of optima. Basically, the value function is nearly measurable, as is the graph of the arg max correspondence and the set on which the optimum is achieved.

Theorem 7.9.19 *Let $(\Omega, \mathcal{F})$ be a measurable space, $(H, \mathcal{H})$ an analytic subset of a csm space with the trace σ-field, $u : \Omega \times H \to [-\infty, +\infty]$ an $\mathcal{F} \otimes \mathcal{H}$-measurable function, and $S : \Omega \rightrightarrows H$ a correspondence with graph $\mathrm{gr}\ S \in \mathcal{F} \otimes \mathcal{H}$. Then*

1. *$v(\omega) := \sup\{u(\omega, h) : (\omega, h) \in \mathrm{gr}\ S\}$ satisfies $\{v > r\} \in \mathbf{A}(\mathcal{F})$ for all $r \in [-\infty, +\infty]$, so that v is $\mathcal{F}^U$-measurable;*

2. *the graph of the arg max correspondence, $M := \{(\omega, h) \in S : u(\omega, h) = v(\omega)\}$ belongs to $\sigma(\mathbf{A}(\mathcal{F})) \otimes \mathcal{H}$; and*

3. *the event that u achieves its supremum, $I = \{\omega : (\exists h)[[(\omega, h) \in S] \wedge [u(\omega, h) = v(h)]]\} \in \mathcal{F}^U$.*

Proof. For (1), $\{v > r\} = \mathrm{proj}_\Omega(\mathrm{gr}\ S \cap \{u > r\})$, so that $\{v > r\} \in \mathbf{A}(\mathcal{F})$. For (2), $\{v > r\} \in \mathbf{A}(\mathcal{F})$ implies that v is $\sigma(\mathbf{A}(\mathcal{F}))$-measurable. Since u is $\mathcal{F} \otimes \mathcal{H}$-measurable, it is also $\sigma(\mathbf{A}(\mathcal{F})) \otimes \mathcal{H}$-measurable. Hence, $M = \{(\omega, h) : v(\omega) = u(\omega, h)\}$ also belongs to $\sigma(\mathbf{A}(\mathcal{F})) \otimes \mathcal{H}$. For (3), $I = \mathrm{proj}_\Omega(M)$, which belongs to $\mathbf{A}(\sigma(\mathbf{A}(\mathcal{F})))$, a subset of $\mathcal{F}^U$. ■

Exercise 7.9.20 Suppose in Theorem 7.9.19 that $\mathrm{gr}\ S \in \mathbf{A}(\mathcal{F} \otimes \mathcal{H})$. Show that all that has to change in the results is that M belongs to $\mathbf{A}(\sigma(\mathbf{A}(\mathcal{F})) \otimes \mathcal{H})$, but the measurability of v and I remain unchanged.

In econometric examples where these issues arise, H is a space of parameters over which one is optimizing an objective function, such as a likelihood function, and the set of parameters over which one is optimizing, $S(\omega) = \{h : (\omega, h) \in S\}$, is determined by the data. In game theory examples where these issues are of concern, H is a space of actions, $S(\omega)$ is the set of actions available when ω occurs, and $u(\omega, \cdot)$ is an agent's utility function.

7.10 ◆ Convergence in Distribution and Skorohod's Theorem

Let $(X, (X_n)_{n \in \mathbb{N}}) \subset L^0(\Omega, \mathcal{F}, P)$ be random variables. We have seen almost everywhere convergence, L^1 convergence, and convergence in probability:

- $X_n \to X$ a.e. is $P(X_n \to X) = 1$,
- if $(X, (X_n)_{n \in \mathbb{N}}) \subset L^1$, then L^1 convergence is $E |X_n - X| \to 0$, and
- $\alpha(X_n, X) \to 0$, equivalently $(\forall \epsilon > 0)[P(|X_n - X| > \epsilon) \to 0]$ is convergence in probability.

The following gathers the relevant facts proved earlier.

1. If $X_n \to X$ a.e., then $\alpha(X_n, X) \to 0$, and if $\alpha(X_n, X) \to 0$, then there exists a subsequence $X_{n'}$ such that $X_{n'} \to X$ a.e. [The first statement is immediate; the proof of the second uses Borel-Cantelli.]

2. If $(X, (X_n)_{n \in \mathbb{N}}) \subset L^1$, then $E |X_n - X| \to 0$ iff $\alpha(X_n, X) \to 0$ and the collection $(X, (X_n)_{n \in \mathbb{N}})$ is uniformly integrable. [The proofs of these results involve a small-events continuity argument.]

In particular, either $X_n \to X$ a.e. or $E |X_n - X| \to 0$ implies that $X_n \to X$ in probability.

The random variables X and $(X_n)_{n \in \mathbb{N}}$ have distributions on $\mathbb{R}$ given by their cdf's, $F(r) := P(X \leq r)$ and $F_n(r) := P(X_n \leq r)$. This section is a brief investigation of how convergence of random variables is related to the convergence of their distributions.[16]

7.10.a Convergence in Distribution

Recall that F_n converges weakly to F, written $F_n \to_{w*} F$, if for all x that are continuity points of F, $F_n(x) \to F(x)$. One way to remember this is that the most useful form of convergence of distributions/cdf's is weak convergence, and random variables converge in distributions iff their distributions converge is this useful sense.

Definition 7.10.1 X_n *converges to* X *in* ***distribution***, *written* $X_n \to_{\mathcal{D}} X$, *if* $F_n \to_{w*} F$.

One can have $X_n \to_{\mathcal{D}} X$ without having a.e. convergence, convergence in probability, or L^1 convergence,

Example 7.10.2 *With* $(\Omega, \mathcal{F}, P) = ((0, 1], \mathcal{B}, \lambda)$, *let* $X_n(\omega) = \omega$ *for all* $n \in \mathbb{N}$ *and let* $X(\omega) = 1 - \omega$. *We have* $F_n \equiv F$ *even though* $P(X_n \to X) = 0$, $E |X_n - X| \equiv \frac{1}{2}$, *and* $\alpha(X_n, X) \equiv \frac{1}{3}$.

Theorem 7.10.3 $[\alpha(X_n, X) \to 0] \Rightarrow [X_n \to_{\mathcal{D}} X]$.

In particular, if either $X_n \to X$ a.e. or $E |X_n - X| \to 0$, then $X_n \to_{\mathcal{D}} X$.

---

16. Chapter 9 includes much more extensive coverage of these topics.

Proof.   Let x be a continuity point of F and pick $\epsilon > 0$. Pick δ small enough that $[x - \delta < r < x + \delta] \Rightarrow [F(x) - \epsilon < F(r) < F(x) + \epsilon]$. Let $\eta = \min\{\epsilon, \delta\}$. For n large enough that $\alpha(X_n, X) < \eta$, $P(|X_n - X| \geq \delta) < \epsilon$, that is, $P(|X_n - X| < \delta) > 1 - \epsilon$, so that $F(x) - \epsilon < F_n(x) < F(x) + \epsilon$.   ■

7.10.b   Skorohod's Representation Theorem

Despite the thoroughness with which Example 7.10.2 demonstrates that $[X_n \to_{\mathcal{D}} X] \nRightarrow [\alpha(X_n, X) \to 0]$, there is a kind of converse, a representation result due to Skorohod.

Theorem 7.10.4 (Skorohod)   *If $(X, (X_n)_{n \in \mathbb{N}}) \subset L^0(\Omega, \mathcal{F}, P)$ and $X_n \to_{\mathcal{D}} X$, then there exists $(X', (X'_n)_{n \in \mathbb{N}}) \subset L^0((0, 1], \mathcal{B}, \lambda)$ with $P(X \in A) = P(X' \in A)$, $P(X_n \in A) = P(X'_n \in A)$, and $X'_n \to X'$ a.e.*

In other words, convergence in distribution implies a.e. convergence of a *different* sequence of random variables defined on a different probability space, and the elements of different sequences have the same distributions as those of the original sequence. The following is a useful device for the present proof and for other purposes.

Definition 7.10.5   *The **probability integral transform of a cdf** G is the function $X_G : (0, 1) \to \mathbb{R}$ defined by $X_G(r) = \inf\{x \in \mathbb{R} : G(x) \geq r\}$.*

The probability integral transform is a "pseudoinverse." If G is strictly increasing and continuous, then X_G is the inverse of G. However, if G has jumps or flat spots, it has no inverse: if G has a jump at x, then for all $r \in [G(x-), G(x)]$, $X_G(r) = x$; if G puts no mass on an interval (a, b) but puts mass on every $(a - \epsilon, a]$ interval, then for $r = G(a)$, $X_G(r) = a$, and X_G has an upward jump at r. In any of these cases, X_G is nondecreasing.

Exercise 7.10.6   Show that for all $x \in \mathbb{R}$, $\lambda(\{r : X_G(r) \leq x\}) = G(x)$, that is, that X_G is a random variable having cdf G.

Proof of Theorem 7.10.4.   Let F be the cdf of X, F_n the cdf of X_n, $n \in \mathbb{N}$. Set $X' = X_F$, $X'_n = X_{F_n}$. We must show that $\lambda(\{r : X'_n(r) \to X'(r)\}) = 1$. Since X' is nondecreasing, it has at most countably many jumps of size $1/m$ for any m, hence has at most countably many discontinuities. As λ of a countable set is 0, showing that $X'_n(r) \to X'(r)$ for the continuity points of X' completes the proof.

Let r be any continuity point of X'. We show that $X'(r) \leq \liminf_n X'_n(r)$ and $\limsup_n X'_n(r) \leq X'(r)$.

Pick $\epsilon > 0$. Since the discontinuities of F are at most countable, there exists an $x \in (X'(r) - \epsilon, X'(r))$ that is a continuity point of F. Since $F_n(x) \to F(x)$, for large n, $F_n(x) < r$, so that $X'(r) - \epsilon < x < X'_n(r)$. As ϵ was arbitrary, $X'(r) \leq \liminf_n X'_n(r)$. The proof for $\limsup_n X'_n(r) \leq X'(r)$ is essentially identical.   ■

7.10.c   Some Corollaries to Skorohod's Theorem

Corollary 7.10.7   *If $g : \mathbb{R} \to \mathbb{R}$ is continuous, $X_n \to_{\mathcal{D}} X$, then $g(X_n) \to_{\mathcal{D}} g(X)$.*

Proof. Let X', $(X'_n)_{n \in \mathbb{N}}$ be the sequence guaranteed by Skorohod's representation theorem. By the continuity of g, $g(X'_n) \to g(X')$ a.e., and a.e. convergence implies convergence in distribution. ∎

That proof is very easy. It is perhaps worth thinking about the difficulties one would have in proving the result without Skorohod's theorem: Let $Y = g(X)$. Since g can be many-to-one, F_Y, the cdf of Y, can have discontinuities even if the cdf of X does not; if y is a continuity point of F_Y, one has to show that $P(Y_n \in g^{-1}(-\infty, y)) \to P(Y \in g^{-1}(-\infty, y))$; the set $g^{-1}(-\infty, y)$ is not an interval unless g is monotonic.

Notation 7.10.8 *For a function $h : \mathbb{R} \to \mathbb{R}$, D_h denotes the discontinuity points of h.*

Lemma 7.10.9 *D_h is measurable whether or not h is measurable.*

Proof. $D_h = \cup_\epsilon \cap_\delta A_{\epsilon,\delta}$, where the union and intersection range of strictly positive rationals and $A_{\epsilon,\delta}$ is the open set $\{x \in \mathbb{R} : \exists y, z \in B_\delta(x), \ |h(y) - h(z)| \geq \epsilon\}$. ∎

Corollary 7.10.10 *If $h : \mathbb{R} \to \mathbb{R}$ is measurable, $X_n \to_{\mathcal{D}} X$, and $P(X \in D_h) = 0$, then $h(X_n) \to_{\mathcal{D}} h(X)$.*

Proof. Let X', $(X'_n)_{n \in \mathbb{N}}$ be the sequence guaranteed by Skorohod's representation theorem. Outside of the null set of ω such that $X' \in D_h$, $h(X'_n(\omega)) \to h(X'(\omega))$. ∎

Again, that proof is easy, and alternative proofs might not be.

Corollary 7.10.11 *If $X_n \to_{\mathcal{D}} a$ and g is continuous at a, then $g(X_n) \to_{\mathcal{D}} g(a)$.*

So far, the function h or g has remained constant.

Corollary 7.10.12 *If $f_n \in C(\mathbb{R})$ converges uniformly to $f \in C(\mathbb{R})$ on compact sets and $X_n \to_{\mathcal{D}} X$, then $f(X_n) \to_{\mathcal{D}} f(X)$.*

Proof. First suppose that $P(X \in K) = 1$ for some compact K. Since f is continuous on K, it is uniformly continuous. Hence, for any $\epsilon > 0$ there is a $\delta > 0$ such that $[|X'_n - X'|(\omega) < \delta] \Rightarrow [|f(X'_n) - f(X')|(\omega) < \epsilon]$, and a.e. convergence implies convergence in distribution. To finish the proof, consider the sequence of compact sets, $K_m = [-m, +m]$, noting that $P(X \in K_m) \uparrow 1$. ∎

In particular, $(a_n X_n + b_n) \to_{\mathcal{D}} (aX + b)$ if $(a_n, b_n) \to (a, b)$ and $X_n \to_{\mathcal{D}} X$. This kind of result turns out to be useful when one is rescaling estimators.

7.10.d More on Weak Convergence

Skorohod's representation theorem also makes it easier to give some alternative characterizations of $F_n \to_{w^*} F$. We need a few more pieces.

Start with a cdf F. From Theorem 7.6.4 and Carathéodory's extension theorem, there is a unique probability P_F on the Borel subsets of $\mathbb{R}$ with $P_F((-\infty, r]) = F(r)$. Further, the identity function $r \mapsto r$ has the distribution with cdf F.

Definition 7.10.13   *For $f \in C_b(\mathbb{R})$ and a cdf F, $\int f(r)\, dF(r)$ is defined as $\int f(\omega)\, dP_F(\omega)$, where $(\Omega, \mathcal{F}, P) = (\mathbb{R}, \mathcal{B}_\mathbb{R}, P_F)$.*

Definition 7.10.14   *For a probability P on $(\mathbb{R}, \mathcal{B}_\mathbb{R})$, $A \in \mathcal{B}_\mathbb{R}$ is a P-**continuity set** if $P(\partial A) = 0$.*

Finally, recall that Lemma 6.2.21 contains an alternative characterization of the Levy metric, d_L, for example, $\rho(F, G) = \inf\{\epsilon > 0 : (\forall x)\, [\, [P_F((-\infty, x]) \le P_G((-\infty, x]^\epsilon)] \wedge [P_G((-\infty, x]) \le P_F((-\infty, x]^\epsilon)]\,]\}$.

Theorem 7.10.15   *Let $(X, (X_n)_{n \in \mathbb{N}}) \subset L^0(\Omega, \mathcal{F}, P)$, let $(F, (F_n)_{n \in \mathbb{N}})$ be the associated cdf's, and let $(P, (P_n)_{n \in \mathbb{N}})$ be the associated probability distributions on $\mathbb{R}$. The following are equivalent.*

1. $X_n \to_\mathcal{D} X$.
2. $F_n \to_{w^*} F$.
3. $\rho(F_n, F) \to 0$.
4. *For all $f \in C_b(\mathbb{R})$, $\int f\, dF_n \to \int f\, dF$.*
5. *For all P-continuity sets, $P_n(A) \to P(A)$.*

Proof.   (1) $\Leftrightarrow$ (2) by definition. That (2) $\Leftrightarrow$ (3) was the point of §6.3.b, specifically Theorem 6.3.9.

(1) $\Rightarrow$ (4). Since $\int f\, dF_n = \int_\Omega f(X_n(\omega))\, dP$ and $\int f\, dF = \int_\Omega f(X(\omega))\, dP$, and f is continuous and bounded, which follows from Skorohod's representation theorem and dominated convergence.

(1) $\Rightarrow$ (5). If A is a P-continuity set, then $D_h = \partial A$ when $h = 1_A$. Apply Corollary 7.10.10 and dominated convergence.

(5) $\Rightarrow$ (2). If $A = (-\infty, x]$, then $\partial A = \{x\}$ and x is a continuity point of F iff $P(\{x\}) = 0$.

(4) $\Rightarrow$ (2). Pick $\epsilon > 0$. For $x \in \mathbb{R}$, define $f(r) = \max\{0, 1 - \frac{1}{\epsilon} d(r, (-\infty, x])\}$ so that

$$F_n(x) \le \int f\, dF_n \quad \text{and} \quad \int f\, dF \le F(x + \epsilon).$$

Since $f \in C_b(\mathbb{R})$, $\limsup_n F_n(x) \le F(x + \epsilon)$. Since F is right continuous, $\limsup_n F_n(x) \le F(x)$.

Define $g(r) = f(x + \epsilon)$; that is, shift the function f to the left by ϵ. We have

$$F(x - \epsilon) \le \int f\, dF \quad \text{and} \quad \int f\, dF_n \le F_n(x).$$

As $g \in C_b(\mathbb{R})$, $F(x - \epsilon) \le \liminf_n F_n(x)$. As F has limits from the left, $F(x-) \le \liminf_n F_n(x)$.

At continuity points of F, $F(x-) = F(x)$, so that $F_n(x) \to F(x)$.   ∎

7.11 ◆ Complements and Extras

7.11.a More on Convergence in Probability

The Ky Fan metric for $X, Y \in L^0(\Omega, \mathcal{F}, P)$ is $\alpha(X, Y) = \inf\{\epsilon \geq 0 : P(\{|X - Y| \geq \epsilon\}) < \epsilon$. It is translation invariant, and there are several equivalent metrics.

Exercise 7.11.1 Show that $\alpha(\cdot, \cdot)$ is, like the norm distance on L^1, translation invariant; that is, for all $X, Y, Z \in L^0$, $\alpha(X + Z, Y + Z) = \alpha(X, Y)$ and for all $X, Y, Z \in L^1$, $d_1(X + Z, Y + Z) = d_1(X, Y)$.

Exercise 7.11.2 (Different Ways to Metrize Convergence in Probability) Show that $e(X, Y) = E(1 \wedge |X - Y|)$ and $r(X, Y) = \inf\{\delta > 0 : \delta + P(|X - Y| > \delta)\}$ are translation invariant pseudometrics on $L^0(\Omega, \mathcal{F}, P)$. Further, show that the following are equivalent:

1. $\forall \epsilon > 0, P(|X_n - X| > \epsilon) \to 0$,
2. $\alpha(X_n, X) \to 0$,
3. $e(X_n, X) \to 0$, and
4. $r(X_n, X) \to 0$.

Exercise 7.11.3 Let $L_0((\Omega, \mathcal{F}, P); M)$ be the set of measurable functions from $(\Omega, \mathcal{F}, P)$ to the metric space M. Show the following:

1. If M is separable, then $\{d(X, Y) > \epsilon\} \in \mathcal{F}$.
2. If M is separable, then $\alpha(X, Y) = \inf\{\epsilon \geq 0 : P(\{d(X, Y) > \epsilon\}) \leq \epsilon\}$ is a metric on $L_0((\Omega, \mathcal{F}, P); M)$ and $\alpha(X_n, X) \to 0$ iff $P(\{d(X_n, X) > \epsilon\}) \to 0$ for all $\epsilon > 0$.
3. If M is complete and separable, then $L_0((\Omega, \mathcal{F}, P); M)$ is complete.
4. If $(\Omega, \mathcal{F}, P)$ is countably generated and M is complete and separable, then $L_0((\Omega, \mathcal{F}, P); M)$ is a complete separable metric space.

7.11.b A Lower Hemicontinuity Result for Sub-σ-Fields

If we have a sequence of random variables X_n in $L^0(\Omega, \mathcal{F}, P)$ and $\alpha(X_n, X) \to 0$, we might expect that $\sigma(X_n)$ converges to $\sigma(X)$ in some reasonable fashion. The following dashes these rosy expectations.

Example 7.11.4 Let $(\Omega, \mathcal{F}, P) = ((0, 1], \mathcal{B}, \lambda)$, $X_n(\omega) = \omega/n$, and $X(\omega) = 0$. $\alpha(X_n, X) < 1/n \downarrow 0$, and set $\sigma(X_n) = \mathcal{B}$ for all $n \in \mathbb{N}$ and $\sigma(X) = \{\emptyset, (0, 1]\}$.

What happens is an implosion: all but two of the sets in the sequence of σ-fields disappear from the "limit" σ-field. We see this kind of implosion in the tail σ-field, P can be nonatomic on each $\mathcal{F}_{M+}$ but have a single atom on $\cap_M \mathcal{F}_{M+}$.

Theorem 7.11.5 For X_n, X in $L^0(\Omega, \mathcal{F}, P)$, if $\alpha(X_n, X) \to 0$ and $A \in \sigma(X)$, then $(\forall \epsilon > 0)(\exists N)(\forall n \geq N)(\exists A_n \in \sigma(X_n))[\rho_P(A, A_n) < \epsilon]$.

Exercise 7.11.6 Fill in the details in the following sketch of a proof of Theorem 7.11.5.

1. $\mathfrak{X}_\partial := \{A \in \mathcal{B}_\mathbb{R} : P(X \in \partial A) = 0\}$ is a field of subsets of $\mathbb{R}$ that generates $\mathcal{B}_\mathbb{R}$. [The field part should be straightforward. To see that it generates $\mathcal{B}_\mathbb{R}$, start by showing that for any $x \in \mathbb{R}$, there are at most countably many $r > 0$ such that $P(\{X \in \partial B_r(x)\}) > 0$.]

2. Let $\mathcal{G}_\partial$ be the set of functions from $\mathbb{R}$ to the two-point set $\{0, 1\}$ of the form $g(x) = 1_A(x)$ for some $A \in \mathfrak{X}_\partial$. Show that for any $g \in \mathcal{G}_\partial$, $\alpha(g(X_n), g(X)) \to 0$.

3. The field of sets of the form $\{(g \circ X)^{-1}(r) : r = 0, 1, \ g \in \mathcal{G}_\partial\}$ generates $\sigma(X)$.

4. Use the ρ_P-denseness of fields in the σ-fields that they generated to complete the proof.

7.11.c Jensen's Inequality and Risk Aversion

We begin with Jensen's inequality for concave functions of random variables taking values in $\mathbb{R}^1$.

Theorem 7.11.7 (Jensen) *If $f : C \to \mathbb{R}$ is a concave function from a convex $C \subset \mathbb{R}$ and $X : \Omega \to C$ is integrable, then $E\, f(X) \le f(E\, X)$. For a convex f, the inequality is reversed.*

Proof. If $P(X = x) = 1$ for an $x \in \text{extr}(C)$, then the inequality is satisfied as an equality. In all other cases, $E\, X \in \text{int}(C)$.

Since f is concave, it is supdifferentiable at $E\, X$; that is, there exists a real number, v, such that for all $x \in C$,

$$f(x) \le f(E\, X) + v(x - E\, X). \tag{7.45}$$

Therefore, for all ω, $f(X(\omega)) \le f(E\, X) + v(X(\omega) - E\, X)$. Monotonicity of the integral yields

$$E\, f(X) \le f(E\, X) + E\, v(X - E\, X). \tag{7.46}$$

Linearity of the integral yields $E\, v(X - E\, X) = v E\, (X - E\, X) = 0$. Substituting $-f$ for f reverses the inequalities and covers the convex case. ∎

Exercise 7.11.8 $f(X)$ need not be integrable in general, but with X integrable and f concave, show that $E\, f(X) = -\infty$ is the only nonintegrable possibility.

For C a convex subset of $\mathbb{R}$, let $L^1(C)$ denote the set of $X \in L^1(\Omega, \mathcal{F}, P)$ with $P(\{X \in C\}) = 1$.

Definition 7.11.9 *A complete transitive preference relation, $\succsim$, on $L^1(C)$ is* **monotonic** *if $[X \ge Y \text{ a.e.}] \Rightarrow [X \succsim Y]$, and it is* **strictly monotonic** *if $[[X \ge Y \text{ a.e.}] \wedge [P(X > Y) > 0]] \Rightarrow [X \succ Y]$.*

Monotonicity captures the idea that more is better. For example, if $X = a$ and $Y = b$ are two degenerate random variables, then $X \succsim Y$ iff $a \ge b$.

Definition 7.11.10 *A complete transitive preference relation, $\succsim$, on $L^1(C)$ is* **risk averse** *if for all $X \in L^1(C)$, $E\, X 1_\Omega \succsim X$, and it is* **strictly risk averse** *if*

*E $X1_\Omega \succ X$ for all X that are not a.e. constant. **Risk-loving** and **strictly risk-loving** preferences reverse the preference orderings.*

Risk aversion captures the idea that receiving the average for sure beats facing randomness.

Example 7.11.11 *Suppose that a person with strictly monotonic risk-averse preferences $\succsim$ has a choice between a certain income of W and a risky income given by a random variable $X \in L^1(\Omega, \mathcal{F}, P)$. We say that X is **acceptable** if $X \succsim W$. If X is acceptable, then $E X \geq W$. If not, $W > E X$, then $W > E X1_\Omega$ a.e., and $E X1_\Omega \succsim X$. Put another way, the only acceptable risks must have expected payoffs higher than the payoff for not taking the risk.*

The previous example can be turned around.

Exercise 7.11.12 Suppose that a person faces a risky income given by a random variable $X \in L^1(\Omega, \mathcal{F}, P)$. A **complete insurance contract** is one that gives the person a certain income W. An acceptable complete insurance contract is one with $W \succsim X$. Show that if a person is risk averse, then the set of acceptable complete insurance contracts includes the interval $[E X, \infty)$.

Definition 7.11.13 *A preference relation $\succsim$ on $L^1(C)$ has an **expected utility representation** if it can be represented by a utility function $U : L^1(C) \to \mathbb{R}$ of the form $U(X) = \int u(X) \, dP$ for some $u : C \to \mathbb{R}$.*

For expected utility preferences, the restriction that $U(X) \in \mathbb{R}$ for all $X \in L^1(C)$ can only be met if u is bounded.

Exercise 7.11.14 Suppose that $(\Omega, \mathcal{F}, P)$ is nonatomic and let $E_1, E_2, \ldots$ be a partition of Ω with $P(E_n) = 1/2^n$. Suppose also that u is unbounded below; that is, for each $n \in \mathbb{N}$, there exists an x_n such that $u(x_n) \leq -2^n$. Pick $x_0 \in C$. Show that the random variables

$$X_m = \left(\sum_{n \leq m} x_n \cdot 1_{E_n} \right) + x_0 \cdot 1_{\cup_{n > m} E_n}$$

have $U(X_m) \downarrow -\infty$; that $X_n \to X_\infty$ a.e. for some random variable X_∞; and that $\omega \mapsto u(X_\infty(\omega))$ is not integrable.

Preferences on $L^1(C)$ are preferences over functions from Ω to C, that is, over subsets of $\Omega \times C$. By contrast, an important aspect of expected utility preferences is that ω does not enter except through the values of $X(\omega)$.

Exercise 7.11.15 Show that if X and Y in $L^1(C)$ have the same distribution, that is, $P(X \leq x) = P(Y \leq r)$ for all r, and the preference relation $\succsim$ has an expected utility representation, then $X \sim Y$. [Change of variables.]

Now suppose that C is a compact convex subset of $\mathbb{R}$, that is, $C = [a, b]$ for some $a, b \in \mathbb{R}$, and that $(\Omega, \mathcal{F}, P)$ is nonatomic. Let $\Delta(C)$ denote the set of probability distributions on C. Since C is a csm space, for each $\mu \in \Delta(C)$, there is at least one $X \in L^1(C)$ such that $\mu(A) = P(X \in A)$ for all measurable $A \subset C$. For expected utility preferences, change of variables tells us that $U : \Delta(A) \to \mathbb{R}$ can be defined by $U(\mu) = \int_C u(x) \, d\mu(x)$.

Exercise 7.11.16 Suppose that preferences on $\Delta(C)$, C a compact convex subset of $\mathbb{R}$, can be represented by $U(\mu) = \int u(x)\, d\mu(x)$ for some concave $u : C \to \mathbb{R}$. Show that the preferences are risk averse. If u is convex, show that the preferences are risk loving. [Jensen's inequality.]

The previous result has a converse.

Exercise 7.11.17 Suppose that $C = [a, b] \subset \mathbb{R}$ and that preferences on $\Delta(C)$ can be represented by $U(\mu) = \int u(x)\, d\mu(x)$ for some bounded $u : C \to \mathbb{R}$. Show that if the preferences are risk averse, then u is concave, and if risk loving, then u is convex.

Exercise 7.11.18 If $(\Omega, \mathcal{F}, P)$ is nonatomic, C is convex, and expected utility preferences $\succsim$ on $L^1(C)$ represented by $U(X) = \int u(X)\, dP$ satisy $E\, X 1_\Omega \succsim X$ for all $X \in L^1(C)$ taking on two values, then $u : C \to \mathbb{R}$ is concave. [Let $X = x1_E + y1_{E^c}$ for $E \in \mathcal{F}$ with $P(E) = \alpha$.]

Exercise 7.11.19 If C is compact and expected utility preferences on $L^1(C)$ can be represented by $U(X) = \int u(X)\, dP$, where $u : C \to \mathbb{R}$ is continuous, then $U : L^1(C) \to \mathbb{R}$ is α-continuous, and if $\mu_n \to_{w*} \mu$, then $U(\mu_n) \to U(\mu)$.

Risk-averse expected utility preferences can be discontinuous on $L^1(C)$ if the convex C does not contain its left boundary point.

Exercise 7.11.20 If $C \subset \mathbb{R}$ is the interval $(a, b|$, $a \geq -\infty$, then there exists a concave $u : C \to \mathbb{R}$ and an $X \in L^1(C)$ such that $\int u(X)\, dP = -\infty$. Further, if $a_n \downarrow a$, then $X_n := X \cdot 1_{\{X \in (a, a_n]\}} \to 0$ a.e. but $U(X_n) \not\to U(0)$.

7.11.d Differences of Opinion and the "No Bet" Theorem

As Mark Twain said in 1895, "It is not best that we should all think alike; it is differences of opinion that make horse races." In modeling uncertainty, differences of opinion between persons 1 and 2 can arise from two nonexclusive sources: differences of probability measures, P_1 and P_2, on $(\Omega, \mathcal{F})$, and differential information, modeled as sub-σ-fields, for example, $\mathcal{X}_1 = \sigma(S_1)$ and $\mathcal{X}_2 = \sigma(S_2)$, where S_1 and S_2 are 1 and 2's private signals. It is somewhat surprising that differential information cannot make bets palatable if people share an understanding of how the world works, that is, if $P_1 = P_2$. The formal result is the "no bet" theorem, but before giving it, here is some intuition.

Two people are considering betting against each other about a random outcome, $Y(\omega)$, for example, which horse will win the next race. Before the horses race, the two people get different information, $S_1(\omega)$ and $S_2(\omega)$, about how the various horses in the race are performing, say from watching morning training sessions, reading their previous results, talking to horse racing enthusiasts, or from some other source(s).

1. Suppose that person 1, on the basis of his information, is willing to take a $100 even-money bet that the horse named BabyJane will win the next race.

2. Suppose that person 2, on the basis of her information, is willing to take a $100 even-money bet that BabyJane will not win.

The question is, "Will the two shake hands or otherwise commit themselves to this bet?" Provided that the two people have the same understanding of how the world works, that is, provided that $P_1 = P_2$, the answer is, "no!"

The information that the other person is willing to take an even-money bet against you means that his information tells him that he is more likely to win than you are. After more argument, it turns out that both people using this information leads to the result that there is no bet that both would be willing to take. Perhaps this is not too surprising; after all, $P_1 = P_2$ requires that both have the same set of conditional probabilities $P_1(A \mid B) = P_2(A \mid B)$ no matter how complicated the sets A and B.

Here is the formal setting. A **bet** is a $\mathcal{G}$-measurable function $B : \Omega \to \mathbb{R}$, $\mathcal{G} \subset \mathcal{F}$, for example, $\mathcal{G} = \sigma(Y)$. The interpretation is that if 1 and 2 take the bet, then person 1 will have $W_1 + B(\omega)$ and person 2 will have $W_2 - B(\omega)$, where W_1 and W_2 are the nonstochastic initial wealths of the two people. The expected utility functions, u_1 and u_2, are weakly concave, increasing functions from $\mathbb{R}$ to $\mathbb{R}$. A bet is acceptable if both are willing to accept it and at least one is strictly willing to accept it. For risk-neutral or risk-averse people, a necessary condition for acceptability is that both people view the bet as having a positive expected payoff.

Definition 7.11.21 *A $\mathcal{G}$-measurable bet $B : \Omega \to \mathbb{R}$ is **potentially acceptable** if*

$$\int B \, dP_1 > 0 \quad \text{and} \quad \int (-B) \, dP_2 \geq 0 \quad \text{or if}$$

$$\int B \, dP_1 \geq 0 \quad \text{and} \quad \int (-B) \, dP_2 > 0. \tag{7.47}$$

Theorem 7.11.22 (No Bet) *$P_1 = P_2$ on $\mathcal{G}$, that is, $P_1(A) = P_2(A)$ for all $A \in \mathcal{G}$, iff there is no $\mathcal{G}$-measurable potentially acceptable bet.*

Before giving the proof, let us consider some special cases:

1. If $\mathcal{G} = \sigma(Y)$, then the bet depends only on the publicly observable Y, and if both people agree about the distribution of Y, then there is no potentially acceptable bet.

2. If $\mathcal{G} = \sigma((\mathcal{X}_1 \cap \mathcal{X}_2), \sigma(Y))$, then the bet must depend on the publicly observable value of Y and those aspects of each other's private information that both share. Agreement about the joint distribution of the mutually observable information and Y means that there are no potentially acceptable bets.

3. If $\mathcal{G} = \sigma(\mathcal{X}_1, \mathcal{X}_2, \sigma(Y))$, then private information cannot deliver a potentially acceptable bet when both people agree about the process that generates each other's information and the outcome Y.

Proof of the "No Bet" Theorem. If $P_1 = P_2$ on $\mathcal{G}$, then $[\int B \, dP_1 > 0] \Leftrightarrow [\int (-B) \, dP_2 < 0]$, so no bet can be potentially acceptable.

If $P_1 \neq P_2$ on $\mathcal{G}$, then there exists $A \in \mathcal{G}$ such that $P_1(A) > P_2(A)$ (if one has the reverse inequality, consider A^c). Define the $\mathcal{G}$-measurable bet B by $B(\omega) = 1_A(\omega) - P_2(A)$. In this case, $\int B \, dP_1 = P_1(A) - P_2(A) > 0$ and $\int (-B) \, dP_2 = -(P_2(A) - P_2(A)) = 0$. ∎

People bet all the time. There are two reasons that naturally appear in the models considered here: first, the wealths W_i are not stochastically independent of $\mathcal{G}$, in which case betting is a form of insurance; second, the utility of money depends on the outcome Y, more generally, that the utility is **state dependent**.

Exercise 7.11.23 (Bets as Insurance)   Suppose that $P_1 = P_2 = P$ and $P(Y = 1) = P(Y = 2) = \frac{1}{2}$ and that if $Y(\omega) = 1$, then $(W_1, W_2)(\omega) = (1, 0)$ and that if $Y(\omega) = 2$, then $(W_1, W_2)(\omega) = (0, 1)$. Show that if u_1 and u_2 are strictly concave increasing expected utility functions, there are acceptable bets making both strictly better off. Show that this is not true if u_1 and u_2 are linear.

Exercise 7.11.24 (State-Dependent Preferences)   Suppose that $Y = 1$ if person 1's horse wins the race and $Y = 2$ if person 2's horse wins. Suppose also that $u_i(i, b) = 2\sqrt{b}$ and $u_i(j, b) = \sqrt{b}$ for $i \neq j$, i, $j \in \{1, 2\}$. This captures the idea that money is sweeter when your horse wins than when it loses. Suppose that $W_1 = W_2 = 100$ and that $P_1 = P_2 = P$, where $P(Y = 1) = P(Y = 2) = \frac{1}{2}$. Show that both would happily accept an even-money bet of 44 on their own horse.

Despite the strength of the assumption that $P_1(A \mid B) = P_2(A \mid B)$ for all A and B, no matter how complicated, the assumption is widely accepted by economists. The reason is that if they are not equal, then a **free lunch** exists.

1. If $P_1(A) > P_2(A)$, then for some small $x > 0$, the bet $B_x(\omega) = [1_A(\omega) - P_2(A)] - x$ gives 1 strictly positive expected returns and the bet $-B_x(\omega) = x - [1_A(\omega) - P_2(A)]$ gives 2 strictly positive expected returns.

2. Strictly positive expected returns are worth some amount, $r > 0$, to both people. Thus, both 1 and 2 would be willing to pay me r to take B_x.

3. I sign legally binding contracts with both of them specifying the bet and take their money, r.

4. If A happens, then person 2 owes me 1 unit of account, she meets her obligation, and I give this to person 1, satisfying my legal obligation.

5. If A^c happens, then person 1 owes me 1 unit of account, she meets her obligation, and I give this to person 2, satisfying my legal obligation.

6. At the end of the day, I am, with probability 1, r richer and the two of them are, jointly, r poorer.

This means that if $P_1 \neq P_2$, then I can get something for nothing. Assuming that both people are risk averse complicates the logic, but does not change the conclusion: $P_1 \neq P_2$ implies the existence of a free lunch for someone. What does change the logic is the existence of costs to finding, making, and enforcing the bets.

7.11.e   Ambiguity and λ-Systems

We have insisted that the appropriate domain for probabilities is a field in all contexts. When modeling choice under uncertainty, this may not be perfectly sensible.

Example 7.11.25 *The* 100 *balls in an urn may be red, blue, yellow, or white. I know that the number of R plus the number of B is equal to 40, R + B = 40. I also know that the number of B plus the number of Y is equal to 50, B + Y = 50. From these, some additions and subtractions yield* $10 \le W \le 50$, $B = W - 10$, $R = 50 - W$, *and* $Y = 60 - W$.

Draw a ball at random from the urn. $P(R \cup B) = 4/10$ *and* $P(B \cup Y) = 5/10$. *The distribution is* **ambiguous***, meaning that our information does not completely specify all of P if we insist that P is defined on a field—if we insist that the* $P(E \cap F)$ *is known when* $P(E)$ *and* $P(F)$ *are known, then* $P(B)$ *must be known. Here it is someplace between* 0 *and* 4/10.

The set of events for which the probability is known in this example is far from being a field.

Exercise 7.11.26 For probabilities P, Q on a σ-field $\mathcal{F}$, show that $\mathcal{A} = \{E \in \mathcal{F} : P(E) = Q(E)\}$ is a λ-system. [Go back to Dynkin's π-λ theorem 7.3.28 if you have forgotten the relevant definitions.]

Perhaps λ-systems are the appropriate domain for probabilities, at least sometimes.

A bet is a function $B : \Omega \to \mathbb{R}$. One of the widely used classes of utility functions for modeling choice between bets when there is ambiguity uses utility functions of the form

$$U(B) = \alpha \min\{\int u \, dP : P \in S\} + (1 - \alpha) \max\{\int u \, dQ : Q \in S\}, \quad (7.48)$$

where $0 \le \alpha \le 1$ and S is a set of probability distributions. Larger sets S correspond to more ambiguity and smaller sets to less ambiguity, and larger α correspond to disliking ambiguity more. An easy extension of Exercise 7.11.26 shows that the set of E on which all P, $Q \in S$ agree is a λ-system.

One could decrease S in Example 7.11.25 by, say, giving information such as $W \ge 20$ and could increase S by replacing the information that there are 100 balls by the information that the total number of balls is between 100 and 200 and leaving the rest unchanged. The relations between choice under ambiguity and information is a fascinating and open field of study.

7.11.f Countable Partitions and σ-Fields

In the case that $\mathcal{X}_1$ and $\mathcal{X}_2$ are finite or generated by a countable partition of Ω, there are informative partition-based comparisons between $\mathcal{X}_1 \cap \mathcal{X}_2$ and $\sigma(\mathcal{X}_1 \cup \mathcal{X}_2)$.

For countable partitions $\mathcal{E}_1$, $\mathcal{E}_2 \subset \mathcal{F}$ of Ω, define $\mathcal{E}_1 \succeq \mathcal{E}_2$ if $\mathcal{E}_1$ is **finer than** $\mathcal{E}_2$ or $\mathcal{E}_2$ is **coarser than** $\mathcal{E}_1$, that is, if $[E \in \mathcal{E}_2] \Rightarrow (\exists \{E_a : a \in A\} \subset \mathcal{E}_1)[E = \cup_a E_a]$.

Exercise 7.11.27 Let $\mathfrak{E}$ denote the set of all countable measurable partitions of Ω. Show that the partial order $\succeq$ "finer than" is transitive and that $(\mathfrak{E}, \succeq)$ is a lattice.

$\mathcal{E}_1 \vee \mathcal{E}_2$ is the coarsest partition finer than both $\mathcal{E}_1$ and $\mathcal{E}_2$, and it is known as the **coarsest common refinement**. $\mathcal{E}_1 \wedge \mathcal{E}_2$ is the finest partition coarser than both $\mathcal{E}_1$ and $\mathcal{E}_2$, and it is known as the **finest common coarsening**.

Exercise 7.11.28 If $\mathcal{X}_i = \sigma(\mathcal{E}_i)$, $\mathcal{E}_i$ a countable measurable partition of Ω, $i = 1, 2$, then $\sigma(\mathcal{X}_1 \cup \mathcal{X}_2) = \sigma(\mathcal{E}_1 \vee \mathcal{E}_2)$ and $\sigma(\mathcal{X}_1 \cap \mathcal{X}_2) = \sigma(\mathcal{E}_1 \wedge \mathcal{E}_2)$.

7.11.g The Paucity of the Uncountable Product σ-Field

Consider the space $\Omega = \mathbb{R}^{[0,1]}$, the set of all functions from $[0, 1]$ to $\mathbb{R}$. Since Ω is a set of functions, we use f, g, h to denote typical members. For each $r \in [0, 1]$, the canonical projection mapping is $\text{proj}_r(f) = f(r)$.

Definition 7.11.29 *The **product field** on $\mathbb{R}^{[0,1]}$ is $\mathcal{F}^\circ$, the smallest field containing $\{\text{proj}_r^{-1}(A) : A \in \mathcal{B}\}$, and the **product σ-field** is $\mathcal{F} = \sigma(\mathcal{F}^\circ)$.*

Given a finite subset $T = \{r_1, \ldots, r_I\}$ of $[0, 1]$ and a finite collection $A = \{A_{r_1}, \ldots, A_{r_I}\} \subset \mathcal{B}$ of Borel subsets of $\mathbb{R}$, the associated **cylinder set** is $C_{T,A} = \{f \in \mathbb{R}^{[0,1]} : f(r_i) \in A_i, \ i = 1, \ldots, I\} = \text{proj}_T(f) = (f(r))_{r \in T}$.

Exercise 7.11.30 Show that the class of cylinder sets is a field. Show that the same is true if we substitute any field $\mathcal{B}^\circ \subset \mathcal{B}$ that generates $\mathcal{B}$ into the definitions.

From Carathéodory's extension theorem, we know that any countably additive probability on $\mathcal{F}^\circ$ has a unique countably additive extension to $\mathcal{F}$. A sufficient condition for countable additivity on $\mathcal{F}^\circ$ is **Kolmogorov's consistency condition**.

Definition 7.11.31 *Suppose that we have a collection $\{P_T : T \in \mathcal{P}_F([0, 1])\}$ of countably additive probabilities on $\mathbb{R}^T$. It satisfies **Kolmogorov's consistency condition** if for all finite $T \subset S \subset [0, 1]$, the marginal distribution of P_S on $\mathbb{R}^T$ is equal to P_T.*

Such a collection gives all of the joint distributions of a continuous-time stochastic process with time set $[0, 1]$. Specifically, $P_T(\times_{r \in T} A_r)$ is the probability that the process takes its value in A_r at all of the times $r \in T$. Given a collection satisfying Kolmogorov's consistency condition, define P on $\mathcal{F}^\circ$ by $P(C_{T,A}) = P_T(\times_{r \in T} A_r)$. The proof of the following is another application of compact classes.

Theorem 7.11.32 (Kolmogorov) *P is countably additive on $\mathcal{F}^\circ$.*

What this means is that if you have consistently specified the joint distribution that you want from a stochastic process $(\omega, t) \mapsto X(\omega, t)$, then you can use $(\mathbb{R}^{[0,1]}, \mathcal{F}, P)$ as the probability space and define $X(\omega, t) = \varphi(t, \omega)$, where φ is the evaluation mapping, $\varphi(t, f) = f(t) = \text{proj}_t(f)$.

This is nice, but a bit more limited than it seems. The essential problem is that most of the useful subsets of functions does not belong to $\mathcal{F}$.

Theorem 7.11.33 *$A \in \mathcal{F}$ iff $A \in \sigma(\{\text{proj}_s^{-1}(\mathcal{B}) : s \in S\})$ for some countable $S \subset [0, 1]$.*

Proof. For any countable $S \subset [0, 1]$, let $\mathcal{F}_S$ denote $\sigma(\{\text{proj}_s^{-1}(\mathcal{B}) : s \in S\})$. Check that $\cup_S \mathcal{F}_S$ is the smallest σ-field containing the cylinder sets. ∎

Thus, the set of continuous functions do not belong to $\mathcal{F}$.

Exercise 7.11.34 Show that $C([0, 1]) \notin \mathcal{F}$.

Let $J \subset \mathbb{R}^{[0,1]}$ be the set of pure jump paths, that is, the set of functions that are right continuous and both right and left constant. More specifically, $f \in J$ iff it is right continuous; for all $t \in (0, 1]$, there exists an $\epsilon > 0$ such that f is constant on $(t - \epsilon, t)$; and for all $t \in [0, 1)$, there exists an $\epsilon > 0$ such that f is constant on $[t, t + \epsilon)$. Poisson processes take values only in J, in fact, in the subset of J taking on only integer values.

Exercise 7.11.35 Show that $J \notin \mathcal{F}$, and that the subset of J taking on only integer values is also not in $\mathcal{F}$.

7.12 ◆ Appendix on Lebesgue Integration

A leading example of the probabilities we developed in this chapter is λ, the uniform distribution on $((0, 1], \mathcal{B})$. We use $\mathcal{B}^\circ$, the set of finite disjoint unions of interval subsets of $[0, 1]$. For an interval $|a, b| \in \mathcal{B}^\circ$, we define $\lambda(|a, b|) = (b - a)$. For any finite disjoint union of intervals, I_n, we define $\lambda(\cup_{n \leq N} I_n) = \sum_{n \leq N} \lambda(I_n)$. This makes $\lambda : \mathcal{B}^\circ \to [0, 1]$ a finitely additive probability on $\mathcal{B}^\circ$.

To extend λ to a larger class of sets, we show that $\mathcal{B}^\circ$ is ρ_λ-dense in $\mathcal{B}$ and identify $(\mathcal{B}, \rho_\lambda)$ with the metric completion of the space $(\mathcal{B}^\circ, \rho_\lambda)$. This approach uses Cauchy sequences $A_n \in \mathcal{B}^\circ$ and looks for representation sets A with $\lambda(A_n \Delta A) \to 0$. The classical approach to defining the measure of A, $\lambda(A)$, is one involving sandwiching. One finds $F \subset A \subset G$, where we can assign a measure to F and G and $(G \setminus F) \downarrow \emptyset$. This appendix fills in the details of this more classical construction.

7.12.a Overview

To sandwich a set A, we begin by approximating it from the outside, asking for as tight a cover of *countable* unions of sets in $\mathcal{B}^\circ$ as is possible. A bit more formally, for any $A \subset [0, 1]$, we define the outer measure of A, $m^*(A)$, as the infimum of the measure of all such covers.

To get the sandwiching, we then approximate the set A^c in the same way, defining $m^*(A^c)$, the outer measure of A^c. The sandwich picture arises because if $A \subset G_n$ and $A^c \subset H_n$, then $F_n \subset A \subset G_n$, where $F_n := H_n^c$. This means that A is sandwiched between two sets to which we can assign measures.

If $m^*(A) + m^*(A^c) = 1$, we say that A is Lebesgue measurable. For Lebesgue measurable sets, we define $\lambda(A) = m^*(A)$. Given our work on nonmeasurable sets, this seems like a sensible approach.

Our last steps involve showing that the set of A with $m^*(A) + m^*(A^c) = 1$ is the completed σ-field containing $\mathcal{B}^\circ$ and that λ is countably additive on this σ-field.

7.12.b Outer and Inner Measure

Definition 7.12.1 *For each set $A \subset [0, 1]$, the **outer measure of** A is defined by*

$$m^*(A) = \inf\{\sum_n \lambda(I_n) : A \subset \cup_n I_n, \ I_n \in \mathcal{B}^\circ\}. \tag{7.49}$$

*The **inner measure** is $1 - m^*(A^c)$.*

The outer measure is the least overestimate of the length of a given set A using countable collections of intervals. Here are some of the basic properties of m^*:

1. Since $[0, 1]$ is an interval containing any A, we know that $m^*(A) \leq 1$.

2. Restricted to $\mathcal{B}^\circ$, $m^* = \lambda$.

3. Since $\mathcal{B}^\circ$ is a field, the unions $\cup_n I_n$, in (7.49) can be taken to be disjoint.

Bear in mind that m^* applies to *all* subsets of $[0, 1]$. Another basic property is monotonicity, as follows.

Theorem 7.12.2 *If $A \subset B$, then $m^*(A) \leq m^*(B)$.*

Proof. If $A \subset B \subset \cup_n I_n$, then $A \subset \cup_n I_n$. ∎

Here is a step on the way to countable additivity.

Theorem 7.12.3 *m^* is **countably subadditive**; that is, for any collection A_n of subsets of $[0, 1]$, $m^*(\cup_n A_n) \leq \sum_n m^*(A_n)$. In particular, $m^*(A) + m^*(A^c) \geq 1$ for all $A \subset [0, 1]$.*

Proof. For each A_n, let $\cup_m I_{n,m}$ be a countable collection of intervals with

$$m^*(A_n) \leq \sum_m \lambda(I_{n,m}) \leq m^*(A_n) + \frac{\epsilon}{2^n}.$$

Since $\cup_n A_n \subset \cup_{n,m} I_{n,m}$, monotonicity implies $m^*(\cup_{n \in \mathbb{N}} A_n) \leq \sum_{n \in \mathbb{N}} m^*(A_n) + \epsilon$. The last part follows from $A \cup A^c = [0, 1]$, so that $m^*(A) + m^*(A^c) \geq m^*([0, 1]) = 1$. ∎

Corollary 7.12.4 *If A is a countable set, then $m^*(A) = 0$.*

Proof. Let $A = \{a_n : n \in \mathbb{N}\}$ be a countable set and pick $\epsilon > 0$. For each $n \in \mathbb{N}$, let $I_n(\epsilon)$ be the interval $|a_n - \epsilon/2^n, a + \epsilon/2^n|$. $A \subset \cup_n I_n(\epsilon)$, and $\sum_n \lambda(I_n) = \sum_n \epsilon/2^n = \epsilon$. ∎

If $A = \mathbb{Q} \cap [0, 1]$, the countable cover by the intervals $I_n(\epsilon)$ just used is rather difficult to visualize—it is open and dense, its complement is closed and nowhere dense, and it has "length" smaller than ϵ. This yields (yet another) proof of the uncountability of $\mathbb{R}$.

Corollary 7.12.5 *If $0 \leq a < b \leq 1$, then any set containing the interval (a, b) is uncountable.*

Proof. m^* agrees with λ on $\mathcal{B}^\circ$ and $\lambda((a, b)) = (b - a) > 0$. ∎

The converse is not true. The Cantor set, C, of Example 7.9.8 is an uncountable set with $m^*(C) = 0$.

Notation 7.12.6 *$(\mathcal{B}^\circ, \bigcup^c)$ is defined as the class of all countable unions of sets in $\mathcal{B}^\circ$.*

Theorem 7.6.4 (p. 412) tells us that λ is countably additive on $\mathcal{B}^\circ$.

Lemma 7.12.7 *For any $A \in (\mathcal{B}^\circ, \bigcup^c)$, $m^*(A) = \sum_n \lambda(I_n)$, where $\cup_n I_n$ expresses A as a disjoint union of elements on $\mathcal{B}^\circ$.*

Proof. Let $A = \cup_n I_n$ and $A = \cup_m I'_m$ be two different expressions of A as a countable disjoint union of elements of $\mathcal{B}^\circ$. $p = \sum_n \lambda(I_n)$ and $p_N = \sum_{n \leq N} \lambda(I_n)$, so that $p_N \uparrow p$. Define p' and p'_M in a parallel fashion for the disjoint sequence I'_m. By monotonicity, it is sufficient to show that $p = p'$.

For any fixed m, $I'_m \setminus (\cup_{n \leq M} I_n) \downarrow \emptyset$. By the countable additivity of λ, $\lambda(I'_m \setminus (\cup_{n \leq M} I_n)) \downarrow 0$. Thus, for all M, $p'_M \leq p$, which implies that $p' \leq p$. If we reverse the roles of the I'_m and the I_n, $p \leq p'$. ∎

Corollary 7.12.8 *For any* $A \in (\mathcal{B}^\circ, \bigcup^c)$, $m^*(A) + m^*(A^c) = 1$.

Proof. If $A \subset \cup_n I_n$, $\{I_n : n \in \mathbb{N}\}$ a disjoint collection, and $A^c \subset \cup_m I'_m$, $\{I'_m : m \in \mathbb{N}\}$ a disjoint collection, $\sum_n \lambda(I_n) + \sum_n \lambda(I'_m) \geq 1$ because there may be overlap between $\cup_n I_n$ and $\cup_m I'_m$. For $A \in (\mathcal{B}^\circ, \bigcup^c)$, there need be no overlap, so that $(\cup_n I_n) \cup (\cup_m I_m)$ expresses $[0, 1]$ as a disjoint union of elements in $\mathcal{B}^\circ$, and λ is countably additive on $\mathcal{B}^\circ$. ∎

7.12.c Lebesgue Measurable Sets

While the outer measure has the advantage that it is defined for every subset of $[0, 1]$, we have only proved that it is countably subadditive, not countably additive. In §7.8.b we showed that it cannot be countably additive on $\mathcal{P}([0, 1])$. In order to satisfy countable additivity, we have to restrict the domain of the function m^* to some suitable subset.

Definition 7.12.9 *A set* $A \subset [0, 1]$ *is* **Lebesgue measurable** *or* $\mathcal{L}$-**measurable** *if* $m^*(A) + m^*(A^c) = 1$, *where* $\mathcal{L}$ *denotes the set of* $\mathcal{L}$-*measurable sets.*

Corollary 7.12.8 showed that $(\mathcal{B}^\circ, \bigcup^c) \subset \mathcal{L}$.

Our aim is to show that $\mathcal{L}$ is a σ-field containing $\mathcal{B}^\circ$ and that m^* is countably additive on it. Since $m^*(A) + m^*(A^c) = m^*(A^c) + m^*(A)$, $\mathcal{L}$ is closed under complementation.

Lemma 7.12.10 *If* $m^*(A) = 0$, *then* A *is* $\mathcal{L}$-*measurable.*

Proof. Suppose that $A \subset \cup_n I_n$, that the I_n are disjoint, and that $\sum_n \lambda(I_n) < \epsilon$. $\cap_n I_n^c \subset A^c$ and $\cup_n I_n$ is $\mathcal{L}$-measurable. Hence, by monotonicity, $1 = m^*([0, 1]) \geq m^*(A^c) > 1 - \epsilon$. ∎

Exercise 7.12.11 Show that if $m^*(E) = 0$, then $m^*(E \cup A) = m^*(A)$.

Lemma 7.12.12 m^* *is finitely additive on* $\mathcal{L}$; *that is, if* $E_1, E_2 \in \mathcal{L}$ *and* $E_1 \cap E_2 = \emptyset$, *then* $m^*(E_1 \cup E_2) = m^*(E_1) + m^*(E_2)$.

Proof. Immediate. ∎

Lemma 7.12.13 *If* $E_1, E_2 \in \mathcal{L}$, *then* $E_1 \cup E_2 \in \mathcal{L}$.

Proof. We know that $m^*(E_1 \cup E_2) + m^*([E_1 \cup E_2]^c) \geq 1$. If we show that $1 \geq m^*(E_1 \cup E_2) + m^*([E_1 \cup E_2]^c)$, then we have shown that $m^*(E_1 \cup E_2) + m^*([E_1 \cup E_2]^c) = 1$, completing the proof. Now,

$1 = m^*(E_1) + m^*(E_1^c)$ (because $E_1 \in \mathcal{L}$)

$\geq m^*(E_1) + m^*(E_1^c \cap E_2) + m^*(E_1^c \cap E_2^c)$ (by the monotonicity of m^*)

$= m^*(E_1) + m^*(E_1^c \cap E_2) + m^*([E_1 \cup E_2]^c)$ (by DeMorgan)

$= m^*(E_1 \cup E_2) + m^*([E_1 \cup E_2]^c)$ (by finite additivity of m^*).   ∎

If we combine what we have so far, $\mathcal{L}$ is a field containing $\mathcal{B}^\circ$ and m^* is finitely additive on $\mathcal{L}$.

Theorem 7.12.14 $\mathcal{L}$ *is a σ-field containing $\mathcal{B}^\circ$ and m^* is countably additive on $\mathcal{L}$.*

Proof.   Let E_n be a sequence of disjoint sets in $\mathcal{L}$. We wish to show that $m^*(\cup_n E_n) = \sum_n m^*(E_n)$. By countable subadditivity, we know that $m^*(\cup_n E_n) \leq \sum_n m^*(E_n)$. We show that for every $\epsilon > 0$,

$$\sum_n m^*(E_n) - \epsilon < m^*(\cup_n E_n) \leq \sum_n m^*(E_n). \qquad (7.50)$$

Since $\cup_{n \leq N} E_n \in \mathcal{L}$ for all N, $\sum_{n \leq N} m^*(E_n) \leq 1$, so that $\sum_n m^*(E_n) \leq 1$. Therefore, for any $\epsilon > 0$, we can pick N large enough that

$$\sum_n m^*(E_n) - \epsilon < \sum_{n \leq N} m^*(E_n) \leq \sum_n m^*(E_n). \qquad (7.51)$$

By monotonicity, $\sum_{n \leq N} m^*(E_n) \leq m^*(\cup_n E_n)$, which yields (7.50).

Finally, $m^*(\cup_n E_n) + m^*([\cup_n E_n]^c) = 1$ because $m^*(\cup_{n \leq N} E_n) + m^*([\cup_{n \leq N} E_n]^c) = 1$ for each N.   ∎

The set function $\lambda : \mathcal{L} \to [0, 1]$ obtained by restricting the functions m^* to the $\mathcal{L}$ is called the **Lebesgue measure**. We now have a second proof that the Lebesgue measure is a countably additive probability on the completion of $\mathcal{B}$.

A final comment: The approach given here—to use inner-outer approximation methods of extending the Lebesgue measure from a particular field of sets to the σ-field that it generates—does generalize. With some further complications, one can prove Carathéodory's extension theorem using this approach. The full proof using inner and outer approximation seems at least as complicated as the approach we took here, completing the metric space $(\mathcal{F}^\circ, d_P)$.

7.13 ◆ Bibliography

P. Billingsley's *Probability and Measure* (3rd Ed., New York: Wiley, 1995) is one of our favorite textbooks and is the model for what we have tried to do in this book—develop mathematics hand in hand with its applications. For pure measure theory, C. Dellacherie and P. A. Meyer's advanced monograph, *Probabilities and Potential* (translated and prepared by J. P. Wilson; Amsterdam: North-Holland, 1982) is hard to beat. R. M. Dudley's *Real Analysis and Probability* (Cambridge: Cambridge University Press, 2002) covers a huge amount of mathematics, with applications in probability theory having the place of honor. W. Feller's two-volume classic, *An Introduction to Probability Theory and Its Applications* (New York: Wiley, 1971), contains the best probabilistic intuitions we have ever seen.

The $L^p(\Omega, \mathcal{F}, P)$ and ℓ^p Spaces, $p \in [1, \infty]$

A complete normed vector space is called a **Banach space**. The classical Banach spaces are the spaces of measurable functions $L^p(\Omega, \mathcal{F}, P)$, $p \in [1, \infty]$, and the sequence spaces ℓ^p. These spaces are used extensively in econometrics as spaces of random variables, especially in regression theory. In economic theory, nonatomic probability spaces are a frequently used model of large populations, and measurable functions become the strategies or consumptions of the people in the population.

We begin this chapter with an overview of the uses of Banach spaces in econometrics and game theory and then cover the basic facts about $L^p(\Omega, \mathcal{F}, P)$ and ℓ^p with an eye toward later applications. With this in place, we deal with $L^2(\Omega, \mathcal{F}, P)$, which is the most useful of the L^p spaces, in more detail. After developing the requisite geometric tools, we discuss the theory of linear, nonlinear, and nonparametric regression.

We then turn back to further developing the structure of $L^p(\Omega, \mathcal{F}, P)$, investigating signed measures, vector measures, and densities in more detail. Of particular note are the Jordan and the Hahn decompositions of finite signed measures, the Radon-Nikodym theorem, and the Lebesgue decomposition of probabilities on $\mathbb{R}$. These lead to Lyapunov's theorem, which delivers the basic tool that allows us to prove the existence of equilibrium for games and for models of exchange economies with nonatomic spaces of agents. Nonatomic probability spaces are also useful models of infinitely divisible sets of goods that must be shared among a finite number of people, and we briefly investigate the main results in what are called "games of fair division."

We devote the next two sections to convex subsets of the dual spaces of $L^p(\Omega, \mathcal{F}, P)$ and to proving the basic separation result for these spaces, the Hahn-Banach theorem, which allows us to define and investigate basic properties of the weak and the weak* topologies. We also cover a separable version of Alaoglu's theorem, which gives a useful compactness criterion. We end the chapter with a brief discussion about the major themes in classical parametric statistics.

8.1 ♦ Some Uses in Statistics and Econometrics

The most prevalent uses of the $L^p(\Omega, \mathcal{F}, P)$ spaces have already appeared: they are spaces of random variables satisfying moment conditions.

8.1.a The Classical Parametric Statistical Model

The classical statistical paradigm assumes the existence of a **parametric model**, which is, by assumption, a "true description" of the data. Elaborating the meaning of these terms is our first task.

Definition 8.1.1 *A **parametric statistical model** is a collection, $\{\mu_\theta : \theta \in \Theta\}$, of probability distributions on a metric space M, where for $\theta \neq \theta'$, $\mu_\theta \neq \mu_{\theta'}$. The nonempty set Θ is called the **parameter space** of the model.*

In applications, the metric space M is almost exclusively $(\mathbb{R}^k)^n$, where k is the number of random quantities that are observed and n is the number of instances in which they are observed. Usually, the parameter space, Θ, is a compact subset of $\mathbb{R}^m$ for some m, and the mapping $\theta \mapsto \mu_\theta(A)$ is continuous, either for all measurable A or for a broad class of measurable A. Crucial to our progress is the notion of an induced distribution or an image law (compare with Theorem 7.6.5 (p. 412)). The following recalls and relabels the definition of image laws.

Definition 8.1.2 *For an M-valued random variable, X, the **induced distribution** or **image law** of X is given by $\mu_X(A) - P(X^{-1}(A))$ for $A \in \mathcal{B}_M$.*

The assumption that the model is a "true description" or that the model is "correctly specified" is that we observe a random variable X taking values in M and

> Correct specification: $(\exists \theta^\circ \in \Theta)[\mu_X = \mu_{\theta^\circ}]$, that is, $\mu_X \in \{\mu_\theta : \theta \in \Theta\}$.

After observing X, we try to pick $\theta \in \Theta$ on the basis of which μ_θ best fits the observation X. This is most familiar in the iid variant of the classical statistical model, in which the "correct specification" assumption is that the data, $X_1, \ldots, X_n$, are iid μ_{θ° for some $\theta^\circ \in \Theta$. That is,

> Correct specification (iid): Each μ_θ is a distribution on $\mathbb{R}^k$, $X = (X_1, \ldots, X_n)$ is an iid collection of random variables in $\mathbb{R}^k$, and there exists a $\theta^\circ \in \Theta$ such that for all $i \leq n$, $\mu_{X_i} = \mu_{\theta^\circ}$.

To pick $\widehat{\theta}$, our best guess for the unknown value θ°, we use the **empirical distribution** of the data, that is, the probability distribution $\mu_{X,n}(A) := \frac{1}{n}\#\{i : X_i \in A\}$. We then solve some variant of the problem

$$\min_\theta g(\mu_{X,n}, \theta) \tag{8.1}$$

for $\widehat{\theta}_n = \widehat{\theta}(\mu_{X,n})$. If $g(\cdot, \cdot)$ is sensibly chosen, then as $\mu_{X,n} \to \mu_X$; if the model is correctly specified, some variant of the theorem of the maximum will guarantee that the random variable $\widehat{\theta}(\mu_{X,n})$ converges to θ° in probability or almost everywhere.

Definition 8.1.3 *If a sequence of estimators converges to $\theta°$ in probability, then it is a **consistent sequence of estimators**.*

Consistency is a minimal standard that estimators are held to because, if it fails, no number of data will reliably reveal the true value of θ even if the model is correctly specified.

If $\widehat{\theta}$ is a good estimator of θ, then we know that the data are, to a good degree of approximation, distributed $\mu_{\widehat{\theta}}$. If future observations have the same distribution as $X_1, \ldots, X_n$, we have learned something about the world. Sometimes, this extra understanding is useful for determining how to improve economic policies.

8.1.b Parametric Regression Analysis

In regression analysis, one typically starts with data, that is, with a set of observations, $(X_i, Y_i)_{i=1}^n$, $X_i \in \mathbb{R}^\ell$, $Y_i \in \mathbb{R}$. These are $n \cdot (\ell + 1)$ random variables. For example, each $X_i(\omega)$ might be an ℓ-vector of observable characteristics of a randomly chosen person, i. The ℓ-vector might consist, for example, of a list that includes years of schooling, GPA, IQ test scores, parental income, parental profession, number of siblings, gender, ethnicity, marital status, whether or not the spouse has a paying job, the number of magazine subscriptions, age, and the number of children. In this case, $Y_i(\omega)$ might be the number of months out of the last twelve that the randomly chosen person was fully employed, or his take-home pay for the last year, or some other variable we are interested in.

Since the list of the person's characteristics is so drastically incomplete a description of the randomly chosen person's life and history, for any function, $f : \mathbb{R}^\ell \to \mathbb{R}^1, \epsilon_i(\omega) := Y_i(\omega) - f(X_i(\omega))$, has a strictly positive variance. We often wish to find the function f that minimizes this variance. If we find this function f, we learn a great deal about the statistical determinants of Y_i. Added to this, we would like to know about the variance of ϵ_i conditional on $X_i = x$, because this tells how good a job f is doing for different values of the X_i.

In these contexts, we are typically interested in finding the **theoretical regression function**, $f(x) = E(Y \mid X = x)$, from the data. Later we carefully develop the properties of the function $E(Y \mid X)$, called the **conditional expectation of Y given X**. In particular, we show that the conditional expectation of Y given X is the f that solves the problem

$$\min_{\{f : f(X) \in L^2(\Omega, \mathcal{F}, P)\}} \int (Y(\omega) - f(X(\omega)))^2 \, dP(\omega). \tag{8.2}$$

Let μ_X be the distribution on $\mathbb{R}^\ell$ induced by the random vector X and $\mu_{(X,Y)}$ be the distribution on $\mathbb{R}^\ell \times \mathbb{R}^1$ induced by the random vector (X, Y). By the change of variable theorem 7.3.13 (p. 391), (8.2) is the same as

$$\min_{\{f : f(X) \in L^2(\mathbb{R}^\ell, \mathcal{B}^\ell, \mu_X)\}} \int (y - f(x))^2 \, d\mu_{X,Y}(x, y). \tag{8.3}$$

The data are $(X_i, Y_i)_{i=1}^n$, $X_i \in \mathbb{R}^\ell$, $Y_i \in \mathbb{R}$. The hope/assumption is that the **empirical distributions**, $\mu_{X,n}(A) := \frac{1}{n}\#\{i : X_i \in A\}$, $A \in \mathcal{B}^\ell$, and $\mu_{(X,Y),n}(B) := \frac{1}{n}\#\{i : (X_i, Y_i) \in B\}$, $B \in \mathcal{B}^{\ell+1}$, are close to the true distributions, μ_X and $\mu_{X,Y}$.

In parametric regression analysis, the set Θ parameterizes a set of functions; for example, $\theta = (\theta_0, \theta_1)$ and $f(x : \theta_0, \theta_1) = \theta_0 + \theta_1 x$ parameterizes the set of affine functions from $\mathbb{R}$ to $\mathbb{R}$. For parametric regressions, we have the following:

Correct specification (for conditional means): $\exists \theta^\circ \in \Theta$, $E(Y \mid X = x) = f(x : \theta^\circ)$.

Assuming a parameterized class of functions, $x \mapsto f(x : \theta)$, $\theta \in \Theta$, we solve the problem

$$\min_{\theta \in \Theta} \int (y - f(x : \theta))^2 \, d\mu_{(X,Y),n}(x, y) = \min_{\theta \in \Theta} \frac{1}{n} \sum_i (Y_i - f(X_i : \theta))^2,$$

$$(8.4)$$

which yields a function, $f(\cdot : \widehat{\theta})$, based on the chosen parametric model and the observable data. If the model is correctly specified, then the theorem of the maximum and $\mu_{(X,Y),n}$ being close to $\mu_{(X,Y)}$ should tell us that $\widehat{\theta}$ is close to the solution of

$$\min_{\theta \in \Theta} \int (y - f(x : \theta))^2 \, d\mu_{(X,Y)}(x, y), \qquad (8.5)$$

which is a function based on the unknown $\mu_{(X,Y)}$.

8.1.c Nonparametric Regression Theory

The correct specification assumption for parametric regression is essentially always violated in the contexts that economists face. It is a matter of context-specific judgment, about which reasonable people may sometimes disagree, whether or not the violation is sufficiently important to significantly weaken, or even vitiate, the analysis.[1] An alternative to making the parametric regression assumption is to allow the set of functions over which we are minimizing in (8.4) to become richer and richer as we gather more data.

One version of becoming richer and richer as we gather more data is **local averaging**, which has many variants. A simple variant picks, for every x, a small subset, $E(x)$, containing x, for example, $E(x) = B_\epsilon(x)$. The main constraint on $E(x)$ is that $\mu_{X,n}(E(x)) > 0$, in which case we locally approximate the true regression function, $f(x) = E(Y \mid X = x)$ by

$$\widehat{f}_n(x) := \frac{1}{\#\{i : X_i \in E(x)\}} \sum_{i : X_i \in E(x)} Y_i. \qquad (8.6)$$

The set $E(x)$ "slides around" as we move x, and the smaller the set $E(x)$, the more detail the function $\widehat{f}_n$ can have.

As n grows, the size of the $E(x)$ necessary to guarantee that $\mu_{X,n}(E(x)) > 0$ shrinks. One tries to arrange that the diameter of $E(x)$ goes to 0 as $n \to \infty$ in such a fashion that $\#\{i : X_i \in E(x)\} \to \infty$. In this case, one expects that $\widehat{f}_n(x) \to f(x)$ for a set of x having μ_X-probability 1.

---

1. To vitiate in the sense of corrupting by "carelessness, arbitrary changes, or the introduction of foreign elements" (*Oxford English Dictionary*).

Another version of becoming richer and richer as we gather more data is **series expansions**, and it, too, has many variants. In a simple variant, one picks a set $\{e_k : k \in \mathbb{N}\}$ with a dense span and approximates the true regression function by

$$\widehat{f}_n(x) := \sum_{k \leq K(n)} \beta_k e_k(x). \tag{8.7}$$

Here, $K(n) \uparrow \infty$ as $n \uparrow$, but in such a fashion that $K(n)/n \to 0$. The idea is that when there are a great many data so that $K(n)$ is large, the function $\widehat{f}_n$ can have more and more detail.

The main trade-off between parametric and nonparametric regression is between interpretability and accuracy.

1. If one uses the parametric class of functions $f(x : \theta_0, \theta_1) = \theta_0 + \theta_1 x$ and arrives at an estimate, $\widehat{\theta}_1$ of θ_1, the interpretation that forcefully suggests itself is that $\widehat{\theta}_1$ is the marginal effect of X on Y. Since arguments using marginal effects abound in economics, this has the advantage of being an object about which economists feel they know a great deal.

2. If one uses a nonparametric class of functions to get at the marginal effect of X on Y, one might evaluate $\partial \widehat{f}_n(x)/\partial x$ at the data points $X_1, \ldots, X_n$, and then present the average, $\frac{1}{n} \sum_{i=1}^{n} \frac{\partial \widehat{f}_n(X_i)}{\partial x}$, as an estimate of the marginal effect. However, if the partial derivative of the estimated function $\widehat{f}_n$ is variable, this interpretation feels strained, partly because the average loses any information to be found in the nonconstancy of the marginal effect.

8.2 ◆ Some Uses in Economic Theory

Infinite-dimensional spaces are often used as models of both commodities and the different actions taken by large numbers of people.

8.2.a Models of Commodities

Commodities can be consumed at sequences of times, they can be random, and they can be both random and consumed at different points in time.

8.2.a.1 Commodities over Time
There are situations in which the best models of commodities are infinite dimensional. For example, we often use a model of an economy that operates at an infinite sequence of points in time as a tool to help us think about economies operating into the indefinite future. In such models, a consumption bundle is given by a vector $x = (x_1, x_2, \ldots)$, where each $x_t \in \mathbb{R}^k_+$. Typically, one assumes that x belongs to a space such as $\ell^{\infty,k} = \{x \in (\mathbb{R}^k)^{\mathbb{N}} : \sup_t \|x_t\| < \infty\}$ or $\ell^{p,k} = \{x \in (\mathbb{R}^k)^{\mathbb{N}} : \sum_t \|x_t\|^p < \infty\}$.

We think that the market value of this consumption bundle should be a continuous linear function of x. This means that we have to understand the structure of spaces of sequences and of continuous linear functions on these sequence spaces.

As usual, the space of continuous linear functions is called the dual space. The difference between how an element of the dual space treats x_s and x_t tells us about the marginal rate of substitution between consumption at times s and t.

8.2.a.2 Random Commodities

When consumption depends on the random state, ω, commodities are often modeled as points in $L^{2,k}(\Omega, \mathcal{F}, P)$, the set of random variables taking values in $\mathbb{R}^k$ and having a finite variance-covariance matrix. The market value of a consumption bundle, X, should be a continuous linear function of X, that is, an element of the dual space. The Hahn-Banach theorem yields separation results for convex sets using elements of the dual space. The difference between how an element of the dual space treats consumption at different values of ω tells us about the marginal rate of substitution between consumption in different states.

8.2.a.3 Random Commodities over Time

The previous two examples are often combined. Consumption at time t is a random variable, X_t, and the market value of a sequence of random variables $(X_1, X_2, \ldots)$ should be a continuous linear function of the sequence. When the time set is continuous, for example, $[0, T]$, we often use $X(t, \omega) \in L^2(\Omega \times [0, T], \mathcal{F} \otimes \mathcal{B}_T, P \times \lambda_T)$ to model random consumption streams, where $\mathcal{B}_T$ is the Borel σ-field of subsets of $[0, T]$, $\mathcal{F} \otimes \mathcal{B}_T$ is the product σ-field, λ_T is the uniform distribution on $[0, T]$, and $P \times \lambda_T(E \times B)$ is defined (using Fubini's theorem) by $P(E) \cdot \lambda_T(B)$. Again, the market value of X should be an element of the dual space. Elements of the dual space treat consumption at different points in time and different states differently, and the difference tells us about the marginal rate of substitution between consumption at different time-state pairs. Typically, the information available to someone in the economy about what has happened up until t is some sub-σ-field of $\mathcal{F}_t := \sigma(\{X(s, \cdot) : s \leq t\})$.

8.2.b Large Populations and Their Choices

There is another type of use for the probability spaces, $(\Omega, \mathcal{F}, P)$. They can be utilized to model large populations of people—populations sufficiently large that we think that we should ignore the effect of any single individual. This captures the notions of perfect competition and anonymity.

8.2.b.1 Perfect Competition

With measure space $(\Omega, \mathcal{F})$, we interpret the points $\omega \in \Omega$ as consumers and $\mathcal{F}$ as possible groups of people. The space $L^{1,k}(\Omega, \mathcal{F}, P)$ of integrable $\mathbb{R}^k$-valued random vectors can represent the set of endowments and consumptions. In the case of a finite Ω having N elements and the uniform distribution $U(\omega) = 1/N$, if person ω consumes the vector $X(\omega)$, then the total consumption of a group A is $\sum_{\omega \in A} X(\omega)$ and the average consumption is $\frac{\#A}{N} \sum_{\omega \in A} X(\omega)$, which is exactly equal to $\int_A X \, dU$.

When N is huge, we can think of replacing our model of the population, $(\{1, \ldots, N\}, \mathcal{P}(\Omega), U)$, with $(\Omega, \mathcal{F}, P)$, where P is nonatomic. This captures the sense that any one individual ω is a negligible part of the economy. Further, groups of people, $A \in \mathcal{F}$, can always be subdivided into two or more parts of any proportional size.

Individual endowments are modeled as an integrable $\omega \mapsto Y(\omega) \in \mathbb{R}_+^k$. The feasible allocations become the set of $\omega \mapsto X(\omega) \in \mathbb{R}^k$ such that $E\ X \le E\ Y$ in $\mathbb{R}^k$. Since any integrable Y is nearly simple, there are many ways to subdivide people with any given endowment into subgroups. We will see that this ability to subdivide groups into arbitrarily small parts, all with essentially the same endowments, means that competition is perfect.

8.2.b.2 Anonymity

We interpret the points $\omega \in \Omega$ as players in a game, each choosing among different actions, $a \in A$, A a finite set. When person ω picks $s_\omega \in A$ with $s : \Omega \to A$ measurable, the population average distribution of actions is given by $\mu_s(a) = P(\{\omega : s_\omega = a\})$. If person ω picks at random, $s_\omega \in \Delta(A)$, $\Delta(A) := \{p \in \mathbb{R}_+^A : \sum_{a \in A} p(a) = 1\}$ being the set of probability distributions on A, then the population average choice is $\mu_s(a) = \int_\Omega s_\omega(a)\,dP(\omega)$. Here, strategies are a subset of the measurable mappings from Ω to $\mathbb{R}^A$.

Each person ω has the expected utility function $(\omega, b, \mu_s) \mapsto u(\omega, b, \mu_s)$, where b is his own choice of action and μ_s is the population average choice. Explicitly, if $\omega \mapsto s_\omega \in \Delta(A)$ gives the choices of each $\omega \in \Omega$ and person ω picks b, then person ω's expected utility is $\int_A u(\omega, b, a)\,d\mu_s(a)$. In the case of a finite A, this reduces to the sum $\sum_{a \in A} u(\omega, b, a)\mu_s(a)$.

When the space of players is a nonatomic probability space, we have a sense of anonymity, each person's utility depending on her own preferences, her own actions, and the average of what other people are doing. Each ω is anonymous in that his choice of action does not affect the average. For example, driving on a crowded freeway, what matters is the pair of exits/entrances one drives between and the proportion of the population that has chosen to drive between the other exit/entrance pairs. The anonymity captures the notion that who exactly is doing the driving does not affect utility, just the volume of traffic—that any individual is lost in the crowd.

8.3 ◆ The Basics of $L^p(\Omega, \mathcal{F}, P)$ and ℓ^p

The L^p spaces are spaces of random variables and the ℓ^p spaces are spaces of sequences, $p \in [1, \infty]$. The cases $p \in [1, \infty)$ and $p = \infty$ are quite different. A **Banach space** is a complete normed vector space. We soon show that L^p and ℓ^p, $p \in [1, \infty]$ are Banach spaces.

Definition 8.3.1 (L^p Spaces) *For $p \in [1, \infty)$, $L^p = L^p(\Omega, \mathcal{F}, P)$ is the set of $\mathbb{R}$-valued random variables with norm*

$$\|X\|_p := \left(\int |X|^p \, dP \right)^{1/p} < \infty, \tag{8.8}$$

*and $L^\infty = L^\infty(\Omega, \mathcal{F}, P)$ is the set of random variables with **essential supremum** norm, defined by*

$$\|X\|_\infty := \operatorname{ess\,sup} |X| := \sup\{B : P(|X| < B) < 1\} < \infty. \tag{8.9}$$

For ℓ^p spaces, read as "little ell pee spaces," summation replaces integration in the definitions. Given the close connection between integration and summation, it is, perhaps, not surprising that the L^p and ℓ^p spaces share many properties.

Definition 8.3.2 (ℓ^p Spaces) *For $p \in [1, \infty)$, ℓ^p is the set of $\mathbb{R}$-valued sequences with norm*

$$\|x\|_p := \left(\sum_t |x_t|^p\right)^{1/p} < \infty, \tag{8.10}$$

and ℓ^∞ is the set of sequences with norm

$$\|x\|_\infty := \sup_t |x_t| < \infty. \tag{8.11}$$

8.3.a Summary of Basic Results

The first result is that the L^p spaces are Banach spaces. There are two parts to this result, the norm (or triangle) inequality and completeness. The norm inequality for the $\|\cdot\|_p$, $p \in [1, \infty)$, comes from the Minkowski-Riesz inequality, Theorem 8.3.4. This is a direct consequence of Jensen's inequality, Theorem 8.3.3, which is more broadly useful. The completeness is Theorem 8.3.8, and the proof relies on the Borel-Cantelli lemma and Fatou's lemma.

An immediate implication of the norm inequality is that Banach spaces are locally convex. This is the property that if U_x is a neighborhood of x, then there is an open convex G_x with $x \in G_x \subset U_x$. Many of the properties of Banach spaces flow from this simple geometric fact.

Our next result is Theorem 8.3.13, which shows that for $p \in [1, \infty)$, ℓ^p is a special case of an $L^p(\Omega, \mathcal{F}, P)$. By "is a special case" we mean that there is an order-preserving linear isometry between ℓ^p and $L^p(\Omega, \mathcal{F}, P)$ for a wide range of probability spaces. Therefore, results about L^p spaces give us results about ℓ^p spaces, at least for $p \in [1, \infty)$.

A good way to understand the structure of a space is to examine simpler subsets that are dense. In the $L^p(\Omega, \mathcal{F}, P)$ spaces, $p \in [1, \infty)$, the simple functions are always dense. If $(\Omega, \mathcal{F}) = (M, \mathcal{B}_M)$ is a metric space with its Borel σ-field, then Lusin's theorem, 8.3.25, shows that the continuous functions are dense in a rather striking fashion.

The best kind of complete metric space is a separable one: ℓ^p is separable iff $p \in [1, \infty)$; for $p \in [1, \infty)$, L^p is separable iff there is a countable field $\mathcal{F}^\circ$ generating $\mathcal{F}$; and $L^\infty(\Omega, \mathcal{F}, P)$ is separable iff P is purely atomic on $\mathcal{F}$ and has only finitely many atoms, that is, iff L^∞ is finite dimensional.

The partial order $\leq$ for $\mathbb{R}^k$ gives us a very useful lattice structure. The partial orders $X \leq Y$ iff $P(X \leq Y) = 1$ and $x \leq y$ iff $(\forall t)[x_t \leq y_t]$ are similarly useful in L^p and ℓ^p. A difference between the finite-dimensional case and the infinite-dimensional case is that $\{y : x \leq y\}$ has empty interior ℓ^p and typically has empty interior in L^p, $p \in [1, \infty)$.

The next result is the boundedness principle, which says that a linear function from one normed space to another is continuous iff it is bounded on norm-bounded sets. We use this extensively in the study of the duals of normed spaces in §8.8. Here we utilize the boundedness principle to show that any finite-dimensional subspace of a normed vector space is linearly homeomorphic to $\mathbb{R}^k$. In other words, our infinite-dimensional normed spaces "look just like" $\mathbb{R}^k$ when we focus on finite-dimensional linear (or affine) subspaces.

In the penultimate part of this section we characterize compact subsets of normed spaces—they are closed and bounded and satisfy approximate flatness. In L^p, the approximate flatness has a useful reformulation that uses conditional expectations.

In the last part of this section we investigate the different notions of a basis, the Hamel basis that allows only finite combinations of basis elements, and the Schauder basis, that allows countable sums of basis elements. Hamel bases are useful for theoretical investigations and Schauder bases for applied work. For infinite-dimensional normed spaces, the Hamel bases lead directly to the existence of discontinuous linear functions, which are weird. They also allow us to show that in any infinite-dimensional normed space, there are two disjoint dense convex sets whose union is all of the space, which is also weird.

8.3.b $L^p(\Omega, \mathcal{F}, P)$ and ℓ^p Are Banach Spaces

A complete normed vector space is called a **Banach** space. We now show that $\| \cdot \|_p$ is a norm, and that the associated metric, $d_p(X_1, X_2) = \|X_1 - X_2\|_p$, makes L^p a complete metric space. The norm inequality that we must show is $\|X + Y\|_p \leq \|X\|_p + \|Y\|_p$, because with normed spaces, we use the metric $d(X, Y) = \|X - Y\|_p$. Setting $X = R - S$ and $Y = S - T$ yields $\|X + Y\|_p \leq \|X\|_p + \|Y\|_p$ iff $\|S - T\|_p \leq \|R - S\|_p + \|S - T\|_p$, which is the triangle inequality.

The proof of the requisite norm inequality is a fairly easy implication of **Jensen's inequality**, as are many other useful and interesting facts in probability and expected utility theory,

Theorem 8.3.3 (Jensen) *If C is a convex subset of $\mathbb{R}^k$, $f : C \to \mathbb{R}$ is concave, and $X : \Omega \to C$ is an integrable random vector, then $E\, f(X) \leq f(E\, X)$. If f is convex, then $E\, f(X) \geq f(E\, X)$.*

If C has empty interior, we can work with its relative interior as in Definition 5.5.15 (p. 197).

Proof. Suppose first that $E\, X \in \text{int}(C)$. Since f is supdifferentiable at $E\, X$, there exists a $\mathbf{v} \in \mathbb{R}^k$ satisfying, for all $\mathbf{x} \in C$, $f(\mathbf{x}) \leq f(E\, X) + \mathbf{v} \cdot (\mathbf{x} - E\, X)$. Therefore, for all ω,

$$f(X(\omega)) \leq f(E\, X) + \mathbf{v} \cdot (X(\omega) - E\, X). \tag{8.12}$$

By monotonicity of the integral,

$$E\, f(X) \leq f(E\, X) + E\, \mathbf{v} \cdot (X - E\, X). \tag{8.13}$$

Since the integral is linear, $E\, \mathbf{v} \cdot (X - E\, X) = \mathbf{v} \cdot (E\, X - E\, X) = 0$. For convexity, reverse the inequalities.

If $E\, X \in \text{extr}(C)$, then $P(X = x) = 1$ for some $x \in \text{extr}(C)$ and Jensen's inequality is satisfied as an equality. If $E\, X \in \partial(C) \setminus \text{extr}(C)$, then there is an affine subset, A, with $P(X \in (A \cap C)) = 1$. In this case, $E\, X \in \text{rel int}(A \cap C)$, so that $f_{|(A \cap C)}$ is supdifferentiable at $E\, X$, and the previous argument works. ∎

Here is the promised norm inequality.

Theorem 8.3.4 (Minkowski–Riesz) *For $X, Y \in L^p(\Omega, \mathcal{F}, P)$, $p \in [1, \infty]$, $\|X + Y\|_p \leq \|X\|_p + \|Y\|_p$.*

Proof. For $p = 1, \infty$, this is easy. For $1 < p < \infty$, we start with the observation that for all ω, $|X + Y| \leq |X| + |Y|$, so that $\|X + Y\|_p \leq \|\, |X| + |Y| \,\|_p$. Now consider the concave function $f_p(u, v) = (u^{1/p} + v^{1/p})^p$ defined on $\mathbb{R}^2_+$. Using Jensen's inequality and applying it to the random variables $U = |X|^p$ and $V = |Y|^p$, we have $E\, f_p(U, V) \leq f_p(E\, U, E\, V)$; that is,

$$E\, \left((|X|^p)^{1/p} + (|Y|^p)^{1/p} \right)^p \leq \left((E\, |X|^p)^{1/p} + (E\, |Y|^p)^{1/p} \right)^p, \quad \text{or}$$

$$E\, (|X| + |Y|)^p \leq \left(\|X\|_p + \|Y\|_p \right)^p.$$

Taking the pth root on both sides gives us $\|X + Y\|_p \leq \|\, |X| + |Y| \,\|_p \leq \|X\|_p + \|Y\|_p$. ∎

Exercise 8.3.5 If $\Omega = \{1, \ldots, k\}$, $\mathcal{F} = \mathcal{P}(\Omega)$, and $P(\{\omega\}) = 1/k$, derive the Minkowski inequality for $\mathbb{R}^k$; that is, show that for all $p \in [1, \infty]$, $\left| \sum_i |x_i - y_i|^p \right|^{1/p} \leq \left| \sum_i |x_i|^p \right|^{1/p} + \left| \sum_i |y_i|^p \right|^{1/p}$.

We have already seen that if $X \in L^p$, then $X \in L^{p'}$ for all $1 \leq p < p'$. That is, if $E\, |X|^p$ is finite, then $E\, |X|^{p'}$ is also finite. Jensen's inequality yields the following.

Lemma 8.3.6 *For $1 \leq p < p'$ and $X \in L^p(\Omega, \mathcal{F}, P)$, $\|X\|_p \leq \|X\|_{p'}$.*

If we allow ∞ as an integral, this result would hold without the need to assume that $E\, |X|^p < \infty$, that is, without assuming that $X \in L^p$. To see how this works, you have to go through Jensen's inequality allowing infinite integrals.

Proof. Suppose first that $1 = p < p'$. The function $f_p(x) = x^{p'}$ is convex on $\mathbb{R}_+$. Applying Jensen's inequality to the random variable $|X|$, we have $E\, |X|^{p'} \geq (E\, |X|)^{p'}$. Taking roots on both sides, we have $(E\, |X|^{p'})^{1/p'} \geq E\, |X|$, that is, $\|X\|_{p'} \geq \|X\|_1 = \|X\|_p$. For general p, we use the function $f_r(x) = x^r, r = p/p'$. ∎

Here is a basic relation between convergence in the L^p spaces and convergence in L^0.

Theorem 8.3.7 *If $1 \leq p < p' \leq \infty$, then $\left[\|X_n - X\|_{p'} \to 0 \right] \Rightarrow \left[\|X_n - X\|_p \to 0 \right] \Rightarrow \left[\alpha(X_n, X) \to 0 \right]$, and there is a subsequence X_{n_k} with $X_{n_k} \to X$ a.e.*

Proof. By the previous lemma, $\|X_n - X\|_p \leq \|X_n - X\|_{p'}$ and $\|X_n - X\|_{p'} \to 0$. For all $\epsilon > 0$, $P(\{|X_n - X| \geq \epsilon\}) \leq \frac{E\, |X_n - X|}{\epsilon} \to 0$, so $\alpha(X_n, X) \to 0$, and α-convergence implies the existence of the requisite subsequence. ∎

Knowing that $L^p(\Omega, \mathcal{F}, P)$ is a normed space, we now show that it is complete, which means that it is a **Banach space**, that is, a complete normed vector space.

Theorem 8.3.8 (Completeness of L^p)   *For $p \in [1, \infty]$, $L^p(\Omega, \mathcal{F}, P)$ is a complete metric space.*

Proof.   We consider $p \in [1, \infty)$, and the case $p = \infty$ is left as an exercise. Let X_n be a Cauchy sequence. Pick a strictly increasing sequence N_k such that for all $m, n \geq N_k$, $\|X_n - X_m\|_p < 1/4^k$ and consider the subsequence X_{N_k}, temporarily relabeled as X_k.

We first find a candidate limit $X(\omega)$. Note that $\|X_n - X_m\|_p < 1/4^k$ is the same as $E\,|X_n - X_m|^p < (1/4^k)^p$. Let $A_k = \{\omega : |X_k(\omega) - X_{k+1}(\omega)| \geq 1/2^k\}$. Combining with Markov's inequality gives

$$P(A_k) \leq \frac{E\,|X_k - X_{k+1}|^p}{(1/2^k)^p} \leq \frac{(1/4^k)^p}{(1/2^k)^p} = \frac{1}{2^{kp}}.$$

Since $\sum_k 1/(2^{kp}) < \infty$, $P([A_k \text{ i.o.}]) = 0$, which means that $P([A_k^c \text{ a.a.}]) = 1$, and for all $\omega \in B := [A_k^c \text{ a.a.}]$, $X_k(\omega)$ is Cauchy. Define $X(\omega) = 1_B(\omega) \lim_k X_k(\omega)$.

We now show that $\|X_n - X\|_p \to 0$. Pick arbitrary $\epsilon > 0$ (but < 1) and M such that for all $m, n \geq M$, $\left(\|X_m - X_n\|_p\right)^p < \epsilon$. For any $m \geq M$,

$$\epsilon \geq \liminf_k \|X_m - X_{N_k}\|_p^p$$

$$= \liminf_k E\,|X_m - X_{N_k}|^p \geq E\,\liminf_k |X_m - X_{N_k}|^p,$$

where the last inequality is from Fatou's lemma (p. 410). Since $X_{N_k}(\omega) \to X(\omega)$ for all $\omega \in B$, $\liminf_k |X_m - X_{N_k}|^p(\omega) = |X_m - X|^p(\omega)$. If we combine for all $m \geq M$, $\epsilon \geq E\,|X_m - X|^p$, so that $\|X_m - X\|_p \leq \epsilon$.   ■

Exercise 8.3.9   Show that $L^\infty(\Omega, \mathcal{F}, P)$ and ℓ^∞ are complete metric spaces.

8.3.c   Local Convexity

When we draw diagrams of open balls, we inevitably make them convex.

Lemma 8.3.10   *In any normed space, $B_\epsilon(x) = \{y : \|x - y\| < \epsilon\}$ is a convex set.*

Proof. $\|x - (\alpha y + (1 - \alpha)z)\| = \|\alpha(x - y) + (1 - \alpha)(x - z)\| \leq \|\alpha(x - y)\| + \|(1 - \alpha)(x - z)\| = \alpha\|x - y\| + (1 - \alpha)\|x - z\|$.   ■

This means that if U_x is an open subset of a normed space containing x, then there is an open convex G_x with $x \in G_x \subset U_x$; simply take G_x to be $B_\delta(x)$ for a sufficiently small δ. It is the failure of this property that makes us avoid the L^p spaces with $p \in (0, 1)$.

Example 8.3.11   *Let $(\Omega, \mathcal{F}, P) = ((0, 1], \mathcal{B}, \lambda)$, $X = 1_{(0, \frac{1}{2}]}$, $Y = 1_{(0, \frac{1}{2}]}$, so that $X + Y = 1_\Omega$. For $p \in (0, 1)$, $\left(\int |X + Y|^p \, dP\right)^{1/p} = 1$, $\left(\int |X|^p \, dP\right)^{1/p} = \left(\int |Y|^p \, dP\right)^{1/p} = \left(\frac{1}{2}\right)^{1/p}$, and because $1/p > 1$, $\left(\frac{1}{2}\right)^{1/p} < \frac{1}{2}$. Therefore, $\|X + Y\|_p = 1 > \|X\|_p + \|Y\|_p$, violating the essential triangle inequality for norm distances.*

8.3.d ℓ^p as a Special Case of $L^p(\Omega, \mathcal{F}, P)$ for $p \in [1, \infty)$

Here we focus on $p \in [1, \infty)$; the case $p = \infty$ takes one into very deep mathematical areas. We now show that any result that we prove for $L^p(\Omega, \mathcal{F}, P)$ gives us essentially the same result for ℓ^p. We define the order $\leq$ on $L^p(\Omega, \mathcal{F}, P)$ by $X \leq Y$ if $P(X \leq Y) = 1$ and the order $\leq$ or ℓ^p by $x \leq y$ if $x_t \leq y_t$ for all t.

Definition 8.3.12   *A bijection* $T : \ell^p \leftrightarrow L^p(\Omega, \mathcal{F}, P)$ *is **order preserving** if* $[x \leq y] \Leftrightarrow [T(x) \leq T(y)]$.

Order-preserving maps preserve lattice structures, that is, $T(x \vee y) = T(x) \vee T(y)$ and $T(x \wedge y) = T(x) \wedge T(y)$.

Theorem 8.3.13   *If* $(\Omega, \mathcal{F}, P) = (\mathbb{N}, \mathcal{P}(\mathbb{N}), \mu)$, *where* $\mu(\{n\}) > 0$ *for all* $n \in \mathbb{N}$, *then for all* $p \in [1, \infty)$, *there is an order-preserving linear isometric bijection* $T : \ell^p \leftrightarrow L^p(\Omega, \mathcal{F}, P)$.

Proof.   We take $\mu(\{t\}) = 1/2^t$, but for any other μ the proof is essentially the same. Fix arbitrary $p \in [1, \infty)$. For any $x = (x_1, x_2, x_3, \ldots) \in \ell^p$, define $T(x) = (2^{1/p}x_1, 2^{2/p}x_2, 2^{2/p}x_3, \ldots)$. $\|x\|_p = \left(\int |T(x)|^p \, d\mu \right)^{1/p}$ because

$$\sum_t |x_t|^p = \sum_t \frac{|2^{\frac{t}{p}}|^p |x_t|^p}{2^t} = \int |T(x)|^p \, d\mu.$$

This shows that T is an isometric bijection. T is clearly linear and order preserving. ∎

Corollary 8.3.14   *If* $(\Omega, \mathcal{F}, P)$ *is nonatomic, then for all* $p \in [1, \infty)$, *there is an order-preserving linear isometric bijection between* ℓ^p *and a closed linear subspace of* $L^p(\Omega, \mathcal{F}, P)$.

Proof.   Take a countable partition $\mathcal{E} = \{E_1, E_2, E_3, \ldots\}$ of Ω with $P(E_n) = \mu(\{n\})$. The linear subspace is $L^2(\Omega, \sigma(\mathcal{E}), P)$ and the order-preserving linear isometric bijection is $T(x) = \sum_t 2^{t/p} x_t 1_{E_t}$. ∎

Exercise 8.3.15   Theorem 8.3.13 shows that for any $p \in [1, \infty)$, $L^p(\Omega, \mathcal{F}, P)$ being a Banach space implies that ℓ^p is a Banach space. Using T^{-1} from the proof, give a direct proof that ℓ^p is a Banach space.

8.3.e   Several Common Dense Subsets

We prove three classes of results: every measurable function in $L^p(\Omega, \mathcal{F}, P)$ is nearly simple; the approximations are almost uniform; and every measurable function on a metric space is nearly continuous. To avoid trivializing this part of the analysis, we suppose, in our study of dense subsets of $L^p(\Omega, \mathcal{F}, P)$, that P is not purely atomic with a finite number of atoms in $(\Omega, \mathcal{F}, P)$.

8.3.e.1   The Denseness of Nearly Simple Functions

For any $\infty \geq p > p' \geq 1$, Lemma 8.3.6 implies that $L^p \subset L^{p'}$. We now show that L^p is a dense infinite-dimensional vector subspace of $L^{p'}$.

Lemma 8.3.16   *For any $\infty \geq p > p' \geq 1$, L^p is a $\|\cdot\|_{p'}$-dense vector subspace of $L^{p'}$.*

Proof.   Since a dense subset of a dense set is dense, it is sufficient to show that L^∞ is dense in $L^{p'}$. Define $X_m = X \cdot 1_{|X| \leq m} \in L^\infty$. If $X \in L^{p'}$, then $\|X_m - X\|_{p'} \to 0$ by dominated convergence.   ∎

For $\mathcal{F}^\circ$ a field of sets, $M_s(\mathcal{F}^\circ)$ denotes the set of simple $\mathcal{F}^\circ$-measurable functions.

Lemma 8.3.17   *If $\mathcal{F}^\circ$ is a field of sets generating $\mathcal{F}$, then $M_s(\mathcal{F}^\circ)$ is dense in L^p, $p \in [1, \infty)$, but not necessarily in L^∞.*

Proof.   The simple $\mathcal{F}$-measurable functions are dense, so it is sufficient to show that $M_s(\mathcal{F}^\circ)$ is dense in the simple functions. Pick arbitrary simple $\mathcal{F}$-measurable $X = \sum_{k \leq K} \beta_k 1_{E_k}$ and $\epsilon > 0$. Choose $\{E'_k : k \leq K\} \subset \mathcal{F}^\circ$ to partition Ω and satisfy $\sum_{k \leq K} P(E'_k \Delta E_k) < \epsilon^p / 2B$, where $B := \max_{k \leq K} |\beta_k|^p$. Set $Y = \sum_{k \leq K} \beta_k 1_{E'_k}$. $\int |X - Y|^p \, dP \leq \sum_{k \leq K} 2B 1_{E'_k \Delta E_k} < \epsilon^p$, so that $\|X - Y\|_p < \epsilon$.

To see that $M_s(\mathcal{F}^\circ)$ need not be dense in L^∞, take $\mathcal{F}^\circ$ to be the class of finite disjoint unions of intervals and $X \in L^\infty((0, 1], \mathcal{B}, \lambda)$ to be the indicator of a countable number of intervals at a positive distance from each other with nonempty interior.   ∎

8.3.e.2 Nearly Uniform Convergence

We now prove that almost everywhere convergence is almost uniform, that is, uniform except on arbitrarily small sets. The same is not true for convergence in probability.

Definition 8.3.18   *A sequence X_n in $L^0(\Omega, \mathcal{F}, P)$ converges **almost uniformly** to X if for every $\epsilon > 0$, there exists $A \in \mathcal{F}$ with $P(A) > 1 - \epsilon$ and $\sup_{\omega \in A} |X_n(\omega) - X(\omega)| \to 0$.*

Theorem 8.3.19 (Egoroff)   *For X_n, $X \in L^0$, if $X_n \to X$ a.e., then X_n converges almost uniformly to X.*

Proof.   For $m, n \in \mathbb{N}$, define $A_{m,n} = \cap_{k \geq n} \{\omega : |X_k(\omega) - X(\omega)| \leq 1/m\}$. For each fixed m, $P(A_{m,n}) \uparrow 1$ as $n \uparrow$. Therefore, we can pick $n(m)$ such that $P(A_{m,n(m)}) > 1 - \epsilon/2^m$. Define $A = \cap_m A_{m,n(m)}$ so that $P(A) > 1 - \epsilon$ and $\sup_{\omega \in A} |X_n(\omega) - X(\omega)| \to 0$.   ∎

Exercise 8.3.20   We have seen that $\alpha(X_n, X) \to 0$ need not imply that $X_n \to X$ a.e. Show that Egoroff's theorem is false if we substitute "$\alpha(X_n, X) \to 0$" for "$X_n \to X$ a.e."

Exercise 8.3.21   Show that Egoroff's theorem is true if X_n, X take values in an arbitrary metric space.

8.3.e.3 The Denseness of $C_b(M)$

We first give an easy version of Lusin's theorem, followed by the real Lusin's theorem. We use the idea of the essential infimum and essential supremum, which is tightly related to the definition of the L^∞-norm. The difference between the essential infimum/supremum of a measurable function and the infimum/supremum of the function is that the former ignores what happens on sets of measure 0, hence is applicable to a.e. equivalence classes of random variables.

Definition 8.3.22   *For $X \in L^\infty(\Omega, \mathcal{F}, P)$, the **essential infimum of** X is defined by* ess inf $X = \inf\{r \in \mathbb{R} : P(\{X \geq r\}) > 0\}$ *and the **essential supremum of** X is defined by* ess sup $X = \sup\{r \in \mathbb{R} : P(\{X \leq r\}) < 1\}$.

   $P(X \in [\text{ess inf } X, \text{ess sup } X]) = 1$, no smaller interval has this property, and $\|X\|_\infty = \text{ess sup } |X|$.

Theorem 8.3.23 (Easy Lusin)   *If $(M, \mathcal{B}_M, \mu)$ is a metric space, its Borel σ-field, and a countably additive probability and $g \in L^\infty(M, \mathcal{M}_B, \mu)$, then for all $\epsilon > 0$, there exists a continuous f such that $\mu(|f - g| < \epsilon) > 1 - \epsilon$ and $\forall x \in M$, ess inf $g \leq f(x) \leq$ ess sup g.*

Proof.   Since g is bounded, by scaling and shifting, there is no loss in assuming that ess inf $g = 0$ and ess sup $g = 1$. Pick n large enough that $g_n := \sum_{j=0}^{2^n} \frac{j}{2^n} 1_{E_j}$ is everywhere within ϵ of g, where $E_j = \{g \in [\frac{j}{2^n}, \frac{(j+1)}{2^n})\}$. Choose closed $F_j \subset E_j$ such that $\sum_j P(E_j \setminus F_j) < \epsilon$, $j \in \{0, \ldots, 2^n\}$. The F_j are closed and, being subsets of disjoint sets, are disjoint. By the Tietze-Urysohn extension theorem (p. 330), for each j there is a continuous function f_j taking values in $[0, j/2^n]$, taking the value $j/2^n$ on F_j, and taking the value 0 on F_k, $k \neq j$. Set $f = \max_{j=0,\ldots,2^n} f_j$.   ∎

Corollary 8.3.24   *$C_b(M)$ is dense in $L^p(M, \mathcal{B}_M, \mu)$, $p \in [1, \infty)$.*

Proof.   By easy Lusin, $C_b(M)$ is $\|\cdot\|_p$-dense in $L^\infty(M, \mathcal{B}_M, \mu)$ and L^∞ is $\|\cdot\|_p$-dense in L^p.   ∎

   In the proof of the easy version of Lusin's theorem, we approximated g on a closed set by functions that were constant on the sets F_j. Using the completeness of $C_b(F)$ for any $F \subset M$, we arrive at the following, which gives a striking property of measurable functions on metric spaces.

Theorem 8.3.25 (Lusin)   *If $(M, \mathcal{B}_M, \mu)$ is a metric space, its Borel σ-field, and a countably additive probability and $g : M \to \mathbb{R}$ measurable, then for all $\epsilon > 0$, there exists a closed set F such that $\mu(F) > 1 - \epsilon$ and $g_{|F}$ is continuous and bounded.*

   This strengthens easy Lusin because $g_{|F} \in C_b(F)$ can be continuously extended to all of M.

Proof.   It is sufficient to prove the result for bounded g since $g_m := g \cdot 1_{\{|g| \leq m\}}$ converges to g a.e. By the easy version of Lusin's theorem, for every $n \in \mathbb{N}$ there is a closed set F_n with $\mu(F_n) > 1 - \epsilon/2^n$ and a continuous f_n such that $\sup_{x \in F_n} |f_n(x) - g(x)| < 1/2^n$. Let F be the closed set $\cap_n F_n$ so that $\mu(F) > 1 - \epsilon$. Restricted to F, the sequence of functions f_n is Cauchy in $C_b(F)$; hence, it has a continuous limit we call f. For all $x \in F$ and all $n \in \mathbb{N}$, $|f(x) - g(x)| < 1/2^n$, so $g_{|F}$ is equal to f.   ∎

8.3.f Separability and Nonseparability in $L^p(\Omega, \mathcal{F}, P)$

The cases $p = \infty$ and $p \in [1, \infty)$ are very different. L^∞ is separable iff P is purely atomic and has only finitely many atoms, which is in turn true iff L^∞ is finite dimensional. For $p \in [1, \infty)$, L^p is separable iff the metric space $(\mathcal{F}, \rho_P)$ is separable, which in turn depends on satisfying the following.

Definition 8.3.26 *A σ-field $\mathcal{F}$ is* **countably generated** *if there is a countable collection of events $\mathcal{E} = \{E_n : n \in \mathbb{N}\}$ such that $\rho_P(\mathcal{F}, \sigma(\mathcal{E})) = 0$.*

The Borel σ-field of any separable metric space is countably generated. Simply take $\mathcal{E}$ to the set of open balls with rational radius around each point in a countable dense set.

8.3.f.1 Separability of $L^p(\Omega, \mathcal{F}, P)$, $p \in [1, \infty)$

Theorem 8.3.27 *$\mathcal{F}$ is countably generated iff $(\mathcal{F}, \rho_P)$ is separable iff for all $p \in [1, \infty)$, L^p is separable.*

Proof. Since the smallest field containing a countable collection of sets is countable, the first two are equivalent. If $\mathcal{F}^\circ$ is a countable field generating $\mathcal{F}$, then for all $p \in [1, \infty)$, a countable dense class is the set of simple $\mathcal{F}^\circ$-measurable functions taking on only rational values. If L^p is separable for all $p \in [1, \infty)$, pick a countable dense subset of the closed set $\mathcal{I} = \{1_E : E \in \mathcal{F}\}$, and let $\mathcal{F}^\circ$ denote the smallest field containing it. Since $\|1_E - 1_F\|_p = \rho_P(E, F)$ for all $p \in [1, \infty)$, $\mathcal{F}^\circ$ is ρ_P-dense in $\mathcal{F}$. ∎

8.3.f.2 L^∞ and Atoms

It is impossible to arrange for L^∞ to be separable when $(\Omega, \mathcal{F}, P)$ is infinite.

Example 8.3.28 *Let $(\Omega, \mathcal{F}, P) = ((0, 1], \mathcal{B}, \lambda)$. For $0 < a \le 1$, define $X_a(\omega) = 1_{(0,a]}(\omega)$. The collection $\{X_a : 0 < a \le 1\}$ is an uncountable subset of L^∞ at norm distance 1 from each other.*

Exercise 8.3.29 Suppose that $\Omega = \{\omega_n : n \in \mathbb{N}\}$ is countable, $\mathcal{F}$ is the set of all subsets of Ω, and for each ω_n, $P(\omega_n) > 0$. Show that $L^\infty(\Omega, \mathcal{F}, P)$ is not separable. [Think about indicators of subsets of Ω.]

A partition $\mathcal{E} \subset \mathcal{F}$ of Ω is **nonnull** if for $E \in \mathcal{E}$, $P(E) > 0$. Nonnull partitions are either finite or countably infinite. To see why, let $\mathcal{E}$ be a nonnull partition; there are at most k elements of $\mathcal{E}$ having mass greater than $1/k$, hence at most countably many elements in $\mathcal{E}$.

Definition 8.3.30 *$(\Omega, \mathcal{F}, P)$ is* **essentially finite** *if Ω can be partitioned into finitely many atoms of P.*

Theorem 8.3.31 *$L^\infty(\Omega, \mathcal{F}, P)$ is separable iff $(\Omega, \mathcal{F}, P)$ is essentially finite.*

Proof. Suppose that $(\Omega, \mathcal{F}, P)$ is not essentially finite, that is, that there exists a countable nonnull partition of Ω, $\mathcal{E} = \{E_n : n \in \mathbb{N}\} \subset \mathcal{F}$ with $P(E_n) > 0$ for all $n \in \mathbb{N}$. For each $A \subset \mathbb{N}$, define $X_A = \sum_{n \in A} 1_{E_n}$. This is an uncountable collection of elements of L^∞ at $\| \cdot \|_\infty$-distance 1 from each other.

Now suppose that $(\Omega, \mathcal{F}, P)$ is essentially finite and let $\{A_k : k \le K\}$ be the set of nonnull atoms. Up to a.e. equivalence, every element of L^∞ is of the form $\sum_{k \le K} \beta_k 1_{A_k}$. Restricting the β_k to be rational gives a countable dense set.   ∎

8.3.f.3 Separability of ℓ^p, $p \in [1, \infty]$

Theorem 8.3.32   *ℓ^p is separable iff $p \in [1, \infty)$.*

Proof.   For any $p \in [1, \infty)$, the set of vectors with at most finitely many nonzero components is dense. If the nonzero components are rational, we have a countable dense set. Taking $\{1_A : A \subset \mathbb{N}\}$ gives an uncountable subset of ℓ^∞, all at distance 1 from each other.   ∎

8.3.g The Nonnegative Orthant

The **nonnegative orthant** is $L_+^p := \{X \in L^p : X \ge 0\}$, and the **positive orthant** is $L_{++}^p := \{X \in L^p : X > 0\}$. The nonnegative orthant is a closed convex set that spans L^p because $X = X^+ - X^-$ and $X^+, X^- \in L_+^p$. When goods/consumption belong to L^p, then L_+^p is the natural consumption set. One might guess that L_{++}^p is the interior of L_+^p. One would be, mostly, wrong.

Lemma 8.3.33   *For $p \in [1, \infty)$, $\mathrm{int}(L_+^p) = \emptyset$ iff $(\Omega, \mathcal{F}, P)$ is not essentially finite.*

Proof.   If $(\Omega, \mathcal{F}, P)$ is essentially finite, then every $X \in L^p$ is a finite sum of the form $\sum_{k \le K} \beta_k 1_{E_k}$, the E_k forming a K-element nonnull partition that cannot be refined. $X > 0$ iff each $\beta_k > 0$. Set $\beta = \min_k \beta_k > 0$. The set of Y of the form $\sum_k \gamma_k 1_{E_k}$ with $|\beta_k - \gamma_k|^p < \beta$ is an open subset of L_{++}^p containing X.
   If $(\Omega, \mathcal{F}, P)$ is not essentially finite, then there is a nonnull partition $\mathcal{E} = \{E_1, E_2, \ldots\}$ with $P(E_m) \downarrow 0$. Pick arbitrary $X \ge 0$ and define $X_m = X - (X + 1) \cdot 1_{E_m}$ so that $\|X - X_m\|_p \to 0$. On E_m, $X_m = -1$, so that $X_m \notin L_+^p$, yet $X_m \to_{L^p} X$, so that $\mathrm{int}(L_+^p) = \emptyset$.   ∎

Exercise 8.3.34   Show that $\mathrm{int}(L_+^\infty) \ne \emptyset$.

The lack of an interior means that several arguments in general equilibrium theory become much harder when L^p, $p \in [1, \infty)$, is used as the consumption space. For example, let $(X_i)_{i \in I}$, $X_i \in L_+^p$, be an allocation to each of I people and suppose that it is Pareto optimal. Let $B_i = \{Y_i \in L_+^p : Y_i \succ X_i\}$ be the set of points that i thinks are strictly better than X_i. If preferences are continuous, then each B_i is open *relative* to L_+^p, but is not open if $(\Omega, \mathcal{F}, P)$ is not essentially finite. For finite-dimensional rather than infinite-dimensional consumption spaces, the second welfare theorem says that there exists a point in the dual space that strictly separates $\sum_i X_i$ from $\sum_i B_i$. This requires that $\sum_i B_i$ have nonempty interior.
   For equilibrium existence arguments with finite-dimensional consumption spaces, one typically assumes that not all of the indifference curves cut the axes. This guarantees that as p_i becomes small, the excess demand for good i becomes strictly positive, a crucial part of the fixed-point arguments. An implication of Lemma 8.3.33 is that for any $X > 0$, there are indifferent $Y \in (L_+^p \setminus L_{++}^p)$, at least when the utility function is monotonic, that is, if $[X > X'] \Rightarrow [u(X) > u(X')]$.

Lemma 8.3.35   *If $(\Omega, \mathcal{F}, P)$ is not essentially finite, $u : L^p_+ \to \mathbb{R}$ is monotonic and continuous, $p \in [1, \infty)$, and $X \in L^p_{++}$, then there exists $Y \in (L^p_+ \setminus L^p_{++})$ such that $u(X) = u(Y)$.*

Proof.   The Tietze extension theorem 6.7.25 (p. 330) tells us that any continuous bounded function on a closed set has a continuous extension to the whole space. Taking a bounded monotonic transformation if necessary, we can assume that $u : L^p \to \mathbb{R}$ is continuous.

Pick arbitrary $\gamma \in (0, 1)$. $0 < (1 - \gamma)X < X < (1 + \gamma)X$, so that, by monotonicity, $u((1 - \gamma)X) < u(X) < u((1 + \gamma)X)$. By continuity, $u((1 - \gamma)[X - X1_E]) < u(X) < u((1 + \gamma)[X - X1_E])$ if $P(E)$ is sufficiently small. The functions $R := (1 - \gamma)[X - X1_E]$ and $S := (1 + \gamma)[X - X1_E])$ are equal to 0 on the set E and are otherwise strictly positive. By the intermediate value theorem, for some $\alpha \in (0, 1)$, $u(\alpha R + (1 - \alpha)S) = u(X)$. ∎

Exercise 8.3.36   For $p \in [1, \infty)$, we know that there is an order-preserving linear isometry between ℓ^p and $L^p(\mathbb{N}, \mathcal{P}(\mathbb{N}), \mu)$, and this tells us that $\text{int}(\ell^p_+) = \emptyset$ iff $p \in [1, \infty)$. Give a direct proof.

8.3.h   The Boundedness Principle and Finite-Dimensional Subspaces

The boundedness principle is the result that a linear function from one normed space to another is continuous iff it is bounded. This implies that finite-dimensional subspaces of normed spaces are linearly homeomorphic to $\mathbb{R}^k$.

Notation Alert 8.3.A   *A function from one vector space to another is a function, but it is often called an **operator** simply to distinguish the fact that it has a special class of domain and range.*

8.3.h.1   Continuous Linear Operators from One Normed Space to Another
We saw in §7.5.a that the linear function $X \mapsto \int X \, dP$ is not continuous on the metric space (L^1, α). The discontinuity seemed extreme—for every $r > 0$ and every $\epsilon > 0$, there is an $X \in L^1((0, 1], \mathcal{B}, \lambda)$ with $\alpha(X, 0) < \epsilon$ and $\int X \, dP > r$. This kind of extreme behavior turns out to be typical of discontinuous linear functions. In the following, we use the same symbol, $\| \cdot \|$, for the norm on two different spaces, letting context determine which norm it must be. Furthermore, we do not assume that the norms are complete.

Theorem 8.3.37   *Let $(\mathfrak{X}, \| \cdot \|)$ and $(\mathfrak{Y}, \| \cdot \|)$ be two normed vector spaces. For a linear operator $T : \mathfrak{X} \to \mathfrak{Y}$, the following are equivalent.*

   1. T is continuous.

   2. T is continuous at 0.

   3. $\sup_{\|x\| \leq 1} \|T(x)\| < \infty$.

   4. For some $M \in \mathbb{R}$, $\|T(x)\| \leq M \cdot \|x\|$ for all $x \in \mathfrak{X}$.

At work in the equivalence of (1) and (2) is the **translation invariance** of norm distances, which is the observation that the distance between x and y, $\|x - y\|$, is equal to the distance between $x + z$ and $y + z$, $\|(x + z) - (y + z)\|$. We call the equivalence of (1) and (4) the **boundedness principle**, which says that a linear operator is continuous iff it is bounded on bounded sets.

Proof. (1) ⇔ (2): If T is continuous, it is continuous at 0. Suppose that T is continuous at 0 and pick arbitary $x \in \mathfrak{X}$. We must show that T is continuous at x. Since T is continuous at 0, $[\|y_n\| \to 0] \Rightarrow [\|T(y_n)\| \to 0]$. Choose x_n in $\mathfrak{X}$ with $\|x_n - x\| \to 0$. Set $y_n = x_n - x$ so that $x_n = x + y_n$ and $\|y_n\| \to 0$. By linearity $T(x_n) = T(x) + T(y_n)$. By translation invariance, $\|T(x_n) - T(x)\| = \|T(y_n)\| \to 0$, so T is continuous at x.

(2) ⇒ (4): Pick $\epsilon > 0$ so that $[\|y\| < \epsilon] \Rightarrow [\|T(y)\| < 1]$. We claim that for $M = 2/\epsilon$, $\|T(x)\| \leq M \cdot \|x\|$ for all $x \in \mathfrak{X}$. To see why, choose arbitrary $x \neq 0$ and set $y = \epsilon x / 2\|x\|$ so that $\|y\| < \epsilon$. By linearity and choice of ϵ, $T(y) = \frac{\epsilon}{2\|x\|} T(x) < 1$, so that $T(x) < \frac{2}{\epsilon}\|x\|$.

(4) ⇒ (2): If $\|x_n\| \to 0$, then $\|T(x_n)\| \leq M\|x_n\| \to 0$.

(3) ⇔ (4): Set $M = \sup_{\|x\| \leq 1} \|T(x)\|$. If $x = 0$, linearity implies that $T(x) = 0$. For any $x \neq 0$, $\|T(x)\| = \|x\|\|T(\frac{x}{\|x\|})\| \leq M\|x\|$. Finally, $\|T(x - y)\| \leq M \cdot \|x - y\|$ implies that T is continuous. ∎

8.3.h.2 Finite-Dimensional Subsets

The first and most basic result that we cover is that k-dimensional subspaces of L^p and ℓ^p are linearly homeomorphic to $\mathbb{R}^k$. If $f_1, \ldots, f_k$ is a linearly independent collection of functions in $L^p(\Omega, \mathcal{F}, P)$, then $L := \mathbf{span}(\{f_1, \ldots, f_k\})$ is the set of functions of the form $\omega \mapsto \sum_{i=1}^k \beta_i f_i(\omega)$. The linear homeomorphism in question is $(\beta_1, \ldots, \beta_k) \leftrightarrow \sum_{i=1}^k \beta_i f_i$.

Theorem 8.3.38 *If $\{x_1, \ldots, x_k\}$ is linearly independent in the normed space $\mathfrak{X}$, then*

1. *$L = \mathbf{span}(x_1, \ldots, x_k)$ is closed and complete, and*

2. *the mapping $T(\beta_1, \ldots, \beta_k) := \sum_{i=1}^k \beta_i x_i$ is a linear homeomorphism of $\mathbb{R}^k$ and L.*

Proof. We first show that T is a linear homeomorphism. The linearity is clear. Further, since the x_i are linearly independent, T is a bijection of $\mathbb{R}^k$ and L. To show that T is a linear homeomorphism, we must show that T and T^{-1} are continuous.

T is continuous: Define $\beta = (\beta_1, \ldots, \beta_k)$ and $T(\beta) = \sum_{i=1}^k \beta_i x_i$. By the linearity of T and the basic properties of a norm, $\|T(\beta)\| \leq \sum |\beta_i|\|x_i\|$, so that T is continuous.

T^{-1} is continuous: If not, there exist a sequence $x_n \in L$ and an $\epsilon > 0$ with $\|x_n\| \to 0$ and $\|T^{-1}(x_n)\| \geq \epsilon$. Defining $y_n \in \mathbb{R}^k$ as $T^{-1}(x_n)/\|T^{-1}(x_n)\|$ and taking a subsequence if necessary yields $y_n \to y^\circ$ for some $y^\circ = (\beta_1^\circ, \ldots, \beta_k^\circ) \in \mathbb{R}^k$ with $\|y^\circ\| = 1$. By the first step, T is continuous, so $T(y_n) = x_n \to T(y^\circ)$. Since $x_n \to 0$, $T(y^\circ) = 0$, that is, $\sum_i \beta_i^\circ x_i = 0$. Since the x_i are linearly independent, this means that $\beta_i^\circ = 0$ for all i, contradicting $\|y^\circ\| = 1$.

Finally, since T and T^{-1} are continuous, the boundedness principle implies that T induces a bijection between Cauchy sequences in $\mathbb{R}^k$ and Cauchy sequences in L. ∎

8.3.i Compactness

For all normed spaces, compactness is tightly related to a condition called approximate flatness. When we have the additional structure of an L^p, the approximately flat sets have an interpretation in terms of conditional expectations.

8.3.i.1 Approximate Flatness in Normed Spaces

Since the finite-dimensional $L \subset \mathfrak{X}$ is complete, it is a Banach space. Thus, we have proved the following.

Corollary 8.3.39 *Every finite-dimensional Banach space is linearly homeomorphic to $\mathbb{R}^k$ and every closed and bounded subset of a finite-dimensional Banach space is compact.*

The closed unit ball in the Banach space $C([0, 1])$ is closed and bounded, but not compact, so the finite dimensionality is crucial for the compactness. Being bounded and "nearly" finite dimensional is a characterization of compact subsets of all Banach spaces. As usual, $A^\epsilon = \cup_{x \in A} B_\epsilon(x)$ is the ϵ-ball around the set A.

Definition 8.3.40 *A subset F of a normed space $\mathfrak{X}$ is **approximately flat** if for all $\epsilon > 0$ there exists a finite-dimensional linear subspace, $W \subset \mathfrak{X}$, such that $F \subset W^\epsilon$.*

Since Banach spaces are complete, a subset of a Banach space is complete iff it is closed.

Theorem 8.3.41 *A closed and bounded subset of a Banach space $\mathfrak{X}$ is compact iff it is approximately flat.*

Proof. Suppose that K is closed, bounded by B, and approximately flat. As it is closed, it is complete. For compactness, it is therefore sufficient to show that K is totally bounded. Pick $\epsilon > 0$. Let W be a finite-dimensional subspace of $\mathfrak{X}$ with $K \subset W^{\epsilon/2}$ and W_B the set of points in W with norm less than or equal to B. Since W is linearly homeomorphic to $\mathbb{R}^k$ for some k, W_B is compact, hence has a finite $\epsilon/2$-net. This provides an ϵ-net for K.

If K is compact, there is a finite ϵ-net, and we take W to be the span of the ϵ-net.   ∎

8.3.i.2 Compactness and Conditional Expectations

Since the simple functions are dense in the normed spaces $L^p(\Omega, \mathcal{F}, P)$, we can give a bit more structure to the finite-dimensional sets that approximately contain a compact set. Toward this end, we introduce conditional expectations with respect to a nonnull partition of Ω. Conditional expectations with respect to more general σ-fields are the main construct in regression analysis, the subject of §8.4.

Definition 8.3.42 *For any nonnull partition $\mathcal{E}$ of Ω and $X \in L^1$, define the **conditional expectation of X given $\mathcal{E}$** as the function*

$$E(X \mid \mathcal{E})(\omega) = X_{\mathcal{E}}(\omega) = \sum_{E_k \in \mathcal{E}} \left(\frac{1}{P(E_k)} \int_{E_k} X \, dP \right) \cdot 1_{E_k}(\omega). \quad (8.14)$$

From introductory probability, the reader should recognize $\frac{1}{P(E_k)} \int_{E_k} X \, dP$ as $E(X \mid E_k)$, the conditional expectation of X given the set E_k. The function

$\omega \mapsto X_{\mathcal{E}}(\omega)$ takes different values on different E_k's in $\mathcal{E}$, that is, *the conditional expectation is itself a random variable*.

For two nonnull partitions $\mathcal{E}'$ and $\mathcal{E}$, $\mathcal{E}'$ refines $\mathcal{E}$, written $\mathcal{E}' \succ \mathcal{E}$, if every $E \in \mathcal{E}$ is a union of elements of $\mathcal{E}'$. Intuitively, if $\mathcal{E}' \succ \mathcal{E}$, then $X_{\mathcal{E}'}$ is a better approximation to $X_{\mathcal{E}}$. We see this idea at work in the following lemma.

Lemma 8.3.43   *For any $X \in L^p(\Omega, \mathcal{F}, P)$, $p \in [1, \infty)$, any partition $\mathcal{E} = \{E_k : k \leq K\} \subset \mathcal{F}$ of Ω, and any simple $\mathcal{E}$-measurable function $\sum_{k \leq K} \beta_k 1_{E_k}$,*

$$\int |X - \sum_{k \leq K} \beta_k 1_{E_k}|^p \, dP \geq \int |X_{\mathcal{E}} - \sum_{k \leq K} \beta_k 1_{E_k}|^p \, dP, \qquad (8.15)$$

equivalently $\|X - \sum_k \beta_k 1_{E_k}\|_p \geq \|X_{\mathcal{E}} - \sum_k \beta_k 1_{E_k}\|_p$.

Proof.   For any nonnull $A \in \mathcal{F}$, define the countably additive probability Q_A by $Q_A(B) = P(A \cap B)/P(B)$ for $B \in \mathcal{F}$. For any $X \in L^1(\Omega, \mathcal{F}, Q_A)$, Jensen's inequality and the convexity of the function $r \mapsto |r - \beta|^p$ imply that for any $\beta \in \mathbb{R}$, $\int |X - \beta|^p \, dQ_A \geq |\int (X - \beta) \, dQ_A|^p$. Applying this to each nonnull E_k in turn and adding up the results gives us $\int |X - \sum_{k \leq K} \beta_k 1_{E_k}|^p \, dP \geq \int |X_{\mathcal{E}} - \sum_{k \leq K} \beta_k 1_{E_k}|^p \, dP$.   ∎

Theorem 8.3.44 (Compactness in $L^p(\Omega, \mathcal{F}, P)$)   *A closed and bounded $K \subset L^p(\Omega, \mathcal{F}, P)$, $p \in [1, \infty)$, is compact iff for every $\epsilon > 0$, there exists a finite nonnull partition, $\mathcal{E}$, such that for all $\mathcal{E}' \succ \mathcal{E}$, $\sup_{X \in K} \|X - X_{\mathcal{E}'}\|_p < \epsilon$.*

Proof.   Suppose that K is compact. It has a finite $\epsilon/4$-net, $\mathcal{X} = \{X_1, \ldots, X_k\}$, for any $\epsilon > 0$. For each $X_i \in \mathcal{X}$, pick a simple $\sum_j \gamma_j 1_{A_j}$ such that $\|X_i - \sum_j \gamma_j 1_{A_j}\|_p < \epsilon/4$. Let $\mathcal{E}$ be the coarsest partition of Ω making each of these simple functions measurable. For each $X_i \in \mathcal{X}$, there is an $\mathcal{E}$-measurable simple function $Y_i = \sum_{k(i)} \beta_{k(i)} 1_{E_{k(i)}}$ with $\|X_i - \sum_{k(i)} \beta_{k(i)} 1_{E_{k(i)}}\|_p < \epsilon/4$. For any $X \in K$ and $X_i \in \mathcal{X}$,

$$\|X - X_{\mathcal{E}}\|_p \leq \|X - X_i\|_p + \|X_i - Y_i\|_p + \|Y_i - X_{i,\mathcal{E}}\|_p + \|X_{i,\mathcal{E}} - X_{\mathcal{E}}\|_p.$$

$$(8.16)$$

By construction, we can pick X_i such that both of the first two terms are less than $\epsilon/4$. Applying the previous lemma twice and noting that $Y_i = Y_{i,\mathcal{E}}$, the last two terms are also less than $\epsilon/4$. Hence, $\sup_{X \in K} \|X - X_{\mathcal{E}}\|_p < \epsilon$. Finally, if $\mathcal{E}' \succ \mathcal{E}$, the last three terms can be made even smaller.

Now suppose that there exists a finite nonnull partition, $\mathcal{E}$, such that for all $\mathcal{E}' \succ \mathcal{E}$, $\sup_{X \in K} \|X - X_{\mathcal{E}'}\|_p < \epsilon/2$. It is sufficient to find a finite ϵ-net. Let B satisfy $\|X\|_p \leq B$ for all $X \in K$ and set $m = \min\{P(E) : E \in \mathcal{E}\}$. Let C be a finite subset of $[-B/m, +B/m]$ with the property that for all $\beta \in [-B/m, +B/m]$ there exists a $\gamma \in C$ such that $|\beta - \gamma| < \epsilon/2$. The set $\{\sum_{E_k \in \mathcal{E}} \gamma_k 1_{E_k} : \gamma_k \in C\}$ is the requisite finite ϵ-net.   ∎

8.3.j Bases and Discontinuous Linear Functions

For the most part, we do not encounter discontinuous linear functions in economics. However, they are worth studying because they tell us how different infinite-dimensional vector spaces are from $\mathbb{R}^k$ and something about the extra assumptions we need in order to guarantee a useful dual space.

8.3.j.1 Discontinuous Linear Functions on Incomplete Normed Spaces

The boundedness principle shows that for a linear function on a normed space to be discontinuous, it must be unbounded on neighborhoods of 0; equivalently, arbitrarily small points in the domain must map to points bounded away from 0. Here is a fairly easy way to arrange this.

Example 8.3.45 *Give the space $L^2((0, 1], \mathcal{B}, \lambda)$ the $\|\cdot\|_1$ norm. Note that $(L^2, \|\cdot\|_1)$ is not complete. Pick a small $\delta > 0$ and let $r = (1 - \delta)/2$. The function $Y(\omega) = (1/\omega)^r$ belongs to L^2. Define $T : L^2 \to \mathbb{R}$ by $T(X) = \int (X \cdot Y) \, d\lambda$. (The Cauchy-Schwarz inequality, Theorem 8.4.5, just below shows that T is well defined on all of L^2, and it is certainly well defined on the dense subset L^∞). Consider the sequence $X_n = \beta_n 1_{(0,1/n]}$ with $\beta_n = n^{1-r}$. $\|X_n\|_1 = n^{-r} \to 0$, but $T(X_n) = 1/(1-r) \not\to 0$.*

The problem is that $\|X_n\|_1$ can go to 0 without $\|X_n\|_2$ going to 0. That is, it is easier to converge to 0 in the L^1-norm than in the L^2-norm. More sequences going to 0 means more possibilities for linear functions to be discontinuous, and that is what is behind the foregoing example. What the example does not demonstrate is a discontinuous linear function on an L^p space that has its own norm.

8.3.j.2 Hamel Bases and Discontinuous Linear Functions

For ease of reference, we repeat the definition of **linear independence**: an infinite set is linearly independent if all finite subsets are linearly independent; and **Zorn's lemma**: if A is a partially ordered set such that each chain in A has an upper bound in A, then A has a maximal element. Also, recall that the span of a set is the set of all *finite* linear combinations of elements in the set.

Definition 8.3.46 *A linearly independent H in a vector space $\mathfrak{X}$ is a **Hamel basis** if* **span** $(H) = \mathfrak{X}$.

Lemma 8.3.47 *Every vector space $\mathfrak{X}$ has a Hamel basis, and if H is a Hamel basis for $\mathfrak{X}$, then every nonzero $x \in \mathfrak{X}$ has a unique representation as a finite linear combination of elements of H with nonzero coefficients.*

Proof. Let $\mathfrak{I}$ denote the class of linearly independent subsets of $\mathfrak{X}$. Define the partial order $\preceq$ on $\mathfrak{I}$ by $B \preceq B'$ if $B \subset B'$ so that $(\mathfrak{I}, \preceq)$ is a partially ordered set. Let $C = \{B_\alpha : \alpha \in A\}$ be a chain in $\mathfrak{I}$. $B := \cup_\alpha B_\alpha$ is an upper bound for C.

We show that $B \in \mathfrak{I}$, that is, that B is a linearly independent subset of $\mathfrak{X}$. If not, there is a finite $F \subset B$ that is linearly dependent. Since F is finite, it belongs to B_α for some α. This contradicts the assumption that $B_\alpha \in \mathfrak{I}$.

From Zorn's lemma, we conclude that there exists a maximal linearly independent subset of $\mathfrak{X}$ that we call H. We show that **span**$(H) = \mathfrak{X}$. If not, there exists an $x \in \mathfrak{X}$ that cannot be expressed as a finite linear combination of elements of

H. That means that $H \cup \{x\} \supsetneq H$ is a linearly independent set, contradicting the maximality of H.

Finally, suppose there exists $x \in \mathcal{X}$ that can be expressed as two different finite sums of elements of H, $x = \sum_{i \in I(x)} \alpha_i h_i$ and $x = \sum_{j \in J(x)} \beta_j h_j$, $I(x)$, $J(x) \subset H$ finite, with none of the α_i or β_j being equal to 0. Now, since $x - x = 0$,

$$\sum_{i \in I(x)} \alpha_i h_i - \sum_{j \in J(x)} \beta_j h_j = 0. \tag{8.17}$$

If $I(x) = J(x)$, then by the linear independence of H, the representations are not different. One the other hand, if $I(x) \neq J(x)$, since none of the α_i nor the β_j are 0, the equality in (8.17) does not hold.   ■

Corollary 8.3.48   *If $\mathcal{X}$ is an infinite-dimensional normed space, then there exists a linear function $L : \mathcal{X} \to \mathbb{R}$ that is not continuous.*

Proof.   A Hamel basis, H, for $\mathcal{X}$ must contain a sequence of linearly independent vectors $S = \{h_n : n \in \mathbb{N}\}$. Dividing by $\|h_n\|$ if necessary, we can assume that each $\|h_n\| = 1$. Define $L(h_n) = n$ for each $h_n \in S$ and $L(x) = 0$ for each $x \in (H \setminus S)$. For $x = \sum_a \beta_a h_a$, $h_a \in H$, $L(x) = \sum_a \beta_a L(h_a)$ defines a linear function from $\mathcal{X}$ to $\mathbb{R}$. In particular, $L(0) = 0$. Consider the sequence $x_n = h_n/n \to 0$ in $\mathcal{X}$. For each n, $L(x_n) = n/n = 1$ and $1 \not\to 0$, so that L is not continuous.   ■

As they are linear even though they are discontinuous, the sets $L^{-1}((-\infty, r])$ and $L^{-1}((r, +\infty))$ are a partition of $\mathcal{X}$ into two disjoint nonempty convex sets, but this does not fully capture the extent of the weirdness. The following two results make our heads hurt.

Lemma 8.3.49   *A nonzero linear $L : \mathcal{X} \to \mathbb{R}$ is discontinuous iff $L^{-1}(0)$ is a dense convex subset of $\mathcal{X}$.*

Proof.   Suppose first that the nonzero L is continuous. Since $\{0\}$ is a closed subset of $\mathbb{R}$, $L^{-1}(0)$ is a closed subset of $\mathcal{X}$. Thus, $L^{-1}(0)$ is dense iff it is equal to $\mathcal{X}$, which implies that L is identically equal to 0.

If L is discontinuous, then $L(U)$ is unbounded, where $U = \{x \in \mathcal{X} : \|x\| < 1\}$ is the unit ball. Since $x \in U$ iff $-x \in U$ and L is linear, $L(U) = \mathbb{R}$, yielding, for every $\epsilon > 0$, $L(\epsilon \cdot U) = \mathbb{R}$. Therefore, for any $y \in \mathcal{X}$, $L(y + \epsilon U) = \mathbb{R}$, so that $L^{-1}(0) \cap B_\epsilon(y) \neq \emptyset$.   ■

Exercise 8.3.50   Generalizing the previous lemma, show that for a discontinuous linear $L : \mathcal{X} \to \mathbb{R}$, $L^{-1}((-\infty, 0])$ and $L^{-1}((0, +\infty))$ are disjoint convex dense sets that partition $\mathcal{X}$.

Lemma 8.3.51   *Every infinite-dimensional normed vector space can be partitioned into two dense convex sets.*

Proof.   The foregoing exercise delivers this, but an alternate argument is instructive. The axiom of choice is equivalent to the well-ordering lemma, 2.11.3, which says that every set can be well ordered. Well-order a Hamel basis H for an infinite-dimensional normed vector space $\mathcal{X}$. For any $x \in \mathcal{X}$, let $\sum_{h_i \in I(x)} \beta_i h_i$, $I(x) \subset H$, be its unique representation with nonzero β_i in terms of the Hamel basis. Define

x to belong to E^+ if the last h_i in $I(x)$ has $\beta_i > 0$ and to belong to E^- if the last $h_i \in I(x)$ has $\beta_i < 0$.

E^+ and E^- partition $\mathfrak{X}$ into two disjoint nonempty convex sets. Suppose that for some x and some $\epsilon > 0$, $B_\epsilon(x) \subset E^+$. Let $x = \sum_{h_i \in I(x)} \beta_i h_i$ be the Hamel representation of x and let h_j be the last h_i in $I(x)$. Let $x' = \beta'_j h_j + \sum_{i \neq j} \beta_i h_x$. For $\beta'_j > 0$ sufficiently close to β_j, the translation $U := B_\epsilon(x) - \{x'\}$ is an open neighborhood of 0. Every $y \in \mathfrak{X}$ is therefore a scalar multiple of an element in U, which implies that $E^+ = \mathfrak{X}$.　■

Exercise 8.3.52　Show that any Hamel basis of an infinite-dimensional Banach space must be uncountable. Show that completeness is crucial by considering the incomplete normed vector space consisting of sequences in $\mathbb{R}$ that are nonzero at most finitely often, with the sup norm.

8.3.j.3 Schauder Bases
The Hamel bases just defined are helpful for showing how strange the L^p spaces can be when compared with $\mathbb{R}^k$. They are not otherwise very useful, but the following can be utilized.

Definition 8.3.53　*For a separable Banach space, $\mathfrak{X}$, a countable $S = \{s_n : n \in \mathbb{N}\} \subset \mathfrak{X}$ is a **Schauder basis** if every $v \in \mathfrak{X}$ has a unique sequence r_n in $\mathbb{R}$ such that $\|v - \sum_{i \leq n} r_i s_i\| \to 0$.*

Exercise 8.3.54　If S is a Schauder basis, then it is linearly independent.

Just as in $\mathbb{R}$, the infinite sum, $\sum_i r_i s_i$, is defined as $\lim_n \sum_{i \leq n} r_i s_i$. For this limit to exist, the sums must be Cauchy. Since every v is the limit of such a sequence, the span of S, that is, the set of finite linear combinations of elements of S, is dense.

Exercise 8.3.55　The span of S is a complete metric space iff $\mathfrak{X}$ is finite dimensional.

There are separable Banach spaces that do not have Schauder bases. However, the classical Banach spaces, $L^p(\Omega, \mathcal{F}, P)$ and ℓ^p, $p \in [1, \infty)$, do have Schauder bases when they are separable. As with many results, this is far easier to demonstrate in $L^2(\Omega, \mathcal{F}, P)$, and we will give the construction in §8.4.b.5

8.4 ◆ Regression Analysis

We begin with an overview of how regression is connected to conditional expectations. Hilbert spaces are a special class of Banach spaces, and $L^2(\Omega, \mathcal{F}, P)$ is a Hilbert space. The formal development of conditional expectations in $L^2(\Omega, \mathcal{F}, P)$ comes after the requisite geometry of Hilbert spaces. Of particular importance are orthogonal projections, which we use to define conditional expectations in $L^2(\Omega, \mathcal{F}, P)$. With these tools in hand, we first examine linear regression and the subset of nonlinear regressions most often used for finding conditional means and then turn to nonparametric regression.

8.4.a Overview

The data, $(X_i, Y_i)_{i=1}^n$, are assumed to come from some distribution, μ, on $\mathbb{R}^{K+1}$, $X_i \in \mathbb{R}^K$, $Y_i \in \mathbb{R}$. The **empirical distribution** of the data is the probability $\mu_n(A) := \frac{1}{n} \#\{i : (X_i, Y_i) \in A\}$. By "come from some distribution" we mean that $\mu_n \to \mu$ as $n \to \infty$ in some fashion that may vary depending on context.

Pick a small subset, $E(x) \subset \mathbb{R}^K$ containing x, for example, $E(x) = B_\epsilon(x)$, with $\mu_n(E(x) \times \mathbb{R}) > 0$. The empirical expectation of Y_i conditional on $X_i \in E(x)$ is

$$\widehat{f}_n(x) := \frac{1}{\#\{i : X_i \in E\}} \sum_{i : X_i \in E} Y_i. \tag{8.18}$$

This is called "local averaging" because one takes the average value of the Y_i in some local neighborhood of x. It "ought to be true" that if we send the diameter of $E(x)$ to 0 and $n \to \infty$ in such a fashion that $\#\{i : X_i \in E(x)\} \to \infty$, then $\widehat{f}_n(x) \to f(x)$, where f is the target function $E(Y \mid X = x)$, for a set of x having probability 1. In this section we examine this problem, not only for the estimating sequence $\widehat{f}_n$ just described, but for many other estimating sequences.

8.4.b Some Geometry of Hilbert Spaces

Regression analysis uses the notation X and Y for very specific kinds of random variables. To avoid trampling on those conventions, we use f, g, and h to denote elements of the Hilbert space, $\mathbb{H}$, which may be $L^2(\Omega, \mathcal{F}, P)$, or ℓ^2, or any one of the other Hilbert spaces.

8.4.b.1 Inner Products

Hilbert spaces are Banach spaces where the norm is given by an inner product. The leading example is $\mathbb{R}^k$ where $\|\mathbf{x}\|_2 = \sqrt{\mathbf{x} \cdot \mathbf{x}}$.

Definition 8.4.1 *An **inner product on a vector space** $\mathfrak{X}$ is a function $(f, g) \mapsto \langle f, g \rangle$ from $\mathfrak{X} \times \mathfrak{X}$ to $\mathbb{R}$ with the properties: $\forall \alpha, \beta \in \mathbb{R}$, $\forall f, g, h \in \mathfrak{X}$,*

1. $\langle f, g \rangle = \langle g, f \rangle$ (symmetry),

2. $\langle \alpha f + \beta g, h \rangle = \alpha \langle f, h \rangle + \beta \langle g, h \rangle$ (linearity),

3. $\langle f, f \rangle \geq 0$ with equality iff $f = 0$.

*The norm associated with an inner product is given by $\|f\| = \sqrt{\langle f, f \rangle}$. If the norm is complete, $\mathfrak{X}$ is a special kind of Banach space called a **Hilbert space**. Hilbert spaces will be denoted $\mathbb{H}$.*

Combining symmetry and linearity makes the mapping $\langle \cdot, \cdot \rangle$ **bilinear**, that is, linear in each of its arguments when the other is fixed. To be really explicit here, for all f, g, $h \in \mathfrak{X}$ and all $\alpha, \beta \in \mathbb{R}$, $\langle \alpha f + \beta g, h \rangle = \alpha \langle f, h \rangle + \beta \langle g, h \rangle$ and $\langle f, \alpha g + \beta h \rangle = \alpha \langle f, g \rangle + \beta \langle g, h \rangle$. Thus, Hilbert spaces are complete normed vector spaces equipped with a bilinear inner product that gives their norm.

For f, $g \in L^2(\Omega, \mathcal{F}, P)$, define $\langle f, g \rangle = \int f(\omega) g(\omega) \, dP(\omega)$. It is clear that for $f \in L^2(\Omega, \mathcal{F}, P)$, $\|f\|_2 = \sqrt{\langle f, f \rangle}$, but not clear that $\langle f, g \rangle$ is well defined for all f, $g \in L^2$. We saw the following in the previous chapter.

Lemma 8.4.2 *For $f, g \in L^2(\Omega, \mathcal{F}, P)$, $\int |fg|\, dP \leq \frac{1}{2}(\int f^2\, dP + \int g^2\, dP)$*
$< \infty$.

Proof. For all ω, $(f(\omega) \pm g(\omega))^2 \geq 0$, so that $f^2(\omega) + g^2(\omega) \geq \mp 2 f(\omega) g(\omega)$.
Thus, for all ω, $|f(\omega) g(\omega)| \leq \frac{1}{2}(f^2(\omega) + g^2(\omega))$. ∎

Exercise 8.4.3 Show that ℓ^2 is a Hilbert space with the inner product $\langle f, g \rangle = \sum_n f_n g_n$.

There are many inner products on $\mathbb{R}^k$.

Exercise 8.4.4 Show the following.

1. For any symmetric positive definite $k \times k$ matrix, Σ, $(\mathbf{x}, \mathbf{y}) \mapsto \mathbf{x}^T \Sigma \mathbf{y}$ is an inner product on $\mathbb{R}^k$. With the associated norm, $\| \cdot \|_\Sigma$, the unit ball is $\{\mathbf{x} \in \mathbb{R}^k : \mathbf{x}^T \Sigma \mathbf{x} < 1\}$, which is ellipsoidal. [First diagonalize Σ.]

2. The usual norm on $\mathbb{R}^k$ is $\|\mathbf{x}\|_2 = \| \cdot \|_I$, where I is the $k \times k$ identity matrix.

3. If $L : \mathbb{R}^k \to \mathbb{R}^k$ is continuous, linear, and invertible, represented by a matrix A, then one can define the L-norm by $\|\mathbf{x}\|_L = \|Lx\|_2$. The inner product that gives this norm can be expressed in matrix form as before by setting $\Sigma = \frac{1}{2}A^T A + \frac{1}{2}A A^T$.

8.4.b.2 A Basic Inequality
The following is a special case of Hölder's inequality, Theorem 8.8.4.

Theorem 8.4.5 (Cauchy–Schwarz) *For $f, g \in \mathbb{H}$, $\mathbb{H}$ a Hilbert space, $|\langle f, g \rangle|$*
$\leq \|f\| \|g\|$.

Proof. If $f = 0$ or $g = 0$, the inequality is immediate. If $f, g \neq 0$, expand $\left\| \frac{f}{\|f\|} + \frac{g}{\|g\|} \right\| \geq 0$ and $\left\| \frac{f}{\|f\|} - \frac{g}{\|g\|} \right\| \geq 0$. After rearrangement, these yield $\|f\| \|g\|$ $\geq -\langle f, g \rangle$ and $\|f\| \|g\| \geq \langle f, g \rangle$. Combining gives us $|\langle f, g \rangle| \leq \|f\| \|g\|$. ∎

Exercise 8.4.6 For $f, g \in \mathbb{H}$, use Cauchy-Schwarz to prove the norm inequality $\|f + g\| \leq \|f\| + \|g\|$.

In $\mathbb{R}^k$, the geometry of inner products in the plane gives more information than is contained in the Cauchy-Schwarz inequality. It tells us that $\langle \mathbf{x}, \mathbf{y} \rangle = \cos(\theta) \|\mathbf{x}\|_2 \|\mathbf{y}\|_2$. Consider the two-dimensional subspaces of $\mathbb{H}$ spanned by f and g. We know that this is linearly homeomorphic to the plane $\mathbb{R}^2$ (Theorem 8.3.38). After we develop orthonormal bases, we will show that the linear homeomorphism is in fact an isometry.

8.4.b.3 Orthogonal Projection onto Closed Linear Subspaces
In $\mathbb{R}^k$, every linear subspace is closed. However, the same is not true for general Hilbert spaces, and this has implications for separation using continuous linear functions.[2]

Exercise 8.4.7 For each $n \in \mathbb{N}$ let $f_{k,n}$ be the indicator of $(k/2^n, (k+1)/2^n]$ in $L^2((0, 1], \mathcal{B}, \lambda)$ and let $D = \{f_{k,n} : 0 \leq k \leq 2^n - 1, \ n \in \mathbb{N}\}$.

---

2. If it has been a while, review the basic separation results of Chapter 5; §5.3 contains the basic definitions.

1. Show that the set $V = \mathbf{span}(D)$ is a dense linear subspace, but that it is not closed as it fails to contain $f(\omega) = \omega$.

2. Enumerate the set D as $\{g_1, g_2, \ldots\}$. Let $E^+ \subset V$ be the set of finite sums $x = \sum_{k \in I(x)} \beta_k$ having $\beta_K > 0$ for the largest $K \in I(x)$. Show that E^+ and its complement, E^-, are dense convex subsets of V.

3. Show that for any nonzero continuous linear L on $L^2((0, 1], \mathcal{B}, \lambda)$, $L(E^+) = L(E^-) = \mathbb{R}$. [This is an extreme failure of continuous linear functions to separate convex sets.]

4. Show that if nonempty C_1, C_2 are convex subsets of $\mathbb{R}^k$ with $C_1 \cup C_2$ dense in $\mathbb{R}^k$, then there exists a continuous linear function $L : \mathbb{R}^k \to \mathbb{R}$ and $r \in \mathbb{R}$ with $L(\text{int}(C_1)) = (-\infty, r)$ and $L(\text{int}(C_2)) = (r, +\infty)$.

Definition 8.4.8   *f and g in a Hilbert space $\mathbb{H}$ are **orthogonal**, written $f \perp g$, if $\langle f, g \rangle = 0$. For $S \subset \mathbb{H}$, $f \perp S$ if for all $g \in S$, $f \perp g$.*

In $L^2(\Omega, \mathcal{F}, P)$, $\text{Cov}(f, g) = \langle f - Ef, g - Eg \rangle$ and people sometimes abuse the term orthogonal by saying that f and g are orthogonal if $\text{Cov}(f, g) = 0$. We are careful to avoid that particular piece of abuse.

Theorem 8.4.9 (Geometry of Orthogonal Projection)   *Let $\mathbb{H}_0$ be a closed vector subspace of a Hilbert space, $\mathbb{H}$. For every $f \in \mathbb{H}$, there is a unique $h_0 \in \mathbb{H}_0$ such that $(f - h_0) \perp \mathbb{H}_0$. The point h_0 solves the problem $\min\{\|f - h\| : h \in \mathbb{H}_0\}$.*

If $\mathbb{H}_0$ was the nonclosed set V of Exercise 8.4.7, the conclusions of this theorem would fail rather completely. We have separate terminology for the h_0 guaranteed by this theorem.

Definition 8.4.10   *For $f \in \mathbb{H}$ and $\mathbb{H}_0$ a closed vector subspace of $\mathbb{H}$, **the orthogonal projection of f onto** $\mathbb{H}_0$ is defined as the unique point $h_0 \in \mathbb{H}_0$ such that $(f - h_0) \perp \mathbb{H}_0$. The point h_0 is denoted $h_0 = \text{proj}_{\mathbb{H}_0}(f)$.*

Proof of Theorem 8.4.9.   Let h_n be a sequence in $\mathbb{H}_0$ such that $\|f - h_n\| \to \delta := \inf\{\|f - h\| : h \in \mathbb{H}_0\}$. We now show that h_n is a Cauchy sequence. Since $\mathbb{H}_0$ is complete (being a closed subset of a complete space), this gives the existence of h_0 minimizing the distance. Uniqueness and perpendicularity are separate arguments.

1. By canceling cross-terms, it is easy to show that for all $g, g' \in \mathbb{H}$, $\|g + g'\|^2 + \|g - g'\|^2 = 2\|g\|^2 + 2\|g'\|^2$. (This is sometimes called the **parallelogram law**—think of 0, g, g', and $g + g'$ as the vertices of a parallelogram.)

2. Setting $g = f - h_n$ and $g' = f - h_m$, we have

$$4 \left\| f - \frac{h_n + h_m}{2} \right\|^2 + \left\| h_n - h_m \right\|^2 = 2 \left\| f - h_n \right\|^2 + 2 \left\| f - h_m \right\|^2.$$

Since $\mathbb{H}_0$ is a vector subspace, it is convex and $\frac{h_n + h_m}{2} \in \mathbb{H}_0$, so the first term is $\geq 4\delta^2$. The right-hand side of the inequality converges to $4\delta^2$, which means that h_n is a Cauchy sequence in the complete space $\mathbb{H}_0$.

3. Let $h_0 = \lim_n h_n$. By continuity, $\|f - h_0\| = \delta$, so that h_0 solves the problem $\min\{\|f - h\| : h \in \mathbb{H}_0\}$.

4. $(f - h_0) \perp \mathbb{H}_0$. Pick arbitrary $g \in \mathbb{H}_0$; for $t \in \mathbb{R}$, we know that $\| f - (h_0 + tg)\|^2 \geq \delta^2$ and $\| f - (h_0 + tg)\|^2 = \| f - h_0\|^2 + 2t \langle f - h_0, g\rangle + t^2 \|g\|^2$, so that for all t,

$$\delta^2 + 2t\langle f - h_0, g\rangle + t^2\|g\| \geq \delta^2, \text{ or } 2t\langle f - h_0, g\rangle + t^2\|g\| \geq 0. \quad (8.19)$$

only $\langle f - h_0, g\rangle = 0$ is consistent with (8.19) for all small positive and negative t.

5. Finally, if h_0 and h_1 are both solutions to $\min\{\| f - h\| : h \in \mathbb{H}_0\}$, then $(f - h_0) \perp (h_0 - h_1)$ and $(f - h_1) \perp (h_0 - h_1)$, equivalently

$$0 = \langle f, (h_0 - h_1)\rangle - \langle h_0, (h_0 - h_1)\rangle = \langle f, (h_0 - h_1)\rangle - \langle h_1, (h_0 - h_1)\rangle,$$

which implies that $\langle h_0, (h_0 - h_1)\rangle - \langle h_1, (h_0 - h_1)\rangle = 0$; that is, $\|h_0 - h_1\| = 0$, so that $h_0 = h_1$. ∎

The next result shows that $\text{proj}_{\mathbb{H}_0}$ is nonexpansive. By the boundedness principle, it must be continuous.

Corollary 8.4.11 *Let h_0 the projection of h onto $\mathbb{H}_0$, a closed linear subspace of $\mathbb{H}$. Then $\|h_0\| \leq \|h\|$ with equality iff $h = h_0$. In particular, orthogonal projection is continuous.*

Proof. $\|h\| = \|(h - h_0) + h_0\| = \langle (h - h_0) + h_0, (h - h_0) + h_0\rangle = \|h_0 - h\| + 2\langle (h - h_0), (h - h_0)\rangle + \|h_0\|$. Since $\langle (h - h_0), h_0\rangle = 0$, $\|h\| = \|h - h_0\| + \|h_0\|$. Thus, $\|h_0\| \leq \|h\|$ with equality iff $h = h_0$. ∎

8.4.b.4 Characterizing Continuous Linear Functions on $\mathbb{H}$

We now show that Hilbert spaces are self-dual, that is, $\mathbb{H}^*$ can be identified with $\mathbb{H}$ itself. We have already seen this result in the special case that $\mathbb{H} = \mathbb{R}^k$.

Exercise 8.4.12 For any $h_0 \in \mathbb{H}$, the function $T(h) := \langle h, h_0\rangle$ is continuous and linear. In particular, $T^{-1}(0)$ is a closed linear subspace of $\mathbb{H}$.

Recall that the norm of a continuous linear function, T, from one normed space to another is $\|T\| = \sup_{\|x\| \leq 1} \|T(x)\|$.

Theorem 8.4.13 (Riesz-Fréchet) *If $T : \mathbb{H} \to \mathbb{R}$ is a continuous linear function, then there exists a unique $h_0 \in \mathbb{H}$ such that $T(h) = \langle h, h_0\rangle$ and $\|T\| = \|h\|$.*

Proof. Let $\mathbb{H}_0 = \{h : T(h) = 0\}$. This is a closed linear subspace of $\mathbb{H}$. If $T \equiv 0$, then $\mathbb{H}_0 = \mathbb{H}$ and $h_0 = 0$. The remaining case is when there exists h_2 such that $T(h_2) \neq 0$.

For existence, set $h_3 = h_2 - \text{proj}_{\mathbb{H}_0}(h_2)$ and $h_0 = \frac{T(h_3)}{\|h_3\|^2}h_3$. For uniqueness, suppose that for all h, $\langle h, h_0\rangle = \langle h, h_1\rangle$. Then, taking $h = (h_0 - h_1)$, we have $\langle (h_0 - h_1), h_0\rangle = \langle (h_0 - h_1), h_1\rangle$, so that $\langle (h_0 - h_1), (h_0 - h_1)\rangle = 0$.

Check that $T(h) \equiv \langle h, h_0\rangle$ and, by considering $T(h_0)$, show that $\|T\| = \|h_0\|$.

∎

8.4.b.5 Orthonormal Bases in $\mathbb{H}$

In $\mathbb{R}^k$, a set $\{\mathbf{x}_1, \ldots, \mathbf{x}_m\}$ is **orthonormal** if it is **ortho**gonal, $\mathbf{x}_j \perp \mathbf{x}_k$ for all $j \neq k$, and **normal**ized, $\|\mathbf{x}_j\| = 1$ for $j = 1, \ldots, m$.

Definition 8.4.14 $S \subset \mathbb{H}$ is **orthonormal** if for all $f \neq g \in S$, $f \perp g$ and for all $f \in S$, $\|f\| = 1$. If S is also a Schauder basis, then S is an **orthonormal basis**.

Theorem 8.4.15 If $\mathbb{H}$ is separable, then it has an orthonormal Schauder basis.

The proof proceeds by inductively applying the Gram-Schmidt procedure (p. 180) from $\mathbb{R}^k$ to this infinite-dimensional context.

Proof. Enumerate a countable dense set $E = \{f_n : n \in \mathbb{N}\} \subset \mathbb{H}$. Without loss, for all n, $f_n \neq 0$, and for all $n \neq m$, $f_n \neq f_m$. We inductively construct an orthonormal $S = \{g_1, g_2, \ldots\}$ from E.

Let $g_1 = f_1/\|f_1\|$ and remove f_1 from E. Given that $g_1, \ldots, g_n$ have been defined, there are two cases: the first f_k left in E satisfies $f_k \in H_n := \mathbf{span}(g_1, \ldots, g_n)$ and $f_k \notin H_n$. If $f_k \in \mathbf{span}(g_1, \ldots, g_n)$, remove it from E and check f_{k+1}. If $f_k \notin H_n$, let f'_k be its orthogonal projection onto H_n, define $g_{n+1} = (f_k - f'_k)/\|f_k - f'_k\|$, remove f_k from E, and proceed with $g_1, \ldots, g_{n+1}$ as above.

If this process stops with a finite set $\{g_1, \ldots, g_n\}$, then $\mathbb{H}$ is finite dimensional, n must be the dimension, and $\{g_1, \ldots, g_n\}$ is an orthonormal basis. Suppose for the rest of the proof that $\mathbb{H}$ is infinite dimensional.

At each step, $\{g_1, \ldots, g_n\}$ is orthonormal, which implies that $S = \{g_n : n \in \mathbb{N}\}$ is orthonormal. All that is left to show is that S is a Schauder basis. First observe that $\mathbf{span}(S) = \mathbf{span}(E)$, so that $\mathbf{span}(S)$ is dense in $\mathbb{H}$. This makes it plausible that every $h \in \mathbb{H}$ can be approximated by at least one sequence $\sum_n \beta_n g_n$. To prove it, pick $h \in \mathbb{H}$ and let $h_n = \sum_n \beta_n g_n$ be the projection of h onto H_n. Since H_n is perpendicular to g_{n+1}, the projection of h_{n+1} onto H_n is h_n, so the sequence β_n is well defined. Since $\cup_n H_n = \mathbf{span}(S)$ is dense, $h_n \to h$.

Now suppose that h can be approximated by another sequence, $\sum_n \gamma_n g_n$. $\|\sum_n \beta_n g_n - \sum_n \gamma_n g_n\| = \|\sum_n (\beta_n - \gamma_n) g_n\| = \langle \sum_n (\beta_n - \gamma_n) g_n, \sum_n (\beta_n - \gamma_n) g_n \rangle^{1/2}$. Now, $\langle \sum_n (\beta_n - \gamma_n) g_n, \sum_n (\beta_n - \gamma_n) \rangle = \sum_n (\beta_n - \gamma_n)^2$ because the g_n are orthonormal. Thus, $\beta_n = \gamma_n$ for all n, and g is a Schauder basis. ∎

Let W be a finite-dimensional subspace of $\mathbb{H}$ and let $g_1, \ldots, g_k$ be an orthonormal basis for W. $\sum_i \beta_i g_i \leftrightarrow (\beta_1, \ldots, \beta_k)$ is not only a linear homeomorphism between W and $\mathbb{R}^k$, but it is also an isometry, which means that we again have $\langle f, g \rangle = \cos(\theta) \|f\| \|g\|$.

8.4.c Conditional Expectations as Projections

Projection in $L^2(\Omega, \mathcal{F}, P)$ has a special name when the closed linear subspace is

$$L^2(\mathcal{G}) := \{f \in L^2(\Omega, \mathcal{F}, P) : f \text{ is } \mathcal{G}\text{-measurable}\},$$

$\mathcal{G}$ a sub-σ-field of $\mathcal{F}$. When $\mathcal{G} = \sigma(X)$, we write $L^2(\sigma(X))$ as $L^2(X)$. By Doob's theorem, 7.2.45 (p. 383), $g \in L^2(X)$ iff there exists a measurable f such that $g = f(X)$ and $E g^2 < \infty$.

Exercise 8.4.16 $L^2(\mathcal{G})$ is a closed vector subspace of $L^2(\Omega, \mathcal{F}, P)$.

Definition 8.4.17 For $Y \in L^2(\Omega, \mathcal{F}, P)$, the **conditional expectation of Y given** $\mathcal{G}$, written as $E(Y \mid \mathcal{G})$, is defined by $E(Y \mid \mathcal{G}) = \mathrm{proj}_{L^2(\mathcal{G})}(Y)$.

Thus, $E(Y \mid \mathcal{G})$ is the unique solution to the problem

$$\min_{f \in L^2(\mathcal{G})} \|Y - f\|_2. \tag{8.20}$$

When $\mathcal{G} = \sigma(X)$, Doob's theorem, 7.2.45, shows that this can be rewritten as

$$\min_{\{f: f(X) \in L^2(\Omega, \mathcal{F}, P)\}} \|Y - f(X)\|_2. \tag{8.21}$$

We now work through some informative special cases, and we suggest reviewing Definition 8.3.42 (p. 470).

Example 8.4.18 *Here are four conditional expectations given finite σ-fields.*

1. *Let $\mathcal{G} = \{\emptyset, \Omega\}$. For any Y, $E(Y \mid \mathcal{G}) = EY$. To see why, consider the problem $\min_{r \in \mathbb{R}} E(Y - r \, 1_\Omega)^2$. The term being minimized is $E(Y - r \, 1_\Omega) \cdot (Y - r \, 1_\Omega) = EY^2 - 2r \, EY + r^2$, which is minimized at $r = EY$.*

2. *Let $\mathcal{G} = \{\emptyset, A, A^c, \Omega\}$ and $Y = 1_B$. $E(Y \mid \mathcal{G}) = P(B \mid A) \cdot 1_A + P(B \mid A^c) \cdot 1_{A^c}$. To see this, minimize over β_1, β_2, $E(Y - (\beta_1 \cdot 1_A + \beta_2 \cdot 1_{A^c}))^2$.*

3. *Let $\mathcal{G} = \{\emptyset, A, A^c, \Omega\}$ and $Y \in L^2(\Omega, \mathcal{F}, P)$. $E(Y \mid \mathcal{G}) = E(Y \mid A) \cdot 1_A + E(Y \mid A^c) \cdot 1_{A^c}$, where $E(Y \mid A) := \frac{1}{P(A)} \int_A Y \, dP$ and $E(Y \mid A^c) := \frac{1}{P(A^c)} \int_{A^c} Y \, dP$.*

4. *More generally, let $\mathcal{E} = \{A_1, \ldots, A_n\}$ be a finite partition of Ω, $\mathcal{G} = \sigma(\mathcal{E})$, $Y \in L^2(\Omega, \mathcal{F}, P)$. $E(Y \mid \mathcal{G}) = \sum_i E(Y \mid A_i) \cdot 1_{A_i}$.*

The following is immediate from orthogonal projection, but extremely useful nonetheless—it is often used as the definition of the conditional expectation.

Theorem 8.4.19 $E(Y \mid \mathcal{G})$ *is the unique element of $L^2(\mathcal{G})$ with $\int_A Y \, dP = \int_A E(Y \mid \mathcal{G}) \, dP$ for all $A \in \mathcal{G}$,*

Proof. We must show that if R is a $\mathcal{G}$-measurable random variable with finite variance that has the property that for all $A \in \mathcal{G}$, $\int_A Y \, dP = \int_A R \, dP$, then $R = E(Y \mid \mathcal{G})$.

$\int_A Y \, dP = \int_A R \, dP$ can be rewritten as $\int_A (Y - R) \, dP = 0$, equivalently as $1_A \perp (Y - R)$. Thus, we must show that if $1_A \perp (Y - R)$ for all $A \in \mathcal{G}$, then $(Y - R) \perp L^2(\mathcal{G})$.

If $1_A \perp (Y - R)$ for all $A \in \mathcal{G}$, then $S \perp (Y - R)$ for all simple $S \in L^2(\mathcal{G})$. Since $X \mapsto \langle X, (Y - R) \rangle$ is continuous, $\langle \cdot, (Y - R) \rangle$ being 0 on a dense subset of $L^2(\mathcal{G})$ implies that it is 0 on the whole set. ∎

A special case of theorem 8.4.19 takes $A = \Omega$.

Corollary 8.4.20 (Iterated Expectations) *For any $Y \in L^2(\Omega, \mathcal{F}, P)$ and sub-σ-field $\mathcal{G}$, $EY = E(E(Y \mid \mathcal{G}))$.*

Exercise 8.4.21 Let $(\Omega, \mathcal{F}, P) = ((0, 1]^2, \mathcal{B}^2, \lambda^2)$, $Y = 1_{(0, \frac{1}{2}] \times (0, \frac{1}{2}]}$, and $X(\omega_1, \omega_2) = \omega_1 + \omega_2$. Find $E(Y \mid \sigma(X))$ and verify that your answer satisfies the condition given in Theorem 8.4.19.

Theorem 8.4.22   *The mapping $Y \mapsto E(Y \mid \mathcal{G})$ from L^2 to itself is a nonexpansive, nonnegative mapping, that is, $[Y \geq 0] \Rightarrow [E(Y \mid \mathcal{G}) \geq 0]$ and linear; for all $X, Y \in L^2$ and $\alpha, \beta \in \mathbb{R}$, $E(\alpha X + \beta Y \mid \mathcal{G}) = \alpha E(X \mid \mathcal{G}) + \beta E(Y \mid \mathcal{G})$.*

Proof.   The nonexpansiveness is Corollary 8.4.11. Linearity is immediate from properties of projection. Suppose that $Y \geq 0$ and let $A = \{E(Y \mid \mathcal{G}) < 0\}$. If $P(A) > 0$, then $\int_A E(Y \mid \mathcal{G}) \, dP < 0$, but we know that $\int_A E(Y \mid \mathcal{G}) \, dP = \int_A Y \, dP$ and $\int_A Y \, dP \geq 0$ because $Y \geq 0$ a.e.   ■

8.4.d Linear and Nonlinear Regression

The conditional expectation of Y given X is the primary object of interest in regression analysis, both linear and nonlinear. Throughout, the convention is that $X(\omega)$ is a row vector. We begin with linear regression. The class of nonlinear regression analyses that we consider here is, from a theoretical point of view, nearly indistinguishable.

Notation 8.4.23   *The **conditional expectation of** Y **given** X is written $E(Y \mid X)$ and defined by $E(Y \mid \sigma(X))$.*

8.4.d.1 Correct Specification of the Affine Model
Let $X_0(\omega) \equiv 1$ be the random "variable" that always takes the value 1. In linear regression, the object is to find the function $\omega \mapsto \sum_{k=0}^{K} \beta_k X_k(\omega)$ that does the best job of approximating the random variable $Y(\omega)$. Let $\widetilde{X}$ be the $1 \times (K + 1)$ random vector $(X_0, X_1, \ldots, X_K)$. Let $A(\widetilde{X}) = \{\widetilde{X}\beta : \beta \in \mathbb{R}^{K+1}\}$ be the closed linear subspace of affine functions of $\widetilde{X}$. The problem becomes

$$\min_{f \in A(\widetilde{X})} \|Y - f\|_2, \tag{8.22}$$

and the solution to this must be the projection of Y onto $A(\widetilde{X})$. Let $\widetilde{X}\beta^*$ be the solution to (8.22). Since $A(\widetilde{X}) \subset L^2(X)$, the minimum in (8.22) is greater than or equal to

$$\min_{f \in L^2(X)} \|Y - f\|_2, \tag{8.23}$$

with equality iff $E(Y \mid X) = \widetilde{X}\beta^*$. This explains the potency of the correct specification assumption.

Definition 8.4.24   *The affine model is **correctly specified** for the conditional mean $E(Y \mid X) = \widetilde{X}\beta^\circ$ for some β°.*

8.4.d.2 The Ideal Regression Coefficients
It is important to bear in mind that there is a unique point in $A(\widetilde{X})$ solving the problem in (8.22) whether or not the affine model is correctly specified. The ideal regression coefficients are the ones that specify that point. They are ideal in the sense that they depend on the unknown joint distribution of (X, Y) rather than on the data.

Notation Alert 8.4.A *Before launching into linear regression, we absorb the constant random variable, $X_0(\omega) \equiv 1$, into the X vector and renumber. This means that $K + 1$ above becomes K below and that X replaces $\widetilde{X}$.*

Since squaring is an increasing transformation on the positive numbers, (8.22) is equivalent to

$$\min_{\beta \in \mathbb{R}^K} \langle Y - X\beta, Y - X\beta \rangle. \tag{8.24}$$

Expanding the objective function yields $\langle Y, Y \rangle - 2\langle Y, X\beta \rangle + \langle X\beta, X\beta \rangle$. The first term does not depend on β.

Let $(E\, X^T X)$ be the $K \times K$ matrix with (i, j)th entry $E\, X_i X_j$ (which exists because $X \in L^2$). Inspection shows that $\langle X\beta, X\beta \rangle = \beta^T E\, X^T X \beta$. With $(E\, YX)$ being a K-dimensional row vector, we are interested in solving the problem

$$\min_{\beta \in \mathbb{R}^K} (E\, YX)\beta + \beta^T (E\, X^T X)\beta. \tag{8.25}$$

This function is strictly convex iff $(E\, X^T X)$ is positive definite.

Lemma 8.4.25 *$(E\, X^T X)$ is positive definite iff the random variables X_i, $i = 1, \ldots, K$ are linearly independent.*

Proof. The X_i are linearly independent if $[\, \|\sum_i \gamma_i X_i\| > 0] \Leftrightarrow [\gamma \neq 0]\,]$. Now, $\|\sum_i \gamma_i X_i\|^2$ is equal to $E\, (\sum_i \gamma_i X_i)^2$, and this is $\gamma^T (E\, X^T X)\gamma$. ∎

Theorem 8.4.26 *If the X_i are linearly independent, then $\beta^* = (E\, X^T X)^{-1}(E\, X^T Y)$ is the unique solution to $\min_{\beta \in \mathbb{R}^K} \|Y - X\beta\|$.*

Proof. Since the objective function is continuously differentiable and strictly convex, the unique solution occurs where the derivative is equal to 0. The derivative is $2(E\, X^T X)\beta - 2(E\, X^T Y)$ is equal to 0 iff $\beta = (E\, X^T X)^{-1}(E\, X^T Y)$. ∎

The β^* in Theorem 8.4.26 is the vector of **ideal regression coefficients**. The vector is ideal in the sense that we have to know (several aspects of) the joint distribution of Y and X in order to calculate it. If we knew the joint distribution, there would be no need to look for data and run regressions.

In order to facilitate the comparison of the ideal regression coefficients and data-based regression coefficients, we make a change of variables. We change $(\Omega, \mathcal{F}, P)$ for $(\mathbb{R}^{K+1}, \mathcal{B}^{K+1}, \mu)$, where $\mu(A) := P((X, Y) \in A)$ for all measurable $A \subset \mathbb{R}^{K+1}$. This has the effect of making $X_i(\omega) = \text{proj}_i(\omega)$ for $\omega \in \mathbb{R}^{K+1}$ and $i \in \{1, \ldots, K\}$ and of making $Y(\omega) = \text{proj}_{K+1}(\omega)$.

It is important to be clear that $f : \mathbb{R}^K \to \mathbb{R}$ satisfies $f(X) = E(Y \mid X)$ for $X, Y \in L^2(\Omega, \mathcal{F}, P)$ iff $f(X) = E(Y \mid X)$ for $X, Y \in L^2(\mathbb{R}^{K+1}, \mathcal{B}^{K+1}, \mu)$. Taking $g(\mathbf{x}, y) = |y - f(\mathbf{x})|^2$ in the following lemma, which is basically a restatement of what change of variables means, should make this clear.

Lemma 8.4.27 *If $g : \mathbb{R}^{K+1} \to \mathbb{R}$ is measurable, then for any measurable $B \subset \mathbb{R}$, $P(\{\omega : g(X(\omega), Y(\omega)) \in B\}) = \mu(\{(\mathbf{x}, y) : g(\mathbf{x}, y) \in B\})$, $\mathbf{x} \in \mathbb{R}^K$, $y \in \mathbb{R}^K$.*

Proof. By a familiar argument, it is sufficient to show that this is true for all sets B of the form $(-\infty, r]$. Since $\mu(A) := P((X, Y) \in A)$ for all measurable $A \subset \mathbb{R}^{K+1}$, the equality holds for all simple g. Suppose first that $g \geq 0$ and take a sequence of simple measurable functions $g_n \uparrow g$ for all $(\mathbf{x}, y) \in \mathbb{R}^{K+1}$. For all n, $P(\{\omega : g_n(X(\omega), Y(\omega)) \le r\}) = \mu(\{(\mathbf{x}, y) : g_n(\mathbf{x}, y) \le r\})$, so that $P(\{\omega : g(X(\omega), Y(\omega)) \le r\}) = \mu(\{(\mathbf{x}, y) : g(\mathbf{x}, y) \le r\})$. The proof for $g \le 0$ is parallel, and the proof for general g uses $g = g^+ - g^-$. ∎

8.4.d.3 Data–Based Regression Coefficients

Suppose that for each $n \in \mathbb{N}$, we have data, $(X_i, Y_i)_{i=1}^n$, and, as before, we define $\mu_n(A) := \frac{1}{n} \#\{i : (X_i, Y_i) \in A\}$ to be the **empirical distribution** of the data. If we believe that μ_n is a good approximation to a distribution, μ, on $\mathbb{R}^{K+1}$ that we care about, then it makes sense to consider the problem

$$\min_{\beta \in \mathbb{R}^K} \int_{\mathbb{R}^{K+1}} |y - \mathbf{x}\beta|^2 \, d\mu_n(\mathbf{x}, y) \tag{8.26}$$

as an approximation to the ideal regression,

$$\min_{\beta \in \mathbb{R}^K} \int_{\mathbb{R}^{K+1}} |y - \mathbf{x}\beta|^2 \, d\mu(\mathbf{x}, y). \tag{8.27}$$

When Theorem 8.4.26 is used, what probability distribution on what probability space it is applied to is irrelevant. The only thing that is required for a unique solution is that the X_i be linearly independent. To examine the solutions to (8.26), arrange the data and the vector β as follows:

$$\mathbf{Y} = \begin{bmatrix} Y_1 \\ Y_2 \\ \vdots \\ \vdots \\ Y_n \end{bmatrix}, \quad \mathbf{X} = \begin{bmatrix} X_{1,1} = 1 & X_{1,2} & X_{1,3} & \cdots & X_{1,K} \\ X_{2,1} = 1 & X_{2,2} & X_{2,3} & \cdots & X_{2,K} \\ \vdots & \vdots & \vdots & \vdots & \vdots \\ \vdots & \vdots & \vdots & \vdots & \vdots \\ X_{n,1} = 1 & X_{n,2} & X_{n,3} & \cdots & X_{n,K} \end{bmatrix}, \quad \text{and } \beta = \begin{bmatrix} \beta_1 \\ \beta_2 \\ \vdots \\ \beta_K \end{bmatrix}.$$

Lemma 8.4.28 *If the X_i are linearly independent under the distribution μ_n, then the solution to (8.26) is $\beta_n^* = (\mathbf{X}^T \mathbf{X})^{-1} \mathbf{X}^T \mathbf{Y}$.*

Proof. By the definition of μ_n, the (k, j)th entry of $(\int X^T X \, d\mu_n)$ is equal to $\frac{1}{n} X_k X_j$, so that $(\int X^T Y \, d\mu_n) = \frac{1}{n}(\mathbf{X}^T \mathbf{X})$. Basic linear algebra of inverses implies $(\int X^T Y \, d\mu_n)^{-1} = n(\mathbf{X}^T \mathbf{X})^{-1}$. In the same way, $(\int X^T Y \, d\mu_n) = \frac{1}{n} \mathbf{X}^T \mathbf{Y}$. Combining gives us $(\int X^T X \, d\mu_n)^{-1}(\int X^T Y \, d\mu_n) = \frac{n}{n}(\mathbf{X}^T \mathbf{X})^{-1} \mathbf{X}^T \mathbf{Y}$. ∎

The $\beta_n^* = \beta_n^*((X_i, Y_i)_{i=1}^n)$ are the **data-based regression coefficients**.

Corollary 8.4.29 *If $\int X^T X \, d\mu_n \to \int X^T X \, d\mu$, $\int X^T Y \, d\mu_n \to \int X^T Y \, d\mu$ and the X_i are linearly independent under μ, then $\beta_n^* \to \beta$ where β^* is the vector of ideal regression coefficients.*

Proof. The mapping from matrices to their inverses is continuous at all invertible matrices, which implies that $(\int X^T X \, d\mu_n)^{-1} \to (\int X^T X \, d\mu)^{-1}$. Since determinants are continuous functions of matrices, for all large n, $(\int X^T X \, d\mu_n)$ is positive definite, which means that the X_i are linearly independent under μ_n for large n, which in turn implies that the sequence β_n^* is well defined, at least for large n. Since multiplication is continuous, the observation that $\int X^T Y \, d\mu_n \to \int X^T Y \, d\mu$ completes the proof. ∎

8.4.d.4 Data–Generating Processes

Until now, the data have been given to us and we have not asked anything about where they came from. We assume they are realizations of random variables.

Definition 8.4.30 *A data-generating process (DGP) is a set of random variables* $\omega \mapsto (X_i(\omega), Y_i(\omega))_{i \in \mathbb{N}}$.

By change of variables, we could as well say that a DGP is a probability distribution on the product space $(\mathbb{R}^{K+1})^{\mathbb{N}}$, the X_i and Y_i being the appropriate projections. Since $(\mathbb{R}^{K+1})^{\mathbb{N}}$ is a complete separable metric space, there would be no loss in assuming that $(\Omega, \mathcal{F}, P) = ((0, 1], \mathcal{B}, \lambda)$.

Since the data $(X_i, Y_i)_{i=1}^n$ are a function of ω, so is the empirical distribution μ_n, that is, the μ_n are random distributions. Therefore, the $\int X^T X \, d\mu_n$ are random variables, $\int X^T Y \, d\mu_n$ are random variables, and the β_n^* are random variables.[3]

Theorem 8.4.31 *Suppose that* μ *is a distribution on* $\mathbb{R}^{K+1}$ *for which the* X_i *are linearly independent and let* β^* *be the associated vector of ideal regression coefficients. If the random variables* $(\int X^T X \, d\mu_n)$ *converge in probability to the constant* $(\int X^T X \, d\mu)$ *and the random variables* $(\int X^T Y \, d\mu_n)$ *converge in probability to the constant* $(\int X^T Y \, d\mu)$, *then* $\beta_n^* \to \beta^*$ *in probability.*

Proof. Pick $\epsilon > 0$ and let G be an open neighborhood in $\mathbb{R}^{K \cdot K} \times \mathbb{R}^K$ with the property that for all $(\mathbf{M}, \mathbf{v}) \in G$, $d(\mathbf{M}\mathbf{v}, \beta^*) < \epsilon$. Pick $\delta > 0$ such that $B := B_\delta(\int X^T X \, d\mu, \int X^T Y \, d\mu) \subset G$. By assumption, $P((\int X^T X \, d\mu_n, \int X^T Y \, d\mu_n) \in B) \to 1$. ∎

Some observations are in order.

1. The conclusion that $\beta_n^* \to \beta^*$ in probability is called the **consistency** of the sequence β_n^*.

2. There are a number of stationarity assumptions on the DGP that guarantee the existence of a unique μ with $\mu_n \to \mu$ in a variety of senses. The strongest of these is the assumption that the vectors (X_i, Y_i) are iid with distribution μ. However, as long as the $\mu_n(\omega)$ deliver convergent integrals, β_n^* will be consistent.

3. If the affine model is correctly specified for μ, then β^* is the unique vector with $X\beta^* = E(Y \mid X)$. This means that the sequence of estimated functions $\widehat{f_n}(\mathbf{x}) := \mathbf{x}\beta_n^*$ has the property that $\|\widehat{f_n}(X) - E(Y \mid X)\| = \|\beta_n^* - \beta^*\|_\Sigma$, where Σ is the matrix $E \, X^T X$, and this number goes to 0 in probability. This gives the sense in which the estimated function goes to the true regression function.

4. The β_n^* may be consistent for many different μ'. Specifically, there are many $\mu \neq \mu'$ having the same relevant integrals, that is, many $\mu \neq \mu'$ with $\int X^T X \, d\mu = \int X^T X \, d\mu'$ and $\int X^T Y \, d\mu = \int X^T Y \, d\mu'$. The β_n^* would

---

3. The proof of the next theorem is essentially a simple application of the theorem of the maximum 6.1.31 (p. 268), one with $x_n \to x$ being $\mu_n \to \mu$, a constant opportunity set, $\Gamma(x) = \Gamma(x')$, and a unique maximum for each x. What makes it particularly simple is that the optimum is a fixed function of the x_n.

converge to the same value if the "true distribution" were any one of these μ'.

5. The theorem assumed that there was a fixed nonrandom pair $\mathbf{M} = \int X^T X \, d\mu$ and $\mathbf{v} = \int X^T Y \, d\mu$. We have seen some examples, and there are many more, of DGPs in which the influence of early events does not disappear. In this case, we could have random variables $\omega \mapsto (\mathbf{M}(\omega), \mathbf{v}(\omega))$ and the $\beta_n^*(\omega)$ would converge to some random variable $\beta^*(\omega)$.

8.4.d.5 Extension to Some Nonlinear Regressions

Many of the nonlinear regressions used for estimating conditional means involve a nonlinear function $f : \mathbb{R}^K \to \mathbb{R}^M$ and then solve

$$\min_{\beta \in \mathbb{R}^M} \|Y - f(X)\beta\|_2. \tag{8.28}$$

Instead of minimizing over the closed linear subspace of affine functions of X, this minimizes over the closed linear subspace of linear functions of $f(X) \in \mathbb{R}^M$. All of the definitions we used for linear regression are just as sensible here (after minor modification), and results we proved are as true. See §6.5.c for more details.

8.4.e Residuals, Orthogonality, and Independence

The theoretical residual, a random variable, is defined as $R = (Y - E(Y \mid X))$. They contain the random variability of Y around its conditional mean given X. We know that for any function h with $h(X) \in L^2$, $E \, R \cdot h(X) = 0$ because R is orthogonal to L^2. This result leads to tests of whether or not parametric models are correctly specified for the conditional mean.

Exercise 8.4.32 Show that $E \, (Y \mid X) = f(X; \theta^\circ)$ for some $\theta^\circ \in \Theta$ in the parametric model $\{f(\cdot; \theta) : \theta \in \Theta\}$ iff $E(Y - f(X; \theta^\circ))h(x) = 0$ for all functions h with $h(X) \in L^2$.

There are many tests of correct specification that pick functions h and examine whether or not the empirical version of $E \, Rh(X)$ is equal to 0. The empirical version uses the estimator, $\widehat{\theta}_n = \widehat{\theta}(X_1, \ldots, X_n)$, to calculate a data-based version of R as $\{R_i = Y_i - f(X_i; \widehat{\theta}_n) : i \leq n\}$. If the parametric model is correctly specified for the conditional mean and $\widehat{\theta}_n$ is close enough to θ°, then the random variables $\{R_i h(X_i) : i \leq n\}$ should have mean 0. Techniques for checking whether or not a collection has mean 0 are well known. For example, if the function h is bounded, then $R_i h(X_i)$ should have finite variance, one can estimate it in the usual fashion, and test whether or not the average of the random variables $\{R_i h(X_i) : i \leq n\}$ is within an acceptable number of standard deviations of 0.

We now examine the difference between the orthogonality property and independence. Recall that the random variables/vectors S_1 and S_2 are independent, $S_1 \perp\!\!\!\perp S_2$, if for all g, h with $g(S_1), h(S_2) \in L^2$, $E \, g(S_1)h(S_2) = E \, g(S_1) \cdot E \, h(S_2)$.

Exercise 8.4.33 Show that S_1 and S_2 are independent iff for all g, h with $g(S_1), h(S_2) \in L^2$ and $E \, g(S_1) = 0$, $E \, g(S_1)h(S_2) = 0$.

Taking $g(r) = r$ in the previous exercise, we see that independence of R and X implies orthogonality between R and $L^2(X)$. The reverse is not true. What makes this possible is that the value of X may contain information about the distribution of R above and beyond its conditional mean being 0. Heteroskedasticity is a classical example of this.

Exercise 8.4.34 Suppose that $X \sim U[0, 1]$, $e \sim U[-1, +1]$, $X \perp\!\!\!\perp e$, and $Y = X \cdot e$. Show that $E(Y \mid X) = 0$, that $R = Y - E(Y \mid X)$ is orthogonal to $L^2(X)$, but that R and X are not independent.

Sometimes it matters that the variability of $Y - E(Y \mid X)$ is a function of X. To see why, consider what happens when it is close to 0 for some values, x_a, of X, and very high for other values, x_b. If we have observations X_i around x_a, then, if the parametric model is correctly specified, then (nearly) the only determinant of the value of Y_i is $f(X_i; \theta^\circ)$. By contrast, observations X_i around x_b tell us very little about how Y_i depends on θ°.

8.4.f Nonparametric Regression

The previous analysis of linear and nonlinear regression achieved consistency under the relatively weak assumption that some cross-moments of the μ_n converged to the corresponding moments for μ. In case of correct specification, to find the conditional expectation, one need only search a small subspace of $L^2(X)$ in solving the minimization problem, a set determined by the regression coefficients.

The whole motivation to do nonparametric regression is that one is unwilling to accept the assumption that the parametric model is correctly specified. Crudely, this means that we substitute "regression function" for "regression coefficient" and replicate the approach to linear regression. This requires a drastic increase in the size of the space we are searching over, hence requires a much stronger assumption about how μ_n converges to μ.

The substitute for the parametric assumption that there exists $\theta^\circ \in \Theta$ such that $f(X; \theta^\circ) = E(Y \mid X)$ is the nonparametric assumption that there exists a measurable function f such that $f^*(X) = E(Y \mid X)$. By Doob's theorem,

$$f^* = \arg \min_{f \in L^2(X)} \int |Y - f| \, d\mu, \tag{8.29}$$

so the assumption is only that $Y \in L^2$. Essentially, we replace the parameterized set $\{f(\cdot; \theta) : \theta \in \Theta\}$ with all functions on $\mathbb{R}^K$.

8.4.f.1 The Ideal Regression Functions

Fix the probability space $(\mathbb{R}^{K+1}, \mathcal{B}^{K+1}, \mu)$. As before, X is the square integrable random variable $\text{proj}_{1,\dots,K}$ and Y is the square integrable random variable proj_{K+1}.

The unit ball in $L^2(X)$ is the set $U = \{f \in L^2(X) : \|f\| < 1\}$, its closure is $\overline{U}$, and $\partial U = \{f \in L^2(X) : \|f\| = 1\}$ is its boundary. As in $\mathbb{R}^k$, a **cone in** $L^2(X)$ is a set $F \subset L^2(X)$ with $F = \mathbb{R}_+ \cdot F$. Replacing $\mathbb{R}_+$ with $\mathbb{R}$ allows for multiplication by both negative and positive scalars, giving **two-way cones**.

Definition 8.4.35 *A set $C \subset L^2(X)$ is a **two-way cone** if $C = \mathbb{R} \cdot C$, and it is a **compactly generated two-way cone (cgtwc)** if there exists a compact $E \subset \overline{U}$, $0 \notin E$, such that $C = \mathbb{R} \cdot E$.*

A frequently seen example of a compactly generated two-way cone (cgtwc) is a finite-dimensional subspace, W, where E can be taken to be the compact set $W \cap \partial U$. In this case, C is convex. Even though two-way cones are not compact, they contain linear subspaces and have basic existence properties.

Lemma 8.4.36 *If C is a cgtwc, then $\min_{f \in C} \|Y - f\|$ has a compact set of solutions, and if C is convex, the solution is unique.*

Proof.   It is easy to show that any cgtwc is closed. Further, any closed and norm-bounded subset is approximately flat. Take the span of an ϵ-net for the compact set $B \cdot (\overline{U} \cap C)$, where B is the norm bound. Therefore, any closed norm-bounded subset of a cgtwc is compact.

Pick any $f \in C$, define $r = \|Y - f\|$, let $C' = \{g \in L^2(X) : \|Y - g\| \le r\}$, and note that $C' \cap C$ is a closed norm-bounded subset of C. This means that the continuous function $h(f) = \|Y - f\|$ achieves its minimum on $C' \cap C$ and that this must be the minimum over all of C. The set of solutions is closed, hence compact.

Suppose now that C is convex. If $f \ne f'$ both solve the minimization problem, then $h := \frac{1}{2}f + \frac{1}{2}f'$ belongs to C and achieves a strictly lower minimum because $\|Y - f\| \le \|Y - h\| + \|h - f\|$, and $\|h - f\| > 0$. ∎

For nonparametric regression, instead finding the best-fitting function in a fixed set, one lets the set grow as n grows.

Definition 8.4.37 *Let C_n be a sequence of cgtwc's in $L^2(X)$. The associated **sequence of ideal regression functions** is $f_n^* := \arg\min_{f \in C_n} \|Y - f\|$, and the sequence C_n is **consistent** if $d(f, C_n) \to 0$ for all $f \in L^2(X)$.*

By the previous lemma, the sets f_n^* are compact and nonempty, and they are singletons when the C_n are convex. We now show that consistency of the C_n guarantees consistency of the ideal regression functions.

Lemma 8.4.38 *If C_n is a sequence of cgtwc's in $L^2(X)$ with $d(f^*, C_n) \to 0$, then $d_H(f_n^*, \{f^*\}) \to 0$, where $d_H(\cdot, \cdot)$ is the Hausdorff metric.*

Proof. $d_H(f_n^*, \{f^*\}) = d(f^*, C_n) \to 0$. ∎

8.4.f.2 Data–Based Regression Functions

The easiest kind of nonparametric regression is called series expansions. Let $\mathcal{E} = \{e_k : k \in \mathbb{N}\}$ be a sequence of linearly independent functions in $L^2(X)$ with **span**$(\mathcal{E})$ dense. Pick a sequence $\kappa(n) \uparrow \infty$ with $\kappa(n)/n \to 0$. This gives rise to a sequence $C_{\kappa(n)} = \mathbf{span}\{e_k : k \le \kappa(n)\}$ of cgtwc's with $d(f^*, C_{\kappa(n)}) \to 0$.

Definition 8.4.39 *For a sequence of cgtwc's $C_{\kappa(n)} = \mathbf{span}\{e_k : k \le \kappa(n)\}$ with $d(f^*, C_{\kappa(n)}) \to 0$, with μ_n being the empirical distribution of the data, the associated **data-based regression functions** are*

$$\widehat{f_n^*} := \arg\min_{f \in C_{\kappa(n)}} \int (y - f(\mathbf{x}))^2 \, d\widehat{\mu}_n(\mathbf{x}, y). \tag{8.30}$$

We now have three related problems:

1. $f^* = \arg\min_{f \in L^2(X)} \int (y - f) \, d\mu$, the target function,
2. $f_n^* = \arg\min_{f \in C_{\kappa(n)}} \int (y - f(\mathbf{x})) \, d\mu$, the ideal regression functions, and

3. $\widehat{f}_n^* = \arg\min_{f \in C_{\kappa(n)}} \int (y - f(\mathbf{x})) \, d\mu_n$, the data-based regression functions.

We are interested in the distance, or **total error**, $\|\widehat{f}_n^* - f^*\|$, which satisfies

$$\|\widehat{f}_n^* - f^*\| \leq \epsilon_n + a_n := \underbrace{\|\widehat{f}_n^* - f_n^*\|}_{\text{estimation error}} + \underbrace{\|f_n^* - f^*\|}_{\text{approx. error}}. \qquad (8.31)$$

The first term, ϵ_n, is called estimation error because it is the error associated with using the data-based estimate, μ_n, of the true distribution, μ. The second term is called approximation error because it has to do with how well the set C_n approximates the true regression function, f^*.

The larger is $C_{\kappa(n)}$, the smaller is a_n. The trade-off is that larger $C_{\kappa(n)}$ can lead to overfitting, which shows up as a larger ϵ_n. In order for the estimation error to go to 0, we need a condition on the DGP that guarantees that the μ_n approximate μ.

Definition 8.4.40 *The DGP is **locally convergent** if for some $\delta > 0$ and all f such that $\|f(X) - f^*(X)\| < \delta$, $\int |y - f(\mathbf{x})|^2 \, d\mu_n \to \int |y - f(\mathbf{x})|^2 \, d\mu$.*

A strong sufficient condition for the DGP to be locally convergent is that the vectors (X_i, Y_i) are iid. In the case of linear regression, instead of convergence of the integral of all f in a local neighborhood, we need only the convergence of a finite set of functions.

The trade-off may appear a bit more complicated than it is. In the iid case, we can show fairly easily that if $\kappa(n)/n \to 0$, there always exists a dense set, $D \subset L^2(X)$, with the property that if $f^* \in D$, then $\limsup(\epsilon_n + a_n)/\sqrt{n \log(n)} < \infty$, which is essentially the parametric rate one learns about in first-year econometrics.[4]

Some examples might help.

8.4.f.3 Series Expansions

Fourier series, wavelets, splines, and the various polynomial schemes specify a countable set $E = \{e_k : k \in \mathbb{N}\} \subset \partial U$ with the property that $\overline{\mathbf{sp}} \, E = L^2(X)$. What differs among the various expansions are the details of the functions e_k. The data-based regression functions, $\widehat{f}_n^*$, are functions of the form

$$\widehat{f}_n^*(x) = \sum_{k \leq \kappa(n)} \widehat{\beta}_k e_k(x). \qquad (8.32)$$

The estimators $\widehat{f}_n$ belong to $C_{\kappa(n)} := \mathbf{span}\{e_1, \ldots, e_{\kappa(n)}\}$. As this is a finite-dimension subspace of $L^2(X)$, it is a compactly generated two-way cone.

Since $\overline{\mathbf{sp}} \, E = L^2(X)$, having $\lim_n \kappa(n) = \infty$ guarantees that the $\widehat{f}_n$ can approximate any function; hence the assumptions of Lemma 8.4.38 are satisfied and the approximation error, a_n, goes to 0. To avoid overfitting and its implied biases, not letting $\kappa(n)$ go to infinity too quickly, for example, $\kappa(n)/n \to 0$ guarantees that

---

4. For the cognoscenti, $1/\sqrt{n \log(n)}$ is a tiny bit slower than the parametric rate of convergence, and what may be a bit surprising is that K, the dimensionality of X, does not appear. This is because the analysis is carried out in the space $L^2(X)$ from the first, and only properties of $L^2(X)$ are used. If one starts with the assumption that f^* was a Lipschitz function of X, one can recover, if one wishes, the curse of dimensionality.

the estimation error, ϵ_n, also goes to 0. If $\kappa(n) \to \infty$ is regarded a sequence of parameters to be estimated, for example, by cross-validation as in Definition 6.5.10 (p. 310) and $\kappa(n)$ depends on both ω and f^*, then $C_{\kappa(n)} = C_{\kappa(n)}(\omega, f^*)$.

8.4.f.4 Kernel and Locally Weighted Regression Estimators

Kernel estimators for functions on a compact domain typically begin with a function $K : \mathbb{R} \to \mathbb{R}$, supported (i.e. nonzero) only on $[-1, +1]$, having its maximum at 0 and satisfying $\int_{-1}^{+1} K(u)\, du = 1$, $\int_{-1}^{+1} u K(u)\, du = 0$, and $\int_{-1}^{+1} u^2 K(u)\, du \neq 0$. Univariate kernel regression functions are (often) of the form

$$\widehat{f}_n^*(x) = \sum_i^n \widehat{\beta}_i g(x; X_i, h_n) = \sum_i^n \widehat{\beta}_i K((x - X_i)/h_n). \tag{8.33}$$

Here $C_{\kappa(n)}(\omega, f^*) = \mathbf{span}\{K((x - X_i(\omega))/h_n : i = 1, \dots, n\}$.

Taking the kernel function to be the indicator of the interval $(-\frac{1}{2}, +\frac{1}{2})$ recovers the local averaging estimator of (8.18), the one with which we opened this section. More typically, the kernel function, $K(\cdot)$, is assumed to be smooth and to have all of its derivatives, $K^{(\alpha)}$, satisfy $\lim_{|u| \to 1} K^{(\alpha)}(u) = 0$.

The n-data points, X_i, $i = 1, \dots, n$, and the window-size or bandwidth parameter h_n, defines n nonzero functions, $g(\cdot; \theta_{i,n})$, $\theta_{i,n} = (X_i, h_n)$. The estimator, $\widehat{f}_n^*$, belongs to the span of these n functions. As established earlier, the span of a finite set of nonzero functions is a compactly generated two-way cone.

The considerations for choosing the window sizes, $h_n \to 0$, parallel those for choosing the $\kappa(n) \to \infty$ in the series expansions. They can be chosen, either deterministically or by cross-validation, so that $h_n \to 0$, to guarantee that the kernel estimators can approximate any function, driving the approximation error, a_n, to 0, but not too quickly, so that the estimation error, ϵ_n, goes to 0, avoiding overfitting.

The considerations for multivariate kernel regression functions are almost entirely analogous.

8.4.f.5 Artificial Neural Networks

Single hidden layer feedforward (slff) estimators with activation function $G : \mathbb{R} \to \mathbb{R}$ often have $E \subset L^2(X)$ defined as $E = \{\mathbf{x} \mapsto G(\gamma'\tilde{\mathbf{x}}) : \gamma \in \Gamma\}$. Here $\mathbf{x} \in \mathbb{R}^K$, $\tilde{\mathbf{x}}' = (1, \mathbf{x}')' \in \mathbb{R}^{K+1}$, and Γ is a compact subset of $\mathbb{R}^{n+1}$ with nonempty interior. The slff data-based regression functions are of the form

$$\widehat{f}_n^*(\mathbf{x}) = \sum_{k \leq \kappa(n)} \widehat{\beta}_k G(\widehat{\gamma}_k' \tilde{\mathbf{x}}), \tag{8.34}$$

where the $\widehat{\gamma}_k$ belongs to Γ. Specifically, $C_{\kappa(n)} = \{\sum_{k \leq \kappa(n)} \beta_k c_k : c_k \in E\}$ is the compactly generated two-way cone of slff estimators. This is typically not a convex cone, which makes the numerical problem of finding $\widehat{f}_n^*(\mathbf{x})$ more challenging.

As before, $\kappa(n) \to \infty$, $\kappa(n)/n \to 0$, and $\overline{\mathbf{sp}}\, E = L^2(X)$ are needed for the total error to go to 0. An easy sufficient condition on G that guarantees $\overline{\mathbf{sp}}\, E = L^2(X)$ is that G be analytic[5] and nonpolynomial. Also as before, $\kappa(n)$ may be regarded as a parameter and estimated by cross-validation.

---

5. Here the word "analytic" is being used in a different sense than in analytic functions from measure theory. G should be everywhere equal to its infinite Taylor series expansion.

8.5 ◆ Signed Measures, Vector Measures, and Densities

So far, we have looked at countably additive probability measures, $P : \mathcal{F} \to [0, 1]$. In this section we study countably additive vector measures, $\nu : \mathcal{F} \to \mathbb{R}^k$. Until the end of this section, our main focus is on the case where $k = 1$. The theory of integration of random vectors has $E\, X := (E\, X_1, \dots, E\, X_k)$, so the integration of $\mathbb{R}$-valued functions immediately delivers the integration of $\mathbb{R}^k$-valued functions. For vector measures, the relation between $k = 1$ and general k is essentially the same.

There are two results that we use in important ways later: the Radon-Nikodym theorem, 8.5.18, which delivers the existence of densities in a wide variety of situations and allows us to characterize the dual spaces of $L^p(\Omega, \mathcal{F}, P)$ for $p \in [1, \infty)$, and Lyapunov's theorem, 8.5.36, which allows us to prove the existence of Walrasian equilibria in exchange economies where the set of people is modeled as a nonatomic probability space.

8.5.a The Main Themes

Definition 8.5.1 *A **finite signed vector measure** is a bounded countably additive function $\nu : \mathcal{F} \to \mathbb{R}^k$, where countable additivity means that for all disjoint sequences E_n in $\mathcal{F}$, the sequence $\nu(E_n)$ is absolutely summable and $\nu(\cup_n E_n) = \sum_n \nu(E_n)$. ν is **nonnegative** if $\nu(E) \geq 0$ for all E and **nonpositive** if $\nu(E) \leq 0$ for all E.*

Notation 8.5.2 *If we say that ν is a "finite signed measure" without including the word "vector," we mean that $\nu(E) \in \mathbb{R}^1$.*

Example 8.5.3 *For probabilities P, Q and α, $\beta \geq 0$, $\nu(E) := \alpha P(E) - \beta Q(E)$ is a finite signed measure with αP nonnegative and $-\beta Q$ nonpositive. If there exists a measurable A such that $P(A) = 1$, $Q(B) = 1$, and $A \cap B = \emptyset$, then P and Q are **mutually singular**. If P and Q are mutually singular, then $\nu(A') \geq 0$ for all $A' \subset A$, $\nu(B') \leq 0$ for all $B' \subset B$, and (A, B) is the **Hahn decomposition** of ν. $\nu^+(E) := \nu(E \cap A)$ and $\nu^-(E) := \nu(E \cap B)$ are called the **Jordan decomposition** of ν, and $|\nu|(E) := \nu^+(E) - \nu^-(E)$ is called the **total variation of** ν.*

If ν has a density with respect to a probability, then it is easy to find a Hahn-Jordan decomposition and the total variation.

Example 8.5.4 *For $f \in L^1(\Omega, \mathcal{F}, P)$, $\nu_f(E) := \int_A f\, dP$ is a finite signed measure and f is called the **density of ν with respect to** P. To see that ν_f is a signed measure, note that the sequence $|\nu(E_n)|$ is dominated by the sequence $\int_{E_n} |f|\, dP$, which is absolutely summable, and $\nu(\cup_n E_n) = \sum_n \nu(E_n)$ is a consequence of dominated convergence. The signed measure ν_f can be divided into a positive and a negative part by setting $A = \{f > 0\}$ and $B = \{f < 0\}$, and $|\nu|(E) = \int_E |f|\, dP$ is total variation of ν.*

Taken together, the results in this section show that all finite signed measures have a Hahn-Jordan decomposition; that is, they are all of the form $\nu = \alpha P - \beta Q$

for mutually singular probabilities P, Q, and α, $\beta \geq 0$. Along the way, we prove the Radon-Nikodym theorem, which tells us that all signed measures have a density with respect to a probability.

Exercise 8.5.5 For the probability space $(\Omega, \mathcal{F}, P) = ((0, 1], \mathcal{B}, \lambda)$, define $\nu(E) = \int_E (x - \frac{1}{2}) \, d\lambda(x)$.

1. Find E, $F \in \mathcal{B}$, $E \subset F$ such that $\nu(E) < 0$ but $\nu(F) > 0$.
2. Find the Hahn-Jordan decomposition of ν.
3. Find the total variation of ν.

Exercise 8.5.6 Let $\Omega = \{1, 2, 3, \ldots\}$, $\mathcal{F} = \mathcal{P}(\Omega)$. For $E \in \mathcal{F}$, define $\nu(E) = \sum_{n \in E} \left(\frac{-1}{2} \right)^n$.

1. Show that ν is a signed measure.
2. Give $E \subset F$ such that $\nu(E) > 0$ but $\nu(F) < 0$.
3. Express ν as $\alpha P - \beta Q$ and $\alpha' P' - \beta' Q'$ for two different pairs of probabilities P, Q and P', Q'.
4. Give the Hahn-Jordan decomposition of ν.
5. Give the total variation of ν.

8.5.b Absolute Continuity and the Radon–Nikodym Theorem

Definition 8.5.7 *A nonnegative ν is **absolutely continuous with respect to** P, or P **dominates** ν, written $\nu \ll P$, if $[P(E) = 0] \Rightarrow [\nu(E) = 0]$.*

There is an alternative ϵ-δ definition of absolute continuity.

Lemma 8.5.8 *For a nonnegative ν, $\nu \ll P$ iff $(\forall \epsilon > 0)(\exists \delta > 0) \, [\, [P(A) < \delta] \Rightarrow [\nu(A) < \epsilon] \,]$.*

Proof. Suppose that $\nu \ll P$ but that there exists $\epsilon > 0$ and A_n with $P(A_n) \leq 1/2^n$ such that $\nu(A_n) \geq \epsilon$. We know that $P([A_n \text{ i.o.}]) = 0$ and $B_m \downarrow [A_n \text{ i.o.}]$, where $B_m := \cup_{n \geq m} A_n$. Countable additivity of ν implies that $\nu([A_n \text{ i.o.}]) \geq \epsilon$. On the other hand, if $P(E) = 0$, then $\nu(E) < \epsilon$ for all $\epsilon > 0$. ∎

Exercise 8.5.9 For nonnegative ν, show that if $\nu \ll P$, then $A \mapsto \nu(A)$ is ρ_P-continuous (where ρ_P is the probability metric $\rho_P(E, F) = P(E \Delta F)$).

In this section, we will see that all finite signed measures have densities with respect to a probability; hence, using their densities, $\mathcal{F}$, they can be divided into a positive and a negative part. In terms of the $\alpha P - \beta Q$ representation, $\{f = 0\}$ is the set on which P and Q cancel each other out.

Theorem 8.5.10 (Radon–Nikodym for Nonnegative Measures) *If ν is a nonnegative measure on $(\Omega, \mathcal{F})$ and $\nu \ll P$, then ν has a density with respect to P; that is, there exists an integrable $f : \Omega \to \mathbb{R}_+$ such that $\nu(A) = \int_A f \, dP$ for all $A \in \mathcal{F}$ and f is unique up to a set of P-measure 0.*

Proof. Define the probability η by $\eta(A) = \nu(A)/\nu(\Omega)$ and the probability Q by $Q(A) = \frac{1}{2}\eta(A) + \frac{1}{2}P(A)$. It is sufficient to show that η has a density with respect to P. Consider the Hilbert space $L^2 = L^2(\Omega, \mathcal{F}, Q)$, Define the linear function $\Phi : L^2 \to \mathbb{R}$ by $\Phi(g) = \frac{1}{2} \cdot \int g \, dP$. Φ is continuous because $|\Phi(g_n) - \Phi(g)| = \frac{1}{2} |\int (g_n - g) \, dP| \le \frac{1}{2} \int |g_n - g| \, dP \le \int |g_n - g| \, d(\frac{1}{2}\eta + \frac{1}{2}P) \le \|g_n - g\|_2$.

By Riesz-Fréchet for L^2 (Theorem 8.4.13, p. 478), there exists an $h_0 \in L^2$ such that for all $g \in L^2$, $\Phi(g) = \int g \cdot h_0 \, d(\frac{1}{2}\nu + \frac{1}{2}P)$. Since $\Phi(g) = \frac{1}{2} \int g \, dP$,

$$\tfrac{1}{2} \int g \, dP = \tfrac{1}{2} \int g \cdot h_0 \, d\nu + \tfrac{1}{2} \int g \cdot h_0 \, dP, \text{ that is, } \int g(1 - h_0) \, dP = \int g \cdot h_0 \, d\nu$$

(8.35)

for all $g \in L^2$. Let $B = \{h_0 \le 0\}$, $1_B \in L^2$, and $1_B \le 1_B(1 - h_0)$. We observe that $P(B) = 0$ because $P(B) = \int 1_B \, dP \le \int 1_B(1 - h_0) \, dP = \int 1_B \cdot h_0 \, d\eta \le 0$. Since $\nu \ll P$, $\eta(B) = 0$. Define $f = 1_{B^c} \frac{1-h_0}{h_0}$. To check $\eta(A) = P_f(A) = \int_A f \, dP$ for any $A \in \mathcal{F}$, $\eta(A) = \int 1_A \, d\eta = \int 1_{B^c} \frac{1}{h_0} 1_A h_0 \, d\eta = \int 1_{B^c} \frac{1}{h_0} 1_A (1 - h_0) \, dP = \int_A f \, dP$, which follows by taking $g = 1_{B^c} \frac{1}{h_0} 1_A$ in (8.35). ∎

Exercise 8.5.11 What about uniqueness in the previous proof?

The Radon-Nikodym theorem delivers the existence of conditional expectations for all $Y \in L^1(\Omega, \mathcal{F}, P)$, not just for $Y \in L^2(\Omega, \mathcal{F}, P)$.

Corollary 8.5.12 *If $Y \in L^1(\Omega, \mathcal{F}, P)$ and $\mathcal{G}$ is a sub-σ-field, then there exists $R \in L^1(\mathcal{G})$ with $\int_A Y \, dP = \int_A R \, dP$ for all $A \in \mathcal{G}$.*

Proof. For $A \in \mathcal{G}$, define $\nu^+(A) = \int_A Y^+ \, dP$ and $\nu^-(A) = \int_A Y^- \, dP$. The ν^+ and ν^- are nonnegative measures on $\mathcal{G}$, both absolutely continuous with respect to P. Let f^+ and f^- be their densities with respect to P so that $\int_A f^+ \, dP = \int_A Y^+ \, dP$ and $\int_A f^- \, dP = \int_A Y^- \, dP$. Letting $f = f^+ - f^-$ yields $\int_A f \, dP = \int_A Y \, dP$. ∎

There is an alternative method for finding $E(Y \mid \mathcal{G})$ for $Y \in L^1(\Omega, \mathcal{F}, P)$.

Example 8.5.13 *For $Y \in L^2(\Omega, \mathcal{F}, P)$, we defined $E(Y \mid \mathcal{G})$ as a projection in L^2. Truncating $Y^+ \in L^1(\Omega, \mathcal{F}, P)$ above by $n \to \infty$ gives a nondecreasing sequence $E(Y_n^+ \mid \mathcal{G})$ with bounded expectation, so $E(Y_n^+ \mid \mathcal{G}) \uparrow R^+$ for some R^+ in $L^1(\mathcal{G})$. Dominated convergence implies that $\int_A Y^+ \, dP = \int_A R^+ \, dP$ for all $A \in \mathcal{G}$. Repeating this process with Y^- gives R^- with the parallel properties. Defining $R = R^+ - R^-$ gives $R \in L^1(\mathcal{G})$ with $\int_A Y \, dP = \int_A R \, dP$ for all $A \in \mathcal{G}$.*

8.5.c The Jordan and the Hahn Decompositions of Finite Signed Measures

For a function $f : X \to \mathbb{R}$, $f^+ = \frac{1}{2}(|f| + f)$ and $f^- = \frac{1}{2}(|f| - f)$, so that $|f| = f^+ + f^-$. The Jordan decomposition of a finite signed measure defines $|\nu|$ and uses this logic to define the Jordan decomposition of ν into a positive and a negative part. The Radon-Nikodym theorem then delivers the Hahn decomposition.

Definition 8.5.14 *If $\nu : \mathcal{F} \to \mathbb{R}$ is a finite signed measure, then the **absolute variation of ν** or **total variation of ν** is denoted, for $E \in \mathcal{F}$, by $|\nu|(E)$ and defined by $|\nu|(E) := \sup \sum_i |\nu(F_i)|$, where the supremum is taken over finite disjoint* *collections of measurable subsets of E. The **positive part** of ν is $\nu^+ := \frac{1}{2}(|\nu| + \nu)$ and the **negative part** is $\nu^- := \frac{1}{2}(|\nu| - \nu)$.*

The following implies that ν^+ and ν^- are finite, countably additive nonnegative measures.

Lemma 8.5.15 *$|\nu|(\Omega) \leq 2 \sup_{E \in \mathcal{F}} |\nu(E)|$ and $|\nu|$ is a nonnegative countably additive measure.*

Proof. Let $M = \sup_{E \in \mathcal{F}} |\nu(E)|$. For any finite disjoint set of F_i,

$$\sum_i |\nu(F_i)| = \sum_{i:\nu(F_i) \geq 0} \nu(F_i) - \sum_{j:\nu(F_j)<0} \nu(F_i) \leq 2M.$$

$|\nu|$ is easily seen to be finitely additive.[6] Let E_n be a disjoint sequence in $\mathcal{F}$, $E = \cup_n E_n$. By additivity, $\forall m$,

$$|\nu|(E) \geq |\nu|(\cup_{n \leq m} E_n) = \sum_{n \leq m} |\nu|(E_n) \tag{8.36}$$

so that $|\nu|(E) \geq \sum_n |\nu|(E_n)$. To show that $|\nu|(E) \leq \sum_n |\nu|(E_n)$, let $\{F_i : i \leq I\}$ be a finite disjoint collection of measurable subsets of $E = \cup_n E_n$:

$$\sum_{i \leq I} |\nu(F_i)| = \sum_{i \leq I} |\nu(\cup_n (E_n \cap F_i))| \ (\text{because } E = \cup_n E_n)$$

$$= \sum_{i \leq I} |\sum_n \nu(E_n \cap F_i)| \ (\text{because } \nu \text{ is countably additive})$$

$$\leq \sum_n \sum_{i \leq I} |\nu(E_n \cap F_i)| \ (\text{because } \nu(E_n \cap F_i) < 0 \text{ is possible})$$

$$\leq \sum_n |\nu|(E_n) \ (\text{by definition of } |\nu|(E_n)).$$

Taking the supremum on the left-hand side yields $|\nu|(E) \leq \sum_n |\nu|(E_n)$. ∎

Definition 8.5.16 *A finite signed measure ν is **absolutely continuous with respect to** P or P **dominates** ν, written $\nu \ll P$, if $|\nu| \ll P$, that is, if $[P(E) = 0] \Rightarrow [|\nu|(E) = 0]$.*

Exercise 8.5.17 Show that $|\nu| \ll P$ iff $[P(E) = 0] \Rightarrow [|\nu(E)| = 0]$, give an alternative ϵ-δ definition of $\nu \ll P$, and show that if $\nu \ll P$, then $A \mapsto \nu(A)$ is ρ_P-continuous.

Theorem 8.5.18 (Radon–Nikodym for Signed Measures) *If ν is a finite signed measure that is absolutely continuous with respect to P, then it has a density with respect to P.*

Proof. ν^+ and ν^- are nonnegative, hence have densities f^+ and f^-. The density of ν is $f := f^+ - f^-$. ∎

---

6. Indeed, $|\nu|$ is the smallest finitely additive measure satisfying $|\nu|(E) \geq |\nu(E)|$ for all measurable E.

Theorem 8.5.19 (Hahn Decomposition) *If ν is a finite signed measure, then there exist measurable A, B that partition Ω with the property that for all measurable E, $\nu^+(E) = \nu(E \cap A)$ and $\nu^-(E) = \nu(E \cap B)$.*

Proof. If ν is the 0 measure, any partition will do. Otherwise, define the probability P by $P(E) = \frac{|\nu|(E)}{|\nu|(\Omega)}$ so that $|\nu| \ll P$. Let f be the derivative of ν with respect to P and set $A = \{f \geq 0\}$ and $B = \{f < 0\}$. ∎

Exercise 8.5.20 Show that for any $E \in \mathcal{F}$, $\nu^+(E) = \sup\{\nu(F) : F \subset E\}$ and $\nu^-(E) = \inf\{\nu(F) : F \subset E\}$.

Finally, to see how this all connects back to $\nu = \alpha P - \beta Q$ with P and Q mutually singular, suppose that ν takes on both negative and positive values and set $\alpha = \nu^+(\Omega)$, $P(E) = \frac{1}{\alpha}\nu^+(E)$, $\beta = \nu^-(\Omega)$, and $Q(E) = \frac{1}{\beta}\nu^-(E)$.

8.5.d The Density of One Signed Measure with Respect to Another

So far, we have only allowed integration with respect to finite signed measures that are probabilities.

Definition 8.5.21 *If $\nu = \alpha P$ for $\alpha \in \mathbb{R}$ and P a probability, then $\int g \, d\nu :=$ $\alpha \int g \, dP$, and g is ν-integrable if $\alpha = 0$ or if g is P-integrable. If $\nu = \nu^+ - \nu^-$ is the Jordan decomposition of a finite signed measure ν, then $\int g \, d\nu := \int g \, d\nu^+ - \int g \, d\nu^-$ so long as both integrals are finite; otherwise g is nonintegrable. For $p \in [1, \infty)$, the space $L^p(\Omega, \mathcal{F}, \nu)$ is the space of measurable functions with norm $\|g\|_{p,\nu} := \left(\int |g|^p \, d|\nu|\right)^{1/p} < \infty$.*

To define the norm, we integrate against the total variation $|\nu|$ rather than against ν to avoid the norm being 0 while $g \neq 0$. Since all that is involved is rescaling a probability, the $L^p(\Omega, \mathcal{F}, \nu)$ spaces have the same properties as the spaces $L^p(\Omega, \mathcal{F}, P)$.

Exercise 8.5.22 Give an equivalent formulation of $\int g \, d\nu$ using densities.

So far, we have only allowed densities with respect to probabilities.

Definition 8.5.23 *For finite signed measures ν and η, η **dominates** ν, written, $\nu \ll \eta$, if $[|\eta|(E) = 0] \Rightarrow [|\nu|(E) = 0]$, η and ν are **mutually absolutely continuous**, written $\eta \sim \nu$, if $\nu \ll \eta$ and $\eta \ll \nu$, and η and ν are **mutually singular** if there exists measurable A, B such that $|\eta|(A) = |\eta|(\Omega)$, $|\nu|(B) = |\nu|(B)$, and $A \cap B = \emptyset$.*

Exercise 8.5.24 Show that $\ll$ is transitive and that $\sim$ is an equivalence relation.

Exercise 8.5.25 Show that if $\nu \ll \eta$, then ν has a density with respect to η. [This involves rescaling and the Radon-Nikodym theorem.]

Notation 8.5.26 *If f is the density of ν with respect to η, we write $f = \frac{d\nu}{d\eta}$.*

The reason for this notation is the chain rule: if $\nu \ll \eta$ and f is the density, then $\int g \, d\nu = \int gf \, d\eta = \int g \frac{d\nu}{d\eta} \, d\eta$. The next two exercises show that we can remember this fact by canceling the $d\eta$'s.

Exercise 8.5.27   Show that if $\nu \sim \eta$, f is the density of ν with respect to η, and g is the density of η with respect to ν, then $fg = 1$.

Exercise 8.5.28   Show that if $\nu \ll \eta \ll \gamma$, $f = \frac{d\nu}{d\eta}$, and $g = \frac{d\eta}{d\gamma}$, then $fg = \frac{d\nu}{d\gamma}$.

8.5.e   The Lebesgue Decomposition of Probabilities on $\mathbb{R}$

Borel probabilities on $\mathbb{R}$ are economists' most frequently used set of probabilities. From Definition 7.5.22 (p. 406), the **Lebesgue measure of a set** E is $\sum_{n=-\infty}^{\infty} P_n(E)$, where P_n is the uniform distribution on the interval $(n, n+1]$. It is possible that the Lebesgue measure is ∞. If it is 0, we say that E is a **Lebesgue null set**.

Definition 8.5.29   *A finite nonnegative measure μ on $(\mathbb{R}, \mathcal{B}_\mathbb{R})$ is **absolutely continuous with respect to the Lebesgue measure** if $\mu(E) = 0$ for every E having Lebesgue measure 0, it is **purely atomic** if there is some countable E with $\mu(E) = \mu(\mathbb{R})$, and it is **singular with respect to the Lebesgue measure** if it has no atoms and there exists a measurable E having Lebesgue measure 0 with $\mu(E) = \mu(\mathbb{R})$.*

We have seen a number of singular probabilities.

1. In Example 7.9.8, we defined the Cantor set as the image of $X = \{0, 2\}^{\mathbb{N}}$ under the one-to-one mapping $r(x) = \sum_n \frac{x_n}{3^n}$, $x = (x_1, x_2, \ldots) \in \{0, 2\}^{\mathbb{N}}$. Giving X the necessarily nonatomic uniform distribution, the cdf of the one-to-one random variable $x \mapsto r(x)$ has a continuous cdf, F_r, and the associated probability on $\mathbb{R}$ is singular with respect to Lebesgue measure.

2. If X_n is a set of iid random variables with $P(X_n = 1) = p$, $P(X_n = 0) = 1 - p$, $0 < p < 1$, and $p \neq \frac{1}{2}$, then the random variable $R := \sum_n \frac{X_n}{2^n}$ has a continuous strictly increasing cdf on the interval $[0, 1]$, and the associated probability is singular with respect to Lebesgue measure. (To see why, note that the strong law of large numbers says that with probability 1, the dyadic expansion of $R(\omega)$ has, in the limit, p of its terms being equal to 1. It also says that if we pick $x \in (0, 1]$ at random according to the uniform distribution, then the proportion of 1's in the dyadic expansion of x is, in the limit, $\frac{1}{2}$ on a set of x having probability 1.)

From the Radon-Nikodym theorem applied to each of the intervals $(n, n+1]$ with the uniform distribution, if μ is absolutely continuous with respect to Lebesgue measure, then it has a density $f \geq 0$, defined on all of $\mathbb{R}$, such that $\mu(E) = \int_E f(x)\, dx$. Here is the historic root of the name "absolute continuity" for $\nu \ll P$.

Theorem 8.5.30   *A probability μ is absolutely continuous with respect to Lebesgue iff the cdf $F_\mu(r) = \mu((-\infty, r])$ is an absolutely continuous nondecreasing function; that is, for all $\epsilon > 0$, there exists $\delta > 0$ such that for all r, $F_\mu(r + \delta) - F_\mu(r) < \epsilon$.*

The proof uses a variant of the argument we find in Scheffé's theorem, 7.5.29.

Proof.   Suppose that μ is absolutely continuous with respect to Lebesgue measure and let f be its density, so that $F_\mu(r) = \int_{(-\infty, r]} f(x)\, dx$. We must show that for every $\epsilon > 0$, there exists $\delta > 0$ such that for all r, $F_\mu(r + \delta) - F_\mu(r) < \epsilon$. Pick M with $\int_{[-M, +M]} f(x) \, dx > 1 - \epsilon/2$, so that F_μ can vary by at most $\epsilon/2$ on $[-M, +M]^c$. Let Q_M be the uniform distribution on $[-M, +M]$, so that for any g, $\int_{[-M, +M]} g(x) \, dQ_M(x) = \frac{1}{2M} \int_{-M}^{+M} g(x) \, dx$. By dominated convergence, there exists $\eta > 0$ such that $[Q_M((a, b]) < \eta] \Rightarrow [\int_{(a,b]} f(x) \, dQ_M(x) < \epsilon/2]$. Pick $\delta > 0$ such that $Q_M((a, a + \delta]) < \eta$, for example, $\delta = \eta/2M$.

Now suppose that $F = F_\mu$ is absolutely continuous. To begin with, suppose that $F(0) = 0$ and $F(1) = 1$, that is, that $\mu([0, 1]) = 1$, and let P be the uniform distribution on $[0, 1]$. If $F(x + \delta) - F(x) < \epsilon$ for all x, then for all $A, B \in \mathcal{B}^\circ$, the set of finite disjoint unions of interval subsets of $[0, 1]$, $[P(A \triangle B) < \delta] \Rightarrow [\mu(A \triangle B) < \epsilon]$. This means that $A \mapsto \mu(A)$ is ρ_P-uniformly continuous on $\mathcal{B}^\circ$, a dense subset of $\mathcal{B}$. Therefore, $A \mapsto \mu(A)$ is ρ_P-uniformly continuous on $\mathcal{B}$, so that $[P(E) < \delta] \Rightarrow [\mu(E) < \epsilon]$. The extension to noncompactly supported μ follows the usual pattern. ∎

Exercise 8.5.31 Show that ess sup $f = L$ in the previous theorem iff F_μ has Lipschitz constant L.

Theorem 8.5.32 (Lebesgue Decomposition) *If μ is a probability on $\mathbb{R}$, then μ has a unique expression as the sum of three nonnegative measures, $\mu = \mu_a + \mu_{ac} + \mu_s$, where μ_a is purely atomic, μ_{ac} is absolutely continuous with respect to Lebesgue measure, and μ_s is singular with respect to Lebesgue measure.*

Combined with the previous theorem, this means that any probability on $\mathbb{R}$ has three parts: atoms, which are easy to understand; a density with respect to Lebesgue measure, which is easy to understand; and a weird part. In this context, "singular" is a mathematical synonym for "weird."

Proof. The uniqueness is immediate. Let A be the countable set (possibly empty) of atoms of μ, define $\mu_a(E) = \mu(A \cap E)$, and define the nonatomic part of μ as $\mu_{na} = \mu - \mu_a$.

Partially order the class $\mathcal{N}$ of Lebesgue null sets by $A \preceq B$ if $\mu_{na}(A) \le \mu_{na}(B)$. Let $\mathcal{C}$ be a chain of Lebesgue null sets in this partial order and set $\delta = \sup\{\mu_{na}(A) : A \in \mathcal{C}\}$. To show that $\mathcal{C}$ has an upper bound in $\mathcal{N}$, let A_n be a sequence in $\mathcal{C}$ with $\mu_{na}(A_n) \uparrow \delta$ and note that $A := \cup_n A_n$ is an upper bound. By Zorn's lemma, $\mathcal{N}$ has a maximal element, which we call S. Define $\mu_s(A) = \mu_{na}(A \cap S)$ and $\mu_{ac} = \mu_{na} - \mu_s$. Since S is a Lebesgue null set, μ_s is singular with respect to Lebesgue measure. If μ_{na} is not absolutely continuous, then there exists a Lebesgue null set E with $\mu_{na}(E) > 0$. Setting $S' = S \cup E$ shows that this contradicts the maximality of S. ∎

8.5.f Nonatomic Vector Measures

The following generalizes probabilities being nonatomic.

Definition 8.5.33 *A measurable A is an **atom of a vector measure** v if for all measurable $A' \subset A$, $v(A') = 0$ or $v(A') = v(A)$. The v is **nonatomic** if it has no atoms.*

The Radon-Nikodym theorem allows us to characterize nonatomic vector measures in a useful fashion.

Lemma 8.5.34   *A vector measure $v : \mathcal{F} \to \mathbb{R}^k$ is nonatomic iff there is a nonatomic probability P on $\mathcal{F}$ and a measurable $f : \Omega \to \mathbb{R}^k$ such that $v(E) = \int_E f \, dP$ for all measurable E.*

Proof. $v(E) = (v_1(E), \ldots, v_k(E)) \in \mathbb{R}^k$.   Define $P(E) = \sum_i |v_i|(E) / \sum_i |v_i|(\Omega)$. Each v_i satisfies $v_i \ll P$, so each has a density f_i with respect to P. For measurable E, $v(E) = \int f \, dP$.   ∎

Lemma 8.5.35   *If $k = 1$ and v is nonatomic, then $\{v(E) : E \in \mathcal{F}\}$ is the interval $[v^-(\Omega), v^+(\Omega)]$.*

Proof.   Let (A, B) be the Hahn decomposition of v, use Theorem 7.6.23(4) on the sets A and B, and appeal to the intermediate value theorem.   ∎

Lyapunov's theorem is the name of the generalization of this result to the case of general k. It has very useful implications for models of exchange economies in which the set of agents is taken to be a nonatomic probability space.

Theorem 8.5.36 (Lyapunov)   *If v is a nonatomic vector measure, then $\{v(E) : E \in \mathcal{F}\}$ is a compact convex set.*

Proof.   By Lemma 8.5.34, it is sufficient to show that for nonatomic P and integrable $f : \Omega \to \mathbb{R}^k_+$, the set $R(f) := \{\int 1_E(\omega) \cdot f(\omega) \, dP(\omega) : E \in \mathcal{F}\}$ is compact and convex. We prove this first for simple f and then take limits. Note that $R(f)$ must be bounded, indeed must be a subset of $\times_{i=k}^k [-\int |f_i|, \, dP + \int |f_i| \, dP]$.

Step 1: f is an indicator function. If $f = \mathbf{x} \cdot 1_A$, $\mathbf{x} \in \mathbb{R}^k$, then for any E, $\int_E f \, dP = \mathbf{x} \cdot P(E \cap A)$. Since the set of possible $P(E \cap A)$ is the set $[0, P(A)]$, $R(1_E) = [0, \mathbf{x} \cdot P(A)]$.

Step 2: f is simple. Suppose that $f = \sum_k \mathbf{x}_k 1_{A_k}$ and that the A_k form a partition of Ω. Any measurable E is the disjoint union of the sets $E \cap A_k$. Since P is nonatomic, the set of possible $P(E \cap A_k)$ is the set $[0, P(A_k)]$. This allows us to express $R(f)$ as the necessarily compact and convex sum of sets, $\sum_k [0, \mathbf{x}_k \cdot P(A_k)]$.

Step 3: Taking limits. Let S be the subset of $g : \Omega \to \mathbb{R}^k$ with each $g_i(\omega)$ belonging to the interval $[-|f_i|(\omega), +|f_i|(\omega)]$, that is, the set vector dominated by $|f|$. Give S the metric $d_1(f, g) = \|f - g\|_1$ and give the compact convex subsets of $\mathbb{R}^k$ the Hausdorff metric. The mapping $f \mapsto R(f)$ maps into a complete metric space and is uniformly continuous on the dense set of simple functions in S, hence has a unique uniformly continuous extension to all of S.   ∎

There is a proof that only takes a couple of lines, provided one has the Krein-Milman theorem for compact convex subsets of topological vector spaces and Alaoglu's theorem on the weak* compactness of norm-bounded subsets of a Banach space. We give these theorems and the proof of Lyapunov's theorem in Chapter 10. At the end of this chapter, there is an alternative proof, which is, essentially, based on no more than the intermediate value theorem.

Exercise 8.5.37 Taking $\mathcal{G} = f^{-1}(\mathcal{B}^k)$, show that there is no loss in assuming that $(\Omega, \mathcal{F}, P)$ is countably generated in Lyapunov's theorem. Going further, show that there is no loss in assuming that $(\Omega, \mathcal{F}, P) = ((0, 1], \mathcal{B}, \lambda)$.

Exercise 8.5.38 Let $f(\omega) = (\omega, 2\omega)$ for $\omega \in ((0, 1], \mathcal{B}, \lambda)$. Give $R(f) \subset \mathbb{R}^2$.

Exercise 8.5.39 Let $f_1(\omega) = 1_{(0, \frac{1}{2}]}(\omega)$, $f_2(\omega) = \omega$ for $\omega \in ((0, 1], \mathcal{B}, \lambda)$. Give $R(f) \subset \mathbb{R}^2$. [Start with E being an interval, proceed to $E \in \mathcal{B}^\circ$.]

8.6 ◆ Measure Space Exchange Economies

The nonatomic probability spaces are especially useful as models of populations so large that any individual has a negligible effect on the aggregate.

8.6.a The Basics

We interpret each ω as an agent in an economy or as a player in a game. We assume that for all $\omega \in \Omega$, $\{\omega\} \in \mathcal{F}$. A subset $E \in \mathcal{F}$ is a **coalition** of people, and we interpret $P(E)$ as the proportion of the people in the economy that belong to the coalition E. Nonatomicity neatly captures the notion that each player is insignificant when compared to the whole, since $P(\{\omega\})$ must be equal to 0. This is the notion that every person is a very small proportion of the economy. Indeed, for any finite set E, $P(E) = 0$.

In the finite-person exchange economies we discussed in Chapter 1, people take prices as given and choose their most preferred consumption in the set of affordable bundles. They do this and do not take into account their own effect on the prices. This seems a reasonable approximation when there are many people. It is an exact statement when we model people as points in a nonatomic probability space. We now turn to the definition of an exchange economy with a probability space as our model of the set of people in the economy.

All of the assumptions in the following are in force for the rest of this section.

Definition 8.6.1 *An **assignment** is a measurable function* $\mathbf{x} : \Omega \to \mathbb{R}_+^k$.

The interpretation is that person ω receives $\mathbf{x}(\omega)$ if the assignment is $\mathbf{x}$.

Definition 8.6.2 *The **endowment** or **initial assigment** is an integrable assigment function* $\mathbf{i} : \Omega \to \mathbb{R}_+^k$ *such that* $\int \mathbf{i} \, dP > 0$. *A **trade** or **final assigment** is an assignment* $\mathbf{x}$ *such that* $\int \mathbf{x} \, dP = \int \mathbf{i} \, dP$.

Note that changing the final assignment of any finite set of people leaves the two integrals unchanged because any finite set of people adds up to 0 as a proportion of the entire economy.

Definition 8.6.3 *Preferences*, $\succ$, *are a mapping* $\omega \mapsto \succ_\omega$ *from* Ω *to the binary relations*.

Here $\succ_\omega$ is interpreted as a strict preference relation. The relevant assumptions on $\succ$ are:

1. monotonicity, for all ω, $[x > y] \Rightarrow [x \succ_\omega y]$,
2. continuity, for all ω and $y \in \mathbb{R}^k_+$, $\{x : x \succ_\omega y\}$ and $\{x : y \succ_\omega x\}$ are open subsets of $\mathbb{R}^k_+$, and
3. measurability, for any assignments $\mathbf{x}$, $\mathbf{y}$, $\{\omega : \mathbf{x}(\omega) \succ_\omega \mathbf{y}(\omega)\} \in \mathcal{F}$.

Unlike what we have done in the past, we do *not* assume that the preferences are complete. One story behind incomplete preferences is that we inhabit different social roles at different times and that our preferences reflect the role we are playing. We sometimes think of ourselves as members of a family and choose, for instance, sensible, nutritious food, and we sometimes think of ourselves as being as we were years ago, feed the kids Twinkies and sup on nachos and beer. In this case, the preference $\succ$ need not be complete, and we define the demand set for person ω as the set of consumption vectors that are not strictly beaten by any affordable consumption vector, that is, $\mathbf{x}^*(\omega)(p, \mathbf{i}(\omega)) = \{y \in \mathbb{R}^k_+ : \forall z \in \mathbb{R}^k_+, \ [p \cdot z \leq p \cdot \mathbf{i}(\omega)] \Rightarrow \neg[z \succ y]\}$. When the preferences are complete, this is the usual demand set, but it allows many more kinds of behavior.

Exercise 8.6.4   Define $(u, v) : \mathbb{R}^k_+ \to \mathbb{R}^2$ by

$$(u(\mathbf{x}), v(\mathbf{x})) = (u(x_1, x_2), v(x_1, x_2)) = (x_1 \cdot x_2, \min\{x_1, x_2\}),$$

and $\mathbf{x} \succ_\omega \mathbf{y}$ if $(u(\mathbf{x}), v(\mathbf{x})) > (u(\mathbf{y}), v(\mathbf{y}))$.

1. Find $\mathbf{x}^*(\omega)((8, 6), (11, 11))$ and $\mathbf{x}^*(\omega)((8, 5), (10, 10))$.
2. Show that no rational preference relation can give rise to these demand sets.

Gathering these together leads to the following.

Definition 8.6.5   *An economy $\mathcal{E}$ is defined as $((\Omega, \mathcal{F}, P), \mathbf{i}, \succ)$, where $(\Omega, \mathcal{F}, P)$ is nonatomic, $\mathbf{i}$ is an endowment, and preferences $\succ$ are monotonic, continuous, and measurable.*

Definition 8.6.6   *A **coalition** is a set $E \in \mathcal{F}$. It is a **null coalition** if $P(E) = 0$. An allocation $\mathbf{y}$ **dominates** $\mathbf{x}$ **via** E if $\mathbf{y}(\omega) \succ_\omega \mathbf{x}(\omega)$ for each $\omega \in E$ and $\int_E \mathbf{y} \, dP = \int_E \mathbf{x} \, dP$.*

The idea is that the coalition E, if it had $\mathbf{x}$, could secede from society, take its holdings, $\int_E \mathbf{x} \, dP$, are reallocate them among themselves, $\int_E \mathbf{y} \, dP = \int_E \mathbf{x} \, dP$, and make everyone better off, $\mathbf{y}(\omega) \succ_\omega \mathbf{x}(\omega)$ for each $\omega \in E$.

Definition 8.6.7   *The **core** of $\mathcal{E}$ is the set of final allocations that are not dominated by any nonnull coalition.*

There is a strong element of kindergarden emotional logic at work here—if we don't like what's going on, we'll take our toys and play by ourselves. At a more sophisticated level, core allocations go after collective action logic in a fashion that is very strong in one way and perhaps not as strong as one would like in another.

1. The collective action logic argument for the core is strong in the sense that something is in the core if it is not possible to gather *any* group of people who can successfully object, and checking through all possible groups of people is a difficult problem.

2. The collective action logic is not so strong in the sense that the presumption is that the group will completely secede from society and only trade among themselves.

So, we check many groups, but we only allow them the option of complete secession.

Definition 8.6.8 *A **price** is a vector $\mathbf{p} \neq 0$ in $\mathbb{R}_+^k$. An **equilibrium for** $\mathcal{E}$ is a price vector–final allocation pair, $(\mathbf{p}, \mathbf{x})$, such that for almost all ω, $\mathbf{x}(\omega)$ is maximal for ω in ω's budget set $B(\omega) = \{\mathbf{x} \in \mathbb{R}_+^k : \mathbf{p} \cdot \mathbf{x} \leq \mathbf{x} \cdot \mathbf{i}(\omega)\}$. A final allocation $\mathbf{x}$ is an **equilibrium allocation** if there exists a price vector $\mathbf{p}$ such that $(\mathbf{p}, \mathbf{x})$ is an equilibrium.*

8.6.b The Results

There are two results that we look at.

Theorem 8.6.9 *Under the assumptions given earlier, an equilibrium exists.*

This is comforting; our theory is about something. The core equivalence theorem is as follows.

Theorem 8.6.10 (Aumann's Core Equivalence) *The core and the set of equilibrium allocations are identical.*

This is really quite amazing. It says that if trades are coordinated by prices, then there is no group that can reorganize its own holdings so as to make itself better off. In combination with the first result, it says that our models of the economy can be so organized.

It is clearly too much to conclude that this is true about the real world. As we will see in examples of games with a continuum of agents, when one group's actions affect the welfare of other groups, coordination problems arise and equilibria no longer have this property. In this case, coordination can improve everyone's welfare and is what we mean by collective action in such cases. However, we often go looking for coordination mechanisms that replicate what prices would do if they were available. We think that it is generally the case that the actions of large groups affect our own welfare, but are not willing to completely abandon the coordination insight contained in the core equivalence theorem. In some settings, we think that prices work wonderfully.

8.6.c Proof of Equilibrium Existence

Full presentation and understanding of the results that go into these proofs requires another book. We point out what we are not proving as we pass by.

The demand correspondence is $\omega \mapsto \mathbf{x}^*(\mathbf{p})(\omega)$. If this is a singleton-valued correspondence for all ω, as it would be if preferences were rational and strictly convex, then **mean demand** is $\mathbf{t}(\mathbf{p}) = \int \mathbf{x}^*(\mathbf{p})(\omega) \, dP(\omega)$. If this is not a singleton-valued correspondence, then demand at prices $\mathbf{p}$ is any $\gamma(\omega) \in \mathbf{x}^*(\mathbf{p})(\omega)$, that is, where $\gamma(\cdot)$ is a selection from $\mathbf{x}^*(\mathbf{p})(\cdot)$. In this case, mean demand is defined as the integral of the correspondence.

Definition 8.6.11 *A nonempty valued correspondence* $\Psi : \Omega \twoheadrightarrow \mathbb{R}^k$ *is **integrable** if there exists an integrable function* $\psi : \Omega \to \mathbb{R}^k$ *with* $P(\{\omega : \psi(\omega) \in \Psi(\omega)\}) = 1$, *in which case the **integral of a correspondence** $\Psi : \Omega \twoheadrightarrow \mathbb{R}^k$ is denoted* $\int \Psi \, dP$ *and defined as the set of all integrals of functions* $\psi : \Omega \to \mathbb{R}^k$ *such that* $P(\{\omega : \psi(\omega) \in \Psi(\omega)\}) = 1$.

We write $P(\{\omega : \psi(\omega) \in \Psi(\omega)\}) = 1$ as $\psi \in \Psi$ a.e. Some conditions on Ψ are necessary for the existence of any measurable ψ with $\psi \in \Psi$ a.e. The proof of the following is at least as involved as the proof of the Borel isomorphism theorem, and we do not give it here.

Theorem 8.6.12 *If* Ψ *is nonempty valued and the graph of* Ψ *is an analytic subset of* $\Omega \times \mathbb{R}^k$, *then there exists universally measurable* ψ *with* $\psi \in \Psi$ *a.e.*

One can easily express $\{\omega : \psi(\omega) \in \Psi(\omega)\}$ as a projection, so one expects it to be, at best, analytic, hence universally measurable [see Definition 7.9.3 (p. 430)]. Note that the integral of a correspondence is, in general, a set rather than a point.

Exercise 8.6.13 If $\Psi(\omega)$ is integrable and a.e. convex, then $\int \Psi \, dP$ is convex.

Exercise 8.6.14 Find the integrals of the following correspondences.

1. $\Gamma(\omega) = [0, 2]$ when $(\Omega, \mathcal{F}, P)$ is nonatomic.
2. $\Gamma(\omega) = \{0, 2\}$ when $(\Omega, \mathcal{F}, P)$ is nonatomic.
3. $\Gamma(\omega) = \{-2\omega, +2\omega\}$ when $(\Omega, \mathcal{F}, P) = ((0, 1], \mathcal{B}, \lambda)$.
4. $\Gamma(\omega) = \{-2\omega, +2\omega\}$ when $(\Omega, \mathcal{F}, P) = (\{1, \ldots, n\}, \mathcal{P}(\Omega), U_n)$, where U_n is the uniform distribution on $\{1, \ldots, n\}$.
5. $\Gamma(\omega) = [-2\omega, +2\omega]$ when $(\Omega, \mathcal{F}, P) = ((0, 1], \mathcal{B}, \lambda)$.
6. $\Gamma(\omega) = [-2\omega, +2\omega]$ when $(\Omega, \mathcal{F}, P) = (\{1, \ldots, n\}, \mathcal{P}(\Omega), U_n)$, where U_n is the uniform distribution on $\{1, \ldots, n\}$.

Excess demand is given by $\mathbf{e}(\mathbf{p}) = \mathbf{t}(\mathbf{p}) - \int \mathbf{i}(\omega) \, dP(\omega)$ and, again, this may be a set because $\mathbf{t}(\mathbf{p})$ may be a set. For our fixed-point arguments to work, this should be a compact convex set and it should depend upper hemicontinuously on p.

Lemma 8.6.15 *For each* ω, *the demand correspondence* $\mathbf{p} \mapsto \mathbf{x}^*(\mathbf{p})(\omega)$ *from* $\mathbb{R}^k_{++}$ *to* $\mathbb{R}^k_+$ *is nonempty valued, compact valued, upper hemicontinuous, homogeneous of degree 0, and satisfies* $\mathbf{p} \cdot \mathbf{y} = \mathbf{p} \cdot \mathbf{i}(\omega)$ *for all* $\mathbf{y} \in \mathbf{x}^*(\mathbf{p})(\omega)$.

Exercise 8.6.16 Prove Lemma 8.6.15.

The mean demand is defined as the set of integrals of selections from the correspondence $\omega \mapsto \mathbf{x}^*(\mathbf{p})(\omega)$. The existence of measurable selections, $\gamma(\cdot)$ from $\mathbf{x}^*(\mathbf{p})(\cdot)$ depends on the graph being analytic.

Theorem 8.6.17 *There exist measurable selections from the correspondence* $\omega \mapsto \mathbf{x}^*(\mathbf{p})(\omega)$.

Given that $\mathbf{p} \mapsto \mathbf{t}(\mathbf{p})$ is nonempty valued, one would like it to be well behaved.

Lemma 8.6.18 $\mathbf{p} \mapsto \mathbf{t}(\mathbf{p})$ *is compact valued and upper hemicontinuous, and for all* $\mathbf{y} \in \mathbf{t}(\mathbf{p})$, $\mathbf{p} \cdot \mathbf{y} = \mathbf{p} \cdot \int \mathbf{i}(\omega) \, dP(\omega)$.

This is what one expects—for each person ω, the demand set may explode but not implode at $\mathbf{p}$, so the set of selections may explode but not implode at $\mathbf{p}$. The proof is an exercise in keeping this insight firmly in place. The following is, perhaps, unexpected.

Lemma 8.6.19 $\mathbf{p} \mapsto \mathbf{t}(\mathbf{p})$ *is convex valued.*

We prove the following.

Theorem 8.6.20 *If $(\Omega, \mathcal{F}, P)$ is nonatomic and $\Psi : \Omega \twoheadrightarrow \mathbb{R}^k$ is integrable, then $\int \Psi \, dP$ is convex.*

Proof. Let $\mathbf{z}_1 = \int \psi_1 \, dP$ and $\mathbf{z}_2 = \int \psi_2 \, dP$ with $\psi_1, \psi_2 \in \Psi$ a.e. Let $\psi = (\psi_1, \psi_2)$ and define the vector measure $P_\psi(E) = (\int_E \psi_1 \, dP, \int_E \psi_2 \, dP)$. Since P is nonatomic, Lyapunov's theorem tells us that $R := \{P_\psi(E) : E \in \mathcal{F}\}$ is a compact convex set. Since $\emptyset, \Omega \in \mathcal{F}$, $(0, 0) \in R$ and $(\mathbf{z}_1, \mathbf{z}_2) \in R$. Pick $E \in \mathcal{F}$ such that $P_\psi(E) = (\alpha \mathbf{z}_1, \alpha \mathbf{z}_2)$. The function $\eta = \psi_1 1_E + \psi_2 1_{E^c}$ is a.e. in Ψ and $\int \eta \, dP = \alpha \mathbf{z}_1 + (1 - \alpha)\mathbf{z}_2$. ∎

$\Delta := \{\mathbf{p} \in \mathbb{R}^k : \sum_i p_i = 1\}$ is the price simplex.

Lemma 8.6.21 *If $\mathbf{p}_n \to \mathbf{p}$, where $\mathbf{p}_n \in \Delta$ and $\mathbf{p} \in \Delta^\circ$, then $\|\mathbf{t}^*(\mathbf{p}_n)\| \to \infty$.*

Proof. Standard implication of monotonicity. ∎

We now have an excess demand correspondence satisfying the conditions satisfied by the excess demand correspondence from a finite economy with convex preferences. We now give the **excess demand theorem**, which is the crucial result for proofs of the existence of Walrasian equilibria.

Theorem 8.6.22 (Excess Demand) *Let S be a closed convex subset of $\mathbb{R}^k_{++}$ and B a convex compact subset of $\mathbb{R}^k$. If $g : S \Rightarrow \mathbb{R}^k$ is an upper hemicontinuous convex-valued correspondence from S to B such that for every $\mathbf{p} \in S$ and every $\mathbf{z} \in g(\mathbf{p})$, $\mathbf{p} \cdot \mathbf{z} \leq 0$, then there exists $\mathbf{p}^* \in S$ and $\mathbf{z}^* \in g(\mathbf{p}^*)$ such that for all $\mathbf{p} \in S$, $\mathbf{p} \cdot \mathbf{z}^* \leq 0$.*

Proof. Define the correspondence $f : B \Rightarrow S$ by $f(\mathbf{z}) = \{\mathbf{p} \in S : \mathbf{p} \cdot \mathbf{z} = \max_{\mathbf{p}' \in S} \mathbf{p}' \cdot \mathbf{z}\}$ so that f is uhc and convex valued.

Define the uhc convex-valued $h : (S \times B) \Rightarrow (S \times B)$ by $h(\mathbf{p}, \mathbf{z}) = f(\mathbf{z}) \times g(\mathbf{p})$. By Kakutani's theorem, there exists $(\mathbf{p}^*, \mathbf{z}^*) \in f(\mathbf{z}^*) \times g(\mathbf{p}^*)$.

By the definition of f, $\mathbf{p} \cdot \mathbf{z}^* \leq \mathbf{p}^* \cdot \mathbf{z}^*$ for all $\mathbf{p} \in S$. $\mathbf{z}^* \in g(\mathbf{p}^*)$ implies $\mathbf{p}^* \cdot \mathbf{z}^* \leq 0$. Combining, we get that for all $\mathbf{p} \in S$, $\mathbf{p} \cdot \mathbf{z}^* \leq 0$. ∎

Recall that to show equilibrium existence, it is sufficient to show that for some $\mathbf{p}^* \in \mathbb{R}^k_{++}$, $0 \in \mathbf{e}(\mathbf{p}^*) = (\mathbf{t}(\mathbf{p}) - \int \mathbf{i}(\omega) \, dP(\omega))$.

Proof of Equilibrium Existence. For $n \geq k$, define $S_n = \{\mathbf{p} \in \Delta : p_i \geq \frac{1}{n}\}$ so that $\Delta^\circ := \text{rel int}(\Delta) = \cup_n S_n$. On each S_n, the total excess demand correspondence, $\mathbf{e}(\cdot)$, satisfies the assumptions of the excess demand theorem. Therefore, there exists a sequence $\mathbf{p}_n^* \in S_n$ and a sequence $\mathbf{z}_n^* \in \mathbf{e}(\mathbf{p}_n^*)$ such that $\mathbf{p} \cdot \mathbf{z}_n^* \leq 0$ for all $\mathbf{p} \in S_n$.

Taking a subsequence if necessary, we get $\mathbf{p}_n^* \to \mathbf{p}^*$ for some $\mathbf{p}^* \in \Delta$. Set $\mathbf{p}^\circ = (1/k, \cdots, 1/k)$. Since $\mathbf{e}(\cdot)$ is bounded from below and $\mathbf{p}^\circ \mathbf{z}_n^* \leq 0$, $\mathbf{z}_n^*$ is bounded, so if we take a further subsequence if necessary, it converges to some $\mathbf{z}^*$. If $\mathbf{p}^* \in \Delta \setminus \Delta^\circ$, then the sequence $\mathbf{z}_n^*$ is unbounded, so $\mathbf{p}^* \in \Delta^\circ$. Since $\mathbf{e}(\cdot)$ is uhc, $\mathbf{z}^* \in \mathbf{e}(\mathbf{p}^*)$.

Now, $\mathbf{p}^* \cdot \mathbf{z}^* = 0$ by monotonicity of the preferences. By the excess demand theorem, for all $\mathbf{p} \in \Delta$, $\mathbf{p} \cdot \mathbf{z}^* \leq 0$, hence $\mathbf{z}^* \leq 0$. But since $\mathbf{p}^* \gg 0$, $\mathbf{p}^* \cdot \mathbf{z}^* = 0$ implies that $\mathbf{z}^* = 0$, that is, $0 \in \mathbf{e}(\mathbf{p}^*)$. ∎

8.6.d Proof of the Easy Part of Core Equivalence

The easy part is that every equilibrium is a core allocation.

Lemma 8.6.23 *If* $(\mathbf{p}, \mathbf{x})$ *is an equilibrium, then* $\mathbf{x}$ *is in the core.*

The proof mimics the finite proof.

Proof. If not, by contradiction, there is an allocation, $\mathbf{y}$, and nonnull coalition, E, such that $\mathbf{y}$ blocks $\mathbf{x}$ via E. This means that $\int_E \mathbf{y} \, dP = \int_E \mathbf{x} \, dP$ and for all $\omega \in E$, $\mathbf{y}(\omega) \succ \mathbf{x}(\omega)$. By monotonicity of preferences, $\mathbf{p} \cdot \int_E \mathbf{x} \, dP = \mathbf{p} \cdot \int_E \mathbf{i} \, dP$.

By the definition of an equilibrium, this implies that for all $\omega \in E$, $\mathbf{p} \cdot \mathbf{y}(\omega) > \mathbf{p} \cdot \mathbf{x}(\omega)$, and since $\mathbf{p} > 0$, this means that $\mathbf{y}$ is not feasible. ∎

The second welfare theorem for economies with finite sets of agents tells us that every equilibrium is Pareto optimal. It requires much stronger assumptions than the first welfare theorem and a careful use of the supporting hyperplanes. Being in the core is a much more stringent condition than being Pareto optimal, and it is not generally true that equilibria are in the core in models with finitely many agents. The proofs that equilibrium allocations are in the core when the space of agents is nonatomic is very involved. Very.

8.7 ◆ Measure Space Games

There are two kinds of uses of nonatomic measure spaces, $(\Omega, \mathcal{F}, P)$, in game theory. The first is as a model of a population of anonymous individuals, each of whom cares about his own action and the population average of the others' actions. As $(\Omega, \mathcal{F}, P)$ is nonatomic, the actions of any single individual do not affect the population average, hence everyone is effectively anonymous. The second is as a model of resources that have to be shared and that can be divided into arbitrarily small pieces and reallocated, and we call such models "games of (fair) division."

8.7.a Anonymous Games

The basic ingredients for this class of games are a nonatomic probability space modeling the people in the game, a set of actions for the people, and a set of utilities for the people depending on own actions and the population average of others' actions.

Definition 8.7.1 $[(\Omega, \mathcal{F}, P), A, u]$ *is an* ***anonymous game*** *if*

1. $(\Omega, \mathcal{F}, P)$ is a nonatomic probability space (of agents),

2. A is a compact metric space (of actions), and

3. $u : \Omega \times A \times \Delta(A) \to \mathbb{R}$ *is a jointly measurable (utility) function, continuous in $A \times \Delta(A)$ for every ω.*

A **strategy** *is a measurable mapping $\omega \mapsto s_\omega$ from Ω to A. The **population average choice associated with a strategy** s is $\mu_s(B) := P(\{\omega : s_\omega \in B\})$ for B in the class $\mathcal{B}_A$ of Borel measurable subsets of A. A strategy is an **equilibrium** if $P(\{\omega : (\forall a \in A)[u(\omega, s_\omega, \mu_s) \geq u(\omega, a, \mu_s)]\}) = 1$.*

In particular, when A is finite, the utility functions, $u : \Omega \times A \times \Delta(A) \to \mathbb{R}$, where $\Delta(A) = \{p \in \mathbb{R}_+^A : \sum_a p(a) = 1\}$.

By contrast with finite games, the strategies here are pure. In the terminology developed for finite games, a strategy is an equilibrium if it is almost everywhere a mutual best response. If we complete the σ-field, we can delete the adjective "almost." We begin with examples of anonymous games in which coordination is not achieved through prices, and then turn to equilibrium existence.

8.7.a.1 Examples of Anonymous Games

Example 8.7.2 (Mandeville's Fable of the Bees #1) *For a strategy s, each person ω in the set of people Ω chooses an action $a_\omega^*(s) \in [0, M]$ to solve*

$$\max_{a \in \mathbb{R}_+} \ u(\omega, a, E\,s) - c_\omega a,$$

where $c_\omega > 0$, $u(\omega, \cdot, \cdot)$ is monotonic in both arguments and $E\,s := \int s_\omega \, dP(\omega)$. We assume that $\frac{\partial^2 u(\omega, \cdot, \cdot)}{\partial a_\omega \partial E\,s} > 0$, which means that an increase in the average activity level in the economy increases the marginal reward/utility of person ω's activity levels and we assume that $\frac{\partial^2 u}{\partial a^2} < 0$. This implies that a unique solution $a_\omega^*(E\,s)$ exists and increases with $E\,s$. As it is a singleton-valued upper hemicontinuous correspondence, $a^*(\cdot)$ is a continuous increasing function from $[0, M]$ to $[0, M]$. We also assume that the mapping $\omega \mapsto a_\omega^*$ is measurable; for example, $\mathcal{F}$ is the set of universally measurable subsets and $\omega \mapsto u(\omega, \cdot, \cdot) \in C([0, M] \times [0, M])$ is measurable.

For $r \in [0, M]$, define

$$\alpha(r) = \int a_\omega^*(r) \, dP(\omega).$$

By dominated convergence, $\alpha(\cdot)$ is continuous, and since each $a_\omega^*(\cdot)$ is increasing, so is $\alpha(\cdot)$, and under the fairly mild conditions given in §7.5.d, $\alpha(\cdot)$ is even continuously differentiable, $\frac{d\alpha(a)}{d\bar{a}} = \int_\Omega \frac{\partial a_\omega^*}{\partial \bar{a}} \, dP(\omega) > 0$.
Either continuity or increasingness is sufficient to guarantee the existence of a fixed point. Any fixed point, $\bar{a}$, is an equilibrium average (or aggregate) level of activity in which person ω plays $a_\omega^*(\bar{a})$. If $\bar{a}^\dagger > \bar{a}^\circ$ are two fixed points, then the equilibrium with activity level $\bar{a}^\dagger$ is unanimously preferred to the equilibrium with activity level $\bar{a}^\circ$, even though both involve each $\omega \in \Omega$ playing strict best responses to what others are choosing.[7]

---

7. Mandeville wrote, in a poem called "The Grumbling Hive," of how

Exercise 8.7.3   Suppose that $\overline{a}^\dagger$ and $\overline{a}^\circ$ are both equilibrium aggregate levels of activity in the **Fable of the Bees #1**. Show that if $\overline{a}^\dagger > \overline{a}^\circ$, then everyone strictly prefers the $\overline{a}^\dagger$ equilibrium to the $\overline{a}^\circ$ equilibrium.

Exercise 8.7.4   Suppose that $u(\omega, a_\omega, \overline{a}) = 2\sqrt{a_\omega \overline{a}}$ and that $c_\omega \equiv 1$. Show that $a_\omega^* \equiv 0$ is one equilibrium for Fable #1. Show that $a_\omega^*(\overline{a}) = \overline{a}$ for all ω. From this, show that any $\overline{a}$ can be an equilibrium.

The next version of the fable involves the level of present activity affecting the rate of return on savings.

Example 8.7.5 (Fable of the Bees #2)   *With the same set of people as above, each person $\omega \in \Omega$ picks present demand, d_ω, and savings for future demand to maximize*

$$u(\omega, d_\omega, (1+r)a_\omega) \text{ subject to } d_\omega + a_\omega = m_\omega, d_\omega, s_\omega \geq 0,$$

where

$$r = r(\overline{d}), \qquad r'(d) > 0, \qquad \overline{d} = \int_\Omega d_\omega \, dP(\omega),$$

and the mapping $\omega \mapsto u(\omega, \cdot, \cdot)$ is measurable. In other words, the more people spend now, the higher the level of economic activity, $\overline{d}$, which leads to a higher return on capital, $r(\overline{d})$, which means more to spend next period for each unit saved. If we believed that there was an aggregate production function, it would have to involve higher activity by one group increasing the marginal product of other groups in order for $r'(d) > 0$.

For any given $\overline{d}$, denote by (d_ω^*, s_ω^*) the solution to the problem

$$\max \; u(\omega, d_\omega, (1+r(\overline{d}))a_\omega) \quad \text{subject to} \quad d_\omega + a_\omega = m_\omega, d_\omega, s_\omega \geq 0.$$

The Root of Evil, Avarice,
That damnéd ill-naturéd baneful Vice,
Was Slave to Prodigality,
That noble Sin; whilst Luxury
Employéd a Million of the Poor,
And odious Pride a Million more:
Envy itself, and Vanity,
Were Ministers of Industry;
Their darling Folly, Fickleness,
In Diet, Furniture and Dress,
That strange ridiculous Vice, was made
The very Wheel that turnéd the Trade.

One interpretation is that it is the high activity levels of others, their avarice, prodigality, odious pride, and vanity, that make employment and hard work so rewarding. Later in the poem, when a vengeful deity showed the population how bad these sins really are, total economic collapse soon followed.

For many reasonable specifications of $u(\omega, \cdot, \cdot)$, $d_\omega^*(\cdot)$ is increasing in $\overline{d}$; for even more specifications,

$$\delta(\overline{d}) := \int_\Omega d_\omega^* \, dP(\omega)$$

is increasing in $\overline{d}$. Any $\overline{d}$ such that $\delta(\overline{d}) = \overline{d}$ is an equilibrium aggregate level of demand activity, and it can be arranged that there are many equilibria. Equilibria with higher $\overline{d}$'s are strictly preferred to those with lower.[8]

Both these fables can be recast as stories about people not internalizing the external effects of their own actions. Another class of external effect that peoples' actions have on others arises when there is congestion.

Example 8.7.6 (Fable of the Commute) *With $(\Omega, \mathcal{F}, P)$ a nonatomic probability space, each person ω picks an action $a_\omega \in A = \{0, 1, 2\}$, where $a_\omega = 0$ means staying at home; $a_\omega = 1$ means commuting during the first hour of morning commute time; and $a_\omega = 2$ means commuting during the second hour of morning commute time. If $s : \Omega \to \{0, 1, 2\}$ is a strategy, then the delays in time 1 and 2 are $d_1(s) = f(P(s^{-1}(1))$ and $d_2(s) = f(P(s^{-1}(2))$, where $f : \mathbb{R}_+ \to \mathbb{R}$ is a strictly increasing convex function with $f(0) = 0$. Person ω's utility if she chooses action a_ω and everyone else acts according to s is*

$$u(\omega, a_\omega, s) = \begin{cases} v_\omega(0) & \text{if } a_\omega = 0, \\ v_\omega(1) - r_{\omega,1} d_1(s) & \text{if } a_\omega = 1, \text{ and} \\ v_\omega(2) - r_{\omega,2} d_2(s) & \text{if } a_\omega = 2. \end{cases}$$

Here, $(v_\omega(0), v_\omega(1), v_\omega(2)) \in \mathbb{R}_+^3$ give, respectively, person ω's baseline utilities for staying home, arriving at work during the first hour, and arriving during the second hour. In a similar fashion, $(r_{\omega,1}, r_{\omega,2}) \in \mathbb{R}_{++}^2$ gives person ω's disutility of time delay.

A pure strategy equilibrium is a function $s : \Omega \to \{0, 1, 2\}$ such that for each ω, $u(\omega, s(\omega), s) \geq u(\omega, A, s)$. Suppose (heroically) that the values can be measured in units of money that are transferable from one person to another. An equilibrium is likely to be inefficient if there is a group of people commuting during, say, the first hour that have high ratios of $r_{\omega,1} d_1(s)/r_{\omega,2} d_2(s)$ and a group having low values of this ratio. Those with the high ratio would be willing to bribe the low-ratio ones to switch commute times, making everyone better off. If the commute times were priced and choices took into account those prices as well as the delays, the problem could be alleviated. This is the rationale for toll roads with charges that vary with the time of day. Whether or not there is actually an increase in some measure of

---

8. As Keynes put it,

> A diminished propensity to consume to-day can only be accommodated to the public advantage if an increased propensity to consume is expected to exist some day. We are reminded of "The Fable of the Bees," the gay of tomorrow are absolutely indispensable to provide a raison d'être for the grave of to-day.

social welfare will depend on any number of other equilibrium phenomena, for example, if the cost of collecting tolls outweighs the benefits or if people vote for bonds to build roads that reduce the function f because they hate the idea of toll roads once they have tried them.

8.7.a.2 Equilibrium Existence for Anonymous Games

Definition 8.7.7 $\Gamma = ((\Omega, \mathcal{F}, P), A, u)$ is a **finite action anonymous game** if $(\Omega, \mathcal{F}, P)$ is a nonatomic probability space, A is a finite set, $\omega \mapsto u(\omega, \cdot, \cdot)$ is a measurable function from Ω to $C(A \times \Delta(A))$, and $\int \|u(\omega, \cdot, \cdot)\|_\infty \, dP < \infty$.

Theorem 8.7.8 *Every finite-action anonymous game has an equilibrium.*

Proof. For $\mu \in \Delta(A)$ and $S \subset A$, $S \neq \emptyset$, let

$$E_S(\mu) = \{\omega : (\forall a \in S, \forall b \notin S)[u(\omega, a, \mu)$$
$$\geq u(\omega, A, \mu) \wedge u(\omega, a, \mu) > u(\omega, b, \mu)]\},$$

that is, $E_S(\mu)$ is the set of ω that has the set S being the set of best responses to the population average play being μ.

Identify any S with the set of point masses, $\{\mathbf{e}_s : s \in S\} \subset \Delta(A)$. For each μ, define $\Psi_\mu : \Omega \twoheadrightarrow \Delta(A)$ by $\Psi_\mu(\omega) = \sum_{S \subset A} S \cdot 1_{E_S(\mu)}(\omega)$. By Theorem 8.6.20, $\int \Psi_\mu \, dP$ is convex, and it is easily seen to be compact as well (e.g., because Ψ takes on only finite many values).

As $\mu_n \to \mu$, each ω's set of optimal choices changes upper hemicontinuously, which implies that $\mu \mapsto \int \Psi_\mu \, dP$ is upper hemicontinuous. Therefore, by Kakutani's theorem, there exists $\mu^* \in \int \Psi_\mu \, dP$. By the definition of the integral of a correspondence, any selection ψ with $\int \psi \, dP = \mu^*$ is an equilibrium. ■

A striking aspect of the previous result is that, unlike games with finite sets of players, mixed strategies were not required. This is due to the convexity provided by Lyapunov's theorem for finite-dimensional vector measures. The intuition is that if everyone in a set E has two or more optimal actions, splitting them into groups of appropriate size playing the different pure strategies replaces randomization.

The result for the existence of pure strategy equilibria does not generalize to measure space games with compact sets of actions because Lyapunov's theorem does not generalize to infinite-dimension vector measures. One can substitute mixed strategies for the pure strategies and prove equilibrium existence with the generalization of Kakutani's theorem 10.8.2 to infinite-dimensional vector spaces. One can also use a better class of measure space, the Loeb spaces of §11.3, for which Lyapunov's theorem does hold for more general vector measures.

8.7.b Games of Fair Division

A parent dividing a last piece of cake between two squabbling siblings can easily find a fair division by letting one child divide the cake into two pieces while giving the second child the first pick. The child doing the division can guarantee that he receives a piece that he views as fair by paying scrupulous attention to dividing the cake into two pieces that he views as being equally attractive. Further, if he knows that the two of them view different aspects of the cake differently, for example, one has a weakness for extra icing not shared by the other, he can divide the cake in such a fashion that they both end up with pieces that they regard as being better than the other's piece.

To generalize this to $n > 2$ children, suppose that we continuously increase the size being offered and everyone can yell "stop!" whenever he pleases, stopping the increase and receiving the piece designated at the instant he yells. If there is a tie, that is, if two or more yell at the same instant, randomly allocate the piece to one of them, and continue as before with $n - 1$ children. Again, everyone has a strategy that guarantees her at least $1/n$ of what she thinks is the value of the whole cake, but it seems that everyone ought to be able to do better than that. We are going to briefly investigate this question.

Definition 8.7.9 *A **division problem** is a measurable space, $(\Omega, \mathcal{F})$ and a finite set, $\{u_i : i \in I\}$, of nonnegative countably additive measures, $u : \mathcal{F} \to \mathbb{R}_+^I$, that denotes the associated vector measure. A **division** is a measurable partition, $\mathcal{E} = \{E_i : i \in I\}$.*

The idea is that $u_i(E_i)$ is i's utility if he receives E_i and that i receiving ω means that $j \neq i$ cannot receive it. The countable additivity is the generalization of the continuity of preferences in the size of the piece of pie one receives.

Theorem 8.7.10 *For any collection of strictly positive weights, λ_i, $i = 1, \ldots, n$, there is a partition, $\{E_1, \ldots, E_n\}$, of Ω, that maximizes, over all possible partitions, $\sum_i \lambda_i u_i(E_i)$.*

Proof. By the Radon-Nikodym theorem, there is no loss in assuming the existence of a probability, $P : \mathcal{F} \to [0, 1]$, and a vector function, $g : \Omega \to \mathbb{R}^I$, such that $u(E) = P_g(E) := \int g \, dP$.

Define $f_i(\omega) = \lambda_i \cdot g_i(\omega)$ and $f(\omega) = \max_i f_i(\omega)$. The basic idea is to give person i the little piece $\{\omega\}$ if $f_i(\omega) = f(\omega)$, breaking any ties in favor of the person with the lower number. Specifically, let $E_1 = \{\omega : f_1(\omega) = f(\omega)\}$, and given that E_i has been defined, define $E_{i+1} = \{f_{i+1} = f\} \setminus \cup_{j \leq i} E_j$.

For this partition, $\sum_i \lambda_i u_i(E_i) = \sum_i \int_{E_i} f_i \, dP = \sum_i \int_{E_i} f \, dP = \int_\Omega f \, dP$.

We now show that for any other partition, $\mathcal{B} = \{B_i : i \in I\}$, $\sum_i \lambda_i u_i(B_i) \leq \int_\Omega f \, dP$. This follows from the observation that $\sum_i \lambda_i u_i(B_i) = \sum_i \int_{B_i} f_i \, dP \leq \sum_i \int_{E_i} f \, dP = \int_\Omega f \, dP$. ■

Corollary 8.7.11 *If each u_i is nonatomic, then the set of Pareto-optimal partitions is given by the E_i of Theorem 8.7.10.*

Proof. It is sufficient to show that R'', the set of vectors of utilities from partitions, is compact and convex. This is easiest to prove when $I = 2$, but the same idea applies more generally.

If u_1 and u_2 are nonatomic, then Lyapunov's theorem tells us that $R = \{(u_1(E), u_2(E)) : E \in \mathcal{F}\}$ is a compact convex subset of $\mathbb{R}^2$. However, if 1 has E, then 2 cannot have it.

Let $u_1^* = u_1(\Omega)$ and $u_2^* = u_2(\Omega)$ be 1's and 2's highest achievable utility levels. Define the affine mapping $(x_1, x_2) \mapsto (x_1, x_2, u_1^* - x_1, u_2^* - x_2)$ from $\mathbb{R}^2$ to $\mathbb{R}^4$. The image, R', of R under this mapping is compact and convex. The projection mapping $(y_1, y_2, y_3, y_4) \mapsto (y_1, y_4)$ from $\mathbb{R}^4$ to $\mathbb{R}^2$ is linear. Let R'' be the necessarily compact and convex projection of R'. R'' is the set of possible utility profiles for the two people. ∎

Another aspect of nonatomicity is that even division is always possible.

Theorem 8.7.12 *If each u_i is nonatomic, there exists a partition $\{E_1, \dots, E_n\}$ such that for each i, $u_i(E_i) - \frac{1}{n}u_i(\Omega)$.*

Proof. As above, let $P_g(E) = u(E)$ for $g : \Omega \to \mathbb{R}^I$. Let E_1 satisfy $P_g(E_1) = \frac{1}{n}P_g(\Omega)$ and $E_2 \subset (\Omega \setminus E_1)$ satisfy $P_g(E_2) = \frac{1}{n-1}P_g(\Omega \setminus E_1) = \frac{1}{n}P_g(\Omega)$. Continue inductively. ∎

In fact, a small change in this last proof gives arbitrary divisions.

Exercise 8.7.13 If $\alpha_i > 0$, $\sum_i \alpha_i = 1$, and the u_i are nonatomic, show that there is a partition with $u_i(E_i) = \alpha_i u_i(\Omega)$. Further, show that if everyone's tastes are exactly the same, that is, if $u_i = u_j$ for all i, j, then this gives the entire set of achievable utilities.

Theorem 8.7.14 *If each u_i is nonatomic and $u_i \neq u_j$ for some i, j, there exists a partition $\{E_1, \dots, E_n\}$ such that for each i, $u_i(E_i) > \alpha_i u_i(\Omega)$, where $\sum_i \alpha_i = 1$.*

Exercise 8.7.15 Prove Theorem 8.7.14.

8.8 ◆ Dual Spaces: Representations and Separation

For linear functions, T, from one normed space, $\mathfrak{X}$, to another, for example, from $L^p(\Omega, \mathcal{F}, P)$ to $\mathbb{R}$, the boundedness principle says that continuity of T is equivalent to $\|T(B)\| < \infty$ for all norm-bounded sets B and that the norm of T is defined as $\|T\|^* := \sup_{\|f\| \leq 1} \|T(f)\|$. $\mathfrak{X}^*$ denotes the set of continuous linear $\mathbb{R}$-valued functions on the normed space $\mathfrak{X}$. This section has three about normed spaces and Banach spaces.

- First, Theorem 8.8.2 shows that $(\mathfrak{X}^*, \|\cdot\|^*)$ is a Banach space whether or not the normed space $\mathfrak{X}$ is complete.
- Second, the Riesz-Frechet theorem, 8.8.7, identifies $(L^p)^*$, $p \in [1, \infty)$. Up to linear isometry it is L^q, where p and q are a **dual pair of indexes**; that is, $\frac{1}{p} + \frac{1}{q} = 1$ and we define $\frac{1}{\infty} = 0$.
- Third, the Hahn-Banach theorem, 8.8.10, shows that disjoint convex sets, one of which has an interior, can be separated by continuous linear functions.

Unlike $\mathbb{R}^k$, being disjoint and convex is not sufficient for separation by continuous linear functions. Lemma 8.3.51 (p. 473) and Exercise 8.4.7 (p. 476) demonstrate the existence of disjoint dense convex sets E_1 and E_2 with the property that $L(E_1) = L(E_2) = \mathbb{R}$ for any $L \in \mathfrak{X}^*$.

8.8.a The Completeness of Dual Spaces

For the Banach space $\mathbb{R}^k$, a function, $T : \mathbb{R}^k \to \mathbb{R}$, is linear and continuous iff it can be represented as the inner product, $T(\mathbf{x}) = \mathbf{x} \cdot \mathbf{y} = \sum_i x_i y_i$, for some $\mathbf{y} \in \mathbb{R}^k$, and further $\|T\|^* = \|y\|$. This shows that $\mathbb{R}^k$ is **self-dual**, that is, that up to linear isometry, it is its own dual space.

Definition 8.8.1 $\mathfrak{X}^*$, **dual** *of a normed space* $\mathfrak{X}$, *is the set of continuous linear functions on* $\mathfrak{X}$. *The norm on* $\mathfrak{X}^*$ *is defined by* $\|T\|^* := \sup_{\|f\| \le 1} \|T(f)\|$.

We care about dual spaces for several reasons. One basic reason is that the market value of a bundle of goods $\mathbf{x}$ at prices $\mathbf{p}$ is $\mathbf{p} \cdot \mathbf{x}$. This means that the existence result for a competitive equilibrium tells us about the existence of a point in the dual space. The proof of the second fundamental theorem of welfare economics uses the separating hyperplane theorem, 5.7.1 (p. 209), another existence result for a point in the dual space. More generally, the convex analysis that gives us the properties of the basic constructs of neoclassical microeconomics (profit functions, supply and demand functions for firms, cost functions, indirect utility functions, Hicksian demand functions for consumers) uses the dual space of the space of commodities.

Theorem 8.8.2 $(\mathfrak{X}^*, \|\cdot\|^*)$ *is a Banach space.*

Proof.   The argument for the norm inequality, $\|T_1 + T_2\|^* \le \|T_1\|^* + \|T_2\|^*$, is a familiar fact about optima. Specifically, for any $f \in \mathfrak{X}$, $|T_1(f) + T_2(f)| \le |T_1(f)| + |T_2(f)|$, so that

$$\sup_{\|f\| \le 1} |T_1(f) + T_2(f)| \le \sup_{\|f\| \le 1} \left(|T_1(f)| + |T_2(f)|\right) \tag{8.37}$$

and

$$\sup_{\|f\| \le 1} \left(|T_1(f)| + |T_2(f)|\right) \le \sup_{\|f_1\| \le 1} |T_1(f_1)| + \sup_{\|f_2\| \le 1} |T_2(f_2)|.$$

To show that $(\mathfrak{X}^*, \|\cdot\|^*)$ is complete, let T_n be a Cauchy sequence. Consider the restrictions of the T_n to the unit ball, $\{f : \|f\| \le 1\}$. Since this is a Cauchy sequence in the complete space $C_b(U)$, it has a continuous limit, which we call T_0. Now, T_0 is defined only on U. However, if we define $T(f) := \|f\| T_0(f/\|f\|)$, we have a linear function. Since T is bounded on U it must be continuous on all of $\mathfrak{X}$.

If $\|T\|^* = 0$, then $|T(f)| = \|f\| T(f/\|f\|) = 0$ for all nonzero $f \in \mathfrak{X}$. If $T(f) \ne 0$ for some $f \ne 0$, then $\|T\|^* \ge |T(f)|/\|f\| > 0$.   ∎

8.8.b Representing the Dual Space of $L^p(\Omega, \mathcal{F}, P)$, $p \in [1, \infty)$

The dual spaces of L^p are different than the dual space of $\mathbb{R}^k$ in (at least) two fundamental ways. First, as we have seen, there are discontinuous linear functions. Second, L^p spaces are rarely their own dual spaces.

Definition 8.8.3 (p, q), $p, q \in [1, \infty]$ *are a **dual pair of indexes** if* $\frac{1}{p} + \frac{1}{q} = 1$, *where we define* $\frac{1}{\infty} = 0$.

The easiest and most useful dual pair is $(p, q) = (2, 2)$, an extreme case is $(p, q) = (1, \infty)$, and an example of a seldom used case is $(p, q) = (3, \frac{3}{2})$. The basic result that we now state is that for $p \in [1, \infty)$ and (p, q) a dual pair, the set of all continuous linear functions on $L^p(\Omega, \mathcal{F}, P)$ is, up to isometry, $L^q(\Omega, \mathcal{F}, P)$, that is, $(L^p)^* = L^q$. This requires defining the analogue of the inner product.

The dual space for L^∞ is a fairly complicated beast, strictly containing L^1. It leads us to one of the few times that we care about topological spaces that are not metric spaces. This sort of space can be analyzed using material in Chapter 11. The requisite analogue of the inner product for L^p spaces is $\langle X, Y \rangle := E\, XY = \int_\Omega X(\omega) Y(\omega) \, dP(\omega)$. Our first job is to make sure that the inner product is well defined.

Theorem 8.8.4 (Hölder)   *If (p, q) are a dual pair of indexes, $X \in L^p$ and $Y \in L^q$, then $|E\, XY| \leq E\, |XY| \leq \|X\|_p \|Y\|_q$. Further, the inequality is tight; that is, there exists $X \in L^p$ such that $E\, |XY| = \|X\|_p \|Y\|_q$.*

Proof.   If $p = \infty, q = \infty$, or $Y = 0$, this is immediate, so suppose that $1 < p, q < \infty$ and $Y \neq 0$.

Set $\alpha = \frac{1}{p}$ so that $(1 - \alpha) = \frac{1}{q}$, and consider the concave function $g_\alpha(u, v) = u^\alpha v^{1-\alpha}$ for $(u, v) \in \mathbb{R}_+^2$. Applying Jensen's inequality (Theorem 8.3.3) to g_α and random variables $U \geq 0$ and $V \geq 0$ gives

$$E\, U^\alpha V^{1-\alpha} \leq (E\, U)^\alpha (E\, V)^{1-\alpha}.$$

With $U = |X|^p$ and $V = |Y|^q$, this yields $E\, (|XY|) \leq \|X\|_p \|Y\|_q$.

If $Y = 0$, the inequality is tight for all $X \in L^p$. We show that it is tight when $\|Y\|_q = 1$ and that, by rescaling, this is sufficient. For $p \in (1, \infty)$, we must fiddle with the exponents carefully. Note that $\frac{1}{p} + \frac{1}{q} = 1$ iff $\frac{1}{p} = \frac{q-1}{q} = 1$ iff $\frac{q}{p} = (q - 1)$. Pick $Y \in L^q$ with $E\, |Y|^q = 1$ so that $\|Y\|_q = 1$. Set $X = |Y|^{q/p} = |Y|^{(q-1)}$. This gives $|X|^p = |Y|^q$, so that $E\, |X|^p = 1 = \|X\|_p$. Now, $|XY| = |X|\, |Y| = |Y|^{(q-1)}|Y| = |Y|^q$. Therefore, $E\, |XY| = 1 = \|X\|_p \|Y\|_q$. Finally, scaling Y by any $r \neq 0$ gives $E\, |XrY| = |r| \cdot E\, |XY| = |r|\, \|X\|_p \|Y\|_q$, so that tightness holds for all $Y \in L^q$.   ∎

Exercise 8.8.5   Prove that Hölder's inequality really is "immediate" if $p = \infty$ or $q = \infty$, and show that in these cases it is also tight.

Hölder's inequality can be restated using dual-space norms.

Theorem 8.8.6   *For any dual pair of indexes (p, q), $p \in [1, \infty)$ and $Y \in L^q$, the mapping $T_Y(X) = \langle X, Y \rangle$ from L^p to $\mathbb{R}$ is continuous, linear, and has norm $\|T\| = \|Y\|_q$. Further, if $Y \neq Y'$, then $T_Y \neq T_{Y'}$.*

Proof.   The linearity of T_Y comes from the linearity of the integral. The continuity comes from the boundedness principle and Hölder's inequality, which gives $|T_Y(X)| \leq \|X\|_p \|Y\|_q$. The tightness of Hölder's inequality shows that $\|T_Y\| = \|Y\|_q$. For uniqueness, set $E = \{Y > Y'\}$ and $F = \{Y' > Y\}$. If $P(E) > 0$, then we set $X = 1_E \in L^p$, and if $P(E) = 0$, then $P(F) > 0$ and we set $X = 1_F$. In either case, $T_Y(X) \neq T_{Y'}(X)$.   ∎

Theorem 8.8.7 (Riesz-Fréchet) *For any dual pair of indexes (p, q), $p \in [1, \infty)$, if $T : L^p(\Omega, \mathcal{F}, P) \to \mathbb{R}$ is continuous and linear, then there exists a unique $Y \in L^q(\Omega, \mathcal{F}, P)$ such that $T(X) = \langle X, Y \rangle$, and $\|T\| = \|Y\|_q$.*

Proof. Define $\nu(E) = T(1_E)$. It is immediate that ν is a countably additive finite signed measure and that $\nu \ll P$. Radon-Nikodym implies that there exists a unique Y such that $\int_E Y(\omega) \, dP(\omega)$ for each $E \in \mathcal{F}$. By the linearity of T, for any simple function X, $T(X) = \int X(\omega) Y(\omega) \, dP(\omega)$. Since the simple functions are dense and T is continuous, for all X, $T(X) = \int X(\omega) Y(\omega) \, dP(\omega)$.

All that remains is to show that $Y \in L^q$. For this, note that $\|T\| = \sup\{|T(X)| : \|X\|_p \le 1\}$ and that for any measurable Y, $\int |Y|^q \, dP = \sup\{| \int XY \, dP|^q : \|X\|_p \le 1\}$ so that $\|Y\|_q = \|T\|$. ∎

Exercises 8.8.20 and 8.8.21 contain information about two different parts of $(L^\infty)^*$. One of the parts has a familiar feel and the other involves a purely finitely additive measure.

Our next result shows that the continuous linear functions are a rich enough class to separate points from convex sets having nonempty interior.

8.8.c Separation: The Hahn-Banach Theorem

If one had a peculiar turn of mind, one could spend time debating which result in functional analysis is the most important. Many who have that peculiar turn of mind would agree with you if you chose the Hahn-Banach theorem.

8.8.c.1 Statement and Proof

We now generalize Definition 5.7.15 from $\mathbb{R}^k$ to any vector space $\mathfrak{X}$. We are not assuming that we have a norm on $\mathfrak{X}$.

Definition 8.8.8 *A function $m : \mathfrak{X} \to \mathbb{R}_+$, $\mathfrak{X}$ a vector space, is called **sublinear** if:*

1. *it is **homogeneous of degree** 1, that is, for any $\alpha \ge 0$, $m(\alpha \mathbf{x}) = \alpha m(\mathbf{x})$, and*

2. *it **goes up sublinearly**, for any $\mathbf{x}, \mathbf{y} \in \mathbb{R}^\ell$, $m(\mathbf{x} + \mathbf{y}) \le m(\mathbf{x}) + m(\mathbf{y})$.*

Norms are special sublinear functions.

Exercise 8.8.9 Show that a sublinear function is a norm iff $[m(x) = 0] \Leftrightarrow [x = 0]$.

Just to be clear, we do not assume that $\mathfrak{X}$ has a norm in the following.

Theorem 8.8.10 (Hahn-Banach) *If V_0 is a vector subspace of the vector space $\mathfrak{X}$, $T_0 : V_0 \to \mathbb{R}$ is linear, $m : \mathfrak{X} \to \mathbb{R}$ is sublinear, and for all $x_0 \in V_0$, $T_0(x_0) \le m(x_0)$, then there exists a linear function, $T : \mathfrak{X} \to \mathbb{R}$ such that $T(x_0) = T_0(x_0)$ and for all $x_0 \in V_0$ and all $x \in \mathfrak{X}$, $T(x) \le m(x)$.*

Proof. An extension of (V_0, T_0) is a pair, (V_a, T_a), $V_a \supset V_0$ a subspace containing V_0, and $T_a \supset T_0$ a linear function (recalling that one identifies functions with their graphs). The set of extensions is partially ordered by inclusion, and any chain $\{(V_a, T_a) : a \in C\}$ has an upper bound $(V_C, T_C) := (\cup_a V_a, \cup_a T_a)$ that is an extension. By Zorn's lemma, there exists a maximal extension, (V, T). Our goal is to show that $V = \mathfrak{X}$.

Suppose that there exists $z \in \mathfrak{X} \setminus V$ and let $V_z = \{x + \alpha z : x \in V, \alpha \in \mathbb{R}\}$. Any linear extension, T_z of T to V_z, is of the form $T_z(x + \alpha z) = T(x) + \alpha c$ for some $c \in \mathbb{R}$. If we show that there exists a $c \in \mathbb{R}$ such that for all $x \in V$,

$$T(x) + \alpha c \leq m(x + \alpha z), \tag{8.38}$$

then (V, T) was not maximal. We first show that (8.38) can be satisfied for all $x \in V$ when $\alpha = \pm 1$, that is,

$$T(x) + c \leq m(x + z), \quad \text{and} \quad T(x) - c \leq m(x - z). \tag{8.39}$$

For any $x, y \in V$,

$$T(x) - T(y) = T(x - y) \leq m(x - y) \tag{8.40}$$
$$= m((x - z + z - y)) \leq m(x - z) + m(z - y),$$

which, after putting the x terms on the left and the y terms on the right, yields, for all $x, y \in V$,

$$T(x) - m(x - z) \leq T(y) + m(z - y). \tag{8.41}$$

Since the inequality holds for all $x, y \in V$, and x only appears on the left-hand side and y only appears on the right-hand side, we have

$$a := \sup_{x \in V}\{T(x) - m(x - z)\} \leq b := \inf_{y \in V}\{T(y) + m(z - y)\}. \tag{8.42}$$

We claim that picking any $c \in [a, b]$ makes (8.39) hold. To check this, $a \leq c \leq b$ and (8.42) yield

$$(\forall y \in V)[c \leq T(y) + m(z - y)], \text{ and}(\forall x \in V)[T(x) - m(x - z) \leq c], \text{ that is,}$$
$$\tag{8.43}$$

$$(\forall y \in V)[c - T(y) \leq m(z - y)], \text{ and}(\forall x \in V)[T(x) - c \leq m(x - z)], \text{ that is,}$$
$$\tag{8.44}$$

$$(\forall y \in V)[T(y) + c \leq m(y + z)], \text{ and}(\forall x \in V)[T(x) - c \leq m(x - z)], \tag{8.45}$$

where the last line, which is equivalent to (8.39), comes from the previous one by noting that $y \in V$ iff $-y \in V$ and $-T(y) = T(-y)$. Finally, (8.38) is immediate for $\alpha = 0$ and in the remaining cases comes from considering $\alpha > 0$ in $T_z(x \pm \alpha z) = T_z(\alpha(\frac{x}{\alpha} \pm z)) = \alpha T_z(\frac{x}{\alpha} \pm z) = T(x) \pm \alpha c$.   ∎

8.8.c.2 Corollaries
We love the Hahn-Banach theorem because of its corollaries.

Corollary 8.8.11 (Supporting Hyperplane) *If K is a convex subset of a normed vector space $\mathfrak{X}$, int$(K) \neq \emptyset$, and $x^\circ \notin$ int(K), then there exists a $T \in \mathfrak{X}^*$ such that $T(\text{int}(K)) < T(x^\circ)$.*

Proof. By translation, there is no loss in assuming that $0 \in \text{int}(K)$. Let $V_0 = \text{span}(\{x^\circ\})$ and define $m_K(y) = \inf\{\lambda \geq 0 : y \in \lambda K\}$. m_K is the Minkowski gauge of the set K, and since $x^\circ \notin K$, $\beta := m_K(x^\circ) \geq 1$. For $r \in \mathbb{R}$, define $T_0(rx^\circ) = r\beta$ so that $T_0(x_0) \leq m_K(x_0)$ for all $x_0 \in V_0$. By the Hahn-Banach theorem, T_0 has an extension, $T : \mathfrak{X} \to \mathbb{R}$, with $T(x) \leq m_K(x)$ for all $x \in \mathfrak{X}$. We have to show that T is continuous and that $T(\text{int}(K)) < T(x^\circ)$.

T is continuous: Since $0 \in \text{int}(K)$, there exists $\delta > 0$ such that $B_\delta(0) \subset K$. This yields $m_K(x) \leq \frac{1}{\delta}\|x\|$, which in turn implies that for all $x \in \mathfrak{X}$, $|T(x)| \leq \frac{1}{\delta}\|x\|$. By the boundedness principle, T is continuous.

$T(\text{int}(K)) < T(x^\circ)$: Since $m_K(y) \leq 1$ for all $y \in K$, we know that for all $y \in K$, $T(y) \leq T(x^\circ) = 1$. For $y \in \text{int}(K)$, there exists some $\epsilon > 0$ such that $y + \epsilon x^\circ \in K$. Now, $T(y + \epsilon x^\circ) = T(y) + \epsilon T(x^\circ)$ and $T(x^\circ) = 1$. Thus, $T(y) + \epsilon \leq T(x^\circ)$. ∎

The set $T^{-1}(1)$ is the closed supporting hyperplane if $x \in \partial K$.

Exercise 8.8.12 (Strict Separation of a Point and a Disjoint Closed Set)
If $x \notin \text{cl}(K)$, K a convex subset of a normed space, $\mathfrak{X}$, then there exists a $T \in \mathfrak{X}^*$ and an $\epsilon > 0$ such that $T(x) \geq T(K) + \epsilon$. [Translate so that $x = 0$, find $\delta > 0$ so that $B_\delta(0) \cap \text{cl}(K) = \emptyset$, and take m to be the Minkowski gauge of $B_{\delta/2}(0)$.]

Corollary 8.8.13 *If K_1 is a compact convex subset of a normed space $\mathfrak{X}$ and K_2 is a disjoint closed convex subset, then K_1 and K_2 can be strictly separated.*

Proof. $K := K_1 - K_2$ is closed and $0 \notin K$. ∎

The following is a direct generalization of the result for $\mathbb{R}^k$.

Corollary 8.8.14 *For every $x \neq 0$ in a normed space $\mathfrak{X}$, there exists $T \in \mathfrak{X}^*$ with $T(x) = 1$, and every closed convex subset of $\mathfrak{X}$ is the intersection of the closed hyperplanes that contain it.*

Proof. For $x \neq 0$, let $K = B_r(0)$, where $r = \|x\|$, and apply the supporting hyperplane theorem. The second part is a direct result of Exercise 8.8.12. ∎

In the middle of the proof of Corollary 8.8.11, we saw that a linear function on a normed space that is dominated by a sublinear function that dominates a multiple of the norm is necessarily continuous. This is worth recording separately.

Corollary 8.8.15 *If $\mathfrak{X}$ is a normed space, V_0 is a vector subspace, and $T_0 : V_0 \to \mathbb{R}$ is continuous with $B = \sup_{\{x_0 \in V_0, \|x_0\| \leq 1\}} |T_0(x_0)| < \infty$, then there exists $T \in \mathfrak{X}^*$ with $T_{|V_0} = T_0$ and $\|T\| = B$.*

Proof. $B \cdot \|x\|$ is sublinear and dominates T_0 on V_0. ∎

8.8.c.3 Skeletons in the Closet
Having seen a number of results showing that normed vector spaces behave just like $\mathbb{R}^k$, it is time to strive for a more balanced perspective. Prices are supposed to be positive continuous linear functions, and we are now going to examine the case where the commodity space is a normed vector space. Compare the following with the commodity space $\mathbb{R}^k_+$.

Example 8.8.16 *A price vector $p \in \ell^\infty_{++}$ has **bounded marginal rates of substitution** if $\sup\{\frac{x_s}{x_t} : s, t \in \mathbb{N}\} < \infty$. In $\mathbb{R}^k$, any strictly positive price vector has bounded marginal rates of substitution. In ℓ^2, no strictly positive price vector has bounded marginal rates of substitution.*

Example 8.8.17 *If the commodity space is ℓ^∞, representing infinite streams of rewards, the endowment of each of two people is a vector w with $w_t \geq r > 0$ a.e., $u_1(x)$ is the $\|\cdot\|_\infty$-continuous "patient" utility function $u_1(x) = \liminf_t x_t$, and $u_2(x)$ is the $\|\cdot\|_\infty$-continuous "impatient" utility function $u_2(x) = \sum_t x_t y_t$ for some $y = (y_1, y_2, \ldots) \in \ell^1_{++}$, then there is no Pareto-optimal allocation, hence no equilibrium.*

We know that for $p \in [1, \infty)$, $T \in (L^p)^*$ is of the form $T(X) = \int XY \, dP$ for some $Y \in L^q$, where (p, q) is a dual pair. Here is part of what goes wrong when $p = \infty$.

Corollary 8.8.18 *For $(\Omega, \mathcal{F}, P) = ([0, 1], \mathcal{B}, \lambda)$, there exists $T \in (L^\infty)^*$ that cannot be represented by $T(X) = \int XY \, d\lambda$ for some $Y \in L^1$, that is, $(L^\infty)^* \not\supseteq L^1$.*

Proof. Let V_0 be the vector subspace of L^∞ consisting of the set of a.e. equivalence classes of $C([0, 1])$. For $X \in V_0$, define $T_0(X) = \lim_{\epsilon \downarrow 0} \frac{1}{\epsilon} \int_{[0, \epsilon)} X \, d\lambda$, that is, $T_0(X) = f(0)$, where X belongs to the equivalence class of f. It is clear that $\|T_0\| = 1$; hence it has a continuous linear extension, T, to all of L^∞. Suppose that $T(X) = \int XY \, d\lambda$ for some $Y \in L^1$. Let $X_n(\omega) = \max\{1 - n\omega, 0\}$, so that $T_0(X_n) = T(X_n) = 1$ for all $n \in \mathbb{N}$. By dominated convergence, $\int X_n Y \, d\lambda \to 0$, so $T(X) \neq \int XY \, d\lambda$. ∎

Exercise 8.8.19 Let $c = \{x \in \ell^\infty : \lim_t x_t \text{ exists}\}$. Show that c is a vector subspace of ℓ^∞ and give a continuous linear function on c that does not have the form $T(x) = \sum_t x_t y_t$ for some $y \in \ell^1$ when extended to ℓ^∞.

There is at least one way to answer the question "What is $(\ell^\infty)^*$ anyway?" The space ℓ^∞ is a nonseparable algebra of bounded functions separating points. If we generalize the separable compact embeddings in metric spaces, it is possible to embed $\mathbb{N}$ as a dense subset of a nonmetric compact space, $\beta\mathbb{N}$, known as the Stone-Čech compactification of $\mathbb{N}$, in such a fashion that there is an isometry between $x \in \ell^\infty$ and $C(\beta\mathbb{N})$, the set of continuous functions on $\beta\mathbb{N}$.

A similar construction is available for $L^\infty(\Omega, \mathcal{F}, P)$, though there is the conceptual hurdle that elements of L^∞ are equivalence classes of functions, not functions. Here is some information about two different part of $(L^\infty)^*$.

Exercise 8.8.20 A continuous linear $T : L^\infty((\Omega, \mathcal{F}, P)) \to \mathbb{R}$ is of the form $T(X) = \int XY \, dP$ for some $Y \in L^1(\Omega, \mathcal{F}, P)$ iff the set function $\nu(E) := T(1_E)$ is countably additive. [One direction is dominated convergence and the other is the Radon-Nikodym theorem.]

Exercise 8.8.21 Let F_n be a sequence of sets in $\mathcal{F}$ with $P(F_n) > 0$ and $F_n \downarrow \emptyset$. Define $T_0(1_E) = 1$ if there exists an n such that $F_n \subset E$ a.e. and $T_0(1_E) = 0$ otherwise. Extend T_0 to the simple functions by linearity and check that it is continuous and nonnegative and that $\nu_0(E) := T_0(1_E)$ is finitely additive. By the Hahn-Banach theorem, T_0 has a continuous extension, T, to all of L^∞. By construction $\nu(E) := T(1_E)$ is a purely finitely additive probability since $\nu(F_n) \equiv 1$ but $F_n \downarrow \emptyset$.

Another aspect of the Hahn-Banach theorem is the lack of uniqueness of the extension.

Example 8.8.22 *Let $B((0, 1])$ be the set of bounded, not necessarily measurable functions on $(0, 1]$ with the sup norm. Let V_0 be the set of simple $\mathcal{B}^\circ$-measurable functions. Define $T_0 : V_0 \to \mathbb{R}$ by $T(X) = \int X \, d\lambda$. T_0 is sup norm continuous, so it has at least one sup norm continuous extension from V_0 to $B((0, 1])$.*

1. *Let $V_1 = \{X + \alpha 1_\mathbb{Q} : X \in V_0, \alpha \in \mathbb{R}\}$. For any $c \in [0, 1]$, $T_1(X + \alpha 1_\mathbb{Q}) = T_0(X) + \alpha c$ is a continuous linear extension of T_0. Thus it is possible to sup norm continuously extend Lebesgue measure on $\mathcal{B}^\circ$ in such a fashion as to assign any mass in $[0, 1]$ to $\mathbb{Q}$.*

2. *Let $V_2 \supset V_0$ be the set of Borel measurable simple functions and define $T_2 : V_2 \to \mathbb{R}$ by $T_2(X) = \int X \, d\lambda$. T_2 extends T_0 and has continuous linear extension to $B((0, 1])$. For any H with outer measure 1 and inner measure 0, $T_2(1_H)$ can be any number $c \in [0, 1]$.*

Example 8.8.23 *Let $ca(K)$ denote the set of finite signed Borel measures on a compact K metric space, for example, $K \subset \mathbb{R}^k$ with the norm $\|\nu\| = \max_{E \in \mathcal{B}_K} |\nu|(E)$. $ca(K)$ is sometimes used as a commodity space modeling attributes. K is a set of attributes that goods may or may not have, and attributes are continuously adjustable by changes in the production process. If a good ν is a nonnegative measure on K, the interpretation is that the good has $\nu(E)$ of the attributes in the set E.*

1. *Let $V_0 = \{\nu \in ca(K, \mathcal{B}_K) : |\nu|(F) = |\nu|(K) \text{ for some finite set } F\}$. The continuous linear functions on V_0 are of the form $T_0(\nu) = \sum_{x \in F} u(x)\nu(\{x\})$ for some bounded, possibly nonmeasurable $u : K \to \mathbb{R}$. If T is an extension of T_0, and ν is nonatomic, $T(\nu)$ need have nothing to do with u. More generally, ν and η are mutually singular, even if their supports are dense, in which case $T(\nu)$ and $T(\eta)$ need not be at all related.*

2. *If we ask that nearby points in K be close substitutes, the situation is somewhat less distressing. Letting $\{f_n : n \in \mathbb{N}\}$ be a countable dense subset of the unit ball in $C(K)$, one can define the metric $d(\nu_1, \nu_2) = \sum_n \frac{1}{2^n} |\int f \, d\nu_1 - \int f \, d\nu_2|$ and ask about linear functions that are continuous with respect to this metric. We will see that such linear functions are of the form $T(\nu) = \int g \, d\nu$ for some $g \in C(K)$. The idea is that for $x \in K$, $g(x)$ is the price of the attribute x.*

8.8.d A Primer on Welfare Economics in Vector Spaces

In finite-dimensional problems, under certain convexity conditions on preferences, we know that we can decentralize a Pareto-optimal allocation through a competitive price system. We now sketch some of the sufficient conditions under which such a result holds in infinite-dimensional commodity spaces such as that of an Arrow-Debreu complete forward markets environment.

The equilibrium concept we use to study this issue is called a valuation equilibrium[9] because the equilibrium is characterized by a linear functional that assigns a value to each commodity bundle. In general, the valuation functional that we are interested in may not be representable as a price system; that is, there may be no price in the dual space such that the value of a commodity bundle is the inner product of the price vector and the quantity vector.

The specific environment is one in which there is trade in a complete set of state contingent claims that take place at $t = 0$. There are I consumers. The ith consumer has a consumption set $X_i \subset S$, where S is the space of infinite sequences of vectors in $\mathbb{R}^k$. The consumer evaluates bundles according to a utility function $u_i : X_i \to \mathbb{R}$. There are J firms. The jth firm chooses among points in a production set $Y_j \subset S$, where the negative components in $y_t \in \mathbb{R}^k$ denote inputs at time t and the positive components denote outputs. If, for example, it takes two periods for an input of time and raw materials to produce a machine that will make final goods one period after it is finished, then Y_j must have interactions between periods t and $t + 3$. These kinds of models allow us to analyze **capital theory**, that is, the theory of produced means of production.

The aggregate production possibilities set is $Y = \sum_{j=1}^{J} Y_j$. An $(I + J)$-tuple $[\{x_i\}_{i=1}^{I}, \{y_j\}_{j=1}^{J}]$ describing the consumption x_i of each consumer and the production y_j of each firm is an **allocation** for this economy. An allocation is **feasible** if $x_i \in X_i$, $y_j \in Y_j$, and $\sum_i x_i - \sum_j y_j = 0$. An allocation $[\{x_i\}, \{y_j\}]$ is **Pareto optimal** if the usual conditions apply, if it is feasible, and if there is no other feasible allocation $[\{x_i'\}, \{y_j'\}]$ with $u_i(x_i') \geq u_i(x_i)$ $\forall i$ and $u_j(x_j') > u_j(x_i)$ for some j.

A feasible allocation $[\{x_i^*\}, \{y_j^*\}]$ together with a bounded linear functional $\phi : S \to \mathbb{R}$ is a **competitive equilibrium** if: (i) $[\{x_i^*\}, \{y_j^*\}]$ is feasible (market clearing); (ii) for each i, $x \in X_i$, and $\phi(x) \leq \phi(x_i^*)$ (i.e., bundle x costs less than bundle x_i^*) implies $u_i(x) \leq u_i(x_i^*)$ (given the price system ϕ, the allocation x_i^* is utility maximizing); and (iii) for each j, $y \in Y_j$ implies $\phi(y) \leq \phi(y_j^*)$ (the allocation y_j^* is profit maximizing).

The **second welfare theorem** can be stated as follows. Suppose: (i) X_i is convex; (ii) convexity of preferences (i.e., if $u_i(x) > u_i(x')$, then $u_i(\theta x + (1 - \theta)x') > u(x')$ for $\theta \in (0, 1)$); (iii) $u_i : X_i \to \mathbb{R}$ is continuous; and (iv) Y is convex and has an interior point (this last assumption is not something that is needed in finite-dimensional spaces). Let $[(x_i^*), (y_j^*)]$ be a Pareto-optimal allocation. Then there exists a continuous linear functional $\phi : S \to \mathbb{R}$ (not identically zero) that satisfies conditions (ii) and (iii) of the definition of a competitive equilibrium. In both finite-dimensional spaces and infinite-dimensional spaces, the proof can be done using the Hahn-Banach theorem, but in both cases, it is just a thinly disguised version of a separation result.

There are issues about which commodity space $S = \{x = \{x_0, x_1, \ldots\} : x_t \in \mathbb{R}^k, \forall t, \text{ and } \|x\|_p < \infty\}$ to choose to satisfy the assumptions of the second welfare

---

9. See G. Debreu, "Valuation equilibrium and Pareto optimum," *Proceedings of the National Academy of Sciences of the USA* 40, 588–92 (1954).

theorem. In particular, if we choose ℓ_p with $p < \infty$, then $x \in \ell_p$ only if the series $\sum_{t=0}^{\infty} |x_t|^p$ converges, which requires $\lim_{t \to \infty} x_t = 0$. Furthermore, the production set may not have an interior point. However, these two conditions are satisfied if we choose ℓ_∞. Unfortunately, even with ℓ_∞, the valuation function $\phi(x)$ may not be representable in the more familiar form of an inner product (i.e., ℓ_1). That is, we may not have $\phi(x) = \sum_{t=0}^{\infty} p_t x_t$ for some sequence of price vectors $\{p_t\}$. To handle this issue, we have to impose additional requirements on preferences and technologies to ensure the existence of an inner product representation.

Let $x^T \in S$ denote the truncated sequence $(x_0, x_1, \ldots, x_T, 0, 0, \ldots)$. One useful strategy is to decompose $\phi(x)$ into a "well-behaved part," which we call $\psi(x) = \lim_{T \to \infty} \phi(x^T)$, and the remainder. It is possible to show that this sequence is bounded above by $\|\phi\| \, \|x\|_\infty$ and hence continuous. Furthermore, for each time period, we can use standard finite-dimensional analysis to show that there exists a bounded linear functional $\psi_t(x_t)$ and can write $\psi(x) = \sum_{t=0}^{\infty} \psi_t(x_t) = \sum_{t=0}^{\infty} p_t \cdot x_t = \sum_{t=0}^{\infty} \sum_{k=1}^{K} p_{t,k} x_{t,k}$. However, to get this to work, we need two additional assumptions: (i) truncated feasible sequences are feasible for all T sufficiently large; and (ii) a continuity requirement on preferences such that sufficiently distant consumption is not that important (i.e., for each i, if $x, x' \in X_i$ and $u_i(x) > u_i(x')$, then $u_i(x^T) > u_i(x')$ for all T sufficiently large). In that case it is possible to show that if $[\{x_i^*\}, \{y_j^*\}, \phi]$ is a competitive equilibrium, then $[\{x_i^*\}, \{y_j^*\}, \psi]$ is also a competitive equilibrium.

8.9 ◆ Weak Convergence in $L^p(\Omega, \mathcal{F}, P)$, $p \in [1, \infty)$

In $\mathbb{R}^k$, $\|\mathbf{x}_n - \mathbf{x}\| \to 0$ iff for all $\mathbf{y}$ in the dual of $\mathbb{R}^k$, $\mathbf{x}_n \cdot \mathbf{y} \to \mathbf{x} \cdot \mathbf{y}$. To see why, take $\mathbf{y} = \mathbf{e}_i$, $i = 1, \ldots, k$. This is emphatically *not* true in the L^p spaces.

8.9.a Weak Convergence Is Different

There are three main ways in which weak convergence is different.

1. It is different than norm convergence: Example 8.9.2 gives a weakly convergent sequence of points that are all at norm distance $\frac{1}{2}$ from each other.

2. It is different than finite-dimensional convergence: $\|X_n - X\|_p \to 0$ and $\langle X_n, Y \rangle \to \langle X, Y \rangle$ fail to be equivalent iff L^p is infinite dimensional.

3. It is different than any kind of convergence we have seen before: Example 8.9.7 shows that there is no metric that gives weak convergence.

Definition 8.9.1 *For a dual pair (p, q), $p \in [1, \infty)$, we say that a sequence X_n in L^p **converges weakly** to X, written $X_n \to_w X$, if for all $Y \in L^q$, $\langle X_n, Y \rangle \to \langle X, Y \rangle$. A set $K \subset L^p$ is **weakly closed** if it contains the limits of all weakly convergent sequences X_n in K, it is **weakly open** if it is the complement of a weakly closed set, and it is **weakly sequentially compact** if every sequence in K has a weakly convergent subsequence.*

It should be clear that an arbitrary intersection of weakly closed sets is closed; hence, by DeMorgan's rules, an arbitrary union of open sets is open. It should also be clear that $\emptyset$ and L^p are both weakly closed, and being complements of each other, they are both weakly open. Therefore the class of weakly open sets is a topology [see Definition 4.1.14 (p. 110)]. We will soon see, in Example 8.9.7, that there is no metric that gives this topology. However, Theorem 8.9.6 shows that there is a metric that gives the weakly open subsets of norm-bounded sets.

It is worth the time to get a full understanding of the following example showing that weak and norm convergence are very different, partly because it is a good example of proving something for simple functions and then using the fact that every measurable function is nearly simple.

Example 8.9.2 *For $(\Omega, \mathcal{F}, P) = ((0, 1], \mathcal{B}, \lambda)$, define $X_n(\omega) - 1$ if ω belongs to an even dyadic interval of order n and $X_n(\omega) = 0$ otherwise. For all $m \neq n$, $\|X_n - X_m\|_2 = \frac{1}{2}$, so that X_n is not a convergent sequence in $L^2((0, 1], \mathcal{B}, \lambda)$. Pick an arbitrary Y in the dual of $L^2((0, 1], \mathcal{B}, \lambda)$. We claim that*

$$\langle X_n, Y \rangle \to \langle \tfrac{1}{2}, Y \rangle. \tag{8.46}$$

*That is, this sequence X_n of "sawtooth" functions, sometimes known as **Rademacher functions**, that bounce up and down more and more quickly, converge weakly to the function that is constant at $\frac{1}{2}$. The intuitive reason is that as n becomes large, the X_n spend, roughly, half of any interval $(a, b]$ being equal to 1 and half being equal to 0. In other words, over the interval $(a, b]$, the X_n are nearly equal to $\frac{1}{2}$ on average.*

The essential idea of the argument is that: (a) the result is true for a dense class of simple functions, (b) every $Y \in L^2$ is nearly one of these simple functions, and (c) this is close enough.

(a). For $m \in \mathbb{N}$, let $\mathcal{B}(m)$ denote the smallest field containing all intervals of the form $(k/2^m, (k + 1)/2^m]$, $k \leq 2^m - 1$. Then $\mathcal{B}° = \cup_m \mathcal{B}(m)$ is the smallest field containing all dyadic intervals and $\mathcal{B} = \sigma(\mathcal{B}°)$. Since $\mathcal{B}°$ is a field generating $\mathcal{B}$, the simple $\mathcal{B}°$-measurable functions are dense in L^2.

Let R be a simple $\mathcal{B}°$-measurable function, that is, $R = \sum_{k \leq K} \beta_k 1_{B_k}$ for $B_k \in \mathcal{B}°$. Since K is finite, there exists an M such that each B_k belongs to $\mathcal{B}(M)$. For all $n > M$, $\langle X_n, R \rangle = \langle \frac{1}{2}, R \rangle$.

(b) and (c). Pick a simple, $\mathcal{B}°$-measurable R with $\|Y - R\|_2 \leq \epsilon$. By Hölder, for large n,

$$|\langle X_n, Y \rangle - \tfrac{1}{2}E\,R| = |\langle X_n, Y \rangle - \langle X_n, R \rangle|$$

$$= |\langle X_n, Y - R \rangle| \leq \|X_n\|_2 \cdot \|Y - R\|_2 < \epsilon/2. \tag{8.47}$$

Now, $\frac{1}{2}E\,Y$ is almost equal to $\frac{1}{2}E\,R$, specifically $|E\,\frac{1}{2}Y - E\,\frac{1}{2}R| \leq E\,|Y - R|_1 \leq \|Y - R\|_2 < \epsilon/2$. Combining yields $|\langle X_n, Y \rangle - \frac{1}{2}E\,Y| = |\langle X_n, Y \rangle - \langle \frac{1}{2}, Y \rangle| < \epsilon$.

8.9.b A Metric for Weak Convergence on Bounded Sets

The following metric is equivalent to $X_n \to_w X$ only in the special case that the sequence of X_n has a uniformly bounded norm and that we are working in an

$L^p(\Omega, \mathcal{F}, P)$ space with a separable dual space, that is, $p \in (1, \infty)$ and $\mathcal{F}$ is countably generated. Fortunately, this happy situation is typical.

Definition 8.9.3 *If $p, q \in (1, \infty)$, we define the **bounded weak convergence metric** on a separable $L^p(\Omega, \mathcal{F}, P)$ by $\rho_p(X_1, X_2) = \sum_m \frac{1}{2^m} |\langle X_1, Y_m \rangle - \langle X_2, Y_m \rangle| = \sum_m \frac{1}{2^m} |\langle X_1 - X_2, Y_m \rangle|$, where Y_m is a countable dense subset of $B_1(0) \subset L^q$, the unit ball in L^q.*

The other properties of a metric being clear, all that must be shown is the following.

Lemma 8.9.4 *If $\rho_p(X_1, X_2) = 0$, then $X_1 = X_2$.*

Proof. Since $\rho_p(X_1, X_2) = \rho_p(X_1 - X_2, 0)$, it is sufficient to show that if $\rho_p(X, 0) = 0$, then $X = 0$. The mapping $X \mapsto \rho_p(X, 0)$ is continuous (by Hölder), so it is sufficient to show that it is true for simple X. For fixed $X \in L^p$, Hölder shows that the mapping $Y \mapsto \langle X, Y \rangle$ is continuous on L^q. If $\langle X, Y_n \rangle = 0$ for all Y_n in a dense subset of $B_1(0)$, then $\langle X, Y \rangle = 0$ for all $Y \in B_1(0)$. Note that $X = \sum_{k \leq K} \beta_k 1_{E_k}$ and that $1_{E_k} \in B_1(0)$. Hence $\beta_k = 0$ for all $k \leq K$. ∎

Example 8.9.2 gave a sequence X_n with norm $\frac{1}{2}$ converging weakly to 0, so weak convergence does not imply norm convergence. However, norm convergence does imply weak convergence.

Exercise 8.9.5 *If $\|X_n - X\|_p \to 0$ for $1 < p < \infty$, L^p separable, then $\rho_p(X_n, X) \to 0$.* [Use Hölder's inequality.]

For norm-bounded sets, weak convergence and the bounded weak convergence metric are equivalent (which is the reason for the name of the metric).

Theorem 8.9.6 *If $X_n \to_{w^*} X$, then $\rho_p(X_n, X) \to 0$, and if there exists a B such that $\|X\|_p$, $\sup_n \|X_n\|_p \leq B$, then $\rho_p(X_n, X) \to 0$ implies that $X_n \to_{w^*} X$.*

Proof. If $\langle Y, X_n \rangle \to \langle Y, X \rangle$ for all Y, then it converges uniformly for all of the finite sets $\{Y_1, \ldots, Y_M\}$, which implies that $\rho_p(X_n, X) \to 0$.

Suppose now that $\|X\| \leq B$, $\sup_n \|X_n\|_p < B$, and $\rho_p(X_n, X) \to 0$. Pick $Y \in L^q$; we want to show that $\langle X_n, Y \rangle \to \langle X, Y \rangle$. By linearity, it is sufficient to consider the case where $\|Y\|_q = 1$. We must show that for large n, $|\langle (X_n - X), Y \rangle| < \epsilon$. Pick Y_m in the dense subset of $B_1(0)$ so that $\|Y_m - Y\|_q < \epsilon/8B$. Since $\rho_p(X_n, X) \to 0$, there exists an N such that for all $n \geq N$, $|\langle (X_n - X), Y_m \rangle| < \epsilon/2$. Therefore,

$$|\langle (X_n - X), (Y_m - Y) \rangle| \leq \|X_n - X\|_p \|Y_m - Y\|_q < 2B \cdot (2\epsilon/8B) = \epsilon/2.$$

(8.48)

Combining, for large n, we get $|\langle (X_n - X), Y \rangle| < \epsilon$. ∎

8.9.c The Nonmetrizability of Weak Convergence

A subtle problem arises here—when we do not have bounded sets, there is *no* metric that captures weak convergence. Happily, this ingenious perversity, revealed in the following example, does not matter very much to economists, and we can be almost exclusively happy with metric spaces. The example uses arguments about orthonormal sets from our coverage of the Hilbert space L^2 in regression analysis.

Example 8.9.7 (von Neumann) [↑*Example 8.9.2*] *For $n \in \mathbb{N}$, let $X'_n = 2(X_n - \frac{1}{2})$, so that $\|X'_n\|_2 = 1$ and for $n \neq m$, $\langle X'_m, X'_n \rangle = 0$. Let $A = \{X'_m + m \cdot X'_n : m, n \in \mathbb{N}, n > m\}$. There are two claims: First, for every $\epsilon > 0$ and every $Y \in L^2$, there exists $\mathbb{R}_k \in A$ such that $|\langle Y, \mathbb{R}_k \rangle| < \epsilon$, and second, there is no sequence $\mathbb{R}_k$ in A such that $|\langle Y, \mathbb{R}_k \rangle| \to 0$ for all $Y \in L^2$. This means that there can be no metric for weak convergence.*

For the first claim, we know that $|\langle Y, X'_m \rangle| \to 0$ and wish to show that $\lim_m \inf_{m<n} |\langle Y, (X'_m + m X'_n) \rangle| = 0$. Pick $\epsilon > 0$ and m such that $|\langle Y, X'_m \rangle| < \epsilon/2$. Now choose $n > m$ such that $|\langle Y, X'_n \rangle| < \epsilon/2m$. For this m, n we have $|\langle Y, (X'_m + m X'_n) \rangle| \leq |\langle Y, X'_m \rangle| + m |\langle Y, X'_n \rangle| < \epsilon/2 + m(\epsilon/2m) = \epsilon$.

For the second claim, pick an arbitrary sequence $R_k = X'_{m_k} + m_k X'_{n_k}$ in A. We show that there exists a $Y \in L^2$ such that $\langle Y, R_k \rangle \not\to 0$.

If m_k is bounded, then along a subsequence it is constant, say at M. For $Y = X'_M$, $|\langle Y, R_k \rangle| \geq 1$.

Suppose now that m_k is unbounded, so that n_k is also unbounded. Without changing the labeling, take a subsequence so that both m_k and n_k are strictly increasing and $m_k > k^2$. Let $Y = \sum_k \left(\frac{1}{k}\right)^2 X'_{n_k}$. We now make two subsidiary claims: first, that $Y \in L^2$, and second, that $\langle Y, R_k \rangle \geq \frac{m_k}{k}$. Since $m_k > k^2$, we have $\frac{m_k}{k^2} \geq 1$. Therefore, these subsidiary claims establish that $\langle Y, R_k \rangle \not\to 0$,

For the first subsidiary claim, let $Y_K = \sum_{k \leq K} \left(\frac{1}{k}\right)^2 X'_{n_k}$. For $K_1 \leq K_2$,

$$\left\| Y_{K_1} - Y_{K_2} \right\|_2 = \left\| \sum_{K_1 < k \leq K_2} \left(\frac{1}{k}\right)^2 X'_{n_k} \right\|_2 = \sum_{K_1 < k \leq K_2} \frac{1}{k^2}.$$

Since $\sum_k \frac{1}{k^2} < \infty$, this means that Y_K is a Cauchy sequence, so that Y is well defined. Finally, for the second subsidiary claim, since $\|Y_K - Y\|_2 \to 0$, $\langle Y, R_k \rangle = \lim_{K \to \infty} \langle Y_K, R_k \rangle \geq \frac{m_k}{k}$.

8.9.d   Weak Compactness in $L^p(\Omega, \mathcal{F}, P)$

We covered compactness using the norm distance, known as norm compactness, earlier. Further, since $[\|X_n - X\|_p \to 0] \Rightarrow [X_n \to_{w^*} X]$, any norm compact set is ρ_p-compact. The reverse is not true. Remember that ρ_p is only defined if $\mathcal{F}$ is countably generated. The following is a special case of Alaoglu's theorem in which the Banach space is separable and is its own second dual, $((L^p)^*)^* = (L^q)^* = L^p$. We discuss the general version in Chapter 10.

Theorem 8.9.8 (Weak Compactness)   *For $p \in (1, \infty)$, if K is norm bounded and weakly closed, then it is ρ_p-compact.*

Proof.   Suppose that $\|K\|_p \leq B$. The space $M = \times_{m=1}^\infty [-B, B]$ with the metric $d_M(x, y) = \sum_m \frac{1}{2^m} |x_m - y_m|$ is compact. For each m and $X \in K$, define $r_m(X) =$

$\langle X, Y_m \rangle$. The mapping $X \leftrightarrow (r_1(X), r_2(X), \ldots)$ from K to its range in M is an isometry. Any closed subset of a compact space is compact. ■

Thus, the closure of the unit ball in a separable $L^p(\Omega, \mathcal{F}, P)$, $p \in (1, \infty)$, is always weakly compact. We have seen many examples of infinite-dimensional norm-bounded sets that fail to be compact. There is a good reason that we can find such examples.

Exercise 8.9.9 Show that the closure of the unit ball in $L^p(\Omega, \mathcal{F}, P)$ is norm compact iff $(\Omega, \mathcal{F}, P)$ is essentially finite, that is, iff $L^p(\Omega, \mathcal{F}, P)$ is finite dimensional. [If $(\Omega, \mathcal{F}, P)$ is not essentially finite, then Ω has a countable nonnull partition $\{E_n : n \in \mathbb{N}\}$. Consider sequences of the form $X_n = \beta_n 1_{E_n}$. Choose the β_n carefully.]

8.10 ◆ Optimization of Nonlinear Operators

In this chapter we have dealt with linear operators and functionals. There are many problems in economics that involve nonlinear operators. For instance, contraction operators (as in many dynamic programming problems) are typically nonlinear. In this section, we show how variational methods and fixed-point theory (in the form of dynamic programming) can be used to prove the existence of an optimum of a nonlinear operator.

8.10.a Overview

We know that if $f : K \to \mathbb{R}$ is upper semicontinuous and K is compact, then f achieves its optimum on K; that is, there exists an $x^* \in K$ such that $f(x^*) \geq f(K)$ [see either of the equivalent definitions, 6.11.30 or 4.10.29, and Lemma 6.11.33 (p. 349)]. Working with K a subset of one of the separable $L^p(\Omega, \mathcal{F}, P)$ spaces with $p \in (1, \infty)$, we now have two metrics to choose between, the norm metric and the bounded weak metric. There is a trade-off to be made.

The essentials of the trade-off come from the observation that more sequences converge weakly than converge strongly; equivalently, it is easier for a sequence to converge weakly than to converge strongly. Here are the basic consequences of this observation.

Theorem 8.10.1 *If $L^p(\Omega, \mathcal{F}, P)$ is separable, $p \in (1, \infty)$, then*

   1. if $K \subset L^p$ is weakly closed, then it is norm closed;

   2. if $K \subset L^p$ is norm compact, then it is weakly sequentially compact;

   3. if $f : K \to \mathbb{R}$ is weakly continuous, then it is norm continuous; and

   4. if $f : K \to \mathbb{R}$ is weakly upper semicontinuous, then it is norm upper semi-continuous.

Proof. Throughout, X_n is a sequence in K. The crucial observation is that $[\|X_n - X\|_p \to 0] \Rightarrow [X_n \to_{w^*} X]$. To see why this is true, note that for every $Y \in L^q$, (p, q) a dual pair, $|\langle X_n, Y \rangle - \langle X_n, Y \rangle| = |\langle (X_n - X), Y \rangle| \leq \|X_n - X\|_p \|Y\|_q$ by Hölder's inequality, Theorem 8.8.4.

Suppose that K is weakly closed, that is, contains the weak limits of all sequences in K. Then K must contain all of its strong limits because $[\|X_n - X\|_p \to 0] \Rightarrow [X_n \to_{w^*} X]$.

Suppose that K is norm compact, that is, every sequence contains a subsequence that norm converges. The subsequence must weakly converge because $[\|X_n - X\|_p \to 0] \Rightarrow [X_n \to_{w^*} X]$.

If f is weakly continuous, then $[X_n \to_{w^*} X] \Rightarrow [f(X_n) \to f(X)]$. Therefore f is norm continuous because $[\|X_n - X\|_p \to 0] \Rightarrow [X_n \to_{w^*} X]$.

If f is weakly semicontinuous, then $[X_n \to_{w^*} X] \Rightarrow [f(X) \geq \limsup_n f(X_n)]$. Therefore f is norm semicontinuous because $[\|X_n - X\|_p \to 0] \Rightarrow [X_n \to_{w^*} X]$. ∎

Thus, it is easier for a set to be weakly compact but harder for a function to be weakly upper semicontinuous. The most useful way of making the trade-off for economic theory involves using norm continuous concave functions on weakly compact sets because it turns out that norm continuous concave functions are actually weakly upper semicontinuous.

8.10.b An Existence Theorem

We wish to show the following.

Theorem 8.10.2 *If K is a nonempty convex norm-closed norm-bounded subset of $L^p(\Omega, \mathcal{F}, P)$, $p \in (1, \infty)$, and $f : K \to \mathbb{R}$ is norm continuous and concave, then there exists $X^* \in K$ such that $f(X^*) \geq f(K)$.*

The crucial piece of the proof is the following lemma, which is another corollary of the Hahn-Banach theorem.

Lemma 8.10.3 *If K is norm closed and convex, then it is weakly closed.*

Proof. Let X_n be a sequence in K and suppose that for all $Y \in L^q$, (p, q) a dual pair, $\langle X_n, Y \rangle \to \langle X, Y \rangle$. We wish to show that $X \in K$. If it is not, then, since K is closed and convex and $X \notin K$, X and K can be strictly separated by a continuous linear function. By the Riesz representation theorem, this means that there exists a $Y^\circ \in L^q$ and an $\epsilon > 0$ such that $\langle X, Y^\circ \rangle > \langle K, Y^\circ \rangle + \epsilon$, contradicting $\langle X_n, Y^\circ \rangle \to \langle X, Y^\circ \rangle$. ∎

Proof of Theorem 8.10.2. *The proof is an exercise in putting together several disparate pieces.*

- *Since K is norm closed and convex, it is weakly closed.*
- *Since K is norm bounded and weakly closed, it is compact in the bounded weak convergence metric.*
- *Since f is continuous and concave, for every a, $\{X : f(X) \geq a\}$ is norm closed and convex, hence weakly closed.*

Therefore f is weakly upper semicontinuous on a compact set and so achieves its maximum.

The conclusions of the theorem do not apply to all Banach spaces, and the following two examples should help demonstrate why. In the first example, everything works, whereas in the second, things fall apart.

Example 8.10.4 *For $p \in (1, \infty)$, define $T : L^p((0, 1], \mathcal{B}, \lambda) \to \mathbb{R}$ by $T(X) = \int_0^{\frac{1}{2}} X(t) \, dt - \int_{\frac{1}{2}}^1 X(t) \, dt$. Letting $Y = 1_{(0, \frac{1}{2}]} - 1_{(\frac{1}{2}, 1]}$ yields $T(X) = \langle X, Y \rangle$, and since $Y \in L^\infty$, T is continuous and linear, hence concave. Let K be the closure of the unit ball in $L^p((0, 1], \mathcal{B}, \lambda)$. K is a nonempty convex norm-closed norm-bounded subset of the L^p, and T achieves its maximum on K at the point $X^* = Y$.*

Example 8.10.5 *Define $T : C([0, 1]) \to \mathbb{R}$ by $T(f) = \int_0^{\frac{1}{2}} f(t) \, dt - \int_{\frac{1}{2}}^1 f(t) \, dt$ and let K be the closure of the unit ball in $C([0, 1])$. T is bounded, hence continuous, and linear, hence concave. K is a nonempty convex norm-closed norm-bounded subset of the Banach space $C([0, 1])$, yet T fails to achieve its maximum on K.*

There is more to this last example. We will see in Chapter 9 that $(C([0, 1]))^*$, the dual space of $C([0, 1])$, is $ca([0, 1])$, the set of countably additive finite signed Borel measures on $[0, 1]$, where the linear functions are of the form $\langle f, \nu \rangle = \int f \, d\nu$. Thus, a sequence f_n in $C([0, 1])$ converges weakly to a function f iff $\int f_n \, d\mu \to \int f \, d\mu$ for all μ. By dominated convergence, for a uniformly bounded sequence of f_n, this is equivalent to $f_n(t) \to f(t)$ for all t. However, this does not mean that the limit function, f, is continuous. All that is required is that f belong to the second dual space, $((C([0, 1]))^*)^* = (ca([0, 1]))^*$, which contains, for example, all of the bounded measurable functions. In particular, it contains the function $f = 1_{(0, \frac{1}{2}]} - 1_{(\frac{1}{2}, 1]}$, which solves the maximization problem and is at norm distance at least 1 from every element of $C([0, 1])$. By contrast, for the L^p spaces of the existence result, the second dual space, $((L^p)^*)^*$, is equal to L^p.

The following is a typical application of Theorem 8.10.2. Recall that $L^{p,k}(\Omega, \mathcal{F}, P)$ is the set of $\mathbb{R}^k$-valued random variables having $\int \|X\|^p \, dP < \infty$.

Theorem 8.10.6 *For each i in a finite set I, let $X_i = L_+^{p,k}$, $p \in (1, \infty)$, be a consumption set an $w_i \in X_i$ an endowment. If each $u_i : X_i \to \mathbb{R}$ is norm continuous and concave, then there exists a Pareto-optimal allocation.*

Proof. Let $w = \sum_i w_i$ be the aggregate endowment. The set of feasible allocations is $F := \{(x_i)_{i \in I} \in \times_{i \in I} X_i : \sum_i x_i \leq w\}$, which is norm bounded (because $0 \leq x_i$ for each $i \in I$), norm closed, and convex. Since each u_i is norm continuous and concave, so is $U((x_i)_{i \in I}) := \sum_i \alpha_i u_i(x_i)$ for nonnegative α_i. U achieves its maximum on F, and if the α_i are strictly positive, any such maximum is Pareto optimal. ∎

8.10.c Dynamic Programming with a Compact Set of State Variables

An important and frequently used example of operators is dynamic programming. In infinite-horizon problems, dynamic programming turns the problem of finding an infinite sequence (or plan) describing the evolution of a vector of (endogenous) state variables into simply choosing a single vector value for the state variables and finding the solution to a functional equation.[10]

A dynamic programming problem is a quadruple (X, Γ, r, β) about which we make a number of assumptions. X denotes the set of possible values of (endogenous) state variables with typical element x, and $\Gamma : X \twoheadrightarrow X$ denotes the constraint correspondence describing feasible values for the endogenous state variable. For this subsection, we make the following assumptions:

Assumption 1: $X \subset \mathbb{R}^n$ is compact and convex and $\Gamma(x)$ is nonempty, compact-valued, and continuous.

In the next subsection, we examine some delicate issues that arise when we drop the compactness assumption, making the next assumption:

Assumption 1': $X \subset \mathbb{R}^n$ is convex and $\Gamma(x)$ is nonempty, compact-valued, and continuous.

Let $G = \{(x, y) \in X \times X : y \in \Gamma(x)\}$ denote the graph of Γ. The instantaneous returns or per-period objective function is $r : G \rightarrow \mathbb{R}$, and we will always make this last assumption:

Assumption 2: $r : G \rightarrow \mathbb{R}$ is a continuous function.

Finally, $\beta \in (0, 1)$ denotes the discount factor.

Definition 8.10.7   *A **plan**, a **feasible plan**, the **k-period objective function**, and the **objective function** are defined by*

1. *a sequence $x = (x_t)_{t=0}^{\infty}$ is a **plan**;*

2. *a plan is **feasible from** x° if $x_0 = x^{\circ}$ and for all t, $x_{t+1} \in \Gamma(x_t)$;*

3. *for any feasible x, the **k-period objective function** is $\varphi_k(x) = \sum_{t=0}^{k} \beta^t r(x_t, x_{t+1})$; and*

4. *for any feasible x, the **objective function** is $\varphi(x) = \sum_{t=0}^{\infty} \beta^t r(x_t, x_{t+1})$.*

The set of feasible plans from x° is denoted $F(x^{\circ})$.

Recall that $\mathbb{Z}_+ = \{0, 1, \ldots\}$.

Lemma 8.10.8   *For all $x^{\circ} \in X$, $F(x^{\circ})$ is a nonempty compact subset of $X^{\mathbb{Z}_+}$ with the product topology.*

Proof.   The nonemptiness of $F(x^{\circ})$ follows from (the axiom of choice and) $\Gamma : X \twoheadrightarrow X$ being nonempty valued for all x. Since Γ is continuous and the continuous image of a compact set under a continuous correspondence is compact (by Lemma 6.1.26), $\text{proj}_{0,\ldots,t}(F(x^{\circ}))$ is easily seen to be a closed, hence compact, subset of $\times_{s=0}^{t} X$. This in turn implies that $F(x^{\circ})$ is a closed, hence compact, subset of $X^{\mathbb{Z}_+}$.   ∎

We are interested in solving the **sequence problem starting from** x°, $SP(x^{\circ})$, which is defined by

$$SP(x^{\circ}) : \max_{x \in F(x^{\circ})} \varphi(x), \tag{8.49}$$

---

10. For a more general and detailed treatment of dynamic programming than our next two examples, see *Recursive Methods in Economic Dynamics* by N. L. Stokey and R. E. Lucas, Jr., with E. C. Prescott (Cambridge, Mass.: Harvard University Press, 1989).

and which has the **value function** defined by

$$v(x^\circ) = \max_{x \in F(x^\circ)} \varphi(x). \qquad (8.50)$$

We have only one result, but it has several parts.

Theorem 8.10.9 *Under Assumptions 1 and 2, for all $x^\circ \in X$, [(1)]*

1. *$SP(x^\circ)$ has a compact and nonempty set of solutions, S;*

2. *for all $\epsilon > 0$, for large k, any solution, x_k^* in S_k, the set of solutions to the k-period problem,*

$$SP_k(x^\circ): \max_{x \in F(x^\circ)} \varphi_k(x), \qquad (8.51)$$

is within ϵ of S, $d(x_k^*, S) < \epsilon$;

3. *the value functions, v_k, for the k-period problem, are continuous and converge uniformly to the continuous function v;*

4. *v is the unique solution to the functional equation (FE),*

$$v(x^\circ) = \max_{y \in \Gamma(x^\circ)} r(x^\circ, y) + \beta v(y); \quad \text{and} \qquad (8.52)$$

5. *any plan satisfying*

$$x_0 = x^\circ \quad \text{and} \ (\forall t) \left[x_{t+1} \in \arg\max_{y \in \Gamma(x_t)} [r(x_t, y) + \beta v(y)] \right] \qquad (8.53)$$

solves the problem $SP(x^\circ)$.

Notation 8.10.10 *If $r(\cdot, \cdot)$ and $v(\cdot)$ are smooth and the solutions to (8.52) are interior, then setting the derivatives equal to 0,*

$$D_y[r(x^\circ, y) + \beta v(y)] = 0, \qquad (8.54)$$

*yields what are called the **Euler equations**. The arg max correspondence, $x^\circ \mapsto y^*(x^\circ)$, is called the **policy function**.*

Proof. The first three results come fairly directly from the theorem of the maximum.

Give $C(X^{\mathbb{Z}_+})$ the sup norm, give $\Theta := X \times C(X^{\mathbb{Z}_+})$ the product topology, and let $\theta^\circ = (x^\circ, \varphi^\circ)$ be a typical element. Define the correspondence $\theta \twoheadrightarrow \Psi(\theta)$ from Θ to $X^{\mathbb{Z}_+}$ by $\Psi(x^\circ, \varphi^\circ) = F(x^\circ)$. Define $U : X^{\mathbb{Z}_+} \times \Theta \to \mathbb{R}$ by $U(x, (x^\circ, \varphi^\circ)) = \varphi^\circ(x)$.

The correspondence Ψ is continuous, compact valued, and nonempty valued and U is continuous. By the theorem of the maximum, $(x^\circ, \varphi^\circ) \mapsto v(x^\circ, \varphi^\circ)$ is continuous, and the arg max correspondence, $(x^\circ, \varphi^\circ) \mapsto G(x^\circ, \varphi^\circ)$, is nonempty valued, compact valued, and upper hemicontinuous. Furthermore, the collection of functions $v_k(\cdot) := v(\cdot, \varphi_k)$ is uniformly equicontinuous because U and Ψ are uniformly continuous. Therefore the v_k converge uniformly to $v(\cdot, \varphi)$.

The argument that v solves the FE (8.52) and that following any plan with $x_0 = x^\circ$ and $(\forall t)\left[x_{t+1} \in \arg\max_{y \in \Gamma(x_t)}[r(x_t, y) + \beta v(y)]\right]$ is optimal is straightforward. We first show that $v(x^\circ)$ is the value to following any such policy and that

$v(x^\circ) \geq \max_{y \in \Gamma(x^\circ)} r(x^\circ, y) + \beta v(y)$, and then show that a strict inequality leads to a contradiction.

1. To see that $v(x^\circ) \geq \max_{y \in \Gamma(x^\circ)} r(x^\circ, y) + \beta v(y)$, y', solve $\max_{y \in \Gamma(x^\circ)} r(x^\circ, y) + \beta v(y)$ and let $x' \in X^{\mathbb{Z}_+}$ be an optimal solution starting from $y' \in X$. Any solution starting from x° must do at least as well as the feasible solution (y', x'). By induction, this is what the plan does.

2. By the same logic, if the inequality is strict, then there is a feasible plan starting at x° doing better than any other feasible plan starting at x°.

Finally, to show that v is the unique solution to the FE, we prove a lemma, due to Blackwell, that shows that the FE is a contraction mapping on the complete space $C(X)$.   ∎

Lemma 8.10.11 (Blackwell's Sufficient Conditions for a Contraction)
Suppose that

1. *X is a nonempty set;*

2. *$(\mathfrak{B}(X), \| \cdot \|_\infty)$ is the Banach space of bounded functions on X with the sup norm; and*

3. *$T : \mathfrak{B}(X) \to \mathfrak{B}(X)$ satisfies*
   1. *(monotonicity) for all $f, g \in \mathfrak{B}(X)$, $[f \leq g] \Rightarrow [T(f) \leq T(g)]$, and*
   2. *(discounting) $\exists \rho \in (0, 1)$ such that for all $a \geq 0$, $[T(f + a)](x) \leq (Tf)(x) + \rho a$.*

Then T is a contraction with modulus ρ.

Proof.   For any $f, g \in \mathfrak{B}$, $f \leq g + \|f - g\|_\infty$. By monotonicity and discounting, respectively, we have the two inequalities,

$$T(f) \leq T\left(g + \|f - g\|_\infty\right) \leq T(g) + \rho \|f - g\|_\infty.$$

Reversing the roles of f and g gives $T(g) \leq T(f) + \rho \|f - g\|_\infty$. Since the inequalities hold at all x, $\|T(f) - T(g)\|_\infty \leq \rho \|f - g\|_\infty$.   ∎

Completing the Proof of Theorem 8.10.9.   *The mapping $v \mapsto T(v)$ defined by $T(v)(x^\circ) = \max_{y \in \Gamma(x^\circ)}[r(x^\circ, y) + \beta v(y)]$ is clearly monotonic and discounts with $\beta \in (0, 1)$. Further, $C(X)$ is a closed vector subspace of $\mathfrak{B}(X)$ and $T : C(X) \to C(X)$. Hence, by Banach's contraction mapping theorem, T has a unique fixed point in $C(X)$.*

Under additional assumptions on the reward function, we get more information about the value function. At work is the same kind of argument that completed the proof of Theorem 8.10.9—if a contraction mapping takes a closed set into itself, then the fixed point must be in the closed set.

Assumption 3. For each y, $r(x, y)$ is strictly increasing in x.

Assumption 4. Γ is monotone in the sense that $x \leq \widehat{x}$ implies $\Gamma(x) \subset \Gamma(\widehat{x})$.

Theorem 8.10.12   *Let $\mathfrak{I}(X)$ denote the set of bounded continuous strictly increasing functions. Under Assumptions 1–4, $v \in \mathfrak{I}(X)$.*

Proof. It is sufficient to show that T takes $\mathrm{cl}(\mathcal{I}(X))$ to $\mathcal{I}(X)$. $v \in \mathrm{cl}(\mathcal{I}(X))$ iff v is continuous and nondecreasing. We show that for any $x_1 > x_0$, $(Tv)(x_1) > (Tv)(x_0)$. This follows from the following sequence: $(Tv)(x_1) = \max_{y \in \Gamma(x_1)}[r(x_1, y) + \beta v(y)] = r(x_1, y_1^*) + \beta v(y_1^*) \geq r(x_1, y_0^*) + \beta v(y_0^*) > r(x_0, y_0^*) + \beta v(y_0^*) = \max_{y \in \Gamma(x_x)}[r(x_0, y) + \beta v(y)] = (Tv)(x_0)$. ∎

Sometimes, the following two additional assumptions can reasonably be made:
Assumption 5. G, the graph of Γ, is a convex set.
Assumption 6. r is strictly concave.

Theorem 8.10.13 *Let $\mathcal{K}(X)$ denote the set of bounded continuous strictly increasing concave functions. Under Assumptions 1–6, $v \in \mathcal{K}(X)$.*

Proof. Since $\mathcal{K}(X) \subset \mathcal{C}_b(X)$ this follows from the closedness property of the contraction mapping theorem (CMT), provided we show that for $x_\alpha = \alpha x_0 + (1 - \alpha)x_1$ with $x_0 \neq x_1$ and $\alpha \in (0, 1)$, $(Tv)(x_\alpha) > \alpha(Tv)(x_0) + (1 - \alpha)(Tv)(x_1)$. Let $y_i^* \in \Gamma(x_i)$ attain $(Tv)(x_i)$ and let $y_\alpha = \alpha y_0^* + (1 - \alpha)y_1^*$. Then

$$(Tv)(x_\alpha) = \max_{y \in \Gamma(x_\alpha)} r(x_\alpha, y) + \beta v(y)$$

$$\geq r(x_\alpha, y_\alpha) + \beta v(y_\alpha)$$

$$> \alpha \left[r(x_0, y_0^*) + \beta v(y_0^*) \right] + (1 - \alpha) \left[r(x_1, y_1^*) + \beta v(y_1^*) \right]$$

$$= \alpha(Tv)(x_0) + (1 - \alpha)(Tv)(x_1),$$

where the (first) weak inequality follows since $y_a \in \Gamma(x_\alpha)$ is feasible from x_α by Assumption 5 but not necessarily optimal and the second strict inequality follows from Assumption 6. ∎

8.11 ◆ A Simple Case of Parametric Estimation

In this section we cover some of the major themes that show up in parametric estimation: the classical statistical model, linear and nonlinear estimators, bias of estimators, best linear unbiased estimators, the Cramér-Rao lower bound and best unbiased estimators, better MSE estimators, Bayes estimators, power functions, hypothesis testing, the Neyman-Pearson lemma, conditional probabilities and causal arguments, and selection/misspecification problems.

To be so ambitious in scope in a mere section of a single chapter requires a rather drastic sharpening of focus. We do all of the above in the context of the simplest parametric estimation model that we know of—trying to estimate the probability of success from a sequence of independent trials.

8.11.a The Model

In the simplest variant of the model, we suppose that there are n iid random variables, data, $(X_1(\omega), \ldots, X_n(\omega)) = (X_1, \ldots, X_n)(\omega)$ with $P(X_i = 1) = \theta$, and $P(X_i = 0) = 1 - \theta$ for some true value θ° in $\Theta = [0, 1]$, abbreviated "$X_1, \ldots, X_n$

is iid Bern(θ) for some $\theta \in \Theta$." Here "Bern" is short for Bernoulli, who first studied these problems systematically. The parameter[11] of interest is $\theta°$, and that is what we would like the data to tell us about.

The idea is that there are n independent trials of some process that may succeed or fail. Examples include, but are hardly limited to: a vaccine that may protect only a proportion, $\theta°$, of the vaccinated; a job training program that leads to a proportion, $\theta°$, finding jobs; a pre-kindergarden care program that helps lead to some proportion, $\theta°$, of enrolled children graduating from high school; a set of building codes that leads to some proportion, θ, of new buildings remaining standing in a strong earthquake; or a monitoring and prosecution policy by the S.E.C. that leads to the management of some proportion, $\theta°$, of firms breaching fiduciary duty. We record $X_i = 1$ if the ith trial is a "success" and $X_i = 0$ otherwise. The word "success" is in quotes because the 0 and 1 labels are arbitrary. We would have, in all essential ways, the same problem if we had defined $Y_i(\omega) := 1 - X_i(\omega)$ and tried to infer the parameter $\gamma° := 1 - \theta°$.

We are interested in estimating the true value of θ, the probability of success in this process. We are also often interested in comparing one process with another to see which has the higher success rate. For example: we might be interested in comparing the survival rates of the vaccinated and the nonvaccinated or the survival rates of those vaccinated with the new vaccine to those vaccinated with the old vaccine; in the comparison of a job training program that teaches literacy and other job skills and one that teaches how to interview; in comparing the effects of a having and not having a prekindergarden program on the likelihood of death before adulthood; in comparing a building code that requires strong and rigid frames for buildings to one that requires flexible frames; or in comparing the rate of corporate fraud under a monitoring and prosecution policy that uses only self-reported statistics and negotiated settlements with corporate criminals to one that audits and prosecutes criminals.

All of these comparisons are modeled as being based on a comparison of one set of data, $X_1, \ldots, X_n$, that is iid θ_X, to another, independent data set, $Y_1, \ldots, Y_m$, that is iid θ_Y.

Definition 8.11.1 *A function of $X_1, \ldots, X_n$ is called a **statistic**, and an **estimator** is a statistic $(X_1, \ldots, X_n) \mapsto \hat{\theta}(X_1, \ldots, X_n)$ from observations to Θ.*

Our present focus is finding and evaluating different elements of the class of random variables known as estimators.

Example 8.11.2 *Consider the following, bearing in mind that $\sum_{i \leq n} X_i$ is simply the count of the number of successes in the data and that $\frac{1}{n} \sum_{i \leq n} X_i$ is the proportion of successes in the data.*

---

11. From the *Oxford English Dictionary*, the word "parameter" comes from a combination of the Greek words for "beside" or "subsidiary to," and "measure." It was first used mathematically in the 1850s, and is now understood as "A quantity which is constant (as distinct from the ordinary variables) in a particular case considered, but which varies in different cases; esp. a constant occurring in the equation of a curve or surface, by the variation of which the equation is made to represent a family of such curves or surfaces."

1. *Our favorite estimator is $\widehat{\theta}(X_1, \ldots, X_n) \equiv \frac{1}{7}$. It has a really minimal amount of variance, namely 0. Other than that redeeming feature, unless θ° is very close to $\frac{1}{7}$, it is, intuitively, a pretty poor estimator.*

2. *If we are pretty sure that $\theta^\circ = \frac{1}{2}$, say, one appealing class of estimators is $\widehat{\theta}(X_1, \ldots, X_n) = \alpha_n \frac{1}{2} + (1 - \alpha_n) \frac{1}{n} \sum_i X_i$ with $\alpha_n \to 0$ as $n \to \infty$.*

3. *If we are completely, 100% sure that $\theta^\circ \in [\frac{1}{2}, 1]$, a sensible estimator might be $\widehat{\theta}(X_1, \ldots, X_n) = \max\{\frac{1}{2}, \frac{1}{n} \sum_i X_i\}$.*

4. *If we think that the iid assumption is violated in a fashion that makes the first n' observations more likely to be informative, an appealing class of estimators is $\widehat{\theta}(X_1, \ldots, X_n) = \alpha \frac{1}{n'} \sum_{i \le n'} X_i + (1 - \alpha) \frac{1}{n-n'} \sum_{n' < j \le n} X_j$.*

5. *If we had the inside knowledge of a powerful diety, we could use $\widehat{\theta} \equiv \theta^\circ$. This is not a statistic because it takes different values depending on something other than the data.*

For this particular estimation problem, estimating the probability of success, the one rather obvious estimator that has almost all of the optimality properties that have been studied is $\widehat{\theta}(\omega) = \frac{1}{n} \sum_{i \le n} X_i(\omega)$, the sample proportion of successes. Using this estimator for different processes and comparing the estimates allows us to statistically decide, that is, to guess, which process has the better success rate. You should be very careful not to overgeneralize the results discussed here: it is not generally true that the average is the best estimator.

8.11.a.1 Maximum Likelihood Estimators

From elementary statistics, the likelihood of observing $(X_1, \ldots, X_n) = (x_1, \ldots, x_n)$, also written $\mathbf{X} = \mathbf{x}$, when θ is the true success rate is

$$P(\{\omega : \mathbf{X}(\omega) = \mathbf{x}\}) = L(x_1, \ldots, x_n \mid \theta) = \Pi_i \theta^{x_i} (1 - \theta)^{1 - x_i}.$$

A **maximum likelihood estimator (MLE)** of θ solves the problem

$$\max_{\theta \in [0,1]} \Pi_i \theta^{x_i} (1 - \theta)^{1 - x_i}.$$

If we take logs and let $\mathcal{L}(\mathbf{x} \mid \theta) = \log(L(\mathbf{x} \mid \theta))$, the MLE solves the problem

$$\max_{\theta \in [0,1]} \mathcal{L}(\mathbf{x} \mid \theta) = \log(\Pi_i \theta^{x_i} (1 - \theta)^{1 - x_i})$$

$$= \sum_i [x_i \log(\theta) + (1 - x_i) \log(1 - \theta)].$$

As it is the sum of concave functions, $\theta \mapsto \mathcal{L}(\mathbf{x} \mid \theta)$ is concave. If the solution is in $(0, 1)$, the first-order conditions characterize it and $\partial \mathcal{L}(\mathbf{x} \mid \widehat{\theta}_{MLE})/\partial \theta = 0$ iff $\widehat{\theta}_{MLE}(x) = \frac{1}{n} \sum_{i \le n} x_i$.

Note that we have, at different times, written $\widehat{\theta}_{MLE}$ as depending on two different objects: on $\mathbf{x}$, the realized value of the random vector $\mathbf{X}$, and on ω, the point in the probability space on which the random vector is defined. This was done on purpose. The estimator is a random variable, that is, a function of ω, but it only depends on ω through the realized value of $\mathbf{X}$; that is, it is a function of $\mathbf{x}$.

8.11.a.2 Best Linear Unbiased Estimators

Perhaps because of the psychological bias against the word "biased," we often focus on unbiased estimators.

Definition 8.11.3   *If $\Theta \subset \mathbb{R}^k$ for some k, then $\widehat{\theta}$ is **unbiased** if for all $\theta \in \Theta$, $\int \widehat{\theta}(x)\, d\mu_\theta(x) = \theta$.*

This is often written as $\forall \theta$, $E_\theta \widehat{\theta}_{MLE} = \theta$, where we understand "$E_\theta$" to be "expectation when the true value of the parameter is θ," that is, E_θ is integration against μ_θ.

Example 8.11.4   *The random variable $\widehat{\theta}_{MLE}$ is an **unbiased estimator**. That is, for all θ, if $X_1, \ldots, X_n$ are iid with $P(\{\omega : X_i = 1\}) = \theta$ and $P(\{\omega : X_i = 0\}) = 1 - \theta$, then $E\,\widehat{\theta}_{MLE} = \theta$.*

The statistic $\widehat{\theta}_{MLE}$ is also a linear function of the data, being of the form $\sum_i x_i y_i$ with each $y_i = \frac{1}{n}$.

Definition 8.11.5   *An unbiased linear estimator, $\widehat{\theta}_{\mathbf{y}} = \mathbf{xy}$, is **more efficient** than another unbiased estimator, $\widehat{\theta}_{\mathbf{z}} = \mathbf{xz}$, if for all θ, $\mathrm{Var}_\theta(\widehat{\theta}_{\mathbf{y}}) \leq \mathrm{Var}_\theta(\widehat{\theta}_{\mathbf{z}})$, where $\mathrm{Var}_\theta(\widehat{\theta}) := \int (\widehat{\theta}(x) - \theta)^2\, d\mu_\theta$ is the variance of $\widehat{\theta}$ when the data are distributed according to μ_θ. A linear estimator, $\widehat{\theta}_{\mathbf{y}}$, is a **best linear unbiased estimator** (**BLUE**), if for all θ and all unbiased linear estimators, $\widehat{\theta}_{\mathbf{z}}$, $\mathrm{Var}_\theta(\widehat{\theta}_{\mathbf{y}}) \leq \mathrm{Var}_\theta(\widehat{\theta}_{\mathbf{z}})$.*

A BLUE is a most efficient linear estimator. If we think that variance is a good measure of the size of the spread of a random variable around its mean, then being better is the requirement that the estimator is less widely spread around the true value.

Lemma 8.11.6   *If $(X_1, \ldots, X_n)$ is iid Bern(θ), then $\widehat{\theta} = \frac{1}{n} \sum_{i \leq n} X_i$ is the unique BLUE of θ.*

Proof.   For any linear unbiased estimator $\tilde{\theta}$, the variance is $E\,(\tilde{\theta} - \theta)^2 = \theta(1 - \theta) \cdot (\sum_i y_i^2)$. Solving the problem $\min \sum_i y_i^2$ subject to $\sum_i y_1 = 1$ gives the BLUE. This is an easy application of the Kuhn-Tucker theorem, and the answer is that $y_i = \frac{1}{n}$ gives the BLUE of θ.   ∎

8.11.a.3 Fisher's Information Inequality

It is possible that there is a nonlinear function of the data that gives a better unbiased estimator of θ, that is, one with a lower variance no matter what θ is. The **Cramér-Rao lower bound**, sometimes known as **the information inequality**, tells us that this is not possible. This is a very strong statement: there is no unbiased estimator with lower variance.[12]

Theorem 8.11.7 (Cramér-Rao Lower Bound for Bernoulli Data)   *If the data are iid Bern(θ) for some $\theta \in (0, 1)$, then for any unbiased estimator $\widehat{\theta}$,*

$$\mathrm{Var}_\theta(\widehat{\theta}) \geq \frac{1}{E_\theta(D_\theta \log f(\mathbf{X} \mid \theta))^2}.$$

---

12. If you want a biased estimator with truly minimal variance, we suggest $\widehat{\theta} = 1/7$.

Exercise 8.11.8 Show that for $\widehat{\theta} = \frac{1}{n} \sum_i X_i$, $\mathrm{Var}_\theta(\widehat{\theta}) = 1/(E_\theta(D_\theta \log f(\mathbf{X} \mid \theta))^2)$. [If you get stuck, consult (8.57).]

The proof passes through a strengthened form of the Cauchy-Schwarz inequality, here applied to random variables with the norm $\|X\|_2 = [\int_\Omega X^2 \, dP]^{1/2} < \infty$. Recall that the covariance of two random variables in defined as $\mathrm{Cov}(X, Y) = E(X - E\,X)(Y - E\,Y) = E\,XY - E\,X E\,Y$, and the variance is defined as $\mathrm{Var}(X) = \mathrm{Cov}(X, X)$.

Lemma 8.11.9 $\mathrm{Cov}(X, Y)^2 \leq \mathrm{Var}(X)\,\mathrm{Var}(Y)$, *with equality iff* $(X - E\,X)(\omega)$ *is, with probability* 1, *a linear function of* $(Y - E\,Y)(\omega)$.

The Cauchy-Schwarz inequality delivers $\mathrm{Cov}(X, Y)^2 \leq \mathrm{Var}(X)\,\mathrm{Var}(Y)$, but what is new is the characterization of the mean 0 pairs (X, Y) for which there is equality.

Proof. It is sufficient to show that this is true if $E\,X = E\,Y = 0$. For any a, b, $(a - b)^2 \geq 0$ with equality iff $a = b$, that is,

$$\tfrac{1}{2}a^2 + \tfrac{1}{2}b^2 \geq ab \quad \text{or} \quad a^2 + b^2 \geq 2ab \text{ with equality iff } a = b.$$

Now, setting $a(\omega) = X/\|X\|_2$ and $b(\omega) = Y/\|Y\|_2$, for all $\omega \in \Omega$, we have

$$\frac{1}{2}\frac{X^2(\omega)}{\|X\|_2^2} + \frac{1}{2}\frac{Y^2(\omega)}{\|Y\|_2^2} \geq \frac{X(\omega)Y(\omega)}{\|X\|_2\,\|Y\|_2},$$

with equality iff a.e. $X = cY$ a.e., where $c = \|X\|_2/\|Y\|_2$. Taking expectations on each side yields

$$\frac{1}{2}\frac{E\,X^2}{\|X\|_2^2} + \frac{1}{2}\frac{E\,Y^2}{\|Y\|_2^2} \geq \frac{E\,XY}{\|X\|_2\,\|Y\|_2}. \tag{8.55}$$

The left-hand side in (8.55) is equal to 1. Rearranging gives $E\,XY \leq \|X\|_2 \cdot \|Y\|_2$ with equality iff $X = cY$. ■

Proof of Theorem 8.11.7. Let $\mathbf{X}$ be the random vector of data. We take $R = R(\mathbf{X})$ to be an unbiased estimator of θ and $S = S(\mathbf{X}) = D_\theta \log f(\mathbf{X} \mid \theta)$, where $f(\mathbf{X} \mid \theta)$ is the density of $\mathbf{X}$ when θ is the true value. For $\theta \in (0, 1)$,

$$\mathrm{Var}_\theta(R) \geq \frac{\mathrm{Cov}_\theta(R, S)^2}{\mathrm{Var}_\theta(S)},$$

with equality iff $R(\mathbf{X}) - \theta$ is a linear function of $S(\mathbf{X})$ (because we will see that $E_\theta S(\mathbf{X}) = 0$). We now complete the proof by showing that for any unbiased R,

$$\mathrm{Cov}_\theta(R, S)^2 = 1, \quad \text{and} \quad E_\theta(D_\theta \log f(\mathbf{X} \mid \theta)) = 0.$$

Rewriting from above, we have

$$\log f(\mathbf{X} \mid \theta) = \log(\Pi_i \theta^{X_i}(1-\theta)^{1-X_i}), \text{ so that} \tag{8.56}$$

$$S = D_\theta \log f(\mathbf{X} \mid \theta) = \frac{\sum_i X_i}{\theta} - \frac{\sum_i(1-X_i)}{(1-\theta)}, \text{ so that} \tag{8.57}$$

$$E_\theta S = E_\theta D_\theta \log f(\mathbf{X} \mid \theta) = \frac{n\theta}{\theta} - \frac{n(1-\theta)}{(1-\theta)} = 0. \tag{8.58}$$

We now show that $\text{Cov}_\theta(R, S) = 1$. Since $E_\theta S = 0$, this is equivalent to $E\,RS = 1$. $R(\mathbf{X})$ is an unbiased estimator, $E_\theta R(\mathbf{X}) \equiv \theta$, and some manipulation of $dE_\theta R(\mathbf{X})/d\theta \equiv 1$ gives the desired information. We let $\mathfrak{X} = \{0, 1\}^n$ denote the set of all possible strings of 0's and 1's that $(X_1, \ldots, X_n)$ could take on:

$$1 = \frac{d}{d\theta} E_\theta R(\mathbf{X}) = \frac{d}{d\theta} \sum_{\mathbf{x} \in \mathfrak{X}} R(\mathbf{x}) f(\mathbf{x} \mid \theta) \left(= \int_{\mathfrak{X}} R(\mathbf{x}) \, d\mu_\theta(\mathbf{x}) \right) \tag{8.59}$$

$$= \sum_{\mathbf{x} \in \mathfrak{X}} R(\mathbf{x}) \frac{d}{d\theta} f(\mathbf{x} \mid \theta) = \sum_{\mathbf{x} \in \mathfrak{X}} R(\mathbf{x}) \frac{d}{d\theta} f(\mathbf{x} \mid \theta) \frac{f(\mathbf{x} \mid \theta)}{f(\mathbf{x} \mid \theta)} \tag{8.60}$$

$$= E_\theta R(\mathbf{X}) \frac{\frac{d}{d\theta} f(\mathbf{X} \mid \theta)}{f(\mathbf{X} \mid \theta)} = E_\theta R(\mathbf{X}) \frac{d}{d\theta} \log f(\mathbf{X} \mid \theta) \tag{8.61}$$

$$= E_\theta R(\mathbf{X}) S(\mathbf{X}), \tag{8.62}$$

so that $1 = \text{Cov}(R, S)$, as we needed. ∎

No unbiased estimator can have variance smaller than $1/(E_\theta(D_\theta \log f(\mathbf{X} \mid \theta))^2)$, and we have one that has exactly this much variance. The right-hand side of the inequality does not depend on which estimator you choose.

8.11.b More General Considerations: A Short Detour

There are examples in which no estimator exactly satisfies the information inequality. However, in the present case, estimating θ from iid Bern (θ)'s, $\widehat{\theta}_{MLE}$ satisfies the bound, showing that it is not only the BLUE, but is the **best unbiased estimator** (**BUE**).

The quantity $E_\theta(D_\theta \log f(\mathbf{X} \mid \theta))^2$ is sometimes called the **Fisher information** of the sample, and the inequality, in this form, is often called the **information inequality**. It should, after some thought, be intuitive (well, plausible anyway) that having $E_\theta(D_\theta \log f(\mathbf{X} \mid \theta))^2$ large means that the sample tells us a great about the value of the parameter, so that we can estimate it more closely. This is especially true when we think about MLEs when the log likelihood function has a great deal of curvature.

Substituting $\frac{d}{d\theta} \int_{\mathfrak{X}} R(\mathbf{x}) \, d\mu(\mathbf{x})$ for $\frac{d}{d\theta} \sum_{\mathbf{x} \in \mathfrak{X}} R(\mathbf{x}) f(\mathbf{x} \mid \theta)$ in (8.59) suggests how far we can and cannot generalize this result. In particular, the calculations required that $\mathfrak{X}$, the region over which we integrate with respect to μ_θ, cannot depend on θ. Leibniz's rule, given in Theorem 7.5.18 (p. 404) tells us that if $\mathcal{X}$ depends on θ there is an extra term.

Example 8.11.10 (Finding the Upper Bound of the Population) *If $X_1, \ldots, X_n$ is iid $U[0, \theta]$ (uniformly distributed on the interval $[0, \theta]$), $\widehat{\theta}_{MLE}(X_1, \ldots, X_n)$ is the biased estimator $\max\{X_i : i \le n\}$, specifically $E\,\theta\widehat{\theta}_{MLE} = (1 - \frac{1}{n})\theta$. Thus, $\widetilde{\theta} := \frac{n}{n-1}\widehat{\theta}_{MLE}$ is unbiased. Calculation shows that $\mathrm{Var}_\theta(\widetilde{\theta}) < 1/\big(E_\theta(D_\theta \log f(\mathbf{X} \mid \theta))^2\big)$, that it reverses the inequality we found. The reason is close to hand, in (8.59); there should be an extra, strictly positive term, which we call $q_\theta > 0$, coming from an increase in the area over which μ_θ is integrated. Thus, $1 = E_\theta R(\mathbf{X})S(\mathbf{X}) + q_\theta$ in the proof, so that the correct version of the bound is $\mathrm{Var}_\theta(\widetilde{\theta}) \ge (1 - q_\theta)/\big(E_\theta(D_\theta \log f(\mathbf{X} \mid \theta))^2\big)$.*

8.11.c Shrunken Estimators and Mean Squared Error

We now turn to a brief study of the virtues of biased estimators.

Definition 8.11.11 *The **mean squared error** of an estimator $\widehat{\theta}$ is defined as $\mathrm{MSE}_\theta(\widehat{\theta}) = E_\theta(\widehat{\theta} - \theta)^2$.*

Exercise 8.11.12 Consider the four estimators $\widehat{\theta}_{1,n} = \frac{1}{n}\sum_{i \le n} X_i$, $\widehat{\theta}_{2,n} = \frac{2}{n}\sum_{i \le n/2} X_i$, $\widehat{\theta}_{3,n} = X_1$, and $\widehat{\theta}_{4,n} = \frac{1}{7}$. As a function of the true θ, graph MSE_θ for each of these estimators. What happens to these comparisons as $n \to \infty$?

The bias of an estimator $\widehat{\theta}$ is the number $\mathrm{Bias}_\theta(\widehat{\theta}) = E\,(\widehat{\theta} - \theta)$. $\widehat{\theta}$ is an unbiased estimator if for all θ, $\mathrm{Bias}_\theta(\widehat{\theta}) = 0$. Unbiased estimators do not allow us to make any trade-offs between variance and bias; they just set the bias to 0.

Exercise 8.11.13 Show that for all θ, $\mathrm{MSE}_\theta(\widehat{\theta}) = \mathrm{Bias}_\theta^2(\widehat{\theta}) + \mathrm{Var}_\theta(\widehat{\theta})$.

This means that unbiased estimators can often be beaten by a biased estimator for all θ in terms of MSE. An estimator that can be beaten for all θ is called **dominated**.[13]

Example 8.11.14 (Shrunken Estimators) *Suppose that $\widehat{\theta} \in (0, 1)$ is an unbiased estimator of $\theta \in (0, 1)$ based on an iid sample X_i, $i = 1, \ldots, n$ of $\mathrm{Bern}(\theta)$'s. The question to be asked is what multiple of $\widehat{\theta}$ minimizes mean squared error? Define*

$$f(a) = E(a\widehat{\theta} - \theta)^2 = a^2\,\mathrm{Var}(\widehat{\theta}) + \theta^2(a - 1)^2.$$

Let $v = \mathrm{Var}(\widehat{\theta})$, so that $f(a) = a^2 v + \theta^2(a - 1)^2$. As f is a quadratic in a with positive coefficients on a^2, the first-order conditions are sufficient for a minimum. Taking derivatives yields $\frac{1}{2}f'(a) = av + \theta^2(a - 1)$, so that

$$a^* = \frac{\theta^2}{v + \theta^2} = \frac{1}{1 + v^\circ} < 1, \qquad (8.63)$$

---

13. Using a class of estimators that is dominated seems odd. Perhaps we do it because the term "biased" sounds so bad. Perhaps we do it because we do not believe in MSE.

where $v° := \frac{v}{\theta^2}$ (which is known as the standardized variation of $\widehat{\theta}$). Thus, the optimal MSE estimator is given by $a^*\widehat{\theta}$. As $a^* < 1$, these are sometimes called shrunken estimators.

There are two potential problems with implementing shrunken estimators. First, it is often (not always) the case that the optimal shrinkage factor depends on the unknown θ. When this is the case, a reasonable procedure is to use $\widehat{\theta}$ in (8.63). However, this may give an estimator with even worse MSE than the original unbiased estimator that we are trying to improve. Second, when θ is small, the largest possible value for the bias is very small, which means that the a^* will try to eliminate a great deal of the variance by getting very close to 0. Both of these problems appear in the following exercise.

Exercise 8.11.15   Suppose that $X_1, \ldots, X_n$ are iid Bern(θ) and that $\widehat{\theta} = \frac{1}{n} \sum_{i \leq n} X_i$ is the BLUE and the BUE. Show the following.

1. $a^* = a^*(n, \theta) = \frac{n\theta}{n\theta + (1-\theta)}$.

2. For $n > 1$, $\theta \mapsto a^*(\theta)$ is strictly increasing and concave and $a^*(0) = 0$ and $a^*(1) = 1$. Conclude that $a^*(n, \theta) > \theta$ for $\theta \in (0, 1)$ and $n > 1$.

3. $\forall \theta \in (0, 1], \lim_n a^*(n, \theta) = 1$ but $\lim_n \max_{\theta \in [0,1]} |1 - a^*(n, \theta)| = 1$, so that a^* converges pointwise but not uniformly.

4. Show that the MSE of $a^*(2, \widehat{\theta}) \cdot \widehat{\theta}$ is greater than the MSE of $\widehat{\theta}$ when $n = 2$ and $\theta = \frac{1}{2}$. [You can explicitly calculate both of these without resorting to a calculator. Your answer should involve roughly a 22% difference in MSE.]

Here is an example where the optimal shrinkage factor does not depend on the unknown parameter.

Exercise 8.11.16   A random variable Y taking its values in $\mathbb{Z}_+$, the nonnegative integers, has a **Poisson distribution with parameter** λ, $\lambda > 0$, if $P(Y = k) = \frac{\lambda^k e^{-\lambda}}{k!}$. If Y has a Poisson distribution with parameter λ, then $E\, Y = \lambda$ and $\text{Var}(Y) = \lambda$ (the first is easy, and the second not too hard to check directly). Suppose that $Y_1, \ldots, Y_n$ are iid Poisson's with unknown parameter $\lambda > 0$. Show the following.

1. The maximum likelihood estimator of λ is $\widehat{\lambda}_{MLE} = \frac{1}{n} \sum_{i \leq n} Y_i$.

2. The maximum likelihood estimator is unbiased and achieves the Cramér-Rao lower bound.

3. The MSE optimal shrinkage factor for $\widehat{\lambda}_{MLE}$ depends on n but not on the unknown λ.

Poisson distributions do a very good job of describing the number of arrivals at an emergency room in a hospital during a given hour, the number of times a web server is accessed during a given minute, the number of mutations in a given string of DNA after exposure to radiation. The reason is that the Poisson distribution counts the number of instances in which something rare happens when the rare event has many chances of occurring.

Exercise 8.11.17 (Rare Events Distribution)   Let $X_{n,1}, \ldots, X_{n,n}$ be iid Bernoulli with parameter λ/n and let Y be Poisson with parameter λ. Show that for all k, $\lim_n P(\sum_{i \leq n} X_{n,i} = k) = P(Y = k)$.

The central limit theorem shows that the distribution of the sum of a large number of independent random variables, all of which are small, is approximately Gaussian. The previous is the easiest example of the non-Gaussian part of the analysis of the sums of many independent random variables. The averages of the $X_{n,i}$ are small, but they are not all small.

8.11.d The Bayes Estimators

Suppose that before seeing any of the data, we know, we are absolutely sure, we are just dead positive, that the true value of θ belongs to the interval $[\frac{1}{2}, 1)$. Sometimes, even if $\theta = 0.9$, we get a sample with less than half of the X_i being successes. Using $\widehat{\theta} = \frac{1}{n} \sum_{i \leq n} X_i$ seems rather foolish in such a case, and perhaps here the nonlinear estimator $\widetilde{\theta} = \max\{\widehat{\theta}, \frac{1}{2}\}$ is more sensible.

Exercise 8.11.18 Suppose that $(X_1, \ldots, X_n)$ are iid Bern(θ)'s with $\theta \in [\frac{1}{2}, 1)$. Show that $\widetilde{\theta} = \max\{\widehat{\theta}, \frac{1}{2}\}$ is the maximum likelihood estimator of θ subject to the constraint that $\theta \geq \frac{1}{2}$. Show that $\widetilde{\theta}$ is biased.

Knowledge that we have before we see any of the data is called **prior knowledge**. The most detailed form of prior knowledge that is regularly studied involves thinking that the true value of $\theta \in [0, 1]$ is a random variable with a known cdf, $G(\cdot)$. The distribution $G(\cdot)$ is called the **prior distribution**. In this case, we evaluate, say, the MSE of an estimator $\widehat{\theta}(X)$ as

$$\text{MSE}(\widehat{\theta}) = \int_{[0,1]} E_\theta(\widehat{\theta}(X) - \theta)^2 \, dG(\theta).$$

Note that we no longer write MSE_θ; the idea is that θ is random and we take the expected value. When G has a density g, this has the form

$$\text{MSE}(\widehat{\theta}) = \int_{[0,1]} E_\theta(\widehat{\theta}(X) - \theta)^2 \, g(\theta) d\theta.$$

It used to be the case that we had to be really careful about choosing a prior distribution G or a g so that the expression for $\text{MSE}(\widehat{\theta})$ would have a closed form that we could write down. We needed this in order to evaluate different estimators. Progress in numerical techniques has loosened this limitation.

Exercise 8.11.19 Suppose that we know that the true value of θ is in the interval $[\frac{1}{2}, 1]$ and that intervals of equal size in $[\frac{1}{2}, 1]$ are equally likely; that is, our prior distribution is $U[\frac{1}{2}, 1]$, so that $G(r) = 2(r - \frac{1}{2})$ for $r \in [\frac{1}{2}, 1]$. The posterior density as a function of the data is

$$P(\theta \mid \mathbf{x}) = k_x \theta \cdot \Pi_i \theta^{X_i} (1 - \theta)^{1 - X_i}, \; \theta \in [\tfrac{1}{2}, 1],$$

where k_x is some constant chosen to make $\int_{[\frac{1}{2}, 1]} k P(\theta \mid \mathbf{x}) \, d\theta = 1$. Take logarithms and maximize over $[\frac{1}{2}, 1]$ to find the Bayesian MLE estimator, watching out for corner solutions.

One of the fascinating aspects of the study of statistics is the interplay between the ideas implicit in the MLE approach and this Bayesian approach. The basic issues in epistemology appear: How do we know the information in the prior? And how sure of it are we? One way to answer these questions appears when there are a lot of data. In this case, there is a tight relation between Bayesian and MLE estimators.

Suppose that $\theta \in \Theta$, a Bayesian has a prior distribution with density $p(\theta)$, and we observe $X_1, \ldots, X_n$ with density $f(x \mid \theta)$. Then the posterior distribution has a density

$$P(\theta \mid X_1, \ldots, X_n) = kp(\theta)L(X_1, \ldots, X_n \mid \theta)$$

for some constant k. A Bayesian might well solve the problem $\max_\theta P(\theta \mid X_1, \ldots, X_n)$. Taking logarithms gives us

$$\max_\theta[\log p(\theta) + \sum_i \log f(X_i \mid \theta)] = \max_\theta \sum_i [\log f(X_i \mid \theta) + \tfrac{1}{n} \log p(\theta)].$$

You may be able to convince yourself that there are conditions under which the solution to this problem approaches the MLE as $n \uparrow \infty$. We interpret this as saying that the prior distribution becomes irrelevant, as it is eventually swamped by the data. For moderate n, the approximation may not be that good.

8.11.e Classical Hypothesis Testing and the Neyman–Pearson Lemma

At the other end of the methodological spectrum we have classical statistics. Karl Pearson tried to make statistics a process that eliminated the prejudices of the analyst. This was, he believed, the only way to conduct "science." With the benefit of a century of hindsight, we can admire his ambition and be amazed at his naivete. However, despite any naivete, the intellectual structures he built are tremendously powerful and are still the most widely used structures in the analysis of data.

Central to Pearson's methodology is the notion of a hypothesis:

> Hypothesis—A supposition or conjecture put forth to account for known facts; esp. in the sciences, a provisional supposition from which to draw conclusions that shall be in accordance with known facts, and which serves as a starting-point for further investigation by which it may be proved or disproved and the true theory arrived at. (*Oxford English Dictionary*)

Hypothesis testing involves formulating a supposition, or conjecture, or guess, called the "null hypothesis," written H_0, and the "alternative hypothesis," H_1 or H_A, observing data that have different distributions under H_0 and H_1, and then choosing between the two hypotheses on the basis of the data. Under study are the possible processes of picking on the basis of the data. This is formulated as a decision rule, a.k.a. a rejection rule: that is, for some set of possible data points, $\mathbb{X}_r$, we reject H_0, while for all others we accept it.

There are two types of errors that can be committed using a rejection rule, unimaginatively called **Type I** and **Type II** errors:

1. You can reject a null hypothesis even though it is true, this kind of false rejection of a true null hypothesis being called a **Type I** error, and the probability of a Type I error is denoted α.

2. You can accept the null hypothesis even though it is false, this kind of false acceptance of a null hypothesis being called a **Type II** error, and the probability of a Type II error is denoted β.

If you adopt a rejection rule that makes α small, you are very rarely rejecting the null hypothesis. This means that you are running a pretty high risk of making a Type II error, that is, β is fairly large. This works the other way too, as getting β small requires accepting a large α. The Neyman-Pearson lemma concerns a class of situations in which we can find the best possible decision rule for given α.

The essential ingredients are:

1. the basic statistical model, $\mathbf{X} \sim f(\mathbf{x} \mid \theta), \theta \in \Theta$,
2. a null hypothesis, $H_0 : \theta \in \Theta_0, \Theta_0 \subset \Theta$,
3. the alternative, $H_1 : \theta \notin \Theta_0$, and
4. a decision rule, reject H_0 if $\mathbf{x} \in \mathbb{X}_r$ and accept H_0 if $\mathbf{x} \notin \mathbb{X}_r$.

We can then examine the probabilistic properties of the decision rule using the **power function**, $\beta(\theta) = P(\mathbb{X}_r \mid \theta)$. The perfect power function is $\beta(\theta) = 1_{\Theta_0^c}(\theta)$, that is, reject iff the null hypothesis is false.[14] However, the idea behind the basic statistical model is that we do not observe θ directly, but rather that we observe the data $\mathbf{X}$ and that they contain probabilistic information about θ. In econometrics, we do not expect to see perfect power functions very often, as they correspond to having positive proof or disproof of a null hypothesis.

Continuing with our simplest of examples, we suppose that $H_0 : \theta = \theta_0$, $H_1 : \theta = \theta_1, \theta_0 \neq \theta_1$. Note how much structure we have already put on the problem of picking a decision rule. In particular, we have replaced $\Theta = [0, 1]$ by $\Theta = \{\theta_0, \theta_1\}$.

We suppose that $\theta_0 < \theta_1$, while the opposite case just reverses inequalities. There is a pretty strong intuition that the best decision rule is to accept H_0 if $\widehat{\theta}_n < \theta^*$ and to reject otherwise, for some $\theta^* \in (\theta_0, \theta_1)$. Work through why this is true, thinking of the analogy of filling bookcases with the lightest possible set of books.

Exercise 8.11.20 (Neyman-Pearson Lemma) Suppose that $\mathbf{X} = (X_1, \ldots, X_n)$ has the Bernoulli density $f(\mathbf{x} \mid \theta), \theta \in \Theta = \{\theta_0, \theta_1\}$. We have seen that there is typically a trade-off between α, the probability of a Type I error, and β, the probability of a Type II error. Let us suppose that we dislike both types of errors, and in particular that we are trying to devise a test, characterized by its rejection region, $\mathbb{X}_r$, to minimize

$$a \cdot \alpha(\mathbb{X}_r) + b \cdot \beta(\mathbb{X}_r),$$

---

14. If we could, we would sing a bar of "To Dream the Impossible Dream."

where $a, b > 0$, $\alpha(\mathbb{X}_r) = P(\mathbf{X} \in \mathbb{X}_r \mid \theta_0)$, and $\beta(\mathbb{X}_r) = P(\mathbf{X} \notin \mathbb{X}_r \mid \theta_1)$. The idea is that the ratio of a to b specifies our trade-off between the two types of errors: the higher a is relative to b, the lower we want α to be relative to β. This problem asks about tests of the form

$$\mathbb{X}_{a,b} = \left\{\mathbf{x} : af(\mathbf{x} \mid \theta_0) < bf(\mathbf{x} \mid \theta_1)\right\} = \left\{\mathbf{x} : \frac{f(\mathbf{x} \mid \theta_1)}{f(\mathbf{x} \mid \theta_0)} > \frac{a}{b}\right\}.$$

This decision rule is based on the **likelihood ratio**, and likelihood ratio tests appear regularly in statistics.

1. Show that a test of the form $\mathbb{X}_{a,b}$ solves the minimization problem given above. [Hint: Let $\phi(\mathbf{x}) = 1$ if $\mathbf{x} \in \mathbb{X}_r$ and $\phi(x) = 0$ otherwise. Note that $a \cdot \alpha(\mathbb{X}_r) + b \cdot \beta(\mathbb{X}_r) = a \int \phi(\mathbf{x}) f(\mathbf{x} \mid \theta_0) \, d\mathbf{x} + b \int (1 - \phi(\mathbf{x})) f(\mathbf{x} \mid \theta_1) \, d\mathbf{x}$ and that this is in turn equal to $b + \int \phi(\mathbf{x})[af(\mathbf{x} \mid \theta_0) - bf(\mathbf{x} \mid \theta_1)] \, d\mathbf{x}$. The idea is to minimize the last term in this expression by the choice of $\phi(\mathbf{x})$. Which $\mathbf{x}$'s should have $\phi(\mathbf{x}) = 1$?]

2. As a function of a and b, find $\mathbb{X}_{a,b}$ when $(X_1, \ldots, X_n)$ is iid Bern(θ), $\theta \in \Theta = \{\theta_0, \theta_1\} \subset (0, 1)$.

8.11.f Selection and Misspecification Issues

If we randomly pick a set of n unemployed people and *compel* them to take a course on job search techniques, we could set $X_i = 1$ if they have a job 6 months later and $X_i = 0$ otherwise. Here the randomness in ω would have to encompass the selection process and the state of the economy where person i is looking for a job, as well as the other random influences on i's job-holding status in 6 months.

When one studies lab rats or electrons, one can force them to undergo surgery after hours of running on a treadmill or to pass through a magnetic field. When studying people, however, one has some moral qualms about compulsion.[15] If we were to randomly pick an unemployed person, offer him the chance to take a job search class, and then check whether or not he has a job in 6 months, we would have a very different experiment than the one described in the previous paragraph. We might expect that the most motivated people are the ones most likely to take the job search class and also the most likely to have jobs in 6 months. Thus, observing $X_i = 1$ tells us something about both the qualities of the person and the qualities of the job training program.

At work is the selection process itself—those who would choose to take such a course have different characteristics, still random and unknown to the researcher, than those who would not choose to take such a course. In other words, the choice to select oneself into the pool matters. The conditional probability of having a job 6 months later conditional on opting to take the course is probably not the same as the probability of having a job 6 months after job training.

---

15. And there is, fortunately, a well-developed set of human subject rules for research of this kind.

This selection problem is the essential reason that most empirical economics research is so much more difficult than the corresponding research in the physical or biological sciences. We are interested in the θ that describes the effectiveness of the job search course, but we observe random variables described by the θ' that describes the effectiveness of the job search course on those who opt to take it.

There is yet another kind of problem: it is possible, perhaps even likely, that the statistical model is itself misspecified. In this simple case, such a failure would be a failure of the X_i to be independent or to have the same distribution. For example, if the job search happens in an economic boom or in an economic slump, then the X_i are correlated. For another, slightly more subtle example, if some identifiable subpopulation suffers discrimination, positive or negative, then its θ will be lower or higher than others, and the best that can be said for the θ that we estimate using the whole population is that it is an average of some kind, although we do not know what kind of average it is. In thinking about these issues, it is wise to keep in mind the following observation, due to T. Rothenberg: "Every estimator is a best estimate of itself."

Failures to have the same probability of success (i.e., the same distribution) used to happen all the time in clinical trials of new medicines, and it was a selection issue. Doctors would select the patients to receive the new treatment and those to receive the old one. If they selected only the most hopeless cases to receive the new treatment, the research would tend to favor the old treatment. If they (say, had a financial interest in the new treatment being approved they might well) choose to try the new treatment on those having the highest probability of recovery, the research would tend to favor the new treatment.

There are a number of strategies for dealing with the selection problem. Here is a brief sketch of one of them.

Example 8.11.21 (Manski Bounds)   *Let J_i be the event, that is, the set of ω, in which person i takes a job training course. Let A be the event, that is, the set of ω, that a randomly chosen unemployed person would choose to take the job training course if offered and A^c the event that she would not take it if offered. Let $X_i(\omega) = 1$ if person i has a paying job in 6 months and $X_i(\omega) = 0$ otherwise. To evaulate the program, one would like to know $P(X_i = 1 \mid J_i)$ and $P(X_i \mid J_i^c)$.*

If we randomly select a large group of people and randomly give some proportion of them the choice of whether or not to take the job training course, we can estimate $P(A)$ (from the proportion of those offered the course who took it). If we follow everyone in the group for 6 months, and people are roughly the same, we can also estimate $P(X = 1 \mid J \cap A)$ (from the job histories of those offered the training course who took it), $P(X = 1 \mid J^c \cap A^c)$ (from the job histories of those offered the training course who refused it), and $P(X = 1 \mid J^c)$ (from the job histories of those not offered the course). Now

$$P(X = 1 \mid J) = P(X = 1 \mid J \cap A) \cdot P(A) + x,$$

where the unknown x is $P(X = 1 \mid J^c \cap A^c) P(A^c)$, the counterfactual conditional probability of a person having a job in 6 months given that she is the type to refuse a job training course but took it nonetheless.

The selection issue is that $P(X = 1 \mid J)$ is a convex combination of two numbers, and, if being in A is positively associated with being interested in and finding a job, then we expect that x is the smaller of the two. This means that estimating $P(X = 1 \mid J)$ by our estimate of $P(X = 1 \mid J \cap A)$ will bias our results upward.

Now $P(X = 1 \mid J^c) = P(X = 1 \mid A \cup A^c) = P(X = 1 \mid A)P(A) + P(X = 1 \mid A^c)P(A^c)$, and another relevant probability that we do not observe is $y := P(X = 1 \mid A)$. One measure of the size of the effect of the job training program is $D := P(X = 1 \mid J) - P(X = 1 \mid J^c)$. Doing a little bit of algebra, we can express the difference as a sum of two terms, one estimatable and one involving the unobservable x and y. This then gives bounds on the estimated size of D, bounds that become sharper as we make more assumptions about the ranges that contain x and y. If you are reasonably certain that the job training program does not actually harm one's chances of getting a job, then $y = P(X = 1 \mid A) \leq P(X = 1 \mid J \cap A)$ seems reasonable.

8.12 ♦ Complements and Extras

8.12.a Information and Maximization

Jensen's inequality tells us that for a convex $\varphi : \mathbb{R} \to \mathbb{R}$, $\varphi(E\,X) \leq E\,\varphi(X)$. Here is the conditional version of Jensen's inequality.

Theorem 8.12.1 (Conditional Jensen) *If $\varphi : \mathbb{R} \to \mathbb{R}$ is convex and both X and $\varphi(X)$ are integrable, then $\varphi(E\,(X \mid \mathcal{G})) \leq E\,(\varphi(X) \mid \mathcal{G})$ a.e.*

Proof. By the separating hyperplane theorem, $\varphi(x) = \sup\{A_n(x) : n \in \mathbb{N}\}$ for some countable collection of affine functions $A_n(x) = a_n x + b_n$. The random variables $A_n(X)$ are integrable and $A_n(E\,(X \mid \mathcal{G})) = E\,(A_n(X) \mid \mathcal{G}) \leq E\,(\varphi(X) \mid \mathcal{G})$ a.e. Take the supremum over the A_n in the leftmost expression. ∎

Exercise 8.12.2 (Expected Profits) For a technology set Y, the necessarily convex function $\pi(\mathbf{p})$ is defined as $\pi(\mathbf{p}) = \sup\{\mathbf{py} : \mathbf{y} \in Y\}$, and for this problem, we assume that the supremum is always finite, and always achieved at a point (or set of points) $\mathbf{y}^*(\mathbf{p})$. A firm solving the problem max $\mathbf{py}$ subject to $\mathbf{y} \in Y$ faces uncertain prices, that is, $\mathbf{p} = \mathbf{p}(\omega)$ is a random variable. We consider two cases.

(A). The firm has no information about $\mathbf{p}$ but its distribution; that is, they solve max $E\,\mathbf{py}$ subject to $\mathbf{y} \in Y$, and

(B). the firm can pick any $\mathcal{G}$-measurable $\omega \mapsto \mathbf{y}(\omega)$, $\mathcal{G}$ a sub-σ-field of $\mathcal{F}$.

1. Give the solution in case (A).
2. Give the solution in case (B).
3. Compare the profits in the two cases.

Exercise 8.12.3 A decision maker (DM) is picking an $a \in A$ to maximize $\int u(a, Y(\omega))\,dP(\omega)$, where $u(\cdot, \cdot)$ is bounded and measurable and Y is a random variable. We consider two cases.

(A). The DM has no information about Y except its distribution; that is, they solve max $E\, u(a, Y)$ subject to $a \in A$, and

(B). the DM can pick any $\mathcal{G}$-measurable $\omega \mapsto a(\omega)$, $\mathcal{G}$ a sub-σ-field of $\mathcal{F}$.

Show that the DM is at least weakly better off in (B) than in (A).

If we compare the previous two problems, the role of convexity may seem obscure, mattering in the first case, but seeming not to matter in the second, but this is not true. The role of convexity is just better hidden in the second case. The following problem is an introduction to the Blackwell ordering on information structures.

Exercise 8.12.4 Consider the DM facing the problem max $u(a, Y)$ subject to $a \in \{1, 2, 3\}$, where $P(Y = 1) = P(Y = 2) = \frac{1}{2}$. We consider three different information structures for this problem, where X and X' are described further below.

(A). the DM has no information about Y except its distribution, $\mathcal{G} = \{\emptyset, \{1, 2\}\}$;

(B). the DM can choose Y to be any $\sigma(X)$-measurable random variable; and

(C). the DM can choose Y to be any $\sigma(X')$-measurable random variable.

The utilities and the distributions of the random variables are given in the following tables.

$a = 3$	0	40		$X = 2$	0.3	0.8		$X' = 4$	0.1	0.8
$a = 2$	30	30		$X = 1$	0.7	0.2		$X' = 3$	0.9	0.2
$a = 1$	50	0								
	$Y = 1$	$Y = 2$			$Y = 1$	$Y = 2$			$Y = 1$	$Y = 2$

where, for example, $50 = u(a = 1, Y = 1)$, $0.3 = P(X = 2 | Y = 1)$, and so on.

The **posterior distribution of Y given X** is the random function $P(Y \in A \mid X = x)(\omega)$, and the posterior distribution of $Y \mid X'$ is the corresponding $P(Y \in A \mid X' = x')(\omega)$. The posterior takes values in $\Delta(\{1, 2\})$, the set of probability distributions over the set $\{1, 2\}$.

1. Calculate the posterior distributions given $\mathcal{G}$, X, and X' and the probabilities that the posterior distributions take on each of their possible values. Show that the average posterior distribution is $(\frac{1}{2}, \frac{1}{2})$. Using the Law of Iterated Expectations, show that the average posterior is always the prior.

2. Let $(\beta, 1 - \beta) \in \Delta(\{1, 2\})$ be a possible distribution of Y. Give the set of β for which

   1. arg max$_{a \in A} \int u(a, y)\, d\beta(y) = \{1\}$,
   2. arg max$_{a \in A} \int u(a, y)\, d\beta(y) = \{2\}$, and
   3. arg max$_{a \in A} \int u(a, y)\, d\beta(y) = \{3\}$.

3. Graph the function $V : \Delta(\{1, 2\}) \to \mathbb{R}$ given by $V(\beta) = $ max$_{a \in A} \int u(a, y)\, d\beta(y)$ and show that it is convex.

4. Solve the DM's problem in the cases (A), (B), and (C). Show that in each case, the DM's solution $a^*(X)$ or $a^*(X')$ can be had by having the

DM calculate the posterior distribution and then maximize according to the rules you found in (2). Show that this DM prefers (C) to (B) to (A).

5. Consider a new DM facing the problem $\max_{b \in B} \int v(b, y) \, d\beta(y)$, where B is an arbitrary set. Define a new function $W : \Delta(\{1, 2\}) \to \mathbb{R}$ by $W(\beta) = \sup_{b \in B} \int v(b, y) \, d\beta(y)$. Show that W is convex.

6. Let η and η' be the distributions of the posteriors you calculated in 1. To be specific here, η and η' are points in $\Delta(\Delta(\{1, 2\}))$, that is, distributions over distributions. Show that for any convex $f : \Delta(\{1, 2\}) \to \mathbb{R}$, $\int f(\beta) \, d\eta'(\beta) \geq \int f(\beta) \, d\eta(\beta)$.

7. Show that any expected utility-maximizing DM with utility depending on Y and his own actions would prefer the information structure $\sigma(X')$ to the information structure $\sigma(X)$. [Any DM must have a utility function and a set of options, and she has a corresponding $W : \Delta(\{1, 2\}) \to \mathbb{R}$. What property or properties must it have?]

8.12.b   More on Conditional Expectation as a Linear Operator

Corollary 8.12.5   *For any $p \in [1, \infty)$, the mapping $X \mapsto E(X \mid \mathcal{G})$ is a continuous linear mapping with norm 1 from L^p to itself.*

Proof.   The function $\varphi(x) = |x|^p$ is convex, so that $\|E(X \mid \mathcal{G})\|_p \leq \|X\|_p$ is the conditional version of Jensen's inequality and the norm of the mapping is less than or equal to 1. For $X \in L^p(\mathcal{G})$, $E(X \mid \mathcal{G}) = X$, so the norm is exactly 1.   ∎

Exercise 8.12.6   Prove the last corollary for the case $p = \infty$.

Recall from our study of orthogonal projections in $\mathbb{R}^N$ that the projection of **y** onto a linear subspace A gives rise to $\mathbf{y} - \text{proj}_A \mathbf{y}$, which is a linear function projecting **y** onto $A^\perp$.

Exercise 8.12.7   Define the function $T(X) = (X - E(X \mid \mathcal{G}))$. Show that T is a continuous linear mapping from $L^p(\mathcal{F})$ to itself and that $L^p(\mathcal{G}) = T^{-1}(0)$.

8.12.c   Translation Invariant Norms, Seminorms, and Metrics

Exercise 8.12.8   A metric on L^p is **translation invariant** if for all X, Y, Z, $d(X, Y) = d(X + Z, Y + Z)$. Check that for all of the L^p spaces, the norm distance is translation invariant and that for all L^p spaces where it is defined, so is the bounded weak convergence metric.

Exercise 8.12.9   Throughout this problem, (p, q) is a dual pair of indexes and $p \in [1, \infty)$. A **seminorm** on L^p is a symmetric sublinear function, that is, a function $\| \cdot \| : L^p \to \mathbb{R}$ with the properties that

1. for all $\alpha \in \mathbb{R}$ and $X \in L^p$, $\|\alpha X\| = |\alpha| \, \|X\|$, and

2. $\|X_1 + X_2\| \leq \|X_1\| + \|X_2\|$ (the triangle inequality) for all $X_1, X_2 \in L^p$.

The difference between a sublinear function and a seminorm is that a seminorm treats multiplication by $\pm \alpha$ the same, whereas a sublinear function may treat them differently. A seminorm is a norm if $[\|X\| = 0] \Rightarrow [X = 0]$. Thus, a seminorm distance is a pseudometric.

A set K is **balanced** if $K = -K$, for example, $K = \frac{1}{2}(A - A)$ is always balanced. A set K is **absorbent** if $L^p = \cup_{\alpha \in \mathbb{R}} \alpha K$, that is, if for all $X \in L^p$, $X \in \alpha K$ for some $\alpha \geq 0$. Every open neighborhood of 0 is absorbent.

Show the following:

1. Seminorms are translation invariant.

2. If K is a balanced convex set with $0 \in \mathrm{int}(K)$, then the Minkowski gauge of K, $m_K(X) := \inf\{\alpha \geq 0 : X \in \alpha K\}$, is a seminorm.

3. If K is a bounded balanced convex set with $0 \in \mathrm{int}(K)$, then the Minkowski gauge of K is a norm, $m_K(X_n - X) \to 0$ iff $\|X_n - X\|_p \to 0$.

4. For any $Y \in L^q$, the function $\|X\|_Y := |Y \cdot X|$ is a translation invariant seminorm.

5. For any $r > 0$, $\{X : \|X\|_Y < r\}$ is an unbounded open set containing the origin.

6. Express the bounded weak convergence metric using seminorms.

7. Express the norm distance using seminorms. [Use the tightness of Hölder's inequality.]

Exercise 8.12.10 Consider the following alternative to ρ_p—$m_p(X_1, X_2) = \sum_m \frac{1}{2^m} \frac{|Y_m \cdot (X_1 - X_2)|}{1 + |Y_m \cdot (X_1 - X_2)|}$, where Y_m is a dense subset of L^q, not just of the unit ball in L^q. Show that m_p is translation invariant and that if $\|X\|_p \leq B$ and $\sup_n \|X_n\|_p \leq B$, then $\rho_p(X_n, X) \to 0$ iff $m_p(X_n, X) \to 0$.

8.12.d More on Compactness in $L^p(\Omega, \mathcal{F}, P)$

Exercise 8.12.11 Show that in $L^1((0, 1], \mathcal{B}, \lambda)$, the set $\mathcal{C}[a, b]$ of nondecreasing functions taking values in $[a, b]$ is norm compact. Further show that this implies that for all $p \in [1, \infty)$, $\mathcal{C}[a, b]$ is norm compact, hence α-compact. [$\alpha(\cdot, \cdot)$ is the Ky Fan metric.]

If $K \subset L^{p'}(\Omega, \mathcal{F}, P)$ is compact and $1 \leq p < p' \leq \infty$, then K is a compact subset of $L^p(\Omega, \mathcal{F}, P)$ because $\|X\|_{p'} \geq \|X\|_p$; hence $L^{p'}$-convergence of a subsequence implies L^p-convergence.

Exercise 8.12.12 For $1 \leq p < p'$, if $K \subset L^{p'}$ is a compact subset of L^p, then K need not be a compact subset of $L^{p'}$.

Remember that $\rho_P(A, B) = P(A \triangle B) = E |1_A - 1_B|$.

Exercise 8.12.13 If $\mathcal{K} \subset \mathcal{F}$ is a compact subset of $(\mathcal{F}, \rho_P)$, then for all $\epsilon > 0$, there is a finite partition, $\mathcal{E} = \{E_1, \ldots, E_n\}$, of Ω such that $\sup_{A \in \mathcal{K}} \sum_n b_n P(E_n) < \epsilon$, where $b_n = |1 - P(A \mid E_n)|$.

Here is a compactness criterion for (L^0, α). Recall that a set $S \subset C(M)$, M compact, cannot be compact if it is either unbounded or too variable. There are parallel ideas for compact subsets of L^0. It is worth considering what compactness implies for choice under uncertainty as presented in §7.2.c.4, and comparing it with weak compactness.

Theorem 8.12.14 $S \subset L^0(\Omega, \mathcal{F}, P)$ *has compact closure iff for all $\epsilon > 0$, there exists a finite $B \subset \mathbb{R}$ and a finite measurable partition $\mathcal{E} \subset \mathcal{F}$ such that the class of functions $\sum_{E_j \in \mathcal{E}} \beta_j 1_{E_j}$, $\beta_j \in B$, is an ϵ-net for S.*

Proof. ($\Rightarrow$) Take a finite $\epsilon/3$-net, $\{X_i : i \leq I\}$, in S for the compact closure of S. Take N such that for each $i \leq I$, $\alpha((X_i \vee N) \wedge (-N), X_i) < \epsilon/3$. For each $(X_i \vee N) \wedge (-N)$, take a finite partition $\mathcal{E}_i$ of Ω for which there exists $\sum_{E_i \in \mathcal{E}_i} \beta_{E_i} 1_{E_i}$ within α-distance less than $\epsilon/3$ of X_i. Let $\mathcal{E}$ be the coarsest partition of Ω that is finer than all of the $\mathcal{E}_i$ and take $B = \cup_{i \leq i} \{\beta_{E_i} : E_i \in \mathcal{E}_i\}$.

($\Leftarrow$) The closure of S is complete because L^0 is complete. Hence the existence of a finite ϵ-net for every $\epsilon > 0$ implies compactness. ∎

Here is another formulation of compactness in L^0 using conditional expectations.

Exercise 8.12.15 (Compactness in L^0) For $X \in L^0$ and $N \in \mathbb{R}_+$, define $X_N = (X \vee N) \wedge (-N)$. Show that $S \subset L^0$ has compact closure iff $\forall \epsilon > 0$, $\exists N \in \mathbb{R}_+$ and a finite measurable partition $\mathcal{E}$ of Ω such that

1. $\sup_{X \in S} P(\{X \neq X_N\}) < \epsilon$, and
2. $\sup_{X \in S} \alpha(E(X_N \mid \mathcal{E}), X_N) < \epsilon$.

8.12.e Small Subsets of Banach Spaces

One way to say that subsets of a Banach space are small is to ask that they be Baire small, that is, that they be countable unions of nowhere dense sets. From Exercise 6.11.50 (p. 353), this may seem unsatisfactory, as there are Baire small subsets of $(0, 1]$ having Lebesgue measure 1. We introduce a different notion of smallness, called **shyness**, that seems to capture the intuitive notion of smallness better. However, given that we do this for infinite-dimensional spaces, the intuitive notion still has some undesirable properties.

Just as we defined Lebesgue measure on $\mathbb{R}^1$ by pasting together countably many copies of the uniform distribution on interval $(n, n+1]$, we can define Lebesgue measure on $\mathbb{R}^k$ by pasting together countably many copies of the uniform distribution on products of intervals.

Definition 8.12.16 *The **uniform distribution on** $(0, 1]^k$ is the unique countably additive extension of the distribution satisfying $U(\times_{i=1}^k |a_i, b_i|) = \Pi_{i=1}^k (b_i - a_i)$. For $\mathbf{n} = (n_1, \ldots, n_k) \in \mathbb{Z}^k$, the uniform distribution on $I_n := \times_{i=1}^k (n_i, n_i + 1]$ is defined by $U_{\mathbf{n}}(A) = U(A - n)$.*

Since each $U_{\mathbf{n}}$ is countably additive, so is the Lebesgue measure.

In the same way that we define $U_{\mathbf{n}}$, we can define $U_{\mathbf{x}}$ as the uniform distribution on $\times_{i=1}^k (x_i, x_i + 1]$ for any $\mathbf{x} = (x_1, \ldots, x_k) \in \mathbb{R}^k$.

Definition 8.12.17 *The **Lebesgue measure** of a measurable $S \subset \mathbb{R}^k$ is defined as $\sum_n U_n(S)$ and S' is a **Lebesgue null set** if $S' \subset S$, where S is universally measurable and has Lebesgue measure 0.*

Exercise 8.12.18 Show that Lebesgue measure is translation invariant and that the Lebesgue measure of S is equal to the Lebesgue measure of $S + \mathbf{x}$ for all $\mathbf{x} \in \mathbb{R}^k$. Further show that the Lebesgue measure of any set with nonempty interior is strictly positive and that it is finite if the set is bounded.

Exercise 8.12.19 Show that S has Lebesgue measure 0 iff for all $\mathbf{x} \in \mathbb{R}$, $U(S - \mathbf{x}) = 0$.

Definition 8.12.20 *A subset $S' \subset \mathbb{R}^k$ is **shy** if it is a subset of a universally measurable S with $P(S - \mathbf{x}) = 0$ for all $\mathbf{x} \in \mathbb{R}^k$ for some distribution P on $(\mathbb{R}^k, \mathcal{B}^k)$.*

Exercise 8.12.21 Show that S' is shy iff it is a Lebesgue null set [Fubini's theorem].

The reason to introduce shyness in this fashion, rather than just saying "Lebesgue null," is that infinite-dimensional vector spaces do not have a version of Lebesgue measure. Recall that Lebesgue measure is translation invariant and assigns strictly positive probability to every open set.

Example 8.12.22 *If $(\Omega, \mathcal{F}, P)$ is nonatomic, then for every $p \in [1, \infty)$, there are countably many **disjoint** $\epsilon/4$-balls inside every ϵ-ball. By scaling and translating, it is sufficient to show that the unit ball, $B_1(0)$, contains countably many disjoint $1/4$-balls. Since $L^p(\Omega, \mathcal{F}, P)$ contains a linearly isometric copy of ℓ^p, it is sufficient to show that this is true for ℓ^p. For $t \in \mathbb{N}$, let e_t be the unit vector in the tth direction in ℓ^p, that is, the vector of 0's except for a 1 in component t. For $s \neq t$, $\|\frac{1}{2}e_t - \frac{1}{2}e_s\|_p = \frac{1}{2}2^{1/p} > \frac{1}{2}$. This means that the countably infinite collection $\{B_{1/4}(e_t) : t \in \mathbb{N}\}$ is disjoint.*

Corollary 8.12.23 *There is no translation invariant measure on $L^p(\Omega, \mathcal{F}, P)$ that is countably additive, translation invariant, and assigns finite strictly positive mass to every open set.*

Proof. If there were such a measure, it would assign strictly positive and identical mass to each of the countably many disjoint $\epsilon/4$-balls that are subsets of $B_\epsilon(x)$. By countable additivity, the mass of $B_\epsilon(x)$ is therefore infinite. ∎

Definition 8.12.24 *A subset S' of a Banach space $\mathfrak{X}$ is **shy** if it is a subset of a universally measurable S with $\mu(S - x) = 0$ for all $x \in \mathfrak{X}$ for some Borel probability distribution μ on $\mathfrak{X}$.*

Very often, it is possible to take P to be the uniform distribution on a line segment in $\mathfrak{X}$.

Exercise 8.12.25 Show the following.

1. The set $S = \{f \in C([0, 1]) : f(\frac{1}{2}) = 0\}$ is shy. [Let P be the uniform distribution on the set of functions $f(x) \equiv r$, $r \in [0, 1]$. For any $g \in C([0, 1])$, $P(S - g) = 0$ because for any g and any $f \in S$, $(f - g)(\frac{1}{2})$ equals 0 iff $f(x) \equiv g(\frac{1}{2})$, and only one point on the line between $f(x) \equiv 0$ and $f(x) \equiv r$ can satisfy this condition.]

2. The set $S = \{f \in C([0, 1]) : f'(\frac{1}{2})$ exists $\}$ is shy. [Let $g(x) = |x - \frac{1}{2}|^{1/2}$ and P be the uniform distribution on the set of functions $r \cdot g(\cdot)$, $r \in [0, 1]$.]

3. The set of concave functions is shy in $C([0, 1])$.

4. The set of continuous functions is shy in $L^p((0, 1], \mathcal{B}, \lambda)$.

5. The set of simple functions is shy in $L^p((0, 1], \mathcal{B}, \lambda)$.

One cannot always get away with using P's that are supported on line segments in $\mathfrak{X}$. The convolution of two probabilities, μ_1 and μ_2, is the distribution of $X_1 + X_2$, where $X_1 \perp\!\!\!\perp X_2$ and X_1 and X_2 have the distributions μ_1 and μ_2. In showing that the countable union of shy sets is shy, the easiest way to proceed is to take a countable convolution of a sequence of scaled versions of a sequence μ_n.

Here is a pair of useful general results about classes of sets that are / are not shy. The one about open sets is easy and there is a clever trick for the compact sets that uses the completeness of $\mathfrak{X}$ in a crucial way.

Lemma 8.12.26 *If $\mathfrak{X}$ is a separable infinite-dimensional Banach space and $K \subset \mathfrak{X}$ is compact, then K is shy. If $\mathrm{int}(K) \neq \emptyset$, then K is not shy.*

8.12.f Mean–Preserving Spreads

Another way to think about individuals that have risk averse preference is that they dislike what we call mean-preserving spreads. The idea is that one scoops mass out of some middle region and deposits it on outer regions in a fashion that leaves the mean unchanged.

Definition 8.12.27 *Suppose that $X, Y \in L^0(\Omega, \mathcal{F}, P)$ are random variables taking values in $[a, b]$ with probability 1. Y is a **mean-preserving spread** of X if $E\,Y = E\,X$ and there is an interval $|c, d| \subset [a, b]$, $a < c < d < b$, with the properties that $A = \{\omega : X(\omega) = Y(\omega)\}$ and $B = \{\omega : X(\omega) \in |c, d|\} \cap \{\omega : Y(\omega) \notin |c, d|\}$ partition Ω. This implies that for every measurable $E \subset [a, b]$,*

1. $P(\{Y \in E \cap |c, d|\}) \leq P(\{X \in (E \cap |c, d|)\})$, and

2. $P(\{Y \in (E \cap |c, d|^c)\}) \geq P(\{X \in (E \cap |c, d|^c)\})$.

*The distribution G is a **mean-preserving spread** of the distribution F if there exist random variables X having distribution F and Y having distribution G, with Y a mean-preserving spread of X.*

Theorem 8.12.28 *For X, Y having the same mean, if Y is a mean-preserving spread of X, then $E\,u(Y) \leq E\,u(X)$ for all concave $u : [a, b] \to \mathbb{R}$, and if $E\,u(Y) \leq E\,u(X)$ for all Y that are mean-preserving spreads of X, then u is concave.*

Exercise 8.12.29 Suppose that Y is a mean-preserving spread of X and show that $E\,(u(Y) \mid X) \leq E\,(u(X) \mid X)$ a.e.

8.12.g Essentially Countably Generated σ-Fields

The following renames countably generated σ-fields.

Definition 8.12.30 *A σ-field $\mathcal{F}$ is **countably generated** if there is a countable collection of events $\mathcal{E} = \{E_n : n \in \mathbb{N}\}$ such $\mathcal{F} = \sigma(\mathcal{E})$. It is **essentially countably generated** if there is a countable collection of events $\mathcal{E} = \{E_n : n \in \mathbb{N}\}$ such that $\rho_P(\mathcal{F}, \sigma(\mathcal{E})) = 0$.*

As noted, the Borel σ-field of any separable metric space is countably generated; simply take $\mathcal{E}$ to be the set of open balls with rational radius around each point in a countable dense set. It often happens that a σ-field is essentially countably generated even though it is not countably generated.

Lemma 8.12.31 *In a probability space* $(\Omega, \mathcal{F}, P)$, *if* $\mathcal{F}$ *contains an uncountable set* E *with* $P(E) = 0$, *then the completion,* $\mathcal{F}^P$, *is not countably generated.*

The proof of this uses the result that the cardinality of any countably generated σ-field is at most $\mathcal{P}(\mathbb{N})$, whereas the cardinality of the completed σ-field must be at least that of $\mathcal{P}(E)$, which is at least $\mathcal{P}(\mathcal{P}(\mathbb{N}))$. We do not develop the theory of ordinal and cardinal numbers that gives this result, but the interested reader should consult Dugundji (II,9.4, p. 55).[16]

8.12.h An Elementary Proof of Lyapunov's Theorem

Proof of the convexity part of Theorem 8.5.36. Let $P_f(E)$ denote $\int_E f \, dP$. We wish to show that $R := \{P_f(E) : E \in \mathcal{F}\} \subset \mathbb{R}^k$, is convex.

Step 1: We show that it is sufficient to show that for all $E \in \mathcal{F}$, there is a subfamily $\{E_r : r \in [0, 1]\}$ of subsets of E that is nested, that $r < s$ implies that $E_r \subset E_s$, and that $P(E_r) = r \cdot \nu(E)$.

Proof of Step 1: Let $z_1 = \nu(E_1)$ and $z_2 = \nu(E_2)$ be two points in R. Pick $F_1 \subset (E_1 \setminus E_2)$ such that $\nu(F_1) = \alpha \nu(E_1 \setminus E_2)$ and $F_2 \subset (E_2 \setminus E_1)$ such that $\nu(F_2) = (1 - \alpha)\nu(E_2 \setminus E_1)$. Check that $\nu(F_1 \cup (E_1 \cap E_2) \cup F_2) = \alpha z_1 + (1 - \alpha)z_2$.

Step 2: Lemma 8.5.35 tells us that the subfamily result is true for $k = 1$.

Step 3: We now show that if the subfamily result holds for k, then it holds in the following seemingly weaker form for $k + 1$: for arbitrary $E \in \mathcal{F}$ there exists an $E' \subset E$ and a $\beta \in (0, 1)$ such that $\nu(E') = \beta\nu(E)$. From the $(k + 1)$-length vector of functions $(f_1, f_2, \ldots, f_{k+1})$, form the k-length vector of functions, $g := (f_2, \ldots, f_{k+1})$. Two applications of the result to the k-length vector g yields a partition $\{E_1, E_2, E_3\}$ of E with the property that $P_g(E_i) = \frac{1}{3} P_g(E)$, $i = 1, 2, 3$. There are three cases:

1. $P_{f_1}(E_i) = \frac{1}{3} P_{f_1}(E)$ for at least one of the $i \in \{1, 2, 3\}$.

2. $P_{f_1}(E_i) > \frac{1}{3} P_{f_1}(E)$ for two of the $i \in \{1, 2, 3\}$ and $\frac{1}{3} P_{f_1}(E) > P_{f_1}(E_j)$ for the other index $j \in \{1, 2, 3\}$.

3. $P_{f_1}(E_i) < \frac{1}{3} P_{f_1}(E)$ for two of the $i \in \{1, 2, 3\}$ and $\frac{1}{3} P_{f_1}(E) < P_{f_1}(E_j)$ for the other index $j \in \{1, 2, 3\}$.

Case (1) yields the weak form of the result with $\beta = \frac{1}{3}$.

For Case (2), we relabel (if necessary) the E_i so that $P_{f_1}(E_1) \geq P_{f_1}(E_3) > \frac{1}{3} P_{f_1}(E) > P_{f_1}(E_2)$. For each E_i, form the nested subfamily $\{E_{i,s} : s \in [0, 1]\}$ such that $P_g(E_{i,s}) = s \cdot P_g(E_i)$. Define a subfamily $\{E_r : r \in [0, 1]\}$ of subsets of $E = E_1 \cup E_2 \cup E_3$ by

---

16. J. Dugundji (1966) *Topology*, Boston: Allyn and Bacon.

$$E_r = \begin{cases} E_{1,3r} & \text{if } 0 \le r \le \frac{1}{3}, \\ E_1 \cup E_{2,3r-1} & \text{if } \frac{1}{3} < r \le \frac{2}{3}, \\ E_1 \cup E_2 \cup E_{3,3r-2} & \text{if } \frac{2}{3} < r \le 1. \end{cases}$$

By construction, $P_g(E_r) = r \cdot P_g(E)$. If $P_{f_1}(E_1) = 0$, we are in Case (1). Otherwise, define $\varphi(r) = P_{f_1}(E_r)/P_{f_1}(E)$. By dominated convergence, this is a continuous function. Further, $\varphi(\frac{1}{3}) > \frac{1}{3}$ and $\varphi(\frac{2}{3}) < \frac{2}{3}$. By the intermediate value theorem, $\varphi(r) = r$ for some $r \in (\frac{1}{3}, \frac{2}{3})$. For that r, $P_{f_1,\dots,f_{k+1}}(E_r) = r \cdot P_{f_1,\dots,f_{k+1}}(E)$.

For Case (3), we have $\varphi(\frac{1}{3}) < \frac{1}{3}$ and $\varphi(\frac{2}{3}) > \frac{2}{3}$.

Step 4: We now show that if the result holds in its weak form, then it holds in its subfamily form. In the previous step, we showed that for $E \in \mathcal{F}$, there exists an $r^* \in (\frac{1}{3}, \frac{2}{3})$ and an $E_{r^*} \subset E$ such that $\nu(E_{r^*}) = r^* \cdot \nu(E)$. By repeated subdivisions, this implies that there is a dense set, D, of $r \in [0, 1]$ and a corresponding nested set $(E_r)_{r \in D}$ such that $\nu(E_r) = r \cdot \nu(E)$ for all $r \in D$. For $s \notin D$, define $E_s = \cup_{r<s} E_r$. $(E_s)_{s \in [0,1]}$ is the requisite nested class of subsets. ∎

Proof of the compactness part of Theorem 8.5.36.   The boundedness of $R \subset \mathbb{R}^k_+$ comes from the integrability of f. We suppose, without loss, that the f_i are linearly independent.

Step 1: The result is true if $k = 1$.

Step 2: Now suppose that the result is true for $k - 1$. Pick arbitrary $\mathbf{v} \in \mathbb{R}^k$ such that $\mathbf{v} \cdot \mathbf{v} = 1$. We show that arg max$\{\mathbf{v} \cdot r : r \in \text{cl}(R)\} \subset R$. Let $E = \{\omega : \mathbf{v} \cdot f(\omega) > 0\}$, so that $\nu(E) \in R$ solves the problem max$\{\mathbf{v} \cdot r : r \in R\}$, and so solves the problem max$\{\mathbf{v} \cdot r : r \in \text{cl}(R)\}$. Let $Z = \{\omega : \mathbf{v} \cdot f(\omega) = 0\}$, so that $\mathbf{v} \cdot \nu(Z') = 0$ for all $Z' \subset Z$. By the inductive hypothesis, the set $\{\nu(Z') : Z' \subset Z\}$ is closed and convex. Therefore the class of sets $\{\nu(E \cup Z') : Z' \subset Z\}$ is arg max$\{\mathbf{v} \cdot r : r \in \text{cl}(R)\} \subset R$.

Step 3: The mapping $\mathbf{v} \mapsto$ arg max$\{\mathbf{v} \cdot r : r \in \text{cl}(R)\}$ is upper hemicontinuous. The upper hemicontinuous image of a compact set is compact, so the set of extreme points of R is a compact set.

Step 4: R is the convex hull of its extreme points, and the convex hull of any compact set is compact. ∎

8.12.i   The Further Problems of L^p, $p \in (0, 1)$

We saw that if $p \in (0, 1)$, then the triangle inequality fails for the associated attempt to define a norm distance. The same problem appears in $\mathbb{R}^k$, $k \ge 2$.

Example 8.12.32 *For $p \in (0, 1)$ and $\mathbf{x} \in \mathbb{R}^k$, if we define $\|\mathbf{x}\|_p = \left(\sum_{i \le k} |x_i|^p\right)^{1/p}$ and $d_p(\mathbf{x}, \mathbf{y}) = \|\mathbf{x} - \mathbf{y}\|_p$, then $d_p(\cdot, \cdot)$ violates the triangle inequality, as one sees in $\mathbb{R}^2$ by taking $\mathbf{x} = (0, 0)$, $\mathbf{y} = (1, 0)$, and $\mathbf{z} = (1, 0)$, so that $d_p(\mathbf{x}, \mathbf{y}) = d_p(\mathbf{y}, \mathbf{z}) = 1$ and $d_p(\mathbf{x}, \mathbf{z}) = 2^{1/p} > 2$, so that $d_p(\mathbf{x}, \mathbf{z}) > d_p(\mathbf{x}, \mathbf{y}) + d(\mathbf{y}, \mathbf{z})$, violating the triangle inequality.*

The failure of local convexity has implications for linear functions on $L^p((\Omega, \mathcal{F}, P))$ whenever $(\Omega, \mathcal{F}, P)$ is nonatomic.

Example 8.12.33 *Take your favorite $p \in (0, 1)$ and let L^p denote the set of $X \in L^0(\Omega, \mathcal{F}, P)$ with $\int |X|^p \, dP < \infty$, $(\Omega, \mathcal{F}, P)$ nonatomic. Suppose that $T : L^p \to \mathbb{R}$ is linear and that for some simple X, $T(X) \neq 0$. We show that there exists a sequence X_n with $\left(\int |X_n|^p \, dP\right)^{1/p} \to 0$ and $T(X_n) \equiv 1$. Thus, a linear T that is nonzero on some simple function must be discontinuous.*

If $X = \sum_i \beta_i 1_{E_i}$ and the E_i are a partition of Ω, then, by linearity, $T(X) = \sum_i \beta_i T(1_{E_i})$, and for one of the E_i, $T(1_{E_i}) \neq 0$. Using linearity again, one can find γ such that $T(\gamma 1_{E_i}) = 1$. Divide E_i into two equal probability sets, A_1, B_1, so that $T(\gamma 1_{A_1}) = T(\gamma 1_{B_1}) = \frac{1}{2}$ and $T(2\gamma 1_{A_1}) = 1$. Split A_1 into two equal probability sets, A_2 and B_2, so that $T(4\gamma 1_{A_2}) = 1$. Continuing inductively, we have $T(2^n \gamma 1_{A_n}) = 1$. Now, $\left(\int |2^n \gamma 1_{A_n}|^p \, dP\right)^{1/p} = \left[P(E_i)/2^n\right]^{1/p} \cdot (2^n |\gamma|) \to 0$ because $1/p > 1$.

If you have a liking for perversities, you can adapt the foregoing to show that X need not be a simple function.

8.13 ◆ Bibliography

All of the mathematics per se developed here is in *Linear Operators,* Vol. 1 (New York: Interscience, 1958) by N. Dunford and J. T. Schwartz, though their coverage is so much broader that one could get lost in it. The mathematical properties of $L^p(\Omega, \mathcal{F}, P)$ spaces that are so central to economic theory and econometrics are covered, wonderfully well, in a number of places, usually as part of a larger agenda. However, in E. Ward Cheney's *Analysis for Applied Mathematics* (New York: Springer-Verlag, 2001) they occupy a more central place. Other sources that treat the properties of L^p spaces are R. M. Dudley's *Real Analysis and Probability* (Cambridge: Cambridge University Press, 2002), which puts convexity in a central role, and H. L. Royden's *Real Analysis* (3rd Ed., New York: Macmillan, 1988). We have absorbed the relations of these spaces to economic models and econometrics from colleagues and students too numerous to name over the years, though Hal White in particular should be thanked.

On a more specific level, the elementary proof of the convexity part of Lyapunov's theorem is from D. A. Ross, "An elementary proof of Lyapunov's theorem," *American Mathematical Monthly* 112(7), 651–53 (2005), while the elementary compactness argument is based on ideas in L. E. Dubins and E. H. Spanier's "How to cut a cake fairly," *American Mathematical Monthly* 68(1), 1–17 (1961).

CHAPTER 9

Probabilities on Metric Spaces

Economists use probabilities on metric spaces in two main ways: as models of choice under uncertainty and as models of stochastic processes. We begin with an overview of these two kinds of applications and then turn to the most important metric on the space of probabilities, the Prokhorov metric. We first give four characterizations of convergence and a compactness criterion for sets of probabilities. We then turn to the study of probabilities as continuous linear functionals, which we use to develop regular conditional probabilities, and as continuous linear operators, which leads to a proof of the central limit theorem.

Throughout, M is a metric space and $\Delta = \Delta(M)$ is the set of countably additive probabilities on $\mathcal{B}_M$, the Borel σ-field of subsets of M.

9.1 ◆ Choice under Uncertainty

The basic model of choice under uncertainty that economists use can be made to look somewhat like the model for parametric statistics. We start with a set Θ. For each $\theta \in \Theta$, there is a probability distribution, P_θ, over the metric space, M, of consequences. However, instead of trying to estimate the true value of θ, the decision maker picks his favorite P_θ according to some utility function. The set Θ can be quite general.

Example 9.1.1 *An insurance company offers a menu, Θ, of insurance policies against loss, with different premia, p; different deductibles, D; and different proportions, r, of the loss that will be reimbursed. If you choose insurance policy $\theta = (p, D, r)$, then if X is your random income and L your random loss, the random variable that you face is*

$$R_\theta = (X - P) - (D - rL) \cdot 1_{L>D} - L \cdot 1_{L \le D}.$$

If we define $P_\theta \in \Delta(\mathbb{R})$ as the distribution of R_θ, the consumer solves the problem

$$\max_\theta U(P_\theta),$$

where $U : \Delta(\mathbb{R}) \to \mathbb{R}$ is the consumer's utility function over the set Δ of distributions.

Two very sneaky things just happened.

1. We assumed that the consequence space, $\mathbb{R}$, representing money, is all that enters into the consumer's utility. This simplifies many decision problems, but is clearly wrong in any generality. If there are other aspects of the random consequences that matter to the consumer's choice in a fashion that affects the economics of the situation being examined, (for example, if this period's consumption is too low, then the children die), then these aspects should be included in the utility function. For the abstract theory, any separable metric space of consequences will do.

2. We assumed that the consumer knows the distributions of X and L, or at least, that solving the optimization problem in which the consumer knows them captures the important aspects of the consumer's behavior. There are behavioral axioms that make a fairly compelling case for this assumption, and it is certainly a convenient one. If there are aspects of the distributions over consequences that matter to the consumer's choice in such a fashion as to affect the economics of the situation being examined, then these aspects should be included in the model.

Much of the theory of choice under uncertainty uses properties of continuous linear functions $U : \Delta \to \mathbb{R}$. We will start with a study of the metric space $(\Delta(M), \rho)$, where ρ is the **Prokhorov metric**, then turn to continuous linear functions.

9.2 ◆ Stochastic Processes

A process is something that happens over time, and stochastic means random.[1] A distribution over a metric space of time paths is the most common model of a stochastic process. The most common spaces of paths are $\mathbb{R}^T$ for T a discrete subset of $\mathbb{R}$; $C([0, 1])$, the set of continuous paths on the time set $T = [0, 1]$; and $D[0, 1]$, the set of discontinuous jump paths on the time set $T = [0, 1]$.

The easiest kinds of stochastic processes to begin with happen in discrete time, for example, at equally spaced times like $T = \{0, 1, 2, \ldots, \tau\}$ or $T = \{0, 1, 2, \ldots\}$. Here, we can take $\Omega = M = \mathbb{R}^T$ with the product metric and the corresponding Borel σ-field, $\mathcal{B}_M$. If μ is a distribution on $\mathcal{B}_M$, then for each $t \in T$, the mapping $\text{proj}_t(\omega)$ is the value of the process at time t if ω is drawn and $\text{proj}_t(\mu)$ is the distribution of the process at time t. If the process takes values in $\mathbb{R}^k$ at times in T, then we can take $\Omega = M = (\mathbb{R}^k)^T$.

If the time set is $T = [0, 1]$ and the stochastic process moves continuously over time, then we can use $C([0, 1])$. If $\Omega = C([0, 1])$ and if ω is drawn according to μ, a distribution on Ω, then at time t, the stochastic process takes on the value $X(t, \omega) := \text{proj}_t(\omega) = \omega(t)$, and $\text{proj}_t(\mu) \in \Delta(\mathbb{R})$ is the distribution. If $0 \leq t_1 < t_2 < \cdots < t_K \leq 1$, then $\text{proj}_{t_1,\ldots,t_K} : \Omega \to \mathbb{R}^K$ gives the joint distribution of the stochastic process at the time points $\{t_1, t_2, \ldots, t_K\}$, and $\text{proj}_{t_1,\ldots,t_K}(\mu) \in \Delta(\mathbb{R}^K)$ is the distribution.

---

1. "Stochastic" is derived from the Greek root for aiming at something, or a guess, while "random" comes from Old French, to run at a fast gallop, with the implication of being out of control.

In order for $\text{proj}_{t_1,\ldots,t_K}(\mu)$ to be a distribution, the projection mapping must be measurable. In fact it is continuous, hence measurable, and the projection mappings generate the Borel σ-field.

Exercise 9.2.1 Show that for all finite $S \subset [0, 1]$, $\text{proj}_S : C([0, 1]) \to \mathbb{R}^S$ is continuous, hence measurable. With $\mathcal{B}$ being the Borel σ-field generated by the sup norm topology on $C([0, 1])$, show that

$$\mathcal{B} = \sigma(\{\text{proj}_S^{-1}(A) : S \text{ finite and } A \text{ a Borel subset of } \mathbb{R}^S\}).$$

From introductory statistics, you learned that the sum of independent Gaussian distributions is another Gaussian distribution. A fancier way to say this is that the Gaussian distribution is **infinitely divisible**. That is, for every Gaussian distribution, Q, and every $k \in \mathbb{N}$, there are k iid random variables, $Y_1, \ldots, Y_k$, such that the distribution of $\sum Y_i$ is Q. It turns out that this means that there is a distribution, μ, on $C([0, 1])$, called a **Brownian motion** or a **Weiner measure**, with the property that the increments, $Y_1 := [X(\cdot, \frac{1}{k}) - X(\cdot, 0)]$, $Y_2 := [X(\cdot, \frac{2}{k}) - X(\cdot, \frac{1}{k})], \ldots,$ $Y_k := [X(\cdot, 1) - X(\cdot, \frac{k-1}{k})]$, are iid Gaussians with mean 0 and variance $\frac{1}{k}$.

There are many more infinitely divisible distributions and they can all be reached in this fashion—a continuous-time stochastic process over the time interval $[0, 1]$, having independent increments. The simplest example is the Poisson distribution, and it shows why we need spaces of times paths that allow jumps.

Definition 9.2.2 *A random variable X has the **Poisson distribution with parameter** r, written $X \sim Poisson(r)$, if $P(X = m) = e^{-r}\frac{r^m}{m!}$, $m = 0, 1, \ldots$.*

Exercise 9.2.3 If X and Y are independent Poisson random variables with parameters r and s, then $X + Y$ is a Poisson random variable with parameter $r + s$.

Let $D[0, 1]$ denote the set of functions mapping $[0, 1] \to \mathbb{R}$ that are right-continuous and have limits from the left, and give $D[0, 1]$ the sup norm metric.

Exercise 9.2.4 Show that the metric space $D[0, 1]$ is not separable, and that

$$\mathcal{B} \supset \sigma(\{\text{proj}_S^{-1}(A) : S \text{ finite and } A \text{ a Borel subset of } \mathbb{R}^S\}).$$

Further, every element of $D[0, 1]$ is bounded on $[0, 1]$, though it may or may not attain its supremum and/or infimum.

It turns out that for every $r > 0$, there is a distribution, μ_r, and a separable measurable set, E_r, with the properties that $\mu_r(E_r) = 1$, and that the increments, $Y_1 := [X(\cdot, \frac{1}{k}) - X(\cdot, 0)]$, $Y_2 := [X(\cdot, \frac{2}{k}) - X(\cdot, \frac{1}{k})], \ldots, Y_k := [X(\cdot, 1) - X(\cdot, \frac{k-1}{k})]$, are iid Poissons with parameter $\frac{r}{k}$.

9.3 ◆ The Metric Space $(\Delta(M), \rho)$

We define a metric, ρ, on $\Delta(M)$ that is useful for studying both choice under uncertainty and the convergence of stochastic processes. Earlier, in §6.3, we defined the Levy metric on $\Delta(\mathbb{R})$ and showed that this made $\Delta(\mathbb{R})$ into a complete separable metric space. For probabilities on general metric spaces, we have the

Prokhorov metric, which is equivalent to the Levy metric when $M = \mathbb{R}$. Recall that $A^\epsilon := \cup_{x \in A} B_\epsilon(x)$ is the ϵ-ball around the set A.

Definition 9.3.1 *The **Prokhorov metric** on $\Delta(M)$ is*

$$\rho(P, Q) := \inf\{\epsilon > 0 : \forall F \quad \text{closed},$$

$$P(F) \le Q(F^\epsilon) + \epsilon \text{ and } Q(F) \le P(F^\epsilon) + \epsilon\}.$$

The comparison with the Levy metric is instructive. The Levy metric restricts the closed F to be sets of the form $(-\infty, x]$.

When we have enough background in place, we will prove the following.

Theorem 9.3.2 $\rho(P_n, P) \to 0$ *iff for all* $f \in C_b(M)$, $\int f \, dP_n \to \int f \, dP$. *Further, (M, d) is complete (respectively separable, respectively compact) iff $(\Delta(M), \rho)$ is as well.*

9.3.a A Detour through Naming ρ-Convergence

We begin with the following.

Notation 9.3.3 *If P, P_n is a sequence of probabilities and $\int f \, dP_n \to \int f \, dP$ for all $f \in C_b(M)$, we write $P_n \to_{w^*} P$ and say that P_n **converges weak*** to P. If X, X_n is a sequence of random variables and $E\, f(X_n) \to E\, f(X)$ for all $f \in C_b(M)$, we write $X_n \to_{\mathcal{D}} X$ and say that X_n **converges in distribution to** P.*

It is well worth keeping this clear—random variables converging in distribution is the same as their distributions converging weak*.

In this notation, Theorem 9.3.2 says that $\rho(P_n, P) \to 0$ iff $P_n \to_{w^*} P$. There is some confusion about why this kind of convergence is called either "weak* convergence" or the less correct but more commonly used "weak convergence." The roots of this fancy name come from the following considerations.

$\mathfrak{X} = C_b(M)$ is a Banach space, and we show that if M is complete and separable, then $\mathfrak{X}^*$ is $ca(M)$, the set of countably additive finite signed Borel measures on M. The dual norm on $ca(M)$ is

$$\|\nu\|^* = \sup_{f \in C_b(M), \|f\| \le 1} \left| \int f \, d\nu \right|.$$

Since continuous functions do a good job approximating indicator functions, one can show that

$$\|\nu\|^* = \sup_{E \in \mathcal{F}} \left| \int 1_E \, d\nu - \int 1_{E^c} \nu \right|,$$

which yields the following.

Lemma 9.3.4 *For $P, Q \in \Delta(M) \subset ca(M)$, $\|P - Q\|^* = \sup\{| \int f \, d(P - Q)| : f \in C_b(M), \|f\| = 1\} = 2 \cdot \sup_{A \in \mathcal{B}} |P(A) - Q(A)|$.*

The quantity $\sup_{A \in \mathcal{B}} |P(A) - Q(A)|$ is known as the variation norm of the signed measure $P - Q$, and it showed up in a crucial way in the middle of the Hahn-Jordan decomposition theorem. In terms of the variation norm, the lemma says that $\|P - Q\|^* = 2 \cdot \|P - Q\|_{vn}$.

It is difficult to converge in the norm metric; probabilities supported on the rationals cannot converge to a continuous supported probability.[2] For choice models, it is easy for objective functions to be continuous in the norm metric and difficult for sets of distributions to be compact.

There are two weaker kinds of convergence, weak convergence and the weaker weak* convergence:

ν_α converges weakly to ν in $\mathfrak{X}^*$ if $|T(\nu_\alpha) - T(\nu)| \to 0$ for all $T \in \mathfrak{X}^{**}$,

ν_α converges weak* to ν in $\mathfrak{X}^*$ if $|t(\nu_\alpha) - t(\nu)| \to 0$ for all $t \in \mathfrak{X} \subset \mathfrak{X}^{**}$.

When we put these together, strong convergence implies weak convergence (because we are using norm continuous linear functions), and weak convergence implies weak* convergence.

Now, $\mathfrak{X}^{**}$ contains the linear functions of the form $T(\nu) = \int h\, d\nu$ for all bounded measurable h. Once again, if a sequence of probabilities is supported on the rationals, then it cannot converge to a continuous probability; simply take $h = 1_{\mathbb{Q}}$.

Hence the name "weak*" and why the name "weak" is not correct. However, "weak" is becoming correct because words mean what people use them to mean, and many people use weak to mean weak*. Go figure.

Exercise 9.3.5   Let P_n be the uniform distribution on $\{\frac{k}{n} : k = 1, 2, \ldots, n\}$ and let $P = \lambda$ be the uniform distribution on $[0, 1]$. Show that $P_n \to_{w*} P$ but $\|P_n - P\|_{vn} \equiv 1$.

Variation norm closeness is even stronger than the last exercise suggests. Roughly, two probabilities are variation norm close iff they are almost equal.

Exercise 9.3.6   Show that $\|P_n - P\|_{vn} \to 0$ iff $P_n = \epsilon_n Q_n + (1 - \epsilon_n)P$ for some sequence $\epsilon_n \to 0$ and some sequence of probabilities Q_n.

It is now time to turn from fascinating etymological issues back to the substance of this chapter.

9.3.b   Basic Properties of the Prokhorov Metric

As a first step, we have to establish that $\rho(\cdot, \cdot)$ is a metric and that we only ever have to establish one of the inequalities in its definition.

Lemma 9.3.7   *For all P, Q, $\rho(P, Q) = \inf\{\epsilon > 0 : \forall F \text{closed}, P(F) \leq Q(F^\epsilon) + \epsilon\}$ and ρ is a metric.*

---

2. Scheffé's Theorem 7.5.29 gives conditions on sequences of densities that deliver norm convergence.

Proof. We first show that for all P, Q and $\epsilon > 0$, $P(F) \leq Q(F^\epsilon) + \epsilon$ for all closed F iff $Q(F) \leq P(F^\epsilon) + \epsilon$ for all closed F. Toward this end, define $C(\epsilon) = \{F : F \neq \emptyset \text{ is closed, and } (F^\epsilon)^c \neq \emptyset\}$. Since $P(\emptyset) = Q(\emptyset) = 0$ and $P(M) = Q(M) = 1$,

$P(F) \leq Q(F^\epsilon) + \epsilon$ for all closed F iff $P(F) \leq Q(F^\epsilon) + \epsilon$ for all $F \in C(\epsilon)$, and

$Q(F) \leq P(F^\epsilon) + \epsilon$ for all closed F iff $Q(F) \leq P(F^\epsilon) + \epsilon$ for all $F \in C(\epsilon)$.

For any closed F,

$$P(F) \leq Q(F^\epsilon) + \epsilon \text{ iff } (1 - P(F)) + \epsilon \geq (1 - Q(F^\epsilon)) \text{ iff}$$

$$Q((F^\epsilon)^c) \leq P(F^c) + \epsilon.$$

Finally, note that $((F^\epsilon)^c)^\epsilon = F^c$ and that $F \in C(\epsilon)$ iff $(F^\epsilon)^c \in C(\epsilon)$.

We now turn to showing that ρ is a metric. By the symmetry of the definition, $\rho(P, Q) = \rho(Q, P)$. If $\rho(P, Q) = 0$, then for all closed F and all $n \in \mathbb{N}$, $P(F) \leq Q(F^{1/n}) + 1/n$. Since $F^{1/n} \downarrow F$, this gives $P(F) \leq Q(F)$. Interchanging P and Q shows that $Q(F) \leq P(F)$. Therefore P and Q agree on closed sets. Since the closed sets generate $\mathcal{B}_M$, a (by now standard) good sets argument completes the proof that $P = Q$.

For the triangle inequality, if $\rho(P, Q) < r$ and $\rho(Q, R) < s$, then for any closed F,

$$P(F) \leq Q(F^r) + r \leq R((F^r)^s) + r + s \leq R(F^{r+s}) + r + s,$$

so that $\rho(P, R) \leq r + s$. As this is true for all $r > \rho(P, Q)$ and $s > \rho(Q, R)$, $\rho(P, R) \leq \rho(P, Q) + \rho(Q, R)$. ■

Exercise 9.3.8 The last proof used $(F^r)^s \subset F^{r+s}$. Show that $(F^r)^s = F^{r+s}$.

The next exercise asks you to show that the Prokhorov metric is a generalization of the Levy metric.

Exercise 9.3.9 In $\Delta(\mathbb{R})$, show that $\rho(P_n, P) \to 0$ iff $d_L(F_n, F) \to 0$ [where $d_L(\cdot, \cdot)$ is the Levy metric, and F_n and F are the cdf's of P_n and P].

9.3.c Finitely Supported Probabilities

A particularly important class of probabilities includes the **point masses** and the **finitely supported probabilities**.

Definition 9.3.10 *Point mass on* x *is the probability denoted/defined by* $\delta_x(A) = 1_A(x)$. *A finitely supported probability is a finite convex combination of point masses, that is, a finite sum of the form* $\sum \alpha_x \delta_x$, *where the* $\alpha_x > 0$ *and* $\sum \alpha_x = 1$,

Exercise 9.3.11 Show that $\rho(\delta_x, \delta_y) = \min\{1, d(x, y)\}$. From this show that:

1. $\rho(\delta_{x_n}, \delta_x) \to 0$ iff $d(x_n, x) \to 0$;
2. $x \leftrightarrow \delta_x$ is a homeomorphism;
3. $\{\delta_x : x \in M\}$ is a closed subset of $\Delta(M)$;

    4. if $(\Delta(M), \rho)$ is complete, then (M, d) is complete;

    5. if $(\Delta(M), \rho)$ is separable, then (M, d) is separable; and

    6. if $(\Delta(M), \rho)$ is compact, then (M, d) is compact.

For choice theory, the last exercise has a simple, but useful, implication.

Lemma 9.3.12   *If $U : \Delta(M) \to \mathbb{R}$ is continuous, then $u : M \to \mathbb{R}$ defined by $u(x) = U(\delta_x)$ is continuous.*

Proof.   Let $x_n \to x$, so that $\delta_{x_n} \to \delta_x$. By the continuity of U, $u(x_n) = U(\delta_{x_n}) \to U(\delta_x) =: u(x)$.   ∎

This is important when we study functions $U : \Delta(M) \to \mathbb{R}$ that are not only continuous, but are also *linear*. To see why, let $\mu = \sum \alpha_x \delta_x$ be a finitely supported probability. If U is linear, then $U(\mu) = \sum \alpha_x U(\delta_x) = \sum \alpha_x u(x) = \int u(x) \, d\mu(x)$. Thus, at least for the finitely supported probabilities, a continuous linear $U : \Delta(M) \to \mathbb{R}$ is the integral of a function $u : M \to \mathbb{R}$. The function u is called the **von Neumann-Morgenstern** utility function, and it is important to keep clear that u need not be linear.

Example 9.3.13   *If $M = \mathbb{R}$ and u is concave, then we know that $u(E\,X) \ge E\,u(X)$ when X is a random variable (Jensen's inequality, Theorem 8.3.3, p. 460). Let μ be the distribution of X, that is, $\mu(A) = P(X^{-1}(A))$. Preferences represented by $U(\mu) = \int_M u(x) \, d\mu(x)$ with a concave u favor receiving the average of a random variable for sure, $u(E\,X)$, to facing randomness, $E\,u(X)$.*

Definition 9.3.14   *A utility function $U : \Delta(M) \to \mathbb{R}$ is a **continuous expected utility function** if there exists $u \in C_b(M)$ such that $U(\mu) = \int u \, d\mu$.*

One can obviously take increasing transformations of U and leave the preferences unchanged. One cannot, in general, take increasing nonlinear transformations of the von Neumann-Morgenstern utility function u and leave the preference relation unchanged, though one can take affine transformations—for $r > 0$, $\int (r \cdot u + s) \, d\mu \ge \int (r \cdot u + s) \, d\nu$ iff $\int u \, d\mu \ge \int u \, d\nu$.

Our proximate aim is to prove the following.

Theorem 9.3.15   *If M is separable, then U is a continuous expected utility function iff U is continuous and linear.*

To prove this, we show that the finitely supported probabilities are ρ-dense and that $\rho(\mu_n, \mu) \to 0$ implies that $\int u \, d\mu_n \to \int u \, d\mu$ for all continuous u. Rather than try to prove these assertions in some minimal way, we prove a number of related results that will be useful in the long run. These involve understanding Prokhorov convergence in much more detail.

9.3.d   Equivalent Formulations of Convergence in the Prokhorov Metric

The crucial fact about convergence in the Prokhorov metric is that it is equivalent to the convergence of integrals of continuous bounded functions. There are also some other, very useful equivalences.

Recall that the boundary of a set A is $\partial A = \mathrm{cl}(A) \cap \mathrm{cl}(A^c) = \mathrm{cl}(A) \setminus \mathrm{int}(A)$. We say that A is a P-**continuity set** if $P(\partial A) = 0$. Let $U_b(M) \subset C_b(M)$ be the set of uniformly continuous bounded functions on M and $Lip_b(M) \subset U_b(M)$ the set of bounded Lipschitz functions. The following is called a *portmanteau* theorem because it covers so many contingencies.

Theorem 9.3.16 (Portmanteau)   *The following are equivalent:*

(1) $\rho(P_n, P) \to 0$,

(2) $\int f \, dP_n \to \int f \, dP$ for all $f \in C_b(M)$,

(2′) $\int f \, dP_n \to \int f \, dP$ for all $f \in U_b(M)$,

(2″) $\int f \, dP_n \to \int f \, dP$ for all $f \in Lip_b(M)$,

(3) $\limsup_n P_n(F) \le P(F)$ for all closed F,

(4) $\liminf_n P_n(G) \ge P(G)$ for all open G, and

(5) $\lim_n P_n(A) = P(A)$ for all P-continuity sets A.

The last three conditions have a common intuition that can be seen using point masses.

Exercise 9.3.17   Give an example of an open set, G, and a sequence, P_n, of point masses at x_n, with $x_n \to x$ such that one has a strict inequality in (4) in the previous result. Taking complements, give an example of a strict inequality in (3). Show that the G and F in your examples are *not* P-continuity sets.

Proof.   The structure of the proof is (3) ⇔ (4), (3) ⇒ (5) ⇒ (2) ⇒ (2′) ⇒ (2″) ⇒ (3) plus (1) ⇒ (3) and (3) ⇒ (1).

(3) ⇔ (4). Take complements.

[(3) ∧ (4)] ⇒ (5). Pick an arbitrary P-continuity set A. (3) ∧ (4) implies that for any P-continuity set A,

$$\liminf_n P_n(\mathrm{int}(A)) \ge P(\mathrm{int}(A))$$

$$= P(\mathrm{cl}(A)) \ge \limsup_n P_n(\mathrm{cl}(A)) \ge \liminf_n P_n(\mathrm{int}(A)),$$

where the last inequality comes from $P_n(\mathrm{cl}(A)) \ge P_n(\mathrm{int}(A))$. Since $\mathrm{int}(A) \subset A \subset \mathrm{cl}(A)$, $P_n(A) \to P(A)$ because $P_n(\mathrm{int}(A)) \le P_n(A) \le P_n(\mathrm{cl}(A))$.

(5) ⇒ (2). Pick an arbitrary $f \in C_b(M)$ and suppose that for all P-continuity sets A, $P_n(A) \to P(A)$. We must show that $\int f \, dP_n \to \int f \, dP$. Pick $\epsilon > 0$.

For all but countably many $r \in \mathbb{R}$, $P(f^{-1}(r)) = 0$. Hence, we can find points $a_1 < b_1 = a_2 < b_2 = \cdots = a_K < b_K$ with $a_1 < f(x) < b_K$, $b_k - a_k < \epsilon/2$, and $P(f^{-1}(a_k)) = P(f^{-1}(b_k)) = 0$, $k = 1, \ldots, K$. In particular, each $f^{-1}((a_k, b_k])$ is a P-continuity set. Since $\sum a_k 1_{f^{-1}((a_k, b_k])}(x) < f(x) \le \sum b_k 1_{f^{-1}((a_k, b_k])}(x)$,

$$A := \sum a_k P(f^{-1}(a_k, b_k]) < \int f(x) \, dP(x) \le \sum b_k P(f^{-1}(a_k, b_k]) =: B,$$

and for each P_n,

$$A_n := \sum a_k P_n(f^{-1}(a_k, b_k]) < \int f(x)\, dP_n(x) \le \sum b_k P_n(f^{-1}(a_k, b_k]) =: B_n.$$

Since $b_k - a_k < \epsilon/2$, $B - A < \epsilon/2$ and $B_n - A_n < \epsilon/2$. As each $f^{-1}((a_k, b_k])$ is a P-continuity set, for large n, $|A - A_n| < \epsilon/2$ and $|B - B_n| < \epsilon/2$. Therefore, for large n, $|\int f\, dP_n - \int f\, dP| < \epsilon$.

$(2) \Rightarrow (2') \Rightarrow (2'')$ is clear because $Lip_b(M) \subset U_b(M) \subset C_b(M)$.

$(2'') \Rightarrow (3)$. Suppose that $(2'')$ holds, that F is closed, and that δ is an arbitrary strictly positive number. Shrink wrapping implies that there exists an $\epsilon > 0$ such that $P(F^\epsilon) < P(F) + \delta$. The function $f(x) := \max\{0, 1 - \frac{1}{\epsilon} d(x, F)\}$ has Lipschitz constant $L = \frac{1}{\epsilon}$ because $|f(x) - f(y)| \le \frac{1}{\epsilon}|d(x, F) - d(y, F)| \le L d(x, y)$. Now, since $(2'')$ holds,

$$P_n(F) = \int_F f\, dP_n \le \int f\, dP_n \to \int f\, dP = \int_{F^\epsilon} f\, dP \le P(F^\epsilon) < P(F) + \delta.$$

Therefore, $\limsup_n P_n(F) \le P(F) + \delta$ for arbitrary $\delta > 0$.

$(1) \Rightarrow (3)$. Suppose that (1) holds, that F is closed, and that δ is an arbitrary strictly positive number. Since $F^{1/m} \downarrow F$, by shrink wrapping, there exists an $\epsilon > 0$ such that $P(F^\epsilon) < P(F) + \delta$. Without loss, $\epsilon < \delta$. Since (1) holds, for large n, $P_n(F) \le P(F^\epsilon) + \epsilon$. Combining yields

$$P_n(F) \le P(F^\epsilon) + \epsilon < P(F) + \delta + \epsilon < P(F) + 2\delta.$$

As δ was arbitrary, $\limsup_n P_n(F) \le P(F)$.

$(3) \Rightarrow (1)$. Suppose that (3) holds and pick $\epsilon > 0$. It is sufficient to show that for large enough n, for all F, $P_n(F) \le P(F^\epsilon) + \epsilon$.

Let $\{x_m : m \in \mathbb{N}\}$ be a countable dense subset of M. Pick $\delta < \epsilon/3$ such that for each x_m, $P(\partial B_{\delta/2}(x_m)) = 0$. This can be done because the set of r such that $P(\partial B_r(x_m)) > 0$ is at most countable for each x_m. Define $A_0 = \emptyset$ and $A_{m+1} = B_{\delta/2}(x_m) \setminus \cup_{k \le m} A_k$. This gives a disjoint cover of M by sets of diameter less than or equal to δ. Pick M large enough that $P(\cup_{m \le M} A_m) > 1 - \delta$. Let $\mathcal{A} = \sigma(\{A_m : m \le M\})$ and note that for all $A \in \mathcal{A}$, $P(\partial A) = 0$. Pick N large enough that for all $n \ge N$ and for all $A \in \mathcal{A}$, $|P_n(A) - P(A)| < \delta$.

For each closed F, define $A_F \in \mathcal{A}$ to be $\cup\{A_m : m \le M, \ A_m \cap F \ne \emptyset\}$. Thus, $F \subset A_F \cup B$, $B := (\cup_{m \le M} A_m)^c$, and A_F and B are disjoint elements of $\mathcal{A}$. Since the diameter of the A_m is less than δ, $F \subset A_F \subset F^\delta$. Combining everything, we get

$$P_n(F) \le P_n(A) + P_n(B) \le P(A) + P(B) + 2\delta \le P(A) + 3\delta \le P(F^\delta) + 3\delta.$$

Since $\delta < 3\epsilon$, and $F^\delta \subset F^{3\delta} \subset F^\epsilon$, $P_n(F) \le P(F^\epsilon) + \epsilon$ for all closed F. ∎

Exercise 9.3.18 Give an example of an $f \in C_b(M)$ such that all of the following inclusions are proper: $f^{-1}(a, b) \subset f^{-1}(a, b] \subset f^{-1}[a, b] \subset \mathrm{cl}(f^{-1}(a, b))$. Show that if $P(f^{-1}(a)) = P(f^{-1}(b)) = 0$, then all four sets are P-continuity sets.

Exercise 9.3.19 Fill in the details for the following alternative proof that $(3) \Rightarrow (2)$. Suppose that (3) holds and that f is a continuous function. First, $\limsup_n \int f \, dP_n \leq \int f \, dP$ by the following arguments:

1. Rescaling and shifting if necessary, assuming $0 < f(x) < 1$ is without loss.
2. For $m \in \mathbb{N}$ and $k \in \{0, \dots, 2^m\}$, let $F_{k,m}$ be the closed set $\{\frac{k}{2^m} \leq f\}$ and $E_{k,m}$ the measurable set $\{\frac{k}{2^m} \leq f < \frac{k+1}{2^m}\}$.
3. Show that for any probability Q, this yields

$$\sum_{k=1}^{2^m} \frac{k-1}{2^m} Q(E_{k,m}) \leq \int f \, dQ < \sum_{k=1}^{2^m} \frac{k}{2^m} Q(E_{k,m}).$$

4. Using $Q(E_{k,m}) = Q(F_{k-1,m}) - Q(F_{k,m})$, rearrange both sums to give

$$\frac{1}{2^m} \sum_{k=1}^{2^m} Q(F_{k,m}) \leq \int f \, dQ < \frac{1}{2^m} + \frac{1}{2^m} \sum_{k=1}^{2^m} Q(F_{k,m}). \quad (9.1)$$

5. Letting $Q = P_n$ and using the rightmost inequality in (9.1), show that

$$\limsup_n \int f \, dP_n \leq \frac{1}{2^m} + \frac{1}{2^m} \sum_{k=1}^{2^m} \limsup_n P_n(F_{k,m}).$$

6. Using (3), show that $\limsup_n P_n(F_{k,m}) \leq P(F_{k,m})$ for each k, m and, using the leftmost inequality in (9.1), that

$$\limsup_n \int f \, dP_n \leq \frac{1}{2^m} + \int f \, dP.$$

7. Show $\limsup_n \int f \, dP_n \leq \int f \, dP$.

Adding a constant, conclude from the foregoing that $\limsup_n \int (-f) \, dP_n \leq \int (-f) \, dP$ and we have $\liminf_n \int f \, dP_n \geq \int f \, dP$. Finish the proof.

9.3.e Characteristic and Moment-Generating Functions

One intuition for the Prokhorov metric is that knowing that $|\int f \, dP - \int f \, dQ|$ is small, or zero, for a rich class of bounded continuous functions f tells us that P and Q are close, or identical. If we know this for a class of f that is dense, or that has a dense span, then we come to the same conclusions. We can understand characteristic functions and moment-generating functions this way. Here we sketch the theory of both of these classes of functions for sets of compactly supported probabilities.

Definition 9.3.20 *A probability P is **compactly supported** if $P(K) = 1$ for some compact K. For a given compact K, $\Delta(K)$ denotes the set of probabilities with $P(K) = 1$.*

Theorem 9.3.21 *Suppose that $K \subset \mathbb{R}$ is compact and $H \subset C(K)$ has the property that* **span**(H) *is dense in $C(K)$. Then $\rho(P_n, P) \to 0$ iff $\int h \, dP_n \to \int h \, dP$ for all $h \in H$.*

Proof. If $\rho(P_n, P) \to 0$, then $\int f \, dP_n \to \int f \, dP$ for all $f \in C(K)$, hence for all $h \in H \subset C(K)$.

Now suppose that $\int h \, dP_n \to \int h \, dP$ for all $h \in H$, where **span**(H) is dense in $C(K)$. Pick arbitrary $f \in C(K)$ and $\epsilon > 0$. Since **span**(H) is dense in $C(K)$, there is a finite sum of the form $g = \sum_k \beta_k h_k \in$ **span**(H) with $d_\infty(f, g) < \epsilon/3$. For each of the finitely many h_k, pick N_k such that for all $n \geq N_k$, $|\int h \, dP_n \to \int h \, dP| < \epsilon/3|\beta_k|$. For all $n \geq \max N_k$, $|\int f \, dP_n \quad \int f \, dP|$ is equal to

$$\left| \int f \, dP_n - \int g \, dP_n + \int g \, dP_n - \int g \, dP + \int g \, dP - \int f \, dP \right|$$

$$\leq \left| \int f \, dP_n - \int g \, dP_n \right| + \left| \int g \, dP_n - \int g \, dP \right| + \left| \int g \, dP - \int f \, dP \right|$$

$$\leq \frac{\epsilon}{3} + \frac{\epsilon}{3} + \frac{\epsilon}{3},$$

where the first and third inequalities arise from $\max_x |f(x) - g(x)| < \epsilon/3$. ■

The following turns weak* convergence of probabilities into pointwise convergence of continuous functions.

Example 9.3.22 (Moment–Generating Functions) *By the Stone-Weierstrass theorem, the span of the set of functions $x \mapsto h_t(x) = e^{tx}$ is dense in $C(K)$ for any compact $K \subset \mathbb{R}$. The **moment-generating function** for a $P \in \Delta(K)$ is $m_P(t) := \int_K e^{tx} \, dP(x)$. By dominated convergence, the function $m_P(\cdot)$ is a continuous function of t. By Theorem 9.3.21, for $P_n, P \in \Delta(K)$, $\rho(P_n, P) \to 0$ iff for all $t \in \mathbb{R}$, $m_{P_n}(t) \to m_P(t)$.*

Exercise 9.3.23 Carefully prove that for $P \in \Delta(K)$, the nth derivative of m_P evaluated at 0 satisfies $m_P^{(n)} = \int x^n \, dP(x)$.

Example 9.3.24 (Characteristic Functions) *By the Stone-Weierstrass theorem, the span of the set of functions $x \mapsto \cos(tx)$ and $x \mapsto \sin(tx)$, $t \in \mathbb{R}$, is dense in $C(K)$ for any compact $K \subset \mathbb{R}$. The **characteristic function** for a $P \in \Delta(K)$ is $\phi_P(t) := \int_K (\cos(tx), \sin(tx)) \, dP(x)$. By dominated convergence, the function $\phi_P(\cdot)$ is a continuous $\mathbb{R}^2$-valued function of t. By Theorem 9.3.21, for $P_n, P \in \Delta(K)$, $\rho(P_n, P) \to 0$ iff for all $t \in \mathbb{R}$, $\phi_{P_n}(t) \to \phi_P(t)$.*

Exercise 9.3.25 Find and carefully prove the relation between of ϕ_P evaluated at 0 and $\int x^n \, dP(x)$ for $P \in \Delta(K)$.

There are two reasons why characteristic functions are so widely used. First, the restriction to compactly supported sets of probabilities is not needed, though this is significantly more difficult to prove. The second involves how characteristic functions turn convolution of probabilities into the addition of functions.

Example 9.3.26 *Euler's formula is $e^{itx} = \cos(tx) + i \sin(tx)$ and one can derive this by using the definition of $e^x = \sum_{k=0}^{\infty} \frac{x^k}{k!}$, absolutely convergent series,* *and the rules of complex multiplication, which yield $i^2 = -1$. This allows us to reformulate characteristic functions $\phi_P(t) = \int e^{itx} \, dP(x)$, or, if X is a random variable having distribution P, as $\phi_X(t) := E \, e^{itX}$*

 If $X \perp\!\!\!\perp Y$, X with distribution P and Y with distribution Q, then $P * Q$ denotes the distribution of $X + Y$ and is called the **convolution** *of P and Q. Convolution is a moderately complicated operation to study directly. However, independence directly delivers $\phi_{P*Q}(t) = \phi_P(t) \cdot \phi_Q(t)$, equivalently, $\phi_{X+Y}(t) = \phi_X(t) \cdot \phi_Y(t)$.*

9.4 ◆ Two Useful Implications

The portmanteau theorem is the starting point for almost everything that one can say about weak* convergence. We now use it to examine the behavior of the continuous images of a convergent sequence of probabilities and to look at approximation by finitely supported probabilities.

9.4.a The Continuous Mapping Theorem

The composition of continuous functions being continuous tells us that if $\rho(P_n, P) \to 0$, that is, $P_n \to_{w*} P$ in $\Delta(M)$, and $f : M \to M'$ is continuous, then $f(P_n) \to_{w*} f(P)$ in $\Delta(M')$. This result is called the **continuous mapping theorem**. It can be substantially generalized, but this requires a background result on the measurability of a set of discontinuities.

 For a function $h : M \to M'$, not necessarily measurable, let D_h denote its set of discontinuities, that is, the set of $x \in M$ such that there exists a sequence $x_n \to x$ with $h(x_n) \not\to h(x)$.

Exercise 9.4.1   Let $h = 1_A$, where $A \subset M$ may or may not be measurable. Show that $D_h = \partial A$, that is, $D_h \in \mathcal{B}_M$.

 The result of the exercise is more general.

Lemma 9.4.2 *If $h : M \to M'$ is a function, then $D_h \in \mathcal{B}_M$.*

Proof. Define $A_{\epsilon,\delta} = \{x \in M : \exists y, z \in B_\delta(x), \; d'(h(y), h(z)) \geq \epsilon\}$. $A_{\epsilon,\delta}$ is open and $D_h = \cup_\epsilon \cap_\delta A_{\epsilon,\delta}$, where the union and intersection range over strictly positive rationals, hence are countable. ∎

Theorem 9.4.3 (Continuous Mapping) *If $P_n \to_{w*} P$ in $\Delta(M)$, $h : M \to M'$ is a measurable function, and $P(D_h) = 0$, then $h(P_n) \to_{w*} h(P)$ in $\Delta(M')$.*

Proof. Let F be a closed subset of M'. It is sufficient to show that $\limsup_n h(P_n)(F) \leq h(P)(F)$, equivalently $\limsup_n P_n(h^{-1}(F)) \leq P(h^{-1}(F))$. Since $\mathrm{cl}(h^{-1}(F)) \subset D_h \cup h^{-1}(F)$, $P(\mathrm{cl}(h^{-1}(F))) = 0 + P(h^{-1}(F))$.

 Since $P_n \to_{w*} P$, $\limsup_n P_n(\mathrm{cl}(h^{-1}(F)) \leq P(\mathrm{cl}(h^{-1}(F))$. Since $P_n(h^{-1}(F) \leq P_n(\mathrm{cl}(h^{-1}(F))$ and $P(\mathrm{cl}(h^{-1}(F)) = P(h^{-1}(F))$, we have $\limsup_n P_n(h^{-1}(F))$ $\leq P(h^{-1}(F))$. ∎

9.4.b Approximation by Finitely Supported Probabilities

Finding a space M in which the finitely supported probabilities are not dense is much harder than the following suggests, touching as it does on deep problems in the foundations of mathematics. We steer clear of the deep waters.

Theorem 9.4.4 *If M is separable, then the finitely supported probabilities are dense in $\Delta(M)$.*

The "disjointification" in the following proof is quite a useful general device.

Proof. Pick arbitrary $P \in \wedge(M)$. We must show that there is a finitely supported Q with $\rho(P, Q) < \epsilon$.

Choose $0 < \delta < \epsilon/2$. Let $\{x_n : n \in \mathbb{N}\}$ be a countable dense subset of M and define $B_n = B_\delta(x_n)$ so that $M \subset \cup_n B_n$. Disjointify the B_n; that is, define $A_1 = B_1$ and $A_{k+1} = B_{k+1} \setminus (\cup_{m \leq k} A_k)$. Since $\delta < \epsilon/2$, the diameter of each A_k is less than $\epsilon/2$.

Pick K such that $P(\cup_{k \leq K} A_k) > 1 - \delta$ and define $\mathcal{A} = \sigma\{A_k : k \leq K\}$. For each nonempty A_k, $k \leq K$, choose $x_k \in A_k$, and if $A_{K+1} := (\cup_{k \leq K} A_k)^c$ is nonempty, pick $x_{K+1} \in A_{K+1}$. Define the finitely supported Q by $Q(x_k) = P(A_k)$, $k = 1, \ldots, (K + 1)$.

We now show that $\rho(P, Q) < \epsilon$. For any closed F, define $A_F = \cup\{A_k : k \leq K, \text{ and } A_k \cap F \neq \emptyset\}$. Since the diameter of each A_k, $k \leq K$, is less than $\epsilon/2$, $(F \cap A_F) \subset A_F \subset F^\epsilon$. As $P(A_{K+1}) < \epsilon$, $Q(F) \leq Q(A_F) + \epsilon \leq P(A_F) + \epsilon \leq P(F^\epsilon) + \epsilon$, completing the argument since F was an arbitrary closed set. ∎

Corollary 9.4.5 *If (M, d) is separable, then $(\Delta(M), \rho)$ is separable.*

Exercise 9.4.6 Prove Corollary 9.4.5. [If $X := \{x_n : n \in \mathbb{N}\}$ is dense in M, show that the countable set $P = \{\sum q_n \delta_{x_n} : q_n \in \mathbb{Q}\}$ of finitely supported probabilities is dense in the finitely supported probabilities, hence dense in $\Delta(M)$.]

9.5 ◆ Expected Utility Preferences

We are now in a position to prove Theorem 9.3.15, which says that for separable M, $U : \Delta(M) \to \mathbb{R}$ is a continuous expected utility function iff U is continuous and linear.

Proof of Theorem 9.3.15. If $U(\mu) = \int u \, d\mu$ for $u \in C_b(M)$, then U is clearly linear and, by the portmanteau theorem, is continuous.

Now suppose that $U : \Delta(M) \to \mathbb{R}$ is continuous and linear. For $x \in M$, define $u(x) = U(\delta_x)$. By linearity, $U(\mu) = \int u(x) \, d\mu(x)$ for any finitely supported μ. For ν that is not finitely supported, let μ_n be a sequence of finitely supported probabilities converging to ν. By continuity, $U(\nu) = \lim_n U(\mu_n) = \lim_n \int u \, d\mu_n = \int u \, d\nu$ by the portmanteau theorem. ∎

The boundedness of the von Neumann-Morgenstern utility function u is necessary for U to be defined on all of $\Delta(M)$.

Example 9.5.1 (The St. Petersburg "Paradox") *For a while, there was a theory that the correct way to decide between random variables $X, Y \in \mathbb{R}$ is* *by their expectations. If μ_X and μ_Y are the distributions induced by X and Y, respectively, this involves ranking distributions by $\int x \, d\mu_X(x)$ and $\int x \, d\mu_Y(x)$, and the function being integrated, $u(x) = x$, is unbounded on $\mathbb{R}$. The problem is that this means that some random variables cannot be ranked, for example, if they have infinite or undefined expectations. This leads to all kinds of problems. For instance, when $E\, X = \infty$, it is hard to answer the question "How much is the random variable X worth to you?" For another example, if $X \geq 0$ and $E\, X = \infty$, then $Y := X + X^2 + 10$ is always strictly greater than X, sometimes by a huge amount, but $E\, X = E\, Y = \infty$, so they are indifferent.*

More generally, suppose that $u : M \to \mathbb{R}$ is unbounded. Then there is a countably supported μ such that $|\int u \, d\mu| = \infty$. To see why, suppose than $\sup_{x \in M} u(x) = \infty$. Pick a sequence x_n such that $u(x_n) \geq 2^n$ and $u(x_{n+1}) > u(x_n)$. Define $\mu = \sum \frac{1}{2^n} \delta_{x_n}$. Note that $\int u(x) \, d\mu(x) \geq \sum_n 2^n \cdot \frac{1}{2^n} = \infty$. Comparing the probability μ to the unambiguously better probability $\nu = \sum \frac{1}{2^n} \delta_{x_{n+1}}$ is impossible using the expected value of u because $\int u \, d\nu = \int u \, d\mu = \infty$. (Similar constructions can be made if $\inf_{x \in M} u(x) = -\infty$.)

Recall that the only bounded concave functions on all of $\mathbb{R}$ are constant. This means that if $U : \Delta(\mathbb{R}) \to \mathbb{R}$ is continuous, linear, and nonconstant, then it cannot be represented as the integral of a concave function. If we believe in monotonic preferences, then only nondecreasing u's make sense. Within this class, the problem of nonrepresentability by a concave u goes away only if we substitute an interval bounded below for $\mathbb{R}$.

9.5.a Stochastic Dominance and Risk Aversion

Throughout this subsection, random variables, X, Y, and Z have cdf's, F_X, F_Y, and F_Z, and cdf's F and G have corresponding probabilities P_F and P_G. The following pair of partial orders is attempting to get at the idea of one random variable being indubitably better than another random variable.

Definition 9.5.2 (First Attempt) *Suppose that X, Y, and Z are $\mathbb{R}$-valued random variables and that $Z = X + Y$. If $Y \leq 0$, then X **first-order stochastically dominates** Z, written $X \succeq_{FOSD} Z$. If X, Y, and Z are also integrable and $E\,(Y \mid X) = 0$, then X **second-order stochastically dominates** Z, written $X \succeq_{SOSD} Z$.*

In the first case, the difference between X and Z is the addition of a penalty, making Z worse for anyone who likes more. In the second case, the difference is the addition of a penalty that is 0 on average, but adds noise, making Z worse for anyone who likes more certainty. In terms of L^2 spaces, if X, Y, and Z have finite variance, then $E\,(Y \mid X) = 0$ is the same as $Y \perp L^2(X)$ and this says that Z is riskier than X if it is equal to X plus something perpendicular to all functions of X.

As sensible as these definitions look, they do have a serious shortcoming.

Exercise 9.5.3 On the probability space $([0, 1], \mathcal{B}, \lambda)$, let $X(\omega) = \omega + \frac{1}{2}$, $Y(\omega) \equiv -\frac{1}{2}$, $Z(\omega) = X(\omega) + Y(\omega)$ and $Z'(\omega) = 1 - \omega$. Show that $X \succ_{FOSD} Z$ and that Z and Z' have the same distribution, but yet $X \not\succeq_{FOSD} Z'$. Give a similar example for second-order stochastic dominance. [Having the probability space be $[0, 1]^2$ might make things easier.]

If the ω do not enter into preferences, a very strong but frequently made assumption, then all that matters about a random variable is its distribution, that is, its cdf. Further, Skorohod's representation theorem, 7.10.4, tells us that we have a great deal of freedom in choosing the probability space. In the following, there is no requirement that the random variables be defined on the same probability space, and the equivalence comes from the probability integral transforms of X, Y, and Z.

Definition 9.5.4 (First- and Second-Order Stochastic Dominance) *Suppose that X, Y, and Z are integrable $\mathbb{R}$-valued random variables, with cdf's F_X, F_Y, and F_Z. We say that $X \succeq_{FOSD} Z$ or $F_X \succeq_{FOSD} F_Z$ if the cdf's satisfy, for all r, $F_X(r) \leq F_Z(r)$, equivalently if there are random variables X', Y', and Z', defined on a common probability space satisfying first-order stochastic dominance as in Definition 9.5.2 and having the same cdf's. We say that $X \succeq_{SOSD} Z$ or $F_X \succeq F_Z$ if there are random variables, X', Y', and Z', defined on a common probability space satisfying second-order stochastic dominance as in Definition 9.5.2.*

Definition 9.5.5 *Preferences, $\succeq$, on $\Delta(\mathbb{R})$ **respect FOSD** if $[F_X \succeq_{FOSD} F_Z] \Rightarrow [F_X \succeq F_Z]$. Preferences, $\succeq$, on any subset of integrable elements of $\Delta(\mathbb{R})$ **respect SOSD** if whenever X and Y have the same expectation, $[F_X \succeq_{SOSD} F_Z] \Rightarrow [F_X \succeq F_Z]$.*

The following shows that these really are only partial orders.

Exercise 9.5.6 Draw the simplex $\Delta(C)$, $C = \{10, 20, 30\}$, as a two-dimensional isoceles triangle. Find the subset of $\Delta(C)$ that first-order stochastically dominates $F = (\frac{1}{3}, \frac{1}{3}, \frac{1}{3})$. Find the subset of $\Delta(C)$ that must be preferred to F when preferences satisfy both first- and second-order stochastic dominance.

Remember that boundedness of the von Neumann-Morgenstern u is needed for expected utility preferences to be defined over all of $\Delta(\mathbb{R})$, that only half intervals $|a, \infty)$ or bounded intervals $|a, b|$ support bounded strictly concave functions, and that $\mathbb{R}$ is also an interval subset of $\mathbb{R}$. In particular, the following shows that expected utility preferences over $\Delta(\mathbb{R})$ cannot respect SOSD except by being indifferent to everything.

Theorem 9.5.7 *Expected utility preferences over $\Delta(A)$, A an interval subset of $\mathbb{R}$, respect FOSD iff the von Neumann-Morgenstern utility function, u, is non-decreasing and they respect SOSD iff u is concave.*

Proof. FOSD: If $u : \mathbb{R} \to \mathbb{R}$ is nondecreasing and $Z = X + Y$, $Y \leq 0$, then for all ω, $u(Z(\omega)) \leq u(X(\omega))$ so that $E\, u(Z) \leq E\, u(X)$. If u is not nondecreasing, then there exists $r < s$ such that $u(r) > u(s)$. Taking $X \equiv s$ and $Y \equiv (s - r)$ gives $E\, u(Z) > E\, u(X)$.

SOSD: Suppose that u is concave. Let $f(x) = E\, (u(X + Y) \mid X = x)$ so that $E\, f(X) = E\, u(X + Y)$. Since u is concave and $E\, (Y \mid X = x) = 0$, Jensen's inequality implies that $f(x) \leq u(x)$ for all x. Therefore, $E\, u(X + Y) \leq E\, u(X)$. Finally, suppose that the preferences respect SOSD. Pick $x < y \in \mathbb{R}$ and $\alpha \in (0, 1)$.

Set $X \equiv r := \alpha x + (1 - \alpha)y$. If $Y = r - x$ with probability α and $Y = y - x$ with probability $(1 - \alpha)$, then $E(Y \mid X) = 0$. Since preferences satisfy SOSD, $E\, u(X) \geq E\, u(X + Y)$, that is, $u(\alpha x + (1 - \alpha)y) \geq \alpha u(x) + (1 - \alpha)u(y)$. ∎

One could therefore have started with the following alternate definitions of FOSD and SOSD: $X \succeq_{FOSD} Y$ iff $E\, f(X) \geq E\, f(Y)$ for all bounded nondecreasing f and $X \succeq_{SOSD} Y$ iff $E\, f(X) \geq E\, f(Y)$ for all concave f.

9.5.b Mean-Preserving Spreads

As convenient as both definitions of SOSD may be, they are still somewhat vague. We now turn to a remedy for this vagueness.

The cdf G is a mean-preserving spread of the cdf F if you can get to G from F by taking mass out of P_F from a bounded interval, $|a, b|$, and giving it to P_G anyplace in the complement of $|a, b|$ in a fashion that preserves the mean. Visually, this can be thought of as scooping mass from $|a, b|$ and spreading it outside of $|a, b|$ in a balanced fashion. This certainly seems like a way to make a distribution riskier, and it is most easily understood by examples. Writing the probability integral transforms in the following cases is informative.

Example 9.5.8 *If $P_F = \delta_0$ and $P_G = \frac{1}{2}\delta_{-2} + \frac{1}{2}\delta_{+2}$, we can take $|a, b| = [0, 0]$, scoop out all of the mass, and put half of it above and half below, equally spaced around 0. Clearly, G is riskier than F, $F \succ_{SOSD} G$.*

If $P_F = U[0, 1]$ and $G = \frac{1}{2}\delta_0 + \frac{1}{2}\delta_1$, we can take $|a, b| = (0, 1)$ and scoop out all of the mass. Again, G is clearly riskier than F, $F \succ_{SOSD} G$.

Definition 9.5.9 *P_G is a **mean-preserving spread (mps)** of P_F if they have the same mean and there is a bounded interval $|a, b|$ such that for all $E \subset |a, b|$, $P_F(E) \geq P_G(E)$, while for all $E \subset |a, b|^c$, $P_F(E) \leq P_G(E)$.*

Exercise 9.5.10 For this problem, $P_F = U[0, 1]$. For each of the following G's or P_G's, determine whether or not G is an mps of F or the reverse and whether or not G second-order stochastically dominates F or the reverse. If one is an mps of the other, give the interval $|a, b|$ and the amount of mass that has been scooped.

1. $G(x) = 2x$ for $0 \leq x \leq \frac{1}{4}$ and $G(x) = 2(x - \frac{3}{4})$ for $\frac{3}{4} \leq x \leq 1$.
2. $P_G = \delta_{\frac{1}{2}}$.
3. $P_G = \frac{1}{2}U + \frac{1}{2}\delta_0$.
4. $P_G = \frac{1}{3}U + \frac{1}{3}\delta_{\frac{1}{3}} + \frac{1}{3}\delta_{\frac{2}{3}}$.
5. $P_G = \frac{1}{3}U + \frac{1}{3}\delta_0 + \frac{1}{3}\delta_1$.

If F_1 is an mps of F_0 and F_2 is an mps of F_1, then F_2 ought to be riskier than F_0. One might expect that the same ought to hold true for the limit of a sequence of mps's. However, it is not true.

Example 9.5.11 (Müller) *Let $F_0 = \delta_0$, $F_1 = (\frac{1}{2})^1 \delta_{-(2^1)} + \frac{1}{2}^1 \delta_{+(2^1)}$, $F_2 = (\frac{1}{2})\delta_0 + (\frac{1}{2})^2 \delta_{-(2^2)} + (\frac{1}{2})^2 \delta_{+(2^2)}$, and $F_n = (1 - \frac{1}{2}^{n-1})\delta_0 + (\frac{1}{2})^n \delta_{-(2^n)} + (\frac{1}{2})^n \delta_{+(2^n)}$. To get to F_1 from F_0, scoop all of the mass in $(-1, +1)$ and split it evenly on ± 2. To get from F_1 to F_2 takes two mps's: first, scoop all of the mass in*

$(1, 3)$ *and split it evenly between* 0 *and* 4*; second, scoop all of the mass in* $(-3, -1)$ *and split it evenly between* -4 *and* 0. *In general, to get from* F_n *to* F_{n+1}, *take the masses at* $\pm 2^n$ *and split them in half, putting half on* 0 *and half twice as far away from* 0.

This gives a sequence of increasingly risky F_n. *However,* $F_n \to_{w^*} F_0$. *In other words, a sequence of mps's of the certain outcome* 0 *becomes riskier and riskier, converging back to the certain outcome* 0. *This means that no continuous* $U :$ $\Delta(\mathbb{R}) \to \mathbb{R}$ *can nontrivially respect SOSD, because if it did,* $U(F_0) > U(F_1) >$ $U(F_2) > \cdots$ *but* $U(F_n) \to U(F_0)$, *which is blatantly ridiculous.*

When the random variables are bounded, the basic result is given in the following.

Theorem 9.5.12 *Let X and Z be random variables taking values in an interval* $[a, b]$. *The following statements are equivalent:*

1. *$X \succeq_{SOSD} Z$, and*
2. *$F_Z = \lim_n F_n$, where $F_0 = F_X$, and for all $n \geq 1$, F_n is carried on $[a, b]$ and F_n is an mps of F_{n-1}.*

The proof is moderately involved, even when the random variables are simple and defined on a nonatomic space.

9.6 ◆ The Riesz Representation Theorem for $\Delta(M)$, M Compact

This section is devoted to showing that countably additive finite signed Borel measures on a compact M and continuous linear functions on $C(M)$ are one and the same, which is known as the Riesz representation theorem. In the following section we do the work necessary to extend this result to complete separable metric spaces M.

9.6.a Probabilities and Integrals of Continuous Functions

We first record the following easy observation, which shows that probabilities are determined by their integrals against bounded continuous functions. The essential intuition for this section is contained in the proof, which shows that for the purposes of analyzing probabilities, the indicators of closed and/or open sets are nearly continuous functions.

Lemma 9.6.1 *If M is an arbitrary metric space, then for μ, $\nu \in \Delta(M)$, $\mu = \nu$ iff $\int f \, d\mu = \int f \, d\nu$ for all $f \in C_b(M)$.*

Proof. If $\mu = \nu$, then clearly, $\int f \, d\mu = \int f \, d\nu$ for all $f \in C_b(M)$.

Now suppose that $\int f \, d\mu = \int f \, d\nu$ for all $f \in C_b(M)$. Since $\mathcal{B}_M$ is generated by the closed sets, it is sufficient to show that $\mu(F) = \nu(F)$ for all closed sets F. For every closed F, there is a sequence, f_n, of uniformly bounded continuous functions, for example, $f_n(x) = \max\{0, 1 - n \cdot d(x, F)\}$, with the property that for all $x \in M$, $f_n(x) \downarrow 1_F(x)$. By assumption, $\int f_n \, d\mu = \int f_n \, d\nu$. By dominated convergence, $\int f_n \, d\mu \downarrow \mu(F)$ and $\int f_n \, d\nu \downarrow \nu(F)$. ■

Exercise 9.6.2 For a given closed F, the proof gave a sequence, f_n, of continuous functions with the property that $f_n(x) \downarrow 1_F(x)$ for all x. Give an alternate proof that, for a given open G, constructs a sequence of uniformly bounded continuous functions, g_n, with the property that $g_n(x) \uparrow 1_G(x)$ for all x.

9.6.b The Riesz Representation Theorem

For any $\mu \in \Delta(M)$, the mapping $L_\mu(f) := \int f \, d\mu$ from $C(M)$ to $\mathbb{R}$ is continuous, linear, and positive, that is, if $f \geq 0$, then $\int f \, d\mu \geq 0$ and $\int 1 \, d\mu = 1$. Let us make these virtues into a definition.

Definition 9.6.3 *A continuous linear $L : C(M) \to \mathbb{R}$ is **probabilistic** if it is **positive**, for all $f \geq 0$, $L(f) \geq 0$, and sends 1_M to 1, that is, $L(1_M) = 1$.*

The following is another representation theorem. It tells us, in the compact case for now, that we can think of probabilities as functions on the class of measurable sets or as functions on the class of continuous functions.

Theorem 9.6.4 (Riesz) *For compact (M, d), a positive $L : C(M) \to \mathbb{R}$ is continuous, linear, and probabilistic iff there is a $\mu_L \in \Delta(M)$ such that for all $f \in C(M)$, $L(f) = \int f \, d\mu_L$. Further, the mapping $L \leftrightarrow \mu_L$ is a bijection.*

The proof relies on compact classes and Lemma 7.6.16 (p. 416).

Proof. By Lemma 9.6.1, the mapping $L \mapsto \mu_L$ is one-to-one and the mapping $\mu \mapsto L_\mu$ is one-to-one. Therefore, the first part of the theorem implies the second part.

If $L_\mu(f) = \int f \, d\mu$, then L_μ is continuous, linear, and positive and sends 1_M to 1. Now suppose that L is continuous, linear, and probabilistic.

For closed $C \subset M$, define $\mu_L(C) = \inf\{L(f) : 1_C \leq f\}$, and for $E \in \mathcal{B}_M$, define $\mu_L(E) = \sup\{\mu_L(C) : C \subset E\}$. For disjoint closed C_1 and C_2, there exists an $\epsilon > 0$ such that $C_1^\epsilon \cap C_2^\epsilon = \emptyset$, and continuous functions f_1 and f_2 such that $1_{C_1} \leq f_1 \leq 1_{C_1^\epsilon}$ and $1_{C_2} \leq f_2 \leq 1_{C_2^\epsilon}$. Since L is linear, this implies that $\mu_L(C_1 \cup C_2) = \mu_L(C_1) + \mu_L(C_2)$. From this, μ_L is finitely additive on $\mathcal{B}_M$. By Lemma 7.6.16, it is countably additive.

Having constructed μ_L, we must show that it represents L, that is, we must show that $L(f) = \int f \, d\mu_L$ for all $f \in C(M)$. Since $L(1_\Omega) = 1$ and L is linear, we can shift and rescale f so that it takes values in $[0, 1]$. In this case, we show that $L(f) \geq \int f \, d\mu_L$. Since this implies $L(-f) \geq \int -f \, d\mu_L$, after shifting and rescaling it also implies that $L(f) = \int f \, d\mu_L$ and completes the proof.

For $k = 0, 1, \ldots, 2^n$, define $G_k = \{f > \frac{k}{2^n}\}$, so that $\emptyset = G_{2^n} \subset G_{2^n-1} \subset \cdots \subset G_1 \subset G_0$. For each $k \geq 1$, define the continuous $g_k : [0, 1] \to [0, 1]$ to be equal to 0 on $[0, \frac{k-1}{2^n}]$ and equal to 1 on $[\frac{k}{2^n}, 1]$, and by linear interpolation (with slope 2^n) on the remaining interval, $[\frac{k}{2^n}, \frac{k-1}{2^n}]$. Since $\frac{1}{2^n} \sum_k g_k(r) \equiv r$, $f(x) \equiv \frac{1}{2^n} \sum_k g_k(f(x))$. Define $f_k(x) = g_k(f(x))$.

Since L is linear, $L(f) = \frac{1}{2^n} \sum_k L(f_k)$. As $f_k \geq 1_{G_k}$, $L(f_k) \geq \mu_L(G_k)$. The following string of (in)equalities shows that $L(f) \geq \int_M f \, d\mu_L - \frac{1}{2^n} \mu_L(G_0)$.

$$L(f) = \tfrac{1}{2^n} \sum_k L(f_k) \geq \tfrac{1}{2^n} \left[\mu_L(G_1) + \mu_L(G_2) + \cdots + \mu_L(G_{2^n}) \right] \qquad (9.2)$$

$$= \left[\left(\frac{1}{2^n} - \frac{1-1}{2^n} \right) \mu_L(G_1) + \left(\frac{2}{2^n} - \frac{2-1}{2^n} \right) \mu_L(G_2) \right.$$

$$\left. + \cdots + \left(\frac{2^n}{2^n} - \frac{2^n - 1}{2^n} \right) \mu_L(G_{2^n}) \right]$$

$$= \tfrac{1}{2^n} \mu_L(G_1 \setminus G_2) + \tfrac{2}{2^n} \mu_L(G_2 \setminus G_3) + \cdots + \tfrac{2^n - 1}{2^n} \mu_L(G_{2^n - 1} \setminus G_{2^n})$$

$$= \sum_{k=1}^{2^n - 1} \tfrac{k+1}{2^n} \mu_L(G_k \setminus G_{k+1}) - \tfrac{1}{2^n} \mu_L(G_1)$$

$$\geq \sum_{k=1}^{2^n - 1} \int_{G_k \setminus G_{k+1}} f \, d\mu_L - \tfrac{1}{2^n} \mu_L(G_1)$$

$$\geq \int_M f \, d\mu_L - \tfrac{1}{2^n}.$$

Since n was arbitrary, this completes the proof.   ■

Bearing in mind the Hahn-Jordan decomposition, we see that slight variations on the same proof show that any continuous linear function on $C(M)$, probabilistic or not, has a representation as an integral against a countably additive finite signed measure. Since this chapter is concerned with probabilities, we do not go into the details.

The compactness in this result is essential, as without it, there must be extra restrictions on the continuous linear L's. This leads us to study a new class of measure spaces.

9.7 ◆ Polish Measure Spaces and Polish Metric Spaces

There is a broadly useful class of the measurable spaces for which topological kinds of arguments are often possible.

Definition 9.7.1   *We say that a metric space is **Polish** if it is complete and separable.*

There is a related class of measurable spaces, the ones that "look like" Borel measurable subsets of Polish metric spaces. The following gathers together some previous material.

Definition 9.7.2   *Let X and Y be two nonempty sets, d_X and d_Y metrics on X and Y, and $\mathcal{X}$ and $\mathcal{Y}$ σ-fields. A one-to-one onto function $\varphi : X \leftrightarrow Y$ is*

1. *an **isometry** if for all $x, x' \in X$, $d_X(x, x') = d_Y(\varphi(x), \varphi(x'))$,*

2. *a **homeomorphism** if it is continuous and has a continuous inverse, and*

3. *a **measurable isomorphism** if it is measurable and has a measurable inverse.*

We have seen that isometries are best understood as renamings of points. If we have the freedom to change to equivalent metrics, homeomorphisms are isometries.

Exercise 9.7.3 If (X, d_X) and (Y, d_Y) are metric spaces and $f : X \to Y$ is one-to-one, onto, continuous, and has a continuous inverse, then $\rho_X(x, x') := d_Y(f(x), f(x'))$ is equivalent to d_X and $\rho_Y(y, y') := d_X(f^{-1}(y), f^{-1}(y'))$ is equivalent to d_Y.

There is a very similar result for measurable isomorphisms. It requires the following.

Definition 9.7.4 *A field or a σ-field $\mathcal{X}$ on a set X is **Hausdorff (or separated)** if for any two points $x \neq x'$, there exists an $E \in \mathcal{X}$ such that $1_E(x) \neq 1_E(x')$.*

Notation Alert 9.7.A *Hausdorff topological spaces are spaces that have topologies with a similar kind of separation property. There is a distinction between a separated σ-field and a separable, that is, countably generated σ-field.*

Exercise 9.7.5 Show that σ-field $\mathcal{X}$ is Hausdorff iff any field $\mathcal{X}^\circ$ generating it is Hausdorff.

For countably generated $\mathcal{X}$, there is little loss in thinking of a measurable isomorphism as an isometry.

Example 9.7.6 (The Blackwell Metric) *Suppose that $\varphi : X \leftrightarrow Y$ is a measurable isomorphism and that $\mathcal{X}$ is a countably generated Hausdorff σ-field. Let $\mathcal{X}^\circ = \{E_n : n \in \mathbb{N}\}$ be an enumeration of a field generating $\mathcal{X}$, so that $\{\varphi(E_n) : n \in \mathbb{N}\}$ is a field generating $\mathcal{Y}$. The **Blackwell metric** on X is defined by*

$$d_X(x, x') = \frac{1}{\min\{n : 1_{E_n}(x) \neq 1_{E_n}(x')\}}, \tag{9.3}$$

and the corresponding Blackwell metric on Y is defined as

$$d_Y(y, y') = \frac{1}{\min\{n : 1_{\varphi(E)_n}(x) \neq 1_{\varphi(E)_n}(x')\}}. \tag{9.4}$$

If $\mathcal{B}_X$ is the Borel σ-field generated by the metric d_X, then $\mathcal{B}_X = \mathcal{X}$. Further, φ is an isometry between the metric spaces (X, d_X) and (Y, d_Y).

Exercise 9.7.7 Show that the sets $E_n \in \mathcal{X}^\circ$ are **clopen**, that is, both closed and open. Further show that the mapping $x \leftrightarrow (1_{E_n}(x))_{n \in \mathbb{N}}$ is a homeomorphism between (X, d_X) and its image in the compact product metric space $\{0, 1\}^{\mathbb{N}}$.

The image of X in the compact metric space $\{0, 1\}^{\mathbb{N}}$ under the embedding just given need not be a measurable set. When it is, we have the special class of measurable spaces under study here.

Definition 9.7.8 *A measurable space $(X, \mathcal{X})$ is **Polish** if it is measurably isomorphic to a Borel subset of a compact metric space.[3]*

---

3. These are so named in honor of the group of Polish mathematicians who developed their properties before World War II. Sometimes Polish measurable spaces are called **standard** measurable

This definition seems a bit loose. After all, there are many compact metric spaces and many measurable isomorphisms. This generality is a bit illusory. Theorems 6.6.40 and 6.6.42 show that a metric space (M, d) is separable iff it is homeomorphic to some $E \subset [0, 1]^{\mathbb{N}}$ and that it is compact iff it is homeomorphic to some closed (hence compact) $E \subset [0, 1]^{\mathbb{N}}$. Thus, we could have defined a measure space to be Polish if its σ-field is Borel with respect to some metric for which the space is homeomorphic to a Borel subset of $[0, 1]^{\mathbb{N}}$.

Exercise 9.7.9   Show that $(X, \mathcal{X})$ is a Polish measurable space iff $\mathcal{X}$ is the Borel σ-field for a metric d_X making X homeomorphic with a measurable subset of a compact metric space.

9.8 ◆ The Riesz Representation Theorem for Polish Metric Spaces

For compact (M, d), $C(M) = C_b(M)$, and the Riesz representation theorem tells us that any continuous linear $L : C(M) \to \mathbb{R}$ can be represented as $L(f) = \int_M f(x) \, d\eta_L(x)$ for a unique countably additive measure, η_L, on the Borel σ-field of M. When (M, d) is not compact, there are continuous linear $L : C_b(M) \to \mathbb{R}$ that are not representable as integration against any countably additive measure. The aim of this section is to prove Theorem 9.8.2 and to understand why the additional "continuity from above" condition is needed.

Definition 9.8.1   *A positive $L : C_b(M) \to \mathbb{R}$ is **continuous from above** if for any decreasing sequence $f_1 \geq f_2 \geq \cdots$ in $C_b(M)$ with $f_n(x) \downarrow 0$ for all $x \in M$, $L(f_n) \downarrow 0$.*

Recall that the countable additivity of probabilities is also a "continuity from above" condition. Specifically, P is countably additive iff for all $E_n \downarrow \emptyset$, $P(E_n) \downarrow 0$. In terms of integrals, this is the condition that if $1_{E_n}(\omega) \downarrow 0$ for all ω, then $\int 1_{E_n} \, dP \downarrow 0$.

Theorem 9.8.2   *For Polish (M, d), a positive $L : C_b(M) \to \mathbb{R}$ is continuous, linear, probabilistic, and continuous from above iff there is a $\mu_L \in \Delta(M)$ such that for all $f \in C(M)$, $L(f) = \int f \, d\mu_L$. Further, the mapping $L \leftrightarrow \mu_L$ is a bijection.*

Compared to the similar result for compact M, we have weakened the assumptions on M and strengthened the assumptions on L. We should understand why we have to strengthen the assumptions on L.

---

spaces. Aside from the historical aptness, fewer concepts in mathematics are described with the adjective "Polish" than the adjective "standard."

9.8.a A Purely Finitely Additive Probability on (0, 1]

We work in the metric space $M = (0, 1]$ with the usual (Euclidean) metric $d(x, y) = |x - y|$. (M, d) is separable, but not complete—any $x_n \downarrow 0$ is Cauchy, and, while it converges to 0, it does not converge to any point in M. The restriction of any $f \in C([0, 1])$ to $(0, 1]$ gives an element of $C_b(M)$. However, $g(x) = \sin(1/x)$ belongs to $C_b(M)$ but has no continuous extension to $[0, 1]$.

Let $\mathcal{M}^\circ$ denote the field, not the σ-field, of finite unions of interval subsets of M and $\mathcal{M} = \sigma(\mathcal{M}^\circ)$ denote the Borel σ-field on M. We want to show that there are positive continuous linear probabilistic $L : C_b(M) \to \mathbb{R}$ that cannot be represented as integrals of some countably additive probability μ on $\mathcal{M}$. To do this, we show that there exist purely finitely additive probabilities on $\mathcal{M}$.

Let $x_n \downarrow 0$ be a strictly decreasing sequence in $(0, 1]$. For each $n \in \mathbb{N}$, let $X_n = \{x_m : m \geq n\}$ be a terminal set for the sequence, so that $\cap_n X_n = \emptyset$. For $E \in \mathcal{M}^\circ$, define $\eta_{fa}(E) = 1$ if $X_n \subset E$ for some n and $\eta_{fa}(E) = 0$ otherwise. The function $\eta_{fa} : \mathcal{M}^\circ \to \{0, 1\}$ is finitely additive on $\mathcal{M}^\circ$. It is not countably additive and cannot have a countably additive extension to $\mathcal{M}$ because $(0, 1/m] \downarrow \emptyset$, but $\eta_{fa}((0, 1/m]) \equiv 1 \not\downarrow 0$.

The problem with extending η_{fa} to all of $\mathcal{M}$ is that it is not clear what probability should be assigned to the sets $\{x_m : m$ is odd $\}$ and $\{x_m : m$ is even $\}$. The answer is that one of the sets should have mass 1 and the other should have mass 0. This choice has implications for the probability of $\{x_m : m$ is prime $\}$, $\{x_m : m$ is a multiple of 24 $\}$, and so on. The Hahn-Banach theorem shows that it is possible to make all of these choices in a consistent fashion.

Lemma 9.8.3 *There exists an extension of η_{fa} from $\mathcal{M}^\circ$ to $\mathcal{M}$.*

Proof. Let V_0 denote the set of simple $\mathcal{B}^\circ$-measurable functions, $X = \sum_i \beta_i 1_{E_i}$, $E_i \in \mathcal{B}^\circ$. Define $L_0 : V_0 \to \mathbb{R}$ by $L_0(X) = \sum_i \beta_i \eta_{fa}(E_i)$. L_0 is sup norm continuous on V_0, hence has a sup norm continuous extension, L, to the set of all bounded functions on $(0, 1]$. The requisite extension of η_{fa} is $\eta(E) = L(1_E)$. ∎

The η cannot be countably additive because $\eta((0, 1/m]) \not\to 0$. Despite this, the following shows that η behaves somewhat like a countably additive point mass on 0.

Exercise 9.8.4 Show that for $f \in C([0, 1])$, $T(f) = \int_M f(x) \, d\eta(x) = f(0)$. Further show that for $g \in C((0, 1])$, $T(g) \in [\liminf_n g(x_n), \limsup_n g(x_n)]$, where $x_n \downarrow 0$ is the sequence used in defining η_{fa}.

One way to understand this is that η is putting mass on a new point, someplace "just to the left of 0" but not negative, and the integral gives the limit behavior of g at 0 along a sequence having η mass 1.

Exercise 9.8.5 Show that the mapping $g \mapsto T(g)$ from $C_b(M)$ to $\mathbb{R}$ is positive, linear, continuous, and probabilistic, but not continuous from above.

9.8.b Topological Completeness

The proof of the Riesz representation theorem for Polish spaces requires a bit of background on Polish spaces. We have observed that completeness is *not* a topological concept; a metric space can be complete in one metric and not complete in another equivalent metric.

Definition 9.8.6 *A metric space* (M, d) *is **topologically complete** if there is a metric* ρ *that is equivalent to* d *such that the metric space* (M, ρ) *is complete.*

Exercise 9.8.7 As observed, the space $(0, 1]$ with the metric $d(x, y) = |x - y|$ is not complete. Show that the metric $\rho(x, y) := d(x, y) + |\frac{1}{x} - \frac{1}{y}|$ is equivalent to d and that the metric space $((0, 1], \rho)$ is complete.

Topological completeness survives under homeomorphism.

Exercise 9.8.8 If $\varphi : X \leftrightarrow Y$ is a homeomorphism between the metric spaces (X, d_X) and (Y, d_Y) and (X, d_X) is topologically complete, then so is (Y, d_Y).

Definition 9.8.9 $E \subset M$ *is a* G_δ *set if* $E = \cap_{m \in \mathbb{N}} G_m$, *where each* G_m *is open.*

In words, a G_δ is a countable intersection of open sets. Any open set, G, is a G_δ; take $G_m \equiv G$. By shrink wrapping, any closed set, F, is a G_δ; to see why, take $G_m = F^{1/m}$.

Exercise 9.8.10 If G is an open proper subset of a complete metric space, (M, d), show that $\rho(x, y) := d(x, y) + |\frac{1}{d(x,G^c)} - \frac{1}{d(y,G^c)}|$ is equivalent to d on G and makes (G, ρ) complete.

Theorem 9.8.11 *If* (M, d) *is a complete metric space, then* $E \subset M$ *is topologically complete iff* E *is a* G_δ.

Proof. Suppose first that $E = \cap_k U_k$, each U_k open. Set $G_m = \cap_{k \leq m} U_k$, so that $G_m \downarrow E$. For $x, y \in E$, define

$$\rho_m(x, y) = \min \left\{ 1, d(x, y) + \left| \frac{1}{d(x, G_m^c)} - \frac{1}{d(y, G_m^c)} \right| \right\}, \qquad (9.5)$$

and define $\rho(x, y) = \sum_m \frac{1}{2^m} \rho_m(x, y)$. It is clear that ρ is equivalent to d. If x_n is a ρ-Cauchy sequence in E, then x_n is a ρ_m-Cauchy sequence, hence converges to some $y_m \in G_m$. If $y_m \neq y_{m'}$ for $m \neq m'$, then x_n is not ρ-Cauchy. Let x denote the common value of the y_m so that $\rho(x_n, x) \to 0$.

Now suppose that (E, ρ) is complete and ρ is equivalent to d on E. Since ρ is equivalent to d, every ρ-open ball, $B^\rho(x, \frac{1}{k})$, is of the form $U_k(x) \cap E$ for some open $U_k(x) \subset M$. Define $G_m(x) = E^{1/m} \cap \cap_{k \leq m} U_k(x)$ and $G_m = \cup_{x \in E} G_m(x)$. We show that $E = \cap_m G_m$.

The first observation is that E is a d-dense subset of $\cap_m G_m$. Now pick $x \in \cap_m G_m$. We must show that $x \in E$. Since E is d-dense, there is a sequence x_n in E such that $d(x_n, x) \to 0$. Since $x \in \cap_m G_m$, for each m, $x \in G_m$, for some $x'_m \in E$, x belongs to the open set $G_m(x'_m)$. This in turn implies that for some N and for all $n \geq N$, $x_n \in G_m(x'_m)$. By construction, this means that for all $n, n' \geq N$, $\rho(x_n, x_{n'}) < \frac{2}{m}$. Since m is arbitrary, x_n is a ρ-Cauchy sequence in E. Since (E, ρ) is complete, $x \in E$. ∎

Exercise 9.8.12 Something like the product metric appeared in the first step of the previous proof. This is not an accident. Show that if each (M_k, d_k) is a topologically complete metric space, then so is $M = \times_{k \in \mathbb{N}} M_k$ with the product metric. [In words, the countable product of topologically complete metric spaces is a topologically complete metric space.]

Here is the corollary that we use to study measures on Polish spaces as linear functionals.

Corollary 9.8.13   *(M, d) is a complete separable metric space iff M is homeomorphic to a G_δ in $[0, 1]^{\mathbb{N}}$.*

Proof. $[0, 1]^{\mathbb{N}}$ is compact, hence complete. The homeomorphic image of M is a G_δ iff M is topologically complete.   ■

9.8.c The Riesz Theorem

Recall that we wish to prove that for a Polish metric space (M, d), a positive $L : C_b(M) \to \mathbb{R}$ is continuous, linear, probabilistic, and continuous from above iff there is a $\mu_L \in \Delta(M)$ such that for all $f \in C(M)$, $L(f) = \int f \, d\mu_L$. Further, the mapping $L \leftrightarrow \mu_L$ is a bijection.

Proof of Theorem 9.8.2.   If $L(f) = \int f \, d\mu$, then L is continuous, linear, probabilistic, and, by dominated convergence, continuous from above.

Now suppose that $L : C_b(M) \to \mathbb{R}$ is continuous, linear, probabilistic, and continuous from above. Let $K = [0, 1]^{\mathbb{N}}$ with the product metric. Let $\varphi : M \leftrightarrow E$ be a homeomorphism between M and $E = \cap_m G_m$, a G_δ in K. For any $f \in C(K)$, the function $f \circ \varphi$ belongs to $C_b(M)$, so that $\mathcal{L}(f) := L(f \circ \varphi)$ is a continuous linear probabilistic function on $C([0, 1]^{\mathbb{N}})$. By the Riesz theorem for compact spaces, there is a unique probability μ_L on K such that $\mathcal{L}(f) = \int_K f \, d\mu_L$. We show that $\mu_L(E) = 1$. It is sufficient to demonstrate that $\mu_L(G_m^c) = 0$.

Let $f_n(y) = \max\{0, 1 - n \cdot d(x, G_m^c)\}$, so that $f_n(y) \downarrow 1_{G_m^c}(y)$ for all $y \in K$. For all $x \in M$, $(f_n \circ \varphi)(x) \downarrow 0$. By continuity from above, $L(f_n \circ \varphi) \downarrow 0$. Therefore, $\int f_n \, d\mu_L \downarrow 0$, so that $\mu_L(G_m^c) = 0$.   ■

9.9 ◆ Compactness in $\Delta(M)$

There are two important results here. The first one, Theorem 9.9.2, is that $\Delta(M)$ is compact iff M is compact. In particular, M is compact iff every $S \subset \Delta(M)$ has a compact closure. The second result, Theorem 9.9.6, characterizes the subsets of $\Delta(M)$ that have compact closure when M is Polish, but need not be compact. It says that $S \subset \Delta(M)$ has compact closure iff S is nearly supported by a compact set; that is, iff for every $\epsilon > 0$ there is a compact $K_\epsilon \subset M$ such that for all $\mu \in S$, $\mu(K_\epsilon) > 1 - \epsilon$, a property called **tightness**.

When M is a normed space, there is a class of results called "central limit theorems." These are statements of the form that the distribution, μ_n, of the sum of n independent small random variables is, for very large n, approximately Gaussian. This is, arguably, the most important and useful result in statistics. The classical proofs use tightness as follows:

1. One shows that for any function f in a rich class, $F \subset C(M)$, and for any subsequence, $\mu_{n'}$, there is a constant, r_f, depending only on f, such that $\int f \, d\mu_{n'} \to r_f$.

2. One shows that the set $\{\mu_n : n \in \mathbb{N}\}$ is tight.

3. One shows that the class F determines a probability, that is, the mapping $f \mapsto r_f$ is of the form $r_f = \int f \, d\mu$ for some unique μ.

4. Using the second result, we know that for any subsequence, $\mu_{n'}$, there is a further subsequence, $\mu_{n''}$, converging to some ν.

5. By continuity, $\int f \, d\nu = r_f$, and since F determines probabilities, $\nu = \mu$.

6. Finally, since any subsequence has a further subsequence converging to μ, the whole sequence must converge to μ.

9.9.a $\Delta(M)$ Is Compact iff M Is Compact

We know that $C_b(M)$ is separable iff M is compact. For compact M, let g_k be a countable dense subset of the unit ball in the separable Banach space $C(M)$. Define the metric $e(P, Q) = \sum_k \frac{1}{2^k} | \int g_k \, dP - \int g_k \, dQ |$.

Lemma 9.9.1   *For compact M, $\rho(P_n, P) \to 0$ iff $e(P_n, P) \to 0$.*

Thus, $\rho(P_n, P) \to 0$ iff for all g_k in a dense set, $\int g_k \, dP_n \to \int g_k \, dP$. This should be compared with Lemma 9.6.1, which showed that a probability is determined by its integrals against all bounded continuous functions. This lemma shows that when M is compact, one can substitute a countable set of functions. Later, we leverage this fact to a proof that a countable set of functions can be used whenever M is separable but not necessarily compact.

Proof.   If $\rho(P_n, P) \to 0$, then for all g_k, $| \int g_k \, dP_n - \int g_k \, dP | \to 0$. Suppose $e(P_n, P) \to 0$, let $f \in C(M)$, and pick $\epsilon > 0$. Since $\{g_k : k \in \mathbb{N}\}$ is dense, there exists $g_k \in B_\epsilon(f - \epsilon)$ and $g_K \in B_\epsilon(f + \epsilon)$ so that $g_k(x) < f(x) < g_K(x)$ and $\max_x (g_K(x) - g_k(x)) < 4 \cdot \epsilon$. For large n, $| \int f \, dP_n - \int f \, dP | < 6 \cdot \epsilon$.   ∎

Theorem 9.9.2 (Compactness of $\Delta(M)$)   *$\Delta(M)$ is compact iff M is compact.*

Proof.   If $\Delta(M)$ is compact, then so is the closed subset, $\{\delta_x : x \in M\}$. This set is homeomorphic to M, so M is compact.

Now suppose that M is compact and let $G = \{g_k : k \in \mathbb{N}\}$ be a countable dense subset of $C(M)$ that is also a vector space over the rationals; that is, for rationals $q_1, q_2 \in \mathbb{Q}$ and $g_1, g_2 \in G$, $q_1 g_1 + q_2 g_n \in G$. For each k, define I_k as the compact interval $[\min_{x \in M} g_k(x), \max_{x \in M} g_k(x)]$. The mapping $P \mapsto \varphi(P) := \left(\int g_1 \, dP, \int g_2 \, dP, \ldots \right)$ embeds $\Delta(M)$ in the compact product space $\times_k I_k$. Let P_n be a sequence in $\Delta(M)$ and take a subsequence, still labeled P_n, such that $\varphi(P_n) \to L$ for some $L \in \times_k I_k$. We chose the notation "L" advisedly, as defining $L(g_k) = \operatorname{proj}_k(L)$ gives a continuous linear function from $\{g_k : k \in \mathbb{N}\}$ to $\mathbb{R}$. It is immediate that L has a unique continuous extension from $\{g_k : k \in \mathbb{N}\}$ to $C(M)$ and that the extension is probabilistic, which means that it corresponds to a probability $P = P_L \in \Delta(M)$. By construction, $e(P_n, P) \to 0$.   ■

When $\Delta(M)$ is compact, we now show that continuous linear functions on $\Delta(M)$ have a very useful structure—the integral of some continuous function $u : M \to \mathbb{R}$.

Corollary 9.9.3   *For compact $\Delta(M)$, $U : \Delta(M) \to \mathbb{R}$ is continuous and linear iff $U(P) = \int u \, dP$ for some $u \in C(M)$.*

In choice theory, the function $u \in C(M)$ is called the **von Neumann-Morgenstern** utility function.

Proof.   Suppose first that $u \in C(M)$ and that $U(P) = \int u \, dP$. Linearity is clear, and if $P_n \to P$, since u is uniformly continuous, $\int u \, dP_n \to \int u \, dP$.

Now suppose that U is continuous and linear. Since $\Delta(M)$ is compact, U is uniformly continuous, so $u(x) := U(\delta_x)$ is uniformly continuous. Since U is linear, for any finitely supported $P = \sum_x \alpha_x \delta_x$, any finitely supported $P = \sum_x \alpha_x u(x) = \int u \, dP$. Thus $P \mapsto \int u \, dP$ is a uniformly continuous function defined on the dense set of finitely supported probabilities. Since $V(P) := \int u \, dP$ is continuous, hence uniformly continuous on the compact set $\Delta(M)$, and V agrees with U on the dense set, $V = U$.   ■

9.9.b   Tightness and Compact Subsets of $\Delta(M)$

A subset S of $\Delta(\mathbb{R})$ has compact closure in the Levy metric iff for every $\epsilon > 0$, there is a $B \in \mathbb{R}$ with every cdf in S satisfying $F(-B) < \epsilon$ and $F(B) > 1 - \epsilon$.

Definition 9.9.4   *$S \subset \Delta(M)$ is **tight** if for all $\epsilon > 0$, there exists a compact K_ϵ with $\mu(K_\epsilon) > 1 - \epsilon$ for all $\mu \in S$,*

Example 9.9.5   *If M is a complete separable metric space, then any $S = \{\mu\}$ is tight by Theorem 7.6.26, (p. 421). If M is compact, then we can take $K_\epsilon = M$, and any subset of $\Delta(M)$ has compact closure because any closed subset of a compact set is compact.*

Theorem 9.9.6 (Prokhorov)   *If $S \subset \Delta(M)$ is tight, then it has compact closure. If M is a complete separable metric space and $S \subset \Delta(M)$ has compact closure, then S is tight.*

The really useful direction is that tightness implies compactness.

Proof.   For the first half, we suppose that S is tight, and for the second half, we suppose that S has compact closure.

1. First half: Suppose that S is tight.

We cover the cases $M = \mathbb{R}^k$, $M = \mathbb{R}^{\mathbb{N}}$, M is σ-compact, and general M in turn, reducing each case to the previous one.

Case 1: S a tight subset of $\Delta(\mathbb{R}^k)$. The case of $\mathbb{R}^1$ was Theorem 6.2.53, which used the Levy metric. In $\mathbb{R}^k$, everything, including the Levy metric, $d_L^k(\cdot, \cdot)$, is more difficult to visualize.

A cdf for $\mu \in \Delta(\mathbb{R}^k)$ is the function $F_\mu(a) = \mu\{y : y \le a\}$. For a cdf F and $x \in \mathbb{R}^k$, define $H_F(x) = \inf\{r > 0 : r > F(x - \sqrt{k}\tilde{r})\}$, where $\tilde{r}$ is the vector $(r, \dots, r)' \in \mathbb{R}^k$. $H_F(x)$ is monotonic and has a Lipschitz constant that is at most 1. For two cdf's, F and G on $\mathbb{R}^k$, define the Levy metric $d_L^k(F, G) = \sup_{x \in \mathbb{R}^k} |H_F(x) - H_G(x)|$. It is tedious to check, but true, that $d_L^k(F_n, F) \to 0$ iff the corresponding probabilities converge weakly.

Since S is tight, we can pick a compact set K such that $\mu(K) > 1 - \epsilon$ for all $\mu \in S$. Expanding K if necessary, we can assume that it is of the form $K = [-B, +B]^k$, with $F_\mu(-\tilde{B}) < \epsilon$ and $F_\mu(+\tilde{B}) > 1 - \epsilon$ for all $\mu \in S$. Restricted to K, the set of functions H_μ is bounded and equicontinuous, hence has a finite ϵ-net, which is a $3 \cdot \epsilon$-net for S.

Case 2: S a tight subset of $\Delta(\mathbb{R}^{\mathbb{N}})$. For any finite $\{1, \dots, k\} \subset \mathbb{N}$, $\text{proj}_{\{1,\dots,k\}} : \mathbb{R}^{\mathbb{N}} \to \mathbb{R}^k$ is continuous. Since the continuous image of a compact set is compact, the tightness of S implies the tightness of $\text{proj}_{\{1,\dots,k\}}(S)$ in $\Delta(\mathbb{R}^k)$. Thus, for each k, there is a subsequence, $(\mu_{k,n})_{n \in \mathbb{N}}$, such that $\text{proj}_{\{1,\dots,k\}}(\mu_{k,n})$ converges to a measure $\nu_k \in \Delta(\mathbb{R}^k)$. By diagonalization (Theorem 6.2.62, p. 292), there is a subsequence, n', of all of the subsequences such that for all $k \in \mathbb{N}$, $\text{proj}_{\{1,\dots,k\}}(\mu_{n'})$ converges to a measure $\nu_k \in \Delta(\mathbb{R}^k)$.

For $k > \ell \ge 1$, $\text{proj}_{\{1,\dots,\ell\}} \circ \text{proj}_{\{1,\dots,k\}} = \text{proj}_{\{1,\dots,\ell\}}$. This means that $\mu_\ell = \text{proj}_{\{1,\dots,\ell\}}(\mu_k)$. This in turn implies that the collection $(\mu_k)_{k \in \mathbb{N}}$ defines a finitely additive probability, μ, on the field, not the σ-field, $\mathcal{B}^\circ = \cup_k \text{proj}_{\{1,\dots,k\}}^{-1}(\mathcal{B}^k)$, where $\mathcal{B}^k$ is the Borel σ-field on $\mathbb{R}^k$. The Borel σ-field on $\mathbb{R}^{\mathbb{N}}$ is generated by $\mathcal{B}^\circ$, so, by Carathéordory's extension theorem, it is sufficient to show that μ is countably additive. By Lemma 7.6.16, it is enough to find a compact class $\mathcal{C} \subset \mathcal{B}^\circ$ such that for all $E \in \mathcal{B}^\circ$, $\mu(E) = \sup\{\mu(C) : C \subset E, \ C \in \mathcal{C}\}$. The class $\mathcal{C} = \cup_k \text{proj}_{\{1,\dots,k\}}^{-1}(C_k)$, C_k a nonempty compact subset of $\mathbb{R}^k$, satisfies these requirements.

Case 3: M is σ-compact. Any σ-compact set is separable, hence homeomorphic to a subset of $[0, 1]^{\mathbb{N}}$, a compact subset of $\mathbb{R}^{\mathbb{N}}$. Since the homeomorphic image of a compact set is compact, M is homeomorphic to a countable union of compact sets, hence homeomorphic to a measurable set. Since weak* convergence is preserved under homeomorphism (by the continuous mapping theorem), we can, and do so without change of notation, replace M with its homeomorphic image $M \subset \mathbb{R}^{\mathbb{N}}$.

Every $\mu \in S$ has the property that $\mu(M) = 1$. It can be extended to $\mathbb{R}^{\mathbb{N}}$ by defining $\mu^{ext}(A) = \mu(A \cap M)$ for any measurable $A \subset \mathbb{R}^{\mathbb{N}}$. The set S^{ext} is tight, so by the previous step, any sequence, μ_n, has a convergent subsequence converging to a limit $\mu \in \Delta(\mathbb{R}^{\mathbb{N}})$. All that must be shown is that $\mu(M) = 1$. Let $K \subset M$ be a compact set with $\mu(K) > 1 - \epsilon$ for all $\mu \in S^{ext}$. Since $\mu(K) \ge \limsup_n \mu_n(K)$, $\mu(K) \ge 1 - \epsilon$. Sending ϵ to 0 completes the proof.

Case 4: General M. Take K_n such that $\mu(K_n) > 1 - \frac{1}{n}$ for all $\mu \in S$ and set $M' = \cup_n K_n$. M' is a σ-compact subset of M and the restrictions of $\mu \in S$ to M' form a tight set. Apply the previous step and note that any limit must assign mass 1 to M'.

2. Second half: Suppose that S has compact closure.

We first show that for every $\epsilon, \delta > 0$, there is a finite set of open balls $\{B_\delta(x_i) : i \le n\}$ such that for all $\mu \in S$, $\mu(\cup_{i \le n} B_\delta(x_i)) > 1 - \epsilon$. If not, then for some $\epsilon, \delta > 0$, for every finite set $\{B_\delta(x_i) : i \le n\}$, there is a $\mu_n \in S$ such that $\mu(\cup_{i \le n} B_\delta(x_i)) \le$

$1 - \epsilon$. If we take a subsequence and relabel if necessary, the compactness of S implies that $\mu_n \to_{w^*} \mu$ for some $\mu \in \text{cl}(S)$. Since M is separable, it can be covered by a sequence of open balls, $B_i := B_\delta(x_i)$, so that $C_m := \cup_{i \leq m} B_i \uparrow M$. For any m, C_m is open, so that $\mu(C_m) \leq \liminf_n \mu_n(C_m) \leq 1 - \epsilon$. This contradicts the countable additivity of μ because $C_m \uparrow M$.

Given that for every $\epsilon, \delta > 0$, there is a finite set $\{B_\delta(x_i) : i \leq n\}$ such that for all $\mu \in S$, $\mu(\cup_{i \leq n} B_\delta(x_i)) > 1 - \epsilon$, the proof of Theorem 7.6.26 (p. 421) can be replicated. ∎

Exercise 9.9.7 We saw that every cdf $F : \mathbb{R} \to [0, 1]$ on $\mathbb{R}^1$ corresponds to a unique probability $\mu \in \Delta(\mathbb{R})$. Show that the same is true for cdf's on $\mathbb{R}^k$.

The following generalizes one of the steps in the previous proof.

Exercise 9.9.8 Show that if $f : M \to M'$ is a continuous mapping between two metric spaces and S is a tight subset of $\Delta(M)$, then $f(S)$ is a tight subset of $\Delta(M')$.

9.10 ◆ An Operator Proof of the Central Limit Theorem

Random variables can be nowhere equal yet have the same distribution, so convergence of their distributions may have nothing to do with convergence of the functions $\omega \mapsto X_n(\omega)$. The limits of the distributions of sums of small iid random variables can take many forms. The most common limit is Gaussian, the next most common the Poisson, and a peculiar limit distribution is the Cauchy distribution.

Example 9.10.1 (Poisson Limits) *For $n \in \mathbb{N}$, let $X_{n,i}$, $i = 1, \ldots, n$, be iid Bernoulli random variables with probability $P(X_{n,i} = 1) = r/n$. It is easy to show, in any of a variety of ways, that the distribution of $Y_n := \sum_{i \leq n} X_{n,i}$ converges to a Poisson distribution with parameter r.*

Example 9.10.2 (Cauchy Limits) *The standard Cauchy distribution has density $f(x) = \frac{1}{\pi(1+x^2)}$. Using characteristic functions, we can show fairly easily that if X_i, $i = 1, 2, \ldots$ is an iid collection of Cauchy random variables then for all n, the distribution of $\frac{1}{n} \sum_{i \leq n} X_i$ is that of a standard Cauchy.*

In this section, we go through an operator theoretic proof of the simplest version of the Gaussian central limit theorem and indicate how it extends to a less simple version. Here Z is a standard Gaussian random variable, that is, $P(Z \leq r) = \int_{-\infty}^r \phi(x)\, dx$, where $\phi(x) = \frac{1}{\sqrt{2\pi}} e^{-\frac{1}{2}x^2}$.

Theorem 9.10.3 (Simplest Central Limit Theorem) *If X_n is a sequence of iid random variables with $E\, X_n = 0$ and $\text{Var}(X_n) = 1$, then $S_n := \frac{1}{\sqrt{n}} \sum_{k \leq n} X_k \to_{w^*} Z$*

The less simple version of this uses independence but allows the X_n to have different distributions so long as the probability that any individual X_n contributes significantly to the sum is negligible.

The proofs we give here regard probabilities as operators, each taking a particular Banach space to itself.

9.10.a The Banach Space $C([-\infty, +\infty])$

We have already seen that a probability on a compact metric space M can be understood as linear operator mapping the Banach space $C(M)$ to the Banach space $\mathbb{R}$. Here, probabilities map the Banach space $C([-\infty, +\infty])$ to itself.

Exercise 9.10.4 If $f \in C([-\infty, +\infty])$, then $g := f_{|(-\infty,+\infty)} \in C(\mathbb{R})$ is bounded, $\lim_{x \to -\infty} g(x)$ exists, and $\lim_{x \to +\infty} g(x)$ exists. Conversely, if $g \in C(\mathbb{R})$ is bounded, $\lim_{x \to -\infty} g(x)$ exists and $\lim_{x \to +\infty} g(x)$ exists, then defining $f(x) = g(x)$ for $x \in \mathbb{R}$, $f(-\infty) = \lim_{x \to -\infty} g(x)$, and $f(+\infty) = \lim_{x \to +\infty} g(x)$ gives a function $f \subset C([-\infty, +\infty])$.

Owing to this last exercise, we are not explicit about the difference between functions on $[-\infty, +\infty]$ and their restrictions to $(-\infty, +\infty)$. $C^s(-\infty, +\infty)$ denotes the set of **smooth** functions in $C([-\infty, +\infty])$, that is, the set of functions that have derivatives of all orders on $(-\infty, +\infty)$, where the derivatives also belong to $C([-\infty, +\infty])$.

Lemma 9.10.5 $C^s(-\infty, +\infty)$ is dense in $C([-\infty, +\infty])$.

Proof. $C^s(-\infty, +\infty)$ is an algebra of continuous functions that separates points, and $[-\infty, +\infty]$ is compact. ∎

We have seen many linear functions from one normed space to another, and the conditional expectation operator maps a space to a subset of itself. Here the operators map $C([-\infty, +\infty])$ to itself. Following convention, we use the word "operator" instead of "function" when the domain and/or range of a function is itself a set of functions.

Notation 9.10.6 *For this section, when an operator, T, is linear, we use "Tu" instead of "$T(u)$" to keep the number of brackets under control.*

The following specializes the general definition of bounded operators to bounded operators from a space to itself.

Definition 9.10.7 *For a Banach space $\mathfrak{X}$, a linear operator $T : \mathfrak{X} \to \mathfrak{X}$ is* **bounded** *if for some $a \in \mathbb{R}$, for all $x \in \mathfrak{X}$, $\|Tx\| \le a\|x\|$. The norm of an operator is denoted $\|T\|$ and defined as the infimum of the set of $a \in \mathbb{R}$ such that for all $x \in \mathfrak{X}$, $\|Tx\| \le a\|x\|$. If $\mathfrak{X}$ is partially ordered by $\ge$, we say that an operator is* **positive** *if $[x \ge 0] \Rightarrow [Tx \ge 0]$.*

In $\mathfrak{X} = L^2(\Omega, \mathcal{F}, P)$ and $\mathcal{G} \subset \mathcal{F}$, $Tf := E(f \mid \mathcal{G})$ is a bounded positive operator with norm 1. The boundedness principle from the previous chapter says that an operator is continuous iff it is bounded.

There are two norms floating around here, the norm on $\mathfrak{X}$ and the norm on the operators from $\mathfrak{X}$ to itself. Keeping them straight is important.

Exercise 9.10.8 Show that $\|T\| = \sup\{\frac{\|Tu\|}{\|u\|} : u \ne 0\}$. Further, show that there is no loss in restricting u to be in a dense set.

Despite this result, there is a good reason for using the word "bounded" instead of "continuous." Sometimes we use different topologies, for example, weak or weak* topologies, on the domain or the range of an operator. An operator could be continuous in one of these without being bounded. In other words, "bounded" is shorthand for "continuous when both the domain and the range have their norm topologies." Thus, if we have, say, three distinct topologies on $\mathfrak{X}$, there are nine possible combinations of topologies on the domain and the range, which leads, in principle, to nine different classes of "continuous" operators.[4]

There are, roughly, as many different classes of continuous operators as there are notions of convergence of operators. For our present purposes, the following, pointwise convergence is enough.

Definition 9.10.9   *A sequence of operators T_n converges to T, written $T_n \to T$, if for all $x \in \mathfrak{X}$, $\|T_n x - T x\| \to 0$.*

The following gives a brief taste of different notions of the convergence of operators.

Example 9.10.10   *Let $\mathfrak{X} = L^2(\Omega, \mathcal{F}, P)$, $E_n : n \in \mathbb{N}\}$ a countable subset of $\mathcal{F}$, $\mathcal{G} = \sigma(\{E_n \in \mathcal{F} : n \in \mathbb{N}\})$, and $\mathcal{G}_m = \sigma(\{E_n \in \mathcal{F} : n \leq m\})$. Define $T_m f = E(f \mid \mathcal{G}_m)$, and $T f = E(f \mid \mathcal{G})$. Then $T_m \to T$, but, in general, $\sup\{\|T_m f - T f\| : \|f\| \leq 1\} \nrightarrow 0$, for example, if $\{1_{E_n} : n \in \mathbb{N}\}$ is a sequence of independent events with $P(E_n) = \frac{1}{2}$. However, for every compact $K \subset L^2(\Omega, \mathcal{F}, P)$, $\sup\{\|T_m f - T f\| : f \in K\} \to 0$.*

In general, linear operators from $\mathbb{R}^k$ to itself do not commute, that is, for $k \times k$ matrices A and B, in general $AB \neq BA$. Having commuting operators makes it possible to do algebra with them.

Definition 9.10.11   *Two operators T and U from $\mathfrak{X}$ to itself commute if for all $x \in \mathfrak{X}$, $TUx = UTx$, written $TU = UT$.*

Exercise 9.10.12   Show that if A and B are symmetric $k \times k$ matrices, then the linear mappings they represent do not necessarily commute. Give several different classes of $k \times k$ matrices that do commute.

9.10.b   Convolution Operators

The word "convoluted" is often used as a synonym for "tricky and complicated." Despite initial appearances, that is not what is going on.

Definition 9.10.13   *For a function $f \in C([-\infty, +\infty])$ and a probability $\mu \in \Delta(\mathbb{R})$, we define the convolution of μ and f, $\mu * f$, as the function $g \in C([-\infty, +\infty])$ given by*

$$g(t) = (\mu * f)(t) = \int f(t - x) \, d\mu(x).$$

For any $\mu \in \Delta(\mathbb{R})$, the operator $(\mu * f)$ is denoted by T_μ so that $T_\mu f := (\mu * f)$.

Equivalently, if X is a random variable with distribution μ, then $T_\mu f(t) = E f(t - X)$.

Exercise 9.10.14   The foregoing definition assumed that $(\mu * f) \in C([-\infty, +\infty])$. Show that this is true. [Dominated convergence is part of the argument.]

---

4. The mind boggles.

Exercise 9.10.15   Show that any T_μ is a bounded positive operator with norm 1. Further, show that this class of operators commutes, that is, if $T = T_\mu$ and $U = T_\nu$, then $TU = UT$.

Theorem 9.10.16 $\mu_n \to \mu$ in $\Delta(\mathbb{R})$ iff $T_{\mu_n} \to T_\mu$.

Proof.   Since $x \mapsto f(t - x)$ is continuous and bounded, $\mu_n \to \mu$ implies that the sequence of continuous function $g_n := \mu_n * f$ converges pointwise to the continuous function $g := \mu * f$. Since the sequence g_n is equicontinuous, this pointwise convergence is in fact uniform.

Now suppose that for all $f \in C([-\infty, +\infty])$, $\max_{t \in [-\infty, +\infty]} |\mu_n * f(t) - \mu * f(t)| \to 0$. In particular, this means that for all $f \in C([-\infty, +\infty])$, $\int f d\mu_n \to \int f d\mu$, implying that μ_n converges to a probability on $[-\infty, +\infty]$. We must show that it puts mass 0 on the points $\pm\infty$. For this it is sufficient to show that for all $\varepsilon > 0$ there exists $t > 0$ such that $\mu_n((2t, +\infty)) < \varepsilon$ a.a. and $\mu_n((-\infty, -2t)) < \varepsilon$ a.a., that is, that the sequence μ_n is tight. Suppose that the sequence is not tight; that is, suppose that for some $\varepsilon^\circ > 0$ and for all t, $\mu_n((2t, +\infty)) \geq \varepsilon^\circ$ i.o. or $\mu_n((-\infty, -2t)) \geq \varepsilon^\circ$ i.o. We derive a contradiction from for all t, $\mu_n((2t, +\infty)) \geq \varepsilon^\circ$ i.o., the remaining case is parallel.

Take $f \in C([-\infty, +\infty])$ to be any continuous, strictly monotonic cdf. As f is a cdf, for any $\eta > 0$, we can pick $t^\circ > 0$ such that for all $t > t^\circ$, $f(-t) < \eta$ and $\mu * f(t) > 1 - \varepsilon^\circ/3$. Now for any $t > t^\circ$, $\mu_n * f(t) = \int_{\{X_n > 2t\}} f(t - X_n(\omega)) dP(\omega) + \int_{\{X_n \leq 2t\}} f(t - X_n(\omega)) dP(\omega) \leq \eta \cdot \varepsilon^\circ + 1 \cdot (1 - \varepsilon^\circ)$, and for appropriate choice of η, this is less than $1 - 2\varepsilon^\circ/3$, which contradicts $\max_t |\mu_n * f(t) - \mu * f(t)| \to 0$. ∎

We could have used $C^s(-\infty, +\infty)$ instead of $C([-\infty, +\infty])$, because it is dense.

Corollary 9.10.17 $\mu_n \to \mu$ in $\Delta(\mathbb{R})$ iff for all $\varphi \in C^s(-\infty, +\infty)$, $\|T_{\mu_n}\varphi - T_\mu\varphi\| \to 0$.

Exercise 9.10.18   Prove the last corollary.

9.10.c The Simplest Central Limit Theorem

There are two lemmas that go into the proof. Throughout, all operators belong to the set $\{T_\mu : \mu \in \Delta(\mathbb{R})\}$ and we define $T^n = T \cdots T$ where we apply T n times.

Lemma 9.10.19   *For all T_1, T_2, U_1, U_2, and all $f \in C([-\infty, +\infty])$, $\|T_1 T_2 f - U_1 U_2 f\| \leq \|T_1 f - U_1 f\| + \|T_2 f - U_2 f\|$, so that $\|T^n f - U^n f\| \leq n \cdot \|Tf - Uf\|$.*

Proof. $(T_1 T_2 - U_1 U_2) = T_1 T_2 - U_1 T_2 + U_1 T_2 - U_1 U_2$. Regrouping and using commutativity yields $(T_1 T_2 - U_1 U_2) = (T_1 - U_1)T_2 + (T_2 - U_2)U_1$. By the triangle inequality, $\|(T_1 T_2 - U_1 U_2)f\| \leq \|(T_1 - U_1)T_2 f\| + \|(T_2 - U_2)U_1 f\|$. Finally, since the norm of T_2 and U_1 are 1, $\|(T_1 - U_1)T_2 f\| + \|(T_2 - U_2)U_1 f\| \leq \|(T_1 - U_1)f\| + \|(T_2 - U_2)f\|$. The last part follows by induction. ∎

The following exercise is a special case of the second lemma we need.

Exercise 9.10.20   Let X be a random variable with distribution μ, $E\,X = 0$, and $E\,X^2 = \mathrm{Var}(X) = 1$, and define $Y_n = -X/\sqrt{n}$ so that $E\,Y = 0$, $E\,Y^2 = \mathrm{Var}(Y) = 1/n$. Show that for $f \in C^s[-\infty, +\infty]$ and $t_0 \in (-\infty, +\infty)$,

$$n[E\,f(t_0 + Y_n) - f(t_0)] \to \tfrac{1}{2} f''(t_0).$$

[Since f is smooth, it is three times continuously differentiable. Its second-order Taylor expansion is $f(t_0 + y) = f(t_0) + f'(t_0)y + \tfrac{1}{2} f''(t_0) y^2 + o(y^2)$. Therefore, $n[E\,f(t_0 + Y_n) - f(t_0)] = n[(f(t_0) - f(t_0)) + f'(t_0)E\,Y_n + \tfrac{1}{2} f''(t_0)\,\mathrm{Var}(Y_n) + o(\mathrm{Var}(Y_n))] = \tfrac{1}{2} f''(t_0) + o(1)$.]

The pointwise convergence result in the previous lemma is in fact uniform.

Lemma 9.10.21   If $f \in C^s(-\infty, +\infty)$, $X \sim \mu$ with $E\,X = 0$, $E\,X^2 = 1$, and T_n is the operator associated with μ_n, the distribution of $X/\sqrt{n}$, then

$$\sup_{t \in (-\infty, +\infty)} |n[T_n f - f](t) - \tfrac{1}{2} f''(t)| \to 0. \qquad (9.6)$$

Proof.   Since $f \in C^s[-\infty, +\infty]$, its third derivative is bounded; hence the o term in Exercise 9.10.20 goes to 0 uniformly.   ■

We are now in a position to prove the following.

Theorem 9.10.3   If X_n is a sequence of iid random variables with $E\,X_n = 0$ and $\mathrm{Var}(X_n) = 1$, then $S_n := \frac{1}{\sqrt{n}} \sum_{k \le n} X_k \to_{w^*} Z$

Proof.   Let Z have a Gaussian distribution, let U and U_n be the operators associated with (the distributions of) Z and $Z/\sqrt{n}$, and let T_n be the operator associated with the (distribution of the) random variable $X_i/\sqrt{n}$. We wish to show that for every $f \in C^s(-\infty, +\infty)$,

$$\|T_n^n f - Uf\| \to 0.$$

Since $U_n^n = U$ (from your introductory statistics class), from Lemma 9.10.19, we have

$$\|T_n^n f - Uf\| = \|T_n^n f - U_n^n f\| \le n\|T_n f - U_n f\|.$$

Now, $(T_n f - U_n f) = (T_n f - f + f - U_n f)$, so

$$n\|T_n f - U_n f\| \le n\|T_n f - f\| + n\|U_n f - f\|,$$

and by Lemma 9.10.21, the right-hand side goes to 0.   ■

When the X_n have mean 0 but infinite variances, the central limit theorem will still hold. One has to find norming constants other than $1/\sqrt{n}$ for Lemma 9.10.21.

Exercise 9.10.22   Proving and using a variant of the two lemmas presented here, show that if X_n is an iid sequence of random variables with mean 0, then $\frac{1}{n} \sum_{i \le n} X_n \to_{w^*} 0$.

Here is Lindeberg's version of the central limit theorem; the sufficient conditions given here are actually necessary.

Theorem 9.10.23 (Lindeberg) *If X_n is an independent collection of random variables with distributions P_n, $E\ X_n = 0$, $\mathrm{Var}(X_n) = \sigma_n^2$, and $s_n^2 := \sum_{k \le n} \sigma_k^2$ and for all $t > 0$,*

$$\frac{1}{s_n^2} \sum_{k \le n} E\ X_k^2 1_{|X_k| < ts_n} \to 1, \tag{9.7}$$

then $\frac{1}{s_n} \sum_{k \le n} X_k \to_{w^*} Z$.

Lindeberg's condition (9.7) asks that in the limit, all of the contributions to the total variance made by the X_k be made by the parts of X_k with X_k/s_n vanishingly small. By contrast, in the Poisson limit case, Example 9.10.1, a great deal of the action happens a long way from 0 and the limits in (9.7) are 0.

Previously, s_n was $\sqrt{n}$, but here it can go to ∞ at very different rates. If we let $T_{k,n}$ and $U_{k,n}$ be the operator associated with the random variables X_k/s_n and any Z/s_n, respectively, the proof must show that $\sum_{k \le n} \|T_{k,n} f - U_{k,n} f\| \to 0$. The inequality in (9.6) splits into n parts, each having weight σ_k^2/s_n^2, and the Lindeberg condition is what makes everything add up correctly.

9.11 ◆ Regular Conditional Probabilities

One of the best properties of Polish measure spaces is that conditional probabilities on them have such a nice form. The technical term for "nice" is "regular." In this section we introduce regular conditional probabilities. We show that they need not exist, but that they do if we work with Polish measure spaces.

9.11.a Definition and Examples

Suppose that $(\Omega, \mathcal{F}, P)$ is a probability space, $X : \Omega \to M$ is measurable, M is finite, and for $A \subset M$, $\mu(A) := P(X^{-1}(A))$. Then, there is a mapping $x \mapsto P_x$, from M to probabilities on $(\Omega, \mathcal{F})$ with the property that for all integrable g, $E\ (g \mid \sigma(X))(\omega) = \int g(\omega')\, dP_{X(\omega)}(\omega')$. That is, P_x is the conditional probability on $(\Omega, \mathcal{F})$ given that $X = x$.

Exercise 9.11.1 For $x \in M$, M a finite set, with $\mu(x) > 0$, show that the requisite $x \mapsto P_x$ is $\frac{P(X^{-1}(x) \cap E)}{P(X^{-1}(x))}$, while if $\mu(x) = 0$, then one can take P_x to be any probability.

The situation is more complicated when M is infinite. Throughout this section, $(M, \mathcal{M})$ is a nonempty set and a σ-field of subsets with $\{x\} \in \mathcal{M}$ for all $x \in M$. For the best results, we require that $(M, \mathcal{M})$ be a separable metric space and $\mathcal{M}$ its Borel σ-field.

We are interested in measurable mappings $x \mapsto P_x$ from M to $\Delta = \Delta(\mathcal{F})$, the set of probabilities on $\mathcal{F}$. For this, we must introduce a σ-field on Δ.

Notation 9.11.2 *For a measurable space $(\Omega, \mathcal{F})$, the associated measure space of probabilities is $(\Delta, \mathcal{D})$, where $\mathcal{D}$ is the smallest σ-field containing the sets $\{Q : Q(E) \le r\}$, $E \in \mathcal{F}$, $r \in \mathbb{R}$.*

Exercise 9.11.3 If $(\Omega, \mathcal{F})$ is a Polish measure space, d is any metric for which $\mathcal{F}$ is the Borel σ-field and ρ is the Prokhorov metric associated with d, then $\mathcal{D}$ is the Borel σ-field $\sigma(\tau_d)$. [Starting with Ω compact might be the easiest approach.]

We defined conditional expectations $E(Y \mid X)$ some time ago.

Definition 9.11.4 *The **conditional probability of a set** A **given** X is $E(1_A \mid X)$. A realization of the random variable $P(A \mid X)$ is called a **posterior probability** of A.*

Exercise 9.11.5 Show that the average of posterior distributions is the prior distribution, that is, that for all measurable A, $E(P(A \mid X)) = P(A)$. [Iterated expectations.]

We know that for any set A, $P(A \mid X)$ is uniquely defined up to a set of measure 0. We know that for any countable collection of sets, A_n, $P(A_n \mid X)$ is uniquely defined up to a set of measure 0. What we do not know is that there is a $\sigma(X)$-measurable mapping from the range of X to $\Delta(M)$ such that for all measurable sets A, $x \mapsto P_x$ is a version of $P(A \mid X)$. For this to be true, we would need $\omega \mapsto P_{X(\omega)}(A)$ to be a.e. equal to $P(A \mid X)(\omega)$ for all measurable sets A simultaneously.

We have good news and bad news. The bad news is that we cannot always find such a mapping $x \mapsto P_x$. The good news is that we can find such a mapping in all the cases that we care about.

Definition 9.11.6 *Given a random variable $X : \Omega \to M$ on a probability space $(\Omega, \mathcal{F}, P)$, a **regular conditional probability (rcp) for** X is a measurable function $x \mapsto P_x$ from M to Δ with the property that for all $E \in \mathcal{F}$, the function $\omega \mapsto P_{X(\omega)}(E)$ is a version of $P(E \mid \sigma(X))$. An rcp is **a.e. proper**, or satisfies the **support condition**, if for μ-a.e., $x \in M$, $P_x(X^{-1}(x)) = 1$, where $\mu(A) := P(X \in A)$ defines the image law of P in M.*

You have seen rcp's before.

Example 9.11.7 *Let $(\Omega, \mathcal{F}) = (\mathbb{R}^2, \mathcal{B}^2)$ and suppose that P has a strictly positive density, $f : \mathbb{R}^2 \to \mathbb{R}$, so that $P(E) = \int_E f(\omega_1, \omega_2) \, d\omega_1 d\omega_2$. Define $X(\omega_1, \omega_2) = \omega_1$. For each $x \in \mathbb{R}$, $X^{-1}(x) = \{(x, \omega_2) : \omega_2 \in \mathbb{R}\}$. Define $P_x \in \Delta$ as the probability on the line $X^{-1}(x)$ with density $y \mapsto \phi(y)$ given by $\phi(y) = \kappa_x f(x, y)$, where κ_x is a constant chosen to make $\phi(\cdot)$ integrate to 1. By construction, P_x satisfies the support condition. To verify that $x \mapsto P_x$ is measurable, it is sufficient to check that for all a and r, $\{x : P_x((-\infty, a]) \leq r\} \in \mathcal{B}$.*

Exercise 9.11.8 Verify that $x \mapsto P_x$ is measurable in the previous example.

It is tempting, but very, very wrong, to think that one can always rescale densities along sets $X^{-1}(x)$ to find the conditional probabilities.

Exercise 9.11.9 (Borel's Paradox) Suppose that $(\Omega, \mathcal{F}) = (\mathbb{R}^2, \mathcal{B}^2)$ and suppose that P is the distribution of two independent standard normal random variables. The density for P is $f_P(\omega_1, \omega_2) = \kappa e^{-\frac{1}{2}(\omega_1^2 + \omega_2^2)}$, where κ is a constant involving π. Define $X(\omega_1, \omega_2) = \frac{\omega_2}{\omega_1}$ if $\omega_1 \neq 0$ and $X(0, \omega_2) = 7$.

Let $E = \{(\omega_1, \omega_2) : \omega_2 < 1\}$ and consider the bet $r1_E - s1_{E^c}$ that pays $r > 0$ if E occurs and costs $s > 0$ if E^c occurs. Suppose that $r, s > 0$ are chosen so that a risk-averse expected utility maximizer is indifferent between taking the bet and not taking the bet.

Show the following:

1. Restricted to any line $X^{-1}(x)$ in a set of x having μ-mass 1, the scaled density is that of a standard normal.

2. Define Q_x to be the scaled density that you just found and show that $x \mapsto Q_x$ is measurable and satisfies the support condition.

3. Show that if $x \mapsto Q_x$ is the rcp for X, then on a set of x having μ-mass 1, after hearing that $X = x$, the expected utility maximizer strictly prefers not to have taken the bet. Conclude that $x \mapsto Q_x$ is *not* the rcp for X.

What goes wrong in the previous exercise is that the radial lines $X^{-1}(x)$ do not do a good job of representing the information contained in the random variable X. Heuristically, sets of the form $X^{-1}(|a, b|)$ with $|a - b|$ small are narrow two-way cones with a point at the origin. If one knew that X had taken its value in some small interval, $|a, b|$, this is the set one would focus on. As one moves further away from the origin, the width of the two-way cone increases linearly. This means that the densities for P_x ought to be of the form $\kappa |r| e^{-\frac{1}{2}r^2}$, where r is the radial distance from the origin and κ is chosen to make the integral equal to 1. This heuristic also works in Example 9.11.7, where the sets $X^{-1}(|a, b|)$ with $|a - b|$ small are narrow strips. Behind the validity of this heuristic is the Martingale convergence theorem, so if you ever have to find a specific rcp, do not be shy about using it. As always, be sure to check your answer in other ways too.

9.11.b Existence of rcp's for Polish Measure Spaces

It turns out that if $(\Omega, \mathcal{F}, P)$ is Polish, rcp's exist. We begin with the following special case. Since Polish measure spaces are measurably isomorphic to Borel subsets of a compact metric space, replacing the compact metric space in the following is easy.

Theorem 9.11.10   *If Ω is a compact metric space, $\mathcal{F}$ is the associated Borel σ-field, P is a probability on $\mathcal{F}$, $(M, \mathcal{M})$ is a measure space, and $X : \Omega \to M$ is measurable, then an rcp for X exists.*

The idea of the proof is to produce, for each x a set having μ-mass 1, a continuous probabilistic positive linear function, L_x, defined on a rich countable subclass of the bounded measurable functions on Ω. Since Ω is compact, this countable subclass can be taken to be a class of continuous functions. If we let P_x be the probability associated with L_x, what is left is the rather lengthy verification that $\omega \mapsto \int g \, dP_{X(\omega)}$ is indeed a version of $E(g \mid X)$.

Proof.   Let $\{f_n : n \in \mathbb{N}\}$ be a countable dense subset of $C(\Omega)$ and let $\mathcal{A}_{\mathbb{Q}}$ denote the (necessarily countable) set of all polynomials in the f_n having rational coefficients. The algebra $\mathcal{A}_{\mathbb{Q}}$ is dense in $C(A)$, and for any $\alpha, \beta \in \mathbb{Q}$, $f, g \in \mathcal{A}_{\mathbb{Q}}$ and $\alpha f + \beta g \in \mathcal{A}_{\mathbb{Q}}$. (In other, fancier words, $\mathcal{A}_{\mathbb{Q}}$ is a vector space over $\mathbb{Q}$.)

1. *Producing the linear functions L_x*. For each $f \in \mathcal{A}_{\mathbb{Q}}$, define a finite signed measure on $\mathcal{F}$ by $Q_f(E) = \int_E f \, dP$ and an associated image measure $\mu_f(A) = Q_f(X^{-1}(A))$ on $\mathcal{M}$. Since each μ_f is absolutely continuous with respect to μ, it has a Radon-Nikodym derivative, d_f.

For $\alpha, \beta \in \mathbb{Q}$ and $f, g \in \mathcal{A}_{\mathbb{Q}}$, since $Q_{\alpha f + \beta g}(E) = \alpha Q_f(E) + \beta Q_g(E)$, we expect that for $x \in M$, $d_{\alpha f + \beta g}(x) = \alpha d_f(x) + \beta d_g(x)$. Put in other terms, we expect that for $x \in M$, the function $L_x(f) := d_f(x)$ is probabilistic, and positive, linear, and continuous, at least on $\mathcal{A}_{\mathbb{Q}}$. The problem is that each density d_f is unique only up to a null set. We identify four null sets: B_{prob} having to do with being probabilistic, B_{pos} with positivity, B_{lin} with $\mathbb{Q}$-linearity, and B_{cont} with continuity.

Probabilistic: $\mu(\{x : d_1(x) \neq 1\}) = 0$, let $B_{prob} = \{d_1 \neq 1\}$.

A positivity null set: For $f, g \in \mathcal{A}_{\mathbb{Q}}$, $f \geq g$, for all $E \in \mathcal{F}$, $Q_f(E) \leq Q_g(E)$. Therefore, $\mu(B_{f,g}) = 0$, where $B_{f,g} = \{x : d_f(x) > d_g(x)\}$. Define the μ-null set $B_{pos} = \cup B_{f,g}$, where the countable union is taken over $f, g \in \mathcal{A}_{\mathbb{Q}}$.

A $\mathbb{Q}$-linearity null set: Let $B_{\alpha,\beta,f,g} = \{x \in M : d_{\alpha f + \beta g}(x) \neq \alpha d_f(x) + \beta d_g(x)\}$. We show that $\mu(B_{\alpha,\beta,f,g}) = 0$. If it is strictly positive, then either the set where $d_{\alpha f + \beta g}(x) > \alpha d_f(x) + \beta d_g(x)\}$ or the set where $d_{\alpha f + \beta g}(x) < \alpha d_f(x) + \beta d_g(x)\}$ has strictly positive μ-mass. Let A be one of those sets having strictly positive mass and set $E = X^{-1}(A)$. It is immediate that $Q_{\alpha f + \beta g}(E) \neq \alpha Q_f(E) + \beta Q_g(E)$, a contradiction. Define the μ-null set $B_{lin} = \cup B_{\alpha,\beta,f,g}$ where the countable union is taken over $f, g \in \mathcal{A}_{\mathbb{Q}}$ and $\alpha, \beta \in \mathbb{Q}$.

A continuity null set: Let U be the unit ball in $\mathcal{A}_{\mathbb{Q}}$. We show that $\mu(\cup_N B_N) = 1$, where $B_N := \{x \in M : \forall f \in U, \ |d_f(x)| \leq N\}$. For all $f \in U$ and $E \in \mathcal{F}$, $|Q_f(E)| \leq Q_{|f|}(E) = \int_E |f| \, dP \leq \int_E 1 \, dP = P(E)$, which yields $\mu(\{x : |d_f(x)| \leq d_1(x)\}) = 1$. Therefore, for any $N \in \mathbb{N}$, $\{x \in M : \forall f \in U, \ |d_f(x)| > N\} \subset \{x : d_1(x) > N\}$, and since $d_1(\cdot)$ is integrable, $\mu(d_1(x) > N) \downarrow 0$, implying that $\mu(\cup_N B_N) = 1$. Let $B_{cont} = (\cup_N B_N)^c$.

For each x outside of the four null sets, the mapping $L_f(x) := d_f(x)$ is probalistic; bounded on $\mathcal{A}_{\mathbb{Q}}$, hence continuous on $C(\Omega)$ because it is $\mathbb{Q}$-linear on $\mathcal{A}_{\mathbb{Q}}$, hence linear on $C(\Omega)$; and positive on $\mathcal{A}_{\mathbb{Q}}$, hence positive on $C(\Omega)$ by continuity.

Since Ω is compact, for each x outside of the four null sets, there is a unique probability P_x on Ω defined by $\int_\Omega f(\omega) \, dP_x(\omega) = L_x(f) = d_f(x)$. For other x, define $P_x = P^\circ$, your favorite probability on Ω.

2. *Verifying that we have a version of the conditional expectation*. We show that for all bounded measurable g, $\omega \mapsto \int_\Omega g(\omega') \, dP_{X(\omega)}(\omega')$ is measurable and is a version of $\omega \mapsto E(g \mid X)(\omega)$. Since the continuous functions are dense, for the "version" part of the proof, it is sufficient to prove that it is true for all $g \in \mathcal{A}_{\mathbb{Q}}$.

Measurability: For any $f \in \mathcal{A}_{\mathbb{Q}}$, $\int f \, dP_{X(\omega)} = d_f(X(\omega))$, which is a measurable function (being the composition of measurable functions). Since $\mathcal{A}_{\mathbb{Q}}$ is dense in $C(\Omega)$, $\omega \mapsto \int g \, dP_{X(\omega)}$ is measurable for all $g \in C(\Omega)$. Finally, the vector space of bounded measurable g's such that $\omega \mapsto \int g \, dP_{X(\omega)}$ is measurable contains the continuous functions and is closed under bounded monotone limits (by dominated convergence). Therefore, for all closed F, $\omega \mapsto \int 1_F \, dP_{X(\omega)}$ is measurable (by considering the functions $g_n = \max\{0, 1 - nd(\omega, F)\}$.) For any measurable $E \in \mathcal{F}$, there exists a sequence of closed $F_n \uparrow E' \subset E$ with $P(E \setminus F_n) \downarrow 0$. Therefore for any measurable E, $\omega \mapsto \int 1_F \, dP_{X(\omega)}$ is measurable (being the limit of a monotonic sequence of measurable functions). This implies that for any simple function, g, $\omega \mapsto \int g \, dP_{X(\omega)}$ is measurable. Finally, since every bounded measurable function is uniformly close to a simple function, $\omega \mapsto \int g \, dP_{X(\omega)}$ is measurable for all bounded measurable g.

Conditional expectation: By the measurability just established, the function $\omega \mapsto \int_{\Omega} g(\omega') \, dP_{X(\omega)}(\omega')$ is measurable and depends on ω only through $X(\omega)$; hence it is $\sigma(X)$ measurable and thus a function of the form $h(X)$. We must show that for all $f \in \mathcal{A}_{\mathbb{Q}}$, $\int_{\Omega} f(\omega) \cdot h(X(\omega)) \, dP(\omega) = \int_{\Omega} f(\omega) \cdot g(\omega) \, dP(\omega)$:

$$\int_{\Omega} f(\omega) \cdot h(X(\omega)) \, dP(\omega)$$

$$= \int_{\Omega} f(\omega) \cdot \left[\int_{\Omega} g(\omega') \, dP_{X(\omega)}(\omega') \right] dP(\omega) = \tag{9.8}$$

$$\int_{\Omega} f(\omega) d_g(X(\omega)) \, dP(\omega) = \int_{\Omega} d_g(X(\omega)) \, dQ_f(\omega). \tag{9.9}$$

We now change variables and begin integrating over M,

$$\int_{\Omega} d_g(X(\omega)) \, dQ_f(\omega) = \int_M d_g(x) \, d\mu_f(x) = \int_M d_g(x) d_f(x) \, d\mu(x) = \tag{9.10}$$

$$\int_M d\mu_{f \cdot g}(x) = \int_{\Omega} f \cdot g \, dP, \tag{9.11}$$

where the last two equalities follow from the definition of μ_f and μ_g and the observation that for any $f, g \in C(\Omega)$, $d_f \cdot d_g = d_{f \cdot g}$ μ-a.e. ∎

We soon show that a.e. properness, a.k.a. the support condition, can be guaranteed μ-a.e. when $(M, \mathcal{M})$ is a separable metric space with its Borel σ-field. It might seem that there is something mysteriously virtuous about metrics as the basis for σ-fields for probability spaces. It has to do with two aspects of the implied measure theoretic structure.

Definition 9.11.11   *For a measure space $(X, \mathcal{X})$ and a sub-σ-field $\mathcal{G} \subset \mathcal{X}$, the $\mathcal{G}$-atoms of X are the sets $At^{\mathcal{G}}(\omega) = \cap\{B : \omega \in B \in \mathcal{G}\}$.*

Exercise 9.11.12   If $\mathcal{G} = \sigma(B_1, B_2, \ldots)$ is countably generated, then for all ω, $At^{\mathcal{G}}(\omega) = \cap\{B_n : \omega \in B_n\}$, which is necessarily measurable.

Lemma 9.11.13   *$(M, \mathcal{M})$ is a separable metric space with its Borel σ-field iff $\mathcal{M}$ is countably generated and all points are atoms, that is, for all $x \in M$, $At^{\mathcal{M}}(x) = \{x\}$.*

Proof.   If countably generated and its points are atoms, then the Blackwell metric (9.3) (p. 570) will work. ∎

Exercise 9.11.14   Provide an alternate proof of the foregoing lemma based on the function $\varphi(\omega) := \sum_n \frac{1}{10^n} 1_{B_n}(x)$, where $\mathcal{M} = \sigma(B_1, B_2, \ldots)$ by observing that this gives a measurable isomorphism between M and its image in $\mathbb{R}$ and then doing the right thing.

Theorem 9.11.15 *If the conditions of Theorem 9.11.10 hold and $(M, \mathcal{M})$ is a separable metric space, then there exists a proper rcp.*

Proof. Take any rcp, $x \mapsto P_x$. We must show that $\mu(\{x : P_x(X^{-1}(x)) = 1\}) = 1$. For any $B \in \mathcal{B}$, $P(X \in B \mid X) = 1_{X^{-1}(B)}$ P-a.e., that is, $\mu(\{x : P_x(B) = 1_B(x)\}) = 1$. Let $\mathcal{M} = \sigma(B_1, B_2, \ldots)$ and define $At^n(x) = \cap\{B_m : \omega \in B_m, \ m \le n\}$, so that $A_n := \{x : P_x(X^{-1}(At^n(x))) = 1\}$ has μ-mass 1. For all $x \in A := \cap_n A_n$, $\mu(\{x : P_x(X^{-1}(x)) = 1\}) = 1$ because $At^n(x) \downarrow At^{\mathcal{M}}(x) = \{x\}$. ∎

There is a special case of these last two results, the existence of rcp's and the existence of proper rcp's, which happens often enough that it ought to have its own statement.

Theorem 9.11.16 *If $(\Omega, \mathcal{F}, P)$ is Polish and $\mathcal{G}$ is a sub-σ-field of $\mathcal{F}$, then there is a $\mathcal{G}$-measurable mapping $\omega \mapsto P_\omega^{\mathcal{G}}$ such that $P_\omega^{\mathcal{G}}(A)$ is a version of $P(A \mid \mathcal{G})(\omega)$ for all $A \in \mathcal{F}$ and $\int f(\omega') \, dP_\omega^{\mathcal{G}}(\omega')$ is a version of $E(f \mid \mathcal{G})(\omega)$ for all bounded measurable f. Further, if $\mathcal{G}$ is countably generated, the rcp is proper, that is, $P(\{\omega : P_\omega^{\mathcal{G}}(At^{\mathcal{G}}(\omega)) = 1\}) = 1$.*

Proof. For existence of an rcp, take $(M, \mathcal{M}) = (\Omega, \mathcal{G})$ and use Theorem 9.11.10. For properness, define $\omega \sim_{\mathcal{G}} \omega'$ if $At^{\mathcal{G}}(\omega) = At^{\mathcal{G}}(\omega')$, let $\widetilde{\Omega}$ be the set of $\sim_{\mathcal{G}}$-equivalence classes, take $X(\omega)$ to be ω's equivalence class, and define $\widetilde{\mathcal{G}} = \{X(A) : A \in \mathcal{G}\}$. The measure space $(\widetilde{\Omega}, \widetilde{\mathcal{G}})$ is countably generated and its points are atoms. Use Theorem 9.11.15. ∎

For Polish measure spaces, there is nearly no loss in assuming that sub-σ-fields are countably generated. This comes from the observation that with the metric $d_P(A, B) = P(A \triangle B)$, $(\mathcal{F}, d)$ is a Polish metric space (when we identify sets that are a.e. equal). However, this may change the atoms.

Exercise 9.11.17 Let $(\Omega, \mathcal{F}, P) = ([0, 1], \mathcal{B}, \lambda)$ and let $\mathcal{G}$ be the smallest σ-field containing the sets $\{\omega\}$. Show that $\mathcal{G} = \{E \subset [0, 1] : E \text{ or } E^c \text{ is countable}\}$, that for all ω, $At^{\mathcal{G}}(\omega) = At^{\mathcal{B}}(\omega)$, that any $\mathcal{G}$-measurable random variable is a.e. constant, and that $d_P(\mathcal{G}, \{\emptyset, \Omega\}) = 0$. [These last two mean that $\mathcal{G}$ is a very poor σ-field indeed.] In particular, if $\omega \mapsto P_\omega^{\mathcal{G}}$ is an rcp, then it is a.e. constant and assigns mass 0 to every $\mathcal{G}$-atom.

To generalize the existence results from compact metric spaces to Polish measure spaces, we observe that every Polish measure space is measurably isomorphic to a measurable subset of the compact metric space $[0, 1]^{\mathbb{N}}$. We use the previous result to find an rcp with the image distribution in $[0, 1]^{\mathbb{N}}$ and then use the measurable isomorphism to pull it back to the original space, yielding the following.

Corollary 9.11.18 *If $(\Omega, \mathcal{F})$ is a Polish measure space, P is a probability on $\mathcal{F}$, $(M, \mathcal{M})$ is a measure space, and $X : \Omega \to M$ is measurable, then an rcp for X exists, and if $(M, \mathcal{M})$ is a separable metric space and its Borel σ-field, the rcp is a.e. proper.*

9.12 ◆ Conditional Probabilities from Maximization

Some random ω in a probability space $(\Omega, \mathcal{F}, P)$ will happen. The decision maker (DM) will observe a signal, $X(\omega)$, that has information about ω in it. After observing the signal, the DM will choose an action $a \in A$. The DM's utility if he chooses a and ω happens is $u(a, \omega)$, and the DM maximizes expected utility.

There are two ways to solve this problem.

1. Make a complete contingent plan, that is decide ahead of time what action, $a^*(x)$, to take if $X(\omega) = x$ for each and every possible value of x.

2. "Cross the bridge when you get to it," that is, decide what to do only after observing the value of x.

The planning approach feels like overkill. These two approaches are intimately linked by regular conditional probabilities.

Here is the easiest case: $X(\omega)$ is a simple function $\sum_i x_i 1_{E_i}$, where the E_i partition Ω, $\mathcal{X}$ is the set of possible x_i's and the set of actions, A, is a finite set. The first approach looks at the problem

$$\max_{a:\mathcal{X}\to A} \int u(a(X(\omega)), \omega) \, dP(\omega). \tag{9.12}$$

Let us rearrange the objective function,

$$\int u(a(X(\omega)), \omega) \, dP(\omega) = \sum_i \int_{E_i} u(a(x_i), \omega) \, dP(\omega)$$

$$= \sum_i P(E_i) \frac{1}{P(E_i)} \int_{E_i} u(a(x_i), \omega) \, dP(\omega). \tag{9.13}$$

Define $P_{x_i}(A) = \frac{1}{P(E_i)} P(A \cap E_i)$.

Exercise 9.12.1   Show that $x_i \mapsto P_{x_i}(\cdot)$ is a regular conditional probability for X. In particular, show that for any measurable $A \subset \Omega$, $\sum_i P(E_i) P_{x_i}(A) = P(A)$, that is, the average of the posterior distributions is the prior.

If we take careful note of the subscripts x_i on the P_{x_i}, we see that this means that (9.12) is equivalent to

$$\max \sum_i P(E_i) \int u(a(x_i), \omega) \, dP_{x_i}(\omega). \tag{9.14}$$

The objective function is a convex combination of the numbers $\int u(a(x_i), \omega) \, dP_{x_i}(\omega)$.

Here is the conclusion that this rearrangement inexorably leads us to.

Lemma 9.12.2   *A plan for the DM maximizes expected utility iff for each value of x_i, the plan solves the problem $\max_{a \in A} \int u(a, \omega) \, dP_{x_i}(\omega)$.*

This says that the "cross that bridge when you get to it" strategy solves the maximization problem, but only if it is a specific form of the strategy. After seeing x_i, the DM should choose a to maximize expected utility using the regular conditional probability P_{x_i}. This is the sense in which conditional probabilities arise from maximization.

The finiteness of the foregoing is not needed if rcp's exist. We assume that $(\Omega, \mathcal{F})$ is a Polish measure space, that X takes values in a Polish metric space, that actions are chosen in a Polish space, and that for all distributions Q on $(\Omega, \mathcal{F})$, there exists an $a^*(Q) \in A$ that solves the problem

$$\max_{a \in A} \int u(a, \omega) \, dQ(\omega) = \max_{a \in A} \langle u(a, \cdot), Q \rangle. \tag{9.15}$$

Theorem 9.12.3 *If $\omega \mapsto P_{\omega}^{\mathcal{G}}$ is an rcp for $\mathcal{G}$, then a plan a solves* $\max\{\int u(a(\omega), \omega) \, dP(\omega) : a(\cdot)$ *is $\mathcal{G}$-measurable\} iff with probability 1, $a(\omega) = a^*(P_{\omega}^{\mathcal{G}})$.*

Exercise 9.12.4 Prove Theorem 9.12.3, but be sure not to work very hard.

This brings us to Blackwell's work on what he called "experiments." Let Δ denote the set of probabilities on Ω so that Δ is also a Polish space (with the Prokhorov metric). An information structure with an rcp gives rise to a mapping $\omega \mapsto P_{\omega}$ from Ω to Δ. Let β be the induced distribution, that is, $\beta \in \Delta(\Delta)$. The set of possible information structures is the set of β in $\Delta(\Delta)$ that average to the prior distribution.

Definition 9.12.5 *The **Blackwell partial order on information structures** is defined by $\beta \succsim \beta'$ if $\int V(Q) \, d\beta(Q) \geq \int V(Q) \, d\beta'(Q)$ for all convex $V : \Delta \to \mathbb{R}$.*

Compare the following with Exercise 8.12.4.

Theorem 9.12.6 *Every decision maker likes β better than β' iff $\beta \succsim \beta'$ in the Blackwell partial order, that is, iff the distributions of priors under β is unambiguously more dispersed under β than under β'.*

There is good intuition for this at the extremes. The most dispersed set of probabilities is the set of point masses. This corresponds to being sure of the exact value of ω with probability 1 before having to choose an action. The least dispersed set is having β be point mass on the prior distribution, that is, having X be completely uninformative.

The essential intuition for the proof is that every convex V is the upper envelope of a set of linear functions on Δ, and each of these is of the form $Q \mapsto \langle u, Q \rangle$. Arrange the function $a \mapsto u(\omega, a)$ so that the set of $\{u(\cdot, a) : a \in A\}$ is the requisite enveloping set.

9.13 ◆ Nonexistence of rcp's

We begin with a positive result on the role of atoms in identifying σ-fields. With this in hand we suppose that we can show that a very peculiar probability space exists. If it does, then rcp's fail to exist in general. The rest of this section is devoted to showing that the peculiar probability space exists.

9.13.a   The Positive Role of Atoms

Atoms do not, in general, determine σ-fields, as was seen in Exercise 9.11.17. However, the following result, which we do not prove, tells that they do for a large and useful class of measure space models of randomness. To interpret the following, which is one of many results known as "Blackwell's theorem," you should know that every Polish measure space is a Blackwell measure space (and that Blackwell himself called such spaces Lusin spaces).

Theorem 9.13.1 (Blackwell)   *If $(\Omega, \mathcal{F})$ is a Blackwell measure space and $\mathcal{S}$ is a countably generated sub σ field of $\mathcal{F}$, then for any sub-σ-field $\mathcal{G}$, $\mathcal{G} \subset \mathcal{S}$ iff for all ω, each $At^{\mathcal{G}}(\omega)$ is a union of $\mathcal{S}$-atoms; that is, for all ω, $At^{\mathcal{G}}(\omega) = \cup\{At^{\mathcal{S}}(\omega') : \omega' \in At^{\mathcal{G}}(\omega)\}$.*

When the randomness takes on only finitely many values, partitions and σ-fields can be identified. Blackwell's theorem says that the same is true for a large and useful class of measure space models of randomness.

Exercise 9.13.2   Using Blackwell's theorem, show that two countably generated sub-σ-fields of a Polish measure space, $(\Omega, \mathcal{F})$, are equal iff their atoms are equal. Further, if P is any probability on $\mathcal{F}$, show that for any sub-σ-fields, $\mathcal{G}_1, \mathcal{G}_2, d_P(\mathcal{G}_1, \mathcal{G}_2) = 0$ iff there exist countably generated sub-σ-fields, $\mathcal{G}_1', \mathcal{G}_2'$, with $d_P(\mathcal{G}_1', \mathcal{G}_1) = d_P(\mathcal{G}_2', \mathcal{G}_2) = 0$, where $\mathcal{G}_1'$ and $\mathcal{G}_2'$ have the same atoms.

9.13.b   A Spectacular Failure of Atoms

For some measure spaces, the conclusion of Blackwell's theorem fails spectacularly. In the next subsection we show how to construct a probability space, $(\Omega, \mathcal{F}, P)$, that is countably generated and contains a set $H \in \mathcal{F}$ and a countably generated sub-σ-field $\mathcal{B}$ with $At^{\mathcal{B}}(\omega) \equiv At^{\mathcal{F}}(\omega) \equiv \{\omega\}$, even though $P(H) = \frac{1}{2}$ and $H \perp\!\!\!\perp \mathcal{B}$, so that $d_P(\mathcal{F}, \mathcal{B}) = \frac{1}{2}$, even though they have exactly the same atoms. Further, rcp's need not exist.

Exercise 9.13.3   Suppose that $(\Omega, \mathcal{F}, P)$ is countably generated and contains a set $H \in \mathcal{F}$ with $P(H) = \frac{1}{2}$ and a countably generated sub-σ-field $\mathcal{B}$ with $At^{\mathcal{B}}(\omega) \equiv At^{\mathcal{F}}(\omega) \equiv \{\omega\}$ satisfying $H \perp\!\!\!\perp \mathcal{B}$. Show that if $\omega \mapsto P_{\omega}^{\mathcal{B}}$ is an rcp, then for all $A \in \mathcal{F}$, $P_{\omega}^{\mathcal{B}}(B) = 1_B(\omega)$ a.e. Show that $H \perp\!\!\!\perp \mathcal{B}$ implies that if $\omega \mapsto P_{\omega}^{\mathcal{B}}$ is an rcp, then $P_{\omega}^{\mathcal{B}}(H) = \frac{1}{2}$ a.e. Conclude that an rcp for $\mathcal{B}$ cannot exist.

We now show that the situation in Exercise 9.13.3 can arise. This takes quite a bit of work.

9.13.c   Nonmeasurable Sets

One of the basic properties of a Polish measure space $(\Omega, \mathcal{F}, P)$ is the ability to tightly sandwich all measurable sets from within by a compact set and from without by open sets. Specifcally, for any $E \in \mathcal{F}$ and $\epsilon > 0$, there is a compact K and an open G with $K \subset E \subset G$ and $P(G \setminus K) < \epsilon$. This is what leads to the ability of continuous functions to approximate the indicators of measurable functions, hence to approximate simple functions, hence to approximate measurable functions. This is what is behind continuous linear functions on the space of continuous functions determining probabilities.

Recall that the Lusin's theorem, 8.3.25 (p. 465), tells us that for any $\epsilon > 0$ and for any bounded measurable function, g, on a metric space, there is a continuous function, f, that has the same bounds and is equal to g with probability at least $1 - \epsilon$. The verbal shorthand for this result is that "every measurable function is nearly continuous."

Here, we start with the Polish space $([0, 1), \mathcal{B}, \lambda)$ and show how, using the axiom of choice, to prove the existence of an $H \subset [0, 1)$ such that for all $E \in \mathcal{B}$, compact, open or other, if $E \subset H$, then $\lambda(E) = 0$ and if $H \subset E$, then $\lambda(E) = 1$. This means that the tightest possible sandwiching would have the property that $E \subset H \subset F$ with $\lambda(F \setminus E) = 1$. We then extend λ to $\widehat{\lambda}$ on $\sigma(\{H\}, \mathcal{B})$, so that $H \perp\!\!\!\perp \mathcal{B}$ and $\widehat{\lambda}(H) = \frac{1}{2}$.

Pick your favorite *irrational* number $\theta \in \mathbb{R}$. Let $A = A(\theta) = \{m + n\theta : m, n \in \mathbb{Z}\}$, $B = B(\theta) = \{m + 2n\theta : m, n \in \mathbb{Z}\}$, and $C = C(\theta) = \{m + (2n + 1)\theta : m, n \in \mathbb{Z}\}$. Define the equivalence relation $\sim_A$ by $x \sim_A y$ if $x - y \in A$. Using the axiom of choice, we let H_0 contain exactly one element of each equivalence class so that $\mathbb{R} = H_0 + A$. The nonmeasurable subset is $H = (H_0 + B) \cap [0, 1)$ and its complement is $H^c = (H_0 + C)$. The rest of this subsection is aimed at showing how this works.

Exercise 9.13.4   Show that $m_1 + n_1\theta = m_2 + n_2\theta, m_1, m_2, n_1, n_2 \in \mathbb{Z}$, iff $m_1 = m_2$ and $n_1 = n_2$.

Note that $A \cap [0, 1) = [A \cap [m, m + 1)] - m$; that is, A is a countable union of translates of the same subset of $[0, 1)$, and the same is true for B and C.

Lemma 9.13.5 $A \cap [0, 1)$, $B \cap [0, 1)$, *and* $C \cap [0, 1)$ *are dense in* $[0, 1)$.

Proof.   We prove this statement for the set A. For each $i \in \mathbb{Z}$, let m_i be the unique m such that $m + i\theta \in [0, 1)$ and define $r_i = m_i + i\theta$. For any $i \neq j, \theta_i \neq \theta_j$, because if they are equal, then $m_i + i\theta = m_j + j\theta$, implying that $\theta = (m_j - m_i)/(i - j)$, that is, that θ is rational. Any open $G \subset [0, 1)$ contains an interval of length ϵ for some $\epsilon > 0$. If $\epsilon < 1/k$, then at least one pair of the numbers $r_1, \ldots, r_{k+1}$ must satisfy $|\theta_i - \theta_j| < \epsilon$. The numbers $m \cdot (\theta_i - \theta_j)$ belong to A and, translated back to $[0, 1)$, one of them must be in any interval of length ϵ.   ∎

Exercise 9.13.6   Show that $B(\theta) \cap [0, 1) = A(2\theta) \cap [0, 1)$ so that $B \cap [0, 1)$ is dense in $[0, 1)$. Show that C is also dense.

Notation 9.13.7   *For* $x \in \mathbb{R}$, $\lfloor x \rfloor$ *is the integer part of* x, *that is,* $\max\{n \in \mathbb{Z} : n \leq x\}$. *Thus, in the foregoing proof,* $\theta_i = i\theta - \lfloor i\theta \rfloor$.

Example 9.13.8 (Addition on the Circle)   *For* $r, s \in [0, 1)$, *we define addition and subtraction* mod(1) *in* $[0, 1)$ *by* $r \oplus s = (r + s) - \lfloor (r + s) \rfloor$ *and* $r \ominus s = (r - s) - \lfloor (r - s) \rfloor$. *One can imagine traveling around the circle on the x-y plane and arriving back at the same place. Changing units if need be, one starts at point 0 (which is $(1, 0)$ in the x-y plane), travels around a circle 1 unit of distance, and returns to the point 0. We now measure distance in $[0, 1)$ in a different fashion, as the shortest arc distance between two points. For example, the* *distance between* 0.9 *and* 0.1 *is* 0.2, *because one travels from* 0.9 *forward to* 0.1. *For* $x \leq y \in [0, 1)$, *define* $\xi(x, y) = \min\{y - x, (1 - y) + x\}$, *and for* $r, s \in [0, 1)$, *define* $d_o(r, s) = \xi(\min\{r, s\}, \max\{r, s\})$.

Exercise 9.13.9   Show that $([0, 1), d_o)$ is a compact metric space and that its Borel σ-field, $\mathcal{B}_d$, is the same as the usual one, $\mathcal{B}$.

Exercise 9.13.10   Show that the uniform distribution on $\mathcal{B}_d$ is translation invariant. That is, for all $E \in \mathcal{B}_d$ and all $r \in [0, 1)$, $\lambda(E \oplus r) = \lambda(E)$. [Think good sets.]

Define $x \sim_A y$ if $x - y \in A$. Each equivalence class contains an element in any interval subset of $[0, 1)$ because A is dense. Using the axiom of choice, we let $H_0 \subset [0, 1)$ be a set containing exactly one element of each equivance class.

Lemma 9.13.11   *If* $E \subset H_0$ *is measurable, then* $\lambda(E) = 0$. *Further,* H_0 *is not measurable.*

Proof.   For each distinct pair $a_1, a_2 \in A$, $\lfloor H_0 + a_1 \rfloor$ and $\lfloor H_0 + a_2 \rfloor$ are disjoint; if they are not, then r is rational. Thus, if a measurable $E \subset H_0$ has strictly positive measure, then the countable disjoint union $\cup_{a \in A} E \oplus a$ has infinite measure, since, by translation invariance, each set $E \oplus a$ has the same measure as E. Further, if H_0 is measurable, then $[0, 1)$ is a countable union of disjoint, measure 0 sets.   ∎

Exercise 9.13.12   We have seen that H_0 can be chosen to be a subset of (a, b) for any $0 \leq a < b \leq 1$. Show that H_0 can be chosen to be a dense subset of $[0, 1)$.

Recall that $B = B(\theta) = \{m + 2n\theta : m, n \in \mathbb{Z}\}$, $C = C(\theta) = \{m + (2n + 1)\theta : m, n \in \mathbb{Z}\}$, and that these disjoint sets are dense. Define $H = \lfloor H_0 + B \rfloor$ so that $H^c = \lfloor H_0 + C \rfloor$. Here is the antisandwiching result.

Theorem 9.13.13   *If* $E \subset H \subset F$ *and both* E *and* F *are measurable, then* $\lambda(E) = 0$ *and* $\lambda(F) = 1$.

All of the proofs of this are fairly elaborate because they involve subtle properties of measurable subsets of the real line. The following is the most direct one that we know. Another, which builds on ideas in the following lemma, is sketched in the next section.

Measurable subsets of $[0, 1)$ are "nearly" finite unions of intervals. The next lemma is immediate for any finite union of intervals and gives another sense in which measurable sets behave like finite unions of intervals.

Lemma 9.13.14   *If* $E \subset [0, 1)$ *and* $\lambda(E) > 0$, *then for all* $\alpha \in [0, 1)$, *there is an open interval,* $(c - r, c + r) \subset [0, 1)$, *such that* $\frac{\lambda(E \cap (c-r, c+r))}{\lambda((c-r, c+r))} \geq \alpha$.

Proof.   By sandwiching for Polish spaces, there is an open G such that $\alpha\lambda(G) \leq \lambda(E)$. G is a.e. the countable disjoint union of intervals, $U_n = (c_n - r_n, c_n + r_n)$, implying that $\alpha \sum_n \lambda(G_n) \leq \sum_n \lambda(E \cap G_n)$, which in turn means that at least one of the intervals satisfies $\lambda(E \cap (c - r, c + r)) \geq \alpha\lambda(c - r, c + r)$.   ∎

Again, measurable subsets of $[0, 1)$ are "nearly" finite unions of intervals, and the following is immediate for any finite union of intervals.

Lemma 9.13.15 *If $\lambda(E) > 0$, then $E \ominus E$ contains an open interval around 0 in the metric space $([0, 1), d_o)$.*

Proof. Pick an interval $(c - r, c + r)$ such that $\lambda(E \cap (c - r, c + r)) \geq \frac{3}{4}\lambda((c - r, c + r))$. The requisite interval is $(0 \ominus \frac{1}{2}r, 0 \oplus \frac{1}{2}r)$. To see why, pick any $x \in (0 \ominus \frac{1}{2}r, 0 \oplus \frac{1}{2}r)$. It is sufficient to show that $[E \cap (c - r, c + r)]$ and $[E \cap (c - r, c + r)] \oplus x$ are not disjoint. If they were, then by translation invariance, $\lambda([E \cap (c - r, c + r)] \cup [E \cap (c - r, c + r)] \oplus x) \geq \frac{3}{2}\lambda(c - r, c + r)$, contradicting $[E \cap (c - r, c + r)] \cup [(E \cap (c - r, c + r)) \oplus x] \subset U \cup [U \oplus x]$ and the observation that the interval $U \cup [U \oplus x]$ has length less than $\frac{3}{4}r$. ∎

Proof of Theorem 9.13.13. Suppose that $E \subset H$ and E is measurable. If $\lambda(E) > 0$, then by Lemma 9.13.15, $E \ominus E$ contains an open interval. Since C is dense, $E \ominus E$ must contain an element of C, a contradiction. In exactly the same fashion, if $F \subset H^c$ and F is measurable, then $\lambda(F) = 0$. When we combine, if measurable sets E, F satisfy $E \subset H \subset F$, then $\lambda(E) = 0$ and $\lambda(F) = 1$. ∎

9.13.d Returning to rcp's

At long long last, we give the measure space that showed us that rcp's need not exist.

Example 9.13.16 *We start with the Polish space $([0, 1), \mathcal{B}, \lambda)$ and let $\mathcal{F} = \sigma(\mathcal{B}, \{H\}) = \{[H \cap E_1] \cup [H^c \cap E_2] : E_1, E_2 \in \mathcal{B}\}$, where H is the nonmeasurable subset of $[0, 1)$ having the properties of Theorem 9.13.13. We define $\widehat{\lambda}([H \cap E_1] \cup [H^c \cap E_2]) = \frac{1}{2}\lambda(E_1) + \frac{1}{2}\lambda(E_2)$. Taking $E_1 = \Omega$, $E_2 = \emptyset$, we see that $\widehat{\lambda}(H) = \frac{1}{2}$. Further, for any $E \in \mathcal{B}$, $\widehat{\lambda}(H \cap E) = \widehat{\lambda}(H) \cdot \widehat{\lambda}(E)$, so that $H \perp\!\!\!\perp \mathcal{B}$. Since $\mathcal{B} \subset \mathcal{F}$ and points are the $\mathcal{B}$-atoms, points must also be $\mathcal{F}$-atoms.*

9.14 ◆ Bibliography

Historically, the material in this chapter was first pulled together in textbook form in K. R. Parthasarathy's *Probability Measures on Metric Spaces* (New York: Academic Press, 1967) and P. Billingsley's *Convergence of Probability Measures* (New York: Wiley, 1968). D. Pollard's *Convergence of Stochastic Processes* (New York: Springer-Verlag, 1984) is an updated version of this approach that leads one more directly into current research using probability theory in econometrics.

CHAPTER 10

Infinite-Dimensional Convex Analysis

As we saw in Chapter 5, convexity plays a crucial role throughout the parts of economic theory that use $\mathbb{R}^\ell$. In Chapter 8, convexity reappeared in a central role for infinite-dimensional normed vector spaces. The main result related to convexity was the Hahn-Banach theorem.

We now turn to convexity in yet more general vector spaces and, again, the Hahn-Banach theorem plays a central role. For this, we require topologies that are sometimes not metric, and we begin with a brief coverage of the requisite topological concepts. The basic wisdom here is that well-behaved topologies have almost the same useful properties as metric topologies, but the proofs are a bit harder since they must be based on more general, more fundamental properties, Theorems 10.2.13 and 10.2.16.

We then turn to the basic geometry of convexity and how it interacts with topology, especially with continuous linear functions.

Throughout: V is a vector space. When V is also a metric space, we assume that each $B_r(x)$ is a convex set; if we write $\|x\|$ for $x \in V$, we have assumed that there is a norm on V and that the metric is $d(x, y) = \|x - y\|$; these explicitly include the cases $V = \mathbb{R}^\ell$ and $V = \mathbb{H}$; except in clearly marked examples, V is complete, hence a Banach space if it is normed; $\mathbb{H}$ always denotes a Hilbert space, that is, a Banach space with norm given by an inner product.

The metric spaces are the ones for which the topology, that is, the class of open subsets, can be given by a metric. We care about nonmetric topological spaces as they have arisen in the weak and the weak* topologies; the so-called continuum law of large numbers; the resolution of the paradoxes that arise from failures of countable additivity; and in representation theorems in stochastic process theory.

10.1 ◆ Topological Spaces

Metric spaces are special cases of topological spaces. It is rare that economists need more generality than metric spaces, but it does sometimes happen. Most of our intuitions for metric spaces apply to the nonmetric topological spaces that we care about. The hard part is knowing when they do and when they do not. To help with this, we look at some rather weird nonmetric spaces to get an idea of where the pitfalls might be.

10.1.a Some Reminders

Since it has been a while, some reminders: a **metric on** M, M nonempty, is a function $d : M \times M \to \mathbb{R}_+$ such that for all $x, y, z \in M$, $d(x, y) = d(y, x)$, $d(x, y) = 0$ iff $x = y$, and $d(x, y) + d(y, z) \geq d(x, z)$, the **triangle inequality**. A **metric space** is a pair (M, d) where M is nonempty and d is a metric on M. For a metric space (M, d), the **metric topology** $\tau_d \subset \mathcal{P}(M)$ is the class of open sets, that is, $\tau_d = \{G \subset M : \forall x \in G \, \exists \epsilon > 0 \, B_\epsilon(x) \subset G\}$. The metric topology has the following properties:

1. $\emptyset, M \in \tau_d$,
2. τ_d is closed under arbitrary intersections,
3. τ_d is closed under finite intersections, and
4. for all $x, y \in M$, there exist $G_x, G_y \in \tau_d$ such that $x \in G_x$, $y \in G_y$, and $G_x \cap G_y = \emptyset$.

The first three properties of metric topologies are the definition of a class of sets being a topology. The fourth is a separation property, which is almost always also true about the topologies that we care about, but to see what it means, we give examples of topologies failing it.

10.1.b Topologies

We begin with the following.

Definition 10.1.1 *A pair (X, τ) is a topological space if X is nonempty and $\tau \subset \mathcal{P}(X)$ satisfies*

1. *$\emptyset, X \in \tau$,*
2. *τ is closed under arbitrary intersections, and*
3. *τ is closed under finite intersections,*

*in which case, τ is called a **topology on** X. Any $G \in \tau$ is called a τ-**open set** (or simply an **open set**). A set $F \subset X$ is τ-**closed** (or simply closed) if $X \setminus F$ is open. If $x \in G \in \tau$, then G is an **open neighborhood** of x, and if U contains an open neighborhood of x, then it is a **neighborhood** of x.*

Notation Alert 10.1.A *Neighborhoods need not be open, but open neighborhoods must be open.*

If we have a metric on M and $E \subset M$, then restricting the metric to $E \times E$ gives another metric space, and the open sets on E are of the form $G \cap E$, where G is an open subset of M. This idea translates directly.

Definition 10.1.2   *For a topological space (X, τ) and nonempty $E \subset X$, we define the topological space $(E, \tau_{|E})$ by setting $\tau_{|E} = \{G \cap E : G \in \tau\}$. $\tau_{|E}$ is called the **relative topology on** E and $(E, \tau_{|E})$ is called a **topological subspace of** X.*

Since arbitrary unions of open sets are open, arbitrary intersections of closed sets are closed. This means that the shrink-wrapping characterization of closed subsets of metric spaces carries over directly.

Definition 10.1.3   *For $C \subset X$, the **closure of** C, written $\operatorname{cl}(C)$, is defined as $\bigcap \{F : C \subset F, \ F \text{ closed}\}$. The **interior of** C, written $\operatorname{int}(C)$, is defined as $\bigcup \{G : G \subset C, \ G \text{ open}\}$.*

Thus, the closure of C is the smallest closed set containing C and the interior of C is the largest open set contained in C.

Definition 10.1.4   *A set K is **compact** if every open cover has a finite subcover.*

Lemma 10.1.5   *If (X, τ) is compact and F is a closed subset of X, then F is compact.*

Proof.   Let $\mathcal{G}$ be an open cover of a closed $F \subset X$. $\mathcal{G} \cup \{F^c\}$ is an open cover of X, hence has a finite subcover. Take F^c out of the finite subcover of X to find a finite subcover of F.   ∎

10.1.c   Pseudometrics and the Failure to Separate

Recall that a pseudometric only fails to be a metric by allowing $d(x, y) = 0$ for some $x \neq y$. Any pseudometric space is also a topological space. The reverse is not true.

Definition 10.1.6   *A topological space (X, τ) is **metrizable (respectively pseudometrizable)** if there exists a metric d (respectively a pseudometric) on X such that $\tau = \tau_d$.*

Not all topological spaces are metrizable. As we saw in Example 8.9.7, the weak topology on L^p spaces is a topology and is not metrizable. Throughout the following examples, X is a nonempty set containing at least two elements.

Example 10.1.7   *The class $\tau = \mathcal{P}(X)$ is called the **discrete topology**. It is the largest possible topology since every topology is a subset of $\mathcal{P}(X)$. As we saw in §4.2, it can be metrized by $d(x, x) = 0$ and $d(x, y) = 1$ for $x \neq y$.*

Example 10.1.8   *The class $\tau_0 = \{\emptyset, X\}$ is called the **trivial topology** on X. If τ is any other topology on X, then $\tau_0 \subset \tau$, so that the trivial topology is the smallest possible topology. This topology is not metrizable because it fails the first separation property. However, it is **pseudometrizable** by $d(x, y) \equiv 0$.*

Example 10.1.9   *Suppose that $X = \{a, b\}$. $\tau = \{\emptyset, \{a\}, X\}$ is called the **Sierpinski topology**. It is not metrizable or pseudometrizable.*

Example 10.1.10   *Suppose that X is an infinite set. $\tau := \{G : X \setminus G \text{ is finite}\}$ is called the **finite complement topology**.*

Exercise 10.1.11 Show that the finite complement topology is not metrizable and that the closure of a set C is

$$\operatorname{cl}(C) = \begin{cases} C \text{ if } C \text{ is finite,} \\ X \text{ if } C \text{ is infinite.} \end{cases}$$

10.1.d Continuity

Continuity is defined in what ought to look like an obvious generalization of metric space continuity.

Definition 10.1.12 *For topological spaces (X, τ_X) and (Y, τ_Y), a function $f : X \to Y$ is τ_Y/τ_X continuous, or simply continuous, if $f^{-1}(\tau_Y) \subset \tau_X$. When $(Y, \tau_Y) = (\mathbb{R}, \tau_E)$, the set of continuous functions is denoted $C(X)$ and the set of continuous bounded functions is denoted $C_b(X)$.*

Since $f^{-1}(Y \setminus A) = X \setminus f^{-1}(A)$, we immediately have the following.

Lemma 10.1.13 *$f : X \to Y$ is continuous iff $f^{-1}(F)$ is a τ_X-closed subset of X for all τ_Y-closed $F \subset Y$.*

Lemma 10.1.14 *If $f : X \to Y$ is continuous and X is compact, then $f(X)$ is compact.*

Proof. f^{-1} of an open cover of $f(X)$ is an open cover of X, which has a finite subcover, which gives a finite subcover in the range space. ■

Exercise 10.1.15 (A Fixed–Set Result) Let (X, τ) be a compact Hausdorff space and $f : X \to X$ a continuous function. Show that there exists a nonempty closed $F \subset X$ such that $f(F) = F$. [For metric spaces, one could start from any closed F_0, define $F_{n+1} = f(F_n)$, and pull out a convergent subsequence. What happens if you start with $F_0 = X$ in this case? Think of the finite intersection property.]

Definition 10.1.16 *If $f : (X, \tau_X) \leftrightarrow (Y, \tau_Y)$ is continuous, one-to-one and onto and has a continuous inverse, then f is a **homeomorphism** and the spaces (X, τ_X) and (Y, τ_Y) are **homeomorphic**.*

Isometries are a way of saying that two spaces are the same except for a change of names. Homeomorphisms have very similar properties because $\tau_X = f^{-1}(\tau_Y)$ and $\tau_Y = f(\tau_X)$.

Exercise 10.1.17 If f is a homeomorphism between the metric spaces (X, d_X) and (Y, d_Y), then there is a metric, ρ_Y, on Y that is equivalent to d_Y for which f is an isometry. Give the metric ρ_Y in the case $X = \mathbb{R}$, $Y = (0, 1)$, and $f(x) = e^x/(1 + e^x)$.

The larger τ_X, the more functions are continuous. Here are the extreme cases.

Example 10.1.18 *If $f : X \to Y$ is constant, then it is continuous for all τ_Y and all τ_X. If τ_X is the discrete topology, then for any τ_Y, any $f : X \to Y$ is continuous.*

More generally, we have the following results.

Exercise 10.1.19 Show that:

1. If $f : X \to Y$ is τ_Y/τ_X continuous and $\tau'_X \supset \tau_X$ is another topology on X, then $f : X \to Y$ is τ_Y/τ'_X continuous.

2. If $\tau''_X = f^{-1}(\tau_Y)$, then $f : X \to Y$ is τ_Y/τ''_X continuous.

3. If d and ρ are metrics on X, then $[\tau_d \subset \tau_\rho] \Leftrightarrow [[\rho(x_n, x) \to 0] \Rightarrow [d(x_n, x) \to 0]]$.

Constant functions are always continuous. Sometimes they are the only continuous functions.

Exercise 10.1.20 Let $X = \{x, y\}$ and let τ be the Sierpinski topology. Show that $f : X \to \mathbb{R}$ is continuous iff it is constant.

Here is another case where having a small topology cripples the set of continuous $\mathbb{R}$-valued functions.

Exercise 10.1.21 Let X be an infinite set and let τ be the finite complement topology. Show that $f : X \to \mathbb{R}$ is continuous iff it has finite range and for exactly one r in the range of f, $f^{-1}(r)$ is infinite.

10.1.e Induced Topologies

One way to define a topology is by specifying that we want a particular function to be continuous.

Example 10.1.22 *Let $f : X \to \mathbb{R}$ and define $\tau_f = f^{-1}(\tau_\mathbb{R})$. This is the smallest topology making f continuous, and τ_f is metrizable, for example, by $d(x, y) = |f(x) - f(y)|$, iff f is invertible; otherwise it is pseudometrizable.*

Sometimes we want a particular class of functions on X to be continuous. We did this systematically for sup norm separable classes of functions using compact embeddings in §6.7.d. Here we generalize this idea by giving X what we call an **induced topology**. The first step toward defining induced topologies is the following.

Lemma 10.1.23 *If $\{\tau_\alpha : \alpha \in A\}$ is a class of topologies on X, then $\tau := \cap_\alpha \tau_\alpha$ is a topology.*

Definition 10.1.24 *Let $\mathcal{S} \subset \mathcal{P}(X)$. The **topology generated by** $\mathcal{S}$ is defined by $\tau(\mathcal{S}) = \cap\{\tau : \mathcal{S} \subset \tau, \ \tau \text{ is a topology on } X\}$.*

The topology generated by S is well defined because $\mathcal{S} \subset \mathcal{P}(X)$ and $\mathcal{P}(X)$ is a topology (indeed, $\mathcal{P}(X)$ is the discrete topology). The topology generated by $\mathcal{S}$ is another instance of shrink wrapping—it is the smallest topology containing $\mathcal{S}$.

Lemma 10.1.25 *$G \in \tau(\mathcal{S})$ iff $G = \emptyset$, $G = X$, or G can be expressed as a union of finite intersections of sets in $\mathcal{S}$.*

Proof. If τ is a topology containing S, then τ must contain all unions of finite intersections of elements of $\mathcal{S}$. Therefore it is sufficient to show that τ', the class of sets consisting of ∅, X and all unions of finite intersections of elements of S, is in fact a topology. We arranged τ' to contain ∅ and X. Since a finite intersection of finite intersections is a finite intersection and a union of unions is a union, τ' is a topology. ∎

Definition 10.1.26   *Let* $\mathbb{F} = \{f_\alpha : \alpha \in A\}$ *be a class of functions mapping* X *to* $\mathbb{R}$. *The **topology on** X **induced by** $\mathbb{F}$ is denoted $\tau(\mathbb{F})$ and defined as $\tau(S)$ where $S = \{f_\alpha^{-1}(\tau_\mathbb{R}) : \alpha \in A\}$.*

Another way to put this is to say that $\tau(\mathbb{F})$ is the smallest topology on X making each $f_\alpha \in \mathbb{F}$ continuous.

Exercise 10.1.27   Let (M, d) be a metric space and let $\mathbb{F} = C_b(M)$. Show that $\tau_d = \tau(\mathbb{F})$.

Exercise 10.1.28   Let (M, d) be compact metric space and let $\mathbb{F} = \{f_n : n \in \mathbb{N}\}$ be a countable dense subset of $C(M)$. Show that $\tau_d = \tau(\mathbb{F})$ and that $\rho(x, y) := \sum_n \frac{1}{2^n} \min\{|f_n(x) - f_n(y)|, 1\}$ is equivalent to d.

Exercise 10.1.29   If X is infinite and has the finite complement topology, characterize the topology $\tau(C_b(X))$.

Exercise 10.1.30   If $X = \{a, b\}$ has the Sierpinski topology, characterize the topology $\tau(C_b(X))$.

Let us turn away from the last two perverse examples and return to metric spaces.

Example 10.1.31   *Let* $M = \times_{a \in A} M_a$ *be a countable product of metric spaces. The product metric topology on* M *is the smallest topology making each of the projections continuous.*

The following generalizes the product metric topology to arbitrary Cartesian products of topological spaces. Perhaps the easiest way to proceed is to review the definition of countable products of metric spaces and check that this definition agrees with the metrizable product topology defined there.

Definition 10.1.32   *Let* $(X_\alpha, \tau_\alpha)_{\alpha \in A}$ *be a collection of topological spaces. Let* $X = \Pi_{\alpha \in A} X_\alpha$ *be the Cartesian product of the* X_α's. *For each* $\alpha \in A$, *define the canonical projection map by* $\text{proj}_\alpha(x) = x_\alpha$ *and let* $\mathbb{P}$ *denote the set of projections. The **product topology on** X is $\tau(\mathbb{P})$, that is, the weakest topology that makes each projection mapping continuous.*

Recall that it is the axiom of choice that guarantees that $X \neq \emptyset$.

Notation Alert 10.1.B   *The possibility of having uncountable products is what led us to change from the Cartesian product notation* $\times_{\alpha \in A} X_\alpha$ *to* $\Pi_{\alpha \in A} X_\alpha$.

If (X, d_X) and (Y, d_Y) are two metric spaces, then one metric giving the product topology is $d((x, y), (x', y')) = \max\{d_X(x, x'), d_Y(y, y')\}$. Here is the general version of this observation.

Exercise 10.1.33   If (X, τ_X) and (Y, τ_Y) are two topological spaces, then $G \subset (X \times Y)$ is open in the product topology iff for all $(x, y) \in G$, there exist $G_x \in \tau_X$ and $G_y \in \tau_Y$ such that $(x, y) \in (G_x \times G_y) \subset (X \times Y)$.

In the case of a countable product of metric spaces, the class of cylinder sets with open base sets played a crucial role in §6.6.c. Here is the generalization of that role.

Exercise 10.1.34 Let $X = \Pi_{\alpha \in A} X_\alpha$ be the product of a collection of topological spaces, (X_α, τ_α), with the product topology. Show that $G \subset X$ is open iff for all $x \in G$ there is a finite collection $A_F \subset A$ and for each $a \in A_F$ an open $G_a \in \tau_a$ such that $x \in \text{proj}_{A_F}^{-1}(\times_{a \in A_F} G_a)$ with $\text{proj}_{A_F}^{-1}(\times_{a \in A_F} G_a) \subset G$.

The finiteness is not special to product spaces.

Exercise 10.1.35 Let $x \in G \in \tau(\mathbb{F})$, where $\tau(\mathbb{F})$ is an induced topology on X. Then there exists a finite set of $f \in \mathbb{F}$ and a finite collection G_f of open sets such that $x \in \cap_f f^{-1}(G_f) \subset G$.

10.1.f Two Separation Axioms

The trivial, the Sierpinski, and the finite-complement topology were rather pitifully small, and as a result, had very few continuous functions. They all violate both of the following conditions, the second of which guarantees that we have as many continuous functions as we need.

Definition 10.1.36 *A topological space (X, τ)*

1. *is **Hausdorff, or** T_2, if for all $x \neq y$, there exists U, $V \in \tau$ such that $x \in U$, $y \in V$, and $U \cap V = \emptyset$, and*

2. *is **normal, or** T_4, if it is Hausdorff and for all closed sets F, F' with $F \cap F' = \emptyset$, there exists U, $V \in \tau$ such that $F \subset U$, $F' \subset V$, and $U \cap V = \emptyset$.*

Every metric space is Hausdorff—to see why, take $G_x = B_\epsilon(x)$ and $G_y = B_\epsilon(y)$ for $0 < \epsilon \le d(x, y)/2$.

Exercise 10.1.37 Show that every metric space is normal.

Pseudometric spaces that are not metric provide a large class of non-Hausdorff topological spaces.

Exercise 10.1.38 Let $\mathbb{F} = \{f_\alpha : \alpha \in A\}$ be a class of functions mapping X to $\mathbb{R}$. $\mathbb{F}$ **separates points** if for all $x \neq y \in X$, there exists an $f \in \mathbb{F}$ such that $f(x) \neq f(y)$. Show that $\tau(\mathbb{F})$ is Hausdorff iff $\mathbb{F}$ separates points.

Exercise 10.1.39 Show that if (X, τ) is Hausdorff, then every finite $F \subset X$ is closed.

Metric spaces are not the only normal ones.

Lemma 10.1.40 *If (X, τ) is compact and Hausdorff, then it is normal.*

Proof. Fix an arbitrary $x \in F$. We first prove the existence of an open set $G'(x)$ containing F' and an open G_x containing x that is disjoint from $G'(x)$.

For each $x' \in F'$, let $G_x(x')$ and $G'_{x'}$ be disjoint open sets containing x and x', respectively. The collection $\{G'_{x'} : x' \in F'\}$ is an open cover of F'. F' is compact because it is a closed subset of a compact space; hence there is a finite subcover,

$\{G'_{x'_i} : i = 1, \ldots, I\}$. Define $G'(x) = \cup_i G'_{x'_i}$ and $G_x = \cap_i G_x(x'_i)$. Since G_x is a finite intersection of open sets, it is open. By construction, it is disjoint from $G'(x)$.

We now complete the proof by showing the existence of a pair of disjoint open sets, $G \supset F$ and $G' \supset F'$. The collection $\{G_x : x \in F\}$ is an open covering of F. Take a finite subcovering, $\{G_{x_j} : j = 1, \ldots J\}$. Define $G = \cup_j G_{x_j}$ and $G' = \cap_j G'(x_j)$. ∎

Urysohn's lemma shows that normal spaces have a rich set of continuous functions, in particular, that they approximate the indicators of closed sets. This ties continuous functions to probabilities on metric spaces, and Chapter 9 should have convinced you that this is an important idea.

Lemma 10.1.41 (Urysohn) *If (X, τ) is normal, $F \subset G$, F closed, and G open, then there exists an $f \in C(X)$ such that $f : X \to [0, 1]$ and $\forall x \in F$, $f(x) = 1$ and for all $y \in G^c$, $f(y) = 0$.*

Exercise 10.1.42 Fill in the details in the following sketch of a proof of Urysohn's lemma. The first time through, metric space intuitions are useful.

1. In any Hausdorff space, points are closed; that is, $\{x\}$ is a closed set for all $x \in X$.

2. For each $x \in F$, show that we can pick a disjoint pair of open sets U_x, V_x, such that $x \in U_x$ and $G^c \subset V_x$.

3. Define $G_0 = \cup_{x \in F} U_x$. Show that $F \subset G_0 \subset G$.

For dyadic rationals $q = k/2^n$ strictly between 0 and 1; that is, with $k \in \{1, \ldots, 2^n - 1\}$, we inductively define a collection of closed and open sets $G_q \subset F_q$ with the property that $q < q'$ implies that

$$G_0 \supset F_q \supset G_q \supset F_{q'} \supset G_{q'} \supset F_1 := F. \tag{10.1}$$

For any $x \in G_0$, the function f is defined as the supremum of the dyadic rationals, $k/2^n$, such that $x \in F_{k/2^n}$, and for $x \in G_0^c$, f is set equal to 0. This yields $f(x) = 1$ for $x \in F_1 = F$ and $f(x) = 0$ for all $x \in G^c$, and all that is left is to show that f is in fact continuous.

4. The initial step: Show that there exists an open set $G_{1/2}$ with closure $F_{1/2}$ such that $F_1 \subset G_{1/2} \subset F_{1/2} \subset G_0$.

5. The inductive step: Suppose that for all dyadic rationals, $1 > q_1 > \cdots > q_{2^n - 1} > 0$ of order n, we have open and closed sets with $F_1 \subset G_{q_1} \subset F_{q_1} \subset \cdots \subset G_{q_{2^n - 1}} \subset F_{q_{2^n - 1}} \subset G_0$. Show that we can define open and closed G's and F's for the dyadic rationals of order $n + 1$ that are not of order n so that (10.1) holds for all dyadic rationals of order $n + 1$.

6. Define $f(x) = \sup\{q : x \in F_q\}$. Show that $\{x : f(x) > r\} = \cup_{q > r} G_q$ is open and that $\{x : f(x) < r\} = \cup_{q < r} F_q^c$ is also open, so that f is continuous.

10.2 ◆ Locally Convex Topological Vector Spaces

We work with vector spaces over $\mathbb{R}$. In this way, we do our part to help perpetuate economists' irrational fears of complex numbers. The following relabels Definition 4.3.1 as being part of the present chapter.

Definition 10.2.1   *A set $\mathfrak{X}$ is a **vector space (vs)** (over $\mathbb{R}$) if for all x, y, $z \in \mathfrak{X}$ and all α, $\beta \in \mathbb{R}$ the mappings $(x, y) \mapsto x + y$ from $\mathfrak{X} \times \mathfrak{X}$ to $\mathfrak{X}$ and $(r, x) \mapsto rx$ from $\mathbb{R} \times \mathfrak{X}$ to $\mathfrak{X}$ satisfy $x + y = y + x$; $(x + y) + z = x + (y + z)$; the point $0 \in \mathfrak{X}$ is defined as $0z$ for $0 \in \mathbb{R}$, is called the "origin," and satisfies $y + (-1)y = 0$ and $0 + x = x$; $(\alpha\beta)x = \alpha(\beta x)$; and $(\alpha + \beta)x = \alpha x + \beta x$.*

If we have a topology τ on $\mathfrak{X}$, then we always give $\mathfrak{X} \times \mathbb{R}$ and $\mathfrak{X} \times \mathfrak{X}$ the corresponding product topologies. What distinguishes vector space topologies on $\mathfrak{X}$ from other topologies is that addition and multiplication by constants are continuous.

Definition 10.2.2   *A $(\mathfrak{X}, \tau)$ is called a **topological vector space (tvs)** if $\mathfrak{X}$ is a vector space and the mappings $(x, r) \mapsto rx$ from $\mathbb{R} \times \mathfrak{X}$ to $\mathfrak{X}$ and $(x, y) \mapsto x + y$ from $\mathfrak{X} \times \mathfrak{X}$ to $\mathfrak{X}$ are continuous. A tvs is a **locally convex tvs (lctvs)** if for all $x \in \mathfrak{X}$ and all open neighborhoods $x \in U_x \in \tau$, there is an open convex G_x with $x \in G_x \subset U_x$.*

10.2.a   Examples

Seminorms appear in all of the following. This is not an accident.

Definition 10.2.3   *For a vector space $\mathfrak{X}$, a function $m : \mathfrak{X} \to \mathbb{R}_+$ is a **seminorm** if $m(x + y) \leq m(x) + m(y)$ and $m(\alpha x) = |\alpha| m(x)$ for all x, $y \in \mathfrak{X}$ and $\alpha \in \mathbb{R}$. A seminorm is a **norm** if $[m(x) = 0] \Rightarrow [f = 0]$.*

A seminorm is convex, that is $m(\alpha x + (1 - \alpha)y) \leq m(\alpha x) + m((1 - \alpha)y) = \alpha m(x) + (1 - \alpha)m(y)$. Further, since $m(x) \leq m(y) + m(x - y)$ and $m(y) \leq m(x) + m(y - x)$, $|m(x) - m(y)| \leq m(x - y)$. The only difference between a seminorm and a sublinear function is that a seminorm is symmetric, that is, $m(x) = m(-x)$.

Example 10.2.4   *For (X, τ) a topological space, $C_b(X)$ denotes the set of continuous functions with the sup norm metric and the associated topology τ_∞. $(C_b(X), \tau_\infty)$ is a locally convex topological vector space (because it is a Banach space.)*

The class of functions in the next example has proved useful in the generalizations of the central limit theorem that tell us about the possible limit distributions of sums of small independent random variables. The simplest example of sums of small independent random variables that do not have a Gaussian limit has for each $n \in \mathbb{N}$, $\{X_{n,i} : i = 1, \ldots n\}$, an iid collection of Bernoullis with $P(X_{n,i} = 1) = \mu/n$ and defines $Y_n := \sum_i X_{n,i}$. Y_n is a sum of small independent random variables and $Y_n \to_{w^*} Y$, where Y is a Poisson with parameter μ. The important part of the $X_{n,i}$ is violating the Lindeberg conditions for the Gaussian central limit theorem, and any $X_{n,i}$ that satisfies the Lindeberg conditions eventually has no effect when integrated against any element in the following.

Example 10.2.5   *Let $\mathfrak{X}$ denote the set of bounded continuous functions, $f : \mathbb{R} \to \mathbb{R}$ for which there exists some $\epsilon > 0$ with $f(B_\epsilon(0)) = 0$. With the sup norm, this is an incomplete normed space, hence a lctvs. For each $n \in \mathbb{N}$, define $m_n(f) = \sup\{|f(x)| : |x| \geq 1/n\}$; τ is the topology induced by the collection $\{m_n : n \in \mathbb{N}\}$. Each of the m_n is a seminorm.*

Exercise 10.2.6   For $\mathfrak{X} = C(\mathbb{R})$ and $r \in \mathbb{R}$, define $\mathrm{proj}_r : \mathfrak{X} \to \mathbb{R}$ by $\mathrm{proj}_r(f) = f(r)$. Show that the collection $\mathbb{F} = \{\mathrm{proj}_r : r \in \mathbb{R}\}$ separates points and that $(C(\mathbb{R}), \tau(\mathbb{F}))$ is a lctvs. For each $r \in \mathbb{R}$, $m_r(f) := |f(r)|$ is a seminorm. [This is called **topology of pointwise convergence** and is very different than measure theoretic a.e. convergence.]

Exercise 10.2.7   For $\mathfrak{X} = C(\mathbb{R})$ and K a compact subset of $\mathbb{R}$, define $d_K(f, g) = \max_{x \in K} |f(x) - g(x)|$. Show that d_K is not a metric. Setting $K_n = [-n, +n]$ and $d(f, g) := \sum_{n \in \mathbb{N}} \frac{1}{2^n} \min\{1, d_{K_n}(f, g)\}$ makes $(\mathfrak{X}, \tau_d)$ into a locally convex topological vector space. Show the following.

1. $\tau_d \not\supseteq \tau(\mathbb{F})$, where $\tau(\mathbb{F})$ is the locally convex topology from the previous problem.

2. There is no norm on $C(\mathbb{R})$ giving this topology. [Suppose there were such a norm, $\|\cdot\|$; let U be the open unit ball in this norm. Show that there must exist an $n \in \mathbb{N}$ and an $\epsilon > 0$ with the property that $\{f : d_{K_n}(0, f) < \epsilon\}$. By scale invariance of everything we are using here, $\{f : d_{K_n}(0, f) < \epsilon/n\} \subset \frac{1}{n} U$ for all $n \in \mathbb{N}$. If $\|\cdot\|$ is to be a norm, then $\cap_n \frac{1}{n} U = \{0\}$. Conclude that $[d_{K_n}(0, f) = 0] \Rightarrow [f = 0]$, which is ridiculous.]

3. For each K_n, $m_n(f) := \max_{x \in K_n} |f(x)|$ is a seminorm and τ_d is the topology induced by the collection of seminorms m_n.

Exercise 10.2.8   Let $C^{(m)}(K)$ denote the set of functions f on a compact $K \subset \mathbb{R}^\ell$ with the property that there is an open neighborhood of K such that f is m times continuous differentiable on that open neighborhood. For $\alpha \in \{0, \ldots, m\}^\ell$ with $|\alpha| := \sum_i \alpha_i \leq m$, define $\|f\|_\alpha = \max_{x \in K} |D^\alpha f|$ and $\|f\|_m = \sum_{|\alpha| \leq q} \|f\|_\alpha$. Show that $\|\cdot\|_m$ is a norm on $C^{(m)}(K)$, and that each $\|\cdot\|_\alpha$ is a seminorm.

10.2.b Absorbent Convex Sets and Seminorms

If U is an open neighborhood of 0 in a normed vector space $\mathfrak{X}$, then $\mathfrak{X} = \cup_{\alpha > 0} \alpha U$. Further, U contains an open neighborhood, G, of 0, namely some $B_\delta(0)$, with the property that $G = -G$.

For any tvs, if $G = T^{-1}(B_\epsilon(0))$ for a continuous linear $T : \mathfrak{X} \to \mathbb{R}$, then G is an open neighborhood of 0, $\mathfrak{X} = \cup_{\alpha > 0} \alpha G$, and $G = -G$.

The following names these two properties.

Definition 10.2.9   *For a vector space $\mathfrak{X}$, a convex set K is **balanced** if $K = -K$. A balanced convex set K is **absorbent** if $\mathfrak{X} = \cup_{\alpha > 0} \alpha K$.*

Any balanced convex set must contain 0.

Definition 10.2.10   *The **Minkowski gauge of an absorbent set** is $m_K(x) = \inf\{\alpha \geq 0 : x \in \alpha K\}$.*

Lemma 10.2.11   *If m is a seminorm, then the sets $\{x : m(x) \leq 1\}$ and $\{x : m(x) < 1\}$ are balanced convex absorbent sets and the Minkowski gauge m_K of a balanced convex absorbent set K is a seminorm with $\{x : m(x) < 1\} \subset K \subset \{x : m(x) \leq 1\}$.*

Proof.   The first statement is immediate. K being absorbent guarantees that $m_K(x) \in \mathbb{R}_+$. For the last part, suppose that $x \in rK$ and $y \in sK$ for $r, s > 0$. Then $x + y \in rK + sK$, and since $rK + sK = (r+s)K$, $m_K(x+y) \leq r + s$.   ∎

10.2.c   Weak Topologies

As we have seen, normed vector spaces are lctvs's with their norm topologies. They are also lctvs's with their weak topologies.

Definition 10.2.12   *Let $\mathfrak{X}$ be a normed vector space and $\mathfrak{X}^*$ its dual. The **weak topology on** $\mathfrak{X}$, denoted τ_w, is the weakest topology making each $T \in \mathfrak{X}^*$ continuous; that is, it is $\tau(\mathfrak{X}^*)$, the topology on $\mathfrak{X}$ induced by $\mathfrak{X}^*$.*

The Hausdorff part of the following comes from the Hahn-Banach theorem.

Theorem 10.2.13   *If $\mathfrak{X}$ is a normed vector space, then $(\mathfrak{X}, \tau_w)$ is a Hausdorff lctvs.*

Proof.   If $x \neq y$, then there exists $T \in \mathfrak{X}^*$ and $\epsilon > 0$ such that $r = T(x) > T(y) + 3\epsilon$. The disjoint open sets in τ_w can be taken to be $G_x = T^{-1}(r - \epsilon, \infty)$ and $G_y = T^{-1}(-\infty, r - 2\epsilon)$.

If $x \in G \in \tau(\mathfrak{X}^*)$, then there exists a finite collection $\{T_i : i = 1, \ldots, I\} \subset \mathfrak{X}^*$ and a finite collection of open sets $G_i \subset \mathbb{R}$, such that $x \in G_x := \cap_i T_i^{-1}(G_i) \subset G$. Since $T_i(x) \in G_i$ and G_i is open, there is no loss in assuming that G_i is an interval, (a_i, b_i). As each T_i is linear, $T_i^{-1}((a_i, b_i))$ is convex. Since the intersection of convex sets is convex, G_x is convex.   ∎

Weak topologies are also rich enough to be useful when $(\mathfrak{X}, \tau)$ is locally convex and Hausdorff. To show this, we look first at translation and dilation. Wherever one is in $\mathbb{R}^k$, and at whatever scale one looks at the space, what one sees is the same in all of the essential ways. The following is the general statement of this idea of invariance under translation and dilation.

Lemma 10.2.14   *If $(\mathfrak{X}, \tau)$ is a topological vector space, then for all $y \in \mathfrak{X}$ and all $r \neq 0$, the mapping $x \mapsto (rx + y)$ is a homeomorphism of $\mathfrak{X}$ with itself.*

Proof. $x \mapsto (rx + y)$ is the composition of continuous functions, hence continuous. Its inverse is the mapping $z \mapsto (\frac{1}{r}z - \frac{1}{r}y)$, and as $r \neq 0$, it is also a composition of continuous functions.   ∎

The reason that it is crucial is that it means that for all $x \in \mathfrak{X}$, the **neighborhood basis at** x, $\tau_x := \{G_x \in \tau : x \in G_x\}$, is the translate of the neighborhood basis at 0; that is, $G \in \tau_x$ iff there exists $G_0 \in \tau_0$ such that $G = G_0 + x$.

Lemma 10.2.15 *If* $(\mathfrak{X}, \tau)$ *is a lctvs, then every neighborhood of* 0 *is absorbent and contains a balanced convex neighborhood of* 0 *(which must itself be absorbent).*

Proof. Let U be a neighborhood of 0. For any $x \in \mathfrak{X}$, the mapping $\alpha \mapsto \alpha x$ is continuous; hence there exists $\epsilon > 0$ such that $[|\alpha| < \epsilon] \Rightarrow [\alpha x \in U]$ so that U is absorbent.

The function $h(r, x) = rx$ is a dilation, hence is continuous at $r = 0$ and $x = 0$. Therefore there exists a neighborhood $V \in \tau$ of 0 and an $\epsilon > 0$ such that $|\alpha| \leq \epsilon$ and $x \in V$ imply that $\alpha x \in U$. The set $W = \cap_{|s| \geq 1} sU$ is convex and balanced and contains V, so is a neighborhood of 0 and a subset of U. ∎

Theorem 10.2.16 *If* $(\mathfrak{X}, \tau)$ *is a Hausdorff lctvs, then* $(\mathfrak{X}, \tau_w)$ *is also a Hausdorff lctvs.*

Proof. Let τ' be the topology induced by the Minkowski gauges of the balanced convex neighborhoods of the origin. Since $(\mathfrak{X}, \tau)$ is Hausdorff, $\{0\} = \cap\{U : U$ is a neighborhood of $0\}$. By the previous lemma, the same is true if we use only the convex balanced neighborhoods of 0; hence τ' is Hausdorff. By the Hahn-Banach theorem, the continuous functions dominated by the Minkowski gauges of the convex balanced neighborhoods of 0 separate every point from 0. Being dominated by a seminorm, they are all τ'-continuous, hence τ-continuous. ∎

10.2.d The Generalized Boundedness Principle

The following is the generalization of the boundedness principle for linear functions on normed spaces. It means that any continuous linear function on a tvs is bounded on some open neighborhood of the origin.

Exercise 10.2.17 For any tvs $(\mathfrak{X}, \tau)$, a linear $T : \mathfrak{X} \to \mathbb{R}$ is continuous iff for every $\epsilon > 0$, $T^{-1}(B_\epsilon(0))$ is an open neighborhood of $0 \in \mathfrak{X}$.

10.3 ◆ The Dual Space and Separation

It is here that some of the differences between finite- and infinite-dimensional lctvs's first become glaringly apparent.

Definition 10.3.1 *If* $(\mathfrak{X}, \tau)$ *is a tvs, then* $\mathfrak{X}^*(\tau) = \mathfrak{X}^*$ *denotes the space of* τ-*continuous linear functions on* $\mathfrak{X}$ *and is called the* ***topological dual of*** $(\mathfrak{X}, \tau)$ *or the* ***dual of*** $(\mathfrak{X}, \tau)$. *When the topology is understood, we simply call* $\mathfrak{X}^*$ *the* ***dual of*** $\mathfrak{X}$.

We are interested in the possibilities of separating convex sets using continuous linear functions. We begin with a number of examples to see what is involved.

10.3.a Examples without a Norm

We begin with the space, $C(\mathbb{N})$, of $\mathbb{R}$-valued sequences, which is not a normed space. What we do here is a special set of applications of the Hahn-Jordan decomposition and the Riesz representation theorem, but since we work with the really simple metric space $\mathbb{N}$, we do not have to pull out the heavy machinery to see what is going on.

Example 10.3.2   *Give $\mathbb{N}$ the discrete topology so that $K \subset \mathbb{N}$ is compact iff it is finite. Let $C(\mathbb{N}) = \mathbb{R}^{\mathbb{N}}$ be the space of $\mathbb{R}$-valued sequences. For $a \in \mathbb{N}$ and $x \in \mathbb{R}^{\mathbb{N}}$, define $\mathrm{proj}_a(x) = x(a)$ so that*

$$x = (x(1), x(2), x(3), \ldots) = (\mathrm{proj}_1(x), \mathrm{proj}_2(x), \mathrm{proj}_3(x), \ldots).$$

Give $\mathbb{R}^{\mathbb{N}}$ the topology $\tau = \tau(\{\mathrm{proj}_a : a \in \mathbb{N}\})$. From earlier work with this sequence space, we know that τ can be metrized by $\rho(x, y) = \sum_a \frac{1}{2^a} \min\{1, |x(a) - y(a)|\}$, and it easy to show that $\rho(x_n, x) \to 0$ iff for every compact $K \subset \mathbb{N}$, $\max_{a \in K} |x_n(a) - x(a)| \to 0$. $(\mathbb{R}^{\mathbb{N}}, \tau)$ is a lctvs.

Definition 10.3.3   *A **positive measure** on $\mathbb{N}$ is a countably additive function $g : \mathcal{P}(\mathbb{N}) \to \mathbb{R}_+$. A **signed measure** on $\mathbb{N}$ is a function $\eta(E) = g_+(E) - g_-(E)$, where g_+ and g_- are positive measures.*

Letting $E_+ = \{a \in \mathbb{N} : \eta(a) > 0\}$, $E_0 = \{a \in \mathbb{N} : \eta(a) = 0\}$, and $E_- = \{a \in \mathbb{N} : \eta(a) < 0\}$ gives a partition of $\mathbb{N}$. We use this partition to define the positive and the negative part of η by $\eta_+(E) = \eta(E \cap E_+)$, $\eta_-(E) = \eta(E \cap E_-)$. The **variation** of η is the positive measure $|\eta|(E) := \eta_+(E) - \eta_-(E)$, and the **variation norm** of η is defined as $|\eta|(\mathbb{N})$. We say that η is **finitely supported** if both E_+ and E_- are finite.

Exercise 10.3.4   If $\eta_+ \neq 0$ and $\eta_- \neq 0$, then η has a unique representation of the form $\eta(E) = \alpha \mu(E) - \beta \nu(E)$, where μ and ν are probabilities with disjoint support and $\alpha, \beta > 0$.

The following bilinear function plays a role throughout this chapter.

Notation 10.3.5   *For $x \in \mathfrak{X}$ and $T \in \mathfrak{X}^*$, we write $\langle x, T \rangle$ or $\langle T, x \rangle$ for $T(x)$. The mapping $(x, T) \mapsto \langle x, T \rangle$ from $\mathfrak{X} \times \mathfrak{X}^*$ to $\mathbb{R}$ is **bilinear** because for all $x, y \in \mathfrak{X}$, $T, T' \in \mathfrak{X}^*$ and $\alpha, \beta \in \mathbb{R}$, $\langle \alpha x + \beta y, T \rangle = \alpha \langle x, T \rangle + \beta \langle y, T \rangle$ and $\langle x, \alpha T + \beta T' \rangle = \alpha \langle x, T \rangle + \beta \langle x, T' \rangle$.*

Example 10.3.6   *For any finitely supported η, the mapping $T_\eta(x) := \sum_{a \in \mathbb{N}} x(a) \eta(a)$ is continuous and linear and is denoted $\langle x, \eta \rangle$. Thus, every T_η belongs to $(\mathbb{R}^{\mathbb{N}})^*$. The finite support aspect is crucial. Let $\mu(\{a\}) = \frac{1}{2^a}$, $x_n = (2, 2^2, \ldots, 2^n, 0, 0, 0, \ldots)$, $x = (2, 2^2, \ldots, 2^n, 2^{n+1}, \ldots)$, $y_n = x_n - x_{n-1}$, and $y = (0, 0, 0, \ldots)$. $\langle x, \mu \rangle$ is well defined on the dense vector subspace of $\mathbb{R}^{\mathbb{N}}$ given by $W = \{x \in \mathbb{R}^{\mathbb{N}} : \#\{a : |x(a)| > 0\} < \infty\}$. Note that $x_n, y_n, y \in W$ but $x \notin W$. Further, $\rho(x_n, x) \to 0$, $\rho(y_n, y) \to 0$. Finally, $\langle x_n, \mu \rangle = n$, $\langle x, \mu \rangle$ is undefined, $\langle y_n, \mu \rangle \equiv 1$, and $\langle y, \mu \rangle = 0$, so that $T_\mu(\cdot)$ is discontinuous on the subspace W, where it is well defined.*

Exercise 10.3.7   Suppose that $T \in (\mathbb{R}^{\mathbb{N}})^*$. For $n \in \mathbb{N}$, define $\eta_T(a) = T(1_{\{a\}}(\cdot))$ and $\eta_T(E) = \sum_{a \in E} \eta(a)$. Show that η_T is finitely supported, that $T(x) = \langle x, \eta \rangle$, and that the mapping $T \mapsto \eta_T$ is one-to-one, onto, and invertible.

Combined with the previous example, this yields the following.

Lemma 10.3.8 *The dual space of $C(\mathbb{N})$ is the set of finitely supported finite signed measures on $\mathbb{N}$.*

Knowing the dual space, let us examine the extent to which it can separate points from convex sets. First we examine some of the difficulties.

Example 10.3.9 *Let $W = \{x \in \mathbb{R}_+^{\mathbb{N}} : 1 \le \#\{a : x(a) > 0\} < \infty\}$. That is, W is the convex set of nonnegative sequences that have at least one and at most finitely many strictly positive elements. Note that $\operatorname{aff}(W)$ is the dense vector subspace of $\mathbb{R}^{\mathbb{N}}$ given by $D = \{x \in \mathbb{R}^{\mathbb{N}} : \#\{a : x(a) \ne 0\} < \infty\}$. Further, $0 \in \partial W$. However, no $T \in (\mathbb{R}^{\mathbb{N}})^*$ can strictly separate 0 from W. Given a finitely supported η, pick x so that $x(a) = 1$ for some a with $\eta(a) = 0$, and note that $T_\eta(x) = T_\eta(0) = 0$.*

Linear subspaces are convex. When they are dense, they cannot be separated from anything, at least not when using continuous linear functions.[1]

Example 10.3.10 *Continuing the previous example, we let $x = (1, \frac{1}{2}, \frac{1}{3}, \ldots)$. Since $x \notin D$ and D is dense, $x \in \partial D$. No nonzero $T \in (\mathbb{R}^{\mathbb{N}})^*$ can separate x from D at all because $T(D) = \mathbb{R}$.*

Exercise 10.3.11 Let $h : \mathcal{P}(\mathbb{N}) \to \mathbb{R}$ be countably additive, that is, for any sequence E_n of disjoint subsets of $\mathbb{N}$, $\{|h(E_n)| : n \in \mathbb{N}\}$ is absolutely summable and $h(\cup_n E_n) = \sum_n E_n$. Show that h is a finite signed measure as defined above.

Finally, let us see that it is possible to use elements of the dual space to strictly separate points from **open** convex sets in $\mathbb{R}^{\mathbb{N}}$.

Example 10.3.12 *Let C be an open convex set containing the origin and x a point not in C. By the definition of τ (or of ρ), there exists an N such that for all $n \ge N$, $\operatorname{proj}_n(C) = \mathbb{R}$. Let N' be the minimal such N. If $x \notin C$, then $x^\circ := \operatorname{proj}_{1,\ldots,N'-1}(x)$ is not an element of $\operatorname{proj}_{1,\ldots,N'-1}(C)$. By the separation theorems for $\mathbb{R}^k$ (with $\ell = N' - 1$), there exists a nonzero $y \in \mathbb{R}^{N'-1}$ such that $x^\circ y > cy$ for all $c \in \operatorname{proj}_{1,\ldots,N'-1}(C)$. Define $\eta(a) = y_a$ for $a \le N' - 1$; L_η then strictly separates x from C.*

10.3.b Related Examples with a Norm

There are many interesting normed vector subspaces of $\mathbb{R}^{\mathbb{N}}$, and we have seen and studied them several times before. The next reviews some salient facts.

Example 10.3.13 *Define $\ell^\infty = \{x \in \mathbb{R}^{\mathbb{N}} : \sup_{a \in \mathbb{N}} |x(a)| < \infty\}$, $c = \{x \in \mathbb{R}^{\mathbb{N}} : \lim_{a \to \infty} x(a) \text{ exists}\}$, and $c_0 = \{x \in \mathbb{R}^{\mathbb{N}} : \lim_{a \to \infty} x(a) = 0\}$. Since $\ell^\infty = C_b(\mathbb{N})$, it is a lctvs with the sup norm topology. Both c and c_0 are closed vector subspaces of ℓ^∞, hence are themselves lctvs's.*

---

1. In $\mathbb{R}^k$, the only dense linear subspace is $\mathbb{R}^k$, so the statement is not surprising in this case.

The norm topology and the relative $\tau = \tau(\{\mathrm{proj}_a : a \in \mathbb{N}\})$ topology are quite different. In working the following, keep in mind that c_0 is the smallest of the vector subspaces and compare with Example 10.3.12.

Exercise 10.3.14 c_0 is τ-dense in $\mathbb{R}^{\mathbb{N}}$ and any τ-open neighborhood of 0 in c_0 contains elements of an arbitrarily large norm.

To characterize c^*, the dual space of c, we add an ideal point, ∞, to $\mathbb{N}$, define $\overline{\mathbb{N}} = \mathbb{N} \cup \{\infty\}$, and extend each $x \in c$ to $\overline{\mathbb{N}}$ by defining $x(\infty) = \lim_a x(a)$. A **finite signed measure on** $\overline{\mathbb{N}}$ is a countably additive function $\eta : \mathcal{P}(\overline{\mathbb{N}}) \to \mathbb{R}$.

Exercise 10.3.15 $T \in c^*$ iff there exists a finite signed measure, η, on $\overline{\mathbb{N}}$ such that $T(x) = \langle x, \eta \rangle := \sum_{a \in \overline{\mathbb{N}}} x(a)\eta(a)$.

Example 10.3.16   *The dual norm on c^* is defined as $\|\eta\| = \sup\{|\langle x, \eta \rangle| : \|x\| \le 1\}$. If we could pick any bounded x mapping $\overline{\mathbb{N}}$ to $\mathbb{R}$, we would set $E_+ = \{a \in \overline{\mathbb{N}} : \eta(a) > 0\}$ and $E_- = \{a \in \mathbb{N} : \eta(a) < 0\}$ and pick $x(a) = 1_{E_+}(a) - 1_{E_-}(a)$, thus delivering $\|\eta\| = \eta(E_+) - \eta(E_-) = |\eta|(\overline{\mathbb{N}})$, which explains the names given to $|\eta|$ and $|\eta|(\overline{\mathbb{N}})$ above. By noting that $|\eta|(\{N, N+1, \ldots\}) \downarrow 0$ as $N \uparrow$, it can be shown that there exists $x \in c$ with $\|x\| \le 1$ for which the $|\langle x, \eta \rangle|$ comes arbitrarily close to $|\eta|(\overline{\mathbb{N}})$.*

The bilinear function, $(x, T) \mapsto \langle x, T \rangle$ from $\mathfrak{X} \times \mathfrak{X}^*$ to $\mathbb{R}$, is used to define some extremely useful topologies, the **weak topology** on $\mathfrak{X}$, and the **weak* topology** on $\mathfrak{X}^*$ (pronounced as the "weak star" topology).

Definition 10.3.17   *The **weak topology** on $\mathfrak{X}$ is the weakest topology making all of the mappings $x \mapsto \langle x, T \rangle$, $T \in \mathfrak{X}^*$, continuous. The **weak* topology**, τ_{w^*}, on $\mathfrak{X}^*$ is the weakest topology making all the mappings $T \mapsto \langle x, T \rangle$, $x \in \mathfrak{X}$, continuous.*

Example 10.3.18   *When $\mathfrak{X} = C(M)$ with the sup norm, M a compact metric space, $\mathfrak{X}^*$ is $ca(M)$, the set of finite signed countably additive Borel measures on M, each of which has a unique representation as $\alpha P - \beta Q$ with P, Q mutually singular probabilities. The weak* topology on $\Delta(M)$ is the restriction of the weak* topology on $ca(M) = C(M)^*$ to $\Delta(M)$.*

Thus, when one uses the linear functions in the dual of a space, one gets the weak topology. When one uses evaluation at points in the domain of the dual space to define linear functions, one gets the weak* topology. When $\mathfrak{X} = (\mathfrak{X}^*)^*$, as happens when $\mathfrak{X} = L^p(\Omega, \mathcal{F}, P)$ with $p \in (1, \infty)$, the weak and the weak* topology are identical.

Definition 10.3.19   *If $\mathfrak{X}$ is a Banach space with $\mathfrak{X} = (\mathfrak{X}^*)^*$, we say that $\mathfrak{X}$ is **reflexive**.*

In the notation of induced topologies, τ_{w^*} is $\tau(\mathbb{F})$, where $\mathbb{F} = \{T \mapsto \langle x, T \rangle : x \in \mathfrak{X}\}$. The weak* topology is, in general, smaller (weaker) than the dual norm topology, but not in the case of $\mathbb{R}^k$.

Exercise 10.3.20   In reflexive spaces, the weak and the weak* topology are the same. In $\mathbb{R}^k$, the norm topology and the weak* topology are the same.

In general, it is difficult to converge in the norm topology and relatively easy to converge in the weak* topology. Since we have not (yet) said what convergence is about in topological spaces, we make statements about the size of the topologies.

Exercise 10.3.21 Show the following:

1. The set of mappings $\mathbb{F} = \{\eta \mapsto \langle x, \eta \rangle : x \in c\}$ separates points in c^*, that is, for every $\eta \neq \eta'$ in c^*, there exists an $x \in c$ with $\langle x, \eta \rangle \neq \langle x, \eta' \rangle$.

2. If $\mathbb{F}' = \{\eta \mapsto \langle x, \eta \rangle : \|x\| \leq 1, \ x \in C\}$, then $\tau(\mathbb{F}') = \tau(\mathbb{F})$.

3. Since the norm topology on c^* is defined by $d(\eta, \eta') = \|\eta - \eta'\| = \sup\{|\langle x, (\eta - \eta') \rangle| : x \in \mathbb{F}'\}$, the norm topology is larger (i.e., finer) than τ_{w^*}, that is, $\tau_{w^*} \subset \tau_d$.

4. Point mass on a point $a \in \overline{\mathbb{N}}$ is defined by $\delta_a(E) = 1_E(a)$. Any weak* neighborhood of δ_∞ contains the set $\{\delta_n : n \geq N\}$ for some N, but no norm ball of radius less than 1 contains any δ_n, $n \in \mathbb{N}$, that is, $\tau_{w^*} \subsetneqq \tau_d$.

If W is a (necessarily closed) vector subspace of $\mathbb{R}^k$ and T_W is a continuous linear function on W, then T_W has many continuous linear extensions, T, to all of $\mathbb{R}^k$. To construct them, note that there is a unique representation of any $\mathbf{x} \in \mathbb{R}^k$ as $\mathbf{x} = \mathbf{y} + \mathbf{z}$, where $\mathbf{y} = \text{proj}_W(\mathbf{x})$ and $\mathbf{z} = \mathbf{x} - \mathbf{y}$. Since $\mathbf{z} \in W^\perp$ and this is itself a (necessarily closed) linear subspace of $\mathbb{R}^k$, the set of continuous linear extensions of T_W is of the form $T(\mathbf{x}) = T_W(\mathbf{y}) + T_{W^\perp}(\mathbf{z})$, where $T_{W^\perp}$ is a continuous linear function on $W^\perp$.

Since c_0 is a proper vector subspace of c, any element of c^* is also an element of c_0^*. Any element of c_0^* has many continuous linear extensions to a function on c.

Exercise 10.3.22 Show that c_0^* is the set of signed measures on $\mathbb{N}$ and that $\eta \in c_0^*$ has a one-dimensional set of continuous linear extensions to c. [Hint: For $E \subset \overline{\mathbb{N}}$ containing ∞ and $r \in \mathbb{R}$, the extension $\widetilde{\eta}_r$ can be defined by $\widetilde{\eta}(E) = \eta(E \setminus \{\infty\}) + r$, and for E not containing ∞, $\widetilde{\eta}_r(E) = \eta(E)$.]

Since c is a closed linear subspace of ℓ^∞, any element of $(\ell^\infty)^*$ is an element of c^*. However, as ℓ^∞ is so much larger than c, there is much more than a single-dimensional set of extensions of an element of c^* to ℓ^∞. The new elements are finite signed measures on $\mathbb{N}$ that fail to be countably additive.

10.4 ◆ Filterbases, Filters, and Ultrafilters

Filterbases, filters, and ultrafilters are the crucial tools for talking about convergence in general topological spaces. We discuss them on their own because they have an important role to play in Chapter 11 as a way to show the existence of a signed measure on $\mathbb{N}$ that fails to be countably additive.

Definition 10.4.1 *A **filterbase in** X is a nonempty collection $\mathfrak{F}^b \subset \mathcal{P}(X)$ of nonempty sets such that for all $F, G \in \mathfrak{F}^b$, there exists an $H \in \mathfrak{F}^b$ such that $H \subset (F \cap G)$.*

Neighborhood bases are the classic filterbases.

Example 10.4.2 *In a topological space (X, τ), $\tau_x = \{G \in \tau : x \in G\}$ is a filterbase for any $x \in X$. In a metric space (M, d), $\{B(x, \epsilon) : \epsilon > 0\}$ and $\{B(x, \frac{1}{n}) : n \in \mathbb{N}\}$ are filterbases that are smaller than τ_x.*

Definition 10.4.3 *A **filter** in X is a nonempty collection $\mathfrak{F}^\circ \subset \mathcal{P}(X)$ of nonempty sets such that for all $F \in \mathfrak{F}^\circ$ and $G \supset F$, $G \in \mathfrak{F}^\circ$, and for all $F, G \in \mathfrak{F}^\circ$, $F \cap G \in \mathfrak{F}^\circ$. A filter $\mathfrak{F}$ on X is called an **ultrafilter** if for all $E \subset X$, either $E \in \mathfrak{F}$ or $(X \setminus E) \in \mathfrak{F}$.*

Example 10.4.4 *In a metric space (M, d), $\{G \in \tau_d : x \in G\}$ is not a filter, but is close to being one. $\mathfrak{F}^\circ = \{E \in \mathcal{P}(M) : \exists G \in \tau_d, \ x \in G \subset E\}$ is a filter. It is not, in general, an ultrafilter. In considering the set $E = \mathbb{Q}$ in the metric space $\mathbb{R}$, we see that neither E nor E^c contain open neighborhoods of any $x \in \mathbb{R}$.*

There are ultrafilters that are easy to characterize.

Example 10.4.5 *$\mathfrak{F}_x = \{E \in \mathcal{P}(X) : x \in E\}$ is called a **point ultrafilter**. Each point ultrafilter is associated with a countably additive point mass probability on $\mathcal{P}(X)$ defined by $\mu(E) = 1$ if $x \in E$ and by $\mu(E) = 0$ otherwise.*

Exercise 10.4.6 If X is finite, then the only kind of ultrafilters on X are the point ultrafilters. If (X, τ) has discrete topology, then $\mathfrak{F}_x$ is equal to τ_x, the neighborhood basis at x.

Example 10.4.7 *$\mathfrak{F}^\circ = \{E \in \mathcal{P}(X) : E^c \text{ is finite}\}$ is a filter, but is not an ultrafilter if X is infinite. To see why, let $E = \{\text{odds}\}$ so that both E and E^c are infinite.*

Filters that are not ultrafilters can be expanded. Suppose that $\mathfrak{F}^\circ$ is a filter that is not an ultrafilter so that we can pick $E \in \mathcal{P}(X)$ such that $E, E^c \notin \mathfrak{F}^\circ$. Since $E \notin \mathfrak{F}^\circ$, for all $F \in \mathfrak{F}^\circ$, $E \not\supset F$. Since $E^c \notin \mathfrak{F}^\circ$, for all $F \in \mathfrak{F}^\circ$, $E^c \not\supset F$. Thus, every $F \in \mathfrak{F}^\circ$ intersects both E and E^c. Pick either E or E^c, add it, and all of its intersections with elements of $\mathfrak{F}^\circ$ to arrive at a new ultrafilter. One could make an inductive argument along these lines to show that ultrafilters exist—keep adding either E or E^c until no more sets remain. Formalizing that argument requires what is called **transfinite** induction. Instead, we use Zorn's lemma to prove that ultrafilters exist.

Lemma 10.4.8 *Every filter is contained within an ultrafilter.*

Proof. Partially order the class of filters by inclusion. Any chain in this partially ordered class has a maximal element in the class of filters, namely the union of the filters in the chain. By Zorn's lemma, there is a filter, $\mathfrak{F}$, that is maximal with respect to set inclusion. If $\mathfrak{F}$ is not an ultrafilter, then there exists E such that $E, E^c \notin \mathfrak{F}$. By the arguments preceding the statement of this lemma, $\mathfrak{F}$ can be expanded to a filter containing one or the other, contradicting the maximality of $\mathfrak{F}$. ∎

Definition 10.4.9 *A filter or an ultrafilter with $\bigcap\{E : E \in \mathfrak{F}\} = \emptyset$ is called **free**.*

Example 10.4.10 *When X is infinite, the canonical free filter is the class of subsets having finite complements, and any free ultrafilter contains this filter.*

Nonpoint ultrafilters can be a bit difficult to visualize.

Example 10.4.11   *Let $X = \mathbb{N}$ and let $\mathfrak{F}°$ be the class of subsets of $\mathbb{N}$ with finite complements. This is a free filter, hence is contained in a free ultrafilter, $\mathfrak{F} \supsetneq \mathfrak{F}°$.*

Either the evens or the odds belong to $\mathfrak{F}$. If the evens belong, then either the set of multiples of, say, 24 belongs to $\mathfrak{F}$ or its complement belongs. If the odds belong, then either the set of primes belongs to $\mathfrak{F}$ or its complement belongs. And so on and so forth.

Exercise 10.4.12   If X is infinite and $\mathfrak{F}$ is an ultrafilter, then either $\mathfrak{F}$ is free or it is a point ultrafilter.

In Example 10.4.5, point ultrafilters were associated with point masses. Free ultrafilters correspond to finitely additive measures that resemble point masses. One can understand these measures as point masses on "ideal" points added to X.

Example 10.4.13   *Let $\mathfrak{F}$ be a free ultrafilter on $\mathbb{N}$ and define $\mu : \mathcal{P}(\mathbb{N}) \to \{0, 1\}$ by $\mu(E) = 1$ if $E \in \mathfrak{F}$ and $\mu(E) = 0$ otherwise. For this μ, for any finite $F \subset \mathbb{N}$, $\mu(F) = 0$ because, being free, $\mathfrak{F}$ must contain F^c. $\mu(\cdot)$ is finitely additive because for any finite partition $E_1, \ldots, E_M$ of $\mathbb{N}$, exactly one of the $E_m \in \mathfrak{F}$. $\mu(\cdot)$ is not countably additive since $\mu(\{1, \ldots, n\}) = 0$ for all $n \in \mathbb{N}$, but $\{1, \ldots, n\} \uparrow \mathbb{N}$, $\mu(\mathbb{N}) = 1$, and $0 \not\to 1$.*

We have defined integrals for countably additive probabilities. For probabilities that are finitely but not necessarily countably additive, the definition requires uniform approximation.

Definition 10.4.14   *If μ is a finitely additive probability on $\mathbb{N}$ and $x(a) = \sum_i \beta_i 1_{E_i}(a)$ is a simple function on $\mathbb{N}$, its integral is $\int x(a) \, d\mu(a) = \sum_i \beta_i \mu(E_i)$. For any $x \in \ell^\infty$ and any $\epsilon > 0$, there is a sandwiching pair of simple functions $x^+ \geq x \geq x_-$ such that $\|x^+ - x_-\| < \epsilon$, so that $|\int x^+ \, d\mu - \int x_- \, d\mu| < \epsilon$. The **integral of x with respect to a finitely additive** μ is denoted $\int x \, d\mu$ and defined as the unique number in these intervals as $\epsilon \downarrow 0$.*

Example 10.4.15   *For any pair of finitely additive probabilities, μ and ν, and any $\alpha, \beta \geq 0$, define $\eta(E) = \alpha\mu(E) - \beta\nu(E)$. The mapping $x \mapsto \int x(a) \, d\eta(a)$ is continuous and linear. If $x \in c$ and μ is the finitely additive probability associated with any free ultrafilter, $\mathbb{F}$, then $\int x(a) \, d\mu(a) = x(\infty)$. For any infinite $E \subset \mathbb{N}$ with an infinite complement, $x = 1_E \notin c$ and $\int x \, d\mu = 1$ or $\int (1 - x) \, d\mu = 1$. This means that specifying an extension of an element of c^* to all of ℓ^∞ requires infinitely many choices.*

10.5 ◆ Bases, Subbases, Nets, and Convergence

In specifying the topology generated by a class of sets, we are specifying a class of sets that we want open and then finding the smallest topology with that property. We begin by figuring out a bit more about that process, expanding on our earlier work on induced topologies.

10.5.a   Bases and Subbases for a Topology

For any class of sets $\mathfrak{T} \subset \mathcal{P}(X)$, $\mathfrak{T}^{\cap f}$ denotes the class of finite intersections of sets in the class $\mathfrak{T}$ and $\mathfrak{T}^{\cup a}$ denotes the class of arbitrary unions of sets in the class $\mathfrak{T}$.

Lemma 10.5.1 *If $\mathfrak{T} \subset \mathcal{P}(X)$ is a class of sets that contains $\emptyset$ and X and is closed under finite intersection, then $\mathfrak{T}^{\cup a}$ is a topology.*

Proof. $\mathfrak{T}^{\cup a}$ contains $\emptyset$ and X and is closed under arbitrary unions, so all that is left to show is that it is closed under finite intersections. Let $G_1, G_2 \in \mathfrak{T}^{\cup a}$ so that $G_1 = \cup_{\alpha_1 \in A_1} T_{\alpha_1}$ and $G_2 = \cup_{\alpha_2 \in A_2} T_{\alpha_2}$ for A_1, A_2 arbitrary and $T_{\alpha_i} \in \mathfrak{T}$, $i = 1, 2$. Consider the set $G := \cup_{(\alpha_1, \alpha_2) \in (A_1 \times A_2)} (T_{\alpha_1} \cap T_{\alpha_2})$. It is clear that $G \in \mathfrak{T}^{\cup a}$. We claim that $G = G_1 \cap G_2$.

Note that $x \in G$ iff $\exists \alpha_1 \in A_1$, $\alpha_2 \in A_2$, such that $x \in (T_{\alpha_1} \cap T_{\alpha_2})$. On the other hand, $x \in G_1 \cap G_2$ iff $(\exists \alpha_1 \in A_1)[x \in T_{\alpha_1}]$ and $(\exists \alpha_2 \in A_2)[x \in T_{\alpha_2}]$. Induction completes the proof. ∎

Theorem 10.5.2 *For a class $\mathcal{S} \subset \mathcal{P}(X)$ containing $\emptyset$ and X, $\tau(\mathcal{S}) = (\mathcal{S}^{\cap f})^{\cup a}$.*

Proof. Since $\tau(\mathcal{S})$ is a topology containing $\mathcal{S}$ and topologies are closed under both finite intersections and arbitrary unions, $(\mathcal{S}^{\cap f})^{\cup a} \subset \tau(\mathcal{S})$.

Since $\tau(\mathcal{S})$ is the smallest topology containing $\mathcal{S}$ and $\mathcal{S} \subset (\mathcal{S}^{\cap f})^{\cup a}$, to show equality, it is sufficient to show that $(\mathcal{S}^{\cap f})^{\cup a}$ is itself a topology. By assumption, it contains $\emptyset$ and X, and by definition, $(\mathcal{S}^{\cap f})$ is closed under finite intersection. By the previous lemma, $(\mathcal{S}^{\cap f})^{\cup a}$ is a topology. ∎

We put these facts into a definition.

Definition 10.5.3 *A class of sets $\mathcal{S} \subset \mathcal{P}(X)$ is a **basis for** τ if every $G \in \tau$ can be expressed as a union of sets in τ. It is a **subbasis for** τ if every $G \in \tau$ can be expressed as a union of sets, each of which is a finite intersection of elements of $\mathcal{S}$.*

We know that τ is always a subbasis for itself, but that is not really interesting. We use $\tau_{\mathbb{R}}$ for the topology on $\mathbb{R}$ associated with the usual metric $d(x, y) = |x - y|$.

Example 10.5.4 $\mathcal{S} = \{\emptyset\} \cup \{(a, b) : a < b \in \mathbb{R}\}$ *is a basis for* $\tau_{\mathbb{R}}$. *More generally, for any metric space,* $\mathcal{S} = \{\emptyset\} \cup \{B_\epsilon(x) : x \in X, \ \epsilon > 0\}$ *is a basis for* τ_d. $\mathcal{S} = \{(-\infty, b) : b \in \mathbb{R}\} \cup \{(a, +\infty) : a \in \mathbb{R}\}$ *is a subbasis for* $\tau_{\mathbb{R}}$.

Exercise 10.5.5 Show that $\mathcal{S}$ is a basis for τ iff for all $G \in \tau$ and all $x \in G$, there exists $B \in \mathcal{S}$ such that $x \in B \subset G$.

Point-set topology is about behavior at very small scales. One way to get at this is through the idea of a local basis.

Definition 10.5.6 *For a topological space (X, τ) and $x \in X$, the **neighborhood system at** x is the collection $\tau_x = \{G \in \tau : x \in G\}$. $\mathcal{S}_x \subset \tau$ is a **local basis at** x if for every $G_x \in \tau_x$, there exists a $B_x \in \mathcal{S}_x$ such that $x \in B_x \subset G_x$.*

Exercise 10.5.7 Show that $\mathcal{S}$ is a basis for τ iff it is a local basis at x for all $x \in X$.

Definition 10.5.8 *A topology τ is **first countable** if for every $x \in X$, there is a countable local basis at x. It is **second countable** if there is a countable basis for τ.*

Example 10.5.9 *If τ is second countable, it must be first countable. If X is uncountable and τ is the discrete topology, then τ is first countable but not second countable.*

Lemma 10.5.10 *Every metric space is first countable, and if it is separable, then it it is also second countable.*

Proof. The collection $\{B_q(x) : q \in \mathbb{Q}_{++}\}$ is a countable local basis at x. If $C = \{c_n : n \in \mathbb{N}\}$ is a countable dense subset of M, then $\{B_q(c) : c \in C, \ q \in \mathbb{Q}_{++}\}$ is a countable basis for τ_d. ∎

Exercise 10.5.11 Check that the class of sets $\{[a, b) : a < b \in \mathbb{R}\}$ is the basis for a topology, τ_L, called the **lower-order topology**. Show that:

1. τ_L is first countable but not second countable.
2. Any cdf $F : \mathbb{R} \to [0, 1]$ is τ_L-continuous (when $[0, 1]$ has the usual metric).
3. If x_n is a decreasing sequence, $x_n \downarrow x$, then x_n converges to x in the lower-order topology, that is, x_n is a.a. in any neighborhood of x.
4. If x_n is an increasing sequence, $x_n \uparrow x$, then x_n does *not* converge to x in the lower-order topology.
5. If $E \subset \mathbb{R}$ is a nonempty interval (a, b), then there are τ_L-open coverings of E that have no finite subcovering.
6. Give an infinite compact subset of $(\mathbb{R}, \tau_L)$.

10.5.b Nets and Convergence

For metric spaces, we could use sequences to talk about closed sets, hence about open sets, hence about continuity and compactness. Nets generalize sequences to topological spaces. We begin by recalling the important properties that we are going to generalize.

A sequence in a space X is a mapping $n \mapsto x_n$ from $\mathbb{N}$ to X. Sequences do two useful things: converge and accumulate. The following recapitulates previous material.

Definition 10.5.12 *A sequence x_n in a metric space (M, d) **converges to** x, written $x_n \to x$, if for all $B_\epsilon(x)$, there exists an N such that for all $n \geq N$, $x_n \in B_\epsilon(x)$. It **accumulates at** x, written $x_n \rightsquigarrow x$, if for all $\epsilon > 0$, for all N, there exists an $n > N$ such $x_n \in B_\epsilon(x)$.*

Since the open balls are a basis for the metric topology, we can equivalently say that $x_n \to x$ if for all open neighborhoods G_x of x, $x_n \in G_x$ a.a., and $x_n \rightsquigarrow x$ if for all open neighborhoods G_x of x, $x_n \in G_x$ i.o.

The integers are a special directed set. A pair $(D, \precsim)$ is **preordered** if D is a nonempty set and $\precsim$ is a reflexive transitive binary relation on D.

Definition 10.5.13   *A preordered set $(D, \precsim)$ is a **directed set** if for all $a, b \in D$, there exists a $c \in D$ such that $a \precsim c$ and $b \precsim c$. A **net** in X is a function $d \mapsto x_d$ from some directed set to X. For $C \subset X$, a net in C is a net in X with $x_d \in C$ for all $d \in D$.*

> *1. A net $d \mapsto x_d$ **converges to** x, written $x_d \to x$, if for all open neighborhoods G_x of x, there exists an $a \in D$ such that for all $b \succsim a$, $x_b \in G_x$.*
>
> *2. A net $d \mapsto x_d$ **accumulates at** x, written $x_d \leadsto x$, if for all open neighborhoods G_x of x and for all $a \in D$, there exists $b \succsim a$ such that $x_b \in G_x$.*

With the directed set $(\mathbb{N}, \leq)$, these replicate the definitions of convergence and accumulation for sequences—take $c = \max\{a, b\}$ in the definition.

There were several arguments in our study of metric spaces where we took $\epsilon_n \downarrow 0$ and took an $x_n \in B_{\epsilon_n}(x)$. This gave us a sequence $x_n \to x$. Taking a neighborhood basis and selecting from each element gives us a canonical way to give a convergent directed set.

Example 10.5.14   *Let $\mathcal{G} \subset \tau_x$ be a b-basis for τ at x, and define the preorder $G \precsim G'$ if $G \supset G'$. With this preorder, going "higher" or "further" in the directed set $\mathcal{G}$ corresponds to taking smaller and smaller neighborhoods of x. For each $G \in \mathcal{G}$, pick a point $x_G \in G$. The net $G \mapsto x_G$ converges to x, that is, $x_G \to x$.*

We use a variant of this device in proving the following.

Theorem 10.5.15   *(X, τ) is a Hausdorff space iff every convergent net in X converges to exactly one point.*

This generalizes the result that convergent sequences in metric spaces can have only one limit.

Proof.   Suppose that (X, τ) is Hausdorff and that $x_d \to x$ and $x_d \to y$, $x \neq y$. Let G_x and G_y be disjoint neighborhoods of x and y. There exists a such that for all $b \succsim a$, $x_b \in G_x$, and there exists a' such that for all $b' \succsim a'$, $x_{b'} \in G_y$. Let $c \succsim a$ and $c \succsim a'$. We have $x_c \in G_x \cap G_y$, a contradiction.

Now suppose that (X, τ) is not Hausdorff. This means that there exists $x \neq y$ such that for all $G_x \in \tau_x$ and $G_y \in \tau_y$, there exists $z \in (G_x \cap G_y)$. Consider the directed set $D = \tau_x \times \tau_y$ with the partial order $(G_x, G_y) \precsim (G'_x, G'_y)$ if $G_x \cap G_y \supset G'_x \cap G'_y$. To define the net, map each $(G_x, G_y) \in D$ to some $z \in G_x \cap G_y$. This gives a net that converges to both x and y.   ∎

We saw that a subset E of a metric space is closed iff there is a sequence x_n in E such that $x_n \to x$. One part of proving this involves looking at the sequence of neighborhoods $B_{1/n}(x)$ and picking $x_n \in B_{1/n}(x)$. We generalize this idea in the proof of the following.

Theorem 10.5.16   *For $C \subset X$, (X, τ) a Hausdorff topological space, $x \in \mathrm{cl}(C)$ iff there is a net in C converging to x. If τ is first countable, then $x \in \mathrm{cl}(C)$ iff there is a sequence in C converging to x.*

Proof.   Let $D_x = \tau_x = \{G_x \in \tau : x \in G_x\}$ be the class of open neighborhoods of x. For $a, b \in D_x$, define $a \precsim b$ if $b \subset a$. That is, as one goes "further along" the net, one has smaller sets. Since the class of open sets is closed under finite intersection, each $(D_x, \precsim)$ is a directed set.

For Hausdorff spaces, $x \in \text{cl}(C)$ iff for $G \in D_x$, $G \cap C \neq \emptyset$. Define a net by picking an $x_G \in G \cap C$ for each $G \in D_x$. It is immediate that $x_G \to x$.

Now suppose that there exists a net $d \mapsto x_d$ in C that converges to x. Pick an open neighborhood G_x of x. We must show that $G_x \cap C \neq \emptyset$. Since $x_d \to x$, we know that there exists a such that $x_a \in G_x$. Since $x_a \in C$, $G_x \cap C \neq \emptyset$. For the first countable part of the result, replace τ_x with a countable local basis, G_n, at x and pick $x_n \in \cap_{m \leq n} G_m$. ∎

10.5.c Two Applications

We have investigated, in some detail, weak* convergence in two contexts: the Prokhorov topology on probabilities on metric spaces and weak convergence in separable L^p spaces, $p \in [1, \infty)$. The first was metrizable and the second was not, but the definition of the second was given in terms of sequences.

10.5.c.1 Bases and Subbases for the Prokhorov Topology on Probabilities

We have talked about Prokhorov or weak* convergence of a sequence of probabilities on a metric space M, and have talked about weak convergence of a sequence in L^p. Weak* convergence is convergence in ρ, the Prokhorov metric, so there is a metric topology, τ_ρ, on the space of probabilities.

It is time to turn these observations into topology. The following is a first step that is more broadly useful.

Theorem 10.5.17 *Let $\mathcal{S} = \{G_\alpha : \alpha \in A\} \subset \mathcal{P}(X)$ satisfy the condition that for each $\alpha, \beta \in A$ and $x \in G_\alpha \cap G_\beta$, there exists $\gamma \in A$ such that $x \in G_\gamma \subset G_\alpha \cap G_\beta$. Then $\{\emptyset\} \cup \{X\} \cup \mathcal{S}$ is a basis for $\tau(\mathcal{S})$.*

Exercise 10.5.18 Prove Theorem 10.5.17.

For $P \in \Delta(M)$, $f_1, \dots, f_K \in C_b(M)$, and $\epsilon_1, \dots, \epsilon_K > 0$, define

$$G(P : (f_1, \dots, f_K); (\epsilon_1, \dots, \epsilon_K))$$

$$= \{Q \in \Delta(M) : |\textstyle\int f_k \, dQ - \int f_k \, dP| < \epsilon_k, \quad k = 1, \dots, K\}.$$

Exercise 10.5.19 Using (the portmanteau) Theorem 9.3.16, show that every $G(P : (f_1, \dots, f_K); (\epsilon_1, \dots, \epsilon_K))$ is τ_ρ-open; that is, for every $Q \in G(P : (f_1, \dots, f_K); (\epsilon_1, \dots, \epsilon_K))$, there exists $\delta > 0$ such that $[\rho(Q, Q') < \delta] \Rightarrow [Q' \in G(P : (f_1, \dots, f_K); (\epsilon_1, \dots, \epsilon_K))]$.

Lemma 10.5.20 *If $\mathcal{S}_*$ is the class of sets of the form $G(P : (f_1, \dots, f_K); (\epsilon_1, \dots, \epsilon_K))$, then it satisfies the condition in Theorem 10.5.17.*

On the basis of this result, we use τ_* to denote the topology generated by $\mathcal{S}_*$.

Proof. [Sketch] Take $Q \in G(P : f; \epsilon) \cap G(P' : g; \delta)$, set $r_f = |\int f \, dQ - \int f \, dP|$, $r_g = |\int g \, dQ - \int g \, dP'|$, and define $\eta = \min\{\epsilon - r_f, \delta - r_g\}$. First show that $\eta > 0$, and then demonstrate that $G(Q : f, g; \eta, \eta) \subset G(P : f; \epsilon) \cap G(P' : g; \delta)$. Induction completes the proof. ∎

Exercise 10.5.21 Give the missing details in the previous proof.

Theorem 10.5.22 *Suppose that* $\mathcal{S}$ *and* $\mathcal{T}$ *are bases for topologies* $\tau_{\mathcal{S}}$ *and* $\tau_{\mathcal{T}}$ *and that*

> 1. *for all* $G_{\mathcal{S}} \in \tau_{\mathcal{S}}$ *and* $x \in G_{\mathcal{S}}$, *there exists* $G_{\mathcal{T}} \in \mathcal{T}$ *such that* $x \in G_{\mathcal{T}} \subset G_{\mathcal{S}}$, *and that*
>
> 2. *for all* $G_{\mathcal{T}} \in \tau_{\mathcal{T}}$ *and* $x \in G_{\mathcal{T}}$, *there exists* $G_{\mathcal{S}} \in \mathcal{S}$ *such that* $x \in G_{\mathcal{S}} \subset G_{\mathcal{T}}$.

Then $\tau_{\mathcal{S}} = \tau_{\mathcal{T}}$. *Further, if* $\tau_{\mathcal{S}} = \tau_{\mathcal{T}}$, *then the previous two conditions hold.*

When $\tau_{\mathcal{S}} = \tau_{\mathcal{T}}$, we say that the bases $\mathcal{S}$ and $\mathcal{T}$ are **equivalent**. Thus, two bases are equivalent iff the conditions in the previous theorem hold.

Exercise 10.5.23 Let $\mathcal{T}_Q$ be the set of Prokhorov ϵ-balls around Q, and $\mathcal{T} := \{\mathcal{T}_Q : Q \in \Delta(M)\}$. Apply the previous theorem to $\mathcal{S}_*$ and $\mathcal{T}$ to conclude that the Prokhorov metric topology and τ_* are the same. [You will have to go back into the details of the proof of the portmanteau theorem for this.]

The point of the following is that the Prokhorov topology is an induced topology, namely, the weakest topology making all integrals against bounded continuous functions into continuous mappings.

Exercise 10.5.24 An alternative basis for the Prokhorov topology is the class of set $L_f^{-1}(\tau_{\mathbb{R}})$ where, for $f \in C_b(M)$, $L_f(P) := \int f \, dP$.

10.5.c.2 Weak Topologies on L^p, $p \in (1, \infty)$

For $p \in (1, \infty)$, the Banach spaces $L^p(\Omega, \mathcal{F}, P)$ are reflexive. This means that the weak and the weak* topologies on L^p are the same. Examples 8.9.2 and 8.9.7 showed that there is no metric giving weak convergence of sequences in $L^p = L^p(\Omega, \mathcal{F}, P)$, $(\Omega, \mathcal{F}, P)$ a nonatomic probability space. There, weak convergence of X_n in L^p was defined by $\langle X_n, Y \rangle \to \langle X, Y \rangle$ for all $Y \in L^q$. We can now see that this really does define a topology.

Exercise 10.5.25 Show that if $(\Omega, \mathcal{F}, P)$ is separable, that is, $\mathcal{F}$ is generated by a countable class of sets $\{E_n : n \in \mathbb{N}\}$) and $p \in (1, \infty)$, then the weak topology on L^p is first and second countable.

10.6 ◆ Compactness

It is worthwhile reviewing the arguments for two results in metric spaces: (1) K is compact iff every sequence in K has a convergent subsequence; and (2) the product $\times_{a \in \mathbb{N}} K_a$ of compact metric spaces is compact in the product topology.

The argument for (1) runs as follows. Let $\{x_n : n \in \mathbb{N}\}$ be a sequence in K and define $F_n = \text{cl}\{x_m : m \geq n\}$. This class of closed sets has the finite intersection property; hence, by compactness, its intersection contains some $x \in K$. Some subsequence of x_n must therefore converge to x.

The argument for (2) runs as follows. If x_n is a sequence in $\times_{a \in \mathbb{N}} K_a$, then for each a, $x_n(a)$ has a convergent subsequence. Using diagonalization, we construct a subsequence of all of the subsequences; we have a subsequence of x_n that converges in the product topology.

In this section, we generalize these arguments to compact topological spaces.

10.6.a Terminal Sets

The sets $\{x_m : n \leq m\}$ used just above are called **terminal sets**.

Definition 10.6.1 *Let $d \mapsto x_d$ be a net in X. A **terminal set for the net** is a set of the form $T_a := \{x_b : a \lesssim b\}$, $a \in D$.*

Convergence and accumulation can be phrased in terms of terminal sets, $x_d \to x$ iff $\forall G_x, \exists T_a \subset G_x$, and $x_d \rightsquigarrow x$ iff $\forall G_x, \forall T_a, T_a \cap G_x \neq \emptyset$.

We characterize continuous mappings between metric spaces using sequences.

Theorem 10.6.2 *For topological spaces (X, τ_X) and (Y, τ_Y), $f : X \to Y$ is continuous iff for all $x \in X$ and all nets $x_d \to x$, $f(x_d) \to f(x)$.*

Exercise 10.6.3 Prove the foregoing theorem. [Reviewing Theorem 4.8.5 might be worthwhile.]

10.6.b Subsequences and Subordinate Filterbases

To get at the generalization of subsequences, which is crucial for compactness, we once again need **filterbases**, **filters**, and **ultrafilters**. These are tied to our present needs through terminal sets. Recall that a filterbase in X is a nonempty collection $\mathfrak{F}^\circ \subset \mathcal{P}(X)$ of nonempty sets such that for all $F, G \in \mathfrak{F}^\circ$, there exists an $H \in \mathfrak{F}^\circ$ such that $H \subset (F \cap G)$. It is very important to bear in mind that the empty set cannot belong to a filterbase.

Example 10.6.4 *The class of terminal sets of a net is a filterbase.*

Exercise 10.6.5 For all $x \in X$, the class $\tau_x = \{G_x \in \tau : x \in G_x\}$ is a filterbase.

We get to filters by including supersets of the elements of filterbases. Here is the canonical example in topology.

Exercise 10.6.6 Let $a \mapsto x_a$ be a net converging to x and let $\{T_a : a \in A\}$ be the associated collection of terminal sets preordered by set inclusion. Let $\mathfrak{F} = \{H \subset X : \exists a \in D, \ T_a \subset H\}$, and preorder $\mathfrak{F}$ by inclusion. For each $A \in \mathfrak{F}$, pick $x_A \in A$. Show that the net $A \mapsto x_A$ converges to x.

Definition 10.6.7 *A **filter in** X is a nonempty collection $\mathfrak{F} \subset \mathcal{P}(X)$ of nonempty sets such that for all $F \in \mathfrak{F}$ and $G \supset F$, $G \in \mathfrak{F}$ and for all $F, G \in \mathfrak{F}$, $F \cap G \in \mathfrak{F}$.*

Again, the empty set does not belong to any filter.

Exercise 10.6.8 Show that every filter is a filterbase, that the class of sets defined in Exercise 10.6.6 is a filter, and that for any filterbase $\mathfrak{F}^\circ$, the class of sets $\mathfrak{F} := \{H \subset X : H \supset G \text{ for some } G \in \mathfrak{F}^\circ\}$ is a filter.

Filterbases can converge.

Definition 10.6.9 *A filterbase $\mathfrak{F}^\circ$ **converges to** $x \in X$, written $\mathfrak{F}^\circ \to x$, if for all open neighborhoods G_x of x, $\exists G_\alpha \in \mathfrak{F}^\circ$, $G_\alpha \subset G_x$. It **accumulates at** x written $\mathfrak{F}^\circ \rightsquigarrow x$, if for all open neighborhoods G_x of x, $\forall G_\alpha \in \mathfrak{F}^\circ$, $G_\alpha \cap G_x \neq \emptyset$.*

Here is the notion of subsequences.

Definition 10.6.10 *The filterbase $\mathfrak{F}°$ **subordinates** the filterbase $\mathfrak{H}°$, equivalently the filterbase $\mathfrak{H}°$ is **subordinate to** the filterbase $\mathfrak{F}°$, if for all $F \in \mathfrak{F}°$ there exists $H \in \mathfrak{H}°$ such that $H \subset F$.*

Exercise 10.6.11 Let x_n be a sequence in X and x_{n_k} a subsequence. Let $\mathfrak{F}°$ be the filterbase of terminal sets of the sequence and $\mathfrak{H}°$ the filterbase of terminal sets of the subsequence. Show that $\mathfrak{F}°$ subordinates $\mathfrak{H}°$.

We expanded filterbases to get filters. When we can do no more expansion, we have an ultrafilter.

Definition 10.6.12 *A filter $\mathfrak{F}$ on X is called an **ultrafilter** if for all $E \subset X$, either $E \in \mathfrak{F}$ or $(X \setminus E) \in \mathfrak{F}$.*

Again, ultrafilters cannot contain the empty set.

Example 10.6.13 $\mathfrak{F}_x := \{A \subset X : x \in A\}$ *is an ultrafilter, called a **point ultrafilter**.*

Exercise 10.6.14 A filter $\mathfrak{F}$ is an ultrafilter iff for all filters $\mathfrak{G} \supset \mathfrak{F}$, $\mathfrak{G} = \mathfrak{F}$.

As noted above, Zorn's lemma and the previous exercise deliver the following.

Exercise 10.6.15 Review the proof that every filter is contained in some ultrafilter. Then show that X is finite iff every ultrafilter on X is a point ultrafilter.

Example 10.6.16 *Let X be an infinite set and let $\mathfrak{F}$ denote the class of sets with finite complements. This is a filter, and no ultrafilter containing it is a point ultrafilter.*

10.6.c Compactness

We have given many characterizations of compactness for metric spaces. One of the most frequently used is that every sequence has a convergent subsequence. Theorem 10.6.21 provides the generalization of this result to topological spaces.

Definition 10.6.17 *An **open covering** of X is a collection $\mathcal{U} \subset \tau$ such that $X \subset \cup\{G : G \in \mathcal{U}\}$. A **finite subcovering** of X is a finite subset $\mathcal{V} \subset \mathcal{U}$ such that $X \subset \cup\{G : G \in \mathcal{V}\}$. X is **compact** if every open covering of X has a finite subcovering.*

Here are three fairly direct generalizations of metric space results that we proved earlier.

Theorem 10.6.18 *If (X, τ) is compact, then every continuous $f : X \to \mathbb{R}$ achieves its maximum.*

Proof. The open covering $\{f^{-1}((-n, +n)) : n \in \mathbb{N}\}$ must have a finite subcovering, so f is bounded. Let $r = \sup_{x \in X} f(x)$. If no x satisfies $f(x) = r$, then $\{f^{-1}((-\infty, r - \frac{1}{n})) : n \in \mathbb{N}\}$ is an open covering of X. If it has a finite subcovering, then r is not the supremum. ∎

Lemma 10.6.19 *If F is a closed subset of X and (X, τ) is compact, then F is compact.*

Proof. Let $\mathcal{U} = \{G_\alpha \cap F : \alpha \in A\}$ be an open covering of F so that $\{G_\alpha : \alpha \in A\} \cup \{\{X \setminus F\}\}$ is an open covering of X. Take a finite subcover. Omitting $X \setminus F$ if necessary, we have a finite subcover of F.   ■

Exercise 10.6.20   If (X, τ) is compact and $f : X \to Y$ is continuous, then $f(X)$ is a compact subset of Y.

The following theorem is not, so far as we can tell, a direct generalization of any of the common metric space results. Ultrafilters are something like "really small" subsequences.

Theorem 10.6.21 (X, τ) *is compact iff every ultrafilter in X converges.*

Proof.   Let (X, τ) be compact and let $\mathfrak{F}$ be an ultrafilter. Suppose that $\mathfrak{F}$ is not convergent. For all $x \in X$, there exists an open neighborhood $G_x \notin \mathfrak{F}$. Since $\mathfrak{F}$ is an ultrafilter, this means that $F_x := (X \setminus G_x) \in \mathfrak{F}$. Take a finite subcover, $G_{x_1}, \ldots, G_{x_K}$ so that $\cap_{k=1}^{K} F_{x_k} = \emptyset$. However, being a filter, $\mathfrak{F}$ is closed under finite intersection, so this implies that $\emptyset \in \mathfrak{F}$, a contradiction.

Now suppose that (X, τ) is not compact. Let $\mathfrak{O}$ be an open cover of X with no finite subcover. Let $\mathfrak{F}^\circ$ denote the set of complements of finite unions of elements on $\mathfrak{O}$. $\mathfrak{F}^\circ$ is a filterbase, and no ultrafilter containing it can converge.   ■

Exercise 10.6.22   Translate the language of the previous proof into filterbases, filters, and ultrafilters generated by the terminal sets of sequences in a compact metric space.

10.6.d The Compactness of Product Spaces

We now turn to Tychonoff's theorem, which says that the product of compact spaces is compact in the product topology.

Exercise 10.6.23   A basis for the product topology is the class of sets of the form $G(x : \alpha_1, \ldots, \alpha_K; G_1, \ldots, G_K)$, where the α_k belong to A and the G_k belong to τ_{α_k}.

Exercise 10.6.24   If each (X_α, τ_α) is Hausdorff, then so is $X := \Pi_{\alpha \in A} X_\alpha$ with the product topology.

Exercise 10.6.25   Verify that the previous definition agrees with the metrizable product topology defined above for the countable products of metric spaces.

Theorem 10.6.26 (Tychonoff)   *If each (X_α, τ_α) is compact, then $X := \Pi_{\alpha \in A} X_\alpha$ is compact in the product topology.*

Proof.   If $\mathfrak{F}$ is an ultrafilter in X, then $\mathrm{proj}_\alpha(\mathfrak{F})$ converges to some x_α in each X_α. Let $x = (x_\alpha)_{\alpha \in A}$. By Exercise 10.6.23, $\mathfrak{F} \to x$.   ■

Four comments:

1. The converse is immediate; if $X := \Pi_{\alpha \in A} X_\alpha$ is compact in the product topology, then each (X_α, τ_α) is compact because it is the image of a compact space under the continuous mapping proj_α.

2. In metric spaces, $C(M)$ is separable iff M is compact. If X is, say, $X = [0, 1]^{[0,1]}$, Tychonoff's theorem tells us that it is compact, but each of the uncountably many mappings $\text{proj}_t : X \to [0, 1]$ is continuous and each is at sup norm distance 1 from all of the others.

3. In particular, the last observation implies that (X, τ) is not metrizable.

4. As we have assumed that each space is Hausdorff, the x_α in the previous proof is unique.

The following is relevant to Example 7.6.30 (p. 423).

Exercise 10.6.27 (Existence of Nonatomic Purely Finitely Additive Probabilities)   For each $n \in \mathbb{N}$ and $A \subset \mathbb{N}$, define $\mu_n(A) = \frac{1}{n}\#(A \cap \{1, \ldots, n\})$ so that μ_n is a point in the compact space $[0, 1]^{\mathcal{P}(\mathbb{N})}$. Consider the filterbase of sets of the form $E_N := \{\mu_n : n \geq N\}$. It is contained in some ultrafilter, which converges to a point μ. Show that μ is a purely finitely additive nonatomic probability.

A point that is sometimes overlooked is that the identity function from X to X need not be a homeomorphism if the range and domain are given different topologies. This shows up in the proof of the following fact, which is a useful piece of information about compact Hausdorff spaces.

Lemma 10.6.28   *If (X, τ) is a compact Hausdorff space and τ' and τ'' are two topologies satisfying $\tau' \subsetneqq \tau \subsetneqq \tau''$, then (X, τ') is not Hausdorff and (X, τ'') is not compact.*

Proof.   Define $f(x) = x$. If (X, τ'') is compact, then $f : (X, \tau'') \to (X, \tau)$ is continuous and $f(F)$ is closed; hence f is a homeomorphism, which contradicts $\tau \subsetneqq \tau''$. If (X, τ') is Hausdorff, then $f : (X, \tau) \to (X, \tau')$ is continuous and $f(F)$ is closed; hence f is a homeomorphism, which contradicts $\tau \subsetneqq \tau''$.   ∎

Exercise 10.6.29   Suppose that (M, d) is a compact metric space. Show that there is no metric d' on M with $\tau_{d'} \subsetneqq \tau_d$, and there is no metric d'' on M with $\tau_d \subsetneqq \tau_{d''}$ such that (M, d'') is compact. [This is a direct implication of the previous result, and it is informative to give a more direct argument using sequences.]

Corollary 10.6.30   *If (M, d) is a metric space and ρ is the Prokhorov metric on $\Delta(M)$, then (M, d) is compact iff $(\Delta(M), \rho)$ is compact.*

Proof.   Suppose first that M is compact. The Prokhorov topology is $\tau(\{L_f : f \in C(M)\})$, where $L_f(\mu) := \int f \, d\mu$. Let $X = \Pi_{f \in C(M)} f(M)$ be the compact product space. It is easy to show that $\{(L_f(\mu) : \mu \in \Delta(M)\}$ is a closed, hence compact, subset of X and that $\mu \leftrightarrow (L_f(\mu))_{p \in C(M)}$ is a homeomorphism.

Suppose now that $(\Delta(M), \rho)$ is compact. It is easy to show that $F = \{\delta_x : x \in M\}$ is closed, hence compact. Further, $x \leftrightarrow \delta_x$ is a homeomorphism between M and F.   ∎

10.7 ◆ Compactness in Topological Vector Spaces

We now present two results, a very useful weak* compactness criterion for subsets of the dual of a normed space, known as Alaoglu's theorem, and the Krein-Milman theorem, which tells us that the closed convex hull of a compact convex set in a topological vector space is the closed convex hull of its extreme points.

10.7.a A Criterion for Weak* Compactness

Theorem 10.7.1 (Alaoglu) *If $\mathfrak{X}$ is a normed vector space, then the closed unit ball of $\mathfrak{X}^*$ is weak* compact.*

Proof. Let U be the closed unit ball in $\mathfrak{X}^*$. For each $x \in \mathfrak{X}$, $I_x := \langle x, U \rangle$ is a compact interval. By Tychonoff's theorem, $X := \Pi_{x \in \mathfrak{X}} I_x$ is compact in the product topology. We embed each $x^* \in U$ in X as the point with $\text{proj}_x(x^*) = \langle x, x^* \rangle \in I_x$. By the definition of the weak* topology and the product topology, the embedding is a homeomorphism. ■

Here are some useful fairly immediate consequences.

1. If $\mathfrak{X}$ is a reflexive Banach space, then any weakly closed norm-bounded set is weakly compact. [Revisit Theorem 8.10.1 for this.]

2. If M is a compact metric space, then variation norm-bounded subsets of $ca(M)$ have weak* compact closures. In particular, $\Delta(M)$ is weak* compact.

3. If $\mathfrak{X}$ is a normed space, K is a norm-bounded, norm-closed convex subset of $\mathfrak{X}^*$ and $f : \mathfrak{X}^* \to \mathbb{R}$ is norm continuous and concave, then f achieves its maximum on K.

10.7.b Convex Hulls of Compact Sets

For a, b in a vector space $\mathfrak{X}$, $[a, b]$ denotes the line segment joining, that is, a and b, that is, $[a, b] = \{\alpha a + (1 - \alpha)b : \alpha \in [0, 1]\}$. The definition of extreme points of sets is the same in general vector spaces as it is in $\mathbb{R}^k$—x is an extreme point of a set K if $[x \in [a, b] \subset K] \Rightarrow [x \in \{a, b\}]$. Put another way, x is an extreme point of K iff x is at one end or the other of every line segment in K that contains x.

Example 10.7.2 *Let $K \subset L^\infty(\Omega, \mathcal{F}, P)$ be the set of functions with $0 \leq$ ess inf X and ess sup $X \leq 1$. X is an extreme point of K iff $X = 1_E$ for some $E \in \mathcal{F}$. If $T : L^\infty \to \mathbb{R}$ is continuous and linear, then it achieves its maximum on K. If $T(X) = \int XY \, dP$ for some $Y \in L^1$, then the optimum is 1_E, where $E = \{Y \geq 0\}$.*

Exercise 10.7.3 Show that the only extreme points of the closed unit ball, $cl(U)$, in $C([0, 1])$ are the constants $f \equiv 1$ and $g \equiv -1$. Also show that the set $Z = \{f \in cl(U) : \int f(t) \, dt = 0\}$ is closed and convex and has no extreme points.

Lemma 10.7.4 *For (M, d) a metric space, the extreme points of $\Delta(M)$ are the point masses, $\delta_x(E) = 1_E(x)$.*

Proof. It is clear that every point mass is an extreme point. If $\mu \in \Delta(M)$ is not a point mass, the support of μ is defined as $\cap\{F : F$ is closed and $\mu(F) = 1\}$. If $F = \{x\}$, then μ is a point mass, hence F contains at least two points, $x \neq y$. For every

$\epsilon > 0$, $\mu(B_\epsilon(x)) > 0$ and $\mu(B_\epsilon(y)) > 0$. For ϵ small enough that $B_\epsilon(x)$ and $B_\epsilon(y)$ are disjoint, we have a set E with $0 < \mu(E) < 1$ and $0 < \mu(E^c) < 1$. We can pick $\epsilon, \delta > 0$ but small enough that $\nu_1(A) = (1 + \epsilon)\mu(A \cap E) + (1 - \delta)\mu(A \cap E^c)$ and $\nu_2(A) = (1 - \epsilon)\mu(A \cap E) + (1 + \delta)\mu(A \cap E^c)$ are probabilities. $\mu = \frac{1}{2}\nu_1 + \frac{1}{2}\nu_2$.   ∎

Definition 10.7.5 *For K a subset of a vector space $\mathfrak{X}$, the* **convex hull** *of K is* $\mathrm{co}(K) := \cap\{C : K \subset C, \ C \text{ convex}\}$ *and the* **closed convex hull** *of K is* $\overline{\mathrm{co}}(K) := \cap\{C : K \subset C, \ C \text{ convex and closed}\}$.

Recall that in $\mathbb{R}^k$, the convex hull of a compact set is compact, hence closed. This is another aspect of finite dimensionality that gets lost in larger vector spaces.

Example 10.7.6 *The set K of point masses on $[0, 1]$ is weak* compact, $\mathrm{co}(K)$ is the set of finitely supported probabilities on $[0, 1]$, and $\overline{\mathrm{co}}(K) = \Delta([0, 1])$ is the set of all probabilities on $[0, 1]$.*

Theorem 10.7.7 (Krein–Milman) *If $(\mathfrak{X}, \tau)$ is a Hausdorff locally convex topological vector space and K is a compact convex subset of $\mathfrak{X}$, then K is the closed convex hull of its extreme points.*

Proof.   The first and hardest step is to show that there exists at least one extreme point.

Let us define a nonempty compact convex $C \subset K$ to be an extreme *set* if for all $x, y \in K$, if $\frac{1}{2}x + \frac{1}{2}y \in C$, then both $x \in C$ and $y \in C$. A standard application of Zorn's lemma shows that there exists a smallest extreme set, which we call X. If X contains only one point, it is an extreme point. Suppose that X contains two or more points, $x_1 \neq x_2$. Since we are working in a Hausdorff locally convex space, there is a continuous linear $T : \mathfrak{X} \to \mathbb{R}$ with $T(x_1) < T(x_2)$. Let $a = \min_{x \in K} T(x)$, which exists by compactness and continuity, and let $Y = \{x \in X : T(x) = a\}$. Now, Y is itself an extreme set [by the same arguments used in the proof of Krein-Milman for $\mathbb{R}^k$ (Theorem 5.6.29)]. However, Y is a nonempty strict subset of X, contradicting the minimality of X.

Now let A be the necessarily nonempty closed convex hull of the extreme points of K. Since K is convex, $A \subset K$. If the subset relation is strict, we can pick $z \in (K \setminus A)$. There is a continuous linear T such that $b := \min_{x \in K} T(x) > T(z)$. If $F = \{x \in X : T(x) = b\}$, then F is compact and convex, and by the first step, has an extreme point, v, which must be an extreme point of X (again by the finite-dimensional arguments of Theorem 5.6.29). But this means that $v \in K$ because K is the convex hull of all of the extreme points of X.   ∎

10.7.c Convexification Again

Recall the we proved that the range of a nonatomic vector measure is convex. By Radon-Nikodym, this involved showing that if $(\Omega, \mathcal{F}, P)$ is nonatomic and that $f : \Omega \to \mathbb{R}^k$, then $R(f) := \{\int_E f \, dP : E \in \mathcal{F}\}$ is compact and convex. We now give the Lindenstrauss proof that represents $R(f)$ as the continuous linear image of a compact convex set.

Proof of Theorem 8.5.36. Let $B = \{g \in L^\infty(\Omega, \mathcal{F}, P) : 0 \le \text{ess inf } g \le \text{ess sup } g \le 1\}$. We know that B is convex and weak* compact (by Alaoglu's theorem) and that its extreme points are the functions 1_E with $E \in \mathcal{F}$. Therefore, the set $S(f) = \{\int fg \, dP : g \in B\}$ is a compact convex hull of $R(f)$. To complete the proof it is sufficient to show that for any $\mathbf{x} \in S(f)$, there is an extreme point of B with $\int fg \, dP = \mathbf{x}$.

Let $B_{\mathbf{x}} = \{g \in B : \int fg \, dP = \mathbf{x}\}$ and let g be an extreme point of this convex closed, hence compact, set and suppose that $P(\{\omega : g(\omega) \notin \{0, 1\}\}) > 0$. Then for some $\epsilon > 0$, $P(E) > 0$, where $E = \{\epsilon \le g \le 1 - \epsilon\}$. Partition E into $k + 1$ equally sized pieces and adjust g up and down on the pieces so as to maintain the integral against f. This can be done because we have $k + 1$ unknowns and k equalities to satisfy. ∎

10.8 ◆ Fixed Points

We have seen a number of fixed-point theorems: Tarski's for monotone functions from a lattice to itself; Banach's for contraction mappings from a complete metric space to itself; Brouwer's for continuous functions from a compact convex subset of $\mathbb{R}^\ell$ to itself; and Kakutani's for an upper hemicontinuous closed-graph convex-valued correspondence from a compact convex set of $\mathbb{R}^\ell$ to itself. In this section we generalize the last one to locally convex topological vector spaces, but first we generalize Brouwer's theorem to normed spaces.

Theorem 10.8.1 (Schauder for Normed Spaces) *If K is a nonempty compact convex subset of a normed vector space $\mathfrak{X}$ and $f : K \to K$ is a continuous function, then f has a fixed point.*

Proof. The essential idea of the proof is that for every $\epsilon > 0$, there is a continuous map, g_ϵ, from K to a finite-dimensional subset of K that is within ϵ of K. The function $g_\epsilon \circ f$ maps from a finite-dimensional compact convex set to itself, hence has a fixed point. Taking limits as $\epsilon \to 0$ completes the proof.

For any finite subset $N = \{x_1, \ldots, x_n\}$ of $\mathfrak{X}$ and any $\epsilon > 0$, let N^ϵ be the open ball around N, that is, $N^\epsilon = \cup_i B_\epsilon(x_i)$. Define $\alpha_i : N^\epsilon \to \mathbb{R}_+$ by $\alpha_i(x) = \max\{0, \epsilon - \|x - x_i\|\}$. Thus, $\alpha_i(x_i) = \epsilon$, if x is closer to x_j than any of the other x_i, $\alpha_j(x) > \alpha_i(x)$, and if x is more than ϵ away from x_i, then $\alpha_i(x) = 0$. Each α_i is continuous, yielding the continuity of the function $g_\epsilon : N^\epsilon \to \text{co}(N)$ defined by

$$g_\epsilon(x) = \frac{\sum_i \alpha_i(x) c_i}{\sum_i \alpha_i(x)},$$

and is called a Schauder projection. For our purposes, the important observation is that for all $x \in N^\epsilon$, $\|x - g_\epsilon(x)\| < \epsilon$. To see why this is true, note that

$$\|x - g_\epsilon(x)\| = \left\| \frac{\sum_i \alpha_i(x)x}{\sum_i \alpha_i(x)} - \frac{\sum_i \alpha_i(x)c_i}{\sum_i \alpha_i(x)} \right\| \tag{10.2}$$

$$= \frac{1}{\sum_i \alpha_i(x)} \left\| \sum_i \alpha_i(x)(x - c_i) \right\|$$

$$\leq \frac{\sum_i \alpha_i(x)\|x - c_i\|}{\sum_i \alpha_i(x)} < \epsilon,$$

where the last inequality comes from the observation that if x is more than ϵ away from c_i, $\alpha_i(x) = 0$.

Let $\epsilon_n \to 0$. Since K is compact, it has a finite ϵ-net, $N_n = \{x_1, \ldots, x_{m(n)}\}$, so that $K \subset N_n^{\epsilon_n}$ and $d_H(\mathrm{co}(N_n), K) < \epsilon$. Let g_n be the associated Schauder projection with $\epsilon = \epsilon_n$.

Restricted to $\mathrm{co}(N_n)$, the mappings $g_n \circ f$ are continuous from a finite-dimensional compact convex subset of $\mathfrak{X}$ to itself. Since finite-dimensional subspaces are linearly homeomorphic to $\mathbb{R}^k$, Brouwer's theorem applies, and $g_n \circ f$ has a fixed point x_n. Taking a convergent sequence if necessary give us $x_n \to x^*$ for some $x^* \in K$. Since $\|f(x_n) - g_n(f(x_n))\| < \epsilon_n$, $\|f(x_n) - x_n\| \to 0$, so that $f(x^*) = x^*$.   ∎

There are useful compactness criteria for the two normed spaces we have used most often.

1. In $C(M)$, M compact, we need K to be bounded and equicontinuous.
2. In $L^p(\Omega, \mathcal{F}, P)$, $p \in [1, \infty]$, we need every $X \in K$ to be nearly equal to its conditional expectation given a fine enough partition of Ω.

One strategy for proving Kakutani's fixed-point theorem from Brouwer's theorem in $\mathbb{R}^k$ is to show that, for every $\epsilon > 0$, one can approximate the graph of a nonempty convex-valued upper hemicontinuous correspondence by an open set at an ϵ distance in such a fashion as to allow a continuous selection. The limits of fixed points of the continuous extensions end up being fixed points of the correspondence. The same strategy works here, and we record the following without proof.

Theorem 10.8.2 *If K is a nonempty compact convex subset of a normed vector space and Ψ is a nonempty compact convex-valued upper hemicontinuous correspondence from K to itself, then Ψ has a fixed point.*

Exercise 10.8.3 (Advanced)   We showed that measure space games with finite sets of actions have equilibria. What extra assumptions or changes in the model and proof are needed to show that measure space games with compact sets of strategies have equilibria? [Lyapunov's theorem does not hold for functions taking values in general vector spaces.]

As we saw in §8.10, it is easier for a set to be compact in the weak topology that in the norm topology because there are fewer weakly open covers than there are norm open covers, which means that the need for finite subcovers is less pressing. Moreover, it is more difficult for a function to be continuous because there are fewer weakly open sets than there are norm open sets. Thus, the compactness assumption in the following may be easier to satisfy, but satisfying the continuity requirement is more difficult.

Theorem 10.8.4 (Schauder for Locally Convex Spaces) *If K is a nonempty compact convex subset of a Hausdorff locally convex topological vector space $(\mathfrak{X}, \tau)$ and $f : K \to K$ is a τ-continuous function, then f has a fixed point.*

Proof.   We briefly sketch how to extend Schauder projections to the present setting. Let G be a balanced convex neighborhood of the origin and let m_G be its Minkowski gauge. For a finite set $N = \{x_1, \ldots, x_n\}$, $\beta_i(x) = \max\{0, 1 - m_G(x - x_i)\}$ and

$$g_U(x) = \frac{\sum_i \beta_i(x) c_i}{\sum_i \beta_i(x)}.$$

(The set U in the normed case was $B_\epsilon(0)$; here it is more general.) It is easy to show that $x - g_U(x) \in U$ and that g_U is continuous. As U shrinks, the Hausdorff assumption delivers the convergence of the fixed points of $g_U \circ f$.   ∎

10.9 ◆ Bibliography

A. P. Robertson and W. Robertson's *Topological Vector Spaces* (2nd Ed., Cambridge: Cambridge University Press, 1973) lays out, with surprising economy and readability, the basics of topology for locally convex vector spaces. *Infinite Dimensional Analysis: A Hitchhiker's Guide* (3rd Ed., New York: Springer-Verlag, 2006) by C. D. Aliprantis and K. C. Border covers a broad range of vector space topics relevant for general equilibrium analysis beyond what we touch. N. L. Carothers's *A Short Course on Banach Space Theory* (Cambridge: Cambridge University Press, 2005) gives a lovely coverage of the major classical topics in Banach space theory.

CHAPTER 11

Expanded Spaces

We constructed the space $\mathbb{R}$ by completing $\mathbb{Q}$, defining $\mathbb{R}$ as the set of Cauchy equivalence classes of Cauchy sequences. This significantly expanded $\mathbb{Q}$ and gave us a model of quantities with many useful properties. Sometimes, however, the model is inadequate for our purposes and we have to add more elements to $\mathbb{R}$. The most useful class of these expansions goes by the misleading name of "nonstandard," and this part of real analysis is often called "nonstandard analysis."

Given a set G, we construct $^*G$ (read "star G") the expanded version of G, as follows:

1. consider $\mathcal{S}(G)$, the set of all sequences of elements of G,
2. define a very fine equivalence relation, $\sim$, on $\mathcal{S}(G)$, and
3. define $^*G$ as the set of $\sim$-equivalence classes.

By comparison, we construct $\mathbb{R}$ from $\mathbb{Q}$ as follows:

1. consider $\mathfrak{C}(\mathbb{Q})$, the set of Cauchy sequences of elements of $\mathbb{Q}$,
2. define the Cauchy equivalence relation, $\sim_{\mathfrak{C}}$, on $\mathfrak{C}(\mathbb{Q})$, and
3. define $\mathbb{R}$ as the set of $\sim_{\mathfrak{C}}$-equivalence classes.

The two main differences are the larger set of sequences and the finer equivalence relation. Unless G is finite, the result is that $^*G$ is much larger than G. The construction of $^*G$ works with any set G.

When $G = \mathbb{R}$, the most useful new elements of $^*\mathbb{R}$ are the **infinitesimals**, which are the df's, dx's, and dt's of your calculus class. A useful choice of G is $\mathcal{P}_F(A)$, the set of finite subsets of a set A. Elements of $^*\mathcal{P}_F(A)$ are called the **star-finite** or **hyperfinite** subsets of A.

We spend a fair amount of time on the basics of $^*\mathbb{R}$, going over continuity and differentiability using infinitesimals. With enough familiarity, the sequence construction begins to recede and the new elements become as intuitive as the old ones. This familiarity greatly facilitates the study of star-finite sets.

11.1 ◆ The Basics of $^*\mathbb{R}$

The constructions and ideas in this section have much broader applicability.

11.1.a A First Expansion of $\mathbb{R}$

The following shows the essential idea that we wish to make operational.

Example 11.1.1 *Consider the sequences $\vec{0} := (0, 0, 0, 0, \ldots)$, $dx := (1, \frac{1}{2}, \frac{1}{3}, \frac{1}{4}, \ldots)$, and $\vec{\epsilon} := (\epsilon, \epsilon, \epsilon, \epsilon, \ldots)$ for some strictly positive ϵ. The first two, $\vec{0}$ and dx, are Cauchy equivalent.*

Let us, temporarily, use "aa" for "almost all integers," by which we mean "for all but finitely many." Thus, (a_n) $^{aa}\!<$ (b_n) iff $\{n \in \mathbb{N} : a_n < b_n\}$ contains all but at most finitely many integers. With this, we have for every $\epsilon \in \mathbb{R}_{++}$,

$$\vec{0} \; ^{aa}\!< \; dx \; ^{aa}\!< \; \vec{\epsilon}.$$

*This represents dx being a new kind of number, which is strictly positive yet smaller than any of our usual positive numbers. This is the idea of an **infinitesimal**.*

Since the finite union of finite sets is finite, defining (a_n) $^{aa}\!=$ (b_n) in a parallel fashion gives an equivalence relation on $\mathbb{R}^{\mathbb{N}}$. Let $^{aa}\mathbb{R}$ denote the set of equivalence classes. Since $\vec{r} = (r, r, r, r, \ldots)$ embeds $\mathbb{R}$ in $^{aa}\mathbb{R}$ and $dx \notin \, ^{aa}\mathbb{R}$, we have expanded $\mathbb{R}$ to contain infinitesimals.

The expansion to $^{aa}\mathbb{R}$ is too big.

Exercise 11.1.2 Show that $^{aa}\!\leq$ is transitive but not complete. [Compare $\vec{0}$ and $(-1, 1, -1, 1, -1, 1, \ldots)$.]

This means that $^{aa}\mathbb{R}$ does not have one of the basic properties that we would like our model of quantities to have—we cannot compare sizes. This is a symptom of a more general problem with $^{aa}\mathbb{R}$, which is that the basic properties of relations on $\mathbb{R}$ do not carry over. We want it all—infinitesimals and preservation of the basic properties of relations.

11.1.b Constructing $^*\mathbb{R}$

The problem with $^{aa}\!=$ is that it is too fine an equivalence relation. The equivalence relation that we want, $\sim$, is based on a purely finitely additive, $\{0, 1\}$-valued measure μ on $\mathcal{P}(\mathbb{N})$, the set of all subsets of $\mathbb{N}$. μ should have the following properties: $\mu(\mathbb{N}) = 1$ and $\mu(A) = 0$ for all finite $A \in \mathcal{P}(\mathbb{N})$. The following is immediate.

Lemma 11.1.3 *If $\mu : \mathcal{P}(\mathbb{N}) \to \{0, 1\}$ is finitely additive, $\mu(\mathbb{N}) = 1$, and $\mu(A) = 0$ for all finite $A \in \mathcal{P}(\mathbb{N})$, then*

1. *for $A \subset \mathbb{N}$, either $\mu(A) = 1$ or $\mu(A^c) = 1$, but not both,*
2. *if $A = \cap\{A_k : 1 \leq k \leq K\}$ and $\mu(A_k) = 1$ for some k, then $\mu(A) = 1$, and*

3. *if $\{A_k : 1 \leq k \leq K\}$ is a finite partition of $\mathbb{N}$, then $\mu(A_k) = 1$ for one and only one of the A_k.*

The existence of a μ with such properties is a direct implication of Zorn's lemma, which is in turn equivalent to the axiom of choice. The details are in §11.5. Here is *$\mathbb{R}$.

Definition 11.1.4 *$(a_n) \sim (b_n)$ in $\mathbb{R}^{\mathbb{N}}$ if $\mu(\{n \in \mathbb{N} : a_n = b_n\}) = 1$. The $\sim$-equivalence class of (a_n) is written $\langle a_n \rangle$. *$\mathbb{R} = \mathbb{R}^{\mathbb{N}}/\sim$ is the set of $\sim$-equivalence classes. For any relation $R \subset \mathbb{R} \times \mathbb{R}$ on $\mathbb{R}$, we define *R by $\langle a_n \rangle$*$R\langle b_n \rangle$ iff $\mu(\{n \subset \mathbb{N} : a_n R b_n\}) = 1$.*

Note that $(a_n) \sim (b_n)$ is the same as (a_n) *$= (b_n)$.

The completeness of *$\leq$ arises from the property that for $A \subset \mathbb{N}$, either $\mu(A) = 1$ or $\mu(A^c) = 1$, but not both. The infinitesimals belong to *$\mathbb{R}$ because $\mu(A) = 1$ for all A with a finite complement.

Exercise 11.1.5 Using Lemma 11.1.3, show that $\sim$ is an equivalence relation, that *$\leq$ is transitive and complete on *$\mathbb{R}$, and that for all $\epsilon \in \mathbb{R}_{++}$, $\langle \vec{0} \rangle$ *$< \langle dx \rangle$ *$< \langle \vec{\epsilon} \rangle$.

In principle, the equivalence relation $\sim$, hence *$\mathbb{R}$, depends on the choice of μ, and there are many μ having the given properties (see Exercises 11.5.6 and 11.5.7). However, nothing we do depends on μ except through the basic properties listed in Lemma 11.1.3.

11.1.c Infinitesimals, Standard Parts, and Nonstandard Numbers

An intuitive picture of the part of *$\mathbb{R}$ that is very near to $\mathbb{R}$ can be had by imagining a cloud of new points, $r \pm dx$, around each $r \in \mathbb{R}$, where dx is the equivalence class of a sequence converging to 0. Out beyond the points near $\mathbb{R}$ are the infinite numbers, for example, the equivalence class of the sequence $(1, 2, 3, 4, 5, \ldots)$. We now give some definitions and results that make these intuitive pictures useful.

The first point about the intuitive picture is that every $r \in \mathbb{R}$ also belongs to *$\mathbb{R}$. Specifically, $\mathbb{R}$ is embedded in *$\mathbb{R}$ by $r \mapsto \langle (r, r, r, r, \ldots) \rangle$. This is directly parallel to how we embedded $\mathbb{Q}$ in $\mathbb{R}$.

The second point is that functions are relations and, from Definition 11.1.4, we know how to extend relations. Specifically, if $f : \mathbb{R} \to \mathbb{R}$ so that the graph of f is a subset of $\mathbb{R} \times \mathbb{R}$, then *$f(\langle x_n \rangle) = \langle f(x_n) \rangle$. In particular, if f is the absolute value function, $f(r) = |r|$, then *$|\langle x_n \rangle| = \langle |x_n| \rangle$.

Notation Alert 11.1.A *It is a pain to write "*$|\langle x_n \rangle|$," because of the "*," because of the "$\langle \cdot \rangle$," and because of the sequence notation, "x_n." The Cauchy equivalence classes of rationals, $[q_n]$, became, more simply, r's. The same happens here. Rather than write "*$|\langle x_n \rangle|$" for the absolute value of an element $\langle x_n \rangle \in$ *$\mathbb{R}$, we write $|x|$ for $x \in$ *$\mathbb{R}$ and leave it to context to clarify what we mean. For example, for $r \in \mathbb{R}$, we write r rather than $\langle (r, r, r, r, \ldots) \rangle$.*

The third point is the definition of an infinitesimal, from which come the near-standard numbers.

Definition 11.1.6 $x \in {}^*\mathbb{R}$ is **infinitesimal**, written $x \simeq 0$, if for every $\epsilon \in \mathbb{R}_{++}$, $0 \leq |x| < \epsilon$.

Note that we also did not write the more correct "$0 \, {}^* \leq |x| \, {}^* < \epsilon$." Mostly, we use nonzero infinitesimals, but it is convenient for some results to allow 0 as well.

Exercise 11.1.7 Let (x_n) be a sequence in $\mathbb{R}$ with $x_n \neq 0$ for all n. Show that either $x := \langle x_n \rangle < 0$ or $x > 0$.

In terms of the sequence construction, x is infinitesimal iff it is the equivalence class of a sequence going to 0. This gives us the cloud of new points around 0. To get the cloud of new points around $r \neq 0$, we have to see how addition works in ${}^*\mathbb{R}$. You should not be surprised to find out that it works just the same way as in $\mathbb{R}$.

The functions $(x, y) \mapsto x + y$, $(x, y) \mapsto x - y$, $(x, y) \mapsto x \cdot y$, and when $y \neq 0$, $(x, y) \mapsto x/y$, from $\mathbb{R}^2$ to $\mathbb{R}$ also have obvious extensions. Specifically, for $x, y \in {}^*\mathbb{R}$, $x = \langle x_n \rangle$, $y = \langle y_n \rangle$, $x + y := \langle x_n + y_n \rangle$, and so forth. This directly parallels how we extended addition of rational numbers to addition of real numbers.

Definition 11.1.8 $x \in {}^*\mathbb{R}$ is **near-standard** if there exists $r \in \mathbb{R}$ such that $r - x \simeq 0$, it is **infinite** if $|x| \geq M$ for all $M \in \mathbb{N}$, and $ns({}^*\mathbb{R})$ denotes the near-standard points.

Exercise 11.1.9 In terms of sequences, x is near-standard iff it is the equivalence class of a convergent sequence and it is infinite iff it is the equivalence class of a sequence going either to $+\infty$ or to $-\infty$. Show that a nonzero $x \in {}^*\mathbb{R}$ is infinite iff $1/x$ is infinitesimal.

The last exercise proves the fourth point about the intuitive picture.

Lemma 11.1.10 An $x \in {}^*\mathbb{R}$ is either near-standard or infinite.

The following arises from the continuity of addition.

Lemma 11.1.11 If $x, y \in {}^*\mathbb{R}$ are infinitesimal, then so is $x + y$.

Exercise 11.1.12 Prove Lemma 11.1.11.

In light of the last two exercises, the following reiterates that a convergent sequence can have only one limit.

Theorem 11.1.13 If $x \in {}^*\mathbb{R}$ is near-standard, then there is a unique $r \in \mathbb{R}$ such that $r - x \simeq 0$.

Proof. If x is near-standard, then there exists at least one $r \in \mathbb{R}$ such that $r - x \simeq 0$. Suppose that $r' \in \mathbb{R}$, $r' \neq r$, and $r' - x \simeq 0$. In terms of sequences, $|r - x_n| + |x_n - r'| \geq |r - r'|$ for all n. Hence, if $r' \neq r$, then $|r - x| + |x - r'| \geq |r - r'|$. This says that the sum of two infinitesimals is greater than $|r - r'| \in \mathbb{R}_{++}$, contradicting Lemma 11.1.11. ∎

Theorem 11.1.13 shows that the following is well defined.

Definition 11.1.14 The mapping $x \mapsto st(x)$ assigns to each $x \in ns({}^*\mathbb{R})$ the unique $r \in \mathbb{R}$ such that $x - r \simeq 0$. It is called the **standard part mapping**.

You should suspect that the standard part mapping has many similarities to taking limits of sequences.

Combining, we have now shown that the cloud of points in *ℝ around an $r \in \mathbb{R}$ is of the form $r \pm dx$ for some $dx \simeq 0$ and that everything else in *ℝ is infinite. Now we should do something useful with these pictures.

11.1.d Using Infinitesimals

To start, we use infinitesimals to re-prove things that we already know. This is by way of starting gently. We rely, at first, on intuitions gained from limit arguments.

Lemma 11.1.15 $f : \mathbb{R} \to \mathbb{R}$ *is continuous iff for all near-standard $x \in$ *ℝ and all infinitesimals dx, $f(x + dx) \simeq f(x)$, equivalently for all near-standard x, $f(\mathrm{st}(x)) = \mathrm{st}(f(x))$.*

In words, this lemma says that if you move an infinitesimal amount in the domain, you move at most an infinitesimal amount in the range. This is the core intuitive idea of continuity. [1] Also, note that we did not write $^*f(x + dx) \simeq {}^*f(x)$, though perhaps we should have. The idea is that the context, evaluating a function at points in *ℝ, should have made it clear that we must be using $^*f$, the extension of f from ℝ to *ℝ.

Proof.   Suppose that f is continuous and pick arbitrary $\epsilon \in \mathbb{R}_{++}$. To show that $f(\mathrm{st}(x)) = \mathrm{st}(f(x))$, it is sufficient to show that $|f(r) - \mathrm{st}(f(x))| < \epsilon$, where $r = \mathrm{st}(x)$. Since f is continuous at r, there exists a $\delta > 0$ such that $[|r - y| < \delta] \Rightarrow [|f(r) - f(y)| < \epsilon/2]$. Since $|r - x| \simeq 0$, it is less than δ. Therefore $|^*f(r) - {}^*f(x)|^* < \epsilon/2$, which implies that $|f(r) - \mathrm{st}\, f(x)| < \epsilon$.

Now suppose that f is not continuous at some $r \in \mathbb{R}$. This means that there exists $\epsilon > 0$ such that for all $\delta > 0$, there exists $y \in \mathbb{R}$ such that $|r - y| < \delta$ and $|f(r) - f(y)| \geq \epsilon$. Take z to be the equivalence class of a sequence $y_n \to r$ with $|f(r) - f(y_n)| \geq \epsilon$ and observe that the $\mathrm{st}(z) = r$. ■

For $X \subset \mathbb{R}$, $^*X := \{\langle x_n \rangle : \mu(\{n \in \mathbb{N} : x_n \in X\}) = 1\}$.

Theorem 11.1.16 $X \subset \mathbb{R}$ *is closed iff $X = \mathrm{st}(^*X \cap ns(^*\mathbb{R}))$, and a closed $X \subset \mathbb{R}$ is compact iff every $x \in {}^*X$ is near-standard.*

Proof.   For the first part, observe that $x \in \mathrm{st}(^*X \cap ns(^*\mathbb{R}))$ iff it is the equivalence class of a sequence in X that converges to x. For the second part, observe that a set X is bounded iff $^*X$ contains no infinite elements, that is, no equivalence classes of sequences going either to $+\infty$ or to $-\infty$. ■

---

1. Working before a rigorous model of the real numbers had been given, Augustin Cauchy's *Cours d'Analyse de l'École Royal Polytechnique, 1.re Partie. Analyse Algébrique* (De l'Imprimerie Royale, 1821) was one of the most successful early attempts to put real analysis on a firm mathematical foundation. For Cauchy, a function is continuous if "an infinitely small increment of the variable always produces an infinitely small increment of the function itself." [The translation and history are from M. Kline's *Mathematical Thought from Ancient to Modern Times* (New York: Oxford University Press, 1972).] Eventually this definition was replaced by the well-known ϵ–δ definition because of difficulties in giving a firm logical foundation to "infinitely small increments," a difficulty only solved in the 1960s by A. Robinson; see his *Non-standard Analysis* (Rev. Ed., Amsterdam: North-Holland, 1974 (1980 printing)).

Internal sets, introduced in the following, are crucial to what follows. The class of internal sets called the star-finite ones is especially important.

Definition 11.1.17   *The equivalence class, $X = \langle X_n \rangle$, of a sequence X_n of subsets of $\mathbb{R}$ is called an **internal set**. If the X_n can be taken to be equal to each other, then (the necessarily internal) set X is called a **standard set**. If each of the X_n can be taken to be finite, then the internal X is called **star-finite** or *-**finite**.*

To reiterate, every standard set is internal, but the reverse is not true.

Now, $x = \langle x_n \rangle$ belongs to an internal set $X = \langle X_n \rangle$ iff $\mu(\{n \in \mathbb{N} : x_n \in X_n\}) = 1$. A really useful example to keep in mind takes $X_n = \{k/2^n : k = 0, 1, \ldots, 2^n\}$ and $X = \langle X_n \rangle$. Note that the Hausdorff distance between X and *$[0, 1]$ is $\langle 1/2^n \rangle \simeq 0$. That is, X is a *-finite set that is infinitely close to $[0, 1]$.

If X is a compact subset of $\mathbb{R}$, then for every $n \in \mathbb{N}$, there exists a finite subset, $X_n \subset X$, such that the Hausdorff distance $d_H(X_n, X) < 1/n$. Thus, if X is compact, then there is a star-finite subset, $X' \subset {}^*X$, with $d_H(X', X) \simeq 0$.

Theorem 11.1.18   *If $f : X \to \mathbb{R}$ is continuous and X is compact, then there exists an $x^\circ \in X$ such that $f(x^\circ) \geq f(x)$ for all $x \in X$.*

Proof.   Let X' be a star-finite subset of *X with $d_H(X', X) \simeq 0$ so that $X' = \langle X_n \rangle$ for some sequence of finite sets with $d_H(X_n, X) \to 0$. As X' is star-finite, there exists an $x' \in X'$ that solves $\max_{y \in X'} f(y)$ (because we can take x' to be the equivalence class of any sequence of x_n that solves the problem $\max_{y \in X_n} f(y)$). Set $x^* = \operatorname{st}(x')$. If there exists an $x^\circ \in X$ such that $f(x^\circ) > f(x^*) + 2\epsilon$ for some $\epsilon > 0$, then there exists a δ-ball, $B_\delta(x^0)$, on which f is greater than $f(x^*) + \epsilon$. Since $d_H(X', X) \simeq 0$, $X' \cap {}^*B_\delta(x^0) \neq \emptyset$, which contradicts the definition of x'. ∎

With the following in hand, we can prove the chain rule, $\frac{df(g(x))}{dx} = f'(g)g'(x)$, by cancellation, $\frac{df}{dg} \cdot \frac{dg}{dx} = \frac{df}{dx}$.

Definition 11.1.19   *A function $f : \mathbb{R} \to \mathbb{R}$ is **differentiable at x** if there exists an $r \in \mathbb{R}$ such that for all nonzero infinitesimals dx, $r = \operatorname{st}\left(\frac{f(x+dx)-f(x)}{dx} \right)$. With an infinitesimal loss of precision, this is written $f'(x) = \frac{f(x+dx)-f(x)}{dx}$.*

For the novel applications, we need some extra tools, ones that will begin to submerge the sequence construction.

11.2 ◆ Superstructures, Transfer, Spillover, and Saturation

It is time to return to §2.13, where we imposed limitations on the objects we are willing to call sets. In all of the intervening work, we have not had any regrets about imposing these limitations, and we do not regret it here either.

11.2.a Superstructures

Recall, we started with a set S, agreed that S is a set, and required that none of the elements of S contain any elements. Then we had

Definition 2.13.1.   Given a set S, define a sequence of sets $V_m(S)$ by

$$V_0(S) = S, \quad \text{and} \quad V_{m+1}(S) = V_m(S) \cup \mathcal{P}(V_m(S)).$$

The **superstructure over** S is the union $V(S) = \cup_{m \geq 0} V_n(S)$, and S is the **base set**. For any $x \in V(S)$, the **rank of** x is the smallest m such that $x \in V_m(S)$.

   With the right base set, S, a superstructure contains all of the mathematical objects we care about. For example, ordered pairs, (a, b), are defined as $\{\{a\}, \{a, b\}\}$. $\{a\} \in V_1(S)$ and $\{a, b\} \in V_1(S)$, so that the set of all ordered pairs of elements of $\mathbb{R}$ belongs to $V_2(S)$. A function $f : \mathbb{R} \to \mathbb{R}$ is a set of ordered pairs with the property that each first element occurs exactly once. Hence, every function $f : \mathbb{R} \to \mathbb{R}$ belongs to $V_3(S)$. Indeed, for any set $E \subset \mathbb{R}$, every $f : E \to \mathbb{R}$ belongs to $V_3(S)$. Therefore every set of continuous functions on some domain $E \subset \mathbb{R}$ belongs to $V_4(S)$. And so on and so forth.
   We define $V_m(^*S)$ inductively and set $V(^*S) = \cup_{m \geq 0} V_m(^*S)$. The only delicate part is keeping track of membership relations.

1. Let G_n be a sequence in $V_0(S)$. Thus, G_n is a sequence of elements, x_n, in S.
    a. Define $x_n \sim x_n'$ if $\mu(\{n \in \mathbb{N} : x_n = x_n'\}) = 1$ and $V_0(^*S)$ as the set of equivalence classes in $(V_0(S))^{\mathbb{N}}$. (For instance, if $S = \mathbb{R}$, then $V_0(^*S)$ is $^*\mathbb{R}$.)
    b. If $x \in V_0(^*S)$ is the equivalence class of a constant sequence, it is called **standard**; otherwise it is called **nonstandard**. (For instance, elements of $\mathbb{R}$ are standard, while infinitesimals, reals plus infinitesimals, and infinite numbers are nonstandard.)
2. Now let G_n be a sequence in $V_1(S)$ that is not a sequence in $V_0(S)$. Thus, G_n is a sequence of subsets of S.
    a. Define $G_n \sim G_n'$ if $\mu(\{n \in \mathbb{N} : G_n = G_n'\}) = 1$ and $V_1(^*S)$ as the union of the class of equivalence classes in $(V_1(S))^{\mathbb{N}}$ and $V_0(^*S)$. (For instance, if $S = \mathbb{R}$, then $V_1(^*S)$ is the class of internal subsets of $^*\mathbb{R}$).
    b. An element $x = \langle x_n \rangle$ of $V_0(^*S)$ belongs to an internal set, that is, to an equivalence class $G = \langle G_n \rangle$ in $V_1(^*S) \setminus {}^*V_0(^*S)$ if $\mu(\{n \in \mathbb{N} : x_n \in G_n\}) = 1$. This is written $x \, {}^* \in G$ or $x \in G$.
    c. A $G \in V_1(^*S)$ is called **internal**, and if G is the equivalence class of a constant sequence, then it is called **standard**. (For instance, $^*[0, 1]$, the equivalence class of $([0, 1], [0, 1], [0, 1], [0, 1], \ldots)$, is a standard set, whereas the equivalence class of $(\{0, 1\}, \{0, \frac{1}{2}, 1\}, \{0, \frac{1}{4}, \frac{1}{2}, \frac{3}{4}, 1\}, \ldots)$ is a nonstandard set.)

3. Given that $V_{n-1}(*S)$ has been defined, let G_n be a sequence in $V_n(S)$ that is not a sequence in $V_{n-1}(S)$.

   a. Define $G_n \sim G'_n$ if $\mu(\{n \in \mathbb{N} : G_n = G'_n\}) = 1$ and $V_n(*S)$ as the union of the class of equivalence classes in $(V_n(S))^{\mathbb{N}}$ and $V_{n-1}(S)$.

   b. An element $x = \langle x_n \rangle$ of $V_{n-1}(*S)$ belongs to an equivalence class $G = \langle G_n \rangle$ in $V_n(*S)$ if $\mu(\{n \in \mathbb{N} : x_n \in G_n\}) = 1$. This is written $x *\in G$ or $x \in G$.

   c. A $G \in V_n(*S)$ is called **internal**, and if G is the equivalence class of a constant sequence, then it is called **standard**.

4. Finally, $V(*S)$ is defined as $\cup_{m \geq 0} V_m(*S)$ and is called a **nonstandard expansion** of $V(S)$ or a **nonstandard extension** of $V(S)$.

It is useful to see some examples of what has been constructed. When F is infinite, we always get new nonstandard elements, and the expansion is proper.

Example 11.2.1 *If $F \in V(S)$ is finite, then $*F = \{*x : x \in F\}$ because for any sequence x_n in $F = \{f_1, \ldots, f_K\}$, there is a unique k such that $\mu(\{n \in \mathbb{N} : x_n = f_k\}) = 1$. If $F \in V(S)$ is not finite, then it contains a countable set of distinct points. Taking the equivalence class of a sequence enumerating these points gives a $y \in *F$ such for all $x \in F$, $y \neq *x$.*

A more subtle aspect of the nonstandard expansion is that internal sets are either finite or uncountable.

Lemma 11.2.2 *If A is an internal set, then it is either finite or uncountable.*

Be careful about what this is *not* saying. There is *not* a "*" in front of "finite" or "uncountable."

Example 11.2.3 *The infinite star-finite set, $A = \langle A_n \rangle$, $A_n = \{k/2^n : 0 \leq k \leq 2^n\}$, has the property that the many-to-one mapping $x \mapsto \mathrm{st}(x)$ maps onto $[0, 1]$, so that A, though $*$-finite, is uncountable. The $*$-countable set, $*\mathbb{Q}$, contains A, hence is uncountable. The $*$-uncountable set, $*\mathbb{R}$, contains $*\mathbb{Q}$, hence is also uncountable.*

Proof of Lemma 11.2.2. Either $\exists M \in \mathbb{N}$ such that $\mu(\{n \in \mathbb{N} : \#A_n \leq M\}) = 1$ or else $\forall M \in \mathbb{N}$, $\mu(\{n \in \mathbb{N} : \#A_n > M\}) = 1$. In the first case, $\exists m \leq M$ such that $\mu(\{n \in \mathbb{N} : \#A_n = m\}) = 1$. Let f_n be a bijection between each of these A_n and the set $\{1, \ldots, m\}$. Then $f = \langle f_n \rangle$ is a bijection between A and $*\{1, \ldots, m\} = \{1, \ldots, m\}$.

Now suppose that $\forall M \in \mathbb{N}$ and $\mu(\{n \in \mathbb{N} : \#A_n > M\}) = 1$. Enumerate the rationals in $[0, 1]$ as $\{q_i : i \in \mathbb{N}\}$. Define $K_n = \min\{n, \#A_n\}$, set $B_n = \{q_1, \ldots, q_{K_n}\}$, and let f_n be a (potentially many-to-one) mapping from A_n onto B_n. The internal function $f = \langle f_n \rangle$ maps A on $B = \langle B_n \rangle$. $\mathrm{st}(B)$ is closed and contains all of the rationals in $[0, 1]$, so that B is uncountable. ∎

Internal sets are like compact sets: they have the finite intersection property.

Theorem 11.2.4 *If $(A_i)_{i \in \mathbb{N}}$ is a sequence of internal sets of bounded rank and $\cap_{i \leq I} A_i \neq \emptyset$ for all $I \in \mathbb{N}$, then $\cap_{i \in \mathbb{N}} A_i \neq \emptyset$.*

Proof.   Each A_i is of the form $\langle A_{i,n} \rangle$. Since $A_1 \neq \emptyset$, there is no loss in assuming that for all $n \in \mathbb{N}$, $A_{1,n} \neq \emptyset$. In tabular form,

$$A_1 = \langle A_{1,1}, A_{1,2}, A_{1,3}, A_{1,4}, A_{1,5}, \ldots$$

$$A_2 = \langle A_{2,1}, A_{2,2}, A_{2,3}, A_{2,4}, A_{2,5}, \ldots$$

$$A_3 = \langle A_{3,1}, A_{3,2}, A_{3,3}, A_{3,4}, A_{3,5}, \ldots$$

$$A_4 = \langle A_{4,1}, A_{4,2}, A_{4,3}, A_{4,4}, A_{4,5}, \ldots$$

$$A_5 = \langle A_{5,1}, A_{5,2}, A_{5,3}, A_{5,4}, A_{5,5}, \ldots$$

$$\vdots \quad \vdots \quad \vdots \quad \vdots$$

Now, for all I, by assumption, $\mu(\{n \in \mathbb{N} : \cap_{i \leq I} A_{i,n} \neq \emptyset\}) = 1$. We now produce a sequence of I_n that goes to ∞ in a very useful way.

1. Define $I_1 = \max\{I \leq 1 : \cap_{i \leq I} A_{i,1} \neq \emptyset\}$, so that $I_1 = 1$.
2. Define $I_2 = \max\{I \leq 2 : \cap_{i \leq I} A_{i,2} \neq \emptyset\}$.
3. Define $I_3 = \max\{I \leq 3 : \cap_{i \leq I} A_{i,3} \neq \emptyset\}$.

More generally, define $I_n = \max\{I \leq n : \cap_{i \leq I} A_{i,n} \neq \emptyset\}$. In words, for the nth column in the table above, one looks down the column, no further than the nth entry, looking for the maximum depth for which the intersection of the sets down to that depth is nonempty.

For each $n \in \mathbb{N}$, pick $x_n \in \cap_{i \leq I_n} A_{i,n}$ and define $x = \langle x_n \rangle$. We have to show that for all I, $x \in \cap_{i \leq I} A_i$. This follows from the observations that for all I, $\mu(\{n \in \mathbb{N} : n \geq I\}) = 1$ and $\mu(\{n \in \mathbb{N} : \cap_{i \leq I} A_{i,n} \neq \emptyset\}) = 1$. ∎

An immediate (DeMorgan's laws) corollary is the following.

Corollary 11.2.5   *If $(B_i)_{i \in \mathbb{N}}$ is a sequence of internal sets, then $B = \cup_{i \in \mathbb{N}} B_i$ is internal iff $B = \cup_{i \leq I} B_i$ for some $I \in \mathbb{N}$.*

Proof.   Suppose that B is internal so that each $A_i := B \setminus B_i$ is internal. Since $\cap_{i \in \mathbb{N}} A_i = \emptyset$, there exists an $I \in \mathbb{N}$ such that $\cap_{i \leq I} A_i = \emptyset$ so that $B = \cup_{i \leq I} B_i$. ∎

This means that classes of internal sets can be fields, but they cannot be σ-fields. This will be crucial when we start doing probability theory with these expanded spaces.

11.2.b   Examples of Expanded Spaces

Function spaces and measure spaces are crucial to economics. Their expanded versions contain useful limit elements. Here are some examples.

Example 11.2.6   *Suppose that S contains $\mathbb{R}$ so that $V_4(S)$ contains $C([0, 1])$, the set of continuous $\mathbb{R}$-valued functions on $[0, 1]$. $^*C([0, 1])$ is the set of *-continuous functions. Since $^*C([0, 1])$ includes the equivalence class of the sequence of functions $f_n(t) = \sin(nt)$, it contains nonstandard functions with infinite slopes. Since it includes the equivalence class of the sequence of functions $g_n(t) \equiv n$, it contains* *nonstandard functions taking on only infinite values. As it includes the equivalence class of the sequence of functions $h_n(t) = (2t)^n$, it contains nonstandard functions taking on both near-standard and infinite values.*

The idea of near-standard generalizes to metric spaces, for example, $C([0, 1])$.

Example 11.2.7 *An $f \in {}^*C([0, 1])$ is **near-standard** if there is a standard function g such that*

$$\max_t \ |f(t) - g(t)| \simeq 0$$

Note that we used the function on functions, "max," which belongs to $V_6(S)$, that the relevant t^*- belong to $^*[0, 1]$, that both f and "$|\cdot|$" are functions from $\mathbb{R}$ to $\mathbb{R}$, and that "$-$" is a function from $\mathbb{R} \times \mathbb{R}$ to $\mathbb{R}$. Hence, if we were keeping track of all of the *'s, the previous maximum would be written

$$^*\max_{t \ ^*\in \ ^*[0,1]} \ ^*|f(t) \ ^* - \ ^*g(t)| \simeq 0.$$

The six *'s rather clutter that up, so we prefer the previous version.

If $n \in {}^*\mathbb{N} \setminus \mathbb{N}$ is an infinite integer, then $f(t) = \sin(nt)$ is not near-standard, though $h(t) = t^2 + \frac{1}{\sqrt{n}} \sin(nt)$ is near-standard. [Writing $h(\cdot)$ with all of the requisite *'s is really messy.]

The star-finite spaces make particularly convenient probability spaces. The next example gives a model of independent, idiosyncratic shocks to each member of a large, indeed uncountable, population. These average out to something deterministic.

Example 11.2.8 *For an infinite integer $m \in {}^*\mathbb{N}$, let $n = m!$, so that n can be evenly divided into k equally sized parts for all $k \in \mathbb{N}$. Consider the star-finite space $\Omega = \{-1, +1\}^n$, which has 2^n points, and let $\mathcal{F}^\circ$ denote the class of internal subsets of Ω. Let P be the uniform distribution on Ω; that is, for any $A \in \mathcal{F}^\circ$, define $P(A) \in {}^*[0, 1]$ by $P(A) = \frac{\#A}{\#\Omega} = \frac{\#A}{2^n}$.*

For $i \in \{1, 2, \ldots, n\}$, let X_i be the random variable $X_i(\omega) = \text{proj}_i(\omega)$, so that $P(X_i = -1) = P(X_i = +1) = \frac{1}{2}$, $E\,X_i = 0$, and $\text{Var}(X_i) = 1$. For any internal $S \subset \{1, 2, \ldots, n\}$, the collection of random variables, $\{X_i : i \in S\}$ is *-iid ± 1 with probability $\frac{1}{2}$.

If $\#S$ is an internal infinite subset of $\{1, 2, \ldots, n\}$, then the average of the random variables is infinitely close to 0 with probability infinitely close to 1. To see why, calculate the *-mean and *-variance of the random variable $Y_S(\omega) :=$ $\frac{1}{\#S} \sum_{i \in S} X_i(\omega)$. $E\,Y_S = 0$ and $\text{Var}(Y_S) = \left(\frac{1}{\#S}\right)^2 \sum_{i \in S} \text{Var}(X_i) = \frac{\#S}{(\#S)^2} \simeq 0$. By Chebyshev's inequality, for any standard $\epsilon > 0$, $P(\{\omega : |Y_S(\omega)| > \epsilon\}) \simeq 0$.

We think of $I = \{1, 2, \ldots, n\}$ as an infinite population, and we have seen that it is uncountable. We can also think of X_i, $i \in I$, as the shock to person i's wealth. The last calculation showed that, averaged over any internal infinite subset of the population, the shock is 0. That is, independent idiosyncratic shocks average out.

Independence is not needed. Any condition that makes $\left(\frac{1}{\#S}\right)^2 \text{Var}(\sum_{i \in S} X_i) \simeq 0$ will do. This allows, for example, for each person's wealth shock to depend on, say, the shock to the wealth of finitely many of the person's business contacts.

It turns out that the $\mathcal{F}^\circ$ is a field, not a σ-field, and that $\text{st}(P): \mathcal{F}^\circ \to [0, 1]$ has a unique countably additive extension to $\mathcal{F} := \sigma(\mathcal{F}^\circ)$. This extension is denoted $L(P)$ in honor of Peter Loeb, its discoverer, and the probability space $(\Omega, \mathcal{F}, L(P))$ is called a **Loeb space**.

The following use of Loeb spaces is due to Robert Anderson.

Example 11.2.9   [↑*Examples 6.2.23, 11.2.8*] *Let n and $(\Omega, \mathcal{F}^\circ, P)$ be as in the previous example. Let $T \subset {}^*[0, 1]$ be the set of points $T = \{\frac{k}{n} : k = 0, 1, \ldots, n\}$, so that T is an internal subset of ${}^*[0, 1]$ with $d_H(T, [0, 1]) \simeq 0$. Define $dt = \frac{1}{n}$ so that $dt \simeq 0$, $\sqrt{dt} \simeq 0$, and $\frac{\sqrt{dt}}{dt} = \frac{1}{\sqrt{dt}}$ is infinite.*

For each ω and $\frac{k}{n}$, define $X(\omega, \frac{k}{n}) = \sum_{i \leq k} X_i(\omega) \cdot \sqrt{dt}$. For each ω and each $\tau \in {}^*[0, 1] \setminus T$, define $X(\omega, \tau)$ by linear interpolation. This makes $X(\cdot, \cdot)$ a random walk over time intervals of length dt with increments of size $\pm\sqrt{dt}$.

More specifically, each $X(\omega, \cdot) \in {}^*C([0, 1])$, $X(\omega, 0) \equiv 0$, and $P(\{\omega : X(\omega, \frac{1}{n}) = +\sqrt{dt}\}) = P(\{\omega : X(\omega, \frac{1}{n}) = -\sqrt{dt}\}) = \frac{1}{2}$. Define the $(k + 1)$th increment of $X(\cdot, \cdot)$ by $\Delta_{k+1}(\omega) = X(\omega, \frac{k+1}{n}) - X(\omega, \frac{k}{n})$. We have $\Delta_{k+1}(\omega) \in \{-\sqrt{dt}, +\sqrt{dt}\}$ for all ω, and $P(\{\omega : \Delta_{k+1}(\omega) = -\sqrt{dt}\}) = P(\{\omega : \Delta_{k+1}(\omega) = +\sqrt{dt}\}) = \frac{1}{2}$. Also, $X(\omega, \cdot)$ is a straight line with infinite slope $\pm\frac{1}{\sqrt{dt}}$ over every interval $(\frac{k}{n}, \frac{k+1}{n})$.

Using an inequality a bit more sophisticated than Chebyshev's, Anderson showed that there is a set $A \in \mathcal{F}$ such that $L(P)(A) = 1$, and for all $\omega \in A$, $X(\omega, \cdot)$ is near-standard in $C([0, 1])$. Further, for any standard $0 \leq t_1 < t_2 < \cdots < t_K \leq 1$, the random variables $\text{st}(X(\cdot, t_2)) - \text{st}(X(\cdot, t_1))$, $\text{st}(X(\cdot, t_3)) - \text{st}(X(\cdot, t_2))$, $\ldots$, $\text{st}(X(\cdot, t_K)) - \text{st}(X(\cdot, t_{K-1}))$ are independent and Gaussian and have variances $(t_2 - t_1), \ldots, (t_K - t_{K-1})$. This means that $(\omega, t) \mapsto \text{st}(X(\omega, t))$ is a Brownian motion.

The Gaussian distribution is **infinitely divisible**. That is, for every Gaussian distribution, Q, and every $k \in \mathbb{N}$, there are k iid random variables, $Y_1, \ldots, Y_k$, such that the distribution of $\sum Y_i$ is Q. From the construction just given one can see that this is true. Note that $X(\cdot, 1)$ has a standard Gaussian distribution, and that for all ω, $X(\omega, 1) = [X(\omega, \frac{1}{k}) - X(\omega, 0)] + [X(\omega, \frac{2}{k}) - X(\omega, \frac{1}{k})] + \cdots + [X(\omega, 1) - X(\omega, \frac{k-1}{k})]$.

There are many more infinitely divisible distributions and they can all be reached in this fashion—a continuous-time stochastic process over the interval $[0, 1]$, having independent increments. The following shows that the Poisson distribution is infinitely divisible and why it is sometimes called the "rare events" distribution.

Example 11.2.10   *As before, let $n = m!$ for some $m \in {}^*\mathbb{N} \setminus \mathbb{N}$. Unlike before, let $\Omega = \{0, 1\}^n$. Define P on the internal subsets of Ω so that the collections $\{X_i : i \in S\}$ are iid with $P(X_i = 1) = \frac{\lambda}{n}$ for some standard $\lambda > 0$. Define $X(\omega, \frac{k}{n}) = \sum_{i \leq k} X_i$. To calculate $P(\{\omega : X(\omega, 1) = 0\})$, note that $X(\omega, 1) = 0$ iff $X_i = 0$ for $i = 1, 2, \ldots, n$, and that this event happens with probability $(1 - \frac{\lambda}{n})^n \simeq e^{-\lambda}$. In a similar fashion, $P(X(\cdot, 1) = 1) = n \cdot \frac{\lambda}{n}(1 - \frac{\lambda}{n})^{n-1} \simeq \lambda e^{-\lambda}$. Continuing in this fashion, we find that $X(\cdot, 1)$ is distributed as a Poisson with parameter λ. Expressing it as the sum $X(\omega, 1) = [X(\omega, \frac{1}{k}) - X(\omega, 0)] + [X(\omega, \frac{2}{k}) - X(\omega, \frac{1}{k})] + \cdots + [X(\omega, 1) - X(\omega, \frac{k-1}{k})]$ shows that the distribution is infinitely divisible. It* *is called the rare events distribution because, over any small interval of time dt, the probability that an event happens is infinitesimal, $dt \cdot \lambda$.*

11.2.c Transfer and Spillover

The transfer principle is one of the basic tools for manipulating nonstandard objects. It leads directly to the spillover principle, which is one of the basic tools for manipulating nonstandard objects that come from metric spaces. Both transfer and spillover make it easier to recognize whether or not a set is internal, that is, whether or not it is the equivalence class of a sequence of sets, without needing the explicit sequence construction.

Roughly, the transfer principle says that a formal statement that uses only quantifiers ranging over objects in $V(S)$ is true iff the same formal statement is true when the quantifiers range over the $*$'d versions of the objects. Since the sets that we care about are of the form $A = \{x \in X : P(x)\}$ for some statement $P(\cdot)$—indeed, this is what we mean by a statement—this means that putting $*$'s all over the statement $P(\cdot)$ gives us an internal set.

Heuristically, this involves writing out a true formal statement and "putting $*$'s in front of everything in sight," which is fun the first several times that you do it. For example, a formal statement of the "sets that are bounded above have a least upper bound" property of the real line is as follows:

> For all $A \in \mathcal{P}(\mathbb{R})$, if there exists a $B \in \mathbb{R}$ such that for all $a \in A$, $a \le B$, then there exists an $x \in \mathbb{R}$ such that for all $a \in A$, $a \le x$, and if $x' \in \mathbb{R}$ and $x' < x$, then there exists an $a' \in A$ with $x' < a'$.

By transfer, the $*$'d version of this statement is true, that is, the following is true:

> For all $A * \in *\mathcal{P}(\mathbb{R})$, if there exists a $B * \in *\mathbb{R}$ such that for all $a * \in A$, $a * \le B$, then there exists an $x * \in *\mathbb{R}$ such that for all $a * \in A$, $a * \le x$, and if $x' * \in *\mathbb{R}$ and $x' * < x$, then there exists an $a' * \in A$ with $x' * < a'$.

Internal sets occur as soon as a quantifier ranges over a class of sets. In this example, they occur at the very beginning of the $*$'d statement, "for all $A * \in *\mathcal{P}(\mathbb{R})$," that is, for all internal subsets of $*\mathbb{R}$. The **internal definition principle** is a fancy name for the idea that sets defined in this way are internal. This follows from the observations that all of the logical statements that we make correspond to sets, we are working with a superstructure after all, and that we know how to put $*$'s in front of sets. Thus, $\{x \in X : P(x)\}$ corresponds to a subset, A, of X and $\{x \in *X : *P(x)\}$ corresponds to the internal set $*A$.

The convention is that we say "by transfer" whenever we want to indicate that we could have written out a formal version of the statement and put $*$'s all over the place, but we are too lazy and/or too afraid of boring you to death.[2]

---

2. A proof of the transfer principle is beyond the scope of this chapter because we would have to be formal about logical languages; see Hurd and Loeb (1985, Ch. 2) and/or Lindstrøm (1988, Ch. IV).

The transfer principle also helps us show that not all sets are internal.

Example 11.2.11   *The set of all infinitesimals,* $st^{-1}(0)$, *is **not** internal. To show this using sequence arguments, we must show that no sequence,* A_n, *of subsets of* $\mathbb{R}$ *have the property that* $\langle A_n \rangle = st^{-1}(0)$. *Here is the transfer argument:* $st^{-1}(0)$ *is bounded above, for example, by 1; if* $st^{-1}(0)$ *is internal, then it has a least upper bound,* x; *if* x *is not infinitesimal, then* $\frac{1}{2}x$ *is also an upper bound for* $st^{-1}(0)$; *if* x *is infinitesimal, then so is* $2x$, *but* $x < 2x \in st^{-1}(0)$.

This leads us to various versions of the spillover (or overspill) principle:

1. If an internal subset of $^*\mathbb{R}_+$ contains arbitrarily large infinitesimals, then it must contain standard numbers.

2. If an internal subset of $^*\mathbb{R}_+$ contains arbitrarily small noninfinitesimals, then it must contain infinitesimals.

3. If an internal subset of $^*\mathbb{R}_+$ contains arbitrarily large standard numbers, then it must contain infinite numbers.

4. If an internal subset of $^*\mathbb{R}_+$ contains arbitrarily small infinite numbers, then it must contain standard numbers.

The name "spillover" has to do with spilling over the "boundaries" between nonstandard and standard numbers. The proof is an illustration of the use of transfer.

Lemma 11.2.12 (Spillover)   *If* S *is an internal subset of* $^*\mathbb{R}_+$ *and for all standard* $\delta > 0$ *there exists an* $s \in S$ *such that* $s < \delta$, *then* S *contains an infinitesimal.*

Proof.   The set S is bounded below. By transfer, it must have an infimum, x, in $^*\mathbb{R}_+$. By assumption, $x < \delta$ for every standard $\delta > 0$ so that x is infinitesimal. Now, x is the infimum of S. By transfer, x being the infimum of S implies that there must be an $s \in S$ between x and $2x$. Such an s must be infinitesimal.   ∎

Exercise 11.2.13   Use transfer to prove the other versions of the spillover principle.

There are also integer versions of overspill.

Exercise 11.2.14   If S is an internal subset of $^*\mathbb{N}$ containing arbitrarily large standard integers, then it contains an infinite integer.

11.2.d   Spillover in Metric Spaces

Spillover is particularly useful in metric spaces. Throughout the following, we assume that the base set, S, contains M and $\mathbb{R}$ and that (M, d) is a metric space.

Definition 11.2.15   *An* $x \in {}^*M$ *is **near-standard** if there exists a* $y \in M$ *such that* $d(x, y) \simeq 0$, *and the (necessarily unique)* y *is called the **standard part of** x, denoted* $y = st(x)$. *The set of near-standard points in* M *is denoted* $ns(M)$.

In general, $ns(M)$ is not internal, for example, spillover implies that $ns(\mathbb{R})$ is not internal.

Exercise 11.2.16 If $x \in ns(M)$, then there is a unique $y \in M$ such that $d(x, y) \simeq 0$. [If $d(y, x) \simeq 0$, $d(y', x) \simeq 0$ and $d(y, y')$ is a strictly positive standard number, then . . . .]

Lemma 11.2.17 *If E is an internal subset of* $^*M$*, then* $\mathrm{st}(E \cap ns(M))$ *is closed.*

Proof. Pick $x \notin F := \mathrm{st}(E \cap ns(M))$. Consider the internal set $S = \{\delta > 0 : B_\delta(x) \cap E = \emptyset\} \subset {}^*\mathbb{R}_+$. Since $x \notin F$, S contains arbitrarily large infinitesimals. By spillover, it contains a strictly positive standard number, ϵ. $B_{\epsilon/2}(x) \cap F = \emptyset$, so F is closed. ∎

Exercise 11.2.18 $E \subset M$ is closed iff $E = \mathrm{st}(^*E \cap ns(M))$.

Sometimes, $ns(M)$ is internal, for example, when M is compact.

Theorem 11.2.19 (Robinson) $E \subset M$ *is compact iff* $\mathrm{st}(x) \in E$ *for every* $x \in {}^*E$.

Proof. Suppose that E is compact but for some $x \in {}^*E$, $y = \mathrm{st}(x) \notin E$. There is a finite cover of E by ϵ-balls, $B_\epsilon(e_1), \ldots, B_\epsilon(e_K)$, such that $y \notin \cup_k B_\epsilon(e_k)$. Since $E \subset \cup_{k \leq K} B_\epsilon(e_k)$, transfer tells us that $^*E \subset \cup_{k \leq K} {}^*B_\epsilon(e_k)$, which contradicts $x \in {}^*E$.

If E is not compact, then there is an infinite cover, $\{B_{\epsilon_n}(x_n) : n \in \mathbb{N}\}$ of E by open balls having no finite subcover. Therefore, the internal set $S = \{N \in {}^*\mathbb{N} : \cup_{n \leq N} B_{\epsilon_n}(x_n) \not\supseteq {}^*E\}$ contains arbitrarily large standard integers. By spillover, it contains an infinite integer, M. Take $x \in {}^*E$ in the complement of $\cup_{n \leq M} B_{\epsilon_n}(x_n)$. $\mathrm{st}(x)$ cannot be in $\cup_{n \in \mathbb{N}} B_{\epsilon_n}(x_n)$ because every $n \in \mathbb{N}$ belongs to $\{1, 2, \ldots, M - 1, M\}$. ∎

One of the amusing things to do with nonstandard tools is to re-prove known results.

Theorem 6.2.61 *(Arzela-Ascoli).* Let M be a compact space. $K \subset C(M)$ has compact closure iff K is bounded and equicontinuous.

Proof. Consider the internal set $S = \{\|f\| : f \in {}^*K\}$. If K is not uniformly bounded, then S contains arbitrarily large standard numbers, hence contains an infinite number. $f \in {}^*K$ cannot be near-standard if $\|f\|$ is infinite. Consider, for any standard $\epsilon > 0$, the internal set $S(\epsilon) = \{\delta > 0 : (\exists x \neq x')[d(x, x') < \delta \wedge (\exists f \in {}^*K)[|f(x) - f(x')| \geq \epsilon]\}$. If K is not uniformly equicontinuous, $S(\epsilon)$ contains arbitrarily small standard $\delta > 0$ for some standard $\epsilon > 0$. By spillover, it contains infinitesimal δ's. Let x, and x', and f correspond to an infinitesimal δ. $f(\mathrm{st}(x)) = f(\mathrm{st}(x'))$ is a set with radius at least ϵ, so that f is not near-standard.

Now suppose that K is bounded and uniformly equicontinuous and pick $f \in {}^*K$. Since M is compact, a function is continuous iff its graph is a compact subset of $M \times \mathbb{R}$. F, the standard part of the graph of f, is a closed subset of $M \times [-B, +B]$, where B is the uniform bound on K; hence it is compact. Thus, it is sufficient to show that F is the graph of a function, that is, that $F(x)$ contains at most one point. If not, then K is not uniformly equicontinuous. ∎

The following is the interesting compactness part of Theorem 6.6.32 about the properties of countable products of metric spaces. Recall that A is a countably infinite set in the following, hence in a 1-1 relation with $\mathbb{N}$, and that the metric on $M := \times_{a \in A} M_a$ is defined as $d(x, x') = \sum_{a \in A} 2^{-a} \min\{d_a(x_a, x'_a), 1\}$. The standard proof of the following uses diagonalization.

Theorem 11.2.20   *If each (M_a, d_a) is compact, then M is compact.*

Proof.   Pick $x \in {}^*M$, $x = (x_a)_{a \in {}^*A}$. Since each M_a is compact, $a \in A$, $y_a := \mathrm{st}(x_a)$ is well defined. Set $y = (y_a)_{a \in A}$. We have to show that $d(x, y) \simeq 0$. Set $d_a^1(x_a, x'_a) - \min\{d_a(x_a, x'_a), 1\}$, and

$$d(x, y) = \sum_{a \in {}^*A} 2^{-a} d_a^1(x_a, y_a)$$

$$= \sum_{a \leq N} 2^{-a} d_a^1(x_a, y_a) + \sum_{a > N} 2^{-a} d_a^1(x_a, y_a)$$

for every $N \in {}^*\mathbb{N}$. By spillover, there is an infinite N such that the first term is infinitesimal, and for any infinite N, the second term is infinitesimal.   ■

For a metric space, (M, d), the set ${}^*M$ is a set of equivalence classes of all sequences. Contained in the set of all sequences are the Cauchy sequences. Intuitively, taking the standard part corresponds to taking limits. If we put these together, it seems that the right equivalence relation on the right subset of ${}^*M$ ought to be the completion of (M, d). We need a more broadly useful technical result to show this to be true.

Lemma 11.2.21 (Internal Extensions of External Sequences)   *If $(A_i)_{i \in \mathbb{N}}$ is a sequence of sets in $V_m({}^*S)$ for some m, then there is an internal sequence, $(A_i)_{i \in {}^*\mathbb{N}}$, extending $(A_i)_{i \in \mathbb{N}}$.*

Proof.   We have a function $g : \mathbb{N} \to V_m({}^*S)$ defined by $g(i) = A_i$. We must extend this to an internal function, $f : {}^*\mathbb{N} \to V_m({}^*S)$, that agrees with g for all $i \in \mathbb{N}$.

Each internal A_i is the equivalence class of a sequence $A_{i,n}$, $A_i = \langle A_{i,n} \rangle$. For each $n \in \mathbb{N}$, define a function $f_n : \mathbb{N} \to V_m(S)$ by $f_n(i) = A_{i,n}$. The requisite internal function $f = \langle f_n \rangle$. For each $i \in \mathbb{N}$, $f(i) = \langle f_n(i) \rangle = \langle A_{i,n} \rangle = A_i = g(i)$.   ■

Exercise 11.2.22   Give another proof based directly on Theorem 11.2.4. [Let B_i be the internal set of internal functions f that agree with g on the first i terms and show that B_i has the finite intersection property.]

Definition 11.2.23   *A point $x \in {}^*M$ is **pre-near-standard** if for all standard $\epsilon \in \mathbb{R}_{++}$, there exists a $y \in M$ such that ${}^*d(x, y) < \epsilon$. The set of pre-near-standard points in ${}^*M$ is denoted $\mathrm{pns}({}^*M)$.*

For $x, y \in {}^*M$, define $x \approx y$ if $d(x, y) \simeq 0$. Let $\widehat{M} = \mathrm{pns}({}^*M)/\approx$ denote the set of $\approx$-equivalence classes of $\mathrm{pns}({}^*M)$ with typical elements $\hat{x}$ and $\hat{y}$. Define the metric $\widehat{d}(\hat{x}, \hat{y}) = \mathrm{st}(d(x, y))$ for any $x \in \hat{x}$ and any $y \in \hat{y}$. Here is yet another construction of the completion of (M, d).

Theorem 11.2.24 $(\widehat{M}, \widehat{d})$ *is a complete metric space and M is dense in it.*

Before proving this, note that $pns(^*\mathbb{R})$ is the set of all near-standard points, and that for any near-standard x, $\mathrm{st}(x)$ is the unique element of $\mathbb{R}$ in the equivalence class of x. This is another way of getting at the observation that $\mathbb{R}$, the completion of the rationals, is itself complete.

Exercise 11.2.25   Prove Theorem 11.2.24. [You must verify that $(\widehat{M}, \widehat{d})$ is a complete metric space and that M is dense. Verifying that it is a metric space and that M is dense ought to be very easy; completeness is less easy, but not too bad. For completeness, let $\hat{x}_n$ be a Cauchy sequence, pick an $x_n \in \hat{x}_n$, use Lemma 11.2.21 to extend it, and use overspill to find a sufficiently small infinite M so that $\widehat{d}(\hat{x}_n, \hat{y})$, where $\hat{y}$ is the equivalence class containing x_M.]

11.3 ◆ Loeb Spaces

Loeb spaces are a special kind of probability space. As they are based on expanded sets that contain representations of the limits of all kinds of sequences, these new probability spaces are often very convenient tools. We first define Loeb spaces, then turn to measurable functions on them, and finally give a couple of applications.

11.3.a Loeb Spaces from Internal Probability Spaces

A probability space is a triple, $(\Omega, \mathcal{F}, P)$, where Ω is a nonempty set, $\mathcal{F}$ is a σ-field of subsets of Ω, and $P : \mathcal{F} \to [0, 1]$ is a countably additive probability. Putting *'s in front of all of these gives an **internal probability space**. In the following definition, be very careful about distinguishing between situations where a union or sum ranges over $\mathbb{N}$ and a situation where it ranges over $^*\mathbb{N}$.

Definition 11.3.1 $(\Omega, \mathcal{F}^\circ, Q)$ *is an **internal probability space** if Ω is a nonempty internal set, $\mathcal{F}^\circ$ is an internal $^*\sigma$-field (i.e., $\mathcal{F}^\circ$ is internal if $\emptyset, \Omega \in \mathcal{F}^\circ$, $[A \in \mathcal{F}^\circ] \Rightarrow [A^c \in \mathcal{F}^\circ]$, and $[\{A_n : n \in {}^*\mathbb{N}\} \subset \mathcal{F}^\circ] \Rightarrow [\cup_{n \in {}^*\mathbb{N}} A_n \in \mathcal{F}^\circ])$ and $Q : \mathcal{F}^\circ \to {}^*[0, 1]$ is a *-countably additive probability (if $\{A_n : n \in {}^*\mathbb{N}\}$ is a *-countable collection of disjoint subsets of $\mathcal{F}^\circ$, then $Q(\cup_{n \in {}^*\mathbb{N}} A_n) = \sum_{n \in {}^*\mathbb{N}} Q(A_n)$).*

There are two observations that we should make:

1. There are three differences between a probability space and an internal probability space:

    1. the Ω must be internal, which may or may not be the case in general,
    2. as we will see $\mathcal{F}^\circ$ is a field, but is not a σ-field unless it is finite, and
    3. Q takes its values in $^*[0, 1]$ rather than in $[0, 1]$.

2. Since $^*\mathbb{N}$ is an uncountable set, $\mathcal{F}^\circ$ is closed under some uncountable disjoint unions. Since $E \mapsto \mathrm{st}(Q(E))$ takes values in $[0, 1]$, the distinction between countable and uncountable disjoint unions had better not matter very much.

We now show that $\mathcal{F}^\circ$ is a σ-field iff it is finite. This is in preparation for an application of Carathéodory's extension theorem (7.6.2, p. 412), which you should review if it has faded from your memory. Though an internal $*\sigma$-field is closed under some uncountable unions, it is not closed under countable unions unless those countable unions reduce to finite unions. This is Corollary 11.2.5.

Notation Alert 11.3.A *We start taking the standard part of a function evaluated at a point in its internal domain, for example, $\mathrm{st}(Q(E))$ for the function $Q : \mathcal{F}^\circ \to *[0, 1]$. We use a raised $\circ$ in front of the function to denote this, that is, we write $^\circ Q : \mathcal{F}^\circ \to [0, 1]$ for the function $E \mapsto \mathrm{st}(Q(E))$. This is clearly a case where, in trying to be faithful to all of the literature we draw on, we end up using the same symbol in too many ways.*

Theorem 11.3.2 (Loeb) *If $(\Omega, \mathcal{F}^\circ, Q)$ is an internal probability space, then $^\circ Q$ has a unique countably additive extension, $L(Q)$, to $\mathcal{F} = \sigma(\mathcal{F}^\circ)$ and to its completion, $L(\mathcal{F})$, with respect to $L(Q)$.*

$L(Q)$ is called a **Loeb measure** and $(\Omega, \mathcal{F}, L(Q))$ and $(\Omega, L(\mathcal{F}), L(Q))$ are both called **Loeb spaces**. When Ω is star-finite, integration is, up to an infinitesimal, $*$summation. This makes many problems much simpler.

Proof. Since a finite sum of infinitesimals is infinitesimal, $^\circ Q$ is a finitely additive probability on $\mathcal{F}^\circ$. To check that $^\circ Q$ is countably additive on $\mathcal{F}^\circ$, let $\{E_i : i \in \mathbb{N}\}$ be a sequence of disjoint elements of $\mathcal{F}^\circ$. We must check that if $\cup_{i \in \mathbb{N}} E_i \in \mathcal{F}^\circ$, then $^\circ Q(\cup_{i \in \mathbb{N}} E_i) = \sum_{i \in \mathbb{N}} {}^\circ Q(E_i)$. But $\cup_{i \in \mathbb{N}} E_i \in \mathcal{F}^\circ$ only when the union is actually finite. ∎

This result has a slippery feel to it, and there is at least one slightly more intuitive way to get at it. The proof of Carathéodory's extension theorem involved completing the metric space $(\mathcal{F}^\circ, d) = (\mathcal{F}^\circ, d_{\circ Q})$, where $d(E, F) = d_{\circ Q}(E, F) := {}^\circ Q(E \Delta F)$. The fact that $\mathcal{F}^\circ$ is internal ought to mean that it contains representations of the limits of Cauchy sequences in $(\mathcal{F}^\circ, d)$. It does. Up to null sets, there is no loss in restricting our attention to internal sets.

Theorem 11.3.3 *If $E \in L(\mathcal{F})$, then there exists an internal $E' \in \mathcal{F}^\circ$ such that $L(Q)(E \Delta E') = 0$.*

Proof. Pick $E \in L(\mathcal{F})$ and let $(E_n)_{n \in \mathbb{N}}$ be a d-Cauchy sequence in $(\mathcal{F}^\circ, d)$ that converges to E. Extend the sequence to an internal sequence $(E_n)_{n \in *\mathbb{N}}$. Since $(E_n)_{n \in \mathbb{N}}$ is Cauchy, for each $m \in \mathbb{N}$, there is an $N_m \in \mathbb{N}$ such that $d(E_n, E_{N_m}) < \frac{1}{m}$ for all finite $n \geq N_m$. By overspill, there is an infinite H_m such that for all $n \in \{N_m, \dots, H_m\}$, $d(E_n, E_{N_m}) < \frac{1}{m}$. The collection $A_m = \{N_m, \dots, H_m\}$, $m \in \mathbb{N}$, has the finite intersection property, hence has nonempty intersection (by Theorem 11.2.4). Pick $H \in \cap_{m \in \mathbb{N}} A_m$ and set $E' = E_H$. Observe that $d(E, E_H) \simeq 0$, that is, $L(Q)(E \Delta E') = 0$. ∎

11.3.b Measurable Functions Taking Values in Measure Spaces

If $E \in L(\mathcal{F})$, then 1_E is $L(P)$-a.e. equal to $1_{E'}$ for an internal E'. We should expect that the same is true for measurable functions in general. Before we turn to these results, we should look at the next example, which shows why we need $\mathcal{F} = \sigma(\mathcal{F}^\circ)$ for the measurability of interesting functions and how little it matters.

Example 11.3.4 *Let $N \in {}^*\mathbb{N} \setminus \mathbb{N}$ be an infinite integer. Take $\Omega \subset {}^*[0, 1]$ to be the set $\{k/n : k = 0, 1, \ldots, N\}$, let $\mathcal{F}^\circ$ be the set of internal subsets of Ω, and for all $E \in \mathcal{F}^\circ$, set $P(E) = \frac{\#E}{N}$. Both P and $L(P)$ represent the uniform distribution on Ω, where Ω is a model of the numbers between 0 and 1 that is rich enough to represent every conceivable measurement. This uniform distribution yields the usual uniform distribution, λ, after we take standard parts.*

Observe that for all $s \in {}^*\mathbb{R}$, $\{\omega \in \Omega : \omega < s\} \in \mathcal{F}^\circ$. Therefore, $\mathrm{st} : \Omega \to [0, 1]$ is measurable because $\mathrm{st}^{-1}((-\infty, r)) = \cup_n \{\omega : \omega < r - \frac{1}{n}\} \in \mathcal{F} = \sigma(\mathcal{F}^\circ)$.

1. *We have to go to the σ-field because $\mathrm{st}^{-1}((-\infty, r))$ is **not** internal (unless r is outside of $[0, 1)$).*

2. *It matters very little because $L(P)(\{\omega : \omega \leq s\} \Delta\, \mathrm{st}^{-1}((-\infty, r)) = 0$ for any $s \in \Omega$ with $s \simeq r$.*

For $a, b \in [0, 1]$, $L(P)(\mathrm{st}^{-1}((a, b]) = b - a$. Since $L(P)$ is countably additive, this means that $L(P)(\mathrm{st}^{-1}(B)) = \lambda(B)$ for any measurable B. The uniform distribution on $[0, 1]$ is the image law of the random variable $\mathrm{st}(\cdot)$ on $(\Omega, \mathcal{F}, L(P))$.

Recall that any distribution, ν, on any complete separable metric space, M, is the image of the uniform distribution on $[0, 1]$, $\nu = f(\lambda)$. Since the composition of measurable functions is measurable, $\nu = f(\mathrm{st}(L(P)))$ shows that any distribution on any complete separable metric space is the image of the nonstandard uniform distribution.

Returning to more general measurable $\mathbb{R}$-valued functions, we have two general results: (1) every measurable function is nearly internal; and (2) taking the standard part of an internal function gives a measurable function. We call the first **lifting**: we lift a measurable function taking values in the smaller space $\mathbb{R}$ up to an internal function taking values in the larger space ${}^*\mathbb{R}$. We call the second **pushing down**: we take an internal function and push it down from taking values in the large space, ${}^*\mathbb{R}$, to taking values in the smaller space $\mathbb{R}$.

11.3.b.1 Lifting
The obvious place to start is with simple functions.

Lemma 11.3.5 *If $f : \Omega \to \mathbb{R}$ is a simple $L(\mathcal{F})$-measurable function, then it is a.e. equal to a simple internal function.*

Proof. $f = \sum_{i \leq I} \beta_i 1_{E_i}$, where ($I$ is finite and) the E_i are disjoint. Take E_i' internal with $L(P)(E_i \Delta E_i') = 0$, disjointify the E_i' if necessary, and check that the internal function $\sum_i \beta_i 1_{E_i'}$ is a.e. equal to f. ∎

Definition 11.3.6 *If g is an internal function $L(P)$-a.e. infinitesimally close to an $L(\mathcal{F})$-measurable function f, then g is called a **lifting** of f.*

In these terms, the foregoing lemma says that simple measurable functions have simple liftings. The following arises from the usual kinds of approximation arguments in measure theory—every measurable function is the limit of simple functions.

Theorem 11.3.7 *If $f : \Omega \to \mathbb{R}$ is $L(\mathcal{F})$-measurable, then it is has a lifting.*

Proof.   Let f_n be a sequence of simple $L(\mathcal{F})$-measurable functions with $f_n(\omega) \to f(\omega)$ and $|f_n|(\omega) \uparrow |f|(\omega)$ for all ω. Let g_n be a corresponding sequence of internal functions a.e. equal to the f_n. Extend the external sequence, $(g_n)_{n\in\mathbb{N}}$, to an internal sequence $(g_n)_{n\in *\mathbb{N}}$ of *-simple functions. For a sufficiently small infinite N, g_N is the requisite internal function.   ∎

Note that the g_N of the last proof is *-simple. In measure theory, many intuitions and results are found from the observation that every measurable function is nearly simple. Here, when we work in Loeb spaces, up to a null set and infinitesimal differences in value, the last result shows that this is exactly true.

Exercise 11.3.8   The last proof asserted that "For a sufficiently small infinite N, g_N is the requisite internal function." Prove this statement.

11.3.b.2 Pushing Down

Starting with an internal function, $\omega \mapsto g(\omega)$, from Ω to *$\mathbb{R}$, we push it down by considering the mapping $\omega \mapsto {}^\circ g(\omega)$ from Ω to $\mathbb{R}$. We would like the pushed-down function to be measurable. There is a small problem.

Example 11.3.9   *The internal constant function $g(\omega) \equiv x$, x infinite, has the property that ${}^\circ g(\omega)$ is not defined.*

Probably the easiest way to handle this problem is to define our way around it. If $x > 0$ is infinite, define st$(x) = +\infty$; if $x < 0$ is infinite, define st$(x) = -\infty$. Now the pushed-down version of an internal function takes values in $[-\infty, +\infty]$ rather than $\mathbb{R}$. Giving $[-\infty, +\infty]$ its Borel σ-field, we have the following.

Lemma 11.3.10   *If $g : \Omega \to$ *$\mathbb{R}$ is internal, then ${}^\circ g$ is $\mathcal{F}$-measurable.*

Exercise 11.3.11   Prove the lemma. [Note ${}^\circ g^{-1}(+\infty) = \cap_{n\in\mathbb{N}}\{\omega : g(\omega) > n\}$, and so on.]

11.3.c   Integration as Summation on Star-Finite Spaces

For this subsection, we assume that Ω is an infinite star-finite set and that $\mathcal{F}^\circ$ is the class of internal subsets of Ω. The simplest example was the uniform distribution given earlier.

Exercise 11.3.12   Show that if $\max_{\omega\in\Omega} P(\omega) \simeq 0$, then $(\Omega, L(\mathcal{F}), L(P))$ is nonatomic. [It might be easier to start by assuming that the $P(\omega)$ are all equal.]

There is one basic result. It has many implications, but we only use a few of them.

Theorem 11.3.13   *If f is integrable and g is a lifting of f or if g is internal and $f = {}^\circ g$, then $\int_\Omega f \, dL(P)(\omega) \simeq \sum_\omega g(\omega) P(\omega)$.*

Proof.   The first part is clearly true if f is simple and the usual limit arguments cover the rest. The same kind of argument works for the second part.   ∎

11.3.d A Continuum Law of Large Numbers

If I is an infinite, hence uncountable, star-finite set and U is the uniform distribution on the class of all internal subsets of I, $U(E) = \frac{\#E}{\#I}$, one has a very good model of an infinite population. From the sequence construction, we can think of this model as a 'limit' of a sequence of larger and larger finite population models. We examine the average wealth shock to the population when each $i \in I$ receives the stochastic shock $X_i \in \mathbb{R}$.

The joint distribution of the populations' shocks is a probability P on $\Omega := \mathbb{R}^I$, with $X_i(\omega) = \text{proj}_i(\omega)$. There are two other notations for the shock that i receives if ω occurs and both are based on seeing $\mathbb{R}^I$ as the space of (internal) functions from I to $\mathbb{R}$: $\omega(i)$ and $\varphi(i, \omega) := \omega(i)$, where φ is called the **evaluation mapping**. If we suppose that each $E\ X_i$ exists, then the function $f(i) = E\ X_i$ is internal.

We calculate the average value of the shock to the population and give conditions under which it is constant. Constant average shock corresponds to there being 0 aggregate uncertainty from the individual shocks. Having a model containing nontrivial individual risk but no aggregate uncertainty is useful for some of the thought experiments that economists enjoy.

If one picks i at random according to U and independently picks ω at random according to P, then the expected value of the shock is $E\ \varphi(\cdot, \cdot)$, which is

$$E\ \varphi = \sum_{i \in I} \left(\int X_i(\omega)\, dP(\omega) \right) U(i)$$

$$= \sum_{i \in I} f(i)U(i) = \frac{1}{\#I} \sum_{i \in I} f(i) \simeq \int_I {}^\circ f(i)\, dLU(i),$$

where the last equality requires integrability of ${}^\circ f$.

If the shocks average out over the population, one has $\text{Var}(\varphi) \simeq 0$. If the X_i are uncorrelated, then $\text{Var}(\varphi) = \left(\frac{1}{\#I} \right)^2 \sum_i \text{Var}(X_i)$.

Exercise 11.3.14 Using Chebyshev's inequality, show that if the X_i are iid and have finite variance, then $\text{Var}(\varphi) \simeq 0$. Show that this implies that $\text{Var}({}^\circ \varphi) = 0$.

The idea that the wealth shocks to a rich person, i, and a poor person i', have the same distribution is ludicrous, and this may render many of the thought experiments in which we want 0 aggregate uncertainty useless. Fortunately, the iid assumption is not needed.

Exercise 11.3.15 Let $v(i) = \text{Var}(X_i)$. Suppose that the X_i have 0 correlation and that both ${}^\circ f$ and ${}^\circ v$ are $L(U)$-integrable. Show that $\text{Var}({}^\circ \varphi) = 0$. [Chebyshev again.]

A negative shock to the wealth of a person who is remodeling his home or starting a business has a negative effect on everyone he has hired or might hire. Thus, 0 correlation among the shocks to different individuals is far too strong an assumption. There are a huge number of ways to incorporate failures of independence that retain the result that there is no aggregate uncertainty from the shocks, that is, that $\text{Var}({}^\circ \varphi) = 0$. The easy ones require some variant of the idea that the shocks to any individual do not propagate too far through the economy.

Exercise 11.3.16 For any $r \in {}^*[0, 1]$ with $0 < \mathrm{st}(r) < 1$ and any infinite N, N^r is infinite and $\frac{N^r}{N} \simeq 0$. For given infinite N, find the set of $r \simeq 0$ such that N^r is finite and find the set of $r \simeq 1$ such that $\mathrm{st}(\frac{N^r}{N}) > 0$.

Example 11.3.17 *Throughout this example, we use $\sqrt{n}$ to mean the integer part of the square root of n. Suppose that the population, I, can be numbered as $I = \{1, 2, \dots, \#I\}$ in such a fashion that the correlation between X_i and X_j satisfies $|\rho(i, j)| \le \alpha^{|i-j|}$ for some α with $\mathrm{st}(\alpha) < 1$. The idea is to add up the shocks of large blocks of people and to ignore the shocks occurring to the people between the blocks. Choosing the block size appropriately makes this work.*

Block the population into an alternating sequence of larger and smaller sets, $I = \{L_1, S_1, L_2, S_2, \dots\}$, with the large sets containing $\sqrt{\#I}$ people and the smaller sets containing $\sqrt{\sqrt{\#I}}$. Define $Y_k = \sum_{i \in L_k} X_i$. The standard part of the correlation between Y_k and $Y_{k'}$, $k' \ne k$, is 0, as is the proportion of the economy belonging to $\cup_k S_k$. From this and some easy calculations, $\mathrm{Var}({}^\circ \varphi) = 0$.

The previous example can be extended to the case where individuals have a bounded number, $B \in \mathbb{N}$, of nearest neighbors, the distance between individuals is given by the smallest number of nearest-neighbor links that join people, and the correlation satisfies $\rho(i, j) \le \alpha^{d(i,j)}$. Essentially, the large one-dimensional blocks of peoples' positions in the example become B-dimensional hypercubes of peoples' positions and the small blocks become padding around the hypercubes. The length of the sides of the cubes and the width of the paddings become the appropriate powers of I, and the same basic idea works.

11.3.e Lyapunov's Theorem for Measure Space Games

Definition 11.3.18 $[(\Omega, \mathcal{F}, P), A, u]$ *is a* **compact Loeb space game** *if*

1. *$(\Omega, \mathcal{F}, P)$ is a nonatomic Loeb space (of agents),*

2. *A is a compact metric space (of actions), and*

3. *$u : \Omega \times A \times \Delta(A) \to \mathbb{R}$ is a jointly measurable (utility) function, continuous in $A \times \Delta(A)$ for every ω.*

A **strategy** *is a measurable mapping $\omega \mapsto s_\omega$ from Ω to A, the* **population average choice associated with a strategy** *s is $\mu_s(B) := P(\{\omega : s_\omega \in B\})$ for B in the class $\mathcal{B}_A$ of Borel measurable subsets of A. A strategy is an* **equilibrium** *if $P(\{\omega : (\forall a \in A)[u(\omega, s_\omega, \mu_s) \ge u(\omega, a, \mu_s)]\}) = 1$.*

Theorem 11.3.19 *Every compact Loeb space game has an equilibrium.*

Proof. There are two steps: (1) showing that the game has an equilibrium in mixed strategies; and (2) then showing that the equilibrium strategies can be purified, which involves proving a version of Lyapunov's theorem for a function taking values in the infinite-dimensional set $\Delta(A)$.

Step 1. For $\mu \in \Delta(A)$, let $Br_\omega(\mu) \subset \Delta(A)$ denote ω's set of mixed strategy best responses to the population average choice μ. As each $u(\omega, \cdot, \cdot) : A \times \Delta(A) \to \mathbb{R}$ is continuous, for each ω, the correspondence $\omega \mapsto Br_\omega(\mu)$ is nonempty valued, compact-valued, and convex-valued, and upper hemicontinuous. Routine limit arguments show that $\Psi(\mu) := \int Br_\omega \, dP(\omega)$ is also nonempty valued, compact-valued, and convex-valued, and upper hemicontinuous. By Theorem 10.8.2, Ψ has a fixed point, μ^*, provided that we can give $\Delta(A)$ a norm that makes it compact and convex. If we have a fixed point μ^*, then, by definition of the integral of a correspondence, there exists an a.e. selection, f^*, from $\omega \mapsto Br_\omega$ that integrates to μ^*, and f^* is an equilibrium.

To construct the requisite norm, enumerate a dense subset of the unit ball in $C(A)$ as $\{f_n : n \in \mathbb{N}\}$ and consider the linear homeomorphism $\mu \leftrightarrow \varphi(\mu) := (\frac{1}{2^1}\langle \mu, f_1 \rangle, \frac{1}{2^2}\langle \mu, f_2 \rangle, \frac{1}{2^3}\langle \mu, f_3 \rangle, \ldots)$ from $\Delta(A)$ to a compact convex subset of ℓ^2. Define the norm of μ to be $\|\varphi(\mu)\|_2$.

Step 2. We claim that it is sufficient to show that for every measurable $f : \Omega \to \Delta(A)$, there exists a measurable $g : \Omega \to A$ such that for all $E \in \mathcal{F}$ with $P(E) > 0$ and for all open $B \subset A$,

$$\int_E f(\omega)(B) \, dP(\omega) = \int_E 1_{g(\omega) \in B} \, dP(\omega). \tag{11.1}$$

Suppose we have such a g^* for the equilibrium f^*, define $E = \{\omega : g^*(\omega) \notin Br_\omega\}$, and suppose that $P(E) > 0$. For some open B, $E_B := \{\omega \in E : Br_\omega \cap B = \emptyset\}$ has $P(E_B) > 0$, contradicting (11.1). The existence of such a g for each f is the conclusion of the next lemma. ■

Exercise 11.3.20 Show that for every $\mu \in \Delta(A)$, for every internal $E \in \mathcal{F}$ and internal $f : \Omega \to \Delta(A)$, there exists an internal $g : \Omega \to A$ such that for all open B, $\int_E f(\omega)(B) \, dP(\omega) = \int_E 1_{g(\omega) \in B} \, dP(\omega)$. [Hint: Take A_F to be a dense star-finite subset of A; consider g taking values in A_F.]

Lemma 11.3.21 (Loeb Space Lyapunov) *For every measurable $f : \Omega \to \Delta(A)$, there exists a measurable $g : \Omega \to A$ such that for all $E \in \mathcal{F}$ with $P(E) > 0$ and for all open $B \subset A$,*

$$\int_E f(\omega)(B) \, dP(\omega) = \int_E 1_{g(\omega) \in B} \, dP(\omega). \tag{11.2}$$

Proof. Let $\mathcal{B}_\tau = \{B_n : n \in \mathbb{N}\}$ be a countable basis for the metric topology on A. Since every open B can be expressed as an increasing union of elements of $\mathcal{B}_\tau$, by dominated convergence, it is sufficient to prove the result for every open $B \in \mathcal{B}_\tau$. We now appeal to overspill in the following way.

For $\epsilon \in \mathbb{R}_{++}$, let $\mathfrak{E}_F(\epsilon)$ denote the set of finite internal partitions of Ω with $|\mathcal{E}| < \epsilon$, that is, $\max\{P(E) : E \in \mathcal{E}\} < \epsilon$. For $r > 0$, let $\mathcal{B}_\tau(r) = \{B_n : n \leq r\}$. Define

$$S = \{\epsilon > 0 : (\exists \mathcal{E} \in \mathfrak{E}_F(\epsilon))(\forall E \in \sigma(\mathcal{E}))(\forall B \in \mathcal{B}_\tau(1/\epsilon))(\exists g : \Omega \to A)$$

$$\left[\left| \int_E f(\omega)(B) \, dP(\omega) - \int_E 1_{g(\omega) \in B} \, dP(\omega) \right| < \epsilon \right].$$

S is internal, and, by the previous exercise, contains arbitrarily small $\epsilon \in \mathbb{R}_{++}$. By overspill, S contains a strictly positive infinitesimal. Take any internal g associated with that positive infinitesimal. ■

11.4 ◆ Saturation, Star-Finite Maximization Models, and Compactification

We start with some easy observations about some optimization problems. As easy as they are, they contain the keys to understanding star-finite maximization models. The statements about the existence of optima for star-finite choice problems come from transfer of the existence of optima for finite choice problems.

1. Consider the problem $\max_{x \in [0,1]} u(x)$, where $u(x) = x \cdot 1_{[0,1)}(x)$.

    a. This does not have a solution because, for every $x < 1$, there is an $x' \in (x, 1)$. For every $\epsilon > 0$, $x = 1 - \epsilon$ is an ϵ-optimum, as it gives utility within ϵ of $\sup_{x \in [0,1]} u(x)$.

    b. For any star-finite $A \subset {}^*[0, 1]$, the problem $\max_{x \in A} u(x)$ does have a solution, x°, and if $d_H(A, {}^*[0, 1]) \simeq 0$, then $^\circ u(x^\circ) = \sup_{x \in [0,1]} u(x)$.

2. Consider the problem $\max_{x \in [0,1]} v(x)$, where $v(x) = 1_{\mathbb{Q}}(x)$.

    a. This has infinitely many solutions, namely the set $Q \cap [0, 1]$.

    b. For any star-finite $A \subset {}^*[0, 1]$, the problem $\max_{x \in A} v(x)$ does have a solution, x°. If A contains no rationals, then even if $d_H(A, {}^*[0, 1]) \simeq 0$, $^\circ v(x^\circ) = 0 < \max_{x \in [0,1]} v(x)$.

3. Consider the problem $\max_{x \in [0,1]} w(x)$, where $w(x) = \frac{1}{x} \cdot 1_{(0,1]}(x)$.

    a. This has neither an optimum nor an ϵ-optimum.

    b. For any star-finite $A \subset {}^*[0, 1]$, the problem $\max_{x \in A} v(x)$ does have a solution, x°. If $d_H(A, {}^*[0, 1]) \simeq 0$, then $x^\circ \simeq 0$ and $w(x^\circ)$ is infinite.

This section gives a brief overview of the interpretations of the solutions to star-finite versions of optimization problems. The examples show that we want bounded utility functions and that the criterion that the star-finite set be infinitely close to the choice set is not sufficient. This leads us to our first topic: saturation.

11.4.a   Saturation and Star-Finite Sets

Let $A_n = \{\frac{k}{2^n} : k = 0, \ldots, 2^n\}$ and let $A = \langle A_n \rangle$. For any dyadic rational $q \in [0, 1]$, $q \in A$. This means that A is a star-finite set that contains every element of a countably infinite dense subset of $[0, 1]$. For many approximation arguments, this is sufficient, but for some it is not. Upper semicontinuous functions and their relation to maximization problems were considered in Definition 6.11.30 and Lemma 6.11.33.

Example 11.4.1 *If $f : [0, 1] \to \mathbb{R}$ is continuous and $x \in A = \langle A_n \rangle$ solves $\max_{a \in A} f(a)$, then $\mathrm{st}(x)$ solves $\max_{r \in [0,1]} f(r)$. If g is the upper semicontinuous function $g(r) = 1$ if $r = \pi/4$ and $g(r) = 0$ otherwise, the same is not true.*

We used sequences to define $V({}^*S)$. This means that for any countable set, E in $V(S)$, there is a star-finite A in $V({}^*S)$, $A \subset {}^*E$, such that for all $e \in E$, $e \in A$. It would be convenient if the same were true for all $E \in V(S)$.

This would require using something other than $\mathbb{N}$ and an equivalence relation on $V_m(S)^{\mathbb{N}}$ defined using subsets of $\mathbb{N}$. One way to get at this is to replace $\mathbb{N}$ with the much(!) larger set $\mathfrak{N} = \mathcal{P}_F(V(S))$, partially order it by inclusion, and consider generalized sequences, that is, elements of $V_m(S)^{\mathfrak{N}}$. It is possible, but not particularly easy, to show that there exists a purely finitely additive, $\{0, 1\}$-valued probability μ on $\mathcal{P}(\mathfrak{N})$ so that the μ equivalence classes in $V_m(S)^{\mathfrak{N}}$ achieve the following.

Definition 11.4.2 *A nonstandard extension $V(^*S)$ is **polysaturated** if for all $E \in V(S)$, there exists an $A \in {}^*\mathcal{P}_F(E)$ such that for all $e \in E$, $e \in A$. Such an A is called **exhaustive for** E or simply **exhaustive**.*

Notation Alert 11.4.A *We did not write $E \subset A$. This is because E is not internal (when it is infinite) and the subset relation has only been defined for internal sets. To some extent we are being overly picky here, but the hazard of forgetting to check whether or not a set is internal makes the pickiness worthwhile.*

Substituting for $[0, 1]$ an exhaustive star-finite subset of $^*[0, 1]$ loses none of the points in $[0, 1]$ and gains many new ones.

11.4.b Maximization over Exhaustive Star-Finite Sets

Returning to the Example 11.4.1, we let A be an exhaustive star-finite subset of $^*[0, 1]$. If x solves $\max_{a \in A} g(a)$, g upper semicontinuous, then $\mathrm{st}(x)$ solves $\max_{r \in [0,1]} g(r)$. Here is as much as can be generally said.

Theorem 11.4.3 *If $f : X \to \mathbb{R}$ is a function with a supremum by $\overline{f}$ and A is an exhaustive star-finite subset of $^*X$, then*

1. *$\max_{x \in A} f(x)$ has a solution, $x^\diamond$, which solves $\max_{x \in A} {}^\circ f(x)$,*
2. *$f(x^\diamond) \le \overline{f}$,*
3. *$f(x^\diamond) \simeq \overline{f}$,*
4. *for every standard $\epsilon > 0$, $x^\diamond \in {}^*E(\epsilon)$, where $E(\epsilon) = \{x \in X : f(x) > \overline{f} - \epsilon\}$.*

This result says that there are two equivalent ways to interpret the solutions to exhaustive star-finite versions of optimization problems: (1) we can either measure the objective function on the nonstandard scale and optimize to within any standard $\delta > 0$; or (2) we can take the standard part of the objective function and optimize exactly. Either way, the solution $x^\diamond$ gives a representation of the limit of approximate solutions.

Proof. The first and second statements follow from transfer. By the definition of supremum, for every $\epsilon \in \mathbb{R}_{++}$, $E(\epsilon) = \{x \in X : f(x) > \overline{f} - \epsilon\} \ne \emptyset$. Since A is exhaustive, for every standard $\epsilon > 0$, $A \cap {}^*E(\epsilon) \ne \emptyset$, which implies the last two statements. ∎

From now on, we assume that we are using a polysaturated expansion. This means that, if we need it, we can assume the existence of exhaustive sets. This does not mean that sequence intuitions go away. Rather, we can use them, and we can use a bit more. Sometimes, the extra that we can use can be thought of as a kind of anchoring.

Definition 11.4.4 *Let F denote a finite subset of $E \in V(S)$. The sequence of approximations A_n is **anchored at** F if $F \subseteq A_n$ for all sufficiently large $n \in \mathbb{N}$.*

Heuristically, an exhaustive star-finite A can profitably be thought of as a sequence of approximations that is anchored at every finite F.

11.4.c Compactifications

If we can embed a set X in a 1-1 fashion as a dense subset of a compact space $\widehat{X}$ with a function $\iota : X \to \widehat{X}$, then the pair (ι, X) **compactifies** X. We call $\widehat{X}$ a **compactification** of X. We construct compactifications of X with the property that every bounded function $f : X \to \mathbb{R}$ has a unique continuous extension, $\widehat{f} : \widehat{X} \to \mathbb{R}$, determined by the condition that $f(x) = \widehat{f}(\iota(x))$.

Here is the construction:

1. Let $B(X)$ denote the set of all bounded functions $g : X \to \mathbb{R}$.
2. For each g, let I_g denote the interval $[\inf g(X), \sup g(X)]$.
3. Let Y be the compact space $\times_{g \in B(X)} I_g$.
4. We embed X in Y by associating to each $x \in X$ the infinite length vector
   $$\iota(x) := (g(x))_{g \in B(X)}.$$
5. The mapping ι is 1-1, hence invertible on the image, $\iota(X)$.
6. Let $\widehat{X}$ denote the closure of $\iota(X)$ in Y.

As it is a closed subset of the compact space, Y, $\widehat{X}$ is compact. Hence $(\iota, \widehat{X})$ is a compactification of X.

For any $f \in B(X)$, the function $\text{proj}_f : Y \to \mathbb{R}$ is continuous. Hence its restriction to $\widehat{X}$, denoted $\widehat{f}$, is both unique and continuous and is determined by the condition that $f(x) = \widehat{f}(\iota(x))$.

Since $\widehat{f}$ is continuous and $\widehat{X}$ is compact, the problem

$$\max_{\widehat{x} \in \widehat{X}} \widehat{f}(\widehat{x})$$

has a solution, $\widehat{x}^\circ$. We call the maximization problem a **compactified version of the original problem**.

For any $E \subset X$, the set $\text{cl}(\iota(E))$ is closed, hence compact. It is also open, which is just plain weird. Sets that are both open and closed are called **clopen** sets, which is a matchingly weird adjective.

Exercise 11.4.5 Show that $\text{cl}(\iota(E))$ is open. [Let $f(x) = 1_E(x) \in B(X)$ and consider the open set $\widehat{f}^{-1}((0.9, 1.1))$.]

The following leads to the link between compactifications and exhaustive star-finite versions of maximization problems. The proof is an exercise.

Theorem 11.4.6 *If $f : X \to \mathbb{R}$ is a bounded function, then $\widehat{x}^\circ$ solves the problem $\max_{\widehat{x} \in \widehat{X}} \widehat{f}(\widehat{x})$ iff $\widehat{x}^\circ \in \cap_{\epsilon > 0} \text{cl}(\iota(E(\epsilon)))$, $E(\epsilon) = \{x \in X : f(x) > \overline{f} - \epsilon\}$.*

11.4.d Star-Finite Constructions of Compactifications

Let A be an exhaustive star-finite subset of $^*X$. Define the equivalence relation $\approx$ on A by $a \approx b$ if for all $g \in B(X)$, $g(a) \simeq g(b)$, and define $\widetilde{X} = A/\approx$ as the set of equivalence classes. Let τ be the topology generated by the subbasis $\{°g^{-1}(G) : G \subset \mathbb{R}, G \text{ open}\}$.

The following is moderately tedious to prove, in good part because the notation gets way out of hand.

Theorem 11.4.7 $\widetilde{X}$ and $\widehat{X}$ are homeomorphic; in particular, $\widetilde{X}$ is compact.

As they are homeomorphic, we make no distinction between the two. The following, which may be more than anyone sensible would want to do, shows that exhaustive star-finite maxima can be interpreted as points in a compactified version of the original maximization problem.

Exercise 11.4.8 Show that if A is an exhaustive star-finite subset of X and f a bounded function on X and if x° solves $\max_{x \in A} f(x)$, then $\mathrm{st}(x^\circ)$ solves $\max_{\widehat{x} \in \widehat{X}} \widehat{f}(\widehat{x})$.

These kinds of interpretations become significantly trickier when we look at the exhaustive star-finite versions of interactive decision problems, that is, of games.

11.5 ◆ The Existence of a Purely Finitely Additive $\{0, 1\}$-Valued μ

As a reminder, Zorn's lemma says that: if $\mathcal{A}$ is a partially ordered set such that each chain has an upper bound in $\mathcal{A}$, then $\mathcal{A}$ has a maximal element.

Definition 11.5.1 A collection $\mathcal{U} \subset \mathcal{P}(\mathbb{N})$ is a **filter** if

1. $\emptyset \notin \mathcal{U}$,
2. $[(A, B \in \mathcal{U}) \wedge (A \subset B)] \Rightarrow [B \in \mathcal{U}]$, and
3. $[(A, B \in \mathcal{U})] \Rightarrow [A \cap B \in \mathcal{U}]$.

A filter is an **ultrafilter** if it also satisfies

4. $\forall A \in \mathcal{P}(\mathbb{N})$, $A \in \mathcal{U}$ or $A^c \in \mathcal{U}$.

The last condition is a sort of maximality property for filters.

Exercise 11.5.2 Show that the following are filters, and that the first is maximal, but that the second is not.

1. $\mathcal{U}_7 = \{A \in \mathcal{P}(\mathbb{N}) : 7 \in A\}$.
2. $\mathcal{U}_F = \{A \in \mathcal{P}(\mathbb{N}) : A^c \text{ is finite}\}$.

The filter $\mathcal{U}_7$ is "bound," in this case to 7, because it "converges to" 7 in the following way: for each $A \in \mathcal{U}_7$, pick an $x_A \in A$ so that we have $(x_A)_{A \in \mathcal{U}_7} \in \mathbb{N}^{\mathcal{U}_7}$; define $A \succ B$ if $A \subset B$; define $x_A \to 7$ if for all $\epsilon > 0$, there exists an $A' \in \mathcal{U}_7$ such that for all $A \succ A'$, $x_A \in B_\epsilon(7)$. This is an example of using a partially ordered set instead of $\mathbb{N}$ as the index for a sequence.

The filter $\mathcal{U}_F$ is called "free" because it is not bound to any $n \in \mathbb{N}$.

Exercise 11.5.3   Show that $(x_A)_{A \in \mathcal{U}_F}$ does not converge to any $n \in \mathbb{N}$.

In Example 11.1.1 (p. 628), We defined $(a_n) \;^{aa}\!< (b_n)$ by $\{n \in \mathbb{N} : a_n < b_n\} \in \mathcal{U}_F$. The problem was that for pairs of sequences such as $(a_n) = (0, 1, 0, 1, 0, 1, \ldots)$ and $(b_n) = (1, 0, 1, 0, 1, 0, \ldots)$ we had neither $\{n \in \mathbb{N} : a_n \le b_n\} \in \mathcal{U}_F$ nor $\{n \in \mathbb{N} : b_n > a_n\} \in \mathcal{U}_F$. As a result, $^{aa}\!\le$ is not a complete ordering of elements of $^{aa}\mathbb{R}$. Expanding $\mathcal{U}_F$ to an ultrafilter, $\mathcal{U} \supset \mathcal{U}_F$ solves this problem and keeps the infinitesimals. It keeps the infinitesimals because $\mathcal{U}_F \subset \mathcal{U}$, and it solves the completeness problem because, by the last property in the definition of an ultrafilter, either $\{n \in \mathbb{N} : a_n \le b_n\}$ or its complement $\{n \in \mathbb{N} : b_n > a_n\}$ belongs to $\mathcal{U}$.

Definition 11.5.4   *An ultrafilter containing $\mathcal{U}_F$ is called a **free ultrafilter**.*

Exercise 11.5.5   Show that there is a free ultrafilter. [Partially order $\mathcal{A}$, the set of filters containing $\mathcal{U}_F$ by inclusion, and show that every chain in $\mathcal{A}$ has a maximal element. Now apply Zorn's lemma.]

Exercise 11.5.6   Let $\mathcal{U}$ be an ultrafilter containing $\mathcal{U}_F$. Show that setting $\mu(A) = 1$ if $A \in \mathcal{U}$ and $\mu(A) = 0$ if $A \notin \mathcal{U}$ gives a purely finitely additive, $\{0, 1\}$-valued probability on $\mathcal{P}(\mathbb{N})$ with $m(\mathbb{N}) = 1$ and $m(A) = 0$ for all finite $A \subset \mathbb{N}$. Conversely, show that any such measure determines an ultrafilter containing $\mathcal{U}_F$.

There are many free ultrafilters.

Exercise 11.5.7   Let E be an infinite subset of $\mathbb{N}$ with an infinite complement. Show that there is a free ultrafilter containing E and one containing E^c.

11.6 ◆ Problems and Complements

Exercise 11.6.1   Let A be a star-finite subset of $[0, 1]$, $\mathcal{F}$ its internal subsets, and P a probability on $\mathcal{F}$, that is, $P \in {}^*\!\Delta([0, 1])$. Consider the following two distributions: $\mu = \mathrm{st}(L(P))$, that is, the distribution of the random variable X defined on $(A, \sigma(\mathcal{F}), L(P))$ by $X(a) = \mathrm{st}(a)$, and $\nu = \mathrm{st}(P)$, that is, the standard part of P in the compact metric space $\Delta([0, 1])$. Show that $\mu = \nu$. Generalize to A a star-finite subset of $\mathbb{R}$, being careful about tightness.

Exercise 11.6.2 (Order of Convergence)   In dynamic programming, we sometimes construct a sequence of approximate value functions, V_n, with the property that $d_\infty(V_n, V) \le \beta^n K$ for some constant K and $\beta < 1$. In the Gaussian part of the central limit theorem, we have $n^{-\frac{1}{2}} \sum_{i \le n}(X_i - \mu)$ converging to a nondegenerate random variable while for any $\epsilon > 0$, $n^{-\left(\frac{1}{2} + \epsilon\right)} \sum_{i \le n}(X_i - \mu)$ converges to 0 and for $\epsilon < 0$, it diverges.

We write $x_n = o(y_n)$ if $\limsup_n |\frac{x_n}{y_n}| = 0$, and $x_n = O(y_n)$ if $\limsup_n |\frac{x_n}{y_n}| < \infty$. These give orders of asymptotic convergence.

Let "1" be the sequence $(1, 1, 1, 1, \ldots)$. Show that x is infinitesimal iff x is the equivalence class of a sequence $x_n = o(1)$, and is near-standard iff it is the equivalence class of a sequence $x_n = O(1)$.

Exercise 11.6.3   Let $f : \mathbb{R} \to \mathbb{R}$ be continuous at 0. Show that it is possible that $\frac{f(x)}{x}$ is infinite for all nonzero infinitesimals x. If $\frac{f(x)}{x}$ is near-standard for every infinitesimal x, what can you tell about f?

11.7 ◆ Bibliography

We learned most of the material in this chapter from T. Lindstrøm's "Introduction to nonstandard analysis," Ch. 1 in N. Cutland's *Nonstandard Analysis and Its Applications* (Cambridge: Cambridge University Press, 1988) and A. E. Hurd and P. A. Loeb's *An Introduction to Nonstandard Real Analysis* (Orlando, Fla.: Academic Press, 1985). The nonstandard analysis approach to stochastic process theory that grew out of R. Anderson's article, "A nonstandard representation for Brownian motion and Itô integration," *Israel Journal of Mathematics* 25, 15–46 (1976), can be found in K. D. Stroyan and J. M. Bayod's *Foundations of Infinitesimal Stochastic Analysis* (Amsterdam: North-Holland, 1986).

Index

Page numbers followed by f refer to figures.